Institute of Physics Publishing

Scientific Discovery Through Advanced Computing Conference

SciDAC 2008

Journal of Physics: Conference Series Vol. 125

July 13-17 2008
Seattle, Washington

Volume 1 of 2

Printed from e-media with permission by:

Curran Associates, Inc.
57 Morehouse Lane
Red Hook, NY 12571
www.proceedings.com

ISBN: 978-1-60560-608-8

IOP Publishing
The Public Ledger Building
Suite 929
150 South Independence Mall West
Philadelphia, PA 19106
USA

Phone: (215) 627-0880
Fax: (215) 627-0879

info@ioppubusa.com

Scientific Discovery Through Advanced
Computing Conference
2008

TABLE OF CONTENTS

VOLUME 1

VOLUME 2

SciDAC 2008

Rick Stevens 2008 *J. Phys.: Conf. Ser.* **125** 011001 (1p) doi: 10.1088/1742-6596/125/1/011001 (Help)

Rick Stevens

The fourth annual Scientific Discovery through Advanced Computing (SciDAC) Conference was held June 13–18, 2008, in Seattle, Washington. The SciDAC conference series is the premier communitywide venue for presentation of results from the DOE Office of Science's interdisciplinary computational science program.

Started in 2001 and renewed in 2006, the DOE SciDAC program is the country's – and arguably the world's – most significant interdisciplinary research program supporting the development of advanced scientific computing methods and their application to fundamental and applied areas of science. SciDAC supports computational science across many disciplines, including astrophysics, biology, chemistry, fusion sciences, and nuclear physics. Moreover, the program actively encourages the creation of long-term partnerships among scientists focused on challenging problems and computer scientists and applied mathematicians developing the technology and tools needed to address those problems.

The SciDAC program has played an increasingly important role in scientific research by allowing scientists to create more accurate models of complex processes, simulate problems once thought to be impossible, and analyze the growing amount of data generated by experiments. To help further the research community's ability to tap into the capabilities of current and future supercomputers, Under Secretary for Science, Raymond Orbach, launched the Innovative and Novel Computational Impact on Theory and Experiment (INCITE) program in 2003. The INCITE program was conceived specifically to seek out computationally intensive, large-scale research projects with the potential to significantly advance key areas in science and engineering. The program encourages proposals from universities, other research institutions, and industry. During the first two years of the INCITE program, 10 percent of the resources at NERSC were allocated to INCITE awardees. However, demand for supercomputing resources far exceeded available systems; and in 2003, the Office of Science identified increasing computing capability by a factor of 100 as the second priority on its Facilities of the Future list. The goal was to establish leadership-class computing resources to support open science. As a result of a peer reviewed competition, the first leadership computing facility was established at Oak Ridge National Laboratory in 2004. A second leadership computing facility was established at Argonne National Laboratory in 2006.

This expansion of computational resources led to a corresponding expansion of the INCITE program. In 2008, Argonne, Lawrence Berkeley, Oak Ridge, and Pacific

Northwest national laboratories all provided resources for INCITE. By awarding large blocks of computer time on the DOE leadership computing facilities, the INCITE program enables the largest-scale computations to be pursued. In 2009, INCITE will award over half a billion node-hours of time.

The SciDAC conference celebrates progress in advancing science through large-scale modeling and simulation. Over 350 participants attended this year's talks, poster sessions, and tutorials, spanning the disciplines supported by DOE. While the principal focus was on SciDAC accomplishments, this year's conference also included invited presentations and posters from DOE INCITE awardees. Another new feature in the SciDAC conference series was an electronic theater and video poster session, which provided an opportunity for the community to see over 50 scientific visualizations in a venue equipped with many high-resolution large-format displays.

To highlight the growing international interest in petascale computing, this year's SciDAC conference included a keynote presentation by Herman Lederer from the Max Planck Institut, one of the leaders of DEISA (Distributed European Infrastructure for Supercomputing Applications) project and a member of the PRACE consortium, Europe's main petascale project. We also heard excellent talks from several European groups, including Laurent Gicquel of CERFACS, who spoke on `Large-Eddy Simulations of Turbulent Reacting Flows of Real Burners: Status and Challenges', and Jean-Francois Hamelin from EDF, who presented a talk on `Getting Ready for Petaflop Capacities and Beyond: A Utility Perspective'.

Two other compelling addresses gave attendees a glimpse into the future. Tomas Diaz de la Rubia of Lawrence Livermore National Laboratory spoke on a vision for a fusion/fission hybrid reactor known as the `LIFE Engine' and discussed some of the materials and modeling challenges that need to be overcome to realize the vision for a 1000-year greenhouse-gas-free power source. Dan Reed from Microsoft gave a capstone talk on the convergence of technology, architecture, and infrastructure for cloud computing, data-intensive computing, and exascale computing (1018 flops/sec).

High-performance computing is making rapid strides. The SciDAC community's computational resources are expanding dramatically. In the summer of 2008 the first general purpose petascale system (IBM Cell-based RoadRunner at Los Alamos National Laboratory) was recognized in the top 500 list of fastest machines heralding in the dawning of the petascale era. The DOE's leadership computing facility at Argonne reached number three on the Top 500 and is at the moment the most capable open science machine based on an IBM BG/P system with a peak performance of over 550 teraflops/sec. Later this year Oak Ridge is expected to deploy a 1 petaflops/sec Cray XT system.

And even before the scientific community has had an opportunity to make significant use of petascale systems, the computer science research community is forging ahead with ideas and strategies for development of systems that may by the end of the next decade sustain exascale performance. Several talks addressed barriers to, and strategies for, achieving exascale capabilities. The last day of the conference was devoted to tutorials hosted by Microsoft Research at a new conference facility in Redmond, Washington. Over 90 people attended the tutorials, which covered topics ranging from an introduction to BG/P programming to advanced numerical libraries.

The SciDAC and INCITE programs and the DOE Office of Advanced Scientific Computing Research core program investments in applied mathematics, computer science, and computational and networking facilities provide a nearly optimum framework for advancing computational science for DOE's Office of Science. At a broader level this framework also is benefiting the entire American scientific enterprise. As we look forward, it is clear that computational approaches will play an increasingly significant role in addressing challenging problems in basic science, energy, and environmental research.

It takes many people to organize and support the SciDAC conference, and I would like to thank as many of them as possible. The backbone of the conference is the technical program; and the task of selecting, vetting, and recruiting speakers is the job of the organizing committee. I thank the members of this committee for all the hard work and the many tens of conference calls that enabled a wonderful program to be assembled. This year the following people served on the organizing committee: Jim Ahrens, LANL; David Bader, LLNL; Bryan Barnett, Microsoft; Peter Beckman, ANL; Vincent Chan, GA; Jackie Chen, SNL; Lori Diachin, LLNL; Dan Fay, Microsoft; Ian Foster, ANL; Mark Gordon, Ames; Mohammad Khaleel, PNNL; David Keyes, Columbia University; Bob Lucas, University of Southern California; Tony Mezzacappa, ORNL; Jeff Nichols, ORNL; David Nowak, ANL; Michael Papka, ANL; Thomas Schultess, ORNL; Horst Simon, LBNL; David Skinner, LBNL; Panagiotis Spentzouris, Fermilab; Bob Sugar, UCSB; and Kathy Yelick, LBNL.

I owe a special thanks to Mike Papka and Jim Ahrens for handling the electronic theater. I also thank all those who submitted videos. It was a highly successful experiment.

Behind the scenes an enormous amount of work is required to make a large conference go smoothly. First I thank Cheryl Zidel for her tireless efforts as organizing committee liaison and posters chair and, in general, handling all of my end of the program and keeping me calm. I also thank Gail Pieper for her work in editing the proceedings, Beth Cerny Patino for her work on the Organizing Committee website and electronic theater, and Ken Raffenetti for his work in keeping that website working. Jon Bashor and John Hules did an excellent job in handling conference communications. I thank Caitlin Youngquist for the striking graphic design; Dan Fay for tutorials arrangements; and Lynn Dory, Suzanne Stevenson, Sarah Pebelske and Sarah Zidel for on-site registration and conference support. We all owe Yeen Mankin an extra-special thanks for choosing the hotel, handling contracts, arranging menus, securing venues, and reassuring the chair that everything was under control.

We are pleased to have obtained corporate sponsorship from Cray, IBM, Intel, HP, and SiCortex.

I thank all the speakers and panel presenters. I also thank the former conference chairs Tony Metzzacappa, Bill Tang, and David Keyes, who were never far away for advice and encouragement.

Finally, I offer my thanks to Michael Strayer, without whose leadership, vision, and persistence the SciDAC program would not have come into being and flourished. I am honored to be part of his program and his friend.

Rick Stevens
Seattle, Washington
July 18, 2008

SciDAC 2008

Michael Strayer 2008 *J. Phys.: Conf. Ser.* **125** 011002 (1p) doi: 10.1088/1742-6596/125/1/011002 Help

Michael Strayer

Welcome to Seattle and the 2008 SciDAC Conference. This conference, the fourth in the series, is a continuation of the PI meetings we first began under SciDAC-1. I would like to start by thanking the organizing committee, and Rick Stevens in particular, for organizing this year's meeting.

This morning I would like to look briefly at SciDAC, to give you a brief history of SciDAC and also look ahead to see where we plan to go over the next few years.

I think the best description of SciDAC, at least the simulation part, comes from a quote from Dr Ray Orbach, DOE's Under Secretary for Science and Director of the Office of Science. In an interview that appeared in the SciDAC Review magazine, Dr Orbach said, `SciDAC is unique in the world. There isn't any other program like it anywhere else, and it has the remarkable ability to do science by bringing together physical scientists, mathematicians, applied mathematicians, and computer scientists who recognize that computation is not something you do at the end, but rather it needs to be built into the solution of the very problem that one is addressing'.

Of course, that is extended not just to physical scientists, but also to biological scientists.

This is a theme of computational science, this partnership among disciplines, which goes all the way back to the early 1980s and Ken Wilson. It's a unique thread within the Department of Energy. SciDAC-1, launched around the turn of the millennium, created a new generation of scientific simulation codes. It advocated building out mathematical and computing system software in support of science and a new collaboratory software environment for data. The original concept for SciDAC-1 had topical centers for the execution of the various science codes, but several corrections and adjustments were needed. The ASCR scientific computing infrastructure was also upgraded, providing the hardware facilities for the program.

The computing facility that we had at that time was the big 3 teraflop/s center at NERSC and that had to be shared with the programmatic side supporting research across DOE. At the time, ESnet was just slightly over half a gig per sec of bandwidth; and the science being addressed was accelerator science, climate, chemistry, fusion, astrophysics, materials science, and QCD.

We built out the national collaboratories from the ASCR office, and in addition we built Integrated Software Infrastructure Centers (ISICs). Of these, three were in applied mathematics, four in computer science (including a performance evaluation research center), and four were collaboratories or Grid projects having to do with data management.

For science, there were remarkable breakthroughs in simulation, such as full 3D laboratory scale flame simulation. There were also significant improvements in application codes – from factors of almost 3 to more than 100 – and code improvement as people began to realize they had to integrate mathematics tools and computer science tools into their codes to take advantage of the parallelism of the day. The SciDAC data-mining tool, Sapphire, received a 2006 R&D 100 award. And the community as a whole worked well together and began building a publication record that was substantial.

In 2006, we recompeted the program with similar goals – SciDAC-1 was very successful, and we wanted to continue that success and extend what was happening under SciDAC to the broader science community. We opened up the partnership to all of the Offices of Science and the NSF and the NNSA. The goal was to create comprehensive scientific computing software and the infrastructure for the software to enable scientific discovery in the physical, biological, and environmental sciences and take the simulations to an extreme scale, in this case petascale. We would also build out a new generation of data management tools. What we observed during SciDAC-1 was that the data and the data communities – both experimental data from large experimental facilities and observational data, along with simulation data – were expanding at a rate significantly faster than Moore's law. In the past few weeks, the FastBit indexing technology software tool for data analyses and data mining developed under SciDAC's Scientific Data Management project was recognized with an R&D 100 Award, selected by an independent judging panel and editors of R&D Magazine as one of the 100 most technologically significant products introduced into the marketplace over the past year.

For SciDAC-2 we had nearly 250 proposals requesting a total of slightly over $1 billion in funding. Of course, we had nowhere near $1 billion. The facilities and the science we ended up with were not significantly different from what we had in SciDAC-1. But we had put in place substantially increased facilities for science. When SciDAC-1 was originally executed with the facilities at NERSC, there was significant impact on the resources at NERSC, because not only did we have an expanding portfolio of programmatic science, but we had the SciDAC projects that also needed to run at NERSC. Suddenly, NERSC was incredibly oversubscribed.

With SciDAC-2, we had in place leadership-class computing facilities at Argonne with slightly more than half a petaflop and at Oak Ridge with slightly more than a quarter petaflop with an upgrade planned at the end of this year for a petaflop. And we increased the production computing capacity at NERSC to 104 teraflop/s just so that we would not impact the programmatic research and so that we would have a startup facility for SciDAC. At the end of the summer, NERSC will be at 360 teraflop/s. Both the Oak Ridge system and the principal resource at NERSC are Cray systems; Argonne has a different architecture, an IBM Blue Gene/P. At the same time, ESnet has been built out, and we are on a path where we will have dual rings around the country, from 10 to 40 gigabits per second – a factor of 20 to 80 over what was available during SciDAC-1.

The science areas include accelerator science and simulation, astrophysics, climate modeling and simulation, computational biology, fusion science, high-energy physics, petabyte high-energy/ nuclear physics, materials science and chemistry, nuclear physics, QCD, radiation transport, turbulence, and groundwater reactive transport modeling and simulation. They were supported by new enabling technology centers and university-based institutes to develop an educational thread for the SciDAC program. There were four mathematics projects and four computer science projects; and under data management, we see a significant difference in that we are bringing up new visualization projects to support and sustain data-intensive science.

When we look at the budgets, we see growth in the budget from just under $60 million for SciDAC-1 to just over $80 for SciDAC-2. Part of the growth is due to bringing in NSF and NNSA as new partners, and some of the growth is due to some program offices increasing their investment in SciDAC, while other program offices are constant or have decreased their investment. This is not a reflection of their priorities *per se* but, rather, a reflection of the budget process and the difficult times in Washington during the past two years.

New activities are under way in SciDAC – the annual PI meeting has turned into what I would describe as the premier interdisciplinary computational science meeting, one of the best in the world. Doing interdisciplinary meetings is difficult because people tend to develop a focus for their particular subject area. But this is the fourth in the series; and since the first meeting in San Francisco, these conferences have been remarkably successful.

For SciDAC-2 we also created an outreach magazine, SciDAC Review, which highlights scientific discovery as well as high-performance computing. It's been very successful in telling the non-practitioners what SciDAC and computational science are all about.

The other new instrument in SciDAC-2 is an outreach center. As we go from computing at the terascale to computing at the petascale, we face the problem of narrowing our research community. The number of people who are `literate' enough to compute at the terascale is more than the number of those who can compute at the petascale. To address this problem, we established the SciDAC Outreach Center to bring people into the fold and educate them as to how we do SciDAC, how the teams are composed, and what it really means to compute at scale.

The resources I have mentioned don't come for free. As part of the HECRTF law of 2005, Congress mandated that the Secretary would ensure that leadership-class facilities would be open to everyone across all agencies. So we took Congress at its word, and INCITE is our instrument for making allocations at the leadership-class facilities at Argonne and Oak Ridge, as well as smaller allocations at NERSC. Therefore, the selected proposals are very large projects that are computationally intensive, that compute at scale, and that have a high science impact. An important feature is that INCITE is completely open to anyone – there is no requirement of DOE Office of Science funding, and proposals are rigorously reviewed for both the science and the computational readiness.

In 2008, more than 100 proposals were received, requesting about 600 million processor-hours. We allocated just over a quarter of a billion processor-hours. Astrophysics, materials science, lattice gauge theory, and high energy and nuclear physics were the major areas. These were the teams that were computationally ready for the big machines and that had significant science they could identify. In 2009, there will be a significant increase amount of time to be allocated, over half a billion processor-hours. The deadline is August 11 for new proposals and September 12 for renewals. We anticipate a significant increase in the number of requests this year.

We expect you – as successful SciDAC centers, institutes, or partnerships – to compete for and win INCITE program allocation awards. If you have a successful SciDAC proposal, we believe it will make you successful in the INCITE review. We have the expectation that you will among those most prepared and most ready to use the machines and to compute at scale.

Over the past 18 months, we have assembled a team to look across our computational science portfolio and to judge what are the 10 most significant science accomplishments. The ASCR office, as it goes forward with OMB, the new administration, and Congress, will be judged by the science we have accomplished. All of our proposals – such as for increasing SciDAC, increasing applied mathematics, and so on – are tied to what have we accomplished in science. And so these 10 big accomplishments are key to establishing credibility for new budget requests.

Tony Mezzacappa, who chaired the committee, will also give a presentation on the ranking of these top 10, how they got there, and what the science is all about. Here is the list – numbers 2, 5, 6, 7, 9, and 10 are all SciDAC projects.

Rank	Title
1	Modeling the Molecular Basis of Parkinson's Disease (Tsigelny)
2	Discovery of the Standing Accretion Shock Instability and Pulsar Birth Mechanism in a Core-Collapse Supernova Evolution and Explosion (Blondin)
3	Prediction and Design of Macromolecular Structures and Functions (Baker)
4	Understanding How Lifted Flame Stabilized in a Hot Coflow (Yoo)
5	New Insights from LCF-enabled Advanced Kinetic Simulations of Global Turbulence in

	Fusion Systems (Tang)
6	High Transition Temperature Superconductivity: A High-Temperature Superconductive State and a Pairing Mechanism in 2-D Hubbard Model (Scalapino)
7	PETsc: Providing the Solvers for DOE High-Performance Simulations (Smith)
8	Via Lactea II, A Billion Particle Simulation of the Dark Matter Halo of the Milky Way (Madau)
9	Probing the Properties of Water through Advanced Computing (Galli)
10	First Provably Scalable Maxwell Solver Enables Scalable Electromagnetic Simulations (Kovel)

So, what's the future going to look like for us? The office is putting together an initiative with the community, which we call the E3 Initiative. We're looking for a 10-year horizon for what's going to happen. Through the series of town hall meetings, which many of you participated in, we have produced a document on `Transforming Energy, the Environment and Science through simulations at the eXtreme Scale'; it can be found at http://www.science.doe.gov/ascr/ProgramDocuments/TownHall.pdf .

We sometimes call it the Exascale initiative. Exascale computing is the gold-ring level of computing that seems just out of reach; but if we work hard and stretch, we just might be able to reach it. We envision that there will be a SciDAC-X, working at the extreme scale, with SciDAC teams that will perform and carry out science in the areas that will have a great societal impact, such as alternative fuels and transportation, combustion, climate, fusion science, high-energy physics, advanced fuel cycles, carbon management, and groundwater.

We envision institutes for applied mathematics and computer science that probably will segue into algorithms because, at the extreme scale, we see the distinction between the applied math and the algorithm *per se* and its implementation in computer science as being inseparable. We envision an INCITE-X with multi-petaflop platforms, perhaps even exaflop computing resources. ESnet will be best in class – our 10-year plan calls for having 400 terabits per second capacity available in dual rings around the country, an enormously fast data communications network for moving large amounts of data.

In looking at where we've been and where we are going, we can see that the gigaflops and teraflops era was a regime where we were following Moore's law through advances in clock speed. In the current regime, we're introducing massive parallelism, which I think is exemplified by Intel's announcement of their teraflop chip, where they envision more than a thousand cores on a chip. But in order to reach exascale, extrapolations talk about machines that require 100 megawatts of power in terms of current architectures. It's clearly going to require novel architectures, things we have perhaps not yet envisioned. It is of course an era of challenge. There will be an unpredictable evolution of hardware if we are to reach the exascale; and there will clearly be multilevel heterogeneous parallelism, including multilevel memory hierarchies. We have no idea right now as to the programming models needed to execute at such an extreme scale. We have been incredibly successful at the petascale – we know that already. Managing data and just getting communications to scale is an enormous challenge. And it's not just the extreme scaling. It's the rapid increase in complexity that represents the challenge.

Let me end with a metaphor. In previous meetings we have talked about the road to petascale. Indeed, we have seen in hindsight that it was a road well traveled. But perhaps the road to exascale is not a road at all. Perhaps the metaphor will be akin to scaling the south face of K2. That's clearly not something all of us will be able to do, and probably computing at the exascale is not something all of us will do.

But if we achieve that goal, perhaps the words of Emily Dickinson will best summarize where we will be. Perhaps in her words, looking backward and down, you will say:

I climb the `Hill of Science'
I view the landscape o'er;
Such transcendental prospect
I ne'er beheld before!

DEISA2: Supporting and developing a European high-performance computing ecosystem

Hermann Lederer

Garching Computing Center of the Max Planck Society, Max Planck Institute for Plasma Physics, Garching, Germany

E-mail: lederer@rzg.mpg.de

Abstract. The DEISA Consortium has deployed and operated the Distributed European Infrastructure for Supercomputing Applications. Through the EU FP7 DEISA2 project (funded for three years as of May 2008), the consortium is continuing to support and enhance the distributed high-performance computing infrastructure and its activities and services relevant for applications enabling, operation, and technologies, as these are indispensable for the effective support of computational sciences for high-performance computing (HPC). The service-provisioning model will be extended from one that supports single projects to one supporting virtual European communities. Collaborative activities will also be carried out with new European and other international initiatives. Of strategic importance is cooperation with the PRACE project, which is preparing for the installation of a limited number of leadership-class Tier-0 supercomputers in Europe. The key role and aim of DEISA will be to deliver a turnkey operational solution for a persistent European HPC ecosystem that will integrate national Tier-1 centers and the new Tier-0 centers.

1. Background on DEISA

In spring 2002 the idea emerged to overcome the fragmentation of supercomputing resources in Europe in terms of both system availability and the necessary skills for efficient supercomputing support. The establishment of a distributed European supercomputing infrastructure was proposed. In May 2004 the DEISA project was started as a EU FP6 Integrated Infrastructure Initiative by eight leading European supercomputing centers. In 2006 DEISA was joined by three additional leading centers. Through their joint efforts, DEISA quickly reached production quality, supporting leading-edge capability computing for the European scientific community. DEISA has also helped raise awareness of the need for a persistent European high-performance computing (HPC) infrastructure as recommended by the European Strategy Forum on Research Infrastructures (ESFRI) in its 2006 report [1].

2. DEISA extreme computing projects

The DEISA Extreme Computing Initiative (DECI), launched in 2005, continues to support the most challenging supercomputing projects in Europe that require the special resources and skills of DEISA. A European call for "Extreme Computing" proposals has been published annually in the spring with 40 responses in 2005, 50 in 2006, and over 60 in 2007, currently asking for more than 70 million computing hours. By selecting the most appropriate supercomputer architectures for each project, DEISA is opening up the currently most powerful HPC architectures available in Europe for the most challenging projects. This approach mitigates the rapid performance decay of a single national supercomputer within its short lifetime cycle of typically about five years, as implied by Moore's law.

The booklet *DEISA – Advancing Science in Europe* reports on scientific achievements through DEISA, especially by Extreme Computing projects, but also by the scientific Joint Research Activities [12]. The projects reported cover major areas of science including astroscience, Earth science, engineering, life sciences, materials science, plasma physics, and quantum chromodynamics. As one of the highlights, results from a German/British group were published as the cover story of *Nature* [3]. For almost two decades, physicists have been on the track of membrane-mediated interactions. Simulations in DEISA have now revealed that curvy membranes make proteins attractive.

So far, scientists from 157 research institutes and universities from 15 European countries have benefited from DECI, with collaborators from four other continents: North America, South America, Asia, and Australia.

3. DEISA2 essentials

In EU FP7, the DEISA Consortium continues to support and further develop the distributed HPC infrastructure and its services through the DEISA2 project, which was funded for three years as of May 2008. Activities and services relevant for applications enabling, operation, and technologies will be continued and enhanced, as these are indispensable for the effective support of computational sciences in high-performance computing. The service-provisioning model will be extended from one that supports single projects to one supporting virtual European communities. Collaborative activities also will be carried out with new European and other international initiatives.

Of strategic importance is cooperation with the Partnership for Advanced Computing in Europe (PRACE), which is preparing for the installation of a limited number of leadership-class Tier-0 supercomputers in Europe. The key role and aim will be to deliver a turnkey operational solution for a persistent European HPC eco-system, as suggested by ESFRI. The ecosystem will integrate national Tier-1 centers and the new Tier-0 centers, as illustrated in figure 1.

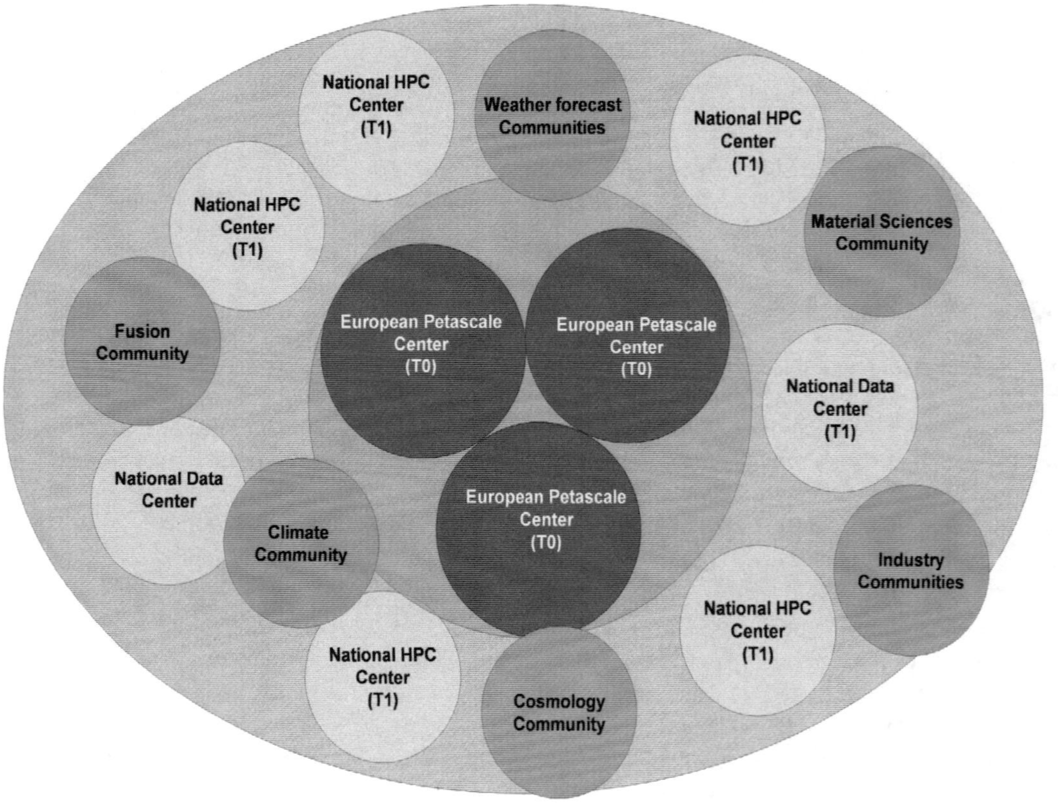

Figure 1. Sketch of the HPC ecosystem integrating national Tier-1 centers, virtual scientific communites, and the new Tier-0 centers.

4. DEISA members and new associate partners

Currently, DEISA has eleven DEISA members from seven countries: BSC (Barcelona, Spain), CINECA (Bologna, Italy), CSC (Espoo, Finland), ECMWF (Reading, UK), EPCC (Edinburgh, UK), FZJ (Juelich, Germany, HLRS (Stuttgart, Germany,) IDRIS (Orsay, France), LRZ (Garching, Germany), RZG (Garching, Germany), and SARA (Amsterdam, The Netherlands). Further centers plan to join DEISA as associate partners, including CEA-CCRT (France), CSCS (Manno, Switzerland), KTH (Stockholm, Sweden), and JSCC (Moscow, Russia).

5. User communities

In DEISA2, the single-project-oriented activities (DECI) will be qualitatively extended for persistent support of virtual science communities. DEISA2 will provide a computational platform for them, offering integration via distributed services and web applications, as well as managing data repositories. Emphasis will be on collaborations with research infrastructure projects established by the ESFRI and on European HPC and Grid projects. The activity reinforces the relations to other European HPC centers; leading international HPC centers in Australia, China, Japan, Russia, and the United States; and leading HPC projects worldwide.

6. Services

Essential services to operate the infrastructure and support its efficient usage have been classified into three categories: operations, technologies, and applications.

- *Operations* refers to operating the infrastructure, including all existing services; adopting approved new services from the Technologies activity; and advancing the operation of the DEISA HPC infrastructure to a turnkey solution for the future European HPC ecosystem by improving the operational model and integrating new sites.
- *Technologies* covers monitoring of technologies already in use in the project, identifying and selecting technologies of relevance for the project, evaluating technologies for preproduction deployment, and planning and designing specific subinfrastructures to upgrade existing services or deliver new ones based on approved technologies.
- *Applications* cover the areas applications enabling and extreme computing projects, environment and user related application support, and benchmarking. *Applications enabling* focuses on enhancing scientific applications from the DEISA Extreme Computing Initiative (DECI), virtual communities, and EU projects. *Environment and user related application support* addresses the maintenance and improvement of the DEISA application environment and interfaces, and DEISA-wide user support in the applications area. *Benchmarking* refers to the provision and maintenance of a European benchmark suite for supercomputers.

Two joint research activities (JRAs) complement the portfolio of service activities. JRA1 (Integrated DEISA Development Environment) aims at an integrated environment for scientific application development, based on a software infrastructure for tools integration, which provides a common user interface across multiple computing platforms. JRA2 (Enhancing Scalability) aims at the enabling of supercomputer applications for the efficient exploitation of current and future supercomputers, to cope with a production infrastructure characterized by an aggressive parallelism on heterogeneous HPC architectures at a European scale.

7. International activities

In support of international science communities traversing political boundaries, DEISA participates in the evaluation and implementation of standards for interoperation.

Acknowledgment

The DEISA Consortium thanks the European Commission for support through contracts RI-508830, RI-031513, and RI-222919.

References

[1] European Strategy Forum on Research Infrastructures http://cordis.europa.eu/esfri/).
[2] *DIESA – Advancing Science in Europe* http://www.prace-project.eu/press-corner/deisa-results/
[3] Reynwar B J, Illya G, Harmandaris V A, Müller M M, Kremer K and Deserno M 2007 Aggregation and vesiculation of membrane proteins by curvature-mediated interactions *Nature* **447**:461–464
[4] The Partnership for Advanced Computing in Europe http://www.prace-project.eu

Journal of Physics: Conference Series **125** (2008) 012001 doi:10.1088/1742-6596/125/1/012001

Multiscale, multiphysics beam dynamics framework design and applications

James F Amundson[1], Douglas Dechow[2], Lois McInnes[3], Boyana Norris[3], Panagiotis Spentzouris[1] and Peter Stoltz[2]

[1] Fermi National Accelerator Laboratory, Batavia, IL 60510, USA
[2] Tech-X Corporation, Boulder, CO 80303, USA
[3] Argonne National Laboratory, Argonne, IL 60439, USA

E-mail: amundson@fnal.gov

Abstract. Modern beam dynamics simulations require nontrivial implementations of multiple physics models. We discuss how component framework design in combination with the Common Component Architecture's component model and implementation eases the process of incorporation of existing state-of-the-art models with newly-developed models. We discuss current developments in componentized beam dynamics software, emphasizing design issues and distribution issues.

1. Introduction

Accelerator-based science is a cornerstone of modern fundamental scientific research. As accelerator-based experiments have progressed, they have required better and better accelerators. "Better" can have several meanings, higher energy and higher luminosity being the two most common. Simulations are important tools for the design of future accelerators, as well as the the optimization of currently running accelerators.

A decade ago, state-of-the-art beam dynamics simulations were highly specialized. One program might have a detailed implementation of higher-order optics but lack or have only trivial implementations of collective effects such as space charge. Another program would implement a fully three-dimensional realization of space charge, but include only a trivial optics component. During the period of SciDAC-1, programs such as MaryLie/IMPACT (ML/I) [1] and Synergia [2] combined state-of-the-art optics codes with IMPACT, a fully three-dimensional, parallel space charge implementation. For the first time, beam dynamics simulations were able to combine detailed space-charge calculations with advanced, nonlinear optics.

While the approach of straightforwardly combining extant codes advanced the capabilities of beam dynamics simulations by enabling studies of the interplay between space charge and non-linear optics effects, extending the same approach to other processes and implementations would be difficult. During SciDAC-2, as part of the Community Petascale Project for Accelerator Science and Simulation (COMPASS) project, we have embarked on the next phase of beam simulation development by transforming our approach into a component-centered framework.

In these proceedings, we discuss the motivations for the component approach and the Common Component Architecture (CCA) implementation of component infrastructure. We then discuss some of the most important issues in component design for beam dynamics. Further, we discuss a solution for the problem of distributing, porting, and building componentized applications.

2. Motivations for the component approach

In this paper we use the word "component" both in the colloquial sense as a well-defined constituent part and in the technical sense as defined by the Common Component Architecture Forum, as discussed in Section 4. For the moment, we restrict ourselves to the general sense of components of being portions of software that can be added and removed from multiple applications. For example, a space charge component is distinct from an application that merely implements space charge in a such a way that it cannot easily be separated from the remainder of the application's implementation.

One goal of the component approach is to increase the availability of physics model implementations for use in simulations. Increased availability arises not only because components written for one application are available for others but also because the existence of well-defined interfaces and other infrastructure makes writing components much easier than it would otherwise be.

Availability of components enables previously impossible scientific applications, namely those that investigate the interplay between multiple physics processes (multiphysics). There are, however, other important advantages of the component approach. Different implementation strategies of the same physical effect allow us to choose the most efficient computational approach for the task at hand. For example, in beam dynamics problems where space charge effects are large, it is important to use a fully three-dimensional, self-consistent space charge implementation. In cases where the effects are relatively small, it may be possible to use a simpler two-dimensional approximation with substantially reduced computational cost.

Another important advantage of the component approach is the improved possibilities for validation and benchmarking. To date, most comparative benchmarks of collective beam dynamics effects have involved comparing entire codes against each other. The different codes typically have independent implementations of accelerator lattice construction and optics, beam generation, and so forth. Benchmarking two or more such codes in their entirety inevitably obscures small, but possibly important, deviations in the various aspects of the code. With component-based simulations, we will be able to perform benchmarks with only a single variable, namely, the component in question.

3. Synergia2 as a component framework

Synergia is a beam dynamics framework that is evolving into a fully component-based simulation tool. The original Synergia [2] was a hybrid code connecting IMPACT, a Fortran 90 three-dimensional space charge application with CHEF, a C++ single-particle beam dynamics library. The user interface for Synergia was controlled by Python, but the Python driver was limited to writing a set of input files to be read by the Fortran 90 application. This approach was successful in allowing the combination of nonlinear optics with space charge, but proved too difficult to extend to other processes.

Synergia has been replaced by Synergia2, a fully Python-driven framework capable of calling the CHEF libraries and the space charge portions of IMPACT as well as a new C++ space charge implementation, Sphyraena, and other modules. Synergia2 was designed for extensibility from the outset. The various modules are all considered on equal footing; new modules are in development. Synergia2 is rapidly becoming a component framework in the colloquial sense described in the previous section. All that is standing in the way for this level of componentization is the formalization of the component interfaces. We discuss some of the relevant design issues in Section 5.

There are still several challenges faced by the Synergia2 componentization effort. The foremost challenge is the difficulty of dealing with cross-language issues. Synergia2 is a Python application with components written in Python, C++, and Fortran 90. Although we have managed to overcome the C++-Python interface issues with little difficulty, the interface to Fortran 90 has proved more problematic. Our current Fortran 90 component implementation was difficult to write, is somewhat fragile, and suffers from portability problems. Furthermore, if other (non-Python) applications want to take advantage of Synergia2 components, they have to deal with their own set of language interoperability problems. We plan to solve these problems with the CCA tools described in the following section.

4. Componentization using the Common Component Architecture

Rapid advances and an increasing diversity in high-performance computing platforms have placed technical computing burdens on the shoulders of scientists who use, as opposed to study, computation. This situation motivates scientists to take advantage of advances in computer science designed to reduce these burdens. Language interoperability is one such problem we have discussed in the context of Synergia2, but many others loom on the horizon, including portability to new platforms and access to profiling information for intelligent optimization.

The CCA Forum [3] is addressing these problems by developing tools for component-based scientific computing in both high-performance parallel and distributed computing contexts. The CCA model and tools [4] provide language-neutral specification of common component interfaces, interoperability for software written in programming languages important to scientific computing, and dynamic composability, all with minimal runtime overhead. The current initiatives, which extend this preliminary work, are motivated by and will be validated through collaborations with a variety of computational science groups, including several SciDAC application teams.

5. Component interface issues

The greatest challenge in componentization is the definition of the components and their interfaces. The components need to be objects that are relevant for multiple implementation of beam dynamics simulations. They also need to have interfaces stripped bare of details that are specific to a particular implementation. Parallelization schemes complicate the issue; two different applications can agree that they both need a certain vector, but they differ in how they expect it to be distributed in a parallel computation.

A beam dynamics application taking on the order of a day on modern, massively parallel hardware would typically include propagating millions of particles through hundreds or thousands of accelerator (beamline) elements interspersed with hundreds of computationally expensive collective-effects calculations, all repeated a few thousand times. Since component calls inevitably have some overhead, calls that happen a few times *per collective effect* would be acceptable, but calls that happen even once *per particle* would be unacceptable.

The elements of beam dynamics that are common to essentially all simulations are the beam bunch and the beamline, as shown in figures 1 and 2. The beam bunch contains a dense array of particles. Since the coordinates of the particles are always relative to a reference particle, the bunch needs to carry the reference also. The beam bunch is the most fundamental component and, fortunately, the most straightforward to define. The beamline is also fundamental but, unfortunately, more problematic. Different beam dynamics implementations have very different ideas about how the elements comprising an accelerator should represented, split, and so forth. As such, it makes sense to have the entire beamline available as a component. One area of commonality betwen models is the representation of action of elements through Taylor maps. We have demonstrated interchanging CHEF and MaryLie beamline components at the map level, even though the beamline models themselves are very different in the two codes.

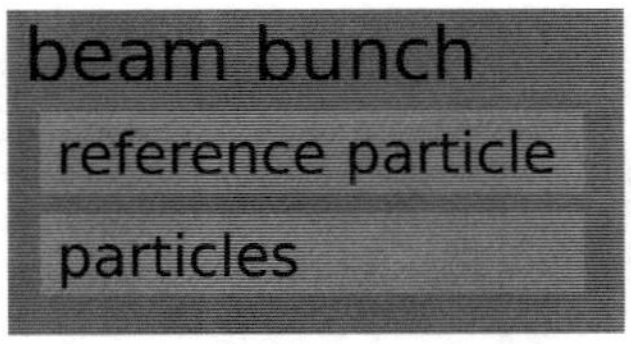

Figure 1. Beam bunch component with constituent reference particle and particles.

Figure 2. Beamline component with constituent elements, each possibly having a map and/or propagator.

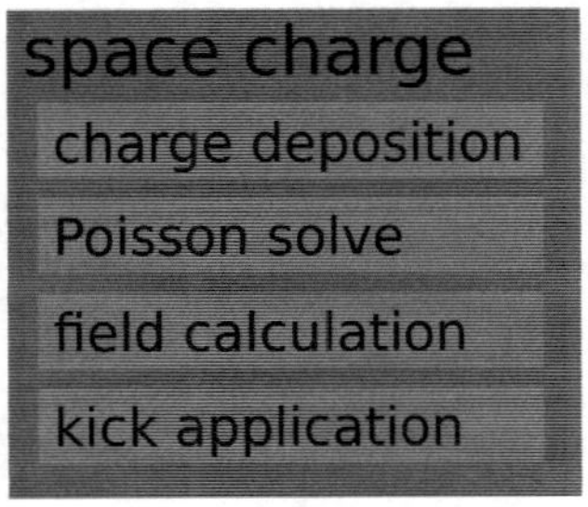

Figure 3. Space charge component with its constituent stages.

The bulk of our work to date on Synergia2 has been in the design and implementation of the space charge module, figure 3. Synergia2 can use the space charge modules from either IMPACT or Sphyraena. Granularity is an issue here. IMPACT and Sphyraena are interchangeable at the overall space charge level, but not at any finer level. The issue is the parallel decomposition. IMPACT uses a two-dimensional decomposition at the particle level, while Sphyraena uses a one-dimensional decomposition at the field level only.

The electron cloud component for Synergia2, Figure 4, is under development. The presence of several constituent stages that overlap with the constituent stages of the space charge component provides an opportunity for nontrivial code reuse. By carefully defining our interfaces, we are able to re-use aspects of the Sphyraena space charge module in the new electron cloud module.

6. Building and distributing componentized applications

While componentized applications represent an advance in capability and maintainability for scientific applications, they also suffer from the same problem facing all multielement software packages: build-time complexity. Figure 5 shows the non-trivial set of interdependent packages required to build Synergia2 and its constituent components. The traditional approach for distributing packages leaves the end user with a long list of dependencies to be built by hand. Our experience shows that expecting users to manually configure and build even a small set of dependent packages creates a substantial barrier to adoption.

We have created the tool Contractor [5] to automatically compile sets of dependent packages. Contractor is a *meta-build* tool; it compiles sets of dependent packages using their own native

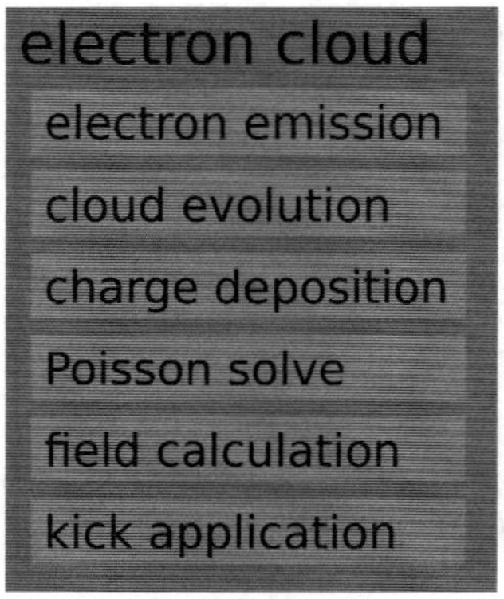

Figure 4. Electron cloud component with its constituent stages.

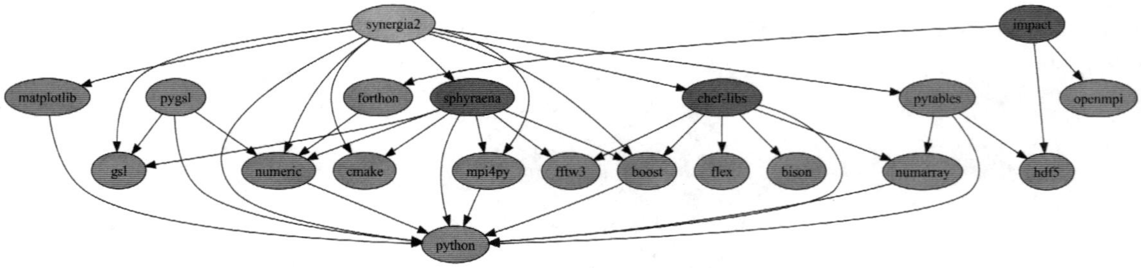

Figure 5. Packages required for building Synergia2, including component packages (blue) and infrastructure (gray).

build packages. In the simplest terms, make is a utility to invoke compilers such as cc and f90, on dependent sets of files. Contractor is a utility to invoke build systems such as GNU Autotools and CMake, on dependent sets of packages. Contractor is flexible enough to invoke arbitrary commands in a build process. It also has explicit support for GNU Autotools, CMake, Make, and the Python Setuptools.

The entire Synergia2 package and all of its dependencies can be compiled with a single Contractor command. The build can be interrupted and restarted. Contractor has a provision for user options, allowing the user to, for example, enable and disable optional packages or use system packages instead of compiling new versions. The state of the system can be queried by the user to see the status of options and the dependency tree itself. Figure 5 was generated from the Synergia2 Contractor description. In addition to being used for Synergia2, Contractor has been used as a meta-build system for CHEF, the CCA software, and other complex systems. We find it to be an invaluable tool for dealing with the build-time complexity arising in componentized applications during both development and distribution phases.

Acknowledgment
This work was supported in part by the U.S. Department of Energy under Contract DE-AC02-06CH11357.

References
[1] Ryne R, Dragt A, Decyk V, Habib S, Neri F, Mottershead C T, Samulyak R and Walstrom P MaryLie/IMPACT manual: A parallel beam dynamics code with space charge based on the MaryLie Lie algebraic beam transport code and the IMPACT parallel PIC code

[2] Amundson J, Spentzouris P, Qiang J and Ryne R 2006 *J. Comp. Phys.* **211** 229–248
[3] Common Component Architecture (CCA) Forum http://www.cca-forum.org
[4] Bernholdt D, *et al.* 2006 *Int. J. High-Perf. Computing Appl., ACTS Collection special issue* **20** 163–202
[5] Contractor Homepage http://home.fnal.gov/ amundson/contractor-www/index.html

Journal of Physics: Conference Series **125** (2008) 012002 doi:10.1088/1742-6596/125/1/012002

Computational studies and optimization of wakefield accelerators

C G R Geddes[1], D L Bruhwiler[2], J R Cary[2,3], W B Mori[4], J-L Vay[1], S F Martins[7], T Katsouleas[6], E Cormier-Michel[1], W M Fawley[1], C Huang[4], X Wang[6], B Cowan[2], V K Decyk[4], E Esarey[1], R A Fonseca[7], W Lu[4], P Messmer[2], P Mullowney[2], K Nakamura[1], K Paul[2], G R Plateau[1], C B Schroeder[1], L O Silva[7], Cs Toth[1], F S Tsung[4], M Tzoufras[4], T Antonsen[5], J Vieira[7] and W P Leemans[1]

[1] Lawrence Berkeley National Laboratory, Berkeley, CA 94720, USA
[2] Tech-X Corporation, Boulder, CO 80303, USA
[3] University of Colorado, Boulder, CO 80309, USA
[4] University of California, Los Angeles, CA 90095, USA
[5] University of Maryland, College Park, MD 20742, USA
[6] University of Southern California, Los Angeles, CA 90089, USA
[7] GoLP/IPFN Instituto Superior Tecnico, Lisbon 1049-001, Portugal

E-mail: cgrgeddes@lbl.gov

Abstract. Laser- and particle beam-driven plasma wakefield accelerators produce accelerating fields thousands of times higher than radio-frequency accelerators, offering compactness and ultrafast bunches to extend the frontiers of high energy physics and to enable laboratory-scale radiation sources. Large-scale kinetic simulations provide essential understanding of accelerator physics to advance beam performance and stability and show and predict the physics behind recent demonstration of narrow energy spread bunches. Benchmarking between codes is establishing validity of the models used and, by testing new reduced models, is extending the reach of simulations to cover upcoming meter-scale multi-GeV experiments. This includes new models that exploit Lorentz boosted simulation frames to speed calculations. Simulations of experiments showed that recently demonstrated plasma gradient injection of electrons can be used as an injector to increase beam quality by orders of magnitude. Simulations are now also modeling accelerator stages of tens of GeV, staging of modules, and new positron sources to design next-generation experiments and to use in applications in high energy physics and light sources.

1. Introduction

Particle accelerators are among the most powerful instruments of scientific discovery. A TeV-class linear collider to extend the energy frontier will require order of 20 km-long conventional radio-frequency (RF) accelerating linacs [1], while machines such as the LCLS will use kilometer-scale linacs to drive undulators for unprecedented x-ray brightness [1]. Greatly increased accelerating gradient is then needed to scale beyond TeV energies and to provide bright, laboratory-scale radiation sources.

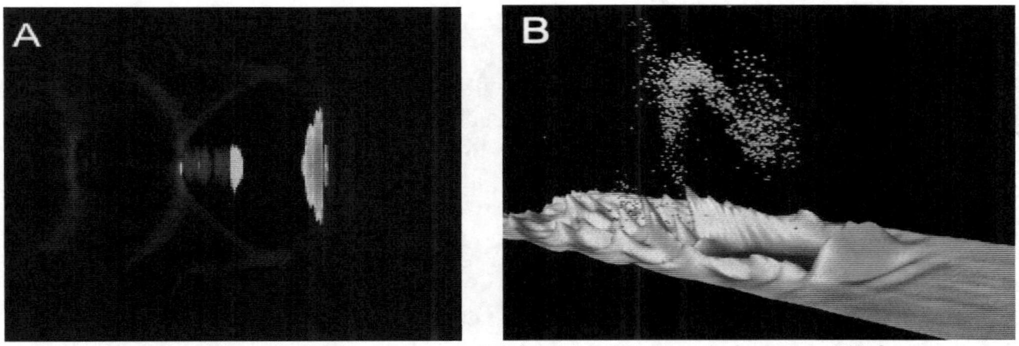

Figure 1. Schematic of an LWFA, from a VORPAL 3D simulation on 4096 processors of a centimeter-scale 0.7 GeV stage (A): the radiation pressure of a laser pulse (red) displaces plasma electrons creating a space charge wave (purple-blue) which accelerates particles, creating a compact bunch (green). Simulations resolve phase space and internal dynamics to optimize WFAs, showing for instance (B) particles in px-x-y space above plasma density in a 100 MeV self-trapped stage (VORPAL 3D).

In laser or particle beam driven plasma wakefield accelerators (LWFAs or PWFAs) [2, 3], the radiation pressure of an intense laser pulse (or space charge of a particle beam) displaces electrons, and the subsequent oscillation of the plasma creates a plasma wave (wake) following the driver (figure 1). The field of the wake is limited by trapping of particles and can be thousands of times that in RF accelerators. The wake period is micrometer scale, producing femtosecond bunches well suited for light sources [4].

Laser wakefield accelerators have produced electron bunches with percent energy spread and low emittance from electrons self-trapped by the wake. Bunches at ~0.1 GeV were produced in a few millimeters using plasma channels [5] or large laser spots [6,7] to extend the interaction length. Using a plasma channel to extend the interaction to centimeter scale, GeV bunches and stable operation at 0.5 GeV were observed [8]. Control over injection has now produced bunches with an order of magnitude lower absolute momentum spread and with stability over several days using plasma ramps [9], and produced bunches with tunable energy [10] using the colliding pulse method [11]. A PWFA has doubled the energy of a fraction of the SLAC beam, achieving energy gain of 42 GeV [12]. Plasma accelerator applications now require development and staging of ≥10 GeV LWFAs, further control over trapping and acceleration to reduce and stabilize momentum spread, and narrow energy spread PWFAs. Plasma scale will increase from centimeter to meter scale, and detailed bunch kinetics will be essential.

Simulations describe nonlinear plasma response, beam trapping, and self-consistent acceleration not accessible to analytic theory. They provide information on internal dynamics to optimize WFAs, and were, for example, essential in modeling the physics of narrow energy spread bunches from self-trapped electrons observed in recent years and evaluating the scaling of these experiments to higher energies as described in [5, 13-16] and references therein, among others.

The VORPAL [17] and OSIRIS [18, 19] codes provide explicit particle-in-cell simulations that resolve the laser wavelength (the shortest major scale) and particle kinetics and were used to model recent experiments. VORPAL now allows fluid description of the wake (reducing noise) with particles for the bunch. Resolving the laser wavelength over the propagation distance ($\sim 10^6$ steps for centimeter-scale GeV simulations) and wake volume ($\sim 10^8$ cells in 3D) drives the computational cost, which is order 10^6 hours for centimeter-scale GeV runs. Envelope simulations decrease cost by resolving the laser envelope but not its fast oscillation, allowing reduced resolution, and quasistatic codes further reduce cost by assuming slow evolution [20]. These assumptions, which require validation in the regime simulated, allow simulation of the same plasma and laser parameters with a 100 to 10000 times savings in CPU time. Envelope models are implemented in VORPAL and quasi-static in QuickPIC [21]. Performing calculations in a boosted frame [22] can also reduce cost, and this method is now being used in WARP[23], VORPAL, and OSIRIS. Optimizing methods for kinetic accuracy are also vital [24].

Here, we describe recent large-scale kinetic simulation of wakefield accelerators by the ComPASS [25] SciDAC project. Section 2 describes a benchmarking and scaling program that confirmed agreement of the codes and is being used to develop the reduced models needed to simulate future large experiments. Section 3 describes modeling of controlled injection in LWFAs towards development of multi-GeV stages with increased stability and reduced momentum spread, and modeling of 10-100 GeV LWFAs. Section 4 describes modeling of PWFA's at 10's of GeV and new positron sources. Finally, in section 5, development of the boosted frame algorithm is described which will be important to simulate m-scale plasmas. PIC algorithm studies were also conducted to optimize kinetic accuracy.

2. Benchmarking and code comparison

Software verification is an important part of the SciDAC mission, and benchmarking exercises are an effective way to achieve this goal in regimes with no analytical results, which is the case for nonlinear 3D plasma wakes. We consider the important example of an intense [I_{peak}~10^{18} W cm^{-2}], ultra-short [τ_{fwhm}=30 fs] Ti-Sapphire [λ_0=0.8 μm] laser pulse entering a uniform density plasma [n_e=1.38x10^{19} cm^{-3}], close to the parameters of recent experiments. We compare 3D simulation results from the time-explicit PIC codes OSIRIS and VORPAL with each other and with the quasi-static code QuickPIC, for several values of a_0, the normalized peak laser field. The rectangular mesh has 512x512x512 cells for the time-explicit case, with 8 macro-particles per cell, for a total of 1.34x10^8 cells and 1.07 x 10^9 particles. The longitudinal grid spacing is 0.04 μm (20 cells per λ_0), and transverse spacing is 0.16 μm.

Figure 2 presents 2D contour plots of the simulated accelerating electric field from each code, in a slice taken from the center of the 3D domain, with longitudinal position along the horizontal axis and transverse position along the vertical axis. The peak accelerating field for trapped electrons is near the middle of the center-left blue region. Acceleration and transverse focusing overlap throughout the right half of this region (roughly), which is where one typically finds trapped, accelerated beams. The lower-right plot in Figure 2 shows lineouts of the accelerating electric field for all three codes along the center of the simulated domain, normalized to the cold nonrelativistic wavebreaking field $E_{wb}=m_e c\omega_0/e$. We note that E_0 is about 10x smaller than E_{wb}. Position is normalized to the central wavenumber of the laser pulse. Agreement between VORPAL and OSIRIS is good, lending confidence in the validity of both codes. The fact that QuickPIC also agreed well gives further confidence as it is based on a completely different reduced algorithm. QuickPIC does not resolve the space/time oscillations driven by the laser pulse but accurately models features on the scale of the

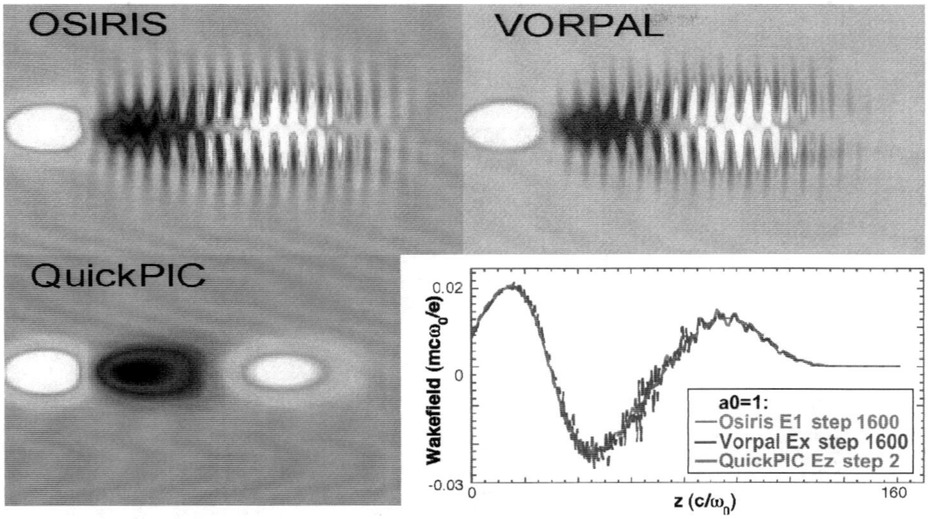

Figure 2. 3D simulation results from OSIRIS, VORPAL, and QuickPIC for an LWFA benchmark problem with a_0=1 show good agreement. The normalized accelerating wakefield is shown as a 2D slice (upper and left), and also overlapping lineouts down the center (lower right).

plasma wavelength so long as there are no self-injected electrons. The benchmarking effort is being extended to include VORPAL fluid and envelope models as well as the quasi-static model in WAKE.

These benchmarking exercises were repeated for a_0=0.5, 1, 2 and 4. Overlapping lineouts like that in figure 2, are shown in figure 3 for all values of a_0. For a_0=1, 2, and 4, these plots include data from OSIRIS and VORPAL simulations using second-order splines for the particle shapes. The higher-order shape, used both for charge-conserving current deposition and force interpolation, filters out high-wavenumber components of the particle currents and forces, which can greatly reduce noise and unphysical effects such as grid heating. Filtering must be done with care, however, as splines and higher-order particle shapes can modify physical effects, like the dispersion relation for plasma waves and the wavelength of the wake. As seen in figure 3, the spline-based particles in OSIRIS and VORPAL agree well with standard first-order shapes and with each other, for a_0=1 and 2, and show only modest differences for a_0=4. QuickPIC also supports higher-order particle shapes.

Extensive scaling of PIC has been performed on large computers to identify areas where current algorithms can be improved to enable scaling of PIC codes to 100,000 processors. Strong scaling studies with OSIRIS, VORPAL, and UPIC have recently been conducted on a billion-particle benchmark simulation with a 512x256x512 grid, using an electromagnetic, relativistic plasma model. The UPIC study is directly applicable to QuickPIC, which depends on this framework. Scaling was carried out on an Opteron-based cluster with Infiniband (ATLAS). The number of processors was varied from 128 to 8192 for UPIC, 4096 for OSIRIS, and 1024 for VORPAL keeping the simulation size fixed. UPIC showed good scaling (92%) for the particle part of the calculation (which typically dominates PIC codes). The fast Fourier transform (FFT) field solve in UPIC scaled well up to 4096 processors, then saturated at 8192. In order to enable this code (and other spectral PIC codes) to scale to 100,000 processors, strategies to improve existing algorithms in the particle manager and the FFT were identified, and a mixed MPI/threaded model has been implemented and is being tested. For OSIRIS both the particle push and field solve scaled well up to the maximum used (4096 processors). Top processing speeds of more than 10 billion particles per second (for an entire step including field solve, etc.) were obtained for UPIC. OSIRIS was found to be 30% slower. VORPAL demonstrated 90% efficiency on up to 1024 processors for the same case, and runs on 4096 processors are in progress.

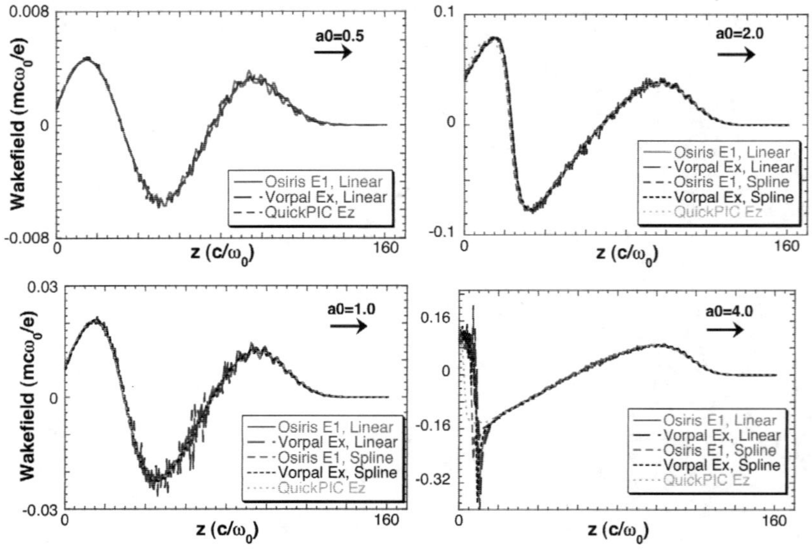

Figure 3. Overlapping lineouts down the center of the simulation domain show good agreement between 3D OSIRIS, VORPAL and QuickPIC simulations of the LWFA benchmark problem for a_0=0.5, 1, 2, and 4, including the use of second-order spline-based particle shapes.

The VORPAL FDTD Maxwell update scaled well to 8192 processors on the Franklin supercomputer at NERSC, demonstrating 95% efficiency with respect to a 4-processor run in weak scaling where the domain was increased to keep equal cells per processor [26]. Tests of I/O scaling on up to 8192 processors also showed that VORPAL's I/O, which uses parallel HDF5, can match that of other optimized applications if the domain sizes are identical on every processor. The restriction appears to come from the HDF5 implementation. For more than 2048 processors, I/O rates begin to decrease, and individual dump files can be used to work around this.

3. Controlled injection, staging, and acceleration

While the energies achieved by LWFAs [8] are sufficient for many radiation source applications [4] and could be staged in series to provide energies needed for high energy physics, improvement of bunch momentum spread and day-to-day accelerator stability are required. Toward this goal, experiments coupled with simulations recently produced stable bunches with longitudinal and transverse momentum spreads an order of magnitude lower than previously observed [9], and show that these bunches can function as injectors to reduce energy spread of high energy LWFAs. The experiments focused a 10 TW laser at the downstream edge of a thin gas jet where density is decreasing, producing bunches with 0.17 (0.02 MeV/c) longitudinal (transverse) momentum spread and with central momenta stable at 0.76±0.02 MeV/c over several days.

Simulation of the experiments [9] using VORPAL explicit PIC showed plasma density gradient control of trapping produced the low energy spread bunches, and indicated that use of such bunches as injectors can greatly improve LWFA bunch quality. In a decreasing gradient, the plasma wavelength λ_p increases, causing the wake fronts to slip behind the laser and decreasing the wake velocity v_ϕ and hence the threshold wake amplitude for trapping and accelerating plasma electrons [27,28]. The simulations showed this modulation of v_ϕ produced trapping without significant modulation of the laser pulse (which is unstable) and at low wake amplitude. It also caused the bunch to quickly outrun the wake structure, producing bunches at ~ 1.5MeV/c, similar to the experiments. The low wake amplitude at trapping also produced low bunch momentum spread, with longitudinal (transverse) momentum spread of 0.2 (0.05 MeV/c), reasonably consistent with the experiments. This low momentum spread, together with the compact bunch diameter of 5μm, indicated a normalized emittance (focusability) of ~0.4 π mm-mrad, approximately an order of magnitude better than other LWFAs. Consistent with observed stability, simulations showed MeV-class bunches over ±10% variation in laser power, plasma density. and plasma length. Simulating focusing through the jet required a large domain, and hence simulations were performed in 2D (with 10^7 cells versus 10^{10} required in 3D), which may explain differences from the experimental energy. Results have now been reproduced by using the VORPAL envelope model, which will enable 3D simulations.

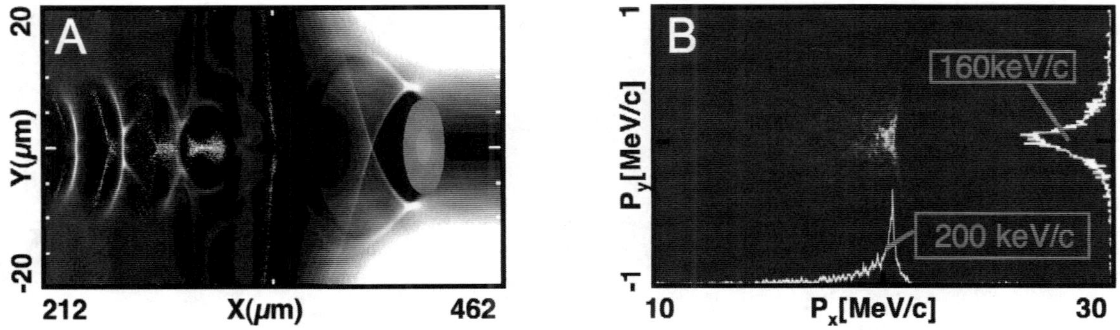

Figure 4. VORPAL simulations merging density gradient injection to an accelerating channel (A) show the laser (red), plasma wake density (grey) and accelerating particles (yellow). Acceleration in the channel (B) produces 20 MeV bunches and preserves the bunch's 0.2 MeV/c momentum spread.

Simulations further showed that merging the ramp into a plasma channel with constant axial density (figure 4) immediately after trapping allows use of these bunches as an injector to improve LWFA bunch quality. This is possible because experiments and simulations show the laser is transmitted through the jet without significant depletion or modulation of the spot, allowing it to drive a wake in the channel. The bunch must also be shorter than λ_p to allow efficient acceleration, and simulations show the trapped bunch meets this condition. The simulated bunch length at the location where the density is correct for emission of THz radiation was benchmarked to THz experimental measurements [29], showing that the simulated length is accurate. This work extended the experimental diagnostics used to compare to simulations to include THz, laser pulse transmission, and bunch transverse momentum in addition to the electron spectrum, increasing the detail and reliability of simulations evaluating injection and postacceleration.

Because the bunch is short compared to the plasma wavelength, it sees a nearly even accelerating field and its momentum spread is nearly preserved as it accelerates in the channel, producing 0.2 MeV/c class momentum spread at high energy. Bunches with 0.2 MeV/c energy spread at energies greater than 20 MeV have so far been demonstrated, limited by computational time with the large domain size. Longer and 3D simulations using the envelope model are in progress to optimize bunch quality, and related simulations [30] indicate this may enable bunches at GeV energies and beyond with < 0.1% energy spread.

Proposed next-generation experiments will use controlled injection coupled with meter-scale plasmas to produce high-quality bunches at ≥10 GeV from a PW class laser. Together with staging, this will be an important step toward high energy physics applications, which could stage multiple 10–100 GeV modules to reach TeV energies. Because meter-scale explicit simulations in multiple dimensions are beyond the capacity of current computers, a combination of Lorentz boosted, envelope, and scaled explicit simulations are being used to detail 10 GeV stage designs (two of which are illustrated in figure 7). Continued simulations of 0.1 to 1 GeV self-trapped stages [15, 31] are validating the codes against additional diagnostics and using this detail to further optimize stage performance. These topics are the subject of upcoming publications; related work was summarized in [15,16] and references therein.

Simulations using QuickPIC have also been used to design high energy, efficient LWFA stages. These designs start from a phenomenological theory [32] that includes the concepts of nonlinear multi-dimensional wake excitation [33], local pump depletion, dephasing, and laser guiding. The simulations show that in this nonlinear blowout regime, a laser can excite a stable wake over distances hundreds of Rayleigh lengths long, as long as its spot size and duration are properly matched: $k_p w_0=\omega_p \tau_L=2(a_0)^{1/2}$. In the simulations a_0 is held fixed at 2, and the plasma density is decreased while the spot size is kept matched. Under these conditions the laser power is equal to the critical power for self-focusing, P_c. A preformed channel is used to keep the leading edge of the laser guided. Stages that provide an average gradient 3.6 GV/m (7.2 GV/m) with a final energy of 100 GeV (25 GeV) were demonstrated (figure 5).

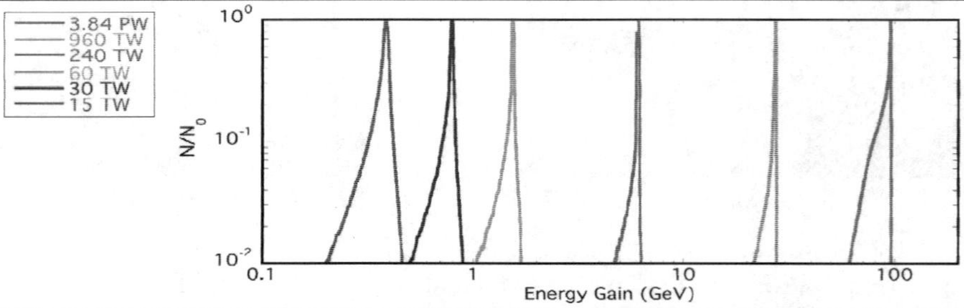

Figure 5. Final electron spectra for a series of LWFA simulations in the nonlinear blowout regime for P/Pc=1. In each case a trailing bunch of electrons was injected. The energy gain is shown, where each color corresponds to a different laser power. Energy gain agreed with scalings (QuickPIC).

4. Optimized PWFA design and positron sources

The theory of Lu et al. [33] can allow one to accurately determine the optimum bunch shapes for a particle driver or the load [34] for either the PWFA or LWFA. As an example, we have conducted QuickPIC simulations of idealized beams for a PWFA stage in which the energy of a 25 GeV beam is doubled to 50 GeV. One could imagine stringing 20 of these stages together to produce a 500 GeV electron beam. The parameters are also relevant to possible two-beam PWFA experiments at SLAC. Based on beam-loading theory [34] in the nonlinear blow-out regime, ideal parameters for the drive beam and the witness beam were obtained, and energy doubling of a 25 GeV beam with energy spread within 1% was demonstrated in the simulation (figure 6). In the simulation, we use 512×512×256 grids with box size of 600×600×270 microns. Both the drive beam and the trailing beam have initial energy of 25 GeV and spot size of 3 microns. After propagating about 0.7 meters in the plasma, the trailing beam has gained approximately 25 GeV in energy. We have also conducted higher resolution (with 2048×2048×256 grids) simulations with and without ion motion and synchrotron radiation loss to investigate their effects on beam acceleration and propagation.

Developing a wakefield-based collider requires schemes for accelerating positrons as well as electrons. One solution is to use a nearly linear wake; in this regime dynamics are similar for electrons and positrons. The nonlinear regime offers higher fields, but there are then significant differences between electrons and positrons. It is found that the positron beam wake is weaker than the electron beam wake motivating schemes to accelerate a positron beam in an electron beam or laser driven wake. Generating a positron beam behind an electron beam is, however, challenging. Simulations of a novel idea to do this were carried out with the code OSIRIS [35]. In order to create a properly phased positron beam, a double-pulse electron beam is collided with a thin foil target embedded within the plasma. This creates a double-pulse positron beam of lesser charge overlapping the electron beam through bremsstrahlung. The first electron pulse drives a weakly nonlinear wake (bubble), and the positrons overlapping this pulse are expelled because they reside in a wake region that focuses electrons but defocuses positrons. Nonlinear wakes have restricted positron focusing phase, but in the weakly nonlinear regime there is a phase of the wake just behind the bubble that has both accelerating and focusing fields for positrons (defocusing for electrons). Placing the second bunch here, the trailing group of electrons will be defocused while the newly created positrons will remain focused and accelerated. Simulations of a possible two-bunch experiment at SLAC show that a large number of positrons (1.2×10^{7}) are injected and accelerated to 6.2 GeV and with relatively narrow energy spread (15%) after 99 cm of plasma.

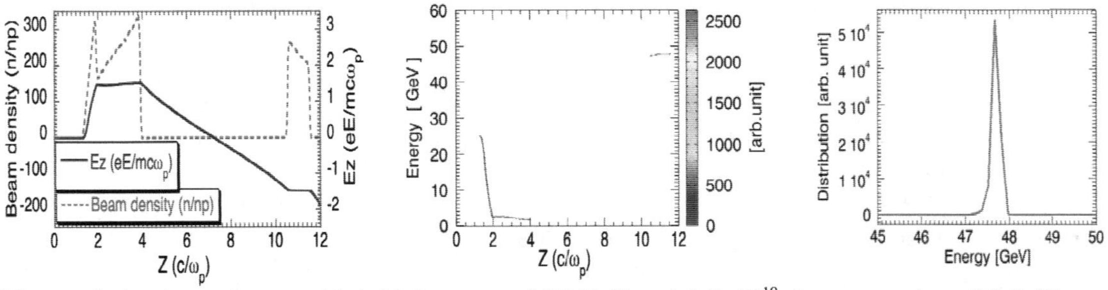

Figure 6. A witness beam with initial energy of 25 GeV and 1.7×10^{10} electrons gains ~25 GeV energy with 0.46% energy spread in a PWFA simulation. The plasma is preionized. Left: the density profile of the drive and witness beam and the longitudinal wake they generated. Middle: Phase space plot of the drive and witness beam, showing energy doubling of the witness beam and depletion of the energy of the drive beam. Right: Energy distribution of the accelerated witness beam (QuickPIC).

5. Boosted frame and PIC kinetic model development

While it is common to model WFAs in the laboratory frame, under certain conditions orders of magnitude speedup can be achieved in modeling relativistic systems by working in a boosted frame [22]. Because of relativistic length contraction and time dilation, the separation of scales between the laser wavelength and the plasma length (which drives the number of timesteps required) is not invariant, and the total number of computer operations can be greatly reduced by choosing a frame for the calculation that is moving at an optimal relativistic velocity.

The boosted frame method is being benchmarked against simulations in the laboratory frame. For example, two-dimensional simulations of coherent spontaneous emission from prebunched beams in linearly polarized undulators agree with standard theory and showed speedups of ~5 orders of magnitude (for γ~250) over standard EM PIC calculations [36]. Standard PIC or other algorithms can be applied in the boosted frame (with corrections as below), so that the fast oscillation is still resolved unlike other reduced models. Compared to standard eikonal and wiggler-period averaging approximations, the boosted frame undulator calculation recovers the "backward wave" emission and also sideband development in a harmonic cascade.

Three-dimensional simulations of electron cloud effects in high-energy physics accelerators reproduced lab frame results with a speedup of 10^{3} but showed the need for new numerical techniques. A new particle pusher was required [37] because errors appeared in the standard Boris push for particles in the boosted frame. The new push may benefit modeling of WFAs also if the improvement is not within EM solve accuracy. The high grid resolution required by standard electromagnetic solvers to accurately model laser group velocity may be aggravated when performing the calculation in the boosted frame due to numerical Cerenkov [38] acting on the plasma traveling near the speed of light. Solvers that are dispersion-free in vacuum [39] along the grid axes for a "magic" time step are being explored [40, 41], but this approach requires mitigation of an even-odd oscillation and potential instability that can develop [40]. Back-scattered radiation is also unresolved in the boosted frame, which is important for many applications.

The Lorentz method has been implemented in all dimensions (1D, 2D, and 3D) in OSIRIS 2.0 [19] and in 1D and 2D in VORPAL [42], dramatically reducing computational resources required to model LWFAs (typically by~γ^2, with γ being that of the boosted frame). While the algorithm needs no change, running in this frame poses particular difficulties. In addition to backscatter and electromagnetic wave considerations, it is important to eliminate round-off errors that can accumulate from the current of the relativistically moving background electrons and ions.

Simulation of laser plasma acceleration setups were reported in one, two, and three dimensions with speedups of respectively x1,500 [42], x150, and x75 [43]. A 1.5 GeV self-injected beam, already fully studied in the laboratory frame [32], was simulated in OSIRIS, and the final bunch energy (1.5 GeV) and injected charge (~0.5 nC) were in reasonable agreement with the laboratory results and theoretical predictions (figure 7A). The simulation was performed with a gamma of 5, giving a total speedup of ~20 times relative to the standard laboratory simulation. A 10 GeV LWFA stage (0.4 m long at a_0=1.6 and n_e=2x10^{17} cm^{-3}) was simulated in VORPAL using lab and boosted frames. Lab frame simulations required ~3x10^7 steps and 5,000 processor hours, and required special techniques (current smoothing, second-order particle shapes) to prevent artificial injection of plasma electrons (i.e. dark current). Boosted frame simulations used γ=27 and obtained a speedup factor of ~1,500 with no artificial dark current problems. Figure 7B shows the boosted-frame longitudinal phase space of test electrons, externally injected with a wide range of phases to sample the wakefields. Energy gain in the lab frame was ~8 GeV, in good agreement with lab frame simulations.

While the above methods extend simulation of high energy stages, improved kinetic accuracy is equally important to accurate accelerator design. Studies of the PIC algorithm in modeling of LWFAs [24] using a local charge conserving deposition code [44] showed that use of high-order spline interpolation for force and current deposition, coupled with smoothing of the current on the grid,

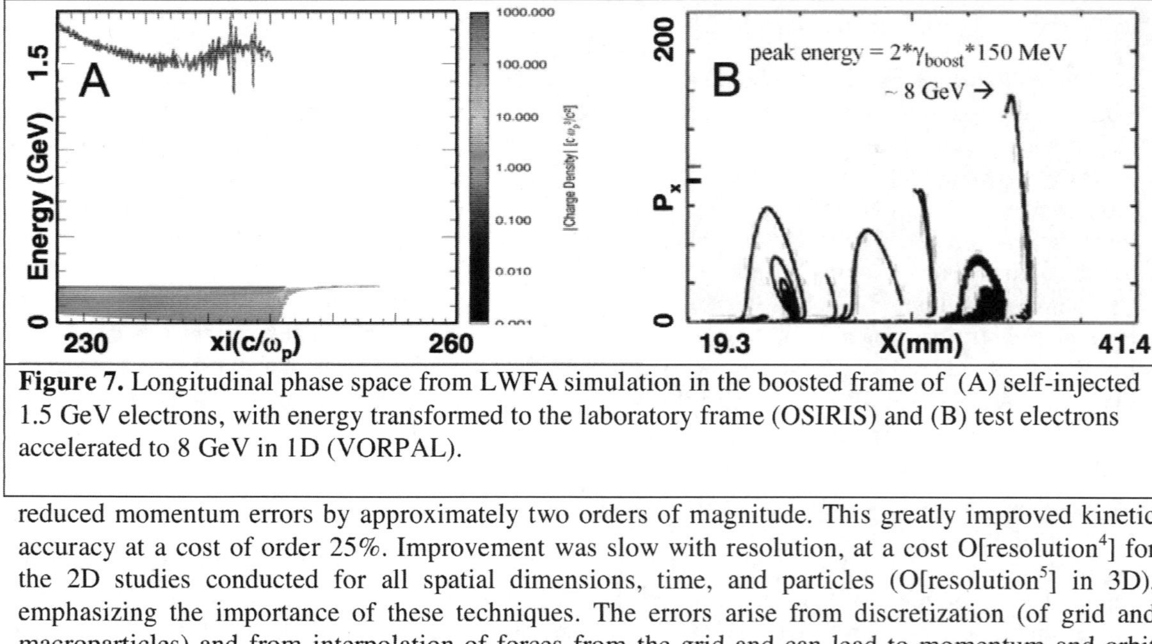

Figure 7. Longitudinal phase space from LWFA simulation in the boosted frame of (A) self-injected 1.5 GeV electrons, with energy transformed to the laboratory frame (OSIRIS) and (B) test electrons accelerated to 8 GeV in 1D (VORPAL).

reduced momentum errors by approximately two orders of magnitude. This greatly improved kinetic accuracy at a cost of order 25%. Improvement was slow with resolution, at a cost O[$resolution^4$] for the 2D studies conducted for all spatial dimensions, time, and particles (O[$resolution^5$] in 3D), emphasizing the importance of these techniques. The errors arise from discretization (of grid and macroparticles) and from interpolation of forces from the grid and can lead to momentum and orbit displacements that result in unphysical trapping, especially important in modeling of dark current free structures for future experiments. Suppression of unphysical trapping for the first few periods of the wake (which are typically modeled) was demonstrated. These methods are implemented in the codes VORPAL and OSIRIS.

6. Conclusions

Large-scale one-to-one simulations, benchmarked to experiments, revealed the physics of formation of narrow energy spread bunches from self-trapping at 0.1 GeV and 1 GeV. They are now providing quantitative understanding for the design of new accelerators. Simulations are being used to develop controlled injection for increased beam quality, accelerator stages for ≥ 10 GeV energies, positron sources, and staging of accelerators to reach high energies. Code development will continue to improve and verify reduced and boosted frame models and to improve the kinetic accuracy of the codes. Inclusion of additional experimental diagnostics will further constrain the codes to increase the detail of information available for accelerator optimization. The new models developed will be used together with scaling to tens of thousands of processors to simulate controlled injection, staging, and meter-scale 10 GeV class experiments with high accuracy and to design next-generation accelerators to push the energy frontier in high energy physics and to develop new light sources.

Acknowledgments
Work supported by the U.S. Department of Energy (DOE) Office of Science, Office of High Energy Physics SciDAC program and under grants including LBL DE-AC02-05CH11231, DE-FG03-95ER40926, DE-FG02-01ER41178, DE-FG02-03ER83857, USC DE-FG03-92ER40745 UCLA DE-FG03-92ER40727, DE-FG02-07ER41500, DE-FG02-03ER54721, TX DE-FC02-07ER41499, DE-FG02-07ER84834, DE-FG02-04ER84097, DE-FG02-04ER41317. Support also comes from the U.S.-LHC Accelerator Research Program (LARP), Office of Nuclear Physics grant DE-FG02-05ER84173, DARPA, from NSF grants 0113907 and 0614001, and from FCT Portugal. This work used the resources of the National Energy Research Scientific Computing Center, which is supported by the DOE Office of Science under Contract No. DE-AC02-05CH11231 as well as the DAWSON Cluster under NSF Phy-0321345 and the LLNL ATLAS facility.

Notes
E Esarey is also at the University of Nevada, Reno; G R Plateau is also at the University of California, Berkeley; W P Leemans is also at Ecole Polytechnique, France.

References

[1] ILC 2008 www.linearcollider.org/cms ; LCLS- www-ssrl.slac.stanford.edu/lcls/
[2] Tajima T and Dawson J M 1979 *Phys. Rev. Lett.* **43:**267
[3] Esarey E, *et al.* 1996 *IEEE Trans. Plasma Sci.* **24:**252
[4] Leemans W P, *et al.* 2005 *IEEE Trans. Plasma Sci.* **33:**8
[5] Geddes C G R, *et al.* 2004 *Nature* **431:**538
[6] Mangles S P D, *et al.* 2004 *Nature* **431:**535
[7] Faure J, *et al.* 2004 *Nature* **431**:541
[8] Leemans W P, *et al.* 2006 *Nature Physics* **2:**696
[9] Geddes C G R, *et al. Phys. Rev. Lett.* **100:**215004.
[10] Faure J, *et al.* 2006 *Nature* **444:**737
[11] Esarey E, *et al.* 1997 *Phys. Rev. Lett.* **79:**2682
[12] Blumenfeld I, *et al.* 2007 *Nature* **445:**741
[13] Pukhov A and Meyer-ter-Vehn J 2002 *Appl. Phys. B* **74:**355
[14] Tsung F S, *et al.* 2004 *Phys. Rev. Lett.* **93:**185002
[15] Geddes C G R, *et al.* 2007 *J. Phys. Conf. Ser.* **78:**12021
[16] Tsung F S, *et al.* 2007 *J. Phys. Conf. Ser.* **78:**12009
[17] Nieter C and Cary J 2004 *J. Comp. Phys.* **196:**448
[18] Hemker R G, *et al.* 1999 *Proc. 1999 PAC Conference*
[19] Fonscea R A, *et al.* 2002 *Lect. Notes Comp. Science* **2329** III-34 (Springer-Verlag)
[20] Mora P, *et al.* 1997 *Phys. Plasmas* **4:**217
[21] Huang C, *et al.* 2006 *J. Comp. Phys.* **217:**658-679
[22] Vay J-L 2007 *Phys. Rev. Lett.* **98:**130405
[23] Grote D P, Friedman A, Vay J-L and Haber I 2005 *AIP Conf. Proc.* **749:**55
[24] Cormier-Michel E, *et al.* 2008 *PRE* (accepted)

[25] Spentzouris P, *et al.* 2008 *Proc. SciDAC 2008* (also compass.fnal.gov)
[26] Cary J, *et al.* 2008 *Proc. SciDAC 2008*
[27 Katsouleas T 1986 *Phys. Rev. A.*, **33**:2056
[28] Bulanov S, *et al.* 1998 *Phys. Rev. E* **58:**R5257
[29] Plateau G, *et al.* 2007 *Proc. Particle Accel. Conf.* (IEEE: Piscataway, NJ)
[30] Andreev N E, *et al.* 2000 *Phys. Rev. ST-AB* **3:**021301
[31] Gonsalves A J, *et al.* 2007 Proc. Particle Accel. Conf. (IEEE: Piscataway, NJ)
[32] Lu W, *et al.* 2007 *Phys. Rev. ST Accel. Beams* **10:**061301 (2007)
[33] Lu W, *et al.* 2006 *Phys. Rev. Lett.* **96:**0165002; Lu W, *et al.* 2006 *Phys. Plasmas.* **13:**056709
[34] Tzoufras M, *et al.* 2008 (submitted for publication); Mtzoufras, Ph.D. UCLA 2008.
[35] Wang X, *et al.* 2008 (submitted for publication)
[36] Fawley W M, *et al.* 2008 (in preparation)
[37] J.-L. Vay, 2008 *Phys. Plasma* 15, 56701 (2008).
[38] review: Greenwood *et al. J. Comp. Phys.* 201, 665 (2004).
[39] Karkkainen M, *et al.* 2006 *Proc. ICAP* Chamonix, France (2006)
[40] Vay J-L 2008 (unpublished)
[41] Cowan B 2008 (unpublished)
[42] Bruhwiler D L 2007 (unpublished)
[43] Martins S F, *et al.* 2007 Mtg. Div. of Plasma Phys. (Orlando, FL)
[44] Ruhl H 2000 Habilitationsschrift (Technische Universitt Darmstadt)

Design and optimization of large accelerator systems through high-fidelity electromagnetic simulations

Cho Ng[1], Volkan Akcelik[1], Arno Candel[1], Sheng Chen[1], Lixin Ge[1], Andreas Kabel[1], Lie-Quan Lee[1], Zenghai Li[1], Ernesto Prudencio[1], Greg Schussman[1], Ravi Uplenchwar[1], Liling Xiao[1], Kwok Ko[1], T. Austin[2], J R Cary[2], S Ovtchinnikov[2], D N Smith[2], G R Werner[2] and L Bellantoni[3]

[1]Stanford Linear Accelerator Center, Menlo Park, CA 94025, USA
[2]TechX Corporation, Boulder, CO 80303, USA
[3]Fermi National Accelerator Laboratory, Batavia, IL 60510-5011, USA

SciDAC CETs/Institutes Collaborators: E Ng, X Li, C Yang (LBNL), L Dianchin (LLNL), K Devine, E Boman (SNL), B Osting, D Keyes (Columbia U.), X Luo, M Shephard (RPI), R Barrett, S Hodson, R Kendall (ORNL), W Gropp (UIUC), Z Bai and K-L Ma (UC Davis)

E-mail: cho@slac.stanford.edu

Abstract. SciDAC-1, with its support for the "Advanced Computing for 21st Century Accelerator Science and Technology" project, witnessed dramatic advances in electromagnetic (EM) simulations for the design and optimization of important accelerators across the Office of Science. In SciDAC2, EM simulations continue to play an important role in the "Community Petascale Project for Accelerator Science and Simulation" (ComPASS), through close collaborations with SciDAC Centers and Insitutes in computational science. Existing codes will be improved and new multi-physics tools will be developed to model large accelerator systems with unprecedented realism and high accuracy using computing resources at petascale. These tools aim at targeting the most challenging problems facing the ComPASS project. Supported by advances in computational science research, they have been successfully applied to the International Linear Collider and the Large Hadron Collider in high energy physics, the JLab 12-GeV Upgrade in nuclear physics, and the Spallation Neutron Source and the Linac Coherent Light Source in basic energy sciences.

1. Introduction

Particle accelerators are important experimental facilities in DOE for discoveries in physical sciences. In the high energy frontier, particle colliders such as the proposed International Linear Collider (ILC) will be used to address many of the most compelling questions about dark matter, dark energy, and the fundamental nature of matter, energy, space, and time. In basic energy sciences research, high-intensity particle beams or high-brightness light sources generated by particle accelerators will probe structures of matter that will benefit science discoveries in biology, chemistry, medical sciences, material sciences and others. Examples of these machines are the Spallation Neutron Source (SNS) and the Linac Coherent Light Source (LCLS). Central to these billion-dollar-class, state-of-the-art accelerators are accelerator structures used to accelerate particle beams to high energies. With the support of SciDAC, advanced electromagnetic simulation tools have been developed for the design and optimization of accelerator structures [1], with the objectives of achieving improved performance, increased reliability and reduced cost of accelerators.

In SciDAC-1, EM modeling [2] in the Advanced Computing for 21st Century Accelerator Science and Technology" (AST) project made dramatic advances in the development of parallel, high-fidelity, and high-accuracy finite-element codes at the Stanford Linear Accelerator Center (SLAC). The application of these codes has tremendous impact on the design, optimization, and analysis of accelerator projects within the Office of Science. For example, AST EM modeling tools were used to simulate beam heating in the PEP-II (a High Energy Physics, HEP, accelerator) interaction region (IR), and the analysis led to a redesign of the IR, resulting in higher luminosity that enabled the discovery of a new elementary particle. The successes of large-scale EM modeling would not have been possible without the computational expertise rendered by ISICs/SAPP partners, especially in the areas of unstructured mesh generation, adaptive mesh refinements, eigensolvers, parallel partition schemes, and visualization [3]. For instance, the finite-element eigensolver Omega3P saw a thousand times speedup through collaborations with Terascale Optimal PDE Simulation (TOPS) center on novel parallel algorithms in eigensolvers and linear solvers.

Following the successes of EM modeling in AST, the ComPASS [4] EM team will continue to strive for advances in accelerator simulation techniques and capabilities that will take advantage of petascale computer hardware in SciDAC-2. The finite-element tools using unstructured grids developed in SciDAC-1 will be augmented to include self-consistent particle-wave interactions and multiphysics analysis that integrates EM, thermal and mechanical effects for accelerator cavity design. In particular, the multiphysics analysis aims at engineering prototyping of accelerator components on computers with unprecedented realism and resolution through petascale modeling, resulting in a fast turnaround time that will revolutionize the current design cycle. The ComPASS EM efforts will also be augmented by the finite-difference time-domain (FDTD) method using a structured mesh approach, which allows for conformal, embedded boundaries for modeling curved surfaces. Collaborations with SciDAC Centers and Institutes in computational science will continue in existing areas and new areas (such as shape determination) to develop scalable numerical algorithms and advanced numerical analysis techniques to facilitate large-scale simulations on petascale computers. It is envisioned that ComPASS EM tools will tackle the most computationally challenging problems in accelerators across the Office of Science, as can be gleaned from the very first applications of ComPASS described in following sections.

2. Parallel electromagnetic codes for high-fidelity simulations

The parallel EM codes [2, 3] developed in the AST project at SLAC were based on the finite-element method using unstructured grids to conform to complicated geometry and high-order basis functions to achieve high solution accuracy, more than that obtained by precise machining of cavities. The suite of finite-element codes include the following:

- Omega3P – Parallel eigensolver for finding normal modes and their damping in accelerating cavities
- S3P – Parallel S-matrix calculation for computing transmission properties of accelerator structures
- T3P – Parallel time-domain solver with beam excitation for wakefield calculations
- Track3P – Parallel particle tracking module for multipacting and dark current simulations
- V3D – Graphic tool for visualizing meshes, fields and particles with unstructured grids

In addition to improving the computational capabilities of these codes, two new codes for treatments of different space-charge effects and a novel multi-physics code with its relevant solvers all based on the finite-element method will be developed in ComPASS. They are

- Gun3P – Parallel particle code for space charge beam optics for electron gun simulations
- Pic3P – Parallel particle-in-cell code for radio-frequency (rf) gun and klystron simulations
- TEM3P – Parallel multiphysics analysis tool including integrated EM, thermal and mechanical effects for cavity design

For the FDTD approach, VORPAL, developed by TechX, uses the Dey-Mittra algorithm [5] for accurate incorporation of boundaries. This is an *embedded boundary* algorithm, which uses a structured mesh throughout most of the simulated region but modifies the cells near the boundary only to obtain improved accuracy. To extract frequencies, we developed an efficient method [6] that combines targeted excitation with filter diagonalization. This allows one to reduce the eigenvalue problem to a small subspace, thus eliminating the need to diagonalize large matrices. Combined with massive parallelism, this provides the ability to cover the scales of cavities from the full size of a multicell cavity down to the tiny holes through which diagnostics are inserted.

3. Collaborations in computational science with SciDAC centers and institutes

Strong collaborations have been established between ComPASS and SciDAC CETs/Institutes. The collaborations include the following R&D activities.

- *Shape Determination & Optimization* (with ITAPS and TOPS) – Obtain cavity deformation from measured mode data by solving a weighted least squares minimization problem
- *Parallel Complex Nonlinear Eigensolver and Linear Solver* (with TOPS) – Develop scalable algorithms for solving large, complex, nonlinear eigenvalue problems to find mode damping for accelerator systems, and for solving highly indefinite systems arising from accelerator modeling
- *Adaptive Mesh Refinement and Meshing Service* [7] (with ITAPS) – Optimize computing resources and increase solution accuracy through adaptive mesh refinement
- *Dynamic Load Balancing* (with CSCAPES [8] and ITAPS) – Implement dynamic partitioning scheme to optimize computational load for particle-field simulations
- *Parallel and Interactive Visualization* (with ISUV) – Visualize electromagnetic fields and particles with large, complex geometries and large aspect ratios
- *Embedded Boundary Methods* (with ITAPS) – Extend the integration of embedded boundary methods with particle-in-cell codes using structured grids

All but the last collaboration are applicable to the finite-element method using unstructured grids. Details on these efforts are described in these proceedings in a separate paper [9]. The importance of the collaborative work will also be illustrated in the following EM applications.

4. Application to accelerator projects in high energy physics

This section describes four applications.

4.1 Simulation of the ILC cryomodule

The determination of wakefield and higher-order-mode (HOM) effects is of utmost importance in the design of the ILC linacs. Previous calculations had been carried out for a single superconducting cavity where the spectrum and the damping factors of HOMs were obtained using the parallel finite-element eigensolver Omega3P [1]. The calculations assumed that the frequencies of the HOMs are below the cutoff frequency of the beampipe. For HOMs above the cutoff, the fields may cover regions more than a single cavity, and therefore calculations involving multiple cavities are required to give a correct account of wakefield effects of these modes. To determine trapped modes and their effects on the wakefield, we modeled the ILC cryomodule, which consists of eight TTF cavities.

In the frequency domain, we used Omega3P to calculate the trapped modes in the cryomodule for the third dipole band, for mode frequencies above the beampipe cutoff frequency. Sixteen dipole modes with high shunt impedances were evaluated and the calculated damping factors agreed reasonably well with the data measured for cryomodules at DESY. This was the first-ever calculation of trapped modes in the cryomodule [10]. Each mode took about one hour with 1500 processors on the Seaborg computer at NERSC. The calculations were facilitated by advances in parallel linear solvers (with TOPS) so the linear systems resulting from the large computational size of the cryomodule could be solved efficiently on the NERSC machine. A new multi-mode waveguide boundary condition was also implemented in Omega3P to properly terminate the modes at the two ends of the cryomodule. The Maxwell system becomes a nonlinear complex eigenvalue problem, which may be solved with a self-consistent iterative algorithm.

In the time domain, we used the finite-element code T3P to drive a Gaussian bunch through the cryomodule. A snapshot of the beam transit in the cryomodule using the visualization tool V3D is shown in figure 1. The Fourier transform of a monitored signal at the beampipe between two cavities shows a sharp peak close to the monopole cutoff frequency of the beampipe. To further determine the nature of this peak, we used Omega3P to solve for modes at frequencies around the peak, and a localized mode was found in the coupler region between two cavities (see figure 2). The shunt impedance and damping factor of this mode indicate that the heat load it generates on the beampipe wall is relatively small and should not be a concern for the ILC linacs.

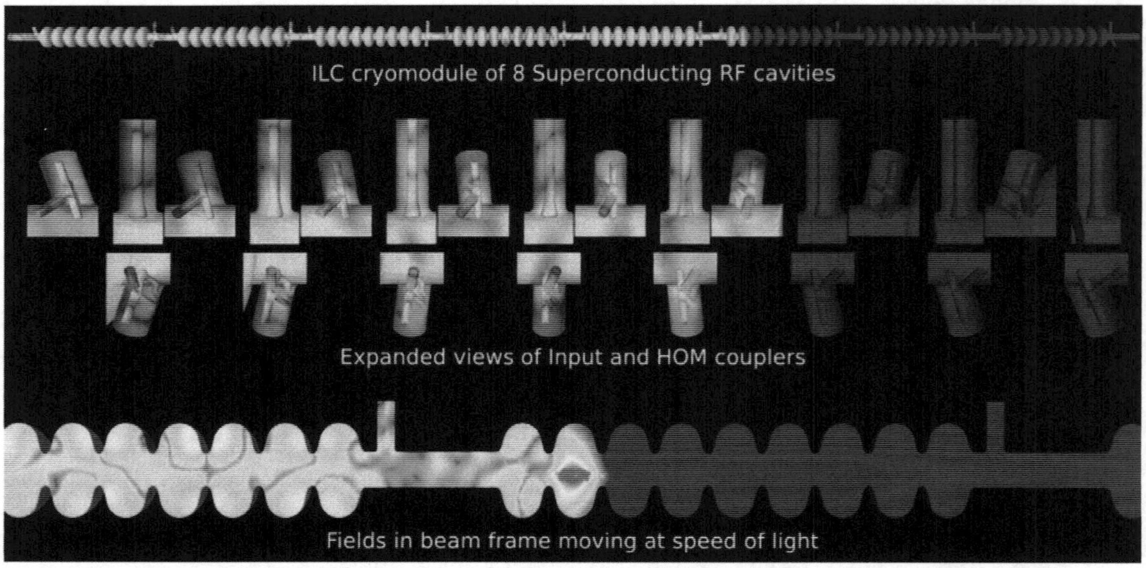

Figure 1. A T3P snapshot of wakefield generated by an electron bunch traversing ILC cryomodule.

Figure 2. A trapped mode in the region between two cavities of the ILC cryomodule using Omega3P.

4.2 Multipacting in ILC TTF-III coupler

The TTF-III power coupler [11] adopted for the ILC baseline cavity design tends to have a long initial high power processing time. The long processing time is believed to result from multipacting in various regions of the coupler. Multipacting is a phenomenon arising from electrons with resonant trajectories hitting a cavity surface at same locations leading to an avalanche of electron currents that will hinder the cavity performance. To understand performance limitations during high power processing and identify problematic regions, SLAC built a flexible high-power coupler test stand to test individual sections of the coupler, including the cold and warm coaxes, the cold and warm bellows, and the cold window. To provide insights for the high power test, detailed numerical simulations of multipacting for these sections were performed using Track3P. Multipacting bands were identified by scanning power levels up to 2 MW. For the cold coax, the simulated multipacting bands agreed very well with measurements.

Multipacting activities were found in the taper region, but they could be suppressed by applying an axial magnetic field above 300 Gauss. Multipacting activity was not apparent at the bellows [10].

4.3 Impedance calculations for the ILC damping rings

The goals of the impedance calculation for the ILC damping rings are to obtain the impedance budget for the vacuum chamber components and to provide a pseudo-Green's function wakefield for beam stability studies. The evaluation of the pseudo-Green's function wakefield necessitates the accurate determination of the short-range wakefield using a bunch length much smaller than the normal bunch length. As the bunch length gets smaller, the computational size becomes larger because the mesh requires refinement to handle the high frequency content of the short bunch.

The conceptual designs of the ILC damping vacuum chamber components were obtained by scaling those in existing machines such as the PEP-II. We used the parallel time-domain code T3P to calculate the pseudo-Green functions of the major vacuum chamber components such as the rf cavity and the beam position monitor (BPM) using a 1 mm bunch (compared with the nominal bunch length of 6 mm). The calculations were carried out on the Bassi computer at NERSC [12].

It is desirable to obtain the pseudo-Green's function wakefield for even smaller bunch lengths for more accurate beam stability studies. However, the computational requirements would become prohibitively high. The difficulty can be alleviated by introducing a computational window moving along with the bunch because the short-range wakefield receives contributions only from fields in the longitudinal region covered by the bunch. The two proposed approaches are to have a moving window with p-refinement and a moving window with h-refinement. In the p-refinement method, high-order finite elements are used in the moving window to improve the solution accuracy, and the region outside the window is represented by lowest order finite elements. This dramatically reduces the problem size and preliminary studies have shown that CPU time can be reduced by an order of magnitude, compared to using basis functions of uniform order. In the h-refinement method, the moving window has a much more refined grid than the region outside the window. Collaboration with RPI (ITAPS) is in progress for moving window with h-refinement to obtain high-quality meshes for computation.

4.4 ILC crab cavity

One of the tasks of the past year has been to carry out verification and validation studies. The first such (verification) study was to recompute the frequency spectrum of the "crab cavity" to be used for the final beam delivery system to enhance the luminosity at the interaction point of the proposed International Linear Collider. Our results differed from previous (obtained with the MAFIA code) by 3 MHz out of 3.9 GHz. This 0.1% error in most fields of physics would not be significant, but cavity frequencies can be measured to parts in 10^5, so it was worth understanding.

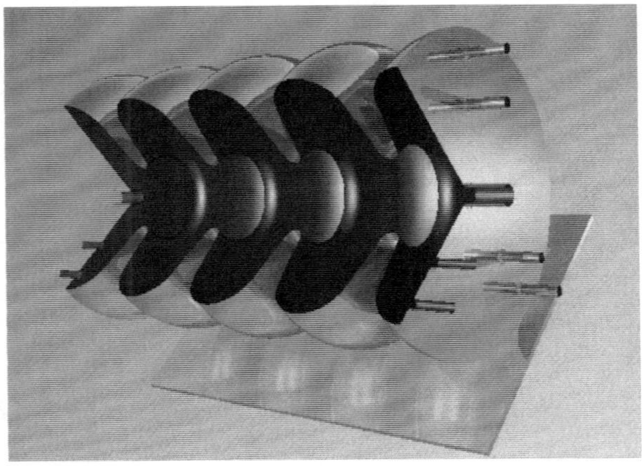

Figure 3. Visualization of the A15 cavity experiment as represented within VORPAL.

Experiments could not act as arbiter in this case, because the superconducting cavities in question cannot be manufactured to sufficiently high accuracy. Instead, we developed computations for the precisely machined A15 cavity experiment shown in figure 3. This had long been a curiosity because previous calculations had differed from experimental results by 5 MHz. After extensive checking, including modeling all of the holes in the cavity, our results also disagreed, but by 2 MHz, which was still unacceptable. Ultimately, we carried out computations for differing values of cavity dimensions and determined that a reduction of the equator radius by 25 mm could account for the observed frequency shift. This led to an as-is measurement of the manufactured cavities, and, indeed, the radius of the equator was found to be smaller by 25 μm, as indicated by the computations. Thus we can say that at this point our cavity calculations have been both verified and validated to within a part in 10^4.

5. Application to accelerator projects in nuclear physics

We focus here on one application: high-Q resonances in JLab high gradient prototype cryomodule [13]. The *Renascence* cryomodule installed in CEBAF in 2007 consists of eight superconducting cavities. Beam-based instability studies have shown a significant beam breakup (BBU) threshold current well below the designed value. Measurements of higher-order modes (HOM) indicate that three of the modes in cavity #5 have external Qs two orders of magnitude higher than similar modes in other cavities (figure 4). It is important to understand the cause of this abnormal behavior so that effective QC measures can be implemented to avoid the BBU problem in the final upgrade design and manufacture.

The abnormality exhibited by cavity #5 could arise from shape deviations of the real cavity from the designed one. However, deviations are unknown in the final cavity installation because of the complicated process of assembly and tuning. The ability to reconstruct the deformed cavity from measured data will provide valuable information for cavity design and performance. Thanks to the support of the Office of Advanced Scientific Computing Research and the collaboration with TOPS scientists, SLAC physicists and computational scientists have developed a shape determination tool to determine the unknown shape deviations using measured cavity rf parameters such as mode frequencies and external Qs by solving an inverse problem [14]. The algorithm is based on the least squares minimization, whose goal is to find the best shape that minimizes the least squares misfit between the measured and simulated models. The shape deviations in the simulated model are parameterized using predefined geometry parameters, and the inversion of the problem quantifies these parameters to determine the reconstructed, deformed shape.

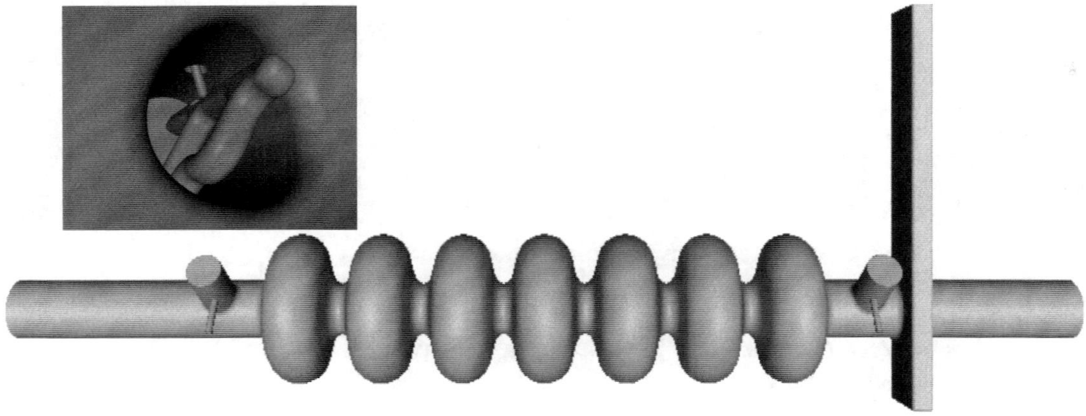

Figure 4. The 7-cell high-gradient cavity model with details of HOM coupler (insert).

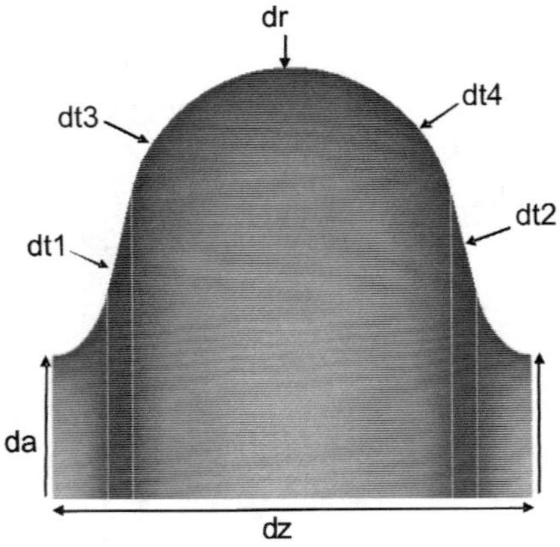

Figure 5. Distortion parameters used in shape determination.

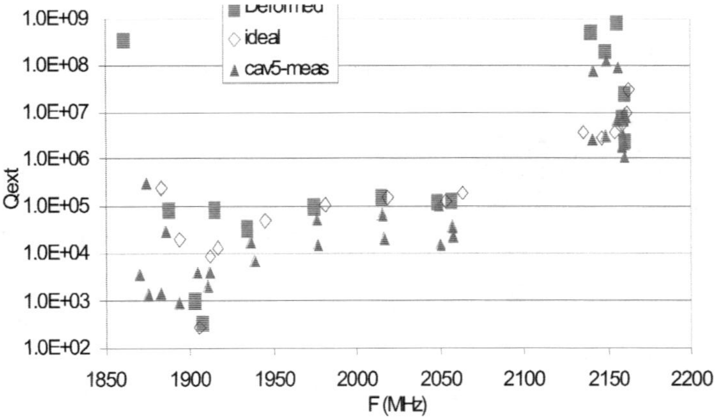

Figure 6. Q_{ext} for ideal and deformed cavities using Omega3P.

The first step of the shape determination is to parameterize each cavity cell with predefined distortion parameters as shown in figure 5. The subsequent goal is to determine the deformed cavity shape by variations of these distortion parameters to achieve the optimal fit to measured cavity rf parameters, which include the 7 monopole mode frequencies, 12 dipole mode frequencies, and 6 external Q (Q_{ext}) values. Figure 6 shows the Q_{ext}'s of dipole modes of the deformed cavity after applying the shape determination tool. Also plotted are the measured data for cavity #5, and those for the ideal designed cavity using Omega3P. The deformed shape not only reproduces the 3 high Q_{ext} modes ($Q_{ext} \sim 10^8$ and frequency ~ 2160 MHz) in the second dipole band, but also the low Q_{ext} modes for the remaining modes in the second dipole band and for the modes in the first band (~ 1850-2050 MHz). The dominant cavity distortion is found to be the distortion in the cell length (dz), and some cells are shorter by 2–3 mm. The total length of the deformed 7-cell cavity is 8.2 mm shorter than that of the ideal design, which is in good agreement with the QC data of cavity #5. Inspections of the field patterns of these three high-Q modes of the deformed cavity show that their field patterns are substantially different from the corresponding modes of the ideal cavity. The reduced couplings of these high-Q modes in the deformed cavity decrease their damping effects, which then leads to the BBU problem. The cell shapes of the deformed cavity from the ideal one is illustrated in figure 7.

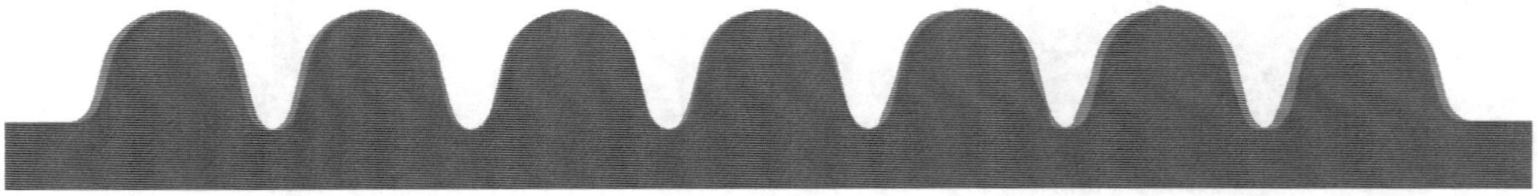

Figure 7. Overlay of distorted cavity (in pink color) on ideal cavity (in green color) (common region in orange color).

6. Application to accelerator projects in basic energy sciences

This section describes a multipacting simulation for HOM coupler in SNS superconducting cavity [15]. During the commissioning and operation of the SNS superconducting linac, abnormal signals were observed at the HOM coupler. These anomalies are believed to be due to the field emission and multipacting electron loadings in the HOM coupler. To understand these issues, the field enhancement and multipacting were simulated using Omega3P and Track3P for the SNS beta=0.81 cavity. The high field gradient in the upstream HOM coupler could cause field emission and multipacting (MP). Multipacting analysis using Track3P shows two multipacting bands in the gradient range up to 20 MV/m. One of the multipacting bands exists in the gap between the enlarged loop head and the cylindrical side wall of the coupler at field gradient levels from 2.8 to 10 MV/m as shown on the left of figure 8. The other MP band exists in the gap between the hook part of the loop and the cylindrical side wall of the coupler as shown on the right of figure 8. No MP activities are found in the notch gap. These MP bands are in good agreement with the experimental observations.

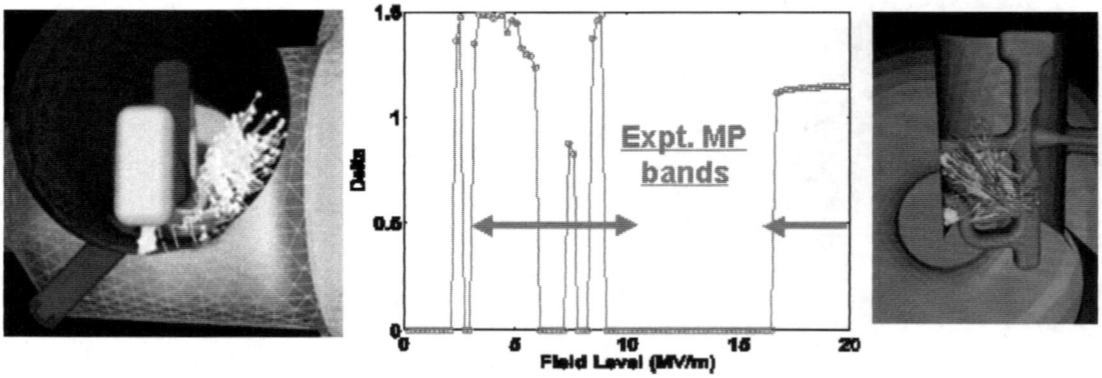

Figure 8. Multipacting activities in the SNS HOM coupler from Track3P simulation.

7. Multiphysics analysis for engineering prototyping

In the design of an accelerator cavity, thermal and mechanical considerations are as important as EM effects. Many failures in accelerator operations are due to heating effects arising from high power or high current operations. At present, the EM, thermal and mechanical calculations are carried out separately with different codes using different meshes. It is highly desirable in the accelerator community to build a modeling package that integrates all these effects. Under ComPASS, a new thermal and a new mechanical solver will be developed and integrated into existing EM codes. Traditionally, thermal and mechanical analysis tools use the finite-element method, and hence it is natural to integrate them into the ComPASS finite-element EM suite. The multiphysics analysis code TEM3P will enable all these calculations to be done within a unified framework using a single model and will provide a complete toolset for engineering prototyping.

To illustrate how TEM3P is used for engineering prototyping, we use the LCLS rf gun as an example [16]. The analysis starts with a CAD model of the rf gun as shown in figure 9. The CAD model contains two regions: the vacuum region (pink) used for EM computation, and the cavity metallic walls with detailed cooling channel layout (yellow) for thermal and mechanical analysis. In an integrated analysis process, the EM eigensolver Omega3P is used to calculate the accelerating mode of the cavity (figure 10a). The rf power loss on the cavity walls is then used as input to the thermal solver to find the temperature distribution inside the cavity wall thickness (figure 10b). The temperature distribution on the cavity wall boundaries is then input to the mechanical solver to calculate the wall distortion and stress distribution (figure 10c). The subsequent distorted cavity due to thermal and mechanical effects is used by the EM eigensolver Omega3P again to determine the frequency shift of the accelerating mode. The analysis process is repeated until an acceptable design is obtained.

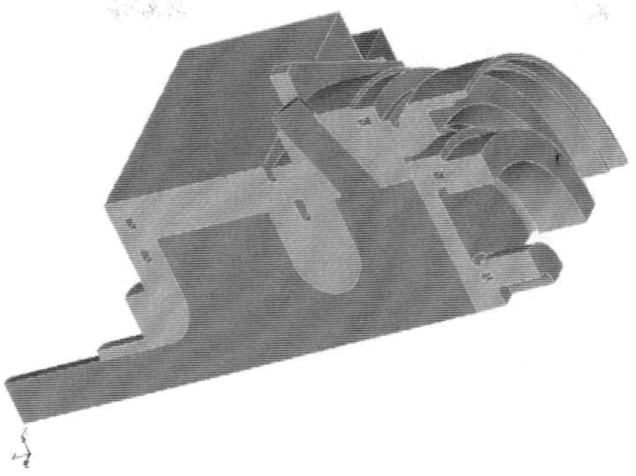

Figure 9. CAD model of LCLS rf gun.

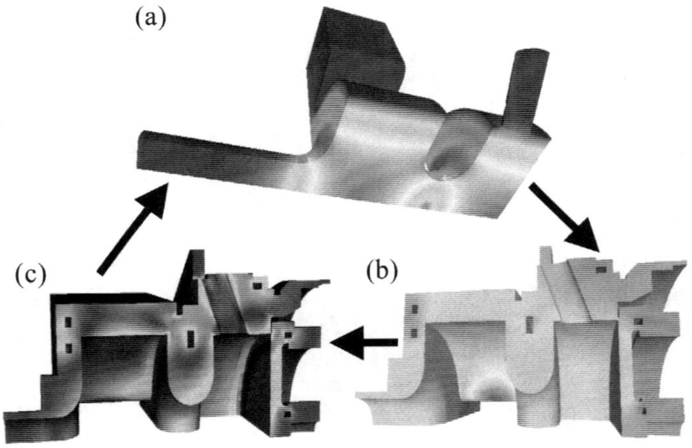

Figure 10. (a) EM field; (b) temperature; (c) mechanical stress distributions from TEM3P.

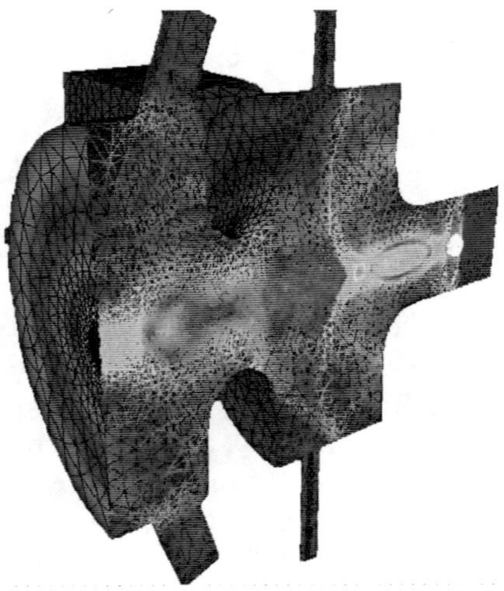

Figure 11. Beam transit in LCLS rf gun using Pic3P.

The complete design of the rf gun also needs to satisfy beam dynamic requirements. The beam emittance of the rf gun can be evaluated with the same mesh using Pic3P in the ComPASS finite-element suite. Pic3P, first of its kind, is a 3D parellel finite-element particle-in-cell code which includes wakefield and retardation effects that are generally ignored in standard space-charge tracking codes. It has been optimized specifically for highly efficient rf gun simulations employing various novel computational techniques [17]. Figure 11 shows a snapshot of the scattered fields excited by the transit of a bunch using Pic3P. Figure 12 shows the beam emittance obtained using Pic3P.

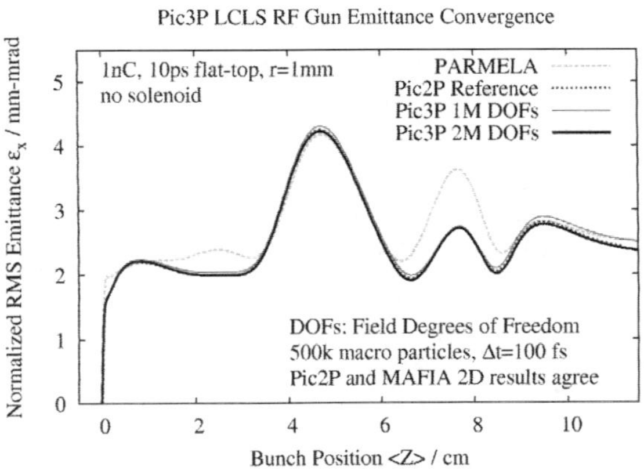

Figure 12. Beam emittance of LCLS rf gun using Pic3P.

8. Summary

Electromagnetic modeling has seen tremendous advances in SciDAC1 AST project, and will continue to play an important role in ComPASS in tackling the most computational challenging accelerator problems in HEP, NP, and BES across the Office of Science. With ASCR and SciDAC CETs/Institutes support, the computational tools – developed through concerted efforts from applied mathematicians, computer scientists and accelerator physicists – have demonstrated their impacts on the design and optimization of accelerators to ensure their performance and reliability while maintaining low costs.

Acknowledgments

We are grateful to H. Wang, F. Marhauser, J. Sekutowicz, C. Reece, and R. Rimmer of TJNAF for initiating the BBU problem in a CEBAF cryomodule and for supplying all the relevant measured data. This work was supported by the U.S. Department of Energy Office of High Energy Physics, Office of Basic Energy Sciences and Office of Advanced Scientific Computing Research under contract DE-AC02-76SF00515 and grants DE-FC02-07ER41499 and DE-FG02-04ER41317. This work used resources of the National Center for Computational Sciences at Oak Ridge National Laboratory and resources of the National Energy Research Scientific Computing Center, which are supported by the Office of Science of the Department of Energy under Contract DE-AC05-00OR22725 and DE-AC02-05CH11231, respectively.

References

[1] Ng C, *et al.* State of the Art in EM Field Computation", Proc. of EPAC2006, Edinburgh, UK, June 26-30, 2006.
[2] Ko K, *et al.* 2005 Impact of SciDAC on accelerator projects across the Office of Science through electromagnetic modeling *J. Physics* **16**:195
[3] Lee L-Q, *et al.* 2006 Advancing computational science research for accelerator design and optimization *Proc. SciDAC 2006 Conference* (Denver, Colorado) June 25-29
[4] Spentzouris P, *et al.* 2008 Community petascale project for accelerator simulation: Advancing computational science for future accelerators and accelerator technologies *Proc. SciDAC 2008*
[5] Dey S and Mittra R 1997 *IEEE Microwave and Guided Wave Lett.* **7**:273
[6] Werner G R and Cary J R 2008 *J. Comput. Phys.* **227**:5200
[7] Shephard M 2008 Curved mesh correction and adaptation tool to improve ComPASS electromagnetic analysis *Proc. SciDAC 2008*
[8] Pothen A 2008 The CSCAPES institute: Recent progress *Proc. SciDAC 2008*
[9] Lee L-Q, et al. 2008 Computational science research in support of petascale electromagnetic modeling *Proc. SciDAC 2008*
[10] Lee L-Q, *et al.* 2007 Multi-cavity trapped mode simulation Presentation at Wake Fest 07 Workshop (Stanford) December 18-20
[11] Ge L, *et al.* 2007 Multipacting simulations of TTF-III coupler components *Proc. PAC2007* (Albuquerque, New Mexico) June 25-29
[12] Ng C 2007 Status and plans for the impedance calculations of the ILC damping rings Presenttion at ILC Damping Rings Mini-Workshop (KEK, Tsukuba, Japan) December 18-20
[13] Kazimi R, *et al.* 2008 Observation and mitigation of multipass BBU in CEBAF *Proc. EPAC2008* (to appear)
[14] Akcelik V, et al. 2008 Maxwell eigenvalue determination for deformed cavities *J. Comput. Phys.* **227**:1722
[15] Li Z, *et al.* 2007 Towards simulations of electromagnetics and beam physics at the petascale *Proc. PAC2007* (Albuquerque, New Mexico, June 25-27)
[16] Akcelik A, *et al.* 2008 Parallel computation of integrated electromagnetic, thermal and structural effects for accelerator cavities *Proc. EPAC2008* (to appear)

[17] Candel A 2008 Load balancing of parallel 3D finite element particle-in-cell code Pic3P Presentation at CSCAPES Workshop (Santa Fe, New Mexico) June 10-13

SciDAC advances in beam dynamics simulation: From light sources to colliders

J Qiang[1], M Borland[1], A Kabel[2], R Li[4], R Ryne[1], E Stern[5], Y Wang[2], H Wasserman[1] and Y Zhang[3]

[1]Lawrence Berkeley National Laboratory, Berkeley, CA 94720, USA
[2]Argonne National Laboratory, Argonne, IL 60439, USA
[3]Stanford Linear Accelerator Center, Menlo Park, CA 94025, USA
[4]Thomas Jefferson National Accelerator Facility, Newport News, VA 23606, USA
[5]Fermi National Accelerator Laboratory, Batavia, IL 60510, USA

E-mail: jqiang@lbl.gov

Abstract. In this paper, we report on progress that has been made in beam dynamics simulation, from light sources to colliders, during the first year of the SciDAC-2 accelerator project "Community Petascale Project for Accelerator Science and Simulation (ComPASS)." Several parallel computational tools for beam dynamics simulation are described. Also presented are number of applications in current and future accelerator facilities (e.g., LCLS, RHIC, Tevatron, LHC, and ELIC).

1. Introduction

Particle accelerators are some of most important tools of scientific discovery. They are widely used in high-energy physics, nuclear physics, and other basic and applied sciences to study the interaction of elementary particles, to probe the internal structure of matter, and to generate high-brightness radiation for research in materials science, chemistry, biology, and other fields. Modern accelerators are complex and expensive devices that may be several kilometers long and may consist of thousands of beamline elements. An accelerator may transport trillions of charged particles that interact electromagnetically among themselves, that interact with fields produced by the accelerator components, and that interact with beam-induced fields. Large-scale beam dynamics simulations on massively parallel computers can help provide understanding of these complex physical phenomena, help minimize design cost, and help optimize machine operation. In this paper, we report on beam dynamics simulations in a variety of accelerators ranging from next generation light sources to high-energy ring colliders that have been studied during the first year of the SciDAC-2 accelerator project.

2. Beam dynamics simulations in light sources

The next generation of accelerator-based light sources will produce unprecedented photon flux, brightness, and temporal resolution. To achieve their full potential requires overcoming significant technological challenges such as producing and transporting high brightness, low-emittance electron beams through the accelerator while maintaining beam quality in the presence of collective effects such as space charge, wakefields, and coherent synchrotron radiation. At Lawrence Berkeley National Laboratory, two parallel particle tracking codes, IMPACT-Z [1] and IMPACT-T [2], have been developed and used to study the electron beam generation from photoinjectors and to simulate beam transport through radio frequency (RF) linacs. Both codes assume a quasi-static model of the beam and calculate space-charge effects self-consistently at each time step together with the external acceleration and focusing fields. There are six Poisson solvers in the IMPACT code suite corresponding to different boundary conditions. These solvers use either a spectral method for closed transverse boundary conditions or a fast Fourier transform-based Green's function convolution method for open transverse boundary conditions. The parallel implementation uses a two-dimensional domain decomposition approach for the three-dimensional computational domain. New features such as a short-range and Coherent Synchrotron Radiation (CSR) wakefield, integrated Green function solver, and particle-field decomposition were added to the code. Figure 1 shows the variation in efficiency of the IMPACT-T and the IMPACT-Z codes executing on increasing numbers of cores of the Cray XT4 computer at NERSC. Both codes still show more than 50% parallel efficiency on one thousand processors.

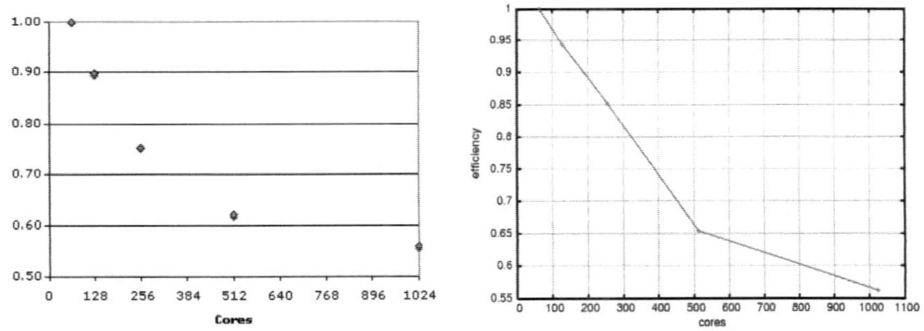

Figure 1. Parallel efficiency of IMPACT-T (left) and IMPACT-Z (right) on Cray XT4.

The IMPACT-T code has been used to help commission the electron generation and transport in the Linac Coherent Light Source (LCLS) photoinjector. Figure 2 shows a simulation of the evolution of rms emittance of the beam with an initial 0.5 mm horizontal and vertical offset through the LCLS photoinjector. The large offset produces a difference between the horizontal and vertical emittance and significantly larger final emittance compared with the nominal design value of 1.2 mm-mrad. The IMPACT-Z code has been used to simulate the beam transport through an RF linac for a next generation FEL light source at LBNL. A large number of macropartices is needed in order to accurately predict the final uncorrelated energy spread of the electron beam due to the presence of the microbunching instability. Figure 3 shows the final energy spread after transport through the FEL linac as a function of the number of macroparticles. It is seen that final energy spread approaches a constant value as the initial number of macroparticles becomes larger than one billion.

At Argonne National Laboratory, the ELEGANT code has been used for design, simulation, and optimization of FEL driver linacs, energy recovery linacs, and storage rings. It also has been used to support for LCLS design and commissioning [3]. The ELEGANT code has recently been completely parallelized. The new version has demonstrated the ability to use 400 million macroparticles on 100 nodes after implementation of parallelized SDDS I/O and elimination of the master node bottleneck in particle management. Figure 4 shows growth of the microbunching instability through a number of dipoles in the Fermi/Elettra FEL linac design.

3. Beam dynamics simulations in colliders

The beam-beam interaction limits the luminosity that can be achieved in high-energy colliders. We have developed a parallel three-dimensional particle-in-cell code, BeamBeam3D, to model beam-beam effect in high-energy ring colliders [4]. This code includes a self-consistent calculation of the electromagnetic forces (beam-beam forces) from two colliding beams (i.e. strong-strong modeling), a linear transfer map model for beam transport between collision points, a stochastic map to treat radiation damping, quantum excitation, an arbitrary orbit separation model, and a single map to account for chromaticity effects. The beam-beam forces are calculated with a parallel implementation of an FFT-based solution to the Poisson equation with a particle-field decomposition to achieve good load balance. It can handle multiple bunches interacting at multiple interaction points (IPs) in arbitrary sequences. The code was extended to include wakefield effects and wire and crab cavity compensation components. Parallel performance of the code was measured on a variety of high-performance computers and showed reasonable scalability up to a thousand processors [5].

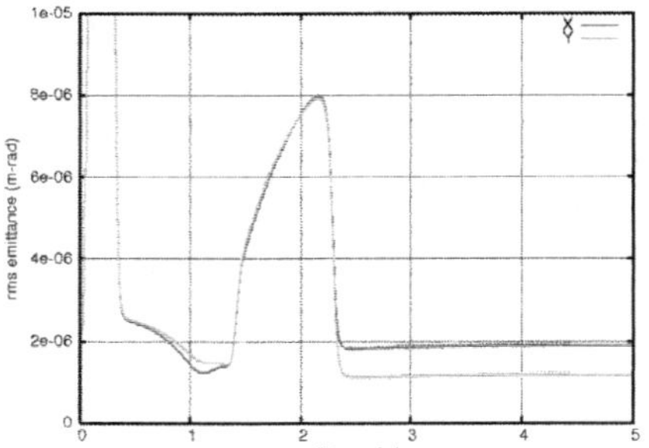

Figure 2. Transverse emittance evolution at LCLS photoinjector with initial 0.5 mm offset from IMPACT-T simulation.

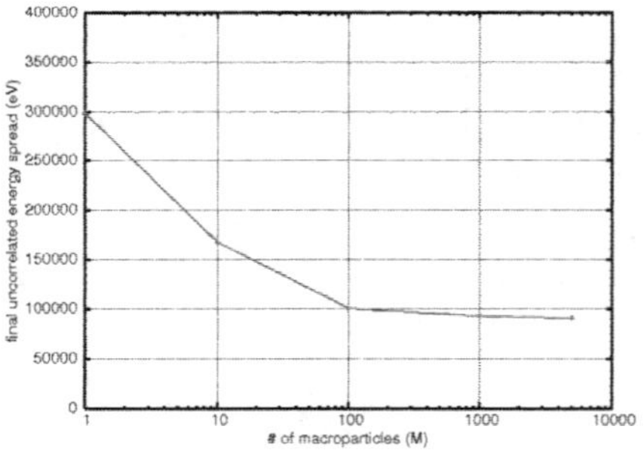

Figure 3. Final energy spread as a function of number of macroparticles from IMPACT-Z simulation.

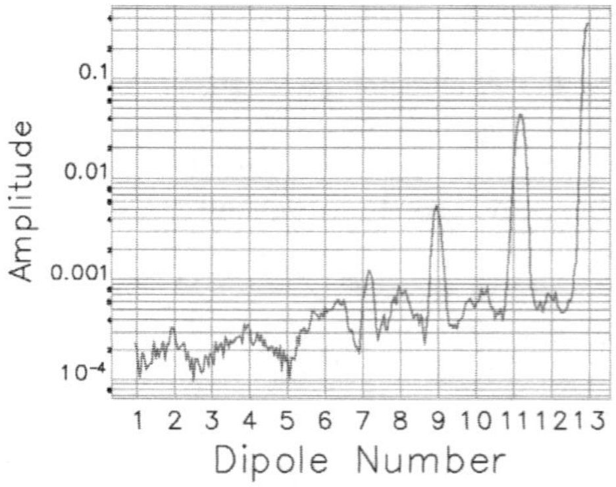

Figure 4. ELEGANT simulation of the growth of microbunching instability in the designed Fermi/Elettra FEL linac.

The BeamBeam3D code has been applied to the study of multibunch coherent beam-beam effects at RHIC. Each beam in the simulation has three bunches coupled via the beam-beam interaction to three bunches of the opposite beam at four interaction points. Figure 5 shows the BeamBeam3D simulation of emittance growth in tune space for the nominal working point near a half integer and for a new working point near an integer. The new working point appears to have a better performance than the nominal one.

BeamBeam3D has been extended and used to study coherent beam-beam effects at the Tevatron. With the commissioning of electron cooling in the Recycler, the head-on beam-beam tune shifts of the two beams become essentially equal. Under these circumstances the coherent beam-beam effects may become an issue. Figure 6 shows the emittance variation among 36 proton bunches interacting with 36 antiproton bunches at both head-on and long-range IPs from a 50000 turn strong-strong Tevatron simulation The pattern of varying emittance growth that develops is characteristic of collective effects produced by the collision.

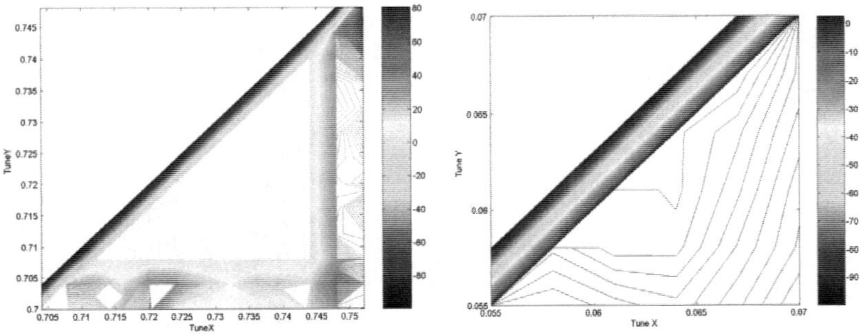

Figure 5. Emittance growth scan in tune space for a nominal working point (left) and for a new working point (right) at RHIC from BeamBeam3D simulation.

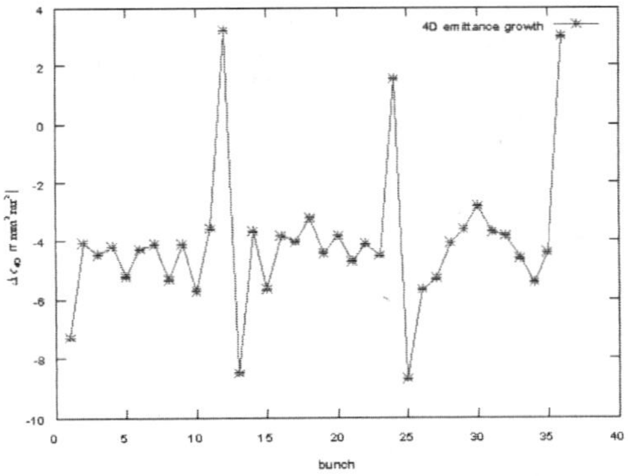

Figure 6. Emittance versus bunch number at the Tevatron from BeamBeam3D simulation.

BeamBeam3D has also been used to study long-range beam-beam effects at the LHC and to study crab cavity compensation of crossing angle collisions at the LHC upgrade. Figure 7 shows the 99.9% emittance growth with and without long-range beam-beam interactions. The simulation includes two head-on collision points (IP 1 and IP5) with 0.3 mrad crossing angle and 64 distributed long-range beam-beam interactions on both sides of the interaction region. The long-range beam-beam interaction has significantly increased the 99.9% emittance growth of the beam which characterizes the tail of the beam. Figure 8 shows the luminosity evolution with and without crab cavities for two beams colliding with 0.3 mrad crossing angle. It is seen that using a crab cavity significantly improves the luminosity at the LHC.

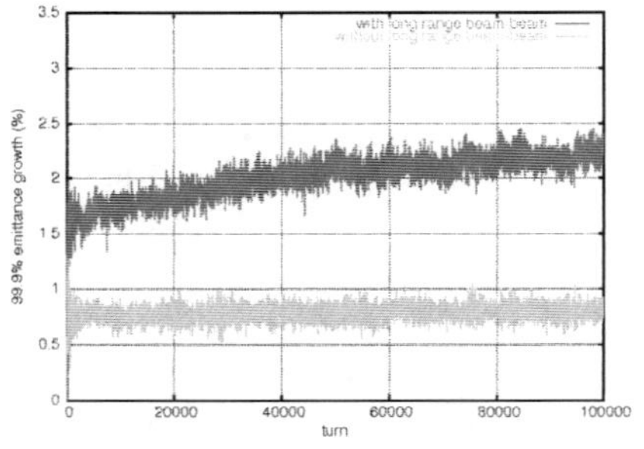

Figure 7. Emittance growth at LHC with and without long-range beam-beam interactions from BeamBeam3D simulation.

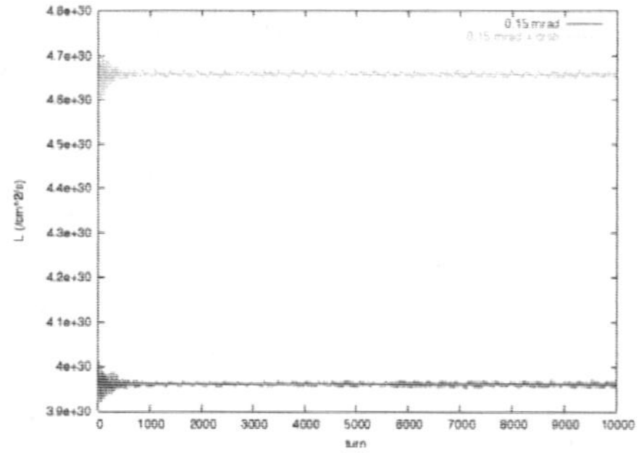

Figure 8. Luminosity evolution at LHC with and without crab cavity compensation from BeamBeam3D simulation.

The BeamBeam3D code has also been used to study the Electron Light Ion Collider (ELIC) proposed at Jefferson Lab. Understanding beam-beam effects at ELIC is crucial for achieving high luminosity. Figure 9 shows the luminosity as a function of electron and proton current using an old working point near the integer and a new working point towards the half integer. The new working point has significantly improved luminosity for all currents.

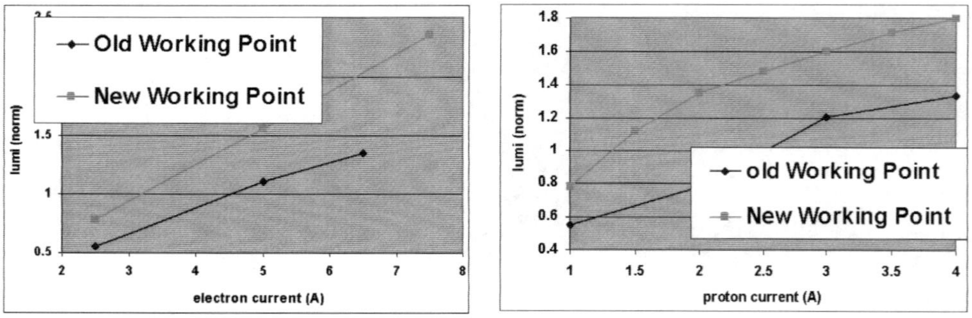

Figure 9. Luminosity as a function of electron (left) and proton (right) current at the JLab proposed ELIC collider from BeamBeam3D simulation.

At SLAC, a particle-tracking framework, PLIBB, has been developed and used to study beam-beam effects at the Tevatron, RHIC, and LHC with emphasis on beam lifetime calculation using a weak-strong model [6]. The physics of the code includes intrabeam scattering, noise sources, common magnetic elements, weak-strong and strong-strong beam-beam interactions, and wire and electron lens compensators. Figure 10 shows the tune space footprint with and without electron lens compensation at LHC. Using an electron lens significantly reduces the size of the tune spread.

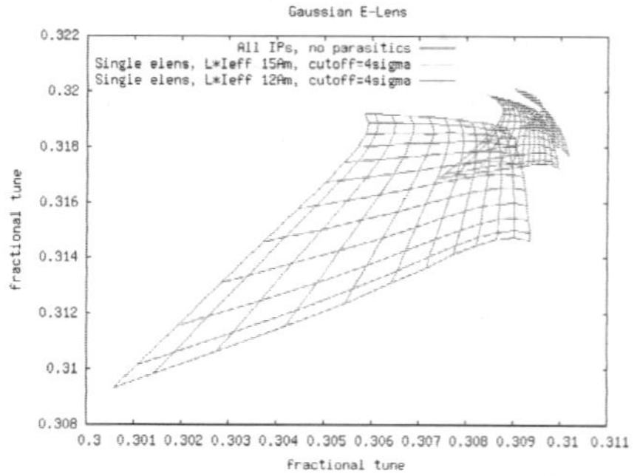

Figure 10. Tune footprint at LHC with/without electron lens from PLIBB simulation.

Acknowledgments
This work used computing resources at the National Energy Research Scientific Computing Center (NERSC), supported by the U.S. Department of Energy under Contract no. DE-AC02-05CH11231.

References
[1] Qiang J, Ryne R D, Habib S and Decyk V 2000 An object-oriented parallel PIC code for beam dynamics simulation in linear accelerators *J. Comp. Phys.* **163:**434–451
[2] Qiang J, Lidia S, Ryne R D and Limborg-Deprey C 2006 A three-dimensional quasi-static model for high brightnees beam dynamics simulation *Phys. Rev. ST Accel. Beams* **9:**044204
[3] Borland M 2000 *APS LS-287*
[4] Qiang J, Furman M A and Ryne R D 2004 A parallel particle-in-cell model for beam-beam interactions in high energy ring colliders *J. Comp. Phys.* **198:**278–294
[5] Shan H, Strohmaier E and Qiang J 2008 Performance analysis of leading HPC architectures with BeamBeam3D *Int. J. High Performance Computing Applications* **22:**21
[6] Kabel A, *et al.* 2003 A parallel code for lifetime simulations in hadron storage rings in the presence of parastic weak-strong beam-beam interactions *Proc. PAC 2003* p. 3542.

Community petascale project for accelerator science and simulation: Advancing computational science for future accelerators and accelerator technologies

Panagiotis Spentzouris[1], John Cary[2], Lois Curfman McInnes[3], Warren Mori[4], Cho Ng[5], Esmond Ng[6] and Robert Ryne[6]

[1]Fermi National Accelerator Laboratory, Batavia, IL 60510, USA
[2]Tech-X Corporation, Boulder, CO 80303, USA
[3]Argonne National Laboratory, Argonne, IL 60439, USA
[4]University of California, Los Angeles Los Angeles, CA 90024, USA
[5]Stanford Linear Accelerator Center, Stanford, CA, USA
[6]Lawrence Berkeley National Laboratory, Berkeley, CA 94720, USA

E-mail: spentz@fnal.gov

Abstract. The design and performance optimization of particle accelerators are essential for the success of the DOE scientific program in the next decade. Particle accelerators are very complex systems whose accurate description involves a large number of degrees of freedom and requires the inclusion of many physics processes. Building on the success of the SciDAC-1 Accelerator Science and Technology project, the SciDAC-2 Community Petascale Project for Accelerator Science and Simulation (ComPASS) is developing a comprehensive set of interoperable components for beam dynamics, electromagnetics, electron cooling, and laser/plasma acceleration modelling. ComPASS is providing accelerator scientists the tools required to enable the necessary accelerator simulation paradigm shift from high-fidelity single physics process modeling (covered under SciDAC1) to high-fidelity multiphysics modeling. Our computational frameworks have been used to model the behavior of a large number of accelerators and accelerator R&D experiments, assisting both their design and performance optimization. As parallel computational applications, the ComPASS codes have been shown to make effective use of thousands of processors.

1. Introduction

Particle accelerators are critical to scientific discovery in the DOE program in America and indeed the world [1]. Of the twenty-eight facilities listed in the DOE report "Facilities for the Future of Science: A Twenty-Year Outlook," fourteen involve accelerators. The development and optimization of accelerators are essential for advancing our understanding of the fundamental properties of matter, energy, space and time and for enabling research in aspects of materials science, chemistry, geosciences, and bioscience.

The High Energy Physics (HEP) program uses accelerators to answer fundamental questions about nature such as the origin of mass and the asymmetry between matter and antimatter and to search for new particles, new symmetries, and possible extra dimensions of space. In the DOE 15-year plan [2] for HEP, the first two action items call for full support of the program of the Large Hadron Collider (LHC) at CERN and for the establishment of leadership in the R&D effort to design and build the proposed International Linear Collider (ILC) on U.S. soil. Even with the current HEP budget difficulties, the recent report of the Particle Physics Project Prioritization Panel (P5) [3] emphasizes in its recommendations the need to maintain leadership in both the energy and the intensity frontier of accelerator science. At the same time, it is imperative to maximize the physics reach of the ongoing DOE/HEP program, and that involves the performance optimization of the Fermilab Tevatron. Furthermore, DOE/HEP is supporting a world-class R&D program to develop new accelerator technologies including laser wakefield and plasma wakefield accelerators, as well as other types of advanced accelerator concepts.

The Nuclear Physics (NP) program uses accelerators to study the properties of nuclear matter and the structure of the nucleus, understand the mechanism of quark confinement, and create and study a heretofore-unknown state of nuclear matter, the quark-gluon plasma. The flagship NP accelerators are CEBAF at Jefferson Lab and RHIC at Brookhaven National Laboratory (BNL). The worldwide nuclear physics community has identified [4] the construction of a polarized electron-ion collider as a long-term objective in the ongoing effort to understand and answer questions in these areas. DOE/NP is considering two approaches for such a facility: e-RHIC [5] and ELIC [6]. DOE/NP is also considering a proposed rare isotope facility that will permit studies of nuclei far from stability that promise to radically improve our understanding of atomic nuclei.

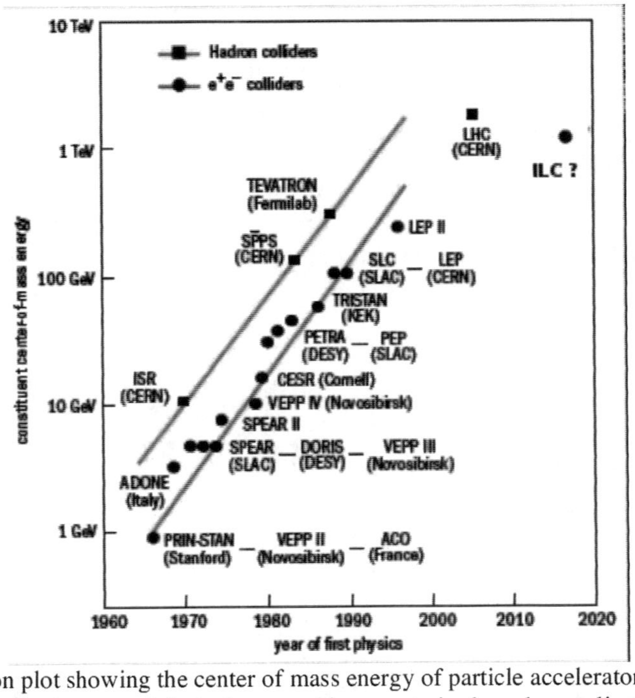

Figure 1. Livingston plot showing the center of mass energy of particle accelerators (lepton and hadron colliders) versus time. In the case of the hadron machines, energies have been adjusted to account for quark and gluon constituents.

The flagship accelerator facilities for the Basic Energy Sciences (BES) program are its spallation neutron sources (SNS, LANSCE) and its synchrotron light sources (ALS, APS, NSLS, SSRL). Over the past decade, the light-source community consensus has been that short-wavelength, free-electron lasers (FELs) will form the basis for the so-called fourth-generation light sources. DOE/BES is building the Linac Coherent Light Source (LCLS) at the Stanford Linear Accelerator Center, which will operate with ultra-short x-ray pulses to shed light on ultra-fast processes in chemistry and biology. LCLS will also further our understanding of the arrangement of atoms in inorganic and organic materials, thus enabling the development of new materials and molecules with desirable properties.

The discussion above clearly shows that accelerators are the primary instruments for scientific discovery at the forefront of basic physics research across all programs of the Office of Science. In order to maintain this forefront position over the past forty years, accelerators have become bigger and much more expensive, with ever-increasing demands for higher energy and intensity (see figure 1), increased complexity, and tighter operational tolerances. The design, cost optimization, and successful operation of modern accelerators require the optimization of many parameters and the understanding and control of many physics processes. In addition, the preceding discussion also shows that the different areas of scientific research have different requirements for the type of particle accelerator that is needed to enable each research activity. Although the basic concepts that guide the design and operation of the different types of particle accelerators are similar, the technologies involved and the critical design and operational parameters vary depending on the application, thus increasing the necessary complexity of high-fidelity modeling tools. The same plurality exists in accelerators used for industrial or medical applications. A representative list of particle accelerator applications in the applied sciences, based on different types of particle beams and accelerator technologies, is shown in table 1. Advances in accelerator technology and careful design have resulted in a more than order of magnitude reduction in the size of accelerators used for medical applications (see figure 2). Advances in accelerator technology triggered by basic research R&D (dielectrics, laser wakefield accelerators) are already discussed as candidates for the next generation of medical accelerators [7].

Table 1. Representative list of accelerator utilization in the applied sciences.

Field	Accelerator type	Topics of study
Atomic physics	Low energy ion beams	Atomic collision processes, study of excited states, electron-ion collisions, electronic stopping power in solids
Condensed matter physics	Synchrotron radiation sources	X-ray studies of crystal structure
	Spallation neutron sources	Neutron scattering studies of metals and crystals, liquids, and amorphous materials
Material science	Ion beams	Proton and X-ray activation analysis of materials; X-ray emission studies; accelerator mass spectrometry
Chemistry and Biology	Synchrotron radiation sources	Chemical bonding studies: dynamics and kinetics; protein and virus crystallography; biological dynamics
Medicine	Proton, electron, and neutron beams	Radiation therapy

Under SciDAC-1, the Accelerator Science and Technology (AST) project produced a powerful suite of parallel simulation tools representing a paradigm shift in computational accelerator science. Using AST codes, simulations that used to take weeks or more now take hours, and simulations once thought impossible are now performed routinely. These codes were applied to major DOE accelerator facilities and future accelerator projects including the Tevatron, PEP-II, LHC, RHIC, NLC, ILC design, SNS, and LCLS [8]. They were also applied to advanced concepts such as plasma-based accelerators, where they played a key role in understanding the physics of doubling the energy of a 42 GeV beam at SLAC and of low-energy-spread beam production in laser wakefield accelerators [9].

Under SciDAC-2, the Community Petascale Project for Accelerator Science and Simulation (ComPASS), the successor of AST, recognizing the increased complexity, precision, and beam intensity requirements of next-generation particle accelerators, changes again the focus of accelerator high-performance simulation software. Our accelerator simulation paradigm is shifting from single-machine, single-component simulations to end-to-end (multistage or complete system), multiphysics simulations. Building on the foundation laid under SciDAC-1, ComPASS is extending terascale capabilities to the petascale and is adding new capabilities in order to deliver a comprehensive, fully integrated accelerator simulation environment. The ComPASS vision is to create a *virtual accelerator modeling* environment for the realistic, inclusive simulation of all relevant beam dynamics effects (single and multi-particle dynamics, realistic geometry and parameters), and a *virtual prototyping* environment for realistic simulation of all relevant accelerator component effects (thermal, mechanical, and electromagnetic properties with accurate geometry description). The ComPASS accelerator modeling applications, discussed in Section 3, target important problems in nearly every accelerator laboratory in the country. The success of these applications depends on the success of the collaboration of ComPASS researchers with accelerator scientists from these laboratories. In addition, the success of the development of the necessary infrastructure that will enable the integration of the high-end capabilities we are building depends on the successful collaboration with ASCR researchers. This collaboration helps ensure that our accelerator tools development uses the most appropriate algorithms and computer science technology for petascale platforms.

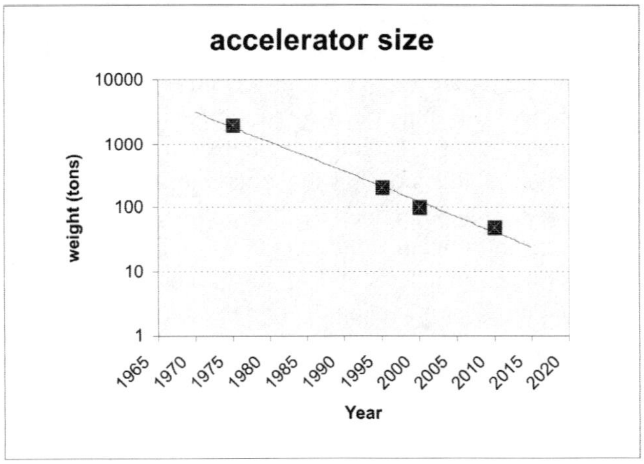

Figure 2. Advances in accelerator technology have resulted in more than order of magnitude reduction in the size of cyclotrons used for medical applications.

2. The ComPASS program

Accelerators rely on electromagnetic fields to provide particle acceleration, and they typically use magnetic fields to guide (bend and focus) the particle beams. Radiofrequency cavities (for acceleration), dipole magnets (for bending), and quadrupole magnets (for focusing) are the most commonly used beamline elements in accelerators. In addition, many more systems are necessary for the operation of a modern accelerator, such as high-order multipoles (to control the nonlinear behavior of the particles), devices for diagnostics (such as beam position monitors), feedback systems, and control systems. A modern accelerator operates like a modern airplane: it requires active damping or it is going to crash! The behavior of the particle beams (the *beam dynamics*) is affected not only by the interaction of the particles with the accelerator elements but also by the interaction of particle ensembles with each other, either via their own fields or via fields created by other particle ensembles. These collective, multiparticle effects in many cases are the limiting performance factor for particle accelerators. All the effects described above combine with operational effects such as misalignments of accelerator components to make the design and operation of an accelerator a very challenging problem. Moreover, the quest for higher and higher energies and particle beam intensities requires continuous R&D for new accelerator techniques and technologies.

2.1. ComPASS computational tools development

Given the size, cost, and complexity of modern accelerators, the research to improve their design and performance becomes as challenging as the basic research and applications they enable. High-fidelity numerical modeling is essential to ensure cost effectiveness and performance optimization. The ComPASS program for computational accelerator physics development is organized so it mirrors the structure of the different accelerator science and technology R&D areas. From the discussion in the previous section, we see that the main ingredients required to compose an accelerator model are tools for electromagnetic field calculations, beam dynamics simulations, and tools that support new accelerator technology R&D. Thus, the main thrust areas of this project are threefold:

- Beam dynamics (BD): Development and applications of tools that model the evolution of beams through beam optics systems, including self-forces and other forces of interaction. A particularly interesting area of BD development is electron cooling (EC), where the beam phase-space is controlled by using the dynamical friction from relativistic copropagating electron beams.
- Advanced acceleration (AA): Development and application of tools that guide the R&D efforts for the realization of new high-gradient acceleration techniques such as plasma or laser wakefield accelerators.
- Electromagnetics (EM): Development and application of tools that model the electromagnetic fields in geometrically complex accelerating cavities and other accelerator components, in order to maximize acceleration while minimizing beam quality degrading effects. Such effects include wakefields, which are generated by a particle bunch as it passes a perturbation in an accelerator structure and can reinteract with the charged particle beam, multipactoring, that is, resonant electron multiplication that builds up an electron avalanche, leading to remarkable power losses and heating of the walls, and so forth.

Of course, these areas are not independent: the dynamics of a particle beam is affected by wakefields that require electromagnetic computations; particle-in-cell methods are common to beam propagation in cavities and advanced acceleration; electron cooling is needed for heavy ion beams because of emittance growth due to intrabeam scattering and beam-beam interactions. Hence, the code development in the three ComPASS areas aims to provide integrated simulation capabilities. This is accomplished by the project's enabling computer science activities in the area of component technologies, such as development of common component interfaces, code refactorization, and development of quality-of-service infrastructure software for our frameworks. In addition, activities in computer science and applied mathematics are common to all areas of development. These activities aim to improve the performance of our simulations and our analysis tools, such as solver development, optimization, and research on new or improved algorithms, data handling, and visualization.

2.2. ComPASS computational tools development

The accelerator modeling codes developed under ComPASS, together with their major applications, are described in the following list:

- Impact-T: A beam dynamics parallel particle tracking code that uses time as the independent variable. This code aims to accurately model low energy beams with large energy spreads, assumes a quasi-static model for the beam, and includes a self-consistent model of space-charge (the effects of the charge of the beam on itself). Applications include modeling of the Advanced Light Source streak camera, Advanced Photon Source upgrade, LCLS injector, the Fermilab A0 experiment, and photoinjectors at BNL, UCLA, and Cornell, the UW-Madison FEL, PSI FEL, SPARC/X, and Fermi/Electra.
- Impact-Z: Beam dynamics simulation in rf linacs using path length as the independent variable. The Impact family of codes (Z and T) share a variety of Poisson solvers, either spectral based or convolution based. Applications for Impact-Z include Berkeley FEL studies, LEDA Halo experiment, ILC damping ring modeling, SNS, RIKEN cyclotron injector, and RIA driver, and JPARC, GSI, CERN SPL, and CSNS linac simulations. The Impact family of codes have been ported to Cray XT4, IBM SP4, and Blue Gene machines.
- MaryLie/IMPACT (ML/I): Beam dynamics framework focusing on ring problems involving high-order optics and space charge, realistic magnetic field profiles, realistic cavity simulations, fitting and optimization. Applications include RHIC upgrade.
- BeamBeam3D: A comprehensive code for modeling beam-beam effects in colliders using a variety of models including strong-strong and weak-strong models, short-range and long-range beam-beam interactions, crab crossing, multiple bunches, multiple interaction points, and impedance effects. Applications include the Fermilab Tevatron, KEKB, RHIC, and the LHC. BeamBeam3D has been ported to desktop MPI PC clusters, Cray XT4, and IBM SP3, SP4, and BG/L machines.
- PLIBB/ Nimzovich: A general-purpose parallel tracking framework for beam-beam simulations. Applications include RHIC wire compensation beam-beam experiments. The code has been used on the NERSC supercomputers.
- Synergia2: An extensible framework for beam dynamics. It is driven by Python, so it can be configured at run time, allowing for arbitrarily complex simulations. The framework includes nonlinear single-particle physics, beam generation and diagnostics, and collective effects modules. The collective effects modules include extensive support for space-charge calculations for a variety of geometries and boundary conditions. These calculations are available both through native Synergia solvers and through an interface to IMPACT. Application focus: Fermilab booster, ILC Damping Ring and Ring to Main Linac accelerator design, and modelling of Fermilab A0 photo injector experiments. The framework is optimized for massively parallel computations utilizing MPI. It is currently ported on desktop PC MPI clusters and Unix supercomputers supporting shared libraries (IBM SP3 and SP4).
- Elegant: Beam dynamics framework with focus on electron beams. Applications include FEL driver linac and energy recovery linac design. The parallel version of elegant is currently developed under ComPASS.
- VORPAL: A framework that can be configured in different ways at run time to solve a variety of problems. It provides models for neutral gases and plasmas interacting with electromagnetic fields and with each other through collisions. Applications include Laser Wake Field Accelerators (LWFA), EM cavity calculations (ILC, JLab), electron cooling (BNL), electron gun modelling and multipactoring (JLab), and modelling and analysis of electron cloud measurements. VORPAL

is running in production on Bassi, Franklin, and Jacquard at NERSC and has been ported to Blue Gene.

- UPIC: A flexible framework for rapid construction of parallel particle-in-cell (PIC) applications. It supports electrostatic, darwin, and EM solvers. Applications focus: QuickPIC, which is based entirely on the UPIC framework and IMPACT solvers. UPIC is strictly Fortran90 compliant, with no dependencies other than MPI and pthreads, so it can be ported to any platform.
- Osiris: Fully explicit PIC code with ionization and coulomb collision packages. Applications of Osiris focus on LWFA and PWFA design and experiment modeling. Osiris runs on all the NERSC machines at LBNL, the ATLAS machine at LLNL, and on os-10 clusters.
- QuickPIC: A quasi-static PIC code based on the UPIC framework. Applications include LWFA and PWFA design and experiment modeling and e-cloud modeling for LHC, ILC, and Fermilab Main Injector applications. QuickPIC is used on the NERSC machines and the UCLA OSX-based cluster.
- Omega3P: Frequency domain eigensolver for cavity mode and damping calculations. Applications include simulations of the ILC cryomodule, LHC crab cavity and collimator, X-band high gradient structure, JLab 12-GeV upgrade superconducting cavity, the BNL RHIC cavity, and the SNS superconducting cavity. The code runs on NERSC and NCCS supercomputers.
- S3P: Frequency domain scattering parameter calculations. S3P was used for ILC superconducting cavity coupler calculations. The code runs on NERSC supercomputers.
- Track3P: Particle tracking for simulation of dark current and multipacting. Applications include modelling the coupler of the ILC superconducting cavity, the SNS superconducting cavity, and BNL RHIC cavity calculations. Track3P runs on NERSC and NCCS supercomputers.
- T3P: Time domain solver for wakefield computations with beam excitation. Applications include the ILC cryomodule and Damping Ring components, CLIC PETS structure, MIT photonic band gap structure. T3P is ported on NERSC and NCCS supercomputers.
- Gun3P: Space-charge trajectory code for beam formation and beam transport. Applications: ILC sheet beam klystron gun. Gun3P runs on NERSC supercomputers.
- Pic3P: Self-consistent particle-in-cell code for rf gun and klystron simulations. Applications include the LCLS rf gun. Pic3P runs on the NERSC supercomputers.
- TEM3P: Framework for integrated electromagnetic, thermal and mechanical analysis for rf cavity design. The framework uses solvers from the 3P family of codes described above. Applications include the LCLS rf gun design. The framework runs on the NERSC supercomputers.
- V3D: Visualization utilities for meshes, field, and particles on unstructured meshes. These utilities are used with the 3P family of codes described above.

Detailed descriptions of the ComPASS codes and their applications can be found in [10] for electromagnetics [11] for advanced acceleration and [12] for beam dynamics. The high-performance-computing (HPC) capabilities of most of these codes were developed under the AST SciDAC project, each targeting specific areas of computational accelerator physics applications, with emphasis on the performance and high fidelity of the calculation. The above list shows not only the wide spectrum of capabilities developed under the SciDAC accelerator modeling project, but also the complexity of the field of computational accelerator physics. Under ComPASS, the emphasis is shifting toward developing the necessary infrastructure for integrating these capabilities into accelerator simulation frameworks, allowing for multi-physics, multi-scale simulations. These activities are in addition to the continuing effort for code and algorithm performance improvements and optimization on the newer HPC platforms.

Journal of Physics: Conference Series **125** (2008) 012005 doi:10.1088/1742-6596/125/1/012005

2.3 Enabling technologies

Software infrastructure for multiphysics accelerator modeling is a central part of ComPASS. Our approach is to develop application components using existing mature physics or algorithmic implementations as the core of each component. Here, "component" is defined as a portion of software implementation that can be added or removed from multiple applications. The most serious challenge in developing simulation components is the definition of their interfaces: the components need to be objects that can be used for multiple implementations of different ComPASS simulations. Thus, the interface definition has to be independent of any details from any particular application implementation, including the parallelization (data distribution) schemes. Our approach builds on AST and ComPASS work on the Synergia2 [13] and MaryLie/IMPACT [14] beam dynamics frameworks and the UPIC framework and is extending to other areas of the project with work such as the TEMP3P electromagnetic simulation framework. Figure 4 shows a schematic of the Synergia2 framework, with the physics components and software infrastructure dependencies.

Use of advanced mathematical techniques, scalable numerical algorithms, and computational tools are also major components of the ComPASS activities. For example, through implementation in VORPAL, SciDAC has supplied the first massively parallel implementation of FDTD electromagnetic computations (see figure 3). While many of the mathematical and computational tools we employ are relatively mature, we need to enhance their capabilities to meet the petascale computational challenges of SciDAC-2. In addition, we need to explore the benefits of employing new techniques and algorithms, and we need to port the new and old implementations to the new petascale capable hardware that is or will be available in the SciDAC-2 era.

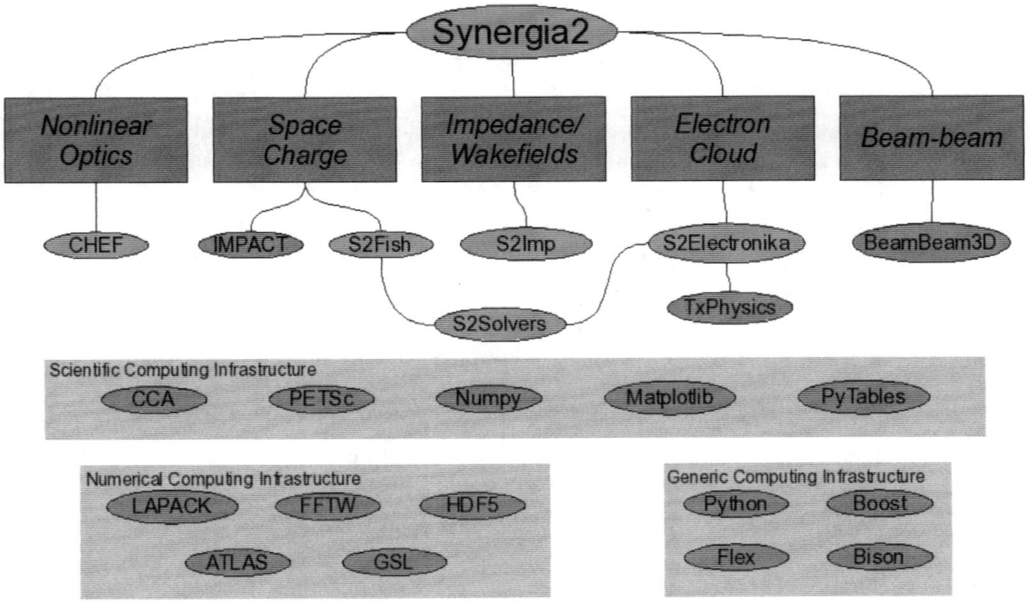

Figure 3. Componentization of the Synergia2 infrastructure. The Synergia2 framework includes many beam dynamics physics modules, both native to Synergia (orange) and reused modules from other ComPASS beam dynamics codes (magenta).

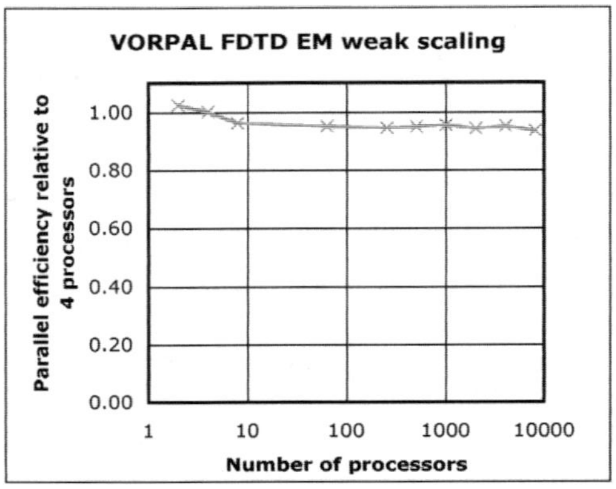

Figure 4. Weak scaling of VORPAL for an 803 domain size relative to 4 processors. VORPAL achieves 95% parallel efficiency out to the limits of the Cray XT4 at NERSC.

Shown below is a list that summarizes the necessary enabling technology activities for the success of the ComPASS program. Note that not all of the listed activities are funded under the ComPASS project, so collaboration and coordination with the SciDAC Centers for Enabling Technologies and SciDAC Institutes is essential.

- Development of scalable parallel eigensolvers, in collaboration with Terascale Optimal PDE Simulation (TOPS) [15]. The success of this activity is essential to enable simulation of complete systems of rf cavities with many millions of degrees of freedom.
- Domain-specific scalable linear solvers development, in collaboration with TOPS. This is another essential activity for detailed modeling of large electromagnetic systems.
- Development of meshing technology for electromagnetic calculations to enable shape adaptation procedure that is essential for the optimal (cost effective) design of rf cavities. This work relies on collaboration with the Interoperable Technologies for Advanced Petascale Simulations (ITAPS) project [16] and TOPS.
- Development of new Poisson solvers that will perform and scale well on petascale platforms, in collaboration with TOPS. Fast and scalable Poisson solvers are essential for petascale applications involving particles in a mean field treatment of space charge.
- Parallel adaptive refinement for finite elements, to improve solution accuracy and reduce computational cost, in collaboration with ITAPS and the Combinatorial Scientific Computing and Petascale Simulations (CSCAPES) project [17].
- Utilization of remote and interactive visualization tools to enable remote collaboration and analysis of large, complex data sets; in collaboration with the Institute for Ultrascale Visualization (ISUV) [18].
- Deployment of performance analysis and optimization tools for ComPASS codes and applications to maximize performance on petascale platforms; work in collaboration with PERI [19].
- Development of embedded boundary methods for PIC simulations of electromagnetic structures, to reduce memory requirement for simulation utilizing a finite differences approach; in collaboration with ITAPS.
- Enhancement of the UPIC framework, to include mesh refinement and optimized preconditioning in reduced PIC and spectral method-based dispersionless solvers. This activity benefits from interactions with the Applied PDE Center (APDEC) [20], TOPS, and PERI.
- Development of "computational quality-of-service" infrastructure for beam dynamics modelling applications. This activity includes the development of interoperable components using the Common Component Architecture (CCA) [21] approach and infrastructure for their effective

utilization in accelerator modelling frameworks; work is in collaboration with TASCS [22], TOPS, and the Performance Engineering Research Institute (PERI).

- Development of high-performance parallel data management and analysis tools for beam dynamics modelling, to provide common parallel particle data representation and utilities necessary for analysis of massive data sets; in collaboration with the Visualization and Analytics Center for Enabling Technologies (VACET) [23].
- Implementation of effective load balancing for particle-field simulations, to improve performance of rf gun simulations and PIC applications in general. This activity depends on collaborating with ITAPS and CSCAPES.

3. Applications

The applications of the new capabilities developed under ComPASS aim to tackle the most computationally challenging problems of near-, medium-, and long-term priorities of HEP, NP, and BES, with the objectives of assisting design and cost optimization and helping improve machine performance. As we discussed in Section 1, the HEP priorities include support for the Fermilab Tevatron and CERN LHC programs, continuation of R&D for a lepton collider (such as the ILC, with exact specifications depending on the LHC results), and support of the high-intensity frontier program at Fermilab (Project-X). The ONP future program focuses on the design of an electron-ion collider to explore the quantum-chromodynamics frontier, and a rare isotope accelerator, to shed light to the laws governing the creation of the elements. The future plans for a large BES facility focus on LCLS, the fourth-generation light source at SLAC. Using these priorities as guidelines for the ComPASS application development program, our activities focus on the following problems: large-scale electromagnetic modeling of an ILC-like rf unit, with realistic cavity shapes and misalignments; assessment of the impact of wakefields on beam dynamics; multiphysics, multibunch modeling of ILC Damping Ring beam dynamics; and Project-X Accumulator and Main Injector rings. We also focus on design optimization of accelerator components with complicated geometries, such as the hybrid rare isotope accelerator rfQ and the ILC crab cavity, which includes couplers with very fine features. For the LHC and LHC upgrade R&D projects (LARP) we focus on beam-beam and electron-cloud simulations to help understand and optimize machine performance. For the Tevatron we focus on multi-bunch beam-beam and impedance effects using exact geometry to help understand the anti-proton current limitations. For the RHIC II proposal, the e-RHIC concept design, the CEBAF upgrade proposal and the ELIC concept design, we focus on three areas: (1) electromagnetic simulation of superconducting rf cavities, with and without self-consistent beam treatment, (2) multiphysics beam dynamics simulations with emphasis on nonlinearities, beam-beam effects, and intrabeam scattering, and (3) electron cooling physics, aiming to quantitatively understand the dynamical friction force on ions moving through electron distributions in the presence of strong external fields. In the area of accelerator based x-ray light sources ComPASS tools are applied to help understand and predict limits on beam brightness, coherent and incoherent undulator radiation, emittance preservation, and microbunching [10, 12]. In addition, ComPASS is assisting the development of advanced accelerator concepts [11]. These technologies have already demonstrated gradients and focusing forces more than 1000 times greater than conventional technology. Under SciDAC-2, we aim to provide real-time or near-real-time feedback between simulation and advanced accelerator experiments.

3.1. ComPASS application example

In the past few years, the Fermilab Tevatron Run II collider program has delivered unprecedented levels of luminosity (see figure 5) to the collider detectors, producing a plethora of interesting new results and

Journal of Physics: Conference Series **125** (2008) 012005 doi:10.1088/1742-6596/125/1/012005

closing in on the discovery of the Higgs boson, the particle that in the Standard Model of particle physics is conjectured to be responsible for the creation of mass. This great performance was achieved as a result of the implementation of a three-part upgrade strategy, which focused on increasing the antiproton production rate, providing a third stage of antiproton cooling (concentrating the beam) with the Recycler storage ring, and increasing the transfer efficiency of antiprotons to the Tevatron (a schematic of the Fermilab accelerator complex is shown in figure 5)

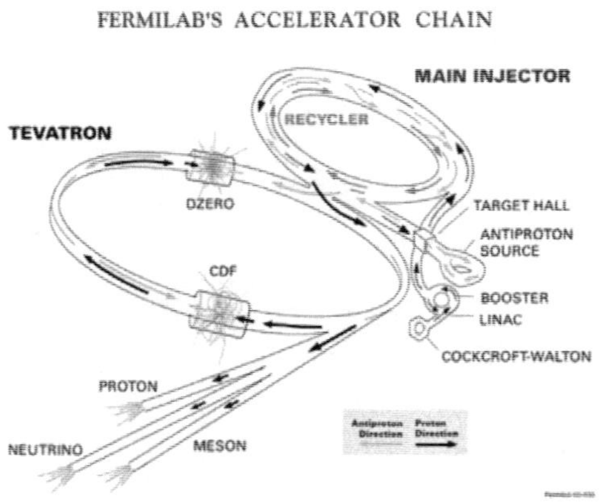

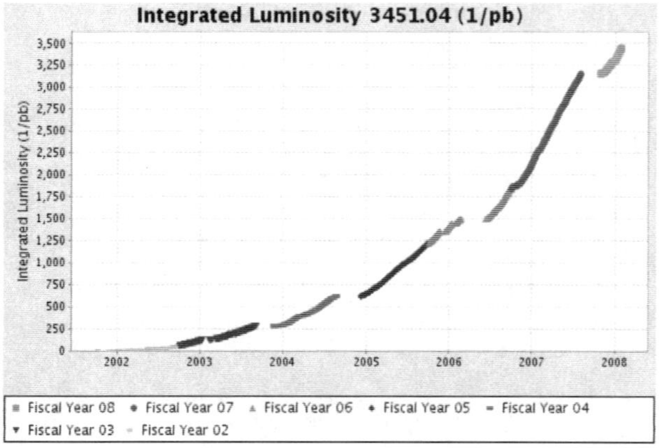

Figure 5. Schematic of the Fermilab accelerator complex (left); Tevatron integrated luminosity versus fiscal year (right).

There were five main areas of improvement:

- *Antiproton production—* antiproton creation and collection rates were increased, decreasing the time it takes to stack them in Fermilab's Accumulator ring. Furthermore, the rate at which those antiprotons are transferred out of the Accumulator was increased. Storing fewer antiprotons in the Accumulator increased efficiency.
- *Slip-stacking protons—* implemented slip-stacking, which effectively doubled the number of protons that the Main Injector ring delivers to the antiproton-producing target. Protons travel around the Main Injector ring in small bunches. Slip-stacking allows extra bunches of protons to be slipped in beside bunches already circulating the Main Injector.
- *Recycler e-cooling* – used the Recycler ring to "cool" antiproton beams using electron cooling thus creating denser antiproton bunches.

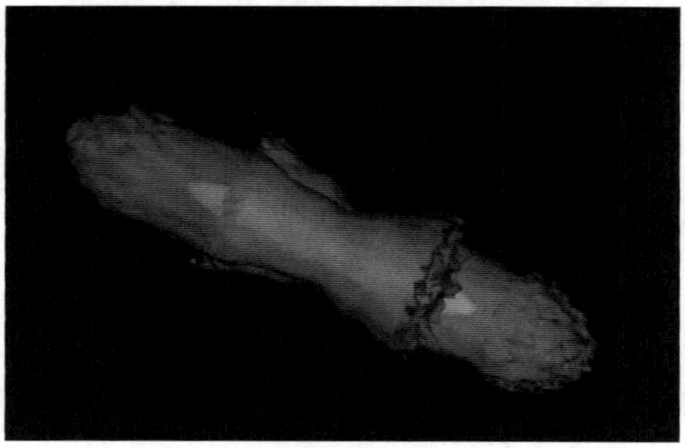

Figure 6. Beam-beam effects for a proton-antiproton head on collision. The bunch charge is 400 times larger than the charge currently used at the Tevatron, to maximize the effect.

Fermilab was the first lab to use electron cooling at high energy.

- *Tevatron beam position monitors* – upgraded the electronics and software for the 240 beam position monitors (BPMs) in the Tevatron, which help prevent the beam drifting too close to a wall and causing losses.
- *Controlling collisions* – controlled premature ("parasitic" or "long range") collisions in the Tevatron waste antiprotons and reduced luminosity at the detectors (see figure 6). Electrostatic separators create an electric field that pulls the protons and antiprotons in opposite directions toward the outside of the beam pipe as they orbit. When Run II began, parasitic collisions claimed 30 to 35 percent of the antiprotons in the beams. With new separators, losses to parasitic collisions have fallen to less than 3 percent.

The ComPASS code BeamBeam3D was used to perform multi-bunch (thirty six proton on thirty six antiproton bunches) Tevatron simulations in an attempt to explain emittance growth patterns in the presence of kicker misfiring. The 4D, transverse phase space, emittance growth of the thirty-six proton bunches is shown in figure 7 as the blue line. This simulation run includes beam-beam interactions at the CDF and D0 detector collision points and one long-range interaction point upstream and downstream of each (these are the most significant points), as well as impedance effects. Beam-beam effects are caused by the two beams interacting with each other through their charge (see figure 6 for an illustration of the effect). Impedance effects are caused by the electromagnetic wake generated by preceding beam bunches interacting with the accelerator chamber. The Tevatron lattice in the simulation was built from analysis of experimental measurements of lattice functions. The three-fold symmetry of the Tevatron filling pattern is evident as is the different behavior of the head bunch of each train: bunches 1, 13, and 25. The first bunch of the train suffers only one parasitic collision around the location of the head-on collision point because there is no preceding bunch from the opposing beam. When the separation of the two beams is artificially increased in the simulation, we obtain the green curve for emittance (Figure 7), which shows no bunch dependence. Figure 8 shows the measured proton intensity loss rate from a Tevatron store [24]. The losses vary significantly along a bunch train, in a systematic fashion that follows the same pattern we see in the simulation with the realistic long-range effects: the losses are lower for bunches at the start of each train nd larger near the end of each train. Since beam emittance is indeed a "figure of merit," where a lower value denotes a higher-quality beam, the simulation is describing very well the loss pattern.

3.2. *Future Work with ComPASS*

It is clear that although the simulations are impressive, the contribution of the ComPASS studies to the Tevatron improvements is a small part of the overall effort. On the other hand, four out of five of the necessary actions for the improved performance could have benefited from studies or designs using ComPASS code capabilities that exist today. The ComPASS vision focuses on using ComPASS codes in a unified simulation environment when a performance improvement problem, such as the Tevatron in the early days of Run II, needs to be tackled. In 2003, AST codes were under development for some of the studies necessary for the Tevatron improvement. Today, the ComPASS code suite is capable of addressing them successfully, and under the ComPASS modelling paradigm, our objective is to be able to perform these studies in the same integrated simulation environment.

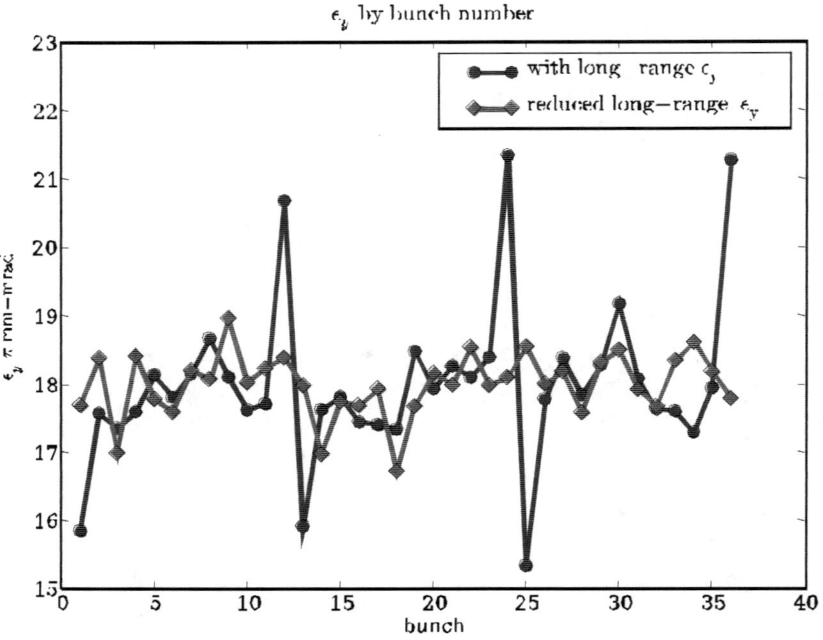

Figure 7. The simulated transverse phase space emittance of the thirty-six proton bunches for the full effect (blue line) and with reduced long-range effects (green line), versus bunch number.

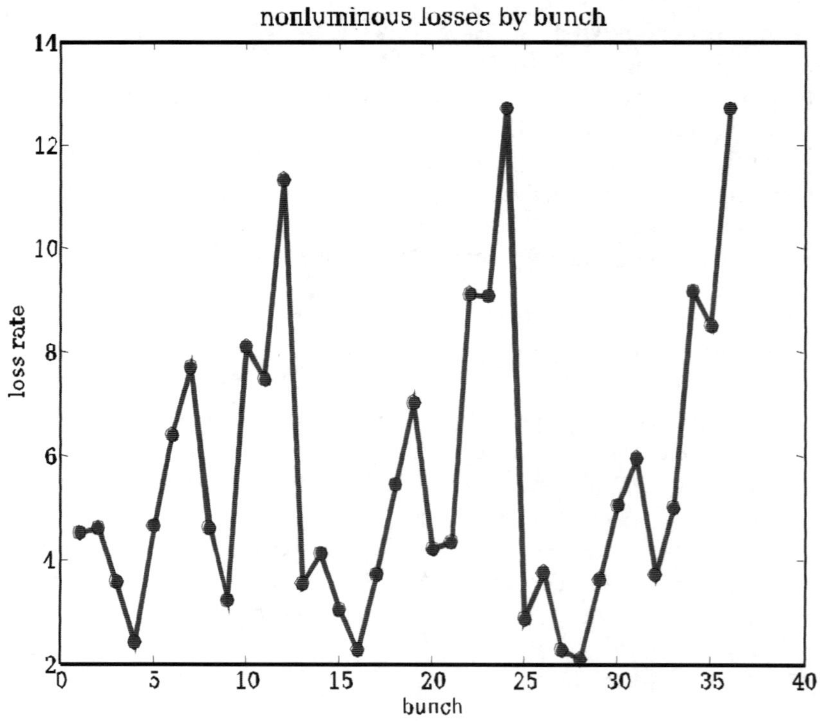

Figure 8. The measured loss rate of the thirty-six proton bunches from a Tevaton store versus bunch number.

4. Summary

The SciDAC Community Petascale Project for Accelerator Science and Simulation (ComPASS) is in the first year of executing its plan to develop the next-generation HPC accelerator modeling tools. ComPASS aims to develop an integrated simulation environment that will utilize existing and new accelerator physics modules with petascale capabilities, by employing modern computing and solver technologies. The ComPASS vision is to deliver to accelerator scientists a virtual ccelerator and virtual prototyping modeling environment, with the necessary multiphysics, multiscale capabilities. The plan for this development includes delivering accelerator modeling applications appropriate for each stage of the ComPASS software evolution. Such applications are already being used to address challenging problems in accelerator design and optimization [10, 11, 12].

The ComPASS organization for software development and applications accounts for the natural domain areas (beam dynamics, electromagnetics, and advanced acceleration), and all areas depend on the enabling technologies activities, such as solvers and component technology, to deliver the desired performance and integrated simulation environment.

The ComPASS applications focus on computationally challenging problems important for design or performance optimization to all major HEP, NP, and BES accelerator facilities. With the cost and complexity of particle accelerators rising, the use of computation to optimize their designs and find improved operating regimes becomes essential, potentially leading to significant cost savings with modest investment.

Acknowledgments

The ComPASS project is supported under the SciDAC program by the U.S. Department of Energy Office of High Energy Physics, Office of Nuclear Physics, Office of Basic Energy Sciences, and Office of Advanced Scientific Computing Research. The simulation work presented here used resources of the National Energy Research Scientific Computing Center and the Fermilab Lattice QCD facility.

References

[1] Abraham S, 2003 Spencer Abraham announces Department of Energy 20-year science facility plan, text available at http://www.er.doe.gov/Sub/Newsroom/News_Releases/DOE-SC/2003/Energy%2020-Year%20Science20Facility%20Plan.htm

[2] NRC Committee on Elementary Particle Physics in the 21st Century 2006 Revealing the hidden nature of space and time: charting the course for elementary particle physics (The National Academies Press)

[3] US Particle Physics: Scientific Opportunities A Strategic Plan for the Next Ten Years, http://www.er.doe.gov/hep/HEPAP/reports/P5_Report%2006022008.pdf .

[4] Deshpande A, Milner R, Venugopalan R and Vogelsang W 2000 *Ann. Rev. of Nuclear and Particle Science* **55:**0506148

[5] e-RHIC website, http://www.bnl.gov/cad/eRHIC/

[6] ELIC website, http://casa.jlab.org/research/elic/elic.shtml

[7] Burr F H 2007 HST.187 Course (Massachusetts General Hospital, Harvard Medical School, Proton Therapy Center)

[8] Ko K, *et al.* 2005 Impact of SciDAC on accelerator projects across the Office of Science through electromagnetic modeling *Proc. SciDAC Conf.* (San Francisco, CA); see also Ryne R 2006 J. *Phys.: Conf. Ser.* **46:**171; Spentzouris P 2006 Accelerator modeling under SciDAC: meeting the challenges of next-generation accelerator design, analysis, and optimization (9th Intern.

Comp. Accel. Physics Conf.) http://bel.gsi.de/icap2006/PAPERS/THMPMP03.PDF; and Qiang J,Lidia S, Ryne R and Limborg-Deprey C 2006 *Phys. Rev. ST/AB* **9:**044204 (2006)

[9] Blumenfeld I, *et al.* 2008 Energy doubling of 42 GeV electrons in a metre scale plasma wakefield accelerator *Nature* (to appear); see also Geddess C G R, *et al.* 2004 *Nature* **431:**538 and Mangles S, *et al.* 2004 *Nature* **431:**535

[10] Ng C 2008 Design and optimization of large accelerator systems through high-fidelity electromagnetic simulations *Proc. SciDAC 2008*

[11] Geddes C 2008 Computational studies and optimization of wakefield accelerators *Proc. SciDAC 2008*

[12] Qiang J 2008 SciDAC advances in beam dynamics simulation: From light sources to colliders *Proc. SciDAC 2008*

[13] Amundson J, *et al.* 2006 Synergia: An accelerator modeling tool with 3-D space charge *J. Comp. Phys.* **211:**229–248; see also Amundson J, *et al.* 2008 Multi-Scale, Multi-Physics Beam Dynamics Framework Design and Applications *Proc. SciDAC 2008*

[14] Ryne R 2006 MaryLie/IMPACT: A Parallel Beam Dynamics Code with Space Charge, http://scidac.nersc.gov/accelerator/marylielayer.html

[15] Keyes D 2007 Towards Optimal Petascale Simulations (TOPS) http://www.scidac.gov/math/TOPS.html

[16] Diachin L 2007 Interoperable Technologies for Advanced Petascale Simulations (ITAPS) www.itaps-scidac.org

[17] Pothen A 2007 Combinatorial Scientific Computing and Petascale Simulations (CSCAPES) http://www.scidac.gov/math/CSCAPES.html

[18] Ma K-L 2007 Institute for Ultrascale Visualization http://www.scidac.gov/viz/ultraviz.html

[19] Lucas R 2007 Performance Engineering Research Institute http://www.peri-scidac.org

[20] Colella P 2007 Applied Partial Differential Equations Center (APDEC) http://www.scidac.gov/math/APDEC.html

[21] Bernholdt D, *et al.* 2006 A Component Architecture for High-Performance Scientific Computing Int J High Performance Computing Applications 20 163–202

[22] Bernholdt 2007 Center for Technology for Advanced Scientific Component Software (TASCS) http://www.scidac.gov/compsci/TASCS.html

[23] Bethel E 2007 Visualization and Analytics Center for Enabling Technologies (VACET) www.vacet.org

[24] Shiltsev V, *et al.* 2005 *Phys. Rev. ST Accel. Beams* **8:**101001

Magnetic field generation by the stationary accretion shock instability

E Endeve[1,2,3]**, C Y Cardall**[1,2]**, R D Budiardja**[2] **and A Mezzacappa**[1]

[1]Physics Division, Oak Ridge National Laboratory, Oak Ridge, TN 37831, USA
[2]Department of Physics and Astronomy, University of Tennessee, Knoxville, TN 37996, USA
[3]Joint Institute for Heavy Ion Research, Oak Ridge National Laboratory, Oak Ridge, TN 37831, USA

E-mail: endevee@ornl.gov

Abstract. By adding a weak magnetic field to a spherically symmetric fluid configuration that caricatures a stalled shock in the post-bounce supernova environment, we explore the capacity of the stationary accretion shock instability (SASI) to generate magnetic fields. The SASI develops upon perturbation of the initial condition, and the ensuing flow generates—*in the absence of rotation*—dynamically significant magnetic fields ($\sim 10^{15}$ G) on a time scale that is relevant for the explosion mechanism of core collapse supernovae. We describe our model, present some recent results, and discuss their potential relevance for supernova models.

1. Introduction

Magnetic fields in the context of core collapse supernovae—the final evolutionary stage of massive stars—have recently attracted considerable interest (see [1, 2], and references therein) due, in part, to the generically observed asphericity of supernova explosions [3].

In particular, the work by [2] demonstrates the ability of magnetic fields to collimate and drive bipolar explosions when simulations of rapidly rotating progenitors are followed for several hundred milliseconds past core bounce: Differential rotation near the surface of the rapidly rotating proto-neutron star (PNS) wind and organize poloidal magnetic fields into "magnetic towers" along the rotation axis, consisting mainly of toroidal fields. Although this mechanism is compelling, current models rely on an initial field strength and topology, and rotation rates that are incompatible with stellar evolution calculations where magnetic effects are taken into account [4].

The magneto-rotational instability (MRI), introduced to core collapse supernova theory by [5], is believed, and often conveniently assumed, to bridge the gap and make the winding mechanism work. Capturing the MRI in numerical simulations requires high spatial resolution and has not been considered in realistic simulations, but the MRI was seen in the work of [1], where a simplified model of stellar collapse was considered. However, the potential impact of the MRI (and magnetic fields) on the core collapse supernova explosion mechanism remains unclear.

A different mechanism that may account for aspherical supernovae is the SASI; a hydrodynamical instability of the stalled supernova shock wave. Idealized two-dimensional and three-dimensional simulations have shown the SASI to have a dramatic impact on the postshock flow [?, 7]: The l=1 (sloshing) mode is seen to dominate 2D simulations, while the m=1 (spiral) mode dominates in 3D and may generate significant pulsar spin. The SASI is also seen to play an important role in the dynamics of more realistic models, e.g., [8, 9, 10, 11].

Given its impact on the postshock flow, it is natural to explore the influence of the SASI on magnetic fields (and vice-versa) in the supernova environment. By extending the axisymmetric model of [?] to include magnetic fields we have initiated an investigation on the evolution of magnetic fields when driven by the SASI: Starting from a spherically symmetric, nonrotating configuration, resembling a stalled supernova shock wave, we find that SASI-induced flows are able to amplify magnetic fields to dynamically significant levels within the time frame a supernova is expected to reach explosive conditions.

2. Physical model and numerical solution

rTo study magnetic field generation by the SASI, and the potential relevance to the core collapse supernova explosion, we adopt a highly idealized description of the post-bounce supernova environment [?]: A steady-state, spherically symmetric accretion shock is located at $r = R_{\rm Sh} = 200$ km from the center of the PNS. Matter at larger radii is falling into the shock at the free-fall speed, $u = \sqrt{2GM/r}$, where G is Newton's constant and the mass of the central object, M, is set to 1.2 $M_\odot$. The accretion rate is set to 0.36 $\rm M_\odot\ s^{-1}$ and is kept constant throughout the simulations. A constant, highly supersonic Mach number of 300 is used to set the pressure in the pre-shock gas. The gravitational potential is given by the point-mass formula: $\Phi = -GM/r$. The Rankine-Hugoniot conditions determine the hydrodynamic state just inside the shock, and the Bernoulli equation is solved for the structure from the shock to the surface of the PNS, which serves as the inner boundary and is located at $r = R_{\rm PNS} = 40$ km. The initial magnetic field is that of a split monopole: $\mathbf{B} = B_r \mathbf{e}_r$, where $B_r = \mathrm{sign}\,(\cos\theta) \times B_0\,(r/R_{\rm PNS})^2$. The strength of the magnetic field at $r = R_{\rm PNS}$, B_0, is set to 10^{10} G in our base model, and θ is the polar angle in the spherical coordinate system. The magnetic field lines are initially parallel to the accreting flow and the configuration is consistent with the steady state initial condition. The system is described by the nonrelativistic ideal magnetohydrodynamic (MHD) equations, which are closed with a polytropic equation of state (EoS) with the ratio of specific heats, γ, set to 4/3; similar to collapsing matter in a supernova.

We have implemented a time-explicit, second-order, semi-discrete central-upwind scheme [12] in our code, GenASiS , for the integration of the time-dependent ideal MHD equations. The equations are formulated in their integral form, appropriate for finite volume shock-capturing methods. We use the HLL formulas given in [13] to compute the numerical fluxes and electric fields, located on faces and edges of a computational cell, respectively. Second-order spatial accuracy in smooth regions, while maintaining a nonoscillatory behavior near shocks and discontinuities, is achieved by linear interpolation via the generalized minmod limiter used in [14]. The method of constrained transport [15] is used to maintain the divergence-free condition on the magnetic field during the evolution of the MHD equations. Second order temporal accuracy is obtained with a two-step Runge-Kutta time step algorithm [14]. The equations are solved on a uniform Cartesian grid ($\Delta x = \Delta y = \Delta z \equiv \Delta l$), and in order to solve axially symmetric problems we have implemented the Cartoon method by [16] for the MHD system.

The results presented in this study have been computed with a spatial resolution given by $\Delta l/R_{\rm PNS} \simeq 0.06$. At this resolution we obtain solutions that are consistent with the MHD equations in the sense that the total energy in the system is conserved at an acceptable level and that the unperturbed model can maintain the steady state of the initial condition indefinitely, without initiating the SASI from numerical noise.

3. Results

In order to initiate the SASI we perturb the initial condition by placing two dense rings in the material ahead of the shock. As the density enhancements pass through the shock, they

Journal of Physics: Conference Series **125** (2008) 012006 doi:10.1088/1742-6596/125/1/012006

introduce nonradial perturbations that grow and initiate the SASI. The $l = 1$ mode quickly manifests itself as an up-down "sloshing" of the global accretion shock configuration. In figure 1 we plot a time series illustrating the evolution of the SASI from the onset of the calculation and well into the nonlinear phase: We show a color plot the magnetic field strength, $|\mathbf{B}|$, at four different times during the evolution. The snapshots shown are taken at 294 ms (a), 336 ms (b), 424 ms (c), and 650 ms (d).

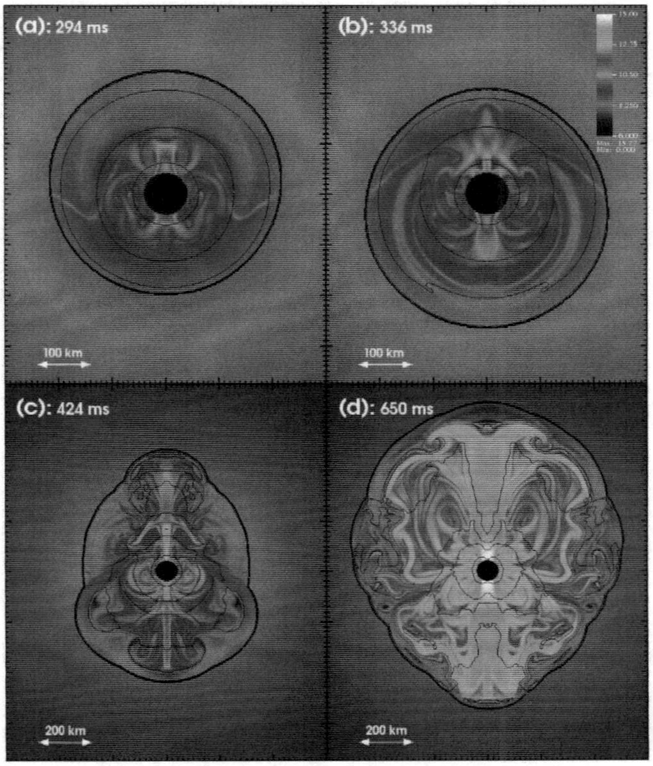

Figure 1. Color plot of the magnitude of the magnetic field, $|\mathbf{B}|$, during the evolution of the SASI. Panel (a) is sampled at 294 ms, panel (b) at 336 ms, while lower panels, (c) and (d), are sampled at 424 ms and 650 ms, respectively. The color-bar in the upper right (valid for all panels) indicates the magnetic field strength on a log scale, extending from 10^6 G to 10^{15} G. Thin lines are contours of constant density, starting with the innermost contour we plot density contours at $\rho = 10^{10}$, 10^9, 3×10^8, and 6×10^7 g cm^{-3}. The thick contour indicates the location of the shock surface. Notice the factor of two difference in the spatial scale used in the upper and lower panels of the plot.

During SASI development the shock starts to deviate from spherical symmetry, the accreting flow hits the shock at an oblique angle, and significant nonradial flows are introduced inside the shocked cavity. The lateral flow carries magnetic field toward the symmetry axis, where the strength of the field increases due to compression: At 294 ms the strongest field is found around the north pole, with $B_{\rm max} \sim 2 \times 10^{12}$ G. At the south pole the field strength is about 3×10^{11} G. Some 40 ms later, at 336 ms, the situation is reversed, and the strongest field, $B_{\rm max} \sim 2 \times 10^{12}$ G, is found around the south pole.

An internal shock forms in the postshock gas around 360 ms, and is connected to the accretion shock in a tripple point [7] (seen as a kink in the shock surface). The internal shock propagates in the north-south direction in panel (c). Ahead of the internal shock, accretion powered flow plunges toward the PNS and the axis of symmetry, causing further field growth. At 424 ms columns of stronger magnetic field are seen in both polar regions, reaching maxima of some 10^{14} G. At 650 ms extended regions of strong magnetic field have formed around both the north and south pole. The strength of the magnetic field in these regions is above 10^{15} G, near the surface of the PNS, and stays above 10^{14} G out to about 100 km beyond the surface of the PNS. The magnetic field is these regions is strong enough to influence the dynamics: From the density contours in panel (c) we see that low density funnels have formed along the polar axis.

The impact of the SASI induced flow on the magnetic field is further illustrated in figure 2 where we plot subsets of the integrated kinetic (red lines) and magnetic (black lines) energies

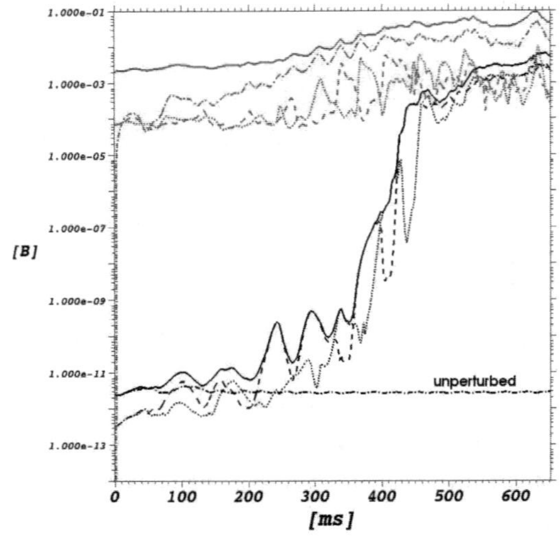

Figure 2. Kinetic (red curves) and magnetic (black curves) energies (see text for detail): $E^{\rm Sh}_{\rm kin}$ (——), $E^{\theta}_{\rm kin}$ (— · —), $E^{\rm N}_{\rm kin}$ (- - - -), and $E^{\rm S}_{\rm kin}$ (· · · · · ·). $E^{\rm Sh}_{\rm mag}$ (——), $E^{\rm N}_{\rm mag}$ (- - - -), and $E^{\rm S}_{\rm mag}$ (· · · · · ·). We also plot $E^{\rm Sh}_{\rm mag}$ for an unperturbed model (— · —). (1 B=10^{51} erg.)

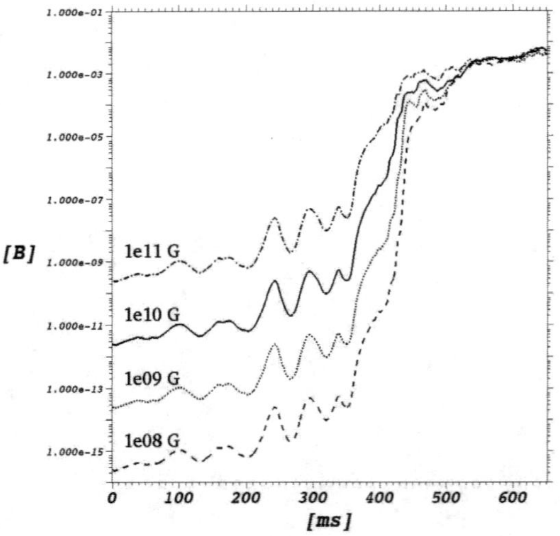

Figure 3. Magnetic energy inside the shocked cavity, $E^{\rm Sh}_{\rm mag}$, for models where we have varied the strength of the initial magnetic field, B_0: 10^8 G (- - - -), 10^9 G (· · · · · ·), 10^{10} G (——), and 10^{11} G (— · —), respectively.

inside the shocked cavity: The solid red line represents the total kinetic energy inside the shock, $E^{\rm Sh}_{\rm kin} = \int_{V_{\rm Sh}} \rho \frac{\mathbf{u}\cdot\mathbf{u}}{2}\, dV$, where $V_{\rm Sh}$ is the total volume encompassed by the shock. The red, dash-dot line represents the kinetic energy of the lateral motion, $E^{\theta}_{\rm kin} = \int_{V_{\rm sh}} \frac{1}{2}\rho u_{\theta}^2\, dV$. The dashed and dotted, red lines represent the kinetic energy around the northern and southern hemispheres, respectively: $E^{\rm N}_{\rm kin} = \int_{V_{\rm N}} \rho \frac{\mathbf{u}\cdot\mathbf{u}}{2}\, dV$ and $E^{\rm S}_{\rm kin} = \int_{V_{\rm S}} \rho \frac{\mathbf{u}\cdot\mathbf{u}}{2}\, dV$. Here $V_{\rm N(S)}$ is the volume bounded by the surface of the PNS, the surface of the accretion shock, and the surface a cylinder with radius $R_{\rm PNS}$, centered on the polar axis in the northern (southern) hemisphere. The solid, black line represents the magnetic energy inside the accretion shock, $E^{\rm Sh}_{\rm mag} = \int_{V_{\rm sh}} \frac{\mathbf{B}\cdot\mathbf{B}}{2\mu_0}\, dV$, while the dashed and dotted lines represent the magnetic energy in cylinders around the north and south pole: $E^{\rm N}_{\rm mag} = \int_{V_{\rm N}} \frac{\mathbf{B}\cdot\mathbf{B}}{2\mu_0}\, dV$ and $E^{\rm S}_{\rm mag} = \int_{V_{\rm S}} \frac{\mathbf{B}\cdot\mathbf{B}}{2\mu_0}\, dV$, respectively. For comparison, we also plot the magnetic energy inside the shock, $E^{\rm Sh}_{\rm mag}$, for an unperturbed model (black, dash-dot line).

The nonradial perturbations lead to the growth of the kinetic energy of the lateral motion inside the shock. This is seen in the rise in $E^{\theta}_{\rm kin}$ for $t < 400$ ms. The sloshing nature of the $l = 1$ mode is also revealed in the alternating pattern of the kinetic energies contained in the cylinders centered around the northern and southern hemispheres. Note the correspondence of rise in magnetic energies with decline in kinetic energies. This persists to about 450 ms, when the flow takes on the character of the $l = 2$ mode. From the evolution of the magnetic energy we can see that the SASI has a quite dramatic impact on the magnetic field. The early evolution is characterized by relatively modest oscillations in the strength of the magnetic field superimposed on a trend of gradual overall increase, but starting at about 360 ms, around the formation of the internal shock, *the magnetic energy increases by six orders of magnitude in roughly 100 ms.* Beyond 450 ms the magnetic energy saturates at a level somewhat below the total kinetic energy inside the shock. The magnetic field is strongly aligned with the symmetry axis, and the magnetic energy is in approximate local equipartition with the kinetic energy in the polar regions.

We have also done calculations with different initial field strengths. In figure 3 we plot the magnetic energy inside the accretion shock, $E_{\rm mag}^{\rm Sh}$, for models where we have varied B_0: 10^8 G (dashed line), 10^9 G (dotted line), 10^{10} G (solid line, the model shown in Figures (1) and (2)), and 10^{11} G (dash-dot line). These calculations show that our result is insensitive to the strength of the initial field: The magnetic energies saturate at the same level for all models.

4. Summary and discussion

By considering an idealized model of a stalled supernova shock wave we have found that the SASI-induced flow beneath the shock is able to generate dynamically significant magnetic fields, in excess of 10^{15} G, on a time scale relevant to the supernova explosion, independent of the initial fields considered. The influence of the SASI on the evolution of magnetic fields in core collapse supernovae will need to be investigated in more realistic models, but our idealized models offer a proof of principle that (1) magnetic fields may be amplified to significant strengths *even in the absence of rotation* and (2) compression may serve as an effective mechanism for rapid field amplification to dynamically significant levels, contrary to conclusions reached in past MHD collapse calculations. Moreover, current 2D models [8, 9, 10, 11] demonstrate that the SASI is common to core collapse supernova models. Our results then suggest that SASI magnetic field amplification and, in particular, the formation of dynamically significant fields may be common as well, and not confined to special cases—for example, to cases of rapid progenitor rotation. The key question now is: What happens in three dimensions? The amplification we document here relies on the continual focusing of fluid flow toward the polar axis, in part an artifact of our assumed axisymmetry. 3D hydrodynamics simulations of the SASI yield a sufficiently different and more complex postshock flow [7]. The consequences of this flow for stellar core magnetic field amplification must be investigated.

Acknowledgments

This work was supported by Oak Ridge National Laboratory (ORNL), managed by UT-Battelle, LLC, for the DoE under contract DE-AC05-00OR22725. This research used resources of the National Center for Computational Sciences at ORNL, which is supported by the Office of Science of the U.S. Department of Energy under Contract No. DEAC05-00OR22725.

References

[1] Shibata M, Liu Y T, Shapiro S L and Stephens B C 2006 *Phys. Rev.* D **74** 104026
[2] Burrows A, Dessart L, Livne E, Ott C D and Murphy J 2007 *ApJ* **664**, 416
[3] Wang L, Howell A D, Höflich P and Wheeler C J 2001 *ApJ* **550** 1030
[4] Heger A, Woosley S E and Spruit H C 2005 *ApJ* **626** 350
[5] Akiyama S, Wheeler C J, Meier D L and Lichtenstadt I 2003 *ApJ* **584** 954
[6] Blondin J M and Mezzacappa A 2007 *Nature* **445**, 58
[7] Blondin J M and Mezzacappa A 2007 *Nature* **445**, 58
[8] Buras R, Janka, T H, Rampp M and Kifonidis K 2006 *A&A* **457** 281
[9] Burrows A, Livne E, Dessart L, Ott C D and Murphy J 2006 *ApJ* **640**, 878
[10] Bruenn S W et al. 2006 *Journ. Phys. Conf. Ser.* **46** 393
[11] Scheck L, Janka, T H, Foglizzio T and Kifonidis K 2008 *A&A* **477** 931
[12] Kurganov A, Noelle S and Petrova G 2001 *SIAM J. Sci. Comput.* **23** 707
[13] Londrillo P and Del Zanna L 2004 *JCP* **195** 17
[14] Kurganov A and Tadmor E 2000 *JCP* **160** 241
[15] Evans C R and Hawley J F 1988 *ApJ* **332** 659
[16] Alcubierre M et al. 2001 *Int. J. Mod. Phys* **10** 273

Journal of Physics: Conference Series **125** (2008) 012007 doi:10.1088/1742-6596/125/1/012007

Multidimensional radiative transfer calculations of the light curves and spectra of Type Ia supernovae

Daniel Kasen[1,2], R C Thomas[3], Fritz Röpke[4], S E Woosley[1]

[1] Department of Physics and Astronomy, University of California, Santa Cruz, CA 95064, USA
[2] Hubble Fellow, University of Callifornia, Santa Cruz, CA 95064, USA
[3] Lawrence Berkeley National Laboratory, Berkeley, CA 94720, USA
[4] Max Plank Institute for Astrophysics, Garching, Germany

E-mail: kasen@ucolick.org

Abstract. The explosion of a white dwarf star in a Type Ia supernova (SN Ia) explosion leads to the burning and ejection of stellar material at a few percent of the speed of light. The spectacle we observe in the months that follow is from the leaking of radiation from this glowing mass of radioactive debris. The modeling of SN Ia light curves and spectra represents a complex problem in time-dependent radiative transfer. Here we discuss numerical methods, in particular Monte Carlo methods, for calculating 3D multi-wavelength radiative transport on massively parallel machines. Our approach involves a newly developed domain decomposition technique in which the memory load is distributed over multiple processors and photon packets are communicated from node to node. We present results for 2-dimensional models that explore white dwarf explosions over a range of explosion paradigms and ignition conditions. These models give insight into how variations in the initial conditions of the explosion affect the light curve we finally observe. We conclude with an outlook (and some initial results) for large scale 3D radiation transport calculations of SNe Ia in an era of petascale computing.

1. Introduction

Among the brightest of stellar explosions are the thermonuclear, or Type Ia, supernovae (SNe Ia). They appear to us as a point source of light, much like a star, but brightening day by day to a peak intrinsic luminosity some billion times greater than that of the sun. We sometimes speak of this transient brilliance as a "supernova explosion," but in fact the explosion had ended long before, having consumed and disrupted the progenitor star in just a few seconds. What we are actually seeing is the stellar debris, burned and ejected at a few percent the speed of light, and glowing under the heat of its own freshly synthesized radioactivity.

SNe Ia can be seen at great distances, billions of light years, and so can be messengers of the universe at earlier ages. Unlike other kinds of supernovae (SNe), the Type Ia events show an impressive regularity in their luminosity and spectral evolution. Empirically, we find that brighter SNe Ia brighten and fade more slowly than dimmer SNe Ia [1], a relation that can be used to infer the intrinsic luminosities from the light curve shape. In this way, SNe Ia are applied as standard candles to measure cosmological distances, probe dark energy and map out the expansion history of the universe. A persistent concern, though, is that the average properties of SNe Ia could evolve over cosmological time (because of demographic shifts in the progenitor star population), which would subject the measurements of cosmological parameters to systematic bias.

A theoretical understanding of SNe Ia, based in numerical simulation, would help secure and refine their utility as cosmological probes. The problem is complex and multifaceted and requires a suite of codes to capture the diverse physics at play, which include stellar evolution, hydrodynamics, turbulent combustion, nucleosynthesis, and radiative transfer. Ideally, we want to simulate all aspects of the phenomenon from first principles. In practice, we still face a number of uncertainties concerning, for example, the exact nature of the progenitor star system, the conditions of ignition, and the physics of the flame propagation that unbinds the star.

Validation of supernova explosion simulations comes ultimately from comparing model predictions to observations of real events. Thus we must model the light curves and spectra of the remnant, a difficult problem in time-dependent radiative transport, and one in which several confounding effects must be considered, for example, radioactive energy deposition, differential expansion, line transport, and deviations from local thermodynamic equilibrium. Realistic explosion models also involve some deviation from spherical symmetry, so the transfer calculations must be carried out in three spatial dimensions as well. This problem in its fullest generality is only beginning to be addressed, and should continue to push the limits of high performance computing for years to come.

The focus here is on radiative transfer in SNe Ia. In Section 2, we recount the first ten seconds of the SNe Ia eruption, followed by a description of the light curve and spectrum formation over the following months (Section 3). In Section 4, we discuss numerical methods for radiation transport, in particular Monte Carlo methods, and strategies from parallelization and domain decomposition (Section 5). We then (Section 6) present results for 2D models that explore stellar explosions over a range of ignition conditions. We end (Section 7) with the outlook and some initial results for large-scale 3D radiation transport calculations of SNe Ia in an era of petascale computing.

2. The supernova eruption

We observe Type Ia SNe exploding in galaxies of all types, even those with old (> 1 Gy) stellar populations. Therefore these events cannot signal the death of a massive star (as in the more common case of core-collapse supernovae) because such stars have lifespans of only ~ 10 million years. We believe that SNe Ia are the disruption of compact white dwarf stars supported by electron degeneracy pressure, that have grown to criticality over longer time-scales. In one favored scenario, the progenitor has a binary companion that, after some extended period of stellar evolution, begins transferring mass to the white dwarf. When the white dwarf mass has grown near the limiting Chandrasekhar value of $\sim 1.4\ M_{\odot}$, the increasing central temperature and density initiate carbon burning in the core.

Carbon burning can continue "simmering" in the white dwarf for several centuries, with the nuclear energy being spirited away by convective flows. Multidimensional models show a complex medium with chaotic fluctuations in entropy that eventually serve as the "sparks" for the explosion [2]. The location of ignition, and whether it happens at one or many points, may profoundly influence the supernova we eventually observe (see Section 6.2). The stochastic nature of this process may be one important reason that not all SNe Ia are exactly alike.

The initial sparks gives rise to a subsonic flame, known as a deflagration, which is modulated by turbulence and fluid instabilities. Because the stellar matter is supported by electron degeneracy pressure, it cannot readily expand in response to the rising temperature, that only promotes further burning. The result is thermonuclear runaway. The details of flame propagation are complex, but within seconds burning has consumed the greater part of the star. The fusion of carbon and oxygen to heavier elements releases sufficient energy ($E \sim 2\times 10^{51}$ ergs) to unbind the white dwarf and eject the stellar material with velocities of order 10,000 km s^{-1}.

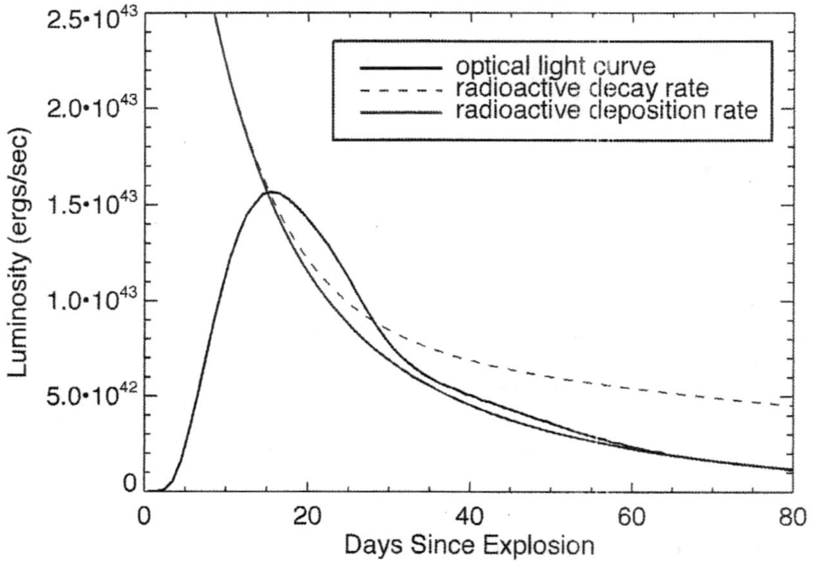

Figure 1. The optical light curve of a Type Ia supernova model (black line) is powered by the decay of radioactive isotopes – this explosion synthesized over 0.6 $M_\odot$ of ^{56}Ni. The total energy generation rate from radioactive decay (dashed red line) tracks the half-lives of the isotopes. After day 20, the ejecta become increasingly transparent to the gamma rays, and only a fraction of the radioactive decay energy is absorbed in the ejecta (solid red line).

Nothing is left behind.

About ten seconds later, hydrodynamical and nucleosynthetic processes abate, and the ejecta reach a state-free expansion. The subsequent evolution is peaceful by comparison: the ejected debris expands, with its radius increasing linearly with time and its density dropping as t^{-3} from geometrical dilution. With no further dynamics, the ejecta structure (i.e., the density profile and chemical stratification) retains a self-similar form, and any asymmetry or substructure that was imparted in the explosion is frozen in.

Throughout the entire process described so far, the stellar material remains so opaque that radiative transport is of little importance. By a day or so after the explosion, however, the densities have declined such that diffusion of optical radiation does become significant on the expansion time-scale. It is ultimately the leaking of photons from the hot mass of ejected debris that we observe as the supernova light curve.

3. Light curves and spectra

The material ejected in the thermonuclear explosion is initially quite hot, around 10^9K, but because of the high velocities it cools rapidly by adiabatic expansion. The thermal energy pool is essentially completely converted to kinetic energy within minutes, well before any appreciable radiation can diffuse to the ejecta surface. If this were the whole story, SNe Ia would lack a luminous display altogether. Remarkably, Nature provides an additional energy reservoir in the form of radioactive isotopes, in particular ^{56}Ni, which are synthesized in abundance during the explosion.

To first order, the peak brightness of a SN Ia is determined by the quantity of ^{56}Ni produced.

A typical event makes around 0.6 $M_\odot$, but for some reason (discussed further in §6.2) different explosions have different yields. The decay chain of nickel (^{56}Ni $\rightarrow^{56}$ Co $\rightarrow^{56}$ Fe) releases $\sim 1 MeV$ gamma rays, which deposit energy in the ejecta primarily through Compton scattering. It is this energy that, through a process of thermalization or excitation/ionization, ultimately re-emerges as the optical/UV radiation we observe.

Figure 1 shows a typical example calculation of a SN Ia light curve. The rate of energy input from radioactive decay follows the half-lives of the isotopes ($\tau_{\rm Ni}$ = 6 days, $\tau_{\rm Co}$ = 78 days). At early times, essentially all of the decay products are absorbed locally, but after $\sim$ 30 days the Compton opacity has dropped such that gamma rays begin to escape the ejecta straightaway. Any absorbed radioactive energy is re-radiated almost immediately in the optical/UV, but at first it takes some weeks for that radiation to diffuse to the ejecta surface. Thus, the visible light rises to a peak in about 20 days, then declines on a similar time-scale. At late times ($t > 80$ days), when the ejecta are optically thin, the light curve tracks directly the exponential decline of radioactive energy deposition.

The time duration of the SN light curve can be considered a measure of the effective diffusion time of photons and hence depends on the total ejecta mass, the rate of expansion, and the opacity of the supernova ejecta. The opacity is complex, being dominated by literally millions of atomic line transitions, each Doppler broadened by the differential expansion of the ejecta and in aggregate blending together into a pseudo-continuum (see figure 2). Transitions from iron group elements (iron, cobalt, nickel) are the most numerous and have their greatest effect at bluer wavelengths. Because the number of optically thick lines depends on the atomic ionization and excitation state, the opacity is a sensitive function of the ejecta thermal and statistical state.

The emergent spectra of SNe Ia are characterized by broad absorption/emission line features superposed on a pseudo-blackbody continuum (figure 3). The line features allow us to diagnose the chemical composition and (by measuring Doppler shifts) the expansion velocity of the ejecta. At early times ($\sim$ 20 days after the explosion) SN Ia spectra are dominated by features of intermediate mass elements (IME) such as silicon, calcium, and sulfur. A few weeks later the opacity has dropped and one peers deeper into the ejecta; the spectrum then evolves to one dominated by iron group features. This leads to a picture in which the iron group elements are concentrated in the ejecta core, which is surrounded by a shell of IME and, perhaps, an outermost clumpy layer of unburned carbon and oxygen.

4. Multidimensional radiative transfer

SN light curves are modeled by post-processing the output of explosion simulations that have been run to the free-expansion phase. At that point, the ejecta structure (i.e., the density, velocity, and chemical composition defined on a spatial mesh) fully determines the emergent light curves, spectra, and spectropolarization. Several codes, using various numerical techniques, have been developed to model aspects of the radiation transport phase in SNe, for example, [3, 4, 5, 6, 7]. We focus here on SEDONA [8], which solves the time-dependent multi-wavelength transfer problem in one, two, or three spatial dimensions.

The full light curve calculation is actually built of several subproblems, with the work flow as follows: (1) calculate the radioactive decay rate and the transport and absorption of emitted gamma rays; (2) determine the material state (i.e., temperature, degree of ionization/excitation) at each spatial point; (3) compute the wavelength-dependent opacity at each spatial point, given the local state; and (4) solve the transport of visible radiation through the rapidly expanding, optically thick ejecta. In fact, the radiation field affects the material state through heating and radiative ionization/excitation, so steps (2) and (4) are coupled together in a nonlinear way.

The radiation field is, in general, a function of 3 spatial dimensions (x, y, z), as well as time t wavelength λ and (because isotropy does not typically hold) of direction as defined by the two propagation angles θ and ϕ. The radiation transfer equation is thus formally a differential

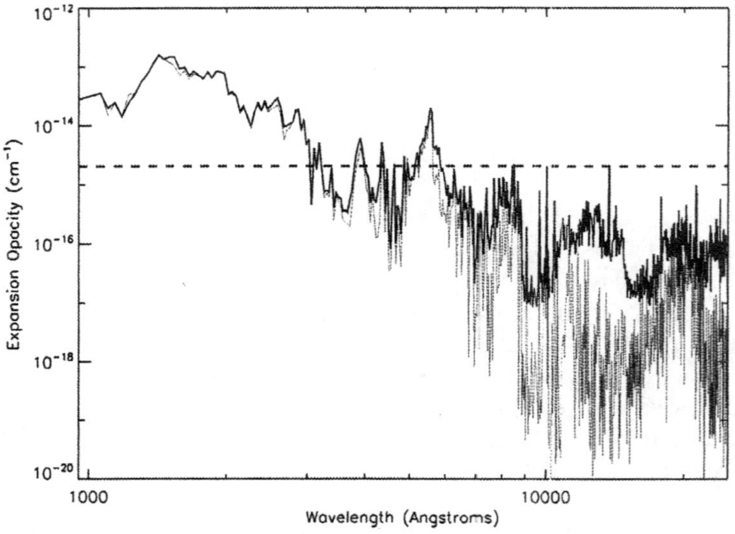

Figure 2. The opacity of SNe Ia is dominated by numerous Doppler-broadened lines that blend together into a pseudo-continuum opacity, here shown for a parcel of cobalt/iron rich gas. The thin red line is calculated by using a restricted atomic linelist of $\sim 500,000$ lines, while the.thick black line is the result by using the full list of ~ 42 million lines. The horizontal dashed line marks the level of electron scattering opacity.

equation involving seven independent variables. Methods for direct numerical solution of the equation using accelerated lambda iteration techniques have recently been implemented in the context of 3D supernovae [9]. More often, approximate schemes are adopted to reduce the dimensionality of the problem by neglecting, for example, the angular dependence (e.g., flux-limited diffusion), the wavelength dependence (grey transport), or time (stationarity approach). Our goal is to address the problem in its full generality.

Given the complexity and high-dimensionality of the problem, Monte Carlo (MC) methods are appealing and the approach adopted in SEDONA. In the MC scenario, packets of radiant energy ("photons") are emitted throughout the ejecta and tracked through randomized scatterings and absorptions until they escape the envelope. Each packet possesses a unique wavelength and polarization Stokes vector, which are updated at each scattering event. Packets passing through a grid cell are tallied to construct estimators of the local radiation field, and the corresponding rates of radiative heating, ionization, and excitation. Packets escaping the grid can be used to construct the emergent spectrum and light curve. All calculated quantities possess statistical noise, which is reduced as the number of propagated packets is increased.

Photon packets diffuse through all dimensions of the problem: stochastically through space, forward in time and – due to absorption and remission processes – back and forth in wavelength space. Because the ejecta is in differential expansion, the packet wavelength is also constantly Doppler shifting with respect to the local co-moving frame, causing packets to sweep through lines one by one. In this case, line scattering and absorption processes can be simplified by adopting the narrow-line limit, or Sobolev approximation, which is appropriate for media in high differential expansion [10]. Here, the ejecta properties are assumed constant over the small spatial region in which the photon Doppler shifts into resonance with a line. A more exact treatment of line transport, which we have also begun implementing, requires very high

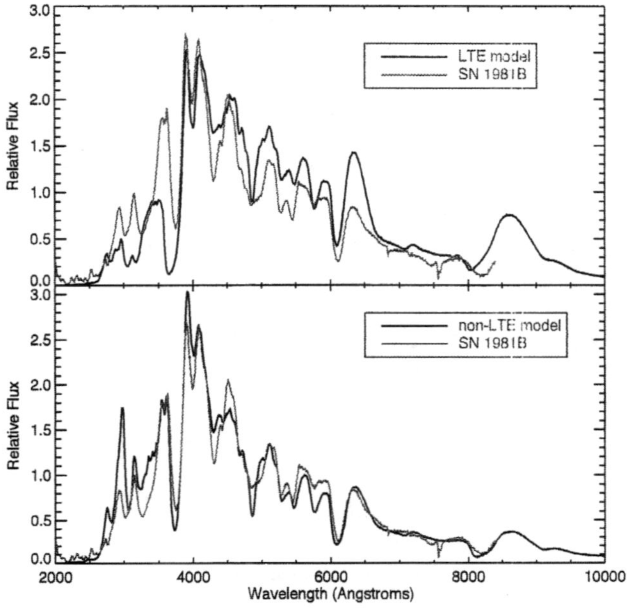

Figure 3. The spectrum of the observed Type Ia SN 1981B compared to the synthetic spectra of a 1-D SN Ia model (black line) computed assuming LTE (*top*) and non-LTE (*bottom*). While LTE is a generally good approximation, it overpredicts the emission component of strong line features and gives too red a color for the pseudo-continuum.

wavelength resolution ($\sim 10^5$ points) in order to resolve individual line profiles. The memory requirements in 3D are therefore extreme but should be accessible using current high performance platforms and new parallelization strategies (see Section 5)

The line opacities and emissivities depend on the ionization and excitation state of the ejecta gas. Under the condition of local thermodynamic equilibrium (LTE) the calculation of these is relatively simple, as they are well-defined functions of the local temperature, density, and composition. Unfortunately, in the tenuous atmospheres of SNe, thermalizing processes (collisions) are rare, and radiative transitions and radioactive excitation can drive the material out of equilibrium. In this case, one must address the full non-LTE problem by solving a set of nonlinear coupled equations describing the rates into and out of each atomic level. Typically, several 100's levels per ion are considered, coupled by 10,000's of line transitions. The matrix must be solved at all spatial points and at all times, making fast algorithms essential.

We have implemented a non-LTE solver into SEDONA that uses tallies of Monte Carlo packets to estimate the rates of radiative transitions (treated in the Sobolev formalism). Figure 3 compares a LTE synthetic spectrum to a non-LTE spectrum for a 1D SN Ia model at maximum light (20 days after explosion). Clearly LTE provides a good qualitative approximation, which justifies its use in our initial multidimensional calculations. In detail, however, the color of the spectrum and the strength of individual line features are sensitive to non-LTE effects. At yet later times (> 70 days after explosion) LTE breaks down altogether.

Realistic opacity calculations require extensive atomic data – we use an atomic linelist from

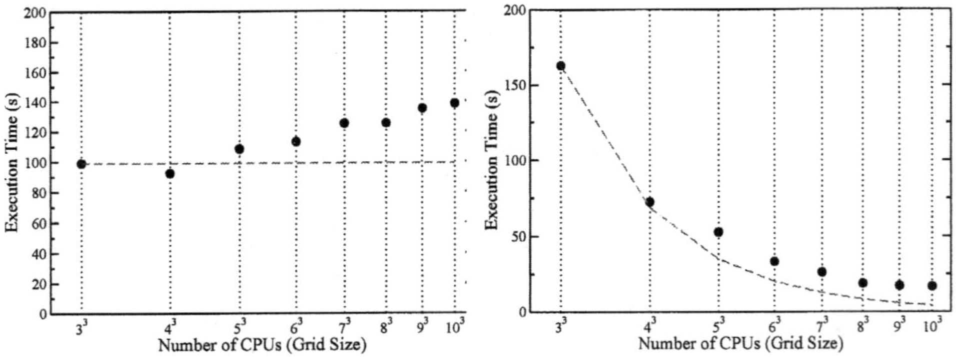

Figure 4. Weak scaling (*left*) and strong scaling (*right*) results for the domain-decomposed Monte Carlo test problem. The blue dashed line represents "perfect" scaling. The points are actual code performance on the Franklin supercomputer at NERSC.

[11], which provides data for over 40 million bound-bound transitions. Even this list is likely incomplete or inaccurate to some degree, especially at longer wavelengths (figure 2). Input from experiments and advancing theoretical calculations will, we hope, improve the atomic data set for astrophysics in the future.

5. Parallelization and domain decomposition strategies

For simple 1D problems, Monte Carlo methods can be computationally inefficient because a very large number of photon packets must be propagated to achieve reasonable noise levels. The MC approach, however, becomes of increasing benefit when facing problems of higher complexity and dimensionality. For 3D multi-wavelength radiation transfer, MC methods could rival or surpass deterministic methods, especially when made to run efficiently and scalably on massively parallel machines.

In problems involving one or two spatial dimensions, where memory requirements are fairly modest, Monte Carlo is "embarrassingly parallel." Each processor holds a copy of the entire grid and propagates packets independently, with the emergent packet statistics being combined at the end of the calculation. This approach, known as "full replication," scales perfectly and without limit, since communication is isolated to a few well-defined, deterministic parts of the program. Unfortunately, it also restricts the grid resolution to the memory available on an individual node, typically only several gigabytes. The requirements for a well-resolved 3D radiative transfer problem far exceed this. A typical calculation would use order 100^3 spatial points and 10^4 to 10^5 wavelengths points, requiring storage of $\sim 10^{11}$ values of opacity, emissivity, and radiation field tallies. At double precision, this necessitates several hundreds of gigabytes to 1 TB of memory.

The situation is complicated by industry trends, which drive new high-performance computing architectures away from massively parallel, shared-memory systems toward systems with fewer CPUs and less memory per node, but faster internode communication. To most effectively use such machines for 3D radiative transfer, we have recently developed and implemented an algorithm for domain-decomposed Monte Carlo (DDMC), in which the model grid is distributed over a large number of CPUs. As packets propagate they "diffuse" from node to node, not unlike the diffusion of photons in the real supernova debris. With DDMC, grid sizes of a few terabytes are accessible, at the expense of a significant increase in communication.

Our specific DDMC algorithm alternates between propagation and communication cycles. The model atmosphere is divided into a number of contiguous spatial "chunks," each assigned to a single CPU. Each chunk is further subdivided into 3D Cartesian mesh of "cells" within which the values of physical variables are defined. Each CPU is responsible for propagating energy packets through its chunk of the model domain. When a packet reaches a chunk boundary, it is buffered, awaiting communication to a neighboring CPU.

When all CPUs have propagated their load of packets to chunk boundaries, the propagation phase ends and communication begins. Our implementation uses the Message Passing Interface (MPI); and to avoid the overheads in passing numerous small messages (packets are less than 100 bytes each), packets are sent in large batches. Distinct MPI calls (both a send and a receive) must be performed at each of the six faces of the chunk cube, accomplished by looping over dimension (x, y, z) and oscillating a send/receive between the even/odd chunk pairs. An alternative approach of "global array" (GA) parallelism is attractive because it is designed to reduce overheads transparently and obviate the two-phase approach, but GA seems to be less universally available than MPI for the present.

If the required memory for the entire grid is less than the aggregate memory available, the grid can be duplicated on further sets of CPUs to recover some of the favorable scaling properties of the "full replication" approach. Another option is to use spare CPUs for load balancing. Chunks representing the dense inner regions of the supernova ejecta typically have a heavier scattering load than do those located near the outer boundaries. Replicating the heavily loaded chunks on additional CPUs can mitigate the latency induced in the propagation phase, but incurs a further communication cost. Hence the load-balancing option may require intelligent tuning.

We have performed both weak and strong scaling tests of our DDMC code with a simple grey opacity test problem (figure 4). To probe weak scaling, we kept the average number of scatterings per CPU constant and enlarged the problem grid (adding chunks in proportion to the number of CPUs). In the strong scaling tests, the same physical volume was progressively gridded at higher resolution, reducing the per-CPU workload. The results (which depend also on the opacity, and hence propagation load, of the particular model) show a weak scaling increase of only 38% out to 1,000 processors. Both the weak and strong scaling results are promising, especially considering the range of grid sizes tested. Moreover, the block of 1000 processors can now be replicated 10 or 100 times over on additional processors with no further losses in inefficiency. Thus DDMC will permit large-scale radiative transfer calculations to be performed effectively on petascale machines.

6. Results for 1D and 2D Type Ia models

Although several aspects of SN Ia explosion process remain uncertain, progress can be made by parameterizing areas of ignorance and comparing model observables to real data. In previous work [13, 14], we studied the diversity of SN Ia light curves using simplified 1-D explosions that bypassed the details of turbulent nuclear combustion. These studies gave insight into how SN Ia light curves depend on ejecta properties and helped clarify the underpinning of the important width-luminosity relation.

In particular, our radiation transfer calculations emphasized the importance of the opacity's dependence on the ionization state. In SNe Ia, gradual cooling of the ejecta promotes the recombination of Fe III to Fe II soon after the light curve peak. Because optical/infrared lines are more numerous for Fe II than for Fe III, this cooling increases the line opacity and enhances the fluorescence of blue photons to longer wavelengths. The effect of wavelength redistribution in Fe II lines explains why the infrared light curves of SNe Ia rise to a "secondary" peak several weeks after the first (Figure 5). In dimmer SN Ia models, which are generally cooler, we find that recombination sets in earlier and the secondary maximum is hastened, a trend also noted in observations [15]. For the same reason, the flux in the blue bands disappears more quickly in the cooler SNe, resulting in a faster declining B-band light curves for dimmer SNe Ia. Such effects are at the heart of the Type Ia width luminosity relation.

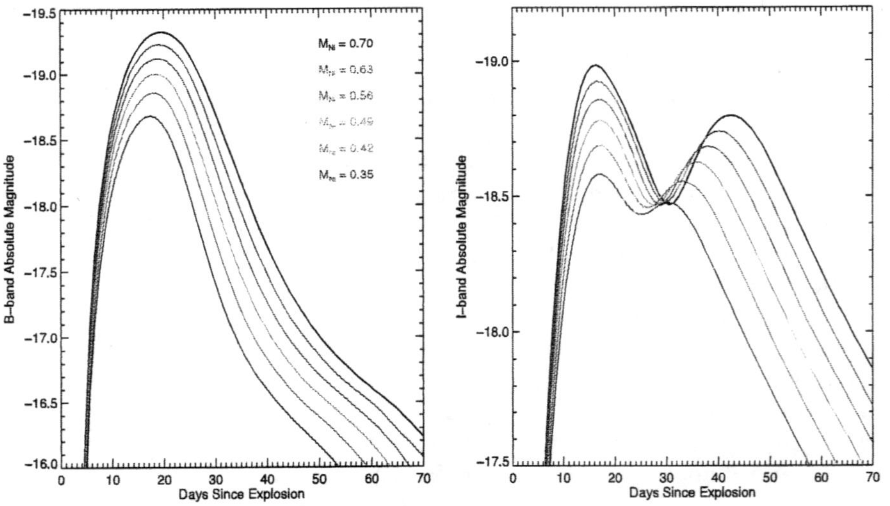

Figure 5. *Left:* Light curves (in the blue wavelength band) of parameterized 1D SN Ia models with ^{56}Ni mass varied between 0.35 and 0.70 $M_{\odot}$. The brighter models decline more slowly than then dimmer ones. *Right:* Near Infrared light curves of the same models. The secondary maximum is due to the redistribution of flux from bluer to redder wavelength, which is enhanced as the ejecta cool and the ionization state lessens.

6.1. 2D deflagration models

While parameterized 1D models are useful for understanding how SN Ia light curves vary with the ejecta properties, they do not constrain the physical origin of such variations. To this end, we have pursued a survey of 2D models based on a more realistic treatment of the explosion physics as captured by the SN Ia code developed by SciDAC collaborators at the Max Planck Institute for Astrophysics. To explore the sensitivity to initial conditions, we parameterized the white dwarf ignition process by inserting randomly distributed "hot spots" around the stellar center. The subsequent propagation of the deflagration flame was tracked under a level-set approach and employed a subgrid-scale model to describe turbulent energy on unresolved scales.

In the deflagration phase, burned ash is buoyant with respect to surrounding cold fuel, resulting in hot rising bubbles subject to shearing and other fluid instabilities. The resulting ejecta structure exhibits irregular plumes of processed material (^{56}Ni and IME) embedded in a substrate of unburned carbon/oxygen (figure 6). The deflagration itself fails to consume all available fuel but burns enough to explode the star. The result is a subenergetic event, with relatively low ^{56}Ni yield (in this particular model only 0.2 $M_{\odot}$).

Postprocessing with our radiative transfer code gives a synthetic spectrum that bears little resemblance to that of observed SNe Ia at maximum light (figure 6). This model failed to synthesize an abundance of high-velocity IME and so does not reproduce the observed blueshifted absorption features from silicon, sulfur, and calcium. The strength of the deflagration phase can be enhanced by increasing the number ignition points (and will be greater yet in full 3D calculations) but even the most vigorous 3-D deflagrations seemingly do not burn enough material to match the properties of the more typical SNe Ia [16].

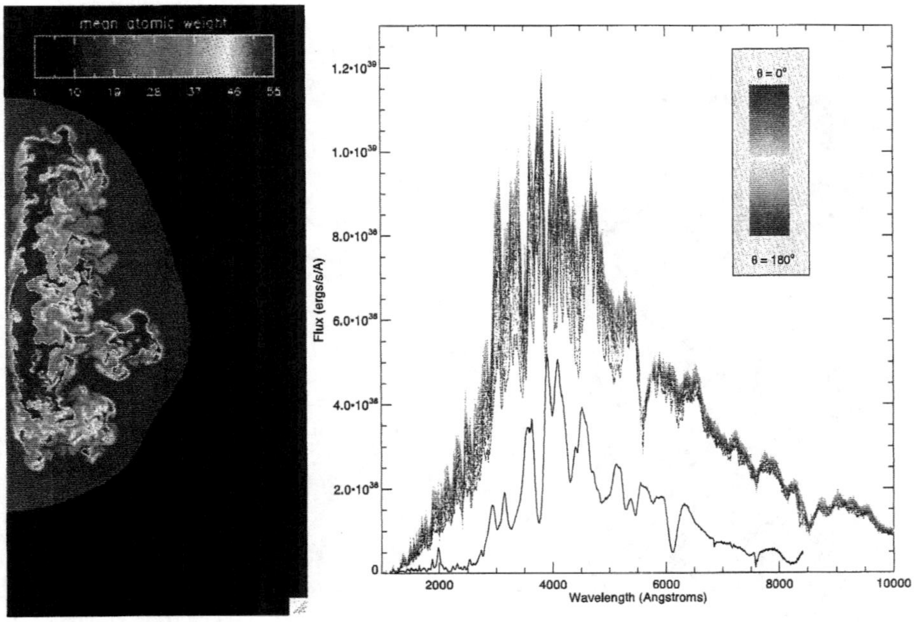

Figure 6. *Left:* Compositional structure, in the coasting phase, of a 2-dimensional pure deflagration model. The color coding shows the mean atomic weight, with red representing ^{56}Ni green intermediate mass elements, and blue unburned C/O. *Right:* Synthetic spectra of the model at maximum light as seen from various viewing angles, compared to the observed spectrum of the Type Ia SN 1981B (solid black line). The model fails to reproduce the depth and blueshift of most of the observed line absorption features, suggesting that pure deflagration explosions do not account for normal SNe Ia. .

6.2. Delayed-detonation models

The speculation is that in real supernova the deflagration flame transitions at some point to a supersonic detonation wave, which sweeps through the star, burning away any remaining fuel, so producing a healthy explosion. The physics of the transition is still not completely understood, although progress is being made in this direction. One possibility is that deflagrating bubbles may break out from the white dwarf surface, initiating large-scale flows that collide and compress the fuel [17]. Another possibility, reconsidered recently in [18], is that a detonation occurs spontaneously as a result of vigorous turbulent mixing of fuel and ash on small scales.

We have introduced a spontaneous detonation transition into our 2D models by simply inserting the detonation by hand once the deflagration flame has reached a density of 10^7 g cm^{-3}, (roughly the conditions when mixing is intensified as a result of burning in the "distributed regime"). The left side of figure 7 shows the final ejecta structure of one such "delayed-detonation" model, which resulted in a stratified compositional structure and 0.64 $M_\odot$ of ^{56}Ni. On the right side of the figure is a time-series of synthetic spectra calculations compared to observations of a typical SNe Ia. The week-by-week evolution follows spectrum formation in progressively deeper layers of the ejecta – at early times ($t < 30$ days) in the outer layers of IME and later in the ^{56}Ni core. The overall good agreement of the model with observations is impressive, especially considering that, apart from the ignition conditions and detonation criterion, the model was not tuned in any way.

The results of figure 7 strongly support some sort of delayed detonation model of SNe Ia.

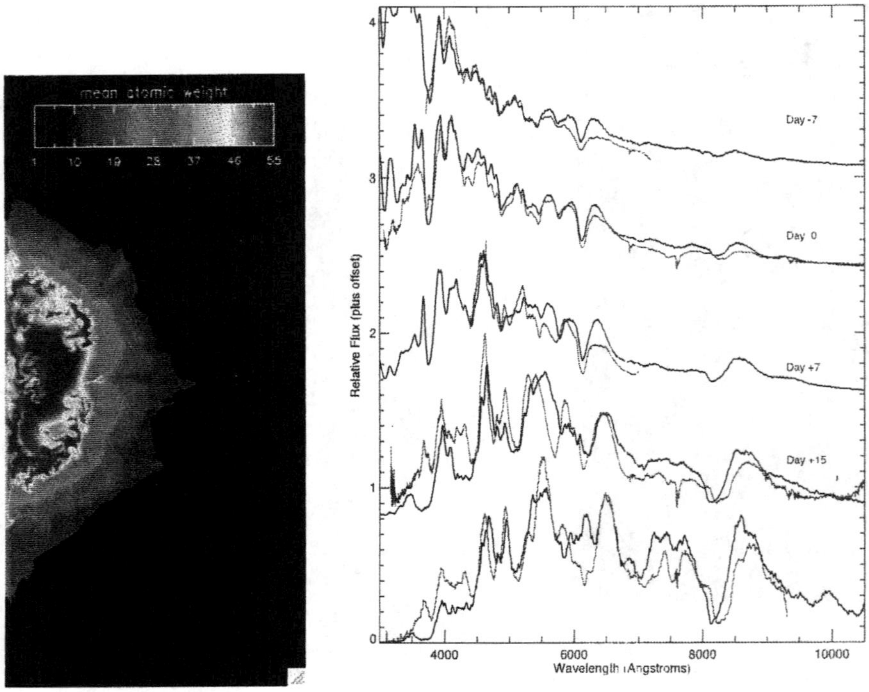

Figure 7. *Left:* Same as figure 6 but for a 2D delayed detonation model that produced 0.64 $M_\odot$ of ^{56}Ni.. *Right:* Angle averaged synthetic spectra of the model (black lines), which compare favorably with the observed week-by-week spectral evolution of the Type Ia SN 1994D (red lines). Such results strongly support some sort of delayed-detonation model for SNe Ia.

Even more interesting, in our specific adopted paradigm we find that the peak brightness of the supernova depends sensitively on the ignition conditions. Increasing the number of ignition "hot-spots" gives (perhaps surprisingly) a *less* luminous event (see figure 8). The reason is that the amount of ^{56}Ni produced by the detonation wave is determined primarily by the density of the material it passes through. Stronger ignition conditions give a stronger deflagration phase, which leads to greater pre-expansion of the white dwarf (hence lower overall densities) at the moment of detonation. A variation of the number of ignition points from $N = 20$, to $N = 150$ leads to a variation in peak brightness of a factor of 2, which is a large part of the luminosity range observed in normal SNe Ia. Inspection of the detailed maximum-light spectra show features that correlate with peak brightness as well. For example, the more luminous models have bluer colors and a weaker ratio of Si II features near 6000 Å.

6.3. Off-center detonation ,odels

The centrally ignited delayed-detonation models discussed so far retain approximate spherical symmetry (see figure 7). Slight asymmetry in the ignition conditions, however, can lead to pronounced global asphericities in the final ejecta structure. Actually, models of the simmering phase of non-rotating white dwarfs find that convection has a dominant dipolar mode, in which case ignition may be expected to occur off-center. We have explored the possibility by distributing the initial hot spots within a cone of opening angle $\pi/4$. The resulting deflagration flame (and subsequent detonation) burned to completion more thoroughly on one side of the

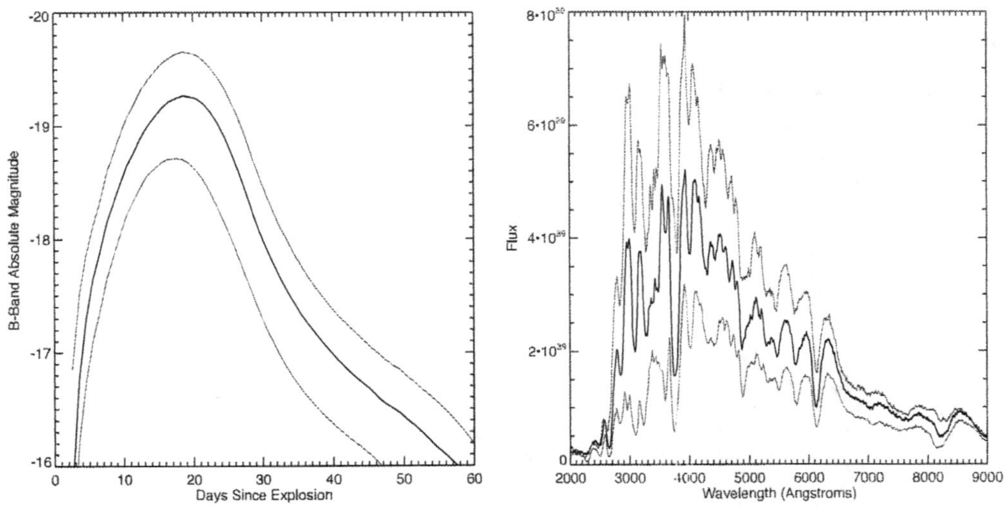

Figure 8. *Left:* Angle averaged light curves (in the blue wavelength band) of 2D SN Ia models with different ignition conditions. The explosions were initiated by randomly distributing 20, 100, and 150 "hot spots," leading to the brightest, middle, and dimmest models, respectively. The greater the number of ignition points, the stronger the deflagration phase and the lower the stellar densities at the time of detonation. *Right:* Angle averaged spectra at maximum light of the same models. The color of the spectra and the strength of some absorption features are found to correlate with luminosity.

star, leading to a predominance of ^{56}Ni on the ignition side ($y > 0$) and of silicon group elements on the other (figure 9). As the deflagration phase in this model was weak (being confined to one side of the star), the ^{56}Ni production was large, $\sim 1.0\ M_{\odot}$. The light curve shape and peak brightness also depend sensitively on the orientation, varying by nearly 25% depending on the viewing angle. When viewed from the ignition side ($\theta = 0°$), the model is as luminous as the brightest SN Ia ever found [19]

Astronomical imaging of the geometry of SNe Ia would clearly give deep insight into the nature of the explosion physics. Unfortunately, virtually all observed SNe are too distant to be directly resolved in the initial phases. In this case, our only direct probe of asphericity is through observations of the polarization of SN light. Electron scattering in the ionized SN envelope will polarize the light, but the net effect will cancel unless there is a preferred direction (i.e., departure from spherical symmetry) in the scaterring medium. Spectropolarimetric observations are typically very challenging, as the light from SNe is dim and quickly fading, and the observed polarization levels are typically small, $P \sim 1\%$. Nevertheless, intrinsic polarization has now been clearly detected in several SNe Ia, both in the lines and the continuum [20, 21, 22], demonstrating that in general SNe Ia aspherical to some extent.

Deriving the detailed ejecta geometry of SNe Ia from polarization observations is a difficult inverse problem which can only really be attempted by comparison with multi-dimensional radiation transfer calculations. Figure 10 shows the polarization spectrum at maximum light of an off-center detonation model. From some orientations, the model resembles observations of particular SNe Ia (e.g., SN 2004dt [23]) with a low continuum polarization but large line polarization peaks at the location of prominent absorption features. Although the agreement with observations is suggestive, the solution may not be unique, and other geometries could make

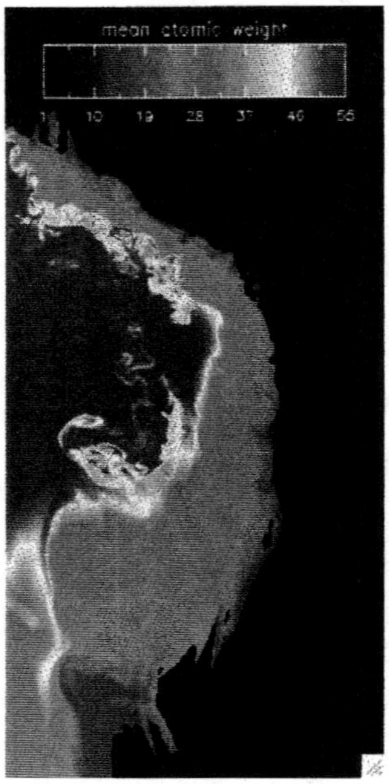

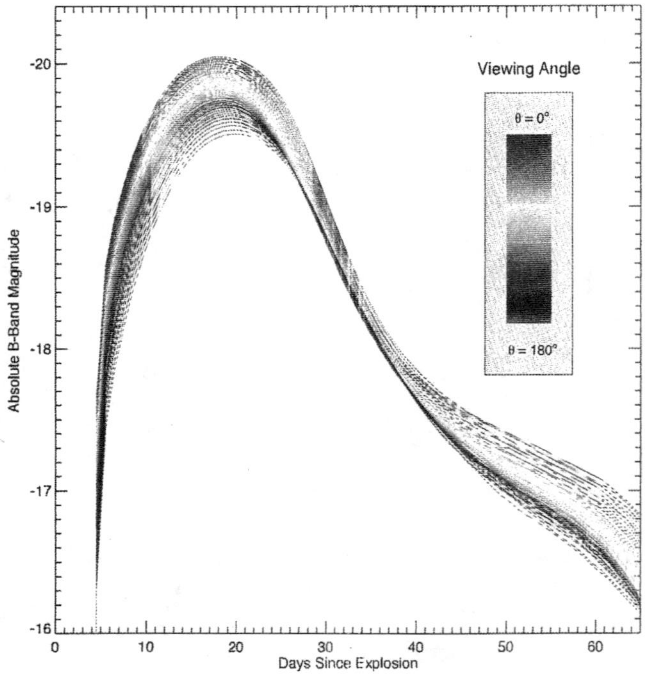

Figure 9. *Left:* Same as figure 6 but for an off-center delayed detonation model. Almost 1.0 $M_{\odot}$of ^{56}Ni was produced, distributed primarily on the ignition side (top half) of the ejecta. *Right:* Light curve at blue wavelengths for the model as a function of viewing angle. The model peak luminosity varies by $\sim 25\%$, being brightest when looking from the ignition side (top down).

similar polarization predictions. The path forward is to obtain spectropolarization observations for a statistical sample of SNe Ia, which probe many different viewing orientations, and compare to the predictions of individual models.

7. Initial results and outlook in 3D

When compared to fully 3D simulations, the 2D explosion models only approximate the ejecta morphology and do not reproduce the correct energetics, ^{56}Ni yields, or detailed sub-structure. Ultimately, the SN Ia problem must be addressed, at all stages, in 3D. The computational expense is then naturally much greater, and a much smaller grid of parameter space can be explored.

While a few full-star 3D SN Ia explosion calculations have been carried out [24, 25, 26, 27], the modeling of synthetic spectra and light curves in 3D is largely unexplored territory. Figure 11 shows one of our milestone calculations – the first high-resolution 3D model of a SN spectrum at maximum light. In this example, an asymmetric geometry was artificially constructed by distorting a spherically symmetric SN Ia model. The transfer calculation, which relied on our newly developed domain-decomposition algorithms, was performed on a grid of 200^3 spatial points and 10^4 wavelength points, amounting to over 1 TB of memory.

Our 3D spectrum run took approximately 10,000 CPU-hours on 1,000 processors on the Cray XT4 machine Jaguar at Oak Ridge National Laboratory. We anticipate that full light

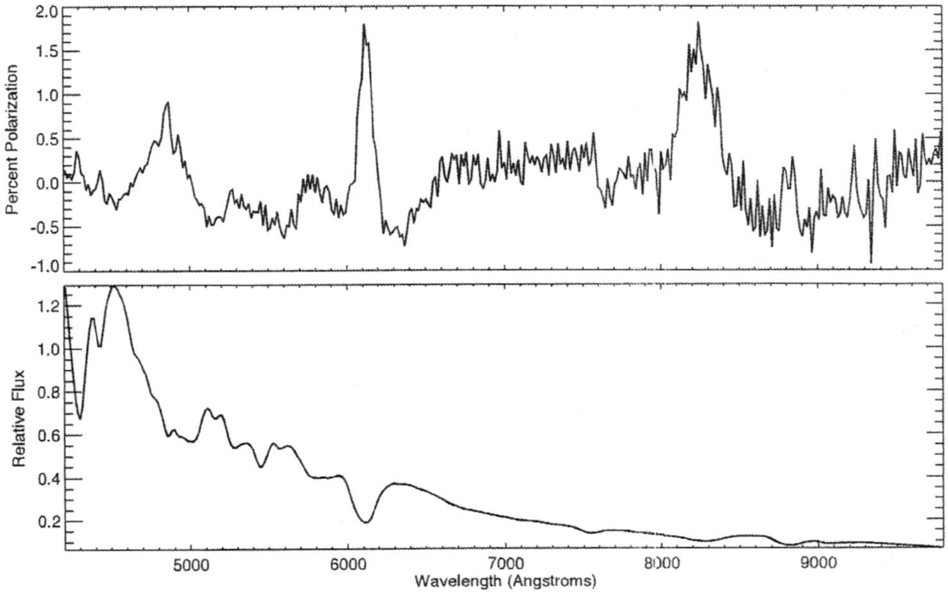

Figure 10. Flux spectrum (*bottom*) and polarization spectrum (*top*) o f an off-center delayed detonation model at maximum light, as seen from one particular viewing angle. The aspherical distribution of IME in the SN ejecta leads to large polarization peaks over line absorption features.

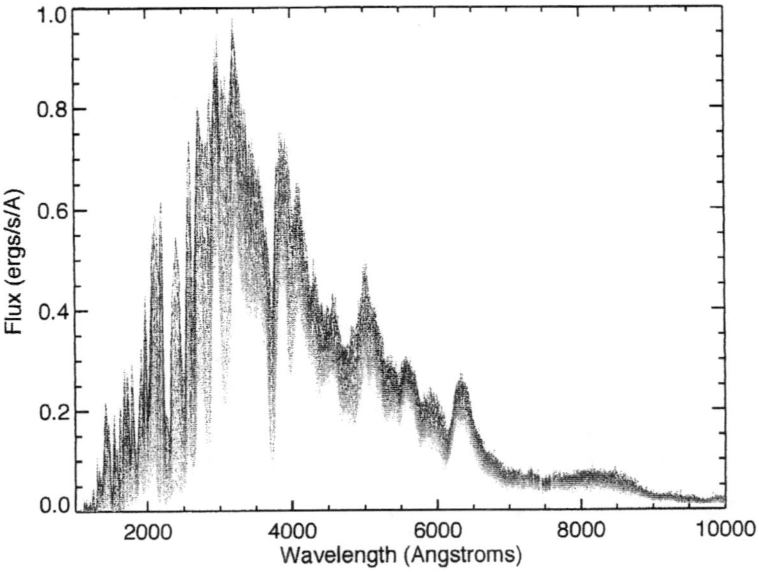

Figure 11. Synthetic spectrum calculation of a model SN Ia calculated in full 3D using the domain decomposition parallelization approach. The underlying model ejecta was constructed artificially by distorting a 1D model.

curve calculations, now under development, will require upwards of 500,000 CPU-hours per model. The introduction of certain Monte Carlo variance reduction techniques may be able to reduce the execution time significantly, but already 3D time-dependent multi-wavelength transfer calculations for supernovae are within our reach.

The ultimate goal is to develop an end-to-end simulation of SNe Ia, from the moments of ignition to the months following explosion, which help us to infer how variations in the SN Ia progenitor system affect the luminosity of the event. This theoretical understanding would be of great help in limiting and controlling the potential evolution of SN Ia properties with redshift, a possible source of systematic error in cosmological studies of dark energy. First-principle calculations of SNe Ia will require our increased understanding at every stage of the event, in particular the precise conditions of ignition, the physics of the deflagration to detonation transition, and the effects of non-LTE on the spectra and light curves. Our SciDAC group has parallel efforts on all these fronts. The advancing numerical methods should continue to make good use of massively parallel machines in the coming era of petascale computing.

Acknowledgments
Support for this work was provided by the DOE SciDAC Program under grant DE FC02-0641438. We are grateful for computer time provided by LLNL on the ATLAS machine, by ORNL through an INCITE award, and by NERSC.

References
[1] Phillips M M, Lira P, Suntzeff N B, Schommer R A, Hamuy M and Maza J 1999 *AJ* **118** 1766–1776
[2] Kuhlen M, Woosley S E and Glatzmaier G A 2006 *ApJ* **640** 407–416 (*Preprint* arXiv:astro-ph/0509367)
[3] Hoeflich P, Mueller E and Khokhlov A 1993 *A&A* **268** 570–590
[4] Hauschildt P H, Baron E and Allard F 1997 *ApJ* **483** 390
[5] Dessart L and Hillier D J 2005 *A&A* **437** 667–685 (*Preprint* arXiv:astro-ph/0504028)
[6] Mazzali P A and Lucy L B 1993 *A&A* **279** 447–456
[7] Tanaka M, Mazzali P A, Maeda K and Nomoto K 2006 *ApJ* **645** 470–479 (*Preprint* arXiv:astro-ph/0603184)
[8] Kasen D, Thomas R C and Nugent P 2006 *ApJ* **651** 366–380 (*Preprint* arXiv:astro-ph/0606111)
[9] Baron E and Hauschildt P H 2007 *A&A* **468** 255–261 (*Preprint* arXiv:astro-ph/0703437)
[10] Sobolev V V 1947 *Moving Envelopes of Stars* (Leningrad: Leningrad State University)
[11] Kurucz R 1993 *CD-ROM 1, Atomic Data for Opacity Calculations* (Cambridge: Smithsonian Astrophysical Observatory)
[12] Kasen D 2006 *ApJ* **649** 939–953 (*Preprint* arXiv:astro-ph/0606449)
[13] Kasen D and Woosley S E 2007 *ApJ* **656** 661–665 (*Preprint* arXiv:astro-ph/0609540)
[14] Woosley S E, Kasen D, Blinnikov S and Sorokina E 2007 *ApJ* **662** 487–503 (*Preprint* arXiv:astro-ph/0609562)
[15] Hamuy M, Phillips M M, Suntzeff N B, Schommer R A, Maza J and Aviles R 1996 *AJ* **112**
[16] Roepke F K 2008 *ArXiv e-prints* **804** (*Preprint* 0804.2147)
[17] Plewa T, Calder A C and Lamb D Q 2004 *ApJ* **612** L37–L40
[18] Woosley S E 2007 *ApJ* **668** 1109–1117 (*Preprint* arXiv:0709.4237)
[19] Howell D A, Sullivan M, Nugent P E, Ellis R S, Conley A J, Le Borgne D, Carlberg R G, Guy J, Balam D, Basa S, Fouchez D, Hook I M, Hsiao E Y, Neill J D, Pain R, Perrett K M and Pritchet C J 2006 *Nature* **443** 308–311 (*Preprint* arXiv:astro-ph/0609616)
[20] Howell D A, Höflich P, Wang L and Wheeler J C 2001 *ApJ* **556** 302–321
[21] Wang L, Baade D, Höflich P, Khokhlov A, Wheeler J C, Kasen D, Nugent P E, Perlmutter S, Fransson C and Lundqvist P 2003 *ApJ* **591** 1110–1128
[22] Chornock R, Filippenko A V, Branch D, Foley R J, Jha S and Li W 2006 *PASP* **118** 722–732 (*Preprint* astro-ph/0603083)
[23] Leonard D C, Li W, Filippenko A V, Foley R J and Chornock R 2005 *ApJ* **632** 450–475
[24] Röpke F K and Hillebrandt W 2005 *A&A* **431** 635–645
[25] Jordan G I, Fisher R, Townsley D, Calder A, Graziani C, Asida S, Lamb D and Truran J 2007 *ArXiv Astrophysics e-prints* (*Preprint* astro-ph/0703573)
[26] Gamezo V N, Khokhlov A M and Oran E S 2005 *ApJ* **623** 337–346 (*Preprint* arXiv:astro-ph/0409598)

[27] Bravo E and García-Senz D 2008 *A&A* **478** 843–853 (*Preprint* arXiv:0712.0510)

The Via Lactea INCITE simulation: Galactic dark matter substructure at high resolution

Michael Kuhlen[1], Jürg Diemand[2], Piero Madau[2] and Marcel Zemp[2]

[1]School of Natural Science, Institute for Advanced Study, Einstein Lane, Princeton, NJ, 08540, USA

[2]Department of Astronomy & Astrophysics, University of California Santa Cruz, Santa Cruz, CA, 95064, USA

E-mail: `mqk@ias.edu`

Abstract. It is a clear and unique prediction of the cold dark matter paradigm of cosmological structure formation that galaxies form hierarchically and are embedded in massive, extended dark halos teeming with self-bound substructure, or "subhalos." The amount and spatial distribution of subhalos around their host provide unique information and clues on the galaxy assembly process and the nature of the dark matter. Here we present results from the *Via Lactea* INCITE simulation, a one billion particle, one million CPU-hour simulation of the formation and evolution of a galactic dark matter halo and its substructure population.

1. Introduction

According to the standard model of cosmology our universe is made up mostly of mysterious dark energy (70%) and cold dark matter (CDM, 25%). Ordinary matter (baryons) makes up only 5% of the mass density in the Universe today. The dark energy component is homogeneous, and its structure formation is dominated by the CDM component. In the past few years cosmological N-body simulations have successfully uncovered how the cold dark matter distribution evolves from almost homogeneous initial conditions near the Big Bang into the present, highly clustered state [1, 2]. CDM halos are built up hierarchically in a long series of mergers of smaller halos. Early low-resolution simulations and simple analytical models found that the end products are smooth, triaxial halos. Higher resolution cosmological N-body simulations later modified this simplistic picture. The merging of progenitors is not always complete: the cores of accreted halos often survive as gravitationally bound subhalos orbiting within a larger host system. CDM halos are not smooth; they exhibit a wealth of substructure on all resolved mass scales [3]. Subhalos are now believed to host cluster galaxies and the satellite galaxies around the Milky Way.

In 2007 we received an Innovative and Novel Computational Impact on Theory and Experiment (INCITE) grant for one million CPU-hours on Oak Ridge National Laboratory's Jaguar Cray XT3 supercomputer, to run a one billion particle simulation of the formation and evolution of a Milky Way scale dark matter halo in an expanding cosmological volume. The resulting simulation, dubbed *Via Lactea Via Lactea II*) has resolved the subhalo population at unprecedented detail and closer to the host halo's center than ever before. (Actually this simulation is the second in a series of simulations of ever-increasing resolution and is referred to in the literature as In the following we describe some of the technical details behind this computational effort, and we present some of the first scientific results that have emerged from this study.

2. Technical details

Measurements of the cosmic microwave background by the Wilkinson Microwave Anisotropy Probe (WMAP) satellite [4] have given us an exquisite view of the state of the universe a mere ∼100,000 years after the big bang. At this time the universe was very simple: a complete statistical description of the primordial temperature fluctuations observed by WMAP requires only six parameters. This simple statistical description of the early universe allows us to generate an initial particle distribution at redshift ~ 100 (only 15 million years after the Big Bang) with small linear density perturbations matching the WMAP measurements. Our simulation then follows the gravitational collapse of these fluctuations into the highly nonlinear clustering regime today.

2.1. The N-body technique

The dark matter density field is sampled using a finite number N of Lagrangian tracer particles, whose evolution is governed by Newton's equations of motion in coordinates co-moving with the expansion of the universe [5]:

$$\frac{\mathrm{d}\vec{x}}{\mathrm{d}t} = \vec{v} \tag{1}$$

$$\frac{\mathrm{d}\vec{v}}{\mathrm{d}t} + 2H(a)\vec{v} = -\frac{1}{a^2}\vec{\nabla}\phi. \tag{2}$$

Here a is the expansion scale factor, $H(a) = \dot{a}/a$ the Hubble constant, and ϕ a modified gravitational potential

$$\phi = \Phi + \frac{1}{2}a\ddot{a}x^2, \tag{3}$$

whose field equation is given by

$$\nabla^2\phi = 4\pi G\rho a^2 - \Lambda a^2 + 3a\ddot{a}, \tag{4}$$

where we are allowing for the contribution of a vacuum energy of density $\rho_v = \Lambda/(8\pi G)$. Using the 2nd Friedman equation

$$\frac{\ddot{a}}{a} = -\frac{4}{3}\pi G\rho_b(t) + \frac{\Lambda}{3}, \tag{5}$$

we reduce equation 4 reduces to

$$\nabla^2\phi = 4\pi G(\rho - \rho_b)a^2. \tag{6}$$

Because the effective gravitational potential is determined by the density contrast $(\rho - \rho_b)$, the computational error on the gravitational potential is set by how accurately the density field is sampled. A larger N will thus result in a more faithful representation of the dark matter dynamics. This has been the driving force leading to ever increasing problem sizes and the use of powerful supercomputers for their solution.

2.2. PKDGRAV*, a hierarchical tree code*

PKDGRAV [6], the code that we used for the *Via Lactea* simulation, is an example of a hierarchical tree code, in which the particles are grouped together hierarchically and the gravitational force on a particle from each group is approximated by a multipole expansion up to low order. The hierarchical grouping is realized in a tree structure, in which the whole computational domain is split up into ever smaller subregions containing successively smaller numbers of particles, down to the leaf nodes of the tree, which contain only a small number of particles. The calculation of the gravitational force felt by each particle proceeds by "walking" the tree, starting with the root node that covers the entire computational domain. A cell-opening criterion is applied to each node to determine whether to use its multipole approximation, ending the descent along this branch, or if the cells needs to be "opened" and its daughter nodes examined.

The use of such a hierarchical multipole expansion requires only $\log N$ interactions per particle, much less than the $N-1$ separate force calculations required with direct summation, and this allows much larger particle numbers to be used. The accuracy of the force calculation is determined by the order to which the multipole expansion is carried out and by how many nodes it is applied to. Increasing the order of the expansion allows greater accuracy but comes at an additional computation cost. PKDGRAV extends the multipole expansion up to fourth order.

PKDGRAV employs a spatially balanced binary tree, a relative of the particle balanced k-d tree. Starting with the root node, every node is recursively split up into two rectangular pieces of equal volume by bisecting the tight bounding box containing all particles in the node. This splitting continues down to the leaf nodes, or "buckets," which are allowed to contain at most 16 particles. Such a tree has been shown to be asymptotically superior to density based decompositions like the k-d tree [7].

The cell opening criterion applied by PKDGRAV is

$$r_{\rm open} = \frac{2}{\sqrt{3}} \frac{b_{\rm max}}{\theta}, \tag{7}$$

where $b_{\rm max}$ is the maximum distance from any point in the cell to the cell's center of mass and θ is a user-specified accuracy parameter typically set to $\theta = 0.7$ for $z < 2$ and $\theta = 0.5$ at higher redshifts. If during the walk a cell is deemed sufficiently far from bucket B_i to allow its multipole approximation, then it is added to B_i's list of particle-multipole interactions. Alternatively, if a particular branch is followed all the way down to its leaf node, then all the particles in that bucket are added to B_i's particle-particle interaction list.

In order to avoid unphysical two-body collisions, the potential is softened for small separations, $|\vec{x}_i - \vec{x}_j| < \epsilon$. This step is accomplished by using a single particle density distribution that is given by a Dirac δ-function convolved with a softening kernel, here taken to be a compensating K_1 kernel [8]. At separations greater than about 2ϵ the potential becomes exactly Newtonian.

2.3. A new dynamical time step criterion

An accurate determination of the gravitational potential is a necessary, but not sufficient, requirement for a reliable solution to the equations of motion of dark matter. Of equal importance is the method used to integrate the equations and the choice of time step. Cosmological simulations typically exhibit a dynamic range of densities spanning many orders of magnitude. Denser particle distributions require shorter time steps for a given accuracy, but using a global time step set by the densest regions in the simulations becomes computationally impractical because the majority of the particles reside in much lower density regions. A more efficient use of computational resources can be achieved by the use of individual time steps that adapt to the local density. PKDGRAV employs an adaptive leapfrog integrator, which allows such adjustable time steps. The time steps are quantized in powers of two of the basic time step T_0.

A widely used time step criterion is based on a rather ad hoc combination of the force softening and the acceleration a given particle feels, $\Delta T \leq 0.2\sqrt{\frac{\epsilon}{|\vec{a}|}}$. This criterion results in a very tight time step distribution and has been shown [9] to produce accurate results at all but the very highest densities. However, this criterion is not directly related to the local dynamical time, and it has been shown [10] that for very high dynamic ranges ($\gtrsim 10^3$) it is not quite

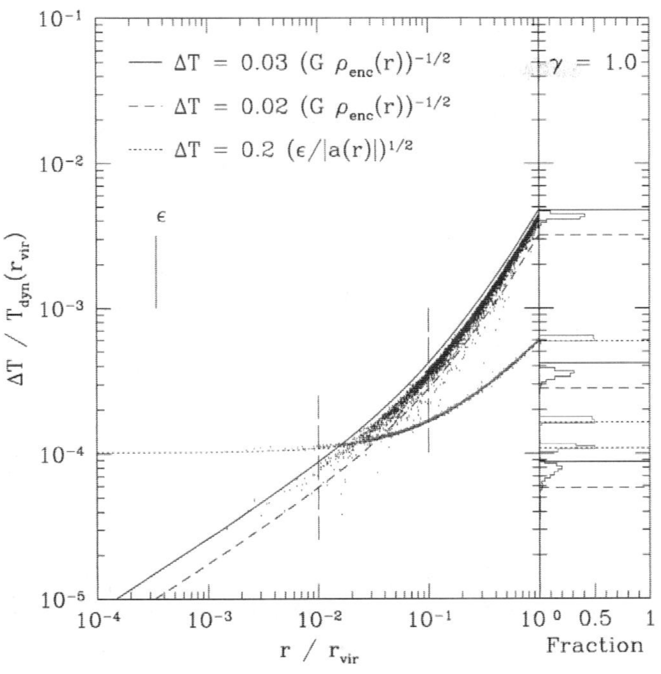

Figure 1. The distribution of time steps for a density profile of slope $\gamma = 1$, for the standard time step $\Delta T \propto \sqrt{\epsilon/|a(r)|}$ (in red) and for the new dynamical time step $\Delta T \propto 1/\sqrt{G\rho(< r)}$ (in blue). From Zemp et al. (2007) [10].

adaptive enough, resulting in overly conservative time steps in the outer low density regions and insufficiently short time steps in the very centers of dark matter halos. Instead, Zemp et al. [10] advocate a time-step criterion that is directly set by the local dynamical time a particle feels,

$$\Delta T \leq \eta \frac{1}{\sqrt{G\rho_{\rm enc}(r)}}, \tag{8}$$

where $\rho_{\rm enc}$ is the enclosed density from the particle's location to the dominant dynamical structure, and η is typically set to 0.02–0.03. The challenge is to quickly determine the dominant structure governing the orbit of a given particle, and Zemp et al. [10] have achieved this by cleverly making use of the hierarchical tree structure used for the force calculation. Figure 1 (taken from [10]) shows a comparison of the time-step distribution using the old and new time-step criteria and reveals that the new time step is more adaptive, with smaller time steps close to the center and larger ones in the outer regions. The most recent version of PKDGRAV contains an implementation of this new algorithm, and it allowed our *Via Lactea* simulation to push to unprecedented central densities at a manageable computational cost.

An unfortunate side effect of the use of individual time steps is that it destroys the symplectic nature of the integrator. This would be highly undesirable for any long-term integration of planetary orbits, for example. In cosmological simulations, however, dark-matter particles typically complete a comparatively much smaller number of orbits over a Hubble time, and so any secular error is not likely to grow significantly during the simulation. In this case the computational advantage of using individual and adaptive time steps outweighs the negative side-effects

Table 1. Simulation parameters.

$L_{\rm box}$ (Mpc)	ϵ (pc)	z_i	$N_{\rm hires}$	$M_{\rm hires}$ ($\rm M_\odot$)	r_{200} (kpc)	M_{200} ($\rm M_\odot$)	$V_{\rm max}$ (km s^{-1})	$r_{V\rm max}$ (kpc)	$N_{\rm sub}$
40.0	40.0	104	1.05×10^9	4.10×10^3	402	1.93×10^{12}	201	60.0	53,653

Box size $L_{\rm box}$, (spline) softening length ϵ, initial redshift z_i, number $N_{\rm hires}$ and mass $M_{\rm hires}$ of high resolution dark matter particles, host halo r_{200} (radius enclosing 200 times the mean density), M_{200} (mass within r_{200}), $V_{\rm max}$ (the peak of the circular velocity curve, $v_{\rm circ} = \sqrt{GM(<r)/r}$) and $r_{V\rm max}$ (the radius at which $V_{\rm max}$ occurs), and number of subhalos within r_{200} $N_{\rm sub}$ at $z = 0$. Force softening lengths ϵ are constant in physical units back to $z = 9$ and constant in co-moving units before.

2.4. Multimass initial conditions

The *Via Lactea* simulation makes use of multi-mass initial conditions [11, 9], which allow the halo of interest to be resolved with very high resolution, while at the same time allowing for the large-scale cosmological tidal field to be represented at much lower resolution. The overall computational volume is (40 Mpc)^3 (one megaparsec, or Mpc, is approximately 3 million light years). But at the end of the simulation 95% of all high-resolution particles are found within the inner (2 Mpc)^3. In total the *Via Lactea* simulation consists of a hierarchy of three particle masses: (16,689,261, 28,645,888, 1,048,772,608) particles with masses (32,768, 64, 1) times the high resolution particle mass of $M_{\rm hires} = 4.10 \times 10^3\,\rm M_\odot$. The remaining parameters of the *Via Lactea* simulation are summarized in table 1.

3. Results

The main scientific driver for our *Via Lactea* simulation is to better characterize the dark matter subhalo population of a Milky Way–scale halo. A visual impression of the enormous abundance of subhalos is given by figure 2, which shows a density-squared projection of the central (566 kpc)^3 cube. Projecting density squared enhances the contrast between individual subhalos and the smooth background. The two insets show progressive zooms into a region of interest. The deepest zoom is a projection of a (40 kpc)^3 cube centered on subhalo with $V_{\rm max} = 21.4$ km s^{-1}* and a tidal radius of 46.8 kpc. The high resolution of *Via Lactea* is able to resolve the second level of the subhalo hierarchy – subhalos of a subhalo.

The overall picture is reminiscent of the "turtles all the way down" image. In fact, in our simulation the clumpiness extends over six orders of magnitude in mass down to our subhalo resolution limit of $\sim 10^5\,\rm M_\odot$. The theoretical expectation is that this clumpiness continues even farther, all the way down to the cut-off in the density fluctuation power spectrum set by the finite velocity dispersion (or temperature) of the cold dark-matter particle [12, 13], some 10 to 15 orders of magnitude below the scales that we currently resolve. An important caveat here is that our simulation does not include the effects of baryons. Interactions with stars or giant molecular clouds in the disk of the Milky Way could destroy some fraction of the subhalos passing through the center. Such baryonic affects are unlikely to affect the subhalo population as a whole, however, since most of the subhalos are not on orbits that carry them through the central disk.

3.1. The dark matter subhalo population

The density profiles of isolated dark matter halos are known to follow a common radial dependence, falling off as r^{-3} in the outer regions, becoming shallower towards the center, and

*$V_{\rm max}$ is the peak of the circular velocity curve ($v_{\rm circ} = \sqrt{GM(<r)/r}$) and is used as a proxy for mass.

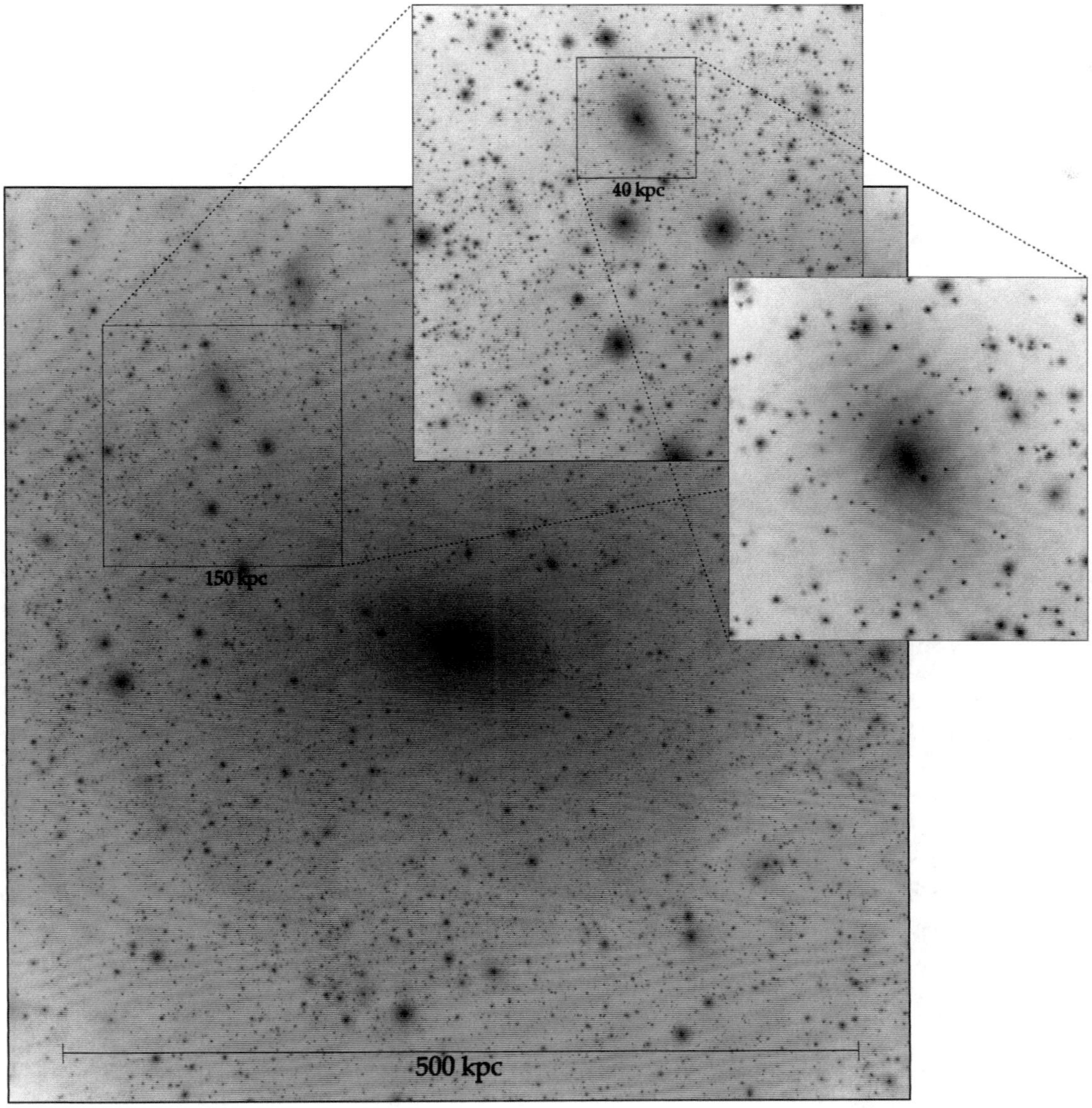

Figure 2. A projection of density squared in a 566 kpc × 566 kpc × 566 kpc region centered on the host halo at $z = 0$. The insets (150 kpc and 40 kpc deep, respectively) show zooms into a region harboring a subhalo exhibiting abundant sub-substructure.

finally approaching a central cusp of $r^{-\gamma}$, with the exact value of γ being somewhat uncertain, but less than 1.5 [14, 15, 16]. The very large number of particles in the *Via Lactea* simulation and the extremely accurate integration of their orbits have allowed a precise determination of the density profile within the inner kpc of the Galactic halo and its satellite halos, see the left panel of figure 3. The host halo density profile is best fit with a central cusp of slope $\gamma = 1.24$, and we find that the inner profiles of subhalos are also consistent with cusps, with slopes scattered around $\gamma = 1.2$. At larger radii the subhalo density profiles generally fall off faster than the host halo (see inset). The explanation is simply that subhalo density profiles are modified by tidal mass loss, which strips material from the outside in but does not affect the inner cusp structure

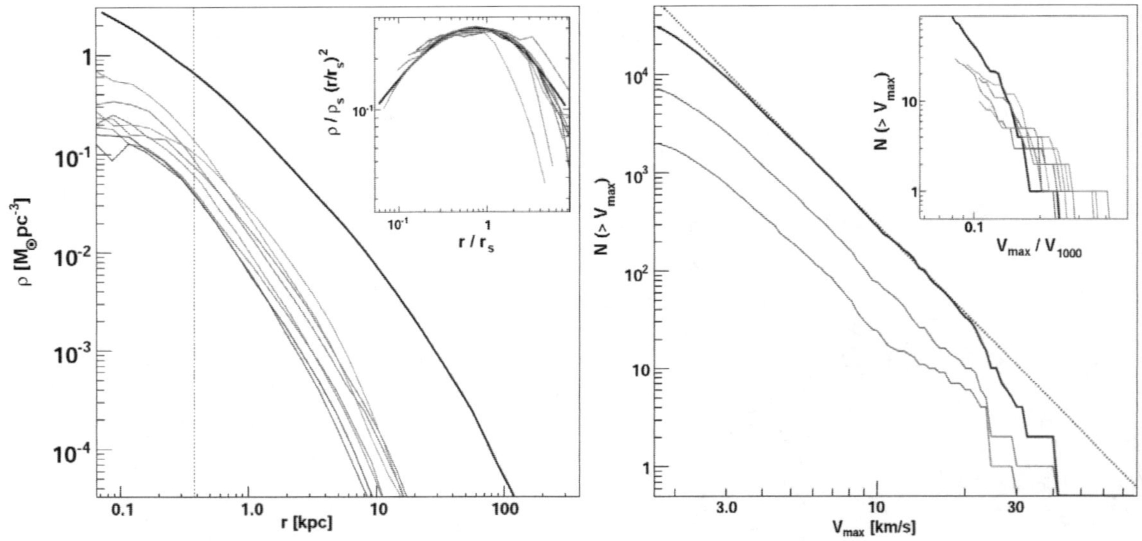

Figure 3. Properties of the subhalo population in *Via Lactea*, from Diemand et al. (2008) [33]. *Left panel:* Density profiles of the host halo (thick black line) and of eight large subhalos (thin colored lines). The profiles remain cuspy down the estimated convergence radius of 380 pc (vertical dotted line). The rescaled profiles in the inset demonstrate that (i) subhalos are quite self-similar in the inner regions and (ii) many subhalos are tidally truncated in the outer region. *Right panel:* The number of subhalos above $V_{\rm max}$ within $r_{200} = 402$ kpc (thick line) and within 100 and 50 kpc of the halo center (thin lines). $V_{\rm max}$ is the peak of the subhalo circular velocity $v_{\rm circ} = \sqrt{GM(<r)/r}$, and serves as a proxy for subhalo mass. The distribution is well fit by a single power-law, $N(>V_{\rm max}) = 0.036(V_{\rm max}/V_{\rm max,host})^{-3}$ (dotted line). Below $V_{\rm max} \simeq 3.5$ km s^{-1} the subhalo abundance is artificially reduced due to numerical limitations. The inset shows the abundance of sub-subhalos in eight large subhalos and is quite similar to the scaled-down version of the host halo.

[17, 18].

Overall *Via Lactea* predicts a remarkably self-similar pattern of dark matter clustering properties. The right panel of figure 3 shows that subhalos appear to be scaled down versions of their host halo, not just in the inner mass distribution, but also in terms of their relative substructure abundance. The overall subhalo abundance as a function of mass, as measured by the cumulative $V_{\rm max}$ function, is well described by a single power law

$$N(>V_{\rm max}) \propto (V_{\rm max}/V_{\rm max,host})^{-3}, \tag{9}$$

independently of distance, within 50, 100, or 402 kpc of the center. Furthermore, for the first time it has become possible to determine the abundance of subhalos at the second level of the substructure hierarchy, and the sub-subhalo $V_{\rm max}$-functions (see inset) appear to be quite similar to the scaled down version of the host halo's subhalo distribution.

3.2. The "missing satellites problem"

It has been known for many years that our host galaxy, the Milky Way, is orbited by a number of so-called dwarf galaxy satellites. The original census of about 11 dwarfs [19] has recently doubled in size by the discovery of ~ 12 additional ultra-faint dwarfs in the Sloan Digital Sky Survey (SDSS) [20, 21, 22, 23, 24, 25, 26, 27]. Many, if not all, of these systems are strongly dark matter dominated in their central regions [28, 29] and are thus thought to be the luminous counterparts to the massive end of the dark matter subhalo population. The apparent discrepancy between the number of observed dwarf galaxies (around 50, after correcting for the incomplete sky coverage of the SDSS) and the vastly larger number of dark matter subhalos predicted by numerical simulations ($\sim$ 50,000 in *Via Lactea*) is well known as the "missing satellites problem."

The solution to this problem could lie in a modification of the standard cold dark matter paradigm of structure formation. For example, *warm* dark matter, with its higher intrinsic velocity dispersion, would lead to a suppression of power at small scales and greatly reduce the number of low mass subhalos seen in CDM simulations. On the other hand, the standard CDM picture might be fine if complicated "gastrophysics," such as radiative or thermal feedback from young galaxies, quasars, or supernovae, prevents star formation in the vast majority of dark matter subhalos. In this case the galactic halo would be populated by an enormous number of invisible dark matter clumps. The resolution of the "missing satellites problem" depends on progress both from the observational side (deeper surveys, more and improved measurements of stellar kinematics in the centers of dwarfs, etc.) and from the theoretical side (better characterization of the properties of dark matter subhalos from simulations, a better understanding of the efficacy of feedback in preventing star formation, etc.). Our *Via Lactea* simulations are one step in this direction [30].

3.3. Indirect detection of dark matter substructure?

An intriguing, if somewhat speculative, possibility for learning more about the dark matter subhalos of our Milky Way arises if the dark matter particle is a WIMP, a weakly interacting massive particle, like the neutralino in the supersymmetric extension to the standard model of particle physics. In that case the dark matter particle would be its own anti-particle and, in regions of sufficiently high density, colliding dark-matter particles would pair-annihilate. The energetic standard model particles that are produced in such annihilations would decay in a particle cascade, resulting in a potentially observable signal of electron-positron pairs, neutrinos, and gamma rays. Space based experiments (such as PAMELA and the upcoming AMS-02, [31]) could detect anti-particles produced in dark matter annihilations within about one kpc, and huge neutrino telescopes built deep into Antarctic ice (e.g. AMANDA-II and its successor IceCube [32]) or in the Mediterranean sea (ANTARES [32]) have also been pursuing such an indirect detection. The strength of such a signal depends on the local dark matter density in the solar neighborhood, and can be boosted by the clumpiness of the dark matter distribution. Extrapolating from the measured subhalo abundance in our simulation, we have estimated a local boost (within 1 kpc) of about 40% [33].

The hope for the gamma-ray signal rests on ground-based atmospheric Cerenkov telescopes (e.g., H.E.S.S., VERITAS, MAGIC, and STACEE, [34]) and on NASA's recently launched Gamma-ray Large Area Space Telescope (GLAST, [35]). We have used the *Via Lactea* simulation to make quantitative predictions for the possible dark-matter annihilation signal that GLAST may detect [36]. By placing fiducial observers at 8 kpc (the distance from the sun to the galactic center) from the host halo center and summing up the annihilation fluxes from all particles in our simulation, we produced all-sky maps of the gamma-ray flux from dark matter annihilations within the Galaxy. An example of such an allsky map is shown in figure 4. Converting these flux maps into expected number counts of gamma-ray photons requires assumptions about the nature of the dark matter particle, namely, its mass, its annihilation cross section, and the gamma-ray spectrum resulting from a single annihilation event. Particle physics theory today does not uniquely predict these quantities, and so we have sampled a region of the theoretically motivated parameter space. Using detailed models of the known astrophysical gamma-ray backgrounds that the dark matter annihilation signal from galactic substructure has to contend with, we determined the expected detection significance for every subhalo in our simulation. Although the

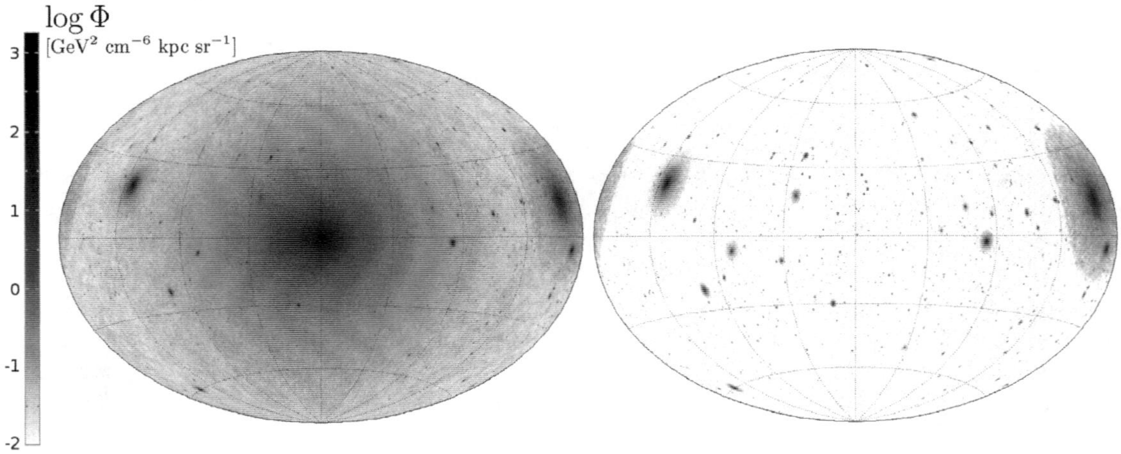

Figure 4. All-sky maps (in a Hammer-Aitoff projection) of the annihilation signal ($\propto \int_{\rm los} \rho^2 c^4 \, d\ell$) from *Via Lactea*, for an observer 8 kpc from the halo center. *Left panel:* the total signal from all particles within r_{200}. *Right panel:* The signal from only the subhalo particles. From Kuhlen et al. (2008) [36].

vast majority of the subhalos are too faint to be detectable, we found that for reasonable choices of the particle physics parameters, GLAST should be able to detect a handful of individual subhalos at more than five sigma significance.

4. Summary

The *Via Lactea* simulation is a one billion particle simulation of the formation and evolution of a Milky Way–scale dark matter halo in an expanding cosmological volume. At a cost of about 1 million cpu hours on ORNL's Cray XT3 supercomputer Jaguar, it provides an unprecedented view of the subhalo population in a dark matter halo like the one harboring our own Milky Way.

PKDGRAV, the code we used for our simulation, is an example of a hierarchical tree code that uses a multipole expansion of the gravitational potential to solve the N-body problem in a fast and accurate manner. In *Via Lactea* we employed for the first time a new dynamical time-step criterion that ensures a highly accurate orbit integration even for particles in very dense environments, while at the same time allowing longer time steps for particles in the outer regions of the halo. This allowed us to follow the central density profile and resolve the subhalo population closer to the host halo center than ever before.

We find that the dark matter clustering properties are remarkably self-similar: isolated halos and subhalos contain about the same relative amount of substructure and both have cuspy inner density profiles. The resolution of the "missing satellites problem," a well-known discrepancy between the number of luminous dwarf galaxies orbiting our Galaxy and the predicted number of dark-matter subhalos, will require either a modification of the standard cold dark-matter paradigm of structure formation or baryonic physics that suppresses star formation in all but a small fraction of subhalos. In the future it may be possible to indirectly detect dark matter substructure, through the measurement of electron-positron pairs, neutrinos, or gamma rays produced in pair-annihilations of dark matter particles in high-density regions such as the centers of subhalos.

References

[1] Diemand J, Moore B and Stadel J, 2005, Nature, 433, 389
[2] Springel V, *et al.* 2005, Nature, 435, 629
[3] Moore B, Governato F, Quinn T, Stadel J and Lake G 1998, ApJ, 499, L5
[4] Spergel D N, *et al.* 2007, ApJS, 170, 377
[5] Peebles P J E 1980, The large-scale structure of the universe (Princeton University Press)
[6] Stadel J G 2001 Ph.D. thesis (Univ. of Washington)
[7] Anderson R 1996 Tree data structures for N-body simulation *FOCS: IEEE Symposium on Foundations of Computer Science* `http://citeseer.ist.psu.edu/177236.html`
[8] Dehnen W 2001, MNRAS, 324, 273
[9] Power C, Navarro J F, Jenkins A, Frenk C S and White S D M, Springel V, Stadel Jand Quinn T 2003, MNRAS, 338, 14
[10] Zemp M, Stadel J, Moore B and Carollo C M 2007, MNRAS, 376, 273
[11] Bertschinger E 2001, ApJS, 137, 1
[12] Green A M, Hofmann S and Schwarz D J 2005, JCAP, 8, 3
[13] Profumo S, Sigurdsoni K and Kamionkowski M 2006 *Phys. Rev. Letters* **97**:031301
[14] Navarro J F, Frenk C S and White S D M 1996, ApJ, 462, 563
[15] Moore B, Ghigna S, Governato F, Lake G, Quinn T, Stadel J and Tozzi P 1999, ApJ, 524, L19
[16] Diemand J, Moore B and Stadel J 2004, MNRAS, 353, 624
[17] Kazantzidis S, Mayer L, Mastropietro C, Diemand J, Stadel J and Moore B 2004 ApJ, 608, 663
[18] Diemand J, Kuhlen M and Madau P 2007, ApJ, 667, 859
[19] Mateo M L 1998, ARA&A, 36, 435
[20] Willman B, *et al.* 2005a, ApJ, 626, L85
[21] Zucker D B, *et al.* 2006a, ApJ, 643, L103
[22] Zucker D B, *et al.* 2006b, ApJ, 650, L41
[23] Belokurov V, *et al.* 2006b, ApJ, 647, L111
[24] Belokurov V, *et al.* 2007b, ApJ, 654, 897
[25] Grillmair C J 2006, ApJ, 645, L37
[26] Sakamoto T and Hasegawa T 2006, ApJ, 653, L29
[27] Walsh S M, Jerjen H and Willman B 2007, ApJ, 662, L83
[28] Strigari L E, Bullock J S, Kaplinghat M, Diemand J, Kuhlen M and Madau P 2007, ApJ, 669, 676
[29] Simon J D and Geha M 2007, ApJ, 670, 313
[30] Madau P, Diemand J and Kuhlen M 2008, ApJ, 679, 1260
[31] `http://pamela.roma2.infn.it/`
`http://ams.cern.ch/`
[32] `http://amanda.uci.edu/`
`http://antares.in2p3.fr/`
`http://icecube.wisc.edu/`
[33] Diemand J, Kuhlen M, Madau P, Zemp M, Moore B, Potter D and Stadel J 2008, ArXiv e-prints, arXiv:0805.1244
[34] `http://www.mpi-hd.mpg.de/hfm/HESS/HESS.html`
`http://veritas.sao.arizona.edu/`
`http://wwwmagic.mppmu.mpg.de/`
`http://www.astro.ucla.edu/ stacee/`
[35] `http://glast.gsfc.nasa.gov/`
[36] Kuhlen M, Diemand J and Madau P 2008, ArXiv e-prints, arXiv:0805.4416

Large-scale simulations of buoyancy-driven turbulent nuclear burning

Dean M Townsley[1,2], Raymond A Bair[3], Anshu Dubey[1], Robert T Fisher[1], Nathan C Hearn[1], Don Q Lamb[1] and Katherine M Riley[3]

[1] Center for Astrophysical Thermonuclear Flashes, University of Chicago, Chicago, IL 60637, USA

[2] Joint Institute for Nuclear Astrophysics, University of Chicago, Chicago, IL 60637,USA

[3] Argonne Leadership Computing Facility, Argonne National Laboratory, Argonne, IL 60439, USA

E-mail: `townsley@uchicago.edu`

Abstract. An critical uncertainty in modeling thermonuclear supernovae is the degree of enhancement of the burning rate by turbulence during the subsonic burning (deflagration) phase. As turbulent combustion in the laboratory is still an active area of research, this remains a challenging problem. A unique feature of turbulent combustion in supernovae is that the driving of the turbulence arises from the strong buoyancy of the burned material. We discuss the large-scale fully three dimensional studies under way. These studies have the goals of characterizing the essential length scales of flame surface structure and thereby developing specific requirements that models of small-scale structure must meet. We discuss some preliminary results of our study concerning the scale-dependence of flame surface structure.

1. Introduction

A fundamental challenge in our understanding of reactive flows is the process by which turbulence wrinkles a combustive flame front, thereby increasing its surface area and effective flame speed. By determining the amount of matter burnt in the initial deflagration phase, and the amount of pre-expansion in the white dwarf progenitor, the process of turbulent combustion plays a central role in our understanding of the explosion mechanism of Type Ia supernovae.

Models of the deflagration phase of Type Ia supernovae typically begin with the ignition of a flame bubble near the center of the progenitor white dwarf. The hot, less dense nuclear ash buoyantly drives the bubble upward through the progenitor. Once the bubble exceeds the fire-polishing scale $\lambda_{\rm fp} \simeq 4\pi S_l^2/Ag$ [1], where S_l is the laminar flame speed, g is the gravity and $A = (\rho_2 - \rho_1)/(\rho_2 + \rho_1)$ is the density contrast, the Sharp-Wheeler growth timescale on the scale of the bubble becomes shorter than the laminar flame crossing time, causing the bubble to become Rayleigh-Taylor unstable and to become turbulent. The turbulent driving on the scale of the bubble establishes the integral scale of the turbulent flow as essentially the bubble radius. On smaller scales, the Gibson scale (defined as the scale at which the flame-crossing timescale is equal to the local eddy turnover time) may possibly establish a lower cutoff to the scales on which the flame surface can be effectively wrinkled. According to this picture, scales intermediate between the flame bubble radius and the Gibson scale are in the inertial regime. Almost immediately after ignition, a substantial inertial range opens up between the flame bubble radius and the fire-polishing scale; *turbulent burning is the dominant reaction mechanism throughout.*

For laboratory combustion, researchers have made some progress in understanding the fundamental physics of turbulent, premixed combustion using a wide variety of theoretical techniques. These techniques can be broadly divided into three classes citepPoinVeyn05, in order of increasing computational cost: Reynolds averaged Navier Stokes (RANS) methods, large eddy simulation (LES) methods, and direct numerical simulation (DNS). Within each of these classes there generally exist several methods that can even make different assumptions about the best way to characterize (or parameterize) turbulent combustion, for example, enhanced flame speed or enhanced flame area. It is important to note that no universal model exists, and existing models have been shown to not necessarily apply in all circumstances. For example, a model that performs satisfactorily in free flow may not work at all well near an igniter or in shear flow near a wall. At the coarsest level of approximation, RANS models resolve only the largest-scale features of the flow and rely on models to describe all the effects of turbulence. While RANS models are generally the most well tested and therefore have the best-understood characterizations, they are limited to steady-state flows and thus cannot be used directly in the inherently nonsteady supernova. At the next-coarsest level of approximation, LES methods resolve some portion of the turbulent cascade itself and use subgrid models to treat the effects of scales beneath the grid scale. In some cases, it has proven valid to forego an explicit model for subgrid effects and instead depend on the corresponding effects of numerical techniques to provide a suitable implicit substitute, leading to implicit LES (ILES) methods [2]. At the highest level of accuracy, DNS fully resolves all scales of interest. DNS simulations require no subgrid model, relying directly on microscopic physical principles instead, making them powerful techniques to verify other RANS or LES turbulent combustion models. The high Reynolds numbers ($\sim 10^{14}$) and likely importance of magnetohydrodynamic effects in the ionized plasma at the smallest scales make informative, true DNS in the stellar context essentially impossible. Hybrid ILES-DNS schemes that resolve the smallest scales of interaction between the flow and burning front are the best available direct verification tests [3; 4].

The physical thickness of the carbon burning stage of the nuclear flame front in the deflagration phase of a Type Ia supernova is 10^{-5} to 10^{4} cm for the densities of interest and is therefore unresolvable in full-star simulations, which typically have linear resolution of several 10^{5} cm. Consequently, the flame front must be modeled in some LES fashion. To date, two primary numerical methodologies have been applied to studies of white dwarf deflagration. One is an LES-level set method, in which the turbulence is treated in an LES model, and the location of the flame front is calculated based on the value of a smooth field defined on the grid and propagated with an advection equation in addition to the hydrodynamics [5]. This method has undergone some evolution ([6] being the most recent) and has been used to study the effect of turbulence on the nuclear burning rate [7] and in many simulations of the deflagration phase of Type Ia supernovae [8–11]. The other method, which is currently implemented in the FLASH code, is an ILES-thickened flame model, in which the turbulent velocity field is treated in an ILES approach, and the flame front is propagated by using an advection-diffusion-reaction (ADR) equation [12]. The ILES/ADR method has been applied to a number of scientific problems, including buoyancy-driven turbulent burning [13], and 2D and 3D simulations of Type Ia models [14–17]. Both of these methods characterize the effects of turbulence on the combustion via an enhanced front propagation speed.

Despite this progress in the development of theoretical and numerical techniques to treat turbulent combustion, a number of key outstanding questions remain, including th following: How does buoyancy-driven turbulent nuclear combustion in a stratified medium differ from turbulent nuclear combustion in a homogeneous isotropic turbulent background? Is it possible to describe the burning rate of a turbulent flame by a single characteristic turbulent timescale?

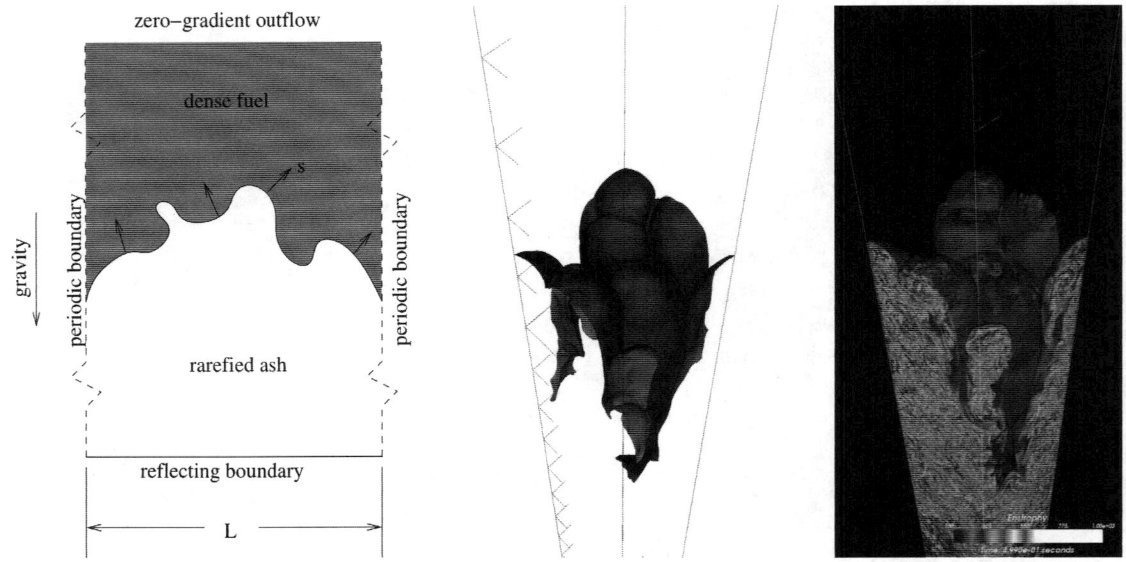

Figure 1. (a) Diagram of simulation configuration for studying a buoyantly-unstable flame front. (b) The flame surface in the fully-developed, self-regulated state. (c) Volume rendering of vorticity magnitude (seconds^{-1}).

If so, what scale dominates the flow – the largest integral scale, the smallest Kolmogorov scale, or some intermediate scale? If it is not possible to describe the turbulent burning rate with a single scale, how can we accurately model the large dynamic range involved in a realistic turbulent combustion flow?

2. Simulation Configuration

Our simulation setup is designed to test the effects of buoyancy and the turbulence that it generates on structure of the flame surface, and therefore the burning rate. (Higher-density) fuel is placed above (lower-density) ash in hydrostatic equilibrium in a square cross-section vertical channel whose dimension in the directions transverse to gravity is L (see figure 1). fhis type of configuration originated with [18], has been used in several studies of buoyancy-driven turbulent burning since, and is similar to many studies of the Rayleigh-Taylor instability; see, for example, [19]. Our implementation is described in detail in [20]. Our simulation is periodic in the transverse direction, closed at the bottom of the channel, and open at the top. The total channel height is much larger than the size of the burning region but still small compared to the pressure scale height. As burning progresses, the expanded ash will force unburned fuel out of the top of the box as the burning surface, which is localized, moves up the channel. We used the 3.0 version of the hydrodynamics code FLASH [21], which implements the Piecewise Parabolic Method for fully compressible, inviscid hydrodynamics with a fully ionized degenerate plasma equation of state on an adaptively block-refined mesh. Simulations of the Rayleigh-Taylor instability with this code have been compared to experiments [22] and a variety of other codes [19]. The reaction, and therefore energy release, is represented by a progress variable, ϕ, which is zero in the fuel and one in the ash. The burning front is propagated bby using an advection-reaction-diffusion scheme [23] with a Kolmogorov Petrovski Piskunov reaction term with the sharpening modification discussed in [15]. The flame width is $\delta \approx 4\Delta$, where Δ is the grid resolution.

3. Metrics for turbulent burning

The complexity of turbulent burning itself and that of the methods used to model it, along with the breadth of work under way on this topic, leads us to consider a variety of metrics to characterize properties of the combustion. Of foremost interest is the scale dependence of the flame surface structure, in particular, determination of the smallest scale of flame surface structure. A quantitative description of this scale dependence is essential for model development in the LES framework, in which a model for subgrid structure must integrate properly with resolved structures. As discussed above, it has been found that in certain circumstances an *implicit* subgrid model may prove sufficient for many characteristics of the combustion. Our large-scale simulations will demonstrate whether this continues to hold at nontrivial Reynolds number as well as giving a wide enough range in resolution to understand why such an ILES approach seems to work and therefore help to quantify its domain of validity. We are using two approaches to measurement of surface structure. The first, described in the next section, uses counts of surface-covering boxes of varying size to quantify the fractal dimension of the flame surface. The second approach will use direct smoothing of the progress variable field to measure the contributions to the total flame surface area which arise from various scales in the flow.

In addition to metrics addressing the flame surface directly, we are applying the most insightful methods available to study the creation, evolution, and decay of the turbulence resulting from buoyancy. This approach is especially important for supernova simulations, as the morphology of the turbulence field is likely to have the largest impact on the spreading of the flame through the star. The nonstationary and nonisotropic nature of our flow makes application of the traditional turbulence workhorse, the energy spectrum, difficult to interpret. For this reason we have come to depend more heavily on structure functions that can be defined appropriately under lateral symmetry. Additionally we are using the ability of the FLASH code to include passive Lagrangian tracer particles in order to measure the Lagrangian structure functions. We expect these to be particularly insightful, as we have found that the turbulence has a markedly different character before and behind the flame surface (see figure 1c).

4. Fractal dimension of the flame surface

As a final point we give brief details about one the techniques being used to measure flame surface characteristics. The burning rate of a flame front depends upon its surface area and is thus sensitive to its complexity (curvature on different scales). Fractal analysis methods provide a measure of the amount of structure in a curve or surface at different length scales and can help our understanding of the evolution and propagation of the flame front [24].

To this end, we have implemented a box counting algorithm in which the simulation domain is divided into a contiguous grid of cubic boxes of width w; those boxes that straddle the flame surface are counted to provide a measure of the surface area at this scale. Over some range of w, we can define a fractal (Minkowski-Bouligand) dimension p such that the number of boxes $N_w \propto w^{-p}$ as $p = -d \log N_w / d \log w$. For a smooth surface, $p \approx 2$, owing to the topological dimension of the surface. If the surface has a rough, complex structure on these scales, $p > 2$.

We find that the flame surface does not have a fixed value of p for all scales, and shows significant time-variation. We can determine an effective dimension p_l for a scale l based on N_w for values of w about l. As anticipated, we find that $p_l \approx 2$ for small l. By determining the range of l for which p_l deviates significantly from 2, as a function of flame speed, we can inform our choice of models for flame propagation.

References

[1] Timmes F X and Woosley S E 1992 *Astrophys. J.* **396** 649–667
[2] Benzi R, Biferale L, Fisher R, Kadanoff L, Lamb D and Toschi F 2008 *Phys. Rev. Lett.* **100**

[3] Bell J B, Day M S, Rendleman C A, Woosley S E and Zingale M 2004 *Astrophys. J.* **608** 883–906
[4] Zingale M, Woosley S E, Rendleman C A, Day M S and Bell J B 2005 *Astrophys. J.* **632** 1021–1034
[5] Reinecke M, Hillebrandt W, Niemeyer J C, Klein R and Gröbl A 1999 *Astron. Astrophys.* **347** 724–733
[6] Schmidt W, Niemeyer J C and Hillebrandt W 2006 *Astron. Astrophys.* **450** 265–281
[7] Schmidt W, Niemeyer J C, Hillebrandt W and Röpke F K 2006 *Astron. Astrophys.* **450** 283–294
[8] Röpke F K, Hillebrandt W, Niemeyer J C and Woosley S E 2006 *Astron. Astrophys.* **448** 1–14
[9] Röpke F K and Niemeyer J C 2007 *Astron. Astrophys.* **464** 683–686
[10] Röpke F K, Hillebrandt W, Schmidt W, Niemeyer J C, Blinnikov S I and Mazzali P A 2007 *Astrophys. J.* **668** 1132–1139
[11] Röpke F K, Woosley S E and Hillebrandt W 2007 *Astrophys. J.* **660** 1344–1356
[12] Khokhlov A M 1995 *Astrophys. J.* **449** 695
[13] Zhang J, Messer O E B, Khokhlov A M and Plewa T 2007 *Astrophys. J.* **656** 347–365
[14] Plewa T 2007 *Astrophys. J.* **657** 942–960
[15] Townsley D M, Calder A C, Asida S M, Seitenzahl I R, Peng F, Vladimirova N, Lamb D Q and Truran J W 2007 *Astrophys. J.* **668** 1118–1131
[16] Jordan G C, Fisher R, Townsley D M, Calder A C, Asida S, Graziani C, Lamb D Q and Truran J W 2008 *Astrophys. J.* **681** 1448
[17] Meakin C A, Seitenzahl I R, Townsley D M, Jordan G C, Truran J and Lamb D Q 2008 *Astrophys. J.* Submitted
[18] Khokhlov A 1993 *Astrophys. J.* **419** L77
[19] Dimonte G *et al.* 2004 *Physics of Fluids* **16** 1668–1693
[20] Townsley D M, Asida S M, Jena T and Lamb D Q 2008 *Astrophys. J.* Submitted
[21] Fryxell B, Olson K, Ricker P, Timmes F X, Zingale M, Lamb D Q, MacNeice P, Rosner R, Truran J W and Tufo H 2000 *Astrophys. J. Suppl. S.* **131** 273–334
[22] Calder A C *et al.* 2002 *Astrophys. J. Suppl. S.* **143** 201–229
[23] Vladimirova N, Weirs G and Ryzhik L 2006 *Combust. Theory Modelling* **10** 727–747
[24] Blinnikov S, Sasorov P and Woosley S 1995 *Space Science Reviews* **74** 299–311

Multidimensional, multiphysics simulations of core-collapse supernovae

O E B Messer[1], S W Bruenn[2], J M Blondin[3], W R Hix[4] and A Mezzacappa[4]

[1] National Center for Computational Sciences, Oak Ridge National Laboratory, Oak Ridge, TN 37831-6008, USA
[2] Department of Physics, Florida Atlantic University, Boca Raton, FL 33431-0991, USA
[3] Department of Physics, North Carolina State University, Raleigh, NC 27695-8202, USA
[4] Physics Division, Oak Ridge National Laboratory, Oak Ridge, TN 37831-6354, USA

Abstract. CHIMERA is a multi-dimensional radiation hydrodynamics code designed to study core-collapse supernovae. The code is made up of three essentially independent parts: a hydrodynamics module, a nuclear burning module, and a neutrino transport solver combined within an operator-split approach. We review the code's architecture and some recently improved implementations used in the code. We also briefly discuss preliminary results obtained with the code in three spatial dimensions.

1. Introduction

The center of a massive star (a star more than 8 or so times the mass of our own Sun) as it nears its demise is composed of iron, nickel, and similar elements. Above this iron core lie concentric layers of successively lighter elements, recapitulating the sequence of nuclear burning that occurred in the star during its lifetime. Unlike prior burning stages, where the ash of one stage became the fuel for its successor, no additional nuclear energy can be released by further fusion in the iron core. No longer can nuclear energy production stave off the inexorable force of gravity. When the iron core grows too massive to be supported against its own self-gravity, the core collapses. This collapse continues until the core reaches densities similar to those of the nucleons in a nucleus, whereupon the hard core of the nuclear interaction renders the core incompressible, halting the collapse. Collision of the supersonically falling overlying layers with this stiffened core produces the bounce shock, which drives these layers outward. However, this bounce shock is sapped of energy by the escape of neutrinos and nuclear dissociation and stalls before it can drive off the envelope of the star; see, for example, [1]. How the shock is revived and proceeds to completely disrupt the star is the central question of supernova theory. A closer examination of some of the known physical processes in the event does provide a short list of probable actors.

After core bounce, $\sim 10^{53}$ erg of energy in the form of neutrinos and antineutrinos of all three types, or flavors, (electron, muon, and tau) is released from the newly formed proto-neutron star (PNS) at the center of the explosion. The supernova explosion energy is $\sim 10^{51}$ erg. Past simulations [2; 3] demonstrate that energy in the form of neutrinos emerging from the PNS can be deposited behind the shock and may revive it. This neutrino reheating is central to modern core-collapse supernova models. However, while a prodigious amount of neutrino energy emerges from the PNS, the neutrinos are weakly coupled to the material directly below the shock, and, in fact, the neutrino heating is very sensitive to the distribution of neutrinos in energy and direction of propagation [4; 5; 6; 7; 8; 9]. Therefore, this ultimately requires neutrino transport algorithms that resolve both the neutrino spectra and the shape of the radiation field at any spatial point in order to compute accurately the neutrino distributions in this region. This renders the core collapse supernova problem a truly multidimensional (space plus neutrino angles and energy), petascale (petaflops, petabytes) problem.

This neutrino heating may also be aided by fluid instabilities (e.g., convection) in the PNS [10; 11; 12; 13; 14], which may boost the luminosity of this central neutrino bulb. Also, convection directly beneath the shock fundamentally alters the nature of neutrino shock reheating [15; 16; 5; 17; 14; 18] relative to the spherically symmetric case, allowing simultaneous downflows that fuel the neutrino luminosities and upflows that bring energy to the shock. A newly discovered instability of the shock wave itself, the Stationary Accretion Shock Instability (SASI), will likely dramatically alter the shock and explosion dynamics [19; 20; 21; 22]. The centrifugal effects of stellar core rotation [17; 14], and other of its effects, will also change supernova dynamics qualitatively and quantitatively, and stellar core magnetic fields, increased perhaps dramatically by compression during collapse, convection (e.g., via a dynamo), and rotation (through wrapping and shear; in the latter case the magnetorotational instability may occur and, if so, would dominate the field evolution), may also play a significant role in driving, and perhaps collimating, core-collapse supernova explosions [23; 24]. The nuclear abundances should be evolved in regions where nuclear statistical equilibrium (NSE) cannot be maintained. This process will enable the potentially observable products of nucleosynthesis to be followed and, most important for the mechanism, the energy released by nuclear burning to be fed back into the computation of the explosion dynamics.

"Ray-by-ray" simulations [25; 26; 14; 18] capture much of the transport realism in multidimensional models, neglecting only contributions from lateral neutrino transport, which will likely be important only below the neutrinospheres in the proto-neutron star where the neutrinos and matter are strongly coupled and the flow may be highly nonspherical. The ray-by-ray approximation allows a substantial amount of physical fidelity in simulations that remain tractable on petascale platforms. The addition of a prescription for nuclear burning to a multidimensional ray-by-ray radiation hydrodynamics scheme allows much of the known physics in the problem to competently addressed.

All of these facts make realistic modeling of the core-collapse supernova mechanism a computational challenge that is nearly unrivaled. Indeed, though the seminal paper of [27] was among the very first examples of computational astrophysics (if not, arguably, computational science in general), the explosion mechanism remains unsure even today, more than 40 years later. The explosion mechanism is apparently inherently multidimensional, with all the attendant difficulties of multidimensional modeling. Furthermore, the mechanism itself must be regarded as energetically marginal. The observed explosion energy of $\approx$1 B (1 bethe [B] = 10^{51} ergs) is only 1 percent of the internal energy residing in the immediate post-collapse core, and the near negative of this in the form of gravitational binding energy. Thus, simulations must be energy conserving to high accuracy if we are to take their outcomes seriously. Ultimately $\approx$300 B in energy is released by the core in neutrinos of all flavors, and their interaction with the stellar core and mantle will either power the explosion itself or play a major role in the explosion dynamics. An inaccurate treatment of neutrino transport can qualitatively change the results of a simulation. Neutrinos interact with matter in a variety of energy-dependent ways, and it is generally accepted that this demands that both the neutrino transport and the interactions receive a full spectral implementation, rather than having the neutrino spectrum prescribed. The angular distribution of the neutrinos is also important to compute accurately. In particular, it affects the neutrino heating, and the latter is primarily determined in a region where the angular distribution can neither be assumed to be isotropic nor radially free streaming. The nuclear abundances should be evolved in regions where NSE cannot be maintained. This will enable the potentially observable products of nucleosynthesis to be followed and, most importantly for the mechanism, the energy released by nuclear burning to be fed back into the computation of the explosion dynamics. Moreover, general relativistic effects must be incorporated as they influence the size of the neutrino heated region, the rate of matter advection through this region, and the neutrino luminosities and RMS energies [28], and can therefore profoundly affect the dynamics. Computational work over the past 40 years has generally consisted of several generations of simulations, each generation containing more physical fidelity than the last, where, as computational power has increased, the treatment of one or more of these myriad physical inputs has been refined. Modern petascale platforms will, for the first time, allow the inclusion of all currently known important physics in supernova simulations. This will mark a watershed moment in supernova simulation, ushering in a new era of predictive capability.

2. Code architecture

Our code, CHIMERA, incorporates most of the aforementioned criteria for realistic core-collapse modeling. At the same time it is efficient enough to evolve progenitors from the onset of collapse to roughly 1-second postbounce, given present state-of-the-art computational resources, such as the Cray XT4. It conserves total energy (gravitational, internal, kinetic, and neutrino) to well within 0.5 B.

CHIMERA can well be described as a "chimera" of three separate, rather mature codes. The codes are tightly coupled in a single executable through a set of interface routines that provide each of the code components access to global variables for use (cf. the FLASH code [29], where each code module "checks out" copies of global variables from a central database structure, uses or evolves them, and then "checks them back in.") The primary code modules are designed to evolve the stellar gas hydrodynamics (VH1), the "ray-by-ray-plus" neutrino transport (MGFLD-TRANS), and the thermonuclear kinetics (XNET). These three "heads" are augmented by a sophisticated equation of state for nuclear matter (e.g., LS-EOS [30]) and a self-gravity solver capable of an approximation to general-relativistic gravity. The hydrodynamics is directionally split, and the ray-by-ray transport and the thermonuclear kinetics solve occur after the radial sweep occurs, when all the necessary data for those modules is local to a processor (see Figure 1). The individual modules are algorithmically coupled in an operator split approach. This approach is well motivated, as the characteristic time scales for each module are widely disparate. Specifically, during the radial sweep of the hydrodynamics, the neutrino transport and the thermonuclear burning are computed along each radial ray, using only data that is local to that ray and, therefore, local to the current process.

The hydrodynamics module in CHIMERA is a modified version of the PPM code VH-1, which has been widely used in astrophysical fluid dynamics simulations and as an important benchmark code for a variety of platforms. VH-1 is a Lagrangian remap implementation of the Piecewise Parabolic Method (PPM) [31]. Being third order in space (for equal zoning) and second order in time, the code is well suited for resolving shocks, composition discontinuities, etc. with modest grid requirements. To avoid the odd-even decoupling and carbuncle phenomenon for shocks aligned parallel to a coordinate axis, we have employed the local oscillation filter method of Sutherland et al. (2003) which subjects only a minimal amount of the computational domain to additional diffusion. We have also found it necessary to incorporate the geometry corrections of [32] in the hydrodynamics module to avoid spurious oscillations along the coordinate axes. Redshift and time dilation corrections are included in both the hydrodynamics and neutrino transport (to be described later). A moving radial grid option, where the radial grid follows the average radial motion of the fluid, makes it possible for the core infall phase to be followed with good resolution. The hydrodynamics equations to be solved can be replaced by the

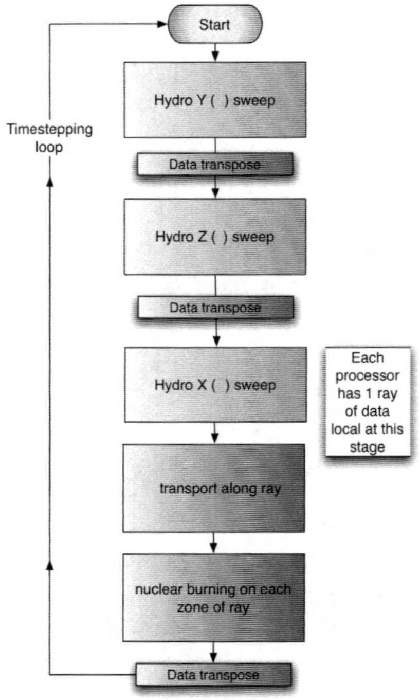

Figure 1. Schematic CHIMERA flowchart.

magnetohydrodynamics equations, enabling the inclusion of stellar magnetic fields in CHIMERA simulations. This change would enable inclusion of *all* known important physical processes in CHIMERA supernova simulations.

The equation of state (EOS) of [30] is currently employed for matter at high densities. The EOS provides a relationship between the matter density and pressure, closing the system of radiation hydrodynamics equations. At densities below about 1.7×108 $\mathrm{g\,cm^{-3}}$ the state of the matter is described by mixture of 4 nuclei (neutrons, protons, helium, and a representative heavy nucleus) in a highly modified version of the EOS described by [33]. For regions not in NSE (at even lower densities), an EOS with a nuclear component consisting of 14 alpha particle nuclei (^{4}He to ^{60}Zn), protons, neutrons, and an ironlike nucleus is used. An electron-positron EOS with arbitrary degeneracy and degree of relativity spans the entire density-temperature regime of interest. The nuclear composition in the non-NSE regions is evolved by the thermonuclear reaction network of [34]. The thermonuclear network is a set of coupled ODEs describing the various nuclear processes that serve to define the isotopic composition of the matter. This is a fully implicit, general-purpose reaction network. Currently, however, we have implemented only a so-called α-network; that is, only reactions linking the 14 alpha nuclei from ^{4}He to ^{60}Zn are used. Data for these reactions is drawn from the REACLIB compilations [35], the world standard for reaction data in simulations of this type. Because the 14 element α reaction network neglects reaction flows involving neutron-rich nuclei, it provides only estimates of the energy generation rates for nuclear burning stages encountered in the supernova [$\pm$50% for oxygen burning and $\pm 10\times$ for silicon burning; 36]. Detailed nucleosynthesis requires evolving 150 or more nuclear isotopes throughout the matter that is ultimately ejected. Simple replacement of the 14-element network with 150 isotopes in the fully implicit Backward Euler integration scheme represents a several hundredfold increase in the computational cost. Given the current ≈5% of execution time required by the alpha network in the current simulations, this cost is prohibitive if implemented naïvely. New thermonuclear kinetic methods can achieve factors of 6–12 decrease in the computational cost of the thermonuclear network during the most computationally expensive supernova burning phases by employing local partial equilibria (termed quasi-equilibrium or QSE) to reduce size of the system of equations which must be integrated [37]. With modest further refinement, these QSE-reduced networks will enable self-consistent nucleosynthesis within the radiation-transport problem at a cost only a few times the current simulations. We have also finished the initial development of a tracer particle tracking module in CHIMERA to allow more detailed postprocessing nucleosynthesis calculations to be performed [38].

Ideally, neutrino transport would be implemented with full multidimensional Boltzmann transport. This important development, however, lies beyond the petascale and will require at least exascale computing to fully realize. We compromise by implementing a "ray-by-ray-plus" approximation (see [14]) for neutrino transport, whereby the lateral effects of neutrinos such as lateral pressure gradients (in optically thick conditions), neutrino advection, and velocity corrections are taken into account, but transport is performed only in the radial direction. Transport is computed by means of multigroup flux-limited diffusion with a flux limiter that has been tuned to reproduce Boltzmann transport results to within a few percent [39]. All $O(v/c)$ observer corrections have been included. The transport solver is fully implicit and solves for four neutrino flavors (types) simultaneously (i.e., ν_e's, $\bar{\nu}_e$'s, ν_μ's and ν_τ's (collectively ν_x's), and $\bar{\nu}_\mu$'s and $\bar{\nu}_\tau$'s (collectively $\bar{\nu}_x$'s)), allowing for neutrino neutrino scattering and pair-exchange, and different neutrino and antineutrino opacities. The PPM technology has been directly applied to both the spatial and energy advection of neutrinos in both the radial and lateral directions. The neutrino opacities employed for the simulations are the standard ones described in [40], with the isoenergetic scattering of nucleons replaced by the more exact formalism of [41], which includes nucleon blocking, recoil, and relativistic effects, and with the addition of nucleon–nucleon bremsstrahlung [42] with the kernel reduced by a factor of five in accordance with the results of [43].

3. Current code development

We have recently implemented a new spatial domain decomposition for the hydrodynamics module used in CHIMERA to enable three-dimensional simulations to be performed. The new spatial domain decomposition divides the computational domain into a series of "pencils" along each coordinate direction. This is in contrast to the original VH-1 "slab" decomposition, where a 2D slice through the computational domain was transposed after each directional sweep. Importantly, the pencil decomposition makes use of a series of data transposes (specifically, using `MPI_Alltoall`) that are performed over row and column subcommunicators. Therefore, although all processes participate in communication roughly simultaneously, producing contention, the actual data movement is significantly less than would be seen an global collective.

Though the preponderance of the work involved in each timestep is performed by the thermonuclear burning and neutrino transport modules, the performance of the hydrodynamics module must also be improved to ensure scalability to tens of thousands of processes, at least. Our choices of ray-by-ray transport and dimensionally split hydrodynamics simplify this task to a large extent but also constrain the methods we use to achieve good scalability. The data transposes between dimensional sweeps and the possibility of overlapping communication and computation in these transposes mark the primary areas of possible improvement.

As these improvements and extensions are added to CHIMERA, the computational infrastructure around the hardware platform will also continue to grow. This growth is inevitable, as the data volumes generated on petascale platforms will dwarf even the most ambitious simulation environments of today. We anticipate each CHIMERA simulation will produce more than 10–100 TB of data per run, the exact number depending strongly on the size of the thermonuclear network used for the nuclear burning and the phase space resolution (20 or more energy groups) of the neutrino transport employed. Robust and straightforward I/O and data workflow tools will be essential to obtain scientific insight from these petascale simulations.

4. Two-dimensional simulations

The current version of CHIMERA has been used to model core-collapse supernovae in two spatial dimensions (for a recent example, see figure 2) [18]. The outcomes in these simulations are promising and have forged a deeper and, in certain important respects, new understanding of the supernova mechanism. We find, as others do, that the shock wave initially stalls and becomes a quasi-stationary accretion shock after propagating out to less than 200 kilometers (the iron core is about 1000 kilometers in size). 200–500 milliseconds then elapse (depending largely on the progenitor), during which time large-scale convective motions develop in the neutrino heated region below the shock, as expected. Distinct, rising high-entropy (hotter) plumes grow and become separated by narrower, low-entropy (cooler) downflows. The rising high-entropy plumes begin to push the shock outward, causing local dome-like distortions. As the evolution continues, these plumes merge and grow, distorting the shock even more. By 140 milliseconds for an 11 solar mass progenitor, or 200 milliseconds for a more massive progenitor, two or three plumes dominate the flow. Significantly, at this time the shock begins to exhibit global distortions of a quasi-oscillatory time-dependent character, which we attribute to the $l = 1$ SASI mode. The shock continues to undulate in a global low-order mode until the oxygen layer, which has been infalling with the matter outside the shock, begins to impinge on one of the extended lobes of the shock. This happens by 180 milliseconds after bounce for an 11 solar mass progenitor, and by 300–500 milliseconds for more massive progenitors. When it does, the energy generated by the oxygen burning of the shock-heated material causes the shock undulations to rapidly grow and a highly asymmetric explosion to occur in the 11 Solar mass case. Simulations beginning with 15 and 20 Solar mass progenitors are ongoing, and the final outcomes are not yet determined but are, nonetheless, promising.

A critical time scale between the stall of the shock and its subsequent rejuvenation is needed in order to explain observed supernova element synthesis, large-scale asymmetry, and remnant neutron star masses. For example, models that explode too quickly [5; 44; 45; 46; 17; 47] do not develop low-mode distortions, which lead to large-scale asymmetries. In our proposed scenario, as briefly described above, this critical time scale is naturally provided by the time required for the oxygen layer to reach the shock. Until that time has elapsed, neutrino energy deposition, aided by convection, supports the shock at a mean radius of several hundred kilometers, but is not able to revive it. The oxygen burning, which commences when the oxygen layer is advected through the shock, provides an additional source of energy. This is sufficient to initiate a shock revival. Moreover, the time to reach the oxygen layer is greatly reduced by the SASI.

5. Three-dimensional rsimulations

The current state of supernova modeling must move to 3D. Virtually all 2D models exhibit large, low-order deviations from spherical symmetry, which appear as sloshing modes (l=1) along the symmetry axis [18]. The work reported in [49] has shown that this mode may not persist in three dimensions, and surprising results can appear in 3D [50]. In 3D simulations, the standing accretion shock instability (SASI) leads to the growth and development of a "spiral mode" with an attendant redistribution of angular momentum below the shock. This spiraling flow beneath the supernova shock is capable of depositing significant angular momentum onto the PNS, causing it to spin up. The predicted spin periods are $\sim$ 50 ms, in the range of the observed spin periods of young pulsars [51; 49; 50]. Moreover, such spins can be generated

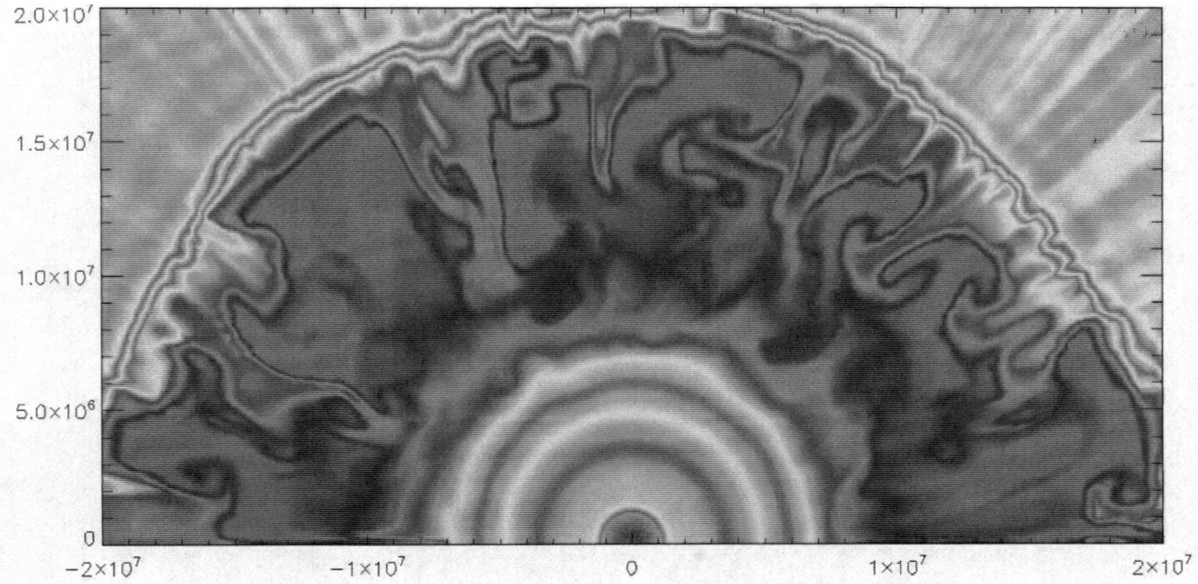

Figure 2. Rendering of the matter entropy in a two-dimensional supernova simulation. The progenitor was the 15 $M_{\odot}$model of [48]. The spatial resolution of the simulation is 256 radial zones by 256 angular zones; 20 energy groups were used to resolve the neutrino spectra.

beginning with nonrotating progenitors, complicating the link between progenitor characteristics and final neutron star properties. These simulations provide specific evidence that core-collapse supernova models must be performed in three dimensions.

No three-dimensional models to date have included all of the following: (1) multifrequency neutrino transport, (2) convection and other fluid instabilities, (3) rotation, (4) nuclear burning, and (5) magnetic fields. In particular, no three-dimensional core-collapse supernova models exist that include at least multifrequency neutrino transport, regardless of other physics included—that is, that satisfy the criterion for realism discussed earlier in this section. (2) and (3) have been included in models without neutrino transport or with gray (frequency-integrated) neutrino transport [17; 51; 49; 52].

We have recently undertaken three-dimensional simulations using CHIMERA. These simulations are evolved through core bounce in one spatial dimension. The profound spherical symmetry seen during core collapse and the absence of symmetry breaking processes like neutrino-driven and PNS convection and the SASI during the infall epoch make this strategy perfectly reasonable. Figure 3 is a snapshot of mater entropy and electron fraction for one of these simulations shortly after the evolution has moved from 1D to 3D (at roughly 19 ms post core bounce). As the evolution proceeds, convection and the SASI will break the apparent spherical symmetry in the matter configuration. These simulations will mark the world's first fully three-dimensional supernova simulations incorporating spectral neutrino transport and detailed nuclear burning capable of modeling realistic nuclear energy release. Though stellar rotation and magnetic fields must be added to future simulations to account for the full panoply of known physics in the problem, these initial simulations mark an important milestone in supernova modeling.

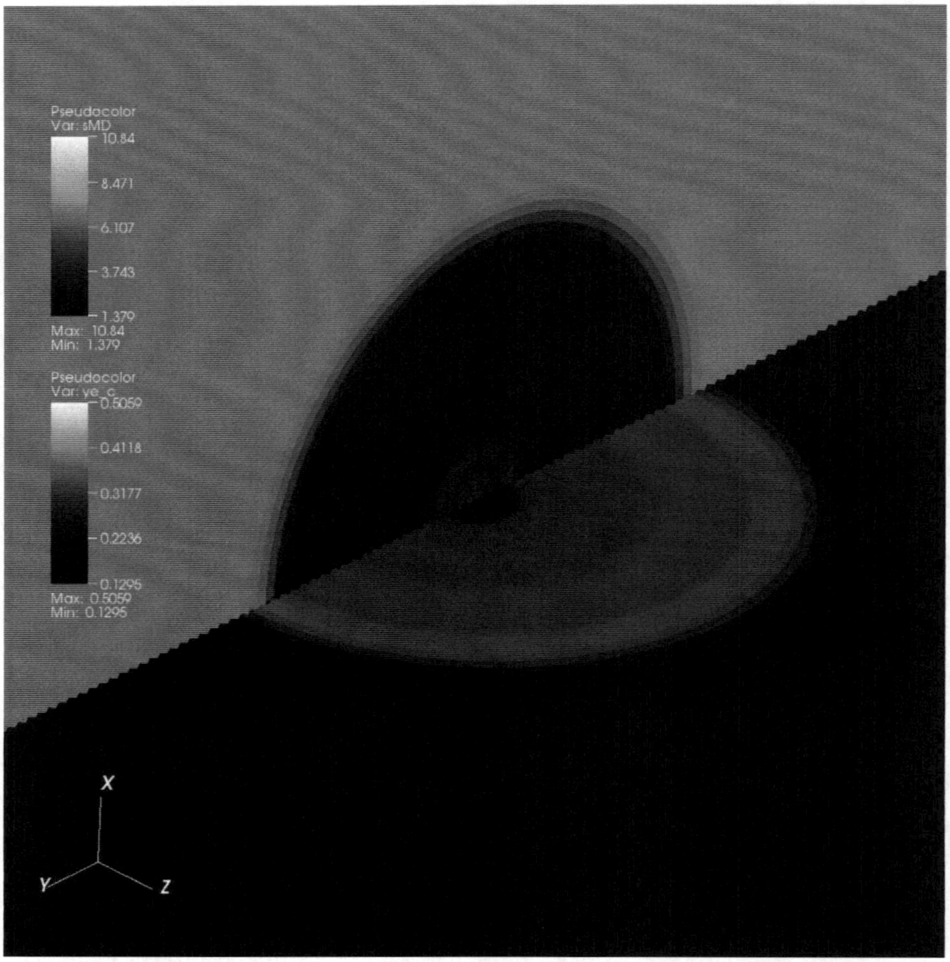

Figure 3. Rendering of the matter entropy (orange colormap) and the electron fraction (blue colormap) in a three-dimensional supernova simulation using CHIMERA. The snapshot is taken roughly 5 ms from the relaxation of spherical symmetry (see text). The spatial resolution of the simulation is 304 radial zones, 76 azimuthal zones, and 152 longitudinal zones; 20 energy groups were used to resolve the neutrino spectra.

Acknowledgments

This work was supported by the NSF PetaApps program under grant numbers OCI-0749242, OCI-0749248, and OCI-0749204, by NASA under grant 07-ATFP07-0011, and by the U. S. Department of Energy under Contract No. DE-AC05-00OR22725 at Oak Ridge National Laboratory. The simulations described here were performed at the National Center for Computational Sciences at ORNL, at the National Energy Research Scientific Computing Center at LBNL, and at the Texas Advanced Computing Center at the University of Texas.

References

[1] Burrows A and Lattimer J M 1985 *ApJ* **299** L19–L22

[2] Wilson J R 1985 *Numerical Astrophysics* ed Centrella J M, LeBlanc J M and Bowers R L (Boston: Jones and Bartlett) pp 422–434

[3] Bethe H A and Wilson J R 1985 *ApJ* **295** 14–23

[4] Burrows A and Goshy J 1993 *ApJ* **416** L75–L78

[5] Janka H T and Müller E 1996 *A&A* **306** 167–198

[6] Messer O E B, Mezzacappa A, Bruenn S W and Guidry M W 1998 *ApJ* **507** 353

[7] Mezzacappa A, Calder A C, Bruenn S W, Blondin J M, Guidry M W, Strayer M R and Umar A S 1998 *ApJ* **495** 911–926

[8] Mezzacappa A, Liebendörfer M, Messer O, Hix W, Thielemann F K and Bruenn S 2001 *Phys. Rev. Lett.* **86** 1935

[9] Janka H T 2001 *A&A* **368** 527–560

[10] Smarr L, Wilson J R, Barton R T and Bowers R L 1981 *ApJ* **246** 515–525

[11] Wilson J R and Mayle R W 1993 *Phys. Rep.* **227** 97–111

[12] Miralles J A, Pons J A and Urpin V A 2002 *ApJ* **574** 356–363

[13] Bruenn S W, Raley E A and Mezzacappa A 2004 *ArXiv Astrophysics e-prints* (*Preprint* astro-ph/0404099)

[14] Buras R, Rampp M, Janka H T and Kifonidis K 2006 *A&A* **447** 1049–1092

[15] Herant M, Benz W, Hix W R, Fryer C L and Colgate S A 1994 *ApJ* **435** 339–361

[16] Burrows A, Hayes J and Fryxell B A 1995 *ApJ* **450** 830–850

[17] Fryer C L and Warren M S 2004 *ApJ* **601** 391–404

[18] Bruenn S W, Dirk C J, Mezzacappa A, Hayes J C, Blondin J M, Hix W R and Messer O E B 2006 *Journal of Physics: Conference Series* **46** 393–402

[19] Blondin J, Mezzacappa A and DeMarino C 2003 *ApJ* **584** 971–980

[20] Janka H T, Buras R, Kitaura Joyanes F S, Marek A, Rampp M and Scheck L 2005 *Nucl. Phys. A* **758** 19–26

[21] Burrows A, Livne E, Dessart L, Ott C D and Murphy J 2006 *ApJ* **640** 878–890

[22] Ohnishi N, Kotake K and Yamada S 2006 *ApJ* **641** 1018–1028

[23] Symbalisty E M D 1984 *ApJ* **285** 729–746

[24] Akiyama S, Wheeler J C, Meier D L and Lichtenstadt I 2003 *ApJ* **584** 954–970

[25] Buras R, Janka H, Keil M T, Raffelt G G and Rampp M 2003 *ApJ* **587** 320–326

[26] Buras R, Rampp M, Janka H T and Kifonidis K 2003 *Phys. Rev. Lett.* **90** 241101–+

[27] Colgate S A and White R H 1966 *ApJ* **143** 626

[28] Bruenn S W, De Nisco K R and Mezzacappa A 2001 *ApJ* **560** 326–338

[29] Fryxell B, Olson K, Ricker P, Timmes F X, Zingale M, Lamb D Q, MacNeice P, Rosner R, Truran J W and Tufo H 2000 *ApJS* **131** 273–334

[30] Lattimer J and Swesty F D 1991 *Nucl. Phys. A* **535** 331–376

[31] Collela P and Woodward P R 1984 *J. Comp. Phys.* **54** 174–201

[32] Blondin J M and Lundqvist P 1993 *ApJ* **405** 337–352

[33] Cooperstein J 1985 *Nucl. Phys. A* **438** 722–739

[34] Hix W R and Thielemann F K 1999 *ApJ* **511** 862–875

[35] Rauscher T, Thielemann F K and Kratz K L 1996 *MmRAS* **67** 851

[36] Timmes F X, Hoffman R D and Woosley S E 2000 *ApJS* **129** 377–398

[37] Hix W R, Parete-Koon S T, Freiburghaus C and Thielemann F K 2007 *ApJ* **667** 476–488

[38] Lee T 2008 *Tracer Particles for Supernova Nucleosynthesis* Ph.D. thesis University of Tennessee, Knoxville

[39] Liebendörfer M, Messer O E B, Mezzacappa A, Bruenn S W, Cardall C Y and Thielemann F K 2004 *ApJS* **150** 263–316
[40] Bruenn S W 1985 *ApJS* **58** 771–841
[41] Reddy S, Prakash M and Lattimer J M 1998 *Phys. Rev. D* **58** 013009
[42] Hannestad S and Raffelt G 1998 *ApJ* **507** 339–352
[43] Hanhart C, Phillips D R and Reddy S 2001 *Phys. Lett. B* **499** 9
[44] Fryer C L 1999 *ApJ* **522** 413–418
[45] Fryer C L and Heger A 2000 *ApJ* **541** 1033–1050
[46] Fryer C L and Warren M S 2002 *ApJ* **574** L65–L68
[47] Kifonidis K, Plewa T, Janka H T and Müller E 2003 *A&A* **408** 621–649
[48] Heger A, Langer N and Woosley S E 2000 *ApJ* **528** 368–396
[49] Blondin J M 2005 *Open Issues in Core Collapse Supernova Theory* ed Mezzacappa A and Fuller G M p 123
[50] Blondin J M and Mezzacappa A 2007 *Nature* **445** 58–60
[51] Blondin J M 2005 *Journal of Physics Conference Series* **16** 370–379
[52] Woosley S and Janka H T 2005 *Nature Physics* **1** 147–154

Optimizing Type Ia supernova follow-up in future dark energy surveys

Peter Nugent, Rollin Thomas and Greg Aldering
Lawrence Berkeley National Laboratory, Berkeley, CA 94720, USA
E-mail: nugent@lbl.gov

Abstract. In the next-generation supernova surveys (Palomar Transient Factory, Dark Energy Survey, LSST, JDEM, etc.) maximizing the cosmological return on each supernova and limiting the systematic uncertainties will be crucial for determining the nature of dark energy. Here we present recent results from the Nearby Supernova Factory, a program that has followed almost 200 SN Ia in the Hubble flow with spectrophotometry. We have used the code SYNAPPS, a combination of the SYNOW spectrum synthesis fitter and the APPSPACK optimization code to automatically perform a highly parameterized fit to observational spectra, to begin the work of identifying the spectral features in a SN Ia that are most important for using these explosions as cosmological probes.

1. Introduction

Observations of high-redshift Type Ia supernovae (SNe Ia) provided the first and to date still the best evidence for the acceleration of the expansion of the universe over the past few billion years, propelled by some kind of mysterious "dark energy," a remarkable result that has shaken the foundations of physics [1–3]. In order to determine whether the dark energy equation of state is consistent with Einstein's cosmological constant (Λ), enormous efforts are under way to accurately populate the SN Ia Hubble diagram with measurements of hundreds of SNe Ia at redshifts $0.2 < z < 0.8$ (e.g., the 5-year CFHT Legacy Survey and the ESSENCE program). Proposed future precision SN Ia cosmology experiments (e.g., DES, JDEM, and LSST) will work at even higher redshifts (up to $z \sim 1.7$) and/or with thousands of SNe Ia to probe the dynamical behavior of the Dark Energy equation of state.

It is crucial to remember that the SN Ia cosmology technique depends on the comparison of the apparent brightnesses of high-redshift SNe Ia to those at low-redshift to measure accurate relative distances. Moreover, our physical understanding of SNe Ia, as well as techniques for standardizing them and correcting for effects such as host galaxy dust extinction, comes from detailed observations of low-redshift SNe Ia. The utility of SNe Ia in the new era of "precision cosmology," where the goal is to discriminate among specific models of the dark energy, depends on a more detailed understanding of SNe Ia and on methods to test for the cosmic evolution of SNe Ia and other systematic effects. To address these issues the Nearby Supernova Factory is discovering and spectrophotometrically following up a few hundred SNe Ia in the nearby smooth Hubble flow ($0.03 < z < 0.08$) [4].

The Nearby Supernova Factory (SNfactory) searches wide-field images obtained by using the QUEST-II CCD camera on the Samuel Oschin 1.2-meter telescope on Mount Palomar [5]. These images are subtracted against older reference images to identify candidates for spectroscopic follow-up. SNfactory also operates the SuperNova Integral Field Spectrograph mounted permanently on the University of Hawaii 2.2-meter telescope on Mauna Kea for dedicated spectroscopic follow-up [6]. Our 30% time share of this facility allows us to obtain high-quality spectral time series of SNe Ia as they evolve (see figure 1). When these are flux calibrated, we can synthesize light curves from these time series in any band (see figure 2).

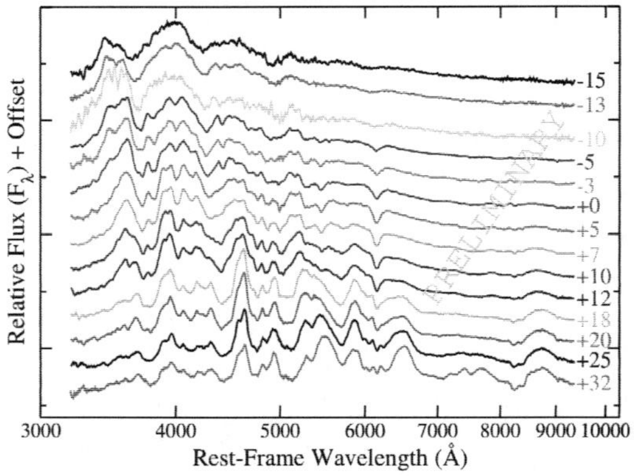

Figure 1. Spectral time series of the SN Ia SNF20070506-006, found by the SNfactory at a redshift of 0.035. Epochs are labeled wrt peak brightness in B-band.

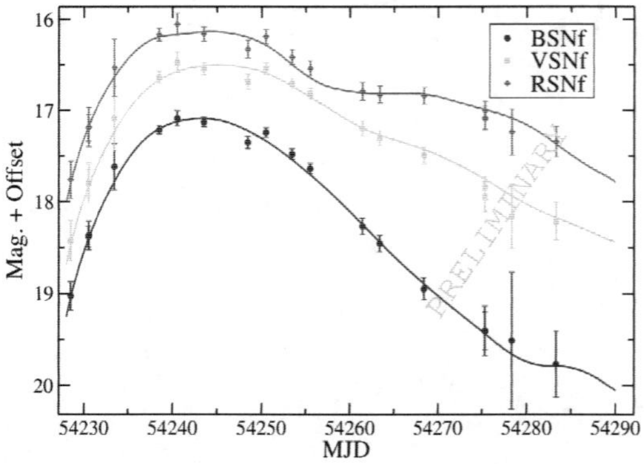

Figure 2. BVR lightcurves of SNF20070506-006 synthesized from the spectrophotometry on the left. Preliminary SALT fits are used to measure the lightcurve shape and colors.

One of the major goals of the SNfactory is to use these smooth Hubble flow SN Ia lightcurves in conjunction with higher redshift data to construct the most accurate SN Ia Hubble diagram yet, due to the augmented low-redshift SN Ia statistics and systematics control of our program. The other major goal is to search for secondary parameters in the spectra of SNe Ia that correlate with luminosity and/or reduce potential systematics in their use as cosmological probes. To achieve these goals, we turn to automated spectrum synthesis.

2. Direct analysis

SN spectra consist of blended, broad absorption and emission features on a thermal or pseudo-thermal continuum [7]. Line features in SN spectra are P Cygni profiles, consisting of a blueshifted absorption component and emission peak centered at the line rest wavelength. P Cygni profiles are typical of rapidly expanding stellar atmospheres, and typical velocities measured from the blue edge of absorption profiles in SN spectra can exceed 20,000 km s^{-1}. SNe evolve on timescales of days to weeks, reaching a peak absolute B magnitudes between -17 and -20 (though there are some noteworthy exceptions [8]).

One approach to the problem of reconciling hydrodynamical stellar explosion models with SN observations is semi-empirical, sometimes referred to as *direct* analysis. This data-driven technique depends on the use of fast, highly parameterized calculations of synthetic SN spectra compared directly to observed ones. A good fit to an observed spectrum constrains explosion models through spectral feature identification and characterization—the presence or absence of particular chemical elements in a SN envelope and corresponding ejecta velocity intervals. Explosion modelers take these results into account when simulating stellar explosions and nucleosynthesis.

The simplest conceptual model of a SN—applicable from a few days to months after explosion—is a sharply defined, continuum-emitting, optically thick photosphere surrounded by an extended line-forming region optically thin to continuum. The most well-known SN direct analysis code is the highly parameterized `SYNOW` code [9], which has been used to study the spectra of SNe of all types. The Sobolev optical depth profile (as a function of velocity) is prescribed for each ion's *reference line*, usually a strong optical line. `SYNOW` uses the full Kurucz line list (more than adequate for direct analysis) and line opacity for other lines of each ion are scaled to the corresponding reference line by assuming Boltzmann excitation of the levels. The source function is that of pure resonance scattering ($S = J$); a more sophisticated source function would require a temperature structure beyond the level of parameterization that is convenient. Since reference line Sobolev optical depths are prescribed by the user individually, no physical constraint is imposed by relative ionizations of chemical elements. Still, this freedom enables one to quickly explore the parameter space of SN spectra to produce basic constraints on the underlying atmosphere.

As an example, figure 3 shows a `SYNOW` fit to the observed spectrum of the Type Ia SN 2006D around 7 days before maximum brightness [10]. This SN revealed the strongest evidence for the presence of unprocessed carbon at photospheric velocities (between 10,000 and 14,000 km s^{-1}). The ion signatures of interest are three C II lines, the strongest of which is the absorption notch around 6300 Å just to the red of the strong Si II absorption at 6100 Å. The identification of carbon with `SYNOW` in the form of a fit to three C II lines simultaneously is an important constraint on explosion models. Certain white dwarf explosion models (delayed detonations) are believed to typically burn all the carbon and oxygen fuel out to a radius of 18,000 km s^{-1} to at least Si-peak elements [11]. In fact, the Si II feature of SN 2006D is fit well by a line profile extending to about 17,000 km s^{-1}, so the Si-peak elements somehow overlap unburned material in velocity space.

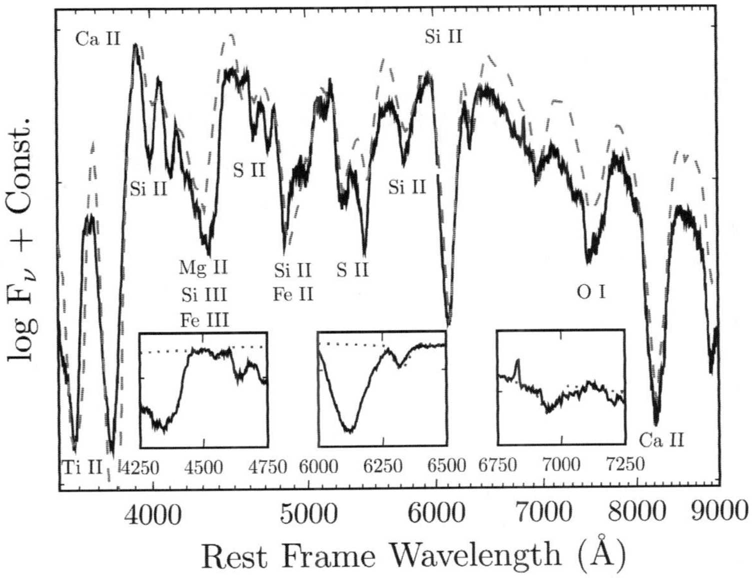

Figure 3. Direct analysis of the Type Ia SN 2006D spectrum at −7 days with respect to peak brightness. The solid black line is the observed spectrum, and the red dashed line is the synthetic `SYNOW` spectrum. The major features are mostly accounted for by various intermediate-mass and iron-peak ions. The insets again show the observed SN spectrum as a solid black line, and a single-ion simultaneous `SYNOW` fit with C II lines λλ 6745, 6580, and 7234 as the blue dotted line.

The SN 2006D results provide an instructive example. While the described constraints do not totally rule out delayed detonations, which are usually expected to completely process carbon at these velocities, they imply that nature allows for unburned material to survive, perhaps in pockets, in SNe Ia. This kind of constraint specifies parameter spaces of composition and energetics that are physically allowed for stellar explosions.

3. SYNAPPS

Until now, successful application of a direct analysis code such as SYNOW to observations has depended on time-consuming, human-supervised, iterative expert use. Guided by experience and rules of thumb, SYNOW users adjust upwards of 50 grouped input parameters to fit SN spectra. This approach has yielded useful results in the past, as discussed, but results may be subjective and homogeneous analysis of a large number of SN spectra produced by a program like the SNfactory and aforementioned future projects with a SN Ia cosmology component would not be feasible.

A typical SYNOW fit requires a human being to interactively manipulate a parameterization containing around 50 or more variables. These include what ions to include, the minimum and maximum velocities where opacity is "turned on," how opacity changes as a function of ejection velocity, and excitation temperature. Note that, in the literature, a SYNOW fit is not actually a "fit" to the observed spectrum in the sense that no figure of merit is applied. The elementary SN model makes use of a sharply defined photosphere emitting a blackbody continuum, when in reality the photosphere has a wavelength-dependent boundary (if it exists at all). For this and other reasons, the theoretical continuum level is offset from the observed spectrum—a minor detail in an analysis where the goal is to simply identify lines and measure the distribution of elements in velocity space. This systematic is a nuisance in the derivation of a robust figure-of-merit, but one route is "local normalization" [12], where the spectrum is smoothed on some physically reasonable wavelength scale and the smoothed "continuum" is divided out. A simple χ^2-minimization approach is then applied.

Each synthetic spectrum calculation takes only a few seconds once the atomic line list has been loaded. Hence, a reasonable approach to such a problem where derivatives are not easily available is a pattern search, such as the familiar downhill simplex search. Promising results have come from coupling a new spectrum synthesizer with the APPSPACK (**a**synchronous **p**arallel **p**attern **s**earch)[13] minimizer package. APPSPACK was designed explicitly to help solve problems with up to about 100 variables on distributed architectures (though serial solutions are supported). The code coupling a direct analysis with the APPSPACK minimizer we call SYNAPPS. SYNAPPS uses a master-worker architecture—the master CPU generates trial points in the parameter space and communicates them to the worker CPUs, which in turn synthesize the resulting spectrum, compute the figure of merit, and report its value back to the master. Depending on various criteria such as sufficient decrease, step size, and contraction factors, the worker CPU evaluates whether the trial point is an improved fit and adjusts the list of trial points accordingly. Convergence is achieved when the trial points have converged to some neighborhood whose size is prescribed by some tolerance. We are also investigating parallelized Newton methods that may possess better convergence properties. Figure 4 includes some preliminary tests with SYNAPPS.

In the left panel of Figure 4, a random (locally normalized) spectrum is shown in which 10 ions are present ($\sim$ 50 parameters), and a S/N ratio of 10 per 2.5 Å bin is applied. The SYNAPPS code producesda successful fit to the test spectrum and finished in less than 2 hours of wall-clock time using 96 Opteron CPUs. As is often the case in optimization problems, the convergence criterion is the subject of some investigation. The best fit near χ^2/dof of 1 was found well before the code completed, indicating that the criterion could be relaxed or the initial guess improved. The input parameters were not exactly recovered by the fit, but this result is not so surprising because the parameterization of optical depth includes some degeneracies. For example, the dependence of line strength on the opacity is roughly logarithmic, so uniformly scaling up an opacity profile can appear degenerate with decreasing its e-folding length.

The right panel of figure 4 includes a SYNAPPS fit to ground-based SN Ia spectroscopy obtained by the SNfactory. The figure shows the locally normalized SNfactory spectrum, along with the standard error. The fit to the spectrum is quite good, reproducing most of the features of the major ions. This fit took only 15 wall-clock minutes with 64 Opteron CPUs.

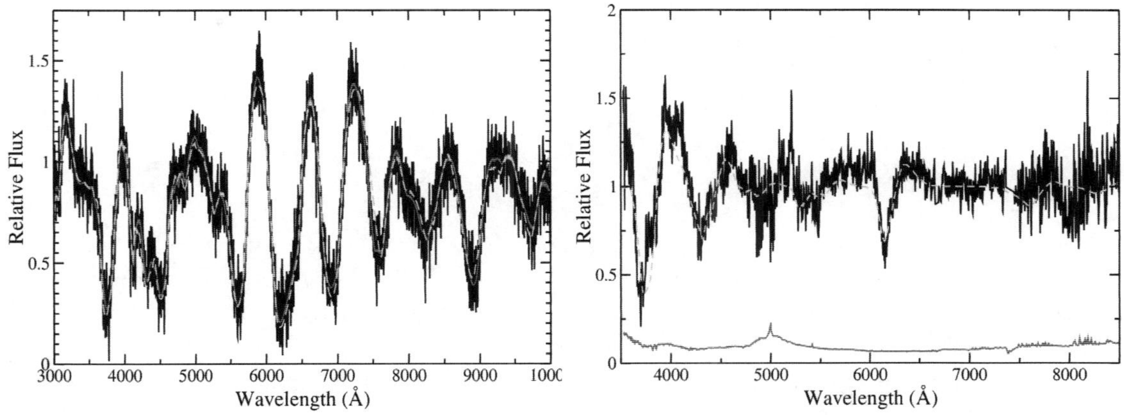

Figure 4. *Left*: Result of a 10-ion test of the SYNAPPS code. A synthetic target spectrum (the solid, noisy, black line) fit (green dashed line) with SYNAPPS using $\sim$ 50 variables with constraints. *Right*: SYNAPPS fit to a continuum-normalized SN spectrum obtained by SNfactory. Both the observed spectrum (solid black line) and standard error (red) are plotted. The fit (green dashed line) reproduces the major features rather well. Ions Ca II, Mg II, Si II&III, O I and S II (IME ions typical of a SN Ia just prior to maximum light) were successfully identified by the code.

A further advance to the code is the recent implementation of a "penalty" function. The purpose of this function is to award the models which fit the minima of weak lines accurately (eg. the C II lines of SN 2006D mentioned above) at the cost of perfecting the fit to a very strong line. This is a generic problem for any χ^2-minimization technique in evaluating the quality of a spectrum synthesis model. We are also working on reducing the level of model parameterization in the automated direct analysis code by replacing its *ad hoc* opacity approach with a more self-consistent calculation. This will enable automated "abundance tomography." That is, the construction of a chemical composition and density model from self-consistent simultaneous fitting of multi-epoch spectroscopy. Again, the SNfactory data set as well as other existing data sets will be used for validation and derivation of SN models.

4. Conclusions

The automated direct analysis and abundance tomography code SYNAPPS will enable, for the first time, large-scale comparative, quantitative, homogeneous analysis of SN spectra. New composition models for SNe will be developed directly from observation for the first time. This capability is in demand now for ground-based projects engaged in the study of SN spectra and SN cosmology and will be in demand for future SN cosmology programs such as the PTF, JDEM, DES, and LSST. The SNfactory will provide more than 100 multiepoch time series of spectra to train and validate this code.

Acknowledgments
We are grateful to the technical and scientific staff of the University of Hawaii 2.2-meter telescope for their assistance in obtaining these data. The authors wish to recognize and acknowledge the very significant cultural role and reverence that the summit of Mauna Kea has always had within the indigenous Hawaiian community. We are most fortunate to have the opportunity to conduct observations from this mountain. This work was supported in part by the Director, Office of Science, Office of High Energy and Nuclear Physics, of the U.S. Department of Energy under Contract No. DE-FG02-92ER40704 This research used resources of the National Energy Research Scientific Computing Center, which is supported by the Office of Science of the U.S. Department of Energy under Contract No. DE-AC02-05CH11231. We also acknowledge support from the U.S. Department of Energy Scientific Discovery through Advanced Computing program under Contract No. DE-FG02-06ER06-04.

References

[1] Riess A G, Filippenko A V, Challis P, Clocchiatti A, Diercks A, Garnavich P M, Gilliland R L, Hogan C J, Jha S, Kirshner R P, Leibundgut B, Phillips M M, Reiss D, Schmidt B P, Schommer R A, Smith R C, Spyromilio J, Stubbs C, Suntzeff N B and Tonry J 1998 *AJ* **116** 1009–1038

[2] Perlmutter S, Aldering G, Goldhaber G, Knop R A, Nugent P, Castro P G, Deustua S, Fabbro S, Goobar A, Groom D E, Hook I M, Kim A G, Kim M Y, Lee J C, Nunes N J, Pain R, Pennypacker C R, Quimby R, Lidman C, Ellis R S, Irwin M, McMahon R G, Ruiz-Lapuente P, Walton N, Schaefer B, Boyle B J, Filippenko A V, Matheson T, Fruchter A S, Panagia N, Newberg H J M, Couch W J and The Supernova Cosmology Project 1999 *ApJ* **517** 565–586

[3] Astier P, Guy J, Regnault N, Pain R, Aubourg E, Balam D, Basa S, Carlberg R G, Fabbro S, Fouchez D, Hook I M, Howell D A, Lafoux H, Neill J D, Palanque-Delabrouille N, Perrett K, Pritchet C J, Rich J, Sullivan M, Taillet R, Aldering G, Antilogus P, Arsenijevic V, Balland C, Baumont S, Bronder J, Courtois H, Ellis R S, Filiol M, Gonçalves A C, Goobar A, Guide D, Hardin D, Lusset V, Lidman C, McMahon R, Mouchet M, Mourao A, Perlmutter S, Ripoche P, Tao C and Walton N 2006 *A&A* **447** 31–48

[4] Aldering G, Adam G, Antilogus P, Astier P, Bacon R, Bongard S, Bonnaud C, Copin Y, Hardin D, Henault F, Howell D A, Lemonnier J P, Levy J M, Loken S C, Nugent P E, Pain R, Pecontal A, Pecontal E, Perlmutter S, Quimby R M, Schahmaneche K, Smadja G and Wood-Vasey W M 2002 *Survey and Other Telescope Technologies and Discoveries. Edited by Tyson, J. Anthony; Wolff, Sidney. Proc. SPIE, Volume 4836, pp 61–72 (2002).* (*Presented at the Society of Photo-Optical Instrumentation Engineers (SPIE) Conference* vol 4836) ed Tyson J A and Wolff S pp 61–72

[5] Baltay C, Rabinowitz D, Andrews P, Bauer A, Ellman N, Emmet W, Hudson R, Hurteau T, Jerke J, Lauer R, Silge J, Szymkowiak A, Adams B, Gebhard M, Musser J, Doyle M, Petrie H, Smith R, Thicksten R and Geary J 2007 *PASP* **119** 1278–1294

[6] Lantz B, Aldering G, Antilogus P, Bonnaud C, Capoani L, Castera A, Copin Y, Dubet D, Gangler E, Henault F, Lemonnier J P, Pain R, Pecontal A, Pecontal E and Smadja G 2004 *Optical Design and Engineering. Edited by Mazuray, Laurent; Rogers, Philip J.; Wartmann, Rolf. Proc. SPIE, Volume 5249, pp 146–155 (2004).* (*Presented at the Society of Photo-Optical Instrumentation Engineers (SPIE) Conference* vol 5249) ed Mazuray L, Rogers P J and Wartmann R pp 146–155

[7] Branch D, Baron E A and Jeffery D J 2003 *Supernovae and Gamma-Ray Bursters* (*Lecture Notes in Physics, Berlin Springer Verlag* vol 598) ed Weiler K pp 47–75

[8] Smith N, Li W, Foley R J, Wheeler J C, Pooley D, Chornock R, Filippenko A V, Silverman J M, Quimby R, Bloom J S and Hansen C 2007 *ApJ* **666** 1116–1128

[9] Fisher A 2000 Ph.D. thesis University of Oklahoma

[10] Thomas R C, Aldering G, Antilogus P, Aragon C, Bailey S, Baltay C, Baron E, Bauer A, Buton C, Bongard S, Copin Y, Gangler E, Gilles S, Kessler R, Loken S, Nugent P, Pain R, Parrent J, Pécontal E, Pereira R, Perlmutter S, Rabinowitz D, Rigaudier G, Runge K, Scalzo R, Smadja G, Wang L and Weaver B A 2007 *ApJL* **654** L53–L56

[11] Marion G H, Höflich P, Wheeler J C, Robinson E L, Gerardy C L and Vacca W D 2006 *ApJ* **645** 1392–1401

[12] Jeffery D J, Ketchum W, Branch D, Baron E, Elmhamdi A and Danziger I J 2007 *ApJS* **171** 493–511

[13] Kolda T 2005 *SIAM J. Optimization* **16** 563

Journal of Physics: Conference Series **125** (2008) 012012 doi:10.1088/1742-6596/125/1/012012

Numerical simulation of low Mach number reacting flows

S E Woosely[1], A J Aspden[2], J B Bell[2], A R Kerstein[3] and V Sankaran[3]

[1]Department of Astronomy and Astrophysics, University of California, Santa Cruz, CA 95064, USA

[2]Center for Computational Science and Engineering, Lawrence Berkeley National Laboratory, Berkeley, CA 94720, USA

[3]Combustion Research Facility, Sandia National Laboratories, Livermore, CA 94550, USA

E-mail: `woosley@ucolick.org`

Abstract. The explosion of a Type Ia supernova (SNIa) begins as a turbulent flame deep within a 1.4 solar-mass white dwarf. Initially the burning happens in the flamelet regime where turbulence serves only to wrinkle and fold an essentially laminar burning front. As the star expands and the flame moves outwards, however, it encounters regions of lower density. At $\sim 2 \times 10^7$g cm^{-3}, the flame transitions to a distributed burning regime. Here individual flamelets are disrupted by turbulent eddies, resulting in a fundamental change in the character of the burning. Detonation does not occur immediately because the turbulently broadened flamelets are still too thin. As the density declines further, however, each flamelet thickens and moves faster until only a few structures are contained within the $\sim$ 10 km integral scale of the SN turbulence. It is here that detonation may occur. We present simulations using both a three-dimensional low Mach number model and a one-dimensional linear eddy model to explore the structure of these flames and quantify their scaling behavior. Our results suggest that detonation may be possible at a density near 1.0×10^7 g cm^{-3}.

1. Introduction

It is widely believed that in order to explain the observations of especially the brightest Type Ia supernovae, nuclear burning that remains subsonic at all times (deflagration) is inadequate [1, 2]. There must be, at late times after the star has already expanded significantly, a transition to a more rapid form of burning, that is, detonation. So far, "delayed detonation" has been introduced into the models as a free parameter, typically an *ad hoc* function of density and turbulent energy [1, 3]. We seek here to understand the physics of the transition and to determine the conditions.

Two previous papers explored the conditions that might lead to a low-density transition to detonation [4] and the nature of turbulent carbon burning resolved in 3D [5]. Here we focus on flames with a fuel of 50% carbon and 50% oxygen at a density of 1×10^7g cm^{-3}, which corresponds to the distributed regime. Through a sequence of simulations, we explore the scaling behavior of flames in this regime and compare the behavior with results using the Linear Eddy Model (LEM [6]) and with Damköhler scaling [7]. After the 1D LEM is verified, calculations are carried out on larger scales. We find a transition in the nature of the burning when the size of an eddy that can burn in a turnover time becomes of order the integral scale [8, 9].

Table 1. Simulation properties - "scaled" values assume Damköhler scaling ($l^{2/3}$) based on case (a) [5].

Case	(a)	(b)	(c)	(d)	(e)
Domain size L (cm)	150	1200	9600	8.0×10^4	3.27×10^6
Integral length scale l (cm)	15	120	960	7680	4.92×10^5
Turbulent intensity u' (cm/s)	2.47×10^5	4.93×10^5	9.86×10^5	1.97×10^6	7.89×10^6
Scaled turbulent speed s_T^* (cm/s)	1.9×10^4	7.6×10^4	3.04×10^5	1.21×10^6	1.95×10^7
Scaled turbulent width l_T^* (cm)	140	560	2240	8960	3.58×10^4
3D-sim turbulent speed s_T (cm/s)	1.9×10^4	7.7×10^4	3.4×10^5	-	-
3D-sim turbulent width l_T (cm)	140	750	3200	-	-
LEM calculated speed s_T^* (cm/s)	1.5×10^4	6.1×10^4	2.3×10^5	8.2×10^5	1.1×10^7

2. Methodologies

The three-dimensional simulations are based on the low Mach number model derived from the compressible flow equations using asymptotic analysis to decompose the pressure into dynamic and thermodynamic components. This analytically removes acoustic waves and therefore the need to resolve them numerically; the equation of state constrains the evolution. The low Mach number equations are discretized by using a fractional-step, projection algorithm and are coupled to a parallel adaptive mesh framework. For details of the derivation of the method for general equations of state and the numerical method, see [10].

The Linear Eddy Model [6] is a numerical technique that captures many of the aspects of 3D turbulence on a 1D grid. A background isotropic turbulence obeying Kolmogorov statistics is assumed and the action of eddies on a field of abundances and temperature is represented by an instantaneous map (a so called triplet map). Eddy locations are random, and size sampling is based on Kolmogorov scaling. The triplet map captures compressive strain and rotational folding effects of eddies and causes no property discontinuities. This approach simulates evolution along a 1D line of sight through a 3D flow.

3. Computations and results

3.1. 3D direct numerical simulations

We initialized the simulations (table 1) with the flat laminar flame in a three-dimensional domain initially filled with homogeneous isotropic turbulence. The flame is oriented so that it propagates downward, although gravitational forces are not included. Periodic boundary conditions were prescribed laterally, a free-slip base, and outflow at the upper boundary. The turbulent velocity field was maintained, following [11], by forcing the momentum equations with a superposition of long wavelength Fourier modes with random amplitudes and phases. The forcing term is scaled by density and so is somewhat reduced in the ash. This approach provides a way to embed the flame in a localized turbulent background, mimicking the much larger inertial range that these flames would experience in a SNIa.

Calculations were carried out for several values of integral scales assuming a constant energy dissipation of $u'^3/l = 10^{15}$ erg g^{-1} s^{-1}. This is on the low end of turbulent intensities expected in an actual supernova [12]. We began with the distributed flame from [5], which has an effective cross-section resolution of 256×256 zones and size 150×150 cm. By demonstrating that we can approximately recover the same turbulent flame speed when the simulation is repeated at a resolution coarsened by a factor of 8 in each direction, we established confidence in simulations with that cell width. Therefore, we can simulate a distributed flame in a 12×12 m domain. This step is then repeated to obtain even larger domains.

Figure 1 shows instantaneous vertical slices of fuel consumption rate (intense burning in red,

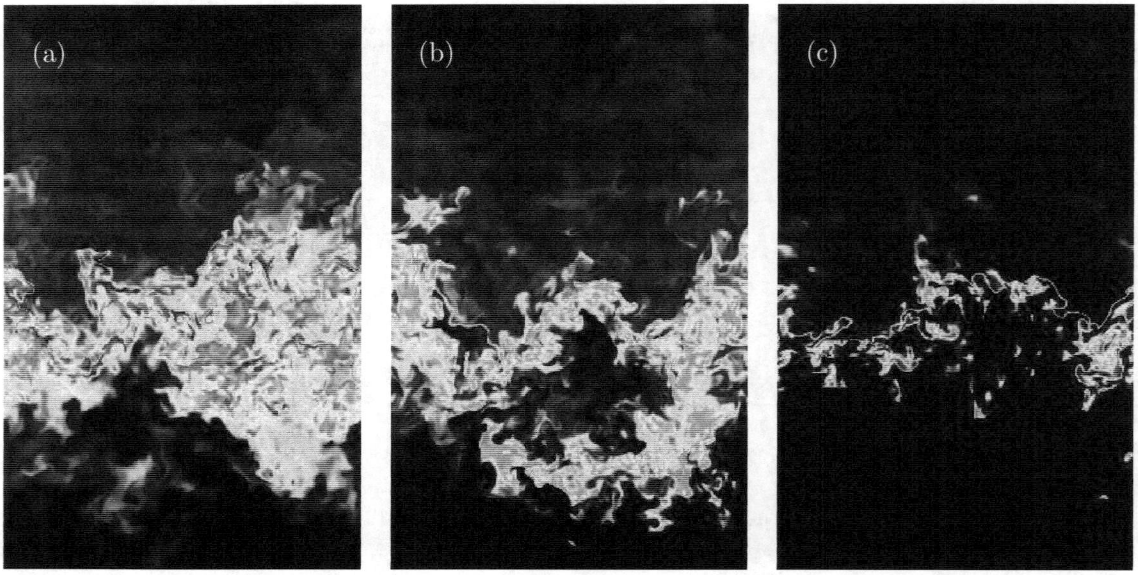

Figure 1. Instantaneous vertical slices of fuel consumption rate (intense burning in red, no burning in blue). The whole domain width is shown, but the height is cropped.

no burning in blue). As the flame burns in increasingly larger integral length scales, there is a decrease in the relative size of the flame structures, with burning occurring in distributed pockets. Figure 2 shows that the flame speed, averaged across the burning volume, is nearly constant after a start up transient. Moreover, taking the $l = 15$ cm case as a base state [5], the steady state speed scales as $l^{2/3}$, the expected scaling for a diffusive flame in which the diffusion coefficient is dominantly due to turbulence ($D \sim u'l$). This scaling must eventually break down because of the restriction $s_T \leq u'$.

3.2. 1D results using LEM

Table 1 shows that LEM is able to reproduce well the bulk speeds and widths of the flames calculated in 3D. The speed also obeys Damköhler scaling but is, over all, a little smaller than the DNS results. Near perfect agreement could have been achieved (but was not) by an adjustment of a parameter in LEM. Heartened by the agreement, we used LEM to explore flame properties on larger length scales than could be done at high resolution in 3D, to 10 km and beyond. Zoning ranged from 0.24 cm for $l = 15$ cm (2048 zones) to 50 cm for $l = 4.5$ km (65536 zones). Until 4.5 km, the $l^{2/3}$ scaling found for smaller integral scales is approximately maintained, but a little below that length scale the flame speed saturates at u'. Above that value, the width and speed are found to scale as $l^{1/3}$; that is, the turbulent flame speed remains pegged to the the overall turbulent speed on the integral scale. As the integral speed approaches this value, the flame front became complex (figure 2) and the burn rate irregular. Variations in integrated burning rate of about a factor of three were observed (the values in table 1 are averages), and the carbon mass fraction in the burning region was about three times less than that in the fuel, corresponding to an overall reduction in burning time scale of a factor of 9.

4. Conclusions

Figure 3 [9] summarizes our conclusions and shows where and how detonation might happen in a SNIa. The Karlovitz number is defined as $\mathrm{Ka} = (l_L/l_G)^{1/2}$, where l_L is the laminar flame thickness (where one exists) and l_G is the Gibson scale. For $\mathrm{Ka} < 1$, one has laminar flames,

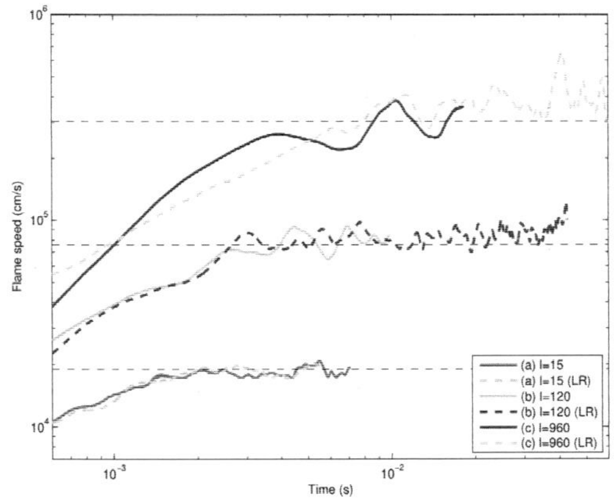

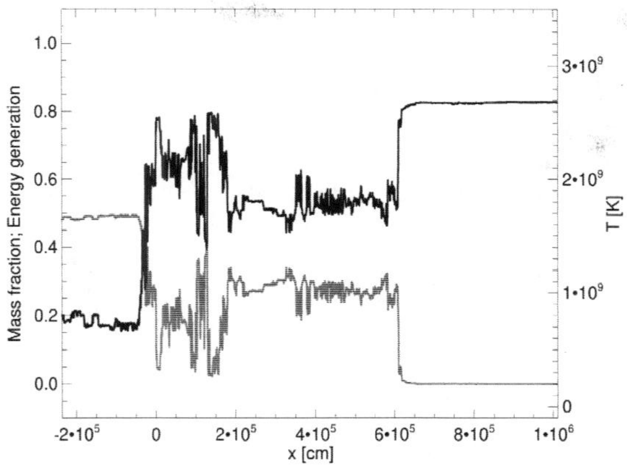

Figure 2. *Left:* Turbulent flame speeds from three-dimensional simulations. Solid lines denote high-resolution simulations (256 zones across;, dashed lines denote low resolution (LR) simulations (32 zones across). Colors denote same cell width. Dashed lines are turbulent flame speeds predicted by Damköhler scaling. *Right:* A calculation using LEM of a single flame for an integral scale of 4.92 km at a density 1×10^7 g cm^{-3}. The turbulent speed on that scale is 78.9 km s^{-1} and the average flame speed and width are also close to these values. Many complex and folded structures are seen, but occasionally surprisingly homogeneous, isothermal regions appear. Red is carbon mass fraction and black is the temperature.

folded and deformed by turbulence, but each still well defined. No detonation is possible in this regime. The overall burning moves at approximately u^*, the characteristic speed of turbulence on the integral scale of the supernova, L^*, and the individual flamelets serve only to "digest" the entrained fuel. The number of such flames contained within the integral scale is approximately $n \approx u^*/s_L$, with s_L the laminar flame speed, and so can be very large. Above Ka $= 10$ (shown on figure 3), turbulence tears individual flamelets and dominates the heat transport. This is the region delimited and studied here in our 3D simulations. For typical turbulence parameters, $L^* = 10$ km and $u^* = 100$ km s^{-1}, we find that the transition to "distributed burning" happens at about 2×10^7 g cm^{-3}.

For lower densities or higher turbulent speeds, Ka > 10; that is, as the supernova expands further, one still has a large collection of flamelets within L, but each flamelet becomes broadened and accelerated by the turbulence. At a given density and turbulent energy dissipation, $\varepsilon = u^{*3}/L^* \sim 10^{15} - 10^{18}$ erg g^{-1} s^{-1} for $u^* = 100$ to 1000 km s^{-1}, there is a characteristic length scale, $\lambda = (\tau_{nuc}^3 \varepsilon)^{1/2}$ [8, 4]. This is the size of an eddy that will burn in a turnover time (a macroscopic analogue, for turbulent flames, of the Gibson length). The characteristic nuclear time, τ_{nuc}, is evaluated in our studies and is $\tau_{nuc} = l_T/s_T$ in table 1. There is only one physical integral scale in the supernova, $L^* \sim 10$ km, but we explored the scaling properties of the turbulent flame speed and width by varying a fictitious integral scale, l, subject to the constraint $\varepsilon =$ constant (table 1), confirming the expected Damköhler scaling, $s_T \propto l^{2/3}$ and $l_T \propto l^{2/3}$ so long as $l < \lambda$. The ratio $(L^*/\lambda)^{2/3} = L^*/(u^* \tau_{nuc})$ is also known as the Damköhler number.

So long as Da $\gg 1$, that is, $\lambda \ll L^*$, there are still many turbulently broadened flamelets, each of width $\sim \lambda$, contained within the integral scale. Stochastic variations in the speeds of individual flamelets will tend to cancel out because $\sqrt{n}$ is large. As Da approaches unity, however, the flamelets become fewer and broader and move faster. Eventually, at Da $= 1$, there is, on the average, only one flame with width L^* and speed u^*. Here our studies with LEM show that the overall flame speed exhibits large fluctuations. The flame structure is very complex, with transient large regions of mixed fuel and ash. "Microexplosions" occur as these pockets burn nearly coherently.

Increasing the turbulence still further, or decreasing the density (hence increasing τ_{nuc} and λ and decreasing Da below 1), does not help. One would then get a single turbulently broadened flame with width $\lambda > L^*$, and speed $< u^*$, similar to those studied here in 3D. Thus, if detonation is to happen, it happens for Da $\approx$ 1 - 10. Our studies with LEM [9] show that the maximum boost from the irregular burning of the flame in this regime corresponds to a reduction in burning time scale of at most a factor of 9. Thus to get supersonic burning, one must also have turbulent speeds no slower than about 10% sonic [12]. Moreover, the region that burns faster than sound must large enough to initiate a self-sustaining detonation [4]. Our calculations show that all these conditions may be simultaneously satisfied at 10^7 g cm^{-3}, though not by a large margin.

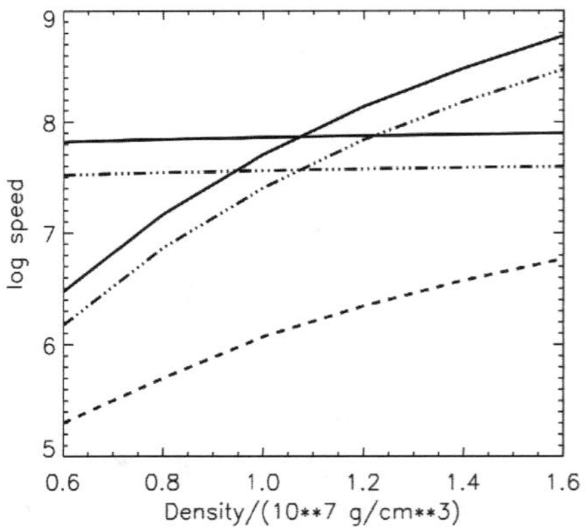

Figure 3. Regions of distributed burning. The y-axis is the log of the turbulent speed on an integral scale, L^*, assumed to be 10 km, and the x-axis is density. The lowest dashed line is Ka = 10. Below this line, one is in the laminar flamelet regime. The solid and dot-dashed slowly rising lines are Da = 1 and 2 respectively. The solid and dot-dashed nearly horizontal lines are sound speed divided by 5 (lower line) and 10. See [9] and text.

Acknowledgments
A. Aspden was supported by a Seaborg Fellowship at LBNL under DOE contract DE-AC02-05CH11231. J. Bell was supported by the SciDAC Program under contract No. DE-AC02-05CH11231. This research used resources of the National Center for Computational Sciences at Oak Ridge National Laboratory, which is supported by the Office of Science of the DOE (DE-AC05-00OR22725). Additional computing resources were provided on the ATLAS Linux Cluster at LLNL as part of a Grand Challenge Project. Work at UCSC was supported by SciDAC (DE-FC02-06ER41438). Sandia National Laboratories are operated by the Sandia Corporation, a Lockheed Martin Company, for the DOE under contract DE-AC04-94AL85000.

References

[1] Hoeflich P and Khokhlov A 1996 *Astrophysical Journal* **457** 500
[2] Kozma C, Fransson C, Hillebrandt W, Travaglio C, Sollerman J, Reinecke M, Röpke F K and Spyromilio J 2005 *Astronomy and Astrophysics* **437** 983
[3] Gamezo V N, Khokhlov A M and Oran E S 2005 *Astrophysical Journal* **623** 337
[4] Woosley S E 2007 *Astrophysical Journal* **668** 1109
[5] Aspden A J, Bell J B, Day M S, Woosely S E and Zingale M 2007 *Submitted to Astrophysical Journal*
[6] Kerstein A 1991 *J. Fluid Mech.* **231** 361
[7] Damköhler G 1940 *Z. Elektrochem* **46** 601 – 652
[8] Kerstein A 2001 *Phys. Rev. E.* **64** 066306
[9] Woosley S E, Kerstein A, Sankaran V and Röpke F 2008 *In preparation*
[10] Bell J B, Day M S, Rendleman C A, Woosley S E and Zingale M A 2004 *Journal of Computational Physics* **195** 677–694
[11] Aspden A J, Nikiforakis N, Dalziel S B and Bell J B 2008 *Submitted for publication*
[12] Röpke F K 2007 *Astrophysical Journal* **668** 1103

Astrophysical applications of the MAESTRO code

M Zingale[1], A S Almgren[2], J B Bell[2], C M Malone[1] and A Nonaka[2]

[1]Department of Physics and Astronomy, Stony Brook University, Stony Brook, NY 11794-3800, USA

[2]Center for Computational Science and Engineering, Lawrence Berkeley National Laboratory, Berkeley, CA, 94720, USA

E-mail: mzingale@mail.astro.sunysb.edu

Abstract. Convective velocities in a white dwarf leading up to a Type Ia supernova or in the accreted layers on a neutron star preceding an X-ray burst are very subsonic. Under these conditions, sound waves can be neglected, but capturing compressibility effects due to nuclear reactions and background stratification is critical to accurately model the flow. We have developed a new algorithm, MAESTRO, based on a low Mach number formulation that exploits the separation of scales between the fluid velocity and the speed of sound. Here, we provide a brief overview of MAESTRO and present the initial astrophysical applications of the algorithm.

1. Introduction

In convectively driven flows, the phenomena of interest evolve on time scales associated with the fluid velocity, which can be orders of magnitude larger than those associated with acoustic wave propagation. Traditional compressible flow solvers typically cannot effectively model these flows. Low Mach number models exploit this separation of scales to derive approximations that allow for large time steps based on fluid velocity rather than sound speed. For large-scale astrophysical applications, a low Mach number model must incorporate a general equation of state [1], must allow for finite-amplitude density and temperature perturbations, and must be able to evolve the base state of the star. We have developed a new low Mach number hydrodynamics code, MAESTRO [2, 3, 4], that accurately and efficiently models these types of astrophysical flows.

In a low Mach number approximation, the pressure is decomposed into a thermodynamic pressure, p_0, and a dynamic pressure, π, such that $\pi/p_0 \sim O(M^2)$, where M is the Mach number of the flow. We define a base state density, ρ_0, by requiring the thermodynamic pressure be in hydrostatic equilibrium, $\partial p_0/\partial r = -\rho_0 g$, with gravitational acceleration, g. The total pressure is replaced by p_0 everywhere except the momentum equation; this decoupling results in sound waves being filtered from the system. This procedure results in the following system of equations:

$$\frac{\partial(\rho X_k)}{\partial t} = -\nabla\cdot(\mathbf{U}\rho X_k) + \rho\dot{\omega}_k \ , \tag{1}$$

$$\frac{\partial(\rho h)}{\partial t} = -\nabla\cdot(\mathbf{U}\rho h) + \frac{Dp_0}{Dt} + \rho H_{\rm nuc} + \rho H_{\rm ext} + \nabla\cdot\kappa\nabla T \ , \tag{2}$$

$$\frac{\partial\mathbf{U}}{\partial t} = -\mathbf{U}\cdot\nabla\mathbf{U} - \frac{1}{\rho}\nabla\pi - \frac{(\rho-\rho_0)}{\rho} g\mathbf{e}_r \ , \tag{3}$$

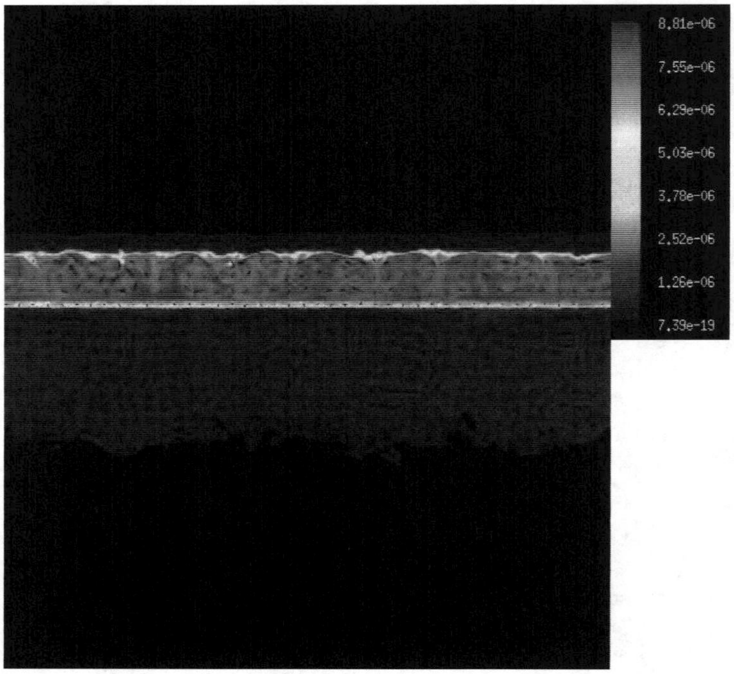

Figure 1. Carbon mass fraction over-plotted with velocity vectors after 0.022 s of evolution for the XRB convection problem. (Note we have zoomed in on the convective region—the total vertical extent of the image is 850 cm.)

$$\nabla \cdot (\beta_0 \mathbf{U}) = \beta_0 \left(S - \frac{1}{\overline{\Gamma}_1 p_0} \frac{\partial p_0}{\partial t} \right) . \tag{4}$$

Here, X_k are the mass fractions of the nuclear species, with production rates, $\dot{\omega}_k$. The total mass density, ρ, is defined as the sum of the partial densities, (ρX_k). The specific enthalpy, h, has heating sources due to reactions, H_{nuc}, (optional) external sources, H_{ext}, and thermal diffusion, where T is the temperature and κ is the thermal conductivity.

The velocity field, $\mathbf{U}$, satisfies a divergence constraint (4) derived from the equation of state. The presence of the densitylike quantity β_0 multiplying $\mathbf{U}$ inside the divergence incorporates the compressibility effects due to the background stratification, similar to the effect of ρ_0 in the divergence constraint in the anelastic approximation. The source term, S, includes the compressibility effects due to heat release and compositional mixing. Finally, $\overline{\Gamma}_1$ is the average of $\Gamma_1 = \partial \log p / \partial \log \rho |_s$ over a layer. Full details of the derivation, the form of β_0 and S, and the solution of this system of equations are given in [4].

As discussed in [3, 4], for the method to accurately represent the physical system, the pressure that follows from the equation of state must remain close to the base state pressure, p_0, as the system evolves. To accommodate this requirement, we evolve the base state in response to the large scale heating and convection on the grid. Extensive comparisons to compressible hydrodynamics methods were shown in [2, 3, 4] and to anelastic hydrodynamics in [2], all showing excellent agreement. Furthermore, we have demonstrated that the algorithm scales well up to 1000 processors [5].

2. X-ray bursts

The standard picture of an x-ray burst (XRB) is the thermonuclear explosion of the accreted H/He layer on the surface of a neutron star (see [6] for a review). As fuel accretes, the temperature and density at the base of the layer increase until conditions are right for a thermonuclear runaway. It is believed that the runaway begins in a small region and propagates subsonically through the fuel layer [7]. Most theoretical work on the subject has been done in one

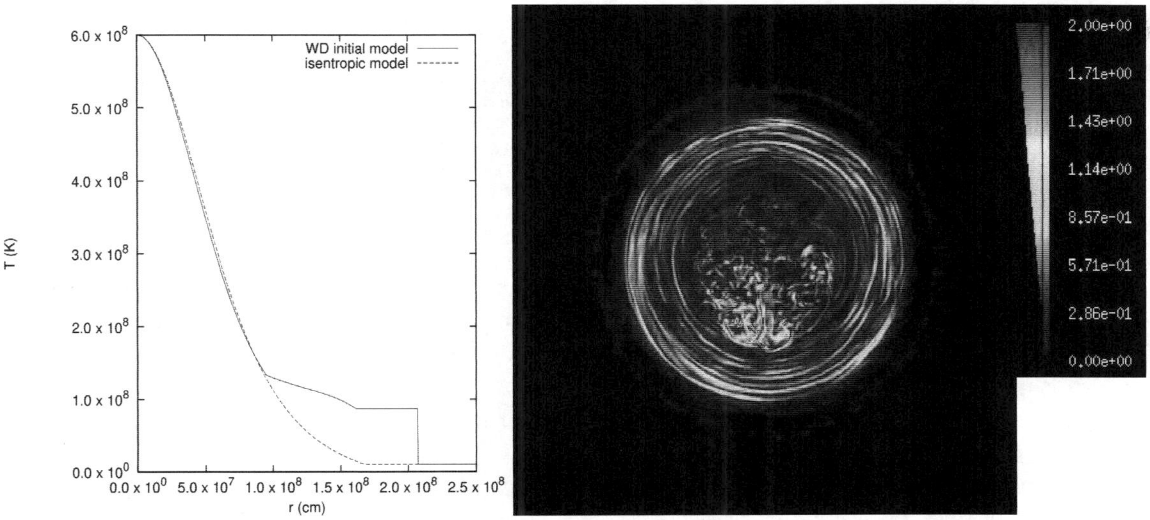

Figure 2. Temperature of the initial model compared to an isentropic model (left) and vorticity (X-Z slice) after 814 s of evolution (right) for the convecting WD problem. (Note that the scale for the vorticity has been reduced from the peak value of 3.27 to enhance detail.)

dimension (see [8, 9] for example). However, these models assume the fuel is burned uniformly over the surface of the star which is unlikely if accretion is not spherically symmetric [10]. Models using the shallow-water approximation [11] have shown the importance of the Coriolis force in localizing the burning front. Previous two-dimensional low Mach number calculations presented in [12] showed the development of convection in the fuel layer. While both MAESTRO and the method used in [12] are based on low Mach number approaches, nontrivial differences between the two have been described in [4]. Our goal for the study of XRBs is to extend the calculations of [12] using our low Mach number method to higher resolution, larger domains, and three dimensions. We will also explore the effect of rotation on localization of the burning.

XRBs are well described by plane-parallel geometries, so the algorithm we use is largely that described in [4], with minor changes. The only significant addition is thermal diffusion, which contributes a term in the enthalpy equation (2) and in S. The treatment of thermal diffusion follows the procedure described in [13] for small-scale astrophysical flames.

To model the XRB we began with a $1.4M_{\odot}$, 10 km neutron star with a thin layer ($\sim$ 5 m) of ^{4}He on the surface. The composition below the accreted layer is set to pure ^{56}Fe. This model is taken as the initial base state for the low Mach number algorithm. We model the upper 10–15 m of the neutron star including its surface; this differs from the model of [12], which did not include the surface. Nuclear burning occurs via 3-α reactions as given in [14] and drives convection throughout the He layer. The simulation we present here had a domain size of 640 cm by 1280 cm on a 512×1024 grid. An initial relative density perturbation of 1.0×10^{-6} was centered just above the helium interface to break the symmetry.

Figure 1 shows the ^{12}C mass fraction over-plotted with velocity vectors after 22 ms of evolution. The sharp interface at the bottom of the red–orange region is the base of the accreted helium layer where 3-α reactions occur most rapidly and drive convection throughout the overlying layers. Additionally there is a large region of convective overshoot below the helium layer, the extent of which seems to have some dependence on the resolution of the model. The formation and persistence of this overshoot region are currently being investigated.

3. White dwarf convection

A Type Ia supernova (SN Ia) is the thermonuclear explosion of a white dwarf that has accreted from a companion to the point where it approaches the Chandrasekhar limit (see [15] for a review). At this point, the temperature is hot enough in the center for carbon fusion reactions to begin, driving convection throughout the star. This convection can last for centuries, increasing the overall temperature of the white dwarf [16]. Eventually, the temperature is high enough that the reactions cannot be quenched by the convective motions of rising plumes. At this point, a burning front is said to have ignited, and it will consume the available carbon/oxygen fuel in seconds. Models of exploding white dwarfs typically begin by seeding one or many hot spots at or near the center of the star and following the subsequent evolution (see for example [17, 18, 19, 20]). These calculations demonstrate that an understanding of the convection leading up to ignition is critical, as this is what determines the number, size, and distribution of the hot spots that seed the burning front. There have been a few investigations of this convection [21, 22, 23], but a lot of work remains.

Extending MAESTRO to handle flows with a radial base state requires a procedure for mapping the one-dimensional base state onto the three-dimensional Cartesian grid, and the inverse operation. At present, we use a base state with finer zoning than the Cartesian cells. The gravitational acceleration in this case is computed based on the base state mass distribution. This variation in g leads to a different evolution equation for the base state velocity than that for plane-parallel simulations. As discussed in [4], the temperature, T, can be derived through the EOS with either h or p_0 as an input in addition to ρ and X_k. For spherical flows, we've found that using $T = T(p_0, \rho, X_k)$ reduces the numerical noise associated with representing a spherical star on a Cartesian grid. This remains a topic of active research, and full details of the spherical algorithm will appear in a subsequent paper. The remainder of the algorithm and input physics follows the description in [4].

To model convection, we begin with a Chandrasekhar mass white dwarf initial model, with a central temperature of 6×10^8 K and a central density of 2.6×10^9 g cm^{-3}. The composition is approximately 0.3 ^{12}C and 0.7 ^{16}O by mass, with some heavier nuclei near the center. This model is based on those explored in [22]. The initial model is taken to be the initial base state and mapped onto the Cartesian grid. For the simulation presented here, we used a domain 5×10^8 cm on a side, with the full star centered on a 384^3 grid. An initial velocity field consisting of a number of random modes of amplitude 10^5 cm/s was seeded at the center of the star. Thermonuclear reactions ($^{12}\mathrm{C} + {}^{12}\mathrm{C} \rightarrow {}^{24}\mathrm{Mg} + \gamma$) begin in the center and drive convection throughout the star.

Figure 2 shows the temperature profile of the initial model compared to a purely isentropic model. The departure of our initial model from the isentropic model at 10^8 cm shows that only the inner part of the star is convectively unstable. The vorticity plot of the right shows this behavior after 13.5 minutes of evolution, with the inner region of the star showing convective behavior. The large descending plume hints at some overall structure to the convective flow, perhaps the onset of a dipole field like that seen in the calculations of [22, 23]. This behavior is still under investigation.

4. Conclusions

We have shown some preliminary results of applying MAESTRO to the study of XRBs and SNe Ia. We will continue to explore the robustness of the results presented here to algorithmic improvements and resolution. These initial results demonstrate that the new algorithm is able to model convective behavior in astrophysical environments. Future work will include the development of an adaptive version of the algorithm, which we will initially apply to the XRB problem to better resolve the base of the burning layer. We will also add rotation, for both the XRB and SNe Ia problems, and turbulent flame models.

Acknowledgments

We thank Stan Woosley for providing the initial model for the convecting WD and many helpful conversations about ignition physics, and Andrew Cumming for providing the initial model for the XRB. We also thank Frank Timmes for providing his equation of state and conductivity routines online. This work was supported by the SciDAC Program of the DOE Office of Mathematics, Information, and Computational Sciences under the U.S. Department of Energy under contract No. DE-AC02-05CH11231 and by a DOE/Office of Nuclear Physics Outstanding Junior Investigator award, grant No. DE-FG02-06ER41448, to Stony Brook.

Computer time for this project was provided through a DOE INCITE at the National Center for Computational Sciences at Oak Ridge National Laboratory, which is supported by the Office of Science of the U.S. Department of Energy under Contract No. DE-AC05-00OR22725, on Livermore Computing's Atlas machine through LLNL's Multiprogrammatic & Institutional Computing Program, and at the National Energy Research Scientific Computing Center, which is supported by the Office of Science of the U.S. Department of Energy under Contract No. DE-AC02-05CH11231.

References

[1] Timmes F X and Swesty F D 2000 *Astrophysical Journal Supplement* **126** 501–516
[2] Almgren A S, Bell J B, Rendleman C A and Zingale M 2006 *Astrophysical Journal* **637** 922–936
[3] Almgren A S, Bell J B, Rendleman C A and Zingale M 2006 *Astrophysical Journal* **649** 927–938
[4] Almgren A S, Bell J B, Nonaka A and Zingale M 2008 *Astrophysical Journal* Accepted for publication
[5] Almgren A S, Bell J B and Zingale M 2007 *Journal of Physics Conference Series* **78** 012085–+
[6] Strohmayer T and Bildsten L 2003 *ArXiv Astrophysics e-prints* (*Preprint* `astro-ph/0301544`)
[7] Bhattacharyya S and Strohmayer T E 2007 *Astrophysical Journal Letters* **666** L85–L88
[8] Taam R E 1980 *Astrophysical Journal* **241** 358–366
[9] Woosley S E, Heger A, Cumming A, Hoffman R D, Pruet J, Rauscher T, Fisker J L, Schatz H, Brown B A and Wiescher M 2004 *Astrophysical Journal Supplement* **151** 75–102
[10] Shara M M 1982 *Astrophysical Journal* **261** 649–660
[11] Spitkovsky A, Levin Y and Ushomirsky G 2002 *Astrophysical Journal* **566** 1018–1038
[12] Lin D J, Bayliss A and Taam R E 2006 *Astrophysical Journal* **653** 545–557
[13] Bell J B, Day M S, Rendleman C A, Woosley S E and Zingale M A 2004 *Journal of Computational Physics* **195** 677–694
[14] Caughlan G R and Fowler W A 1988 *Atomic Data and Nuclear Data Tables* **40** 283–334 see also `http://www.phy.ornl.gov/astrophysics/data/cf88/index.html`
[15] Hillebrandt W and Niemeyer J C 2000 *Annu. Rev. Astron. Astrophys* **38** 191–230
[16] Woosley S E, Wunsch S and Kuhlen M 2004 *Astrophysical Journal* **607** 921–930
[17] García-Senz D and Bravo E 2005 *Astronomy and Astrophysics* **430** 585–602
[18] Gamezo V N, Khokhlov A M and Oran E S 2005 *Astrophysical Journal* **623** 337–346
[19] Plewa T, Calder A C and Lamb D Q 2004 *Astrophysical Journal* **612** L37–L40
[20] Röpke F K, Hillebrandt W, Schmidt W, Niemeyer J C, Blinnikov S I and Mazzali P A 2007 *Astrophysical Journal* **668** 1132–1139
[21] Höflich P and Stein J 2002 *Astrophysical Journal* **568** 779–790
[22] Kuhlen M, Woosley S E and Glatzmaier G A 2006 *Astrophysical Journal* **640** 407–416
[23] Woosley S E, Almgren A, Bell J B, Glatzmaier G, Kasen D, Kerstein A R, Ma H, Nugent P, Röpke F, Sankaran V and Zingale M 2007 *Journal of Physics Conference Series* **78** 012081–+

A molecular approach to understanding complex systems: computational statistical mechanics using state-of-the-art algorithms on terascale computational platforms

Christopher J Mundy[1], Roger Rousseau[1], Alessandro Curioni[2], Shawn M Kathmann[1] and Gregory K. Schenter[1]

[1]Chemical and Materials Sciences Division, Pacific Northwest National Laboratory, Richland, WA 99352
[2] IBM Research Corporation, Zurich Research Lab
CH-8803 Ruschlikon, Switzerland

E-mail: chris.mundy@pnl.gov

Abstract. We present our recent progress toward a molecular picture of reactions in complex fluids and at interfaces. In particular, we highlight the important statistical mechanical methodology as well as the electronic structure methods and algorithms that are used to tackle important unresolved questions in the chemical sciences. Moreover, we demonstrate that using the simulation protocol outlined, in conjunction with current and next generation computer hardware, a quantitative understanding of complex chemical problems in the condensed phase is within reach.

1. Introduction

The abundance of water in nature, its role as a universal solvent, and its importance in many biological processes responsible for sustaining life are the driving forces behind the need for a fundamental understanding of its physical and chemical properties under different conditions and in various environments. Significant advances in our understanding of aqueous phase processes have been made by using molecular simulation approach. This progress was made possible by effectively coupling accurate molecular interactions with statistical mechanical sampling. Iin order to achieve this, the availability of computational resources and development of computational algorithms is essential. Central to the accurate description of the relevant interactions in complex aqueous systems is the underlying potential. Thus, a key to success lies in the choice of interaction potentials and statistical mechanical sampling methods. *Ab initio* interaction potentials provide a direct link to reactivity and spectroscopy but come at a high computational cost. A good compromise between accuracy and computational efficiency is based on the Kohn-Sham (KS) formulation of density functional theory (DFT). The past few years have produced rapid progress toward fast and accurate electronic structure methods. Since the introduction of an electrostatic interaction potential by Bernal and Fowler in 1933 [1], the first computer simulations on water in its liquid state by Barker and Watts [2] in 1969 and Rahman and Stillinger [3] in 1971, the initial parameterization of a pair potential for water from *ab initio* calculations by Clementi and co-workers [4], and the original simulation of liquid water using DFT interaction potentials by Parrinello and Car [5], water has remained an active field of research. To date, however, the overwhelming majority of DFT studies on liquid water have focused on short time dynamics at constant volume to calculate structural properties.

With the recent advent of terascale computing at our national laboratories, we demonstrated that supercomputing hardware is indeed necessary to be able to perform calculations at an unprecedented size and scale using DFT interaction potentials. These calculations have shed new light on the electronic structure of complex chemical systems under a variety of thermodynamic conditions as well as validation of simulation protocol as well as understanding the shortcomings and successes of DFT. Specifically, we have made use of terascale resources in conjunction with open ensembles to examine liquid-vapor coexistence curves for water using DFT interaction potentials [6, 7], quantify complex chemistry under shock loading [8], and understand the important heterogeneous systems such as the aqueous liquid-vapor interface [9]. In the aforementioned studies, it was shown for the first time that DFT could produce a stable liquid-phase, and established simulation protocols [10]. In fact it is these types of calculations that put the continued use of DFT on stable statistical mechanical grounds for the continued understanding of both bulk and heterogeneous processes in aqueous systems. Thus, given the next-generation computing hardware available through the Department of Energy (DOE), we are now in a unique position to use DFT to characterize free energy landscapes of important reactive processes in aqueous and heterogeneous phase interfaces (vapor-liquid/vapor-solid) systems. As stated by Wallqvist and Berne [11] in 1993, "A single model that satisfactorily captures all of the essential experimental features of water has yet to be found." This suggestion infers that *simple* empirical models used to describe the relevant inter- and intramolecular interactions will not suffice to reproduce and predict the chemical and physical properties over wide temperature and pressure ranges.

We have a unique opportunity to go beyond traditional simulation methodology to advance our understanding of reactions in solution and at interfaces. We seek to understand the role of molecular environments on chemical reactions and processes that control the formation of solid phases from materials dissolved in liquids. A major focus is on understanding these processes in aqueous systems. In contrast to gas phase reactions, the constant proximity of solvent molecules can profoundly alter reaction kinetics and thermodynamics. Solvents can influence the course of a reaction in many ways—providing an efficient route for energy transfer, initiating reactions that would not occur in the gas-phase, stabilizing reaction intermediates, donating or accepting electrons, limiting the mass transfer of reacting species, stabilizing transitions states and thus modifying reaction barriers, and moderating collision dynamics. Furthermore, interfaces provide a unique environment where novel chemistry can occur. Our understanding of interfacial processes is still in its infancy – these processes encompass reactions at or near interfaces, ion pairing and clustering, transport to and across interfaces, interfacial structure, and the concomitant chemistry. These ideas are extremely important in separations, biological systems, and catalysis. There are many challenges in the accurate treatment of condensed-phase reactions. Understanding the underlying free energetics requires reliable representations of molecular interaction potentials beyond what current empirical interaction potentials can provide. It is an ongoing research challenge to understand the relation between the level of detail of a molecular potential and the simulated properties. However, the direct application of high-level *ab initio* methods with appropriate statistical mechanical sampling of reaction dynamics in solution remains a computational Grand Challenge due to the large number of relevant degrees of freedom and the sheer volume of phase space to be explored. Furthermore, chemical reactions in solution and at interfaces can occur over a long time scale relative to the length of a molecular dynamics (MD) trajectory and thus are often identified as rare events such that one may never "see" a reaction occur. Methods to bias the sampling of the ensemble can be employed to bring the reactive event into focus during reasonable simulation times. The accuracy of the free energies depends on accurate energetics and the degree of statistical mechanical sampling along the relevant reaction coordinates. Reaction free energies are also influenced by quantum effects like zero-point-energy and tunneling, which can be especially important for processes involving light atoms such as hydrogen. It is the subtle variations in interaction potentials, charge transfer, quantum treatment of nuclear degrees of freedom, and so forth that can alter reaction kinetics by many orders of magnitude.

The concept of sampling complex reaction paths will play a vital role in understanding thermodynamics and kinetics of fundamental processes in the condensed phase. We use potentials of mean force (PMF) to provide reduced representations of the reactive processes, which can simplify the computational costs by focusing on rate-limiting steps. In condensed phase, appropriate reaction paths correspond to minimum free energy paths (MFEP). Sampling high-energy states along the MFEP, such as those near a barrier maximum, can be very difficult because these energy states are, by definition, improbable. One way of sampling these rare event states is to use biased ensembles generated via *metadynamics* [12]. Metadynamics allows one to simultaneously and efficiently sample many complex reaction coordinates to arrive at free energy landscapes and hence deduce barriers and transition states. The aim of metadynamics is to sample the free energy surface $F(\mathbf{s}) = -k_B T \ln Q(\mathbf{s})$ of a set of collective variables $\mathbf{s}$, where Q is the partition function and $k_B T$ is Boltzmann's constant multiplied by temperature. The salient feature of *metadynamics* is that at the beginning of the simulation the self-adapting biasing potential $V(\mathbf{s}, t = 0) = 0$; then, at subsequent times, Gaussian hills are added to yield a potential, $V(\mathbf{s},t)$ that can be shown to converge to the exact free-energy surface under certain limits. Moreover, the extension of metadynamics to incorporate multiple walkers makes this algorithm ideal for further parallelization of the statistical mechanics. This will allow for the development and testing of several reaction coordinates that may give rise to the lowest free energy profile (e.g., coordination number, topological functions). The resulting free energy path will then be further validated by transition path sampling or other saddle-point searches to confirm that reactive moieties represent true transition states.

2. Computational approach

The CPMD (www.cpmd.org) package is written in standard Fortran 77 and solves the Kohn-Sham density functional theory electronic structure equations employing matrix algebra operations (level three BLAS routines) and fast Fourier transform (FFT) either via a Goedecker algorithm or the FFTW package. CPMD draws extensively upon standard LAPACK and BLAS libraries and the FFT library and may be parallelized with MPI or OpenMP or both (depending on which approach best suits a given computational architecture). Moreover, further efficiency is gained for liquids by utilizing the Car-Parrinello molecular dynamics method for integrating the electron equations of motion on the fly. For simulations involving a nudged elastic band (NEB) to obtain free energy surfaces, parallelization between system replicas is, in principal, trivial (e.g., relying on simple communication between neighboring replicas). Hence, parallelization between replicas is essentially linear up to very large numbers. Interprocessor communication is the heaviest within a system replica and dominated by the MPI collective communications of the FFT routines. MPI collective operations may be diminished by subdividing the parallelization of the FFT among groups of communicators (task-groups). Even for systems as small as a few thousand electrons, linear scaling for calculating a wave function may be obtained for 100-1000 CPUs on either an IBM Blue Gene or Cray XT machine. References to the implementation of the electronic structure for CPMD and its performance for high-performance computing applications are given in [13, 14]. Table 1 depicts typical scaling results for CPMD on 32 water molecules on BG/P and BG/L.

Table 1. Benchmarks using CPMD on a 32 water molecule system (96 atoms). The excellent scaling of small number of atoms to large node counts and fast wall clock demonstrates the utility of performing large condensed phase with CPMD on BG architecture. Note the superior scaling of BG/P compared to BG/L.

# of Processors	**1**	**2**	**4**	**8**	**16**	**32**	**64**	**128**	**256**	**512**	**1024**	**2048**
BG/L (co-processor mode) time/steps(s)				10.67	5.23	2.62	1.63	0.94	0.48	0.3	0.17	0.12
BG/P (VNM-4 proc per node) time/step (s)	40.1	19.8	9.02	4.62	2.44	1.54	0.76	0.48	0.24	0.16		

For the CP2K (www.cp2k.berlios.de) code written in Fortran 90 the majority of the calculations of interest also involve the solving Kohn-Sham density functional equations using a localized basis set for the orbitals in addition to an auxiliary plane wave basis set for the density. Hence, CP2K largely employs matrix algebra operations and FFTs as in CPMD. The difference is that the construction of the density matrix in real-space relies on the use of matrix manipulations (both sparse and full) that extensively employ both BLACS and ScaLAPACK. Thus, CP2K requires only a single FFT to construct the Hartree potential using either FFTW or the internal Goedecker libraries. Given that construction of the real-space density scales linear CP2K is able to perform efficient calculations on very large systems scaling up to 512 CPUs on the Cray XT4. Using the method of orbital transformations utilizing only the occupied states [15] provides an extremely efficient algorithm for wavefunction optimization that scales as the number of occupied states squared multiplied by the number of basis functions. This effectively lowers the prefactor of the cubically scaling diagonalization (e.g., going as cube of the number of basis functions). *Metadynamics* or NEB calculations could be implemented efficiently parallelized between system replicas. Hence, the current simulations are in principle capable of being run efficiently on several thousands of CPUs with almost ideal scaling. References to the implementation of the electronic structure algorithms present in CP2K for high-performance computing applications are given in [15, 16].

3. Ongoing research

In recent years, it has become clear that the air-water interface plays a more significant role in gas adsorption and reaction kinetics of many atmospheric processes than had been anticipated. A new conventional wisdom is emerging that implicates surface complexes to explain the initial step in a number of important atmospheric reaction mechanisms [17-20]. Moreover, there is mounting experimental evidence that ions (e.g., Cl^- and Br^-) can reside at the aqueous liquid-vapor interface corroborating results obtained with polarizable empirical interaction potentials [17, 19]. Although there is indirect experimental evidence that places ions in the vicinity of the aqueous liquid-vapor interface, knowledge of the exact concentration excess is still an open question [17, 19]. In order to probe this with molecular simulation, the PMF of a polarizable anion along the surface coordinate can be calculated. To date, however, no analogous calculation has been performed with DFT interaction potentials. Moreover, there is a debate in the recent literature on the proximity of the hydroxide ion at the aqueous liquid-vapor interface [21-25]. Determining the surface activity of a hydroxide ion is an important step toward understanding the complex chemistry that can take place in the vicinity of an interface [25].

We believe that fast electronic structure algorithms, in conjunction with state-of-the-art supercomputer hardware, can provide the path to performing benchmark calculations for these important chemical systems. To this end, the first steps are to provide chemical insight through an appropriate reaction coordinate. Fortunately, pioneering studies using DFT interaction potentials with standard gradient corrected functionals have been used to quantify the bulk phase chemistry of the hydroxide ion[26]. In this study a reasonable reaction coordinate was determined to be the difference in distance ,called d (see figure 1), between the hydrogen involved in the shortest hydrogen bond and the oxygen on the hydroxide and the oxygen on the transferring water molecule [26]. The goal of this ambitious calculation will be to map out the free energy of the hydroxide ion as a function of its distance from the liquid-vapor interface and d. These calculations will be at an unprecedented scale. First, in order to perform the appropriate statistical mechanical sampling, replicas of the liquid-vapor interfacial system must be constructed with the hydroxide ion at some z, (direction normal to the liquid-vapor interface) and d (see figure 2). Each one of these replicas will be simulated for 10-20 ps in order to achieve the appropriate statistics for computing free energies through non-Boltzmann sampling techniques. Together, these calculations will be on the order of 100 times larger than the original pioneering study of the aqueous liquid-vapor interface performed at Lawrence Livermore National Laboratory in 2004 [1]. These simulations will provide true benchmarks for both methodology and accuracy in investigating complex chemical phenomena at interfaces using DFT interaction potentials. We will also use these results to study and validate the free energetics for proton transfer in the bulk and in the vicinity of the interface to make contact with previous experiments and simulations [21. 24. 25].

Conclusions

We have presented our progress on the initial stages of our ongoing research to develop a molecular scale picture of molecular interactions and reactivity at the aqueous liquid-vapor interface using DFT interaction potentials. Our studies make use of scalable electronic structure algorithms in conjunction with statistical mechanical sampling methods that will enable calculations at an unprecedented scale and accuracy. In particular, we have outlined recent progress in our attempt to elucidate the full free energy surface of a reacting species, namely, the hydroxide ion, as a function of its proximity to the interface.

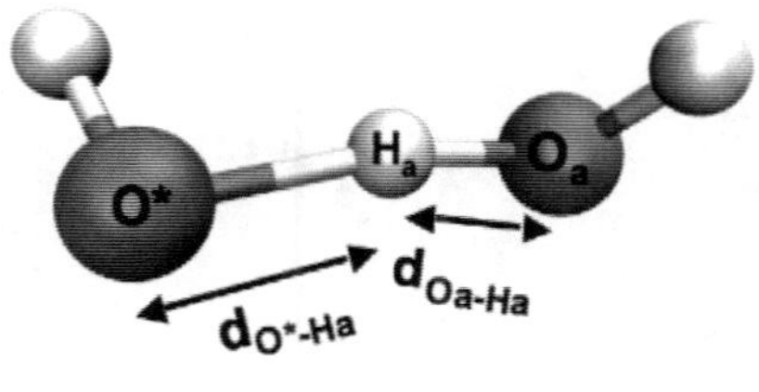

Figure 1. The reaction coordinate $\delta = d_{O^*\text{-}Ha} - d_{Oa\text{-}Ha}$. Here O^* denotes the oxygen on the hydroxide, and O_a and H_a are the oxygen and hydrogen of the transferring water molecule, respectively.

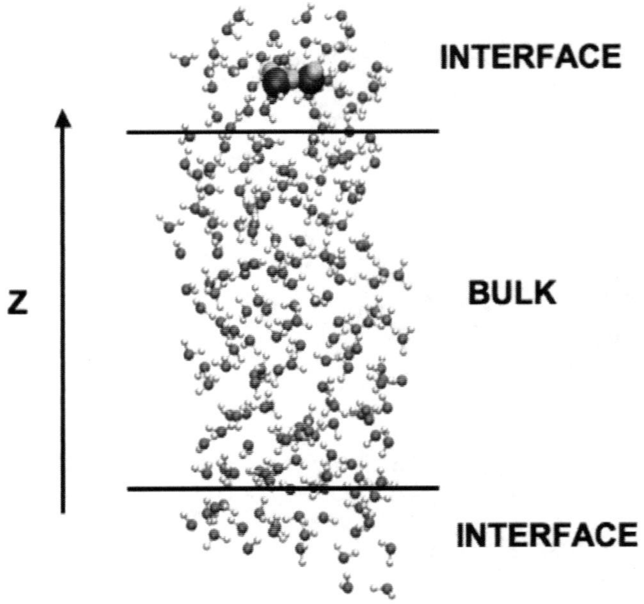

Figure 2. Geometry of the aqueous liquid vapor interface. This is a standard geometry for simulating interfacial systems. *Z* denotes the interfacial coordinate. The two other directions are periodically replicated creating an infinite plane. Thus, one creates two free interfaces (denoted INTERFACE) and a region of bulk (denoted BULK). This geometry contains 215 water molecules (denoted by red oxygen and white hydrogen atoms) and a single hydroxide ion. The complex denoted in blue and green is the constrained value of *Z* and δ (see figure 1). We perform simulations of 8 constrained values of δ and 20 constrained values of *Z* yielding a calculation that is one order or magnitude greater than the original pioneering work on the aqueous liquid-vapor interface performed at LLNL in 2004. Technical details (e.g., exchange and correlation functional, cut-off) are given in [1].

Acknowledgments

This work is supported by the U.S. Department of Energy's (DOE) Office of Basic Energy Sciences Chemical Sciences program. Calculations are currently being performed at National Energy Research Scientific Computing Center (NERSC) on the Cray XT4 (Franklin) under the 2008 re-imbursement program, and through resources afforded by the 2008 INCITE award, namely the Cray XT4 (Jaguar) at Oak Ridge National Laboratory, the IBM Blue Gene/P (Intrepid) at Argonne National Laboratory. We also acknowledge the state-of-the-art facilities at PNNL, namely, the 1500 CPU Linux cluster comprised of 2.33 GHz Clovertown chips (NWice) . CJM acknowledges ongoing collaborations with Prof. Douglas Tobias (UC-Irvine), Dr. I. –F. Will Kuo (LLNL), Dr. Liem Dang (PNNL), and Prof. Dr. J Hutter (U. Zurich).

References

[1] Kuo I F W and Mundy C J 2004 *Science* **303**:658
[2] Bernal J D and Fowler R H 1933 *J. Chemical Physics* **1**:515
[3] Barker J A and Watts R O 1969 *Chemical Physics Letters* **3**:144
[4] Rahman A, Stillinger F H 1971 *J. Chemical Physics* **55**:3336
[5] Popkie H and Kistenma H and Clementi E 1973 *J. Chemical Physics* **59**:1325
[6] Car R and Parrinello M 1985 *Phys. Rev. Letters* **55**:2471
[7] McGrath M J, Siepmann J I, Kuo I F W and Mundy C J 2006 *Molecular Physics* **104**:3619
[8] McGrath M J, Siepmann J I, Kuo I F W, Mundy C J VandeVondele J, Hutter J, Mohamed F and Krack M 2006 *J. Physical Chemistry A* **110**:640
[9] Mundy C J, Curioni A, Goldman N, Kuo I F W, Reed E J, Fried L E and Ianuzzi M 2008 *J. Chemical Physics* **128**
[10] Mundy C J and Kuo I F W 2006 *Chem. Reviews* **106**:1282
[11] Wallqvist A and Berne B J 1993 *J. Physical Chemistry* **97**:13841
[12] Laio A and Parrinello M 2002 *Proc. National Academy of Sciences USA* **99**:12562
[14] Hutter J and Curioni A 2005 *Parallel Computing* **31**
[15] VandeVondele J and Hutter J 2003 *J. Chemical Physics* **118**:4365
[16] VandeVondele J, Krack M, Mohamed F, Parrinello M, Chassaing T and Hutter J 2005 *Computer Physics Comm.* **167**:103
[17] Ghosal S, Hemminger J C, Bluhm H, Mun B S, Hebenstreit E L D, Ketteler G, Ogletree D F, Requejo F G and Salmeron M 2005 *Science* **307**:563
[18] Jungwirth P and Tobias D J 2001 *J. Phys. Chem. B* **105**:10468
[19] Petersen P B, Saykally R J, Mucha M and Jungwirth P 2005 *J. Phys. Chem. B* **109**: 10915.
[20] Tarbuck, T. L.; Richmond, G. L. *J. American Chemical Society* **127**:16806
[21] Petersen P B and Saykally R J 2008 *Chemical Physics Letters* **458**:255.
[22] Beattie J K 2008 *Physical Chemistry Chemical Physics* **10**:330
[23] Vacha R, Buch V, Milet A, Devlin J P and Jungwirth P 2008 *Physical Chemistry Chemical Physics* **10**:332
[24] Vacha R, Buch V, Milet A, Devlin P and Jungwirth P 2007 *Physical Chemistry Chemical Physics* **9**:4736
[25] Buch V, Milet A, Vacha R, Jungwirth P and Devlin J P 2007 *Proc. National Academy of Sciences USA* **104**:7342
[26] Tuckerman M E, Marx D and Parrinello M 2002 *Nature* **417**:925

Building fast, reliable, and adaptive software for computational science

Alistair P Rendell, Joseph Antony, Warren Armstrong, Pete Janes and Rui Yang
Department of Computer Science, Australian National University, Canberra, ACT0200 Australia
E-mail: `Alistair.Rendell@anu.edu.au`

Abstract. Building fast, reliable, and adaptive software is a constant challenge for computational science, especially given recent developments in computer architecture. This paper outlines some of our efforts to address these three issues in the context of computational chemistry. First, a simple linear performance that can be used to model and predict the performance of Hartree-Fock calculations is discussed. Second, the use of interval arithmetic to assess the numerical reliability of the sort of integrals used in electronic structure methods is presented. Third, use of dynamic code modification as part of a framework to support adaptive software is outlined.

1. Introduction

It's an exciting time for computational science. After years of incremental changes to computer architectures we are at the dawn of a new era. Multicore is now on the desktop with processor counts steadily increasing. Graphics cards and special-purpose processors such as the NVIDIA GeForce 8800 GTX and Cell Broadband Engine provide flexible programming environments and astonishing performance for just a few hundred dollars. And if this were not enough, the release of the Sun ROCK processor sees transactional memory move from an academic curiosity into mainstream.

How to harness these new developments and construct code that is fast, reliable, and adaptive to its environment is a real challenge for the computational scientist. In this paper we outline some of our work to address these issues. To build fast software, we are interested in designing performance models that can be used to predict performance and thereby guide program execution. Detailed here is our attempt to use a simple linear performance model to describe and predict the performance of the Gaussian quantum chemistry code [1] as a function of cache usage and memory page placement on a variety of systems. For reliable software we are seeking to include the effects of rounding and truncation errors into scientific computations. To achieve this, we are developing interval versions of standard quantum chemistry methods. To facilitate the development of adaptive software, we are using DynInst [2] to dynamically modify running executables.

2. Building fast software

Central to understanding the performance of any computational application is an appreciation of the costs associated with accessing the various data structures used. Increasingly processors have multiple levels of cache that shadow a memory system that has non-uniform memory access (NUMA) characteristics. Two key issues are therefore understanding application performance as a function of cache usage, and as a function of memory page placement. With these in mind our goal has been to develop simple performance models that capture these effects and can be used to predict performance as a function of algorithmic or system hardware modifications.

Below we outline briefly some of our work that has explored the use of a simple linear performance model to characterize the cache performance of the Gaussian quantum chemistry code [1] on a variety of modern hardware systems. We then show how this model can be combined with cache simulation to study the effect of cache variation on application performance, before discussing how our model can be extended to include the effects of NUMA architectures.

2.1. A simple linear performance model

Recently we proposed that under certain circumstances a simple inear oerformance Model (LPM) may be sufficiently accurate to provide predictive information for use in calculating cache blocking factors for quantum chemistry calculations [3]. In this model the overall performance is predicted as a simple linear combination of instructions issued and cache misses (i.e. requests by the processor to load or store an item from/to memory where that item is not found to be resident in cache):

$$Cycles = \alpha * (I_{Count}) + \beta * (L1_{Misses}) + \gamma * (L2_{Misses}), \tag{1}$$

where I_{Count} is the instruction count, $L1_{Misses}$ the total number of Level 1 cache misses, $L2_{Misses}$ the total number of Level 2 cache misses, and α, β, and γ are penalty factors. Loosely speaking, the value of α reflects the ability of the code to exploit the underlying super-scalar architecture, β the average cost of an L1 cache miss, and γ the average cost of an L2 cache miss.

We have investigated the above in the context of Hartree-Fock (HF) calculations as implemented within the Gaussian quantum chemistry code [1]. Under most circumstances the time required to perform an HF calculation is dominated by formation of the so-called Fock matrix (F). This has the following form:

$$F_{\mu\nu} = H_{\mu\nu} + \sum_{\lambda\sigma}^{N}\sum_{i}^{N_e} C_{\lambda i}C_{\sigma i}^{*}[2(\mu\nu\,|\,\lambda\sigma) - (\mu\lambda\,|\,\nu\sigma)], \tag{2}$$

where $\mu\nu\cdots$ and i are indices for the basis functions and molecular orbitals respectively, N is the total number of basis functions, and N_e is the total number of electrons in the system. In equation 2 each $F_{\mu\nu}$ involves another two-index quantity $H_{\mu\nu}$, and a four-index quantity $(\mu\nu\,|\,\lambda\sigma)$ contracted with a two-index quantity $C_{\lambda i}$. The $(\mu\nu\,|\,\lambda\sigma)$ are the electron repulsion integrals (ERIs), and for most HF calculations evaluation of these integrals and their contraction with C is what dominates the overall computation time.

In the PRISM algorithm [4] used by Gaussian the ERIs are evaluated in batches, where ERIs in the same batch share some common intermediates and involve the same set of operations, the latter make each batch amenable to processing on single instruction multiple data (SIMD) or vector architecture. There is a balance between larger batch sizes causing improved performance because of smaller instruction counts and better exploitation of SIMD architectural features, but too large batches causing the cache to overflow leading to a loss in performance. The Gaussian code uses cache blocking to limit the maximum size of an integral batch and avoid cache overflow.

To illustrate this effect, we show performance data table 1 for an HF calculation as the cache blocking size is varied from 4–1024 KiW. (The blocking factor relates to the size of an internally allocated array within the Gaussian program and is measured in 8-byte words). The test system is taken from a molecular dynamics simulation of a solvated potassium ion equilibrated at 300K and then truncated to contain only the central potassium ion and any water molecules that have atoms within 4Å of it. (In subsequent results the same system will be used but truncated at 8Å and containing 80 water molecules.) This shows cycle times of 5.38×10^{10} with a block size of 4 KiW that reduce to 2.75×10^{10} when the block size is expanded to 64 KiW, but then increases to 4.64×10^{10} as the block size is further increased to 1024 KiW. Between best and worse case scenario this represents a performance difference of nearly a factor of 2. Clearly the goal is to determine the optimal blocking size, but as this may change according to the particular calculation being performed and hardware being used, this should really be done using an adaptive algorithm.

Table 1. Performance of the PRISM integral evaluation code as a function of cache blocking size for a Hartree-Fock calculation on a solvated potassium ion ($K^+(H_2O)_{11}$) using 6-31G* basis on a 2.2 GHz AMD848 Opteron. See text for further details.

Cache Blocking Size (KiW)	**Number of Batches**	**Instruction Count** (x10^{10})	**L1 Cache Misses** (x10^8)	**L2 Cache Misses** (x10^7)	**Floating Point Operations** (x10^9)	**CPU Cycles** (x10^{10})
4	226922	5.38	3.79	1.68	6.24	5.38
16	75729	4.11	3.83	2.35	6.15	2.99
64	20049	3.18	5.43	1.39	6.18	2.75
256	5624	2.91	7.50	4.89	6.14	3.58
1024	2174	2.84	8.82	6.51	6.08	4.64

Using the data in Table 1, we can perform a least squares fit to equation 1 and obtain values for α, β and γ. Sample results for four different computations (molecular system, basis set) obtained on an Opteron system are given in Table 2. The values for α are all positive and suggest that on average between 1.4 and 1.6 instructions are being executed each cycle. For β two values are, however, negative with the unphysical interpretation that it takes negative time to retrieve data from the L1 cache! For γ all values are positive and roughly what we might expect a level 2 cache miss penalty to cost (e.g. if measured using LMBench [5]). On other platforms we found similar problems extracting meaningful values for β and believe this is partly due to the effects of out-of-order execution (although there are also issues identifying exactly which hardware counter(s) to use in order to compute total L1 cache miss). For this reason we chose to remove L1 misses as a parameter in the LPM.

Table 2. LPM Parameters for 5 different HF Calculations on the Opteron.

System	$K^+(H_2O)_{11}$	$K^+(H_2O)_{80}$	$K^+(H_2O)_{11}$	Valinomycin	α-Al_2O_3
Basis	6-31G*	6-31++G(3df,3pd)	6-31G*	3-21G	3-21G*
α	0.7	0.6	0.7	0.7	0.6
β	5.1	2.0	-0.4	-4.5	4.5
γ	320.3	403.1	411.0	367.4	452.6

To access the accuracy of the LPM for a range of different processors, we present in table 3, the average relative errors obtained when using the LPM to fit cycle times for five different HF calculation types on seven different compute platforms. The results show that on average the LPM is able to fit the observed performance to an accuracy of around 5%, and in some cases much better than this.

2.2. *Performance prediction using the LPM*

As demonstrated above the LPM is capable of describing the observed cycle count for Gaussian HF calculations with an accuracy of $\sim 5\%$ across a range of different calculation types and architectures. To obtain this did, however, require that the computation first be run on that hardware recording instructions issued and cache misses. An alternative to using hardware performance counters to obtain this data is to use functional cache simulation; this simulates cache behavior given the load/store memory operations generated by the program. Callgrind [6] is a tool that permits such simulations. It performs dynamic, execution driven cache simulation by using the Valgrind framework for dynamic binary instrumentation

Table 3. Percentage average relative error in fitted cycle time using LPM for a range of different calculations and systems.

	Molecular System and Basis				
	$K^+(H_2O)_{11}$ 6-31G*	$K^+(H_2O)_{80}$ 6-31++G(3df,3pd)	$K^+(H_2O)_{11}$ 6-31G*	Valinomycin 3-21G	α-Al_2O_3 3-21G*
Opteron	3.2	1.0	2.7	1.3	2.0
EM64T	2.7	1.4	2.0	2.7	0.7
Pentium 4	2.3	1.8	2.7	0.5	2.2
Pentium M	2.0	0.8	1.9	1.0	2.6
G5	1.8	1.5	2.7	2.2	2.7
G5-XServe	2.4	1.7	2.2	2.4	3.2

[7]. Thus, by combining the values of α and γ derived by fitting to data obtained on existing hardware, with instruction and cache miss counts given by Callgrind for a "similar" system, we can perform architectural studies. This would include addressing questions such as "What performance effect would we predict for Gaussian HF calculations if we doubled the cache size, or modifying some other characteristic?"

To illustrate this, we show in table 4 the various event counts obtained via hardware performance counters on an Opteron with those given by Callgrind for the original Opteron cache configuration, the same system but with the L1 and L2 cache line size increased and the same system with a larger L2 cache (while also increasing the cache blocking parameter used by Gaussian). (The cache line size determines the number of bytes transferred between main memory and cache when a request is made by the processor to access a data item that is currently not resident in the cache.) By combining these counts with the values for α and γ obtained when running on real hardware, we can predict the overall cycle times on the modified hardware. Not surprising, the results show that decreasing the L1 line size to 32B has no effect, but increasing the L2 line size causes a slight decrease from 2.50×10^{10} to 2.36×10^{10} cycles. Increasing the L2 cache size to 16MB while also adjusting the cache blocking factor is much more significant, resulting in nearly a 20% reduction in the predicted cycle count.

Table 4. Hardware and Callgrind counts for various L1 and L2 cache configurations on AMD Opteron based system system together with LPM predicted cycle counts.

	Hardware	Callgrind			
L1 Size	64KiB	64KiB	64KiB	64KiB	64KiB
L1 Line size	64B	64B	32B	64B	64B
L2 Size	1MB	1MB	1MB	1MB	16MB
L2 Line size	64B	64B	64B	1024B	64B
Blocking Size	64KiW	64KiW	64KiW	64KiW	1MW
ICount	3.28E+10	3.27E+10	3.27E+10	3.27E+10	2.93E+10
L2$Miss	9.72E+06	1.44E+07	1.44E+07	8.90E+06	5.24E+06
Est. Cycles	2.72E+10	2.50E+10	2.50E+10	2.36E+10	2.01E+10

2.3. Extending the LPM for NUMA

The LPM developed above predicts performance based solely on the instruction count and total L2 cache misses. On many multicore shared memory platforms all L2 cache misses are not equal; rather the memory has non-uniform access characteristics. A simple extension to the LPM would be to divide the L2 cache miss penalty (γ) into multiple terms that are each associated with a cache miss to a particular domain within the NUMA system. While in principle this would be easy, in practice hardware counters are currently not able to provide such a breakdown.

We are addressing the above limitation in two ways. First, we have performed a number of studies looking at the effect of memory placement on the performance of the Gaussian application. This work involves strategically placing certain data structures at specific locations within the NUMA architecture and building up a performance model based on the resulting execution times and data from various hardware counters [8, 9]. In [8] we show that even with good cache blocking it is possible to see a difference in runtime of over 40% between for a Gaussian HF calculation that uses good or poor memory placement on 16 cores of a Sunfire X4600 M2 system.

Our second approach is to extend Callgrind to include a NUMA memory model on top of the existing cache model. This work is still in progress, but will allow us to map easily cache misses to particular memory domains and determine which memory pages are best located near a particular process.

3. Building reliable software

To date, the majority of computational science is performed by using 64-bit floating-point arithmetic. This provides $\sim$15 significant figures. On a petaflop machine performing in excess of 10^{15} operations per second it is not at all clear whether this is sufficient accuracy. To address this issue, we have been investigating the use of interval arithmetic to bound errors in computational science applications.

Interval arithmetic is not new; it but was proposed in the early days of computing as a means of bounding rounding errors that occur in floating-point calculations [10]. Formally, an interval $X = [\underline{X}, \overline{X}]$ contains the set of all real numbers between two endpoints corresponding to the lower $\underline{X}$ and upper bound $\overline{X}$, referred to as the infimum and supremum respectively. Thus, an interval may represent, among other things, a bound on the true result of an uncertain quantity, where the source of uncertainty may include a combination of rounding and truncation errors.

The set of basic arithmetic operations $\bullet \in \{+, -, \times, \div\}$ is defined such that uncertainties in X and Y are propagated onto the result:

$$X \bullet Y = \{x \bullet y | x \in X, y \in Y\} \tag{3}$$

where for example $X + Y = [(\underline{X} + \underline{Y})_{\nabla}, (\overline{X} + \overline{Y})_{\Delta}]$, which if implemented using IEEE754 [11] floating-point arithmetic will use downward ∇ and upward rounding Δ for the infimum and supremum respectively.

There exists a duality between expressions in floating-point and interval arithmetic. An interval equivalent of a floating-point expression, referred to as a *natural interval extension*, can be obtained by simply replacing all floating-point arithmetic operations and functions such as exp, sin, cos with their interval equivalents. The interval obtained by this translation contains both the true numerical result and the result that is obtained by the equivalent ordinary floating-point calculation [12]. Thus, if a floating-point calculation is unstable due to large rounding or truncation errors, then its interval extension will return wide interval bounds.

According to the *containment principle* an interval implementation must guarantee that the result it returns will bound the true result of the particular calculation in question, taking into account uncertainties introduced by rounding and truncation errors [13]. Thus if, for example, an expression for $sin(x)$ is computed by evaluating its Taylor series expansion, then the interval returned must include the effects of both the rounding errors associated with the operations used to compute the series, and the truncation error resulting from only evaluating a finite number of terms in an infinite series. One of the most challenging problems faced when writing interval code is to find the interval with the narrowest width, as this places the most precise bounds on the true result.

We are interested in using intervals to place rigorous bounds on the sort of integrals used in quantum chemical calculations, and propagating these through an entire calculation to quantify the errors in the final property being computed. Specifically we have considered the same ERIs as given in Equation 1. The majority of algorithms for computing these rely on the evaluation of the incomplete gamma function [14], denoted here as $F_m(T)$ and defined as follows:

$$F_m(T) \quad = \quad \int_0^1 t^{2m} e^{-Tt^2} dt, \tag{4}$$

where T is a non-negative real value (dependent on the distance between the basis functions) and m is an integer (the value of which depends on the total angular momentum of the functions involved). Values of T range from 0 to over 1,000, while m typically ranges from 0 up to 16. In a typical calculation $F_m(T)$ may be evaluated many millions of times.

In modern quantum chemistry codes exactly how $F_m(T)$ is evaluated depends on both the values of T and m. In general there are two alternative approaches. For a given m if T is less than some threshold (T_F) the value of $F_m(T)$ is based on finite evaluation of an infinite series, while above the threshold an asymptotic approximation is used. In order to speed evaluation of the finite series, interpolation tables based on either Taylor [15, 16] or Chebyshev [17] polynomial approximations are constructed. Interpolation points are generated along the center of each uniformly discretized partition $[X-h, X+h]$ of the domain $[0, T_F]$, with values at each interpolation point computed using the truncated series.

In short, when evaluating $F_m(T)$ errors can result from fundamental approximations made before any computation has begun, such as use of the asymptotic approximation or a particular form of interpolation, as well as from rounding errors associated with the various floating-point operations involved. Using interval arithmetic we have been able to place rigorous error bounds on all of these errors [18]. This is illustrated in table 5 where the average relative interval widths for $F_m(T)$ computed using five different polynomial approximation schemes and two different implementations are given. The results clearly show that the use of ChebyA, a three term interpolation in T gives rise to large interval widths, particularly for large m values. The alternative approaches are all roughly equal, giving about 14 significant figures of accuracy in $F_m(T)$. Alg2, which uses compensated summation to minimize the effect of rounding errors in series evaluation, is able to decrease the interval width by a factor of about 4.

Table 5. Average relative interval width of $F_m(T)$ for various interpolation schemes. Alg2 uses compensated summation to minimize rounding errors in series evaluation. See [18] for full details.

Scheme		m			
		0	4	8	12
ChebyA	Alg1	8.3e-14	4.3e-12	2.2e-11	6.7e-11
	Alg2	3.2e-14	1.6e-12	7.9e-12	2.3e-11
ChebyB	Alg1	1.4e-14	1.7e-14	1.9e-14	2.1e-14
	Alg2	5.7e-15	7.2e-15	8.1e-15	8.9e-15
ChebyC	Alg1	1.5e-14	1.8e-14	2.0e-14	2.1e-14
	Alg2	7.1e-15	8.5e-15	9.3e-15	1.0e-14
TaylorA	Alg1	1.4e-14	1.7e-14	1.8e-14	2.0e-14
	Alg2	4.8e-15	6.2e-15	7.1e-15	8.0e-15
TaylorB	Alg1	1.4e-14	1.7e-14	1.9e-14	2.0e-14
	Alg2	5.6e-15	7.0e-15	7.9e-15	8.7e-15

In other work [19] we have used interval analysis to compare electrostatic energies for a 3D distribution of point charges computed using a pairwise summation with those evaluated using fast multipole methods (FMM). Using intervals, we were able to show how adding terms to the FMM treatment improves accuracy as truncation errors decrease, but then errors begin to increase as rounding errors become significant.

4. Building adaptive software

Traditional methods for program optimization have involved repeated cycles of performance measurement, analysis and tuning. Increasingly, however, tuning is runtime dependent reflecting what else is running on the system at that time. For instance, on a multicore system with a shared cache it may be useful to adjust cache usage dynamically depending on what is happening on other cores. Moreover, there can be cases where the algorithm of choice can change during the course of the computation. For example, in particle simulations it is often advantageous to exploit linear scaling algorithms, but only if the particles involved are well spaced. If they move during the course of the simulation, whether they are well spaced or not can change over time.

While there are a number of possible approaches to achieving dynamic optimization, we have been investigating the use of DynInst [2] to perform dynamic modifications to a running executable. Our basic design goal is to produce a tool that can attach to a running program, inject sensors into it to measure its performance, analyze the results using machine learning techniques, and then affect changes in the execution path of that program [20]. This high-level design is illustrated in figure 1. In principle our goals are similar to what can be achieved using a managed language run on a virtual machine (e.g., Java), but the advantage of our approach is that it is applicable to the large body of existing computational science applications written using C/C++ or Fortran.

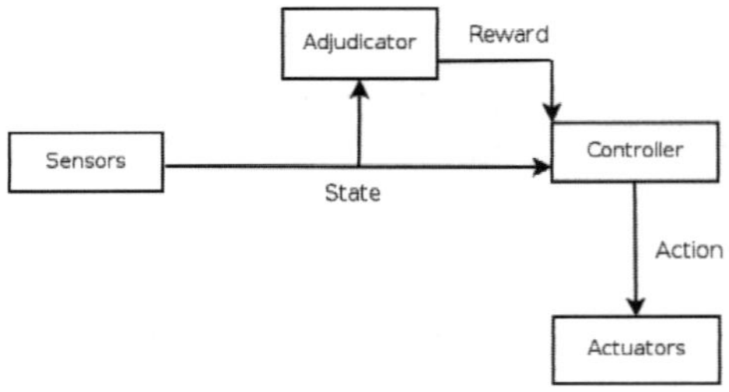

Figure 1. High-level design of adaptive software environment.

As with any optimization that aims to improve performance, it is important to quantify the costs associated with introducing those optimizations. In this case it is important to understand the costs associated with using DynInst to modify a running executable. DynInst provides three main methods for dynamic modification. The first involves inserting fragments of code at designated points - for example, inserting code to measure how many cache misses a function is incurring. The second modification involves manipulating global variables within the address space of the target process, for example, reading them and writing to them. The third modification involves redirecting function calls; for example, if f and g are procedures with identical parameter counts, parameter types, and return types, then any call to f may be replaced by a call to g, while maintaining syntactic validity. (Note that caution must be employed when doing this in the face of interprocedural optimization by the compiler and or the linker.)

The basic mechanism for modifying the behavior of a running binary is known as a *trampoline*. This can be thought of as a diversion of the target program's flow of control. There are two main kinds of trampoline: *base trampolines* and *mini trampolines*. Figure 2 shows how these trampolines interact with each other, and with the target application. The base trampolines are responsible for saving and restoring the state of the CPU, while the mini trampolines contain the inserted code. Only mini trampolines must be explicitly inserted by the user of DynInst; base trampolines are placed automatically where required.

As mentioned above, the base trampoline is responsible for saving and restoring the state of the processor. At a minimum, this state consists of all general-purpose registers and CPU flags. However, if the inserted code is going to make use of floating-point computations, it is also necessary that floating-point state be saved and restored as well. In the general case it is also necessary to guard against the

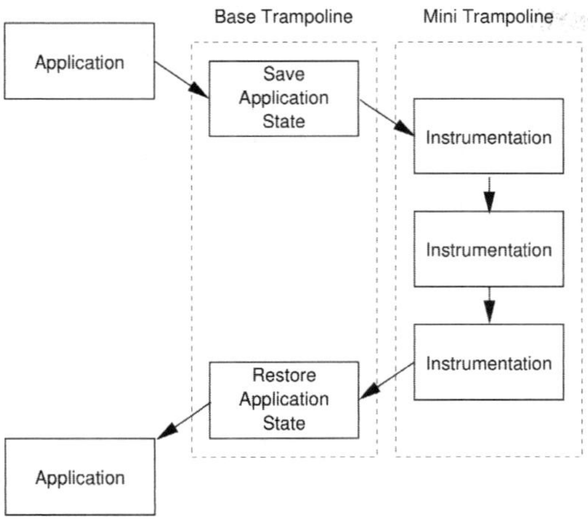

Figure 2. Internal organization of the trampoline mechanism used by DynInst.

possibility that the modified code is inserted into a routine that calls itself, leading to infinite recursion. To handle such situations, DynInst provides guard constructs.

Beyond the one-off cost of modifying the running executable (something that involves stopping the process, making the modifications and then restarting it, and should therefore not be taken lightly), it is of interest to consider the overhead associated with executing the trampolines since this will occur every time that portion of the original code is executed. We have designed a benchmark to measure this overhead [21]. Results are shown in table 6 for a "do-nothing" trampoline located at various points within the target routine. The results show that execution of a trampoline with no checks for recursion or saving of floating point registers adds roughly 110 cycles to the execution time every time this code is executed. Including checks for recursion is relatively cheap, but saving floating point state can be expensive. The latter is true when the code is inserted in the body of a routine, rather than at its head, reflecting the fact that if the trampoline is inserted at the head of a routine floating point state is already saved as part of the calling convention and is therefore not repeated by DynInst.

5. Conclusions and discussion

This paper has outlined some of our efforts to develop fast, reliable and adaptive software for computational science applications. The linear performance model was shown to describe the cache behavior of HF calculations to an accuracy of around 5% across a wide spectrum of hardware. This suggests that over time the intricate details of program execution average out and are dominated by instruction counts and level 2 cache misses. As presented here, the model was parameterized for specific HF calculations, it is of interest to explore its domain of applicability. For example, using values for α and γ obtained by fitting to a specific HF calculation, how well can we predict execution times for a different basis set, molecular system, or method based only on the measured instruction counts and cache misses. Extending the model to include effects of NUMA is a challenge due to lack of hardware counters to breakdown cache misses by memory domain. If page placement was random over the entire NUMA system, using a single γ that was in some way an average would probably suffice. We are investigating these issues by extending the Callgrind tool to include a NUMA memory model.

For the use of interval methods to place rigorous numerical error bounds on computational science, applications support by the Sun compiler of an interval data type for Fortran and a C++ template library has greatly eased our work. Performance of interval codes remains a major issue, not least because they require additional operations to be performed. For example an interval addition requires two floating point addition operations one for the infimum and one for the supremum, while interval multiplication

Table 6. Per-invocation overhead for executing DynInst modified code on an x86 platform as a function of mechanism and location.

	Safety Checks			
	None	Recursion	Floating Point	Recursion & Floating Point
inline trampolines located in loop header				
Wall time (ns)	37±0	39±0	159±0	162±0
CPU cycles	108±0	115±0	476±0	498±3
Instructions	12±0	19±0	20±0	27±0
inline trampolines located in called function				
Wall time (ns)	35±0	38±0	35±0	37±0
CPU cycles	102±0	109±0	102±0	109±0
Instructions	11±0	18±0	11±0	18±0
outline trampolines located in loop header				
Wall time (ns)	37±0	39±0	160±0	168±0
CPU cycles	110±0	116±0	476±0	501±2
Instructions	15±0	24±0	23±0	32±0
outline trampolines located in called function				
Wall time (ns)	35±0	38±0	35±0	38±0
CPU cycles	104±0	112±0	104±0	112±0
Instructions	14±0	23±0	14±0	23±0

can required up to eight (depending on how it is implemented), and in both cases there will be the need to switch rounding modes. Thus our use of intervals to date has been more of a proof of concept rather than a practical large-scale implementation. For carefully rewritten kernels (in developing any tuned kernel care must be taken to ensure that the code still guarantees containment), our work suggests that an interval code will be about one order of magnitude slower than the equivalent floating point code [22, 23]. However, since real estate on the processor die is now less of a bottleneck compared to the rate at which data can be moved on or off the chip, it is possible that future hardware could support native interval operations at a much lower performance penalty. Whether this would be a good option or whether it would be better to move to quadruple precision is partly the rationale for our exploratory work with intervals.

In developing a framework to support adaptive software, use of DynInst to affect change in a running application is an interesting option. Our benchmarking has shown that executing a piece of code that has been modified by DynInst adds between 100 and 500 cycles, depending largely on whether it is necessary to save floating-point state. This factor needs to be considered carefully when using DynInst. What we have not discussed here is the software for deciding when to make modifications to a running code, and what those changes should be. Our current work in this area is focused on use of machine learning techniques, particularly reinforcement learning, in the context of optimal storage formats for sparse matrices [24].

In summary and as mentioned initially, it's an exciting time for computational science. There are many challenges ahead, but also many opportunities.

Acknowledgments
This work was supported in part by Australian Research Council grants Discovery grant DP0558228 and Linkage grant LP0347178 with partners Gaussian Inc. and Sun Microsystems.

References

[1] Frisch M J, Trucks G W, Schlegel H B, Scuseria G E, Robb M A, Cheeseman J R, Montgomery Jr J A, Vreven T, Kudin K N, Burant J C, Millam J M, Iyengar S S, Tomasi J, Barone V, Mennucci B, Cossi M, Scalmani G, Rega N, Petersson G A, Nakatsuji H, Hada M, Ehara M, Toyota K, Fukuda R, Hasegawa J, Ishida M, Nakajima T, Honda Y, Kitao O, Nakai H, Klene M, Li X, Knox J E, Hratchian H P, Cross J B, Bakken V, Adamo C, Jaramillo J, Gomperts R, Stratmann R E, Yazyev O, Austin A J, Cammi R, Pomelli C, Ochterski J W, Ayala P Y, Morokuma K, Voth G A, Salvador P, Dannenberg J J, Zakrzewski V G, Dapprich S, Daniels A D, Strain M C, Farkas O, Malick D K, Rabuck A D, Raghavachari K, Foresman J B, Ortiz J V, Cui Q, Baboul A G, Clifford S, Cioslowski J, Stefanov B B, Liu G, Liashenko A, Piskorz P, Komaromi I, Martin R L, Fox D J, Keith T, Al-Laham M A, Peng C Y, Nanayakkara A, Challacombe M, Gill P M W, Johnson B, Chen W, Wong M W, Gonzalez C and Pople J A Gaussian 03, Revision D.02 Gaussian, Inc., Wallingford, CT, 2004

[2] Buck B R and Hollingsworth J K 2000 *Journal of High Performance Computing Applications* **14** 317 – 329

[3] Antony J, Frisch M and Rendell A P in Press *International Conference on High Performance Scientific Computing, Hanoi, Vietnam March 2006*

[4] Gill P M W, Head-Gordon M and Pople J A 1990 *Journal of Physical Chemistry* **94** 5564–5572

[5] McVoy L and Staelin C 1996 *USENIX Annual Technical Conference* pp 279–294

[6] J Weidendorf KCachegrind – A Profile Visualisation Tool `http://kcachegrind.sourceforge.net/` URL `http://kcachegrind.sourceforge.net/`

[7] Nethercote N and Seward J 2007 *PLDI* ed Ferrante J and McKinley K S (ACM) pp 89–100 ISBN 978-1-59593-633-2

[8] Yang R, Antony J, Janes P P and Rendell A P 2008 *ISPAN* (IEEE Computer Society) pp 31–36

[9] Antony J, Janes P P and Rendell A P 2006 *HiPC* (*Lecture Notes in Computer Science* vol 4297) ed Robert Y, Parashar M, Badrinath R and Prasanna V K (Springer) pp 338–352 ISBN 3-540-68039-X

[10] Moore R E 1959 Automatic error analysis in digital computation Tech. Rep. LMSD-48421 Lockheed Missiles and Space Company URL `interval.louisiana.edu/Moores\_early\_papers/Moore_Lockheed.p%df`

[11] IEEE Computer Society Standards Committee Working group of the Microprocessor Standards Subcommittee and American National Standards Institute 1985 *IEEE standard for binary floating-point arithmetic* ANSI/IEEE Std 754-1985. (Silver Spring, MD: IEEE Computer Society Press)

[12] Caprani O, Madsen K and Nielsen H B 2002 Introduction to interval analysis URL `http://www2.imm.dtu.dk/pubdb/p.php?1462`

[13] Kearfott R B and Kreinovich V 1996 pp 1–22 URL `citeseer.ist.psu.edu/kearfott95applications.html`

[14] P M W Gill 1994 *Advances in Quantum Chemistry* **25** 141–205

[15] McMurchie L E and Davidson E R 1988 *J. Chem. Phys.* **89** 5777–5786 URL `http://link.aip.org/link/?JCP/89/5777/1`

[16] Obara S and Saika A 1986 *J Chem. Phys.* **84** 3963 URL `http://dx.doi.org/`

[17] Gill P M, Johnson B G and Pople J A 1991 *International Journal of Quantum Chemistry* **40** 745–752

[18] Janes P and Rendell A P In Press *2008 IEEE 11th International Conference on Computational Science and Engineering*

[19] Rendell A P, Clarke B, Janes P, Milthorpe J and Yang R 2007 *"International Conference on Computational Science and its Applications (ICCSA), IEEE Conference Proceedings (DOI 10.1109/ICCSA.2007.49)* (IEEE Computer Society) pp 457 – 466

[20] Armstrong W, Christen P, McCreath E and Rendell A P 2006 *Proceedings of the International Workshop on Integrating AI and Data Mining (AIDM 2006), IEEE Publications (ISBN 0-7695-2730-2, Digital Object Identifier: 10.1109/AIDM.2006.9)* pp 18 – 24

[21] Armstrong W and Rendell A P *To be published*

[22] Rendell A P, Clarke B and Milthorpe J 2006 *International Conference on Computational Science (1)* (*Lecture Notes in Computer Science* vol 3991) ed Alexandrov V N, van Albada G D, Sloot P M A and Dongarra J (Springer) pp 218–225 ISBN 3-540-34379-2

[23] Milthorpe J and Rendell A P 2005 *Proceedings of the Third Australian Undergraduate Computing Conference, Canberra 2005* pp 34–41 ISBN 0-9757173-1-6

[24] Armstrong W and Rendell A P 2008 *The Third international Workshop on Automatic Performance Tuning (IWAPT08), (submitted)*

Advanced software for the calculation of thermochemistry, kinetics, and dynamics

R Shepard

Chemical Sciences and Engineering Division, Argonne National Laboratory, Argonne, IL 60439, USA

E-mail: shepard@tcg.anl.gov

Abstract. The Born-Oppenheimer separation of the Schrödinger equation allows the electronic and nuclear motions to be solved in three steps: (1) the solution of the electronic wave function at a discrete set of molecular conformations; (2) the fitting of this discrete set of energy values in order to construct an analytical approximation to the potential energy surface (PES) at all molecular conformations; and (3) the use of this analytical PES to solve for the nuclear motion by using either time-dependent or time-independent formulations to compute molecular energy values, chemical reaction rates, and cumulative reaction probabilities. This project involves the development of technology to address all three of these steps. This report focuses on our recent work with nonlinear parameterization of the electronic wave functions.

1. Introduction

Our work this past year has focused on the use of nonlinear wave function parameters for electronic wave functions, on the computation of spin densities, and on the extension of our nonlinear wave function expansion approach to include spin-orbit interaction.

We have recently proposed a new expansion form for molecular electronic wave functions [1-6]. The wave function is written as a linear combination of graphically contracted functions (GCF), and each GCF in turn is formally equivalent to a linear combination of configuration state functions (CSFs) that comprise the underlying linear expansion space of dimension N_{csf}. The CSF coefficients that define the basis functions are nonlinear functions of a smaller number of variables N_{φ}=N_{csf}. This new approach addresses directly the main bottlenecks of previous electronic structure methods. Much of the computational effort and data storage requirements with this new approach depend only on Log(N_{csf}) rather than N_{csf}. The method is formulated in terms of spin-eigenfunctions using the Graphical Unitary Group Approach (GUGA)[7], and consequently it does not suffer from the spin contamination or spin instability that is often associated with single-reference electronic structure methods. The expansion form is appropriate for both ground and excited states and to closed- and open-shell molecules. The method is not based on the idea of expansion about a "reference" wave function, so its accuracy is not inherently limited because of failures of the Hartree-Fock method or artificially imposed excitation-level restrictions. The exact electronic Hamiltonian operator is used in this method; for example, it does not rely on multipole approximations or exchange-correlation operator approximations (although such approximations may prove useful within an optimization context as preconditioners). Our initial applications to some small molecules are encouraging. As demonstrated previously [4], even a single GCF, N_{α}=1, based on a $\mathbf{6}^6$ Shavitt graph is sufficient to dissociate the N≡N triple bond correctly to ground state ^4S atom fragments, two GCFs are sufficient to approach the full-valence CI electronic energy to within 3 mE_h at all bond lengths, and three GCFs reproduce exactly the full-valence CI energy at all bond lengths. More recently, we have demonstrated [5] convergence of the total energy to within 1 kcal/mole of the full-CI limit (the solution within the full tensor-product basis) with only 13 expansion terms, and to within 1 mE_h of the full-CI limit with only 16 expansion terms, at all bond lengths along the symmetric dissociation curve of H_2O to its atomic fragments $O(^3P)+2H(^2S)$.

2. Wave function optimization

Although the results described in the previous section are encouraging, the most remarkable feature of our new method is the relatively small effort required to construct hamiltonian matrix elements and transition density matrices in the GCF basis[2,3]. Given a wave function expansion of the form

$$|\psi\rangle = \sum_{P}^{N_\alpha} c_P |P\rangle$$

in which the N_α GCFs are denoted $|P\rangle$, each basis function $|P\rangle$ depends on a corresponding set of nonlinear parameters $\boldsymbol{\varphi}^P$ which in turn depend on the set of arc factors $\boldsymbol{\alpha}^P$ associated with the arcs of the Shavitt graph. The wave function thereby depends on the linear expansion coefficients $\mathbf{c}$ and on the nonlinear parameters $\boldsymbol{\varphi}$. Optimization of the expansion coefficients $\mathbf{c}$ to minimize the energy results in the symmetric generalized eigenvalue equation

$$\mathbf{H}\,\mathbf{c} = \mathbf{S}\,\mathbf{c}\,E$$

with $H_{PQ} = \langle P|\hat{H}|Q\rangle$ and $S_{PQ}=\langle P|Q\rangle$. An efficient procedure has been developed to compute Hamiltonian matrix elements and reduced one- and two-particle density matrices for this nonlinear expansion form [2]. The effort required to construct an individual Hamiltonian matrix element H_{PQ} scales as $O(\beta n^4)$ for a wave function expanded in n molecular orbitals. The prefactor β itself scales between N^0 and N^2, for N electrons, depending on the complexity of the underlying Shavitt graph. The corresponding metric matrix element S_{PQ} requires effort that scales as $O(\beta n)$. There is no component of the effort or storage for matrix element computation that scales as N_{csf} , which may be, practically speaking, arbitrarily large. Hamiltonian matrix element timings with our initial implementation of this method are very promising. Wave function expansions that are orders of magnitude larger than can be treated with traditional CI methods require only modest effort with our new method. A matrix element involving GCFs with $n=N=46$ corresponding to an underlying linear expansion space dimension $N_{csf}\approx 5.5\cdot 10^{24}$, or over 9.2 moles of CSFs, requires only a few seconds with our new method [2-4]. The computation of that same Hamiltonian matrix element using traditional full-CI technology is estimated to require about 10^{24} seconds, over a million times the age of the universe. Of course, a traditional full-CI code has the capability of computing $H_{PQ}=\mathbf{x}^{PT}\mathbf{H}\mathbf{x}^Q$ for arbitrary vectors $\mathbf{x}^P$ and $\mathbf{x}^Q$, whereas our new method is restricted to vectors that can be represented in our graphical-based nonlinear expansion form. This restriction affects eventually N_α, the number of GCFs required for convergence to chemical accuracy. These timings show the tremendous potential of our new method. H_{PQ} computation scales approximately the same as the simple Hartree-Fock method, yet the method is capable of approaching the accuracy of full-CI wave functions.

A major challenge presented by the new method is the optimization of the nonlinear parameters. In our previously published algorithm [3], each gradient evaluation requires $O(\beta n^5)$ effort, which is significantly more expensive than an energy evaluation, particularly for larger values of n. Typically, we find that our previous analytic gradient computation is 10 to 100 times faster than a finite difference approximation, and it is 10 to 100 times slower than a single energy computation. In last year's proceedings [8], an example was given of the number of gradient evaluations that are required using various nonlinear optimization methods. For the conjugate-gradient and quasi-newton methods, hundreds or sometimes thousands of iterations might be required, each of which requires an expensive gradient evaluation. These methods along with the substitution and split-optimization methods are based on the construction of the matrices $\mathbf{G}^{[u]}$ and $\mathbf{S}^{[u]}$, which are representations of the Hamiltonian and overlap matrices defined in a basis in which the GCFs are uncontracted at a particular orbital level u. These matrices give a faithful representation of the energy in terms of the arc factors at a single orbital level $\boldsymbol{\alpha}^{[u]}$ within the assumption that the arc factors at all other levels are temporarily held fixed.

$$E\left(\boldsymbol{\alpha}^{[u]}\right)=\frac{\boldsymbol{\alpha}^{[u]T}\mathbf{G}^{[u]}\boldsymbol{\alpha}^{[u]}}{\boldsymbol{\alpha}^{[u]T}\mathbf{S}^{[u]}\boldsymbol{\alpha}^{[u]}}$$

From this expression the analytic gradient and various optimization approaches follow. With our previous algorithm [3,5], as each of the $O(n^4)$ Shavitt-loop types is terminated, resulting in the contributions to H_{PQ} associated with a particular one- or two-electron contribution to the hamiltonian, an immediate reverse (downward) propagation was performed, allowing the contributions to the separate $\mathbf{G}^{[u]}$ matrix elements to be accumulated. Each reverse propagation requires $O(\beta n)$ computational effort, resulting in $O(\beta n^5)$ total effort for all of the $\mathbf{G}^{[u]}$ matrices.

We have recently developed [9] a new algorithm based on *deferred* reverse propagation. This algorithm may be understood by examining a particular sequence of hamiltonian contributions. For simplicity, consider the one-electron contributions h_{pq} for p=1 and q=2,3,4 using the vector and sparse matrix notation introduced in [2].

$$\begin{aligned}H_{PQ} \leftarrow\ & h_{12}\boldsymbol{\gamma}_{(0)}^{PQ}\mathbf{F}_{(1)}^{PQ}(\underline{R})\mathbf{F}_{(2)}^{PQ}(\overline{R})\overline{\boldsymbol{\gamma}}_{(2)}^{PQ}\\
&+h_{13}\boldsymbol{\gamma}_{(0)}^{PQ}\mathbf{F}_{(1)}^{PQ}(\underline{R})\mathbf{F}_{(2)}^{PQ}(R)\mathbf{F}_{(3)}^{PQ}(\overline{R})\overline{\boldsymbol{\gamma}}_{(3)}\\
&+h_{14}\boldsymbol{\gamma}_{(0)}^{PQ}\mathbf{F}_{(1)}^{PQ}(\underline{R})\mathbf{F}_{(2)}^{PQ}(R)\mathbf{F}_{(3)}^{PQ}(R)\mathbf{F}_{(4)}^{PQ}(\overline{R})\overline{\boldsymbol{\gamma}}_{(4)}^{PQ}\end{aligned}$$

In these expressions $\boldsymbol{\gamma}_{(0)}{}^{PQ}$ is a row vector, $\mathbf{F}_{(u)}{}^{PQ}(W)$ are rectangular matrices (treated as a sparse data structure) of segment factors,

$$F_{(u)}^{PQ}(W)_{\mu\mu'}=\alpha_{\mu}^{P}\alpha_{\mu'}^{Q}T^{[u]}(W)_{\mu\mu'}$$

and $\overline{\boldsymbol{\gamma}}_{(u)}^{PQ}$ is a column vector. $T^{[u]}(W)_{\mu\mu'}$ is the Shavitt segment value of the appropriate shape W at orbital level u. Our recursive H_{PQ} algorithm consists of extracting the common factors in each of these terms in left-to-right order.

$$H_{PQ}\leftarrow h_{12}\mathbf{z}_{(1)}(\underline{R})\mathbf{F}_{(2)}^{PQ}(\overline{R})\overline{\boldsymbol{\gamma}}_{(2)}^{PQ}+h_{13}\mathbf{z}_{(2)}\mathbf{F}_{(3)}^{PQ}(\overline{R})\overline{\boldsymbol{\gamma}}_{(3)}+h_{14}\mathbf{z}_{(3)}\mathbf{F}_{(4)}^{PQ}(\overline{R})\overline{\boldsymbol{\gamma}}_{(4)}^{PQ}$$

with the row vectors $\mathbf{z}_{(k)}$ defined with a simple multiplicative recursion as follows.

$$\mathbf{z}_{(1)}=\boldsymbol{\gamma}_{(0)}^{PQ}\mathbf{F}_{(1)}^{PQ}(\underline{R})$$

$$\mathbf{z}_{(k)}=\mathbf{z}_{(k-1)}\mathbf{F}_{(k)}^{PQ}(R)\quad ;k=2\ldots(n-1)$$

In general, this recursive algorithm requires $O(\beta n^2)$ effort for all h_{pq} contributions for a given GCF pair PQ. The construction of the $\mathbf{G}^{[1]}$ matrix with our previous algorithm would consist of three separate contributions from the right-to-left reverse propagation of the above three matrix expressions, or in general $O(\beta n^3)$ effort for n orbitals for a PQ pair. Our new algorithm consists of accumulating these separate contributions recursively according to

$$\overline{\mathbf{z}}_{(4)}=h_{14}\mathbf{F}_{(4)}^{PQ}(\overline{R})\overline{\boldsymbol{\gamma}}_{(4)}^{PQ}$$

$$\overline{\mathbf{z}}_{(k)}=h_{1k}\mathbf{F}_{(k)}^{PQ}(\overline{R})\overline{\boldsymbol{\gamma}}_{(k)}^{PQ}+\mathbf{F}_{(k)}^{PQ}(R)\mathbf{z}_{(k+1)}\quad ;k=(n-1)\ldots 2$$

giving (in principle)

$$H_{PQ}\leftarrow\boldsymbol{\gamma}_{(0)}^{PQ}\mathbf{F}_{(1)}^{PQ}(\underline{R})\overline{\mathbf{z}}_{(2)}$$

from which the $\mathbf{G}^{[1]}$ contributions may be extracted in the single step

$$G_{\mu\mu'}^{[1]}\leftarrow\gamma_{(0)\mu}^{PQ}T^{[1]}(\underline{R})_{\mu\mu'}\overline{z}_{(2)\mu'}$$

over all nonzero Shavitt segment values. This single step replaces the n separate accumulation steps of our previous algorithm. The total effort then requires only $O(\beta n^2)$ for all h_{pq} contributions, reduced from the O(βn^3) effort of the previous algorithm. This deferred reverse propagation algorithm may be extended to the two-electron repulsion contributions to the Hamiltonian operator in a straightforward manner, and the O(βn^5) effort of our previous algorithm is reduced to only O(βn^4).

At the time of writing of this report, we have only partially implemented this algorithm. We hope to have this algorithm completely implemented and to present timing comparisons of both approaches by the time of the SciDAC meeting. A complete description of this new algorithm will be published separately [9]. However, it is clear even at this time that the new algorithm will be a significant improvement over our previous implementation, and it will allow new optimization algorithms to be considered that were impractical previously because of the relatively high costs of the $\mathbf{G}^{[u]}$ matrix constructions. We expect our new algorithm to allow calculations on much larger molecular systems that are associated with larger values of n.

3. Spin-density computation

The GUGA formulation on which our nonlinear wave function expansion method is based is well-suited to the computation of spin-independent properties, including the electronic energy with no external magnetic field as discussed above. However, this approach complicates somewhat the computation of spin-dependent properties (those that depend on the magnetic quantum number M). In particular, the expectation value of such an operator may be written as

$$\langle \Psi; S, M | \hat{A} | \Psi; S, M \rangle = \sum_{p,q} A_{pq}^{(1,0)} D_{qp}^{(1,0;M)} = Tr\left(\mathbf{A}^{(1,0)} \mathbf{D}^{(1,0;M)} \right)$$

with the spin-density matrix defined in terms of the spin-orbital creation and annihilation operators as

$$D_{qp}^{(1,0;M)} = \langle \Psi; S, M | a_{p\alpha}^{\dagger} a_{q\alpha} - a_{p\beta}^{\dagger} a_{q\beta} | \Psi; S, M \rangle .$$

It may be shown that this density matrix is related to the maximal M=S spin-density matrix as

$$\mathbf{D}^{(1,0;M)} = \frac{M}{S} \mathbf{D}^{(1,0;S)} \qquad ;M=-S,-S+1,\ldots,S-1,S$$

and that the maximal spin density matrix is given by

$$(S+1) D_{qp}^{(1,0;S)} = \left(2 - \tfrac{1}{2} N\right) D_{qp}^{(0)} - \sum_k d_{qkkp}$$

with $D_{qp}^{(0)} = \langle \Psi | \hat{E}_{pq} | \Psi \rangle$ and $d_{qkkp} = \langle \Psi | \hat{e}_{pkkq} | \Psi \rangle$, which in turn are both independent of M and may be computed in a straightforward manner. The computation of the full set of d_{qkkp} elements requires O(βn^3) computational effort with a GCF wave function expansion. However, it is not the individual d_{qkkp} elements that are required, but rather the particular summation over the orbital index k indicated above. We have developed an algorithm to compute this summation recursively. Shavitt loop[2] types 4*ab*, 6*ab*, and 8*ab*, along with the two-level loop types 11*ab*, and 12*ab*, and the one-level loop type 14*a* must be included [10]. We demonstrate this recursive algorithm for the simple, but nontrivial, case of the Shavitt loop type 11*b* contributions which would normally require O(βn^2) effort.

Consider the q=4 contributions for k=1…3 for d_{qkkq}. With the matrix notation used in the previous section, the summation of the three contributions is as follows.

$$\begin{aligned} D_{44}^{(1,0;S),PQ} \leftarrow\; & \gamma_{(0)}^{PQ} \mathbf{F}_{(1)}^{PQ}(\underline{RL}) \mathbf{F}_{(2)}^{PQ}(RL) \mathbf{F}_{(3)}^{PQ}(RL) \mathbf{F}_{(4)}^{PQ}(\overline{RL}) \overline{\gamma}_{(4)}^{PQ} \\ & + \gamma_{(1)}^{PQ} \mathbf{F}_{(2)}^{PQ}(\underline{RL}) \mathbf{F}_{(3)}^{PQ}(RL) \mathbf{F}_{(4)}^{PQ}(\overline{RL}) \overline{\gamma}_{(4)} \\ & + \gamma_{(2)}^{PQ} \mathbf{F}_{(3)}^{PQ}(\underline{RL}) \mathbf{F}_{(4)}^{PQ}(\overline{RL}) \overline{\gamma}_{(4)}^{PQ} \end{aligned}$$

The common factors in these terms may be computed using the recursive algorithm

$$\mathbf{z}_{(1)}^{PQ} = \gamma_{(0)}^{PQ} \mathbf{F}_{(1)}^{PQ}(\underline{RL})$$

$$\mathbf{z}_{(k)}^{PQ} = \gamma_{(k-1)}^{PQ} \mathbf{F}_{(k)}^{PQ}(\underline{RL}) + \mathbf{z}_{(k-1)}^{PQ} \mathbf{F}_{(k)}^{PQ}(RL) \qquad ;k = 2\ldots(n-1)$$

giving ultimately for an arbitrary q contribution

$$D_{qq}^{(1,0;S),PQ} \leftarrow \mathbf{z}_{(q-1)}^{PQ}\mathbf{F}_{(q)}^{PQ}(\overline{RL})\overline{\boldsymbol{\gamma}}_{(q)}^{PQ}$$

This recursive procedure allows the row vectors $\mathbf{z}_{(q)}^{PQ}$ to be computed with O(βn) effort for each GCF pair P and Q. The same approach applies also to the 11*a* Shavitt loop type using the corresponding segment factor matrices in the recursion expressions. By using this recursive approach, all O(βn^2) individual contributions from the two-index 11*ab* Shavitt loops to the diagonal elements of the spin-density may be computed with only O(βn) total effort. This same kind of recursion may be applied also to the three-index Shavitt loops 4*ab*, 6*ab*, and 8*ab*. The O(βn^3) individual operator contributions to the off-diagonal spin-density matrix elements may be computed with only O(βn^2) total effort[10].

4. Conclusions

We have presented the latest developments for our recently developed nonlinear wave function expansion form. The energy-based optimization is formulated in terms of a specific kind of uncontraction of each of the graphically contracted basis functions. These orbital-level Hamiltonian matrices give an intuitive representation of the energy as a function of disjoint subsets of the arc factors, they provide for an efficient computation of gradients of the energy with respect to the arc factors, and they allow optimal arc factors to be determined in closed form for subspaces of the full variation problem. Our new analytic gradient and orbital-level Hamiltonian matrix construction algorithm scales approximately as O(βn^4), in contrast to our previous algorithm which scaled approximately as O(βn^5). This new approach will allow applications of our new method to larger molecular systems, and it will allow new optimization algorithms to be considered that were previously impractical. Energy and gradient computations with our new method require relatively little effort compared to other electronic structure methods. Timings for energy and arc factor gradient computations involving expansion spaces of over 10^{24} CSFs have been reported. We have begun to extend our nonlinear expansion method to include spin-orbit interactions and to include the computation of magnetic properties (e.g., NMR and ESR spectra simulation). An algorithm has been presented that allows the reduced spin-density matrix to be computed with only O(βn^2) effort.

References

[1] Shepard R 2005 *J. Phys. Chem. A* **109**:11629
[2] Shepard 2006 *J. Phys. Chem. A* **110**:8880
[3] Shepard R and Minkoff M 2006 *Int. J. Quantum Chem.* **106**:3190-3207
[4] Shepard R, Wagner A F and Gray S K 2006 *J. Physics: Conference Series* **46**:239-243
[5] Shepard R, Minkoff M and Brozell S R 2007 *Int. J. Quantum Chem.* **107**:3203-3218
[6] Brozell S R, Shepard R and Zhang Z 2007 *Int. J. Quantum Chem.* **107**:3191-3202
[7] Shavitt I 1981 *The Graphical Unitary Group Approach and its Application to Direct Configuration Interaction Calculations,* in *The Unitary Group for the Evaluation of Electronic Energy Matrix Elements (Lecture Notes in Chemistry 22)* 51-99 J. Hinze, ed., (Berlin: Springer-Verlag)
[8] Shepard R 2007 *J. Physics: Conference Series* **78**:012067
[9] Shepard R , Gidofalvi G and Hovland P 2008 (*manuscript in preparation*).
[10] Gidofalvi G and Shepard R 2008 (*manuscript in preparation*).

Journal of Physics: Conference Series **125** (2008) 012017 doi:10.1088/1742-6596/125/1/012017

High performance computations using dynamical nucleation theory

Theresa L Windus[1], Shawn M Kathmann[2] and Lonnie D Crosby[1]
[1] Department of Chemistry, Iowa State University and Ames Laboratory, Ames, IA 50014, USA
[2] Chemical and Material Sciences Division, Molecular Interactions and Transformations Group, Pacific Northwest National Laboratory, Richland, WA 99352, USA

E-mail: theresa@fi.ameslab.gov, shawn.kathmann@pnl.gov, lcrosby@iastate.edu

Abstract: Chemists continue to explore the use of very large computations to perform simulations that describe the molecular level physics of critical challenges in science. In this paper, we describe the Dynamical Nucleation Theory Monte Carlo (DNTMC) model – a model for determining molecular scale nucleation rate constants – and its parallel capabilities. The potential for bottlenecks and the challenges to running on future petascale or larger resources are delineated. A "master-slave" solution is proposed to scale to the petascale and will be developed in the NWChem software. In addition, mathematical and data analysis challenges are described.

1. Introduction

Quantum chemistry has a long and distinguished history of using parallel computing (see, e.g., [1]) to solve very difficult chemical problems involving the Schrödinger or Dirac equations [2]. In fact, chemists have often led the charge in using these resources and in developing new algorithms that impact not only their own field, but others as well. With the advent of terascale and petascale computers, however, quantum chemists have been challenged to think beyond the usual limiting box of determining kinetic rates using a few stationary points on the potential energy surface (PES) – the historical approach for over 80 years. The increased size of computational resources allows us to determine properties and chemical reactions that more closely model the real chemistry of complex systems. Toward that end, we present here our work using dynamical nucleation theory (DNT) [3, 4] with *ab initio* potentials to obtain accurate information on the nucleation of small molecular clusters.

Nucleation processes are of critical importance to DOE and the nation; they have been identified in many reports as one of the basic challenges in chemical physics. For example, nucleation is important in controlled crystallization and dissolution of waste forms and fuel formation as well as in gas bubble formation and waste separations in nuclear technology [5]. Nucleation has also been identified as an important process in catalysis (especially in understanding the formation and growth of small particle and bulk catalysts) [6], nanoparticle formation through phase transitions and directed assembly [7], and the release of hydrogen in hydrogen storage devices [8]. In addition, understanding the formation of hydrates of gases such as methane is important for natural gas storage and transportation and for separation processes. In many biological, geological, chemical and material processes, water does not exist in its bulk form, but rather in nanoscopically confined environments. For example, reverse micelles are potentially useful as templates for nanoparticle synthesis [9]. Examining the nucleation rates of contaminants in the air has also been shown to be an effective method for the detection of TNT and DNT in photoinduced nucleation chemical detection [10]. In order to provide species specific identification, the nucleation rate as a function of wavelength is measured and computation in conjunction with experiment will provide a powerful method of detecting chemicals in the air. Furthermore, aerosol and aerosol-ice nucleation is important in the formation of clouds and ultimately in models used for global climate prediction [11].

Since the theoretical aspects of DNT have been presented elsewhere [3, 4], we give only a short introduction to provide the reader with context for the algorithms described below. At the core of DNT is the use of molecular potentials to find the minimum flux surface (and therefore, the evaporation rate) using canonical variational transition state theory [12] for the monomer evaporation from a cluster:

$$N_{i+1} \underset{\beta_i}{\overset{\alpha_{i+1}}{f}} N_i + N_1 \ , \tag{1}$$

where N_1 is a monomer of species N, i is the number of monomers in the cluster, α is the evaporation rate constant and β is the condensation rate constant. Note that a homogeneous cluster of monomers is assumed here in order to simplify the discussion, but is not a requirement.

Through DNT the evaporation rate, $\alpha_i(T, r_{cut})$, for an i-cluster can be written as

$$\alpha_i(T, r_{cut}) = -(2\pi m_1 k_b T)^{-1/2} \left(\frac{dA_i(T,r)}{dr} \right)\Bigg|_{r=r_{cut}} , \tag{2}$$

where the derivative of the Helmholtz fee energy, $A_i(T,r)$, with respect to the radius, r, of an enclosing spherical volume is the central quantity in calculating the rate constant, m_1 is the mass of the monomer, k_b is Boltzman's constant and T is the temperature of interest. The choice of r_{cut} is the variationally optimized spherical dividing surface for the evaporation. At radii less than r_{cut} the molecular i-cluster is intact,and at radii greater than r_{cut} the molecular cluster is an (i-1)-cluster plus a monomer.

The Helmholtz free energy is related to the canonical partition function by

$$Q_i(T,r) = \exp\left(-\frac{A_i(T,r)}{k_b T} \right), \tag{3}$$

which can be evaluated as an integral over configurational space. As examined in detail elsewhere [4], we use a derivative of the canonical ensemble to minimize the number of molecular evaluations that are needed to evaluate the integral:

$$\frac{dQ_i(T,r')}{dr'} = \left(\frac{2\pi m_1 k_b T}{h^2} \right)^{\frac{3i}{2}} \frac{1}{i!} \int \mathbf{dr}^i \exp\left[-\frac{U_i(\mathbf{r}^i)}{k_b T} \right] \sum_k^i \delta\left(r' - |\mathbf{R}_i - \mathbf{r}_k| \right) \prod_{j \neq k}^i \Theta\left(r' - |\mathbf{R}_i - \mathbf{r}_j| \right), \tag{4}$$

where Θ is the Heaviside step function, r' is the enclosing spherical volume centered on the center-of-mass, $\mathbf{R}_i$, of the i-cluster, $\mathbf{r}_j$ and $\mathbf{r}_k$ are monomer coordinates, δ is a delta function that restricts one or more monomers to lie on the surface of the spherical volume, h is Planck's constant and U_i is the i-cluster interaction potential. We note that the minimum in the derivative of the canonical ensemble is also the minimum of the flux that determines r_{cut}. The computational expense lies in the cost of the interaction potential and the extent of the configurational sampling.

In this paper, the parallel algorithms that are used and the current challenges to scale those algorithms to petascale and beyond are discussed. In Section 2, we discuss the current parallel algorithm and implementation. Improvements that will be required to move to the petascale are presented in Section 3. Other improvements to the model that will require mathematical input are presented in Section 4. We present conclusions in Section 5.

2. Current parallel model

The configurational sampling required in (4) is accomplished through a Metropolis Monte Carlo (MC) methodology [13], which is implemented as the DNTMC module in NWChem [14, 15]. Importance sampling is accomplished through the Boltzmann weight of the interaction potential, $\exp\left[-\frac{U_i\left(\mathbf{r}^i\right)}{k_b T}\right]$, where configurations with a lower potential (i.e., a greater interaction potential) are more probable. During the simulation, the monomers are kept rigid to decrease the number of degrees of freedom that must be explored in the MC. Only translations and rotations of the monomers are therefore allowed. Each of these degrees of freedom is uncorrelated and has a probability of ½ in each move, leading to four possible combinations (no moves, only translation, only rotation and both translation and rotation) with a probability of ¼. Each move is then tested for acceptance based on the interaction potential and the temperature of the simulation.

Since the coordinate that is important for this work is r', the distance from the center-of-mass to the center-of-mass of the outermost monomer, the configurational space of interest is projected onto this coordinate and divided into bins to gather the statistics. For example, in water dimer simulations $\mathbf{r}_{\min}$ and $\mathbf{r}_{\max}$ for the simulation are 1.25 and 5.25 Å, respectively. Then, 100 bins are used to evenly divide this space to gather the statistics. So, if a move is accepted (or if the original configuration is to be accepted again), the spherical radius of the cluster is determined from the center of mass and the configuration is added to the appropriate bin. When the system is fully sampled, a probability distribution is obtained that can be used to evaluate the evaporation rate as in (2).

To reduce the statistical noise and to aid in fully exploring the configurational space, we perform multiple MC Markov chains. Since each of these Markov chains is independent of the others, one can run these simulations simultaneously to reduce the time to solution as well as take advantage of the parallel capabilities of the current and future generations of computers. This model lends itself nicely to a multiple level computational model as shown figure 1. This type of algorithm is possible in NWChem through the use of subgroups within the Global Arrays [16] code that is used as the parallel communications model. A similar multilevel parallel algorithm was implemented using classical potentials and Global Arrays [17].

As shown in the figure, the overall DNTMC module coordinates the overall computation in the scope of the world group and creates subgroups for each of the Markov chains to be used in the simulation. Each of the Markov chains will, in turn, use a set of processors to calculate energies. In this manner many subgroups can be used, each with a large number of processors, that allow *ab intio* potentials to be used in the energy computations.

In order for the overall computation to be efficient, the Markov chains are evenly distributed through $\mathbf{r}$ space. Each initial configuration is determined randomly with the only constraint that the molecular cluster lays within the specified $\mathbf{r}$. This allows the initial rotational space to be randomly sampled as well. The overall goal of using these different initial configurations is to get a better sampling of space and to determine the overall probability distribution more efficiently.

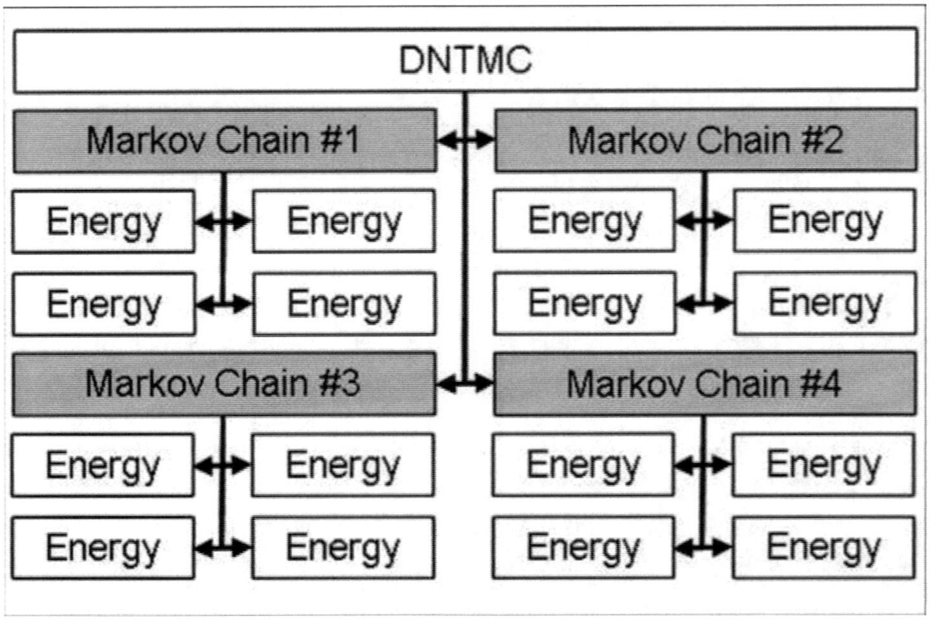

Figure 1. Schematic of the multiple levels of parallelism in the DNTMC module.

The algorithm also includes a specialized group, denoted the zero group, which includes the zero processor on each of the subgroups for the Markov chains. This group is used to gather the statistical results such as the overall probability distribution and the acceptance ratios, determine if the results are converged, write restart information and then continue the calculation. These processes occur after a user specified number of steps in the Markov chain and require synchronization of all of the subgroups before they can occur. Figure 2 shows the overall flow of the DNTMC module. At time zero (represented by the top line in the figure), the module starts in the world group and all processes are involved in the initialization stage that determines the species to be used for the monomers, the number of each monomer species, and the size of the bins to be used in the calculation. At this point, the subgroups and zero group are formed. Each of the subgroups is assigned an initial **r** for the molecular cluster and creates an initial configuration within that **r**. At this point, each of the individual Markov chains starts to compute (represented by the blue arrows in figure 2 and referred to in the text as an MC loop). While the subgroups are computing results, each chain also writes a data file that holds the molecular cluster's coordinates, the energy of the configuration and any other relevant property data (such as dipoles). The practice of writing this data is not normal for MC calculations with classical potentials. However, since the *ab initio* data is so expensive to compute, this data is gathered for future use (described below in Section 4).

After the user specified number of steps in an MC loop have been calculated (usually about 1,000), the groups are synchronized to make sure that all of the calculations are completed. The zero group then communicates results and writes data to the restart file (represented by a yellow arrow in figure 2). The other processes wait in a subgroup synchronization until the zero group is complete (represented by the grey arrow in figure 2). When everyone reaches the subgroup synchronization, the individual Markov chains continue if the calculation has not converged.

While simulations using this model scale very well to several thousand processors (tens of teraflops), it is unclear what the performance will be on petascale computers. It is clear that the synchronization steps with their communication could cause a scaling bottleneck. However, these are very small steps in comparison to the overall computational costs. A much larger concern is the load imbalance that can occur between the individual Markov chains. This can occur since each individual energy calculation can take more or less time depending on how good the starting orbitals are for each calculation. In addition, some of the computations may experience a different load on the processors due to daemons or other processes that run on a sporadic basis.

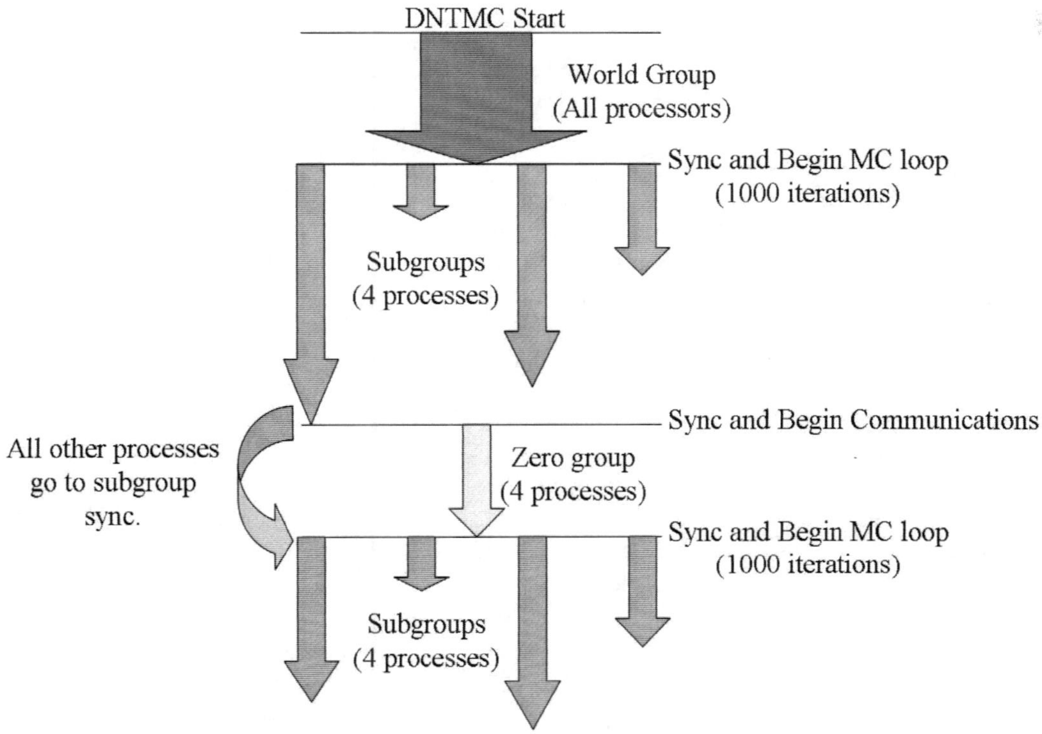

Figure 2. Schematic of the synchronization and work flow in the DNTMC module.

To get an idea of this load imbalance, representative calculations were performed on the mpp2 Itanium system at the Molecular Science Computing Facility at Pacific Northwest National Laboratory [18]. Timers were added to the NWChem code to record the overall time that it took for an MC loop between synchronizations (the time represented by the blue arrows in figure 2). These times were then collected and the minimum time, the maximum time and the average time were calculated to get an overall idea of the load imbalance. As a representative example, a water trimer using the MP2 level of theory [19] and the aug-cc-pVDZ [20] basis set was run for 10 hours. This calculation used a total of 400 processors in 50 subgroups each using 8 processors. While the number of processors used in this computation is relatively small, it is certainly representative of the types of simulations that are performed at a larger scale. The calculation did a total of 3 MC loops, with a minimum loop time of 1.54 hours, a maximum loop time of 3.11 hours and an average time of 2.29 hours.

At a first glance, the results look very poor for the overall load balance with the maximum time being twice that of the minimum time. However, we know from previous experience that long simulations scale very well. By examining the individual timings for each of the MC loops, it becomes obvious that the large load imbalance happens on the first loop. These large imbalances are due to Markov chains that are started in regions where convergence of the calculations are difficult (e.g. very repulsive configurations). This is a property that typically does not show up in classical simulations since the potential function is usually analytical. However, in quantum mechanical calculations, this can be a very large problem, especially at the beginning of a calculation. Later in the simulations after several MC loops, the repulsive geometries (or those that tend to give high energies) are not sampled as often and therefore, the simulations have much better load balance. Further investigation of the load imbalance with progression of the Markov chains will need to be performed.

3. Petascale parallel issues

Several issues must be addressed in scaling this algorithm to the petascale. The first is the individual energy calculations themselves. This is a very complex problem and is not the main point of this paper, so it will not be discussed here. We note, however, that our work with the Common Component Architecture [21] may be of use in this context. As new components are created that can efficiently use larger numbers of processors to calculate energies, they can be directly used by the DNTMC module if it is retrofit to use these components. These components could be specific implementations for a particular architecture or entirely new algorithms that represent the state of the art in chemical accuracy.

The second issue is the load imbalance described in Section 2. Although the load imbalance is small, it will have a large impact on the overall computation at the petascale. This suggests that another implementation may be better suited to get the most performance out of the petascale architectures.

A third issue has to do with the limitations on the number of Markov chains and the number of processors that can be brought to bear on any given molecular configuration. Even with an efficient implementation for individual energy computations, the relatively small sizes of the clusters (up to 10s of monomers) will limit the number of processors that can be brought to bear on the problem. Coupling this with the limited number of Markov chains that can practically be initiated in a given simulation volume and there is a limit on the overall processors that can be used in any simulation. Our best estimates based on current scalability of the energy computations and the practical number of Markov chains would allow approximately 10,000 processors to be used. Of course, the ideal situation would be to examine much larger systems. While this is practical at the level of individual energy computations, the number of configurations (and therefore the computational expense) that must be explored is on the order of $r_{\max}^{3i-3}$. Obviously this makes the number of overall configurations that must be computed very large, but the limitation on the number of Markov chains currently limits the amount of parallelism that can be applied to the problem. Therefore, the overall time to solution is also limited.

One way to avoid some of the limitations on the overall scalability is to change the model to one in which the individual energy computations are independent of any particular Markov chain. For example, one model that would negate this need is to go to a "master-slave" implementation. If one of the nodes (or more depending on the level of parallelism and scalability) generates the configurations associated with many Markov chains, places them in a queue, and then farms them out to the other processors using task management, the tie to the number of Markov chains would be closer to being broken. Of course, since each Markov chain formally needs to get the energy of the current configuration before deciding whether to accept the configuration, the number of Markov chains will determine the number of energy computations to be performed. However, we have found that with a conservative translational and rotational step size, a high acceptance ratio can be maintained (higher than 85% is not uncommon). Hence, it is likely that precomputation of the next step in the chain could also be calculated with little loss. This would at least double the number of energy computations that are needed at any particular time. It is possible that the next step after that one (two steps removed from the current step) could also be computed ahead of time. However, this needs to be examined since one would not want to perform too many unused energy calculations. This master-slave model would require more communication than the current model – although it would be only a very small amount of data (coordinates, energy and any other relevant properties) and the latencies could be hidden through overlap with computation. However, there is likely to be less load imbalance since the slaves could be kept continually busy.

In addition to helping with the load imbalance, this model would negate the need for the zero group. The master node, in addition to determining the configurations from the many Markov chains, could also periodically determine global statistics and convergence. There would be no need for any synchronization since a processor could determine when all of the Markov chains have completed an MC cycle and do the "global" tasks while the rest of the processors continue on the next MC cycle.

An additional consideration in all of this is that the amount of I/O for the results, while not large at any particular point, can get large overall (10s–100s of GB for large simulations). Since many of the proposed petascale computers do not have much or any local disk, it might be possible to pipe that information through one (or a few) of the cores on the master node. This would limit the number of processors trying to access the disk space. The availability of asynchronous I/O would certainly be a boon to this research since the disk files are not read during the simulation, only written. Of course, the *ab initio* energy calculations themselves may require large disk access for both reading and writing, but that will need to be optimized separate from the DNTMC module and is outside the scope of this paper.

As part of an NSF PetaScale project, we are exploring these algorithmic changes to the DNTMC module. One of the challenges is to have an active task manager capability that allows for on-demand results. As part of our exploration, we are examining the use of Adaptive MPI (AMPI) [22] which implements an MPI process as a user-level thread. It tolerates or hides latency by virtualizing the threads, facilitating the migration of the threads and provides dynamic load balancing to distribute work across the algorithm. Thus, it forms an effective way of scheduling a processor (or a group of processors) in the presence of potentially large latencies. Other tools will also be examined, but many of them are not easily made compatible with a Fortran program.

4. Other model issues

In addition to the parallel scalability issues discussed above, several other issues are important to accomplish the overall science. One is the determination of convergence in the simulations. This is a long-standing problem in MC simulations. One of the traditional methods for determining convergence is to allow several Markov chains that were initiated at different configurations to continue running until the probability distribution of each Markov chain is the same as the combined, overall distribution. While this may be practical with classical potentials, it is not practical at all when *ab initio* potentials are used. One of the goals of using many Markov chains is to sample the space more effectively and *not* to calculate the same configurations many times. The ideal would be to calculate very few configurations in many Markov chains and to determine overall convergence based on some global property (such as the evaporation rate constant). Of course the challenge with this is that the properties of a given probability distribution may stay the same for quite a while and suddenly change because a particular configurational space is made available in the Markov chains. So, another traditional method for determining configuration is to double the number of configurations and see if the properties of the distribution change. While not conclusive, if the property does not change significantly, this is an indicator that convergence may be reached. Again, this has the unfortunate side effect of causing many configurations to be computed with possibly no improvement in the overall result (i.e., calculation of too many configurational points).

In our simulations, we have adapted the latter approach for our simulations. While it is not possible, due to the computational expense, to keep doubling the simulation size after about 10 million configurations, we do continue to compute until the variance in the evaporation rate is an order of magnitude lower than the actual rate. Again, while this is not conclusive that we have reached convergence it is a reasonable indicator. We have recently started a collaboration with a statistician to help in studying these convergence issues.

Another issue that must be addressed is the data that is generated from these simulations. With over 10 million *ab initio* determined points, this data is extremely valuable to the chemistry community as a whole. Creating a database out of these points is a challenging task, but one that would be useful in several ways. For example, we do not have any current plans to develop classical potentials from this data; however, it certainly could be used to do this. Making this data available would allow others to take advantage of the computations and benefit a much broader community than just our specific nucleation work. In addition, this data could be mined to determine if local minima and maxima occur on any portion of the potential energy surface. This is important for these simulations since the stationary points and the potential energy surface around them can lead to ergodicity issues when exploring surfaces at low temperature [23]. If the Markov chains become localized and do not exit the minima, the probability distributions will not be representative of the true surface.

Related to this last point is the need to be able to do as few of these large *ab intio* based simulations as possible to obtain the results that we want. We have recently developed an algorithm for performing a simple and computationally inexpensive scaling of high temperature Markov chains to determine lower temperature Markov chains and distributions [23]. In this method, we take high-temperature simulations – for example, 363 K for water – and use a weighting factor to get the results for a lower temperature – for example, 243 K. This weighting factor is

$$w_i(T_1,T_2) = \exp\left[-\frac{U_i}{k_b}\left(\frac{1}{T_2}-\frac{1}{T_1}\right)\right], \tag{5}$$

where $T_1 \geq T_2$ so that all important states at T_2 are thermally accessible at T_1. This scaling is similar to the Weighted Histogram analysis [24]. In our method, however, we weight the individual Markov chains, which allow us to examine properties such as O-O distances and energy distributions in a very straightforward manner. Of course, there are limitations to how far this temperature scaling can be applied and we are currently exploring these limitations.

5. Conclusions

In this paper, we have briefly discussed the DNTMC method, its current implementation, challenges, a possible solution for petascale computing, and other issues associated with convergence, data management, and the reuse of the simulation data. The DNTMC method is just one of a group of methods developed to explore chemical properties using a combination of quantum chemistry and statistical mechanics. These methods hold a significant promise to increase our scientific understanding of the world as well as in the use of petascale computers. However, while these large simulations can now be contemplated, there are still limitations to the parallel scalability and the huge number of computing resources that must be used to obtain solutions. We have presented one possible algorithm to reduce the amount of load imbalance as well as to improve the overall scalability of the computation.

In addition, some fundamental mathematical and data management issues will need to be solved. Convergence has long plagued the field of MC simulations. While we do not have any solutions yet, this clearly is an area of need for expert mathematical and statistical input. Additional input is also needed to explore issues associated with ergodicity and extending the usage of the existing simulations (such as the temperature scaling method described above).

Furthermore, a significant amount of data is being generated from these simulations and methods for storing and mining the information will be required to make the most use out of the data. In addition to the issues discussed previously, we have recently found that examining the individual energy distributions in each of the bins can lead to additional insight into the variation of the configurations at a given **r**. This in turn can give us an idea of the amount of configurational space that is being explored. These types of analysis techniques will need to be developed to exploit the data to its fullest.

Acknowledgments

T. L. Windus and L. D. Crosby gratefully acknowledge Iowa State University for funding and computational resources. This work was also supported by the U.S. Department of Energy (DOE) Office of Basic Energy Sciences, Chemical Sciences program, the U.S. DOE Office of Advanced Scientific Computing Research and the National Science Foundation under award OCI-0749156. This work was performed in part using the Molecular Science Computing Facility (MSCF) in the William R. Wiley Environmental Molecular Sciences Laboratory, a DOE national scientific user facility located at the Pacific Northwest National Laboratory (PNNL). PNNL is operated by Battelle for the U.S. Department of Energy. T. L. Windus dedicates this paper in memory of her mother, Bonnie A. Brown.

References

[1] Ostlund N S 1979 *Int. J. Quant. Chem. Symp.* **13** 15; see also Windus T L, Schmidt M W and Gordon M S 1995 *ACS Symp. Monograph on Parallel Computing in Chemistry* **592** 16; and Harrison R J and Shepard R 1997 *Ann. Rev. Phys. Chem.* **45** 623
[2] Schrödinger E 1926 *Annalen der Physik* **386** 109; Dirac P A M 1926 *Proc. Roy. Soc. London, A* **112** 661
[3] Schenter G K, Kathmann S M and Garrett B C 1999 *J. Chem. Phys.* **110** 7951; Schenter G K, Kathmann S M and Garrett B C 1999 *Phys. Rev. Lett.* **82** 3484
[4] Crosby L D, Kathmann S M and Windus T L 2008 Implementation of dynamical nucleation theory with quantum potentials *J. Comp. Chem.* accepted
[5] Basic Research Needs for Advanced Nuclear Energy Systems http://www.sc.doe.gov/bes/reports/files/ANES_rpt.pdf
[6] Opportunities for Catalysis in the 21st Century http://www.sc.doe.gov/bes/reports/files/OC_rpt.pdf
[7] Nanoscale Science, Engineering and Technology Research Directions http://www.sc.doe.gov/production/bes/reports/files/NSET_rpt.pdf
[8] PNNL Progress within the DOE Center of Excellence for Chemical Hydrogen Storage http://www.hydrogen.energy.gov/pdfs/progress06/iv_b_4b_aardahl.pdf
[9] Tan H-S, Piletic I R, Riter R E, Levinger N E and Fayer M D 2005 *Phys. Rev. Lett.* **94** 174501
[10] Katz J L, Lihavainen H, Rudek M M and Salter B C 2000 *J.Chem.Phys.* **112** 8363
[11] Seinfeld J H, *et al. Our Changing Planet: The U.S. Science Change Program for Fiscal Year 2007 Aerosol Radiative Forcing and Climate Change* (National Academy Press, Washington, D.C.) http://www.usgcrp.gov/usgcrp/Library/ocp2007/ocp2007.pdf
[12] Wigner E 1938 *Trans. Faraday Soc.* **34** 29; Keck J C 1960 *J. Chem. Phys.* **32** 1035; Truhlar D G, Garrett B C and Klippenstein S J 1996 *J. Phys. Chem* **100** 12771
[13] Metropolis N and Ulam S 1949 *J. Am. Stat. Assoc.* **44** 335; Metropolis N, Rosenbluth A W, Rosenbluth M N, Teller A H and Teller E 1953 *J. Chem. Phys.* **21** 1087
[14] Bylaska E J *et al.* 2006 *NWChem, A Computational Chemistry Package for Parallel Computers, Version 5.0* (Pacific Northwest National Laboratory)
[15] Kendall R A, *et al.* 2000 *Computer Phys. Comm.* **128** 260
[16] Nieplocha J, Palmer B, Tipparaju V, Krishnan M, Trease H and Apra E 2006 *Int. J. High Perf. Comput. Appl.* **20** 203
[17] Palmer B J, Kathmann S M, Krishnan M, Tipparaju V and Nieplocha J 2007 *J. Chem. Phys.* **3** 583
[18] Molecular Sciences Computing Facility, http://mscf.emsl.pnl.gov/
[19] Møller C and Plesset M S 1934 *Phys. Rev.* **46** 618; Pople J A, Binkley J S and Seeger R 1976 *Int. J. Quantum Chem.* **S10** 1

[20] Dunning T H, Jr. 1989 *J. Chem. Phys.* **90** 1007; Kendall R A, Dunning T H, Jr. and Harrison R J 1992 *J. Chem. Phys.* **96** 6796

[21] Kenny J P *et al.* 2004 *J. Comp. Chem.* **25** 1717; Alexeev Y *et al.* 2005 *J. Phys.: Conf. Ser.* **16** 536; Bernholdt D E *et al.* 2006 *Int. J. High Performance Comp. Apps.* **20** 163; Janssen C L, Kenny J P, Nielsen I M B, Krishnan M, Gurumoorthi V, Valeev E F and Windus T L 2006 *J. Phys.: Conf. Ser.* **46** 220; Peng F, Wu M S, Sosonkina M, Windus T L, Bentz J, Gordon M S, Kenny J P and Janssen 2007 *Proc. of HPC-GECO/CompFrame 200 7* 101; Kenny J P, Janssen C L, Valeev E F and Windus T L 2008 *J. Comp. Chem.* **29** 562

[22] Bhandarkar M, Kale L V, de Sturler E and Hoeflinger J 2001 Object-based adaptive load balancing for MPI programs *Proc. International Conference on Computational Science*, In Vassil Alexandrov, *et al.* eds. *Lecture Notes in Computer Science* vol. 2074 (Springer Verlag) 108-117; Huang C, Zheng G, Kumar S and Kale L V 2006 *Proc. ACM SIGPLAN Symposium on Principles and Practice of Parallel Programming* 142

[23] Crosby L D and Windus T L 2008 A Temperature Scaling Method for Markov Chains *J. Phys. Chem.* submitted

[24] Ferrenberg A M and Swendsen R H 1988 *Phys. Rev. Lett.* **61** 2635

SciDAC 2008 IOP Publishing
Journal of Physics: Conference Series **125** (2008) 012018 doi:10.1088/1742-6596/125/1/012018

Decadel climate prediction: challenges and opportunities

James W Hurrell
National Center for Atmospheric Research, Climate Analysis Section, Boulder, CO 80307-3000, USA

E-mail: jhurrell@ucar.edu

Abstract. The scientific understanding of climate change is now sufficiently clear to show that climate change from global warming is already upon us, and the rate of change as projected exceeds anything seen in nature in the past 10,000 years. Uncertainties remain, however, especially regarding how climate will change at regional and local scales where the signal of natural variability is large. Addressing many of these uncertainties will require a movement toward high resolution climate system predictions, with a blurring of the distinction between shorter-term predictions and longer-term climate projections. The key is the realization that climate system predictions, regardless of timescale, will require initialization of coupled general circulation models with best estimates of the current observed state of the atmosphere, oceans, cryosphere, and land surface. Formidable challenges exist: for instance, what is the best method of initialization given imperfect observations and systematic errors in models? What effect does initialization have on climate predictions? What predictions should be attempted, and how would they be verified? Despite such challenges, the unrealized predictability that resides in slowly evolving phenomena, such as ocean current systems, is of paramount importance for society to plan and adapt for the next few decades. Moreover, initialized climate predictions will require stronger collaboration with shared knowledge, infrastructure and technical capabilities among those in the weather and climate prediction communities. The potential benefits include improved understanding and predictions on all time scales

1. Introduction

The Earth's climate is changing, and detection and attribution studies consistently find an anthropogenic signal in the climate record of at least the last half-century. Human activities are rapidly changing the composition of the atmosphere through the burning of fossil fuels and changes in land use, such as those associated with agriculture and deforestation.

Many recent observed changes in climate are broadly consistent with increased radiative heating at the Earth's surface. These include (1) an increase in the globally averaged surface temperature of 0.75°C over the past century, including warming of 0.17°C per decade since 1979; (2) the rapid melting of glaciers in nonpolar regions around the world; (3) dramatic decreases in the areal coverage and thickness of Arctic sea ice, especially during spring and summer, and of snow cover over northern continents; (4) decreases in the maximum area covered by seasonally frozen ground in the high latitudes of the Northern Hemisphere (NH); (5) a reduction of about two weeks in the annual duration of northern lake and river ice cover; (6) an increase in global ocean surface temperatures of 0.35°C since 1979, with warming evident at all latitudes over each of the ocean basins and to depths of at least 3000 m; (7) an increase in global sea level of 0.18 cm per year since 1961, with a faster rate (0.31 cm per year) since 1993 due to thermal expansion and contributions from melting glaciers and ice caps; (8) increases in the atmospheric moisture content through increased evaporation of surface moisture; and (9) changes in extremes consistent with the warming, such as an increase in the number of heat waves globally, widespread increases in the numbers of warm nights, and rarer occurrences of cold days, cold nights, and days with frost. These many independent observations, and the physical consistency among them, form the basis for the iconic phrase from the Fourth Assessment Report (AR4) of the Intergovernmental Panel on Climate Change (IPCC) that the "warming of the climate system is unequivocal. "

The indisputable evidence of global warming and the knowledge that surface temperatures will continue to rise over the next several decades under any plausible emission scenario is now a factor in the planning of many organizations and governments. It does not imply, however, that future changes will be uniform around the globe. Regional and seasonal variations in climate will be notable, especially over periods of a few decades or less. Regional differences in land and ocean temperatures arise, for instance, from natural variability and other factors. Natural variability can result from purely internal atmospheric processes as well as from interactions among the different components of the climate system, such as those between the atmosphere and oceans or the atmosphere and land.

One example is the very strong warming of the central and eastern tropical Pacific Ocean that occurs during El Niño events. These events also produce ocean cooling over portions of the subtropics and the tropical western Pacific. Over the Atlantic, average basin-wide warming is imposed on top of strong, natural variability on multidecadal time scales. The level of natural variability, in contrast, is relatively small over the tropical Indian Ocean, where the surface warming has been steady and large over recent decades.

These differences in regional rates of sea surface temperature (SST) change perturb the atmospheric circulation so that some regions become warmer and wetter, while other regions cool and dry out. For instance, the tendency for the most prominent mode of natural atmospheric circulation variability over the NH, the North Atlantic Oscillation (NAO), to remain in a mostly positive index state over the past several decades has contributed to strong warming and wetter-than-normal conditions over much of northern Europe, while colder and drier conditions have prevailed over much of southern Europe and northern Africa. Hurrell et al. [1] find that these pronounced regional changes in climate associated with the NAO have likely been influenced by changes in the distribution of tropical heating induced by regional changes in tropical SSTs. There are many other examples as well. Atmospheric circulation changes have brought cloudier and wetter conditions in recent decades to the southeastern U.S. [2], while warming has been strong over the West (figure 1). Changes in tropical SSTs are likely a key factor in producing the Dust Bowl era of the 1930s over the central U.S. [3, 4], the devastating droughts over the Sahel and other parts of Africa since the 1980s [5, 6], and the widespread drying over subtropical land regions of the NH in more recent years [7]. Multidecadal variations in tropical and Pacific SSTs have been implicated in decadal changes in temperature and rainfall patterns across the Pacific basin [8], while slow changes in Atlantic SSTs have affected regional climate trends over parts of North America and Europe [9, 10, 11], hemispheric temperature anomalies [11] and hurricane activity in the tropical Atlantic and Caribbean [12, 13, 14].

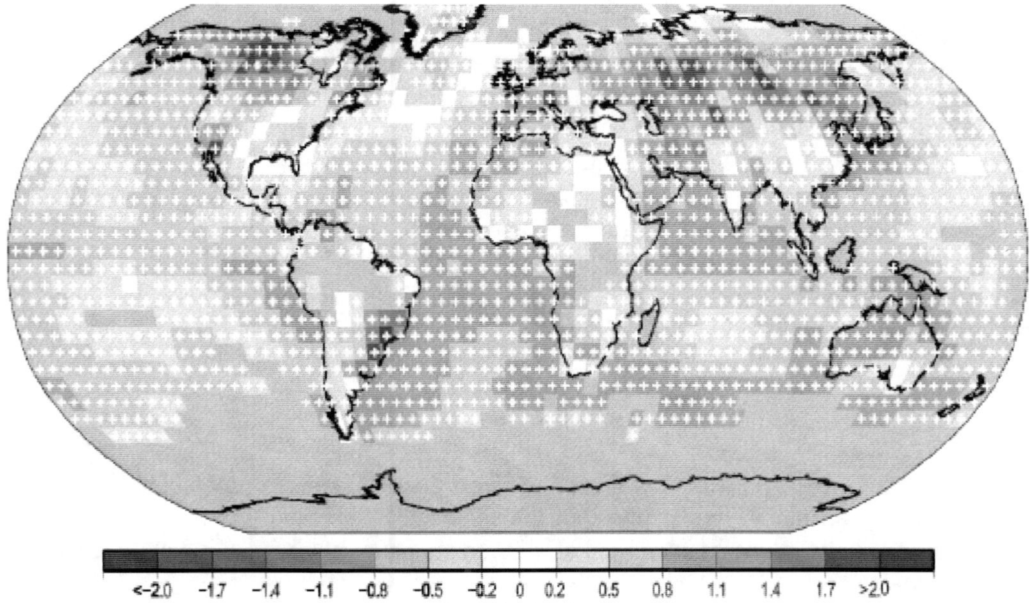

Figure 1. Linear trend of annual temperatures for 1901 to 2005 (°C century^{-1}). Areas in grey have insufficient data to produce reliable trends. Trends significant at the 5% level are indicated by white + marks. The figure is taken from Trenberth et al. [2].

Clearly, therefore, at the regional scales on which most planning decisions are made, future warming will not be smooth. Instead, it will be strongly modulated by natural climate variations, and especially those driven by the slowly varying oceans on a time scale of decades. This nonuniformity of change highlights the challenges of regional climate change that has considerable spatial structure and temporal variability. Moreover, it illustrates the need to predict patterns of SSTs around the globe as accurately as possible. Emerging evidence suggests that this might well be possible, at least for a decade or two in advance. Finding ways to account for these variations through initialized climate predictions will provide society with much better information on what kind of climate change to expect.

2. Climate system prediction and the effect of initial conditions

Good and well-documented reasons exist for the historically different approaches that have been taken toward shorter- and longer-term predictions. For weather prediction, detailed analyses of the observed state of the atmosphere are required, but uncertainties in the initial state grow rapidly over several days. Other components of the climate system are typically fixed at observed values. For climate predictions, the initial state of the atmosphere is less critical, and states separated by a day or so can be substituted. However, the initial states of other climate system components, some of which may not be critical to day-to-day weather prediction, become vital.

Therefore, at least two types of predictions may be possible on longer timescales. The first is a prediction of the internal variability of the climate system based on an initialized state of the ocean, atmosphere, land and cryosphere system. Coupled ocean-atmosphere interactions, for instance, are likely important for understanding the temporal evolution of some extratropical regional modes of climate variability, such as the NAO [15]. Within the tropics, variations in SSTs play a dominant role in regional climate, not only through ENSO, but also through other local modes of coupled variability in the Atlantic and Indian Ocean basins [16]. Moreover, land surface processes, such as the role of soil moisture, as well as changes in both sea ice and land snow cover, are important in establishing the variability of regional climate. Recent research has also indicated that the influence of the stratosphere on the state of the troposphere might also be a significant source of predictability, at least on seasonal timescales.

First attempts at "decadal prediction" with initialized coupled models have shown reduced error growth in globally averaged surface temperature over ten-year periods [17] and improved regional skill relative to predictions made with incomplete knowledge of the ocean state [18]. Decadal-scale predictability in the ocean may arise from fluctuations in gyre and overturning circulations, particularly in the Atlantic. These oceanic fluctuations produce changes in SST that could possibly lead to atmospheric predictability.

In addition to the potential sources of predictability from the initial values of the system, predictability may be derived from changes in radiative forcing. This necessitates the best possible estimates of future emissions of radiatively important pollutants, as well as modeling capabilities to accurately simulate both how these pollutants affect the global energy, carbon and sulfur cycles, and how the climate system subsequently responds to that altered forcing. In this regard, unpredictable volcanic eruptions can be a significant "wild card" to such predictions, although techniques to handle this aspect are under consideration. Despite these difficulties, it is likely that projected changes in radiative forcing are likely an important source of climate predictability, whose importance may grow with the time scale considered.

3. Some challenges

Accurate initial conditions for the global oceans are a major challenge. They could conceivably be provided by existing ocean data assimilation exercises, but salinity reconstructions prior to the ARGO float near-global upper ocean salinity and temperature data remain a significant problem. If model simulations are started prior to the availability of the ocean initial conditions, the model ocean would have to be nudged toward the observed values. How strongly this should be done and what it implies about energy conservation are research issues that are beginning to be explored.

The mass, extent, thickness, and state of sea ice and snow cover are vital at high latitudes. The states of soil moisture and surface vegetation are especially important in understanding and predicting warm season precipitation and temperature anomalies along with other aspects of the land surface. But at this time there is no direct way to provide soil moisture or ground conditions. Any information on systematic changes to the atmosphere (especially its composition and influences from volcanic eruptions) as well as external forcings, such as from changes in the sun, is also needed; otherwise these are specified as fixed at observed values. The potential errors induced by incorrect initial conditions should become less apparent as the simulations evolve but could still be evident through the course of the simulations.

Another significant obstacle is the systematic errors present in current couple climate models, so that the simulated climate is, in some cases, not close to the observed state at any given point in time. All models have systematic errors, some of which are common across the models. There are active research efforts that examine how to initialize the coupled modes of the coupled models given that they do not agree with those of nature. The basic idea is to recognize that the best state estimate for the individual component models may not be the best initial condition for coupled forecasts. Much of the research focuses on how to identify the slow manifold described by the observed estimates and the coupled model and how a mapping between them can be derived.

4. Concluding remarks

Currently, several decadal prediction activities are under way, although all are still in their infancy. The efforts typically fall into one of the following categories: using coupled models for days to decadal prediction or using numerical weather prediction models for seasonal to decadal prediction. Other approaches also are emerging such as the concept of beginning integrations with higher resolution to satisfy weather forecast requirements, then cascading down to lower-resolution versions of the model with consistent physical parameterization schemes. All of these approaches have various scientific issues, as well as advantages and disadvantages, but they all attempt to remove the distinction between weather and climate.

At the heart of this new perspective of a continuum of prediction problems is the realization that all climate system predictions, regardless of timescale, may require initialization of coupled general circulation models with best estimates of the current observed state of the atmosphere, oceans, cryosphere, and land surface. Even though the prediction problem itself is seamless, the best practical approach to it may be described as unified: models aimed at different timescales and phenomena may have large commonality but place emphasis on different aspects of the system. The potential benefits of this commonality, however, are significant and include improved predictions on all timescales and stronger collaboration and shared knowledge, infrastructure and technical capabilities among those in the weather and climate prediction communities.

References

[1] Hurrell, J W, Hoerling M P, Phillips A and Xu T 2004: Twentieth century North Atlantic climate change, Part I: Assessing determinism *Climate Dynamics* **23**:371-389, doi: 10.1007/s00382-004-0432-y

[2] Trenberth, K E, Jones P D, Ambenje P, Bojariu R, Easterling D, Tank A T, Parker D, Rahimzadeh F, Renwick J A, Rusticucci M, Soden B and Zhai P 2007 Observations: surface and atmospheric climate change *Climate Change 2007: The Physical Science Basis*. Contribution of WG 1 to the Fourth Assessment Report of the Intergovernmental Panel on Climate Change, S Solomon, D Qin, M Manning, Z. Chen, M C Marquis, K B Averyt, M Tignor and H L Miller (eds) (Cambridge University Press) 235–336, plus annex online.

[3] Schubert S D, Suarez M J, Pegion P J, Koster R D and Bacmeister J T 2004 Causes of long-term drought in the United States Great Plains *J. Climate* **17**:485-503

[4] Seager R, Kushnir Y, Herweijer C and Naik N 2005 Modeling of tropical forcingof persistent droughts and pluvials over western North America: 1856-2000 *J. Climate* **18**:4068-4091

[5] Giannini A, Saravanan R and Chang P 2003 Oceanic forcing of Sahel rainfall on interannual to interdecadal time scales *Science* **302**:1027-1030

[6] Hoerling M P, Hurrell J W, Eischeid J and Phillips A 2007 Detection and attribution of 20th century Northern and Southern African rainfall change *J. Climate* **19:**3989-4008

[7] Hoerling M P and Kumar A 2003 The perfect ocean for drought *Science* **299**:691-694

[8] Deser C., Phillips A S and Hurrell J W 2004 Pacific interdecadal climate variability: Linkages between the tropics and North Pacific during boreal winter since 1900 *J. Climate* **17**:3109–3124

[9] Enfield D B, Mestas-Nunez A M and Trimble P J 2001 The Atlantic multidecadal oscillation and its relation to rainfall and river flows in the continental U.S. *Geophys. Res. Lett.* **28**(10):2077-2080

[10]Sutton R T, and Hodson D L R 2005 Atlantic Ocean forcing of North American and European summer climate *Science* **309**:115-118, DOI: 10.1126/science.1109496

[11] Zhang R, Delworth T and Held I M 2007 Can the Atlantic Ocean drive the observed multidecadal variability in Northern Hemisphere mean temperature? *Geophys. Res. Lett.* **34**:L02709, doi:10.1029/2006GL028683

[12] Emanuel K 2005 Increasing destructiveness of tropical cyclones over the past 30 years *Nature* **436**:686–688

[13] Webster P J, Holland G J, Curry J A and Chang H-R 2005 Changes in tropical cyclone number, duration and intensity in a warming environment *Science* **309**:1844-1846

[14] Zhang R and Delworth T L 2006 Impact of Atlantic multidecadal oscillations on India/Sahel rainfall and Atlantic hurricanes *Geophys. Res. Lett.* **33**:L17712, doi:10.1029/2006GL026267

[15] Hurrell J W, *et al.* 2006 Atlantic climate variability and predictability: A CLIVAR perspective *J. Climate,* **19:**5100-5121

[16] Xie S P and Carton J A 2004 Tropical Atlantic variability: patterns, mechanisms, and impacts *Earth's climate: The ocean-atmosphere interaction* C Wang, S-P Xie and J A Carton, eds. *Geophysical Monograph* **147:**121-142
[17] Smith D M, *et al.* 2007 Improved surface temperature prediction for the coming decade from a global climate model *Science* **317**:796-799
[18] Keenlyside N S, Latif M, Jungclaus J, Kornblueh L and Roeckner E 2008: Advancing decadal-scale climate prediction in the North Atlantic sector *Nature* **453**:84-88

Generating data ensembles over a model grid from sparse climate point measurements

Meredith Franklin[1,2], V. Rao Kotamarthi[1], Michael L Stein[2] and David R Cook[1]

[1]Argonne National Laboratory, Argonne, IL 60439, USA
[2]University of Chicago Chicago, IL 60637, USA

E-mail: vrkotamarthi@anl.gov

Abstract. Parameterizations in high-end climate models can be evaluated with observed climate and meteorological data. Often this evaluation is achieved by averaging measured quantities over time and space to match the spatial and temporal resolution of a gridded climate model without any regard for the statistical errors in doing so. We present a statistical method to interpolate sparsely located surface measurements into a uniform spatial grid representative of global and regional models of the atmosphere and climate. This method provides estimates of mean values over the entire domain containing the measurement sites as well as the uncertainty in the estimated quantities at the grid locations through multiple simulations to create "data ensembles." We demonstrate this method using measurements of surface sensible heat flux over the southern Great Plains region gathered through the Department of Energy's Atmospheric Radiation Measurement (ARM) program. Application to the numerous other climate variables measured through the ARM program and the development of a software tool to streamline data management and implementation of the statistical models is also discussed.

1. Introduction

Continued improvement in the performance of climate models requires evaluation of their various parameterizations, particularly for processes driven by subgrid features such as orography, soil characteristics, vegetation, and land use. This can be achieved through validation with observed climate-related process measurements over decadal-length scales [1]. Such observations enable validation of models over large spatial scales (regional to continental) and large temporal scales (seasonal).

The Atmospheric Radiation Measurement Program Climate Research Facility (ARM ACRF) hosts a suite of instruments that continually collect data on such measurements at high temporal frequency [2]. While the ARM sites are spaced throughout a region in the U.S. southern Great Plains (SGP) designed to simulate the grid size of a climate model, their spatial scale is not suitable for model evaluation. Sites are widely scattered, with a heavily instrumented central location (figure 1) and ancillary sites arranged to cover different land use and topographic features in the region. The data collected ranges from surface measurements to remote sensing measurements used to generate vertical (3-D) distributions of parameters such as wind profiles.

Measurements are taken on different temporal scales and exhibit varying degrees of spatial heterogeneity. A given quantity, such as surface sensible heat flux, can vary widely across the domain from location to location (figure 2). Much of the variability is due to the variations in topography, local soil conditions, and geographic features. Variables such as temperature may show less deviation over a diurnal cycle at the different SGP site; however, cloud cover, soil moisture, and precipitation show similarly large spatial and temporal deviations.

In an initial sensitivity study, we examined the significance of using ARM data to constrain the Single Column Atmosphere Model (SCAM), which represents a single column of the full Community Climate System Model (CCSM) distributed with the latest version of the full National Center for Atmospheric Research Community Atmosphere Model (CAM3).

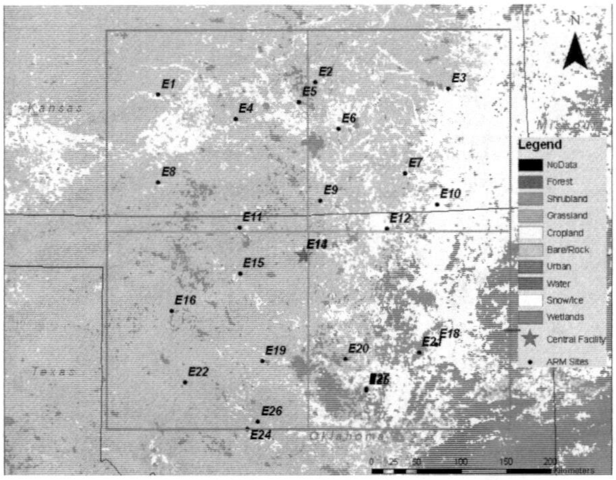

Figure 1. The ACRF SGP site. Around the central facility near Lamont, OK, measurements are made at extended facilities and a few intermediate facilities.

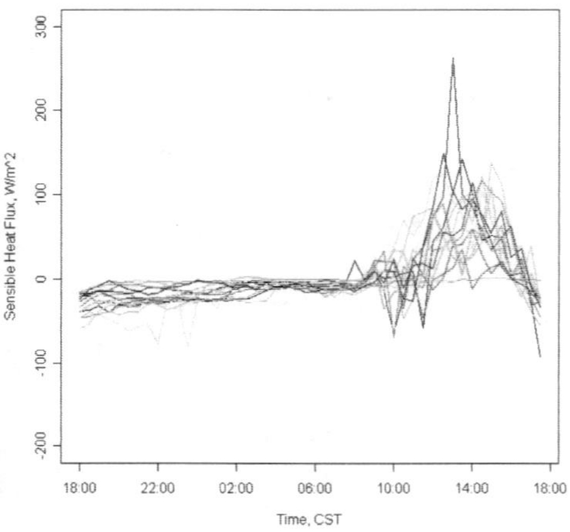

Figure 2. Diurnal variability of surface sensible heat flux recorded at 13 stations within the ARM ACRF SGP site.

The difference in the surface latent heat flux under two identical simulations is shown in figure 3, with the surface latent heat fluxes set to the derived surface latent flux (figure 3a) or constrained with measurements (figure 3b). During daytime hours, the latent heat flux in the unconstrained model is nearly 50% higher than that measured during the ARM 1995 intensive operational period (IOP).

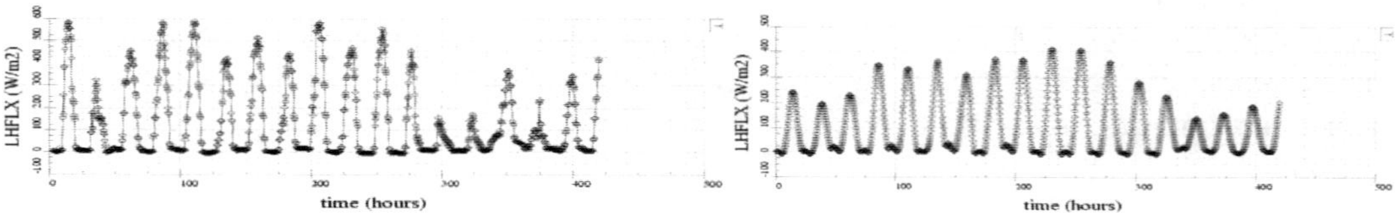

Figure 3. (a) SCAM unconstrained and (b) SCAM constrained with latent heat flux.

The difference in latent heat flux between the constrained and unconstrained model can be attributed to the SCAM's parameterization of planetary boundary layer, which may account for nearly 50% water vapor under different initializations. Thus, a strong constraint on surface heat flux would be a valuable tool for evaluating the model-calculated moisture distribution.

While the 14 ARM Energy Balance Bowen Ratio (EBBR) sites in the SGP make measurements of surface heat flux, to evaluate climate model-calculated quantities, data from the sites must be converted from a point-process to a uniform grid surface by spatio-temporal interpolation. Furthermore, it is important to have a realistic approximation of uncertainty in the interpolated grid surface.

In recent years there has been intensive research in developing statistical methods for spatio-temporal interpolation [3-5] of environmental monitoring data. These methods are based on the construction of covariance matrices that describe an observed space-time process, applied to predict values at unobserved locations and times. Often only a single prediction is generated at any unobserved location, but as Schneider [6] notes, it is extremely important to have multiple predictions in order to quantify uncertainty in the estimates. Furthermore, it is useful to evaluate climate model parameterizations with ensemble runs [7]; this can be facilitated by using "ensembles" of observed data.

We present a statistical approach for spatio-temporal interpolation of ARM data to a uniform spatial grid over the SGP domain. We illustrate our approach with sensible heat flux (h) measurements obtained from the 14 EBBR sites. Finally, by generating multiple realizations of the interpolated grid surface we quantify the uncertainty in our predictions and create data ensembles for evaluating climate model parameterizations.

2. Methods and results

All ARM data require varying degrees of preprocessing before any modeling can be conducted. In certain cases, quality flags have been included in the data product, while in others careful inspection and understanding of the instruments collecting the data are necessary to remove anomalies.

2.1 Data processing and gap-filling of flux data

The 14 SGP EBBR sites (figure 4) produce 30-minute estimates of vertical fluxes of surface sensible and latent heat. The flux estimates are calculated by the Bowen ratio technique, using measured net radiation, soil surface heat flux, and the vertical gradients of temperature and relative humidity. Several known issues with sensible and latent heat flux using the EBBR system were addressed before modeling these data. First, if the measured Bowen ratio, from which h and lambda e are derived, is close to -1, the fluxes become very large: $\lambda e = \frac{-(Q^*+G)}{1+B}$ and $h = B\lambda e$,

where Q* is net radiation, G is average ground heat transfer and B is the Bowen ratio. Second, because of the nature of the EBBR system, the data are reliable only when wind is coming over land that has sufficient fetch. Depending on the land type surrounding the instrument, there may be many instances where the flux data are questionable. A comparison of the data over "good" and "poor" fetch indicated a statistically significant difference in the flux values depending on the wind direction. Moreover, instrument issues can cause short gaps in the data series. After removing data that had any of the above data quality issues, gap filling was necessary to obtain a complete time series for modeling.

As shown in figure 2, sensible heat flux is highly variable by site, and thus we approached filling in missing values on a site-by-site basis. Heat flux is related to net radiation by $Q^* = h + \lambda e + G$, where during daytime $h \propto Q^*$ will be the dominant relationship for sensible heat flux. Thus we use the measured Q* to fill gaps in daytime sensible heat flux. To fill in values for which both h and Q* were missing during the daytime, we empirically calculated the solar radiation received at each EBBR site (insolation) by $R = S_0\varphi \cos z$, where S_0 is the solar constant, phi is the corrected Earth-sun distance, and cosz is the cosine of the zenith angle. We then regressed Q* on R and used this relationship to calculate Q_2*. Missing h for which measured Q* was also missing were filled with Q_2*. To predict the remaining missing h (occurring at night), we regressed available h on Q*, adjusting for diurnal and seasonal patterns with cubic regression spline functions and applied this fitted model to scale Q* and Q_2* to be more representative of nighttime h. A 7-day period at site E2 where missing sensible heat flux measurements were gap-filled with this method is shown in figure 5.

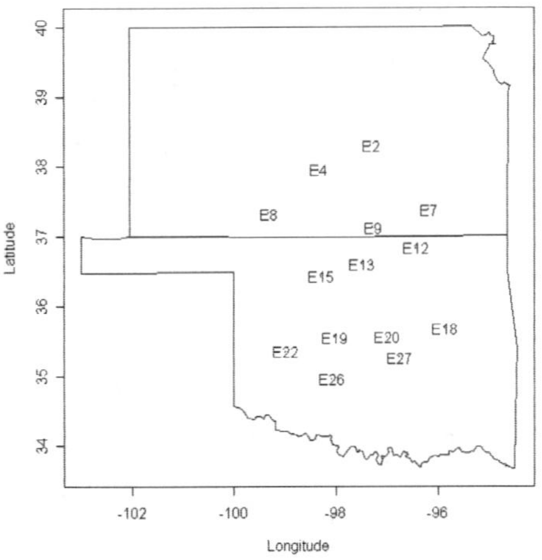

Figure 4. Locations of 14 SGP EBBR sites.

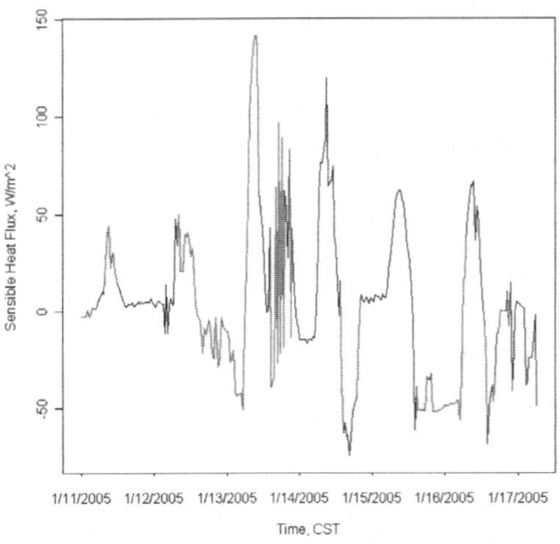

Figure 5. Observed (black) and gap-filled (red) over a 7-day period at EBBR site E2.

2.2 Spatio-temporal interpolation: from point process to model grid

We characterize the space-time process of sensible heat flux over the SGP domain, D, by a nonseparable covariance function:

$$\mathrm{cov}\{Z(s_1,t_1,Z(s_2,t_2)\} = K(s_1 - s_2, t_1 - t_2)$$

for N space-time coordinates $(\mathrm{s}_1,\mathrm{t}_1),\ldots(\mathrm{s}_\mathrm{N},\mathrm{t}_\mathrm{N})$ for all $s \in D, t \geq 0$ where

$$K(s,t) = \int_{-\pi}^{\pi} S_0(\omega)e^{i\omega t}d\omega + \int_{-\pi}^{\pi} S_1(\omega)C(|s|\gamma(\omega))e^{iu's\theta(\omega)+i\omega t}d\omega$$

is spectral in time and has an isotropic covariance function [5, 8].

To fit this model it is necessary to have a stationary Gaussian process. For the sensible heat flux data, this was achieved by removing 15 diurnal and 12 seasonal (monthly) frequencies for each year of data from each site. The time series was then converted to the spectral domain by a discrete fast Fourier transform (DFFT). Other parameters in the model, including the coherence between spectra at the sites ($\gamma(\omega)$ in $C(|s|\gamma(\omega))$, the phase ($\theta(\omega)$ in $e^{iu's\theta(\omega)}$) and $S(\omega)$) were defined with cubic spline basis functions. The isotropic spatial covariance function, C was chosen to be a Matérn function with a smoothness parameter of 1.5 (see [5]). The coordinates of each site were transformed to the UTM zone 14N for ease in computing distances between sites. A grid consisting of 1x1 km grid cells over the entire SGP domain was constructed to serve as the interpolation surface of the ARM data. This fine spatial grid fits within four 2.5-degree Community Climate System Model (CCSM) grid cells.

With all of the parameters initialized, the combined model K(s,t) was fit through optimization to maximize its likelihood. This space-time process was used to predict h at each unobserved grid cell in the 1 km grid (figure 6).

2.3 Data ensembles

We estimated uncertainty in our predictions by computing multiple simulations [8] under the modeling framework described above. These simulations are conditional on the ARM data we observe and the maximum likelihood estimates (MLE) of the spatio-temporal covariance function.

Essentially, the process Z(s*,t*), where s* is the (x, y) location of an unobserved cell in the 1 km grid and t* is any time point (either 30-minute as is observed data or another time scale) was simulated 99 times. This simulation is run for each 1 km grid cell, yielding periodic simulations of the Fourier frequencies at each unobserved grid cell.

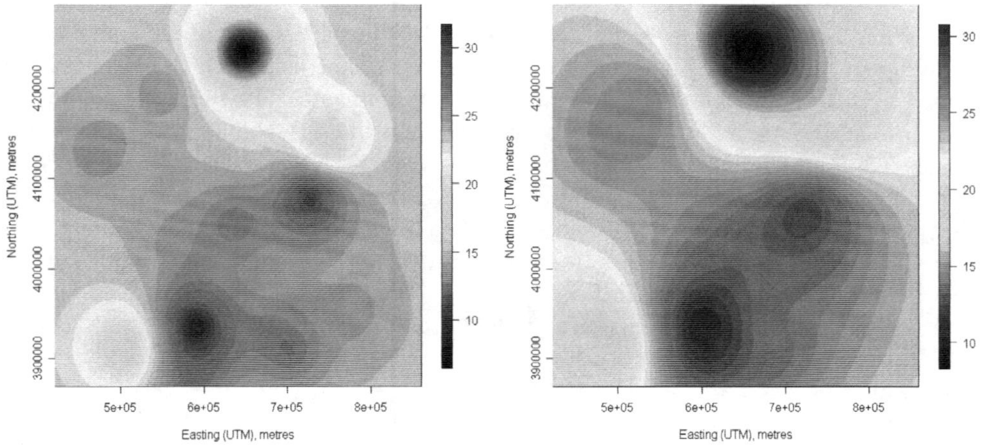

Figure 6. Two conditional simulations of h (W/m^2) into the 1 km gridded surface over the SGP domain (April 29, 2005, 23:30 CST). UTM zone 14N coordinates (meters) were used to meaningfully calculate distances between sites and to create the 1 km grid.

These simulated frequencies were reconverted back to flux values by taking the inverse DFFT and adding back in the diurnal and seasonal cycles. By sampling from the posterior distribution of the maximum likelihood estimates, uncertainty in the parameters may also be taken into account.

Figure 6 is a snapshot the spatial distribution at one time point (April 29, 2005, 23:30 CST) of 2 of the 99 simulations of h over in the 1 km grid.

2.4 Computational issues and software development

The computational aspect of this project, which will be under way by the end of 2008, will deal with the data and computational complexity of this work. While we present an example analysis of sensible heat flux, many other measured parameters are of interest to climate modelers for which we will create gridded surfaces. Furthermore, there is at minimum a 10-year time series for many of these parameters, some at 1-minute intervals. The computational complexity to conduct conditional simulations of the spatio-temporal process to generate data ensembles in the 1 km grid will require both significant processing power as well as a large amount of storage. For example, at the 1 km resolution, there are ~100,000 interpolated locations and 99 simulations of the full time series (17,520 30-minute intervals per year) at each location.

To meet these challenges, we will use component-based software engineering; an approach for managing the complexity of large-scale software systems and increasing productivity. We call this software "a data domain to model domain conversion package" (DMCP).

2.5 DMCP software

The software built for this project will be capable of computing statistical descriptions of the ARM data sets on the basis of the spatio-temporal MLE described above. Furthermore, it will have the flexibility to integrate future data sets, user-defined data dependencies (such as topography and land use) and user-defined spatial and temporal scales for the gridded interpolations. Component-based software engineering (CBSE) is an emerging approach to help manage the complexity of large-scale software systems and increase the productivity of software developers and users. In CBSE, units of software are encapsulated as "components" that interact with other components only through well-defined interfaces, allowing the internal implementation of the component to remain opaque and hiding much of the complexity of the software. These more manageable units of software can be composed together to form applications. When many components are available that conform to standardized interfaces for solvers and other such capabilities, the construction of complex applications is greatly simplified, and adaptation to different problems or computational environments can be as easy as swapping one component implementation with another more suitable to the new situation. This provides a "plug-and-play" environment for the construction of applications. The US Department of Energy has supported a component approach appropriate to tackle the problems we wish to solve — the Common Component Architecture (CCA; http://www.cca-forum.org).

The CCA is a grass-roots effort to bring the benefits of component-based software engineering to high-performance scientific computing. The CCA is specifically designed to preserve the performance of components on the same processor, to support both tightly coupled parallel and distributed computing, and to simplify the incorporation of existing code into the CCA environment. The CCA uses the Babel language http://www.llnl.gov/CASC/components/ ([9] interoperability tool to allow components written in various languages (currently Fortran, C, C++, Java, and Python) to interoperate, providing the necessary support for important scientific languages and data types.

3. Conclusions and future work

The methodology for interpolating sensible heat flux data from 14 ARM EBBR sites and creating data ensembles through multiple simulations is generalizable to all parameters gathered through the ARM program. Care must be taken to pre-process the measurement data before implementing spatio-temporal interpolation and simulation, but with the development of software using CCA, data handling and computational challenges will be addressed. Initial development of the DMCP software is under way, and we expect a first version of the applications to be available by the end of 2008. Next steps also include testing the spatio-temporal process at different spatial and temporal resolutions (i.e., over 5 km or larger grid rather than 1 km grid and at longer time averages such as hourly, every 3 hours, or daily). We will also examine the multivariate processes in space and time by exploring the dependencies of land use and topography in our interpolation. This is particularly relevant for surface flux parameters, are highly dependent upon land and soil characteristics.

Acknowledgements

Work supported by the U.S. Department of Energy, Office of Science, under contract DE-AC02-06CH11357 and the Argonne National Laboratory Environmental Science Division.

References

[1] Hurrell J W, Hack J J, Boville B A, Williamson D L and Kiehl J T 1998 The dynamical simulation of the NCAR Community Climate Model Version 3 *J. Clim.* **11**:1207-1236

[2] Macduff M C and Eagan R C 2004 *ACRF data collection and processing infrastructure* (ARM Technical Report ARM TR-046)
[3] Cressie N and Huang H 1999 Classes of nonseparable spatio-temporal stationary covariance functions *J. Amer. Stat. Assoc.* **94**:1330-1340
[4] Gneiting T 2002 Nonseparable, stationary covariance functions for space-time data *J. Amer. Stat. Assoc.* **97**:590-600
[5] Stein M L 2005 Statistical methods for regular monitoring data *J. Roy. Stat. Soc. B Met.* **67:** 667-687
[6] Schneider T 2006 Analysis of incomplete data: readings from the statistics literature *Bull. Amer. Meteor. Soc.* **87**:1410-1411 [7] Hume T and Jakob C 2005 Ensemble single column modeling (ESCM) in the tropical western Pacific: Forcing data and uncertainty analysis *J. Geophys. Res.* **110**:D13109, doi:10.1029/2004JD005704
[8] Stein M L 2008 Spatial interpolation of high-frequency monitoring data *Ann. Appl. Stat.* (in press)
[9] Dahlgren T, Epperly T and Kumfert G 2003 *Babel Users' Guide, Version 0.8.4* (CASC Technical Publication, Lawrence Livermore National Laboratory)

Journal of Physics: Conference Series **125** (2008) 012020 doi:10.1088/1742-6596/125/1/012020

Aerosol dynamics using the quadrature method of moments: Comparing several quadrature schemes with particle-resolved simulation

Robert McGraw[1], Ling Leng[2], Wei Zhu[2], Nicole Riemer[3] and Matthew West[4]

[1]Environmental Sciences Department, Atmospheric Sciences Division, Brookhaven National Laboratory, Upton, NY 11973-5000, USA

[2]Department of Applied Mathematics and Statistics, Stony Brook University, Stony Brook, NY 11794-3600, USA

[3]Atmospheric Sciences Department, University of Illinois at Urbana-Champaign, Urbana, IL 61801-3070, USA

[4]Department of Mechanical Science and Engineering, University of Illinois at Urbana-Champaign, Urbana, IL 61801-3070, USA

E-mail: rlm@bnl.gov

Abstract. The method of moments (MOM) is a statistically based alternative to sectional and modal methods for aerosol simulation. The MOM is highly efficient as the aerosol distribution is represented by its lower-order moments and only these, not the full distribution itself, are tracked during simulation. Quadrature is introduced to close the moment equations under very general growth laws and to compute aerosol physical and optical properties directly from moments. In this paper the quadrature method of moments (QMOM) is used in a bivariate test tracking of aerosol mixing state. Two aerosol populations, one enriched in soot and the other in sulfate, are allowed to interact through coagulation to form a generally-mixed third particle population. Quadratures of varying complexity (including two candidate schemes for use in climate models) are described and compared with benchmark results obtained by using particle-resolved simulation. Low-order quadratures are found to be highly accurate, and Gauss and Gauss-Radau quadratures appear to give nested lower and upper bounds, respectively, to aerosol mixing rate. These results suggest that the QMOM makes it feasible to represent the generallymixed states of aerosols and track their evolution in climate models.

1. Introduction

Gaussian quadrature provides a systematic method for approximate evaluation of integrals of the form

$$I = \int_0^\infty \sigma(m) f(m) dm \approx \sum_{i=1}^{N} \sigma(m_i) w_i . \qquad (1)$$

Here $\sigma(m)$ is a known kernel function, in this case of particle mass. The weight function $f(m)$ is the aerosol distribution function, and I is some integral property of the distribution. The approximate equality gives the quadrature approximation with abscissas and weights, m_i and w_i, respectively, determined to give correct values for selected moments of the distribution:

$$\mu_k = \int_0^\infty m^k f(m)dm . \tag{2}$$

A key feature of the quadrature method of moments (QMOM) is that one does not need to follow the distribution itself to make the approximation on the right-hand side of equation (1) – only its selected moments [1]. This makes for a highly efficient method. For example, given the $2N$ integral moments for $k = 0$ through $2N$-1, the N quadrature points (N abscissas and N weights in correspondence with these moments) are determined. Efficient algorithms are available do the inversion, and these are well conditioned for small N (here we consider N values of 1, 2, and 3). Furthermore, the approximate equality of equation (1) becomes exact for polynomial kernels of degree less than or equal to $2N$-1, as is most easily seen by substitution into equation (1) and expanding the integral in the moments defined by equation (2).

1.1 Aerosol mixing state

The remarkable efficiency of the QMOM makes this method ideal for use in atmospheric models. The QMOM has been used to represent aerosols in chemical transport models (CTMs) on the subhemispheric and regional scale, wherein several types of aerosols were tracked (e.g., sulfate, dust) using six moments/three quadrature points per type. Later the QMOM was applied to multicomponent, internally mixed populations, with remarkable levels of accuracy demonstrated in comparisons with high-resolution sectional models [2]. Even with chemical resolution capability, the QMOM was limited to internal mixtures and distribution functions defined along a single radius or mass coordinate. Capturing the general mixing states of an evolving multicomponent particle population requires, instead, multidimensional representation. It is here that statistical approaches, most prominently Monte Carlo-based particle resolved simulation and the QMOM, come into their own to offer the greatest advantage over sectional and modal methods. Monte Carlo is best for benchmark simulations, especially for complex multivariate processes in high dimension, whereas the QMOM provides economy of representation and great computational speed.

The QMOM has been applied to bivariate populations of combustion particles undergoing simultaneous coagulation and sintering. Two coordinates, surface area and volume, represent particles of mixed size and nonspherical shape [3]. Calculations were benchmarked bu using a CPU-intensive 2D model of $150 \times 150 = 22500$ sections and by Monte Carlo simulation [4]. The theory of QMOM extension to higher dimensions has been described and illustrated for the general mixing of multicomponent aerosols [5]. Model validation in higher dimension is extremely problematic and often limited to analytic test cases. Particle-resolved (PR) simulation, developed by two of us (Riemer and West), finally provides a platform for benchmarking the QMOM under realistic conditions where other approaches, for example, high-resolution sectional calculations, are impractical to carry out.

Particle-resolved simulation of three populations interacting through coagulation is described in the following section and used to benchmark QMOM accuracy for Gauss and Gauss-Radau quadratures of varying complexity (numbers of tracked moments). We find Gauss quadrature tends to overestimate (and Gauss-Radau quadrature underestimate) coagulation rate, nevertheless yielding nested bounds and very rapid convergence to the PR-simulation result.

2. Calculations

2.1 Interacting particle populations: Description of the model

Aerosol aging is an important atmospheric process that tends to make particles more hydrophilic and thus better able to serve as sites for cloud droplet condensation, thereby affecting cloud properties and resulting in shorter particle lifetimes. Coating of hydrophobic soot particles with sulfate is one such aging process. Here we consider an initial test condition consisting of two distinct particle populations: one enriched in soot (mass composition 90% soot and 10% sulfate) the other in sulfate (90% sulfate and 10% soot). These subsequently interact through coagulation to produce a third, generally mixed, population evolving ultimately to an internally mixed final state. The situation is depicted schematically in figure 1 (I and II are the initial sulfate and soot enriched populations, respectively, and III is the mixed population at some intermediate stage of development). The coagulation kernel is set proportional to the sum of the volumes of the coagulating particles – for simplicity we assume particle densities of unity and use mass and volume interchangeably. The sum kernel allows for a well-known analytic solution, but only for the total population – not for the individual subpopulations. The initial populations are taken to have identical size distributions, differing only in number and composition. The total initial population is set at 100,000 particles. As coagulation proceeds and particle number is reduced, the population is doubled several times through "cloning" to maintain good statistics throughout the PR-simulation. Figure 2 shows the evolution of particle number (corrected for cloning) and comparison with the QMOM result, which is exact for the total population. Agreement supports the excellent statistics achievable with the PR-simulation method.

2.2 Moments and quadratures

Because populations I and II are each of uniform composition, they can be treated by using the univariate moments and quadrature point assignments of equations (1) and (2). Population III, on the other hand, is generally mixed and is treated by tracking bivariate moments of the form

$$\mu_{pq} = \int_0^\infty \int_0^\infty m_1^p m_2^q f(m_1, m_2) dm_1 dm_2 . \tag{3}$$

Particle coordinates m_1 and m_2 are the masses of sulfate and soot, respectively. Quadrature points are easily assigned [5] to give correct values for the lower-order moments listed in figure 1. The blue and green points give two such assignments for population III, each capable of reproducing the particle number, distribution mean, and covariance matrix elements for that population. Figure 2 compares results using 1 point/mode quadrature (red points), tracking number and mass composition for each mode, and a more extensive calculation (blue quadrature points) that tracks 14 moments including the covariance matrix. 1-point quadrature tracking of number and mass, for each of a number of populations, is now in use in the GISS climate model [6] and plans are underway to incorporate the more extensive scheme, tracking four moments for each univariate mode and the covariance matrix for generally mixed populations (the 14 moment case of the test calculation).

Figure 3 shows the effect of using more moments and odd numbers of moments; previous applications of the QMOM all used Gauss quadrature and even numbers (per mode) of moments. Gordon [7] showed in the sixties how to obtain nested upper and lower bounds to the Laplace transform of a positive distribution function using only that functions moments. His approach is reformulated in [8] in the language of Gauss and Gauss-Radau (GR) quadratures. Briefly, the GR scheme places an abscissa at zero (the lower boundary of the distribution function domain) thereby using one less moment. The method is tested here for sequences of 1–6 univariate moments separating the total distribution now into two populations: population I and others (II and III), left panel; population II and others (I and III), right panel. Two separate calculations were performed. It is not clear that the true population decay can be written as the Laplace transforms of any fixed distribution function. Hence, the behavior illustrated in figure 3 may be fortuitous and needs to be explored for other kernels to establish its generality. Nevertheless the calculations clearly show rapid convergence of the quadrature approximation, equation (1), even with low numbers of moments.

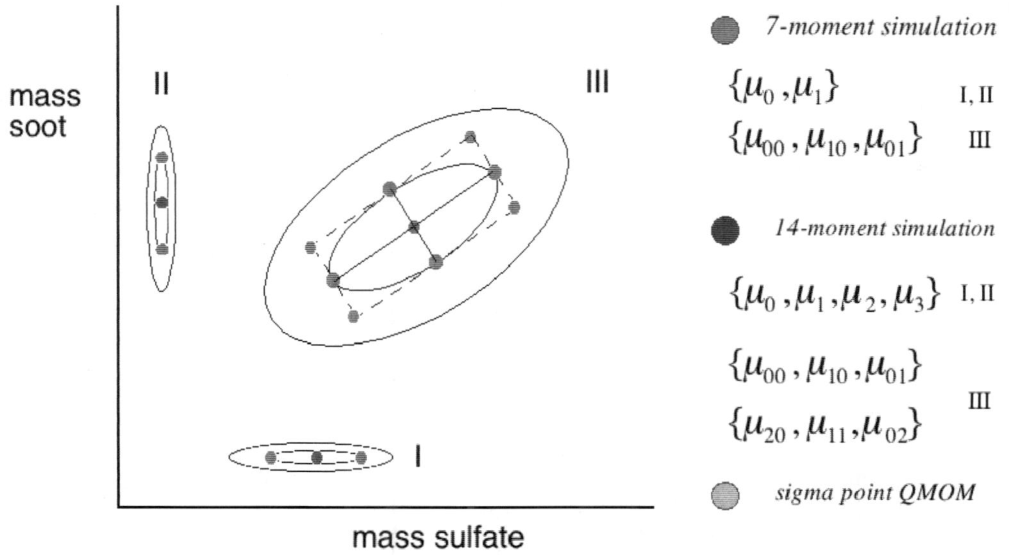

Figure 1. Schematic diagram showing the three particle populations (I, II, and III) and approximate quadrature point locations and population densities (1 and 2σ surfaces).

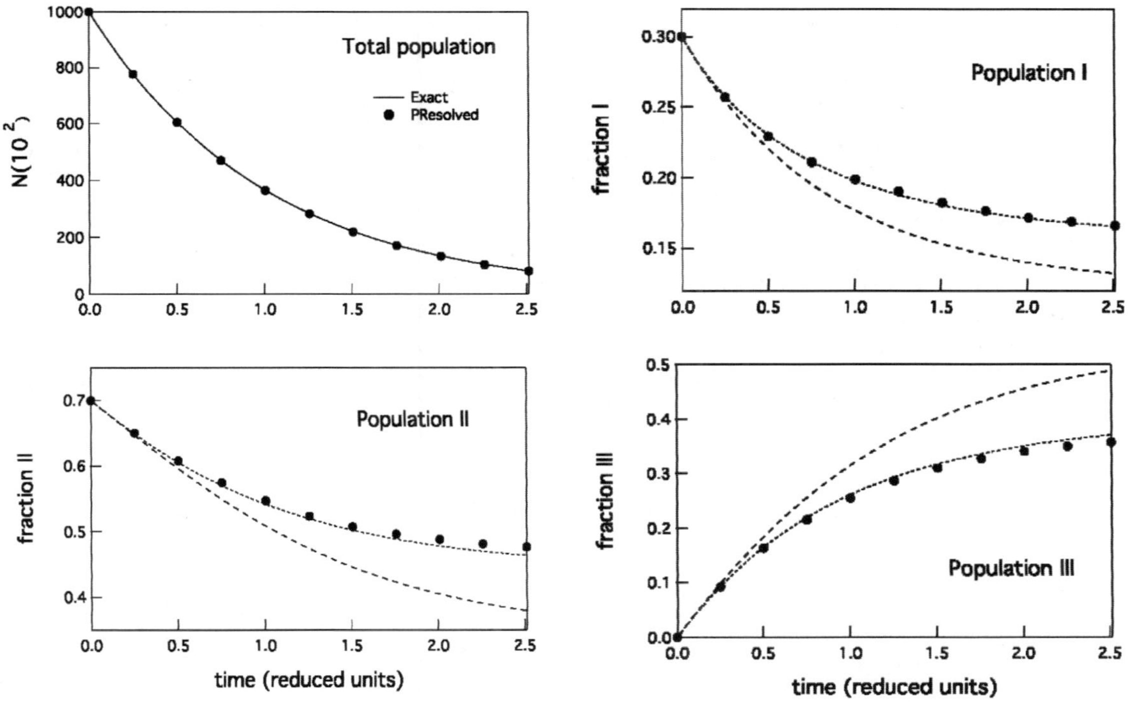

Figure 2. Evolution of total particle number (upper left panel) showing exact QMOM (curve) and particle-resolved simulation (markers) results. Remaining panels show particle number fraction for each population: long-dashed curves, QMOM with 7-moment tracking; short dashed curves, QMOM with 14-moment tracking; markers, particle resolved simulation.

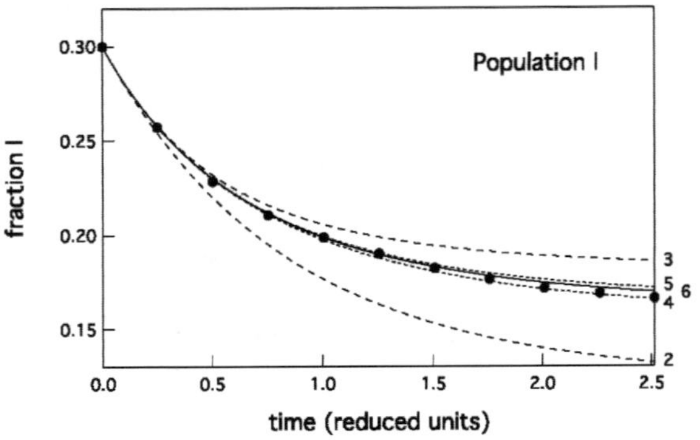

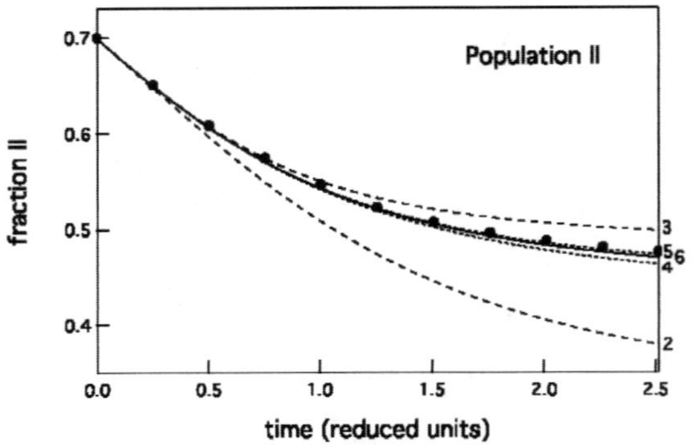

Figure 3. Testing higher-order Gauss and Gauss-Radau quadrature schemes. Curves are labeled by the number of univariate moments for each initial population used in the calculation. Gaussian quadratures use an even number of moments; Gauss-Radau quadratures an odd number. One point Gauss-Radau quadrature (not shown) yields the trivial upper bound – no decay. Note that the use of Gauss-Radau (and Gauss-Lobatto) quadratures in addition to providing bounds extends the capability of QMOM to allow for accurate integrations over boundary-limited domains [9].

A more detailed description of these results will follow in future publications.

References

[1] McGraw R 1997 Description of aerosol dynamics by the quadrature method of moments *J. Aerosol Sci.* **27**:255–265

[2] McGraw R and Wright D L 2003 Chemically resolved aerosol dynamics for internal mixtures by the quadrature method of moments *J. Aerosol Sci.* **34:**189–209

[3] Wright D L, McGraw R and Rosner D E 2001 Bivariate extension of the quadrature method of moments for modeling simultaneous coagulation and sintering of particle populations *J. Coll. Interface Sci.* **236**:242–251

[4] Rosner D E, McGraw R and Tandon P 2003 Multivariate population balances via moment and Monte Carlo simulation methods: An important sol reaction engineering bivariate example and "mixed" moments for the estimation of deposition, scavenging, and optical properties for populations of nonspherical suspended particles *Ind. Eng. Chem.* **42:**2699–2911

[5] Yoon C and McGraw R 2004 Representation of generally mixed aerosols by the quadrature method of moments, I: Statistical foundation *J. Aerosol Sci.* **35:**561–576; II: Aerosol dynamics *J. Aerosol Sci.* **35:**577–598

[6] Bauer S E, Wright D, Koch D, Lewis E R, McGraw R, Chang L-S, Schwartz S E and RuedyR 2008 MATRIX (Multiconfiguration Aerosol Tracker of mIXing state): an aerosol microphysical module for global atmospheric models *Atmos. Chem. Phys. Discuss.* **8:**9931–10003

[7] Gordon R G 1968 Error bounds in equilibrium statistical mechanics *J. Math. Phys.* **9:**65-663

[8] McGraw R 2001 Dynamics of barrier crossing in classical nucleation theory *J. Phys. Chem. B* **105:**11838–11848

[9] Arias-Zugasti M 2006 Application of orthogonal collocation in aerosol science: Fast calculation of the coagulation tensor *J. Aerosol Sci.* **37:**1356–1369

Journal of Physics: Conference Series **125** (2008) 012021 doi:10.1088/1742-6596/125/1/012021

Simulation of global cloudiness

David Randall, and **Celal Konor**
Department of Atmospheric Science, Colorado State University, Fort Collins, Colorado 80523, USA

E-mail: randall@atmos.colostate.edu

Abstract. We present a new model designed to simulate cloudiness over the entire Earth with a grid spacing of less than 5 km. The model uses a geodesic grid. Vertically propagating sound waves are filtered. The three-dimensional vorticity equation is integrated. The code can efficiently use tens of thousands of processors.

1. Introduction

During the 1960s, atmospheric general circulation models (AGCMs) had grid spacings of a few hundred kilometers and up to about 10 layers. The models included parameterizations of moist convection, stratiform clouds, radiation, and the boundary layer. In part because of limitations of computer power, they were typically run for a few simulated months, in perpetual-season mode.

Now, 40 years later, the AGCMs used for numerical weather prediction (NWP) have grid spacings about 10 times finer in the horizontal, implying about 100 times as many grid columns. The number of layers has increased to nearly 100 in some cases. As a result of advances in numerical methods, the time steps of today's high-resolution NWP models are actually larger than those used by the lower-resolution models of the 1960s. The physical parameterizations of today's models are more elaborate than those of the 1960s, but not dramatically so; they require perhaps 10 times more computation than their predecessors. Putting these factors together, we conclude that the most elaborate of today's global NWP models use about 10^4 times as much computer power as the AGCMs of the 1960s.

The AGCMs used for climate simulation today have horizontal grid spacings only slightly finer than those of the 1960s. Similarly, the number of layers has changed from O(10) to perhaps 30 or so -- not a drastic increase. As mentioned above, parameterizations have become more complex, but not drastically so. We conclude that today's AGCMs for climate simulation are only a few hundred times more expensive than their counterparts of the 1960s.

While these modeling changes have been going on, computer power has increased by about a factor of a million – from megaflops to teraflops. A typical laptop computer today is faster, has more memory and disk space, consumes much less power, is more reliable, is considerably easier to use, and costs much, much less than the cutting-edge Cray-1 of the 1970s.

How did we use the millionfold increase in computer power that has been achieved since the 1960s? For NWP models, we have argued above that increases in resolution and other changes have slowed the models down by about a factor of 10^4. The additional factor of 100 needed to go from 10^4 to a million can be accounted for in terms of shorter execution times (enabling operational use on a fixed schedule) and the introduction of forecast ensembles. For the AGCMs used for climate, we have argued that current models are a few hundred times slower than their predecessors. The additional factor of several thousand need to go from a few hundred to a million can be accounted for by much longer simulations – centuries instead of months – and by an increase in the number of simulations performed.

Over the next few decades, computers are expected to speed up by another factor of a million. What are we going to do with that next million? Although there are many worthwhile ways to spend it, we believe that a portion of it will be used to run very high-resolution models for both NWP and climate simulation. The horizontal grid spacings of these models will be in the range 1 to 5 km, and the vertical structure of the atmosphere will be resolved using on the order of 100 layers. We refer to this new generation of global models as "global cloud-resolving models," or GCRMs.

Because they have grid spacings comparable to the native scales of large clouds, GCRMs are free of the parameterizations of deep cumulus convection and stratiform cloudiness needed by conventional AGCMs. GCRMs also have no need for parameterizations of cloud overlap, or gravity-wave drag, because these can be explicitly simulated. GCRMs do still rely on parameterizations of cloud microphysics, turbulence, and radiation, but all three of these parameterization problems become much more tractable with the high resolution of a GCRM. In fact, a strength of GCRMs is that they are ideal vehicles for the implementation of more advanced parameterizations of microphysics, turbulence, and radiation. In addition, the cloud-scale statistics simulated by GCRMs can be compared directly with cloud-scale observations, which greatly simplifies the diagnosis of model deficiencies. Moreover, GCRMs are intrinsically more amenable to modularization than today's AGCMs, simply because the various physical processes represented by the parameterizations are less inter-dependent on the cloud scale.

The transition from today's AGCMs to GCRMs will be possible only after important changes in the formulations of the models. A decrease of the horizontal grid spacing by another factor of ten, together with the introduction of nonhydrostatic dynamics and major changes in parameterizations, will yield global NWP models with grid cells just a few kilometers across, at the coarse end of the "cloud-resolving" range. Taking into account the decrease in time step needed to permit simulation of the birth and death of individual large clouds, we can estimate that these cloud-resolving NWP models of the future will be about 1000 times as expensive as today's high-resolution global NWP models. Such GCRMs will become marginally practical for operational global NWP within the coming decade, which means that now is the time to begin their development.

A GCRM has already been developed at the Frontier Research Center for Global Change (FRCGC), in Japan. The model has been tested on the Earth Simulator, with provocative results [1, 2]. The FRCGC's GCRM, called NICAM (for "Nonhydrostatic Icosahedral Atmospheric Model") is constructed on a geodesic grid based on the icosahedron, similar to the one that we had developed and published before NICAM development began (see Section 2). During the development of NICAM, FRCGC scientists consulted with our team at Colorado State University, especially about the geodesic grid. In January 2008, Dr. Hiroaki Miura of FRCGC began a two-year visit with us at CSU.

At the highest tested resolution, NICAM has cells 3.5 km across. It needs approximately one teraflop-day of computer power (about 10^{17} floating point operations) to simulate one day. A snapshot of just the prognostic (time-stepped) variables fills about 1 TB of memory. Model output can easily exceed 10 PB per simulated year. Because of its high computational cost, NICAM has been run with 3.5 km cell widths for just a few simulated weeks. The model has also been run with coarser grids.

2. Model design

2.1. Introduction

We are developing a new and unique global nonhydrostatic dynamical core, based on a geodesic grid (explained below), and suitable for use with grid spacings ranging from meters to hundreds of kilometers. The core is accurate in the Taylor-series sense, physically consistent, computationally robust, and designed to perform well on modern computer architectures.

Because the core is nonhydrostatic, it can be run with very high spatial resolution (grid spacings of meters), but it can also be using using much coarser resolutions typical of current global numerical weather prediction and climate models (grid spacings on the order of tens or hundreds of kilometers). The core is therefore suitable for a wide range of applications. It is currently (June 2008) at a late stage of development.

We have not simply adapted a conventional nonhydrostatic model for use on a global domain. Our approach has been is to go back to a blank piece of paper and ask, "What is the best way to build a dynamical core, given what we know in 2008?" The model that we are proposing is quite different from anything that has been done before. However, the closest single parallel is the NICAM, the model developed by the Frontier Research Center for Global Change [3-10, 11, 12]. Like NICAM, our global model is nonhydrostatic, and uses a geodesic grid. However, there are many major differences between our model and NICAM, as mentioned briefly in several places below.

2.2. Governing equations

Our GCRM uses what we call the Unified System system of equations derived by Arakawa and Konor (2008). The Unified System is a generalization of the "pseudo-incompressible system" of Durran (1989). It unifies the quasi-hydrostatic system with the anelastic system. Unlike the anelastic system, the Unified System does not use a reference state. As shown by Arakawa and Konor]11], the Unified System filters vertically propagating sound waves, while permitting the Lamb wave and giving accurate phase speeds for both inertia-gravity waves and Rossby waves. For comparison, NICAM uses the fully compressible system.

The Unified System is further modified to solve a parabolic equation, analogous to a diffusion equation, instead of solving an elliptic equation, which is required for any truly anelastic system. The equilibrium solution of the diffusion equation is of course identical to the solution of the elliptic equation. When a large diffusion coefficient is used together with a vertically implicit and horizontally iterative algorithm, there is no need to solve a global elliptic equation, as with the conventional anelastic system. The Unified System system has tested in the model of Jung and Arakawa (hereafter JA) [13], with good results. We are now carrying it forward into the global domain.

Following the approach of JA, our model is based on solution of the three-dimensional vector vorticity equation, rather than the vector momentum equation. The motivation for this choice is that vorticity is at the core of the meteorologically important dynamical processes over a wide range of scales. Large scales are dominated by strongly rotational quasi-geostrophic motion, which is closely related to the vertical component of the vorticity. On small scales, convection arises through the generation of *horizontal* vorticity by buoyancy. Because of the key roles played by vorticity on a wide range of spatial scales, the vector (three-dimensional) vorticity equation represents the meteorologically important dynamical processes much more directly and explicitly than the momentum equation. It is therefore advantageous to formulate the discrete equations of a numerical model by starting from the vector vorticity equation.

It is necessary to predict only two components of the three-dimensional vorticity vector. The third component can then be diagnosed (given a boundary condition), by using the mathematical identity Div(Curl) = 0. In practice, we predict the two components of the horizontal part of the vorticity vector and use the identity to diagnose the vertical component of the vorticity. We also predict the divergence of the three-dimensional velocity. The three-dimensional velocity field can then be reconstructed diagnostically. As shown by JA, when vector vorticity is predicted, the elliptic equation required to filter out sound waves is expressed in terms of the vertical velocity, rather than the pressure as in the usual anelastic models based on the momentum equation. Once this equation has been solved for the vertical velocity, the components of the horizontal velocity can be diagnosed from the horizontal components of the vorticity (given a vertical boundary condition). This makes formulation of the lower boundary condition over irregular terrain significantly simpler. For comparison, NICAM integrates the momentum equation.

We are formulating the first version of the GCRM using height as the vertical coordinate. In later versions of the dynamical core, we may test alternative vertical coordinates. In the first version of the model, the effects of topography will not be included. They will be introduced in the second version.

2.3. Discretization of the sphere

The GCRM will be constructed using the geodesic grid developed by Heikes and Randall [14, 15] and

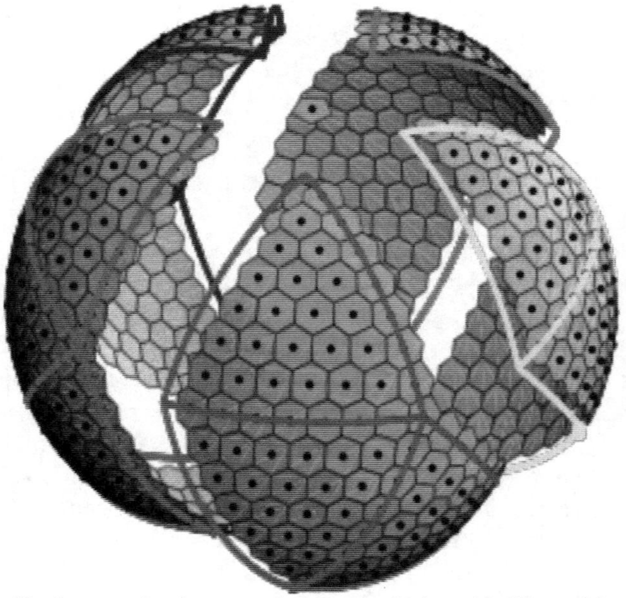

Figure 1. An "exploded" picture of a low-resolution geodesic grid. The grid can be divided into ten logically rectagular panels, as seen in the figure. Memory addressing can be organized using these panels.

extended by Ringler et al. [16] and Ringler and Randall [17, 18]; see also Randall et al. [19]. An "exploded" depiction of a (very low-resolution) geodesic grid is shown in Fig. 1. Models based on geodesic grids are now being used by several major modeling groups around the world, including NOAA's Earth System Research Laboratory in Boulder (http://fim.noaa.gov/), the U.S. Naval Postgraduate School [20], FRCGC in Japan, Richard Peltier's group in Canada [21, 22], the Max Planck Institute for Meteorologie and the Deutsche Wetterdienst in Germany [23-25]), and the University of Reading in the UK [26]. The geodesic grid is well suited to use in high-resolution models. It is also being used in a variety of other applications (e.g., [27]).

An obvious strength of the geodesic grid is that all grid cells on the sphere are very nearly the same size. Variational "tweaking" of the grid [15] can improve the uniformity of the mesh. The largest cells anywhere on the grid are only about 5% larger in area than the smallest cells on the grid (e.g., [19]).

Because of the uniform cell size, computational stability for advection is not a major issue. For example, with a grid cell L km across, and an extreme wind speed of 100 m s^{-1}, the allowed time step for advection is on the order of 10 L seconds. Computational stability for fast waves is discussed below.

A second important property of the geodesic grid is that it is quasi-isotropic. As is well known, only three

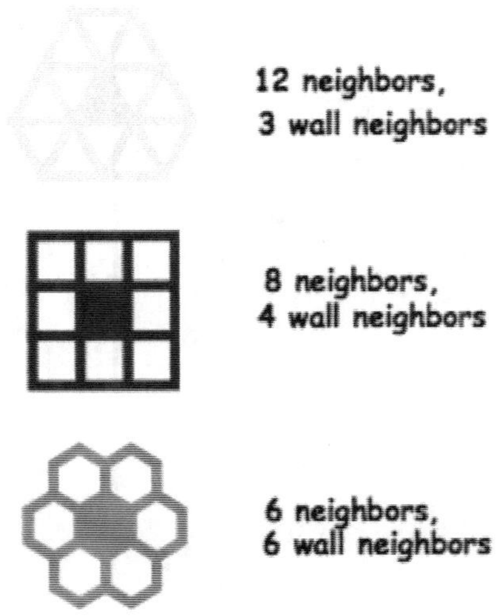

Figure 2. Small "grids" made up of equilateral triangles (top), squares (center), and hexagons (bottom). These are the only regular polygons that tile the plane. The hexagonal grid has the highest symmetry. For example, all neighboring cells of a given hexagonal cell are located across cell walls. In contrast, with either triangles or squares some neighbors are across walls, while others are across corners.

regular polygons tile the plane: equilateral triangles, squares, and hexagons. Figure 2 shows portions of planar grids made up of each of these three possible polygonal elements. On the triangular and square grids, some of the neighbors of a given cell lie directly across cell walls, while others lie across cell vertices. As a result, finite-difference operators constructed on these grids tend to use "wall neighbors" and "vertex neighbors" in different ways. For example, the simplest second-order finite-difference approximation to the gradient, on a square grid, uses only "wall neighbors"; vertex neighbors are ignored. Although it is certainly possible to construct finite-difference operators on square grids (and triangular grids) in which information from all neighboring cells is used (e.g., the Arakawa Jacobian, as discussed by Arakawa [28]), the essential asymmetries of these grids remain and are, unavoidably, manifested in the forms of the finite-difference operators. Hexagonal grids, in contrast, have the property that all neighbors of a given cell lie across cell walls; there are no "vertex neighbors." As a result, finite-difference operators constructed on hexagonal grids treat all neighboring cells in the same way; in this sense, the operators are as symmetrical and isotropic as possible. A geodesic grid on the sphere has twelve pentagonal cells in addition to the many hexagonal cells; nevertheless each cell of the geodesic grid has only wall neighbors; there are no vertex neighbors anywhere on the sphere.

It is sometimes claimed that geodesic grids permit only second-order accurate finite-difference schemes. This is simply not correct. We are currently using third-order-accurate horizontal advection schemes, and we can easily construct schemes of arbitrarily high accuracy (in the Taylor-series sense) by numerically solving straightforward linear algebra problems.

2.4. Spatial differencing

The spatial differencing of the model is a modified version of that described by Jung and Arakawa (2008; hereafter JA). The modifications serve to adapt the scheme from JA's Cartesian grid to the GCRM's

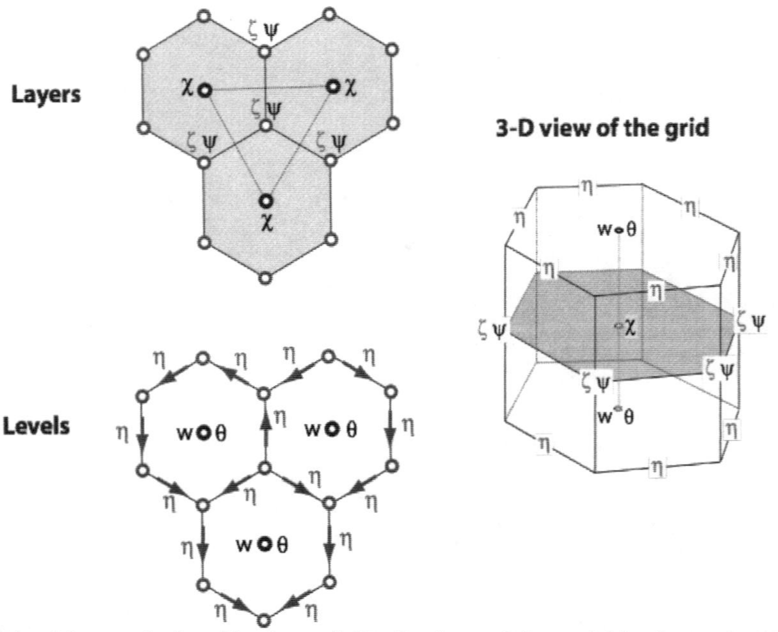

Figure 3. A sketch of the vertical and horizontal distributions of the variables in one hexagonal grid cell of the GCRM. Here eta is the tangential component of the horizontal vorticity vector, defined on the cell walls. The other notation is conventional.

geodesic grid. A sketch showing the horizontal and vertical staggering of the variables in a single hexagonal grid cell is given in figure 3. The horizontal staggering is designed to minimize the computational noise associated with computational modes. The horizontal staggering places vertical velocity, temperature, the divergence of the horizontal wind, and the associated velocity potential all at the centers of hexagonal (and pentagonal) cells. The tangential component of the horizontal vorticity vector is defined on the walls of the hexagons (and pentagons). The vertical component of the vorticity and the associated stream function are defined on the corners of the hexagons (and pentagons).

Again to minimize the noise associated with computational modes, we use the Charney-Phillips vertical staggering, in which temperature and the horizontal components of the vorticity are predicted at the same levels. The moisture variables are also predicted at those levels (for reasons discussed by Konor and Arakawa, 2000), and the vertical velocity is diagnosed there.
For comparison, NICAM uses the Lorenz staggering in the vertical, and an unstaggered grid in the horizontal.

2.5. Time differencing
Because the Unified System filters vertically propagating sound waves, it permits relatively long time steps without the need for time splitting or vertically implicit time-differencing schemes. It thus simplifies the model.

We will test a new explicit horizontal differencing scheme for the wave equation (Konor and Arakawa, 2007). The idea underlying this scheme is very simple. The trapezoidal scheme permits large time steps, but at the cost of using information from all grid points in the entire global domain, thus requiring the solution of large matrix problems that create problems on parallel computer architectures. Far-away points actually contribute very little information to the solution, however. The new scheme mimics the trapezoidal implicit scheme, but approximates it by using information only from relatively nearby points. The maximum Courant number for which computational stability is required is specified in advance; larger required Courant numbers imply larger required stencils. Details are explained by Konor and Arakawa [29]. We will use the new approach in the GCRM to permit large time steps in the presence of the fast-moving Lamb wave.

3. Status

Components of the GCRM are currently being tested on Franklin, a Cray XT-4 at the National Energy Research Supercomputing Center at the Lawrence Berkeley National Laboratory in Berkeley, California. We have already tested a high-resolution "kernel" of the model on 10,000 Franklin processors, with good scaling performance.

We expect to complete coding and preliminary testing of the dynamical cores before the proposed grant begins. Tests will be designed to establish numerical accuracy and stability, conservation properties, and parallel scaling performance. We will also be coupling the cores with our own suite of physical parameterizations, designed for use with cloud-resolving grid spacings.

References

[1] Tomita H, Miura H, Iga S, Nasuno T and Satoh M 2005 A global cloud-resolving simulation: Preliminary results from an aqua planet experiment *Geophys. Res. Lett.* **32**:L08805, doi:10.1029/2005GL022459

[2] Miura H, Tomita H, Nasuno T, Iga S-I, Satoh M and Matsuno T 2005 A climate sensitivity test using a global cloud resolving model under an aquaplanet condition *Geophys. Res. Lett.* **32**,"L19717, doi:10.1029/2005GL023672

[3] Miura H, Satoh M, Nasuno T, Noda A Tand Oouchi K 2007 A Madden–Julian oscillation event simulated using a global cloud-resolving model *Science* **318**:1763–1765

[4] Miura H, Tomita H, Satoh M, Noda, A T, Nasuno T and Iga S 2007 A short-duration global cloud-resolving simulation under a realistic land and sea distribution *Geophys. Res. Lett.* **34**:L02804, doi:10.1029/2006GL027448

[5] Nasuno T, Tomita H, Iga S, Miura H and Satoh M 2008 Convectively coupled equatorial waves simulated on an aquaplanet in a global nonhydrostatic experiment *J. Atmos. Sci.* **65**:1246–1265

[6] Satoh M, Tomita H, Miura H, Iga S and Nasuno T 2005 Development of a global cloud-resolving model—A multi-scale structure of tropical convections *J. Earth Simulator* **3**:11–19

[7] Satoh M, Nasuno T, Miura H, Tomita H, Iga S and Takayabu Y 2007 Precipitation statistics comparison between global cloud resolving simulation with NICAM and TRMM PR data *High Resolution Numerical Modelling of the Atmosphere and Ocean*, K Hamilton and W Ohfuchi, eds. (Springer) 99–109

[8] Satoh M, Matsuno T, Tomita H, Miura H, Nasuno T and Iga S 2008 Nonhydrostatic Icosahedral Atmospheric Model (NICAM) for global cloud-resolving simulations *J. Comput. Phys.* (in press.)

[9] Tomita H, Tsugawa M, Satoh M and Goto, K 2001 Shallow water model on a modified icosahedral geodesic grid by using spring dynamics *J. Comput. Phys.* **174**:579-613

[10] Tomita H, and Satoh M 2004 A new dynamical framework of nonhydrostatic global model using the icosahedral grid *Fluid Dyn. Res.* **34**:357-400

[11] Arakawa A and Konor C S 2008 Unification of the anelastic and quasi-hydrostatic systems of equations *Mon. Wea. Rev.* (under revision)

[12] Durran D R 1989: Improving the anelastic approximation *J. Atmos. Sci.* **46**:1453–1461

[13] Jung J-H and Arakawa A 2008 A three-dimensional anelastic model based on the vorticity equation *Mon. Wea. Rev.* **136**:276-294

[14] Heikes R P and Randall D A 1995 Numerical integration of the shallow water equations on a twisted icosahedral grid, Part I: Basic design and results of tests *Mon. Wea. Rev.* **123**:1862-1880

[15] Heikes R P and Randall D A 1995 Numerical integration of the shallow water equations on a twisted icosahedral grid, Part II: Grid refinement, accuracy and computational performance *Mon. Wea. Rev.* **123**:1881-1887

[16] Ringler T D, Heikes R P and Randall D A 2000 Modeling the atmospheric general circulation using a spherical geodesic grid: A new class of dynamical cores *Mon. Wea. Rev.* **128**:2471-2490

[17] Ringler T D and. Randall, D A 2002 A potential enstrophy and energy conserving numerical scheme for solution of the shallow-water equations on a geodesic grid *Mon. Wea. Rev.* **130**:1397-1410

[18] Ringler T D and Randall D A 2002 The ZM grid: An alternative to the Z grid. *Mon. Wea. Rev.* **130**: 1411-1422

[19] Randall D A, Ringler T D, Heikes R P, Jones P and Baumgardner J 2002 Climate modeling with spherical geodesic grids *Computing in Science and Engr.* **4**:32-41

]20] Giraldo F X 2000 Lagrange-Galerkin methods on spherical geodesic grids: The shallow water equations *J. Comput. Phys.* **160**:336-368

[21] Stuhne G R and Peltier W R 1996 Vortex erosion and amalgamation in a new model of large-scale flow on a sphere *J. Comput. Phys.* **128**:58-81

[22] Stuhne G R and Peltier W R 1999 New icosahedral grid-point discretizations of the shallow water equations on the sphere *J. Comput. Phys.* **144**:23-58

[23] Majewski D, Liermann D, Prohl P, Ritter B, Buchhold M, Hanisch T, Paul G, Wergen W and Baumgardner J 2002 The operational global icosahedral–hexagonal gridpoint model GME: description and Hhgh-resolution tests *Mon. Wea. Rev.* **130**:319–338

[24] Bonaventura L 2004 The ICON project: Development of a unified model using triangular geodesic grid *Proc. ECMWF Annual Seminar on Developments in Numerical Methods for Atmospheric and Ocean Modelling (*ECMWF, Reading)

[25] Bonaventura L and Ringler T D 2005 Analysis of discrete shallow-water models on geodesic Delaunay grids with C-type staggering *Mon. Wea. Rev.* **133**:2351–2373 (under revision)

[26] Thuburn J 1997 A PV-based shallow-water model on a hexagonal-icosahedral grid *Mon. Wea. Rev.* **125:**2328-2347

[27] Simpson J J, Heikes R P and Taflove A 2007 FDTD modeling of a novel ELF radar for major oil deposits using a three-dimensional geodesic grid of the Earth-ionosphere wave guide *IEEE Transaction on Antennas and Propagation* **54**:1734-1741

[28] Arakawa A 1966 Computational design for long-term numerical integration of the equations of fluid motion: Two-dimensional incompressible flow, Part I *J. Comp. Phys.* **1**:119-114

[29] Konor C S and Arakawa A 2000 Choice of a vertical grid in incorporating condensation heating into an isentropic vertical coordinate model *Mon .Wea. Rev.* **128**:3901–3910

[30] Konor C S and Arakawa A 2007 Multipoint explicit differencing (MED) for time integrations of the wave equation *Mon. Wea. Rev.* **135**:3862-3875

WRF nature run

John Michalakes[1], Josh Hacker[1], Richard Loft[1], Michael O McCracken[2], Allan Snavely[2], Nicholas J Wright[2], Tom Spelce[3], Brent Gorda[3] and Robert Walkup[4]

[1]University Corporation for Atmospheric Research (UCAR), Boulder, CO 80307, USA
[2]Performance Modeling and Characterization Lab, San Diego Supercomputer Center, La Jolla, CA 92093, USA
[3]Lawrence Livermore National Laboratory, Livermore, CA 94550, USA
[4]IBM Thomas J. Watson Research Center, Yorktown Heights, NY 10598, USA

E-mail: allans@sdsc.edu

Abstract. The Weather Research and Forecast (WRF) model is a model of the atmosphere for mesoscale research and operational numerical weather prediction (NWP). A petascale problem for WRF is a nature run that provides very high-resolution "truth" against which more coarse simulations or perturbation runs may be com-pared for purposes of studying predictability, stochastic parameterization, and fundamental dynamics. We carried out a nature run involving an idealized high resolution rotating fluid on the hemi-sphere, at a size and resolution never before attempted, and used it to investigate scales that span the k-3 to k-5/3 kinetic energy spectral transition, via simulations. We used up to 15,360 processors of the New York Blue IBM BG/L machine at Stony Brook Uni-versity and Brookhaven National Laboratory. The grid we employed has 4486 by 4486 horizontal grid points and 101 vertical levels (2 billion cells) at 5km resolu-tion; this is 32 times larger than the previously largest 63 million cell 2.5km resolution WRF CONUS benchmark [10]). To solve a problem of this size, we worked through issues of parallel I/O and scalability and employed more processors than have ever been used in a WRF run. We achieved a sustained 3.4 Tflop/s on the New York Blue sys-tem, inputting and then generating an enormous amount of data to pro-duce a scientifically meaningful result. More than 200 GB of data was input to initialize the run, which then generated output datasets of 40 GB each simulated hour. The cost of output was considered a key component of our investigation. Then we ran the same problem on more than 12K processors of the XT4 system at NERSC and achieved 8.8 Tflop/s. Our primary result however is not just scalability and a high Tflop/s number, but capture of atmosphere features never before represented by simulation, and taking an important step towards understanding weather predict-ability at high resolution.

1. Introduction

A fundamental challenge in numerical weather prediction (NWP) is to understand how (or even if) increasingly available computational power can improve weather modeling. An important enabling step toward improving that understanding is to perform a "nature run" to provide a very high-resolution standard against which coarser simu-lations and parameter sweeps may be compared for purposes of studying predictability, stochastic parameteriza-tion, and the underlying physical dynamics.

In this work we carry out a nature run at unprecedented computational scale on one of the world's largest supercomputers. We calculate an idealized high-resolution rotating fluid on the earth's hemisphere to investigate scales that span the wavenumber (k) k^{-3} (large-scale) to $k^{-5/3}$ (small-scale) kinetic energy spectral transition of the observed atmosphere using up to 15K CPUs of the IBM Blue Gene/L (New York Blue) at the New York Center for Computational Sciences (NYCCS), a cooperative effort of Stony Brook University and Brookhaven National Laboratory. Then we set a U.S. performance record of a weather code using the XT4 "Franklin" system at NERSC.

This calculation is neither embarrassingly parallel nor completely floating-point dominated, but memory bandwidth limited in the computational core, latency-bound with respect to interprocessor communication, and very I/O intensive. In these ways it is representative of many scientific calculations, and therefore achieving a high level of performance is challenging. Our primary result, however, is not just the high Tflop/s number or record-setting scalability of an atmosphere simulation, but an important step toward understanding weather predictability at high resolution.

1.1 Science motivation

A nature run that includes planetary, synoptic (barotropic and baroclinic), and near-convective scales in the midlatitudes facilitates a new generation of basic research on predictability and turbulence. It is impossible to study predictability in the real atmosphere, making computer models necessary. The superiority of either increased resolution, or more probabilistic information, can only be established through basic predictability research. A nature run including the transition between the k^{-3} and $k^{-5/3}$ spectral regimes facilitates a new generation of predictability studies that were not previously possible. For example, simple identical-twin experiments on how errors grow within the $k^{-5/3}$ regime and across the transition can now be performed. The hypothesis of enhanced mesoscale predictability near topography with increased resolution of the model can now be rigorously addressed.

It is also difficult to study turbulence in the real atmosphere, and therefore models are attractive here as well. The turbulence community faces several challenges; wave-turbulence interactions occur within the $k^{-5/3}$ regime and across the transition, for example in the jet-stream region of the atmosphere, but wave-wave interactions within the regime and across the transition are but poorly understood.

In the meantime, the growth of computational power is enabling numerical weather prediction model forecasts within the scale region defined by the observed $k^{-5/3}$ scaling in the mesoscale. Yet we have much to learn about how waves and turbulence interact, better understanding of which will affect predictability and optimal subgrid parameterization for predictive calculations within this region and across the observed transition to larger scales. Simply increasing the resolution of operational weather forecasts may not result in improved accuracy unless we can improve understanding of the physics and model parameterizations. The long-term goal of our project is therefore to produce a suite of nature runs, including runs at resolutions achievable only with petascale computing, that can serve as a basis for current and future predictability, turbulence, and parameterization study in a multi-scale environment that spans scales above and below the spectral transition. This work describes our achievement of a milestone in that project.

Previous work of Skamarock et al. [1] showed that, with dedicated computer time on a large machine and using the Weather Research and Forecasting (WRF) model [2], high-resolution nature runs that can produce the appropriate $k^{-5/3}$ spectral slope [3] are enabled. The WRF model includes a moist thermodynamic equation making it appropriate for precipitation processes. WRF is fully nonhydrostatic, so it is appropriate for deep convection and gravity wave breaking. The numerics are stable enough to make additional damping terms, ubiquitous in typical mesoscale models, less necessary. Figure 1, reproduced from that study, encapsulates some of the evidence that the computational model is stable and of high verisimilitude.

Building on that study, the nature run done here contains instances of both stratified and unstratified turbulence, facilitating their study in a rotating fluid on a sphere and in the presence of many other scales. It further allows the study of gravity waves in a realistic environment, including gravity wave breaking. This level of fidelity is unprecedented, and opens new avenues of research to improve NWP.

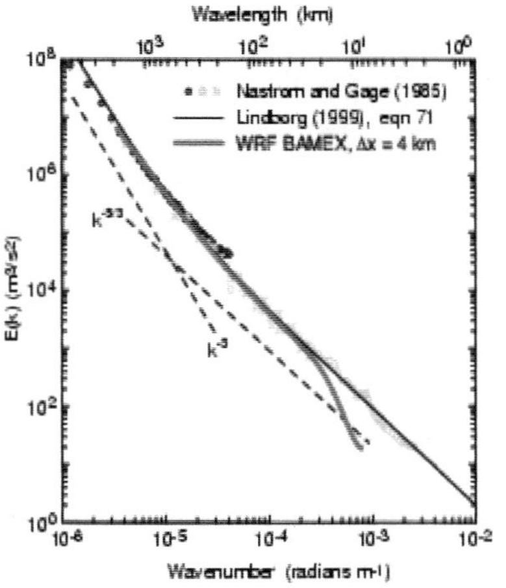

Figure 1. Spectral energy density in the WRF model compared to observations. The red curve is spectra computed from WRF forecasts at 4 km grid spacing, averaged from 3 May 2003 to 14 July 2003. Both the transition of the spectral slope from k^{-3} to $k^{-5/3}$, and the numerical dissipation range are evident. Observations from Nastrom and Gage [4] and Lindborg [5] are shown with points and the solid black curve, respectively (figure courtesy W. C. Skamarock).

1.2 Computational approach

The WRF model [1] is a limited-area model of the atmosphere for mesoscale research and operational NWP. Developed and maintained as a community model, WRF is in widespread use over a range of applications including real-time NWP, tropical cyclone and hurricane research and prediction, regional climate, atmospheric chemistry and air quality, and basic atmospheric research. The WRF model represents the atmosphere as a number of variables of state discretized over regular Cartesian grids. The model solution is computed using an explicit high-order Runge-Kutta time-split integration scheme in the two horizontal dimensions with an implicit solver in the vertical. WRF domains are decomposed over processors in the two horizontal dimensions only; and since the solver is explicit in the horizontal, interprocessor communication is logically nearest-neighbor. Each time-step involves 36 halo exchanges and a total of 144 two-way messages between neighboring processes. The decomposition is two-level: first over distributed memory patches and then again within each patch over shared memory tiles. Thus, WRF exploits hybrid parallel (message passing and multithreaded) architectures such as BG/L, though the runs conducted here were MPI only. Weather prediction codes are by nature I/O (mostly output) intensive, repeatedly writing out a time series of 3D representations of the atmosphere. WRF used Parallel NetCDF [6] as well as direct calls to MPI-IO for these runs.

2. Key aspects of the BlueGene/L architecture for NWP

The Blue Gene/L presents several opportunities and challenges for efficient implementation of NWP simulations. Details of the tightly integrated large-scale system architecture are covered elsewhere [4]. Overall, Stony Brook's BG/L platform has 18K compute nodes (36K CPUs). Here we briefly cover its general architectural aspects here, focusing on those related to our optimizations to WRF.

Each compute node is built from a single CPU ASIC and a set of memory chips. The compute ASIC features two 32-bit superscalar 700 MHz PowerPC 440 cores, with two copies of the PPC floating point unit associated with each core that function as a SIMD-like double FPU [5]. Each node has 512 MB of physical memory.

Achieving high performance requires that the application be fully domain- decomposable into data structures that can fit this relatively modest memory per node. If this can be accomplished, then the network support for scaling is an architectural strength of BG/L, which has five networks; we focus on the 3-D torus, the broadcast/reduction tree and the global interrupt for WRF optimizations. Integration of the network registers into the compute ASIC provides not only fast interprocessor communication but also direct access to network-related hardware performance monitor data. Because of limitations on deadlock-free communication, the MPI implementation uses the tree networks only for global (full-partition) collective operations

3. Computational method

As described in Skamarock et al. [1] the continuous equations solved in WRF are the Euler equations cast in a flux (conservative) form where the vertical coordinate, denoted as η, is defined by a normalized hydrostatic pressure (or mass) following Laprise [6] as

$$\eta = (ph - pht)/\mu \quad (1)$$

where $\mu = phs - pht$ and ph is the hydrostatic component of the pressure, and phs and pht are the values for the dry atmosphere at the surface and top boundaries, respectively. Following common practice we set pht = constant. η decreases monotonically from a value of 1 at the surface to 0 at the upper boundary of the model domain. Using this vertical coordinate, we express the flux form equations as

$$Ut + (\nabla \bullet Vu) + Px(p, \varphi) = FU \quad (2)$$

$$Vt + (\nabla \bullet Vv) + Py(p, \varphi) = FV \quad (3)$$

$$Wt + (\nabla \bullet Vw) + P\eta(p, \mu) = FW \quad (4)$$

$$\Theta t + (\nabla \bullet V\theta) = F\Theta \quad (5)$$

$$\mu t + (\nabla \bullet V) = 0 \quad (6)$$

$$\varphi t + \mu{-}1\ [(V \bullet \nabla\varphi) - gW] = 0 \quad (7)$$

$$(Qm)t + (\nabla \bullet VQm) = FQ \quad (8)$$

where μ(x, y) represents the mass of the dry air per unit area within the column in the model domain at (x, y). Hence the flux form variables are defined as $U = \mu u/m$, $V = \mu v/m$, $W = \mu w/m$, $\Omega = \mu\eta/m$. And m is a map-scale factor that allows mapping of the equations to the sphere (see [7]) and is given as m = (Δx, Δy) distance on the Earth.

The velocities v = (u, v, w) are the physical velocities in the two horizontal and vertical directions, respectively, ω = ˙η is the transformed "vertical" velocity, and θ is the potential temperature. Qm = μqm; Qm= Qv, Qc, Qi ..., represent the mass of water vapor, cloud, rain, ice, etc., and q∗ are their mixing ratios (mass per mass of dry air).

We also define nonconserved variables φ = gz (the geopotential), p (pressure), and α = 1/ρ (the specific volume) that appear in the governing equations. The P's are pressure gradient terms.

4. Data issues

The nature run is data intensive, with a large sum memory footprint. During the New York Blue run we used a 2-billion cell, 4486 by 4486 grid with 101 levels. Horizontal resolution (width of an individual grid cell) was 5km; the time step was 6 seconds. Experimentally, the smallest possible run was 2048 processors, requiring 287 MB/task for WRF data, not counting buffers, executable size, operating system tax, and the like. Each output interval (1 simulation hour), the model generates 40 GB of data, the size of five 3D fields (three components of wind velocity, potential temperature, and mass) and two 2D fields. Normally, a much larger set of fields is output by WRF; however, this was reduced to the barest minimum to reduce cost. On the input side, the size of the dataset read initially to start the model contained 26 3D fields and was over 200 GB.

5. Porting and tuning

To achieve high performance with WRF on BG/L, we overcame two primary hurdles: the size of main memory on BG/L and the nonscalable I/O scheme in WRF.

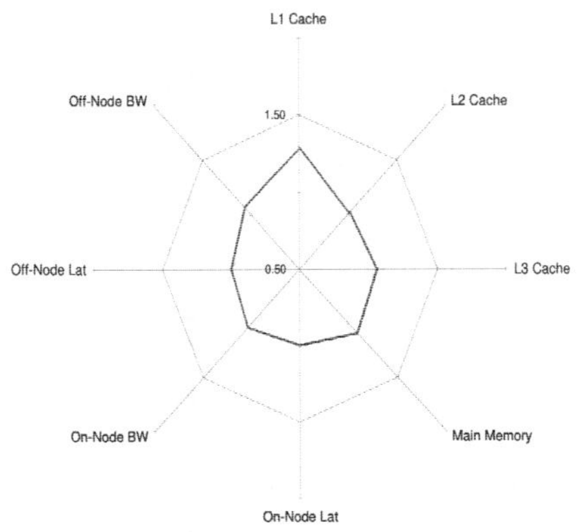

Figure 2. Performance-modeled response of WRF to doubling of processor attributes post-tuning.

Most data structures in WRF scale in memory. The domain decomposition and associated local memory extents used to dynamically allocate state arrays are calculated at run time on each process. However, each processor used to keep the global set of boundary conditions. This had been sufficient up to modest numbers (several hundreds) of processors; but with very large grid sizes on thousands of processors, the memory for arrays that store lateral boundary conditions (LBCs) ballooned out using more memory than the rest of model state combined, quickly exceeding the 512 MB physical memory limit. The solution was to fully decompose all dimensions so that each processor only stores the LBCs used in its calculation. This also involved rewriting the code for performing I/O on LBCs. With this optimization, the full state required by each processor fits in memory even on the very large grid 4486 by 4486, 101 levels, on 15K processors.

The other scaling issue we addressed was also I/O related. WRF, like many parallel applications, historically used a single-reader/single-writer scheme for distributed I/O and thus required large, un-decomposed buffers to be stored on at least one process. Again this quickly exceeded the physical memory of one BG/L node. Support for MPI-IO was added through parallel NetCDF and also direct calls to MPI-IO. Thereby we avoided the need to collect data on a single I/O task. We also encountered the 32-bit addressing limit on Blue Gene/L in the form of inability to define an MPI type large enough write a 3D field (8 GB) in one call to MPI-IO's write routine. Instead, 3D fields were written one level at a time.

Figure 2 shows a performance model developed for WRF using the PMaC methodology [8]. The figure confirms that posttuning WRF is a well-balanced, floating-point intensive code; it is most sensitive to L1 cache bandwidth (i.e., clock speed of processor) for fixed processor counts.

6. Performance measurement and results

Floating-point operations were counted using the APC performance counter library on BG/L. This library accesses the compute node ASIC's hardware performance counters to tracks several events including FPU and some SIMD and load and store operations. Because WRF uses single (32-bit) floating point precision, it was not possible to fully exploit the BG/L "double-hummer" SIMD capability.

Using 15K CPUs of Blue Gene/L at Stony Brook, we achieved 3.39 Tflop/s and set a parallelism record for number of processors running a weather code. The model was run on 6,144 and 15,360 nodes of NY Blue in co-processor mode. Only one of the two processors on each node was in use, an advantage for large- memory, high memory-bandwidth applications such as weather models.

Considering first floating-point rate by itself, the BG/L system delivered 1.49 Tflop/s on 6K processors, and 3.39 Tflop/s on 15K. Thus, scaling relative to the 6K rate was better than 90 percent efficient. Meanwhile, the output bandwidth on the BG/L, rather than degrading with increasing numbers of writers, was seen to scale as well: 242 MB/s on 6K processors and 286 MB/s on 15K. The result was good overall performance, even when the cost of writing 40 GB every simulation hour was considered: WRF still achieves 1.44 Tflop/s on 6K processors (3 percent penalty) and 3.17 Tflop/s on 15K processors (6 percent penalty), even with the cost of model output.

Next, using the tuned code, we set a speed performance record for a U.S. weather model running on the Cray XT4 “Franklin” supercomputer at the Department of Energy’s National Energy Research Computing Center (NERSC) at Lawrence Berkeley National Laboratory. Running on 12,090 processors of this 100 peak teraflops system, we achieved the important milestone of 8.8 teraflops – the fast-est performance of a weather or climate-related application on a U.S. supercomputer.

Initial science results, indicated in figure 3, demonstrate that the proposed suite of nature runs are feasible. In both panels the mid-latitude wave train is obvious, encircling the hemisphere. The top panel shows relative vorticity at approximately 5 km above ground level (model level 30) after 3 hours of simulation (model/earth time). In this figure, reds are large positive (cyclonic) vorticity and dark blues are large negative (anticyclonic) vorticity. We note the multiple scales of flow present, indicative of high resolution, which are not typically seen in simulations of this spatial extent (e.g., global). The 5 km grid spacing (ΔX) in this simulation easily resolves the spectral transition, which will typically reside at scales on the order of $20\Delta X$. Gravity waves related to baroclinic development may be present, suggested by the narrow filaments paralleling the streaks of greatest magnitude. More investigation is required to determine whether they are gravity waves, and then to understand whether they are contributing to the spectral transition from k^{-3} to $k^{-5/3}$.

The bottom panel of figure 3 shows wind speed (in m/s) at approximately jet-stream level (10 km, model level 60), with red indicating maximum wind speeds in jet streaks. In order to reach the long-term scientific objectives of this project, this wave train, the associated baroclinic development, and near-convective scales where gravity waves may break need to be simulated together. This is possible only at high resolution on domains that cover a hemisphere or the globe, so that many instances of waves, turbulence, and their interactions on both sides of the spectral transition are present. We have thus succeeded in opening new lines of investigation, and the path is now open for the next steps.

7. Conclusion

Figure 2 shows example results from the nature run. We set records for performance and scalability of WRF or any other atmospheric simulation. But our primary achievement is not performance or scalability; rather, it is new science to enable improved numerical weather prediction.

We carried out a WRF nature run that provides very high-resolution "truth" against which more coarse simulations or perturbation runs may be compared for purposes of studying predictability, stochastic parameterization, and fundamental dynamics. We studied idealized high resolution rotating fluid on the hemisphere to investigate scales that span the k^{-3} to $k^{-5/3}$ kinetic energy spectral transition of the observed atmosphere using 15K CPUs of BG/L with achieved 3.4 Tflops. Then we set a U.S. performance record of a weather code using the XT4 “Franklin” system at NERSC. We have thereby opened up new avenues of science investigations via simulation.

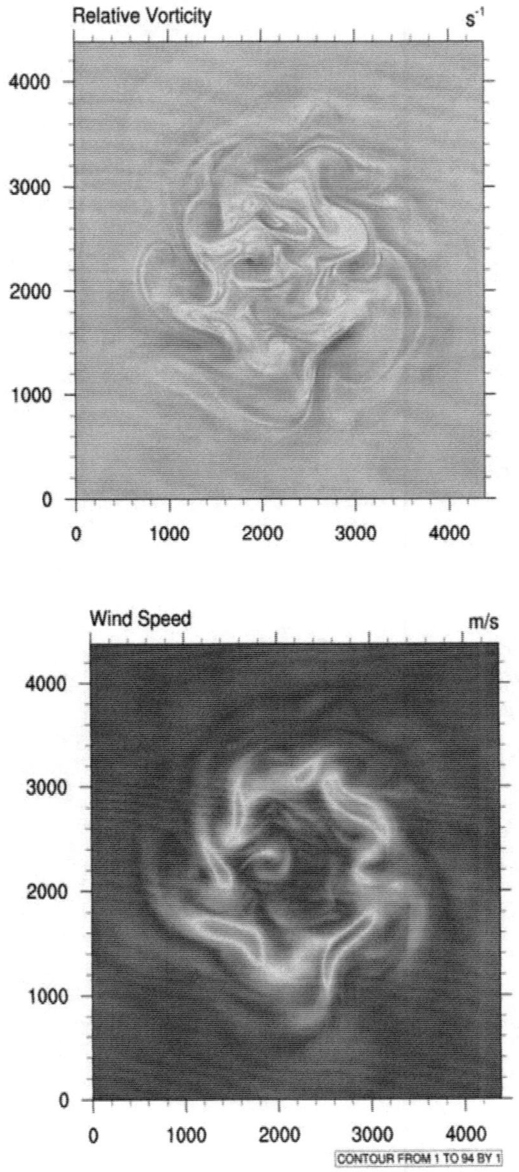

Figure 3. Relative vorticity at approximately 5 km above ground (top), and wind speed near the jet stream at approximately 10 km above ground. The entire midlatitude wave train is simulated at high resolution (ΔX=5 km), and the solutions show the range of scales in the flow.

Acknowledgments

We thank William Skamarock, whose equations and description of WRF's numerical formulation is reprinted with permission in Section 3. This work was sponsored in part by the National Science Foundation via GEO ATM

SGER award #0637994 "Feasibility of Taking the Weather Research and Forecasting (WRF) Model to Petascale" and by the OCI award entitled "The Cyberinfrastructure Evaluation Center", and in part by DOE Office of Science through the SciDAC2 award entitled Performance Engineering Research Institute (PERI). We also thank Andrew Vogelmann, Brian Colle, and Efstratios Efstathiadis of the New York Center for Computational Sciences at Stony Brook University and Brookhaven National Laboratory.

This paper updates results presented in the Gordon Bell Prize track at SC07.

References

[1] Skamarock W, *et al.* 2007 A time-split nonhydrostatic atmospheric model for weather research and forecasting applications *J. Comput. Phys.*

[2] Skamarock W 2004 Evaluating mesoscale NWP models using kinetic energy spectra *Monthly Weather Rev.*

[3] Skamarock W, Klemp J, Dudhia J, Gill D, Barker D and Wang W and Powers J 2005 A description of the advanced research WRF version 2 *NCAR Technical Note*

[4] Nastrom E D and Gage K S 1985 A climatology of aircraft wavenumber spectra observed by commercial aircraft *J. Atmos. Sci.* **42**:950-960

[5] Lindborg E 1999 Can the atmosphere kinetic energy spetrum be explained by two-dimensional turbulence? *J. Fluid Mech.* **388:**259-288

[6] Parallel NetCDF http://www-unix.mcs.anl.gov/parallel-netc

[7] Bachega E L, Chatterjee S, Dockser K, Gunnels J, Gupta M, Gustavson F, Lapkowski C, Liu G, Mendell M, Wait C and Ward T J C 2004 A high-performance SIMD floating point unit design for BlueGene/L: architecture, compilation, and algorithm design *PACT*

[8] Laprise R 1992 The Euler equations of motion with hydrostatic pressure as an independent variable *Mon. Weather Rev.* **120:**197-207

[9] Haltiner G J and Williams R T 1980 *Numerical Weather Prediction and Dynamics Meteorology* 2nd ed. (John Wiley and Sons)

[10] PMaC 1980 htttp:/www.sdsc.edu/pmac

Petascale atmospheric models for the Community Climate System Model: new developments and evaluation of scalable dynamical cores

M A Taylor[1], **J Edwards**[2] **and A St.Cyr**[2]

[1]Sandia National Laboratories, Albuquerque, NM 87185, USA
[2]National Center for Atmospheric Research, Boulder, CO 80303, USA

E-mail: `mataylo@sandia.gov`

Abstract. We present results from the integration and evaluation of the spectral finite-element method into the atmospheric component of the Community Climate System Model (CCSM). This method overcomes the atmospheric scalability bottleneck by allowing the use of a true two-dimensional domain decomposition for the first time in the CCSM. Scalability is demonstrated out to 86,200 processors with an average grid spacing of 0.25° (25 km). We present initial evaluations results using a standardized test problem with the full suite of CCSM atmospheric model forcings and subgrid parametrizations but without the CCSM land, ice, or ocean models. For this realistic setting, the true solution is unknown. Even convergence under mesh refinement is not expected, so we cannot rely on high-resolution reference solutions. Instead we focus on intermodel comparisons and use the Williamson equivalent resolution methodology to evaluate the results.

1. Introduction

Climate change models have been a crucial tool in the development of our understanding of the Earth's climate. This understanding is crucial to determining the response of the Earth's climate to increasing concentrations of greenhouse gases in the atmosphere. Further improvement of these models is needed if they are to be used to guide policy makers in the determination of safe levels of atmospheric greenhouse gas concentrations or to be used to evaluate climate change mitigation and adaptation strategies. Clearly, such next-generation models will require petascale, if not exascale, computing resources. Not withstanding the recent advent of the worlds first petascale computer, effectively using machines with O(100,000) processors for climate modeling remains a challenge because of several scalability bottlenecks present in all modern climate models – the largest bottleneck being created by the numerical methods used in the atmospheric component.

These numerical methods are encapsulated in the *dynamical core* of each atmospheric model. The dynamical core is the component that solves the partial differential equations governing the fluid dynamical aspects of the atmosphere. In addition, atmospheric models contain a suite of subgrid parametrizations for the many physical processes unresolved in an atmospheric model but which drive the dynamics. The physics parametrization suite includes such effects as convection, precipitation, and radiative forcings. Currently, most dynamical cores use latitude-longitude-based grids. These grids create a logically Cartesian orthogonal mesh suitable for a wide array of numerical methods including finite volumes and spectral transform methods. The grid lines cluster at the pole, creating a potentially severe CFL restriction on the time step. There are many successful techniques to handle this *pole problem*; however, most of them substantially degrade parallel scalability by effectively limiting the model to one-dimensional domain decomposition strategies.

Hence there is now a renewed interest in the development and evaluation of scalable dynamical cores tat use quasi-uniform grids for the surface of the sphere and avoid the pole problem. This is evidenced by the recent colloquium [1] held at the National Center for Atmospheric Research (NCAR). Nine modeling groups, close to 40 students, and dozens of keynote speakers met for two weeks to discuss numerical methods, evaluate new and existing dynamical cores, initiate the process of developing idealized test cases to facilitate model intercomparison and to objectively measure the strengths and weaknesses of proposed numerical methods. Data and results from this colloquium are being collected as of this writing. When completed, they will be made available to the community on a special portal to the Earth System Grid [2]. This will represent an unprecedented database of results from many different models all running a suite of standardized test cases at a wide range of resolutions.

Here we describe results from our recent integration of one of these dynamical cores into the CCSM. In particular, we have integrated the High Order Multiscale Modeling Environment (HOMME) [3] into the Community Atmospheric Model (CAM), the atmospheric component of the CCSM. We are motivated by the fact that HOMME is one of the most scalable dynamical cores and by previous work in which the physics parametrizations from CAM were incorporated into a spectral element model [4]. We use an improved formulation of the spectral element method that locally conserves both mass an energy [5, 6] and replaces the local element filter with a more isotropic hyperviscosity term. In addition, we make use of two key features of CAM that allow the HOMME dynamical core to be integrated directly into the full CCSM. The first is the process split dynamics/physics interface from [7], and the second is the infrastructure work to allow for unstructured grids in the CCSM described in [8]. The process split interface allows us to compute the physics parametrizations with a much larger time step as compared to the dynamics time step.

Figure 1. Tiling the surface of the sphere with quadrilaterals. An inscribed cube is projected to the surface of the sphere. The faces of the cubed-sphere are further subdivided to form a quadrilateral grid of the desired resolution. The Gnomonic equal angle projection is used, resulting in a quasi-uniform but nonorthogonal grid.

HOMME is used as a research tool to study several types of quadrilateral element-based dynamical cores. It can be run as a standalone model or as a component of CAM. Here we focus on HOMME's spectral element dynamical core running within CAM, which we refer to as CAM/HOMME. CAM/HOMME models the global circulation of the Earth's atmosphere using the three-dimensional hydrostatic primitive equations. To eliminate the pole problem, CAM/HOMME uses an equal angle cubed-sphere grid [9, 10] to discretize the horizontal directions as shown in figure 1. In the radial direction, CAM/HOMME uses the hybrid η pressure vertical coordinate system [11, 12]. CAM/HOMME uses the *compatible* spectral element formulation from [5, 6], where the compatibility property of the method is used to obtain local conservation of both mass and energy. CAM/HOMME is the first version of CAM that allows for full two-dimensional domain decomposition. It is also the first version of CAM that exactly conserves dry mass without the use of mass fixers, and conserves energy (exactly with exact time integration).

In what follows we first give an overview of the spectral element method and the form of the equations solved by CAM/HOMME. We then present results from aqua planet experiments. These simulations use the full suite of CAM forcings and subgrid physics parametrizations but without the CCSM land, ice, or ocean models. Because of the subgrid parametrizations, conventional mesh-refinement studies are not possible in this setting, so we rely on a resolution sensitivity and model inter-comparison methodology from [13, 14]. We present parallel scalability results that show the two-dimensional domain decomposition will be sufficient for CAM to achieve petascale performance levels at resolutions suitable for climate modeling.

2. Compatible spectral finite elements

The spectral finite element method is a variant of the traditional h–p finite element method. This allows the method to easily handle the nonorthogonal irregular cubed-sphere grids generated by tiling the sphere with quadrilaterals. As with traditional $h-p$ methods, it relies on globally continuous polynomial basis functions. The equations of interest are solved in integral formulation. The unique feature of the spectral element method is that the elements are restricted to quadrilaterals so that the integrals can be approximated by highly accurate Gauss-Lobatto quadrature rules within each element. This allows the construction of compactly supported, globally continuous basis, and trial functions that are orthogonal, leading to a diagonal mass matrix. The diagonal mass matrix means that iterative methods are not needed to invert the mass matrix. Thus, time-dependent problems such as the atmospheric primitive equations can be solved with explicit or semi-implicit methods. It allows the method to remain efficient while retaining the geometric flexibility of unstructured finite elements grids. It also allows the method to run efficiently at very large values of p, which is why it is referred to as a spectral finite element method.

The method has proven accurate and effective for a wide variety of geophysical problems, including global atmospheric circulation modeling [10, 15, 16, 4] ocean modeling [17, 18, 19] and planetary scale seismology [20]. The method has unsurpassed parallel performance; it was used for earthquake modeling by the 2003 Gordon Bell Best Performance winner [20] and for climate modeling by a 2002 Gordon Bell Award honorable mention [21].

The compatible spectral element formulation used in CAM/HOMME locally conserves mass to within round-off error levels. In addition, the forcing terms (such as precipitation) that change the water content of the atmosphere are incorporated into the continuity equation so that CAM/HOMME exactly conserves the atmospheric dry mass, and thus no additional mass fixers such as those used in [12] are needed. The total energy conservation in CAM/HOMME is *semi-discrete*, meaning the conservation is exact (to within round-off error levels) with exact time stepping. With the leapfrog time-stepping scheme used in CAM/HOMME there is no dissipation introduced by the method for linear and quadratic terms such as mass, tracer mass, and internal energy. For cubic terms such as kinetic energy, dissipation is introduced by the time-stepping scheme that is proportional to Δt^2, resulting in an error in the conservation of total energy proportional to Δt^2

For realistic simulations some kinetic energy dissipation is needed in order to obtain a reasonable enstrophy cascade. This is especially true in an scheme such as CAM/HOMME, where the advection operator does not dissipate any kinetic or internal energy. Without additional terms, the only dissipation in the model is a trace amount introduced by the Robert filter in the leapfrog time-stepping scheme. To introduce additional dissipation, we use a ∇^4 hyper-viscosity term modeled after that used in [12] but adapted for the spectral element method. In [6] the energy balance was examined in detail, using a quasi-realistic baroclinic instability test. With average grid spacing of 1° and an explicit time step for the dynamical core of 90 seconds, the dissipation of internal and potential energy was at the round-off error levels as expected. The dissipation of kinetic energy due to the Robert filtered leapfrog time-stepping scheme was on the order of 2×10^{-6} W/m^2, while the kinetic energy dissipation explicitly introduced by the hyper-viscosity operator is much larger, on the order of .02 W/m^2. These values should be compared to the typical atmospheric forcing in a full climate model of around 2 W/m^2.

The spectral element method is capable of refinement in both h (element size) and p (polynomial degree within each element). For the runs presented here, we fix $p = 4$ and adjust the resolution by the choice of the number of quadrilateral elements used to tile the sphere.

3. The hydrostatic primitive equations

In CAM/HOMME, we solve the hydrostatic primitive equations using the hybrid η pressure vertical coordinate system from [11, 12]. The equations, neglecting dissipation and forcing terms are given by

$$\frac{\partial \mathbf{u}}{\partial t} + (\zeta + f)\,\hat{\mathbf{k}}\times\mathbf{u} + \nabla\left(\frac{1}{2}\mathbf{u}^2 + \Phi\right) + \frac{RT}{p}\nabla p + \dot{\eta}\frac{\partial \mathbf{u}}{\partial \eta} = 0 \qquad (1)$$

$$\frac{\partial T}{\partial t} + \mathbf{u}\cdot\nabla T + \dot{\eta}\frac{\partial T}{\partial \eta} - \frac{RT_v}{c_p^*}\frac{\omega}{p} = 0 \qquad (2)$$

$$\frac{\partial}{\partial t}\left(\frac{\partial p}{\partial \eta}\right) + \nabla\cdot\left(\frac{\partial p}{\partial \eta}\mathbf{u}\right) + \frac{\partial}{\partial \eta}\left(\dot{\eta}\frac{\partial p}{\partial \eta}\right) = 0 \qquad (3)$$

$$\frac{\partial}{\partial t}\left(\frac{\partial p}{\partial \eta}q\right) + \nabla\cdot\left(\frac{\partial p}{\partial \eta}q\mathbf{u}\right) + \frac{\partial}{\partial \eta}\left(\dot{\eta}\frac{\partial p}{\partial \eta}q\right) = 0. \qquad (4)$$

There are prognostic equations for the horizontal velocity vector (tangent to the surface of the sphere) $\mathbf{u}$, the temperature T, the quantity $\frac{\partial p}{\partial \eta}$, and specific humidity q. The equations are written with $\mathbf{u}$ and T as the prognostic variables to minimize the number of cubic nonlinearities that appear. The resulting system is not written in conservation form, nor is there a prognostic equation for total energy, as this is not required to obtained energy conservation when using a compatible numerical method. CAM/HOMME also advects two additional tracers, cloud liquid and cloud ice, using equations identical to equation 4.

In the η coordinate model, $\partial p/\partial \eta$ can be thought of as a mass density. The prognostic variables are functions of η and coordinates describing the surface of the sphere. The divergence, gradient and curl operators shown are the standard two-dimensional operators acting on the surface spherical coordinates *only*. The vorticity is denoted by $\zeta = \nabla\times\mathbf{u}$, f is a Coriolis term, R, R_v, c_p, c_{pv} are constants and

$$T_v = \left[1 + \left(\frac{R_v}{R} - 1\right)q\right]T \qquad c_p^* = \left[1 + \left(\frac{c_{pv}}{c_p} - 1\right)q\right]c_p.$$

There are two additional prognostic equations for the geopotential height field Φ and the pressure vertical velocity ω,

$$\Phi = \Phi_s + \int_\eta^1 \frac{T_v}{p}\frac{\partial p}{\partial \eta}\,d\eta \qquad \omega = \mathbf{u}\cdot\nabla p - \int_{\eta_{top}}^{\eta} \nabla\cdot\left(\frac{\partial p}{\partial \eta}\mathbf{u}\right)d\eta$$

where Φ_s is the prescribed surface geopotential height (given at $\eta = 1$) and η_{top} is the η-surface at the top of the model. We note that in the η coordinate system, the pressure is given by

$$p(\eta) = A(\eta)p_0 + B(\eta)p_s$$

for a constant reference pressure p_0. The functions A and B are prescribed to control the spacing of the η-surfaces and $\eta = A(\eta)+B(\eta)$. These functions are chosen to allow the coordinate system to transition from a pure pressure coordinate system near the top of the atmosphere to a terrain following coordinate system near the surface. With this relation, equation 3 can be rewritten as a prognostic equation for surface pressure p_s and a diagnostic equation for $\dot{\eta}\frac{\partial p}{\partial \eta}$,

$$\frac{\partial}{\partial t}p_s + \int_{\eta_{top}}^{1} \nabla\cdot\left(\frac{\partial p}{\partial \eta}\mathbf{u}\right) d\eta = 0 \qquad \dot{\eta}\frac{\partial p}{\partial \eta} = -\frac{\partial p}{\partial t} - \int_{\eta_{top}}^{\eta} \nabla\cdot\left(\frac{\partial p}{\partial \eta}\mathbf{u}\right) d\eta,$$

where $\partial p/\partial t = B(\eta)\partial p_s/\partial t$. The remaining equations are discretized in the form shown in equtions 1, 2 and 4.

4. Aqua planet experiments

When using three-dimensional dynamical cores to simulate idealized configurations without subgrid-scale physics, both short time integrations and climate statistics are observed to converge under mesh refinement [22, 23]. However, physics parametrizations represent subgrid-scale processes and thus generate a significant amount of grid-scale forcings. Under mesh refinement, these forcings will remain at the grid scale, resulting in a term that can be considered as never fully resolved. Thus it is not expected that full atmospheric models will converge under mesh refinement [13]. In order to study the effects of these parametrizations in the simplest possible setting, a suite of aqua planet experiments was proposed in [24]. In these experiments, the model is run with the full physics parametrizations, but the surface boundary conditions are greatly simplified by prescribing a planet covered with water and with a fixed sea surface temperature (SST).

We follow the methodology outlined in [13, 14]. In that work, aqua planet simulations are performed with two dynamical cores in CAM: an Eulerian core based on spherical harmonics (CAM/EUL) and a finite volume core on a latitude/longitude grid (CAM/FV). Williamson shows that some of the sensitivity of the solutions to resolution is in fact due to the large physics time step and can be avoided by using a physics time step of 5 min. In order to further reduce the sensitivity, all simulations are made using the CAM Eulerian dynamical core T85 physics parametrizations tunings. The simulations are started from a previously generated aqua planet state and run for 14 months. The time-averaged statistics presented here are taken from the past 12 months. In this regime, many large scale features are observed to converge under mesh refinement, but many other quantities still show a strong dependence on resolution.

In Table 1 we summarize several global mean quantities related to precipitation, one of the most heavily parametrized processes in CAM. The CAM/EUL data was taken from [13]. First note the large monotonic signal in all the quantities as a function of grid resolution. Compared to these differences, the agreement between CAM/EUL and CAM/HOMME when grouped as shown in the table is remarkable. This initial pairing and the choice of hyperviscosity coefficient were based on the equivalence established between CAM/FV and CAM/EUL in [14]. This suggests that CAM/HOMME at 1.0°, 0.5° and 0.25° resolutions is comparable to CAM/EUL at T42, T85, and T170 resolutions, respectively. However, the table also shows that further tuning of the viscosity coefficients in CAM/HOMME at 1.0° can produce results somewhere between the T85 and T170 CAM/EUL results if the viscosity is reduced. A similar conclusion holds for CAM/HOMME at 0.5°.

Table 1. Global mean quantities as a function of resolution. CAM/HOMME results are presented with a average grid spacing of 1.9, 1 and 0.5 degrees. CAM/EUL results are presented with with a spectral truncation of T42, T85, T170 and T340. Both models were run with the CAM standard 26 vertical levels.

Model/ Resolution	Physics Time Step (minutes)	Hyperviscosity Coefficient	Convective Precipitation (mm/day)	Large-scale Precipitation (mm/day)	Cloud Fraction	Precipitable Water (mm)
EUL T42	5	1×10^{16}	1.71	1.11	0.65	20.21
HOMME 1.9	5	1×10^{16}	1.76	1.14	0.66	20.09
EUL T85	5	1×10^{15}	1.59	1.38	0.60	19.63
HOMME 1.0	5.5	1×10^{15}	1.59	1.43	0.61	19.67
HOMME 1.0	5.5	3×10^{14}	1.45	1.58	0.59	19.71
EUL T170	5	1.5×10^{14}	1.44	1.62	0.55	19.13
HOMME 0.5	5	1.5×10^{14}	1.48	1.62	0.55	19.36
HOMME 0.5	5	5.0×10^{13}	1.39	1.70	0.53	19.18
EUL T340	5	1.5×10^{13}	1.36	1.75	0.50	18.75

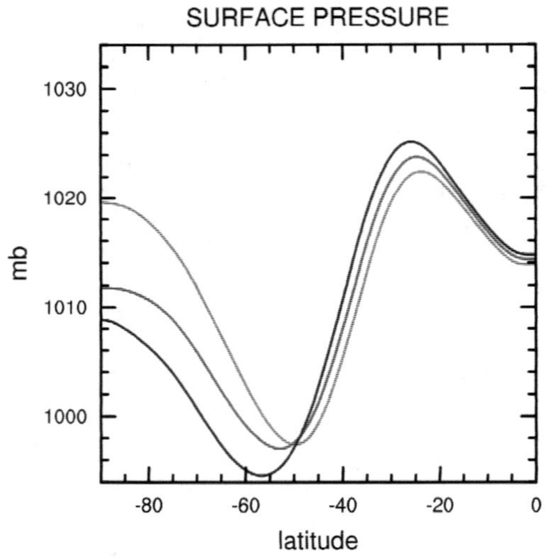

Figure 2. Time-averaged, zonally averaged surface pressure for CAM/HOMME at 1.9°, 1.0°, and 0.5° resolutions plotted respectively in orange, red, and blue.

We now turn to time-averaged, zonally averaged data. We start with the surface pressure, plotted for three different resolutions in figure 2. This should be compared with figure 3 in [13]. As with the global means, there is a strong signal with resolution, but relative to these differences there is good agreement between the CAM/HOMME and CAM/EUL results. When the equivalent resolutions are paired, the pairing is similar to what was found with the global means, except that the 0.5° result as actually quite close to the T340 result.

The time-averaged, zonally averaged total cloud fraction is plotted in figure 3. This should be compared with figure 4 in [13] and figure 2 in [14]. Here again there is a strong signal with resolution, but now the agreement between CAM/HOMME and CAM/EUL is not as clear cut. The decrease in cloud fraction with resolution is quite similar, but CAM/HOMME has consistently fewer clouds at $-15°$ latitude. CAM/HOMME 0.5° does obtain good agreement for the rest of the domain as compared to the T170 and CAM/FV 0.5° results.

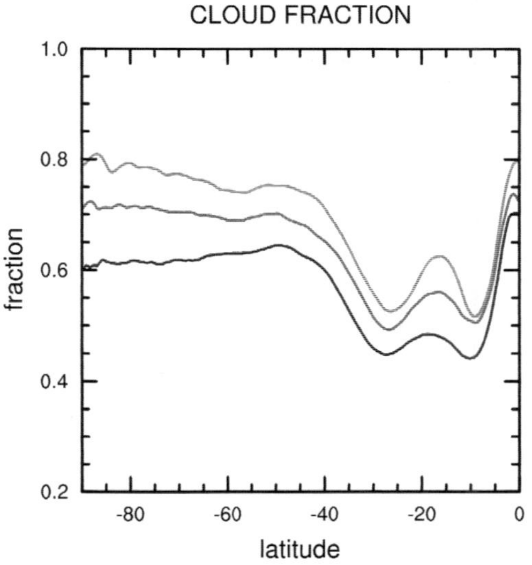

Figure 3. Time averaged, zonally averaged total cloud fraction for CAM/HOMME at 1.9°, 1.0° and 0.5° resolutions plotted respectively in orange, red and blue.

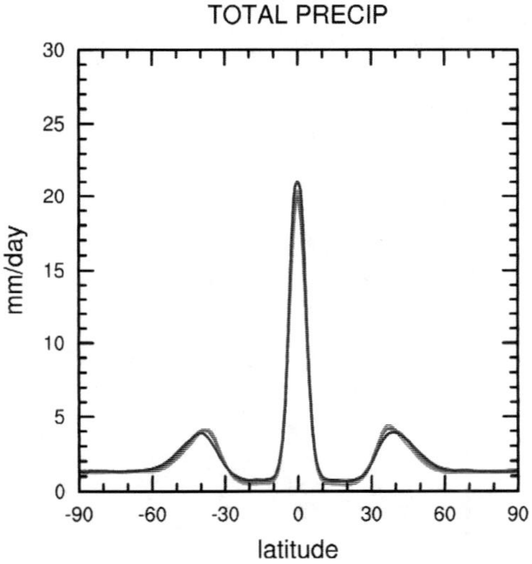

Figure 4. Time averaged, zonally averaged total precipitation for CAM/HOMME at 1.9°, 1.0°, and 0.5° resolutions plotted, respectively in orange, red, and blue.

The time-averaged, zonally averaged total precipitation is plotted in figure 4. This plot should be compared with figure 2 in [13]. The data is consistent with the conclusions drawn in that paper: the larger scales in the equatorial region have converged and the variation in the global averages comes from the remainder of the domain.

In Fig. 5, we plot precipitation probability density functions. This plot should be compared with Fig. 5 in [14]. As with all the other metrics we have examined, there is a strong dependence on resolution, but by choosing different resolution pairs the models can be seen to agree well. The pairings between CAM/HOMME and CAM/EUL continue to hold, except the 1.0° result with a 10 mm bin size appears much closer to the T340 than T170 result.

Grid imprinting is of special concern for methods on non-latitude/longitude grids, as any such grid will have some anisotropy and must contain a few special points. For the cubed-sphere grid, the main concern is anomalous forcing of an $m = 4$ mode caused by the 8 special points at the cube corners that appear in the mid-latitudes. To examine this issue, we plot the time averaged pressure vertical velocity (ω) field from both CAM/HOMME and CAM/EUL in figure 6. This field is very sensitive to grid-scale noise and is susceptible to grid imprinting. The noise in ω

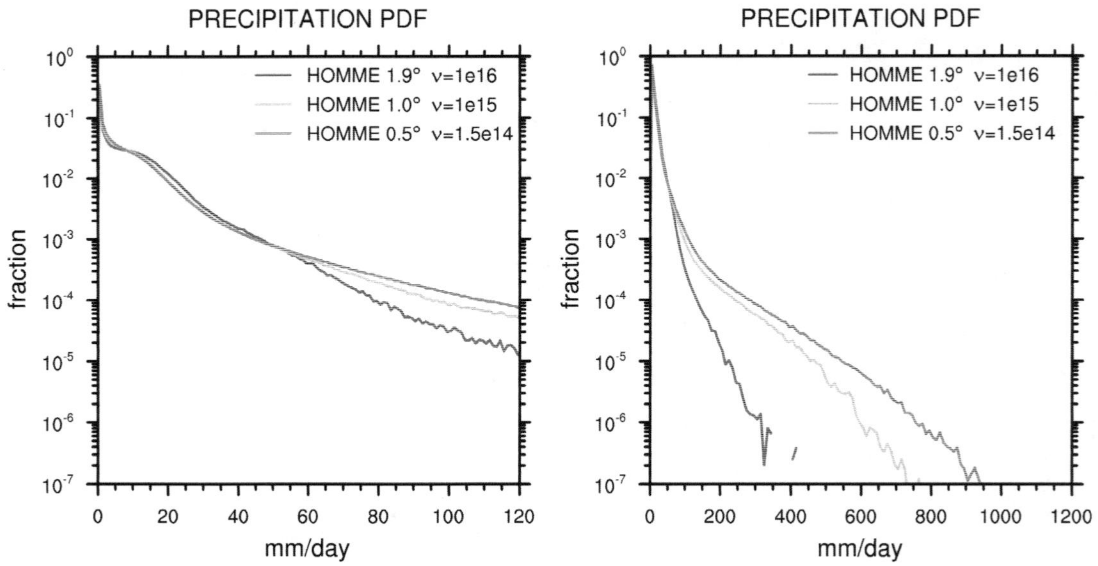

Figure 5. CAM/HOMME probability density function of 6-hour mean precipitation in the equatorial band between ±10 latitude, for 1.9, 1.0 and 0.5 degree resolutions. The left figure uses a bin size of 1 mm, while the right figure uses a bin size of 10 mm.

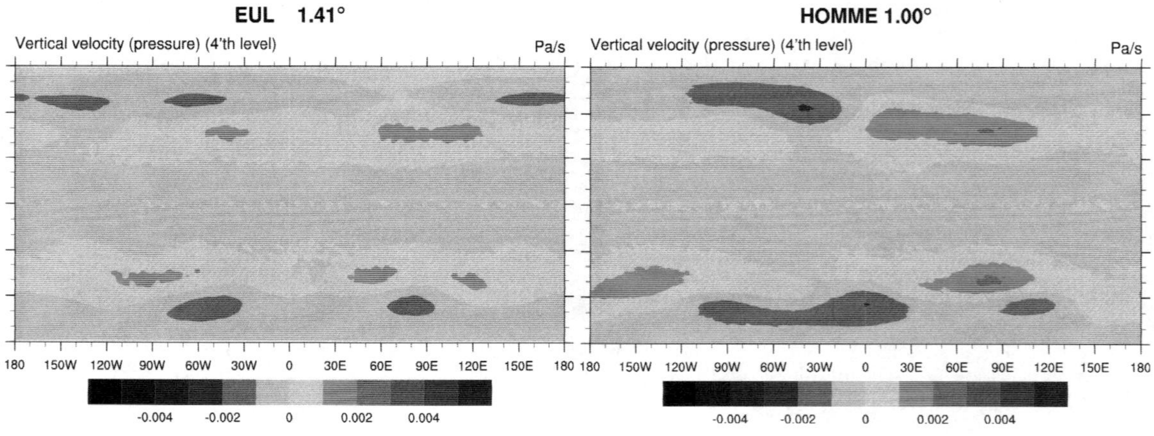

Figure 6. Time-averaged pressure vertical velocity (ω) on an η-surface near the model top for CAM/EUL at T85 and CAM/HOMME at 1.0°.

is not significantly different from that found in the near perfectly isotropic CAM/EUL model. We attribute this to the fact that CAM/HOMME is using relatively high-order (fourth-order) numerics and is an $h-p$ type finite-element method that has a long history of dealing with grids far more unstructured and anisotropic than the cubed sphere.

5. Scalability results

The parallel scalability of CAM/HOMME is demonstrated in Fig. 7. We present benchmark results from the IBM BG/L system installed at Lawrence Livermore National Laboratory (106,496 nodes in a torus interconnect, each with two 700 MHz processors) and *Tbird*, an Intel/Infiniband-based Linux cluster (4480 nodes, each with dual 3.6 GHz Intel EM64T processors) installed at Sandia National Laboratories. The figure shows the integration rate

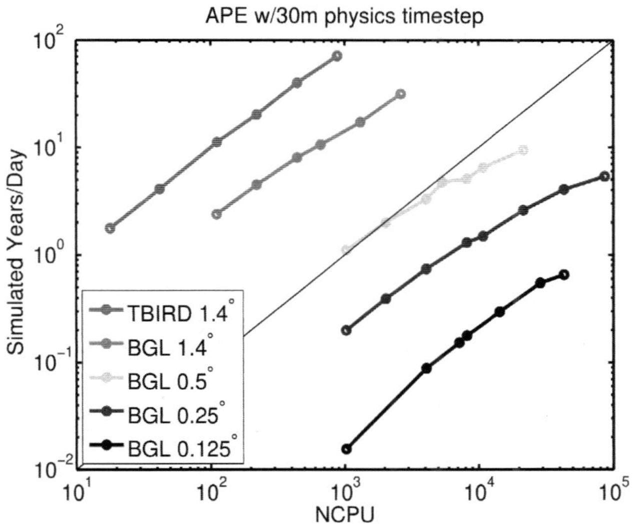

Figure 7. Integration rate measured in simulated years per day is plotted as a function of the number of processors, for several different resolutions. Because of memory constraints, the BG/L results were obtained using only one processor per node, except for the largest run on 86,200 processors that used both processors on each node.

(simulated years per day) obtained as a function of number of processors at several resolutions. Each curve represents a fixed resolution, so that the the total amount of work was kept constant while the processor count was increased. Data from the BG/L runs at 1.4, 0.5 and 0.25 degree resolutions use up to 1536, 21600, and 86400 processors, respectively. These processor counts result in parallel decompositions as fine as one element per processor, the limit of our domain decomposition approach. The thin black line represents perfect parallel scalability. As can be seen in the figure, at moderate processor counts CAM/HOMME obtains close to perfect scalability. Reasonable scalability is obtained even at the limit of one element per processors.

6. Conclusions

The highly scalable dynamical core HOMME has been integrated into CAM, the atmospheric component of the CCSM. Initial results from aqua planet simulations using the full physics subgrid parametrization suite suggest that CAM/HOMME produces results in agreement with other dynamical cores in CAM. In particular, for many quantities, CAM/HOMME at 1° resolution is comparable to CAM/EUL at a resolution between T85 and T170, while CAM/HOMME at 0.5° resolution is comparable to CAM/EUL at a resolution between T170 and T340. The CAM/HOMME sensitivity of all quantities examined to mesh refinement is also similar to CAM/EUL and CAM/FV.

CAM/HOMME is the first non-latitude/longitude dynamical core integrated into CAM, demonstrating the CAM physics/dynamics interface is capable of handling unstructured grids. The parallel scalability of HOMME is maintained within this interface. At 0.25°, CAM/HOMME scales effectively to 86,200 processors, or 1 element per processor. This research suggests that at slightly higher resolutions CAM/HOMME will easily scale to the processor counts expected for upcoming petascale computers. At resolutions of 0.5° and 0.25°, The LLNL BG/L computer is capable of running CAM/HOMME at integration rates of at least 5 simulated years per day, the conventional target for climate integrations.

Acknowledgments

This work supported in part by the DOE/BER SciDAC Project, *Modeling the Earth System.*

References

[1] National Center for Atmospheric Research 2008 NCAR ASP 2008 summer colloquium on numerical techniques for global atmospheric models http://www.wiki.ucar.edu/display/dycores/Home
[2] National Center for Atmospheric Research 2008 ESG-CET gateway prototype for the 2008 NCAR ASP dycore workshop http://dycore.ucar.edu
[3] Thomas S and Loft R 2005 *J. Sci. Comput.* **25** 307–322
[4] Wang H, Tribbia J J, Baer F, Fournier A and Taylor M A 2007 *Monthly Weather Review* To appear
[5] Taylor M A and Fournier A 2007 *In preparation*
[6] Taylor M A, Edwards J, Thomas S and Nair R 2007 *J. Phys. Conf. Ser.* **78**
[7] Williamson D 2002 *Mon. Wea. Rev.* **130** 2024–2041
[8] Worley P H and Drake J B 2005 *International Journal of High Performance Computing Applications* **19** 187 – 201
[9] Rancic M, Purser R and Mesinger F 1996 *Q. J. R. Meteorol. Soc.* **122** 959–982
[10] Taylor M, Tribbia J and Iskandarani M 1997 *J. Comput. Phys.* **130** 92–108
[11] Simmons A J and Burridge B M 1981 *Mon. Wea. Rev.* **109** 758–766
[12] Collins W D, Rasch P J, Boville B A, Hack J J, McCaa J R, Williamson D L, Kiehl J T, Briegleb B, Bitz C, Lin S J, Zhang M and Dai Y 2004 Description of the NCAR Community Atmosphere Model (CAM 3.0) Tech. Rep. TN–464+STR NCAR
[13] Williamson D 2008 *Tellus* To appear
[14] Williamson D 2008 *Tellus* To appear
[15] Giraldo F X 2001 *International Journal for Numerical Methods in Fluids* **35** 869–901
[16] Thomas S and Loft R 2002 *Journal of Scientific Computing* **17** 339–350
[17] Haidvogel D, Curchitser E N, Iskandarani M, Hughes R and Taylor M A 1997 *Atmosphere-Ocean Special* **35** 505–531
[18] Iskandarani M, Haidvogel D, Levin J, Curchitser E N and Edwards C A 2002 *Computing in Science and Engineering* **4**
[19] Molcard A, Pinardi N, Iskandarani M and Haidvogel D 2002 *Dynamics of Atmospheres and Oceans* **35** 97–130
[20] Komatitsch D, Tsuboi S, Ji C and Tromp J 2003 *Proc. ACM / IEEE Supercomputing SC'2003 conference*
[21] Loft R, Thomas S and Dennis J 2001 *Proc. ACM / IEEE Supercomputing SC'2001 conference*
[22] Held I M and Suarez M J 1994 *Bull. Amer. Meteor. Soc.* **75** 1825–1830
[23] Jablonowski C and Williamson D L 2006 *Q. J. R. Meteorol. Soc.* **132** doi:10.1256/qj.06.n
[24] Neale R B and Hoskins B J 2001 *Atmos. Sci. Lett.* **1** doi:10.1006/asle.2000.0019

Journal of Physics: Conference Series **125** (2008) 012024 doi:10.1088/1742-6596/125/1/012024

The use of the Climate-science Computational End Station (CCES) development and grand challenge team for the next IPCC assessment: an operational plan

W M Washington[1], J Drake[2], L Buja[1], D Anderson[3], D Bader[4], R Dickinson[5], D Erickson[2], P Gent[1], S Ghan[6], P Jones[7] and R Jacob[8]

[1]National Center for Atmospheric Research, Boulder, CO 80837, USA
[2]Oak Ridge National Laboratory, Oak Ridge, TN 37831, USA
[3]NASA Headquarters, Washington, DC 20024, USA
[4]Lawrence Livermore National Laboratory, Livermore, CA 94550, USA
[5]Georgia Institute of Technology, Atlanta, GA 30332, USA
[6]Pacific Northwest National Laboratory, Richland, WA 99352, USA
[7]Los Alamos National Laboratory, Los Alamos, NM 87545, USA
[8]Argonne National Laboratory, Argonne, IL 60439, USA

E-mail: wmw@ucar.edu

Abstract. The grand challenge of climate change science is to predict future climates based on scenarios of anthropogenic emissions and other changes resulting from options in energy and development policies. Addressing this challenge requires a Climate Science Computational End Station consisting of a sustained climate model research, development, and application program combined with world-class DOE leadership computing resources to enable advanced computational simulation of the Earth system. This project provides the primary computer allocations for the DOE SciDAC and Climate Change Prediction Program. It builds on the successful interagency collaboration of the National Science and the U.S. Department of Energy in developing and applying the Community Climate System Model (CCSM) for climate change science. It also includes collaboration with the National Aeronautics and Space Administration in carbon data assimilation and university partners with expertise in high-end computational climate research.

1. Introduction

The creation of computational end stations provides the scientific applications software and world-class computational resources needed for researchers in various areas of science. In its fourth year, the Climate Science Computational End Station (CCES) advances climate science through both an aggressive model development activity and an extensive suite of climate model application simulations. Subproject allocations are subject to external scientific review of specific proposed computational experiments. The required computing resources and the scientific goals of the proposed CCES are compatible and consistent with the DOE Office of Science Strategic Plan, the Federal Climate Change Science Program (CCSP) Strategic Plan, and the CCSM Strategic Business Plan for 2004–2008 and with the CCSM Science Plan (http://www.ccsm.ucar.edu).

This multiyear project includes an ambitious simulation plan emphasizing model application with an ongoing core model development that envisions future needs and scales the CCSM for use on petascale computer systems. The CCSM3 is the current state-of-the-science model that accurately simulates global to continental aspects of the physical climate system. Extensive simulations with higher resolution versions of the current model are under investigation to identify and correct bias errors that are unacceptable for future simulations that require subcontinental and smaller-scale accuracy. Using high-resolution ocean and ice models, researchers also plan studies of abrupt climate change, in which large global scale changes occur within a few decades.

The CCSM3 requires the specification of changes to many factors that determine climate change, such as the atmospheric concentrations of greenhouse gases, atmospheric aerosols, and the distribution of

1

vegetation types on the land surface. In the Earth system version of CCSM, these factors result from dynamic physical, chemical and biological processes. These are being added to the current CCSM4 as part of this project, and will be used in future versions to realistically predict future climates. Significant biogeochemical cycles influencing climate change involve carbon, sulfur, and nitrogen cycles. Of particular importance to U.S. energy policy is the correct simulation of the global carbon cycle and its feedbacks to the climate system, including its variability and modulation by ocean and land ecosystems. This continuing model development and extensive testing of the CCSM system to include recent new knowledge about these processes is at the cutting edge of climate science research and is a principal focus of the CCES. We are planning to develop CCSM4 in time for the fifth IPCC assessment.

The CCES *predicts future climates using emissions scenarios and other changes resulting from energy policies options.* We deliver simulations improving the scientific basis, accuracy and fidelity of climate models and perform climate change simulations that directly inform national and international science policy, thereby contributing to the DOE, NSF, and NASA science missions.

2. Resources

The Computational Climate End Station manages resources on the three primary systems that fall under the INCITE program: Cray XT4 Jaguar at NCCS, the Cray XT4 Franklin at NERSC, and the IBM Blue Gene/P at Argonne. The CCES seeks to provide a single source version of the CCSM that runs on all these platforms and that can be used for applied research in climate change. Development time for model extension, validation, and verification, and software scalability forms a significant part of the requested allocation on each platform. The experiments and simulations use all these machines, demonstrating production on cutting-edge high-performance computer (HPC) platforms and taking advantage of the unprecedented computational power for more comprehensive, high resolution climate prediction results. The Earth System Grid infrastructure links climate simulation data and analysis capabilities across the DOE laboratories and NCAR.

3. Brief summaries of subprojects

The CCES is the primary allocation that develops and deploys the Community Climate System Model (CCSM) for use in DOE and NSF climate change simulation experiments. The CCSM is the primary computational instrument. We provide the capability tools and simulation frameworks needed to advance climate-change science and carry out leading climate-science research program in conjunction with our partners and sponsoring agency programs. We execute the high-priority simulations for the DOE that require the high-end computational capability of scalable high-performance computers,. In addition, we provide outreach to the larger research community through simulation products analysis of model results, and workshops.

2.1 Climate change (NCAR/LLNL)

The Climate Change Working Group and the climate modeling community has initiated an experimental design process for the next round of coordinated climate change experiments that will involve atmosphere-ocean general circulation models (AOGCMs) and Earth system models (ESMs) for assessment in the IPCC Fifth Assessment Report in 2013. In consultation with the Integrated Assessment Modeling community that formulates scenarios, a set of climate change experiments and scenarios is proposed that will address both short-term climate change (out to about 2030), with high-resolution coupled models (0.5–1.0 degrees horizontal resolution in the atmosphere) to look at regional climate change and processes involved with producing changes of weather and climate extremes, and long-term climate change (out to 2300) with medium-resolution coupled models (about 1.5–2.0 degrees horizontal resolution in the atmosphere) to quantify feedbacks involved with carbon cycle and to assess climate change adaptation in mitigation scenario experiments. We are planning future climate change simulations to the year 2300, short-term high-resolution climate predictions, coupled ice sheet simulations, and 1000-year biogeochemistry simulations. The simulation data is freely and openly provided to the international climate science community via the Earth System Grid.

2.1 SciDAC biogeochemistry (LLNL/PNNL/ORNL/NCAR/LANL)

The SciDAC Biogeochemistry subproject focuses on the impacts of land use change and dynamic global vegetation on the climate system. This is a follow-on phase to the Carbon-Land Model Intercomparison Project (C-LAMP) and the C4MIP model analysis and evaluation projects. It is hoped that this will be instrumental in defining the biogeochemical/carbon experiments for the IPCC AR5. This will assess the performance using observational datasets of the CLM-CN, CASA, and IBIS models in the CCSM framework to gauge the strength of the feedback between the carbon cycle and the climate system.

2.2 SciDAC Earth system model with sulfur (NASA/ORNL)

As part of the SciDAC CCSM Consortium project, a groundbreaking Earth system model simulation (and model development), is being carried out. The model focuses on dimethyl sulfide interactions to extend and augment the current plans of the CCSM Biogeochemistry working group by focusing on the sulfur cycle and aerosol feedbacks. The simulations will also be used as a testbed for atmospheric chemistry, particularly new aerosol indirect effects.

2.3 NASA carbon assimilation (ORNL)

Our NASA team is pursuing the simulation and assimilation of atmospheric carbon species. Carbon modeling and use of observational data are considered key to the long-term success of the End Station's modeling thrust. This project is performing multiyear data assimilation simulations of atmospheric carbon species at 0.5-degree spatial resolution using the GEOS-5 atmospheric general circulation model. The runs use specified emission boundary fluxes for CO and CO_2.

2.4 Eddy-resolving Parallel Ocean Program (LANL/SIO/NCAR)

The Parallel Ocean Program (POP) ocean model is being used to produce climate simulations with ocean mesoscale eddies (~20 km in size) and small-scale topographic features. These play an important role in determining the structure of important current systems in the ocean and in the eddy transport of heat. Previous simulations have shown that high-resolution (0.1°or less) is required to adequately represent properties of ocean current systems such as that of the Gulf Stream, with its separation at Cape Hatteras and subsequent turn to the northwest near the Grand Banks. At coarser resolutions, it has been impossible to reproduce such features, resulting in significant biases that can strongly distort important feedbacks in the climate system. This version uses a new configuration with a tripole grid and partial bottom cells and also attempts to introduce an improved advection scheme. This high-resolution, standalone, coupled ocean/ice simulation set the stage a fully coupled climate simulation with the eddy-resolving ocean as a high-priority SciDAC project.

2.5 Community Atomosphere Model development (NCAR/LBL/LLNL/PNNL/LANL/ORNL)

High-resolution configurations of the Community Atmosphere Model (CAM3) and prototype CAM4 have been developed to quantify the improvement in simulation fidelity across the wide range of space and time scales observed in the climate system. The development of CAM3 demonstrated a clear breakpoint in simulation quality when horizontal resolution was doubled from a 42-wavenumber truncation (T42) to an 85-wavenumber (T85) truncation. Early results suggest that the large-scale systematic biases in mean climate, which affect the quality of regional performance, are far more related to the parameterized treatment of nonresolvable physics, such as clouds, convection, radiation and boundary layer. Some transient features of the simulation are far more realistic, including the clear presence of tropical storms at the T341 truncation. This project investigates both the spectral Eulerian dynamical core and the prototype CAM4 using the finite-volume dynamical core at resolutions as fine as 0.25 degrees, which should have solution accuracy equivalent to the T341 spectral model.

2.6 CCSM Development (NCAR/LBL/LLNL/LANL/PNNL/ANL/ORNL)

The CCSM Software Engineering Group focus is on CCSM development simulations in four main categories:

1. *Model development toward CCSM version 4.* CCSM Version 3.5 was assembled in June 2007, which reduced several long-standing biases in CCSM 3. The atmosphere model working group is working on incorporating the following new parameterizations: new boundary layer scheme, new shallow convection scheme, revised vertical coordinate, new aerosol microphysics package, and revised gravity wave drag scheme. The land, ocean, and sea ice components have also been upgraded.

2. *Higher-resolution CCSM.* Scientists at LLNL and NCAR have demonstrated that the simulation quality of the finite-volume atmosphere component improves rapidly with increasing resolution. Thus, the development versions of CCSM will be regularly tested at the higher atmosphere resolution as development is progresses toward CCSM 4.

3. *CCSM3.5 with the carbon cycle.* The carbon cycle includes a new land component with fully interactive carbon and nitrogen limitation, an ocean ecosystem component, and several extra fields being transported by the atmosphere component.

4. *First attempts at short-term simulations using a high-resolution version of CCSM 3.5.* A novel type of climate change simulation proposed for the next IPCC report is high-resolution simulations of the near future with regional climate change information. The integrations are proposed to start in 1980, run to 2005 using observed climate forcings, and then proceed to 2030 in predictive mode. Interactive chemistry will be included in some of these simulations in order to include predictive aerosols. This inclusion will enable regional simulations of pollutants near large cities to be made.

3.8 WRF resolution studies (PNNL)

Researchers are assessing the impacts of spatial resolution at cloud-resolving and mesoscale scales using regional to global configurations of the Weather Research and Forecasting (WRF) model. Recent advances in high-performance computing have made it possible to run these model configurations at very high spatial resolution down to the cloud-resolving scale over relatively large domains and long time periods. This project uses WRF to systematically assess the impacts of spatial resolution, with a focus on the model's ability to simulate cloud processes and their interactions with the local to large scale circulation. Taking advantage of the nesting capability and cloud resolving to global modeling capability in WRF, numerical experiments using WRF configured in regional/global domains are also being performed.

3.9 University components (MIT/ORNL and Georgia Institute of Technology)

The magnitude and ubiquity of deforestation in the Amazon basin have become increasingly clear in recent years. Furthermore, there is overwhelming evidence that vegetation is not passively responding to changes in climate and weather but is dynamically evolving with climate and human and natural disturbances to vegetation. Human disturbance of Amazon forests over the next several decades is inevitable. The MIT team is taking the null hypothesis that this disturbance will have no effect on the nonlocal vegetation structure or the spatio-temporal partitioning of energy in the Amazon basin.

The Georgia Tech team is developing and testing a systematic subgrid scaling framework for the land component of the CCSM based on four elements: a complex vegetation tiling representation; an orographic tiling system; a tiling system to describe a distribution of water table parameters that derives a realistic statistical model of wetland; and an extension of our current work on precipitation intensity scaling that incorporates statistical estimation of precipitation intensities based on the physics of the CAM convective parameterization.

3.10 CCSM scaling (Argonne/Sandia National Laboratories)

Researchers at Sandia and Argonne have completed the first scaling tests with the HOMME version of CCSM CAM dynamical core. Results are encouraging: the Blue Gene/P is up to 4 times faster than other Blue Gene systems for a model with full physics and high resolution.

4. Operational plan timelines for IPCC fifth sssessment simulations

Figure 1 shows the various CCSM4 model developments and production steps required to be prepared for the IPCC fifth assessment simulations (AR5). Also shown are the various organizations where the simulations will take place and the areas of programmatic emphasis with time.

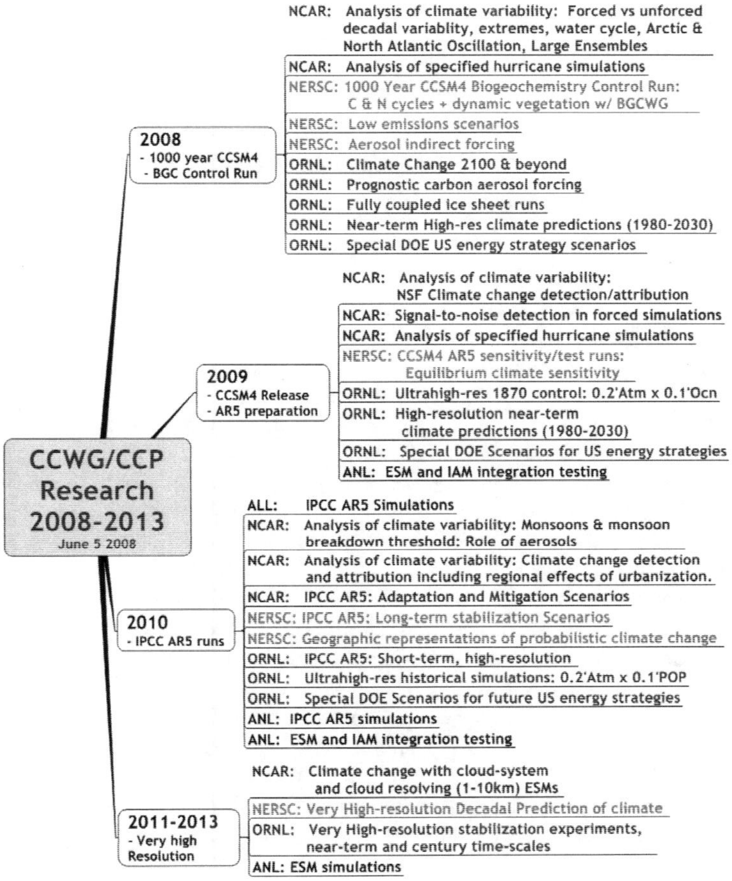

Figure 1. Operational timeline.

CCSM3 simulation of Pacific multi-decadal climate variability: The role of subpolar North Pacific Ocean

Y Zhong and Z Liu
Center for Climatic Research, University of Wisconsin, Madison, WI 53701, USA

E-mail: yafangzhong@wisc.edu

Abstract. Previous analyses of the CCSM3 standard integration have revealed pronounced multidecadal variability in the Pacific climate system. The purpose of the present work is to investigate the physical mechanism underlying the Pacific multidecadal variability (PMV) using specifically designed sensitivity experiments. A novel mechanism is advanced, characterized by a crucial role of the subpolar North Pacific Ocean. The multidecadal signal in ocean temperature and salinity fields is found to originate from the subsurface of the subpolar North Pacific, as result of the wave adjustment to the preceding basin-scale wind curl forcing. The multidecadal signal then ascends to the surface and is amplified through local temperature/salinity convective feedback. Along the southward Oyashio current, the anomaly travels to the Kuroshio Extension (KOE) region and is further intensified through a similar convective feedback in addition to the wind-evaporation-sea surface temperature feedback. The temperature anomaly in the KOE is able to feed back to the large-scale atmospheric circulation, inducing wind curl anomaly over the subpolar region, which in turn generates anomalous oceanic circulation and causes temperature/salinty variability in the subpolar subsurface. Thereby, a closed loop of PMV is established, in the form of a subpolar delayed oscillator.

1. Introduction

The Pacific multideccadal variability (PMV) is one of the most important climate phenomena over the Pacific Ocean [1-3]. However, our knowledge is limited concerning its origin and the physical mechanism because of the paucity and short duration of historical records. Fortunately, the arrival of state-of-art coupled global climate models (CGCMs) provides an unprecedant opportunity for the study of PMV. Recent increasing effort has been witnessed toward improved understanding of PMV using long integration of CGCMs [4, 5]. The CGCMs not only yield sufficient long time series of comprehensive climate variables but also present a powerful tool to help distill responsible physical mechanism.

The origin of PMV is an issue under intense debate. Numerous observational studies projected a tropical origin largely based on a substantiated tropical-North Pacific linkage [3]. In contrast, recent model studies suggested the PMV likely arises from the North Pacific [4, 5]. This discrepancy is attributable partly to a model deficiency that is common to state-of-art CGCMs, that is, a too-weak tropical-North Pacific linkage. Yet, the discrepancy may as likely result from the inconclusiveness (regarding causality) with statistical diagnostics of observational studies.

As a follow-up of [4] using FOAM, Zhong et al. [6] explicitly identified a North Pacific origin for the PMV in CCSM3, using advanced model strategies. The CCSM3 is thus consistent with the FOAM in terms of the origin of PMV. A closer inspection by Zhong et al. revealed a robust tropical-North Pacific linkage in accompany of the North Pacific originated PMV. This model linkage resembles that found in the observation [3]. Since the model PMV is known to arise locally from the North Pacific, the associated tropical-North Pacific linkage indicates more of a climate impact on the tropical Pacific by the PMV rather than a tropical origin. This interpretation of the tropical-North Pacific linkage thus reconciles the two contradictory views regarding the origin of PMV.

The present study is intended for a more thorough investigation of the North Pacific originated PMV in CCSM3. To better understand the responsible mechanism, we perform additional appropriately designed experiments to help isolate the relevant physical processes. The paper is organized as follows. Section 2 introduces the model and the experiment design. A preliminary survey of PMV is undertaken in Section 3 for all experiments. Section 4 is devoted to depicting the life cycle of the PMV. Conclusions are given in Section 5.

2. The model and experiment design

CCSM3 is a state-of-art global climate model [7]. Under the present-day climate conditions, CCSM3 at T31x3 [8] has been integrated for 880 years without flux adjustment later than year 133, showing no apparent climate drift. Analyses of this control integration are based on the past 580 years of simulation.

Aside from the CCSM3 control integration (named CTRL), sensitivity experiments that employ modeling surgery are performed to explore the mechanism of the PMV. The modeling strategy used is partial coupling (PC) and partial blocking (PB) [4] The PC strategy is used to deactivate atmosphere-ocean coupling in specified regions while retaining it elsewhere. In the decoupled regions, the atmosphere "sees" prescribed sea surface temperature (SST) from the seasonal climatology of the CTRL and is therefore not affected by the SST anomalies predicted by the model's ocean component. Meanwhile, the ocean is still driven by heat flux and wind stress anomalies, rather than pure climatological fluxes. The PB strategy is used to block Rossby wave propagation by inserting a sponge wall in a specific longitudinal band of the ocean component of the coupled system. Technically, a restoring boundary of climatological temperature and salinity is inserted in the North Pacific along 180°E with a width of 10°, and from bottom to 100 meters below the surface.

A couple of new experiments (table 1) are specifically designed to assess the significance of Rossby waves in the subtropical/subpolar Pacific for the generation of PMV. In experiment BLKNP, the PB surgery is applied to the latitudinal band 10°N–60°N to block wave propagation from the middle of the basin. In another experiment BLKSTP, the PB is applied to the latitudinal band 10°N–35°N only. For both experiments, the tropical region equatorward of 20° is decoupled using the PC surgery, similar to the experiment PC-ET from previous work of Zhong et al. [6]. A 400-year annual time series is used for the diagnostics.

Table 1. List of the sensitivity experiments.

Case	Experiment Design
CTRL	Standard integration of CCSM3.
PC-ET	PC surgery is applied to the global tropics equatorward of 20°.
BLKNP	PC surgery is applied to the global tropics equatorward of 20°; PB surgery is applied to the North Pacific at 10°–60°N.
BLKSTP	PC surgery is applied to the global tropics equatorward of 20°; PB surgery is applied to the Subtropical North Pacific at 10°–35°N.

3. PMV in sensitivity experiments

Spatial distribution of multidecadal SST variability in the North Pacific is given in figure 1 for new sensitivity experiments in comparison to the CTRL and PC-ET. The BLKSTP shows similar features to the CTRL and PC-ET, exhibiting heaviest loading (over 0.3°C) along the KOE. They also show comparable variability in the western subpolar North Pacific. Differently in the BLKNP (panel c), the activity center in the subpolar North Pacific is missing for the wave blocking there; associated, the other center in the KOE appears considerably weaker.

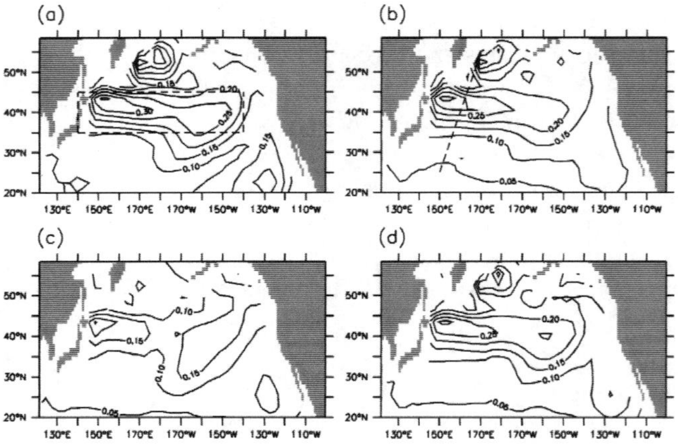

Figure 1. Spatial distribution of standard deviation of 25–80-year band filtered SST in (a) CTRL, (b) PC-ET, (c) BLKNP, and (d) BLKSTP. Contour Interval (CI) is 0.05 °C. The western-central North Pacific (140°E-140°W, 35°N–45°N) is delineated with the dashed rectangular box in (a). The dashed track line in (b) represents the vertical section on which figure 3 is based.

A power spectrum analysis is made for unfiltered SST anomalies averaged over the western-central North Pacific as denoted by the dashed rectangular box in figure 1. The BLKSTP captures the multidecadal peak (quasi-50-year period) that is evident in the CTRL and PC-ET, whereas no similar peak shows up in the BLKNP (figure 2). It is inferable the PMV hinges on the wave propagation in the North Pacific Ocean. As wave propagation is blocked at all latitudes of the North Pacific along the middle of the basin, the PMV is virtually suppressed in the BLKNP. However, as the blocking is applied to the subtropical region only, the PMV is essentially retained in the BLKSTP. Thus the wave propagation at mid-to-high latitudes of the North Pacific is crucial to the generation of the PMV, whereas that at subtropical latitudes is irrelevant at large.

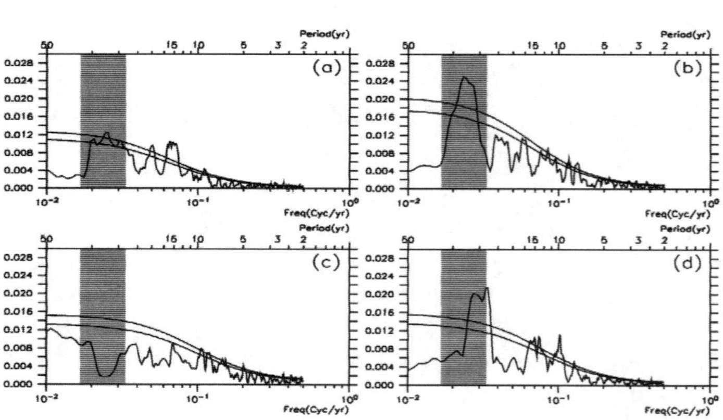

Figure 2. Power spectra of unfiltered SST averaged over the western-central North Pacific in (a) CTRL, (b) PC-ET, (c) BLKNP, and (d) BLKSTP; 95% and 90% confidence levels are indicated. The quasi-50-year frequency band is highlighted with shading.

4. Life cycle of PMV

The evolution of multidecadal temperature and salinity anomalies is shown in figure 3 along the vertical section that crosses both the subpolar and KOE activity centers, as represented by the dashed track line in figure 1b. It is derived as lagged regression of temperature and salinity upon the KOE SST (140°E-170°E, 35°N–45°N), which exhibits the strongest multidecadal signal in figure 1. To show the life cycle more clearly, all data have been 25–80-year band filtered before regression. The multi-decadal signal appears to originate in the subsurface of subpolar North Pacific. A cold anomaly develops first in the subpolar subsurface at lag -12yr (figure 3a2). It grows (figure 3a3-a5), expands to the surface at lag +4yr (figure

3a6), and further intensifies at the surface (see figure 3a1, but with the sign reversed). This cold anomaly from the subpolar region is being advected southward by the Oyashio current and turns the KOE into anomalously coldness eventually (figure 3a3, but with the sign reversed). Similarly, a fresh anomaly emerges in the subpolar subsurface (figure 3b4). It intensifies and upwells to the surface (figure 3b6). At the surface, the fresh anomaly is further amplified while being transported southward to the KOE (figure 3b3, but with the sign reversed). Thus, the multidecadal signal forms in the subsurface of subpolar North Pacific, then ascends to the surface, and travel southward to the KOE afterwards. Herein, three key regions have been identified: the subsurface, surface of the western subpolar North Pacific Ocean, and the KOE. Further insight into the physical interactions among the three key regions is provided in [9. 10]and leads us to a novel mechanism of PMV that is detailed in the next section.

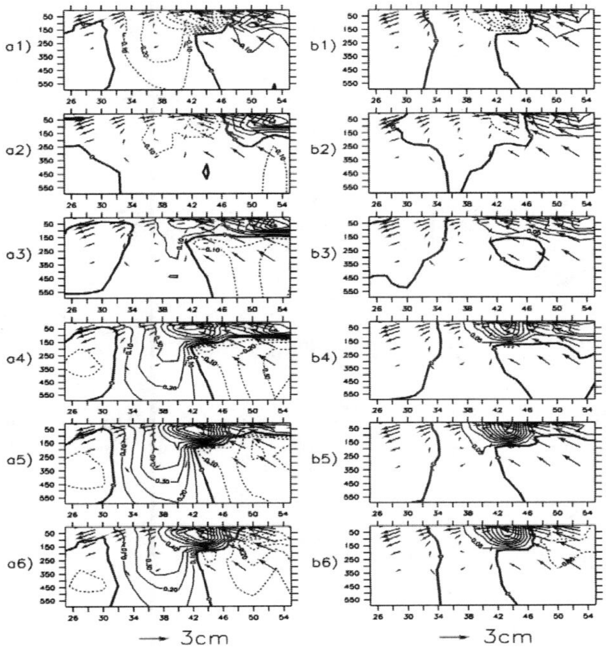

Figure 3. Lagged regression of (a1)-(a6) temperature and (b1)-(b6) salinity upon the normalized SST in the KOE based on output from PC-ET. The vertical section is along the dashed track line in figure 1b. All data are 25-80-year band filtered before regression. Lags are (a1), (b1) -16yr, (a2), (b2) -12yr, (a3), (b3) -8yr, (a4), (b4) -4yr, (a5), (b5) 0yr, (a6), (b6) +4yr respectively. CI = 0.1 °C/°C and 0.05 psu/°C for temperature and salinity, respectively. Also shown in vector is the mean meridional and vertical current with the latter multiplied by 1e4. Unit length of vector represents 3cm/s.

5. Schematic diagram of PMV

The physical processes involved in a multidecadal cycle of PMV are illustrated with the schematic diagram in figure 4. During the mature cold phase of the PMV (figure 4a), the KOE undergoes maximum coldness and freshness. The atmosphere responds to the cold anomaly in the KOE with a cyclonic wind curl anomaly over the subpolar region, which in turn induce Ekman suction and dynamic low pressure anomaly propagating westward across the subpolar North Pacific Ocean in form of first baroclinc Rossby waves. As a result, the subsurface of the western subpolar North Pacific shows substantial warm and saline anomaly owing to the mean halocline stratification (i.e., cold fresh water on top of warm saline water). At the same time, the surface is in a near-normal state ,while the mean upwelling carries the warm saline anomaly from subsurface to surface. After a quarter of cycle (figure 4b), the warm saline anomaly maximizes at surface via local temperature/salinity convective feedback that is preconditioned by the mean halocline stratification. To be more specific, once the warm saline anomaly ascends to the surface, the warm anomaly damps quickly and leaves the saline anomaly behind, which makes the surface water denser and thus weaker stratification. Consequently, convective activity gets more intense and entrains more warm saline water from below. The warm saline anomaly is then advected southward along the

Oyashio current, working to neutralize the cold fresh anomaly in the KOE. Concomitantly, the dynamic low-pressure anomaly proceeds into the western basin of subpolar North Pacific, and the associated geostrophic current starts to diminish heat and salt convergence in the subsurface. It has cooling and freshening effect. As another quarter of cycle goes by (figure 4c), the KOE enters a mature warm saline state, due to the enduring growth via the temperature/salinity convective and WES feedback that follows the initial influx of warm saline anomaly from the subpolar North Pacific. The corresponding anti-cyclonic wind curl anomaly over the subpolar region generates a new wake of Rossby waves producing a cold fresh anomaly in the subsurface. Notably, the cold fresh anomaly is substantially enhanced by the aforementioned cooling and freshening that results from the previous passage of the dynamic low-pressure anomaly. Ascending of the cold fresh anomaly meanwhile, neutralizes the foregoing warm saline anomaly in the surface of subpolar North Pacific, as seen in figure 4b. Thus, a mature warm phase of the PMV (figure 4c) is established, featured by a maximized warm saline anomaly in the KOE and substantial cold fresh anomaly in the subpolar subsurface. The converse occurs during the other half cycle of PMV.

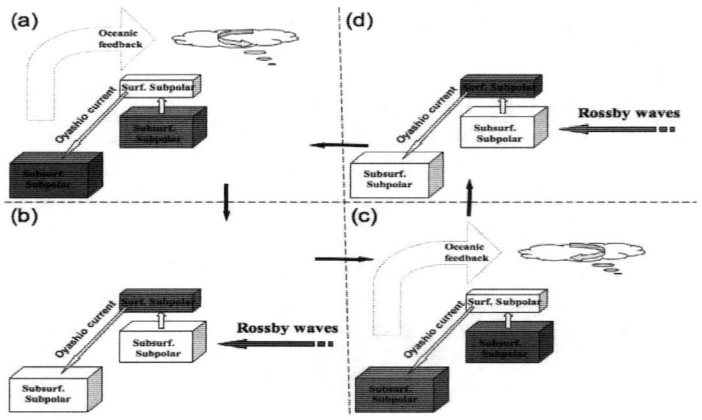

Figure 4. Schematic picture of a full cycle of PMV. Explanation is given in the text.

Acknowledgment
This work is supported by DOE and NOAA.

References

[1] Minobe S 1997 A 50-70 year climatic oscillation over the North Pacific and North America *Geophys. Res. Lett.* **24**:683-686
[2] Mantua N J, Hare S R, Zhang Y, Wallace J M, and Francis R C 1997 A Pacific interdecadal climate oscillation with impacts on salmon production *Bull. Am. Meteorol. Soc.* **78**:1069-1079
[3] Deser C, Phillips A S, and Hurrell J W 2004 Pacific interdecadal climate variability: Linkages between the tropics and the North Pacific during boreal winter since 1900 *J. Clim.* **17**:3109-3124
[4] Wu L, Liu Z, Gallimore R, Jacob R, Lee D, and Zhong Y 2003 Pacific decadal variability: The tropical Pacific mode and the North Pacific mode *J. Clim.* **16**:1101-1120
[5] Latif M, 2006 On North Pacific multi-decadal climate variability *J. Clim.* **19**:2906-2915
[6] Zhong Y, Liu Z and Jacob R 2008 Origin of pacific multi-decadal variability in community climate system model, version 3 (CCSM3): A combined statistical and dynamical assessment *J. Clim.* **21**:114-133
[7] Collins W D and Coauthors 2006 The Community Climate System Model version 3 (CCSM3) *J. Clim.* **19** 2122-2143
[8] Yeager S G, Shields C A, Large W G, and Hack J J 2006 The low-resolution CCSM3 *J. Clim.* **19**: 2545-2566
[9] Zhong Y and Liu Z 2008 A joint statistical and dynamical assessment of atmospheric response to North Pacific Oceanic variability in CCSM3 *J. Clim.* (n press)

[10] Zhong Y, and Liu Z 2008 On the mechanism of Pacific multi-decadal variability in CCSM3: the role of subpolar North Pacific Ocean (in preparation)

Journal of Physics: Conference Series **125** (2008) 012026 doi:10.1088/1742-6596/125/1/012026

Joint Cluster and Non-Negative Least Squares Analysis for Aerosol Mass Spectrum Data

Tianyi Zhang[1], Wei Zhu[1] and Robert McGraw[2]

[1]Department of Applied Mathematics and Statistics, Stony Brook University, Stony Brook, NY 11794-3600
[2]Environmental Sciences Department, Brookhaven National Laboratory, Upton, NY 11973-5000

E-mail: zhu@ams.sunysb.edu

Abstract. Aerosol mass spectrum (AMS) data contain hundreds of mass to charge ratios and their corresponding intensities from air collected through the mass spectrometer. The observations are usually taken sequentially in time to monitor the air composition, quality and temporal change in an area of interest. An important goal of AMS data analysis is to reduce the dimensionality of the original data yielding a small set of representing tracers for various atmospheric and climatic models. In this work, we present an approach to jointly apply the cluster analysis and the non-negative least squares method towards this goal. Application to a relevant study demonstrates the effectiveness of this new approach. Comparisons are made to other relevant multivariate statistical techniques including the principal component analysis and the positive matrix factorization method, and guidelines are provided.

1. Introduction

Atmospheric aerosols are significantly related to environmental issues such as rainfall, air pollution, and climate change. In recent years, quantitative aerosol studies based on mass spectrometry data have evolved rapidly in such field as particle formation research and air pollution analysis (see, e.g., [1]-[6]). Another area of great interest is dimension reduction. The goal is to obtain, from the original high dimensional aerosol mass spectrum (AMS) data, a much smaller set of representative tracers that will elucidate the source and interaction of different aerosol components, and to be plugged in as part of the model variables, to the next generation of climate models.

Several multivariate analysis techniques have been applied individually towards the dimension reduction of AMS data. These include principal component analysis [7] and cluster analysis [8]. The most widely used technique in the atmospheric research community however, is Positive Matrix Factorization (PMF), also known as the non-negative matrix factorization [9]. It has been developed [10] to yield factors with non-negative coefficients which would be more intuitive and interpretable than factors with a mixture of positive and negative coefficients. Plenty of work using PMF have been done [10-14] with two review papers published recently [15, 16]. However, a number of undesirable features inherent in PMF including subjectivity and non-uniqueness [17] have made it clear that better method must be developed to better serve the climate research communities.

In this paper, we propose a joint cluster analysis on variables (VARCLUS) [18-20] and non-negative least squares (NNLS) [21] approach to better achieve the goal of dimension reduction. Each method alone is not new; however, their combination is novel and as shown later, is rather effective in solving the problem at hand. The rest of the paper is arranged as follows. Section 2 provides a brief introduction to the AMS data from Milagro. Section 3 describes the proposed method in details while Section 4 presents the analysis of the Milagro data using the new method. Finally, Section 5 concludes with a comparison between the proposed method and the existing methods, especially the PMF.

2. Aerosol Mass Spectrum Data

2.1 Aerodyne Mass Spectrometer

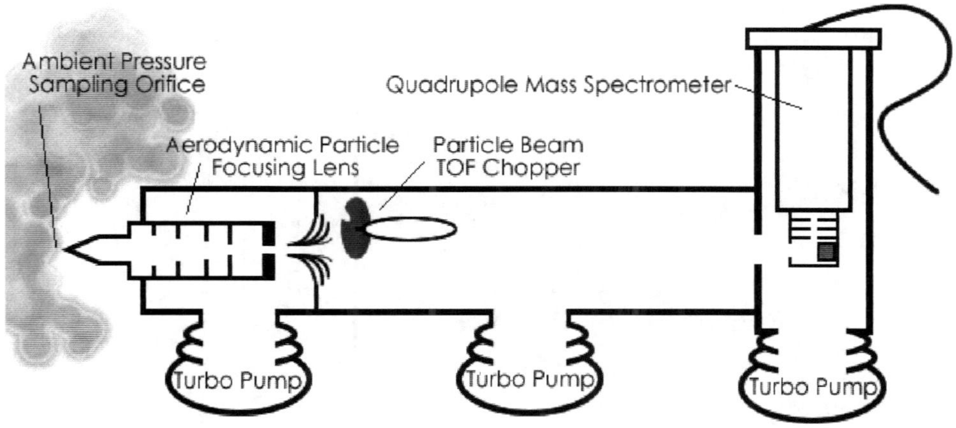

Figure 1. Basic schematic of the Aerodyne aerosol mass spectrometer. (http://cires.colorado.edu/~jjose/ams.html)

The aerosol mass spectrum data analyzed in this work are recorded by the Aerodyne mass spectrometer. Its basic mechanism is illustrated in Figure 1. In summary, the air flows into the orifice. Particles are then focused in the focusing lens and passed through the chopper which is set to either an open or a closed mode. Vaporized and ionized, particles are introduced into the spectrometer sequentially with their weights and frequencies recorded. The instrument alternates in the MS mode and the TOF mode. During the MS mode, the chopper alternates between the open and blocked position every 5 seconds and the particle rates are record, while during the TOF mode mass spectrometer is set to a singe m/z and sampled at a user defined rate. In this work, only data from the MS mode are involved. A full description of the Aerodyne mass spectrometer can be found in [1].

2.2 Data Preprocessing

The data we analyzed were collected on March 19, 2006 from Milagro containing signals with mass to charge ratios (m/z) ranging from 1 to 452 measured every 12 seconds for a total of 943 measurements in time. Two modes of data were obtained -- data recorded when the aerosol beam was unobstructed (MSSopen) and data collected when the beam was blocked by the chopper (MSSClosed). Each mode also has its own baselines, designated as MSSOpenBaseL and MSSClosedBaseL. We used the difference between the open mode and the closed mode to remove the contribution from background gas in the detector. Thus the data file we analyzed in this paper is named MSSDiff and defined as:

MSSDiff = (MSSOpen - MSSOpenBaseL) - (MSSClosed - MSSClosedBaseL)

Data in MSSDiff contain negative values, which are expected for signals that are close to zero. Since MSSDiff contains negative signals, it does not satisfy the requirement to apply PMF directly. Although replacing all negative values with zeros would be a viable approach, it will nonetheless destroy the linearity of the AMS data. Since the proposed joint cluster and NNLS approach would allow negative values, we are able to retain all negative values for the analysis.

Our analysis is focused on the subset organic aerosols ranging in m/z values from 1 to 100. Twenty-eight inorganic aerosols in the range (m/z's 1-11, 14, 16-18, 20-23, 28, 32-36, 39-40, and 47) were removed by setting the corresponding intensities to 0. The dataset was further normalized by dividing by the standard deviation at each m/z value. In addition, two time steps (time = 353, 643 seconds) appeared as outliers and were removed. The processed data are shown in Figure 2.

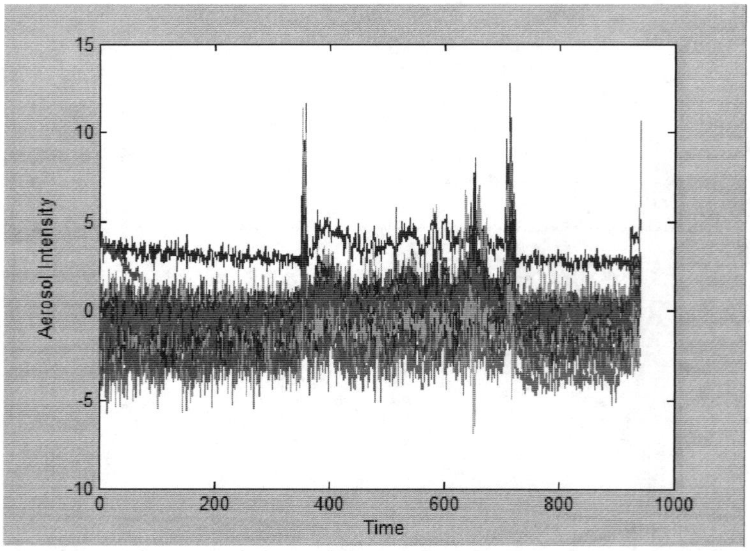

Figure 2. Pre-processed organic aerosol data where each colour represents the time series of a unique m/z value.

3. Method

3.1 Notations and Overview

Let **ORG** represent the pre-processed AMS data with Ni columns for m/z values ranging from 1 to Ni, and Nt rows for Nt temporal measurements. Thus the ith column of the matrix **ORG** is the time series corresponding to m/z value i, and the tth row of **ORG** is the mass spectrum at time point t. Our goal is to find a mass spectrum basis, **MS**, with lower dimension of Nc, such that

$$\mathbf{ORG}=\mathbf{C}*\mathbf{MS}+\mathbf{E} \quad (1)$$

Here, **ORG** is an Nt*Ni matrix (in our work Nt=941, Ni=100). **MS** is an Nc*Ni matrix, Nc<<Nt. Our goal includes two parts: first, we need to decide Nc, the dimension of the basis; next, we need to estimate **MS**.

The above decomposition can be solved by the principal component analysis (PCA), factor analysis (FA) or positive matrix factorization (PMF). PCA and FA could yield a suggestive Nc based on the variation explained by the basis. However, there are two apparent disadvantages to these methods. First, it is hard to interpret the output basis; second, the output basis usually contains negative values that are often unacceptable for practical purposes. The PMF method could avoid the second problem but not the first one. Furthermore, PMF returns different output using different starting matrix. That is, it is not unique [9, 17, 22].

To resolve these issues, we propose a combined cluster analysis on variables (VARCLUS), PCA and non-negative least square (NNLS) approach to better achieve the goal of dimension reduction. In summary, first we use VARCLUS to determine the dimension Nc. Next we perform the PCA to obtain the matrix **C**. Finally we apply the NNLS to estimate the non-negative matrix **MS**. More details are given below.

3.2 VARCLUS and PCA

The goal of VARCLUS is to divide the m/z values into disjoint clusters based on their correlations. The inherent linearity of AMS data usually returns explainable clusters with m/z values belonging to the same aerosol chemical class grouped together. Therefore the number of major aerosol classes can be estimated by the number of major clusters. Since the first principal component for each cluster is a weighted linear combination of all m/z values in the given cluster, with non-negative coefficients, and would explain the most (and often the majority) variation in the given cluster, it is the natural choice of the representing tracer for each cluster. Thus the dimension of basis Nc equals to the number of major clusters. And the basis consists of the first principal components from each major cluster. The matrix C is thus determined.

The VARCLUS procedure follows the following steps. [23, 24]

1. We treat the m/z values as variables, and different time points as observations. The data set is normalized by dividing by the standard deviation at each m/z value.
2. A cluster is chosen for splitting, for which the percentage of variation explained by its cluster component is the smallest. (The whole dataset is in the same cluster at the very beginning.)
3. The chosen cluster is split into two clusters by finding the first two principal components, performing an orthoblique rotation, and assigning each variable to the rotated component with which it has the higher squared correlation.
4. Variables are iteratively reassigned to clusters to maximize the variance accounted for by the cluster components.

Steps 2-4 are repeated until the stop criterion (total percentage of variation accounted for) is satisfied. A hierarchical tree is produced as the final output.

In each cluster, we use the first principal component (PC) to represent the whole cluster. Each PC forms a column of the matrix **C**. The PC's from all major clusters thus form the basis (the tracer set). We denote columns of **ORG** as $\mathbf{MZ_1,\dots,MZ_{100}}$ (each of length 941), while rows of **ORG** as $\mathbf{T_1,\dots T_{941}}$.(each of length 100). Thus, the ith column of **C** is $C_i = \sum_{k\in Cluster\#i} a_k MZ_k$. VARCLUS produces disjoint clusters so that each $\mathbf{C_i}$ is a weighted summation of disjoint subset of the m/z's ($\mathbf{MZ_k}$**'s)**. For practical reasons, clusters that would explain a small amount of variation in the data are either merged to its nearest major clusters, or simply discarded.

3.3 Non-Negative Least Squares

Now that we have obtained the basis dimension Nc and the basis matrix C, our next goal is to find the matrix MS in

$$\underset{941*100}{ORG} = \underset{941*Nc}{C} * \underset{Nc*100}{MS} + E$$

such that we can express each original m/z value as a linear combination of the basis with non-negative coefficients. This will further elucidate the relations between the original AMS data and the newly obtained tracer set. In other words, we need to calculate the non-negative coefficients β in the following equation system:

$$MZ_1 = \beta_{1,1}C_1 + \beta_{1,2}C_2 + \dots + \beta_{1,Nc}C_{Nc} + e_1$$
$$MZ_2 = \beta_{2,1}C_1 + \beta_{2,2}C_2 + \dots + \beta_{2,Nc}C_{Nc} + e_2$$
$$\dots\dots$$
$$MZ_{100} = \beta_{100,1}C_1 + \beta_{100,2}C_2 + \dots + \beta_{100,Nc}C_{Nc} + e_{100}$$

This is achieved through the NNLS algorithm by minimizing $\| ORG - \widehat{ORG} \|^2$, with the constraints that each β is non-negative. The details of the NNLS algorithm can be found in [21]. For better fit, we add constant terms in the above linear equation system as follows.

$$\begin{aligned} MZ_1 &= \beta_{1,0} + \beta_{1,1}C_1 + \beta_{1,2}C_2 + \ldots + \beta_{1,Nc}C_{Nc} + e_1 \\ MZ_2 &= \beta_{2,0} + \beta_{2,1}C_1 + \beta_{2,2}C_2 + \ldots + \beta_{2,Nc}C_{Nc} + e_2 \\ &\ldots\ldots \\ MZ_{100} &= \beta_{100,0} + \beta_{100,1}C_1 + \beta_{100,2}C_2 + \ldots + \beta_{100,Nc}C_{Nc} + e_{100} \end{aligned} \quad (2)$$

In matrix form we have: $\underset{941*100}{ORG} = \underset{941*(Nc+1)}{\widetilde{C}} * \underset{(Nc+1)*100}{\widetilde{MS}}$, where $\widetilde{C}$ is **C** with an extra column of 1, and $\widetilde{MS}$ is **MS** with the added row $(\beta_{1,0}, \beta_{2,0} \ldots, \beta_{100,0})$. Notice that $(\beta_{1,0}, \beta_{2,0} \ldots, \beta_{100,0})$ is allowed to be negative. We used the NNLS command in MATLAB to estimate β.

4. Results

4.1 VARCLUS and NNLS output

The VARCLUS output, a hierarchical tree, is shown in Figure 3. Based on the output tree and the related aerosol information, we have, clearly, 4 major clusters.

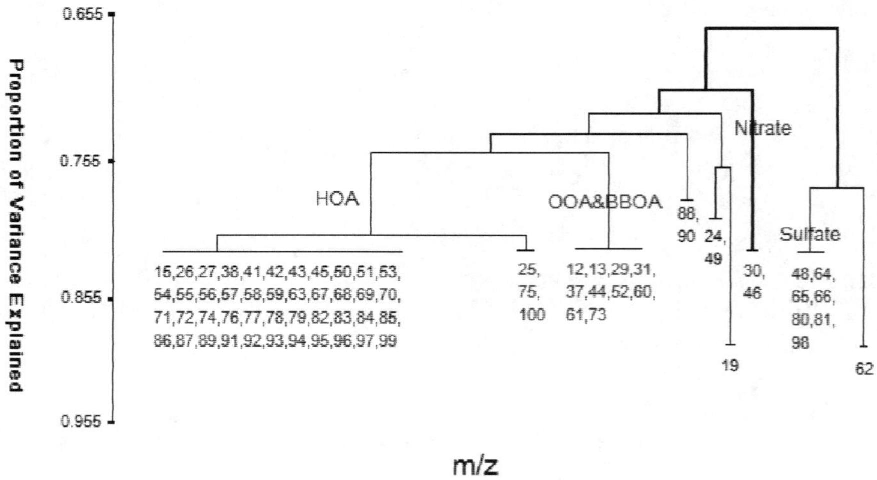

Figure 3. VARCLUS output: the hierarchical tree. From the tree structural, we decided four major clusters. Numbers in blue indicate the m/z values of cluster members while numbers in black are m/z values not belonging to any major cluster. The proportion of variance explained at each step of the binary tree split is displayed along the vertical axis.

Table 1 contains detailed information on these four major clusters. As discussed in Section 3.2, the advantage of using VARCLUS is that the clustering results are usually explainable if there is inherent linearity in the data. The AMS data are inherently linear with organic aerosols usually coming from several main sources. For example, hydrocarbon-like organic aerosols (HOA) mostly come from fossil fuel, while oxygenated organic aerosols (OOA) mostly come from secondary organic aerosols (SOA) [7]. This property supports the VARCLUS data assumptions and in our output we can clearly identify these 4 clusters. All 1st PCs in these clusters represent a high percentage of variation explained with non-negative coefficients. Thus, it is reasonable to use these four 1st PCs as the basis (tracer set) for the given AMS data.

Table 1. VARCLUS and PCA output.
Total variation explained by the basis consisting of four PC's is **71.14%**.

Cluster	Members (m/z values)	Variation Explained	1st PC Variation in Each Cluster	Label
Cluster#1	15,25:27,38,41:43,45,50 51,53:59,63,67:72,74:79, 82:87,89,91:97,99,100	49.72%	76.16%	HOA
Cluster#2	48,64,65,66,80,81,98	9.15%	94.16%	Sulfate
Cluster#3	12,13,29,31,37,44, 52,60,61,73	9.50%	68.36%	OOA and BBOA
Cluster#4	30, 46	2.77%	99.68%	Nitrate

Based on the VARCLUS output, we obtained the matrix **C** mentioned above. The matrix **MS** was estimated using NNLS. The output is shown in Figure 4.

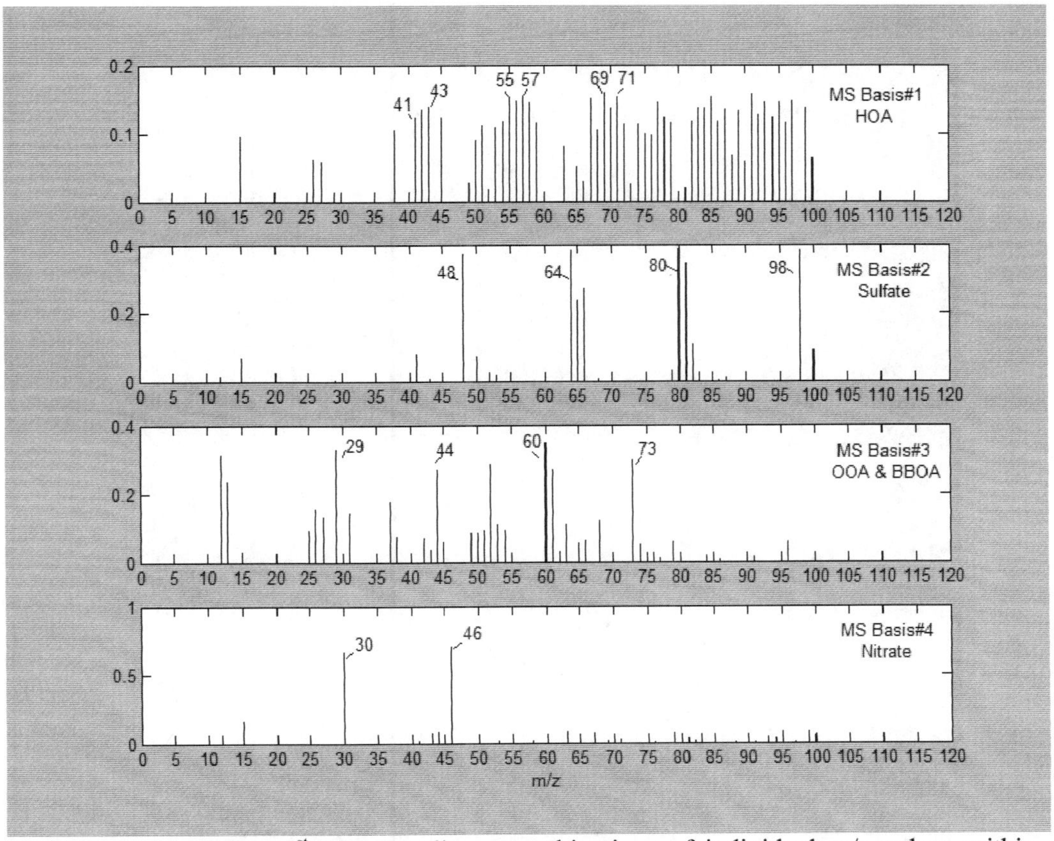

Figure 4. The basis of four 1st PC's are linear combinations of individual m/z values within each cluster. Significant peaks on each 1st PC are labelled.

4.2 Validity of Tracers

We checked the validity of the four tracers from three aspects: (a) whether they are interpretable for the given aerosol study; (b) how much of the total variation they could explain; (c) whether they could represent most of the aerosols in the original AMS data.

The organic aerosol factors from theoretic analysis are listed in Table 2 to check for (a). Compared with Table 1, the VARCLUS and PCA outputs are consistent with these factors for each major cluster containing all the corresponding critical m/z's correctly.

For (b), we found that 71.14% of the total variation is explained by the tracers as shown in Table 1. Figure 4 shows the time series of the four tracers. Compared to the single m/z tracer used by Zhang et al (2005), we found that the new tracers have larger variability as illustrated in Figure 5.

Table 2. Organic aerosol factors. The emphasized m/z's are important signals for the given factor.

Organic aerosol factors	Ionized at 600C	m/z	O: C ratio
HOA hydrocarbon-like organic aerosols	$C_nH_m \rightarrow C_{n-x}H_{m-y}^+$	27,29,**41,43,55,57**,69,71,...	Less Oxidized
OOA2 Oxygenated organic aerosols type II	$C_nH_mO \rightarrow C_2H_3O^+, C_3H_3O^+, R^{\cdot+}$	**43,55**,...	↓
BBOA biomass burning organic aerosols	$R \rightarrow R^{\cdot+}, C_2H_4O_2^+, C_3H_3O_2^+$	**44**,45,...	More Oxidized
OOA1 Oxygenated organic aerosols type I	$C_nH_mO_2 \rightarrow CO_2^+, HCO_2^+, R^{\cdot+}$	**60,73**,...	

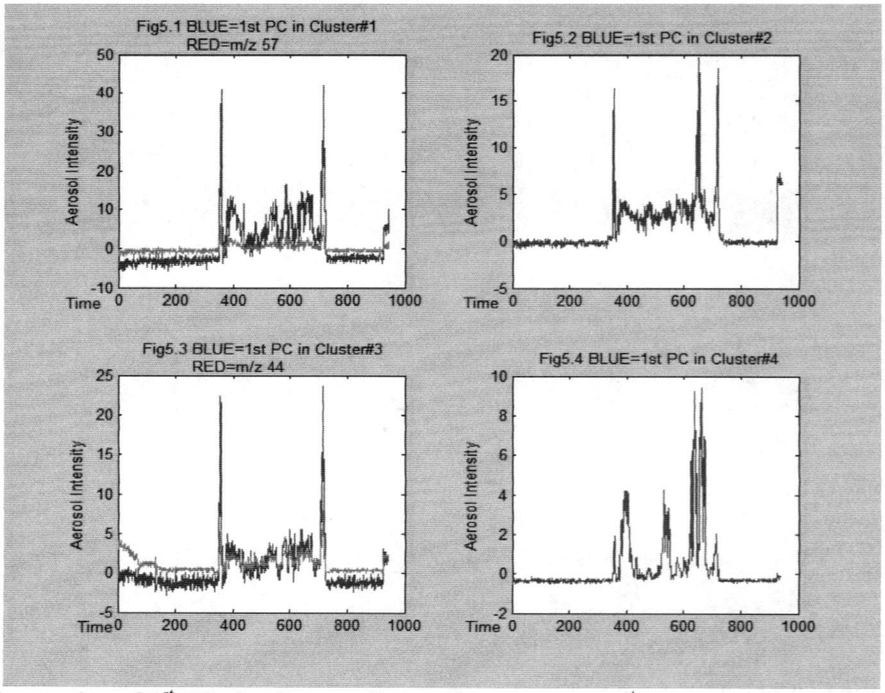

Figure 5. Time series of 1st PCs in clusters. Comparison between 1st PC in cluster#1 (HOA) and m/z 57 is shown in Fig5.1; comparison between 1st PC in Cluster#3 and m/z 44 is shown in Fig5.3. The single m/z values of m/z 44 and m/z 57 were used as tracers for OOA and HOA in Zhang et al (2005).

To check for (c), we need to verify that most of the peaks in the aerosol mass spectrum could be represented by our tracers. We first used the ordinary least squares regression (OLS) for this purpose. For each m/z=i, we fit the equation $MZ_i = \alpha_{i,0} + \alpha_{i,1} * C_1 + \alpha_{i,2} * C_2 + \alpha_{i,3} * C_3 + \alpha_{i,4} * C_4 + e_i$, where C_j represents the 1st PC in the jth major cluster and i=1, 2,…, 100. The goodness of fit for each regression, as evaluated by the coefficient of determinant R^2 for each regression equation, is reported in Figure 6. We found that most R^2 are bigger than 0.8 (the mean of R^2 is 0.81, the standard deviation is 0.23). For the few m/z's with small R^2 such as m/z19, it is because they are not in any major cluster and thus their signals can not be expected to be explained by the tracers. Lastly, stepwise regression was performed. The results support that these four tracers are not reducible.

We therefore conclude that the given tracers are highly satisfactory gauged by the three criteria (a), (b) and (c).

4.3 Error Analysis

To further validate our analysis, especially the NNLS portion of the analysis, we performed an error analysis by examining and comparing the coefficient of determination (R^2) from both OLS and NNLS.

Let $SSE_i = \sum_{j=1}^{941} (ORG_{ji} - \widehat{ORG_{ji}})^2$ be the sum of squared errors (SSE) for m/z=i (the i-th column in **ORG**), and $SST_i = \sum_{j=1}^{941} (ORG_{ji} - \overline{ORG_{ji}})^2$ be the corresponding total sum of squares (SST). The coefficient of determination (at m/z=i) is defined as [25]:

$$R_i^2 = 1 - \frac{SSE_i}{SST_i}$$

R^2 for each m/z using OLS or NNLS were calculated to examine the regression goodness-of-fit. We can see OLS and NNLS give similar R^2 for most m/z's in the given example. R^2=0.36, which corresponds to a multiple correlation of 60%, was used as a threshold to determine outliers. Six outliers were found: they're m/z 19, 24, 49, 62, 88, 90. The reason causing these outliers is quite clear. These are the six m/z's that are not assigned to any of the four major clusters as shown in Figure 7. Hence, all the other m/z's could be well represented by the four tracers (first PC's of the 4 major clusters) with non-negative constraints except for these outsiders.

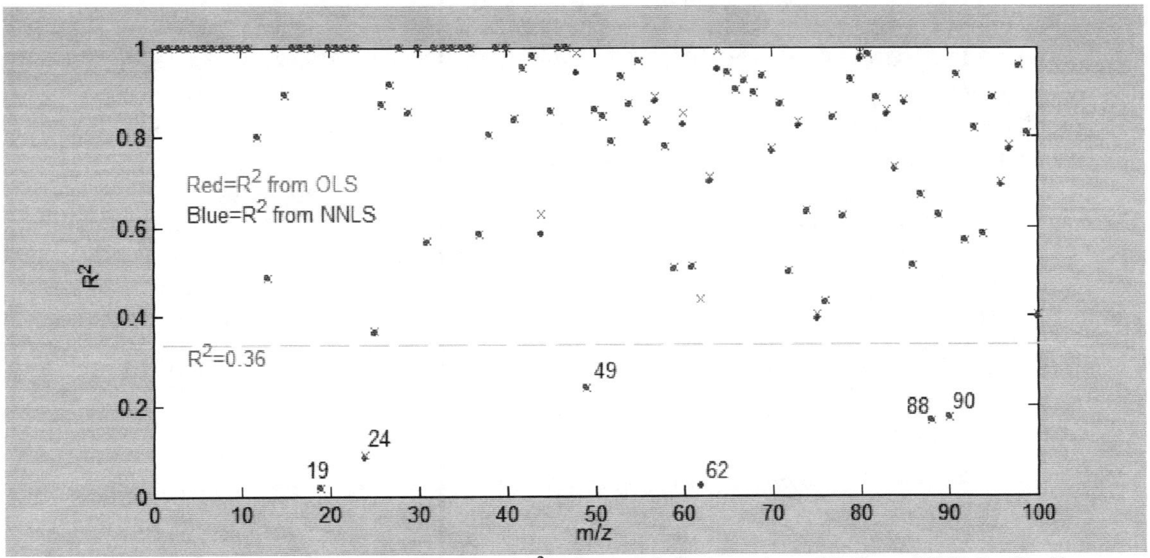

Figure 6. The coefficient of determination (R^2) for each m/z using OLS or NNLS. Outliers are specified.

Figure 7. Outliers are displayed in the hierarchical clustering tree.

5. Discussion

Many commonly used multivariate statistical methods are not suitable in the intended AMS data analysis of dimension reduction and tracer extraction due to negative coefficients of the resulting basis. Both the PMF and the proposed method of joint cluster and NNLS analysis can achieve non-negative tracer coefficients. The major draw-back of the PMF lies in its non-uniqueness and subjectivity.

Generally speaking, PMF is not unique. For example, if X=ABC, where A, B and C are all positive matrix, then we have at least two different PMF's in X=(AB)*C=A*(BC). Different factorizations are called "rotations" in factor analysis. For the proposed joint analysis method, there is little flexibility in deciding the major clusters and thus the basis. Subsequently, given the fixed **C** matrix, the **MS** matrix determined by the NNLS procedure would be unique.

As P. Paatero (See [17]) pointed out "It is unfortunate that introducing a priori information also introduces some subjectivity in the analysis." To apply the PMF analysis, one needs to first decide upon at least two things: the number of factors and the "rotation" parameter. Otherwise one will find the solution non-unique. Both of these two choices have determinant effect on the final output. For our method, we still have to refer to some prior knowledge from the field. However, the "subjectivity" exists in only one step – the determination of major clusters based on the VARCLUS output. Furthermore, even for this step, our decision is based mainly on objective criterion such as the percent variation explained by each cluster as shown in the given study.

There is space yet for us to improve upon in the proposed method. First, since the sources of organic aerosols are complicated, the disjoint clustering method may not perform well for overlapping aerosol groups. Secondly, if the first principal component for a given cluster could only explain a modest amount of variation, one would need to find additional tracer(s) for the given cluster. Perhaps a combined VARCLUS and PMF approach with the PMF done within each cluster would better serve our purposes. Further research is warranted in this area.

References

[1] Jimenez J L, *et al.* 2003 Ambient aerosol sampling using the Aerodyne aerosol mass spectrometer *J Geophysical Research* **108**(D7):8425

[2] Allan J D, *et al.* 2003 Quantitative sampling using an Aerodyne aerosol mass spectrometer, 1: Techniques of data interpretation and error analysis *J. Geophysical Research* **108**(D3) 4090

[3] Allan J D, *et al.* 2003 Quantitative sampling using an Aerodyne aerosol mass spectrometer, 2: Measurements of fine particulate chemical composition in two U.K. cities *J. Geophysical Research* **108**(D3) 4091
[4] Zhang Q, *et al.* 2005 Deconvolution and quantification of hydrocarbon-like and oxygenated organic aerosols based on aerosol mass spectrometry *Environmental Science & Tech.* **39:**4938–4952
[5] Zhang Q, *et al.* 2005 Time- and size-resolved chemical composition of submicron particles in Pittsburgh: Implications for aerosol sources and processes *J. Geophysical Research* **110:**D07S09
[6] DeCarlo P F, *et al.* 2006 A field-deployable high-resolution time-of-flight aerosol mass spectrometer *Analytical Chemistry* **78:**8281–8289
[7] Zhang Q, *et al.* 2007 Ubiquity and dominance of oxygenated species in organic aerosols in anthropogenically-influenced Northern Hemisphere midlatitudes *Geophysical Research Letters* **34:**L13801
[8] Marcolli C, *et al.* 2006 Cluster analysis of the organic peaks in bulk mass spectra obtained during the 2002 New England air quality study with an Aerodyne aerosol mass spectrometer *Atmos. Chem. Phys.* **6:**5649–5666
[9] Lee D D and Seung H S 1999 Learning the parts of objects by non-negative matrix factorization *Nature* **401**(6755): 788–791
[10] Paatero P and Tapper U 1994 Positive matrix factorization: A non-negative factor model with optimal utilization of error estimates of data values *Environmetrics* **5:**111–126
[11] Lee E, *et al.* 1999 Application of positive matrix factorization in source apportionment of particulate pollutants in Hong Kong *Atmos. Environ.* **33:**3201–3212
[12] Ramadan Z, *et al.* 2000 Identification of sources of Phoenix aerosol by positive matrix factorization *J. Air Waste Manag. Assoc.* **50:**1308–1320
[13] Larsen R K and Baker J E 2003 Source apportionment of polycyclic aromatic hydrocarbons in the urban atmosphere: A comparison of three methods *Environ. Sci. Technol.,* **37:**1873–1881
[14] Maykut N N, *et al.* 2003 Source apportionment of PM2.5 at an urban improve site in Seattle, Washington *Environ. Sci. Technol.* **37:**5135–5142
[15] Engel-Cox J and Weber S A 2007 Compilation and assessment of recent positive matrix factorization and UNMIX receptor model studies on fine particulate matter source apportionment for the eastern United States *J. Air Waste Manag. Assoc.* **57:**1307–1316
[16] Reff A, *et al.* 2007 Receptor modeling of ambient particulate matter data using positive matrix factorization: Review of existing methods *J. Air Waste Manag. Assoc.* **57:**146–154
[17] Ulbrich I M, *et al.* 2008 Interpretation of organic components from positive matrix factorization of aerosol mass spectrometric data *Atmos. Chem. Phys. Discuss.* **8:**6729-6791
[18] Harman H H 1976 *Modern Factor Analysis* 3rd ed (University of Chicago Press)
[19] Cattell R B 1965 Factor analysis: An introduction to essentials, II: The role of factor analysis in research *Biometrics* **21**(2):405–435
[20] Rummel R J 1970 *Applied Factor Analysis* (Northewestern Univ. Press: Evanston, Illinois)
[21] Lawson C L and Hanson R J 1974 *Solving Least Squares Problems* (Prentice-Hall) chapter 23
[22] Lee D D and Seung H S 2001 Algorithms for non-negative matrix factorization *Advances in Neural Information Processing Systems 13: Proc. 2000 Conference,* 556–562
[23] SAS Institute Inc 1994 *SAS/STAT User's Guide* Version 6, 4th ed

[24] Sun S 2000 *Multivariate Statistical Analysis and Statistical Software* (Beijing Medical University Press)
[25] http://en.wikipedia.org/wiki/Coefficient_of_determination.

Interaction of turblence and chemistry in a low-swirl burner

J B Bell[1], R K Cheng[2], M S Day[1], V E Beckner[1] and M J Lijewski[1]

[1]Center for Computational Science and Engineering, Lawrence Berkeley National Laboratory, Berkeley, CA, 94720, USA

[2]Environmental Energy Technologies Division, Lawrence Berkeley National Laboratory, Berkeley, CA, 94720, USA

E-mail: jbbell@lbl.gov

Abstract. New combustion systems based on ultra-lean premixed combustion have the potential for dramatically reducing pollutant emissions in transportation systems, heat, and stationary power generation. However, lean premixed flames are highly susceptible to fluid-dynamical combustion instabilities, making robust and reliable systems difficult to design. Low-swirl burners are emerging as an important technology for meeting design requirements in terms of both reliability and emissions for next-generation combustion devices. In this paper, we present simlations of a laboratory-scale low-swirl burner using detailed chemistry and transport without incorporating explicit models for turbulence or turbulence/chemistry interaction. We consider two fuels, methane and hydrogen, each at two turbulent intensities. Here we examine some of the basic properties of the flow field and the flame structure. We focus on the differences in flame behavior for the two fuels, particularly on the hydrogen flame, which burns with a cellular structures.

1. Introduction

Low-swirl burner technology was introduced by Bedat and Cheng [1] as tool for studying the fundamental properties of lean, premixed turbulent flames. Burners based on modifications of the original design have been used by a number of research groups [2–5] and have the potential for use in the design of next-generation, lean premixed combustion systems, including those burning lean hydrogen mixtures, both at atmospheric and elevated pressures [6].

The low-swirl burner concept is extremely simple: premixed fuel (here, lean hydrogen or methane mixtures) exits a pipe with circular cross section (5-10 cm in radius) after passing through a turbulence plate and an annular set of curved vanes. The vanes impart a swirl component to the flow over a narrow layer near the pipe wall. The flow then diverges radially at the pipe exit, and a detached premixed flame propagates freely in the flow field. The turbulence is introduced into the fuel stream to produce wrinkles in the flame and enhances the overall rate of combustion in the device; the flames stabilize where the mean burning speed matches the axial flow velocity. Analysis of experimental velocity measurements for both lean hydrogen-air and methane-air mixtures shows that the mean burning speed of the low-swirl flames is proportional to the intensity of turbulence in the approach flow to the flame [6, 7], and that effective enhancement of the burning speed can exceed a factor of 50 or more. However, because the burner supports ultra-lean mixtures while also lacking a significant postflame flow recirculation, nitrogen-based emissions are extremely low.

Continued development of these types of burners, particularly for practical combustion devices, depends on improving our understanding of basic flame structure, stabilization mechanisms, emissions, and response to changes in fuel. Numerical modeling has the potential to address some of these issues; however, simulation of these types of burners has proven to be difficult because of the large range of spatial and temporal scales in the system; the bulk of the analysis to date has been experimental. The work of Nogenmyr et al. [4], which simulated low-swirl methane flame using LES with a G-equation formulation, was one of the first successful simulations of these types of flames.

The basic issues for these flames become particularly important, and difficult to simulate, for certain advanced fuels. Lean hydrogen-air flames burn in cellular structures—localized regions of intense burning, separated by regions of local extinction. Such a discontinuous flame surface introduces severe difficulties in applying standard turbulence/chemistry interaction models, which are based on the presence of a highly wrinkled but continuous flame surface that propagates locally as an idealized laminar flame structure. The cellular burning patterns additionally make the analysis of experimental data problematic and can lead to significant inaccuracies or misinterpretations.

We consider an alternative approach to the study of these flames based on detailed numerical simulations carried out at the laboratory scale without including turbulence models. Our target low-swirl burner configuration, presented in [2], is under investigation by several groups internationally using matched nozzle hardware and is designed specifically to support validations with computer simulation.

2. Numerical methodology

The simulations presented here are based on a low Mach number formulation [8] of the reacting Navier-Stokes equations that offers a significant performance benefit for these types of turbulent flame simulations [9]. The methodology treats the fluid as a mixture of perfect gases. We use a mixture-averaged model for differential species diffusion, which is critical to capturing the thermodiffusive behavior of lean hydrogen flames (see [10] for a complete discussion of this approximation). We ignore Soret, Dufour, and radiative transport processes. The transport coefficients, thermodynamic relationships, and chemical source terms are obtained from a CHEMKIN-compatible specification. The methane flame with inlet mixture, $\phi = 0.7$, was modeled with DRM-19; the hydrogen flame, $\phi = 0.37$, was based on the hydrogen sub-mechanism of GRI-Mech 2.11. The basic discretization [11] combines a symmetric operator-split treatment of chemistry and transport with a density-weighted approximate projection method to impose the evolution constraint. The resulting integration proceeds on the time scale of the relatively slow advective transport. Faster diffusion and chemistry processes are treated implicitly in time. This integration scheme is embedded in an adaptive mesh refinement algorithm based on a hierarchical system of rectangular grid patches. The data and work are apportioned over a parallel computing system using a coarse-grained load distribution strategy [12]. The complete integration algorithm is second-order accurate in space and time and discretely conserves species mass and enthalpy.

3. Computational setup

We use an experimental characterization of the flow at the nozzle exit to generate a suitable inflow boundary condition for the simulations. Typical profiles for this purpose were provided by Petersson et al. [2] and are shown in figure 1 as a function of radius from the centerline of the fuel pipe. Outside the pipe, a 35 cm/s upward coflow is specified. Turbulent fluctuations in the nozzle were prepared in an auxiliary simulation and added to the mean flows. An initially zero-mean

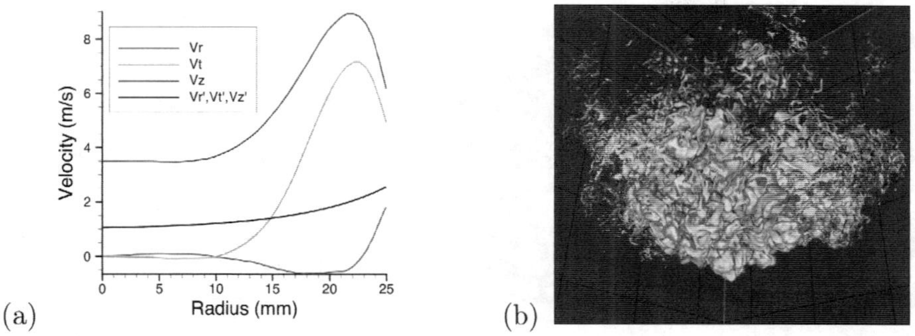

Figure 1: (a) Inlet velocity profiles at the exit plane of the swirl nozzle, adapted from experimental data [2]. (b) Instantaneous snapshot of typical H_2-air flame burning in 3D cellular structures (view angle is looking up from below the exit plane of the nozzle, radius = 25 mm). An isotherm in the flame zone is shown and is colored by the local rate of fuel consumption–the red areas are burning intensely, the blue areas are extinguished.

velocity field, representative of flow through a perforated plate, is evolved until the measured integral length scale was 4 mm, and the fluctuation intensity matched the value measured at the device core. A radial scaling factor was used to shape these synthetic fluctuations to the measured data. For the second, "high-turbulence" case, we simply scaled all inlet profiles at the nozzle by 250%.

The thermal thicknesses of these flames are approximately 600–800 μm; the full half-width maximum of the fuel consumption rate is 250–500 μm. The computational domain for the reacting flow simulations measures $(25\text{cm})^3$. We implicitly assume that the boundaries of this box are sufficiently far from the flame as to not significantly affect its dynamics. The base mesh for the simulation is a uniform grid of 256^3 cells and uses three additional levels of factor-of-two dynamic grid refinement that track regions of high vorticity (turbulence) and reactivity (combustion/flame). In all cases, the flame is contained entirely within the finest level, for an effective resolution of 2048^3. We run the simulations from an arbitrary initial condition until the flame is statistically stabilized without the finest level of resolution. Prior to collecting data for analysis, we add the finest level of resolution and continue the numerical integration until the flame statistics cease to vary in time. At the finest resolution of these computations, our simulation methodology resolves the detailed structure of both the methane and hydrogen flames—the time-dependent location of the peak fuel consumption, the thermal field and major species are numerically converged, and the profiles of most of the flame radicals are well-represented.

4. Simulation results

A typical snapshot of the high-turbulence with lean hydrogen is shown in figure 1(b). The image is of an isotherm in the flame zone (T=1139 K), colored by the local consumption rate of fuel. The red regions indicate burning at up to 16 times the flat laminar flame value. The blue regions indicate local extinction and separate the cellular burning structures in this thermodiffusively unstable flame. In figure 2 we show fuel and OH in a slice through the center of the burner for all four cases. The solution was advanced at both resolutions for 50 coarse time steps after the final level of refinement was added prior to making comparisons. This corresponds to 200 time steps at 244μm and 400 time steps at 122μm. The images demonstrate that the methodology is capable of resolving the overall flame shape as well as the chemical detail of the flame. Similar grid-independence is observed for the other state quantities.

The methane flame shows an increased wrinkling with the increased level of turbulence; however, as shown by the OH profile, combustion along the flame front is relatively insensitive

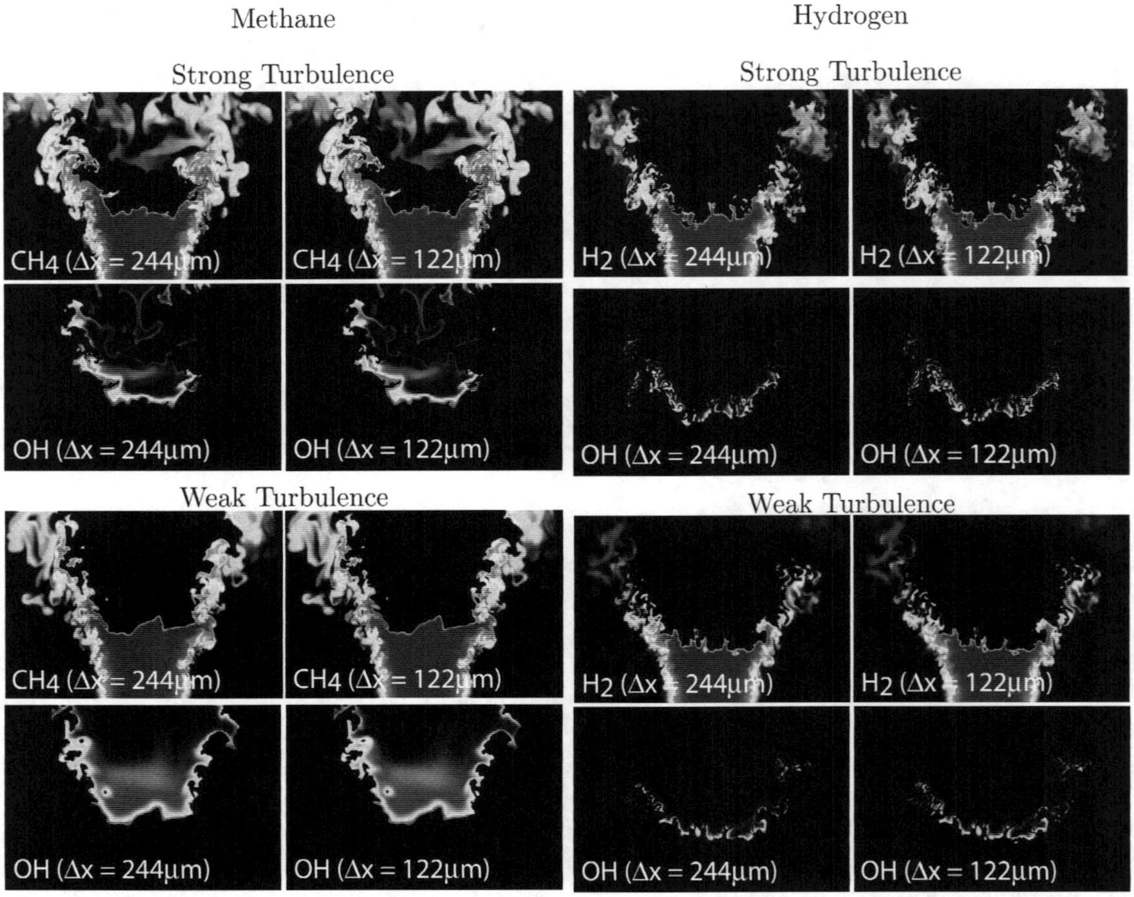

Figure 2: Fuel and OH mole fraction normalized to the peak value in the flat flame, for the four simulations cases. Snapshots taken on slices through centerline of domain indicate robust grid-independent solutions.

to stretch effects as expected for methane flames at $\phi = 0.7$. The high diffusivity of H_2 makes the hydrogen flame at $\phi = 0.37$ thermodiffusively unstable so that it burns with a cellular structure. As a consequence of this cellular burning the wrinkling of the hydrogen flame is considerably more pronounced. Examination of the OH images shows the cellular structure more clearly. The flame front is broken with regions of intense burning separated by regions of local extinction. As in the methane case, increasing turbulence leads to an increase in the wrinkling of the flame and a reduction in the overall size of the features along the flame front.

We next consider the effect of the fuel on the overall turbulent burning rate. In all the flames computed, there is a qualitatively different interaction between the flame and the vortical structures due to the inlet turbulence, and those due to the the breakdown of the mean swirling flow. For that reason, we restrict our consideration to the central core region of the burner. We can estimate the burning speed in this core region by integrating the instantaneous fuel consumption rate over a cylindrical volume with the same radius as the nozzle up to a height of 10 cm. This regions captures all of the core burning. We then normalize this integral by $\rho Y_{F,in} \times A_{in} \times s_L$ where A_{in} is the inflow area of the burner, $\rho Y_{F,in}$ is inflow fuel density and s_L is the laminar flame speed. The resulting data for the four cases are presented in figure 3. (We note that this measure does not represent a traditional definition of the turbulent flame speed. For that type of measurement, we would need the predict the flame brush to correctly normalize the integrals. Here, we do not have sufficient data for that prediction.

Our definition, in effect, assumes that the flame brush is flat.) This figure shows several interesting features. First, the turbulent flame speed enhancement for the hydrogen flame is much larger than for the methane flame. In addition, the response to increasing the level of turbulence is stronger for the hydrogen flame. These characteristics are a result of the thermodiffusive instability of the hdyrogen flame. In addition, we note that extrapolation to zero turbulence gives the laminar flame speed for methane but shows an enhanced speed for hydrogen. This is again because of the thermodiffusive instability; even in the absence of turbulence, the hydrogen flame will burn with a cellular structure at an enhanced flame speed, see [13].

We have presented simulations of an experimental low swirl burner with detailed chemistry and transport without introducing any turbulence models. The simulations demonstrate the level of resolution needed to accurately characterize the flames and shows that the methodology can capture the cellular structure of the hydrogen flame. In future work, we will provide a more detailed analysis of the simulation results focusing on local flame properties and comparison with experimental results. In future simulations, we plan to include nitrogen emissions chemistry to explore the formation of NO_x and to investigate the structure of these types of flames at higher pressures.

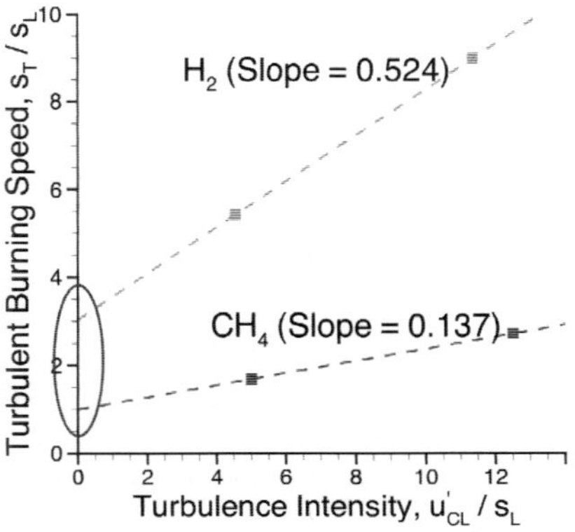

Figure 3: Computed turbulent burning speeds in the central core of the burner.

Acknowledgments

This work was supported by the SciDAC Program of the DOE Office of Mathematics, Information, and Computational Sciences under the U.S. Department of Energy under contract No. DE-AC02-05CH11231. The computations presented here were performed on Franklin, the XT4 at NERSC as part of an INCITE award. The authors thank A. Dreizler and P. Petersson for providing experimental velocity measurements used to define the boundary conditions.

References

[1] Bedat B and Cheng R K 1995 *Combust. Flame* **100** 485–494
[2] Peterson P, Olofsson J, Brackman C, Seyfried H, Zetterberg J, Richter M, Alden M, Linne M, Cheng R, Nauert A, Geyer D and Dreizler A 2007 *Appl. Opt* **46** 3928–3936
[3] Cheng R K 1991 *Combust. Flame* **101** 1–14
[4] Nogenmyr K, Peterson P, Bai X, Nauert A, Olofsson J, Brackman C, Seyfried H, Zetterberg J Li Z S, Richter M, Dreizler A, Linne M and Alden 2007 *Proc. Combust. Inst* **31** 1467–1475 *submitted to LACSEA Feature Issue*
[5] Mansour M and Chen Y C 2007 *Experimental Thermal Fluid Sci. in press*
[6] Cheng R K, Littlejohn D, Strakey P A and Sidwell T 2008 *Proc. Combust. Inst.* 21–46 to appear
[7] Cheng R K and Littlejohn D 2008 *Proceedings of GT2008: ASME Turbo Expo 2008* (Berlin, Germany: ASME) paper GT2008-50504
[8] Rehm R G and Baum H R 1978 *N. B. S. J. Res.* **83** 297–308
[9] Bell J B, Aspden A J, Day M S and Lijewski M J 2007 *Numerical Simulation of Low Mach Number Reacting Flows* (*Journal of Physics Conference Series: SciDAC 2007 (W. Tang, Ed.)* vol 78) (Boston, Massachusettes: Institute of Physics Publishing) p 012004
[10] Ern A and Giovangigli V 1994 *Multicomponent Transport Algorithms* (*Lecture Notes in Physics* vol m24) (Berlin: Springer-Verlag)
[11] Day M S and Bell J B 2000 *Combust. Theory Modelling* **4** 535–556
[12] Rendleman C A, Beckner V E, Lijewski M, Crutchfield W Y and Bell J B 2000 *Computing and Visualization in Science* **3** 147–157
[13] Day M S, Bell J B, Bremer P T, Pascucci V and Beckner V E In preparation

High-fidelity simulations for clean and efficient combustion of alternative fuels

J H Chen, C S Yoo, R Sankaran and J C Oefelein

Reacting Flows Research Department, Sandia National Laboratories, Livermore, CA 94550, USA

E-mail: jhchen@sandia.gov

Abstract. The monolithic nature of transportation technologies offers opportunities for significant improvements in efficiency of 25–50% through strategic technical investments in both advanced fuels and new low-temperature engine concepts. The application of direct numerical simulation (DNS) provides a way to study fundamental issues related to small-scale combustion processes in well-defined canonical configurations, whereas the application of large eddy simulation (LES) provides a formal treatment of the full range of time and length scales that exist in turbulent reacting flows, and thus provides a direct link to experimental studies of relevant combustion devices. In the present study, through DOE INCITE and Oak Ridge National Laboratory 250 Tflop Transition-to-Operations grants in 2008, DNS is performed to understand how a lifted autoignitive flame is stabilized, and LES is performed to understand the high-pressure injection and mixing processes in internal combustion engines. Understanding of these and other fundamental issues is needed to develop robust and reliable ignition and combustion models for the combustion regimes observed under low-temperature combustion engine environments using alternative fuels.

1. DNS of an autoigniting, turbulent-lifted ethylene/air jet flame

In many modern combustion systems such as diesel engines or direct injection stratified gasoline engines and gas turbines, fuel is injected into an environment of hot gases such that a flame may be stabilized through the recirculation of hot air and combustion products. Under such conditions, this process leads to a turbulent lifted flame, and the hot environment supports autoignition as a possible mechanism contributing to the stabilization of the flame base. In addition to autoignition, flame propagation and the role of coherent eddies have been considered as possible mechanisms for stabilization of the lifted flame [1]. In previous 3-D direct numerical simulation (DNS) study of an autoigniting, turbulent-lifted hydrogen/air jet flame, it was found that the lifted flame stabilizes primarily by autoignition that occurs in fuel-lean, hot mixtures in regions of low scalar dissipation (mixing rate) rate [1]. The present DNS study explores the stabilization mechanism for a hydrocarbon fuel with more complex ignition kinetics. In particular, a turbulent lifted ethylene/air jet flame in an autoignitive heated coflow is simulated, and the stabilization mechanism and flame structure are determined. Future DNS studies will examine the effects of renewable fuels (e.g., ethanol and biofuels) on flame stabilization, extinction and reignition, and stratified flame propagation.

An ethylene-air lifted flame in a spatially evolving slot-burner configuration was simulated on the Cray XT system, called Jaguar, at the National Center for Computational Sciences (NCCS).

Jaguar was recently upgraded with quad-core AMD OpteronTM processors, which raised the peak performance of the system from 119 teraflops to 260 teraflops. Prior to the upgrade, Jaguar consisted of 11,508 XT4 nodes each containing a 2.6 GHz dual-core AMD Opteron processor. After the upgrade, Jaguar consists of 7,832 XT4 compute nodes each containing a 2.1 GHz quad-core AMD Opteron processor. More important, while the dual-core opteron processor was, theoretically, able to perform two floating point operations per clock cycle, the quad-core Opteron processor is able to perform four operations. The factor of 2 improvement in performance is due to an increase in the SSE execution width from 64-bit to 128-bit in the new quad-core processor. Vectorization and generation of SSE-instructions are essential to take advantage of the capabilities of the new quad-core processor.

The performance of the Sandia DNS code, S3D, with a detailed ethylene-air mechanism was evaluated using a pressure-wave test, where the propagation of a small amplitude pressure wave through the domain is computed for a short period of time. The fixed problem size contained 30^3 grid points per MPI-thread. Earlier performance studies have shown that S3D can scale up to the entire size of the system on various Office of Science platforms, and show near-ideal weak scaling performance. For brevity, only the single socket performance data is presented here, with a note that the near-ideal weak scaling behavior was observed across the entire system nodes. Table 1 shows a comparison of the time taken to compute the ethylene-air problem on the old XT with 2.6 GHz dual core Opteron and the newer XT with 2.1 GHz quad-core Opteron. For comparison, the table contains performance data for an undersubscribed processor with fewer threads than the number of cores. Practical S3D simulations use all cores in a socket. The initial port to quad-core processor showed a decrease in performance: it took 503 seconds on the quad-core in comparison to 465 seconds on dual-core. Part of this decrease is attributable to the lower clock speed. Clearly, however, vectorization was not as aggressive as desirable. Special effort was devoted to rewriting the chemistry evaluation routine, which was consuming roughly 60% of the overall run time. With assistance from the Cray Supercomputing Center for Excellence, the routine was rewritten to operate on data at multiple grid points using array syntax to facilitate SSE instruction generation. A larger fraction of the floating -point operations are performed through SSE instructions as a result of the code improvement. A decrease in run time of roughly 20% was achieved, and the computation took less time (422 s) on the quad-core processor. The production simulation was then performed using 30,000 Opteron cores on Jaguar consuming 4.5 million CPU-hrs and generated half a terabyte of data every hour.

Table 1. Table 1. S3D performance on quad-core Opteron processor in comparison to dual-core Opteron. Wall-clock time (in seconds) needed to compute 100 time steps is shown. Under MPI mode, the argument to -n denotes the number of MPI-threads, and -N denotes the number of threads per Opteron socket.

Problem Size	Mpi-mode	Dual-core	Quad-core (initial)	Quad-core (vectorized)
$30 \times 30 \times 30$	-n 1 -N 1	404	415	333
$60 \times 30 \times 30$	-n 2 -N 2	465	430	349
$60 \times 60 \times 30$	-n 4 -N 4	-	503	422

The spatially developing turbulent lifted jet flame simulation was performed in a 3-D slot-burner configuration. Fuel issues from a central jet, which consists of 18% ethylene and 82% nitrogen by volume at an inlet temperature of $T_{\rm j} = 550$K. The central jet is surrounded on either side by co-flowing heated air streams at $T_{\rm c} = 1,550K$ and atmospheric pressure. The fuel jet and coflow velocities are specified as $U_{\rm j} = 204$ m/s and $U_{\rm c} = 20$ m/s, and the fuel jet width, H, is 2 mm such that the jet Reynolds number, $Re_{\rm j} = HU_{\rm j}$, is approximately 10,000. The computational domain is 15H $\times$ 20H $\times$ 3H in the streamwise, x, transverse, y, and spanwise, z, directions with

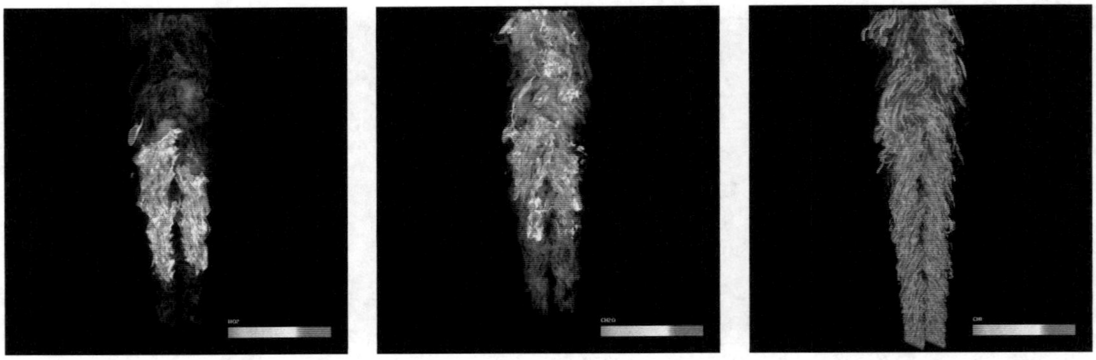

Figure 1. From Left: Y_{HO_2},Y_{CH_2O}, and scalar dissipation rate in a lifted ethylene-air jet flame at Re=10,000.

2025 × 1600 × 400 grids for a total of 1.3 billion grids. A uniform grid spacing of 15 microns is used in the x- and z-directions, while an algebraically stretched mesh is used in the y-direction. The compressible Navier-Stokes, species continuity, and total energy equations were solved by using S3D with a fourth-order Runge-Kutta method for time integration and an eighth-order central differencing scheme for spatial discretization. We adopted a reduced ethylene/air kinetic mechanism which consists of 22 species and 18 global reaction steps (T. Lu and C. K. Law, private communication, 2007). Nonreflecting inflow/outflow boundary conditions [2] were used in the x- and y-directions and periodic boundary conditions were applied in the homogeneous z-direction.

Figure 1 shows volume rendering of hydroperoxy radical (ignition marker) and formaldehyde mass fractions and scalar dissipation rate (local mixing rate) fields. Downstream of the high scalar dissipation region, temperature starts to increase at x/H = 6, following an increase in the heat release rate. Global characteristics of the lifted flame represented by the Favre mean temperature and heat release rate show that, unlike the hydrogen/air lifted jet flame [1], the mean heat release rate starts to increase in the middle of the shear layer. This is attributed not only to the large stoichiometric mixture fraction of the present flame relative to the hydrogen flame but also to differences in the ignition kinetics/transport coupling in the ethylene/air jet flame.

The stabilization mechanism of the present ethylene/air lifted jet flame is further investigated by tracking Lagrangian fluid particles. Tracer particle methods are commonly used to gain fundamental understanding of intermittent flow and flame physics [3]. The particles provide the time history of the turbulence and mixing conditions that a given fluid parcel undergoes while being advected through the domain. In particular, it provides the temporal trajectory of a given localized ignition site as it traverses oncoming turbulence intermittency, thus providing valuable lagrangian velocity-scalar statistics for the validation of models predicting lift-off heights in autoignitive partially-premixed turbulent jet flames.

2. LES of direct-injection processes in internal combustion engines

Direct injection (DI) is a promising option for modern internal-combustion (IC) engines. In conventional diesel and homogeneous charge compression ignition (HCCI) applications, DI lowers soot and NO_x production and improves fuel economy. In hydrogen-fueled engines, DI provides the appropriate energy density required for high efficiency and low NO_x. In order to realize the full benefit of DI, however, the effect of various injection parameters, such as injection timing, duration, pressure, and dilution, must be investigated and optimized under a range of operating conditions. As part of our 2008 INCITE grant, Oefelein has developed a model for high-fidelity

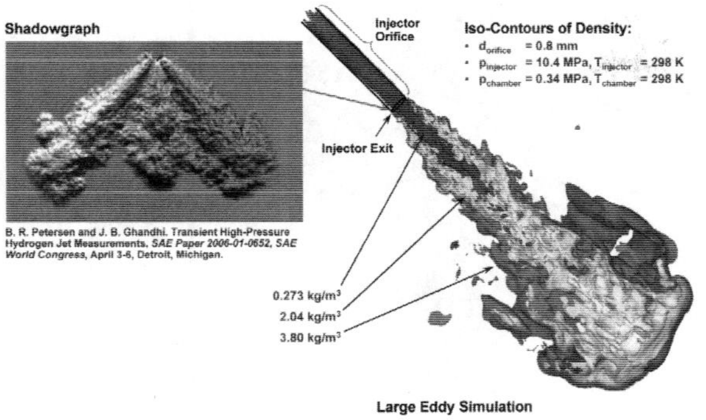

Figure 2. Representative LES of the instantaneous jet structure compared to the corresponding shadowgraph results of Petersen *et al.* [5].

calculations of DI processes using the large eddy simulation (LES) technique and coupled this with an advanced property evaluation scheme designed to handle high-pressure phenomena. The theoretical-numerical framework enables both the canonical cases described here and in-cylinder calculations that will be performed subsequently. Here we focus on high-pressure multi-port gas injectors designed for application in hydrogen-fueled IC-engines. To investigate issues related to high-pressure gas injection processes, we have developed a general model framework that combines a detailed property evaluation scheme for arbitrary gas-liquid hydrocarbon mixtures with the LES technique [4]. In LES, the large energy-containing scales of fluid motion are solved on a computational grid. The small (more universal) subgrid-scale interactions are modeled. This approach allows time-accurate predictions of turbulent flows at extremely high Reynolds numbers. As a step toward systematic validation, we have considered a series of cases using the experimental data acquired by Petersen and Ghandhi [5] for validation. Following is a brief description of simulations enabled by the current INCITE grant.

High-pressure, multiport hydrogen injector studies were conducted in an optically accessible chamber by Petersen and Ghandhi [5]. Single, three, seven, nine, and thirteen port injectors were considered. Jet flow patterns were visualized by using a Schlieren technique. The images were processed to determine the penetration length and angle. The pressure history during injection was also measured to determine the mass flow rate of the injected gas. Injection pressure, chamber density, and chamber gas composition were varied to isolate the effects of pressure, density, and compressibility. As part of our effort to validate the fully coupled theoretical-numerical framework, we have performed a series of high-fidelity LES calculations of the three hole injector. We focus on three cases, where hydrogen is injected into nitrogen at 10.4 MPa, with chamber pressures fixed at atmospheric, 340 kPa and 720 kPa. This set of conditions places the system in the supercritical regime.

A representative result is given in figure 2, which shows the level of qualitative agreement obtained between experimentally observed and simulated results for the case when the chamber pressure is 340 kMPa. On the left is the instantaneous shadowgraph from the experiment. On the right is instantaneous iso-contours of density from the LES solution at the same instant in time. Blue represents the density of the pure hydrogen stream (0.273 kg/m^3), red the density of pure nitrogen (3.80 kg/m^3) and yellow an intermediate value (2.04 kg/m^3). Similar agreement has been obtained for all cases. Comparisons between the shadowgraph data and corresponding LES suggest that the fine-scale turbulent structure, which dominates the spatial evolution of the flow, can be reproduced quite accurately in the simulations. This is a crucial requirement for developing accurate predictive models and provides a way to perform detailed analysis of

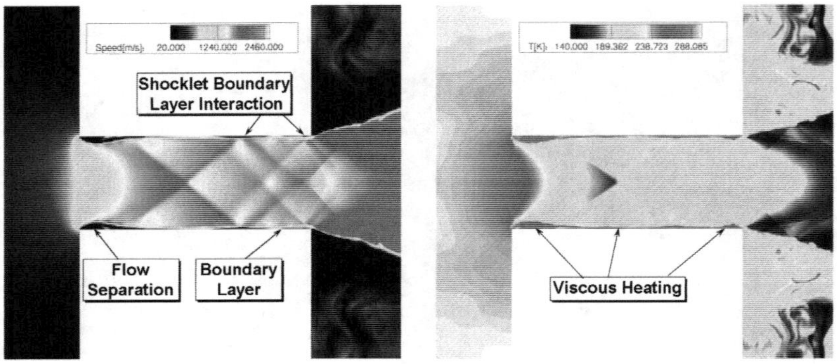

Figure 3. Flow velocity (left) and temperature (right) 18 μs after start of injection.

these types of flow in a manner that is anchored to actual data. In addition to comparisons with the shadowgraph data, we have performed detailed comparisons with the jet penetration data provided from the experiments. In each case, LES results are in good agreement with the experiments.

Having established quantitative agreement with the experiments, we performed detailed analysis of transient processes at the start of injection. The goal is to identify loss mechanisms that degrade the performance of the injector, to investigate the transient evolution of the mass flow rate through respective injector orifices, and to quantify the effect of injection pressure drop on the overall formation and quality of mixture during the injection process (see figure 3). Ongoing studies will include detailed analysis of the turbulent small-scale flow structure and the related scalar mixing processes, with emphasis placed on maximizing engine efficiency and minimizing emissions.

Calculations of the entire set of injectors are in progress. In addition to direct use of the model in a companion set of in-cylinder calculations, results from these studies will be used to systematically progress toward liquid hydrocarbon fuels. We will focus on (1) what is required to rigorously reproduce the essential broadband turbulent dynamics in the simulations using high-fidelity subgrid-scale models, (2) the accuracy that can be achieved through direct comparisons with key experimental data (see www.ca.sandia.gov/ecn), and (3) how the information provided by the high-fidelity models can be used for development of more affordable engineering models.

Acknowledgments

This work was supported by the U. S. Department of Energy, Office of Basic Energy Sciences, Division of Chemical Sciences, Geosciences, and Biosciences. Sandia National Laboratories is a multiprogram laboratory operated by Sandia Corporation, a Lockheed Martin Company, for the United States Department of Energy under contract DE-AC04-94-AL85000. The work at Oak Ridge National Laboratory (ORNL) was supported by and used resources of the National Center for Computational Sciences (NCCS) at ORNL, supported by the DOE Office of Science under contract DE-AC05-00OR22725.

References

[1] C S Yoo R S and Chen J H 2008 *Preprint*

[2] Yoo C S and Im H G 2007 *Comb. Theory and Modelling* **11** 259–286

[3] P Sripakagorn G K and Riley J J 2004 *Combust. Flame* **136** 351–363

[4] Oefelein J C 2006 *Prog. Aerosp. Sci.* **42** 2–37

[5] Petersen B R and Ghandhi J B 2006 *J. Engines* (*SAE Transactions* vol 115) (Detroit, Michigan: SAE International) SAE Paper 2006-01-0652

Large eddy simulations of turbulent reacting flows in real burners: The status and challenges

Laurent Y M Gicquel[1], Gabriel Staffelbach[1], Bénédicte Cuenot[1] and Thierry Poinsot[2]

[1] CERFACS, 31057 Toulouse Cedex 1, France
[2] Institut de Mécanique des Fluides de Toulouse, 31400 Toulouse, France

E-mail: `Laurent.Gicquel@cerfacs.fr`

Abstract. Turbulence is a recurrent and recognised challenge for which the scientific community has not been able to provide reliable methodologies necessary for predictions in complex industrial applications. In fact and throughout the past century, that challenge has been identified as a million dollar achievement simply because of the tremendous impact such a contribution would have on existing industrial processes. Among all industries, the gas-turbine companies are probably the most receptive to new contributions in the field of turbulent reacting flows because of the upcoming new regulations and the existing pressure linked to petroleum consumption and pollutant emissions. From a purely scientific point of view, the treatment of combustion and turbulence yields additional difficulties. However, and because of the advent of high-performane computers providing teraflop capabilities, the fully unsteady, temporally and spatially dependent approach that is large eddy simulations (LES) provides new insights and promises when dealing with these industrial applications. In this document, state-of-the-art LES for complex flows is presented, along with developments and potential challenges that are needed to improve LES in order to yield an efficient and reliable tool for the prediction of real industrial turbulent reacting flows.

1. Introduction

Today, understanding and predicting simple turbulent reacting flows (or, more generally, turbulent flows) remain a great challenge for researchers and current developments heavily rely on computer codes. Major progresses hence depend on the available computing power and the development of new architectures. It impacts not only fundamental work but also industry, which uses numerical tools developed by engineers and scientists. With the recent developments of high-performance computing (HPC) facilities (see the Top 500 list issued on June 2008 and ranking all worldwide supercomputing systems: http://www.top500.org/lists/2008/06#top10.), teraflop capabilities have been reached, and the symbolic pentaflop machine seems within reach. Such computer power increase opens new perspectives to the scientific community and should change the way industry addresses turbulent reacting flows.

Currently, three numerical approaches coexist in the field of turbulent flows. The first approach, named direct numerical simulations (DNS) [1; 2], aims at improving the scientific understanding of turbulence and its coupling with various physical phenomena as encountered in real applications. Among the subjects involving turbulence and investigated by DNS, one notes combustion [3–14] (fuel consumption and pollutant emissions), multi-phase flows [15–20], radiation [21–23], primary and secondary break-up of liquid sheets [24–27], thermo-acoustic instabilities [28; 29], and aero-acoustics [30–33]. With DNS, physicists simulate the governing equations with as much accuracy as possible for ideal or simple configurations. The limit is essentially linked to the computing power available. The highly detailed results are then usually probed to improve the understanding of the physics involved [34–36]. Such databases are also exploited to provide inputs or models for the other set of numerical methods which have less stringent computer costs. Note that DNS is a tool aimed at fundamental issues, and ongoing researchers also benefit from the new computing facilities as illustrated in recent progress [1; 2]. The models constructed from DNS and experimental or theoretical investigations ease the simulation of more complex and realistic flows within industry, which uses other numerical methods.

The approach most routinely used in industry is probably RAS (Reynolds average simulations) and was mainly developed in the 1970s [37–39]. In RAS, one looks only for an estimate of the turbulent flow behavior, namely, the mean solution. With this tool, only the mean quantities (in the sense of time or ensemble average) are available, and all the effects of turbulence on the mean field are modeled. Although computer effective, this approach heavily relies on the models supplied by the scientists. Their range of applicability is theoretically restricted to statistically stationary flows, although industry currently uses RAS beyond this limit to dimension new industrial systems. Here again, the increase of computing power offers an efficient turnover time that allows one to gauge new design points before industrial validation of the final design on expensive experimental test facilities.

In order to extend the domain of validity of the numerical tools, especially in the industrial context, a third approach is currently being developed. This new numerical method, called large eddy simulations (LES) [40; 41], introduces the notion of scale separation that filters out the high frequency flow structures and retains only the large-scale motions that usually depend on the geometry considered. With this separation of scales, the computing effort is increased when compared to RAS, while remaining largely below the inferred computer costs of DNS. The models, still needed with LES, are nonetheless more universal, and more physics is naturally imbedded in the simulations. Indeed, LES offer access to a spatially and temporally dependent flow description, a particularly valuable information in complex configurations. As such and despite the fact that LES are still under development, it opens new perspectives, especially in the industrial context where transient physical phenomena are known to be critical.

The development of LES for an extensive use by industry at the design stage needs to meet several requirements. Code management and scalability are of the foremost importance because these applications aim at running on HPC architectures to bring response times within the industrial requirements or to tackle ever more complex geometries. Systematic model and code validations are necessary to guarantee predictability of the LES tool. In the following, we briefly discuss the state of the art for LES of turbulent reacting flows in complex configurations (Section 2). Issues about modeling, code implementation, validation procedures, and scalability are then discussed in Sections 3, 4, and 5, respectively. Potential perspectives and closing remarks are given in Section 6.

2. LES of turbulent reacting flows in complex geometries: a brief status report

Current HPC facilities provide enough computing power to demonstrate the capacity of advanced LES codes in addressing complex industrial configurations. Although still at the state of demonstration, these simulations [42–45] (figure 1) are usually produced in a specific context where everything is eased to produce the results. Among the necessary conditions usually met for the computations shown on figure 1, one notes the implication of four types of personnel:

- Experts, usually scientists with a strong background on modeling issues specific to the physics simulated by LES

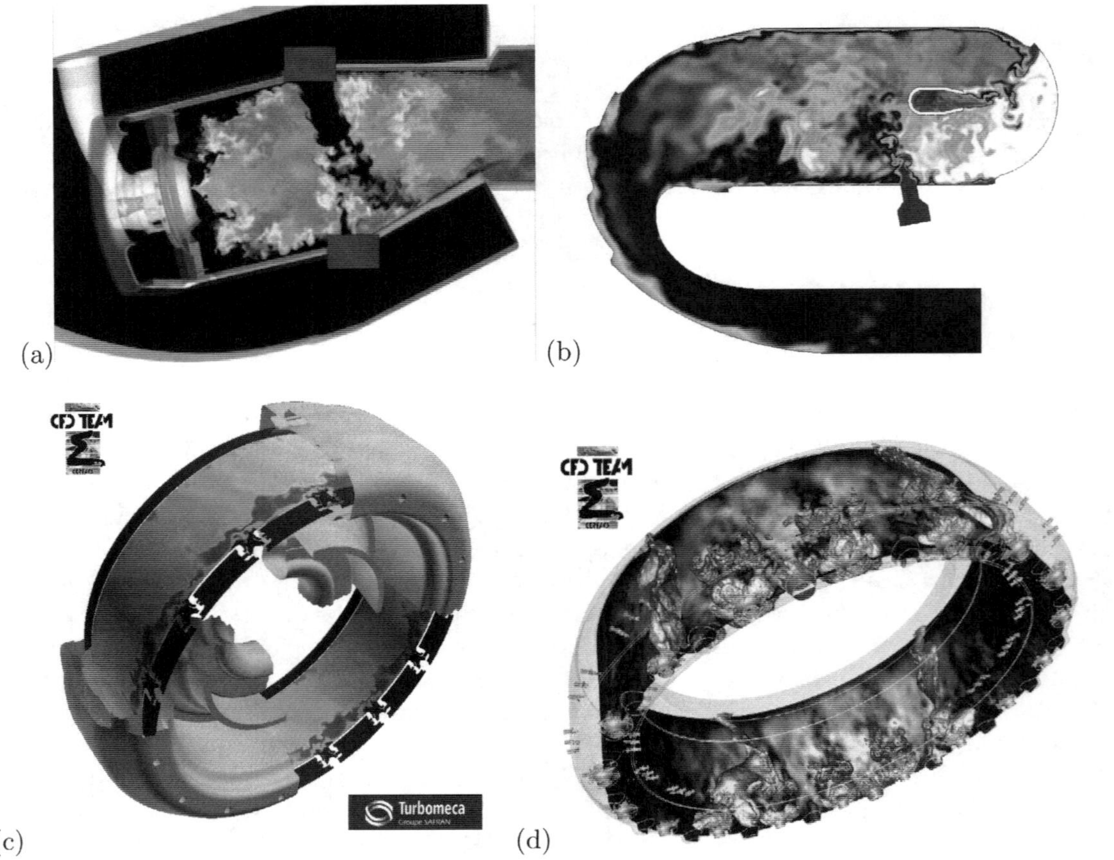

Figure 1. Instantaneous views of state-of-the-art LES applied to real aero-engine combustion chambers: (a) Pratt & Whitney single-sector burner [42], (b) a single-sector Turbomeca helicopter chamber [43], (c) a full Turbomeca chamber [44] and (d) an ignition sequence in a full annular combustion chamber of Turbomeca [45]. All views provide information on the temperature distribution in the burner at one given instant and as obtained by LES.

- Computer scientists with a strong background on programming and HPC and involved in the development of the code devised to produce the simulation
- Computer scientists with a strong background on hardware and HPC, who usually work for the computing center or computer manufacturer hosting the computation
- Engineers from aero-engine companies that provides the real application to be simulated by LES

The final requirement, which may be necessary and which usually dimensions the size and the type of the problem targeted, is accessibility to the computing facility or, more important, access to a large number of processors. For example, the fully transient LES of the ignition phase for a complete helicopter gas-turbine combustion chamber [45] (figure 1 (d)) was achieved thanks to a dedicated access to the top 19 Bull machine (Tera-10 Nova-Scale 5160) from the CEA computing facility located in France, as well as the top 24 IBM (Blue Gene/P) architecture located in Rochester, New York; top 13 Barcelona Super-computing Centre (Blade Centre JS21) IBM, Spain; and top 7 Oak Ridge National Center for Computational Sciences (Cray XT3/XT4) Oak Ridge, Tennessee (classification based on the November 2007 Top 500 list). Note also that the other results shown in figure 1 (a) and (b) treat only a single sector of the full annular combustor and do not necessitate such dedicated accesses. Furthermore, and with the increasing number of ever larger HPC centers (INCITE), full annular chamber computations will rapidly exit the realm of "opportunity" or the need for specific "challenges." It is also interesting to underline that, although all of the reported advanced LES results aimed at being purely demonstrative, the acquired knowledge yielded important scientific contributions with several journal publications in the field of turbulence and combustion. This interesting consequence further emphasizes the importance of such pioneering work for industry but also for the scientific community.

Such computations, although demonstrative and scientifically valid, should be viewed as a first step toward full industrial use of LES. Most of the results are obtained for fully gaseous LES when real engines operate on liquid fuels and chemistry models are simple. In fact, a lot of the fundamental physics present in the real engine combustion chamber is still not understood or modeled in such computations. Technological aspects introduced by the manufacturer to specifically address cooling of the chamber walls, for example, are open issues that only start to be dealt and modeled by LES experts [46–48]. Note also that the validation of these LES simulations in a rigourous scientific way remains a challenge since experimental measurements are scarce for such real burners. Furthermore, issues specific to the LES approach, its analysis and the handling of the huge amount of data generated by such codes are basic questions that need to be addressed to fully capitalize on the information generated by LES in the industrial context. (One snapshot for the instantaneous field of full annular combustor weights about 1.4 Gbytes and a temporal evolution of the main flow structures usually necessitates 4 million iterations.) Although most likely incomplete, part of these questions are presented and discussed in the following section and prior to a discussion on future developments and potential perspectives for HPC LES applications.

3. Specific issues related to advanced LES in complex geometries: the challenges

LES for reactive multispecies mixtures involves the spatial filtering operation, figure 2, which reduces for spatially, temporally invariant, and localized filter functions [49; 50], to

$$\widetilde{f(\mathbf{x},t)} = \frac{1}{\overline{\rho(\mathbf{x},t)}} \int_{-\infty}^{+\infty} \rho(\mathbf{x}',t)\, f(\mathbf{x}',t)\, G(\mathbf{x}'-\mathbf{x})\, d\mathbf{x}', \tag{1}$$

where G denotes the filter function and $\widetilde{f(\mathbf{x},t)}$ is the Favre filtered value of the variable $f(\mathbf{x},t)$ [51]. Note that the quantities of interest have spatial $\mathbf{x}$ (a three-component vector for full $3D$ applications) and temporal t dependencies.

In the mathematical description of gaseous turbulent flows with chemical reactions and species transport, the primary variables are the species volumic mass fractions $\rho_\alpha(\mathbf{x},t)$, the velocity vector $u_i(\mathbf{x},t)$, the total energy $E(\mathbf{x},t) \equiv e_s + 1/2\; u_i u_i$, and the density $\rho(\mathbf{x},t) = \sum_{\alpha=1}^{N} \rho_\alpha(\mathbf{x},t)$. Note that $\rho_\alpha(\mathbf{x},t)$ is linked to the species mass fractions $Y_\alpha(\mathbf{x},t)$; and for a mixture of N species, mass conservation imposes $\sum_{\alpha=1}^{N} Y_\alpha(\mathbf{x},t) = 1$.

The fluid follows the ideal gas law, $p = \rho\, r\, T$ and $e_s = \int_0^T C_p\, dT - p/\rho$, where e_s is the mixture sensible energy, T the temperature, $C_p = \sum_{\alpha=1}^{N} C_{p,\alpha}\, Y_\alpha$ the fluid heat capacity at constant pressure and r is the mixture gas constant, which varies with composition and is obtained by $r = \frac{R}{W} = R \sum_{\alpha=1}^{N} \frac{Y_\alpha}{W_\alpha}$, where $R = 8.314\; \mathtt{kg\, m^2/(s^2K)}$ and W_α is the molecular weight of the species α. The viscous stress tensor, the heat diffusion vector, and the species molecular transport use classical gradient approaches. The fluid viscosity follows Sutherland's law, the heat diffusion coefficient follows Fourier's law, and the species diffusion coefficients are obtained by using a constant species Schmidt number and diffusion velocity corrections for mass conservation [34].

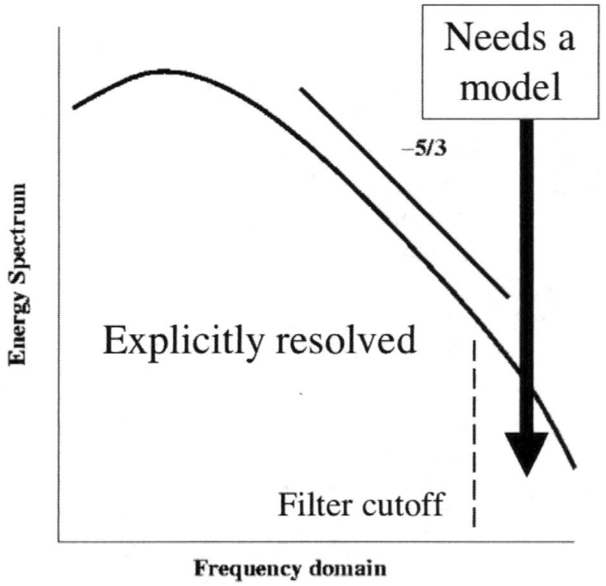

Figure 2. Schematic representation of the LES concept based on the typical energy spectrum of the velocity field for a large Reynolds number, homogeneous, isotropic turbulent flow [52; 53].

As presented above (and only in the context of pure gaseous flows), the mathematical system to be solved is represented by five highly coupled partial differential equations to which you need to add N species transport equations. The resulting number of dependent variables for which one looks for a fully $3D$ and time dependent solution is $N + 5$.

3.1. Modelling the unclosed terms

The application of the filtering operation to the instantaneous set of transport equations with chemical reactions yields the LES equations, which need modeling for the system to be closed [36; 40]. The new unknowns in the momentum transport equation and also called the subgrid-scale (SGS) turbulent velocity tensor is usually addressed through the concept of SGS turbulent viscosity model and the Boussinesq assumption [35; 54–57]. The SGS species flux (SGS mixing of the individual species that appear in the species transport equations), and the SGS energy flux (energy transport equation), are respectively modeled by use of the species SGS turbulent diffusivity along with a turbulent Schmidt number [58–61]. The eddy diffusivity is also used along with a turbulent Prandtl number [58–61]. Note that the performance of the models can be improved through the use of a dynamic formulation [56; 62–65] or more advanced type of closures [66–72].

When dealing with reacting flows, LES is faced with another set of unclosed terms issued by the highly nonlinear expressions involved in the chemical kinetics that represents the production and consumption rates of the various species involved in the chemical reaction. Again, the lack of information below the filter length scale needs to be modeled to properly mimic the interaction between combustion and turbulence. Such models are crucial since combustion usually happens below the filter length scale and a direct use of the filtered mean field (species concentrations, for example) to approximate combustion results in erroneous predictions [34; 35]. Many turbulent/combustion models have been developed to perform LES of turbulent reacting flows [66; 67; 70; 71; 73–78]. The major drawback for these models stems from the fact that

Journal of Physics: Conference Series **125** (2008) 012029 doi:10.1088/1742-6596/125/1/012029

they are usually derived for specific combustion regimes: *i.e.* fully premixed, partially-premixed or diffusion regimes [59]. Off course when dealing with industrial burners, most of the current designs never operate with a given regime of combustion but often exhibit all of the three regimes. Extensive use of these closures in an industrial context needs therefore a lot of caution and exhaustive validations.

3.2. Chemical kinetic schemes

Another open issue lies in the modeling of the chemical kinetics [60]. Indeed, it remains a large area of investigation for chemists: aeronautical fuels and air combustion usually involve thousands of chemical compounds and hundreds of chemical reactions. As such and when known, a detailed kinetic model usually contains too many species to be solved by LES or DNS (figure 3). The induced computer cost would be, in this context, prohibitive and out of reach with today's technologies. Alternative solutions are common, two of which being: kinetic reduction until a limited number of species [60; 79] is obtained to allow the computation (figure 3) or use of tabulated approaches [80; 81] coupled to a progress variable and a Zeldovic'h conserved scalar transport equations to parameterize the combustion table [82]. The induced limitation appears in the need for new kinetic models or tables when the operating point to be simulated is changed: that is, pressure and temperature dependencies of the different rates of productions are lost with these approaches and intermediate species (mostly involved in pollutant) are usually misrepresented, thus also influencing the flame dynamics. Another shortcoming of such approaches is the overall lack of generality of the constructed kinetic or table that will not cover ignition phases, the entire range of premixed and/or diffusion flames.

Detailed kinetics of H_2/O_2

9 species, 38 reactions

Nb	Reaction	B	b	E
1	$H_2 + O_2 \rightleftharpoons 2OH$	1.70E+13	0.000	47780
2	$OH + H_2 \rightleftharpoons H_2O + H$	1.17E+09	1.300	3626
3	$H + O_2 \rightleftharpoons OH + O$	2.00E+14	0.000	16800
4	$O + H_2 \rightleftharpoons OH + H$	1.80E+10	1.000	8826
5[a]	$H + O_2 + M \rightleftharpoons HO_2 + M$	2.10E+18	-1.000	0.
6	$H + O_2 + O_2 \rightleftharpoons HO_2 + O_2$	6.70E+19	-1.420	0.
7	$H + O_2 + N_2 \rightleftharpoons HO_2 + N_2$	6.70E+19	-1.420	0.
8	$OH + HO_2 \rightleftharpoons H_2O + O_2$	5.00E+13	0.000	1000.
9	$H + HO_2 \rightleftharpoons 2OH$	2.50E+14	0.000	1900.
10	$O + HO_2 \rightleftharpoons O_2 + OH$	4.80E+13	0.000	1000.
11	$2OH \rightleftharpoons O + H_2O$	6.00E+18	1.300	0.
12[b]	$H_2 + M \rightleftharpoons H + H + M$	2.23E+12	0.500	92600
13	$O_2 + M \rightleftharpoons O + O + M$	1.85E+11	0.500	95560
14[c]	$H + OH + M \rightleftharpoons H_2O + M$	7.50E+23	-2.600	0.
15	$H + HO_2 \rightleftharpoons H_2 + O_2$	2.50E+13	0.000	700.
16	$HO_2 + HO_2 \rightleftharpoons H_2O_2 + O_2$	2.00E+12	0.000	O.
17	$H_2O_2 + M \rightleftharpoons OH + OH + M$	1.30E+17	0.000	45500
18	$H_2O_2 + H \rightleftharpoons HO_2 + H$	1.60E+12	0.000	3800
19	$H_2O_2 + OH \rightleftharpoons H_2O + HO_2$	1.00E+13	0.000	1800

Reduced kinetics of H_2/O_2

Baurle, 2003: 6 species, 7 reactions

Nb	Reaction	B	b	E
1	$H_2 + O_2 \rightleftharpoons 2OH$	1.70E+13	0.000	48050
2	$OH + H_2 \rightleftharpoons H_2O + H$	2.20E+13	0.000	5155
3	$H + O_2 \rightleftharpoons OH + O$	1.20E+17	- 0.910	16530
4	$O + H_2 \rightleftharpoons OH + H$	5.06E+04	2.670	6293
5[a]	$2 OH \rightleftharpoons H_2O + O$	6.300+12	0.000	1090
6	$H+OH+M \rightleftharpoons H_2O + M$	2.21E+22	- 2.000	0
7	$H+H+M \rightleftharpoons H_2 + M$	7.30+17	-1.000	0

3 species, 1 reaction

Nb	Reaction	B	b	E
1	$H_2 + 1/2\ O_2 \rightarrow H_2O$	2.56E+13	0.000	40000

Figure 3. Typical detailed versus reduced kinetic schemes for the combustion of H_2-O_2.

3.3. Multiphase flows

As shown by the numerous examples published in the past several years [34; 43; 45; 73; 74; 83–91], LES is very accurate in complex turbulent gaseous flows. This accuracy mainly results from the explicit and direct computation of the largest-scale structures of the flow, which are the most difficult to model and contain a significant part of the physics controlling the flame. The introduction of a dispersed liquid fuel raises two kinds of problems:

- The physics of a liquid fuel spray is very complex and is not yet fully understood [46]. The atomisation process of a liquid fuel jet [92–96], the turbulent dispersion of the resulting droplets [97–101], their interaction with walls [102; 103], their evaporation, and their combustion [24] are phenomena occurring for LES at the subgrid scale and therefore requiring accurate modeling.
- The numerical implementation of two-phase flow in LES remains a challenge. The equations for both gaseous and dispersed phases must be solved together at each time step in a strongly coupled manner. This requirement differs from classical RAS, where both phases can be solved in a weakly coupled procedure, bringing first the gas flow to convergence, then calculating the associated dispersed phase and iterating until convergence of both phases.

Attempts to extend RAS formulation to LES of two-phase combustion [86; 104; 105] are all based on a Euler-Lagrange (EL) description of the dispersed phase in which the flow is solved using an Eulerian method and the particles are tracked with a Lagrangian approach. An alternative is the Euler-Euler (EE) description, also called two-fluid approach, in which both the gas and the dispersed phases are solved using an Eulerian formulation. Both strategies have advantages and shortcomings that are further emphasized in the LES context.

Following the individual trajectory of millions of droplets created by standard injectors implies computer resources that are still far beyond the capacities of computers available today and even in the coming years. In order to overcome this issue the stochastic Lagrangian approach is usually introduced, where each particle is only a "numerical" particle, representing in fact a statistically homogeneous group of real particles. This reduces the number of particles to compute but implies modeling for these parcels of particles [106]. Another difficult point is that the topology of the flow in dense zones (e.g., near the injectors) differs from a cloud of droplets and a Lagrangian description may not be adapted in these specific regions. The EE description requires an initial modeling effort much larger than in the EL method [15] and faces difficulties in handling droplet clouds with extended size distributions [107]. Moreover, the resulting set of equations is numerically difficult to handle and requires special care [108].

In the context of LES, just like for the gas phase, modeling issue appears for two-phase flow simulations, either in the EL or EE formulation. The closure problem is here linked to the SGS contribution to the turbulent droplet dispersion. This problem has already been addressed in [109] but is still an open question.

3.4. Numerics

The modeling problems being specifically addressed by LES, a remaining issue discussed here and often disregarded in the context of complex configuration simulations deals with the numerical scheme implemented in the computer code. Indeed, once closed, the system of governing LES transport equations needs to be solved numerically to yield a numerical approximations of the $N + 5$ flow variables. The conventional strategy that is common to any computational fluid dynamics (CFD) code aims at representing the nonlinear mathematical operators by introducing linear discrete operators to ease the temporal integration given an initial solution. Because of the geometrical complexity of target configurations, the scheme should cope with unstructured and/or hybrid meshes. For this specific reason, the finite-volume numerical approach [111] is often retained for CFD codes. Developments and research on the finite-element approaches [111]

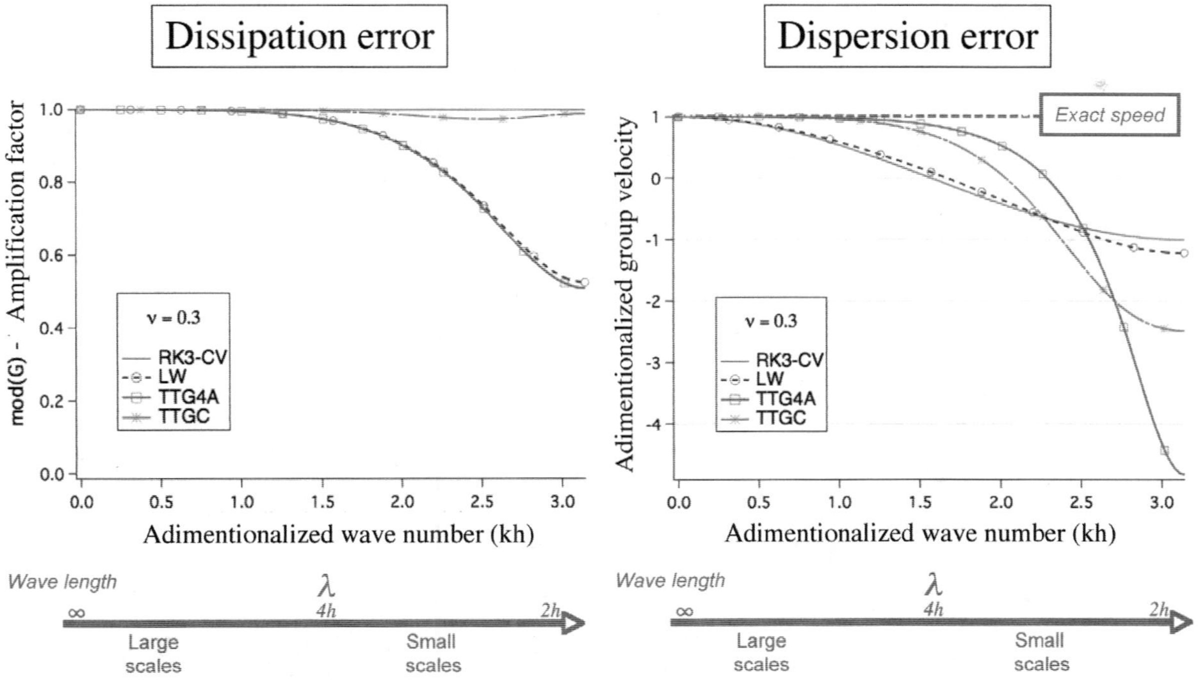

Figure 4. Dissipation and dispersion errors as functions of the wave number and as introduced by conventional numerical schemes used in LES on unstructured meshes: extracted from [110].

for CFD exist but do not seem to be used in advanced LES codes as discussed here. The difficulty is then to derive generic high-order temporal and spatial discrete operators with a decent computer overhead and various cell topologies.

Of course, the restriction of the governing equations onto a discrete set of cells and operations introduces errors. These truncation errors (in time and space) will impact the numerical prediction, and the approximation will differ from the exact solution. For LES, the approach looks for spatially and temporally dependent flow solutions over a range of wavelengths that cover the characteristic size of the geometry all the way to the smallest grid cell. To be accurate, the LES information linked to these wavelengths needs to be correctly propagated in space and in time. Numerics cannot guarantee such constraints especially over such a large range of wave lengths. In fact, and depending on the approach retained, the properties of the schemes are a function of the local mesh size to the wave-length ratio: that is, the number of grid nodes per wave-length. The errors introduced by numeric on a problem of convection are twofold (figure 4):

- Dissipation errors: the simple convection problem the initial solution is simply convected in space at the fixed convective velocity. With numerical scheme based on the notion of up- or downwinding, that property is not guaranteed, and the initial solution is diffused [112]. This lack of conservation is referred to as a dissipative error, and results in the regular decrease of the maximum value as the solution advances in time.
- Dispersion errors: an additional effect introduced by some numerical scheme is a wave-dependent convection velocity. It is traduced by wavelengths that are convected at different velocities and as a function of the wavelength to the grid point number ratio as depicted in figure 4. The initial solution is then deformed as time evolves, in some circumstances with

some information traveling upstream.

The last numerical issue is the well-known Courant-Friedrichs-Lewis (CFL) stability criterion [111], which guarantees the overall stability of the integration. That is to say, the round-off errors or the noise artificially introduced by the discretization will not be amplified by the scheme only if the CFL is respected. Likewise in the context of diffusion dominated physics, schemes are stable under the Fourier criterion [111]. All of these behaviors are usually known issues. However, because of the specific context of LES, these spurious oscillations need to be controlled or at least quantified to properly assess a given SGS model or computer code. In that respect, the coupling and interactions between numeric and models are still not understood by the LES community although recent researches illustrate the impact on LES, its reliability and the understanding of the method [113; 114].

4. LES validation

The previous discussion clearly highlights the need for extensive and comprehensive validation procedures of LES: *i.e.* the physics, the models, the numeric... Such validations are necessary for newly developed LES codes and prior to their application on industrial applications (figure 1). As a matter of fact, all of the recent complex geometry LES results are obtained by research codes with years of development and multiple validations often published in scientific journals.

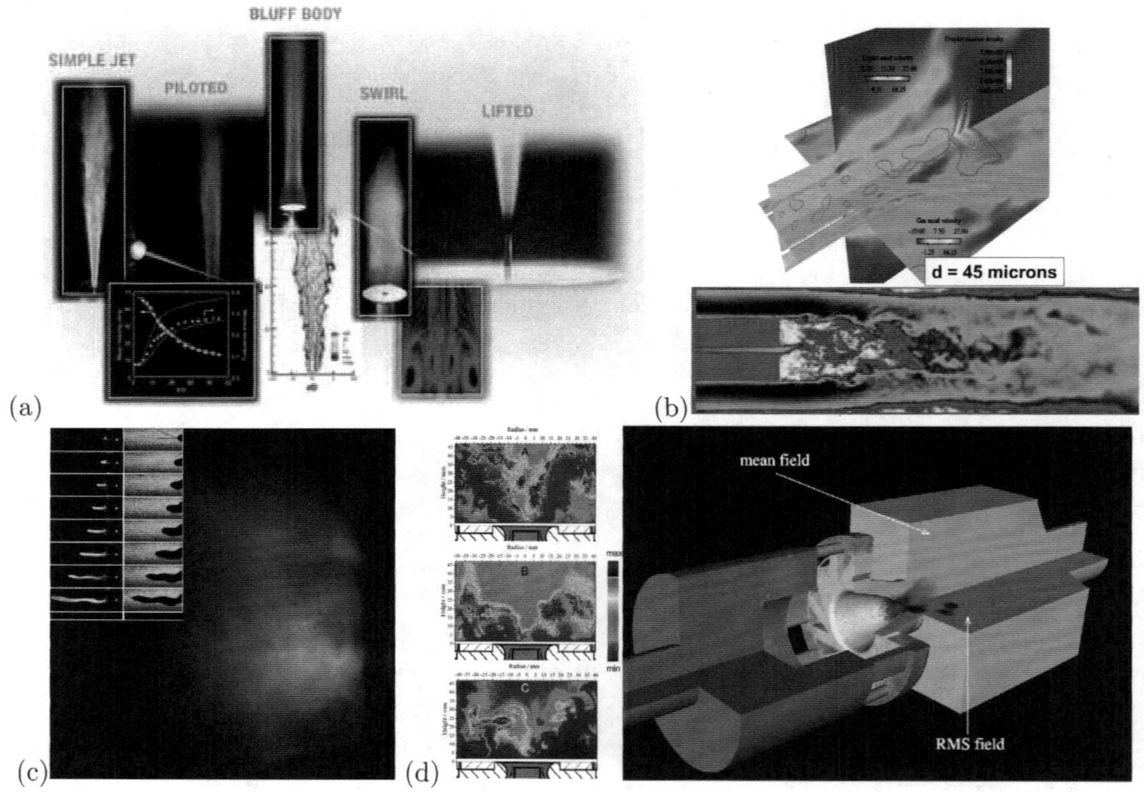

Figure 5. (a) Turbulent nonpremixed flames. (TNF): community covering measurement techniques and numerical validations with combustion. (b) Sommerfeld and Hercules experiments [115] provide data for multiple phase flows. (c) Ignition experiments with fully transient diagnostics [116] and (d) the Preccinsta burner [117] designed for thermo-acoustic instabilities.

From a modeling point of view, LES models should in principle degenerate when the grid resolution meets the DNS context. Likewise, for large filter sizes, RAS type of predictions are expected [34; 35]. For multiple-phase and reacting LES, modeling appears at different levels. The initial modeling of the physics as prescribed by a EL/EE approach or for premixed/diffusion flames does not ease the process. Once the development context fixed, the filtering process of LES is introduced, and the new closure problem appears. Simple analytical test cases are of great help to validate the implemented models and numeric as found in complex codes. *A priori* and *a posteriori* validations of models against DNS predictions are also largely recommended [35; 36], although the Reynolds number dependency of DNS somehow limits a direct extrapolation of the diagnostic to complex industrial flows.

Detailed experiments in increasingly complex configurations are also necessary. However, contrary to the RAS context, these experiments require well-controlled physics, accurate boundary condition description and when possible access to time series for an accurate validation of LES code. Such benchmark test cases (figure 5) are of scientific interest and greatly affect the LES community with a large number of combustion scientists involved. They also constitute a stimulating research environment, especially with the increased difficulty that is unsteadiness. The increase in difficulty of these benchmark test facilities needs to be sustained with more and more physics to be addressed. Statistically stationary turbulent flows are usually first targeted: free and confined jets, recirculating flows, swirl flows, and the like. Fully transient flows, such as ignition sequences, are only starting to be addressed by experimentalists and already raise interesting issues if to be simulated with LES [116].

5. Parallelization and code scalability

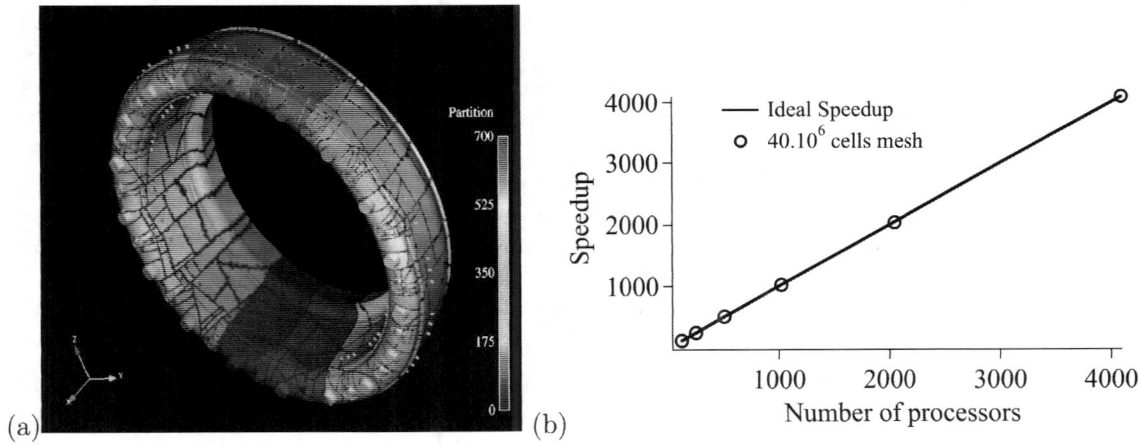

Figure 6. Domain partitioning strategy as implemented for an advanced LES code: (a) view of all the subdomains assigned to each processor (one color corresponds to one part of the domain computed by one processor, blue lines corresponds to surfaces where processors need to exchange information) and (b) scalability test on a HPC system as obtained by the LES code.

LES codes are usually validated on scientific applications with increasing complexity and for which the computing effort does not necessarily justify state-of-the-art HPC. In some circumstances, however, these computer codes and models need to be thought of with an extended range of use and potential end-users in the industrial realm. With this in mind, several new constraints appear. The first belongs to computing sciences and deals with scalability of the computer application (figure 6(b)), that is, the ability to efficiently take advantage of a large number of processors by reducing the number of operations per processor while increasing the number of operations performed in parallel. With scalability, not only is the wall-clock time potentially reduced (the time the end-user has to wait before to have a solution to a problem), but it also provides increased computer memory which allows to tackle large problems.

Contrarily to scientific burners, industry usually treats large and complex geometries for which LES require much larger number of grid nodes and cells. Deploying such grids on HPC systems ultimately necessitates a large number of processors. CFD codes and their algorithmic are heavily affected by such use. Scientific choices are also of importance and should ease scalability. The final aim of the retained strategy for these computer codes is twofold:

- Minimize the time devoted to communications inferred by the dissemination among the processors of pieces of the geometry and which induces the exchange of information between processors with common interfaces
- Maximize the time devoted to the integration of the mathematical system by each processor

For LES, it is easy to show that domain partitioning is efficient onnly on large grids because task parallelization would require the communication of very large three-dimensional datasets at each iteration between all processors. Load balancing among the processors is thus the first issue that provides scalability. With partitioning, the first objective is thus to partition a grid in subdomains that will guarantee a uniform computational effort throughout the processors despite the fact that they will treat different parts of the mathematical problem. Note that depending on the physics, partioning may need to be reinitialized as the computation proceeds as recently developed.

Reduced communications also goes with the choice of the numerical scheme and partly explains why finite volume is preferred to finite difference for example [111]. In the former, interfaces between processors reduce to cell faces (figure 6(a)) and only the fluxes and/or face nodal values are to be communicated from one processor to the next. For the latter, more information is exchanged since some overlap of the different sub-domains is needed. These duplicated cells guarantee coherence of the solutions at the processor interfaces and induce message sizes that scale with the number of points in the overlap region. Note that in the case of high-order schemes, as preferred for LES, the nodes in the overlapping region are large.

The nature of the differential equations used to model the physics is also of importance. For instance, flow dynamics may be addressed in the context of the compressible/incompressible Navier-Stokes equations. The incompressible system needs to find a solution for the pressure field that is obtained by solving a Poisson equation that follows a logarithmic scalability law as the number of grid points increases. The main advantage of that formulation comes from the stability criterion, which relies on the convective velocity. The compressible formalism requires a CFL condition based on the convective plus acoustic speeds (smaller time steps). Scalability of the problem is, however, linear with the number of grid points. Explicit/implicit numerical schemes are other issues that need to be carefully addressed for efficient HPC codes.

Specifically for two-phase flow applications, two methods may be used for LES: (1) task parallelization, in which certain processors compute the gas flow and others compute the droplets flow, and (2) domain partitioning, in which droplets are computed together with the gas flow on geometrical subdomains mapped on parallel processors. With (2) droplets must be exchanged between processors when leaving a sub-domain to enter an adjacent domain. However, codes based on domain partitioning are difficult to optimize on massively parallel architectures when droplets are clustered in one part of the domain (typically, near the fuel injectors). Moreover, the distribution of droplets may change during the computation: for a gas turbine re-ignition sequence, for example, the chamber is filled with droplets when the ignition begins thus ensuring an almost uniform droplet distribution; these droplets then evaporate rapidly during the computation, leaving droplets only in the near injector regions. In order to preserve a high parallel efficiency on thousands of processors, dynamic load-balancing strategies are required that re-decompose the domain during the computation itself. The computer implementation of the EL approach is not well suited to parallel computers: since two different solvers must be coupled, the complexity of the implementation on a parallel computer if improperly done will increase drastically compared to a single-phase code. The EE approach has the important advantage to be straightforward to implement in a numerical tool, and immediately efficient as it allows the use of the same parallel algorithm than for the gas phase [118].

(a) (b)
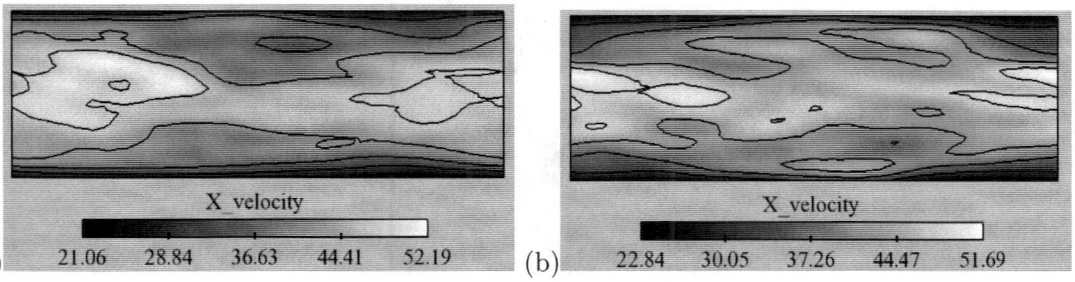

Figure 7. Instantaneous LES snapshots in the mid-channel width showing the axial velocity component in a periodic turbulent channel flow computed using (a) four processors and (b) eight processors. For this test case, initial conditions, mesh resolution, time step, and so forth are identical and the flow fields are obtained after the same number of iterations [119; 120].

HPC LES codes involve huge computer programming skills, and state-of-the-art codes are usually the result of numerous collaborations at the various stages of it developments. Parallel applications bring new issues such as error propagation and impact on LES predictions as illustrated on figure 7. Indeed, partioning is inevitably linked to the number of processors to be used and that number may not be fixed. Subdomains will thus differ as the number of processors varies. Likewise, message ordering among the processors and at interfaces is impossible, so that commutativity errors and round-off errors will differ [119; 120] potentially resulting in different instantaneous LES snapshots. Moreover, LES is intimately linked to the grid resolution, and LES realizations will differ as the mesh resolution is increased (figure 8), further complicating the validation of the LES approach.

6. Conclusions and perspectives

Large eddy simulation is a fully unsteady spatially dependent numerical approach with high potential when applied to industrial flow applications. Current state of the art fully turbulent and reacting use of LES obtained thank to new HPC facilities produce convincing results for aeronautical engines despite the numerous physical issues that are not accounted for in these simulations. Such pioneering results open new research perspectives and applications, not only for the LES community, but also for the entire scientific community that is dealing with combustion. With the benefit from these new computing centres, CFD research should expand, and new subjects linked to fluid mechanics and its coupling to other physical phenomena should appear: radiation, heat transfer, combustion instability, detailed chemistry, multiphase flows, and so on. Fundamental applications also benefiting from HPC such as direct numerical simulations should also help in tackling more and more complex applications with LES. Furthermore, we note that frontier simulations stimulate the scientific community and propagate to the experimentalists with the development of leading-edge diagnostics addressing unsteady flow behaviors.

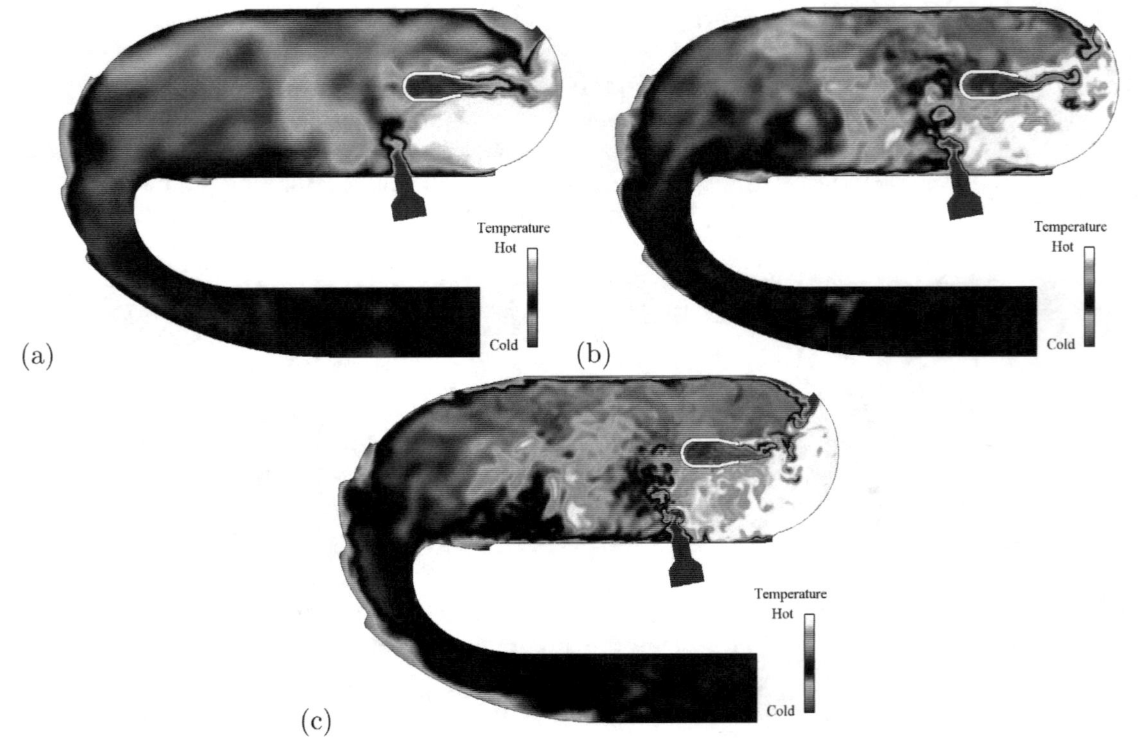

Figure 8. Instantaneous views of the temperature field as obtained by three LES of the same helicopter chamber with different grid resolution [121]: (a) 1,242,086, (b) 10,620,245, and (c) 43,949,682 tetrahedrons.

From a pure computational point of view, application of LES in an industrial context raises new issues that will need to be addressed. Data management and diagnostic will need to adjust to ever bigger databases produced by LES. Code scalability and efficiency will need to be investigated thoughtfully. Moreover, in-depth analysis of grid resolution, error propagation and interaction, and repetitivity of LES predictions with parallel computing raises new questions that will need to addressed before extensive use of LES by industry.

Acknowledgments
We acknowledge the contributions from CERFACS students and colleagues, the SciDAC organization for this invited contribution, and all the HPC contributors without whom most of the new LES applications would not be possible.

References

[1] Moin P and Mahesh K 1998 *Annual Review of Fluid Mechanics* **30** 539–578

[2] Westbrook C, Mizobuchi Y, Poinsot T, Smith P J and Warnatz J 2004 *Proceedings of the Combustion Institute* **30**

[3] Chen J Y and Kollmann W 1994 *Turbulent Reacting Flows* ed Williams P L and FA (Academic Press London) pp 213–308

[4] Baum M, Poinsot T, Haworth D and Darabiha N 1994 *Journal of Fluid Mechanics* **281** 1–32

[5] Im H, Chen J H and Chen J Y 1999 *Combustion and Flame* **118** 204–212

[6] Mizobuchi Y, Tachibana S, Shinjo J, Ogawa S and Takeno T 2002 *Proceedings of the Combustion Institute* **29** 2009 – 2015

[7] Seiser R, Frank J, Liu S, Chen J, Sigurdsson R and Seshadri K 2004 *Proceedings of the Combustion Institute* **30** 423–430

[8] Bell J, Day M, Almgren A, Lijewski M, Rendleman C, Cheng R and Shepherd I 2006 *SciDAC* (*Journal of Physics: Conference Series* vol 46) (Denver, Colorado) pp 1–15

[9] Bell J, Day M, Grcar J, Lijewski M, Driscoll J and Filatyev S 2007 *Proceedings of the Combustion Institute* **31** 1299–1307

[10] Lignell D O, Chen J H, Smith P J, Lu T and Law C K 2007 *Combustion and Flame* **151** 2–28

[11] Woosley S, Almgren A, Bell J, Glatzmaier G, Kasen D, Kerstein A, Ma H, Nugent P, Ropke F, Sankaran V and Zingale M 2007 *SciDAC* Journal of Physics: Conference Series (Boston, Massachusetts)

[12] Mizobuchi Y, Shinjo J, Ogawa S and Takeno T 2004 *Proceedings of the Combustion Institute* **30** 611–619

[13] Hawkes E, Sankaran R, Sutherland J and Chen J 2005 *SciDAC* (*Journal of Physics: Conference Series* vol 16) p 65

[14] Chen J H, Hawkes E R, Sankaran R, Mason S D and Im H G 2006 *Combustion and Flame* **145** 128–144

[15] Février P, Simonin O and Squires K 2005 *Journal of Fluid Mechanics* **533** 1–46

[16] Miller R and Bellan J 1999 *Physics of Fluids* **12**

[17] Réveillon J and Vervisch L 2000 *Combustion and Flame* 75–90

[18] Mashayek F and Taulbee D B 2002 *Num. Heat Transfer, part B* **41** 1–29

[19] Vermorel O, Bédat B, Simonin O and Poinsot T 2003 *Journal of Turbulence* **4, 025**

[20] Domingo P, Vervisch L and Réveillon J 2005 *Combustion and Flame* **140** 172–195

[21] Lee K and Tien C 1985 *Combustion Science and Technology* **43** 167–182

[22] Soufiani A and Djavdan E 1994 *Combustion and Flame* **97** 240–250

[23] Wu Y, Haworth D C, Modest M F and Cuenot B 2005 *Proceedings of the Combustion Institute* **30** 639–646

[24] Gökalp I, Chauveau C, Morin C, Vieille B and Birouk M 2000 *Atomization and Sprays* **10** 475–510

[25] Hopfinger E and Lasheras J 1996 *Physics of Fluids* 1696–1698

[26] Shavit U 2001 *Experiments in Fluids* **31** 550–557

[27] Varga C, Lasheras J and Hopfinger E 2003 *Journal of Fluid Mechanics* **497** 405–434

[28] Lieuwen T, Torres H, Johnson C and Zinn B 2001 *Journal of Engineering for Gas Turbines and Power* **123** 182–189

[29] Truffin K and Poinsot T 2005 *Combustion and Flame* **142** 388–400

[30] Bailly P, Garréton D, Simonin O, Bruel P, Champion M, Deshaies B, Duplantier S and Sanquer S 1996 *Proceedings of the Combustion Institute* **26** 923 – 930

[31] Bogey C and Bailly C 2006 *Theoretical and Computational Fluid Dynamics* **20** 23–40

[32] Freund J B 2001 *Journal of Fluid Mechanics* **438** 277–306

[33] Colonius T and Lele S K 2004 *Progress in Aerospace Sciences* **40** 345–416

[34] Poinsot T and Veynante D 2005 *Theoretical and numerical combustion* (R.T. Edwards, 2nd edition.)

[35] Pope S B 2000 *Turbulent flows* (Cambridge University Press)
[36] Sagaut P 2002 *Large eddy simulation for incompressible flows* (Springer)
[37] Tennekes H and Lumley J L 1972 *A first course in turbulence* (Cambridge: M.I.T. Press)
[38] Lumley J L 1978 *Advances in Applied Mechanics* **18** 123–176
[39] Launder B E, Reece G J and Rodi W 1975 *Journal of Fluid Mechanics* **68** 537–566
[40] Ferziger J 1997 *New tools in turbulence modelling* ed Métais O and Ferziger J (Les Editions de Physique - Springer Verlag) pp 29 – 47
[41] Lesieur M 1997 *New tools in turbulence modelling* ed Métais O and Ferziger J (Les Editions de Physique - Springer Verlag) pp 1 – 28
[42] Moin P and Apte S V 2006 *American Institute of Aeronautics and Astronautics Journal* **44** 698–708
[43] Boudier G, Gicquel L Y M, Poinsot T, Bissières D and Bérat C 2007 *Proceedings of the Combustion Institute* **31** 3075–3082
[44] Staffelbach G, Boudier G, Gicquel L and Poinsot T 2008 *Proceedings of the Combustion Institute* **32** accepted for publication
[45] Boileau M, Staffelbach G, Cuenot B, Poinsot T and Bérat C 2008 *Combustion and Flame* **154** 2–22
[46] Lefebvre A H 1999 *Gas Turbines Combustion* (Taylor & Francis)
[47] James S, Zhu J and Anand M 2006 *American Institute of Aeronautics and Astronautics Journal* **44** 674–686
[48] Mendez S and Nicoud F 2008 *Journal of Fluid Mechanics* **598** 27–65 URL `http://www.cerfacs.fr/ cfdbib/repository/TR_CFD_06_110.pdf`
[49] Aldama A A 1990 *Lecture Notes in Engineering* vol 49 (New York: Springer-Verlag)
[50] Vreman B, Geurts B and H Kuerten H 1994 *Physics of Fluids* **6** 4057–4059
[51] Favre A 1969 *Problems of hydrodynamics and continuum mechanics* (Philadelphia: SIAM) pp 231–266
[52] Kolmogorov A N 1941 *C.R. Acad. Sci. USSR* **30** 301
[53] Kolmogorov A N 1962 *Journal of Fluid Mechanics* **62** 82
[54] Smagorinsky J 1963 *Monthly Weather Review* **91** 99–164
[55] Ducros F, Nicoud F and Poinsot T 1998 *ICFD* (Baines M. J.) pp 293–300
[56] Germano M 1992 *Journal of Fluid Mechanics* **238** 325–336
[57] Vreman A 2004 *Physics of Fluids* **16** 3670
[58] Spalding D B 1978 *17th Symp. (Int.) on Combustion* (The Combustion Institute, Pittsburgh) pp 431–440
[59] Williams F A 1985 *Combustion theory* (Menlo Park, CA: Benjamin Cummings)
[60] Kuo K K 1986 *Principles of Combustion* (New York: John Wiley)
[61] Bell J B, Cheng R K, Day M S and Shepherd I G 2007 *Proceedings of the Combustion Institute* **31** 1309–1317
[62] Ghosal S and Moin P 1995 *Journal of Computational Physics* **118** 24 – 37
[63] Moin P, Squires K D, Cabot W and Lee S 1991 *Physics of Fluids* **A 3** 2746–2757
[64] Lilly D K 1992 *Physics of Fluids* **4** 633–635 URL LES
[65] Meneveau C, Lund T and Cabot W 1996 *Journal of Fluid Mechanics* **319** 353
[66] Colucci P, Jaberi F, Givi P and Pope S 1998 *Physics of Fluids* **10** 499

[67] Jaberi F, Colucci P, James S, Givi P and Pope S 1999 *Journal of Fluid Mechanics* **401** 85–122

[68] Fureby C and Grinstein F F 1999 *AIAA Journal* **37** 544–556

[69] Boyaval S and Dumouchel C 2001 *Proceedings of the 20th ILASS - Europe Meeting 2001*

[70] Gicquel L, Ģivi P, Jaberi F and SB P 2002 *Physics of Fluids* **14** 1196

[71] Sheikhi M, Drozda T, Givi P, Jaberi F and Pope S 2005 *Proceedings of the Combustion Institute* vol 30 pp 549–556

[72] Givi P 2006 *AIAA Journal* **44** 16–23

[73] Colin O, Ducros F, Veynante D and Poinsot T 2000 *Physics of Fluids* **12** 1843–1863

[74] Pitsch H and de la Geneste L D 2002 *Proceedings of the Combustion Institute* **29** 2001–2008

[75] Domingo P, Vervisch L and Bray K 2002 *Combustion Theory and Modelling* **6**

[76] Vervisch L, Hauguel R, Domingo P and Rullaud M 2004 *Journal of Turbulence* **5** 004

[77] Pitsch H 2006 *Annual Review of Fluid Mechanics* **38** 453–482

[78] Raman V, Pitsch H and Fox R O 2005 *Combustion and Flame* **143** 56–78

[79] Miller H P, Mitchell R, Smooke M and Kee R 1982 *19th Symp. (Int.) on Combustion* (The Combustion Institute, Pittsburgh) pp 181–196

[80] Maas U and Pope S B 1992 *Combustion and Flame* **88** 239–264

[81] Gicquel O, Darabiha N and Thévenin D 2000 *Proceedings of the Combustion Institute* **28** 1901–1908

[82] Domingo P, Vervisch L, Payet S and Hauguel R 2005 *Combustion and Flame* **143** 566–586

[83] Caraeni D, Bergstrom C and Fuchs L 2000 *Flow, Turbulence and Combustion* **65** 223–244

[84] Chakravarthy V K and Menon S 2000 *Flow, Turbulence and Combustion* **65** 133–161

[85] Forkel H and Janicka J 2000 *Flow, Turbulence and Combustion* **65** 163–175

[86] Mahesh K, Constantinescu G and Moin P 2004 *Journal of Computational Physics* **197** 215–240 ISSN 0021-9991

[87] Selle L, Lartigue G, Poinsot T, Koch R, Schildmacher K U, Krebs W, Prade B, Kaufmann P and Veynante D 2004 *Combustion and Flame* **137** 489–505

[88] Sommerer Y, Galley D, Poinsot T, Ducruix S, Lacas F and Veynante D 2004 *Journal of Turbulence* **5**

[89] Moureau V, Lartigue G, Sommerer Y, Angelberger C, Colin O and Poinsot T 2005 *Journal of Computational Physics* **202** 710–736

[90] Roux S, Lartigue G, Poinsot T, Meier U and Bérat C 2005 *Combustion and Flame* **141** 40–54

[91] Roux A, Gicquel L Y M, Sommerer Y and Poinsot T J 2007 *Combustion and Flame* **152** 154–176

[92] Weber C and Angrew Z 1931 *Math. Mech.* **11** 136–154

[93] Nukiyama S and Tanasawa Y 1939 *Trans. Soc. Mech. Eng. Japan* **5** 62–67

[94] Lasheras J C and Hopfinger E J 2000 *Annual Review of Fluid Mechanics* **32** 275–308

[95] Reitz R D and Bracco F V 1982 *Physics of Fluids* **25** 1730–1742

[96] Han Z, Parrish S, Farrell P V and Reitz R D 1997 *Atomization and Sprays* **7** 663–684

[97] Apte S V, Mahesh K, Moin P and Oefelein J C 2003 *International Journal of Multiphase Flow* **29** 1311–1331

[98] Alipchenkov V M and Zaichik L I 2006 *Journal of Fluid Dynamics* **41**

[99] Boivin M, Squires K D and Simonin O 2000 *Physics of Fluids* **12** 2080–2090

[100] Oesterlé B and Zaichik L I 2004 *Physics of Fluids* **16** 3374–3384

[101] Marchioli C, Picciotto M and Soldati A 2006 *Journal of Turbulence* **7** 1–12

[102] Benson M, Tanaka T and Eaton J K 2005 *Journal of Fluids Engineering* **127** 250–256

[103] Sommerfeld M and Kussin J 2003 *International Journal of Multiphase Flow* **29** 701–718

[104] Sankaran V and Menon S 2002 *Journal of Turbulence* **3** 011

[105] Apte S V, Gorokhovski M and Moin P 2003 *International Journal of Multiphase Flow* **29** 1503–1522

[106] Moin P 2004 *Progress in Computational Fluid Dynamics* **4** 237–240

[107] Mossa J B 2005 *Extension polydisperse pour la description Euler-Euler des écoulements diphasiques réactifs - TH/CFD/05/74* Phd thesis INP Toulouse URL `http://www.cerfacs.fr/ cfdbib/repository/TH_CFD_05_74.pdf`

[108] Riber E 2007 *Développement de la méthode de simulation aux grandes échelles pour les écoulements diphasiques turbulents* Phd thesis INP Toulouse

[109] Fede P, Simonin O, Villedieu P and Squires K D 2006 *Proceedings of the Summer Program* (Center for Turbulence Research, NASA Ames/Stanford Univ.) pp 247–258

[110] Lamarque N 2007 *Schémas Numériques et conditions limites pour la simulation aux grandes échelles de la combustion diphasique dans les foyer d'hélicoptères* Ph.D. thesis Institut National Polytechnique de Toulouse Toulouse, France

[111] Anderson D A 1984 *Computational Fluid Mechanics and Heat Transfer* (New York: Hemisphere Publishing Corporation)

[112] Colin O and Rudgyard M 2000 *Journal of Computational Physics* **162** 338–371

[113] Boudier G, Staffelbach G, Gicquel L and Poinsot T 2007 *QLES (Quality and reliability of LES) workshop* (Leuven, Belgium: ERCOFTAC)

[114] Meyers J and Sagaut P 2007 *Physics of Fluids* **19** 1–4

[115] Sommerfeld M and Qiu H H 1991 *International Journal of Heat and Fluid Flow* **12**

[116] Ahmed S F, Balachandran R, Marchione T and Mastorakos E 2007 *Combustion and Flame* **151** 366–385

[117] Meier W, Weigand P, Duan X and Giezendanner-Thoben R 2007 *Combustion and Flame* **150** 2–26

[118] Kaufmann A 2004 *Vers la simulation des grandes échelles en formulation Euler/Euler des écoulements réactifs diphasiques* Phd thesis INP Toulouse

[119] Staffelbach G, Senoner J M, Gicquel L and Poinsot T 2008 *8th International meeting on High performance computing for computational science* (ENSEEIHT, Toulouse, France)

[120] Senoner J M, Garcia M, Mendez S, Staffelbach G, Vermorel O and Poinsot T 2008 *AIAA Journal* **in press**

[121] Boudier G, Gicquel L and Poinsot T 2008 *Combustion and Flame* **10.1016/j.combustflame.2008.04.013**

Journal of Physics: Conference Series **125** (2008) 012030 doi:10.1088/1742-6596/125/1/012030

Direct numerical Simulation of turbulent nonpremixed flame extinction by water spray

Hong G Im,[1] Arnaud Trouvé,[2] Paul G Arias[1] and Praveen Narayanan[2]
[1]Department of Mechanical Engineering, University of Michigan, Ann Arbor, MI 48109-2125, USA
[2]Department of Fire Protection Engineering, University of Maryland, College Park, MD20742-3031, USA

E-mail: hgim@umich.edu

Abstract. This paper presents a brief overview of our INCITE 2007 project on the direct numerical simulation of nonpremixed flames subjected to turbulent flows and water spray evaporation. The simulation is a culmination of our recent developments in advanced physical submodels associated with radiative heat transfer and Lagrangian spray dynamics. One of the main objectives is to identify and verify a unified extinction criterion based on the flame weakness factor built on the excess enthalpy variable concept. The results from two-dimensional turbulent ethylene-air flames suggest that the proposed diagnostic tool provides a correct measure of flame weakening. Further work is under way to extend the analysis over a wide range of parametric conditions.

1. Introduction

Recent advances in high-performance computing have enabled the study of three-dimensional, laboratory-scale combustion problems using the direct numerical simulation (DNS) approach [1-3]. The state-of-the-art high-fidelity combustion simulations involve tens of reactive scalar variables, which impart a significant computational overhead in terms of the number of transport equations to be solved as well as the spatial and temporal stiffness arising from the large scale disparities. Furthermore, successful direct simulation of turbulent combustion depends strongly on the fidelity of the embedded physical submodels. The SciDAC project led by a multi-university team [4] has been developing key submodels including thermal radiation based, soot formation, and spray evaporation. A more recent INCITE project [5] has subsequently provided a pilot application of these models to important scientific issues on flame suppression by water spray.

Flame extinction has a significant impact on the performance of non-premixed combustion systems. It determines, in part, the turbulent flame structure and the levels of pollutant emission (NOx, CO, and soot). In engine applications, flame extinction results from the high turbulence intensities associated with the momentum-driven, high-Reynolds number flames and/or the interaction with cold wall surfaces. In fire applications, extinction results from the low combustion intensities associated with the buoyancy-driven, moderate-to-low Reynolds number flames and/or the use of inert gaseous agents or water sprays in fire suppression problems. Laminar diffusion flame theory considers two mechanisms for flame extinction: excessively high values of the rate of fuel-air mixing, that is, insufficient residence times to complete the combustion processes, [6] and excessively low values of the flame temperature, that is, under sluggish combustion conditions that are vulnerable to external cooling. The fast mixing limit is also called kinetic extinction, while the low flame temperature limit is often called radiation extinction [7]. We focus in the following on the kinetic extinction limit in laminar and turbulent flames that are weakened by evaporative cooling from a water spray.

The study of spray-flame interaction has strong scientific and practical relevance, especially in fire suppression technology. Newer designs for water-based fire suppression systems are based on water mist systems [8]. While the conventional sprinkler systems achieve fire suppression by *fuel cooling* and by *flame spread inhibition*, water mist systems achieve suppression by *flame cooling* and have the potential of improved efficiency (reduced water demand) and performance (possible circulation around obstructions and shielded areas). Despite a growing interest in water mist systems, however, a fundamental understanding of the basic physical mechanisms that lead to fire suppression by fine water droplets is still lacking. Key relevant issues include the injection and atomization processes, and overall early liquid water spray dynamics in the vicinity of the injector; the subsequent turbulent dispersion of the water droplets in the fire room and the probability of their transport to the flame location; and the interaction of evaporating water droplets with the turbulent flame, and the resulting decrease in combustion intensity and possible partial/total flame quenching. High-fidelity DNS is well suited to provide valuable information to address these issues.

This paper presents a summary of results from our ongoing investigation of spray-flame interactions. One of the key objectives in this study is to identify a unified extinction criterion that can account for the effects of flame weakening by various heat loss mechanisms. To this end, the concept of the excess enthalpy variable is introduced as a valuable diagnostic that measures deviations from a reference flame conditions resulting from non-adiabatic combustion and non-unity Lewis numbers. The theoretical developments are first evaluated in simulations of spray-flame interactions in the laminar and turbulent counterflow flame simulations.

2. Enthalpy deficit and flame extinction

Following [9], we introduce the excess enthalpy variable, H, that provides a local measure of the relative flame intensity, defined as

$$H = \frac{-(\int_T^{T^{ref}} c_p dT)}{\Delta H_{st}}, \qquad (1)$$

where c_p is the heat capacity of the reactive mixture at constant pressure, T the fluid temperature, T_{ref} the local reference temperature (corresponding to the temperature that would be obtained if combustion was adiabatic and occurring at unity Lewis numbers), and ΔH_{st} the heat of combustion (per unit mass of mixture) obtained under stoichiometric, equilibrium chemistry conditions. With this definition, H is a nondimensional quantity that takes on values of order unity: $H = 0$ under adiabatic and unity-Lewis-number conditions; H is negative in the presence of external heat losses (due to radiative heat transfer [10], convective heat transfer to walls [9], or evaporative cooling) and/or negative Lewis number effects; H is positive in the presence of external heat gains and/or positive Lewis number effects. In the present study, H provides a measure of flame weakening due to cooling by evaporating water droplets, combined with possible flame strengthening/weakening due to non-unity Lewis number effects.

Subsequently, non-premixed flame extinction is predicted to occur according to a modified scalar dissipation rate criterion [9,10]; extinction occurs if:

$$\chi_{st} \geq \chi_{st,ext}^{ref} \times \exp(-T_a(\frac{1}{T_{st}} - \frac{1}{T_{st}^{ref}})) \qquad (2)$$

where χ_{st} is the scalar dissipation rate at the stoichiometric mixture fraction, $\chi_{st,ext}^{ref}$ is the extinction value of χ_{st} for a reference flame with adiabatic condition and unity Lewis number, T_a is an effective activation temperature, T_{st} is the flame temperature at the stoichiometric mixture fraction, and T_{st}^{ref} is the value of flame temperature for a reference flame at the extinction limit. The RHS of equation (2) is the modified extinction scalar dissipation rate, $\chi_{st,ext}$, accounting for the deviations from adiabatic and/or unity Lewis number combustion. The exponential term in $\chi_{st,ext}$ represents the decrease of the critical value that is expected at lower flame temperatures. In other words, flame quenching is more readily achieved in flame configurations that are weakened by heat losses and/or negative nonunity Lewis number effects.

The extinction criterion may alternatively be cast in the form of a flame weakness factor R [9,10]:

$$R = \left(\frac{\chi_{st}}{\chi_{st,ext}^{ref}}\right) \times \exp\left(T_a\left(\frac{1}{T_{st}} - \frac{1}{T_{st}^{ref}}\right)\right) \quad (3)$$

such that extinction occurs when $R \geq 1$. Equation (3) can be rewritten as $R \approx (\chi_{st} / \chi_{st,ext}^{ref}) \times \exp(-\beta H_{st})$, where H_{st} is the value of the excess enthalpy variable at the flame and $\beta = T_a(T_{st}^{ref} - T^{\infty})/(T_{st}^{ref})^2$ is the Zeldovich number with T^{∞} the ambient fluid temperature. The nondimensional activation temperature, T_a, is determined as the sensitivity of the mass burning rate to the adiabatic flame temperature [11]. Note that flame quenching or equivalently large values of the flame weakness factor are promoted by large negative values of the excess enthalpy variable H_{st}, that is, by large values of the flame heat losses.

3. Results and discussion

The current study presents steady counterflow flames that are cooled by fine water droplets that interact with the flame at its vicinity. The flow configuration corresponds to a two-dimensional domain of size 1×2 cm at a strain rate of 440 s^{-1} (40% of the extinction strain-rate). Fuel is injected from the left and oxidizer is injected from the right side boundaries. A reduced reaction mechanism based on directed relation graphs for ethylene-air flame [12] is adopted. Homogeneous turbulence is injected from both vertical boundaries, wrinkling the flame surface. Water droplets are injected at a fixed x-location, but at random y-locations, at the local flow velocity. The droplet injection or loading, defined as a ratio of the latent heat of water droplets injected per second to the integrated flame heat release rate, is set to be 40% of the total flame power, which is expected to weaken the flame significantly enough to cause extinction events [13].

Figure 1 shows an instantaneous image of two-dimensional fields of flame temperature and excess enthalpy variable after the flame-droplet-turbulence interactions have evolved for 40 ms. Droplets are visualized as black dots. The dataset clearly shows variation in the flame thickness due to the action of turbulent eddies. Spray evaporation adds an additional flame weakening effect. The combined turbulence and spray weakening effects are denoted by the excess enthalpy variable isocontours in Figure 1(b), in which two regions of large negative values in the excess enthalpy variable are noticeable: Region A shows a relatively broad flame thickness suggesting that the flame weakening is primarily caused by the spray, while the weakening in Region B is predominantly the action of turbulence straining of the flame.

To further confirm this observation, the spatial variation

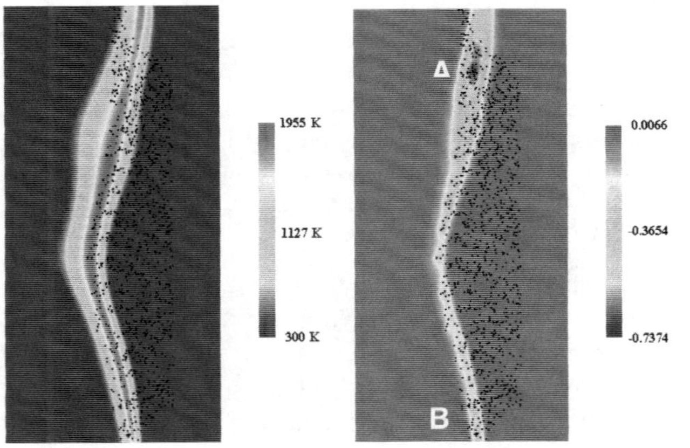

Figure 1. Instantaneous isocontours of (a) temperature and (b) excess enthalpy variable, on which the lagrangian water droplets are overlaid.

along the stoichiometric mixture fraction line (denoted by s) is monitored in terms of temperature, the normalized scalar dissipation rate, $\chi_{st} / \chi_{st,ext}^{ref}$, and the flame weakness factor R defined in equation (3). These are shown in figure 2. The flame temperature is significantly lowered near the zones $1.5 < s < 1.7$ and $0 < s < 0.2$, which correspond to Regions A and B in figure 1, respectively. In the latter (Region B), the large flame weakness factor is primarily due to the enhanced scalar dissipation rate from turbulent eddies, indicating that this is a typical strain-induced flame weakening. On the other hand, flame weakening in Region A is not evidently seen in the scalar dissipation rate, which is actually minimal there, thereby suggesting that the flame weakening is primarily caused by the cooling due to spray evaporation.

As shown in figure 2, the flame diagnosis based on the flame weakness factor appears to provide an unambiguous criterion that accounts for combined effects of flame weakening.

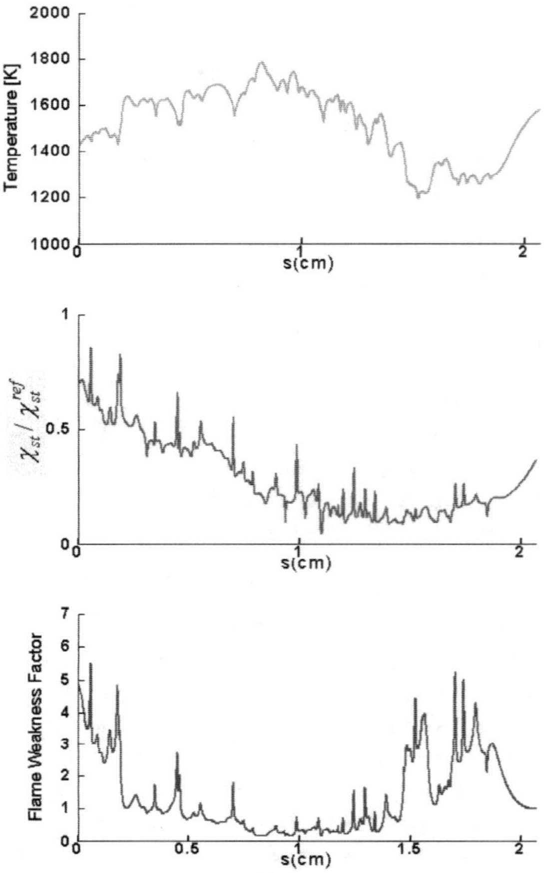

Figure 2. From top to bottom: temperature, normalized scalar dissipation rate, and the flame weakness factor along the stoichiometric line, taken from the data shown in figure 1.

4. Summary and future work

Two-dimensional direct numerical simulations of turbulent counterflow ethylene-air diffusion flames interacting with water spray were conducted. Success in such simulations requires advanced numerical algorithms to deal with strong stiffness arising from detailed chemistry as well as physical submodels associated with radiation and spray dynamics. The main scientific objective of the present study is to identify and verify the unified extinction criterion based on the flame weakness factor built on the excess enthalpy variable concept. Preliminary results are promising but further analysis is needed.

As for future work, more two- and three-dimensional simulations will be undertaken to provide a comprehensive database for parametric studies. Various datasets will be extensively analyzed by using the presented diagnostic tools proposed here, and the results will be synthesized to derive a unified flame weakness metric that accounts for the effects of turbulence and spray evaporation.

Acknowledgments

This work was supported by the U.S. Department of Energy, Office of Basic Energy Science, SciDAC Computational Chemistry Program and the Department of Energy 2007 INCITE (Innovative and Novel Computational Impact on Theory and Experiment) Program. The authors would like to thank Professor Chris Rutland and Dr. Yunliang Wang of University of Wisconsin, Dr. Chun Sang Yoo of Sandia National Laboratories, and Dr. Yi Wang for their contributions in the computational development during the earlier phase of the project.

References

[1] Chen J H and Hawkes E R 2005 Direct numerical simulation of turbulent nonpremixed combustion – fundamental insights towards predictive modeling *Proc. SciDAC 2005 Conference*
[2] Sankaran R, Hawkes E R, Chen J H, Lu T and Law C K 2007 *Proc. Combust. Inst.* **31:**1291
[3] Bell J B, Day M S, Grcar G F, Lijewski M J, Driscoll J F and Filatyev S A 2007 *Proc. Combust. Inst.* **31**:1299
[4] Im H G, Trouvé, A Rutland C J and Chen J H 2005 Terascale high-fidelity simulations of turbulent combustion with detailed chemistry http://www.purl.org/net/tstc
[5] Im H G, Trouvé A and Rutland C J 2007 Direct numerical simulation of turbulent flame quenching by fine water droplets *Innovative and Novel Computational Impact on Theory and Experiment (INCITE),*Department of Energy
[6] Linan A 1974 *Acta Astronautica* **1**:1007
[7] Tien J S 1986 *Comb. Flame* **65**:31
[8] Liu Z and Kim A K 2000 *J. Fire Prot. Engr.* **10**:32
[9] Wang Y and Trouvé A 2006 *Combust. Flame* **144**:461
[10] Narayanan P and Trouvé A 2008 *(Preprint)*
[11] Sun C J, Sung C J, He L and Law C K1999 *Combust. Flame* **118**:108
[12] Lignell D O, Chen J H, Smith P J, Lu T and Law C K 2007 *Combust. Flame* **151**:2
[13] Lentani A M and Chelliah H K1998 *Proc. Combust. Inst.* **27**:2839.

Terascale direct numerical simulations of turbulent combustion – fundamental understanding towards predictive models

D O Lignell, J H Chen, E S Richardson

Reacting Flows Research Department, Sandia National Laboratories, Livermore, CA 94550, USA

E-mail: david@crsim.utah.edu

Abstract.
Advances in high-performance computational capabilities enable scientific simulations with increasingly realistic physical representations. This situation is especially true of turbulent combustion involving multiscale interactions between turbulent flow, complex chemical reaction, and scalar transport. A fundamental understanding of combustion processes is crucial to the development and optimization of next-generation combustion technologies operating with alternative fuels, at higher pressures, and under less stable operating conditions, such as highly dilute, stratified mixtures. Direct numerical simulations (DNS) of turbulent combustion resolving all flow and chemical features in canonical configurations are used to improve fundamental understanding of complex flow processes and to provide a database for the development and validation of combustion models. A description of the DNS solver and its optimization for use in massively parallel simulations is presented. Recent DNS results from a series of three combustion configurations are presented: soot formation and transport in a nonpremixed ethylene jet flame, the effect of fuel stratification in methane Bunsen flames, and extinction and reignition processes in nonpremixed ethylene jet flames.

1. Introduction

The rapid growth in computational capabilities has presented both opportunities and challenges for high-fidelity simulations of turbulent reacting flows. The advent of terascale computing power made it possible to glean fundamental physical insight into fine-grained aero-thermo-chemical interactions in simple laboratory-scale turbulent flames with detailed chemistry using three-dimensional direct numerical simulations [1–5].

1.1. Direct numerical simulation computational approach

In direct numerical simulation (DNS), the instantaneous governing equations are solved without averaging or filtering; all relevant continuum scales are resolved on the grid with no closure models using accurate numerical methods. Such simulations are costly, requiring several million CPU-hours on a terascale computer, up to a billion grid points, and generating hundreds of terabytes of data. It is anticipated that petascale computing (and beyond) will provide access to directly simulating a larger dynamic range of scales ($O(10^4)$) and the transport of greater numbers of species representative of renewable fuels of practical interest, enabling DNS to access relevant parameter regimes.

1.2. Direct numerical simulation of canonical configurations

Next-generation alternative-fuel internal combustion engines and aero-transport and power generation gas turbines will operate in nonconventional, mixed-mode, turbulent combustion under previously unexplored aero-thermo-chemical regimes. Compared to present devices, combustion in next-generation devices is likely to be characterized by higher pressures, lower temperatures, and higher levels of dilution and/or excess air (fuel-lean). Combustion processes in these environments, combined with new physical and chemical fuel properties associated with non-petroleum based-fuels, results in complex interactions that are unknown even at a fundamental level. These unknown processes place new demands on simulation and severely restrict our ability to predict the behavior of these systems from first principles and our ability to optimize them. Therefore, there is an urgent demand for high-fidelity simulation approaches that capture these reacting thermo-fluid interactions and, in particular, capture and discriminate the effects of variations in fuel composition. DNS can uniquely isolate and reveal fundamental causal relationships that are required for physical understanding and to develop predictive models. With petascale computing on the horizon, aero-thermo-chemical interactions are amenable to study by DNS, and here we highlight recent results from terascale DNS focused on soot formation and transport in turbulent ethylene/air jet flames, stratified combustion in turbulent slot jet flames, and extinction and reignition phenomena in turbulent ethylene/air jet flames. For each configuration, a range of physical conditions has been explored. In many instances, the unique simulation benchmarks are being used in a priori development and validation of advanced models for combustion closures used in engineering CFD simulations.

2. DNS performance and nodal optimization

Terascale DNS simulations were performed with the code S3D, which solves the compressible reactive Navier-Stokes, total energy, and species continuity equations for detailed chemistry. Typical DNS currently transport as many as 20–60 species in addition to momentum, energy and density. As such, S3D is data-intensive, and memory constrained on the microprocessor-based supercomputers. S3D is parallelized with a Cartesian domain decomposition, and nearly all the interprocess communication is performed with nearest-neighbor processes. Negligible global communication occurs. As such, weak scaling tests confirm that S3D scales to the full size of current supercomputers, to as many as 30,000 cores on the 250-Tflop Cray XT4 at ORNL. However, it achieves only 10–12% of the peak performance on a nodal basis. A detailed performance analysis of runs on heterogeneous allocations of the Cray XT3 and XT4 at ORNL was performed using TAU [6]. Analysis of execution time of loops and processes distributed among all processes showed performance limitations caused by lower memory bandwidth of the XT3 nodes, causing the faster XT4 nodes to wait in `MPI_wait` statements.

S3D is a data-intensive code that is limited by memory bandwidth. Without careful optimization, scientific codes often achieve as little as 5–10% of peak performance on microprocessor-based systems. Therefore, careful restructuring of S3D's data-intensive kernels was required to reduce its memory footprint and to boost its overall performance. Collaborating with computer scientists of the Performance Engineering Research Institute (PERI) S3D Tiger Team, S3D was reconfigured to reuse data in memory registers improving nodal performance. Rice University's HPCToolkit performance tools [7] were used to study single-processor execution of a pressure wave test on a 50^3 domain. The study used a single node of a Cray XD1 with a 2.2 GHz Opteron 275 processor and DDR 400 memory, which provides 6.4 GB/s of memory bandwidth (as on Jaguar's XT3 nodes). For the model problem S3D achieved 0.305 FLOPs/cycle (15% of peak). The HPCToolkit pinpointed several data-intensive routines in

S3D that did not fully exploit the memory hierarchy. The most important of these was the diffusive flux routine for species, momentum, and energy transport. This routine contains loops over a five-dimensional data structure: three position, one vector direction, and a scalar variable. In the model problem, this loop accounted for 11.3% of the total execution time, and the loop nest achieved only 4% of peak performance. The original code made very poor reuse of data in cache memory. The diffusive flux routine was restructured using Rice University's LoopTool utility [8], which supports source-to-source restructuring of loop nests written in Fortran. Loop reordering and unrolling performed by LoopTool resulted in a speed increase of 2.94 times over the original code, with a 6.8% overall speedup for this change alone.

Further analysis of S3D's node performance with HPCToolkit, tuning loops with LoopTool, and some manual adjustment of procedure argument passing conventions to avoid copying four-dimensional slices of a five-dimensional array yielded an aggregate improvement of 12.7% for the 50^3 problem on the Cray XD1. Future DNS computations will involve increasingly complex combustion chemistry involving up to 60 or more transported species and large numbers of grid points to resolve the turbulence and internal flame structure. Code optimizations will be extended with a focus on efficient evaluation of the chemical reaction source term and spatial transport routines.

3. Turbulent combustion simulations

With the optimized S3D code, we have performed direct numerical simulations of turbulent combustion with detailed chemistry in a wide variety of environments spanning a range of flame regimes and conditions. Here, examples of three configurations are presented:

(i) Soot formation and transport in nonpremixed turbulent ethylene flames
(ii) Fuel/air stratification in methane jet flames
(iii) Flame extinction and reignition in nonpremixed ethylene flames

3.1. Soot formation and transport

Soot formation is an important process in most practical combustion environments. The majority of heat transfer from combustion occurs through radiative emission of soot in the vicinity of high temperature flame zones. Soot is formed on the fuel side of nonpremixed flames as fuel diffuses into the high temperature flame zone where it is pyrolized in the absence of oxidizing species. Large molecular weight species are formed which grow by coagulation and chemical reaction to form primary soot particles which further grow by surface reaction and aggregation. Soot emission occurs when the soot is not fully oxidized as it passes through flames during the burnout of the fuel by oxidizer (e.g., air). This is a particular problem in diesel combustion and results in copious quantities of smoke emission in large-scale fires such as wildfires. Soot emission reduces combustion efficiency through unburned hydrocarbon emission. Emission of soot is also a significant source of small particles in the atmosphere (PM 2.5) causing haze and smog. When breathed, these particles are able to penetrate deeply into the lungs, contributing to health problems including asthma, chronic bronchitis, and even premature death.

A fundamental understanding of soot formation and transport processes is important in order to develop and validate combustion models that can be used to perform engineering equipment design and optimization and fire hazard evaluation. While turbulent combustion is a multiscale physical phenomenon, combustion with soot formation is a special challenge. Experimental measurements of soot in turbulent flames are complicated by the difficulty of noninvasive measurement techniques, such as laser diagnostics, due to the optical thickness of soot. Soot exists as a particulate phase with a particle size distribution that must be measured or modeled in order to accurately describe its formation processes. The composition and reactivity

Journal of Physics: Conference Series **125** (2008) 012031 doi:10.1088/1742-6596/125/1/012031

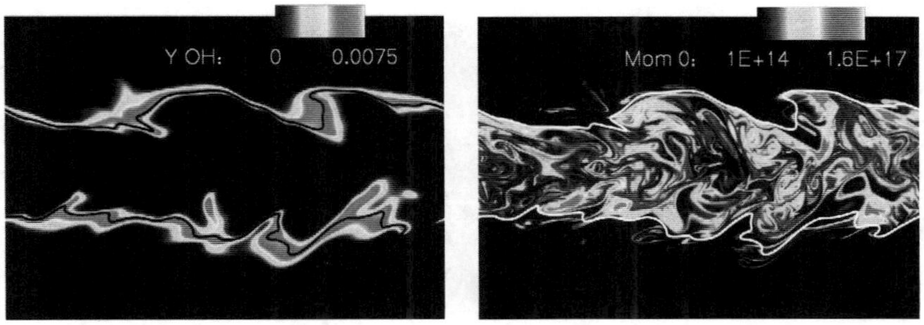

Figure 1. Y_{OH} (left), soot number density (m^{-3}, right) at $41\tau_j$ with stoichiometric contour.

may depend strongly on the soot size and its age in a flame. The basic chemistry involved in soot formation is complicated by the large number of increasingly high molecular weight species involved in its formation and growth. As a particle phase, soot transport is distinct from the gaseous species comprising the diffusion flame. Soot transport occurs primarily through thermophoretic diffusion (diffusion down temperature gradients) as opposed to any significant soot concentration-based driving force.

While many detailed experimental and numerical studies of laminar flames have been performed, experimental turbulent flame data are limited to providing only mean quantities such as mean soot concentration, without simultaneously resolved instantaneous flow and chemical structure information. Simulations of turbulent sooting flames have been limited to RANS and LES simulations in which only large-scale structures are resolved, with fine-scale turbulence-chemistry interactions accounted for through subgrid modeling approaches. Direct numerical simulations have recently been performed with soot formation that reveal the detailed flow-flame structure of turbulent nonpremixed ethylene flames [5, 9]. The application of DNS to soot formation in turbulent combustion provides fundamental insights that are not otherwise available. These insights and the DNS data obtained may be used to develop and validate subgrid-scale models for practical combustion simulations. Here, DNS simulation results are presented of soot formation in a three dimensional nonpremixed ethylene jet flame.

The simulation is a temporally-evolving planar jet with a central C_2H_4/N_2 fuel core with a species mole ratio of 0.342. The fuel core is surrounded by oxidizer with an O_2/N_2 mole ratio of 0.359. These streams result in a stoichiometric mixture fraction ξ_{st} of 0.25, where the mixture fraction (mass fraction of fuel stream material) ξ is a flame coordinate and varies from zero in the oxidizer stream to unity in the fuel stream. The fuel and oxidizer streams flow axially in opposite directions. The boundary conditions are periodic in the streamwise (axial) and spanwise directions, and are nonreflecting outflow in the cross-stream direction. The domain size is $28.8 \times 19.8 \times 10.8$ cm in the streamwise, cross-stream, and spanwise directions. The grid size is 30 μm, resulting in 228 million computational grid points. The initial jet height is 1.8 mm, the velocity difference between the streams is 82 m/s, and the jet evolved for 50 characteristic jet times. The combustion mechanism is a validated reduced mechanism consisting of 19 transported species [9], and the soot model consists of semi-empirical nucleation, growth, oxidation, and coagulation steps, with the particle size distribution represented by transporting the first three mass-moments [5]. The simulation was run on 7920 processors on the Cray Redstorm machine at Sandia National Laboratories at a computational cost of approximately 1.5 million CPU-hours.

Figure 1 shows the OH mass fraction indicative of the flame zone and the soot number density at $41\tau_j$. This figure highlights the qualitative difference between the reactive gaseous species and the soot. The OH mass fraction exists only in the vicinity of the flame zone and has a

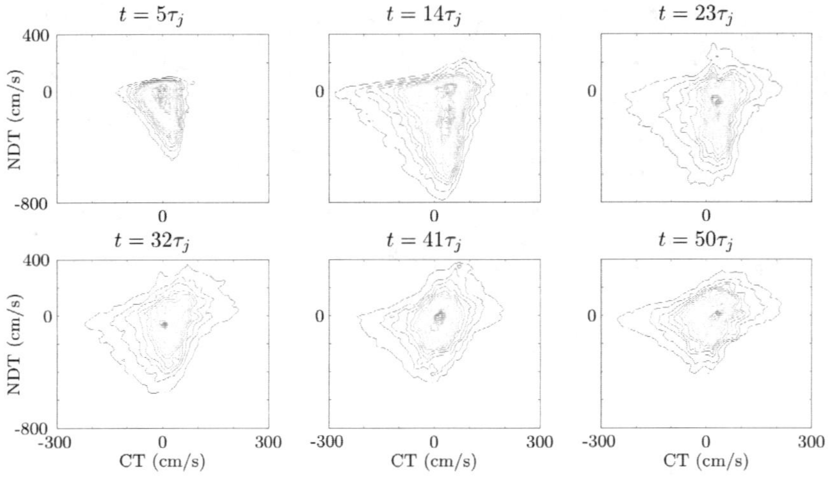

Figure 2. Joint PDFs of normal (NDT) and curvature (CT) components of v_ξ on the flame surface at six times.

relatively high diffusivity, whereas the soot number density, with its low diffusivity, is formed on the fuel side of the flame and convected into the fuel core where it is strained by the turbulence and forms very thin structures.

The transport of soot in the mixture fraction coordinate is of central importance to soot formation in turbulent flames as this dictates the location of the soot relative to the flame zone, hence the temperature, reactivity, radiative emission rate. As soot is a particulate with a low diffusivity, it is primarily convected with the flow and its transport in the mixture fraction (flame) coordinate is determined by the relative velocity between convection and isocontours of the mixture fraction. It was previously shown that this relative flame velocity v_ξ arises from two additive terms: (1) flame curvature, and (2) normal strain [9]:

$$v_\xi = -D_\xi \nabla \cdot \mathbf{n} - \frac{D_\xi}{2}\frac{\partial}{\partial \eta} \ln\left(\rho D_\xi \frac{\partial \xi}{\partial \eta}\right). \tag{1}$$

Here, D is diffusivity, ρ is density, η is the flame normal coordinate, and $\mathbf{n}$ is the flame normal $\left(\frac{\nabla\xi}{|\nabla\xi|}\right)$. Soot concentrations are higher when v_ξ is positive such that soot is convected into the flame, as happens when the center of flame curvature is in the fuel stream. In addition, normal strain tends to move soot away from the flame zone due to the low ξ_{st} relative to the inflection point in a ξ profile across a flame (i.e., as a nominal *S-shape* profile of ξ relaxes diffusively, the stoichiometric value moves away from the higher ξ soot formation zone).

The soot transport is quantified by examining the joint PDF of the curvature and normal diffusion components of v_ξ. Figure 2 shows this joint PDF at six times during the simulation. The curvature term (CT) and normal diffusion term (NDT) are of similar magnitudes throughout the simulation. At each time, the mean of the CT is centered at zero. Conversely, the NDT is strongly biased to negative values at early times (soot transported away from the flame zone), but by the end of the simulation nearly 50% of the NDT is positive (soot convected towards the flame. This motion of the soot is related to the turbulent transport of soot away from its formation zone, along with the bulk mixing of ξ as the jet evolves. Eventually, the fuel core will burn out completely, and the soot will be forced towards the flame zone where it will either burn or be emitted.

Figure 1 shows evidence of soot-flame breakthrough in regions where soot is on the oxidizer side of ξ_{st}. While noticeable levels of soot number density break through the flame, the corresponding soot mass is very low since the soot is oxidized, but the soot particle number is not. However, it is conceivable that under conditions of higher soot loading, significant levels of soot mass may escape. Soot-flame breakthrough is an important problem in environments such as fires in which soot effectively shields surrounding objects from flame radiation.

The DNS data obtained provide not only fundamental understanding of soot-flame interactions but a database by which new turbulent combustion models may be validated. The transport of soot in the ξ (flame) coordinate is of prime importance in modeling turbulent sooting flames. A new conditional moment closure (CMC) [10] model has been proposed to directly account for the differential transport of soot in the ξ coordinate [11, 12]. Figure 3 shows an a priori comparison of the differential transport term in the CMC equation evaluated in its exact, unclosed form from the DNS data, and a modeled, closed form of the term suitable for subgrid-scale modeling. The agreement between the curves is reasonably good and illustrates the utility of DNS data for directly quantifying such models.

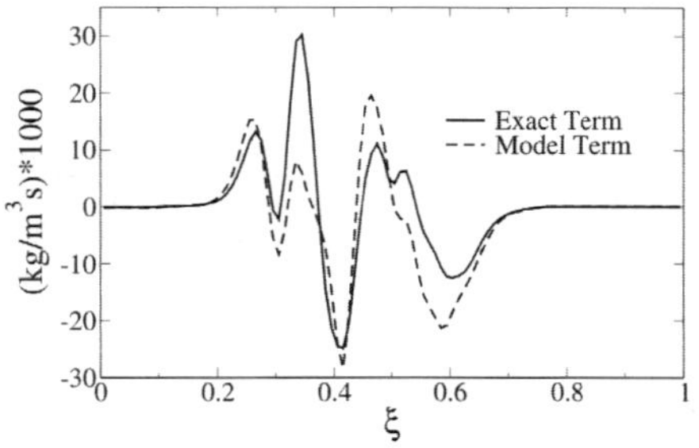

Figure 3. Comparison of modeled and exact differential diffusion terms for a CMC model at $50\tau_j$.

3.2. Stratified methane jet flames

Stratified flames occurs in many practical combustion environments. In direct injection engines, fuel is injected directly into the combustion chamber to generate a highly stratified fuel-air mixture prior to ignition. This is done to reduce fuel consumption, particularly at low-speed light-load operation and is of great interest to both spark-ignition gasoline and compression ignition diesel engine manufacturers. In stratified combustion, the fuel and oxidizer are neither completely mixed nor completely segregated prior to combustion, and therefore traditional modeling approaches based on pure premixed or nonpremixed burning modes are inadequate. Another example of stratified combustion is found in aero-gas turbine engines where fuel and secondary air are injected at different spatial locations, resulting in a nonuniform equivalence ratio at the flame front.

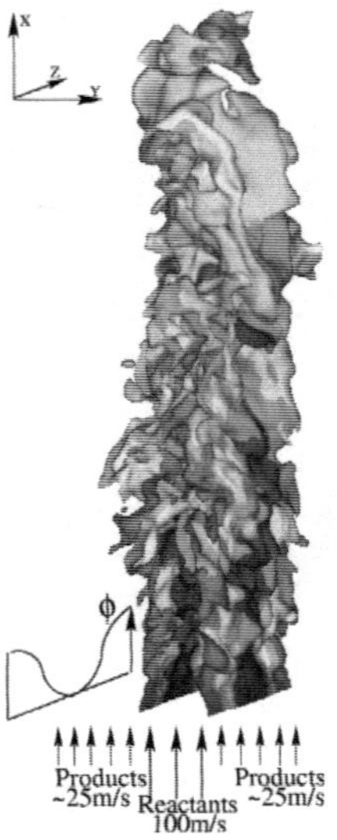

Figure 4. Stratified flame configuration showing the 1200 K isosurface of temperature colored by equivalence ratio from blue at $\phi = 0.41$ to red at $\phi = 1.0$.

Flame response to equivalence ratio gradients has been studied in simulations of laminar premixed flames, [13], and two-dimensional turbulent configurations [14]. Recent developments in experimental diagnostic techniques for stratified flows provide access to fully resolved composition fields at low turbulence levels [15]. A new set of fully resolved three-dimensional direct numerical simulations of turbulent stratified methane Bunsen flames are presented at a jet Reynolds number of 2100. The objective of the study is to determine the combustion efficiency and emissions characteristics (e.g., NO and CO) resulting from a range of initially stratified distributions of fuel and oxidizer. The flame structure and burning velocity under stratified conditions are to be compared to an equivalent uniformly premixed case.

The flame series consists of an existing DNS database of a

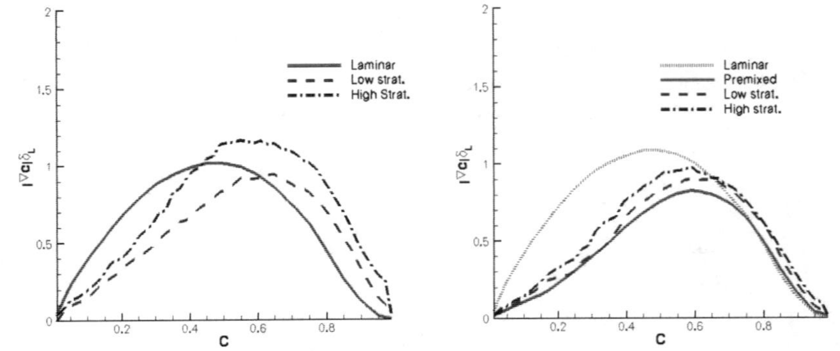

Figure 5. $|\nabla c|\delta_L$ conditionally averaged on progress variable c and equivalence ratio $\phi = 0.5$ (left) and $\phi = 0.7$ (right) compared to the laminar flame profile for three simulations at $\frac{1}{2}$ the flame height.

turbulent premixed Bunsen flame [3], and two new stratified parametric simulations. These two cases differ in their ranges of equivalence ratio (normalized fuel-to-air ratio ϕ) stratification: (1) low stratification, $0.41 < \phi < 1.0$; (2) high stratification, $0.0 < \phi < 1.46$. In the premixed case $\phi = 0.7$), reactants heated to 800K issue from a 1.8 mm wide slot at 100 m/s into a coflow of products moving at 25 m/s. The variation of composition from the jet to the coflow is prescribed by a profile of progress variable (measuring extent of reaction across the flame) and a tabulated premixed flame solution. The stratified cases, shown schematically in figure 4, have a variation of mixture fraction in the periodic spanwise z direction between the equivalence ratio limits given above. The equivalence ratio limits were selected and the coflow velocity profiles calculated to give the same global equivalence ratio as the premixed case. The inlet compositions for the stratified cases are specified as a function of mixture fraction and progress variable with reference to two-dimensional, laminar flame solutions with the corresponding range of equivalence ratios. The premixed and low stratification cases are solved with a nonstiff reduced mechanism for lean methane-air combustion [16], and the high stratification case uses a reduced methane-air mechanism based on GRIMech 3.0 applicable to the full range of stoichiometry encountered [17]. The low-stratification case consists of 260 million grid cells and was performed on approximately 10,000 processors of Franklin at NERSC at a cost of 0.6 million CPU-hours. The high stratification case consists of 300 million grid cells and was run on 25,200 processors on Redstorm at Sandia at a computational cost of 1.2 million CPU-hours.

Turbulent premixed combustion is dependent on flame speed, which depends on flame structure. Hence, a fundamental understanding how a turbulent flow field affects local flame structures is important for developing and testing turbulent premixed and stratified premixed combustion models. Characterizing the turbulence by the velocity fluctuations and length scales evaluated at one-quarter of the domain height [16], and evaluating the laminar flame properties for the range of equivalence ratio in each case, provides a range of Karlovitz numbers ($Ka = (\alpha/S_L l_k)^2$) observed in the simulations. Here, α is the thermal diffusivity, S_L is the laminar flame speed and l_k is the Kolmogorov length scale. Ka is the square of the ratio of the flame thickness to the smallest turbulent eddies. Previous analysis [16] has illustrated that the premixed DNS case, with $Ka = 5.2$, results in turbulent eddies penetrating the flame's preheat layer, but not the reaction zone, placing that case in the thin reaction zones regime [18]. Scaling the Karlovitz number by the ratio of the full width half maximum heat release thickness (δ_H) to the flame's thermal thickness (δ_L), $Ka^* = Ka \cdot (\delta_H/\delta_L)^2$, gives a value of 1.9 for the premixed case, but a range of 17.7-1.0 for the low stratification case. This suggests that the highly turbulent flow will perturb the reaction zone itself in the leaner mixtures of the stratified cases leading to combustion in the broken reaction zones regime.

A progress variable c is defined that varies between zero and unity in the unburnt and burnt gases, respectively. Figure 5 compares the normalized mean progress variable gradient versus c at $\phi = 0.5$, and $\phi = 0.7$. Values of $|\nabla c|\delta_L < 1$ indicate flame thickening relative to the laminar flame, where $|\nabla c|^{-1}$ is a measure of the flame thickness. The peak heat release occurs at $c = 0.65$, and low values of c constitute the flame's preheat zone. Figure 5 shows thickening of the flame's preheat zone relative to the laminar flame. Near the peak heat release zone and at higher c, the turbulence acts to thin the flame $\phi = 0.5$, whereas at $\phi = 0.7$ the turbulent cases are comparable to the laminar flame and to the turbulent premixed flame [16]. These results emphasize the importance of correctly modeling the turbulence-chemistry interactions across the range of equivalence ratios present.

3.3. Nonpremixed flame extinction and reignition

Extinction and reignition phenomena are important in nonpremixed flames under high turbulence conditions. Nonpremixed flames occur at the interface between fuel and oxidizer, and turbulence acts to increase flame surface area and scalar gradients, which enhances mixing rates, hence combustion rates. As turbulence increases, finite rate chemical kinetic effects become important, and flame extinction may occur if heat release rates cannot keep up with diffusive heat losses. This is an important problem in modeling and simulation of combustion devices that push the limits of flame stability [18] to improve combustion efficiencies and reduce pollutant emissions. Flame regions that become extinguished continue to mix with fuel, oxidizer, and combustion products, and may reignite. The mechanism and mode of reignition is a subject of ongoing research. Reignition occurs as turbulent strain rates decrease and fluid parcels may reignite through several possible mechainisms. These include autoignition [19, 20], edge flame propagation [21], flame folding/mixing [22], and premixed or partially premixed flame propagation.

A series of three turbulent flame simulations were performed with varying levels of flame extinction. The flow configuration and chemical mechanism is the same as in the sooting simulation described above, that is, a temporally evolving planar nonpremixed ethylene jet flame. In each case, all flow and geometrical parameters were constant, with the differences being the composition of the fuel and oxidizer streams. These stream compositions were varied to yield a factor of two difference between Cases 1 and 3 in the levels of the stoichiometric steady laminar flamelet extinction scalar dissipation rate (i.e., loosely, the imposed flame strain

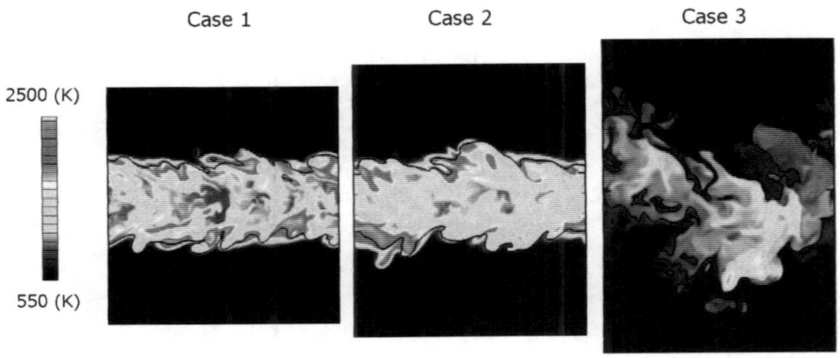

Figure 6. Temperature contours with stoichiometric mixture fraction surface at the point of peak extinction.

of one-dimensional opposed jet diffusion flames at extinction). The streamwise and spanwise domain sizes are 11.5 and 7.7 cm, respectively, while the initial jet width is 0.96 mm. The stream velocity difference is 196 m/s, giving a Reynolds number of approximately 5000 for the three cases. In each case ξ_{st} is 0.17, and the jets evolved for 75–140 characteristic jet times. The three cases were run on Jaguar at NCCS at ORNL, and consisted of 270, 308, and 341 million computational grid cells, the differences being in the cross-stream direction only. The three cases were run on 10,752, 10,752, and 14,112 processors at computational costs of approximately 0.9, 1.2, and 2 million CPU-hours, respectively.

Figure 6 shows axial cuts of the jet for the three cases. Temperature is shown with the stoichiometric contour overlaid. The time of each case corresponds to the maximum level of extinction: 27, 37, $72\tau_j$, for Cases 1, 2, 3, respectively, for which the degree of extinction increases dramatically. The level of extinction is important to the development of the jet and the subsequent reignition of the flame. The presence of a flame reduces the local turbulence intensity through flow dilatation, as well as the increased kinematic viscosity at higher temperatures. This results in higher turbulence intensity as the flame extinguishes, which enhances the flame extinction, and hinders flame reignition. In addition, the total stoichiometric flame surface area is higher in cases with greater amounts of extinction, resulting in higher rates of total heat release during reignition.

The degree of extinction of the flame surface also impacts the mode of reignition. High flame extinction allows significant mixing of fuel and oxidizer in the absence of reaction, allowing reignition to occur in a mixture of fuel, oxidizer, and combustion products. These reignition modes blur the lines between premixed, stratified and nonpremixed combustion. Current efforts are underway to quantify the impact of flame extinction on the jet evolution characteristics, as well as understand the complex reignition processes occurring in the three cases. These investigations make use of flame indicies that quantify the mode of reignition through the alignment of gradients of fuel and oxidizer species. Preliminary results indicate that Case 3 experiences extreme extinction, near the limit of global blowout. This allows significant mixing of the fuel, oxidizer, and products resulting in near complete reignition in a premixed flame mode. Hence, an initially nonpremixed flame reignites through premixed flame propagation. Conversely, Case 1 appears to retain its nonpremixed character through most of the reignition process.

4. Conclusions

The growth in high-performance computing has allowed the simulation of turbulent combustion with realistic chemistry and transport on up to 30,000 processors. The performance of the DNS code S3D has been optimized to make efficient use of available computational resources by using advanced performance and profiling tools to increase use of memory bandwidth resources. A series of results from three recent DNS configurations were presented including nonpremixed flames with soot formation, and extinction and reignition, and stratified methane Bunsen flames. Soot formation in nonpremixed combustion is important to heat transfer, and as an emitted pollutant. Results show the importance of flame structure on transport of soot into and out of reaction zones. Stratified combustion is increasingly important to achieving high efficiency and low pollutant emission. Initial results of the stratified methane Bunsen flames show turbulence-induced thickening of the preheat layer, impacting flame speed and overall burning rate, with high stratification possibly leading to broken flames with disruptions in reaction zones. Flame extinction and reignition phenomena are important in flame stabilization and for flows operating in high-strain environments which maximize combustion rates through high turbulent mixing. Results indicate that the extent of extinction has an important effect on the extent of premixing prior to reignition, hence affecting the reignition mode and flow characteristics.

Acknowledgments
This work was supported by the U. S. Department of Energy, Office of Basic Energy Sciences, Division of Chemical Sciences, Geosciences, and Biosciences. Sandia National Laboratories (SNL) is a multiprogram laboratory operated by Sandia Corporation, a Lockheed Martin Company, for the United States Department of Energy under contract DE-AC04-94-AL85000. Simulations were performed at NERSC at LBNL, NCCS at ORNL, and at SNL. The authors acknowledge the contributions from FYO7 SciDAC PERI S3D Tiger team (J. Mellor-Crummey, N. Wright, D. Bailey, R. Vuduc, S. Shende, J. Larkin, and B. de Supinski) in the profiling and performance optimization of S3D.

References

[1] Pantano C 2004 *J. Fluid Mech.* **514** 231–270

[2] Hawkes E, Sankaran R, Sutherland J and Chen J 2007 *Proc. Combust. Inst.* **31** 1633–1640

[3] Sankaran R, Hawkes E R, Chen J H, Lu T and Law C K 2007 *Proc. Combust. Inst.* **31** 1291–1298

[4] Yoo C S, Chen J H and Sankaran R 2007 *Computational Combustion, ECCOMAS Thematic Conference, Delft, The Netherlands* ed Roekaerts D, Coelho P, Boersma B J and KClaramunt

[5] Lignell D O, Chen J H and Smith P J 2008 *Combust. Flame* In press

[6] Shende S and Malony A D 2006 *Int. J. High Perform. Comput. Appl.* **20** 287–331

[7] Mellor-Crummey J, Tallent N, Fagan M and Odegard J Seattle, WA, 2007 *In Proceedings of the Cray user's group meeting*

[8] A. Qasem, G. Jin, and J. Mellor-Crummey, Improving performance with integrated program transformations, Tech. Rep. TR03-419, Dept. of Computer Science, Rice University, 2003

[9] Lignell D O, Chen J H, Smith P J, Lu T and Law C K 2007 *Combust. Flame* **151** 2–28

[10] Klimenko A Y and Bilger R W 1999 *Prog. Energy Combust. Sci.* **25** 595–687

[11] Hewson J, Ricks A, Tieszen S, Kerstein A and Fox R 2006 *Center for Turbulence Reseach, Proceedings of the Summer Program*

[12] Lignell D O, Hewson J C and Chen J H 2007 *Fall Meeting of the Western States Section of the Combustion Institute* pp 07F–36

[13] Marzouk Y M, Ghoniem A F and Najm H N 2000 *Proc. Combust. Inst.* **28** 1859–1866

[14] Jimenez C, Cuenot B, Poinsot T and Haworth D 2002 *Combust. Flame* **128** 1–21

[15] Barlow R S, Wang G H, Anselmo-Filho P, Sweeney M S and Hochgreb S 2008 *submitted to Proc. of the Combust. Inst.* **32**

[16] Sankaran R, Hawkes E R, Yoo C S, Chen J H, Lu T and Law C K March 2007 *5th Joint US Combustion Meeting of the Combustion Institute, San Diego*

[17] T. Lu and C. K. Law, A CSP-based criterion for the identification of QSS species: a reduced mechanism for methane oxidation with NO chemistry, In preparation (2008)

[18] Peters N 2000 *Turbulent Combustion* (Cambridge, United Kingdom: Cambridge University Press)

[19] Zeldovich Y B 1980 *Combust. Flame* **39** 211–214

[20] Chen J H, Hawkes E R, Sankaran R, Mason S and Im H G 2006 *Combust. Flame* **145** 128–144

[21] Hewson J C and Kerstein A R 2002 *Combust. Sci. Technol.* **174** 35–66

[22] Hawkes E R, Sankaran R and Chen J H 2007 *5th Joint US Combustion Meeting of the Combustion Institute* p B05

High-order AMR computations of reacting flow with adaptive reduction of chemical stiffness

H N Najm[1], J Ray[1], C Safta[1], Y Marzouk[1], M Valorani[3] and D Goussis[4]

[1] Sandia National Laboratories, Livermore, CA 94550, USA
[3] Sapienza, University of Rome, Rome, Italy
[4] National Technical University of Athens, Athens, Greece

E-mail: hnnajm@sandia.gov

Abstract. We outline a high-order adaptive mesh refinement construction for low Mach number reacting flow computations. We illustrate the order of accuracy on a model problem and demonstrate the construction for methane-air ignition in 2D channel flow. We also outline a nonparametric tabulation strategy for computational singular-perturbation time integration of a general chemical system. We demonstrate this adaptive chemistry technique for H_2-O_2 ignition.

1. Introduction

Reacting flows exhibit wide ranges of length and time scales. These lead to significant computational challenges due to associated requirements of fine meshes and small integration time steps. On the other hand, small length scales are typically highly localized in thin flame "sheet" regions, and small time scales are typically "active" only in these regions. This largely universal structure of reacting flows of interest motivates development of associated numerical constructions that target these computational challenges. The present work outlines our recent developments in each of these two areas. Specifically, we discuss (1) high-order adaptive mesh refinement (AMR) computations of reacting flows, largely addressing the spatial length-scale range challenge; and (2) adaptive chemistry tabulation addressing the time-scale range challenge.

We rely on AMR employing structured meshes [1]. Second-order AMR constructions for low Mach number reacting flow have been used in [2–4]. High-order structured AMR constructions have also been under development [5,6]. Our implementation [7] relies on combinations of high-order discretizations, interpolants, and filters, to arrive at a robust high-order AMR construction.

Adaptive chemistry constructions have typically employed either adaptive selection among a set of simplified models [8] or adaptive reduction of stiffness within a detailed chemical model [9]. We use computational singular perturbation (CSP) analysis [10] to identify, save, and reuse local manifold structure for filtering detailed chemical source terms [9]. While model complexity is not addressed, removal of stiffness is highly advantageous, eliminating fast time scales and enabling efficient explicit time integration of detailed chemical models.

This paper presents our recent developments in both the AMR and adaptive chemistry contexts. We present a high-order AMR strategy for low Mach number reacting flow, and demonstrate the construction with ignition in a 2D channel-flow methane-air system. We also present a nonparametric strategy for "tabulation" of low-dimensional manifold structure, and demonstrate its use on the ignition of a H_2-O_2 mixture. These two topics are discussed in the following sections below.

2. Adaptive mesh refinement computations of reacting flow

A fourth-order momentum projection scheme was developed for adaptive mesh computations of reacting flow in the low-Mach number limit. The projection scheme is coupled to an operator-split stiff approach [11] for species and energy transport equations. The velocity-field length scales for reacting laminar flows of interest, while comparable with the temperature/density field length scales, are large compared to those associated with flame radicals. Consequently, we solve the momentum equations on a uniform base mesh that provides sufficient resolution for the density, pressure, and velocity fields, while we use AMR for the scalar transport equations.

The time integration strategy involves (1) momentum projection and (2) scalar update. Step 1 is performed in two substeps: (1a) time advancement of the velocity field neglecting the pressure term, and (1b) projection of the intermediate velocity field on a vector field that satisfies the divergence constraint. An explicit second-order Adams-Bashforth scheme is used for (1a) while spatial derivatives are computed using fourth-order stencils. Figure 1 shows typical spatial convergence rates obtained by the method for manufactured solutions (MMS) [12] for the velocity field. Fourth-order stencils are also used to discretize the variable-coefficient pressure Poisson equation solved in step (1b). The linear system that arises from this discretization is solved using Hypre [13], employing a conjugate-gradient technique coupled with a structured multigrid (SMG) preconditioner. Figure 2 shows typical strong scaling efficiencies. Preconditioning the linear system with SMG improves the parallel efficiency compared to diagonal scaling and, more important, reduces the number of subiterations necessary to obtain convergence by an order of magnitude (not shown). The convergence rates obtained for the pressure equation are also fourth-order, confirming the theoretical order of accuracy.

Step 2 involves Berger-Colella time refinement [1] to advance the scalar field on successively refined levels in the AMR grid hierarchy. In order to compute the convection term in the scalar transport equation, the velocity field is prolonged using sixth-order interpolation stencils from the base level to the rest of the grid hierarchy. This interpolation order is necessary [7] to maintain an overall fourth-order accuracy. The same interpolation order is also used to prolong and restrict scalar fields between adjacent grid levels. The results shown in figure 1 confirm the theoretical order of accuracy of the discretization in the coupled construction.

A reacting channel flow simulation is used to demonstrate the construction. Figure 3 shows a schematic of the flow. The rectangular channel is 3×1 cm^2 and the initial velocity field is set to fully developed channel flow with a mean x-velocity of 2 m/s. The scalar concentration and temperature are initially uniform, air at 800 K, except for a small "hot spot" near the inlet, where temperature is increased by 400K to facilitate the autoignition process. Two levels of refinement, shown in figure 3, are used in addition to the base grid level. The grid cell size is 250 μm for L0, 125 μm for L1, and 62.5 μm for L2.

A one-step, irreversible, global reaction mechanism is used to model the oxidation of methane, with a pre-exponential rate constant of 9.0e23 mol-cm-sec-K, a zero temperature exponent, and an activation energy of 5.0e4 cal/mol. At $t = 0$, the insertion of a diluted fuel plug, 20% CH_4 and 80% N_2 by volume, is initiated at the inflow edge. Hyperbolic tangent profiles are used to create the spatial profiles, employing a $\sim$3 mm-wide plug centered on the channel inlet, and a 1ms time span for plug insertion. The inlet velocity field is maintained at the initial condition throughout the simulation. As fuel is inserted near the left boundary, the hot spot moves downstream ahead of the fuel plug, eventually leading to the ignition of the fuel-oxidizer interface starting at the downstream edge of the fuel plug. At a later time, the flame extends to the sides and trailing region of the fuel plug. The entire available fuel is consumed before it

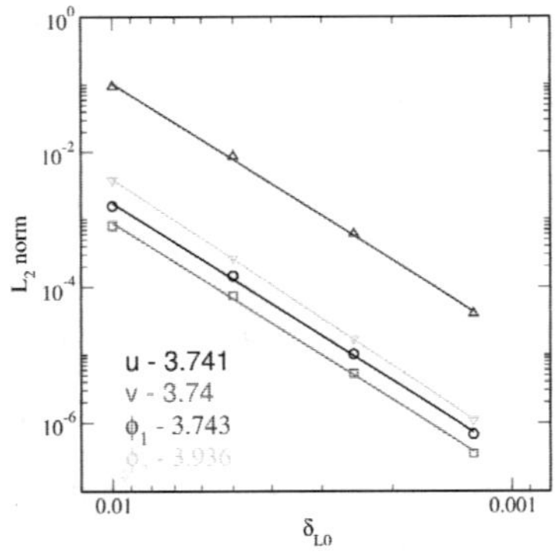

Figure 1. L_2 norms for velocity field components and select scalars on successively refined computational grids. The corresponding convergence rates are also displayed.

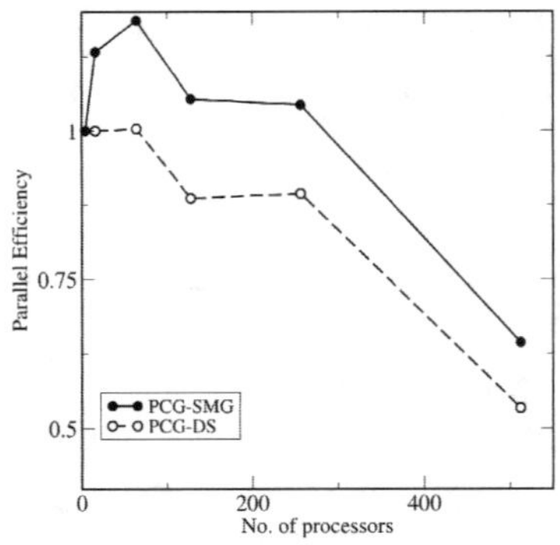

Figure 2. Parallel efficiency (strong scaling) for the solution of the variable coefficient Poisson equation. Results are scaled based on computational times using 4 processors.

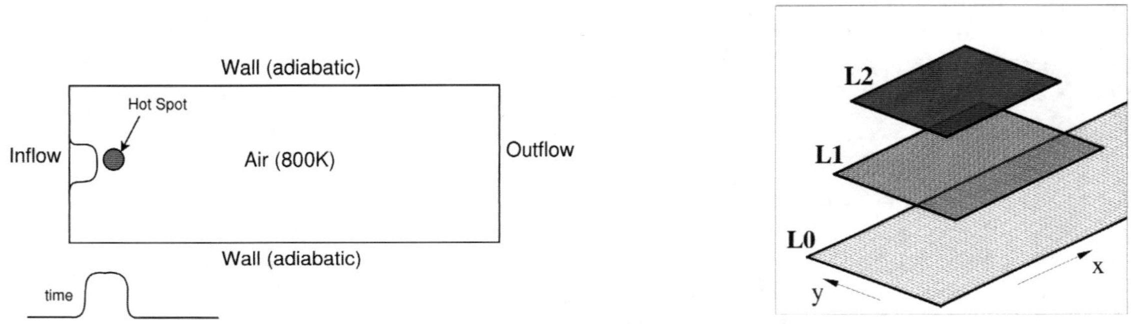

Figure 3. Left: Schematic of the setup for the reacting channel flow. Right: Computational mesh hierarchy. Two levels of refinement, L1 and L2, are used in addition to the base level L0.

reaches the outflow boundary. Figure 4 shows contour lines of the y-component of the velocity field. The heat released during combustion leads to significant transverse velocity gradients, which are accurately resolved on the base mesh. Figure 5 shows a snapshot of CH_4 and CO_2 scalar mass fractions, with finer peak gradients, which are well resovled on the adaptive mesh hierarchy.

3. Adaptive chemistry tabulation

The second component of our reacting flow simulation strategy relies on CSP [14] to identify fast dissipative processes and to decouple them from slower dynamics. Fast dissipative processes define a "slow" manifold, along which the system evolves according to the slow processes. The identification of slow manifolds facilitates a time integration strategy known as explicit time-scale splitting [15]. In this scheme, the ODE system is first advanced according to the *slow* dynamics, then, the CSP "homogeneous correction" [15] is applied in order to return the state to the local slow manifold. This strategy allows the use of explicit time integration schemes, as the slow-dynamics ODE system is non-stiff.

The CSP integrator has been shown to be robust and accurate [9] for large chemical systems. Further, the computational cost of evaluating CSP quantities has been recognized, motivating development of associated *tabulation* schemes [9] to enable re-use and interpolation of these quantities. Previous work has applied CSP-tabulation in simple model problems. We continue

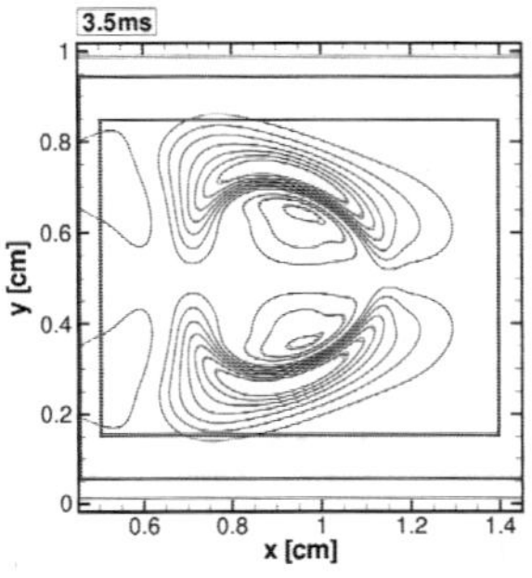

Figure 4. Contour plot of transverse velocity during ignition. The borders of mesh blocks are shown with thick lines.

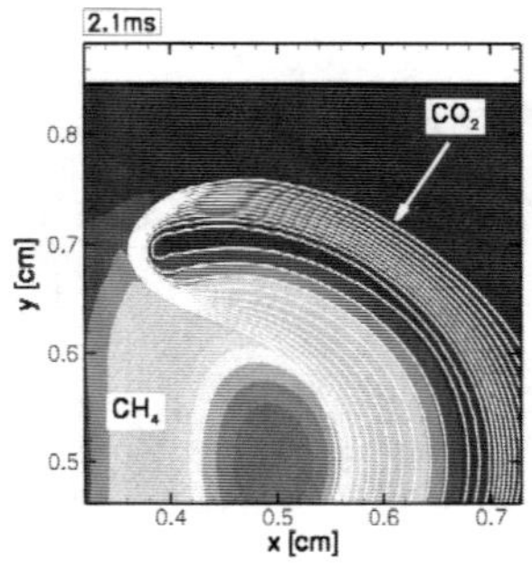

Figure 5. CH_4 (color) and CO_2 (contour lines) mass fraction fields during ignition, all shown in the L2 mesh region.

this development here in the context of detailed H_2-O_2 chemistry.

Design points for our "table" consist of state vectors drawn from a precomputed set of reacting flow simulations, here homogeneous ignitions. CSP analysis is performed on each design point; iterated applications of the homogeneous correction may bring a design point closer to a slow manifold. Each manifold corresponds to a unique choice of M, the number of exhausted modes, and CSP radicals; based on specific error criteria. With an N-dimensional chemical space, each manifold is associated with a corresponding regime of validity—a hypervolume in $\mathbb{R}^N$ defined by the hull of design points that are "attracted" to the manifold.

Regression of CSP quantities requires an efficient representation of each manifold, on the $(N - M)$-dimensional space of "active" species [9]. We adopt the simplest nonparametric scheme available: nearest-neighbor regression. This results in piecewise-constant interpolants, but is easily extended to smoother interpolants via k-nearest-neighbor queries. Efficient nearest-neighbor queries require appropriate spatial data structures. Here, we construct a distinct kd-tree [16] for each manifold, where d, the dimension associated with a tree, is always $N - M$. To further mitigate the impact of dimensionality, we employ the *approximate* nearest-neighbor algorithm of [16], which returns a point within relative error ϵ of the true nearest neighbor. Scaling of per-query cost with dimension becomes increasingly favorable as ϵ increases.

Results from a homogeneous ignition calculation using adaptive CSP tabulation and explicit time-scale splitting are shown in figure 6. The initial condition is an H_2-O_2 mixture at 1000 K, and we use a 9-species chemical mechanism [17]. The right-hand axis shows the number of exhausted modes M identified by the full CSP analysis (in green) and by the tabulation scheme (in red). Brief discrepancies between the two traces are seen near transitions in the value of M. Changes in M correspond to moving from one manifold to another; since the tabulation scheme is dependent entirely on design points to identify the extent of each manifold, sometimes the system state may leave the known regime of validity of one manifold and continue for several steps before entering the known regime of validity of the next manifold. Increasing the number of design points or introducing thin overlap regions tends to counterbalance such events. Note that temperature trajectories computed with tabulation are nearly indistinguishable from those using CSP analysis at each step.

4. Conclusions

We have outlined our recent advances in high-order AMR computations of reacting flow and CSP-based adaptive chemistry time integration. We presented results highlighting the demonstrated accuracy of both constructions. Work is in progress to couple the two techniques to arrive at AMR computations of reacting flow with adaptive chemistry.

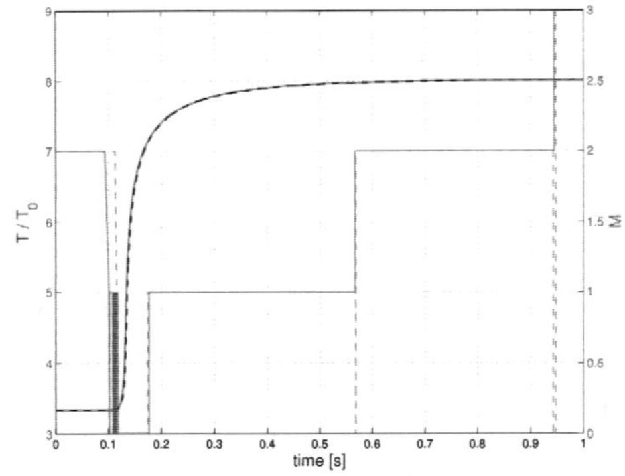

Figure 6. Ignition of a homogeneous H_2-O_2 mixture, showing the number of exhausted CSP modes (M) at each step. Solid-green/dashed-red lines correspond to full/tabulated CSP analysis.

Acknowledgments

This work was supported by the US Department of Energy (DOE), Office of Basic Energy Sciences (BES), SciDAC Computational Chemistry Program. Sandia National Laboratories is a multiprogram laboratory operated by Sandia Corporation, a Lockheed Martin Company, for the United States Department of Energy under contract DE-AC04-94-AL85000. MV acknowledges the support of the Italian Ministry of University and Research (MIUR).

References

[1] Berger M J and Colella P 1989 Local adaptive mesh refinement for shock hydrodynamics *J. Comput. Phys.* **82**:64–84
[2] Bennett B-A V and Smooke M D 1998 Local rectangular refinement with application to axisymmetric laminar flames *Combust. Theory Model.* **2**:221–258
[3] Ray J, Najm H N, Milne R B, Devine K D and Kempka S N 2000 Triple flame structure and dynamics at the stabilization point of an unsteady lifted jet diffusion flame *Proc. Comb. Inst.* **28**:219–226
[4] Day M S and Bell J B 2000 Numerical simulation of laminar reacting flows with complex chemistry *Combust. Theory Modelling* **4**:535–556
[5] Jameson L 2003 AMR vs high order schemes *J. Sci. Comp.* **18(1)**:1–24
[6] Barad M and Colella P A fourth-order accurate local refinement method for Poisson's equation *J. Comp. Phys.* bf 209:1–18
[7] Ray J, Kennedy C A, Lefantzi S and Najm H N 2007 Using high-order methods on adaptively refined block-structured meshes *SIAM J. Sci. Comp.* **29(1)**:139–181
[8] Schwer D A, Lu P and Green Jr. W H An adaptive chemistry approach to modeling complex kinetics in reacting flows. *Combustion and Flame* **133**:451–465
[9] Lee J C, Najm H N, Lefantzi S, Ray J, Frenklach M, Valorani M and Goussis D 2007 A CSP and tabulation based adaptive chemistry model *Combustion Theory and Modeling* **11(1)**:73–102
[10] Lam S H and Goussis D A 1988 Understanding complex chemical kinetics with computational singular perturbation *Proc. Comb. Inst.* **22**:931–941
[11] Najm H N and Knio O M 2005 Modeling Low Mach number reacting flow with detailed chemistry and transport. *J. Sci. Comp.* **25(1)**:263–287
[12] Roache P J 2002 Code verification by the method of manufactured solutions *ASME J. Fluids Engineering* **124**:4–10
[13] Hypre: High performance preconditioners and solvers http://computation.llnl.gov/casc/hypre/software.html
[14] Lam S H 1993 Using CSP to understand complex chemical kinetics *Combustion Science and Technology* **89**:375–404
[15] Valorani M and Goussis D A 2001 Explicit time-scale splitting algorithm for stiff problems: Auto-ignition of gaseous-mixtures behind a steady shock *J. Comput. Phys.* **169**:44–79
[16] Arya S, Mount D M, Netanyahu N S, Silverman R, and Wu A Y 1998 An optimal algorithm for approximate nearest neighbor searching in fixed dimensions *J. ACM* **45(6)**:891–923
[17] Yetter R A, Dryer F L and Rabitz H 1991 A comprehensive reaction mechanism for carbon monoxide/hydrogen/oxygen kinetics *Combust. Sci. Technol.* **79**:97

Science-based model for particle formation from novel fuels

A Violi

University of Michigan, Ann Arbor, MI 48109-2125, USA

E-mail: avioli@umich.edu

Abstract. With the advent of petascale high-performance computing platforms, realistic multiscale modeling can be constructed to incorporate atomic-scale (molecular) information into macroscopic predictions of engineering systems. The overriding theme of the work presented in this paper is developing a multiscale modeling approach for soot formulation where atomistic data is integrated into macroscopic simulations. The prediction of soot formation remains arguably one of the most challenging subjects in combustion science, having an influence over a wide range of applications ranging from combustion efficiency to reducing emissions to slow global warming, to improved heat transfer designs in industrial settings, to predicting the radiation heat transfer from large scale fires. Starting from the fuel structures the new multiscale simulations reveals how chemical changes and transformation can propagate upward in scale to help define the function of the particle structures. In particular, the fuel structure influences the morphology of the nanoparticles, which in turn is critical in determining the overall growth and agglomeration behavior. These simulations make use of a newly proposed combination of molecular dynamics and kinetic Monte Carlo methodologies that will include both chemical reactions and agglomeration processes. The main strength of this approach is the ability to use important atomic-scale information directly into large scale description of the macroscopic phenomena.

1. Introduction

Science and engineering have been revolutionized by the infusion of computational science and simulation in the traditional experimentation-observation-analysis-theory loop and by eliminating the geographic constraints for collaboration and experimentation. New means of computational discovery will augment the traditional discovery-innovation loop with novel computational concepts to aid knowledge discovery, analysis, and experimentation. Petascale computing will allow the prediction of complex system behaviors with unprecedented speed and accuracy. To capitalize on these advances and the enormous potential they hold, we require basic research into the scientific components of modeling and simulation.

The scope of this work is to gain an understanding of the complex phenomena occurring in systems composed of reactive carbonaceous nanoparticles, developing a new paradigm that couples the microscopic information about fuel structures with the description of particle formation (soot) from combustion sources.

The fundamental challenge in predicting soot formation lies in the fact that soot formation is a heterogeneous process at even atomic-length scales, ~$O(10^{-10}m)$, which has an enormous impact on the soot morphology and overall yield. Recent work of Violi et al. [1, 2] has revealed how chemical changes and transformation can propagate upward in scale to help define the function of the particle structures. In particular, the morphology of the nanoparticles is critical in determining the overall growth and agglomeration behavior. Current detailed soot-modeling approaches that have been developed for premixed [3-5] and nonpremixed flames [6, 7] do not account for key morphology dependent nucleation, surface growth, and agglomeration processes and may help explain current discrepancies between observed and predicted amounts of soot formed for even well controlled systems [5, 8, 9]. In addition, the models currently in the literature offer mathematical simplicity but do not yield chemical detail on the high molecular weight species. To help resolve issues such as identification of pathways leading to soot formation, we need to characterize the precursors, in terms of chemical structure andcomponents. Studying relationships between structure and pathways, structure and properties, and structure and reactivity (population of active sites) can lead to a deeper understanding of nanoparticle growth mechanisms. In current models, parameters related to structures are fitted to experimental data.

The new multiscale computational approach can describe the formation of particles following the evolution from fuel, gas-phase species into particles, preserving chemical and physical details of the compounds. The work presented in this paper is designed to help reveal, through multiscale computer simulation, how chemistry (chemical structures or changes at the scale of the molecular building blocks of nanoparticle agglomerates) will influence the emergent structure and function of particle assemblies at significantly large length and time scales. This can be achieved by providing a computational infrastructure to help connect the different scales of chemistry and physics. Carbonaceous nanoparticles will be characterized both chemically and physically, to study their formation and transformation mechanisms in combustion systems.

A major feature of this study is the identification of the chemical specificity of nanoparticles: size, chemical functionalities, and water solubility have a decisive role in establishing the interactions of aerosol with human tissues in the lungs. The same parameters establish their optical properties relevant to direct radiative forcing and to their ability to act as cloud condensation nuclei.

A timely application of this novel computation infrastructure is related to energy and transportation. The automotive industry is facing a challenging goal of further reducing fuel consumption and increasing performance while obtaining cleaner exhaust gas and greater driver comfort. Meeting the challenge of developing and optimizing the combustion processes of bio-diesel fuels in new engines will require a new level of the understanding of the physical and chemical phenomena common to diesel engines and low-temperature combustion engines, which will be strongly affected by the different fuel properties, which in turn affect fuel/air mixture preparation, combustion and emissions. Soot formation is one of the major constraints in engine performance.

After describing the methodology employed to compute chemical and physical properties of nanoparticles in different environments, we present a section on petascale computations and the need to develop a multiscale architecture to study the phenomena involved in soot formation such that the atomistic-level processes (chemistry) can be coupled with the very long length and time-scale processes.

2. Multiscale computational approach: the AMPI code

In high-temperature environments, particle formation from small gas-phase species is a fascinating multiscale problem of nanoparticle growth, both in length scale and in time scale. Starting from simple gas-phase molecules, such as ethylene and acetylene, polycyclic aromatic hydrocarbons including naphthalene and pyrene are formed. The system then undergoes a nucleation process, and the first particles are formed by reactions with the gas-phase species. Coagulation and agglomeration of these nanoparticles lead to particle size of the order of hundreds of nanometers in diameter [10]. The processes involved in the formation of particles exhibit a wide range of time scale, spanning pico- or nanoseconds for intramolecular processes to milliseconds for intermolecular reactions.

For these systems, the coupling between scales can be very strong, and "bidirectional," and microscopic phenomena not only are driven by macroscopic forces but also change these macroscopic forces. In this case, there is a feedback loop between the microscopic and macroscopic scales. For example, particle clusters explicitly contain both micro- and macroscale dimensions in their structure, and the proper modeling of trans-temporal and trans-spatial structures involves crossing this microscopic/macroscopic barrier to incorporate an adequate degree of information transfer between these disparate scales [11]. Because of the small time-step required to integrate the equations of motion, molecular dynamics (MD) – the standard method for dynamical simulations at the atomistic level – is generally limited to nanoseconds. However, the relevant thermally activated processes typically take place over time scales of microseconds or even longer. Although MD simulations provide a powerful tool for investigating complex systems, their substantial computational costs limit their application to investigations on time scales that are less than microseconds and length scales that are significantly less than micrometers [12, 13].

The novel multiscale approach is embedded in the Atomistic Model for Particle Inception (AMPI) code, which combines the MD methodology with kinetic Monte Carlo to allow the extension of the accessible time scales by orders of magnitude relative to direct MD, while retaining full atomistic details of the systems of interest [14].

2.1 The Kinetic Monte Carlo method

Kinetic Monte Carlo (KMC) method is part of a broad class of algorithms that solve problems through the use of random numbers [15, 16]. It was developed for evolving systems dynamically from state to state [17]. The underlying principle in all KMC algorithms is the random selection of a process based on the transition probabilities of all processes, execution of the selected process, and updating the time clock and the transition probabilities [18].

The starting point for a stochastic description of a system is the underlying master equation given by [19]:

$$\frac{dP(\underline{\sigma})}{dt} = \sum_{\underline{\sigma}} G(\underline{\sigma}' \to \underline{\sigma}) P(\underline{\sigma}') - \sum_{\underline{\sigma}} G(\underline{\sigma} \to \underline{\sigma}') P(\underline{\sigma}) \tag{1}$$

with $\underline{\sigma}' \neq \underline{\sigma}$, which gives the evolution of the probability density function $P(\underline{\sigma})$ of observing a state $\underline{\sigma}$, $G(\underline{\sigma} \to \underline{\sigma}')$ is an element of the transition matrix for the transition from state $\underline{\sigma}$ to state $\underline{\sigma}'$ [20]. Atoms or molecules vibrate around locations separated by large free energy barriers and occasionally jump from one location to a nearby one. The momentum degrees of freedom of all atoms and molecules and thermal vibrations are integrated out to compute the microscopic rates used in a stochastic description of a system. The resulting microscopic state variable $\underline{\sigma}$ is a function of only spatial and time coordinates. The atomic jump results in a change in $\underline{\sigma}$ once the jump has occurred. Recently Chatterjee and Vlachos have published a thorough overview of the kinetic Monte Carlo methods and we refer the reader to their paper [21] and citations therein for detailed information on the microscopic and accelerated KMC methodologies.

In summary, in the kinetic Monte Carlo method, the system starts in some state and equations of motion are never directly evolved, but a set of escape pathways connect a state to neighboring states. The key requirement is to specify these escape paths and the rate constant for each one. Given this set of rates, a stochastic algorithm is employed to advance the time and to choose a neighboring state to which the system is moved. The rate constants for the possible moves out of each state are either computed by using first-principles calculations and transition state theory or determined by using the reaction class theory to make a rate catalog [13].

2.2 Molecular dynamics

In the broadest sense, molecular dynamics is concerned with molecular motion and conformational transitions together with local vibrations are the usual subjects of MD. Molecular dynamics alters the intramolecular degrees of freedom in a stepwise fashion, where the steps meaningfully represent the changes in atomic position over time (i.e., velocity) [20, 22]. Newton's equation is used in the molecular dynamics formalism to simulate atomic motion. The rate and direction of motion (velocity) are governed by the forces that the atoms of the system exert on each other as described by Newton's equation. Knowledge of the atomic forces and masses can then be used to solve for the positions of each atom along a series of extremely small time steps (on the order of femtoseconds = 10–15 seconds).

If the potential gives an accurate description of the atomic forces of the system, then the dynamical evolution will be an accurate representation of the real physical system. This explains the popularity of the MD method.

A serious limitation, however, is that accurate integration requires time steps short enough (~10–15 s) to resolve the atomic vibrations. The total simulation time is typically limited to less than one microsecond, while important processes such as diffusion, annihilation, nanoparticle assembly, often take place on much longer time scales.

To summarize, MD simulations follow the dynamical evolution of a system and generate a sequence of points in phase space as function of time corresponding to different conformations of the system and their momenta. This process requires choosing an interatomic potential for the atoms and a set of boundary conditions.

2.3 The AMPI code: synergy between KMC and MD

The AMPI code combines the strengths of KMC and MD methodologies in a unique way. The two modules are placed on an equal footing, and the code alternates between MD and KMC steps during the simulation. This feature represents the peculiarity of the new proposed methodology; the two approaches have been widely used [23] but never been integrated in this way. The novelty of this code lies in the synthesis of methods used to describe reacting systems over long time scales while retaining fully atomistic detail of the system itself.

The algorithm used in the AMPI code includes the following steps [14]:

1. The code reads a set of inputs, such as reaction rates for the KMC module and MD-related quantities.
2. It counts the number of sites available for reactions in the system by identifying each atom as belonging to a specific reaction class, and a rate catalog of possible reactions at a specific time t is constructed.
3. After classifying the reactions as either accepted or rejected, a KMC step is performed, and the structure of the growing species is modified according to the selected reaction. The clock is then incremented in a way that is consistent with the average time for escape from that state, which can be determined easily from the rate constants for the possible escape paths in the list.
4. The code relaxes the newly formed structure using a MD run, allowing for relaxation of the structure towards thermal equilibrium. The potential used to describe hydrocarbon interactions is the adaptive intermolecular reactive bond order (AIREBO) [24]. The potential describes the covalent bonding interactions largely as in Brenner's REBO potential [25].

At this point a new kMC step is performed, and the sequence of steps is repeated. Before performing the next kMC step, the global update method is used in order to re-evaluate all interparticle interactions and all process transition probabilities after a process is executed.

3. Results

The AMPI code is used to understand the physical and chemical processes involved in the formation of nanoparticles in different conditions of temperature, pressure and species concentrations starting from small aromatic compounds. The initial system, used in this example, is composed of 300 gas-phase species at 1500K and 1 atm. The goal is to understand how one of these species, for example benzene evolves into a three dimensional structure of thousands of atoms over long time scales. The three main modules are gas-phase species that can react with the growing species, reaction sites on the growing particle, and reaction rates. For the gas-phase input, the user defines the temperature and the concentrations of the species that contribute to the formation of the nanoparticles. Examples of common gas-phase species for the high-temperature environment are H, OH, H_2, and aliphatic and aromatic species C_xH_y with x and y varying from 1 to 24.

Once the species are defined, it is important to list the possible reactions that can occur over time and describe them in terms of reaction sites, that is, atoms where a particular reaction can occur. The reaction site module governs the definition and counting algorithm of reaction sites, which are capable of undergoing modification (for example, additional reactions). Examples of reaction sites included in the AMPI code are an tom on a 5- or 6-membered ring and an sp^3 carbon.

The third input to the AMPI code is the list of the rate constants that are extracted from smaller length and time scales simulation tools, such as density functional theory (DFT) [26-28], transition state methods, transition state theory (TST) [29], and molecular dynamics [30]. The reactions included in the AMPI code can be broadly classified into three categories.

The unique capability of AMPI code has been succesful in reproducing physical and chemical characteristics of nanoparticles in different environments [14, 31-36]. Nanoparticles have been characterized in terms of chemical structure/components and relationships between structure and pathways, structure and properties, and structure and reactivity population of active sites have been addressed. Results show the comparison between computed properties and experimental data in terms of H/C trends, particle morphology, depolarization ratio, and free radical concentration [2, 14, 19, 34, 35, 37].

The AMPI code also has been successful in identifying reaction mechanisms for the formation of nanoparticles. For example, the AMPI code can identify the role of accessibility to reaction sites on the relative rate of reactions of different molecules. During the site-counting procedure, the code identifies every atom with a local environment that fits the definition of each site listed as input to the code. At the same time, it also checks whether there is sufficient space next to the site for a gas-phase species to penetrate and react. In this way, an atom is considered as a possible site for termination or addition by a gas-

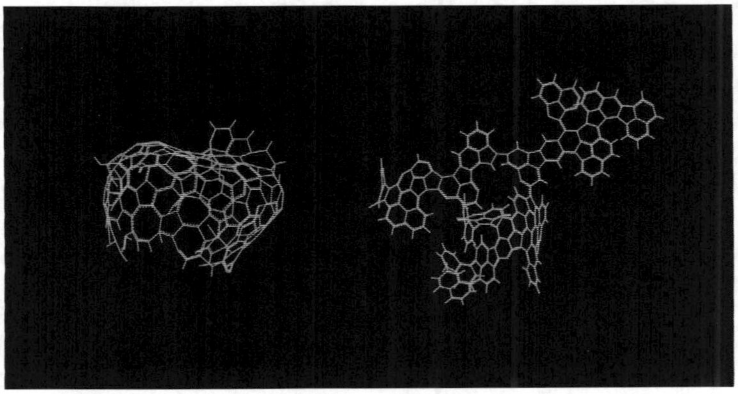

Example of nanoparticles of different morphology produced with the AMPI code.

phase species only if the gas-phase species of interest can be accommodated. The code goes over all atoms in the cluster, including all the possible sites, even the "hidden" ones that may be located inside the structure. The figure on the left shows particles produced from different combustion environments. It was found that the accessibility of the sites on the growing seed molecule decreases with increases in the curvature of the seed particle and size of the reacting molecule. This result can be used to identify new reaction mechanisms for the formation of nanoparticles of specified morphology.

For these systems, the presence of the MD module in the AMPI code is a determinant for a correct representation of the system. Rearrangements reactions are identified by MD leading to structural changes of the particle.

Besides growing by surface addition, the newly formed particles can grow by coalescence/coagulation to produce particles of 20–40 nm [1, 37, 38]. The figure below shows the influence of morphology on particle clustering. The right panel shows particles formed in an aromatic flame – total number of C=189 while the left side presents structures from an aliphatic flame (total number of C atoms is 214). The round particles, characteristic of benzene flames, tend to cluster and exhibit a preferred orientation with the open sections of the particles facing away form each other. The sheetlike particles produced in an acetylene flame, however, show a different behavior. They have more trouble getting together, and they tend to form stacking structures. These results demonstrate that the morphology of the nanoparticles is critical in determining the overall growth and agglomeration behavior.

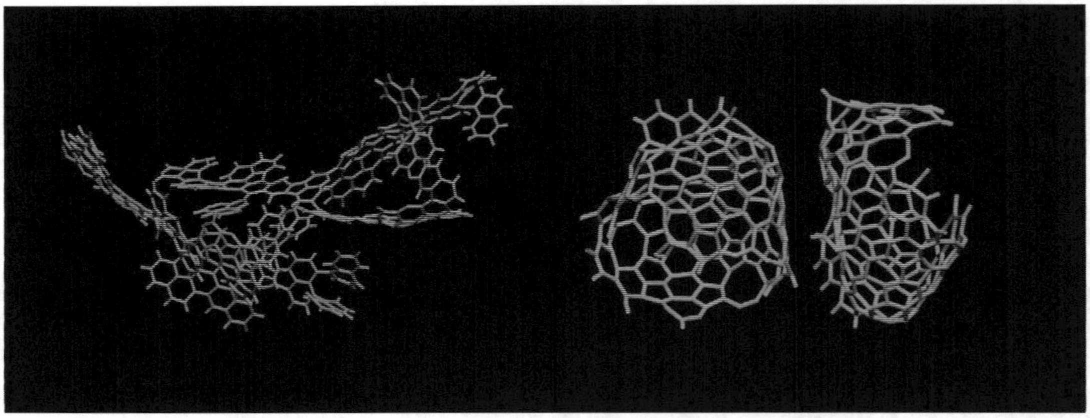

Particles from an aliphatic flame – left panel – and from an aromatic flame on the right. The clustering behavio r of the two systems is different.

The replacement of petroleum-derived fuels by biofuels from renewable resources has gained worldwide interest and is scientifically investigated for its environmental costs and benefits [39]. The chemistry of these new fuels, however, is poorly characterized, particularly under conditions relevant to next generation engine designs. The trend of fuel diversification needs a comprehensive science foundation that must span a range of knowledge from the most fundamental facets of fuel chemical and physical properties to the impact of these properties on the aspects of the performance and emissions of the emerging new engine technologies. The multiscale approach presented in this paper makes it possible to bridge the chemical characteristics of new fuels including reaction pathways and new oxygenated species with the dynamics of particle formation and agglomeration to produce a unified description from fuel properties, such as oxygen content, to particle characteristics, such as morphology and chemical composition.

4. Needs for petascale computing

The work reported in this paper describes a multiscale computational approach that makes it possible to correlate the structures of fuels with the morphology of nanoparticles that is critical in determining the overall growth and agglomeration behavior of higher molecular structures compounds (soot). Specifically, the method reveals how chemical changes and transformation can propagate upward in scale to help define the function of the particle structures. In particular, the morphology of the nanoparticles is critical in determining the overall growth and agglomeration behavior of higher molecular structures compounds (soot). The unique and powerful tool, embedded in the AMPI code, uses information on the chemistry occurring at the atomistic level and bridges it in a multiscale fashion to describe long timescale changes. The code combines the strengths of molecular dynamics and kinetic Monte Carlo and integrates them in a novel way. The synergy of the methods leads to a powerful computational approach that can describe the evolution of reactive systems over long time scales (seconds). To observe a statistically significant number of coagulation events, a large particle ensemble (10^6) has been shown to be effective [2].

With regard to the coagulation process, molecular dynamics simulations of systems composed of particles of different morphologies are used to understand the effect of shape and chemical composition on the clustering assemblies. Current levels of computing allow for roughly 10^4 nanoparticles to be formed using approximately $2x10^6$ atoms over nanosecond time scales. These calculations currently take about 14 days running on 16 processors (where each processor is a circa 2006 commodity processor) using the GROMACS molecular dynamics package. At this level of simulation, soot precursor clusters of approximately 20nm can be formed and studied numerically. For realistic predictions of soot formation, soot clusters of size approaching 1μm are needed, resulting in a factor of 50 increase in computational requirements. Furthermore, the physical dimensions will also be increased accordingly, resulting in a factor of 125,000 more atoms, or roughly 10^{11} atoms in total. With coarse-graining approaches, the number of computational particles will naturally be fewer but will still be on the order of 10^7 nanoparticles. With petascale computing we could begin to approach these requirements.

Acknowledgments

This research is supported by the National Science Foundation (CTS 0553764).

References

[1] Izvekov S, Violi A 2006 A coarse-grained molecular dynamics study of carbon nanoparticle aggregation *J. Chemical Theory and Computation* **2(3):**504-512
[2] Fiedler S L, Izvekov S and Violi A 2007 The effect of temperature on nanoparticle clustering *Carbon.* **45(9):**1786-794
[3] Kazakov A and Frenklach M 1998 Dynamic modeling of soot particle coagulation and aggregation: Implementation with the method of moments and application to high-pressure laminar premixed flames *Combustion and Flame* **114(3-4):**484-501
[4] Singh J, Patterson R I A, Kraft M and Wang H 2006 Numerical simulation and sensitivity analysis of detailed soot particle size distribution in laminar premixed ethylene flames *Combustion and Flame* **145(1-2):**117-127
[5] Morgan N, Kraft M, Balthasar M, Wong D, Frenklach M and Mitchell P 2007 Numerical simulations of soot aggregation in premixed laminar flames *Proc. Combustion Institute* **31:**693-700
[6] Smooke M D, Long M B, Connelly B C, Colket M B and Hall R J 2005 Soot formation in laminar diffusion flames *Combustion and Flame* **143(4):**613-628

[7] Smooke M D, McEnally C S, Pfefferle L D, Hall R J and Colket M 1999 Computational and experimental study of soot formation in a coflow, laminar diffusion flame *Combustion and Flame* **117(1-2):**117-139
[8] Balthasar M, Kraft M and Frenklach M 2005 Kinetic Monte-Carlo simulations of soot particle aggregation *Abstracts of Papers of the Amer. Chem. Soc.* **229:**U860
[9] Balthasar M and Frenklach M 2005 Detailed kinetic modeling of soot aggregate formation in laminar premixed flames *Combustion and Flame* **140(1-2):**130-145
[10] Homann K H and Wagner H G 1968 Chemistry of carbon formation in flames. *Proc. Royal Society of London Series A-Mathematical and Physical Sciences* **307(1489):**141
[11] Wang C Y and Zhang X 2006 Multiscale modeling and related hybrid approaches *Current Opinion in Solid State & Materials Science* **10(1):**2-14
[12] Muller-Plathe F 2002 Coarse-graining in polymer simulation: From the atomistic to the mesoscopic scale and back *Chemphyschem.* **3(9):**754-769
[13] Duan Y and Kollman P A 1998 Pathways to a protein folding intermediate observed in a 1-microsecond simulation in aqueous solution *Science* **282(5389):**740-744
[14] Violi A 2004 Modeling of soot particle inception in aromatic and aliphatic premixed flames *Combustion and Flame* **139(4):**279-287
[15] Metropolis N, Rosenbluth AW, Rosenbluth M N, Teller A H and Teller E 1953 Equation of state calculations by fast computing machines *J. Chem. Phys.* **21(6):**1087-1092
[16] Metropolis N and Ulam S 1949 The Monte Carlo method *J. Amer. Statistical Assoc.* **44(247):**335-341.
[17] Voter A F 2006 Introduction to the kinetic Monte Carlo method. In *Radiation Effects in Solids* E A SKEaK, ed. (Dordrecht, The Netherlands: Springer, NATO Publishing Unit)
[18] Gillespie D T 1976 General method for numerically simulating stochastic time evolution of coupled chemical-reactions *J. Comput. Phys.* **22(4):**403-434
[19] Violi A, Kubota A, Truong T N, Pitz W J, Westbrook C K and Sarofim A F 2003 A fully integrated kinetic Monte Carlo/molecular dynamics approach for the simulation of soot precursor growth Proc. Combustion Institute 29:2343-2349
[20] Allen M P 1989 *Computer Simulation of Liquids* (Oxford: Oxford Science Publications)
[21] Chatterjee A 2007 An overview of spatial microscopic and accelerated kinetic Monte Carlo methods *J.Computer-Aided Materials Design* **14(2):**253-308
[22] Frenkel D 1996 *Understanding Molecular Simulation: From Algorithms to Applications* (San Diego, CA: Academic Press)
[23] McDermott J B, Libanati C, Lamarca C and Klein M T 1990 Quantitative use of model-compound information - Monte-Carlo simulation of the reactions of complex macromolecules *Industrial & Engrg. Chem. Res.* **29(1):**22-29
[24] Stuart S J, Tutein A B and Harrison J A 2000 A reactive potential for hydrocarbons with intermolecular interactions *J. Chem. Phys.* **112(14):**6472-6486
[25] Brenner D W, Shenderova O A, Harrison J A, Stuart S J, Ni B and Sinnott S B 2002 A second-generation reactive empirical bond order (REBO) potential energy expression for hydrocarbons *J. Physics-Condensed Matter* **14(4):**783-802
[26] Hansen E W and Neurock M 2000 First-principles-based Monte Carlo simulation of ethylene hydrogenation kinetics on Pd. *J. Catalysis* **196(2):**241-252
[27] Neurock M and Hansen, E W 1998 First-principles-based molecular simulations of heterogeneous catalytic surface chemistry *Comput. Chem. Eng.* **22:**S1045-S1060
[28] Violi A, Truong TN, Sarofim AF. Kinetics of hydrogen abstraction reactions from polycyclic aromatic hydrocarbons by H atoms. Journal of Physical Chemistry A. 2004 Jun;108(22):4846-4852

[29] Truhlar D G, Garrett B C and Klippenstein S J 1996 Current status of transition-state theory *J. Phys. Chem.* **100(31):**12771-12800

[30] Car R and Parrinello M 1985 Unified approach for molecular-dynamics and density-functional theory Phys. Rev. Letters **55(22):**2471-2474

[31] Violi A, Voth G A and Sarofim A F 2005 The relative roles of acetylene and aromatic precursors during soot particle inception *Proc. Combustion Institute* **30:**1343-1351

[32] Violi A and Venkatnathan A 2006 Combustion-generated nanoparticles produced in a benzene flame: A multiscale approach *J. Chem. Phys.* **125(5)**

[33] Chung S H and Violi A 2007 Insights on the nanoparticle formation process in counterflow diffusion flames *CARBON* in press

[34] Violi A, Sarofim A F and Truong T N 2002 Mechanistic pathways to explain H/C ratio of soot precursors *Combustion Science and Technology* **174(11-12):**205-222

[35] Violi A, Sarofim A F and Voth G A 2004 Kinetic Monte Carlo molecular dynamics approach to model soot inception *Combustion Science and Technology* **176(5-6):**991-1005

[36] Chung S H and Violi A 2007 Insights on the nanoparticle formation process in counterflow diffusion flames *CARBON* in press

[37] Violi A and Izvekov S 2007 Soot primary particle formation from multiscale coarse-grained molecular dynamics simulation *Proc. Combustion Institute* 529-537

[38] Izvekov S, Violi A and Voth G A 2005 Systematic coarse-graining of nanoparticle interactions in molecular dynamics simulation *J. Phys. Chem. B* **109(36):**17019-17024

[39] Hill J, Nelson E, Tilman D, Polasky S and Tiffany D 2006 Environmental, economic, and energetic costs and benefits of biodiesel and ethanol biofuels *Proc. National Academy of Sciences USA* **103(30):**11206-11210

Journal of Physics: Conference Series **125** (2008) 012034 doi:10.1088/1742-6596/125/1/012034

Exploring HPCS Languages in Scientific Computing

Richard F. Barrett, Sadaf R. Alam, Valmor F. de Almeida, David E. Bernholdt, Wael R. Elwasif, Jeffery A. Kuehn, Stephen W. Poole, and Aniruddha G. Shet

Oak Ridge National Laboratory, Oak Ridge, TN

E-mail:
rbarrett,alamsr,dealmeidav,bernholdtde,elwasifwr,kuehn,spoole,shetag@ornl.gov

Abstract.
As computers scale up dramatically to tens and hundreds of thousands of cores, develop deeper computational and memory hierarchies, and increased heterogeneity, developers of scientific software are increasingly challenged to express complex parallel simulations effectively and efficiently. In this paper, we explore the three languages developed under the DARPA High-Productivity Computing Systems (HPCS) program to help address these concerns: Chapel, Fortress, and X10. These languages provide a variety of features not found in currently popular HPC programming environments and make it easier to express powerful computational constructs, leading to new ways of thinking about parallel programming. Though the languages and their implementations are not yet mature enough for a comprehensive evaluation, we discuss some of the important features, and provide examples of how they can be used in scientific computing. We believe that these characteristics will be important to the future of high-performance scientific computing, whether the ultimate language of choice is one of the HPCS languages or something else.

1. Introduction

As high-performance computing reaches the petascale and anticipates reaching the exascale inside of a decade, the nature of the hardware is changing rapidly. High-end systems now routinely encompass tens and even hundreds of thousands of CPU cores. Adding more cores per chip, and the details of how they share resources can be quite important for performance. Heterogeneous processors (LANL's Roadrunner petaflop system [1] includes three different processor architectures), complex memory hierarchies, and heterogeneous interconnects all further complicate the HPC environment. As we build toward exascale, these factors will become even more important.

Unfortunately current programming models are not capable of fully exploiting the enormous performance potential of these architectures [2–4]. Current models are designed to provide access to low level runtime system mechanisms in order to achieve performance. They generally require the programmer to manage every aspect of complex parallel algorithms, including mapping from conceptually global data structures to local storage, coordination of data exchange, and other aspects. As hardware architectures become larger and more complex, low-level approaches to parallel programming become less and less tractable for real applications.

Three new programming languages are being developed, as part of the High-Productivity Computing Systems (HPCS) program, sponsored by the U.S. Defense Advanced Research Projects Agency (DARPA), that are intended to address this issue. Chapel [5, 6], Fortress [7, 8], and X10 [9, 10] represent perhaps the largest concerted investment in the development of new environments for parallel programming in several decades. These "HPCS languages" incorporate the results of past research in parallel programming models with novel ideas and approaches relevant to emerging hardware architectures to produce high-level languages with support for object oriented and generic programming, a broad range of constructs for expressing both task and data parallelism at multiple levels, and a global view of data. We (the authors of this paper) believe that these languages are at least representative of a new generation of parallel programming environments which are appropriate for widespread use in high-end scientific and technical computing, and therefore worthy of deeper examination.

Because these languages are still relatively early in their development, we focus on the general features of the languages and their expressiveness for scientific computing. At present, the details of the language specifications continue to change, and the implementations generally do not provide all of the planned features, nor have they been tuned for performance. Performance is obviously one of the many issues faced by any new programming environment. We believe that the approach taken by these languages offers many potential benefits for performance portability across architectures, with more use of compilers and runtimes to help than is possible with current parallel programming environments. Where possible we point out syntax and semantics we believe could be exploited by the compiler and runtime system for achieving these goals.

We begin with a brief discussion of related efforts, followed by an overview of the new languages. We then explore the use of these languages using computations common to a broad set of scientific applications. Throughout these cases we discuss the opportunities provided by the languages qualified with the challenges faced by the language developers and compiler implementers. We end with some conclusions drawn from our experiences.

2. Background and Related Work

The dominant programming model in high-performance scientific computing couples a sequential programming language (i.e. Fortran) with the MPI message passing library [11, 12]. In this approach, all responsibility for managing parallelism, coordinating data exchange, and implementing global data structures falls on the programmer.

For hierarchical parallelism, of increasing interest as multi-core CPUs become ubiquitous, OpenMP [13] is often added to the mix. OpenMP offers implicit multithreaded parallelism controlled by compiler directives, and targets shared-memory environments.

High Performance Fortran (HPF) [14, 15] was an early attempt at a truly parallel language for high-performance scientific computing. However the first version suffered from a number of limitations (particularly the lack of task parallelism, and limited parallel data structures) and implementation challenges, and interest waned before these could be remedied [16].

More recently, languages such as Co-Array Fortran (CAF, [17]), Unified Parallel C (UPC, [18]), and Titanium [19, 20], collectively known as the partitioned global address space (PGAS) languages have been developed, and have gained some popularity. These languages were designed as *minimal* extensions of their underlying sequential languages (Titanium is based on Java) to provide some parallelism. While they do provide mechanisms to address remote memory via load/store semantics, they do not provide support for sophisticated global data structures. In CAF, global data objects are merely local objects with an index for the processor appended. In UPC, only one-dimensional arrays can be distributed. They also provide little in the way of parallel control structures.

A wide range of research languages developed over the years have also had important influences on the designs of the HPCS languages. The reader is referred to papers from the language teams themselves (see next section) for appropriate details. Of particular significance in the scientific computing context is the ZPL language [21, 22], which demonstrated an approach to describing array operations at a higher level, thus allowing code that is both safer and more flexible for a compiler to map onto the underlying architecture.

3. Overview of Languages

At a high level, the three languages have many similarities, though in detail there are significant differences. All three languages emphasize the expression of parallelism at a high level by the programmer and relying on the compiler/runtime/library infrastructure to produce an optimized implementation for the underlying parallel architecture.

In all three, program execution starts with a single conceptual thread of control, which then generates more parallelism through the use of language constructs (i.e. *not* strictly SPMD). Parallelism is mapped onto a multi-level conceptual model that is *roughly* approximated by the "processes" and "threads" of the traditional MPI-based programming model, for which each language has different terminology. Memory in all three is globally addressable, and data is global and can be distributed. Locality control permits computation and data to be assigned to specific system resources for performance reasons. The base languages are object-oriented and provide generic programming capabilities.

3.1. Chapel

Chapel [5, 6] is being designed by Cray Inc. to support general parallel programming while narrowing the gap between mainstream and HPC languages. Chapel's design builds on concepts from ZPL [21,22], High-Performance Fortran (HPF, [15]), and Cray's multithreaded extensions to C and Fortran [23], while adopting a variety of other useful features from mainstream and academic languages.

A *locale* in Chapel symbolizes a unit of architectural locality on the target machine, containing processing and storage capabilities. A locale's memory is uniformly accessible to computations running on it. Each locale supports a dynamic set of tasks that are created using *begin*, *cobegin*, and *coforall* statements. Tasks are synchronized using synchronization (*sync*) variables that have full/empty semantics, and *atomic* sections that provide transactional memory capabilities. Data and tasks can be mapped to machine resources (locales) using *on* clauses. The mapping may be explicitly specified, or data driven. Chapel supports data parallelism via *domains*, a first-class language concept representing an index set. Domains can be iterated over in parallel using *forall* and *coforall* loops, and are used to declare, resize, and slice arrays. Domains and their arrays may be partitioned across a set of locales using *distributions*, which map from the global view of an aggregate to its implementation on distinct locales.

3.2. Fortress

Fortress [7,8], being developed by Sun Microsystems, Inc., is designed to be an open, growable language. Consequently, it is designed with a small set of core language features, and the majority of concepts are coded in libraries.

Fortress programs are multithreaded; a user may explicitly spawn *threads*, or call *implicitly parallel* constructs that create threads managed by the Fortress language implementation. *Atomic* sections enable synchronization of threads. Fortress *regions* abstractly describe the underlying machine structure and can have an arbitrary hierarchical structure. Thread affinity to particular regions may be specified with *at* expressions, and *distributions* allow management of data locality. Parallelism can be programmed inside libraries as *distributions* and *generators*.

Fortress also provides a variety of novel features targeting the HPCS program's productivity goals, including built-in constructs for managing components and interfaces, expressing tests and contracts, and methods for rendering the code that look like typeset mathematics.

3.3. X10

X10 [9, 10], which is being developed by IBM Corp., is designed to leverage the extensive software ecosystem around the Java language. X10 is defined as a serial subset of Java, extended with additional concurrency, distribution, and locality features.

In X10, a *place* corresponds to a data-coherent processing element, with each place supporting a dynamic set of lightweight *activities*. Activities specify logical parallelism and may be composed in arbitrarily nested ways using *async*, *future*, *foreach*, and *ateach* constructs, and are translated by the X10 compiler/runtime into running threads. An activity executes to completion on the place where it is created, but can launch activities on other places, and detect termination of all such activities via the *finish* statement. *Clocks* enable synchronization of dynamically created activities across places. Activities within a place uniformly and coherently access its memory using *atomic* statements; weaker ordering semantics exist for inter-place data accesses. Similar to Chapel, X10 provides a ZPL-like "array language" to express high-level operations on distributed arrays.

3.4. Language Versions and Status

The versions of the languages available at the time of this writing are shown in Table 1. None of the current implementations fully support the features and capabilities envisioned by their developers. Where possible, we present and discuss code that works today, but given the fact that these languages are still very much under development, it is useful to also discuss what we expect to be possible in future implementations based on our interactions with the developers.

Table 1. Language Versions

Language	Specification	Implementation
Chapel	v0.775 [24]	v0.7 compiler
Fortress	v1.0 [7]	v1.0 interpreter
X10	v1.3 [25]	v1.5 compiler

4. Examples

We illustrate Chapel, Fortress, and X10 using computations common to a broad set of scientific applications. Finite difference stencils, used to solve partial differential equations, illustrate the data parallel model. A mesh sweeping algorithm for solving first-order hyperbolic PDE with linear characteristics illustrates more complex data parallel capabilities and object-oriented features. The Hartree-Fock method, used in quantum chemical simulations, illustrates task parallelism and atomic operations.

4.1. Finite Difference Stencils

A broad range of physical phenomena in science and engineering can be described mathematically using partial differential equations (PDE). Determining the solution of these equations on computers is often accomplished using finite differencing methods, based on stencil computations. Figure 1 illustrates the advantage of the global view which we will exploit. *The local view model requires that the code developer explicitly manage the decomposition of the domain across the parallel processors and the induced inter-process communication sharing requirements; the global view model frees the code developer from these tasks.*

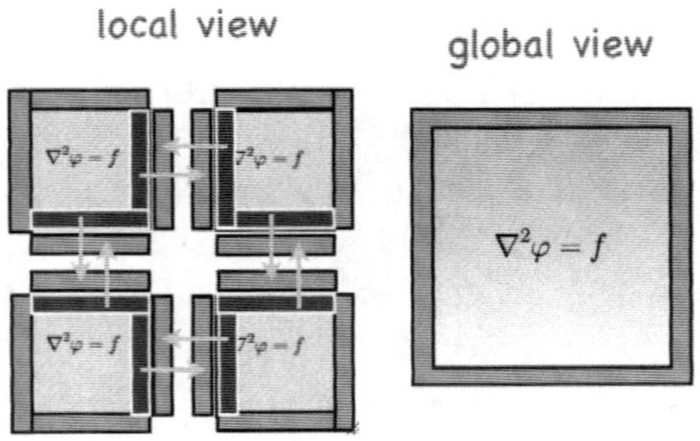

Figure 1. Local-view parallel vs. global-view programing model.

4.1.1. Chapel A 9-point stencil in two dimensions is shown in Code 1. An *arithmetic domain* (`PhysicalSpace`) describes the grid points in the physical space. From it, we derive a second domain (`AllSpace`), which adds space for applying boundary conditions. Arrays are allocated using this boundary space, while iteration is controlled by the physical space. Utilizing these domains in the computation rather than explicit indices provides the compiler flexibility for arranging and operating on data, and also avoids mistakes with loop bounds, a common source of coding errors.

By configuring the stencil computation as a reduction over a set of grid points, we can more clearly express the intent of this computation, conveying to the compiler the intent of the computation within the context of the data structure.

Code 1. 9-point stencil in 2d - Chapel

```
const PhysicalSpace : domain(2) distributed(Block) = [1..m, 1..n],
      AllSpace = PhysicalSpace.expand(1);
const Stencil9pt = [-1..1, -1..1];
const div = 1.0 / 9.0;
var g1, g2 : [AllSpace] real;
forall i in PhysicalSpace do
  g2(i) = ( +reduce [j in Stencil9pt] g1(i+j)) * div;
```

Code 2. 5-point stencil in 2d - Chapel

```
const Stencil:sparse subdomain(Stencil9pt)=
  ((-1,0),(0,-1),(0,0),(0,1),(1,0));
  - OR -
const Stencil:sparse subdomain(Stencil9pt) =
  [(i,j) in Stencil9pt] if (abs(i) + abs(j) < 2) then (ij);
```

Using a *sparse domain* we define a 5-point stencil as a subset of the arithmetic domain 9-point stencil, shown in Code 2 expressed in two different ways. The computation code remains unchanged, illustrating Chapel's polymorphic capabilities. For more on difference stencils using Chapel, see [26, 27].

Code 3. 5-point stencil in 2d - X10

```
public class stencil {
  static final region (:rank==2) Rall = [0:N+1,0:N+1], Rcore = [1:N,1:N];
  static final dist (:rank==2) Dall=(dist(:rank==2)) dist.factory.block(Rall),
                               Dcore=(Dall | Rcore),
                               Dhalo=(Dall - Dcore.region);
final double[Dcore:rect&&rank == 2] gridnew = (double[Dcore:rect&&rank == 2])
  new double[Dcore](point [i,j]) {
    return ((a[i-1,j]+a[i,j-1]+a[i,j]+a[i,j+1]+a[i+1,j])*0.2 );};
```

4.1.2. X10 The X10 implementation is shown in Code 3. Data is distributed across the available *places*, with work-sharing loop constructs performing the computation. The *region* declarations create abstract one dimensional and two dimensional templates, which are sliced up and distributed across the available places by the *dist* declarations. The declared distributions, at this point still abstract objects, are then used when declaring and allocating array objects. The array is allocated and accessed according to its distribution. Included in the array declarations for `gridold` and `gridnew` are pointwise initialization constructs which run in parallel at each X10 place, a fact which is used to update `gridnew` with the 5-point stencil. Array operations such as the difference, absolute value, and sum over the core regions of the old and new grids as well as the update of the core region of the old grid based on the new grid are also performed in parallel.

4.1.3. Fortress A polymorphic implementation using Fortress (instantiated for any `Number` type) of the 5-point stencil computation is shown in Code 4. The loop bound constraints depend on the start values, so the loop iterations are set to `N-3`. Fortress offers multi-dimensional ar-

Code 4. 5-point stencil in 2d - Fortress

```
stfivept[\Elt extends
  Number\](z:Elt,o:Elt) :() = do
  a = array[\Elt\](N,N)
  b = array[\Elt\](N,N)
  for j <- 2#(N-3) do
    for i <- 2#(N-3) do
      b[i,j] :=
      ( a[i-1,j]+a[i,j-1]+a[i,j]+a[i,j+1]+a[i+1,j] ) * 0.2
    end
  end
```

ray data structures called traits, which can be instantiated with any data type. The array elements could be distributed in multiple regions of the memory hierarchy and the region method returns the locality of an array element. In addition, the `for` loop construct in fortress is by default parallel and nondeterministic.

4.2. Hartree-Fock Method

The Hartree-Fock method of quantum chemistry [28], widely used in quantum chemical simulations, combines a task-parallel algorithm with operations on distributed data, and is characterized by irregular and unpredictable parallel tasks as well as distributed data access. A thorough examination of this problem in the HPCS languages appears in [29, 30]. Due to space limitations, we focus here on just one example drawn from this rich computational problem.

The most computationally intensive step in the Hartree-Fock method is construction of the Fock matrix,

$$F_{\mu\nu} \leftarrow D_{\lambda\sigma}\{2(\mu\nu|\lambda\sigma) - (\mu\lambda|\nu\sigma)\}$$

where the indices μ, ν, λ, σ denote the basis functions, and D is the density matrix. $(\mu\nu|\lambda\sigma)$ is a rank-4 tensor representing the two-electron repulsion integrals. Formation of the Fock matrix is an $O(N^4)$ operation for N basis functions.

The scalable parallel implementation of the heart of the Fock matrix construction algorithm is illustrated in Figure 2. The density (input) and Fock (output) matrices are fully-distributed arrays, which are accessed in blocks, sized to match integral blocks. The computationally intensive evaluation of integrals is distributed across the available processes using a four-fold loop over the index space (blocked appropriately). Because the integral blocks are highly non-uniform both in terms of size and computational cost, the implementation must be dynamically load balanced in order to scale.

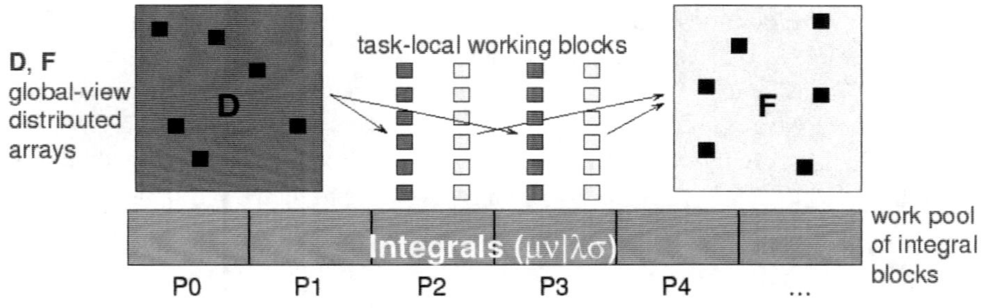

Figure 2. Schematic of the Fock matrix construction algorithm.

In contrast to MPI, the HPCS languages provide a variety of parallel constructs which can be used to implement dynamic load balancing for this algorithm in a number of different ways. One approach makes use of a *task pool* concept. The "pool" is a common work area, or list of tasks. Producers submit tasks to the pool and consumers execute them. In the Fock-build algorithm, the producer would generate the lists of integral blocks that need to be evaluated, and all processes would consume these tasks as they become available (see [29, 30] for details). This approach utilizes the languages' atomic operations to manage the global task pool data structures, and locality-aware task management features to distribute the work. The task pool idea provides a very flexible way of expressing and managing many kinds of parallelism. In addition to irregular problems, task pools can provide flexible management of parallel multitasking and other complex concepts. Task pool implementations can be as sophisticated as necessary to account for data locality and other considerations. As processor (and core) counts grow, one important way of expressing more parallelism is to more carefully examine data dependencies and allow unrelated parallel operations to take place on subsets of the available processes. We believe that the task pool pattern will become increasingly relevant to high-performance scientific computing, and therefore also the language (or library) constructs that support it.

In this paper, we illustrate in more detail another load balancing approach using a shared global counter, which is well-suited to a single large-scale irregular computation. In this case, the four-fold loop indices are mapped onto a linear sequence of computational tasks, which can be represented with a simple integer counter. When a process needs work, it reads the counter to obtain the sequence number

Code 5. Shared counter for dynamic load balancing - X10

```
1 int G = 0;
2 finish ateach(point [p] : dist.factory.unique(place.places)) {
3   int myG, L = 0;
4   future<int> F = future (place.FIRST_PLACE) {read_and_increment_G()};
5   myG = F.force();
6   for(point [iat] : [1:natom])
7     for(point [jat, kat] : [1:iat, 1:iat])
8       for(point [lat] : [1:(kat==iat?jat:kat)]) {
9         if (L == myG) {
10          F = future (place.FIRST_PLACE) {read_and_increment_G()};
11          buildjk_atom4(new blockIndices(...));
12          myG = F.force();
13        }
14        ++L;
15      }
16 }
```

of the next task it should perform, and increments it so that the next reader will get the following task. Clearly this entire operation must be performed atomically, or there is a risk that another process will read the unincremented value and duplicate the task. This is the approach used in the first scalable, dynamically load balanced version of the fully-distributed Hartree-Fock algorithm [28, 31, 32], which used the Global Array (GA) Toolkit [33], which provides the concept of a globally shared counter with an atomic read-and-increment operation, as well as a global view of distributed arrays, and one-sided messaging (get/put semantics), all of which also appear in the HPCS languages.

4.2.1. X10 In Code 5, the *root activity* creates the globally shared counter `G` on the *first place* (line 1). Then it uses the `ateach` construct (line 2) to launch a copy of the Fock-build algorithm (lines 3–16) on each *place*. The `finish` (line 2) causes the root activity to block until the rest of the algorithm completes on every place. Each place iterates over the same sequence of tasks (the four-fold loop, lines 6–8), using `L` to count the tasks. When `L` matches the next task assigned to the place (`myG`), it evaluates that integral block. Assignments (`myG`) are obtained from a remote *atomic* read-and-increment operation on the globally shared counter `G` on the first place (lines 4 and 10). When every place has completed the four-fold loop, all tasks will be evaluated.

X10 requires that remote references to mutable data (in this case the shared counter `G`) be done asynchronously, hence the use of the `future` construct at lines 4 and 10. Separation between spawning the future and forcing it (as in lines 10 and 12) allows computation and communication to be overlapped.

Code 6. Atomic read-and-increment - X10

```
private int read_and_increment_G() {
  int myG;
  atomic myG = G++;
  return myG;
}
```

Code 6 shows how the atomic read-and-increment operation is straightforwardly implemented with an `atomic` section.

4.2.2. Chapel Our Chapel implementation in Code 7 employs the `coforall` statement to create distinct concurrent computations for all the *locales*, with an `on` clause binding one computation to one locale (line 2). Each computation loops over the index space of tasks (lines 4–7).

The shared counter `G` is created on the *first locale* as a *synchronization* variable of the `sync` type, which provides full/empty semantics (line 1). Once written, such a variable cannot be re-written until it is emptied. Likewise, an empty variable cannot be re-read until it is written. Computations attempting to write to a full `sync` variable or read from an empty one will block until another computation changes the variable's state. Taking advantage of these semantics to atomically update the counter in the `readAndIncrementG` method from Code 7, every computation first does a read (line 18) followed immediately by a write of `G` (line 19) to fetch the next task. The processing of a newly assigned task is overlapped with the fetch of the next task inside a `cobegin` statement (Code 7, line 9).

Code 7. Shared counter for dynamic load balancing - Chapel

```
1  var G : sync int = 0;
2  coforall loc in LocaleSpace on Locales(loc) {
3    var (L,lattop,myG) = (0,0,readAndIncrementG());
4    for iat in 1..natom do
5      for (jat, kat) in [1..iat, 1..iat] {
6        lattop = if (kat==iat) then jat else kat;
7        for lat in 1..lattop {
8          if (L == myG) then
9            cobegin {
10             buildjk_atom4(new blockIndices(...));
11             myG = readAndIncrementG();
12           }
13         L += 1;
14       }
15     }
16 }
17 def readAndIncrementG() {
18   const myG : int = G;
19   G = myG + 1;
20   return myG;
21 }
```

4.2.3. Fortress The Fortress version in Code 8 spawns parallel threads using the `for` expression, with the `at` expression denoting the *region* where each thread should run (line 3). (At the time of this writing, since Fortress does not support distinct regions, `numRegs` is used to simulate the different regions.) All threads traverse the iteration space in a serial fashion as mandated by the `seq` multi-generator `for` loop (lines 5–7). The `also` construct (line 11) runs a new task concurrently with update to the shared counter `G`. The `read_and_increment_G` function is implemented as an `atomic` method (lines 18–25).

Code 8. Shared counter for dynamic load balancing - Fortress

```
1  numRegs = 10
2  var G : ZZ32 = 0
3  for reg<-1#numRegs at region(reg) do
4    (L,myG) : ZZ32... := (0,read_and_increment_G())
5    for iat<-seq(1#natom), jat<-seq(1#iat),
6        kat<-seq(1#iat), lat<-seq(1#(if (kat=iat)
7        then jat else kat end)) do
8      if (L = myG) then
9        do
10         buildjk_atom4 blockIndices(...)
11       also do
12         myG := read_and_increment_G()
13       end
14     end
15     L += 1
16   end
17 end
18 read_and_increment_G() : ZZ32 = do
19   var myG : ZZ32
20   atomic do
21     myG := G
22     G += 1
23   end
24   myG
25 end
```

4.3. Mesh sweeping

Methods for solving a first-order, hyperbolic PDE with linear characteristics often employ a sweeping method to traverse the domain along the corresponding characteristic direction. This is accomplished in practice by first, approximating the PDE domain by a polyhedron (polygon if two-dimensional), second, partitioning the polyhedron into a set of non-overlapping sub-polyhedra (sub-polygon), thereby defining a mesh, and third, coloring the mesh representation data so that each color represents a group of mesh elements wherein an exact solution to the original PDE is obtained by simple integration along the characteristic direction. (Figure 3 illustrates the idea.) If all groups of elements are visualized with a continuous colormap, they give rise to a sequence of fronts roughly orthogonal to the characteristic direction. Therefore a colored group of elements is referred to as a wave front.

Although first-order, hyperbolic PDE's with linear characteristics fall short to represent many physical processes of interest, they are at the heart of many iterative schemes used to solve more complex and realistic models. For instance, in neutron transport models used in nuclear reactor simulations, the discretization of the angular variables (known as discrete ordinates) leads to a set of first-order, hyperbolic PDE's in as many directions as there are ordinates; this set is further augmented by the discretization of the neutron energy variable. The overall solution scheme iterates within an outer loop to update the integral terms (scattering and source) in the original model. Another related example is the use of mesh sweeping on radiative transfer problems [34].

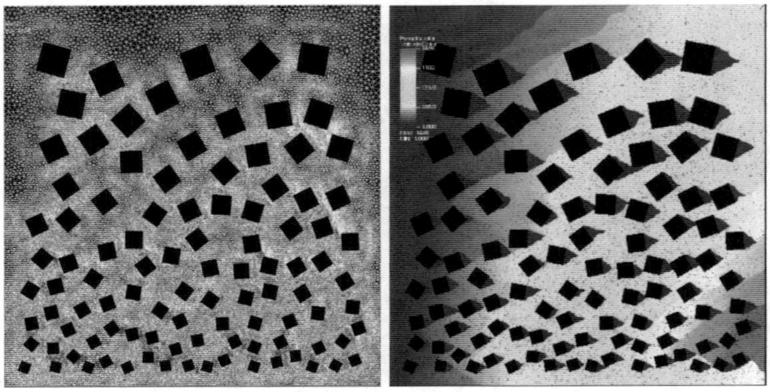

Figure 3. Mesh sweeping
Left: mesh with 123,888 triangles around 100 square particles. Right: wave front coloring of mesh sweeping in the horizontal direction left to right. Blue indicates all elements marked in the first wave front, and red, all elements in the last front.

The most frequent parallel implementation of a mesh sweeping method is obtained by spatially partitioning the graph of the whole mesh (either in parallel or sequentially) and distributing the data across processors explicitly. This requires the time-consuming, error-prone reconstruction of the mesh data on each processor, *e.g.* connectivity, numbering, physical boundary conditions, artificial internal boundary information, etc. by the programmer. Once the data has been distributed, mesh sweeping is applied independently in each processor on the local data, and an outer loop is needed to update interprocessor boundary information when solving the associated PDE problem. This approach no longer guarantees an exact solution of the problem but upon sufficient convergence and multiple full sweeps of the domain, the approximate solution is typically acceptable.

Our implementation of a sweeping algorithm for a two-dimensional (2-D) mesh using Chapel in Code 9 explores the underlying parallel infrastructure of the language to avoid an explicit spatial partitioning of the mesh data. The gains are twofold, no need for mesh partitioning and tedious extra programming to reconstruct local mesh data, and no need for multiple sweeps by the PDE host solver.

For simplicity we do not save the wave fronts in this example code but override them each time a full front has been computed. In line 1, the mesh partition set object, `setT`, is created by instantiation of a finite element partition class whose constructor reads a mesh input data file, `mesh.dat`. The data is then stored in various containers which distribute their corresponding data over Chapel locales. The essential functionality derives from the systematic/nested use of a generic vector container similar to a class template (Code 10). All sweep directions are looped over in parallel starting in line 4. `sweepDir` is a `(real,real)` pair direction from the vector `sweepDirections` of all desired sweep directions spanning the range 0–360°. A mask over all mesh elements is created in line 5 and marked as `blue`, that is, all elements are initially ahead of the current wave front while the null element (line 6) is always behind any front, hence, colored green. This mask is a distributed vector of type `int`, with range `0..setTCardinality`. As the sweeping of the mesh proceeds, the number of blue elements reduces and the while statement in line 7 terminates the process. The core of the look-ahead searching to color the wave front mesh elements as red is in lines 8–20. Before starting the parallel loop over all elements in line 8, the mask never has any red color value, that is, elements are colored either ahead of the front or behind the front. In line 9, only the elements ahead of the front are queried. Initially the color status `setColor` of any given blue element is set to true (line 10) indicating that the element is a candidate for the new wave front. Next, for each edge in the list of all edges, the outward pointing, unit normal vector

Code 9. Look-ahead mesh sweeping - Chapel

```
1 var setT = new FiniteElementPartition("mesh.dat");  //create dist. mesh part. set
2 const setTCardinality:int = setT.GetCardinality();    //# of mesh elements; large
3 enum {blue, green, red};     //color for ahead of front, behind, and on the front
4 forall sweepDir in sweepDirections {        //parallel loop over sweep directions
5  var waveFrontMask = new Vector(int,0..setTCardinality,blue);//distributed vector
6  waveFrontMask(0) = green;                          //null element id points to green
7  while (waveFrontMask.Find(blue)) {                 //keep coloring while blue exists
8   forall fE in setT.GetMembers {
9    if (waveFrontMask(fE.GetId()) == blue) {
10    var setColor:bool = true;
11    forall edge in fE.GetEdges() {
12     const normal = fE.GetEdgeNormal(edge);
13     if (normal^sweepDir < 0.0){                   //if sweep direction enters edge
14      const neMemberId:int = setT.GetNeighborMemberId(fE,edge);
15      if (waveFrontMask(neMemberId) != green) {setColor = false;}
16     }
17     if (setColor == true) {waveFrontMask(fE.GetId()) = red;}
18    }
19   }                                           // end if mesh element ahead of front
20  }                                          // end parallel loop over mesh elements
21  forall i in waveFrontMask {if (i != blue) {i = green;}}
22 }                                                                    // end while
23 }                                       // end parallel loop over sweep directions
```

is computed and compared (via an inner product) to the sweeping direction (line 13). If the sweeping direction points towards the inside of the element then the neighbor element sharing that edge (line 14) is tested for its position relative to the wave front (line 15). If the neighbor is on the front or ahead of the front, then the current element is not selected (line 15). In the event that all edges in the list `fE.GetEdges()` (line 11) of the candidate mesh element, `fE`, fail the test in line 15, then it is colored as a wave front element (line 17). Once all remaining blue elements are searched, the mask color is updated to convert the red elements into green (lines 21).

It is clear from Code 9 that the sweeping algorithm can be expressed almost identically to other sequential object-oriented codes, and it is equally intuitive. However it has the potential to exploit parallelism since all objects eventually use distributed vectors of generic types (Code 10) for their underlying data storage. This implementation of a distributed vector preserves Chapel domain and range variables in the same program unit, and exposes a familiar interface for the user accustomed to object-oriented and generic programming. We have tested this Chapel implementation extensively on various 2-D meshes data sets (Figure 3) which is equaly valid for 3-D meshes upon substitution of `face(s)` for `edge(s)` in Code 9, and implementation of the respective member functions.

Code 10. Sample of a generic vector distributed container - Chapel

```
class Vector { // generic/template class
// Accessors and modifiers
 def GetType() type {return T;}
 def GetSize() {return rng.high-rng.low+1;}
 def GetRange() {return rng;}
 def GetDomain() {return dom;}
 def Resize(n: range = 1..0) {rng = n; dom = [rng];}
 def Clear() {this.Resize();}
 def Find(val:T):bool {return LinearSearch(data,val)(1);}
 def these() var {for i in dom {yield data(i);}}
// Members
 type T;
 var rng:range(int);
 var value:T;
 var dom:domain(1) distributed(Block) = [rng];
 var data:[dom] T = value;
}
```

5. Conclusions

We have attempted to illustrate some of the features in the HPCS languages that make these languages much better aligned with the needs of the next generation of HPC scientific codes and the hardware on which they will run. In contrast to the MPI library+sequential language environment which currently dominates parallel scientific computing, the HPCS languages integrate parallel control structures, distributed data, and other important features such as atomic operations into a comprehensive higher-level expression of the computation that should allow the compiler and the runtime system to effectively perform transformations and optimizations for different underlying hardware platforms.

However, the qualifier *should* in the sentence above is an important one. As we have noted already, these languages are still early in their development, and implementations are not yet mature with respect to capabilities or performance. Performance is obviously on par with expressiveness when it comes to evaluating new programming environments for HPC, and many of the features of the HPCS languages will be challenging to implement efficiently.

For example, the stencil code frees us from dealing with the mechanics of distributing and sharing data, and we expect this to be a manageable task by a compiler due to the regular structure of the grid. The unstructured grid operated on by the sweep algorithm will require advanced partitioning techniques and associated inter-process communication requirements.

Additionally, all three languages anticipate supporting software transactional memory in the form of constructs for synchronizing access to shared data by distributed threads of control. Atomic sections have the potential of easing programming by relieving the programmer from the burden and pitfalls of explicit lock-based schemes. Scalable implementation of transactional memory in distributed memory environments is a challenging area of research. The shared counter schemes of dynamic load balancing will require efficient techniques for doing atomic updates.

On the other hand, the development teams for these languages are all very much aware of the complexities they face, and are sensitive to the experience of High-Performance Fortran, which has been attributed in part to pushing the compiler technology of the time too far [16], and have no wish to repeat it. The developers are confident that the proposed features can be implemented effectively.

Further, while the languages provide constructs suitable for exploiting the heterogeneous and other advanced features of architectures in very flexible ways, effectively managing these capabilities within the context of algorithmic requirements will also be challenging. Though once again, we observe that the higher-level integrated view of the computation provided by the HPCS languages should allow more use of tools and automation, and therefore require less detailed programmer effort to make them work.

In addition to a variety of performance and implementation-related issues, there are many other concerns on the long road to possible widespread of a new language for scientific computing. Large-scale applications and key libraries will not immediately be rewritten in a new language, therefore all three languages plan to provide interoperability with existing languages and libraries. In all languages, the plan is to provide basic interoperability via Babel [35] or a similar mechanism. Parallel data distributions to match those used by important libraries (for example ScaLAPACK) might be predefined to aid users. I/O is clearly extremely important for scientific computing, but has been deferred by the language teams in order to focus on the core of the languages first.

We have found the development teams to be enthusiastic and responsive to our questions, experiences, and suggestions. Feedback related to real scientific applications is particularly valuable at these formative stages. But at the same time, we want to provide two cautions that, given the HPF experience [16], seem very important. First, prospective users should make sure they set their expectations in accord with the stated level of maturity of the languages and implementations. At this stage, for example, most users would find it more productive to begin learning about the new concepts presented in these languages, and thinking about how they might apply to their applications rather than actually trying to write code for the HPCS languages. Second, patience within the user community will be important to allow new languages to reach a level of maturity and quality at which it becomes reasonable to evaluate them as possible alternatives to long-standing approaches, which have thirty or more years of experience and usage behind them.

Although we cannot predict the success of these languages, we believe there is ample justification for the computational scientist to be supportive of the efforts. We look forward to tracking the progress of the HPCS language specifications and their implementations.

Acknowledgements

We thank all three of the language development teams for their support and discussions through this work.

This work has been supported by the Laboratory Directed Research and Development Program of Oak Ridge National Laboratory (ORNL), and the ORNL Postmasters Research Participation Program which is sponsored by ORNL and administered jointly by ORNL and by the Oak Ridge Institute for Science and Education (ORISE). ORNL is managed by UT-Battelle, LLC for the U. S. Department of Energy under Contract No. DE-AC05-00OR22725. ORISE is managed by Oak Ridge Associated Universities for the U. S. Department of Energy under Contract No. DE-AC05-00OR22750.

References

[1] Roadrunner supercomputer, Los Alamos National Laboratory `http://www.lanl.gov/roadrunner`
[2] Alam S, Barrett R, Kuehn J, Roth P and Vetter J 2006 *IEEE International Symposium on Workload Characterization*
[3] Dongarra J, Gannon D, Fox G and Kennedy K 2007 *CTWatchQuarterly* **3**
[4] Post D and Votta L 2005 *Physics Today* **58**
[5] Chamberlain B L, Callahan D and Zima H P 2007 *International Journal of High Performance Computing Applications* **21** 291–312
[6] Chapel home page `http://chapel.cs.washington.edu/`
[7] Allen E, Chase D, Hallett J, Luchangco V, Maessen J W, Ryu S, Jr G L S and Tobin-Hochstadt S 2008 `http://projectfortress.sun.com/Projects/Community/browser/trunk/Specification/fortress.1.0.pdf`
[8] Fortress home page `http://fortress.sunsource.net/`
[9] Charles P, Donawa C, Ebcioglu K, Grothoff C, Kielstra A, von Praun C, Saraswat V and Sarkar V 2005 *Proceedings of Object-Oriented Programming, Systems, Languages, and Applications(OOPSLA)*
[10] X10 home page `http://x10.sf.net/`
[11] Snir M, Otto S, Huss-Lederman S, Walker D and Dongarra J 1998 *MPI: The Complete Reference: Volume 1 - 2nd Edition* (The MIT Press)
[12] Gropp W, Huss-Lederman S, Lumsdaine A, Lusk E, Nitzberg B, Saphir W and Snir M 1998 *MPI: The Complete Reference: Volume 2 - The MPI-2 Extentions* (The MIT Press)
[13] OpenMP home page `http://openmp.org`
[14] High Performance Fortran Forum 1993 High Performance Fortran language specification Tech. Rep. CRPC-TR92225 Center for Research on Parallel Computation, Rice University, Houston, TX
[15] High Performance Fortran Forum 1997 High performance fortran language specification, version 2.0 Tech. rep. Rice University
[16] Kennedy K, Koelbel C and Zima H 2007 *Proceedings of the third ACM SIGPLAN conference on History of programming languages*
[17] Numrich R and Reid J 1998 *ACM Fortran Forum* **17** 1–31
[18] UPC home page `http://upc.gwu.edu`
[19] 2006 The Titanium project home page `http://titanium.cs.berkeley.edu/`
[20] Hilfinger P, Bonachea D, Datta K, Gay D, Graham S, Liblit B, Pike G, Su J and Yelick K 2005 Titanium language reference manual Tech. Rep. UCB/EECS-2005-15 U.C. Berkeley `http://titanium.cs.berkeley.edu/papers/EECS-2005-15.pdf`
[21] Chamberlain B L, Choi S E, Deitz S J and Snyder L 2004 *Proceedings of the IEEE International Workshop on Productivity and Performance in High-End Computing (PPHEC)*
[22] ZPL home page `http://www.cs.washington.edu/research/zpl/`
[23] Cray, Inc. 2005 *Cray MTA-2 Programmer's Guide* s-2320-10
[24] Chapel language specification `http://chapel.cs.washington.edu/spec-0.775.pdf`
[25] X10 language specification `http://x10.cvs.sourceforge.net/x10/x10.man/v1.0/x10.pdf?view=log`
[26] Barrett R, Poole S and Alam S R 2007 Expressing POP from a Global View Using Chapel: Towards a More Productive Ocean Model Tech. Rep. TM-2007/122 Oak Ridge National Laboratory

[27] Barrett R, Alam S and Poole S W 2008 *Proc. 50th Cray User Group Meeting*

[28] Foster I T, Tilson J L, Wagner A F, Shepard R L, Harrison R J, Kendall R A and Littlefield R J 1996 *Journal of Computational Chemistry* **17** 109–123

[29] Shet A G, Elwasif W R, Harrison R J and Bernholdt D E 2008 *2008 IEEE International Parallel and Distributed Processing Symposium*

[30] Shet A G, Elwasif W R, Harrison R J and Bernholdt D E 2008 Programmability of the HPCS languages: A case study with a quantum chemistry kernel (extended version) Technical report ORNL/TM-2008/011 Oak Ridge National Laboratory see http://www.ornl.gov/info/reports/

[31] Harrison R J, Guest M F, Kendall R A, Bernholdt D E, Wong A T, Stave M, Anchell J L, Hess A C, Littlefield R J, Fann G L, Nieplocha J, Thomas G S, Elwood D, Tilson J L, Shepard R L, Wagner A F, Foster I T, Lusk E and Stevens R 1996 *Journal of Computational Chemistry* **17** 124–132

[32] Tilson J L, Minkoff M, Wagner A F, Shepard R, Sutton P, Harrison R J, Kendall R A and Wong A T 1999 *International Journal of High Performance Computing Applications* **13** 291–302

[33] Nieplocha J, Palmer B, Tipparaju V, Krishnan M, Trease H and Apra E 2006 *International Journal of High Performance Computing Applications* **20** 203–231

[34] de Almeida V F 2003 An iterative phase-space explicit discontinuous galerkin method for stellar radiative transfer in extended atmospheres Technical Manuscript ORLN/TM-2003-072 Oak Ridge National Laboratory Tennessee, TN 37831-6181, U.S.A. also available by request to dealmeidav@ornl.gov or at http://dealmeidav.ornl.gov

[35] Lawrence Livermore National Laboratory Babel http://www.llnl.gov/CASC/components/babel.html

Journal of Physics: Conference Series **125** (2008) 012035 doi:10.1088/1742-6596/125/1/012035

Design issues for numerical libraries on scalable multicore architectures

Michael A Heroux

Distinguished Member of the Technical Staff, Sandia National Laboratories, Albuquerque, NM 87185-1320, USA

E-mail: maherou@sandia.gov

Abstract. Future generations of scalable computers will rely on multicore nodes for a significant portion of overall system performance. At present, most applications and libraries cannot exploit multiple cores beyond running addition MPI processes per node. In this paper we discuss important multicore architecture issues, programming models, algorithms requirements and software design related to effective use of scalable multicore computers. In particular, we focus on important issues for library research and development, making recommendations for how to effectively develop libraries for future scalable computer systems.

1. Introduction

Numerical libraries for parallel computers have traditionally had two target machine architectures: highly scalable distributed-memory systems with serial nodes and shared-memory systems with modest scalability. Although shared-memory nodes have been part of large-scale systems for many years, most scalable applications have not explicitly used shared-memory programming on the node, instead relying on the message passing library, MPI in particular, to exploit shared memory for message-passing performance.

The next generations of high-end computer systems will combine scalable node count with shared memory nodes in ways that will likely require most applications and libraries to explicitly managed shared memory resources in addition to performing message passing across nodes, thereby requiring libraries to address both types of parallel resources. In this paper we discuss the architectural features libraries must exploit and support and the implications for library developers and users.

2. Important trends in computer architecture

For more than 15 years scalable computing has been dominated by single-level message-passing applications and computer systems. These are applications that view the computer system as a flat network of equidistant, homogeneous processors and computer systems that are able to behave as such, even though very few were physically configured that way. For a brief period around the year 2000, the parallel computing community thought that systems composed of networked shared memory parallel (SMP) nodes could not be viewed as a flat network and still allow optimal performance [1]. The general impression was that such systems would require explicit shared memory programming underneath message passing in order to achieve best performance, for example, MPI across nodes and OpenMP within a node. However, once message-passing libraries were made "SMP-aware" (such that messages between processors on the same node bypassed the network, thereby reducing network performance bottlenecks), the flat message-passing-only approach – where on a machine with m nodes and k cores on each node, we run an MPI program with $p = m * k$ processes in the traditional single-program-multiple-data (SPMD) model – continued to be the best overall approach to scalable application design and continues to be true even today on most systems.

Despite the success of the single-level, message-passing-only approach, there is growing evidence that future applications and libraries will need to address increasing node complexity by introducing another level of parallelism. There is little doubt that message passing will still be the dominant internode programming model, but there is also much doubt that it will be sufficient for driving all cores on a single node for much longer, at least at a single level. As a result, we must explore new node programming models that can be effectively used in combination with message passing across nodes.

There are many ways to characterize multicore architectures, but for our purposes we consider two classes: homogeneous and heterogeneous. For each class we discuss important hardware and software issues that are impacting library software design and implementation.

2.1. Homogeneous multicore nodes

Homogeneous multicore nodes, for example, Intel and AMD quad-cores and Sun's Niagara2, have identical cores with shared memory (although memory access may be non-uniform). Because of their architectural predecessors, SMP nodes, these nodes have several well-established parallel programming models, such as Posix threads and OpenMP. Furthermore, as long there is sufficient memory bandwidth, message passing is often an effective approach.

Homogeneous multicore nodes are in many ways similar to microprocessor SMP architectures of the past 15 years. The biggest differences are that multicore nodes are now available in every computer and core counts are expected to increase substantially. Previous multiprocessor architectures were primarily available only in servers, and two or four processors were the most common configurations.

2.1.1. Slow growth in single-core Performance For many years, users of commodity microprocessors could realize regular performance improvements in their applications simply due to single processor performance improvements. Many of these improvement were due to increased clock speeds and sophisticated instruction scheduling in hardware. At present, clock speed are staying relatively constant and core architectures are expected to become simpler, if change much at all. The effect is that single core performance is improving but not as rapidly as in the past, and users who want to once again experience substantial performance improvements from one generation of microprocessor design to the next must learn to exploit multiple cores.

Although not an exhaustive list of machines, table 1 shows the performance profile over five years of processors for HPCCG, an unstructured finite-element microapplication. HPCCG performance is a function of both floating point capabilities and memory system performance. Therefore single core performance rates will increase due to memory system improvements, even if floating point rates stagnate. If we restrict HPCCG execution to a single core, however, the vast majority of performance improvement will be missed since parallel execution across multiple cores enables an additional three times performance improvement on four cores (figure 1 red line).

2.1.2. Weak per-core memory vandwidth Although single-core performance growth is slowing, if we can exploit multiple cores, we will continue to see single node performance improve. In addition, however, slow growth in single-core performance, current multicore nodes do not tend to have sufficient memory bandwidth to adequately support all cores when performing memory-intensive computations. This situation implies that memory-bound algorithms will be unable

Table 1. Single-core performance of HPCCG, an unstructured finite-element microapplication, on several machines for a 3D problem with 1 million mesh nodes.

Year	Processor	Speed (GHz)	Cores (per socket)	(MFLOPS/s)
2003	AMD Athlon	1.9	1	178
2004	AMD Opteron	1.6	1	282
2005	Intel Pentium M	2.1	1	310
2006	AMD Opteron	2.2	2	359
2007	Intel Woodcrest	1.9	4	401
2007	AMD Opteron	2.1	4	476
2007	Intel Core Duo	2.3	2	508

to fully use all cores. Figure 1 shows the performance of pHPCCG, a parameterized version of HPCCG that supports arbitrary data types, on one to eight cores of an AMD Barcelona node. On single-core scalable systems, HPCCG scales linearly to more than 10,000 cores. However, in this study, linear speedup was achieved only for single-precision data going from one to two cores. All other data points, single or double precision, show sublinear improvement.

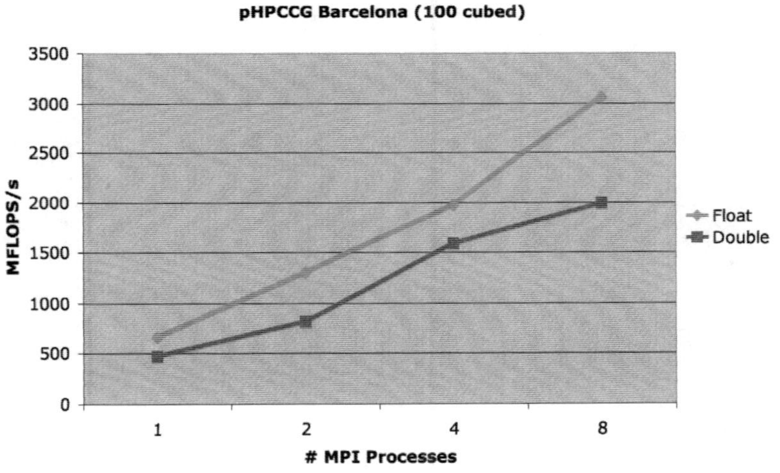

Figure 1. Comparison of single- vs. double-precision performance for pHPCCG, an unstructured implicit finite-element microapplication. Results are for 1M equations per process using 1-8 MPI processes on a dual-chip quadcore AMD Barcelona node.

2.1.3. Multithreading Multithreading is a well-known architectural approach to reducing effective latencies by keeping multiple thread states. If the active thread on a core stalls, the core can attach to another thread that is not stalled. This capability is especially effective for bandwidth-intensive programs and those with irregular memory access patterns. A processor's ability to have multiple outstanding threads can greatly improve performance, often without sophisticated software restructuring, which is otherwise typically required on more traditional microprocessors. The Sun Niagara 2 processor contains eight cores with eight threads per core. Figure 2 shows performance for Trilinos Epetra kernels [2] and indicates that oversubscribing cores with MPI processes can lead to substantial performance improvements. Multithreading will likely be a part of other new node architectures, and exposing a large collection of parallel task can greatly improve core utilization.

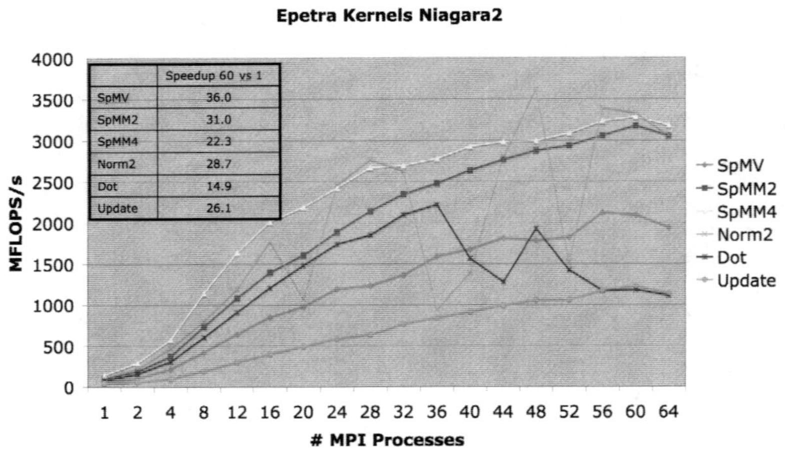

Figure 2. Weak scaling study of Epetra using 1-64 MPI processes on Sun Niagara2. Although there are only eight cores, performance improves due to multi-threading until about 60 processes. Dot product and two-norm performance peak at a speedup of 28.0 and 31.3 using 36 and 48 processes, respectively.

2.1.4. Core count growth When SMP nodes became available about 15 years ago, there was some expectation that processor counts per node would steadily increase on desktop systems. This expectation was not fully realized. At present, we have a similar expectation that core counts on homogeneous multicore systems will steadily increase. If true, we will see tens of cores per node, and programming these cores using a single-level flat MPI model will be difficult. However, there are a number of implementation details that can improve MPI performance for large core counts, which we discuss in Sections 3.1.1 and 3.1.2.

2.2. Heterogeneous multicore nodes

Heterogeneous nodes (e.g., STI Cell, nVidia, and ATI GPUs) have two core types (more in some cases). One processor is used primarily to handle serial and administrative tasks. The other cores are all of of the same type, are very capable of performing floating-point computations, have a communications path to the serial processor and also have their own high-performance memory network. Most heterogeneous multicore nodes have emerged from the multimedia computing field, and each has its own unique programming environment. Although efforts are under way to define a common programming environment for these nodes, at present there is no way to easily write portable code for these nodes.

2.2.1. Native double-precision support Heterogeneous multicore nodes have been of interest to the high performance computing community for many years [7]. However, full support for double precision floating-point arithmetic on these systems has only recently become available. As a result, prior to now, very few scientific and engineering applications could use heterogeneous multicore nodes in a production setting. Some floating point behavior differences still exist, such as lack of support for gradual underflow, but in most situations the degree of double-precision support is sufficient for applications.

2.2.2. Programming interfaces Perhaps the biggest hurdle to using heterogeneous multicore nodes is the lack of a standard, portable programming interface. Each node vendor has its own interface, and some commercial products are available to provide a generic interface, but presently there is no open standard programming interface that works across multiple heterogeneous multicore nodes. Therefore, to use these nodes, we must write our software in the native environment of each node, or provide our own abstract interface with adapters for each node type.

3. Node programming models

Since message passing, specifically MPI, will be the dominant programming model for parallel execution across multicore nodes, the real issue for libraries on future scalable systems is how to program the node. The simplest approach is to use a single-level of MPI processes, and there is some consensus that single-level MPI-only will be a sufficient programming model for most scalable systems for about three to five more years, unless of course an application must be ported to systems such as the Los Alamos Roadrunner [8], which rely on STI Cell processors.

Beyond the three- to five-year time frame, there is no general consensus on how to program multicore nodes. We will likely have many programming models. This issue is probably the single most challenging for library designers, since library software must be embedded into many different software frameworks and cannot generally control its run-time environment.

3.1. MPI under MPI

Although one might assume that a shared-memory multicore node should use a shared memory programming model, there is merit to considering MPI underneath MPI. In this case, we not only use MPI across nodes but also program each multicore node via a local MPI communicator. This approach is attractive because an application and its developers rely on only on a single programming model.

Furthermore, the interoperability of MPI with other parallel programming environments can be an issue at run time. Process scheduling and other run-time resource management problems can lead to difficult-to-debug performance degradation. Even efficient MPI-only execution can be difficult on multicore systems, and we will need to revisit MPI implementations.

3.1.1. Multicore-aware MPI Similar to making MPI SMP-aware 15 years ago, MPI implementations must now become "multicore-aware." In addition to performing shared-memory copies when communicating between shared-memory cores, MPI run-time environments should keep a single copy of the application executable per shared-memory image, especially as core counts increase. This optimization is already available on some systems, for example, Sandia's Redstorm.

MPI process placement and memory allocation are also an issue. Figure 3 shows the before and after performance impact of poor memory placement policy. Since the AMD Barcelona node has a non-uniform memory access (NUMA) design but is logically shared, memory can be allocated nonoptimally for an MPI process. Present run-time environments do not take into account the importance of memory and process affinity. Although there are often system tuning functions that an experienced programmer can use to address these issues, a multicore-aware MPI environment should handle these problems automatically.

3.1.2. Shared-memory MPI Beyond fixing MPI run-time performance so that standard subcommunicators can be used efficiently on multicore nodes, some modest additions to MPI can make shared-memory parallel programming feasible and portable. Specifically, by supporting a shared-memory allocation such that one MPI process can allocate memory that is also visible to other processes on a node, we can implement parallel, fine-grained shared-memory algorithms using MPI. Using MPI barriers to control loop increments and ensure data integrity, MPI can easily be used to implement shared-memory algorithms. Such mechanisms also set us up to consider hybrid run-time environments, which we discuss in Section 3.3.

3.2. Threads under MPI

Perhaps the most natural way to program homogeneous multicore nodes is to use one of several thread-based programming models. Pthreads and OpenMP are the most prevalent threading

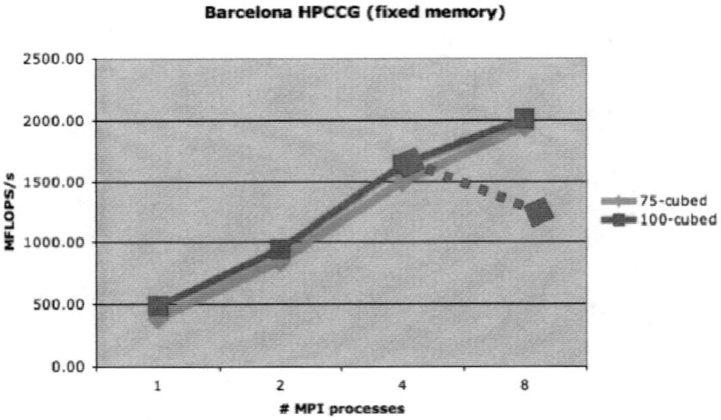

Figure 3. Comparison of performance for HPCCG, an unstructured implicit finite-element microapplication, before and after memory placement policy changes.

models, developed 15-20 years ago for SMP nodes. Threads under MPI are already well-established as a two-level parallel model, used successfully by many applications to extract additional parallelism from their application, for example, shared-memory parallel FFTs within an MPI weather modeling application.

More recently, other threading models have emerged that provide richer run-time semantics that can improve work partitioning and scheduling and integrate into advanced language environments. Perhaps most notable is the Intel Thread Building Blocks library (TBB) [3], which uses advanced features of C++ to support default and user-defined work partitioning, as well as dynamic thread scheduling. TBB is open source and extensible. It also has the potential to be useful on heterogeneous multicore nodes.

3.2.1. Thread management strategies and multiple thread run-time environments One of the more challenging aspects of thread programming is managing, placing, and scheduling threads. Pthreads, OpenMP and similar models provide almost no control over these thread policies, or they give control only at a global level via environment variables or system function calls. To make the situation even more challenging, there is no protocol for how these thread models should interact with each other within the same application. This is a particular problem for libraries that may use one model (e.g., OpenMP) and are called from an application that uses another model. Run-time performance issues are unpredictable and there are very limited, operating system-specific mechanisms to handle thread policies.

Portable performance of thread-based parallelism is probably the single biggest challenge to using threads underneath MPI. When this fact is coupled with the fact that there is no portable programming model for heterogeneous multicore nodes, we realize that the single biggest barrier to shared-memory parallel multicore programming is the lack of a programming model.

3.3. Hybrid run-time environments

A common observation in porting applications to homogeneous multicore-based scalable systems is that the vast majority of computations perform very well using single-level MPI only as the programming model. In fact, for some applications that use third-party solver libraries, all of the application's computation can scale using MPI only, because it is only the solvers that are not scaling well. The solvers are not poorly designed or implemented. They are simply memory bandwidth bound, while the application has good data locality. Figure 4 shows the breakdown of time spent in the Fluid Density Functional Theories program called Tramonto. The setup time is the time spent in Tramonto itself, while the solve time is spent in library kernels. Although Tramontos profile is dramatic, other applications show similar behavior.

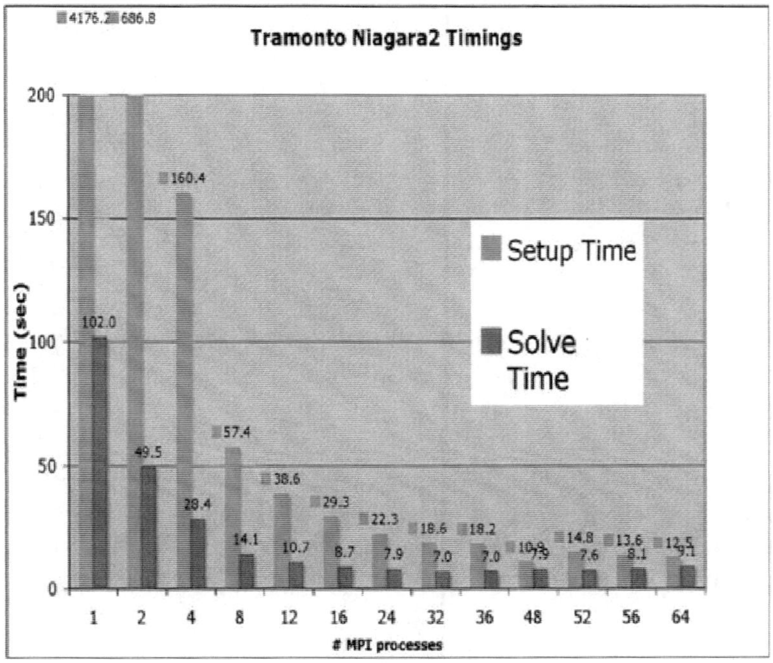

Figure 4. Tramonto is a classical density functional theories code for inhomogeous fluid modeling. Timing results are for 1-64 MPI processes on an 8-core, 8-threads-per-core Niagara2 processor. Setup Time is time spent in Tramonto except for the solver. Solve Time is spent in a Trilinos-based custom preconditioned solver. Note that outside of the solver, Tramonto scales extremely well using MPI-only.

Based on this observation, t becomes attractive to develop data classes that can be used by an application in MPI-only mode such that the data is stored in a way that is accessible to a library via shared-memory addressing on a multicore node. In this way the library can use shared-memory algorithms to improve run-time performance and robustness without requiring any change to the application that is using the library. Although applications will need to become "multicore aware" at some point, by using a hybrid run-time environment we can significantly extend the years during which existing scalable applications can continue using single-level MPI only.

4. Designing libraries for scalable multicore architectures

High-performance libraries have been an important part of scientific and engineering computing for decades, and their importance grows as machine architectures become more complex and as libraries become better designed to integrate easily within an application. Despite the fact that there are many different science and engineering applications, the underlying algorithms used by these applications have remarkable similarities, especially those algorithms that are most challenging to implement well in parallel. It is not too strong a statement to say that libraries are essential to the broad success of scalable high performance computing [4].

Based on the above discussions about multicore trends and programming models, we now discuss what library developers can do to design and develop effective libraries for future scalable systems, as well as the challenges we face going forward.

4.1. Addressing bandwidth challenges

Common to all multicore architectures is bandwidth limited computations for sparse linear algebra and dense vector kernels. Because many applications use these kernels, we must look at algorithms and data structures that can reduce our bandwidth requirements.

4.1.1. Block algorithms Block iterative methods (in contrast to single vector methods) for solving linear systems and eigenvalue problems have attractive performance behavior. Vector updates and dot products become (oddly shaped) dense matrix-matrix multiplies. Sparse matrix vector multiplication (SpMV) becomes multiplication by multiple vectors (SpMM). Figure 2 shows the performance advantage that SpMM has over SpMV. Block methods are also attractive numerically because of robustness properties. Also, as we move toward mathematical optimization, uncertainty quantification and related higher level algorithms, solution of multiple simultaneous systems becomes a basic kernel. As a result, one of the most important ways to address reduce per-core bandwidth is to develop libraries of block iterative methods.

Similarly, blocking of sparse matrix data structures can be very effective. Although automated techniques such as those implemented in OSKI [5] have been shown to be effective for fast SpMV execution, overall efficiency is still best served by problem-defined blocking if it is present in the problem being solved. This is mentioned further in Section 4.3.

4.1.2. Multiprecision data types and mixed-precision algorithms Use of single precision (32-bit) arithmetic has always had some performance and storage advantage over double (64-bit) precision. However, when double precision became widely supported in hardware, most science and engineering applications, with some notable exceptions, moved to using double precision for all computations. The motivation for using double precision in some phases of computation is clear: robust results. However, the motivations for solely using double precision are not well documented in the literature, although are likely due to simplicity in code implementation and maintenance. As observed in Sections 2.1.1 and 2.1.2, there is growing incentive for judicious use of single precision data within double precision applications.

Mixed precision algorithms are well known in the numerical analysis literature. Although the most prevalent of these is iterative refinement [9], mixed-precision approaches are useful in many other settings. For example, using lower precision for grid transfer operators in multigrid and local element stiffness computations in finite-element stiffness matrix calculations can be effective.

The first limitation to developing lower and mixed-precision algorithms is the limited ability to express multiprecision and precision-neutral algorithms in our most common computing languages. Although Fortran 2003 provides some support for generic programming, the only practical approach to implementing multiprecision and precision-neutral algorithms at this point is using templates in C++. In order to provide support for multiprecision algorithms, we need the following:

(i) Generic data classes whose precision can be declared when an instance of the class is created
(ii) The ability to express an algorithm independent of precision
(iii) A rich generic attribute set including the concept of zero and one, and a relative half or double precision

Many existing scalable libraries support 32-bit or 64-bit data types, but often the choice must be made at compile time using the same namespace. Few libraries support the generic expression of an algorithm independent of the floating-point precision, and very few if any support the generic expression of a mixed precision algorithms. Using C++ templates and traits mechanisms, we can provide data classes that support multiple precisions and provide the ability to express generic and multiprecision algorithms.

4.2. Shared-memory algorithms

If we are to benefit from shared-memory multicore nodes beyond making MPI implementations multicore-aware, we will need to develop shared-memory algorithms that are faster, more robust or both when compared to default single-level MPI only execution. There are two primary ways to accomplish this objective:

(i) Exploit fine-grained data parallel algorithms

(ii) Partition work and data for dynamic load balancing

SPMD is not an effective programming model for algorithms that are highly parallel with fine-grained parallel data dependencies or for task pools of work that are difficult to partition across processors *a priori*. For example, via level scheduling algorithms, sparse triangular solves have sufficient parallelism to keep multiple processors active, but the parallel tasks depend on shared-memory access to data that was generated at a previous level. Furthermore, at each level, the parallel tasks vary greatly in the amount of work per task, so static scheduling is typically imbalanced. Figure 5 illustrates level scheduling and shows typical statistics for an unstructured problem. By using shared-memory implementations of sparse triangular solves as a preconditioner or smoother for multilevel preconditioners, we can reduce the number of subdomains in a domain decomposition preconditioner, thereby reducing the scalability complexity of a parallel solver.

$$L = \begin{bmatrix} 1 & 0 & 0 & 0 & 0 \\ l_{21} & 1 & 0 & 0 & 0 \\ l_{31} & 0 & 1 & 0 & 0 \\ 0 & 0 & l_{43} & 1 & 0 \\ l_{51} & 0 & l_{53} & 0 & 1 \end{bmatrix} \quad \text{Solve } Ly = x.$$

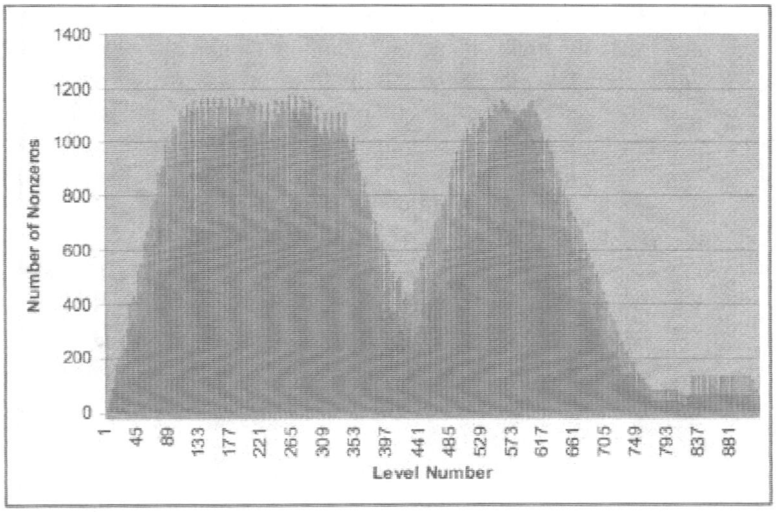

Figure 5. (a) Level scheduling of lower triangular solve: After Row 1 is processed, Rows 2 and 3 can be processed in parallel, followed by Rows 4 and 5 in parallel. (b) Level statistics for a 3D unstructured finite-element matrix.

4.3. Opaque data structures

As mentioned in Section 4.1.1, OSKI and other inspector-executor approaches have become popular for providing opaque data structures that provide a machine-friendly layout of data while hiding the details from the user. This approach has been used very successfully for level-3 dense BLAS in libraries such as ATLAS [6] where data transformations can occur as part of executing the kernel. It is less clear that similar approaches will work in the long run for other library kernels where the computation costs cannot cover the data transformations in real time and must be done as part of a preprocessing step. Although kernel execution time can be faster, production users are reluctant to adopt this degree of opaqueness for a variety of reasons including: duplication of data (the application's copy and the library's), *ad hoc* semantics of updating data, and amortization of the inspector phase costs. Opaque data structures will be necessary for optimal multicore execution. However, we have only started to design high-quality software that can be readily adopted by production application users.

4.4. Node API abstractions

Multicore nodes are evolving rapidly. Although we have mature shared-memory programming models for homogeneous multicore nodes, the run-time environments of these models are often poorly defined and difficult to control within the context of a scalable high-performance application. Furthermore, heterogeneous multicore nodes appear to offer better long-term performance and at present have no portable open standard programming model. Because of these facts, the most important software issue facing scalable libraries at this time is how to develop and define an abstract node programming interface that is compatible with and implementable on current platforms while extensible to future multicore platforms.

An effective approach to solving this problem is to use C++ abstract classes to define a parallel machine model. By describing the parallel machine in terms of attributes and services needed by the library, we have a chance to achieve efficient performance on current architectures and still retain flexibility to adapt to future ones. Currently, no other reasonable approach exists.

5. Conclusions

Multicore nodes present an exciting and challenging environment for scalable application and library developers. For the first time in 15 years we face the real prospect of redesigning much of our software base. Libraries will play a critical role in several ways. They will provide some of the first multicore-optimized software that will enable scalable application performance and portability, and extend the period of time where single-level MPI-only programming still works within the application. Libraries will also provide prototype portable programming interfaces for multicore nodes that can serve as candidates for open standards.

As we move forward, library developers must focus efforts on reducing memory bandwidth requirements by use of new algorithms, lower precision arithmetic and improved data structures. We must also focus on shared-memory algorithms beyond a multicore-aware MPI, enabling spawning of multiple tasks that can be dynamically scheduled. Further, we must develop abstraction layers that support efficient opaque data structures and robust node programming interfaces that allow us to adapt as multicore architectures and programming models evolve.

As a final note, it is clear from the above discussion that advanced object-oriented programming, using C++ as the primary language, is the most effective approach to next-generation library development. Although Fortran 2003 (and soon 2008) contains many attractive features, access to a standard-compliant Fortran 2003 compiler is limited. Furthermore, C++ offers the best combination of access to robust compilers, language interoperability, true object-oriented support and compile-time polymorphism (via templates) for high-performance generic data types and mixed precision algorithms. Adopting C++ does not mean that Fortran should be abandoned, but it does mean that Fortran applications will need to link with C++ easily and portably.

References

[1] Krawezik G 2003 Performance comparison of MPI and three openMP programming styles on shared memory multiprocessors *SPAA '03: Proceedings of the fifteenth annual ACM symposium on Parallel algorithms and architectures* (ACM: New York) 118–127
[2] Heroux M A, Bartlett R A, Howle V E, Hoekstra R J, Hu J J, Kolda T J, Lehoucq R B, Long K R, Pawlowski R P, Phipps, E T, Salinger A G, Thornquist H K, Tuminaro R S, Willenbring J M, Williams A and Stanley K S 2005 An overview of the Trilinos project *ACM Trans. Math. Softw.* **31(3)**:397–423
[3] Pheatt C 2008 Intel®threading building blocks, *J. Comput. Small Coll.* **23(4)**, 298–298
[4] Kothe D and Kendall R 2007 *Computational science requirements for leadership computing* (Oak Ridge National Laboratory ORNL/TM-2007/44)
[5] Vuduc R, Demmel J W' and Yelick K A 2005 OSKI: A library of automatically tuned sparse matrix kernels *J. Physics: Conference Series* **6**: 521–530
[6] ATLAS Homepage 2008 http://math-atlas.sourceforge.net/

[7] General-purpose Graphics Processing Units (GPGPU) Homepage 2008 http://www.gpgpu.org/
[8] Los Alamos Roadrunner Homepage 2008 http://www.lanl.gov/roadrunner/
[9] Buttari A, Dongarra J, Langou J, Langou J, Luszczek P and Kurzak J 2007 Mixed precision iterative refinement techniques for the solution of dense linear systems *Int. J. High Perform. Comput. Appl.* **21(4)**:457–466

Some essential techniques for developing efficient petascale applications

Laxmikant V Kalé

Parallel Programming Laboratory, Computer Science Department, University of Illinois at Urbana-Champaign, Urbana, IL 61810, USA

E-mail: `kale@cs.uiuc.edu`

Abstract. Multiple petaflops-lass machines will appear during the coming year, and many multipetaflops machines are on the anvil. It will be a substantial challenge to make existing parallel CSE applications run efficiently on them, and even more challenging to design new applications that can effectively leverage the large computational power of these machines. Multicore chips and SMP nodes are becoming popular and pose challenges of their own. Further, a new set of challenges in productivity arise, especially if we wish to have a broader set of applications and people to use these machines. Reviewed here is a set of techniques that have proved useful in multiple parallel applications that have scaled to tens of thousands of processors, on machines such as the Blue Gene/L, Blue Gene/P, Cray XT3, and XT4. New challenges and potential solutions for the performance issues are identified. Issues presented by multicore chips and SMP nodes also rre addressed. Also reviewed are some new and old ideas for increasing productivity in parallel programming substantially.

1. Introduction

As I write this paper, the first petascale supercomputer has already been assembled. Many supercomputers with multiple petaFLOPS peak performance are being developed to be deployed within the next few years. Which parallel programming techniques are necessary to harness this unprecedented power for science and engineering applications that could lead to breakthrough discoveries? In this paper, I will present my answer to this question, based on my research group's experience with developing multiple parallel applications that have scaled to tens of thousands of processors.

Multiple trends are contributing to the emergence of powerful parallel machines. The first is the long-standing trend of growth in computer hardware, including processors and memories getting cheaper. Of course, processors have stopped becoming faster in terms of clock frequency, but that is after several decades of continuous improvement on that count as well. As a result, it has been feasible to build very large parallel computers with relatively modest investments. For example, a company in India was able to build a computer with a peak performance of 170 teraflops [1], reportedly for a cost of about $30 million.

The second dramatic trend is an increasing number of cores on each multicore chip. Clock frequencies cannot be increased further because of power dissipation concerns. Yet the feature size keeps decreasing, and the number of transistors per chip keeps going up. Hence, it makes sense to organize all those transistors into multiple processor cores on a chip. Intel's eight-core chip is expected to ship later in 2008; further it is expected that the number of cores per chip will keep doubling every so often (say 18 months) for at least a decade. The ways in which the mainstream desktop market responds to these chips will have significant consequences for the high-performance computing community.

The third trend is the development and incorporation of specialized accelerator chips and acceleratorlike features in mainstream chips. Accelerators, by restricting their domain of application to a few specific patterns of computation, have been able to achieve performance orders of magnitude higher than general-purpose processors. Graphics processors have an extremely specialized function, of computing pixels, given a set of polygons. Yet, they were used by enterprising computational scientists for carrying out computations useful for their applications. But now these accelerators have evolved to a somewhat more general architecture, as seen by NVIDIA's new architecture and the CUDA programming system. IBM's Cell Broadband Engine presents an architecture that is more general purpose than GPGPUs, but still specialized compared with mainstream processors. Each of the eight specialized cores on the chip must work with only a small scratch-pad memory, with explicit DMA for data movements between the scratch-pad and main memory.

At the same time, there exist computational modeling problems that can be solved if one can harness this compute power effectively, which will lead to dramatic benefits for mankind. Breakthroughs are waiting to happen in multiple areas. Such breakthoroughs can improve our understanding of nature: biophysical simulations for understanding the molecular machinery of our bodies, genetic computations to understand evolution, similarities and differences between individuals, and the relationships between the genotype and phenotype. They can improve our ability to predict future: including short-term weather and long-term climate predictions and possibly even global economic predictions. They can lead to better design of engineered artifacts, such as automobile engines, rockets or buildings.

The new generation of computational science and engineering (CSE) applications do not just exploit the larger compute power by "increasing resolution everywhere." Instead, they tend to use techniques such as adaptive and dynamic mesh refinements, sophisticated algorithms such as Barnes-Hut or fast multipole method and are therefore more challenging to parallelize effectively.

To summarize, parallel computers of the petascale era are here and they are likely to have a rich, complex, and possibly heterogeneous structure. High-impact science and engineering applications that need such power exist or are being developed but are challenging to parallelize effectively. The parallel programming challenge before us in the coming era is how to productively and efficiently program the multipetaflops machines. In this paper, I outline some of the lessons I have learned in parallelizing several applications to machines with over 40,000 processors, lessons that I believe will be useful to address this challenge.

1.1. Application experience

I have selected examples from the experience of my research group (Parallel Programming Lab, UIUC – http://charm.cs.uiuc.edu) in parallelizing a few CSE applications that are used by CSE communities in their regular research: NAMD, from our oldest collaboration, is a program for biomolecular modeling [2]; OPENATOM is a Car-Parinello MD program used for simulation of electronic structure (in nanomaterials, as well as biophysics) [3]; ChaNGa, an astronomy code [4]; and RocStar, a code for simulating solid-propellant rockets, such as those in the space shuttle solid rocket booster. Figures 1(a), 1(b), and 1(c) show performance of these applications on three machines: the Blue Gene/L at IBM (up to 32,768 processors), the recent Blue Gene/P at Argonne, and the Cray XT3/4 at Pittsburg and Oak Ridge.

The success of these applications, which are used routinely by scientists, substantiates and leads credence to the lessons presented in the rest of the paper. However, I note that the SciDAC community, as well as the broader HPC community has a significant experience in developing highly scalable codes. So, the "lessons" or dicta presented here should be taken as a

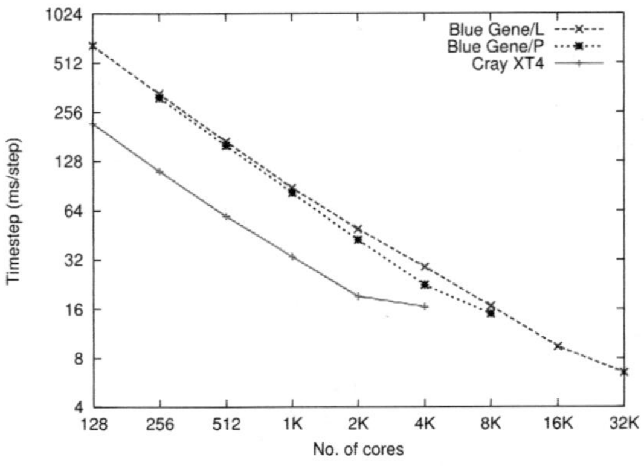

(a) NAMD running STMV (1 million atoms)

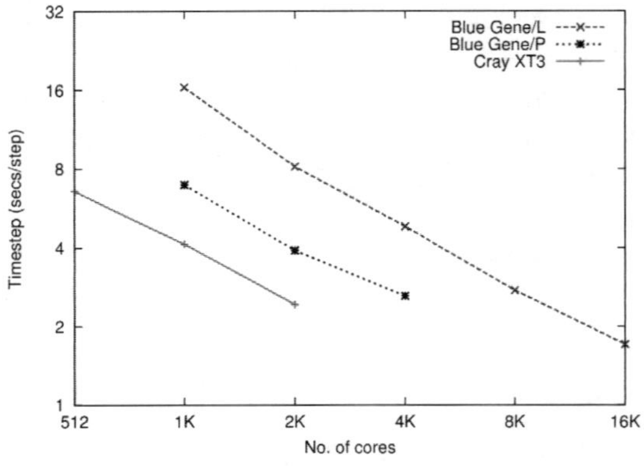

(b) OPENATOM running WATER_256_70Ry (256 water molecules)

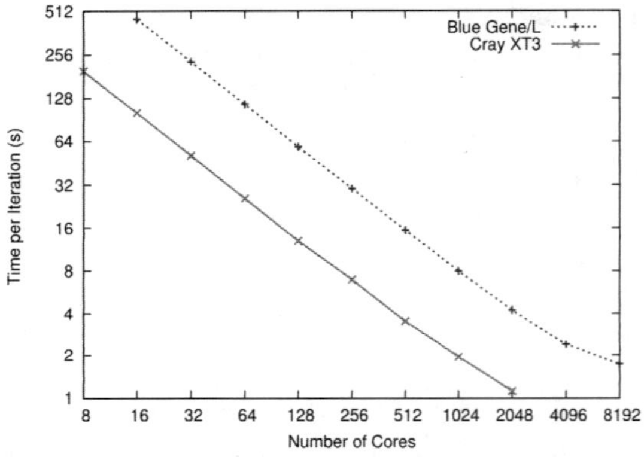

(c) ChaNGa running hwrh_lcdms (16 million particles)

Figure 1. Performance of NAMD, OPENATOM and ChaNGa on Cray XT3, XT4, IBM Blue Gene/L, and Blue Gene/P.

subjective selection based on my experiences and my set of techniques. A union of such lessons and techniques from multiple HPC researchers will be needed to tackle the challenge of petascale application development.

2. Guiding principles

Our computer science research toward the goals mentioned in the introduction involves development of abstractions and techniques that help simplify parallel programming and improve performance of applications developed. We use the following guiding principles to decide which abstractions and techniques to explore.

- **Don't rely on magic (at least until you have learned the trick):** For example, a compilers that automatically parallelizes sequential codes is an attractive research idea, but one that smacks too much of wishful thinking. Parallelizing compilers have achieved close to technical perfection, in that they can extract most of the parallelism *that can*

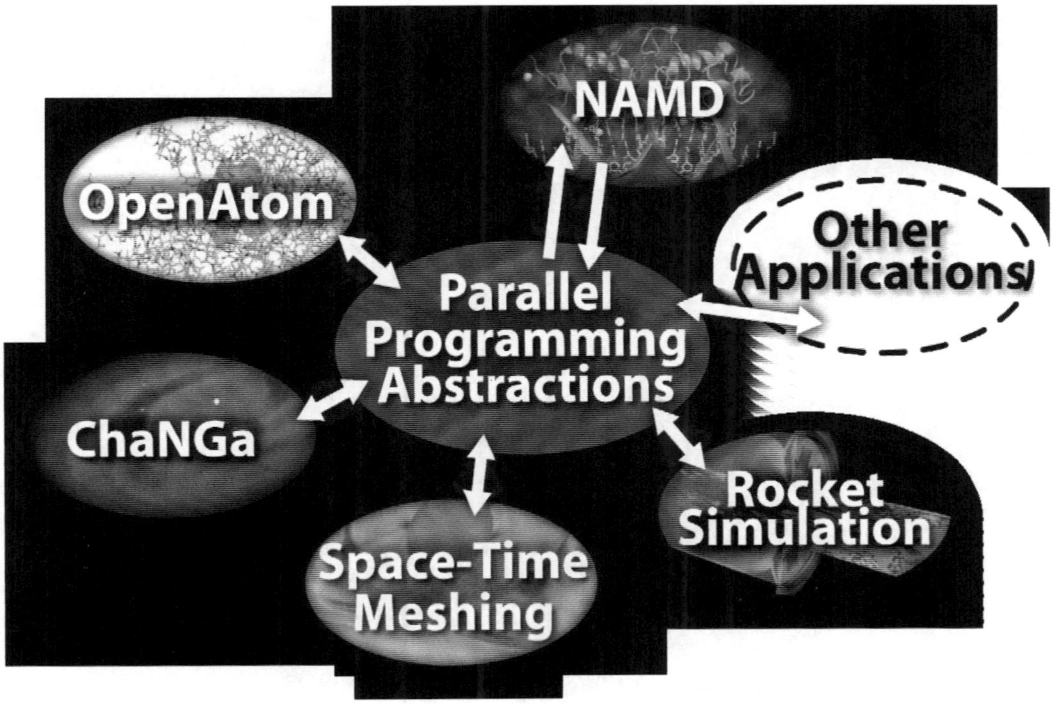

Figure 2. Application-oriented parallel abstractions

be extracted from sequential programs, with a huge intellectual effort by the compiler community. However, they have proved inadequate mostly because sequential programs obscure too much information. In general, it is better to develop abstractions in a bottom-up manner. One aims at developing an abstraction that builds on existing technology, and one that can be seen to be feasible at the outset and can be shown to be useful (in raising the level of abstraction) for realistic applications. Of course, such an organic bottom-up growth of abstractions needs to be controlled with the quest for elegance and simplicity.

- **Seek an optimal division of labor between the system and the programmer:** There are tasks that a human programmer can do much better than any automated system and vice versa. By automating those tasks that humans have most difficulty with, while avoiding automating tasks that are difficult to automate (such as automatic parallelization) but are not that difficult for an application programmer, we will get the most benefit from our research efforts.
- **Design abstractions based solidly on use cases:** As computer scientists, we tend to develop abstractions in an almost "platonic" way. Elegance and beauty may influence us more than relevance to real applications. This tendency may lead to abstractions that one can write papers about but that are not used by the CSE community and thus fail to have an impact on the state of the art. At the other extreme, a computer scientist may help parallelize an individual application, but then the contributions are not transferable to other applications. Instead, one should develop abstractions in the context of multiple full-scale applications. Benchmarks, short algorithms, or kernels don't expose the full complexity of the parallel programming problem, nor does a single application. If the abstractions that are triggered from individual applications are used, tested and honed in the context of multiple applications, they are likely to improve the state-of-art in parallel programming. Figure 2 illustrates this idea, in the context of my group's abstractions and some of the

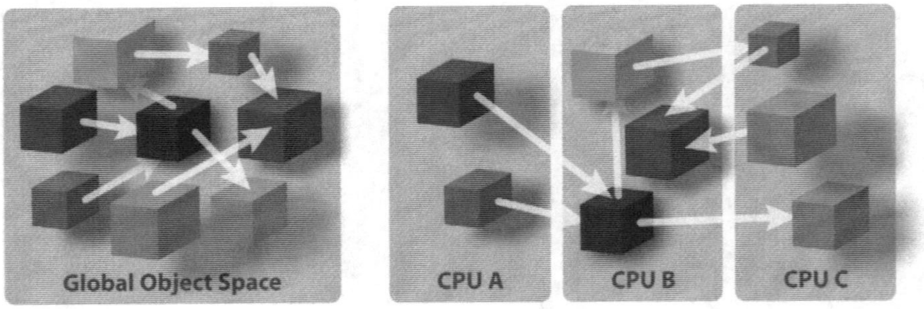

Figure 3. The user is concerned only with objects and communication between them. Actual mapping of objects to processors and underlying messaging is handled by the system.

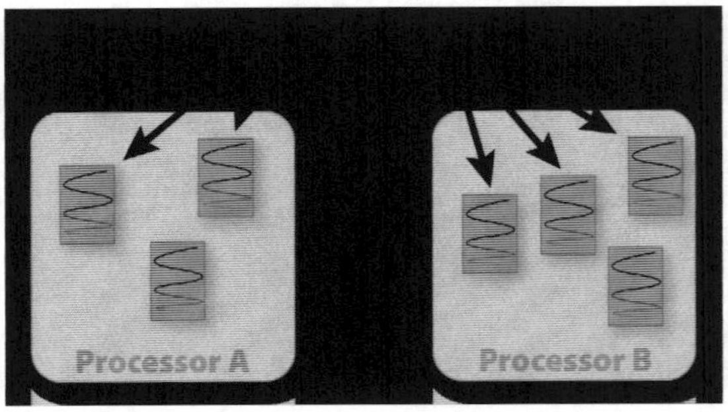

Figure 4. Virtualization for MPI threads using adaptive MPI.

applications developed with this paradigm. We called this approach "application-oriented yet computer science-centered research" [5].

3. Lessons learned

In the following subsections, I will enumerate a few "imperatives" of petascale programming. Each is based on a technique (or set of techniques) that has been proved useful in highly scalable CSE applications, and that I consider crucial to programming the next generation of parallel supercomputers.

3.1. Decouple decomposition from physical processors

Dictum 1: *Use overdecomposition with migratable objects, and leverage its benefits among multiple dimensions via an intelligent runtime system.*

Processor virtualization based on migratable objects was developed in the early 1990s in my group's work on CHARM++. The basic idea is simple: Let the programmer overdecompose the computation and data into logical work-units and data-units, and let an intelligent runtime system (RTS) assign these units to physical processors. Several others have developed similar ideas since early days of parallel computing [6, 7]. The decomposition done by the programmer is independent of the number of processors. Yet, the programmer does exercise grain-size control, ensuring that the overhead is adequately amortized by useful computation in each grain. Since the overhead is independent of the number of processors, the programmer does not have to optimize for the particular number of processors available in their allocation, in most cases.

CHARM++ and AMPI are two systems that embody this strategy. In CHARM++, the data and the computation are decomposed into multiple indexed collections of data-driven migratable objects that communicate with each other via asynchronous method invocations and a few asynchronous collective calls (figure 3). AMPI is an implementation of MPI where each MPI "process" is implemented as a user-level lightweight migratable thread (not to be confused with a POSIX thread; see figure 4). Over the past decade (and more), we have learned that this strategy yields a multitude of rich dividends. The first category of benefits is in the area of better software engineering, whereas the second category consists of benefits arising out of automated program optimizations via intelligent runtime strategies enabled by migratable objects. A longer, albeit somewhat dated, exposition of these benefits can be found in [8, 9]. Here, a brief summary is presented.

Separating the decomposition from the number of physical processors immediately frees one from artificial restrictions like "this program requires a cube of power of two processors". Only the logical entities need to have such a restriction, while the program continues to run on any arbitrary number of processors, with only a small loss of efficiency, if at all – that is, if the number of objects is not evenly divisible by the number of processors, and all the objects have an equal grain size.

Software engineering principles of cohesion and coupling are violated in MPI programs, viewed from a certain perspective: this situation comes about because of the processor-centric organization of an MPI program. Consider a program involving two parallel modules, say a "solids" and a "fluids" module. Typically, in an MPI program, the 17th partition (for example) of the *solids* domain and the 17th partition of the *fluids* domain are glued together on the 17th processor, simply because their respective partitioners called them "17th." Geometrically, they may not be connected, and the corresponding computations may not exchange data with each other. In contrast, with object-based decomposition, one can partition the two domains independently into M and N pieces respectively, express the computation in terms of the $M + N$ objects, and allow the RTS to place together the objects that communicate with each other to the extent possible.

Concurrent composition of parallel modules is well supported by this model: since there are multiple objects on a processor, this model requires a message-driven scheduler on each processor. As a result, if two modules are invoked "simultaneously," idle time in one module can be automatically overlapped with useful computation from the other module.

The second category of benefits arises from the flexibility afforded by migratable objects, which can be exploited by an intelligent runtime system. Probably the most obvious of these is dynamic load balancing, discussed in the next section. In addition, the RTS can shrink and expand the sets of processors allocated to a job by migrating objects accordingly. This can enable more powerful job scheduling where jobs don't need to languish in the queue waiting for the right number of processors to become vacant. An introspective RTS can also observe persistent communication patterns among objects and optimize communication by using appropriate algorithms, say for collective operations such as all-to-all.

Automated checkpointing can be supported now with little extra effort by the programmer, since checkpointing is essentially "migrating" all the objects to the disk. The CHARM++ and AMPI systems provide such automated checkpointing [10] in two separate modes: a traditional file-system based checkpointing, and a memory or local-disk based checkpointing. Migratable objects also enable an innovative fault tolerance scheme that sends only the failed processors back to their checkpoints and parallelizes the recovery by migrating the recovering objects from the failed processor to multiple other processors [11]. As a result, a job can continue to make progress even if the mean time between failures is less than the checkpointing period!

Message-driven scheduling leads to automatic adaptive overlap of communication and computation, as mentioned above. This also makes the programs more tolerant of latencies. This feature has been demonstrated dramatically in our multicluster results [12], where relatively tightly coupled benchmarks with time steps of a few milliseconds were able to tolerate multiple millisecond latencies between clusters located over a hundred miles apart. In addition, the fact that the scheduler can peek ahead at the queue can be exploited to prefetch the objects (and in some extreme cases, such as on IBM cell processors, their code) closer to the processor in the memory hierarchy, which provides a significant performance boost in today's architectures.

In spite of this long list of successes, we believe that only the "low-hanging fruit" has been exploited so far from this idea. Much more sophisticated runtime optimization strategies are yet to be developed. Scheduling challenges, such as those identified in [13] that require making complex trade-offs between parallelism and available memory along with load balancing, are examples of problems that will require further research.

3.2. Automate load balancing

Dictum 2: *Leverage the* principle of persistence *via measurement-based automatic load balancing.*

For most CSE applications, once a computation is expressed in terms of its "natural" objects, the computational loads and communication patterns tend to persist over time. We call this *the principle of persistence.* Note that this is simply an empirical observation, similar to the principle of locality. This principle often holds even when the application behavior is dynamic. Often, either the loads and patterns change slowly over time, or they change abruptly but infrequently. In either case, the recent past becomes a good predictor of near future, and this can be exploited by runtime load-balancing strategies that are enabled by migratable objects.

As an aside, we point out that this methodology leads to a two-level load-balancing: application entities (data and computation) are already partitioned into chunks, and the load balancer only moves the chunks around the processors. As an example, an unstructured-mesh computation is traditionally rebalanced by calling a graph partitioner (such as parMETIS). This is typically a time-consuming process. In contrast, the migratable objects method retains the original overpartitioning and decides to migrate only chunks, say because some chunks have gotten heavier. Since there are thousands of elements in each chunk, this is a much faster process. Moreover, the runtime strategies become domain independent: they apply to unstructured-mesh computations or gravity calculations based on the Barnes-Hut algorithm.

Since the runtime system is in charge of mediating communication and scheduling objects, it can automatically instrument and collect data on how long an object executes and how many bytes and messages are communicated between each communicating pair of objects. A load-balancing strategy, selected from a suite of strategies, then uses this database to create a new mapping of objects to processors, either from scratch or by refining the existing mapping.

Centralized load balancers collect the entire database on one processor and make decisions using a sequential algorithm. Since there are typically only tens (or hundreds) of objects per processor, the sequential bottleneck is not as strong a hurdle as one might expect. Indeed, we have shown that such balancers can be used to several thousand processors effectively. However, for much larger machines, this bottleneck must be eliminated. Fully distributed strategies [14, 15, 16], which we experimented with in early days, do not lead to a good quantity load balance. Instead, hierarchical and hybrid load balancers can combine the benefits of both kinds of strategies.

In our experience (at least currently), a suite of strategies, rather than a single superstrategy, is necessary, catering to differing needs of applications and parallel machines. Some strategies may trade communication costs for better load balance, or vice versa; others may minimize load-balancing time at the cost of quality, which trade-off is desirable for computations where the load changes rapidly.

Recent supercomputers, including the BG/L, BG/P, and Cray XT series, employ a three dimensional torus or grid interconnect, which necessitates another wrinkle for load-balancing strategies. Overall, such interconnects are beneficial because they can maximize link bandwidths

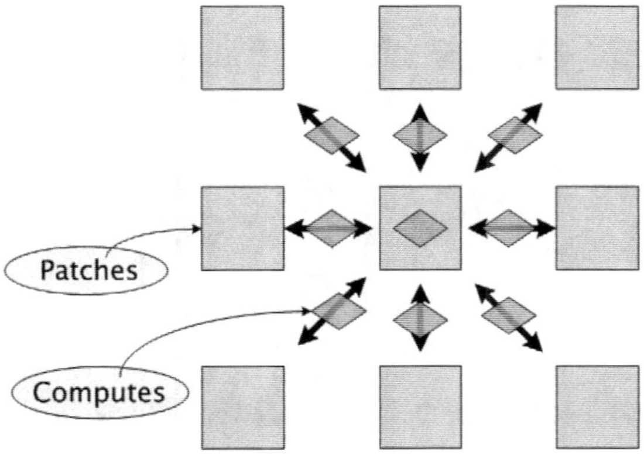

Figure 5. Object-based parallelization for molecular dynamics.

due to shorter wires. Because of wormhole routing and its variants, the communication latency does not increase significantly with the number of hops; however, the fraction of available bandwidth consumed by a message is proportional to the number of hops traveled. Thus, for communication-intensive applications, a load balancer must keep communicating objects on nearby processors to the extent possible.

Automatic load-balancing faces many new challenges with the new generation of machines and applications. In addition to the load, memory footprint variation may have to be considered; heterogeneous processors including accelerators complicate the scenario further. However, it seems clear that for such large machines, an automated system (in the RTS) can do a better job than most humans can productively do.

3.3. Analyze scalability of the algorithm

Dictum 3: *Use asymptotic analysis and particularly isoefficiency analysis, to choose parallel algorithms that will continue to scale to newer generations of machines.*

When one uses a larger machine, one often wants to solve a larger problem. This early insight led to the notion of scaled speedups [17]. Isoefficiency analysis gives a precise definition of when an algorithm can be said to be scalable, and provides a quantification of scalability [18]. Consider an algorithm that is running on P processors with parallel efficiency η. If we increase the number of processors, can we also increase the problem size so that we can regain the same efficiency η? If so, the algorithm is said to be scalable. The rate at which the problem size (as measured by the total amount of computation, for example) must increase to maintain the same efficiency, as a function of the number of processors, quantifies the scalability.

Our early experience in designing the program NAMD illustrates this point. In the mid-1990s, when NAMD was designed, many other biomolecular simulation programs used either static atom decomposition or force decomposition methods. The former was a natural consequence of parallelizing an existing sequential code. Our isoefficiency analysis [19] of both methods showed that the communication to computation ratio will keep on increasing with increasing number of processors, independent of the number of atoms being simulated, thus rendering them nonscalable. This gave a solid basis for our decision to use spatial decomposition (which was also used by a few others, such as [20]). To generate additional parallelism in a flexible manner we hybridized this with force decomposition: a separate migratable object was created for calculating forces between atoms in each neighboring pair of cubes, and it could be placed by the load balancer on a third processor where neither of the interacting cubes are resident (see figure 5). For the machines and molecular system sizes being considered at that time, the non-isoefficient methods were adequate; yet our design, based in part on the isoefficiency analysis,

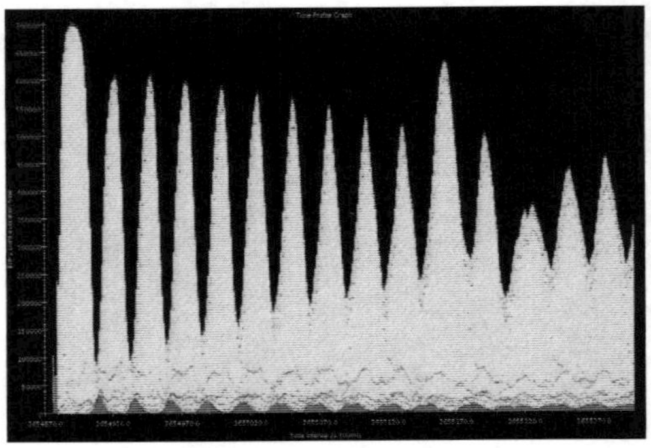

(a) Time Profile View

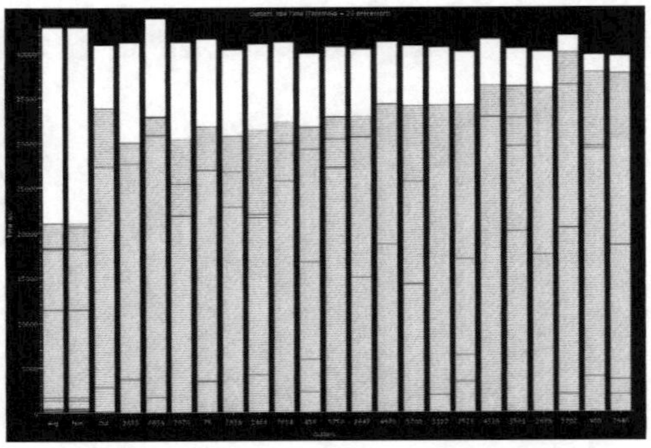

(b) Outliers View

Figure 6. Scalable performance visualization tools.

has survived the test of time for over a dozen years.

3.4. Analyze performance with sophisticated tools

Dictum 4: *Use scalable performance analysis tools.*

Performance of a parallel application on large, modern supercomputers is a function of a large number of interacting factors. With sophisticated multimodule applications, deep communication stacks with adaptive and variable behaviors, and other complexities of a parallel architecture, it is not surprising that even the most well-designed applications require performance debugging (a.k.a. performance tuning). When one is dealing with, say, a hundred thousand processors, running a fine-grained application that synchronizes every few milliseconds, even low-probability imbalances or bottlenecks will occur on some processor. Even on a thousand processors it is possible to view and analyze full traces, albeit barely so. But on larger machines, it becomes well-nigh impossible. A more automated solution is necessary.

Two separate limits are being breached. One is the perceptual limit of how much visualized data can a human brain comprehend; the second is the limit on the volume of trace data that needs to be stored on the file system, and the time needed to store and analyze it. For the former, a well-engineered system that can process, sort, and summarize the relevant performance data in perceptually comprehensible chunks is necessary. For the latter, an automated analysis, carried out in parallel, possibly at the end of the program run while the trace data is still in individual processor memories, will be necessary.

A relatively simple example illustrates this point. While debugging the performance of NAMD on 8,192 processors, which was below expectations, we examined the time-profile view (figure 6(a)), which shows the average processor utilization for a sequence of time intervals. The waxing and waning of utilization with each time step seemed like the signature of load imbalance, yet the dynamic load balancer was reporting that it had managed to bring the load imbalance within a few percent. An outlier analysis was performed to show the 20 most different processors (figure 6(b)). This showed that the overloaded processors had a particular activity on them – an intermediate node of the tree-based multicast over a small set of processors that was assumed to be cheap. A more detailed time-line view of the most overloaded processors and their communicating processors showed the problem clearly (figure 7). Not only were the intermediate modes of the spanning tree taking over 100 μs, which was much longer than expected, but multiple such intermediate nodes were being assigned to the overloaded processors. This went unnoticed by the load balancer because the spanning trees are formed, out of necessity, after the

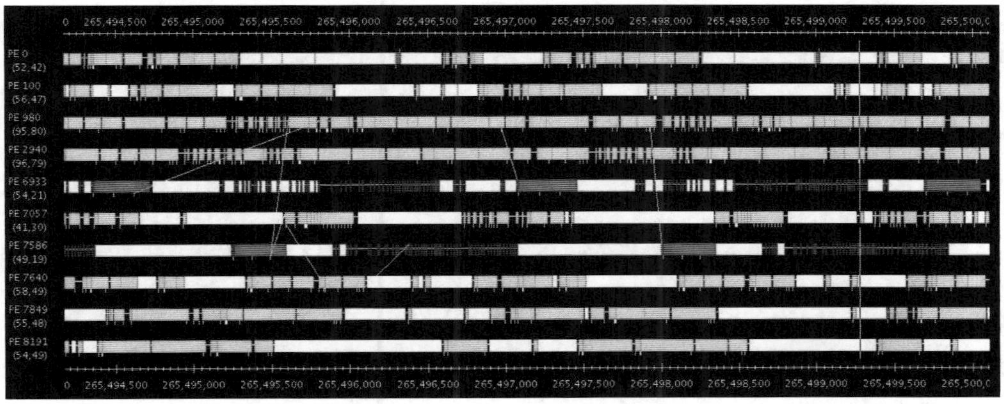

Figure 7. Timeline view of a NAMD run illustrating performance issues.

load-balancing phase, with respect to the new placement of objects. How this chicken-and-egg problem was solved is not pertinent here. The scalable outlier analysis presented a manageable picture to the programmer, which was instrumental in identifying this problem. Our current work [21], as well as that of a few other groups [22, 23, 24], is aimed at carrying out such analysis automatically at runtime, and storing only relevant summaries and a few representative detailed logs.

3.5. Model and predict performance via simulation

Dictum 5: *use an emulation-based program development and performance modeling system to prepare your application for a future parallel machine.*

The NSF-funded Blue Waters system will be available for general use in Illinois in the middle of 2011. Although many details are still not public, it will have at least 200,000 processor cores and will have multiple petaflops peak performance. An application developer who starts tuning (or worse, developing) an application for this machine at that time will have a difficult time getting effective production runs during the first year. Simply doing paper and pencil modeling or scaling it to the other parallel machines available to the NSF community may not be adequate. A similar challenge arises any time a new machine is to be deployed with substantially larger size than or substantially different architecture from the prevailing machines. Further, even when a large machine is deployed, time available on the full machine for performance tuning may be limited and intermittent.

An approach to a solution is provided by the *BigSim* project at Illinois. The idea leverages processor virtualization based on migratable objects mentioned earlier. One first runs the full-fledged application on a (relatively) smaller parallel machine in *emulation mode.* For example, a target application with a million objects (or AMPI threads) meant for a target machine of 100,000 processors can be run on a 10,000 processor machine, with the run-time system providing data structures that make it pretend to be the full-sized machine. During the emulation mode run, trace data is collected that records information about sequential execution blocks (SEBs), messages, and dependencies between them. This trace data is then fed to a parallel discrete event simulator along with architectural parameters of the target machine. The simulator run produces performance traces as if the program has been run on the target machine, which can then be analyzed using scalable performance analysis tools.

This basic outline shows the promise of the approach but hides many technical challenges. Somewhat surprising, there exists a significant class of applications where such a 10-to-1 multiplexing of target processors to emulating processors does not cause a memory footprint problem. For other applications, an out-of-core execution scheme that leverages the peek-ahead capability of message-driven execution is necessary. Predicting execution time for SEBs on the target processors is another challenge. One can use a multiresolution approach here, but at the highest resolution one may need to plug in (and/or interpolate) numbers from cycle-accurate simulators. Similarly, a multiple-resolution approach can be used for modeling communication performance. Significant research is under way toward these objectives and an initial system is in place for application developers.

The technical details of solutions to these challenges are out of scope of this paper [25, 26]. However, it is important to understand the utility of BigSim for petascale application development: (1) the emulation is useful to test the basic scalability authentication data structures; (2) even when the performance prediction is approximate, the performance traces obtained by the simulation are useful for identifying potential performance bottlenecks; (3) application sensitivity to machine parameters (which may not be completely known yet) can be studied by running the simulation with varying machine parameters (for example, one can study how latency tolerant the application is); and (4) run-time strategies such as load balancers can be tuned in an off-line mode by processing the trace data via the simulator. (5) off-line performance tuning using smaller version of the target machine becomes feasible.

4. Challenge of the multicore chip

Multicore chips are now the norm, and the number of cores per chip is expected to grow exponentially for next several years. Multiple such chips are present in individual nodes. Power considerations will keep clock frequencies low, possibly below 4 GHz. How should a petascale application developer adapt to these circumstances?

SMP nodes have been used for many years by the HPC community. However, they are commonly used by setting up a separate MPI process for each processor. Hybrid models, typically using OpenMP within a node and MPI across, have been sporadically used but have not caught on. Experience suggests that a straightforward coding using OpenMP or pthreads within a node leads to worse performance than simply using MPI on each processor. False sharing, cost of locks, and sloppy use of common locks for multiple unrelated critical sections, are some of the reasons. With significant effort, they can be overcome but typically only to equal the MPI performance.

Yet, there are benefits of using shared memory within a node that cannot be ignored, at least for some applications. In gravity calculations, a processor may request a set of particles from a remote processor. Since it may encounter a need for the same particles again during its tree walk, a processor level software cache is maintained. Using shared memory, one can maintain a *node-level* cache and reduce network traffic considerably, by avoiding duplicate requests.

We believe that a programming model that does not have the performance pitfalls of OpenMP or *Pthreads* is needed for exploiting shared memory within a node. CHARM++ and MSA built on top of CHARM++ have some of the features needed in such a model, but clearly further research is needed. For example, CHARM++ objects can access only their own memory, in addition to node-level read-only data. With this pattern, and a careful memory allocation strategy, false sharing can be eliminated completely in user programs, nor are any locks needed in the user program. Of course, the RTS itself must reorganize its own data structures to limit or eliminate false sharing and reduce the penalty of locks and fences. Once the RTS has been tuned, however, all the applications benefit uniformly. Similar benefits can accrue to users of frameworks such as Trilinos [27], which can also insulate the users from dealing with SMP performance issues.

CHARM++ has included an SMP version for many years. However, it was typically not used by application users because of its performance. When we started examining the issue, we found that for messages up to several kilobytes in length, even messaging within the node was slower than using the network to send the message (see the top two curves in figure 8)! The benchmark

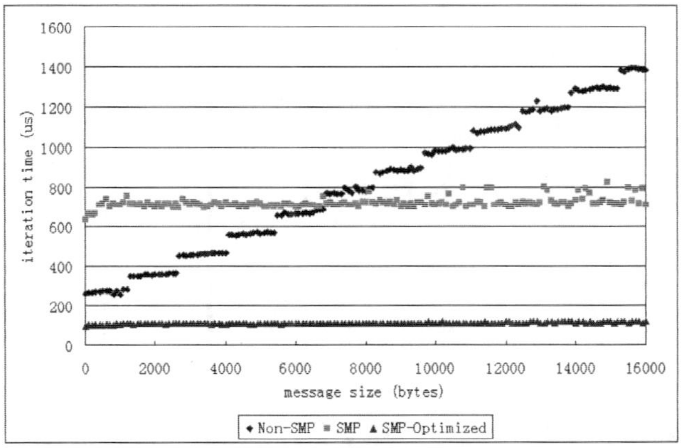

Figure 8. Performance of k-neighbor(multiple pingpong) iteration time.

used to generate the data in figure 8 has each processor core sending and receiving 12 messages in each iteration. With a series of optimizations in the RTS, which are instructive in themselves and will be described elsewhere, the performance was improved dramatically, even for short messages (the bottom curve in Figure 8 shows the performance after recent optimizations). For large messages, of course, the SMP mode performs better because of its avoidance of an extra copy. Note that in CHARM++, the ownership of a message sent to a remote object is transferred to it and therefore no copy at the runtime level is needed.

In addition, the RTS needs to incorporate new versions of load-balancing strategies, communication optimization strategies, and libraries that are "SMP aware" to derive full advantage of SMP nodes. Communicating objects preferably should be placed on the same node if they cannot be placed on the same processor, spanning trees for multicast groups need to be optimized, load balancers need to take multicast targets into account (especially for multicast over a set of objects whose number is significantly smaller than the total number of processors is, which is common in applications such as molecular dynamics) so as to prefer placing them on the same node. Many such strategies are being developed in our research and in frameworks developed by other researchers.

5. Further improvements in oroductivity: new paradigms and interoperability

The task of programming petascale machines for complex multimodule, multiphysics applications that use sophisticated algorithms and dynamic adaptation remains challenging, and can benefit from new programming models. Although systems such as Charm++ automate some aspects (e.g., load balancing), they do not raise the level of abstraction, especially for applications that don't need these features. How should we go about exploring and utilizing new programming models?

Since this is an open area of research, there are many opinions about it. However, two metapoints seem evident: interoperability and need for exploration. The latter point simply means that we should not standardize "too soon," allowing many different approaches to be explored. The need for interoperability arises in part from the same desire. If we facilitate interoperation of multiple modules written in different parallel programming paradigms, it becomes possible to integrate one module based on a radical paradigm into an application consisting of many traditional-paradigm modules. This allows new paradigms to establish a "beachhead" within the HPC landscape. But the long-term reasons for supporting interoperability are deeper: it seems conceivable, or possibly even inevitable, that different paradigms will be suitable for coding different types of algorithms. A system that supports

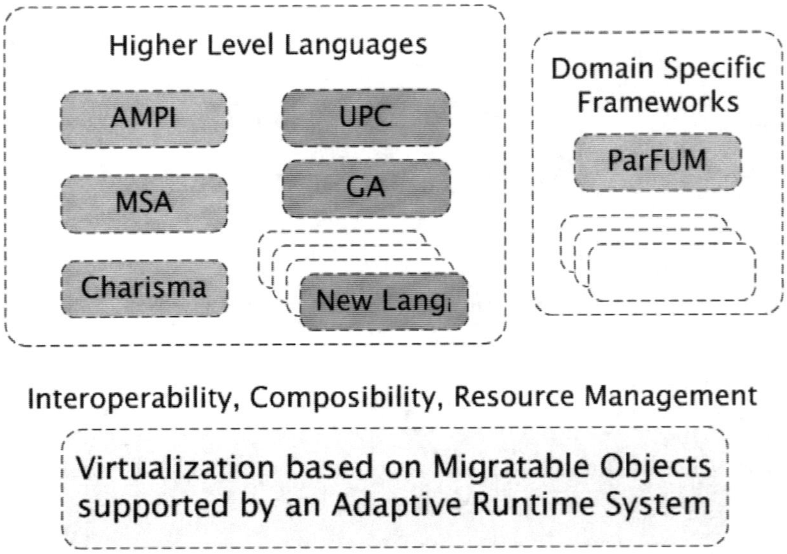

Figure 9. A complete programming solution with multiple interoperable paradigms.

interoperability will therefore enhance productivity by allowing a suitable paradigm for each module.

Interoperability doesn't mean just the ability to call one module from another module written in a different paradigm. It means the ability to concurrently compose multiple patterns and modules. As explained in Section 3.1, this may require an approach based on overdecomposition and message-driven execution.

A somewhat radical idea for raising the level of abstraction is to simplify parallel programming by giving up completeness! One can design a language or paradigm that is simple to use and debug applications with, yet cannot be used for expressing all parallel algorithms. As long as it is useful for expressing a significant subset of parallel programming patterns, it can be used to improve programmer productivity when combined with interoperability. Multiple such paradigms, interoperating with lower-level but complete paradigms such as MPI or Charm++, form a complete parallel programming solution (see figure 9).

In our own work, we have developed two such paradigms. Both are deterministic in the sense that they do not permit expression of programs with race conditions. Multiphase shared arrays (MSA) [28] is a paradigm in which migratable threads can access user-partitioned global data arrays; the restriction is that each array must be accessed only in one of a few access modes, including read-only, exclusive-write, and accumulate. The access modes of an array may change across phases indicated by array-specific synchronization points. Charisma [29] is a language that supports static data flow communication only. The program consists of migratable objects (as in Charm++) that communicate with each other in predetermined patterns. These patterns are expressed in a script like notation, while the sequential code is expressed separately.

Clearly, support for a global address space is a popular abstraction, though it need not be as general as cache-coherent shared memory. Co-array Fortran (CAF), UPC, and GA (Global Arrays) are examples of such paradigms. GA, for example, has been used in several widely used applications, such as NWChem [30]. It is necessary to find a method for supporting these (and related) abstractions with true (concurrently composible) interoperability, migratability and adaptive run-time systems, without losing the ability to leverage hardware support for one-sided (get/put) operations.

An answer along other dimension to the question of raising the level of abstraction is provided by domain-specific frameworks. An unstructured mesh framework, such as Sierra [?] or ParFUM [32], encapsulates commonly needed functionality within that domain into a reusable library, and provides an abstraction that simplifies expression of program modules within such domains. Since many CSE applications are covered by a relatively small number of data structures — unstructured meshes, structured meshes, adaptively refined structured meshes, particles partitioned into uniform bricks/cubes or into hierarchical structures such as oct-trees — frameworks aimed at each of these can enhance productivity. Again, since the entire application may consist of multiple modules only some of which can use such a framework, it is necessary that such frameworks play well in an interoperable framework.

6. Conclusion

I have presented a viewpoint about what techniques are needed for programming upcoming parallel machines with performance of multiple PetaFLOPS. Admittedly, it is somewhat subjective and maybe even a parochial viewpoint, in that it is based on my research group's experience with parallel applications and presents a research agenda that overlaps significantly with my own.

Five techniques were identified and presented in the form of dictums: migratable-objects-based virtualization, load balancing based on principle of persistence, isoefficiency analysis for scalability, scalable performance visualization/analysis, and emulation-based early application development. The paper then discussed challenges presented by multicore chips, and some approaches for overcoming them. It also discussed methods for raising the level of abstraction in petascale programming, via multi-paradigm programming supported by an interoperable RTS that supports concurrent composition, via a collection of specialized, incomplete yet simple parallel programming paradigms, and via domain specific frameworks.

We are poised at a cusp in the history of computational science and engineering. The raw power of upcoming supercomputers is huge; the parallel programming challenge of exploiting this power is enormous; yet, if this challenge can be overcome, the societal benefits arising from science and engineering breakthroughs are even larger. A concerted effort by the HPC community in the next few years will make a difference.

References

[1] Singh S 2000 Tata hopes its supercomputer is a money machine p 18
[2] Bhatele A, Kumar S, Mei C, Phillips J C, Zheng G and Kale L V 2008 Overcoming scaling challenges in biomolecular simulations across multiple platforms *Proc. IEEE International Parallel and Distributed Processing Symposium 2008*
[3] Bohm E, Martyna G J, Bhatele A, Kumar S, Kale L V, Gunnels J A and Tuckerman M E 2008 Fine grained parallelization of the Car-Parrinello ab initio MD method on Blue Gene/L *IBM J. Research and Development: Applications of Massively Parallel Systems* **52**(1/2):159–174
[4] Jetley P, Gioachin F, Mendes C, Kale L V and Quinn T R Massively parallel cosmological simulations with ChaNGa *Proc. IEEE International Parallel and Distributed Processing Symposium 2008*
[5] Kale L V 1994 Application oriented and computer science centered HPCC research 98–105
[6] Fox G C, *et al.* 1998 *Solving Problems on Concurrent Processors* (Prentice-Hall)
[7] Moreira J E and Naik V K 1997 Dynamic resource management on distributed systems using reconfigurable applications *IBM J. Research and Development* **41**(3):303
[8] Kalé J V, Ramkumar B, Sinha A B and Gursoy A 1994 The CHARM parallel programming language and system: Part I – description of language features *Parallel Programming Laboratory Technical Report #95-02*
[9] Kalé J V, Ramkumar B, Sinha A B and Saletore V A 1994 The CHARM parallel programming language and system: Part II – the runtime system *Parallel Programming Laboratory Technical Report #95-03*
[10] Zheng G, Huang C and Kalé L V 2006 Performance evaluation of automatic checkpoint-based fault tolerance for AMPI and CHARM++ *ACM SIGOPS Operating Systems Review: Operating and Runtime Systems for High-end Computing Systems* **40**(2)

[11] Chakravorty S and Kale L V 2007 A fault tolerance protocol with fast fault recovery *Proc. 21st IEEE International Parallel and Distributed Processing Symposium*

[12] Koenig G A and Kale L V 2007 Optimizing distributed application performance using dynamic grid topology-aware load balancing *21st IEEE International Parallel and Distributed Processing Symposium*

[13] Husbands P and Yelick K 2007 Multi-threading and one-sided communication in parallel lu factorization *SC '07: Proc. 2007 ACM/IEEE Conference on Supercomputing* 1–10

[14] Kalé L V, Bhandarkar M and Brunner R 1998 newblock Load balancing in parallel molecular dynamics *Fifth International Symposium on Solving Irregularly Structured Problems in Parallel* vol. **1457** *Lecture Notes in Computer Science* 251–261

[15] Zheng G 2005 *Achieving high performance on extremely large parallel machines: performance prediction and load balancing* Ph.D. thesis (Department of Computer Science, University of Illinois at Urbana-Champaign)

[16] Bhatele A 2007 *Application specific topology aware mapping and load balancing for three dimensional torus topologies* Master's thesis (Dept. of Computer Science, University of Illinois) http://charm.cs.uiuc.edu/papers/BhateleMSThesis07.shtml

[17] Gustafson J L 1990 Fixed time, tiered memory, and superlinear speedup *Proc. Fifth Distributed Memory Computing Conference* **2**:1255 – 1260.

[18] Grama A, Gupta A and Kumar V 1993 Isoefficiency: measuring the scalability of parallel algorithms and architectures *IEEE Parallel & Distributed Technology* **1**(3)

[19] Kalé L, Skeel R, Bhandarkar M, Brunner R, Gursoy A, Krawetz N, Phillips J, Shinozaki A, Varadarajan K and Schulten K 1998 NAMD2: Greater scalability for parallel molecular dynamics *J. Computational Physics*

[20] Plimpton S J and Hendrickson B A 1996 A new parallel method for molecular-dynamics simulation of macromolecular systems *J. Comp. Chem.* **17**:326–337

[21] Lee C-W, Mendes C and Kalé L V 2008 Towards scalable performance analysis and visualization through data reduction *13th International Workshop on High-Level Parallel Programming Models and Supportive Environments* (Miami, Florida)

[22] Malony A and Reed D 1990 Performance analysis, visualization with hyperview *IEEE Software* **5**(3):26

[23] Fowler R J, LeBlanc T J and Mellor-Crummey J-M 1989 An integrated approach to parallel program debugging and performance analysis on large-scale multiprocessors *Proc. ACM SIGPLAN/SIGOPS Workshop on Parallel and Distributed Debugging, published in ACM SIGPLAN Notices* **24**(1):163–173

[24] Shende S and Malony A D 2006 The TAU Parallel Performance System *International J. High Performance Computing Apps.* **20**(2):287–331

[25] Zheng G, Kakulapati G and Kalé L V 2004 Bigsim: A parallel simulator for performance prediction of extremely large parallel machines *18th International Parallel and Distributed Processing Symposium (IPDPS)* 78

[26] Wilmarth T L, Zheng G, Bohm E J, Mehta Y, Choudhury N, Jagadishprasad P and Kale L V 2005 Performance prediction using simulation of large-scale interconnection networks in pose *Proc. Workshop on Principles of Advanced and Distributed Simulation* 109–118

[27] Heroux M A, Bartlett R A, Howle V E, Hoekstra R J, Hu J J, Kolda T G, Lehoucq R B, Long K R, Pawlowski R P, Phipps E T, Salinger A G, Thornquist H K, Tuminaro R S, Willenbring J M, Williams A and Stanley K S 2005 An overview of the Trilinos project *ACM Trans. Math. Softw.* **31**(3):397–423

[28] DeSouza J and Kalé L V 2004 MSA: Multiphase specifically shared arrays *Proc. 17th International Workshop on Languages and Compilers for Parallel Computing* (West Lafayette, Indiana)

[29] Huang C and Kale L V 2006 Charisma: orchestrating migratable parallel objects *Proc. IEEE International Symposium on High Performance Distributed Computing (HPDC)*

[30] Kendall R A, Apra E, Bernholdt D E, Bylaska E J, Dupuis M, Fann G I, Harrison R J, Ju J, Nichols J A, Nieplocha J, Straatsma T P, Windus T L and Wong A T 2000 *A.T. Computer Phys. Comm.*, chapter High Performance Computational Chemistry: an Overview of NWChem a Distributed Parallel Application 260–283

[31] Taylor L M 1999 Sierra - a software framework for developing massively parallel, adaptive, multi-physics, finite element codes (Presentation at the International Conference on Parallel and Distributed Processing Techniques and Applications (PDPTA'99), Las Vegas, Nevada)

[32] Lawlor O, Chakravorty S, Wilmarth T, Choudhury N, Dooley I, Zheng G and Kale L 2005 Parfum: a parallel framework for unstructured meshes for scalable dynamic physics applications *Engineering with Computers* **22**(3-4):215–235

Toward an understanding of the building blocks: constructing programs for high processor count systems

M H Reilly
SiCortex, Inc., Maynard, MA 01754 USA
E-mail: matt.reilly@sicortex.com

Abstract. Technology and industry trends have clearly shown that the future of technical computing lies in exploitation of *more* processors in larger multiprocessor systems. Exploitation of high processor count architectures demands a more thorough understanding of the underlying system dynamics and an accounting for them in the design of high-performance applications. Currently these dynamics are incompletely described by the widely adopted benchmarks and kernel metrics. Systems are most often characterized to allow comparisons and ranking. Often the characterizations are in the form of a scalar measure of some aspect of system performance that is a "not to exceed" number: the maximum possible level of performance that could be attained. While such comparisons typically drive both system design and procurement, more useful characterizations can be used to drive application development and design. This paper explores a few of these measures and presents a few simple examples of their application. The first set of metrics addresses individual processor performance, specifically performance related to memory references. The second set of metrics attempts to describe the behavior of the message-passing system under load and across a range of conditions.

1. Overview

The primary challenge in high performance systems design is not the development of new and powerful architectures, but in bridging the gap between potential system performance and the performance delivered by actual applications: that is, once a system is designed, how can users best make use of its performance features? To some degree, the recent anxiety around the introduction of multi-core processors and the end of the period of dramatic intergenerational improvements in single processor performance is an expression of this very problem. Users have expressed frustration at realizing they are tapping into just a small fraction of their system's potential. In the past this frustration has been ameliorated by the knowledge that a new generation of processors would arrive in 18 months that would still deliver a fraction of its potential, but a potential that had doubled relative to the current generation: the surest software performance upgrade strategy was not tuning, but waiting for new hardware. The era of multicore designs has made that strategy obsolete.

If an application is to tap a greater portion of a multicore-multinode system, the application's design must recognize the key performance characteristics of the underlying system. This is not to say that each application must be tailored to each target hardware platform, but awareness of the characteristics of *some* platform will at least improve performance in one instance. However, few systems are characterized in such a way as to allow "architecture-aware" application design. Instead system characterization is driven most often by benchmark sets developed for pre-procurement comparisons. Where actual *applications* are used as benchmarks, their complex dynamics often obscure the underlying system characteristics to the point that the benchmark result sheds little light on just how a new application could be designed to best use the hardware.

Application design can and should be driven by an understanding of the behavior of the supporting hardware and system platform. Some common contemporary benchmarks are helpful in this effort but do not form a complete set. This paper explores a few new metrics for hardware behavior. First, we describe the SiCortex SC5832, the testbed for the concepts described here. Then we describe a set of metrics that can provide some insight into the design of an application. Then we review a few results in application or kernel design that have taken advantage of some of the new system metrics.

2. The System

The SiCortex SC5832 system comprises a cluster of 972 six-processor computing nodes connected via a high speed parallel point-to-point network. The network (fabric) topology is a Kautz graph that guarantees that the longest path through the network is logarithmic in the number of network nodes. In the case of the SC5832, the network diameter is 6 hops, with each node connecting three input ports and three output ports into the fabric [1], [2].

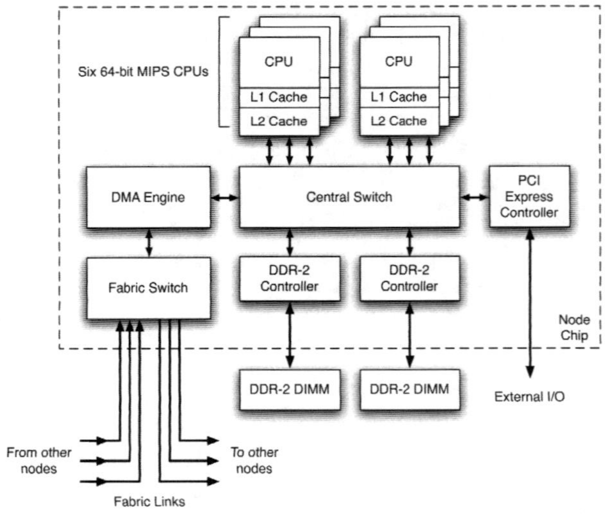

Figure 1. The SiCortex six-processor cluster node.

Figure 1 shows the single-chip, six-processor cluster node. Each processor is a 500 MHz low-power, 64-bit MIPS core and is described in table 1. All caches within the SMP node are coherent. The six processors share access to two independent DDR2 memory controllers, the DMA message handling processor, and a PCI Express controller.

An SC5832 is built from thirty-six modules, each containing twenty-seven cluster nodes, for a total of 972 nodes or 5,832 processors in the system. One node on each module connects to a dual Gigabit Ethernet controller. Three other nodes on each module connect to PCI Express Module slots to support external I/O. The SC5832 provides up to 324 PCI Express Module slots for connections to disk farms or external storage networks.

Table 1. SiCortex Node Chip Characteristics

Peak FP Rate	1000 MFLOP/s
DP Matrix Multiply (DGEMM)	730 MFLOP/s
Stream TRIAD Bandwidth	350 MB/s
Clock Frequency	500 MHz
L1 DCache Size	32 KB
L1 ICache Size	32 KB
L2 Cache Size	256 KB
Power Dissipation	< 1 W

Table 2. HPCC communications characteristics for the SC5832

Random Ring Bandwidth	50 MB/s
Random Ring Latency	4 microseconds
Ping Pong Bandwidth	1.5 GB/s
Ping Pong Latency	1.4 microseconds
PTRANS (Parallel Transpose) Bandwidth	210 GB/s

The 972 nodes in the SC5832 are connected via a logarithmic diameter network described by Kautz almost 40 years ago [3]. The network offers low latency and high bandwidth communications and has the ability to route around failed nodes or links. Each node has three outbound fabric links and three inbound links as shown in figure 1. The peak bandwidth on each link is 1.6 GB/s. Table 2 presents the performance of the communications fabric as measured by the High Performance Computing Challenge benchmarks [4].

3. Memory performance

The STREAM benchmark set, and in particular the STREAM triad metric, is useful in predicting the cost in time of many data intensive operations. However, the STREAM tests stress sequential access to memory. Many large dataset operations exhibit distinctly nonsequential addressing patterns. Notable examples are sort routines and FFT operations.

In the case of sort routines, the address stream can be considered "random." This random sequence can place significant stresses on the virtual memory translation unit, especially when the translation look-aside buffer (TLB) covers a region much smaller than the dataset to be sorted. Further, sort operations often look at a small part of the record. Such patterns exhibit no block level spatial locality, meaning that large cache blocks are often filled on a miss, only to have most of the data fetched into the block ignored. Even in cases where the list to be sorted is dominated by the key field so that several list elements fit in a single cache block, it is unlikely that more than one element in a cache block will be examined before the block is victimized.

In the case of FFT operations, the butterfly or bit-twiddle memory access pattern of the classic Cooley-Tukey algorithm presents a significant challenge to any prefetch logic in a compiler or a processor. For this reason, it is useful to consider two more metrics beyond the STREAM triad bandwidth for a processor. The first is the cost (load to use latency) of an isolated, unpredictable memory reference that misses in all levels of cache. In this case, we assume that the address translation is in the TLB so that the latency is entirely due to the time required to pass through the caches and extract the data from main memory (we consider here only systems with uniform access to local memory). Measuring this operation can be quite difficult, as good compilers and clever reordering hardware can often frustrate attempts to measure real latency. The measurement technique is specific to each processor implementation. Nonetheless, such a number can be quite useful. Typical values are on the order of 100 ns, with some values reported as low as 35 ns.

A second interesting memory performance metric is the cost (again, load to use latency) of a load operation that results in a translation buffer miss where the page table entry misses in the cache *and* the resulting resolved target address also misses in all levels of the cache. Again, this factor can be quite difficult to measure. In the absence of any other knowledge, the cost can often be approximated as the latency of two sequential (non overlapping) references to main memory. Typical values for this metric are on the order of 200nS.

The final memory metric we consider is the "span" of the TLB: the size of the memory region it covers if all TLB entries contain mappings for a nominal sized page. Nominal here is used in the sense of "the page size used to contain most of the data that we are interested in." While almost all modern processors support superpages and other methods of expanding the maximum span of the TLB for specially identified memory regions, it is important for this metric to use the actual page size used in practice. As an example, while the MIPS 5kf processor in SiCortex' Ice9 node chip supports super pages as large as 16 MB, Linux support for variable page sizes is somewhat unsettled, so the dominant page size is 64 KB, making the span of the TLB 6 MB. Datasets larger than 6MB will cause capacity TLB misses. For long random-access sequences over very large databases, this can ensure that many such random accesses result in a TLB miss that causes a cache miss (to fetch the page table entry) *and* a cache miss on behalf of the original load instruction.

4. Communication metrics

4.1. Simple pairwise bandwidth

The HPCC benchmark suite has several interesting communication metrics that can be used to guide program design. However, all are scalar. For example, the ping-pong bandwidth is a measure of the maximum bandwidth available to a pair of communicating processors in an otherwise quiescent system. The PTRANS bandwidth is a good measure of how fast a large dataset can be exchanged among a set of processes. The random ring bandwidth is a good measure of what can be expected for process-to-process transfers of very large messages while the communications fabric is heavily loaded. Unfortunately these scalar measures offer little guidance for optimization as they obscure the relationship between delivered bandwidth and message size.

The ping-pong (zero-contention) and random ring (high contention) tests can be elaborated to allow insight into the dependence of bandwidth on message size.

The zero-contention bandwidth test is performed over a range of message sizes to measure the available transmit bandwidth under no contention. The test calculates the delivered unidirectional bandwidth between nodes of a randomly selected pair from the 5832 processor complex, while all other processors are idle. Figure 2 shows a scatter plot of measured bandwidth vs. message size. Each point in the plot represents a trial for a randomly selected pair and a message size. The peak of 1.4 GB/s is determined by the throughput of the node's DMA engine. The spread over samples at the same message size is caused by differences in the network path length for each randomly chosen pair as it affects the latency of rendezvous response messages, and possibly by interfering low-level system events. Such a plot shows that the fabric becomes more efficient with increasing message size up to about 256 KB and that messages smaller than 2 KB are limited by the fixed cost of initiation. The spread also suggests that not all pairs are alike and even that we should expect system variability to affect delivered bandwidth. We see similar results on a 512-processor segment of the Lonestar Infiniband based cluster; see figure 3.

The random ring bandwidth test can be expanded in a similar manner (though in this case,

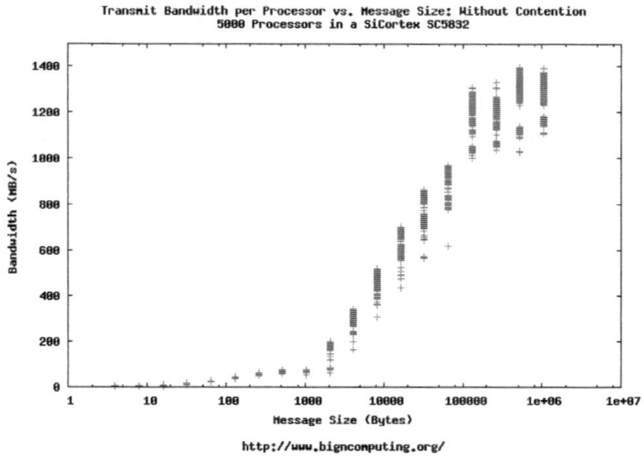

Figure 2. MPI_Send/MPI_Recv delivered bandwidth: no fabric contention,

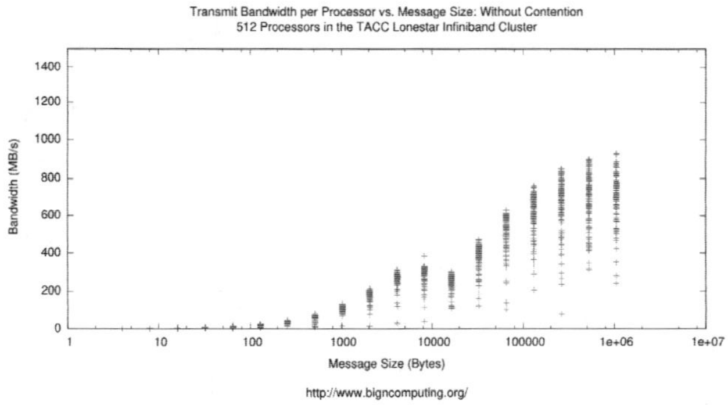

Figure 3. MPI_Send/MPI_Recv delivered bandwidth: no fabric contention.

we do *not* use the random ring communication pattern, as it can confound attempts to measure communications behavior with operating system and benchmark synchronization effects). The available transmit bandwidth under high contention test, shown in figure 4, measures the average unidirectional message bandwidth across 2500 randomly assigned pairs of processors. The measurement is taken while all processors are exchanging messages to maximize fabric utilization.

The first region of the curve, for messages ranging from 4 bytes to 1 KB, is entirely determined by the round-trip delay and flow control overhead for the eager message exchange. Messages larger than 1 KB are transferred via a rendezvous protocol that eliminates the flow control overhead after an initial negotiation. Messages of 128 KB and larger trigger a transition to a multirail transmission scheme that was intended to optimize performance under low contention. Unfortunately the multirail distribution tends to increase network contention and reduce delivered bandwidth.

Note that good performance – in fact, nearly optimal performance – is obtained for message sizes ranging from 2 KB to about 100 KB. Further, the delivered bandwidth for small messages of 256 bytes is high enough to permit consideration of schemes that rely on small data transfers.

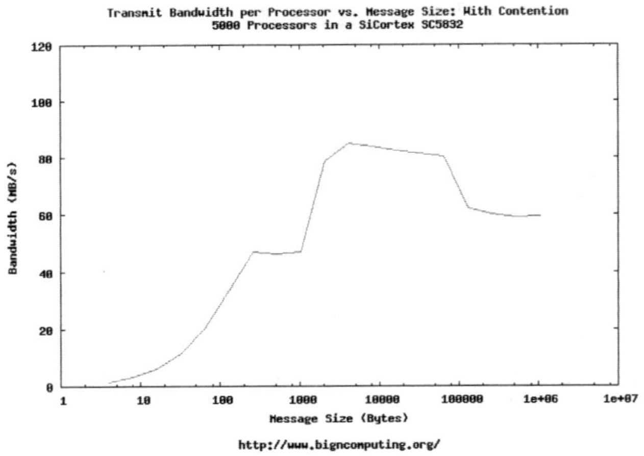

Figure 4. MPI_Send/MPI_Recv delivered bandwidth: high fabric contention.

4.2. Collective metrics

Some applications or kernels rely on collective operations (such as the MPI_Allreduce operators) over the entire group of communicating processors. We have found that characterizing the collective operators relative to vector length and communicator group size has been useful in planning a spanning tree analysis program.

The SC5832 MPI library, based on MPICH, implements the AllReduce collectives using an algorithm requiring $O(\log p)$ message exchanges between the p processes. The cost of the exchanges is $O(l)$, where l is the number of elements in the vector [5]. Figure 5 shows the cost of an integer MIN operation over a range of vector sizes. For larger vectors, the cost is clearly proportional to vector length. For smaller vectors, the initiation cost of the operation dominates, and so we see nearly constant cost for vectors shorter than 128 bytes. The apparent "noisiness" in the result between 64 and 1K bytes may be caused by flow control dynamics in the eager message protocol, though we do not yet have any definitive evidence.

Figure 6 shows the variation in the cost of the AllReduce operation vs. the number of participating processes. The binomial behavior at the foundation of the implementation is clearly apparent in the scatter plot [5].

4.3. Work underneath nonblocking transfers

Frequently algorithms admit to some overlap of communications operations with actual calculation. In the context of MPI operations, it has been useful to know how much work can be fit between a nonblocking MPI message request (e.g., MPI_Irecv, MPI_Isend, MPI_Issend) and the corresponding MPI_Wait completion operation without adversely effecting the communications performance by more than one percent.

This test is more complex than the simple bandwidth measurements described so far. The algorithm for calculating the communication/work overlap for MPI_Isend is shown in figure 7. The "guess" step is driven by the history of s1 values. "Close" is defined such that s1 is within $\pm 1\%$ of s0. Half of the processors in the job execute this timing loop; the other half act as sinks, repeatedly calling MPI_Recv.

Figure 8 shows the result of running the Isend work overlap test on two system configurations, the first using 128 processors on 22 nodes, the second using 4096 processors on 683 nodes. The plot shows the proportion of the communication time (the cost of the MPI_Isend and MPI_Wait

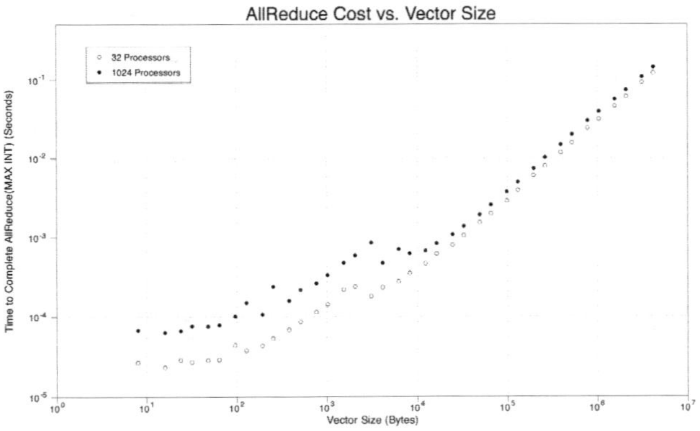

Figure 5. MPI_AllReduce execution time vs. vector size.

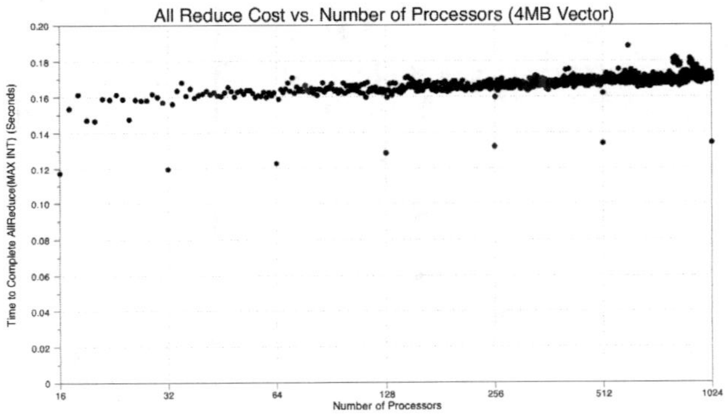

Figure 6. MPI_AllReduce execution time vs. number of processes.

```
double findWorkTime(msg_size)
s0 = Time(MPI_Send(msg_size)) // s0 is the time to complete an MPI_Send operation

until (s1 is "close" to s0) do
    worktime = guess how much work I can do
    s1 = Time (MPI_Isend(msg_size); waste_time(worktime); MPI_Wait())
end
return s1
```

Figure 7. Communication wverlap test for MPI_Isend.

calls plus the time waiting for completion) that could be spent on background calculation. This suggests that applications can perform significant work in the background while waiting for MPI_Isend operations to complete. The noise in the 128-processor test is likely the result of averaging measurements over a small sample size: essentially all measurements of communication latency are subject to variability. Further practice may indicate a need to quantify this variability as well.

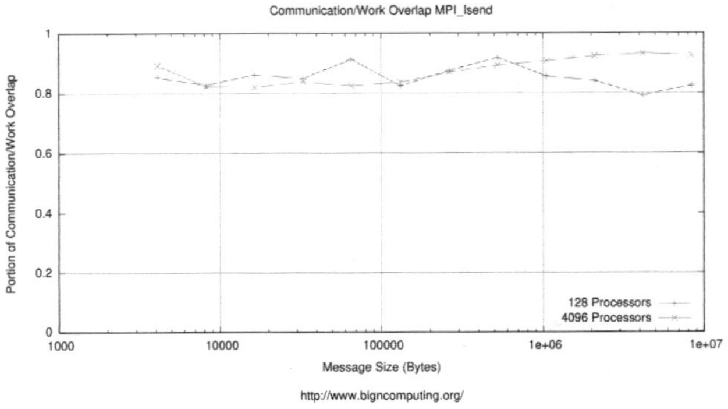

Figure 8. Communication/work overlap: MPI_Isend.

The SiCortex MPICH implementation of MPI_Irecv is such that more computation/communication overlap can be found by calling MPI_Test or MPI_Iprobe at intervals while performing overlapped computation. So, the MPI_Irecv overlap test is modified by breaking the "work time" in figure 7 into 1, 2, 4, or 8 segments. Figure 9 shows the available overlap for 1 and 8 work segments. That is, in the "One Work Segment" case the test called "waste_time" just once and was able to overlap about 77% of the time for a 128 KB message. But if the test called MPI_Test 8 times during the overlap interval, more than 82% of the wait time was available for computation.

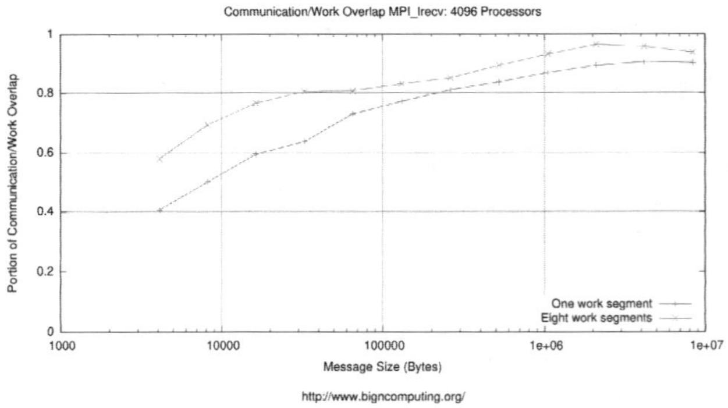

Figure 9. Communication/work overlap: MPI_Irecv for 4096 processors.

5. Experience with the Metrics

Few useful parallel programs are developed as leisure activities. While enjoyable, most parallel development is demand driven and often time constrained. The metrics presented here can assist both in forming algorithmic approaches and in guiding the developer's attention toward areas promising improvement and identifying "good enough" solutions and hopeless cases.

The brief practical experience described here attempts to catalog the influence of some of these metrics on two recent and small development efforts.

5.1. TeraSort

The TeraSort test was inspired by the TeraByte Indy sort benchmark at `http://www.hpl.hp.com/hosted/sortbenchmark/`. TeraSort departs from the original in that it ignores the influence of IO system bandwidth by generating internally all the data to be sorted. This reduces the problem to the sort step alone. The object is to sort 10^9 records of 100 bytes each.

The TeraSort program implements a simple bucket sort to distribute a portion of the ten billion records to each of the cooperating processes. At the start of the test, each of the N processes owns $10^9/N$ records with randomly generated keys. After building a global histogram of the distribution of key values, each process is assigned a contiguous section of the 10 byte keyspace of approximately $10^9/N$ records. Each record is then sent from the process that generated it to the process that is responsible for its range of the keyspace. Once all the records are distributed, each process sorts its portion using the quicksort algorithm.

The initial implementation of the parallel sort took 28.8 seconds to complete on a 5600-processor complement. (This time included 4 seconds to generate the random data.) The sort time was dominated by the exchange step, which took 11.4 seconds, and the local quicksort step, which took 8.78 seconds.

On average, each processor was required to sort about 2M records. The addressing sequence for the quicksort algorithm is somewhat problematic, as little use is made of spatial cache locality, and large record layouts can cause very poor TLB performance. Quicksort suggests that each processor would complete approximately 40M comparisons. Each comparison is between a record that was recently considered and one that was unlikely to have been accessed recently. Thus, over a record list of 2M entries, each of 100B, the second access is likely to result in both a cache miss *and* a TLB miss. (Note the span of the SiCortex SC5832 TLB is only 6 MB.) With a combined cost of about 200nS, the TLB miss traffic accounts for almost all of the 8 second sort time. ($200nS\ per\ TLB\ miss \cdot 1TLB\ miss\ per\ comparison \cdot 40 \cdot 10^6 comparisons = 8Sec$) Clearly, we cannot expect to elminate all of this time, as cache misses are inevitable given the list size and the 256 KB L2 caches. But we can reduce the number of TLB misses by building a shadow copy of the keylist and sorting the shadow copy. The shadow contained only the top 4 bytes of the key and a pointer to the actual record. Comparisons that showed the top 4 bytes as equal would then dereference the record pointer. This condensed the sort list from 200 MB down to 16 MB. While this was still larger than the span of the TLB, the reduction in the TLB miss rate from almost one per compared to about one for every three comparisons was a substantial reduction in cost. Further, with fewer active pages, the TLB miss was more likely to find the page table entry already resident in the L2, further mitigating the cost of a TLB miss.

As a result, with simple reformatting of the list and some small adjustments to the original qsort algorithm, the quicksort phase of the operation was reduced to 3.75 seconds.

Finally, the 11.4 second exchange time was far greater than the expected value. The algorithm had already been designed to exchange relatively large messages: approximately 10 KB. But the communication bandwidth vs. message size graphs encouraged a larger message size at the expense of some local buffer copies. Further, our expectation of a reasonable amount of overlap suggested an implementation that would perform as much of the buffer copy and list reformatting as possible underneath nonblocking receive operations. Finally, a knowledge of the expected exchange bandwidth between nodes (greater than 50 MB/s for the message size in question) suggested a closer look at the communications pattern: 11.4 seconds to perform a 200 MB exchange was clearly far from optimal. Primarly the most significant improvement came from detailed analysis of message exchanges that uncovered an implicit synchronization barrier. The final cost of the exchange, at 8 seconds, still suggests further improvement is possible.

After tuning the application, directed in part by our metrics, the resulting sort was completed in just 19.65 seconds on 5600 processors.

5.2. Three-dimensional FFT

Certain tomography computations depend on a large, three-dimensional FFT calculation over volumes of approximatley 10^9 points. Multidimensional FFTs are composed of single-dimensional FFT operations repeated in each axial dimension. Therefore, a 1040 x 1040 x 1040 point FFT would require about $3 \cdot 10^6$ 1040 point FFTs. (We chose 1040 points as a good fit for the prime-factor FFT. [6])

The primary focus of the effort was to distribute the calculation and data set among 65 processes and optimize the z-axis exchange prior to the FFT scan in the final (third) dimension. This step involved transposing the three-dimensional volume. The graphs of exchange bandwidth under contention vs. message size suggested that not much was to be gained from messages beyond 128 KB or so, which guided the decision to exchange slabs of the volume sized to fit the optimal message size. As a result, the exchange phase accounts for 4 seconds out of the 15.1 seconds taken for the complete 3D operation.

The overlap plots suggest that total operation time could be reduced as much as 20% by overlapping some operations with the exchange, but this has not yet been attempted.

Test programs

The test programs described here are available at `http://www.BigNComputing.org/Big_N_Computing/Test_Programs.html`.

References

[1] Reilly M, Stewart L C, Leonard J and Gingold D 2006 *SiCortex Technical Summary* (SiCortex, Inc.) http://www.sicortex.com/press/sicortex-tech_summary.pdf

[2] Stewart L C and Gingold D 2006 *A New Generation of Cluster Interconnect* (SiCortex, Inc.) http://www.sicortex.com/press/sicortex-cluster_interconnect.pdf

[3] Elspas B, Kautz W H and Turner 1968 *Theory of Cellular Logic Networks and Machines* (Tech. Report Stanford Research Institute AFCRL-68-0668)

[4] High Performance Computing Challenge Benchmarks 2008 http://icl.cs.utk.edu/hpcc

[5] Thakur R, Rabenseifner R and Gropp W 2005 Optimization of collective communication operations in MPICH *International J. High Performance Computer Apps.* **19(1)**:49–66

[6] Temperton C 1985 Implementation of a self-sorting in-place prime factor FFT algorithm *J. Computational Phys.* **58**: 283–299

Journal of Physics: Conference Series **125** (2008) 012038 doi:10.1088/1742-6596/125/1/012038

PERI - Auto-tuning memory-intensive kernels for multicore

Samuel Williams[1,2], **Kaushik Datta**[2], **Jonathan Carter**[1], **Leonid Oliker**[1,2], **John Shalf**[1], **Katherine Yelick**[1,2], **David Bailey**[1]

[1]CRD/NERSC, Lawrence Berkeley National Laboratory, Berkeley, CA 94720, USA
[2]Computer Science Division, University of California at Berkeley, Berkeley, CA 94720, USA

E-mail: `SWWilliams@lbl.gov`, `kdatta@eecs.berkeley.edu`, `JTCarter@lbl.gov`, `LOliker@lbl.gov`, `JShalf@lbl.gov`, `KAYelick@lbl.gov`, `DHBailey@lbl.gov`

Abstract. We present an auto-tuning approach to optimize application performance on emerging multicore architectures. The methodology extends the idea of search-based performance optimizations, popular in linear algebra and FFT libraries, to application-specific computational kernels. Our work applies this strategy to sparse matrix vector multiplication (SpMV), the explicit heat equation PDE on a regular grid (Stencil), and a lattice Boltzmann application (LBMHD). We explore one of the broadest sets of multicore architectures in the high-performance computing literature, including the Intel Xeon Clovertown, AMD Opteron Barcelona, Sun Victoria Falls, and the Sony-Toshiba-IBM (STI) Cell. Rather than hand-tuning each kernel for each system, we develop a code generator for each kernel that allows us identify a highly optimized version for each platform, while amortizing the human programming effort. Results show that our auto-tuned kernel applications often achieve a better than 4× improvement compared with the original code. Additionally, we analyze a Roofline performance model for each platform to reveal hardware bottlenecks and software challenges for future multicore systems and applications.

1. Introduction

The computing revolution toward massive on-chip parallelism is moving forward with relatively little concrete evidence on how to best to use these technologies for real applications [1]. For the foreseeable future, high-performance computing (HPC) machines will almost certainly contain multicore chips, likely tied together into (multisocket) shared memory nodes as the machine building block. As a result, application scientists must fully harness intranode performance in order to effectively leverage the enormous computational potential of emerging multicore-based supercomputers. Thus, understanding the most efficient design and use of these systems, in the context of demanding numerical simulations, is of utmost priority to the HPC community.

In this paper, we present an application-centric approach for producing highly optimized multicore implementations through a study of sparse matrix vector multiplication (SpMV), the explicit heat equation PDE on a regular grid (Stencil), and a Lattice Boltzmann application (LBMHD). Our work uses a novel approach to implementing the computations across one of the broadest sets of multicore platforms in existing HPC literature, including the conventional multicore designs of the dual-socket×quad-core Intel Xeon E5345 (Clovertown) and AMD Opteron 2356 (Barcelona), as well as the hardware multithreaded dual-socket×octal-core Niagara2 – Sun T2+ T5140 (Victoria Falls). In addition, we include the heterogeneous local-store based architecture of the dual-socket×eight-core Sony-Toshiba-IBM (STI) Cell QS20 Blade.

Our work explores a number of important optimization strategies, which we analyze to identify the microarchitecture bottlenecks in each system; this leads to several insights into how to build effective multicore applications, compilers, tools, and hardware. In particular, we discover that, although the original code versions run poorly on all of our superscalar platforms, memory bus bandwidth is often not the limiting factor on most examined systems. Instead, performance is limited by lack of resources for mapping virtual memory pages – limited translation lookaside (TLB) buffer capacity, limited cache capacity and associativity, high memory latency, voluminous cache coherency traffic, and/or poor functional unit scheduling. Although of some these bottlenecks can be ameliorated through code optimizations, the optimizations interact in subtle ways both with each other and the underlying hardware. We therefore create an *auto-tuning* environment for these three codes that searches through a parameter space for a set of optimizations to maximize performance. We believe such application-specific auto-tuners are the most practical near-term approach for obtaining high performance on multicore systems. Additionally, our experience offers concrete challenges to future multicore work on compilers and performance tools.

Results show that our auto-tuned optimizations achieve impressive performance gains – up to 130× speedup compared with the original version. Moreover, our fully optimized implementations sustain the highest fraction of theoretical peak performance on any superscalar platform to date. We also demonstrate that, despite the relatively slow double-precision capabilities, the STI Cell provides considerable advantages in terms of raw performance and power efficiency – at the cost of increased programming complexity. Additionally, we present several key insights into the architectural tradeoffs of emerging multicore designs and their implications on scientific algorithm design.

We facilitate the visualization of the limitations to performance through a roofline model. Attained performance and requisite optimizations were well correlated to those suggested by the model. In fact, the model is a much better indicator of performance than either raw peak flops or bandwidth.

2. Experimental Testbed

Our experimental testbed consists of a diverse selection of multicore system implementations. A summary of key architectural features of our architectural testbed appears in table 1 and is visualized in figure 1. The sustained system power data was obtained by using an in-line digital power meter while the node was under a full computational load. Node power under a computational load can vary dramatically from idle power and from the manufacturer's peak power specifications. Although the node architectures are diverse, they accurately represent building blocks of current and future ultra-scale supercomputing systems. We now present a brief overview of the studied platforms.

2.1. Intel Clovertown

Clovertown is Intel's first foray into the quad-core arena. Reminiscent of Intel's original dual-core designs, two dual-core Xeon chips are paired onto a multichip module (MCM). Each core is based on Intel's Core2 microarchitecture, runs at 2.33 GHz, can fetch and decode four instructions per cycle, execute 6 micro-ops per cycle, and can fully support 128b SSE, for peak double-precision performance of 10.66 GFlop/s per core.

Each Clovertown core includes a 32 KB L1 cache, and each chip (two cores) has a shared 4 MB L2 cache. Each socket has access to a 333MHz quad-pumped front side bus (FSB), delivering a raw bandwidth of 10.66 GB/s. Our study evaluates the Dell PowerEdge 1950 dual-socket platform, which contains two MCMs with dual independent buses. The chipset provides the interface to four fully buffered DDR2-667 DRAM channels that can deliver an aggregate read memory bandwidth of 21.33 GB/s, with a DRAM capacity of 16 GB. The full system has 16 MB of L2 cache and an impressive 85.3 GFlop/s peak performance.

2.2. AMD Barcelona

The Opteron 2356 (Barcelona) is AMD's newest quad-core processor offering. Each core operates at 2.3 GHz, can fetch and decode four x86 instructions per cycle, executes 6 micro-ops per cycle, and fully supports 128b SSE instructions, for peak double-precision performance of 9.2 GFlop/s per core or 36.8 GFlop/s per socket.

Each Opteron core contains a 64KB L1 cache, and a 512 MB L2 victim cache. In addition, each chip instantiates a 2 MB L3 victim cache shared among all four cores. All core prefetched data is placed in the L1 cache of the requesting core, whereas all DRAM prefetched data is placed into the L3. Each socket includes two DDR2-667 memory controllers and a single cache-coherent HyperTransport (HT) link to access the other

Core Architecture	Intel Core2	AMD Barcelona	Sun Niagara2	STI PPE	STI SPE
Type	super scalar out of order	super scalar out of order	MT dual issue†	MT dual issue	SIMD dual issue
Clock (GHz)	2.33	2.30	1.16	3.20	3.20
DP GFlop/s	9.33	9.20	1.16	6.4	1.8
Local Store	–	—	—	—	256KB
per core L1 Data Cache	32KB	64KB	8KB	32KB	–
per core L2 Cache	–	512KB	—	512KB	—
L1 TLB entries	16	32	128	1024	256
Page Size	4KB	4KB	4MB	4KB	4KB

System	Xeon E5345 (Clovertown)	Opteron 2356 (Barcelona)	UltraSparc T5140 T2+ (Victoria Falls)	QS20 Cell Blade	
# Sockets	2	2	2	2	
Cores/Socket	4	4	8	1	8
shared L2/L3 Cache	4×4MB(shared by 2)	2×2MB(shared by 4)	2×4MB(shared by 8)	–	—
DP GFlop/s	74.66	73.6	18.7	12.8	29
DRAM Bandwidth (GB/s)	21.33(read) 10.66(write)	21.33	42.66(read) 21.33(write)	51.2	
DP Flop:Byte Ratio	2.33	3.45	0.29	0.25	0.57
System Power (Watts)§	330	350	610	285‡	
Threading	Pthreads	Pthreads	Pthreads	Pthreads	libspe2.1
Compiler	icc 10.0	gcc 4.1.2	gcc 4.0.4	xlc 8.2	xlc 8.2

Table 1. Architectural summary of evaluated platforms. Top: per core characteristics. Bottom: SMP characteristics. †Each of the two thread groups may issue up to one instruction. §All system power is measured with a digital power meter while under a full computational load. ‡Cell BladeCenter power running SGEMM averaged per blade.

socket's cache and memory; thus delivering 10.66 GB/s per socket, for an aggregate non-uniform memory access (NUMA) memory bandwidth of 21.33 GB/s for the quad-core, dual-socket system examined in our study. Nonuniformity arises from the fact that DRAM is directly attached to each processor. Thus, access to DRAM attached to the other socket comes at the price of lower bandwidth and higher latency. The DRAM capacity of the tested configuration is 16 GB.

2.3. Sun Victoria Falls

The Sun "UltraSparc T2 Plus" dual-socket 8-core processor, referred to as Victoria Falls, presents an interesting departure from mainstream multicore chip design. Rather than depending on four-way superscalar execution, each of the 16 strictly in-order cores supports two groups of four hardware thread contexts (referred to as Chip MultiThreading, or CMT) – providing a total of 64 simultaneous hardware threads per socket. Each core may issue up to one instruction per thread group assuming there is no resource conflict. The CMT approach is designed to tolerate instruction, cache, and DRAM latency through fine-grained multithreading.

Victoria Falls instantiates one floating-point unit (FPU) per core (shared among 8 threads). Our study examines the Sun UltraSparc T5140 with two T2+ processors operating at 1.16 GHz, with a per-core and per-socket peak performance of 1.16 GFlop/s and 9.33 GFlop/s, respectively – no fused-multiply add (FMA) functionality. Each core has access to its own private 8 KB write-through L1 cache but is connected to a shared 4MB L2 cache via a 149 GB/s(read) on-chip crossbar switch. Each of the two sockets is fed by two dual-

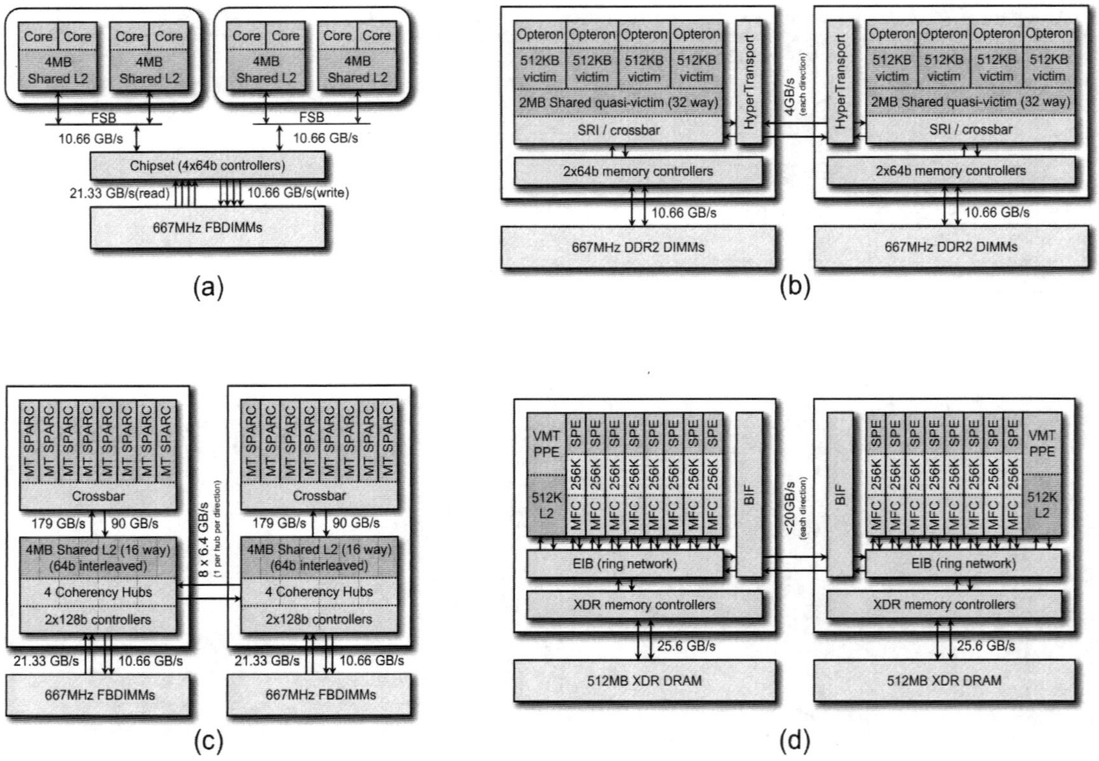

Figure 1. Architectural overview of (a) dual-socket×quad-core Intel Xeon E5345 (Clovertown), (b) dual-socket×quad-core AMD Opteron 2356 (Barcelona), (c) dual-socket×eight-core Sun Niagara2 T2+ T5140 (Victoria Falls), (d) dual-socket×eight-core STI QS20 Cell Blade,

channel 667 MHz FBDIMM memory controllers that deliver an aggregate bandwidth of 32 GB/s (21.33 GB/s for reads, and 10.66 GB/s for writes) to each L2, and a total DRAM capacity of 32 GB. Victoria Falls has no hardware prefetching, and software prefetching places data only in the L2. Although multithreading may hide instruction and cache latency, it may not be able to fully hide DRAM latency.

2.4. *STI Cell*

The Sony Toshiba IBM (STI) Cell processor is the heart of the Sony PlayStation 3 (PS3) video game console, whose aggressive design is intended to meet the demanding computational requirements of video games. Cell adopts a heterogeneous approach to multicore, with one conventional processor core (power processing element, or PPE) to handle OS and control functions, combined with up to eight simpler single instruction multiple data (SIMD) cores (synergistic processing elements, or SPEs) for the computationally intensive work [2].

At first glance, each PPE appears to be a conventional 64b dual-issue, cache-based PowerPC core. In fact, each PPE is two-way vertically multithreaded. Thus, similar to Victoria Falls, multiple thread contexts are maintained in hardware on each core. On any given cycle the core may fetch and issue up to two instructions from one thread or the other. There are no hardware prefetchers and the cores are in-order. Thus one expects the PPEs to be low performing, especially on memory-bound kernels; thus, one should consider their value to be compatibility and productivity, not performance.

The SPEs differ considerably from conventional core architectures due to their use of a disjoint software

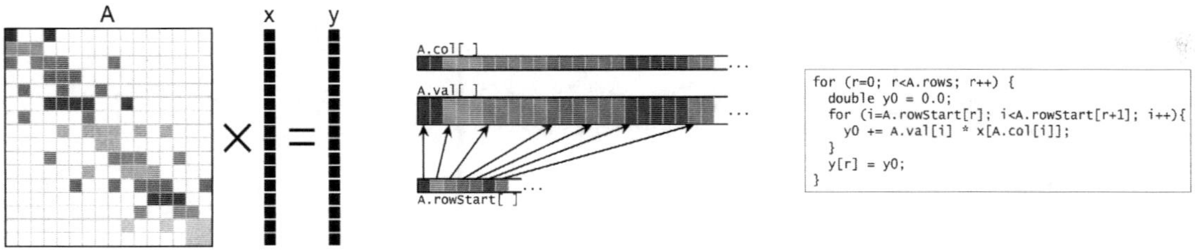

(a) algebra conceptualization **(b)** CSR data structure **(c)** CSR reference code

Figure 2. Sparse Matrix Vector Multiplication (SpMV). (a) visualization of the algebra: $y \leftarrow Ax$, where A is a sparse matrix. (b) Standard compressed sparse row (CSR) representation of the matrix. This structure of arrays implementation is favored on most architectures. (c) The standard implementation of SpMV for a matrix stored in CSR. The outer loop is trivially parallelized without any data dependencies.

controlled local memory instead of the conventional hardware-managed cache hierarchy employed by the PPE. Rather than using prefetch to hide latency, the SPEs have efficient software-controlled DMA engines which asynchronously fetch data from DRAM into the 256KB local store. This approach allows more efficient use of available memory bandwidth than is possible with standard prefetch schemes on conventional cache hierarchies, but also makes the programming model more complex.

Each SPE is a dual issue SIMD architecture that includes a partially pipelined FPU. Although the SPEs can execute two single precision fused-multiply adds (FMAs) per cycle, they can only execute one double-precision FMA SIMD instruction every 7 cycles, for a peak of 1.8 GFlop/s per SPE. This study utilizes the QS20 Cell blade comprised of two sockets with eight SPEs each (29.2 GFlop/s double-precision peak). Each socket has its own dual channel XDR memory controller delivering 25.6 GB/s, with a DRAM capacity of 512 MB per socket (1 GB total). The Cell blade connects the chips with a separate coherent interface delivering up to 20 GB/s, resulting in NUMA characteristics (as with Barcelona and Victoria Falls).

3. Kernels and applications

In this work, we examine the performance of three memory-intensive kernels: SpMV, stencils, and LBMHD. They are all $O(1)$ arithmetic intensity, varying from as little as 0.16 to as high as 1.0 flops per byte. We detail each of them in the following sections.

3.1. Sparse matrix vector multiplication

Sparse matrix vector multiplication (SpMV) dominates the performance of diverse applications in scientific and engineering computing, economic modeling and information retrieval; yet, conventional implementations have historically been relatively poor, running at 10% or less of machine peak on single-core cache-based microprocessor systems [3]. Compared to dense linear algebra kernels, sparse kernels suffer from higher instruction and storage overheads per flop, an inherent lack of instruction- and data-level parallelism in the reference implementations and indirect and irregular memory access patterns. Achieving higher performance on these platforms requires choosing a compact data structure and code transformations that best exploit properties of both the sparse matrix – which may be known only at run time – *and* the underlying machine architecture. This need for optimization and tuning at run time is a major distinction from both the dense case as well as many structured grid codes. We reuse the suite of 14 matrices used in our SC07 [4] paper. These matrices run the gamut of structure, density, and apsect ratio. None should fit in cache.

We consider the SpMV operation $y \leftarrow Ax$, where A is a sparse matrix and x, y are dense vectors. The most common data structure used to store a sparse matrix for SpMV-heavy computations is compressed sparse row (CSR) format, illustrated in figure 2. SpMV has very low arithmetic intensity, performing only 2 flops for every double and integer loaded from DRAM. We define the compulsory miss limited arithmetic intensity to be the

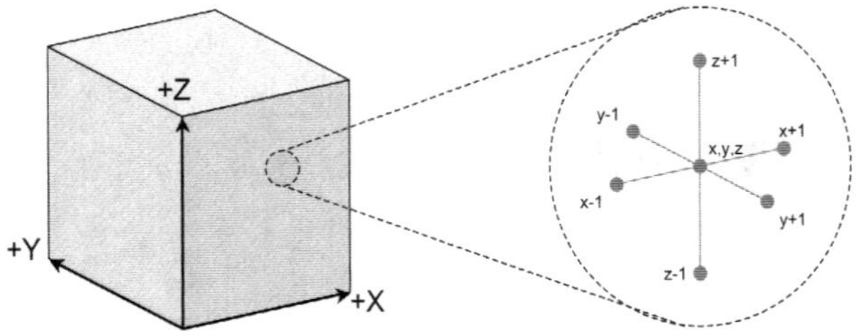


```
Next[x,y,z] =
    C0 * Current[x,y,z] +
    C1 *(
        Current[x+1,y,z] +
        Current[x-1,y,z] +
        Current[x,y+1,z] +
        Current[x,y-1,z] +
        Current[x,y,z+1] +
        Current[x,y,z-1]
       );
```

(a) PDE grid — **(b)** stencil for heat equation PDE — **(c)** inner loop

Figure 3. Visualization of the data structures associated with the heat equation stencil. (a) the 3D temperature grid. (b) the stencil operator performed at each point in the grid. (c) pseudocode for stencil operator.

ratio of the total number of floating-point operations to the compulsory cache misses. True arithmetic intensity includes all cache misses, not just the compulsory ones. With a compulsory miss limited arithmetic intensity of 0.166, one expects most architectures to be heavily memory bound.

For further details on the sparse matrix vector multiplication and our auto-tuning approach, we direct the reader to our SC07 paper [4].

3.2. The heat equation stencil

The second kernel examined in this work is the explicit heat equation partial differential equation (PDE) applied to a uniform 3D grid. PDE solvers constitute a large fraction of scientific applications in such diverse areas as heat diffusion, electromagnetics, and fluid dynamics. These applications are often implemented using iterative finite-difference techniques that sweep over a spatial grid, performing nearest neighbor computations called *stencils*. In a stencil operation, each point in a multidimensional grid is updated with weighted contributions from a subset of its neighbors in both time and space – thereby representing the coefficients of the PDE for that data element. In this work, we examine performance of the 3D heat equation, naïvely expressed as triply nested loops over xyz and shown in figure 3

This seven-point stencil performs a single Jacobi iteration, meaning that the calculation is out-of-place; thus reads and writes occur in two distinct arrays. As a result the compulsory limited arithmetic intensity is 0.33 on write allocate architectures. We use a 256^3 problem when auto-tuning and benchmarking.

For further details on the heat equation and our auto-tuning approach, we direct the reader to our paper submitted to SC08 [5].

3.3. Lattice Boltzmann Magnetohydrodynamics

The final application studied here is Lattice Boltzmann Magnetohydrodynamics (LBMHD) [6]. LBMHD was developed to study homogeneous isotropic turbulence in dissipative magnetohydrodynamics (MHD). MHD is the theory of the macroscopic interaction of electrically conducting fluids with a magnetic field. MHD turbulence plays an important role in many branches of physics [7]: from astrophysical phenomena in stars, accretion discs, and interstellar and intergalactic media to plasma instabilities in magnetic fusion devices. In three-dimensional LBMHD, the 27 velocity momentum lattice is augmented by a 15 (Cartesian vector) velocity magnetic lattice – shown in figure 4. This creates tremendous memory capacity requirements — over 1 KB per point in space. LBM methods iterate on two phases: a *collision()* operator, where the grid is evolved on each timestep, and a *stream()* operator that exchanges data with neighboring processors. Typically, *collision()* constitutes 90% or more of the total run time. Thus, although we parallelize *stream()*, we concentrate on auto-tuning *collision()*. On most architectures, we use a 128^3 problem when auto-tuning and benchmarking. However, this exceeds the Cell Blade's 1 GB memory capacity requirements. As such, we run only a 64^3 problem on Cell.

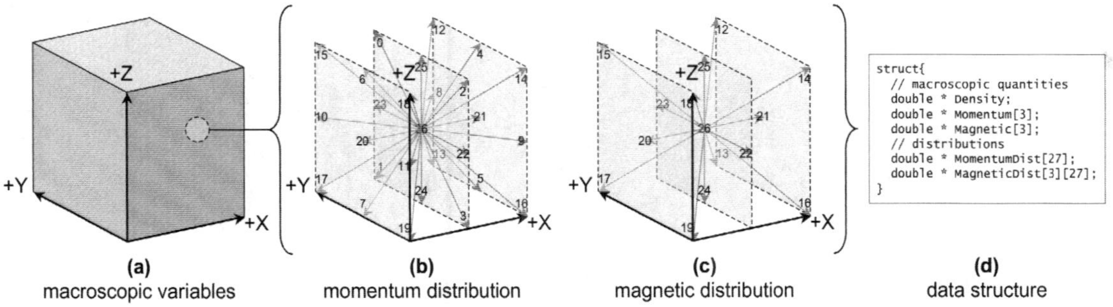

Figure 4. Visualization of the datastructures associated with LBMHD: (a) the 3D macroscopic grid, (b) the D3Q27 momentum scalar velocities, (c) D3Q15 magnetic vector velocities, and (d) C structure of arrays datastructure. Note, each pointer refers t a N^3 grid, and X is the unit stride dimension.

The code is far too complex to duplicate here, although a conceptualization of the lattice method and the data structure itself is shown in figure 4. Nevertheless, the *collision()* operator must read 73 doubles, write 79 doubles, and perform 1300 floating-point operations per lattice update. This results in a compulsory limited arithmetic intensity of about 0.7 on write allocate architectures.

For further details on LBMHD and our auto-tuning approach, we direct the reader to our IPDPS08 paper [8].

4. Roofline Performance Model

The roofline model, as detailed in [9–11] is premised on the belief that the three fundamental components of performance on single program multiple data (SPMD) kernels are communication, computation, and locality In general, communication can be from network, disk, DRAM or even cache. In this work, the communication of interest is from DRAM, and the computation is floating-point operations. The rates associated with these two quantities are peak bandwidth (GB/s) and peak performance (GFlop/s).

Every kernel has an associated arithmetic intensity (AI) – defined as the ratio of floating-point operations to bytes of communication, in this case DRAM bytes. This is essentially a locality metric as cache misses may significantly increase the compulsory memory traffic. One may naïvely estimate performance as the $min(peakGFlop/s, AI \times peakGB/s)$. If one were to plot attainable performance as a function of arithmetic intensity, we would see a rampup in performance followed by a plateau at peak flops much like the slopped rooflines of many houses.

On modern architectures this naïve roofline performance model (guaranteed peak flops or peak bandwidth) is wholly inadequate. Achieving peak in-core flops is premised on fully exploiting every architectural innovation of the past 15 years. For example, to achieve peak flops on an AMD Opteron (Barcelona), one must have a balance between the number of multiplies and adds, fully exploit SSE, unroll by enough to cover the functional unit latency (4 cycles), and ensure that loops are sufficiently long that startup overhead is amortized – see figure 5(a). For instruction bandwidth bound architectures (e.g., Niagara), attainable performance is dictated by the fraction of code that is floating point. Extrapolating to memory bandwidth, for an Opteron to achieve peak bandwidth, one must ensure that the memory accesses are long unit strides, memory allocation must be NUMA-aware, and software prefetching may be needed — figure 5(b). Failing to employ even one of these will diminish the attainable performance and form ceilings below the roofline.

We may now construct a roofline model for each microarchitecture explored in this paper The model is based on microbenchmarks for bandwidth and optimization manuals for computation. The order of optimizations is based on the likelihood of compiler exploitation. For example, fused multiply add (FMA) and multiply/add balance are inherent in linear algebra kernels but nearly impossible to fully exploit in structured grid codes. Thus, this optimization is the lowest of the computational ceilings for SpMV (from sparse linear algebra) but is the roofline for Stencil and LBMHD. Furthermore, compiler unrolling is more likely and easier than SIMDizing a kernel. Similarly, NUMA optimizations are an easier optimization to employ than manually

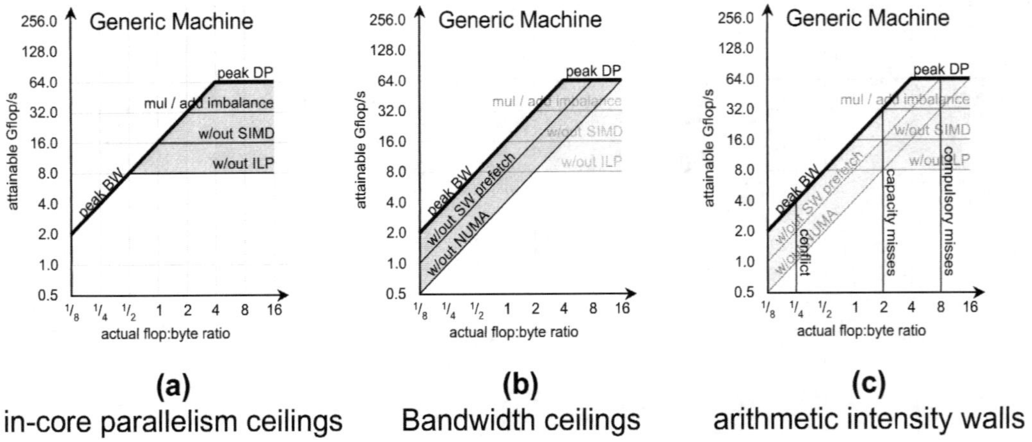

Figure 5. Construction of a DRAM based roofline model for a generic machine. Note the log-log scale (a) Horizontal ceilings are added for 2-way SIMD and 2 cycle functional unit latency. (b) Diagonal ceilings are added for NUMA bandwidth and the fact that software prefetching is required for peak bandwidth. (c) Arithmetic intensity walls are added for the compulsory limit as well as the potential impact of capacity and conflict misses.

software prefetching.

As mentioned, each kernel has an associated flop:compulsory byte arithmetic intensity – vertical lines in figure 5(c). This counts only memory traffic (bytes) from compulsory cache misses. Of course, capacity and conflict misses will significantly reduce this arithmetic intensity — creating a true arithmetic intensity to the left of the compulsory arithmetic intensity. For each kernel, we may scan upward along the specified arithmetic intensity line to determine which optimizations are required to achieve better performance — the ceilings that we must punch through.

5. Optimizations and auto-tuning

Principally, there are three categories of optimizations that must be performed on these memory-intensive SPMD kernels: maximizing memory bandwidth, minimizing memory traffic, and ensuring that there is enough instruction and data level parallelism to keep the machine memory bound.

5.1. Auto-tuning and Code Generation Background

Optimizing compilers are data and algorithm agnostic and rely on heuristics to optimize code. However, it has been shown that not only do search based methods improve both dense linear algebra [12, 13] and spectral methods [14, 15], but data and algorithm foreknowledge allows for complex restructuring of both implementation and data structure [3]. These automatically tuned libraries – auto-tuners — deliver significant performance boosts on existing architectures. They also provide a productive solution to optimizing future architectures.

We know that we will have an extremely large code optimization parameter space to explore. As such, it is imperative to create tools that generate the hundreds of different inner loop variants. We chose to write a parameterized Perl script for each kernel (SpMV, Stencil, LBMHD) capable of producing all the desired variants. The functions are then packed into a pointer to function table that may be indexed by the desired optimizations.

We then create an auto-tuning benchmark for each kernel. It first creates n-1 additional threads that will call a SPMD auto-tuning benchmark functions to create, initialize, and load the datasets. Then, the threads explore the parameter space by applying optimizations in an order appropriate for the given platform. The optimal configuration and performance of these runs is then reported.

The following sections group all the auto-tuned optimizations into the three basic categories and details each of them.

5.2. Threading and parallelization

We selected Pthreads as the threading library for all cache-based microarchitectures. The Cell SPEs use IBM's libspe threading library. On top of Pthreads we implement a shared memory SPMD-like model where all threads execute the same function for all iterations of the kernel. We implemented a flat shared memory barrier for each threading library.

For SpMV, we restricted ourselves to a deterministic 1D decomposition of the matrix (by rows). This ensures that no interthread reductions are necessary. Generally this approach works well when the number of rows is at least an order of magnitude larger than the number of threads.

For LBMHD, we primarily implement a deterministic 1D decomposition of the grid (into slabs or eventually planes). This is acceptable until we have more threads than planes in Z – the least contiguous dimension. At that point we transition, and begin to parallelize in the Y dimension.

The heat equation employs an even more complex parallelization scheme. Not only does it parallelize in all three dimensions, but it auto-tunes to determine the optimal parallelization in each dimension. We simultaneously search for the best cache blocking and thread parallelization.

5.3. Maximizing in-core performance

We must first attempt to ensure that the kernels expected to be memory-bound are, in fact, bound by memory. Essentially, the optimizations detailed here boost in-core performance by ensuring the data and instruction level parallelism inherent in a kernel is expressed in its implementation. The principal optimizations are unrolling, reordering, and explicit SIMDization via intrinsics. This last optimization is a clear departure from portable C code as it requires an optimized SSE implementation, an optimized SPE implementation, an optimized double hummer implementation, etc... for each SIMD ISA. The out-of-the-box SpMV CSR implementation has no instruction-level parallelism (ILP) or data-level parallelism (DLP). Thus these optimizations could not be applied until register blocking (see below) was applied. Nevertheless, the heat equation and LBMHD (after loop restructuring) have ample ILP and DLP. We auto-tune unrolling and reordering in powers of two to determine how much ILP and DLP should be expressed in the implementation.

5.4. Maximizing memory bandwidth

As suggested in the roofline model, many optimizations are required to maximize memory bandwidth. These include limiting the number of memory streams, exploiting the NUMA nature of these machines, and proper use of software and hardware prefetching.

The most basic optimization is to properly collocate data with the threads tasked to process it. We rely on an OS process affinity routine and a first touch policy. We then bind threads to cores and allow each thread, rather than only the master, to initialize its own data. This approach ensures all memory controllers may be engaged and minimizes intersocket communication.

Although SpMV and stencil require only two memory streams per thread, LBMHD generates over 150. In this case, we restructured the loop to avoid TLB misses and also properly engaged the hardware prefetchers [8]. However, this resulted in an unknown quantity – vector length, for which we must auto-tune.

On some architectures, hardware prefetchers can be easily confused. On others, they don't exist. As a result, we instrumented our code generators to produce variants with software prefetches inserted into the inner loops. Our auto-tuning framework then tunes for the appropriate prefetch distance.

5.5. Minimizing memory traffic

When we have optimized our code to the point where it is using nearly all of a machine's sustainable memory bandwidth, the only option for improving performance is to reduce total memory traffic. Basically, this boils down to addressing the 3C's [16] of caches. Essentially any cache miss can be categorized as a conflict, capacity, or compulsory miss. Conflict misses arise due to limited associativity in set associative caches, and capacity misses occur due to limited cache capacity – a bigger cache would eliminate them. However, compulsory misses are ever present. Data starts in DRAM. In order to read it, a cache miss always must occur. In this category we also include data that is hardware prefetched into the cache that would normally generate a compulsory miss if the prefetcher wasn't present.

For all three kernels, the data must start in DRAM, and the result must be returned to DRAM. As a result, for each matrix multiplication or grid sweep we will generate a large number of compulsory misses. We can calculate an upper limit to arithmetic intensity based on the total number of floating-point operations and compulsory cache misses. For SpMV, two floating-point operations are performed for every nonzero (a double and an integer) loaded from DRAM. The heat equation stencil must read every point in the grid (one double each), perform the stencil for each (8 flops), and write the updated point back to DRAM (one double for every point in the grid). LBMHD is similar, but the balance is 73 reads, 1300 flops, and 79 writes per point. On write-allocate cache architectures, each write actually generates 16 bytes of memory traffic – 8 bytes of write-miss fill, and 8 bytes of writeback. Thus we expect flop:compulsory byte ratios of 0.33 and 0.7 respectively for the heat equation and LBMHD.

A flop:byte ratio of 0.166 is not the ultimate limit for SpMV. In fact we may exploit register blocking [3] techniques to improve the flop:byte ratio to 0.25 by eliminating redundant column indices. For the structured grid codes we may exploit an x86 instruction – *movntpd* — and force writes to bypass the cache. The result is that the fill memory traffic may be eliminated. For architectures that allow this optimization, the highest flop:byte ratios of 0.5 for stencil and 1.0 for LBMHD can be achieved.

Conflict misses arise as a result ofthe relatively limited associativity of the caches. Normally, these misses are small, but when power of two problem sizes are coupled with power of two numbers of threads on shared caches with low associativity, the number of conflict misses skyrockets. For SpMV, we address this by skewing the first element of each thread's piece of the matrix to uniformly spaced cache sets. The heat equation has only two arrays total, shared by all threads. As a result, each 3D array is padded in the unit stride dimension to ensure collectively; all the points of the thread's stencils do not hit the same set. The stencil for LBMHD touches one point in each of 150 separate 3D grids. As a result, we pad each grid to ensure that no point within a stencil hits the same set.

Capacity misses occur when the desired working set size exceeds the cache capacity per thread. Depending on how much inherent reuse a kernel has, this will cut the arithmetic intensity in half or more. This result can be devastating on memory-bound kernels. Each dataset for each kernel has a different capacity requirement, and when coupled with the fact that each architecture has a different cache size, it becomes evident that we must auto-tune each to eliminate as many cache misses as possible. For SpMV, capacity misses occur when the working set of the source vector is too big. To rectify this, we cache block the matrix [4] into submatrices known not to touch more data than the cache can hold. A similar technique is applied to the heat equation stencil. Here, to generate solely compulsory misses, we must maintain a working set of 4 planes in the cache. Thus, we auto-tune the loop structure to implicitly block the 3D grid into columns for which this is true. The structure of arrays data structure used in LBMHD was explicitly chosen as it is expected to be free of capacity misses.

5.6. Summary of techniques

Figure 6 visualizes the three general optimizations – maximizing in-core performance, maximizing memory bandwidth, and minimizing memory traffic — by overlaying them on the roofline model for a generic machine. Improving in-core performance is valuable when performance is not bandwidth limited. Conversely, improving bandwidth or increasing arithmetic intensity is valuable when not limited by in-core performance.

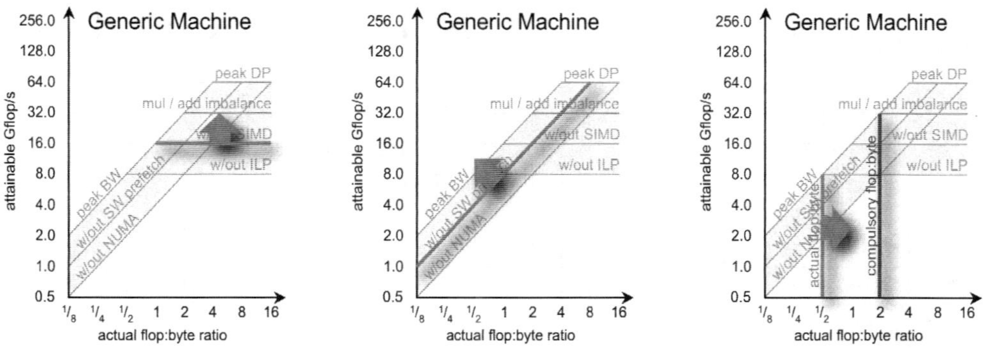

(a) maximizing in-core perf. **(b)** maximizing bandwidth **(c)** minimizing traffic

Figure 6. The three general optimizations associated with the roofline model. (a) Maximizing in-core performance is appropriate when performance at a given arithmetic intensity is not bandwidth limited. (b) Maximizing memory bandwidth is valued on this machine for arithmetic intensities less than 16. (c) Improving arithmetic intensity by minimizing conflict and capacity misses can significantly improve performance when bandwidth limited.

6. Performance results and analysis

Figure 7(a) shows SpMV(suite median), stencil, and LBMHD performance before and after auto-tuning. We have also explicitly shown the benefits of exploiting architecture- or instruction set architecture (ISA)-specific optimizations including SSE for x86 or heterogenity (SPEs) on the Cell blade. On Cell, only the orange, architecture specific, bar represents code running on the SPEs. All other results were obtained on the PPEs. Clearly all architectures benefit from auto-tuning on all kernels, however Barcelona gets much of its performance on stencils and LBMHD from x86 optimized code – perhaps acceptable given the ubiquitous x86 ISA. However, Cell gets almost entirely all of its performance from SIMDized and local store implementations running on the SPEs. Neither of these would be considered productive. In figure 8, we overlay kernel performance onto a roofline model constructed for each architecture. This allows us to visually interpret the fundamental architectural limitations and their impact on kernel performance. For simplicity, we show only SpMV for the dense matrix stored in sparse format. Red diamonds denote untuned performance, where green dots denote auto-tuned performance. The purple regions highlight the expected range of compulsory arithmetic intensities. Clearly, on most architectures, performance is limited by the sustainable memory bandwidth (diagonal lines). We should reiterate that both stencils and LBMHD have disproportionally few multiply instructions. As such, the inherent imbalance between multiplies and adds will limit performance to 50% of peak on all but the SPARC architecture.

A cursory examination of the green bars (out-of-the box pthreads) in figure 7(a) shows that Clovertown, Barcelona, and Victoria Falls all deliver comparable untuned performance for each kernel. The Cell PPE untuned performance is so poor, we can barely see it at this scale. We also see that the benefits of auto-tuning (yellow bars) can significantly improve all machines. When using explicitly SIMDized and auto-tuned code on the cache-based machines, and fully exploiting the heterogeneity of the Cell blade (orange bars), we see that for most kernels, the Cell SPEs deliver considerably better performance than even Barcelona. However, for the most computationally intensive kernel, LBMHD, Cell's weak double-precision implementation is now the bottleneck. Thus, Barcelona has nearly reached parity with Cell.

Simply put, the out-of-the box pthreads code (green bars) represent maximum productivity, while auto-tuned pthreads (yellow bars) makes a one-time sacrifice in productivity for performance portability across most architectures. The orange bars represent an almost complete loss of productivity in favor of great ISA-specific

Figure 7. Comparison of performance (left) and power efficiency (right) across architectures and kernels with increasing optimization. Sustained power was measured under full load with a digital power meter.

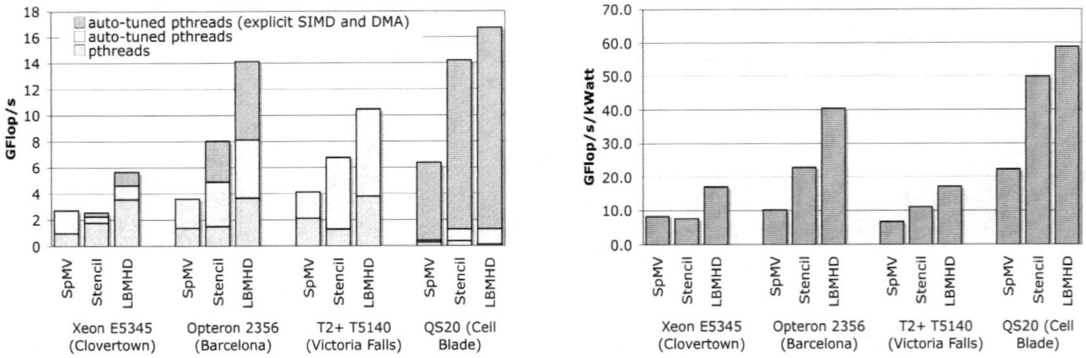

(a) Performance **(b)** Power Efficiency

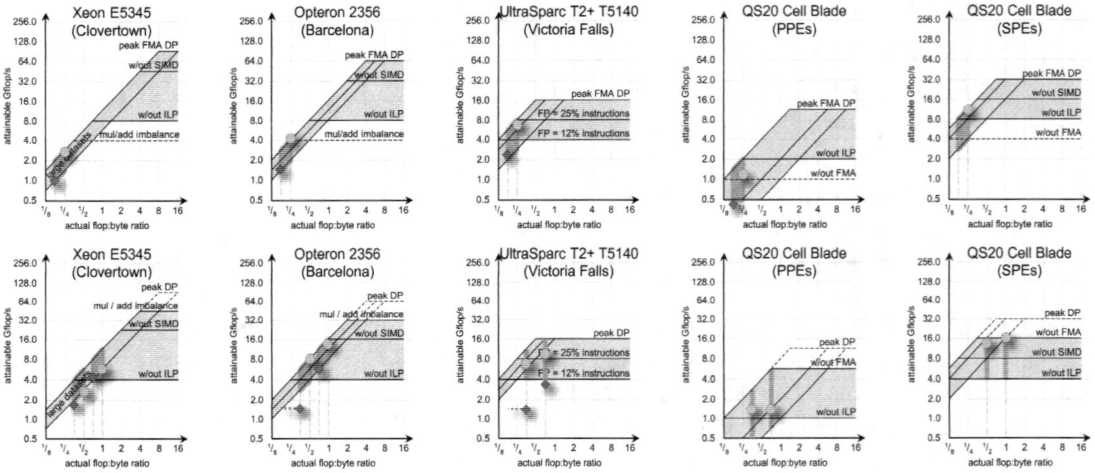

Figure 8. Roofline models for SpMV (top) as well as a combined model (bottom) for both the heat equation($ArithmeticIntensity \leq 0.5$) and LBMHD ($ArithmeticIntensity \leq 1.0$).

auto-tuned performance.

Figure 7(b) shows power efficiency after full optimization. We used a digital power meter to measure sustained system power and quote it in Table 1. Clearly, Victoria Falls' high sustained power – 610 W — significantly reduces its power efficiency, while Cell's relatively low 285 W improves its relative power efficiency.

6.1. Clovertown

Although auto-tuning provided nearly a 3× increase in median performance for SpMV, we see only about a 50% increase in performance for Stencil and LBMHD. Clovertown has no NUMA issues, has numerous hardware prefetchers but relatively poor memory bandwidth, and has giant caches with ample associativity and capacity per thread. As a result, Clovertown started out nearly memory bound, and suffered from few conflict or capacity misses. Thus, auto-tuning provides relatively little benefit on this architecture for these kernels. When referring to the roofline models for the Clovertown in figure 8, we see that Stencil and LBMHD performance seemed to be pegged to the lower bandwidth diagonal associated with large datasets, while the smaller SpMV problems reach the higher optimized bandwidth diagonal. The relatively small snoop filter in the memory controller hub can eliminate the front side bus (FSB) snoop traffic that saps useful bandwidth. Thus, smaller problems will likely see better bandwidth. In either case, it is clear that Clovertown is heavily memory bandwidth starved. In fact scalability studies have shown that quad-core is of little value on SpMV and, surprisingly, dual core is of little value on Stencil and LBMHD. On both Stencil and LBMHD, Clovertown achieves relatively little speedup from explicit SIMDization.

6.2. Barcelona

Although Barcelona has the same number of cores, sockets, and DRAM frequency as Clovertown, we see considerably better performance after auto-tuning. The Opteron is a dual socket NUMA machine. When properly managed, this provides significantly more memory bandwidth per core than the FSB used on the Clovertown. The caches are much smaller making cache blocking a necessity. We also observed that software prefetching could significantly improve memory bandwidth. When bandwidth is maximized and traffic minimized, we must also ensure that in-core performance is not the bottleneck. Explicit SIMDization via SSE intrinsics nearly doubled both LBMHD and Stencil performance. Part of this speedup is from the use of an x86 only cache-bypass instruction, but the other part is from explicitly exploiting the 128b floating-point units. As noted, the order of optimizations in the roofline is based on which is easiest for a compiler to generate. After explicitly SIMDizing the code, perhaps we should have redrawn the roofline explicitly interchanging SIMD and ILP. One should be mindful of this optimization as it is difficult for most programmers to SIMDize even small kernels. As the roofline figure suggests, after auto-tuning, Barcelona is totally memory bound. Auto-tuning provided between a 2.6× and 5.4× speedup.

6.3. Victoria Falls

Victoria Falls is a heavily multithreaded architecture with relatively small shared caches, and even smaller working sets per thread – 1 KB L1 per thread, 64 KB L2 per thread. The more challenging issue is the low associativity of the caches — 16 way L2 for 64 threads. As a result most of the optimizations that showed benefit were ones that addressed the 3C's of caches. Nevertheless, the portable auto-tuned C code for all three kernels runs better on Victoria Falls than any other architecture despite having the lowest peak floating-point capability. When examining the roofline model, we conclude that Victoria Falls is bandwidth bound for SpMV, and possibly Stencil, but has transitioned to being processor bound for LBMHD. Without accurate performance counter data we cannot accurately determine the x-coordinates of the dots and diamonds in the roofline figure. Thus, we cannot conclude with certainty when small changes in arithmetic intensity allow one conclusion or the other. Auto-tuning provided between a 1.9× and 5.2× speedup.

6.4. Cell Blade (PPEs)

Out-of-the box Cell PPE performance is abysmal on all three kernels. Auto-tuning provided significant speedups for Stencil and LBMHD – 3.3× and 10× respectively. Nevertheless, using both threads on both cores delivers far less performance than any other machine. This has serious ramifications as a heterogeneity version of Amdahl's law suggests that the vast majority of the code must be amenable to running on the SPEs to see significant performance speedups. When coupled with the immaturity of the Cell SDK, productivity will suffer greatly.

6.5. Cell Blade (SPEs)

The 3C's model is loosely applicable to local store architectures. First, compulsory misses are universal for all architectures. Second, there are no conflict misses. However, some padding is still required for efficient direct memory access (DMA) operation in much the same way padding is required for SIMDization. Therefore, all capacity misses must be handled in software. Programs whose working set size exceeds the local store capacity must be blocked in software. Thus, the program, rather than the hardware, must find all spatial and temporal locality either when the problem is specified or during execution. The advantage of this approach is that the arithmetic intensity can be calculated easily and exactly. There is no need for performance counters to cope with the uncertainty in the number of conflict or capacity misses. Furthermore, the decoupling of DMA from execution in conjunction with double buffering is a perfect match for the overlapping of communication and computation found in the Roofline model.

Despite the SPE's weak double-precision implementation, we see that their use of DMA ensures good memory bandwidth performance coupled with a minimization of memory traffic. As a result, we see more than a 10× improvement over auto-tuned C code on the PPE. When examining the roofline, we see that SpMV is totally memory bandwidth bound, although the choice of a minimum register blocking of 2×1 means the implementation compulsory misses are higher than the algorithmic compulsory misses. Stencil is likely to be barely compute bound (remember FMAs cannot be exploited), while LBMHD is strongly compute bound. Thus one expects the forthcoming QS22 blades to significantly improve LBMHD performance, slightly improve Stencil performance, but do nothing for SpMV.

7. Summary and Conclusions

The computing industry is moving rapidly away from exponential scaling of clock frequency toward chip multiprocessors in order to better manage trade-offs among performance, energy efficiency, and reliability. Understanding the most effective hardware design choices and code optimizations strategies to enable efficient utilization of these systems is one of the key open questions facing the computational community today.

In this paper we discussed a set of multicore optimizations for Sparse Matrix Vector Multiplication (SpMV), the explicit heat equation PDE on a regular grid (Stencil), and a lattice Boltzmann application (LBMHD). We detailed and presented the results of an auto-tuning approach, which employs a code generator that produces multiple versions of the computational kernels using a set of optimizations with varying parameter settings. The optimizations include TLB and cache blocking, loop unrolling, code reordering, software prefetching, streaming stores, and use of SIMD instructions. The impact of each optimization varies significantly across architecture and kernel, necessitating a machine-dependent approach to automatic tuning. In addition, our detailed analysis reveals the performance bottlenecks for each computation on our evaluated system.

Results show that the Cell processor still offers the highest raw performance and power efficiency for these computations, despite having peak double-precision performance and memory bandwidth that is lower than many of the other platforms in our study. The key architectural feature of Cell is explicit software control of data movement between the local store (cache) and main memory. However, this impressive computational efficiency comes with a high price – a difficult programming environment that is a major departure from conventional programming. However, much of this work must be duplicated on cache-based architectures to achieve good performance. Nevertheless, these performance disparities point to the deficiencies of existing automatically-managed cache hierarchies, even for architectures with sophisticated hardware and software prefetch capabilities. Thus there is considerable room for improvements in the latency tolerance techniques of microprocessor core designs.

Our study has demonstrated that – for the evaluated class of algorithms – auto-tuning is essential in achieving good performance. Moreover, in the highly multithreaded/multicore world, DRAM bandwidth is the preeminent factor in sustained performance. In this world, there is little need for processor designs that emphasize high single thread throughput. Rather, high throughput via large numbers of simpler cores is likely to deliver higher efficiency and vastly simpler implementation. While prior reseach has shown that these design philosophies offer substantial benefits for peak computational rates [17], our work quantifies that this approach can offer significant performance benefits on real scientific applications.

Overall, the auto-tuned codes achieved sustained superscalar performance that is substantially higher than any published results to date, with speedups often in excess of 5× relative to the original code. Auto-tuning amortizes tuning effort across machines by building software to generate tuned code and using computer time rather than human time to search over versions. It can alleviate some of compilation problems with rapidly-changing microarchitectures, since the code generator can produce compiler-friendly versions and can incorporate small amounts of compiler- or machine-specific code. We therefore believe that auto-tuning will be an important tool in making use of multicore-based HPC systems of the future.

Additionally, we analyzed the performance using the roofline model. The roofline makes it clear which architectural paradigms must be exploited to improve performance. Fraction of peak raw bandwidth or flops is an inappropriate metric of performance. The Roofline model presents realistic ceilings to performance. After auto-tuning is applied to the kernels we see that performance is very near the roofline ceilings of each architecture despite potentially low percentage of peak raw flop or bandwidth numbers.

Future work will continue exploring auto-tuning optimization strategies for important numerical kernels on the latest generation of multicore systems, while making these tuning packages publicly available. We hope the roofline model will provide an architecture specific optimization search path that will significantly reduce the time required for auto-tuning.

Acknowledgments

We express our gratitude to Forschungszentrum Jülich for access to their Cell blades, as well Brian Waldecker of AMD for Barcelona access. This work was supported by the ASCR Office in the DOE Office of Science under contract number DE-AC02-05CH11231, by NSF contract CNS-0325873, and by Microsoft and Intel Funding under award #20080469.

References

[1] Asanovic K, *et al.* 2006 The landscape of parallel computing research: A view from Berkeley EECS, *University of California, Berkeley, Report UCB/EECS-2006-183*

[2] Gschwind M 2006 Chip multiprocessing and the cell broadband engine *CF '06: Proc. 3rd Conference on Computing Frontiers* (Ischia,Italy) 1–8

[3] Vuduc R, Demmel J, and Yelick K 2005 OSKI: A library of automatically tuned sparse matrix kernels *Proc. SciDAC 2005, J. of Physics: Conference Series* (Institute of Physics Publishing)

[4] Williams S, Oliker L, Vuduc R, Shalf J, Yelick K, and Demmel J 2007 Optimization of sparse matrix-vector multiplication on emerging multicore platforms *Proc. SC2007: High performance computing, networking, and storage conference*

[5] Datta K, Murphy M, Volkov V, Williams S, Carter J, Oliker L, Patterson D, Shalf J, and Yelick K 2008 *Stencil computation optimization and autotuning on state-of-the-art multicore architectures* Preprint

[6] Macnab A, Vahala G, Vahala L, and Pavlo P 2002 Lattice Boltzmann model for dissipative MHD *Proc. 29th EPS Conference on Controlled Fusion and Plasma Physics* **26B** (Montreux, Switzerland)

[7] Biskamp D 2003 *Magnetohydrodynamic Turbulence* (Cambridge University Press)

[8] Williams S, Carter J, Oliker L,Shalf J, and Yelick K 2008 Lattice Boltzmann simulation optimization on leading multicore platforms *Interational Conf. Parallel and Distributed Computing Systems (IPDPS)* (Miami, Florida)

[9] Williams S 2008 Autotuning performance on multicore computers, Ph.D. thesis (University of California, Berkeley)

[10] Williams S, and Patterson d 2008 The roofline model: An insightful multicore performance model *Preprint*

[11] Williams S, Patterson D, Oliker L, Shalf J, and Yelick K 2008 The roofline model: A pedagogical tool for auto-tuning kernels on multicore architectures *Hot Chips 20: Stanford University, stanford, California, August 24–26, 2008* (IEEE MIcro)

[12] Whaley R C, Petitet A, and DongarraJ 2001 Automated empirical optimizations of software and the ATLAS project *Parallel Computing* **27** **(1)** 3–25

[13] Bilmes J, Asanović K, Chin C W, and Demmel J 1997 Optimizing matrix multiply using PHiPAC: A portable, high-performance, ANSI C coding methodology *Proc. International Conf. Supercomputing* (Vienna, Austria)

[14] Frigo M and Johnson, S G 1998 FFTW: An adaptive software architecture for the FFT *Proc. 1998 IEEE Intl. Conf. Acoustics Speech and Signal Processing vol 3 (IEEE) 1381–1384*

[15] Moura J M F, Johnson J, Johnson R W, Padua D, Prasanna V K, Püschel M, and Veloso M 2000 SPIRAL: Automatic implementation of signal processing algorithms *High Performance Embedded Computing (HPEC)*

[16] Hill M D, and Smith A J 1989 Evaluating associativity in CPU caches *IEEE Trans. Comput.* **38** **(12)** 1612–1630

[17] Sylvester D and Keutzer K 2001 Microarchitectures for systems on a chip in small process geometries *Proc. IEEE* 467–489

Simulation of wave interactions with MHD

D Batchelor[1], G. Abla[2], G. Bateman[3], D Bernholdt[1], L Berry[1], P Bonoli[4], R. Bramley[5], J Breslau[6], M Chance[6], J Chen[6], M Choi[2], W Elwasif[1], G Fu[6], R Harvey[7], E Jaeger[1], S Jardin[6], T Jenkins[10], D Keyes[8], S Klasky[1], S Kruger[9], L Ku[6], V Lynch[1], D McCune[6], J Ramos[4], D Schissel[2], D Schnack[10] and J Wright[4]

[1]Oak Ridge National Laboratory, Oak Ridge, TN 37831, USA

[2]General Atomics, San Diego, CA 92186, USA

[3]Lehigh University, Bethlehem, PA 18015, USA

[4]Plasma Science and Fusion Center, MIT, Cambridge, MA 02139 USA

[5]Indiana University, Bloomington, IN 47405, USA

[6]Princeton Plasma Physics Laboratory, Princeton, NJ 08543, USA

[7]CompX, Del Mar, CA 92014, USA

[8]Columbia University, New York, NY 10027, USA
[9]Tech-X, Boulder, CO 80303, USA

[10]University of Wisconsin, Madison, WI 53706, USA

Abstract. The broad scientific objectives of the SWIM (Simulation of Wave Interaction with MHD) project are twofold: (1) improve our understanding of interactions that both radio frequency (RF) wave and particle sources have on extended-MHD phenomena, and to substantially improve our capability for predicting and optimizing the performance of burning plasmas in devices such as ITER: and (2) develop an integrated computational system for treating multiphysics phenomena with the required flexibility and extensibility to serve as a prototype for the Fusion Simulation Project. The Integrated Plasma Simulator (IPS) has been implemented. Presented here are initial physics results on RF effects on MHD instabilities in tokamaks as well as simulation results for tokamak discharge evolution using the IPS.

E-mail: batchelordb@ornl.gov

1. Introduction

Comprehensive plasma simulations are essential to the development of fusion energy. Not only does simulation serve to advance science by allowing us to evaluate and test basic theory through comparison with experiments, it also has a direct economic benefit by maximizing the productivity of experimental facilities and by supporting design decisions for new facilities. For a device like ITER, such decisions can have large financial consequences. The Center for Simulation of Wave Interactions with Magnetohydrodynamics (SWIM) was begun in 2005 with the scientific objectives of improving our understanding of interactions that both RF wave and particle sources have on extended-magnetohydrodynamics (MHD) phenomena; improving our capability for predicting and optimizing the performance of burning plasmas; developing an integrated computational system for treating multiphysics phenomena with the required flexibility and extensibility to serve as a prototype for the Fusion Simulation Project; addressing mathematics issues related to the multiscale, coupled physics of RF waves and extended MHD; and optimizing the integrated system on high performance computers.

The scientific rationale of the SWIM project is relatively simple. One of the most fundamental observations in fusion devices is that performance is often limited by the appearance of unstable plasma motions that can degrade plasma containment or even terminate the plasma discharge, with the potential for damage to the containment device. These instabilities can be modeled by using extended magnetohydrodynamics. A second observation is that high-power radio frequency

(RF) electromagnetic waves can be used to influence plasma stability—sometimes producing instability and sometimes reducing or eliminating instability. A capability to understand and predict the effects of RF waves on plasma stability is of significant scientific and economic benefit.

2. Structure of the SWIM project

The SWIM Center consists of three elements:

(1) **Development of a computational platform, referred to as the Integrated Plasma Simulator (IPS),** that will allow efficient coupling of the full range of required fusion codes or modules.

(2) **A physics campaign addressing long timescale discharge evolution in the presence of sporadic fast MHD events**. This involves interfacing the IPS to both linear and 3D non-linear extended MHD codes and carrying out a program of research related to use of RF and other driving sources to study and control fast time-scale MHD phenomena such as optimizing burning plasma scenarios and improving the understanding of how RF can be employed to achieve long-time MHD stable discharges and control sawtooth events.

(3) **A physics campaign for modeling the direct interaction of RF and extended MHD for slowly growing modes.** This requires development of new approaches to closure for the fluid equations and the interfacing of RF modules directly with the extended MHD codes and with code modules that implement the fluid closures. The primary physics focus of this campaign is to improve the understanding of how RF can be employed to control neoclassical tearing modes.

3. Design and implementation of the Integrated Plasma Simulator

The computational core of the SWIM project is the Integrated Plasma Simulator (figure 1), a software framework designed and developed not only to support the coupled simulations required by the SWIM science plan but also to explore issues associated with integrated fusion simulation at an even larger scale for future efforts such as the proposed Fusion Simulation Project (FSP). The IPS environment brings together a variety of well-established physics codes and a Plasma State component to manage the exchange of key simulation data, under a SWIM-specific framework written in Python that provides a uniform execution environment, services for data and resource management, and integration with a web portal to facilitate tracking the progress of simulations. It also provides a convenient facility for testing mathematical issues related to multiphysics coupling and time-stepping algorithms.

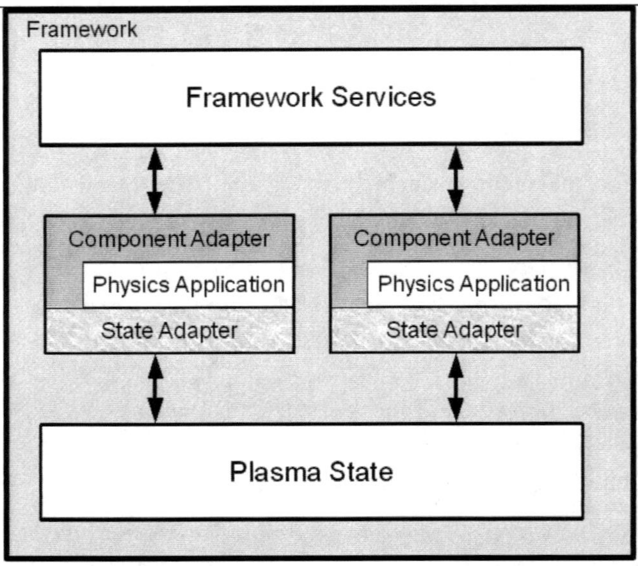

Figure 1. Schematic architecture of the Integrated Plasma Simulator environment (from W. Elwasif 2007).

The physics codes of interest have their own input and output files and formats, so we use small "helper" programs within the component wrapper script to serve as adapters between the Plasma State and the physics code's own input and output files.

The design anticipates the need to shift from file-based to in-memory data exchange eventually, as SWIM's science plan moves to simulations with tighter couplings and exchanging larger volumes of data. The IPS has been designed to accommodate multiple implementations of individual physics components, even within the same simulation run. The framework design imposes few restrictions on the types of simulations that can be carried out, allowing the IPS to be used for studies outside SWIM's physics principle targets of RF interactions with MHD.

An important feature of the IPS is the exchange of simulation data between components using an intermediate Plasma State service that has a simple user interface and a standardized data format. It also provides services to map between different grids when the grid required by the components differs from that on which it is stored in the state. It is implemented as a Fortran 90 module, thus permitting multiple time-step instances to be in memory at once. The Plasma State code is automatically generated from a plasma state specification file allowing for ease and accuracy of modification and extensibility. The SWIM Plasma State service has been adopted by other fusion computing projects as their communication mechanism.

3.1 IPS development and implementation

The framework and Plasma State service are fully implemented and tested on the Jaguar computer at ORNL and on the SGI cluster at PPPL, and several driver components have been developed. Physics components include Equilibrium and Profile Advance (EPA), RF solver, Fokker-Planck solver, MHD Stability, and Visualization. The main EPA component is based on the TSC code. Two RF Ion Cyclotron components based on the TORIC and AORSA codes give us flexibility for comparison and for a wide range of RF physics capabilities. A Fokker Planck solver components based on the CQL3D continuum code and a Monte Carlo RF Fokker Planck solver based on ORBIT–RF from are nearing completion. A Fokker Planck/neutral beam component based on NUBEAM is nearly complete, although NUBEAM can already be used in SWIM simulations through the direct coupling between TSC and TRANSP. The linear MHD component contains the capabilities of equilibrium refinement, flux surface mapping, ballooning stability evaluation using BALLOON, and low-n stability evaluation using DCON, PEST-1, or PEST-2. It will be extended in the near future to include the energetic particle code NOVA-K and ultimately to the nonlinear codes M3D and NIMROD. The Visualization component extracts data from the simulation data structure and presents it in a form for run-time monitoring using the ELVIS system.

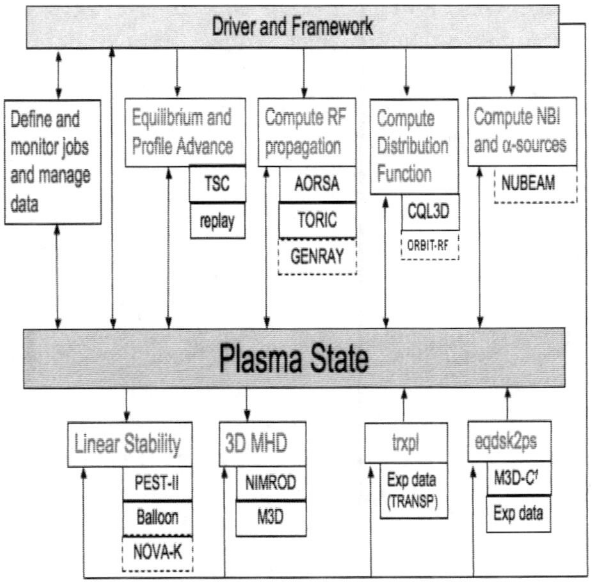

Figure 2. IPS physics components.

3.2 Fast MHD campaign and IPS applications

We have begun using the SWIM framework to simulate the time-dependent evolution of the fast ion distribution during hydrogen minority heating in the Alcator C-Mod device. These simulations will allow us to validate the predictive capability of the RF and MHD components, a prerequisite to investigating sawtooth modification in the presence of energetic particles. The TSC EPA component is being used to advance the plasma profiles and equilibrium for a specific C-Mod discharge, and both the AORSA and CQL3D codes are being used to self-consistently calculate the 3D (r, $v_\perp$, $v_{//}$) fast ion distribution during ICRF minority heating. The simulation is programmed to adjust the OH-coil and other PF coil currents (and thus the loop voltage) to give the experimental values of the plasma current and plasma major radius vs time as well as the line-averaged density and the Z_{eff}.

The simulation used the standard TSC thermal conductivity model, which normally provides a good model for L-mode transport at low power, and the Porcelli sawtooth model was used for sawtooth evolution. Model source terms were used for the electron and ion heating associated with the collisional slowing down of the minority ion tail. We are comparing with the experimental values of sawtooth amplitude and crash time, the loop voltage, the plasma shape, and the variation in the electron temperature during different phases in the heating and sawtooth cycles. As can be seen from figure 3, the overall comparison between the TSC simulation and the experiment is very good.

Using the RF component (AORSA-CQL3D), we have also performed a standalone simulation of the time evolution of the minority tail energy at a time corresponding to 0.91 sec. in the C-Mod discharge simulation of figure 3. Figure 4 compares the minority tail energy from this simulation with the measurement of charge exchange count rate due to energetic minority ions, as detected by a compact neutral particle analyzer (CNPA) diagnostic on C-Mod. The simulated turn-on time of the fast ion tail energy follows the rise time of the CNPA signal quite well for the vertical diagnostic channel viewing at R = 69 cm, corresponding to $\rho \approx 0.1$. The decay time of the energetic tail in the simulation is noticeably longer than the fall time of the CNPA signal, however. This difference may be attributed to the fact that no spatial losses were turned on in the Fokker Planck simulation. Also, the simulation assumed a constant electron and deuteron temperature in time, whereas in reality the background temperatures were evolving during the 60 msec. simulation time, due to the ICRF heating and sawtooth cycle. These latter effects will be captured properly when the model ICRF heating sources used in the TSC simulation are coupled with the AORSA-CQL3D results.

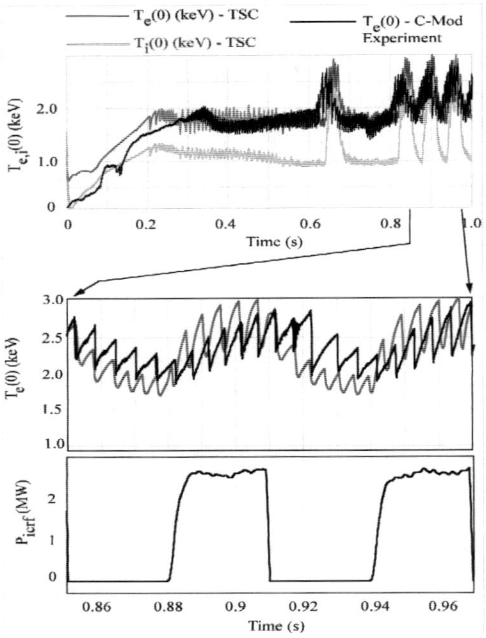

Figure 3. Comparison of experimental central electron temperature (black traces) with TSC simulation. Central electron (ion) temperatures from the simulation are shown in red (green). The bottom graph shows an expanded time period with associated ICRF power pulse.

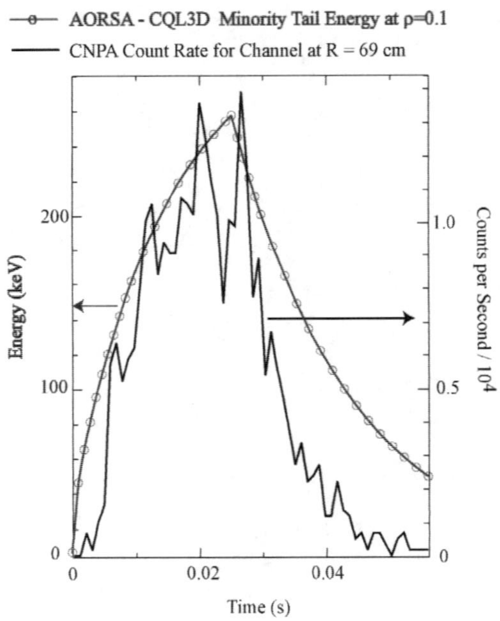

Figure 4. Comparison of minority hydrogen tail temperature (keV) versus time (s) from an AORSA-CQL3D simulation with CNPA diagnostic count rate versus time for a channel at R=69 cm. Plasma parameters correspond to the TSC simulation at 0.91 sec.

3.3 Slow MHD campaign

In the Slow MHD campaign a theoretical approach for including the effects of RF in the fluid closure has been developed, a computational approach to coupling MHD with RF has been developed, and initial numerical studies with reduced models have begun. Figure 5 illustrates the effect of the RF current source on the evolution of a resistive magnetic island. The equilibrium is unstable to a 2/1 tearing mode that is allowed to grow and saturate. An externally specified, axisymmetric RF current source added to the Ohm's law equation in the NIMROD 3D nonlinear MHD code. The current is then gradually in time increased from small amplitude. Comparing the subsequent evolution of the magnetic island with and without the RF current source, we find that the island shrinks in response to the current perturbation. The relative sizes of the magnetic islands in the presence and absence of RF are shown in figure 5. More detailed comparisons are ongoing.

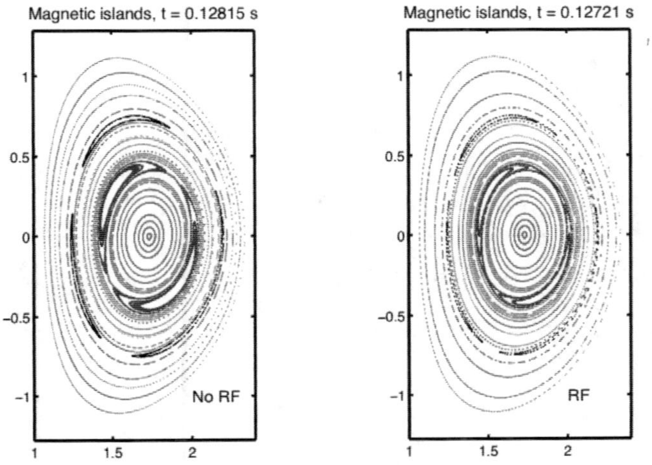

Figure 5. Size of the 2/1 magnetic island near the end of the simulations without (left) and with (right) RF current source.

First results from core-edge parallel composition in the FACETS project

J R Cary[1], J Candy[2], R H Cohen[3], S Krasheninnikov[4], D C McCune[5], D J Estep[6], J Larson[7], A D Malony[8], A. Pankin[9], P H Worley[10], J A Carlsson[1], A H Hakim[1], P Hamill[1], S Kruger[1], M Miah, S Muzsala[1], A Pletzer[1], S Shasharina[1], D Wade-Stein[1], N Wang[1], S. Balay[7], L McInnes[7], H Zhang[7], T Casper[3], L Diachin[3], T Epperly[3], T D Rognlien[3], M R Fahey[10], J Cobb[10], A Morris[8], S Shende[8], G W Hammett[5], K Indireshkumar[5], D Stotler[5], A Yu Pigarov[4]

[1]Tech-X Corporation, Boulder, CO 80303, USA
[2]General Atomics, San Diego, CA 92186, USA
[3]Lawrence Livermore National Laboratory, Livermore, CA 94550, USA
[4]University of California at San Diego, San Diego, CA 92093, USA
[5]Princeton Plasma Physics Laboratory, Princeton, NJ 08543, USA
[6]Colorado State University, Fort Collins, CO 80523, USA
[7]Argonne National Laboratory, Argonne, IL 60439, USA
[8]ParaTools, Inc., Eugene, OR 97405, USA
[9]Lehigh University, Bethlehem, PA 18015, USA
[10]Oak Ridge National Laboratory, Oak Ridge, TN 37831, USA

E-mail: cary@txcorp.com

Abstract. FACETS (Framework Application for Core-Edge Transport Simulations), now in its second year, has achieved its first coupled core-edge transport simulations. In the process, a number of accompanying accomplishments were achieved. These include a new parallel core component, a new wall component, improvements in edge and source components, and the framework for coupling all of this together. These accomplishments were a result of an interdisciplinary collaboration among computational physics, computer scientists, and applied mathematicians on the team.

1. Introduction

The FACETS (Framework Application for Core-Edge Transport Simulations) project [1] has the goal of providing whole-tokamak modeling through coupling separate components for each of the core region, edge region, and wall. A coupling approach is used because (1) there exist computational models for each region, (2) the regions are covered by different approximations, and (3) direct simulation resolving the gyrokinetics time scales in all regions are as too computationally intensive. That is, even when the code is parallelized, computing 1 ms of experimental time using gyrokinetics requires 1 hour of computational time. Thus, computing the full 1000-second discharge would require 10^6 hours.

FACETS has organized multidisciplinary teams for each of the components and for the framework. The computational fusion scientists provide the model and basic component, but in working with applied mathematicians and computers scientists, the implementations are improved. Computer scientists are also involved in interface definition and interlanguage implementation.

2. Core solver including performance analysis

Tokamak transport is dominated by instability-driven turbulence at extremely small spatial scales. The resultant turbulent transport is highly nonlinear and increases rapidly beyond the instability threshold in temperature gradient. A commonly used model of this turbulent transport uses fitted quasilinear theory to model this transport and has shown success in modeling tokamak plasmas [2]. The difficulty of using this model is the length of computation in calculating the fluxes, which is a separate computation at each radial cell, and in advancing the integrated system with these nonlinear, radially dependent fluxes.

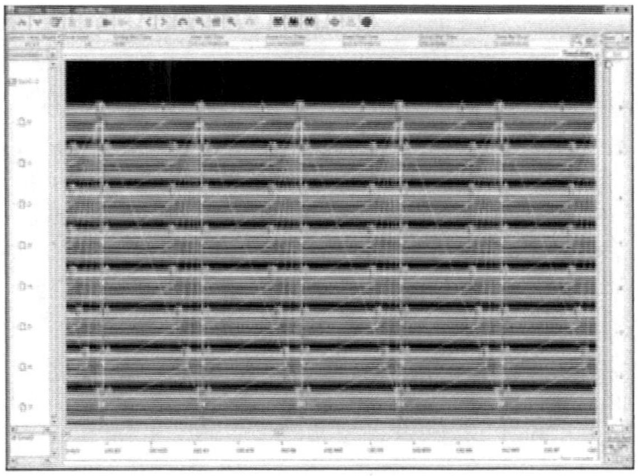

Figure 1. Using the TAU performance system [3] for performance analysis has proven critical in achieving scalable performance.. Shown here is an example of how unbalanced flux calculations (purple) lead to wasted CPU cycles (green) as the time advance requires all of the flux calculations to complete. This lead to a change in the code that allows for more balanced calculations.

To solve these problems, a new core solver was written from scratch using the FACETS infrastructure. As discussed below, the new solver has been verified against established codes, but the new code has allowed us to parallelize the flux calculations and achieve a factor of 30 speedup on 128 processors with a 128-cell grid. The time advance uses a Newton solve with a multilevel method to enlarge the radius of convergence. The overall performance is sensitive to the load balancing of the flux calculations, and performance analysis has proven critical to identifying these problems, as shown in figure 1. As part of an embedded SAP, these flux calculations will be replaced by even more intensive first-principles (gyrokinetic) calculations. The new code has been written to allow us to fully exploit leadership computers as the individual calculations increase in computational cost. By improving the parallelism and algorithms, we hope to perform full 1000-second ITER simulations in less than an hour.

3. Componentization and parallelization of UEDGE

The model for the region from the edge pedestal, through the magnetic separatrix, and to material walls is the existing 2D UEDGE transport code that includes electrons, multi-species ions, and neutrals. UEDGE is the major US-developed edge transport code for modeling the tokamak boundary plasma. Figure 1 shows a typical mesh and electron temperature solution. The complexity of the amount of physics that UEDGE encapsulates has meant that users prefer a python interface for control of UEDGE. While this has been a benefit for users, it has introduced additional interlanguage problems with the C++ framework, described below.

UEDGE has also previously used a Newton-Krylov solver and preconditioning technique, where the latter does not provide any domain-overlap strategy for the parallel UEDGE option that has been revived during this project. Being 2D, edge transport can be a performance-limiting component of the coupled system. To remedy this deficiency, we developed an interface to the nonlinear solvers (SNES) in PETSc. As anticipated, in the serial mode, SNES performance is close to the previous solver. PETSc has allowed the easy exploration of various parallel preconditioning strategies, which show improved convergence efficiency.

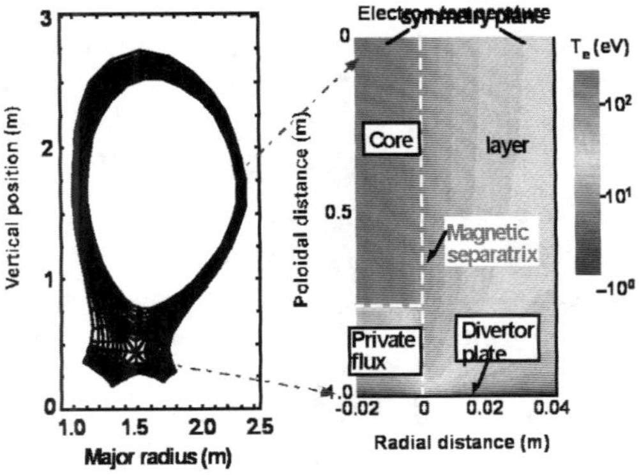

Figure 2. Typical UEDGE mesh and electron temperature showing 2D variation in the outer edge, but 1D variation at the left boundary where core model is coupled.

4. Core sources

Fusion experiments plan to reach ignition temperatures by using external sources: both neutral beam injection and radiofrequency heating. Modeling of these external sources is done by using parallel codes, and can consume considerable resources by itself. The recently parallelized PPPL Monte Carlo package, NUBEAM [2, 3] will be used for providing the core sources from neutral beam injection (NBI) and fusion reactions. The output of NUBEAM includes particle sources (both charged and neutral particles), and momentum and energy source rates along with a wide array of other output including the slowing down fast ion distribution functions. As part of the FACETS project, NUBEAM is undergoing performance analysis and scaling testing. Since the only communications costs in NUBEAM involve the broadcast of initial data and the MPI reductions at the end of deposition and orbiting, good scaling is expected. Recent tests demonstrate excellent scaling up to 128 processors with 1 million particles. Efforts are under way to test this scaling further.

As part of the Simulation of Wave Interactions with MHD (SWIM) SciDAC, common interfaces for all sources were defined, and an API using Fortran was defined. As part of the FACETS project, we have extended this API to include C++ interfaces. The solution of the problem of how to efficiently construct this interface led to a new publication on Fortran90/C++ interoperability. With this new interface, we will be able to leverage the work of the SWIM project and incorporate RF and particle sources into our parallel core solver. Ahead is yet to work out the interfaces for sinks, such as radiation.

5. Framework and coupling

We have used the framework to obtain initial results for the core-edge coupled system. In these simulations the core component is native to FACETS; that is, it is written by using infrastructure provided by the framework itself. The edge component is a non-native component, obtained by wrapping UEDGE using Babel [6]. The framework now provides an explicit coupler component which calls the individual components to advance their internal state (solution) by a given time step. The coupler then exchanges the core-edge boundary data and loops to move the solution forward in time. Current efforts are directed toward the coupled simulation results and the development of an implicit coupler component

6. The wall component

Plasma-surface interactions are complex due to the complex chemistry and plasma physics that occurs on a short timescale. To model these complex interactions, we have developed the Wall and Plasma-Surface Interactions (WallPSI) module, which is the 1-D multiscale, multispecies code for particle and heat transport inside plasma-facing components [7]. The code incorporates new approaches in continuum modeling of hydrogen species in wall materials. The code is being validated against the vast experimental data on hydrogen retention, permeation, and erosion rates for major fusion related materials.

Preliminary results of self-consistent plasma-neutrals-wall modeling were reported for the first time in [7] using WallPSI/Edge1D and showed examples of strong plasma-wall coupling, nonlinear variation of wall hydrogen inventory and recycling coefficient with respect to incident plasma flux, and the featured plasma instabilities. The modeled transitional effects include sawteeth-like oscillations in edge plasma due to switching from wall pumping to gas release occurred in the case without external gas puff/pump feedback, where 15–20% variation in recycling coefficient corresponds to transitions between cold deeply-detached and hot sheath-limited edge plasmas due to thermal plasma instability [8].

Ongoing work on WallPSI integration in FACETS includes development of the infrastructure and plasma-wall interface, implementation of robust nonlinear solvers, and massive parallelization via the FACETS framework based on assigning one CPU to each wall segment of the modeling domain. The coupling of WallPSI to UEDGE (not completed) will be through assigning one instance of WallPSI to each wall facing UEDGE.

6. Steady-state gyrokinetic transport

FACETS has an embedded Scientific Application Partnership (SAP) whose goal to develop a strategy to employ local and global gyrokinetic turbulence simulations in steady state transport calculations. A prototype transport driver has been developed that can run multiple instances of GYRO, a fundamental physics gyrokinetic code, or simpler transport models. This state-of-the-art code computes self-consistent transport using a Newton iteration scheme. Early results show that using this iteration scheme a fairly crude steady state can be obtained in just a few iterations. Ongoing work on the physics of this iteration scheme includes investigating the Newton step-size as well as the GYRO time-averaging length.

Our prototype simulation was a 12-hour simulation run on the Cray XT4 at Oak Ridge National Laboratory using 1284 cores. This simulation used 25 radial grid points over the minor radius: 4 instances of a simpler model were run over the range 0.0375 to 0.15 while 20 instances of GYRO each using 64 processors were run over 0.1875 to 0.9. This could easily have been run with more grid points and with each GYRO instance using more processes, thus the scaling to very high numbers of cores is straightforward. Current efforts are leveraging this embedded SAP to be made available to the new core solver so that this parallelism can be efficiently exploited in a time-dependent simulation.

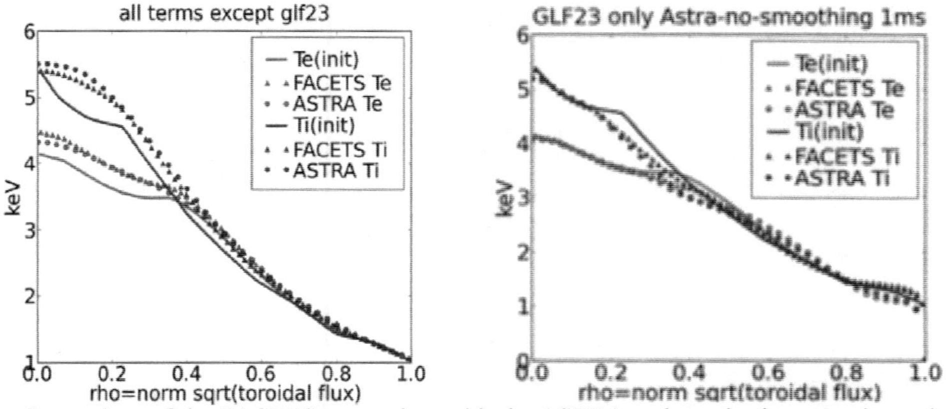

Figure 3. Comparison of the FACETS core solver with the ASTRA code evolved an experimental discharge for 1 microsecond with different terms for energy transport used. On the left, no turbulent transport was considered. The codes agree well except near the core, where the ASTRA code does not conserve energy. The case with turbulent transport also agrees well except for problems near rho of 0.4, believed to be due to time-advance differences.

7. Verification of the core solver

Because the new core solver was written from scratch, verifying the code is critical to ensure its accuracy. To verify the code, the code was compared to the ASTRA transport code, a standard transport code used in Europe. The code was compared both with and without the turbulence transport model described above to isolate problems. The disagreement for the case without the turbulent transport was traced to a lack of energy conservation in the ASTRA code. Disagreements also exist for the case with the turbulent transport, and the differences are believed to be due to different time advance schemes (ASTRA uses an explicit scheme). Efforts are under way to refine these verification efforts by using a steady-state code for comparison.

8. Concluding remarks

FACETS has achieved core-edge coupled simulations in its first year through development of a multicomponent framework that allows separate components to run on different processors. In the process FACETS developed a new, parallel core solver, restored parallelism to the edge component, and implemented a wall model. Applied mathematics and computer science have been an integral part of this project through algorithms, performance analysis, and interlanguage operability.

In its initial phase, FACETS has proceeded with the philosophy of maximally reusing code that is available, while writing new components and infrastructure where warranted. Furthermore, FACETS has worked to provide developed modules to other teams. The sister FMCFM project has defined component interfaces for turbulent transport models, and these are being shared, in particular, with the SWIM project, which has another approach to component integration.

9. Future directions

FACETS has only just started, and so many tasks remain. One of these is to make the core-edge coupling application a robust, user application by improving the initial setup of the edge component. Others include the move to implicit coupling, incorporation of the wall component, bringing in core sources, allowing for dynamic magnetic equilibrium, validation studies, and releasing version 1.0. In addition, we will incorporate the work of the SSGKT SAP program for more fundamental flux calculations.

As well, we will have to continue improvements of our code base. We expect to make further refinements to our interfaces. Non-native components bring a challenge in that they were typically written without the needs of componentization in mind; for example, they often make heavy use of global variables and so are not thread safe. This latter point makes it impossible to make use of mixed parallelism. Another problem is the used of shared objects, as is needed for python-wrapped codes. Shared libraries are not available on the reduced operating systems on the compute nodes of leadership class supercomputers. These challenges require a continual evaluation of component requirements and project needs as we progress. Coupling codes with different mathematical models brings new challenges in ensuring the reliability of the results. We will extend the a posteriori analysis of finite volume methods to time dependent problems using various time integration schemes, including a detailed study of stability issues. Then, we will analyze the effect of coupling through a common boundary, including the interaction with the time integration. We also will analyze the effects of using nonuniform cells on the accuracy of the finite volume method.

Of course, the primary challenge facing FACETS, with its parallel component (MCMD) decomposition model, is effectively distributing the components over the expected 10^5 nodes of a high-performance computer.

Acknowledgments

Work supported by the U.S. Department of Energy (DOE) Office of Science under grants and contracts including DE-FC02-07ER54907 at Tech-X, DE-AC02-06CH11357 at ANL, DE-FC02-07ER54909 at CSU, DE-AC52-07NA27344 at LLNL, DE-AC05-00OR22725 at ORNL, DE-FC02-07ER54910 at ParaTools, DE-AC02-76CH03073 at PPPL, DE-FC0207-ER54908 and DE-FG0204-ER54739 at UCSD. This work used the resources of the National Energy Research Scientific Computing Center, which is supported by the DOE Office of Science under Contract No. DE-AC02-05CH11231, and of the National Center for Computational Sciences at ORNL, which is supported by the DOE Office of Science under Contract No. DE-AC05-00OR22725.

References

[1] https://www.facetsproject.org/facets
[2] Kinsey J E, Staebler G and Waltz R E 2003 *Fus. Sci. Tech.* **44** 763
[3] Shende S and Malony A 2006 The TAU parallel performance system *Int. J. High Perf. Comp. Apps.* **20(2)** 287–331, http://tau.uoregon.edu.
[4] Goldston R J, *et al.* 1981 *J. Comp. Phys.* **43** 61
[5] Pankin A, *et al.* 2004 *Comp. Phys. Comm.* **159** 157
[6] https://computation.llnl.gov/casc/components/babel.html
[7] Pigarov A Yu, *et al.* 2008 Coupled plasma-wall modeling *Proc. 18th Int. Conf. on Plasma Surface Interactions* (Spain)
[8] Krasheninnikov S I *et al.* 2006 *Phys. Plasmas* **13** 094502

Scalable parallel implicit solvers for 3D magnetohydrodynamics

L Chacón
Los Alamos National Laboratory, Los Alamos, NM 87545 USA
E-mail: chacon@lanl.gov

Abstract. The efficient integration of the extended magnetohydrodynamics (XMHD) model is critical for present and future modeling needs of fusion plasmas and is at the core of several interacting SciDAC centers. However, XMHD is a remarkably challenging system of PDEs, featuring a stiff multiscale character both in time and space. Such stiff character has hindered progress toward a truly scalable, efficient, and accurate nonlinear XMHD algorithm. In this paper, we discuss recent algorithmic developments that demonstrate that such an XMHD algorithm is within reach. The approach is based on fully implicit time integration and employs parallel Jacobian-free Newton-Krylov technology, preconditioned with multigrid methods for algorithmic scalability. The algorithm has been tested on a variety of 2D and 3D configurations that demonstrate its excellent algorithmic scalability properties, both serially and in parallel up to 4096 processors and 134 million unknowns.

1. Introduction

With the advent of petascale computing, fundamental scientific understanding will be enabled by the possibility of ever more realistic modeling of complex physical systems. This is, in fact, the ultimate goal of SciDAC. However, hardware power alone is not sufficient to enable such massive computations. Algorithms, as has been recognized in various DOE [1, 2] and NSF [3] reports, are as important, if not more. The emphasis is, or course, on algorithmic scalability, namely, that the computational complexity of a given algorithm scales linearly with the number of unknowns, and inversely proportional to the number of processors.

In this paper, we focus on the development of such an algorithmic capability for the extended magnetohydrodynamic (XMHD) model, which is relevant in modeling various plasmas of interest to DOE such as solar, magnetospheric, and laboratory (e.g., fusion) plasmas. This paper builds on recently published work [4], where a scalable, parallel, fully implicit, fully nonlinear solver for the 3D resistive compressible MHD equations was demonstrated, and augments this work in two fundamental ways. First, we generalize the formulation in the reference (which was limited to small flows) to arbitrary-size plasma flows. This is of interest to enhance the robustness of the solver in the presence of large plasma flows that may appear locally during the simulation (e.g., due to magnetic reconnection processes) and to enable direct-to-steady-state solution procedures. Second, we generalize the formulation in [4] to include electron Hall effects. As a proof of principle, we consider the cold-ion limit to demonstrate the soundness of the approach.

As in [4], we base our nonlinear solver approach on the Newton-Raphson iterative algorithm. Krylov iterative techniques [5], implemented Jacobian-free [6, 7] (i.e., without ever forming and storing the Jacobian matrix) for memory efficiency, are employed for the required algebraic matrix inversions. Here, FGMRES (Flexible Generalized Minimal RESiduals [8, 9]) is employed as the Krylov solver of choice, because of the lack of symmetry in the algebraic system of interest. The flexible character of FGMRES relaxes some of the constraints in the preconditioner step of regular GMRES, which we have found useful in our implementation. In particular, FGMRES allows the preconditioner to change between successive GMRES iterations. For parallelization, we employ the PETSc library [10, 11, 12].

The efficiency and scalability of Krylov methods depend strongly on adequate preconditioning [5]. Here, as described above, we extend previous work on 3D resistive MHD to include arbitrary flows and electron Hall effects. As in [13, 4], a suitable multigrid-based preconditioner is developed around the parabolization concept, whereby a hyperbolic system is reformulated as a parabolic one, which is in turn amenable to a multilevel treatment. The connection between the parabolization procedure and the Schur block decomposition, outlined first in [13], and extended to 3D resistive MHD in [4], is suitably generalized here to accommodate the new physics of interest. We shall show that such generalization is simple conceptually and of straightforward implementation.

The rest of the paper is organized as follows. Section 2 introduces the base model equations. Section 3 introduces the Krylov methods and the specifics of the Jacobian-free implementation. Section 4 discusses the physics-based preconditioner for this particular application is discussed. Section 5 presens serial and parallel efficiency results of the resulting implicit algorithm in various 2D and 3D configurations. We conclude in Section 6.

2. Three-dimensional Hall MHD model

We consider the dimensionless compressible Hall MHD model, given in Alfvénic units (i.e, Alfvén speed $v_A = B_0/\sqrt{\mu_0 \rho_0}$, Alfvén time $\tau_A = L/v_A$, where B_0, ρ_0, and L are the reference magnetic field, density, and length, respectively) by

$$\frac{\partial \rho}{\partial t} + \nabla \cdot (\rho \mathbf{v} - D\nabla\rho) = 0, \tag{1}$$

$$\frac{\partial \mathbf{B}}{\partial t} + \nabla \times \mathbf{E} = 0 \tag{2}$$

$$\frac{\partial (\rho \mathbf{v})}{\partial t} + \nabla \cdot \left[\rho \mathbf{v}\mathbf{v} - \mathbf{B}\mathbf{B} + \mathbb{I}(p + \frac{B^2}{2}) - \rho\nu\nabla\mathbf{v}\right] = 0, \tag{3}$$

$$\frac{\partial T}{\partial t} + \mathbf{v}\cdot\nabla T + (\gamma - 1)\left[T\nabla\cdot\mathbf{v} - \frac{\kappa\nabla^2 T + Q}{\alpha_T \rho}\right] = 0, \tag{4}$$

with $\mathbb{I}$ the identity operator, $p = (1 + \alpha_T)\rho T$ the pressure, $\alpha_T = T_i/T_e$ the ion/electron temperature ratio, and ρ the particle density. In these equations, $\mathbf{v}$ is the plasma velocity; $\mathbf{B}$ is the magnetic field; η and ν are the reciprocals of the Lundquist number and the Reynolds number, respectively; D is a dimensionless particle diffusivity that models cross-field particle diffusion (such particle diffusivity D is not self-consistently considered in equation 3 and hence should be regarded as ad hoc); and $Q = \eta j^2 + \rho\nu\nabla\mathbf{v} : \nabla\mathbf{v}$ contains the Joule and viscous heating sources. Simple closures for the heat flux $-\kappa\nabla T_e$ and the viscous stress tensor $-\rho\nu\nabla\mathbf{v}$ are considered at this stage. More accurate closures (such as parallel electron heat transport and gyro-viscous stresses) will be considered in future work. The electric field $\mathbf{E}$ is given by

$$\mathbf{E} = -\mathbf{u}\times\mathbf{B} + \eta\nabla\times\mathbf{B} + \frac{d_i}{\rho}(\mathbf{j}\times\mathbf{B} - \nabla p_e), \tag{5}$$

where $p_e = \rho T$ is the electron pressure and $d_i = c/(\omega_{pi} L)$ is the ion skin depth, and is a measure of the importance of two-fluid physics for a given simulation.

The system in equations 1–4 is hyperbolic, supporting a variety of linear and dispersive normal modes, depending on the value of kd_i, with k the wavenumber. Linear modes ($kd_i \ll 1$) include fast and slow magnetosonic waves, and the shear Alfvén wave. Dispersive modes ($kd_i \gg 1$) include Whistler and kinetic Alfvén wave. In some applications (notably, in magnetic fusion confinement), dynamical time scales of interest are much slower than those associated with these normal modes. In such contexts, an implicit approach that steps over the normal-mode time scales to resolve the time scales of interest is useful. This is the subject of this paper. However, as we shall see, the hyperbolic character of the MHD model makes the task of developing an optimal, scalable solver difficult.

A word about the discretization of equations 1–4 is in order. Spatially, the system is discretized by using finite volumes, as detailed in [14]. Such spatial discretization has proved to be conservative, solenoidal in the magnetic field to numerical round-off, and remarkably robust in the absence of physical and/or numerical dissipation. Temporally, we employ a θ-scheme, with $\theta = 0.5$ (second-order Crank-Nicolson). This choice results in a set of nonlinear algebraic equations $\mathbf{G}(\mathbf{x}) = \mathbf{0}$, with $\mathbf{x}^T = (\rho, T, \mathbf{B}, \mathbf{v})$, that needs to be inverted every time step. For this, we employ preconditioned Newton-Krylov methods. The next section introduces these methods and some of their properties.

3. Jacobian-free Newton-Krylov methods

We give a brief introduction to Jacobian-free Newton-Krylov methods (JFNK). The motivated reader can find extensive discussions on this approach elsewhere [15]. Newton's method solves the nonlinear system $\mathbf{G}(\mathbf{x}) = \mathbf{0}$ iteratively by inverting linear systems of the form

$$\left.\frac{\partial \mathbf{G}}{\partial \mathbf{x}}\right|_k \delta \mathbf{x}_k = -\mathbf{G}(\mathbf{x}_k),$$

with $\mathbf{x}_{k+1} = \mathbf{x}_k + \delta \mathbf{x}_k$. Nonlinear convergence is achieved when

$$\|\mathbf{G}(\mathbf{x}_k)\|_2 < \epsilon_a + \epsilon_r \|\mathbf{G}(\mathbf{x}_0)\|_2 = \epsilon_t, \tag{6}$$

where $\|\cdot\|_2$ is the ℓ_2-norm (Euclidean norm), $\epsilon_a = \sqrt{N} \times 10^{-15}$ (with N the total number of degrees of freedom) is an absolute tolerance to avoid converging below round-off, ϵ_r is the Newton relative convergence tolerance (set to 10^{-4} in this work), and $\mathbf{G}(\mathbf{x}_0)$ is the initial residual.

Such linear systems are solved iteratively with Krylov methods, which require only matrix-vector products to proceed. Because the linear system matrix is a Jacobian matrix, such matrix-vector products can be implemented Jacobian-free by using the Gateaux derivative

$$\left.\frac{\partial \mathbf{G}}{\partial \mathbf{x}}\right|_k \mathbf{v} = \lim_{\epsilon \to 0} \frac{\mathbf{G}(\mathbf{x}_k + \epsilon \mathbf{v}) - \mathbf{G}(\mathbf{x}_k)}{\epsilon}, \tag{7}$$

where in practice a small but finite ϵ is employed (p. 79 in [15]). Thus, the evaluation of the Jacobian-vector product requires only the function evaluation $\mathbf{G}(\mathbf{x}_k + \epsilon \mathbf{v})$, and there is no need to form or store the Jacobian matrix. This, in turn, allows for a memory-efficient implementation.

An inexact Newton method [16] is used to adjust the convergence tolerance of the Krylov method at every Newton iteration according to the size of the current Newton residual, as follows:

$$\|J_k \delta \mathbf{x}_k + \mathbf{G}(\mathbf{x}_k)\|_2 < \zeta_k \|\mathbf{G}(\mathbf{x}_k)\|_2, \tag{8}$$

where ζ_k is the inexact Newton parameter and $J_k = \left.\frac{\partial \mathbf{G}}{\partial \mathbf{x}}\right|_k$ is the Jacobian matrix. Thus, the convergence tolerance of the Krylov method is loose when the Newton state vector $\mathbf{x}_k$ is far from the nonlinear solution but tightens as $\mathbf{x}_k$ approaches the solution. Hence, the linear solver works the hardest when the Newton state vector is closest to the nonlinear root. Superlinear convergence rates of the inexact Newton method are possible if the sequence of ζ_k is chosen properly (p. 105 in [15]). Here, we employ the same prescription as in [13]:

$$\begin{aligned} \zeta_k^A &= \gamma \left(\frac{\|\mathbf{G}(\mathbf{x}_k)\|_2}{\|\mathbf{G}(\mathbf{x}_{k-1})\|_2} \right)^{\alpha}, \\ \zeta_k^B &= \min[\zeta_{max}, \max(\zeta_k^A, \gamma \zeta_{k-1}^{\alpha})], \\ \zeta_k &= \min[\zeta_{max}, \max(\zeta_k^B, \gamma \frac{\epsilon_t}{\|\mathbf{G}(\mathbf{x}_k)\|_2})], \end{aligned}$$

with $\alpha = 1.5$, $\gamma = 0.9$, and $\zeta_{max} = 0.8$. The convergence tolerance ϵ_t is defined in equation 6. In this prescription, the first step ensures superlinear convergence (for $\alpha > 1$), the second avoids volatile decreases in ζ_k, and the last avoids oversolving in the last Newton iteration.

A further advantage of Krylov methods is that they can be preconditioned by considering the alternate (but equivalent) systems $J_k P_k^{-1} P_k \delta \mathbf{x}_k = -\mathbf{G}_k$ (right preconditioning) or $P_k^{-1} J_k \delta \mathbf{x}_k = -P_k^{-1} \mathbf{G}_k$ (left preconditioning). Such preconditioned systems can be straightforwardly and efficiently implemented in the Krylov algorithm as two consecutive matrix-vector products. A crucial feature of preconditioning is that, while it can substantially improve the convergence properties of the Krylov iteration if $P_k^{-1} \approx J_k^{-1}$, it does not alter the solution of the Jacobian system upon convergence (because the solution $\delta \mathbf{x}_k$ of the preconditioned system is the same as that of the original system). Therefore, one can explore suitable approximations in the preconditioner for efficiency purposes without compromising the accuracy of the converged result. We next discuss our approach to preconditioning.

4. Preconditioning

JFNK requires preconditioning for algorithmic scalability (i.e., with convergence rates independent of the number of unknowns and the number of processors considered). Multigrid methods (MG) have been shown in a variety of applications [17] to produce such optimal JFNK convergence rates, and are at the core of our approach. In particular, our work builds on previous developments in 2D resistive [18] and Hall [13] reduced MHD and 3D resistive MHD [4], where such optimal behavior has been demonstrated. In these references, the key for an effective multigrid implementation was the parabolization of otherwise hyperbolic PDEs. The approach, which was termed "physics-based," aimed at reformulating the semi-discrete (temporally discrete, spatially continuous) set of PDEs into a diagonally dominant set, amenable to classical smoothing (based on stationary iterative techniques) in an MG setting.

MG methods employ a divide-and-conquer approach where multiple grids of varying refinement are employed [19, 20]. The underlying idea is that oscillatory components of the error can be readily attacked at a given grid level (with a so-called smoother), but smooth ones are difficult. In its simplest form (an MG V-cycle), the procedure then involves smoothing the error on a given level and then coarsening (restricting) the smooth components to the next coarse grid level. In the new level, some of the smooth components will appear oscillatory and therefore can be subjected to further smoothing. The process is performed recursively until the coarsest grid level is reached, at which point a direct solve can be performed very cheaply. The solution is then interpolated (prolongated) up the grid hierarchy, until the finest level is reached. While variations of the basic V-cycle are possible to improve the convergence rate of a given MG implementation (p. 47 in [19]), one or several V-cycles are generally enough for preconditioning purposes. As can be understood from the description, the crucial element for a working MG solver is the availability of a smoother. While smoothers can be found fairly easily for diagonally dominant systems (in a point or block sense; p. 96 in [20]), it is remarkably hard otherwise.

Hyperbolic systems (such as MHD) can be shown to be diagonally submissive when time steps larger than the explicit CFL stability constraint are employed (see Ref. [18] for an in-depth explanation of this issue). This, in turn, hinders the task of finding a suitable smoother for MG. However, hyperbolic systems can be conveniently parabolized in an implicit time-stepping setting [18, 13, 4]. The basic idea is to produce a well-conditioned (diagonally dominant) parabolic operator from an ill-conditioned hyperbolic system of equations. Algebraically, the parabolization procedure is equivalent to a Schur complement factorization. For 3D resistive MHD (we will generalize this to Hall MHD in Sec. 4.2), one can succinctly write the Jacobian matrix for the linearized system in terms of the linear updates $\delta\rho$, δT, $\delta\mathbf{B}$, and $\delta\mathbf{v}$ as [4]

$$J\delta\mathbf{x} = \begin{bmatrix} D_\rho & 0 & 0 & U_{\rho\mathbf{v}} \\ 0 & D_T & 0 & U_{T\mathbf{v}} \\ 0 & 0 & D_\mathbf{B} & U_{\mathbf{B}\mathbf{v}} \\ L_{\mathbf{v}\rho} & L_{\mathbf{v}T} & L_{\mathbf{v}\mathbf{B}} & D_\mathbf{v} \end{bmatrix} \begin{pmatrix} \delta\rho \\ \delta T \\ \delta\mathbf{B} \\ \delta\mathbf{v} \end{pmatrix}.$$

The diagonal blocks D_ρ, D_T, $D_\mathbf{B}$, $D_\mathbf{v}$ are given in [4] and contain advection-diffusion terms, which can be readily inverted by using MG if upwinding is employed for the advective terms (only in the preconditioning stage [18, 13]). Off-diagonal blocks L and U (also given in the reference) contain all relevant hyperbolic couplings. The Jacobian matrix has an "arrow" structure, which suggests considering the following 2x2 block structure for analysis purposes:

$$J\delta\mathbf{x} = \begin{bmatrix} M & U \\ L & D_\mathbf{v} \end{bmatrix} \begin{pmatrix} \delta\mathbf{y} \\ \delta\mathbf{v} \end{pmatrix},$$

where $\delta\mathbf{y} = (\delta\rho, \delta T, \delta\mathbf{B})^T$, and

$$M = \begin{pmatrix} D_\rho & 0 & 0 \\ 0 & D_T & 0 \\ 0 & 0 & D_\mathbf{B} \end{pmatrix}.$$

The block M can be easily invertible, since M is block diagonal, and, as stated earlier, the blocks themselves are amenable to MG techniques. The Schur factorization of the inverse of the 2x2 block Jacobian matrix yields

$$\begin{bmatrix} M & U \\ L & D_\mathbf{v} \end{bmatrix}^{-1} = \begin{bmatrix} \mathbb{I} & -M^{-1}U \\ 0 & \mathbb{I} \end{bmatrix} \begin{bmatrix} M^{-1} & 0 \\ 0 & P_{Schur}^{-1} \end{bmatrix} \begin{bmatrix} \mathbb{I} & 0 \\ -LM^{-1} & \mathbb{I} \end{bmatrix},$$

where $P_{Schur} = D_\mathbf{v} - LM^{-1}U$ is the Schur complement, which contains all the information from the off-diagonal blocks L and U. At this point, the MHD system has been effectively parabolized. The Schur factorization translates into the following three-step *exact* inversion algorithm (with $\mathbf{G}$ the nonlinear function residual):

$$\begin{aligned} \text{Predictor} &: \quad \delta\mathbf{y}^* = -M^{-1}\mathbf{G}_\mathbf{y}, \\ \text{Velocity update} &: \quad \delta\mathbf{v} = P_{Schur}^{-1}[-\mathbf{G}_\mathbf{v} - L\delta\mathbf{y}^*], \\ \text{Corrector} &: \quad \delta\mathbf{y} = \delta\mathbf{y}^* - M^{-1}U\delta\mathbf{v}. \end{aligned} \tag{9}$$

We note that, at this point, no approximations have been introduced, and the exact Jacobian inverse requires only finding M^{-1} and P_{Schur}^{-1}. Suitable approximations in the preconditioning strategy will use equation 9 as a starting point.

4.1. Preconditioner for arbitrary plasma flow

As it stands, inverting P_{Schur} is impractical because of the presence of M^{-1}. In [4], the *small flow limit* was considered, in which $M^{-1} \approx \Delta t \mathbb{I}$ (i.e., advection and diffusion were neglected). In this section, we derive a tractable preconditioner formulation that 1) does not require M^{-1} to form P_{Schur}, and 2) is valid for arbitrary flows. Our starting point is the Schur complement, $P_{Schur} = D_{\mathbf{v}} - LM^{-1}U$. The key idea is to find a suitable invertible operator M_* such that $M^{-1}U \approx UM_*^{-1}$, or equivalently $UM_* \approx MU$. This idea is not new: similar considerations have been employed in the context of the implicit integration of incompressible Navier-Stokes (see, e.g., [21] and references therein). Note that M and M_* operate on different spaces, as M_* operates on $\mathbf{v}$-space, and M operates on $\mathbf{y}$-space. If such an operator M_* exists, one straightforwardly finds that

$$P_{Schur}^{-1} \approx P_*^{-1} = [D_{\mathbf{v}} - LUM_*^{-1}]^{-1} = M_*(D_{\mathbf{v}}M_* - LU)^{-1}, \tag{10}$$

which eliminates the need of calculating M^{-1}, does not require the inverse of M_*, and features the *same* operator LU as in the small-flow limit formulation. The resulting P_*^{-1} can be readily implemented in equation (9) by inverting $(D_{\mathbf{v}}M_* - LU)^{-1}$ (using matrix-light multigrid techniques, as described in [4]), followed by an application of M_*. By comparison with the small-flow approach, the arbitrary-flow approach only requires an application of M_* in the velocity update and an additional inversion of M^{-1} in the corrector step, making it slightly more expensive.

There remains to find a suitable operator M_*. For this, we notice that all diagonal blocks in M correspond to the discretization of time-dependent advection-diffusion equations and are therefore similar. Furthermore, the upper blocks U,

$$\begin{aligned} U_{\rho\mathbf{v}}\delta\mathbf{v} &= \theta\nabla\cdot(\rho_0\delta\mathbf{v}), \\ U_{T\mathbf{v}}\delta\mathbf{v} &= \theta[\delta\mathbf{v}\cdot\nabla T_0 + (\gamma-1)T_0\nabla\cdot\delta\mathbf{v}], \\ U_{\mathbf{B}\mathbf{v}}\delta\mathbf{v} &= \theta\nabla\cdot[\delta\mathbf{v}\mathbf{B}_0 - \mathbf{B}_0\delta\mathbf{v}], \end{aligned}$$

correspond to linearized advective terms and are also similar. Given these similarities, we focus on the terms of MU that correspond to the density update $\delta\rho$ (i.e., $D_\rho U_{\rho\mathbf{v}}\delta\mathbf{v}$, with $D_\rho\delta\rho = \frac{\delta\rho}{\Delta t} + \theta[\nabla\cdot(\mathbf{v}_0\delta\rho) - D\nabla^2\delta\rho]$, and $U_{\rho\mathbf{v}}$ defined above) to infer a suitable M_*. After some manipulation, one can derive *exactly* that

$$D_\rho U_{\rho\mathbf{v}}\delta\mathbf{v} = \theta\nabla\cdot\left[\rho_0\left(\frac{\delta\mathbf{v}}{\Delta t} + \theta\frac{\mathbf{v}_0}{\rho_0}\nabla\cdot(\rho_0\delta\mathbf{v}) - \theta\frac{D}{\rho_0}\nabla\nabla\cdot(\rho_0\delta\mathbf{v})\right)\right] = U_{\rho\mathbf{v}}D_*\delta\mathbf{v},$$

with $D_*\delta\mathbf{v} = \frac{\delta\mathbf{v}}{\Delta t} + \theta\frac{\mathbf{v}_0}{\rho_0}\nabla\cdot(\rho_0\delta\mathbf{v}) - \theta\frac{D}{\rho_0}\nabla\nabla\cdot(\rho_0\delta\mathbf{v})$. Inspired by D_*, and for preconditioning purposes, we *define* M_* as

$$M_*\delta\mathbf{v} = \frac{\delta\mathbf{v}}{\Delta t} + \theta\mathbf{v}_0\nabla\cdot\delta\mathbf{v}.$$

Notice we have kept only the flow term (which is anisotropic, and common to all variables) and neglected the diffusion one (which is isotropic, and depends on the variable of interest). This completes the formulation of the arbitrary-flow preconditioner.

4.2. Preconditioner for Hall MHD in the cold-ion limit

The XMHD model in equations 1-5 supports dispersive waves with dispersion relation $\omega \sim k^2$, such as the whistler and kinetic Alfvén waves. In explicit methods, the presence of dispersive waves results in explicit CFL time step limits $\Delta t_{CFL} \sim \Delta x^2$, substantially slowing their numerical integration. In implicit methods, dispersive waves result in poorly conditioned matrices, difficult to treat with iterative methods. In the context of Krylov methods, adequate treatment of dispersive waves puts a premium on preconditioning. We now demonstrate a proof of principle for a preconditioner strategy for XMHD. For this, we focus on the cold ion limit, $T_e \gg T_i$. In this limit, $p \approx p_e$, and therefore the Ohm's law (equation 5) reads:

$$\mathbf{E} \approx -\mathbf{u}\times\mathbf{B} + \eta\nabla\times\mathbf{B} + \frac{d_i}{\rho}(\mathbf{j}\times\mathbf{B} - \nabla p) = -\mathbf{u}\times\mathbf{B} + \eta\nabla\times\mathbf{B} + d_i\left(\partial_t\mathbf{u} + \mathbf{u}\cdot\nabla\mathbf{u} - \nu\nabla^2\mathbf{u}\right). \quad (11)$$

With this simplification in mind, the linearized XMHD model in equations 1–4 has the *same* coupling structure as the resistive MHD model presented earlier,

$$\mathbf{J}\delta\mathbf{x} \approx \begin{bmatrix} D_\rho & 0 & 0 & U_{\mathbf{u}\rho} \\ 0 & D_T & 0 & U_{\mathbf{u}T} \\ 0 & 0 & D_{\mathbf{B}} & U_{\mathbf{uB}} + U_{\mathbf{uB}}^H \\ L_{\rho\mathbf{u}} & L_{T\mathbf{u}} & L_{\mathbf{Bu}} & D_{\mathbf{u}} \end{bmatrix} \begin{pmatrix} \delta\rho \\ \delta T \\ \delta\mathbf{B} \\ \delta\mathbf{u} \end{pmatrix},$$

but with a new contribution $U_{\mathbf{uB}}^H$, coming from the modified Ohm's law in equation 11. Consequently, the modified Schur complement P_* contains a new term $L_{\mathbf{Bu}}U_{\mathbf{uB}}^H \approx \theta d_i \mathbf{B}_0 \times (\nabla \times \nabla \times \delta\mathbf{u})$, which is the whistler wave propagator in the momentum equation in Hall MHD. We have analytical and numerical proof that such term can be smoothed effectively with damped Jacobi or Gauss-Seidel techniques and therefore can be treated effectively with multigrid methods.

5. Numerical results

In what follows, we present results for several test problems aimed at demonstrating some of the advertised properties of the solver. In particular, we employ a 2D magnetic Kelvin-Helmholtz instability (KHI) [14] to demonstrate its performance in the presence of large flows, a 2D tearing mode problem [4] to demonstrate its effectiveness in Hall MHD, and a 3D island coalescence problem [4] to demonstrate its excellent parallel scalability properties. As in [4], the explicit solver employed as a reference to calculate implicit CPU speedup is a second-order predictor-corrector method, which requires two function evaluations per time step. The explicit time step Δt_{CFL} is calculated here as $\pi/(2\omega_{max})$, with ω_{max} the maximum local normal-mode frequency.

We employ multigrid V-cycles to approximate M^{-1} and P_*^{-1} where required. Restriction and prolongation employ second-order splines (of local processor data only). As a smoother, we employ a few passes of weighed Jacobi (p. 10 in [19]; p. 118 in [20]), with weight $\omega_{JB} = 0.7$, for both the restriction and the prolongation steps. In MG terminology, such V-cycle is identified as V(m,n), where the two integers indicate restriction and prolongation smoothing steps, respectively.

5.1. Finite flow: Kelvin-Helmholtz instability

The KHI equilibrium is defined [14] by uniform density and pressure ($\rho_0 = T_0 = 1$), uniform magnetic field in the ignorable direction ($B_{x0} = B_{y0} = 0$, $B_{z0} = 1$), and a sheared velocity profile $v_{y0} = V_0 \tanh[(x - 0.5)/\lambda]$, with $V_0 = 0.5$ and $\lambda = 0.2$. The computation is performed in a 2D rectangular domain with $x \in [0, 1]$ and $y \in [0, 2.5]$, and with $D = 0$ and $\eta = \nu = 10^{-3}$, $\gamma = 5/3$. The boundary conditions are periodic in y and perfect conductor ($B_x = 0$), no stress ($\partial_x v_y = 0$), and impenetrable wall ($v_x = 0$) in x. Homogeneous Neumann boundary conditions are imposed in x for both ρ and T.

The KHI is an ideal instability and therefore features $\gamma \sim \mathcal{O}(1)$. This limits the attainable implicit time step in practice because, for a second-order accurate discretization, accuracy requires $\gamma\Delta t < 1/2$ [18]. However, implicit gains can still be obtained for sufficiently fine grids. Table 1 gives results of a grid convergence study with a fixed implicit time step $\Delta t = 0.1$. For this test, we use one V(3,3) MG cycle. Performance results are averaged over ten time steps. The table shows good performance under grid refinement and moderate gains for sufficiently fine grids, despite the fact that flows are comparable to the Alfvén speed.

Table 1. Serial grid convergence study for the KHI problem with $\Delta t = 0.1$. Results are averaged over ten time steps. The table shows the average number of Newton and FGMRES iterations per time step, the total CPU time, the CPU speedup vs. and explicit code (CPU_{exp}/CPU, with CPU_{exp} the CPU of the explicit approach), and the implicit time step size relative to the explicit CFL limit ($\Delta t/\Delta t_{CFL}$).

N	Newton/Δt	FGMRES/Δt	CPU (s)	CPU_{exp}/CPU	$\Delta t/\Delta t_{CFL}$
32x32	4.0	4.7	29	0.5	17
64x64	4.0	5.0	102	1.2	34
128x128	4.2	5.9	434	2.3	74
256x256	4.9	9.9	2652	3.3	156

Table 2. Serial grid convergence study for Hall MHD using the tearing mode problem with $\Delta t = 1.0$. Results are taken for a single time step. Other quantities reported are defined as in table 1.

N	Newton/Δt	FGMRES/Δt	CPU (s)	CPU_{exp}/CPU	$\Delta t/\Delta t_{CFL}$
32x32	5	32	21.6	0.9	110
64x64	5	9	25.2	9.3	384
128x128	5	8	91.5	37.9	1436
256x256	4	10	467.0	117.0	5660

5.2. Cold-ion Hall MHD: Tearing mode

The tearing mode equilibrium is defined by a uniform density and temperature, no flow, and a force-free magnetic Harris-sheet configuration given by [14, 4]:

$$B_{x0} = 0,\ B_{y0}(x) = \tanh[(x - 0.5)/\lambda],\ B_{z0} = \sqrt{1 - B_{y0}^2(x)},$$

with $\lambda = 0.2$. The domain is rectangular of dimensions with $x \in [0, 1]$ and $y \in [0, 4]$. The computation is performed with $\eta = 10^{-3}$, $\nu = 10^{-3}$, $D = 0$, and $\gamma = 5/3$. The boundary conditions are the same as for the KHI problem. We perturb the equilibrium with a sinusoidal perturbation in ρ, $\delta\rho = 10^{-3}\sin(\frac{2\pi y}{L_y})\cos(\frac{2\pi x}{L_x})$.

We demonstrate the performance of the Hall MHD preconditioner (using the small-flow approximation of [4]) with the tearing mode problem, using $d_i = 0.05$. Grid scaling results are presented in table 2. We employ classical MG V(3,3) cycles (with under-damped Jacobi as a smoother, and a linear tolerance of 10^{-2}) to approximately invert M and P_*. We have fixed the time step to $\Delta t = 1.0$, which spans anywhere from 110 explicit CFL time steps for the coarsest grid to 5660 for the finest. The table reports data for a single implicit time step. (Taking multiple time steps requires including a higher-order dissipation term in the Ohm's law, equation (11), which is left for future work.) Notice from the table that the CPU speedup increases sharply as the grid is refined, reaching a factor of 117 for the finest grid considered.

5.3. Parallel performance: 3D resistive island coalescence

The island coalescence equilibrium is a modification of the Harris sheet equilibrium of the previous section. It features a uniform density, no flow, and magnetic field and pressure given by [4]:

$$\Psi(x, y) = -\lambda \log\left[\cosh\left(\frac{x}{\lambda}\right) + \epsilon\cos\left(\frac{y}{\lambda}\right)\right],$$

$$p(x, y) = p_0 + \frac{(1-\epsilon^2)}{2}\left[\cosh\left(\frac{x}{\lambda}\right) + \epsilon\cos\left(\frac{y}{\lambda}\right)\right]^{-2},$$

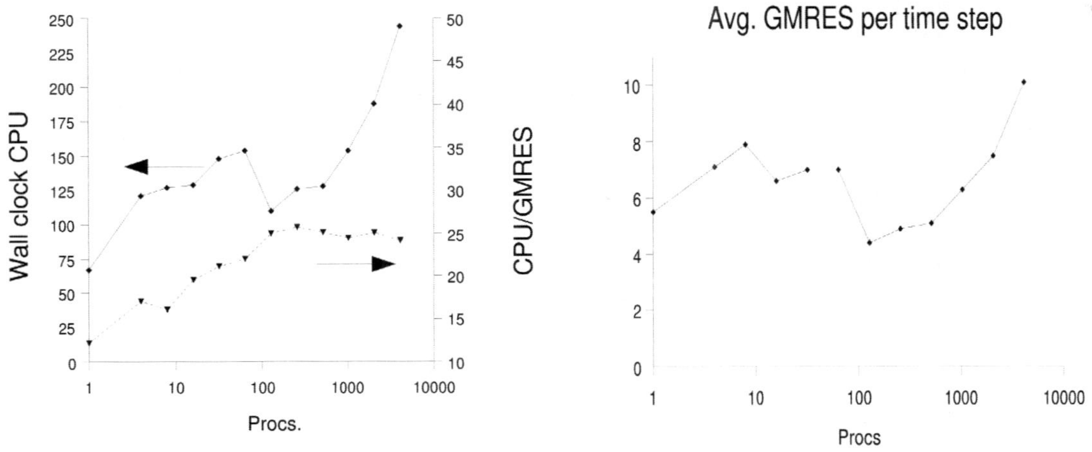

Figure 1. Parallel performance (weak scaling) of the implicit 3D resistive MHD algorithm using the 3D island coalescence problem. For this test, we have considered up to 4096 processors, and 134 million unknowns. Data obtained with a dual-core Cray XT4 (Franklin) at NERSC.

where Ψ is the poloidal flux (from which the magnetic field can be found as $\mathbf{B} = \mathbf{z} \times \nabla\Psi$), and p is the pressure. The domain is square, of dimensions $x, y, z \in [-1, 1]$. The computation is performed with $d_i = 0.0$, $\eta = 10^{-3}$, $\nu = 10^{-3}$, $D = 0$, $\gamma = 5/3$, $p_0 = 1$, $\lambda = L_y/4\pi$, and $\epsilon = 0.2$. The boundary conditions are the same as for the KHI problem. We perturb the equilibrium with a solenoidal magnetic field perturbation of the form $\delta\mathbf{B} = \mathbf{z} \times \nabla\delta\Psi$, with $\delta\Psi = 10^{-3}\sin(\frac{\pi x}{L_x})\cos(\frac{2\pi y}{L_y})\cos(\frac{2\pi z}{L_z})$.

Parallel performance results of the algorithm are given in figure 1. The left figure shows traces of the wall-clock CPU and the CPU per FGMRES iteration. The right figure shows the averaged number of FGMRES per time step (averaged over 10 fixed implicit time steps $\Delta t = 0.1$). These parallel performance data have been obtained in a weak sense, using a base resolution of 16^3 grid points per processor. Such small base resolution is unfavorable from a parallel scaling standpoint because the surface-to-volume ratio (which is a measure of the percentage of points of the local domain that need to be communicated) is fairly large ($\sim$ 37% for $N_p = 16^3$). Nevertheless, the algorithm performs well in parallel up to thousands of processors. From figure 1, left, it is clear that the scaling of the wall-clock CPU is optimal up to 1000 processors, showing some degradation after that. However, such degradation is not due to parallel bottlenecks in the algorithm: the CPU per FGMRES iteration does not increase, and communication costs remain small for all processor numbers considered. Rather, it is a loss of algorithmic performance, manifested in the growth of FGMRES iterations past the 1000-processor level (figure 1, right), which is the root cause of the increase. Such growth is attributed to the lack of a coarse-grid solve (our parallel MG implementation employs a fixed number of levels, determined by the size of the local processor problem), and to the fact that the problem is getting harder as the number of processors increases (N scales linearly with the number of processors n_p, $N \propto n_p$, and the explicit CFL scales as $\Delta t_{CFL} \sim N^{-1/3}$; therefore, $\Delta t/\Delta t_{CFL} \sim N^{1/3} \sim n_p^{1/3}$).

6. Conclusions

In this paper, we have described a fully implicit, scalable approach for 3D compressible Hall MHD. The approach is based on Jacobian-free Newton-Krylov methods. Key to the approach is the availability of suitable preconditioners. In [4], a suitable preconditioner was proposed for 3D compressible resistive MHD in the small-flow limit. Here, we have extended this development in two ways: we have generalized the small-flow preconditioner strategy to deal with arbitrary plasma flows, and we have extended the formulation to deal with Hall MHD in the cold-ion limit. The arbitrary-flow generalization is important to ensure robustness of the solver (as finite flows may develop in nonlinear regimes in simulations of interest, both locally and globally) and to enable direct-to-steady-state solution procedures. The cold-ion Hall MHD development, while still in a proof-of-principle stage, demonstrates that the approach holds promise to deal with the very challenging extended MHD formulation, which is of interest to many SciDAC applications. Moreover, we have presented parallel scalability results for a 3D resistive MHD problem using up to 4096 processors and 134 million unknowns, which demonstrate the potential of the approach to exploit massively parallel environments.

Acknowledgments

We acknowledge useful discussions with D. A. Knoll, J. N. Shadid, and R. Pawlowski. We thank the PETSc team for technical support. The work was funded by the Department of Energy at Los Alamos National Laboratory under contracts No. DE-AC52-06NA25396 and the DOE Office of ASCR program in Applied Mathematical Sciences. This research used resources of the National Energy Research Scientific Computing Center, which is supported by the Office of Science of the U.S. Department of Energy under Contract No. DE-AC02-05CH11231.

References

[1] Colella P, Hou T and Petzold L 2004 Report of the first multiscale mathematics workshop: First steps toward a roadmap http://www.sc.doe.gov/ascr/Research/AM/MultiscaleMathWorkshop1.pdf
[2] Dolbow J, Khaleel M A and Mitchell J 2004 Multiscale mathematics initiative: A roadmap Tech. Rep. PNNL-14966 Pacific Northwest National Laboratory http://www.sc.doe.gov/ascr/Research/AM/MultiscaleMathWorkshop3.pdf
[3] Oden J T, Belytschko T, Fish J, Hughes T J, Johnson C, Keyes D, Laub A, Petzold L, Srolovitz D and Yip S 2006 Simulation-based engineering science: Revolutionizing engineering science through simulation Tech. rep. National Science Foundation
[4] Chacón L 2008 *Phys. Plasmas* **15** 056103
[5] Saad Y 1996 *Iterative Methods for Sparse Linear Systems* (Boston: PWS Publishing Company)
[6] Chan T F and Jackson K R 1984 *SIAM J. Sci. Stat. Comput.* **5** 533–542
[7] Brown P N and Saad Y 1990 *SIAM J. Sci. Stat. Comput.* **11** 450–481
[8] Saad Y and Schultz M 1986 *SIAM J. Sci. Stat. Comput.* **7** 856–869
[9] Saad Y 1992 *SIAM J. Sci. Comput.* **14** 461–469
[10] Balay S, Buschelman K, Eijkhout V, Gropp W D, Kaushik D, Knepley M G, McInnes L C, Smith B F and Zhang H 2004 PETSc users manual Tech. Rep. ANL-95/11 - Revision 2.1.5 Argonne National Laboratory
[11] Balay S, Eijkhout V, Gropp W D, McInnes L C and Smith B F 1997 *Modern Software Tools in Scientific Computing* ed Arge E, Bruaset A M and Langtangen H P (Birkhäuser Press) pp 163–202
[12] Balay S, Buschelman K, Gropp W D, Kaushik D, Knepley M G, McInnes L C, Smith B F and Zhang H 2001 PETSc Web page http://www.mcs.anl.gov/petsc
[13] Chacón L and Knoll D A 2003 *J. Comput. Phys.* **188** 573–592
[14] Chacón L 2004 *Comput. Phys. Comm.* **163** 143–171
[15] Kelley C T 1995 *Iterative methods for linear and nonlinear equations* (Philadelphia: SIAM)
[16] Dembo R, Eisenstat S and Steihaug R 1982 *J. Numer. Anal.* **19** 400
[17] Knoll D A and Keyes D E 2004 *J. Comput. Phys.* **193** 357 – 97
[18] Chacón L, Knoll D A and Finn J M 2002 *J. Comput. Phys.* **178** 15–36
[19] Briggs W L 1987 *A Multigrid Tutorial* (Philadelphia, PA: SIAM)
[20] Wesseling P 1992 *An Introduction to Multigrid Methods* (Chichester: John Wiley & Sons)
[21] Elman H C, Howle V E, Shadid J N and Tuminaro R S 2003 *J. Comput. Phys.* **187** 504 – 23 ISSN 0021-9991

Toward a first-principles integrated simulation of tokamak edge plasmas

C S Chang[1], S Klasky[2], J Cummings[3], R. Samtaney[4], A Shoshani[5], L Sugiyama[6], D Keyes[7], S Ku[1], G Park[1], S Parker[8], N Podhorszki[2], H. Strauss[1], H Abbasi[9], M Adams[7], R Barreto[2], G Bateman[10], K Bennett[11], Y Chen[8], E D'Azevedo[2], C Docan[11], S Ethier[4], E Feibush[4], L Greengard[1], T Hahm[4], F Hinton[12], C Jin[2], A. Khan[13], A Kritz[10], P Krsti[2], T Lao[7], W Lee[4], Z Lin[14], J Lofstead[9], P Mouallem[15], M Nagappan[15], A Pankin[10], M Parashar[11], M Pindzola[16], C Reinhold[2], D Schultz[2], K Schwan[9], D. Silver[11], A Sim[5], D Stotler[4], M Vouk[15], M Wolf[9], H Weitzner[1], P Worley[2], Y Xiao[14], E Yoon[4], D Zorin[1]

[1]New York University, New York, NY 10012, USA
[2]Oak Ridge National Laboratory, Oak Ridge, TN 37381, USA
[3]California Institute of Technology, Pasadena, CA 91125, USA
[4]Princeton Plasma Physics Laboratory, Princeton, NJ 08543, USA
[5]Lawrence Berkeley National Laboratory, Berkeley, CA 94720, USA
[6]Massachusetts Institute of Technology, Cambridge, MA 02139, USA
[7]Columbia University, New York NY 10027, USA
[8]University of Colorado at Boulder, Boulder, CO 80309, USA
[9]Georgia Institute of Technology, Atlanta, GA 30332, USA
[10]Lehigh University, Bethlehem, PA 18015, USA
[11]Rutgers University, New Brunswick, NJ 08901, USA
[12]Hinton Associates
[13]The University of Utah, Salt Lake City, UT 84112, USA
[14]University of California at Irvine, Irvine, CA 92697, USA
[15]North Carolina State University, Raleigh, NC 27695, USA
[16]Auburn University, Auburn, AL 36849, USA

E-mail: cschang@cims.nyu.edu, klasky@ornl.gov

Abstract. Performance of the ITER is anticipated to be highly sensitive to the edge plasma condition. The edge pedestal in ITER needs to be predicted from an integrated simulation of the necessary first-principles, multi-scale physics codes. The mission of the SciDAC Fusion Simulation Project (FSP) Prototype Center for Plasma Edge Simulation (CPES) is to deliver such a code integration framework by (1) building new kinetic codes XGC0 and XGC1, which can simulate the edge pedestal buildup; (2) using and improving the existing MHD codes ELITE, M3D-OMP, M3D-MPP and NIMROD, for study of large-scale edge instabilities called Edge Localized Modes (ELMs); and (3) integrating the codes into a framework using cutting-edge computer science technology. Collaborative effort among physics, computer science, and applied mathematics within CPES has created the first working version of the End-to-end Framework for Fusion Integrated Simulation (EFFIS), which can be used to study the pedestal-ELM cycles.

1. Introduction

The mission of the SciDAC Fusion Simulation Project (FSP) Prototype Center for Plasma Edge Simulation (CPES) is to develop of an integrated predictive plasma edge simulation package that will simulate the edge pedestal buildup and subsequent large-scale edge localize mode (ELM) crash for existing magnetic fusion facilities as well as for next-generation burning plasma experiments such as ITER [1]. Timely progress on this formidable scientific challenge demands a well-coordinated effort involving experts in plasma physics, computer science, and applied mathematics. The performance of ITER depends sensitively on the ability to achieve H-mode operation [2] with a high-edge pressure pedestal while suffering minimal damage from ELM instabilities [3]. Many nonequilibrium physical processes on different spatiotemporal scales in the edge region demand large-scale integrated simulations based on as much first-principles physics as possible using high-performance computers.

The local equilibrium of the edge plasma is nonthermal, for which kinetic models are required. For the first time, a special full-f distribution function (full-f) kinetic transport code suite, consisting of the codes XGC0 and XGC1, has been developed for real tokamak edge geometry. The existence of ion-temperature-gradient (ITG) turbulence in the real geometry pedestal region was demonstrated using the XGC1 code: a particle-in-cell (PIC) approach on a massively parallel, high-performance computing platform (see figure 1). A production run normally uses 10,000–20,000 Cray XT4 processing cores for 20 hours. The resulting ion thermal transport rates are at experimentally estimated level [4]. An effective team effort among the physics, applied mathematics and computer science has been critical for this rapid progress. The XGC1 code benefits from the state-of-the-art applied mathematics tools of the SciDAC Center TOPS and Courant Mathematics and Computing Laboratory (CMCL) and high-performance software engineering from the SciDAC Performance Engineering Research Institute (PERI). XGC1 will be further enhanced to investigate electromagnetic turbulence. XGC0 is a simplified version of XGC1 for carrying out experimental timescale simulation of the kinetic pedestal buildup, which will utilize the turbulent transport information from XGC1. The XGC0 code, using a coarse-grained PIC approach with only a radial electric field solution, shows that the pedestal is routinely built up from neutral ionization resulting from wall-recycling of neutrals and from heat/momentum/particle flux from the core.

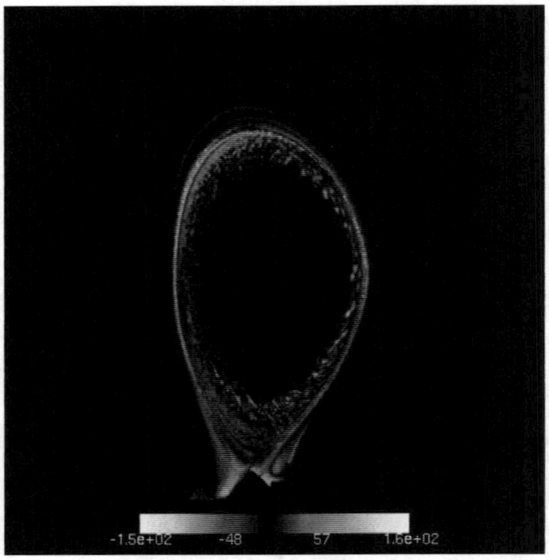

Figure 1. ITG turbulence in a real geometry DIII-D edge with a mild L-mode type pedestal. The temperature gradient is strong enough for ITG instability at the pedestal top but not in the pedestal slope. Still, the ITG grows at both locations.

Our new End-to-end Framework for Fusion Integrated Simulation (EFFIS) has successfully been used to automate integrated simulation of XGC0 kinetic pedestal buildup, Elite [5] stability-boundary check, M3D [6] edge localized mode crash, and back to XGC0 pedestal build up, for multiple ELM cycles. EFFIS uses highly advanced inter-operating computer science tools that include workflow management, fast presentation of image data on a web-based dashboard, speedup of I/O operations, collection and management of provenance metadata, and wide-area data transfer. EFFIS facilitates efficient integration of different fusion codes running on multiple and nonlocal platforms, provides users with real-time dynamic tracking of simulation progress through graphs and images on a dashboard, and has already been used on tens of thousands of processors in secure environments. EFFIS demonstrates the initial aspects of the FSP within the CPES project. Code performance optimization is achieved in collaboration with SciDAC Institute PERI. The performance information is displayed on the dashboard.

2. Pedestal buildup and ELM crash

Even though the full-f 5D gyrokinetic code XGC1 includes equilibrium physics [7], it is prohibitively expensive to use XGC1 to study experimental timescale equilibrium evolution (i.e., pedestal development) together with turbulence physics on the present-day high-performance computers. Use of XGC1 will thus be limited to snapshot studies of critical turbulent transport until faster-than-petascale computers are available. Accumulation of errors in a gyrokinetic code is another issue to be resolved in an experimental timescale simulation. Development of the XGC0 code has been accelerated to fulfill the demand of experimental timescale kinetic pedestal buildup [8]. The XGC0 code, which is a simplified version of the XGC1 code, computes the kinetic equilibrium physics (based on neoclassical physics including orbit loss and wall interaction) in real geometry by solving only the radial electric field E_r in one-dimensional radial cells, instead of the 3D turbulence field in three-dimensional cells. This leads to a reduction in particle number by a factor >1,000. The plasma sheath/presheath solution in the scrape-off region is simplified with the logical sheath algorithm [9]. Since XGC0 is not a turbulence code, it does not perform the gyrokinetic four-point averaging operation, leading to additional significant savings in computing time. As a result, XGC0 is ~1,000 times faster than XGC1. The Lagrangian equations of motion used in XGC0 are identical to those in XGC1.

XGC0 simulates not only the electrons and ions but also the Monte Carlo neutral particles with a wall recycling model. An electron subcycling technique is used to simulate the electron dynamics efficiently. Non-axisymmetric magnetic perturbations are another input, taking advantage of the 3D nature of the particle dynamics. XGC0 incorporates magnetic and limiter geometry information from the g-eqdsk files, as in XGC1. XGC0 accepts the heat, torque, and particle fluxes from the core. It uses a radial profile of anomalous transport coefficients and converts them into radial random and convective walks superimposed on the Lagrangian particle motions. The anomalous transport coefficients can be prescribed analytically, based on experimental validation, or they can be based on simulation results from turbulent transport codes such as XGC1. At the present time, XGC0 is used for pedestal buildup only without any ELM activities. The ELM crash is completely described by the nonlinear resistive MHD code M3D-MPP (cross-verified against NIMROD). Systematic verification of the XGC0 code has been carried out to ensure that the code is accurate to the desired level. XGC0 is a mature code, with validation against experimental data underway.

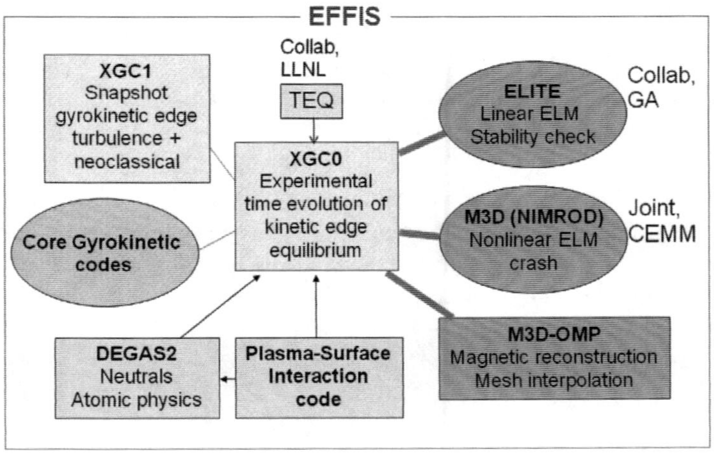

Figure 2. Schematic diagram of code integration in CPES. Thick red lines represent nonlocal data coupling, thin arrow lines show subroutine coupling, and thin red lines indicate manual transport coupling.

The kinetic-dominated pedestal buildup followed by the nonlinear evolution of the MHD-dominated ELM crash is routinely simulated within EFFIS. This is accomplished by data coupling four separate codes: XGC0 for kinetic simulation of the pedestal buildup, M3D-OMP [6] for equilibrium reconstruction and mesh interpolation, ELITE to determine the ideal MHD linear stability of the evolving profiles, and M3D-MPP (without the two-fluid effects) to simulate the nonlinear resistive MHD evolution during an ELM crash when the profiles are deemed unstable by ELITE. (We also use NIMROD for cross-verification.) The density, temperature, and current profiles along with the equilibrium configuration files in g-eqdsk format are transferred among the codes, which employ very different spatial meshes and resolution. The entire process in EFFIS is automated using the Kepler workflow framework, the ADIOS data I/O technology, and the Dashboard run-monitoring, visualization and run-control (see Section 3). Additional necessary integrations are achieved by including them into XGC0 as subroutines (see figure 2).

Below an example of a specific integrated simulation is described. The kinetic equilibrium simulation code XGC0 was run on 128 processing cores on the Cray XT4 at ORNL, the M3D-OMP equilibrium reconstruction and ELITE stability checks were done on one processor of a Linux Opteron cluster called *ewok* at ORNL, and the nonlinear resistive MHD code M3D-MPP was run on 128 cores of *ewok*. During the kinetic simulation, it is assumed that the edge plasma is already in H-mode, where the neoclassical and the neutral atomic physics are important. A small prescribed level of anomalous transport is superimposed on the Lagrangian particle motion as a radial random walk and convection steps. The magnitude is determined by comparison with experimental pedestal shapes. (Turbulent transport information from XGC1 will be integrated into the XGC0's random walk and convective steps in the future.) The edge pedestal builds up in XGC0 from neutral ionization; that is, the density and pressure gradients at the edge get steeper, and the electrical current density becomes higher from the kinetic bootstrap effect (see figure 3 for the pedestal buildup during an example simulation). The pressure and current density profiles are examined periodically by the ELITE code through the end-to-end data coupling network to determine if they are unstable to ideal MHD peeling-ballooning modes. Before the linear stability check, the magnetic equilibrium is automatically reconstructed from the profiles with a free-boundary Grad-Shafranov solver, which is part of the M3D-OMP code. If the profiles are ideal MHD unstable, then the nonlinear resistive MHD code M3D-MPP takes over the simulation of the nonlinear development of the ELMs and the subsequent relaxation. Figure 4 shows a 2D plot of the density perturbation that emerged from the example integration, illustrating the fast expulsion of plasma into the surrounding region by a ``ballooning" type instability. The instability saturates as the density gradient in the plasma edge is reduced below a critical level and the plasma begins to relax back towards the original configuration, but with a broader edge pressure pedestal. Once the flux

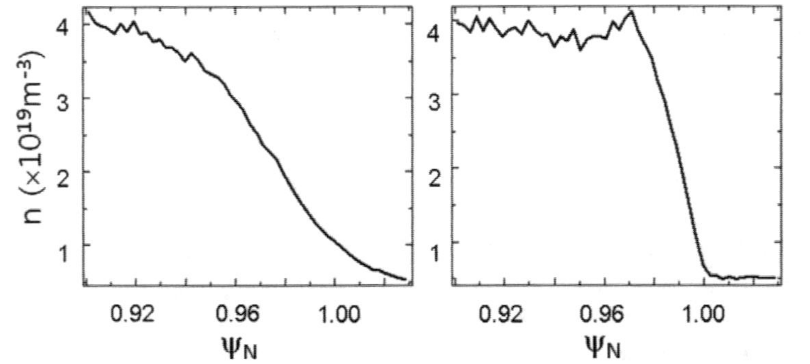

Figure 3. Density profiles at t=0 (left), and t=42 (right) showing the pedestal edge buildup. The times are normalized by the ion transit time.

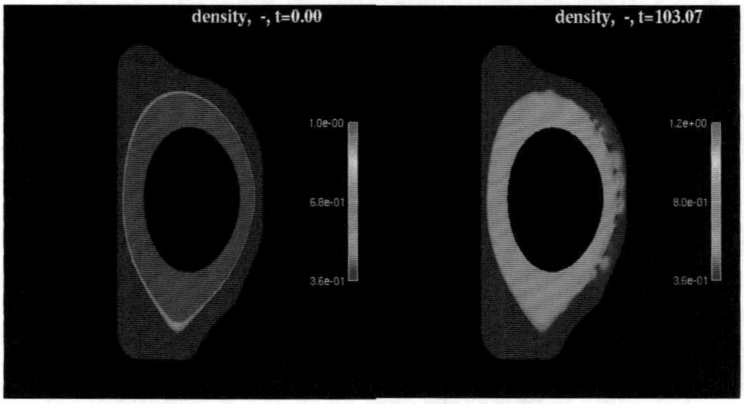

Figure 4. Density images at t=0 (left) and t=103 (right) showing the nonlinear development of the ELMs in the example integrated simulation. The times are normalized by the Alfvén time, and t=0 is at the beginning of the ELM crash.

surfaces have adequately healed, a new equilibrium is computed along with profiles by toroidal averaging and poloidal smoothing of the full 3D fields. The end of the ELM crash is checked using an input parameter specifying a threshold level of density fluctuation. The relaxed plasma profiles and magnetic equilibrium information are then used to initiate a new pedestal buildup in the XGC0 code and automatically repeat the ELM cycle.

3. Computer science in End-to-end Framework for Fusion Integrated Simulation (EFFIS)

In this section several computer science technologies that have been integrated into the single framework EFFIS are described, which facilitates real-time dynamic monitoring of simulations while they are running, code integration of simulations, and dynamic generation of images that help visualize and track the simulations as well as various data movement activities. EFFIS has been successfully performing automated integrated simulation of XGC0 kinetic pedestal buildup, ELITE stability-boundary check, M3D edge localized mode crash, and back to XGC0 pedestal build up, for multiple ELM cycles. EFFIS uses highly advanced interoperating computer science tools that include workflow management, fast presentation of image data on a web-based dashboard, speedup of I/O operations, collection and management of provenance metadata, and wide-area data transfer. EFFIS facilitates efficient integration of different fusion codes running on multiple and nonlocal platforms, and has already been used on tens of thousands of processors in secure environments. CPES researchers recognized that they needed a framework as powerful as EFFIS, since they commonly run simulations on thousands of processors and I/O becomes a bottleneck. Additionally, because the simulations run for a long periods of time, there is a need for easy way to monitor the simulations, which includes invoking analysis and visualization during the monitoring stage.

The EFFIS framework is organized in terms of "foundation technologies" and "enabling technologies (see figure 5). The three foundation technologies are stacked into three layers: the Adaptable I/O System (ADIOS), which is designed to achieve high I/O bandwidth, is at the foundation, the Kepler workflow system is on top of ADIOS, and the Dashboard is on top of the workflow system, indicating that workflows use the ADIOS technology and that the Dashboard displays the results generated by workflows. The enabling technologies interact with these layers to collect provenance and metadata into database repositories and to provide efficient data coupling, advanced visualization, and wide-area data movement. We focus the rest of this section on our Adaptable I/O system, our workflow system, and our Dashboard.

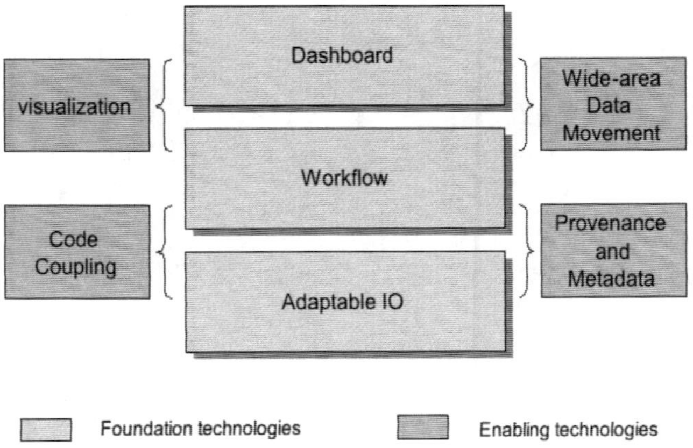

Figure 5. EFFIS foundation and enabling technologies.

One of the main goals of the computer science work in CPES is to facilitate running workflows that will perform dynamic monitoring of simulations and the dynamic analysis of the time steps in order to track the simulations. This required the use and development of special components (called "actors") of the Kepler workflow system, through collaboration with the SDM (Scientific Data Management) center [10]. Such workflows incorporate many other tasks, such as generation of images and movies, data transformation on-the-fly as the data is generated, and combining of multiple small files into a larger file for archiving. The workflows developed not only support "monitoring workflows," but also support "integration workflows" across machines.

A diagram of the Kepler coupling workflow is shown in figure 6. In this figure, the *M3D-OMP* pipeline (top of figure) realizes the loose cyclic coupling between XGC0 and M3D-OMP. From XGC0, a small data file *m3d.in* is transferred to the processing site where M3D-OMP is executed as a batch job. The new magnetic equilibrium is then transferred back to the simulation's directory. The result also triggers the execution of the *ELITE* pipeline, which transfers the p-eqdsk data (plasma density profile) from XGC0 and runs ELITE (also via the job queue). A wrapper script around ELITE performs all the tasks: prepare and submit job and wait for results, check all output files for errors or problems, generate output plots using IDL, and determine whether the XGC0 simulation has become MHD unstable. (All of these subcomponents could also be realized as sub-workflows in Kepler if more detailed control were needed inside the workflow.) The result of the script is a simple "control file" containing the string *stable* or *unstable*, which is checked in the last actor of this sub-workflow. If XGC0 is determined to be MHD unstable, the M3D-MPP code is launched, while the workflow stops XGC0 completely and releases the CPUs until a new start after the M3D-MPP run is finished.

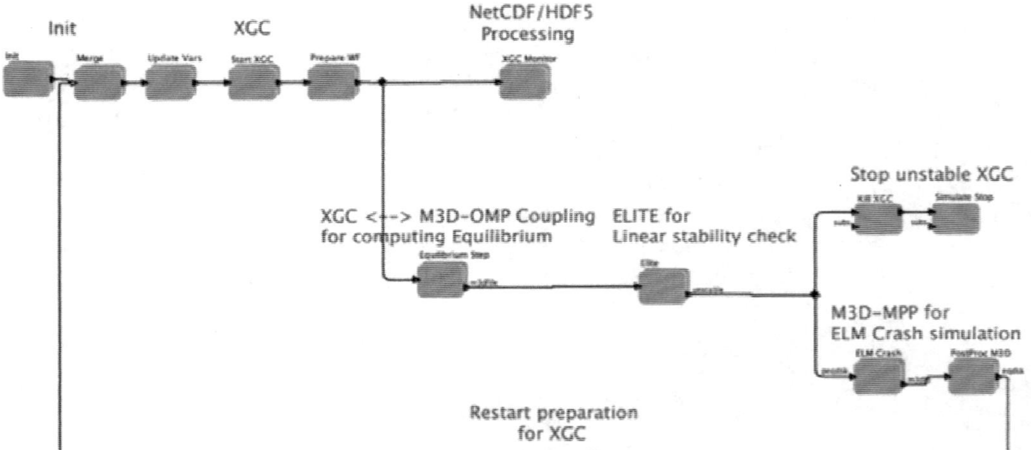

Figure 6. Coupling pipelines performing the M3D-OMP, ELITE, and M3D-MPP steps.

The *M3D-MPP* pipeline is fairly complex and can be run as a stand-alone workflow to execute M3D-MPP and monitor it. This workflow generates images from the HDF5 output of M3D-MPP using AVS/Express, so the structure is similar to the *HDF5 Processing* pipeline that we use in our monitoring workflow. The AVS/Express job is submitted directly from the workflow, not from a script. Thus, the workflow is able to wait for the AVS/Express job to start, perform other tasks while the job is running, and stop the job at the end of the processing. This capability is provided by a set of job-related actors in Kepler that support several job managers (PBS, LoadLeveler, SGE, Condor, and "process fork") and use SSH to connect to remote computing sites. Finally, the coupling workflow completes the cycle by detecting the end of the M3D-MPP simulation, processing the new equilibrium data, and returning it to XGC0 for the next pedestal buildup simulation.

Simulation monitoring and effective code integration are critical tasks in EFFIS. Because of the scarcity and costliness of supercomputing resources, it is very important that any large-scale simulations be well monitored. Such monitoring may include data sorting, filtering, transformation and/or reduction, as well as graphical representations or visualizations and simple remote access to these outputs. Most of these tasks are data-centric and can be automated and thus are amenable to the use of scientific workflows. The main purposes of simulation monitoring are tracking the progress of a run, detecting possible errors in the simulation setup or behavior, and recording selected code diagnostics of interest in real time. Such monitoring can prevent the waste of precious resources and instead focus simulations on key physics insights as identified by the scientist in the setup phase and even while a simulation is in progress. For our workflows, we choose the Kepler scientific workflow management system [11] largely due to its support of advanced computational models such as pipeline-parallel execution. We prefer Kepler over other scientific workflow systems because the majority of them are grid-oriented (where each component must be realized as a job) or web-based (where a component is a web service call). The resources at hand are not part of any grid or web-based software infrastructure. Even if they were, in a grid infrastructure there is no access to a job's output on the fly unless all output is first put onto external storage elements. The quick access to and processing of data and the fast turnaround in the execution of operations are essential for efficient kinetic-MHD code integration and are not achievable using grid-based job submission. Another advantage over many other systems is that Kepler development is community-driven. Thus, the many different projects and teams such as CPES and the SciDAC SDM Center that use Kepler can collaborate to share development tasks and system improvements.

We developed ADIOS [12] in conjunction with Sandia, for the improvement of I/O speed for dumping data from the compute engines. The original idea was to provide fast, scalable and portable I/O for the large data producing codes used in this project. The approach taken is to remove the load from the application codes by permitting generic I/O calls which invoke libraries that use the desired properties of the I/O (type of data, parallelism, etc.) captured in an XML file. This work, the Adaptable I/O System (ADIOS), provides great improvements of the I/O speed, as well as elegant separation of the I/O details from the application code. As a result, the application code does not have to be modified if changes to the I/O properties are desired; the changes are recorded only in the XML file that the I/O libraries use. Since one of the goals under the CPES work was to facilitate code integration, ADIOS also allows us to componentize this layer so that we can move data quickly from different components inside of the CPES framework, and to communicate directly with our integration and monitoring workflows. This allows a CPES scientist to abstract the data movement from CPES components, placing this burden on our EFFIS infrastructure. ADIOS has recently been used in large XGC1 simulations, which have produced over 50 TB in 24 hours on the Cray XT at ORNL on 28,672 cores, writing the data at over 24 GB/sec.

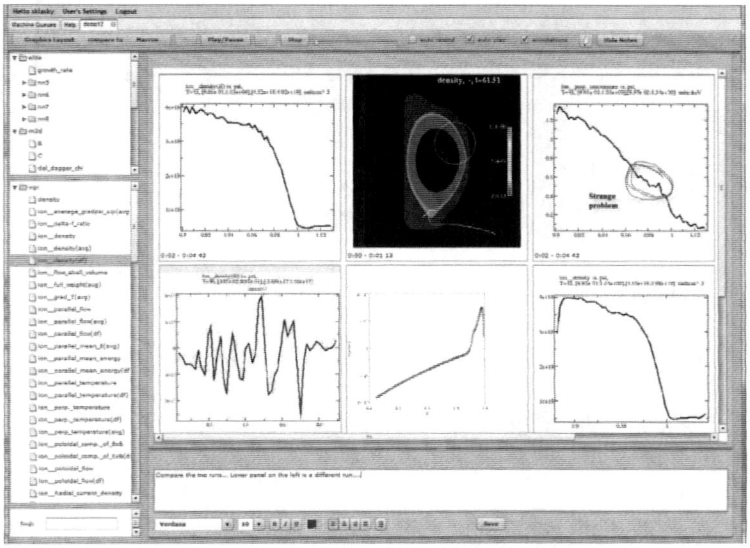

Figure 7. Simulation monitoring via Dashboard.

The eSimMon (electronic Simulation Monitoring [13]) dashboard is a web-based interface that presents several views of simulation resources and results, including an overview of activity on the computers of interest at DOE, a means to monitor data provenance and analytics related to either current or previous runs (or *shots*), and a display of other provenance information and status tools. The purpose of eSimMon is to provide the user a simple "one-stop shopping" access to view the status of simulations (see figure 7), launch new simulations, and analyze or revisit existing runs. It is a lightweight front end for comprehensive end-to-end data and workflow management that provides users with access to large simulation datasets, as well as provenance information and performance records for simulations and their associated workflows. Users can examine the evolution and outputs of their runs or drill down into run-time processes, workflows and system information. More recently, we have added wide-area file movement to the physicist's site through requests made on the Dashboard, using a Storage Resource Manager (SRM) technology used in the SDM center called SRM-lite.

4. Conclusion and discussion

End-to-end Framework for Fusion Integrated Simulation (EFFIS) has been built to perform the pedestal-ELM cycle, aiming to bring together first-principles physics codes and reduced transport model codes running on multiple nonlocal platforms, as well as the physics data information saved on disk. To date, the XGC0 kinetic code for pedestal evolution, M3D-OMP for mesh interpolation and magnetic reconstruction, ELITE for linear ELM stability check, and M3D-MPP for nonlinear resistive ELM crash to relaxation have been integrated in EFFIS. In the future, the XGC1 gyrokinetic turbulence code will be integrated with XGC0 to provide turbulent transport information. M3D-MPP will be improved to perform a more robust two-fluid ELM crash all the way to relaxation. In addition, EFFIS will be used for another urgent ITER problem, namely, prediction of the resistive MHD response to an external resonant magnetic perturbation (RMP), kinetic pedestal behavior, and ELM control.

The physics integrated simulations carried out in EFFIS will be systematically verified and validated. The simulations will be used to analyze edge data in existing tokamaks and to predict edge performance, including the pedestal width and height, and the divertor heat load in ITER. A core turbulence code will be integrated into EFFIS to allow an edge integrated simulation of core turbulent transport. EFFIS will be enhanced to collect and provide system and data provenance information through the Dashboard, as well as additional analytical capabilities. The ADIOS system will be enhanced to include experimental data for comparison with simulation data through the Dashboard.

Acknowledgment

This work is jointly supported by US DOE Office of Fusion Energy Science and Office of Advanced Scientific Computing Research.

References

[1] Aymar R, *et al.* 2001 ITER Joint Central Team and ITER Home Teams *Nucl. Fusion* **41:**1301
[2] Groebner R J, Burrell, K H and Seraydarian R P 1997 *Phys. Rev. Lett.* **64**:3015; Greenwald M, *et al.* 1997 *Nucl. Fusion* **37**:793
[3] Leonard A W, *et al.* 1997 *J. Nucl. Mater.* **241–243**:628
[4] Stacey W M and Groebner R 2007 *Phys. Plasmas* **14**:012501
[5] Snyder P B, *et al.* 2002 *Phys. Plasmas* **9**:2037; Wilson H R, *et al.* 2002 *Phys. Plasmas* **9**:1277
[6] Strauss H, *et al.* 2004 *Comput. Phys. Commun.* **164**:40
[7] Chang C S and Ku S 2008 *Phys. Plasmas* (to be published June)
[8] Park G Y, *et al.* 2007 *J. Phys. Conf. Series* **78**:012087
[9] Parker S E, Procassini R J, Cohen B I and Birdsall C K 1989 *J. Comp. Phys.* **104**:41
[10] Shoshani A, Altintas I, Choudhary A, Critchlow T, Kamath C, Ludäscher B, Nieplocha J, Parker S, Ross R, Samatova N and Vouk M 2007 SDM Center Technologies for Accelerating

Scientific Discoveries *J. Phys. Conf. Series* **78**:012068. See also http://sdmcenter.lbl.gov.

[11] Ludäscher B, Altintas I, Berkley C, Higgins D, Jaeger E, Jones M, Lee E A, Tao J and Zhao Y 2006 Scientific workflow management and the Kepler system *Concurrency and Computation: Practice and Experience* **18**(10):1039-1065

[12] Lofstead J, Klasky S, Abbasi H and Schwan K 2007 Adaptable IO System (ADIOS) for scientific codes *Supercomputing 2007 Petascale Data Storage Workshop*

[13] Klasky S, Barreto R, Kahn A, Parashar M, Podhorszki N, Parker S, Silver D and Vouk M 2008 Collaborative visualization spaces for petascale simulations *2008 International Symposium on Collaborative Technologies and Systems*

Validating simulations of core tokamak turbulence: Current status and future directions

C Holland[1], J Candy[2], R E Waltz[2], A E White[3], G R McKee[4], M W Shafer[4], L Schmitz[3] and G R Tynan[1]

[1]University of California-San Diego, La Jolla, CA 92093, USA
[2]General Atomics, San Diego, CA 92186, USA
[3]University of California-Los Angeles, Los Angeles, CA 90095, USA
[4]University of Wisconsin-Madison, Madison, WI 53706, USA

E-mail: chholland@ucsd.edu

Abstract. Validating predictive models of turbulent transport in magnetically confined plasmas requires comparisons of detailed fluctuation statistics, in addition to net energy flows. Using measurements from new and improved diagnostics on the DIII-D tokamak, we have performed a series of comparisons against predictions from the GYRO code. The development and application of synthetic diagnostics that model the spatial sensitivities of a given experimental fluctuation diagnostic is essential for these comparisons. At r/a = 0.56, we find very good agreement between the predicted and measured energy fluxes and fluctuation power spectra. At r/a = 0.8, however, the simulations underpredict the energy flows by a factor of seven and fluctuation amplitudes by a factor of 3 but successfully reproduce the shapes of the experimentally measured fluctuation power spectra. At both locations significant attenuation in the synthetic power spectra and fluctuation levels is observed relative to "unfiltered" levels. Additional results contrasting local and nonlocal simulation results and convergence in toroidal mode number spacing are presented.

1. Introduction

The development of validated predictive modeling codes has been identified as an essential task for ensuring the success of burning plasma experiments such as ITER [1]. One of the most challenging aspects of this work is the validation of nonlinear simulations of turbulent transport in magnetically confined plasmas. In the absence of large-scale magnetohydrodynamic (MHD) instabilities, the inherent free energy gradients of the confined plasma drive a broad spectrum of nonlinearly interacting small-scale drift-wave instabilities [2] (termed "drift-wave turbulence" hereafter) which in turn drive cross-field flows of particles and energy, limiting the achievable levels of confinement. While these instabilities have received intensive analytic and computational investigation for more than 30 years, only in recent years have simulations matured to the point that they have included what is believed to be the essential physics for making quantitatively accurate predictions of the turbulence and its associated transport. Also essential for validation of these models has been the corresponding maturation of experimental diagnostics able to measure the turbulent fluctuations in relevant parameter ranges, as well as the advances in computing power to make the "full physics" simulations feasible on tractable timescales.

To truly validate a model of drift-wave turbulence, one would like to be able to accurately predict a wide range of fluctuation characteristics, starting with details of the turbulent spectrum such as cross-phases and biphases between different fluctuation fields, to spectral shapes and amplitudes, as well as "higher-order" quantities such as total turbulent flows and fluctuation amplitudes that integrate the lower-order quantities [3]. While there is a long history of comparisons between plasma turbulence measurements and simulation, for large-scale tokamaks these comparisons have been limited by essential physics missing from the simulation (such as equilibrium $\mathbf{E}\times\mathbf{B}$ shear) and by limited relevant fluctuation measurements. In many cases, the comparisons have been limited to comparisons of predicted energy and particle flows against those determined via a power balance analysis of the measured mean plasma profiles (termed the experimental flows hereafter). Taking advantage of the recent advances in code sophistication, available computing power, and experimental measurements, we have begun taking the next step in turbulence model validation, in which predictions of multiple fluctuation spectra as well as transport levels are compared to experimental measurements. A key aspect for this work is the development of synthetic diagnostics that model the spatial sensitivities of the diagnostics used in these comparisons; including theses sensitivities is essential for meaningful quantitative comparisons.

2. Overview of simulation parameters and comparisons of energy flows

The simulations used in this validation exercise were performed with the gyrokinetic turbulence code GYRO [4]. GYRO is a physically comprehensive, nonlinear Eulerian δf initial value code that operates in either a local or nonlocal mode. GYRO was chosen because it is the most physically comprehensive code of its type currently available, able to include all the essential physics necessary for meaningful quantitative comparisons against experiment [5]. This list includes kinetic electrons, realistic electron-ion collisions, electromagnetic effects, equilibrium $\check{E} \times \check{B}$ flow shear, the ability to use experimentally measured density, temperature, rotation and E_r profiles as inputs and a realistic magnetic geometry description via the Miller equilibrium model [6]. In this work, we have focused primarily on local simulations, although we have performed nonlocal ones as well that support the conclusions obtained from the local simulation results. In a local simulation, a $\rho^* = \rho_s / a \to 0$ limit (where $\rho_s = c_s / \Omega_{ci}$ is the ion gyroradius evaluated at the sound speed $c_s = \sqrt{T_e / M_i}$ and a is the minor radius of the plasma) of the underlying gyrokinetic equations [7] is used to self-consistently eliminate spatial variation in the equilibrium density, temperature, and rotation profiles (with only the local gradient values retained), whereas these variations are retained in the nonlocal simulation. GYRO also allows the user to simulate only a finite subsection of the full torus of the tokamak, using N_n toroidal modes with mode number spacing Δn [i.e., toroidal mode number $n = 0, \Delta n, 2\Delta n, \ldots, (N_n - 1)\Delta n$]. Choosing a value of $\Delta n >$ 1 corresponds to simulating a "wedge" of $2\pi / \Delta n$ extent in toroidal angle, allowing for significant computational savings. We show below that this approximation can be used without impact on the predicted energy flows and net fluctuation amplitudes, consistent with the inherent scale separation between the turbulence and the equilibrium scales.

The experiment modeled is a steady, low-power, sawtooth-free L-mode DIII-D discharge, described in greater detail in Refs. [8] and [9]. A power balance analysis conducted with the ONETWO code [10] is used to determine the rates at which ion and electron energy flow across a given magnetic surface, and are denoted by Q_i and Q_e, respectively, in this paper. Note that we reference these net energy flows only to avoid any ambiguities in terms of definitions of fluxes or diffusivity coefficients, and to minimize the impact of statistical and systematic errors in the measured plasma profiles and their gradients. Because the power balance calculations rely on an accurate knowledge of the relevant source terms in order to determine these flows and because the source of particles due to edge recycling is not well constrained, we do not attempt to compare predicted particle fluxes against the power balance results. Although we do not know the particle flux with sufficient accuracy to validate the code predictions, it is expected to be sufficiently small to have a negligible impact on the determination of the energy flows. A comparison of the energy flows predicted by a set of local simulations, as well as an accompanying nonlocal simulation, are shown in figure 1. Each of the local simulations used 16 toroidal modes, generally spanning a range of $k_y \rho_s = [0,1]$, where $k_y = nq/r$ is the "binormal" wavenumber (i.e., perpendicular to the local magnetic field while still lying in the magnetic flux surface); r is the midplane minor radius and $q = rB_\varphi / rB_\theta$ is the safety factor. All simulations are electromagnetic and use the experimentally measured profiles without modification. The local simulations use a fully realistic mass ratio $\sqrt{M/m} = 60$, while the nonlocal simulation uses $\sqrt{M/m} = 40$ for computational savings. Time integration is done with a fourth-order Runge-Kutta algorithm combined with a second-order implicit treatment of linear electron physics. The nonlinear terms are discretized by using the Arakawa algorithm [11]. A standard 128-point velocity-space discretization is used (8 points in pitch angle, 8 points in energy, and 2 signs of velocity). Further details of the grid resolution and timestep h used, as well as local plasma parameters, can be found in table 1. The nonlocal simulation uses 24 toroidal modes, spanning a range of $k_y\rho_s = [0,1.5]$ at $r/a = 0.67$. Because q and ρ_s both vary with r, however, this resolution is also radially dependent, such that it covers the range $k_y\rho_s = [0,3.9]$ at $r/a = 0.38$ and $[0,1.0]$ at $r/a = 0.85$. We also use a nonuniform radial grid spacing for this simulation such that Δr is proportional to the local value of ρ_s, ranging from $\Delta r/\rho_s(r_0) = 0.88$ at $r/a = 0.38$ to $\Delta r/\rho_s(r_0) = 0.28$ at $r/a = 0.85$; here $\rho_s(r_0)$ is the gyroradius evaluated at the reference surface $r_0/a = 0.67$.

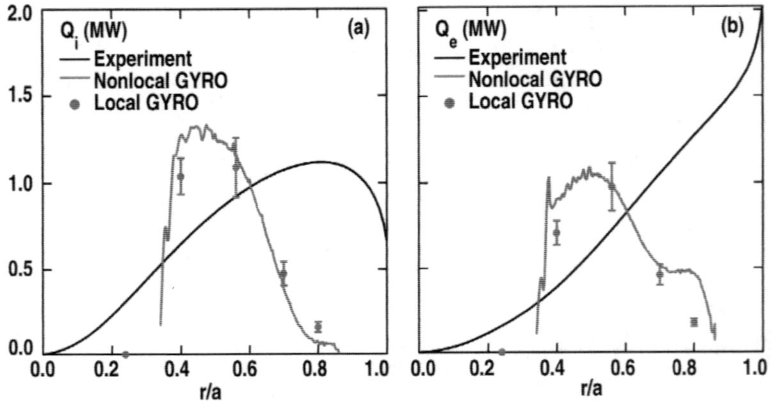

Figure 1. Comparison of ion (Q_i, a) and electron (Q_e, b) energy flows calculated via ONETWO (black) versus those predicted by GYRO (red).

Table 1. Simulation parameters.

	$r/a = 0.24$	$r/a = 0.4$	$r/a = 0.56$	$r/a = 0.7$	$r/a = 0.8$
expt. Q_i (MW)	0.30	0.65	0.93	1.1	1.1
GYRO Q_i (MW)	0	1.0 ± 0.10	1.1 ± 0.17	0.48 ± 0.068	0.16 ± 0.030
expt. Q_e (MW)	0.15	0.40	0.74	1.0	1.3
GYRO Q_e (MW)	0	0.71 ± 0.071	0.97 ± 0.14	0.46 ± 0.060	0.17 ± 0.023
$\Delta r/\rho_s$	0.54	0.54	0.43	0.48	0.40
L_r/ρ_s	109	108	128	120	121
Δn	2	5	8	10	12
hc_s/a	0.02	0.02	0.02	0.01	0.01

As can be seen from figure 1, significant discrepancies exist between the GYRO predicted and experimental heat fluxes at some locations, with very good agreement occurring at $r/a = 0.56$, but a near factor of 7 under prediction at $r/a = 0.8$. Also of interest is that the nonlocal simulation gives results that are generally consistent in magnitude and variation with radius as the local results. The plotted error bars are the standard deviations of the turbulent fluxes, providing some measure of the variation in time at each location; the nonlocal simulation exhibits a similar level of variability. A local simulation at r/a =

0.24 produced no transport, suggesting a complete suppression of the instability by the local $\check{E} \times \check{B}$ shear. A significant possible source of some or all of these differences is due to the effects of "profile stiffness," in which a small change in input temperature or density gradient can lead to a much larger change in the observed energy transport [5]. While the statistical errors in the input experimentally measured profile gradients are actually rather small (generally < 5%–6%) [9], larger systematic profile uncertainties remain possible. The one exception to this limitation is the equilibrium $\check{E} \times \check{B}$ shear rate, which has a statistical uncertainty on the order of 10%–20%. In order to better quantify this effect at $r/a = 0.56$, additional runs artificially varying the input $\check{E} \times \check{B}$ shear rate and ion temperature scale length $a/L_{Ti} = -a\,d(\ln T_i)/dr$ (the primary driving gradient) were performed. The results of these tests are given in table 2. While even better agreement at $r/a = 0.56$ can be achieved by changing a/L_{Ti} and $\check{E} \times \check{B}$ shear rate within the statistical uncertainties, one in general finds the somewhat surprising result that there is perhaps less stiffness at this location than was observed in previous work. We have also investigated the sensitivity of the $r/a = 0.8$ results, first by reducing the $\check{E} \times \check{B}$ shear rate, and then by additionally increasing a/L_{Ti}. The results are shown in table 3; and while a significant proportional response is observed, the magnitude of the mismatch is such that it is hard to reconcile with small changes to the input gradients, even in the presence of stiff transport.

Table 2. Comparisons of energy flows at $r/a = 0.56$.

	Experiment (ONETWO)	GYRO (Base)	GYRO (+20% $\gamma_{E\times B}$)	GYRO (+20% $\gamma_{E\times B}$, −5% a/L_{Ti})	GYRO ($\Delta n = 4$)
Q_i (MW)	0.93	1.1 ± 0.17	0.98 ± 0.16	0.79 ± 0.11	1.1 ± 0.13
Q_e (MW)	0.74	0.97 ± 0.14	0.86 ± 0.12	0.75 ± 0.09	1.0 ± 0.10

Table 3. Comparisons of energy flows at $r/a = 0.8$.

	Experiment (ONETWO)	GYRO (Base)	GYRO (−20% $\gamma_{E\times B}$)	GYRO (−20% $\gamma_{E\times B}$, +10% a/L_{Ti})
Q_i (MW)	1.1	0.16 ±).030	0.22 ± 0.031	0.33 ± 0.044
Q_e (MW)	1.3	0.17 ±).023	0.23 ± 0.026	0.30 ± 0.031

Inclusion of all the necessary physics mechanisms significantly increases the computational costs of these simulations relative to earlier simplified studies, with each local run at $r/a = 0.56$ costing approximately 3000 CPU-hours on either the Jaguar or Franklin Cray XT4 machines, and with the runs at $r/a = 0.8$ costing at least twice as much. While an individual one of these runs may have a small cost relative to so-called hero runs, the need to conduct many such runs in order to quantify sensitivities and test convergences quickly accumulates to significant total computational time requirements. Therefore, it is essential to identify avenues to simplify the simulations where possible, without impacting the physics results. One way in which this can be done is to exploit the inherent scale separation of the small-scale drift-wave turbulence and the equilibrium scale lengths and simulate only a fraction of the full toroidal extent of the plasma. This is done by choosing a toroidal mode number spacing $\Delta n > 1$ and corresponds to simulating a wedge of toroidal extent $\Delta\varphi = 2\pi/\Delta n$. In the local simulations, we have used 16 toroidal modes, with Δn chosen at each location to allow coverage in binormal wavenumber to $k_y\rho_s = 1$.

Previous work demonstrated convergence in Δn for a different set of plasma conditions [12]. We have performed an additional test for this particular set of conditions, in a second simulation performed with 32 toroidal modes at $r/a = 0.56$, using $\Delta n = 4$ rather than 8 as was done in the 16-mode case. The ion and electron energy flows plotted as a function of k_y are shown in figure 2 for these two cases and show an essentially exact match between the two. This can be further quantified by observing that the $\Delta n = 4$ case predicts a $Q_i = 1.1 \pm 0.13$ MW and $Q_e = 1.0 \pm 0.10$ MW, relative to the $\Delta n = 8$ case which predicted $Q_i = 1.1 \pm 0.17$ MW and $Q_e = 0.97 \pm 0.14$ MW (shown in table 2). Furthermore, we show a similar level of agreement in predicted fluctuation levels and spectra in Section 4.

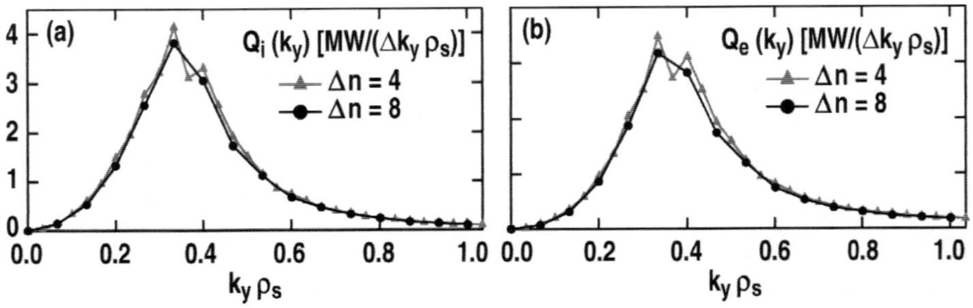

Figure 2. Comparisons of power spectral densities of Q_i (a) and Q_e (b) at $r/a =$ 0.56, calculated using Δn = 4 (red) and Δn = 8 (black).

3. Generation of synthetic diagnostics

While the energy flow comparisons presented in Section 2 are an important initial step in model validation, comparisons against individual fluctuation field amplitudes and spectra represent a more stringent test of the model. Such comparisons can be made by using data from the beam emission spectroscopy [13] (BES) and correlation electron cyclotron emission [8] (CECE) diagnostics on the DIII-D tokamak [14], which provide spatially localized, high time resolution measurements of density and electron temperature fluctuations, respectively. The BES system consists of a 30 measurement channels arranged in a 5x6 array in the (R,Z) plane with roughly a 1 cm spacing in both directions, while the CECE system consists of a pair of channels radially separated by 0.5 cm. Broadly speaking, each diagnostic measures a specific type of radiation from a small but finite region of the plasma, which can be related to local, instantaneous density or electron temperature fluctuation amplitude. To conduct quantitative comparisons against these measurements, one must use synthetic diagnostics that model what one expects the BES and CECE diagnostics to have measured, if they had observed the turbulence predicted by the simulation. For these specific diagnostics, the synthetic diagnostics consist of transforming the GYRO data from a co-rotating plasma reference frame in which the calculations are performed to the lab reference frame, and accounting for the finite spatial integration volume of each diagnostic channel. The transformation from plasma to lab reference frame is easily done by using the formula

$$\delta\hat{X}_l^{LF}(r,\theta,n,t) = \delta\hat{X}_l^{PF}(r,\theta,n,t)e^{-il\Delta n\omega_0 t} \quad , \tag{1}$$

where $\omega_0 = -qcE_r/rB$ is $\check{E} \times \check{B}$ rotation rate of the reference surface r_0, $\delta\hat{X}_l^{PF}$ is the complex amplitude of fluctuations with toroidal mode number $n_l = l\Delta n$ in the plasma frame, and $\delta\hat{X}_l^{LF}$ is the corresponding lab-frame amplitude. When applying this transform, it is generally necessary to interpolate the plasma frame data in time because, while the sampling rate of c_s/a at which the GYRO output is saved is more than sufficient to adequately record the time dynamics in the plasma frame without aliasing, it is insufficient for the Doppler-shifted lab-frame rates.

Accounting for the finite integration volumes of each diagnostic is done via the use of "point spread functions" (PSFs), which model the spatial sensitivities of a given diagnostic channel. To create a synthetic signal δX_{syn} corresponding to the diagnostic channel centered at (R_0, Z_0), we first reconstruct the instantaneous real-space representation of the fluctuation field at a specific toroidal angle φ_0 via

$$\delta X_{GYRO}(R,Z,\varphi_0,t) = \sum_{l=-N_n}^{N_n} \delta\hat{X}_l^{LF}(r,\theta,t)e^{-il\Delta n[\varphi_0+\nu(r,\theta)]} \quad , \tag{2}$$

where $\nu(r,\theta)$ is an eikonal phase function defined by the magnetic geometry model, as is the transformation from (r,θ) coordinate system to the (R,Z) coordinate system. We then convolve δX_{GYRO} with a diagnostic-specific PSF of the form $\psi(R-R_0,Z-Z_0)$, by

$$\delta X_{syn}(R_0,Z_0,\varphi_0,t) = \frac{1}{N_\psi}\iint dRdZ\, \delta X_{\text{GYRO}}(R,Z,\varphi_0,t)\psi(R-R_0,Z-Z_0) \quad , \tag{3}$$

$$N_\psi = \iint dRdZ\, \psi(R-R_0,Z-Z_0) \quad . \tag{4}$$

Further details on how each PSF is calculated can be found in Ref. [9]. This convolution is done for each time t of interest and diagnostic channel to be modeled, generating a set of synthetic timetraces that can then be analyzed in the same fashion as the experimental data. In order to quantify the effects of the PSFs, the "unfiltered" $\delta X_{\text{GYRO}}(R_0,Z_0,\varphi_0,t)$ values for each diagnostic channel are also recorded. Figure 3 illustrates contours of the PSFs for each diagnostic overlaid on the relevant fluctuation fields. The synthetic diagnostic generation algorithm can be summarized as follows:

1. Calculate the point spread function $\psi(R-R_0,Z-Z_0)$ for the diagnostic of interest, and the normalization factor N_ψ.
2. Calculate the phase factor $\nu(r,\theta)$.

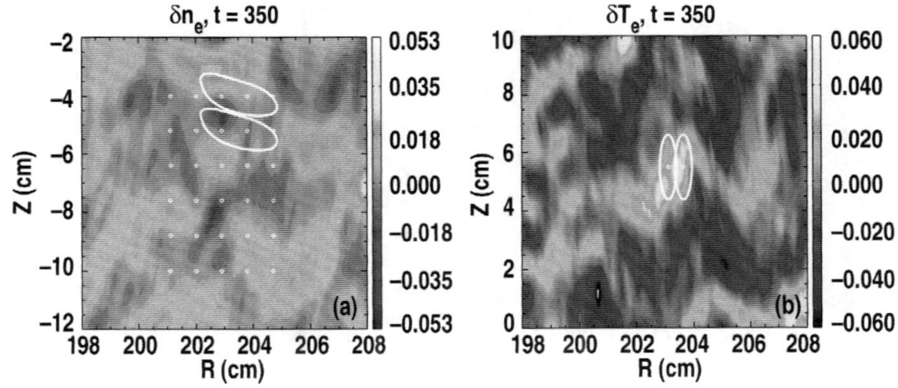

Figure 3. Contours of electron density fluctuations (left) and electron temperature fluctuations (right) in 10 cm x 10 cm regions of the (R,Z) plane from the local simulation of $r/a = 0.56$. Shown are 50% contours of the BES and CECE PSFs overlaid in white, with diamonds indicating the center locations of the individual diagnostic channels.

For each time t of interest, we then

3. Calculate the lab-frame fluctuation field mode amplitudes $\delta\hat{X}_l^{LF}$, interpolating $\delta\hat{X}_l^{PF}$ in time if necessary.
4. Transform the lab-frame mode amplitudes $\delta\hat{X}_l^{LF}$ into a real-space signal δX_{GYRO} at a specified toroidal angle φ_0 via equation (2).

5. Record the unfiltered reference signal $\delta X_{\mathrm{GYRO}}(R_0, Z_0, \varphi_0, t)$.
6. Calculate and record the synthetic signal $\delta X_{\mathrm{syn}}(R_0, Z_0, \varphi_0, t)$ using equation (3).

4. Comparisons of lab-frame fluctuation power spectra

Comparisons of lab-frame power spectra calculated from the unfiltered GYRO output, synthetic results, and experimental measurements are shown for $r/a = 0.56$ (figure 4). To minimize channel-specific noise in the experimental measurements, we compare cross-spectra, defined as $S_{ij}(f) = \left|\langle X_i^*(f) X_j(f)\rangle\right|$, where $X_i(f)$ and $X_j(f)$ correspond to the Fourier transforms of output from two distinct diagnostic channels i and j. In modeling the BES system, we average over all vertically adjacent pairs of channels, while the CECE system uses a pair of radially separated channels. This difference is chosen to minimize the loss of physical signal in the BES measurements, since we expect (and observe) that nearby fluctuations on the same flux surface remain more highly correlated than those on separate ones. In order to generate additional synthetic data realizations, synthetic time traces are generated at four equally spaced toroidal angles in the simulation, and the results averaged together. As seen in figure 4, we find very good agreement in both the shape and magnitude of the synthetic BES system and the experimental measurements, and somewhat weaker agreement in the CECE system. We also observe a clear attenuation in both spectra due to the averaging effects of the PSFs. This attenuation of power is frequency- and diagnostic-dependent and can be understood in terms of the relative spatial anisotropies of each PSF.

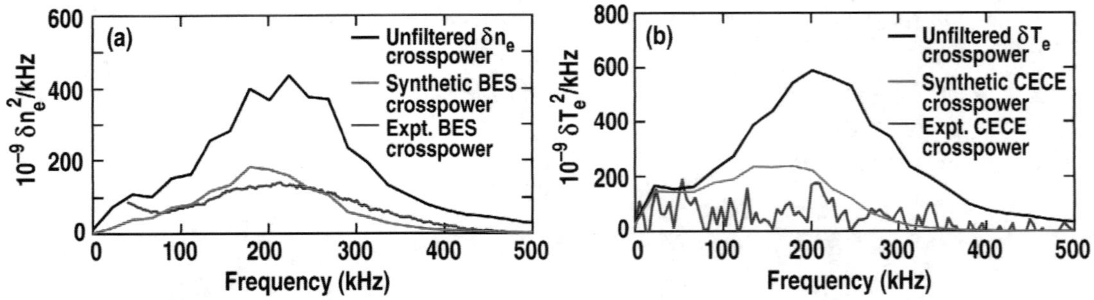

Figure 4. Comparisons of cross-spectra for density (left) and electron temperature (right) fluctuations for the local r/a = 0.56 simulation. The unfiltered GYRO output cross-spectra are shown in black, the synthetic results in red, and experimental measurements in blue.

The first key to understanding this effect is to note that the significant Doppler shift imposed by the transformation to the lab frame yields a near linear relation $f_{lab} = k_y V_y^{ExB}$, where V_y^{ExB} is the local binormal component of the equilibrium $\vec{E} \times \vec{B}$ velocity. For the BES system, the PSF at $r/a = 0.56$ is predominately elongated radially, and thus its major effect will be to filter out higher radial wave numbers at all k_y (and thus lab-frame frequencies), yielding the relatively uniform attenuation observed. There is a small but finite frequency dependence due to the finite poloidal extent of the PSF as well, but this effect is subdominant to the radial averaging. In contrast, the CECE PSF primarily filters higher k_y, leading to the strong frequency dependence observed for those spectra.

Because the BES measurements below 40 kHz are significantly impacted by noise in the diagnostic beam, we compare only the spectra between 40 and 400 kHz (where the signal goes below the noise level of the diagnostic); a similar comparison range is used for the CECE system for consistency. To quantify the agreement, we can calculate the root mean square (RMS) fluctuation level associated with fluctuations within that frequency band via formula

$$\delta X_{RMS}^2 = \int_{40}^{400} df\, S_{ij}(f) \quad . \tag{5}$$

The results are shown in table 4. There is a near-perfect level of agreement in the density fluctuation level (0.55% for the synthetic vs. 0.56% for the experiment), whereas the electron temperature fluctuations are overpredicted by 40%. Because the CECE measurement is so close to the noise level, however, there is a nearly 50% uncertainty in the experimental RMS electron temperature fluctuation level (relative to a 20% uncertainty in the experimental density fluctuation level). We recall that both the ion and electron energy flows were (slightly) over predicted in this simulation and that one might expect a better agreement in the electron temperature fluctuation levels if the input profiles had been adjusted to match the experimental energy flows. How strongly this profile adjustment would affect the density fluctuation levels is uncertain. Also note that the unfiltered and synthetic RMS fluctuation levels calculated from the $\Delta n = 4$ simulation are included in table 5 and are quite close to the $\Delta n = 8$ results. This agreement further highlights the convergence in Δn, and the ability of these wedge simulations to accurately predict integrated quantities such as energy flows and fluctuation amplitudes. A comparison of the unfiltered density and electron temperature fluctuation spectra from these cases is shown in figure 5.

A comparison of lab-frame cross-spectra for $r/a = 0.8$ is shown in figure 6. In contrast to the $r/a = 0.56$ results, the fluctuation levels are significantly underpredicted; net RMS values are shown in table 5. The RMS fluctuation levels are underpredicted by roughly a factor of 3, which is relatively consistent with the factor of 7 underprediction in the energy flows. By consistent, we mean that because the turbulent flows are generally observed to scale with fluctuation amplitude squared, a factor of 7 underprediction in the flows should lead to a generic under prediction of fluctuation levels by $\sqrt{7} = 2.64$, close to the observed factor of 3.

Table 4. Comparisons of RMS fluctuation levels at $r/a = 0.56$.

Level (%)	Unfiltered Cross-Power, 40–400 kHz	Synthetic Cross-Power, 40–400 kHz	Experimental Cross-Power, 40–400 kHz	Unfiltered Cross-Power, $\Delta n = 4$	Synthetic Cross-Power, $\Delta n = 4$
δn	0.90 ± 1.5e-2	0.55 ± 7.7e-3	0.56 ± 0.1	0.98 ± 2.1e-2	0.59 ± 1.1e-2
δT_e	1.1 ± 5.7e-2	0.66 ± 2.4e-2	0.46 ± 0.2	1.1 ± 7.8e-2	0.67 ± 3.2e-2

Table 5. Comparisons of RMS fluctuation levels at $r/a = 0.8$.

Level (%)	Unfiltered Cross-Power, 30-390 kHz	Synthetic Cross-Power, 30-390 kHz	Experimental Cross-Power, 40-400 kHz
δn	0.69 ± 1.9e-2	0.33 ± 6.9e-3	1.1 ± 0.2
δT_e	0.90 ± 7.7e-2	0.50 ± 1.9e-2	1.6 ± 0.2

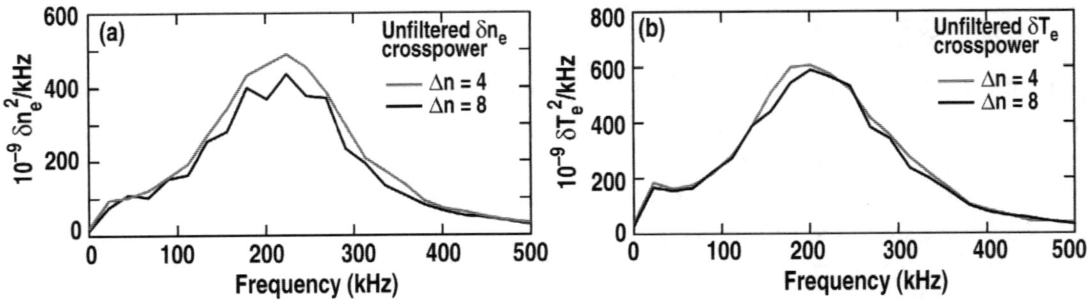

Figure 5. Comparison of unfiltered cross-spectra of density (a) and electron temperature (b) fluctuations at $r/a = 0.56$, for $\Delta n = 4$ (red) and $\Delta n = 8$ (black).

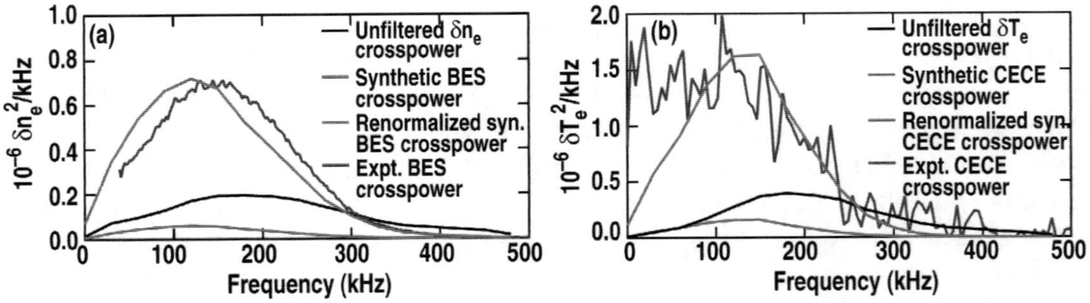

Figure 6. Comparisons of cross-spectra for density (a) and electron temperature (b) fluctuations for the local $r/a = 0.8$ simulation. The unfiltered GYRO output cross-spectra are shown in black, the synthetic results in red, and experimental measurements in blue. The synthetic spectra rescaled to contain the same power in the 40-400 kHz band as the experimental results are shown in purple.

Even more interesting, if we rescale the synthetic fluctuation spectra to contain the same amount of power as the experimental spectra (shown as the purple curves in figure 6), we find very good agreement between the shapes of the synthetic and experimental spectra. Keeping in mind the strong correlation between lab-frame frequency and poloidal mode number, we also observe very good agreement between poloidal correlation functions calculated with the synthetic and experimental data [9]. A stronger frequency dependence observed in the attenuation of the synthetic BES signal can be understood in terms of the PSF at this location exhibiting a more symmetric shape than at $r/a = 0.56$. Current work is focused on understanding why the simulations under predict the fluctuation levels and energy flows at $r/a = 0.8$ so significantly. Some possibilities are discussed in greater detail in [9], including potentially missing long-wavelength modes such as resistive ballooning modes or insufficient range in simulated k_y, neither of which we deem likely due to $\check{E} \times \check{B}$ shear suppression and an inability to increase ion energy transport, respectively. Other candidates include systematic errors in the power balance calculations, the use of an overly simplified electron collision operator in the GYRO simulations, and nonlocal effects coupling the turbulence dynamics at $r/a = 0.8$ to the strong variations in equilibrium profiles and normalized turbulent intensity at larger radii. We also hope to conduct even finer tests of simulation predictions in the near future, by comparing against experimental measurements of fluctuation cross-phases that are crucial components in setting the levels of turbulent flows.

4. Conclusions

Advances in computational models of turbulent plasma transport, coupled with recent increases in available computing power and new and improved fluctuation diagnostics, opens the door to the next level of model validation, simultaneous quantitative comparisons of turbulent fluxes and individual fluctuation field characteristics. The development and application of synthetic diagnostics that model the experimental ones are essential for accurate, meaningful quantitative comparisons between code and experiment. Using newly developed synthetic BES and CECE diagnostics, we have shown that a "full physics" GYRO simulation that accurately predicts the total energy flows Q_i and Q_e at $r/a = 0.56$ also successfully reproduces measured fluctuation characteristics, including fluctuation level and spectral shapes. In a second case in which the energy flows were significantly under predicted at $r/a = 0.8$, the RMS fluctuation amplitudes were likewise underpredicted in a "self-consistent" manner, but the shapes of the fluctuation power spectra were still rather accurately predicted. Understanding the physics underlying the differing levels of success of these simulations in reproducing the experimental observations, and conducting additional validation studies for a broader range of plasma conditions are the next steps for this research and will be presented in future publications.

Acknowledgments
This work supported by the U.S. Department of Energy under DE-FC02-04ER54798, DE-FG02-95ER54309, DE-FG02-07ER54917, DE-FG03-01ER54615 and DE-FG02-89ER53296. We thank T. L. Rhodes, W. A. Peebles, E. J. Doyle, R. Prater, J. C. DeBoo, K. H. Burrell, P. H. Diamond, J. E. Kinsey, G. M. Staebler, D. R. Mikkelsen, R. V. Bravenec, and R. V. Budny for many useful discussions. A. E. White's research was performed under appointment to the Fusion Energy Sciences Fellowship Program administered by Oak Ridge Institute for Science and education under a contract between the U.S. Department of Energy and the Oak Ridge Associated Universities. This research used resources of the National Center for Computational Sciences at Oak Ridge National Laboratory, which is supported by the Office of Science of the Department of Energy under Contract DE-AC05-00OR22725, and the National Energy Research Scientific Computing Center, which is supported by the Office of Science of the U.S. Department of Energy under Contract No. DE-AC02-05CH11231.

References
[1] Shimada M, *et al.* 2007 *Nucl. Fusion* **47** S1
[2] Horton W 1999 *Rev. Mod. Phys.* **71** 735
[3] Terry P W, *et al.* 2008 *Phys. Plasmas* **15** 062503
[4] Candy J and Waltz R E 2003 *J. Comp. Phys.* **186** 545
[5] Candy J and Waltz R E 2003 *Phys. Rev. Lett.* **91** 045001
[6] Miller R L, Chu M S, Greene J M, Lin-Liu Y R, and Waltz R E 1998 *Phys. Plasmas* **5** 973
[7] Brizard A J and Hahm T S 2007 *Rev. Mod. Phys.* **79** 421
[8] White A E *et al.* 2008 *Phys. Plasmas* **15** 056116
[9] Holland C *et al.* 2008 Implementation and application of two synthetic diagnostics for validating simulations of core tokamak turbulence, submitted to *Phys. Plasmas*
[10] St John H E, Taylor T S, Lin-Liu Y R, Turnbull A D 1994 *Plasma Phys. Controlled Nucl. Fusion Res.* **3** 603
[11] Arakawa A 1966 *J. Comput. Phys.* **1** 119
[12] Waltz R E, Austin M E, Burrell K H, and Candy J 2006 *Phys. Plasmas* **13** 052301
[13] McKee G R *et al.* 2007 *Plasma Fusion Res.* **2** S1025
[14] Luxon J L 2002 *Nucl. Fusion* **42** 614

The M3D-C^1 approach to simulating 3D 2-fluid magnetohydrodynamics in magnetic fusion experiments

S. C. Jardin[§1], N. Ferraro[1], X. Luo[2], J. Chen[1], J. Breslau[1], K. E. Jansen[2], and M. S. Shephard[2]
[1]Princeton Plasma Physics Laboratory, P.O. Box 451, Princeton NJ 08543
[2]Rensselaer Polytech Inst, Sci Computat Res Ctr, Troy, NY 12180

Abstract. A new approach for solving the 3D MHD equations in a strongly magnetized toroidal plasma is presented which uses high-order 2D finite elements with C^1 continuity. The vector fields use a physics-based decomposition. An efficient implicit time advance separates the velocity and field advance. ITAPS (SCOREC) adaptivity software and TOPS solvers are used.

1. Introduction

The two-fluid MHD equations for a magnetized fusion-grade plasma are a high-order system of 8 scalar variables that are characterized by a wide range of space and timescales and by extreme anisotropy. The M3D code [1] has proven itself to be an invaluable tool for the simulation and understanding of global nonlinear phenomena in magnetic fusion confinement devices. However, the structure of M3D is not optimal for computing in regimes where two-fluid (2F) effects dominate, or for times that are very long compared to the Alfven transit time. We have built upon many of the favorable features of the M3D approach to construct the M3D-C^1 code [2], which is based on high-order, compact finite elements with C^1 continuity on an unstructured adaptive triangle-based grid. The efficient split-implicit time advance is closely related to the ideal MHD energy principle, and allows time steps several orders of magnitude in excess of the Courant condition based on the Alfven or whistler waves. Previous papers have described this technique applied to the 2F equations in 2D slab geometry. [2,3] In this paper, we discuss a subset of the full method as it is applied to the linearized single fluid MHD equations in 3D toroidal geometry.

2. Numerical method and definitions

We restrict consideration here to solutions linearized about an initial equilibrium with zero velocity, $\mathbf{V}_0 = 0$. The full 2F nonlinear solution, to be presented elsewhere, builds on the method presented here. The split linearized time advance [4,5] can be written as follows. First, introducing the implicit parameter $(0 < \theta < 1)$, the velocity is advanced to the new time level (n+1) using the magnetic field at the half-time level (n+1/2):

$$\left\{\rho - \theta^2(\delta t)^2 L\right\}\mathbf{V}^{n+1} = \left\{\rho - \theta^2(\delta t)^2 L\right\}\mathbf{V}^{n} + \delta t\left\{-\nabla p + \frac{1}{\mu_0}(\nabla\times\mathbf{B})\times\mathbf{B}\right\}^{n+1/2} \quad (1)$$

Here ρ is the plasma density, p is the pressure, and L is the linear ideal MHD operator [6]:

$$L\{\mathbf{V}\} = \frac{1}{\mu_0}\left\{\nabla\times\left[\nabla\times(\mathbf{V}\times\mathbf{B})\right]\right\}\times\mathbf{B} + \frac{1}{\mu_0}(\nabla\times\mathbf{B})\times\left[\nabla\times(\mathbf{V}\times\mathbf{B})\right] + \nabla\left(\mathbf{V}\cdot\nabla p + \gamma p\nabla\cdot\mathbf{V}\right) \quad (2)$$

Next, the pressures, density, and magnetic field are advanced from time level (n+1/2) to (n+3/2) using the velocity at time (n+1). Defining $\theta_m \equiv 1-\theta$, we have:

$$p^{n+3/2} = p^{n+1/2} - \delta t\left[\mathbf{V}^{n+1}\cdot\left(\theta\nabla p^{n+3/2} + \theta_m\nabla p^{n+1/2}\right) - \gamma\left(\theta p^{n+3/2} + \theta_m p^{n+1/2}\right)\nabla\cdot\mathbf{V}^{n+1} + S^{n+1/2}\right] \quad (3)$$

$$\mathbf{B}^{n+3/2} = \mathbf{B}^{n+1/2} + \delta t\nabla\times\left[\mathbf{V}^{n+1}\times\left(\theta\mathbf{B}^{n+3/2} + \theta_m\mathbf{B}^{n+1/2}\right) - \eta\left(\theta\nabla\times\mathbf{B}^{n+3/2} + \theta_m\nabla\times\mathbf{B}^{n+1/2}\right)\right] \quad (4)$$

[§] Corresponding author jardin@pppl.gov

This is unconditionally stable for all δt for implicit parameter $\theta \geq 1/2$. We use a physics-motivated decomposition of the vector fields in toroidal geometry. Using a cylindrical coordinate system (R,φ,Z) with $|\nabla\varphi| = 1/R$, we define the 2D gradient and 2D divergence operators as $\nabla_\perp a \equiv a_R\hat{R} + a_Z\hat{Z}$, $\nabla_\perp \bullet \mathbf{A} \equiv R^{-1}(R\mathbf{A}\bullet\hat{R})_R + (\mathbf{A}\bullet\hat{Z})_Z$. Subscripts denote partial differentiation with respect to *R* and *Z*. The velocity field is represented in terms of the three scalar variables (U,ω,χ) as follows: $\mathbf{V} = R^2\nabla U \times \nabla\varphi + \omega R^2\nabla\varphi + R^{-2}\nabla_\perp\chi$. The magnetic vector potential is given in terms of the two scalar variables (f,ψ) and the constant F_0 (proportional to the current in the toroidal field magnets) as: $\mathbf{A} = R^2\nabla\varphi \times \nabla f + \psi\nabla\varphi - F_0 \ln R\hat{Z}$. Note that the gauge condition implied by this form is: $R^2\nabla_\perp \bullet R^{-2}\mathbf{A} = 0$. The magnetic field and current density are calculated in terms of the vector potential variables as: $\mathbf{B} = \nabla\times\mathbf{A} = \nabla\psi\times\nabla\varphi - \nabla_\perp f' + F\nabla\varphi$ and $\mathbf{J} \equiv \nabla\times\mathbf{B} = \nabla F^* \times \nabla\varphi + R^{-2}\nabla_\perp\psi' - \Delta^*\psi\nabla\varphi$. Here we have defined auxiliary variables $F \equiv F_0 + R^2\nabla\bullet\nabla_\perp f$, $F^* \equiv F_0 + R^2\nabla^2 f = F + f''$, and the operator: $\Delta^*\psi \equiv R^2\nabla_\perp \bullet R^{-2}\nabla\psi$ Primes denote differentiation with respect to the angle φ. Note that **B** and **J** are manifestly divergence free.

3. The scalar equations

Each of the scalar fields is divided into a 2D equilibrium part and a 3D perturbed part. Thus, for example: $\psi(R,\varphi,Z,t) = \psi_0(R,Z) + \psi_1(R,\varphi,Z,t)$. In the work described here, the velocity variables do not have an equilibrium part. All the equilibrium and perturbed variables are expanded in 2D basis functions for the Q_{18} triangular finite element with C^1 continuity [7], which we denote $\nu_i(R,Z)$. To get scalar forms for the momentum equation, we take the weak form of the three following projections of (1) and perform integration by parts as indicated in (5):

$$\begin{aligned}
&\iint d^2R\, \nu_i \nabla\varphi \bullet \nabla_\perp \times R^2(1) & & \iint d^2R\, R^2\nabla_\perp \nu_i \times \nabla\varphi \bullet (1) \\
&\iint d^2R\, \nu_i R^2 \nabla\varphi \bullet (1) & \rightarrow \quad & \iint d^2R\, \nu_i R^2\nabla\varphi\bullet(1) \\
-&\iint d^2R\, \nu_i \nabla_\perp \bullet R^{-2}(1) & & \iint d^2R\, R^{-2}\nabla_\perp \nu_i \bullet (1)
\end{aligned} \quad (5)$$

These projections, combined with the form of the velocity field described above, lead to energy conserving subsets of the full equations [8,9], and a partial decoupling of the part of the velocity field that does not compress the strong toroidal field. [10] The resulting scalar integrands can be written compactly in terms of the inner and Poisson brackets. For any two scalar variables *a* and *b*, we define: $(a,b) \equiv \nabla_\perp a \bullet \nabla_\perp b = a_R b_R + a_Z b_Z$ and $[a,b] \equiv [\nabla a \times \nabla b \bullet \nabla\varphi] = R^{-1}(a_Z b_R - a_R b_Z)$. The partial perturbed energies $\delta W_{ij}(a,b)$ [6] are defined in the Appendix. Thus, for (1) we have:

$$\begin{aligned}
&\left\{\rho R^2(\nu_i,U^{n+1}) - \rho[\nu_i,\chi^{n+1}] - (\theta\delta t)^2\left[\delta W_{11}(\nu_i,U^{n+1}) + \delta W_{12}(\nu_i,\omega^{n+1}) + \delta W_{13}(\nu_i,\chi^{n+1})\right]\right\} \\
&= \left\{\rho R^2(\nu_i,U^n) - \rho[\nu_i,\chi^n] - (\theta\delta t)^2\left[\delta W_{11}(\nu_i,U^n) + \delta W_{12}(\nu_i,\omega^n) + \delta W_{13}(\nu_i,\chi^n)\right]\right\} \\
&+\delta t\left\{\Delta^*\psi\left([\nu_i,\psi] - (\nu_i,f')\right) + \frac{F}{R^2}(\nu_i,\psi') + F[\nu_i,f''] - R^2[p,\nu_i]\right\}^n
\end{aligned} \quad (6a)$$

$$
\begin{aligned}
&\left\{\rho R^2 v_i \omega^{n+1} - (\theta\delta t)^2\left[\delta W_{21}(v_i,U^{n+1}) + \delta W_{22}(v_i,\omega^{n+1}) + \delta W_{23}(v_i,\chi^{n+1})\right]\right\}\\
&=\left\{\rho R^2 v_i \omega^{n} - (\theta\delta t)^2\left[\delta W_{21}(v_i,U^{n}) + \delta W_{22}(v_i,\omega^{n}) + \delta W_{23}(v_i,\chi^{n})\right]\right\}\\
&+\delta t\left\{-\frac{v_i}{R^2}(\psi,\psi') + v_i[f'',\psi] + v_i[F,\psi] + v_i[f',\psi'] - v_i(f',f'') - v_i(f',F) - v_i p'\right\}^n
\end{aligned}
\tag{6b}
$$

$$
\begin{aligned}
&-\frac{\rho}{R^4}(v_i,\chi^{n+1}) - \rho\left[v_i,U^{n+1}\right] - (\theta\delta t)^2\left[\delta W_{31}(v_i,U^{n+1}) + \delta W_{32}(v_i,\omega^{n+1}) + \delta W_{33}(v_i,\chi^{n+1})\right]\\
&= -\frac{\rho}{R^2}(v_i,\chi^{n}) - \rho R^2\left[v_i,U^{n}\right] - (\theta\delta t)^2\left[\delta W_{31}(v_i,U^{n}) + \delta W_{32}(v_i,\omega^{n}) + \delta W_{33}(v_i,\chi^{n})\right]\\
&+\delta t\left\{\Delta^*\psi\frac{1}{R^2}\left[\frac{1}{R^2}(\psi,v_i) + [v_i,f']\right] + \frac{F}{R^4}\left[(v_i,F) + (v_i,f'') - [v_i,\psi']\right] + \frac{1}{R^2}(v_i,p)\right\}^n
\end{aligned}
\tag{6c}
$$

Similarly, taking appropriate projections of (4) gives the time advance integrands for the magnetic field variables:

$$
\begin{aligned}
&-\frac{1}{R^2}(v_i,\psi^{n+3/2}) - \theta\delta t\left\{
\begin{array}{l}
\frac{1}{R^2}\left(v_i,R^2\left[\psi^{n+3/2},U^{n+1}\right]\right) - \frac{F^{n+3/2}}{R^2}\left(v_i,U'^{n+1}\right) + \frac{\omega'^{n+1}}{R^2}\left(v_i,\psi^{n+3/2}\right)\\
+\frac{1}{R^2}\left(v_i,\frac{1}{R^2}\left(\psi^{n+3/2},\chi^{n+1}\right)\right) + \frac{1}{R^4}F^{n+3/2}\left[v_i,\chi'^{n+1}\right]\\
+\frac{\eta}{R^2}\left[\Delta^* v_i\Delta^*\psi^{n+3/2} - \frac{1}{R^2}\left(v_i,\psi''^{n+3/2}\right) - \left[v_i,f'''^{n}\right] - \left[v_i,F'^{n+3/2}\right]\right]
\end{array}\right\}\\
&= -\frac{1}{R^2}(v_i,\psi^{n+1/2}) + \theta_m\delta t\left\{
\begin{array}{l}
\frac{1}{R^2}\left(v_i,R^2\left[\psi^{n+1/2},U^{n+1}\right]\right) - \frac{F^{n+1/2}}{R^2}\left(v_i,U'^{n+1}\right) + \frac{\omega'^{n+1}}{R^2}\left(v_i,\psi^{n+1/2}\right)\\
+\frac{1}{R^2}\left(v_i,\frac{1}{R^2}\left(\psi^{n+1/2},\chi^{n+1}\right)\right) + \frac{1}{R^4}F^{n+1/2}\left[v_i,\chi'^{n+1}\right]\\
+\frac{\eta}{R^2}\left[\Delta^* v_i\Delta^*\psi^{n+1/2} - \frac{1}{R^2}\left(v_i,\psi''^{n+1/2}\right) - \left[v_i,f'''^{n}\right] - \left[v_i,F'^{n+1/2}\right]\right]
\end{array}\right\}
\end{aligned}
\tag{7a}
$$

$$
\begin{aligned}
&\frac{v_i}{R^2}F^{n+3/2} - \theta\delta t\left\{
\begin{array}{l}
F^{n+3/2}\left[v_i,U^{n+1}\right] + \frac{1}{R^4}F^{n+3/2}\left(v_i,\chi^{n+1}\right) - \omega^{n+1}\left[v_i,\psi^{n+3/2}\right]\\
-\frac{\eta}{R^2}\left(v_i,F^{n+3/2}\right) + -\frac{\eta}{R^2}\left(v_i,f''^{n+1/2}\right) + \frac{\eta}{R^2}\left[v_i,\psi'^{n+3/2}\right]
\end{array}\right\}\\
&= \frac{v_i}{R^2}F^{n+1/2} + \theta_m\delta t\left\{
\begin{array}{l}
F^{n+1/2}\left[v_i,U^{n+1}\right] + \frac{1}{R^4}F^{n+1/2}\left(v_i,\chi^{n+1}\right) - \omega^{n+1}\left[v_i,\psi^{n+1/2}\right]\\
-\frac{\eta}{R^2}\left(v_i,F^{n+1/2}\right) + -\frac{\eta}{R^2}\left(v_i,f''^{n+1/2}\right) + \frac{\eta}{R^2}\left[v_i,\psi'^{n+1/2}\right]
\end{array}\right\}
\end{aligned}
\tag{7b}
$$

4. Linearization and solution:

In the benchmarking results presented here, we eliminated the φ dependence by taking all the linearized variables to be complex and varying as exp[$i\ n\ \varphi$], where n is the toroidal mode number. This allows comparison with ideal MHD stability codes such as PEST [11] in the appropriate limit. We form a single matrix to solve for the velocity variables (6abc), and then subsequently form another to solve for the magnetic variables (7ab). For the linearized solution, we use SuperLU_dist [12] to factor the matrices into LU and then perform the backsolves each timestep.

5. Interfaces to support unstructured mesh

ITAPS and SCOREC have coordinated with CEMM and PPPL to support the use of ITAPS mesh adaptivity technologies in M3D-C^1. Four components have been added for applying parallel scalable adaptive simulations as part of these efforts. They are: (1) Input mesh and geometry information using FMDB [13], a flexible distributed mesh database developed in SCOREC. (2) Localization of the ordering of degree-of-freedom (DOFs), distributed vector and matrix objects and operations so as to allow M3D-C^1 to perform all operations needed by the matrix solvers locally, thus hiding the complexity of parallel operations. (3) Access to the distributed matrix storage data structure to assemble matrices for PETSc solvers to effectively apply the PETSc scalability and preconditioning technologies developed by TOPS. (4) Anisotropic size-driven mesh adaptation including solution field transfer during local mesh modifications.

6. Adaptive Meshing

The M3D-C^1 simulations often exhibit high gradients in localized areas of the domain. Using uniform meshes with small mesh size over the entire domain will over-refine the regions beyond the critical local domains with high gradients which can lead to infeasible simulations. Size-driven mesh adaptation [14] based on error indicators is an effective means to resolve this issue by structuring the mesh domains with elements concentrated in the domains needed to improve computational efficiency and accuracy. ITAPS and SCOREC have successfully introduced the size-driven mesh adaptation procedure in M3D-C^1 to account for anisotropic solution characteristics. Interpolation-based error indicators [15] have been employed to compute the mesh size field needed by the mesh adaptation. Anisotropic adaptivity is quite effective in further reducing the number of elements needed for a given level of solution accuracy where the solution gradients are stronger in specific directions as is commonly the case in contained plasma fields.

7. Results of a linear stability application

As part of our verification process, we are computing linear unstable eigenmodes of a fairly simple toroidal configuration that can be evaluated by other codes in the appropriate limits. Using the parameterization described in [7], we defined a near circular cross section toroidal plasma of minor radius 1.0 and major radius 3.0, centered inside a conducting vacuum vessel of dimension 3x3. Defining an equilibrium toroidal field $RB_T = 2.25$, we calculated an equilibrium with plasma current $\mu_0 I_P = 0.7$, central safety factor $q_0 = 1$, central pressure and derivative $p_0 = .030$, $p_1 = -0.5$. The resistivity varied as $p^{-3/2}$, going from 2.5×10^{-6} in the center to 0.415 at the edge and in the region outside the plasma. We show the perturbed values of ψ in Figure 1. These had corresponding growth rates 0.078, 0.1060, and 0.1180, obtained by solving the time-dependent equations for 100 cycles with a time step $\delta t = 2.0$. The mesh used is shown in Figure 1d, superimposed upon colors showing the resistivity contours. It is seen to be fine in the plasma region, continuously varying to coarse in the 'vacuum' region. The growth rates obtained were the same to 3 digits when a uniform mesh was used with the same fineness as used here only in the plasma region, but this required 14,400 vertices as opposed to the 6,245 vertices used here.

8. Summary

The M3D-C^1 code has been extended to 3D in toroidal geometry. This paper describes a subset of the full code that is single fluid MHD linearized about an equilibrium without flow. Future publications will build on the structure and algorithm described here to extend it to a full nonlinear 3D solution of the 2F MHD equations. The implicit time advance splits into two parts: the velocity advance described by (6abc), and the field advance described by (7ab). The magnetic field is manifestly divergence free. Energy conserving subsets of the full equations (reduced MHD) are obtained by solving only (6a) and (7a), or only (6ab) and (7ab). The SCOREC adaptivity software is shown to greatly increase the efficiency of the calculation.

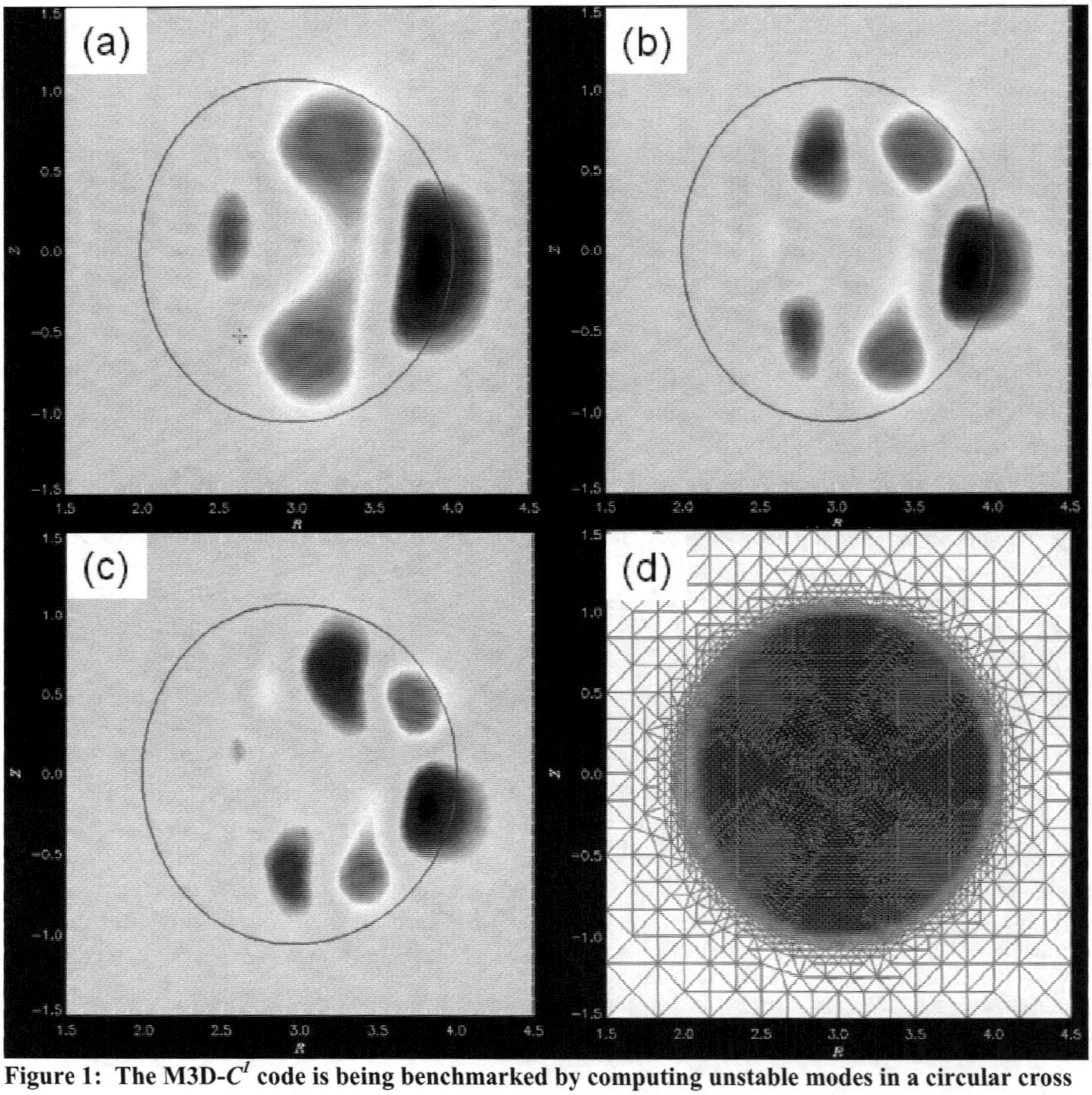

Figure 1: The M3D-C^1 code is being benchmarked by computing unstable modes in a circular cross section toroidal plasma. Shown are eigenmodes of ψ for modes with toroidal mode # (a) n=1, (b) n=2, and (c) n=3. In (d) is the unstructured mesh superimposed on the equilibrium resistivity contours.

Acknowledgment

This work was supported by U.S. DoE contracts DE-FC02-06ER25760 and DE-AC02-76CH03073. The authors benefitted from discussions with Drs. M. Chance, G. Fu, W. Park, H. Strauss, and L. Sugiyama, and from discussions and assistance from Drs. A. Bauer and Y.S. Li.

Journal of Physics: Conference Series **125** (2008) 012044 doi:10.1088/1742-6596/125/1/012044

Appendix

The partial energy terms, $\delta W_{kj}(a_k, b_j)$ are obtained by taking the integral of the inner product of the kth velocity component with the operator L operating on the jth velocity component. For example: $\delta W_{13}(U,\chi) \equiv R^2 \nabla U \times \nabla \varphi \bullet L\{R^{-2}\nabla_{\perp}\chi\}$. In implementing the Galerkin method, each finite element basis function is used in place of the first variable and the second variable is expanded in basis functions. Integration by parts are done so that no more than two derivatives appear on any scalar, consistent with restrictions on C^1 elements. The 9 integrands are as follows:

$$\delta W_{11}(\nu_i,U) = -\frac{1}{R^2}\left(R^2[U,\psi],R^2[\nu_i,\psi]\right) + \Delta^*\psi\left[\nu_i,R^2[U,\psi]\right] - \frac{F}{R^2}\left(U',R^2[\nu_i,\psi]\right)$$
$$+\frac{F}{R^2}\left(R^2[U',\psi],\nu_i\right) + \Delta^*\psi F[\nu_i,U'] + \frac{F^2}{R^2}(U'',\nu_i) + R^2\left[\nu_i,R^2[U,p]\right] + 2R^2\gamma[\nu_i,pU_Z]$$

$$\delta W_{12}(\nu_i,\omega) = \omega'\frac{1}{R^2}\left(\psi,R^2[\nu_i,\psi]\right) - \omega'\Delta^*\psi[\nu_i,\psi] - \frac{F}{R^2}\left(\omega''(\psi,\nu_i)\right) - R^2\gamma[\nu_i,p\omega']$$

$$\delta W_{13}(\nu_i,\chi) = \frac{1}{R^2}\left(R^{-2}(\chi,\psi),R^2[\nu_i,\psi]\right) - \Delta^*\psi\left[\nu_i,R^{-2}(\chi,\psi)\right] + \frac{F}{R^4}\left[R^2[\nu_i,\psi],\chi'\right]$$
$$-\frac{F}{R^4}\left((\chi',\psi),\nu_i\right) + \Delta^*\psi\frac{F}{R^4}(\chi',\nu_i) + \frac{F^2}{R^4}[\chi'',\nu_i] + 2\nu_i\frac{1}{R^2}\left[(\chi,p) + \gamma p\Delta^*\chi\right]$$

$$\delta W_{21}(\nu_i,U) = R^2[\psi,F][U,\nu_i] - \frac{\nu_i}{R^2}\left(R^2[U',\psi],\psi\right) - \nu_i\frac{F}{R^2}(U'',\psi) + \nu_i F[F,U'] - \nu_i R^2[U',p] - 2\nu_i\gamma pU_Z'$$

$$\delta W_{22}(\nu_i,\omega) = -R^2[\nu_i,\psi][\omega,\psi] + \nu_i\frac{\omega''}{R^2}(\psi,\psi) - \nu_i\omega'[F,\psi] + \gamma\nu_i p\omega''$$

$$\delta W_{23}(\nu_i,\chi) = +\frac{\nu_i}{R^2}\left(R^{-2}(\chi',\psi),\psi\right) - \frac{F}{R^4}\left(R^2[\nu_i,\psi],\chi\right) + R^{-2}(\chi,\psi)[F,\nu_i]$$
$$-\frac{\nu_i}{R^4}\left(F[\chi'',\psi]\right) + \nu_i\frac{F}{R^4}(F,\chi') + \left(\nu_i R^{-2}(\chi',p) + \nu_i\gamma pR^{-2}\Delta^*\chi'\right)$$

$$\delta W_{31}(\nu_i,U'') = -\frac{1}{R^2}\left(R^2[U,\psi],R^{-2}(\nu_i,\psi)\right) + \frac{1}{R^4}\Delta^*\psi\left(R^2[U,\psi],\nu_i\right) - \frac{F}{R^4}\left[\nu_i,R^2[U',\psi]\right] - \frac{F}{R^2}\left(U',R^{-2}(\nu_i,\psi)\right)$$
$$+\Delta^*\psi\frac{F}{R^4}(U',\nu_i) - \frac{F^2}{R^4}[\nu_i,U''] + [F,U]F\left[\Delta^*\nu_i\frac{1}{R^2} + \left(\frac{1}{R^2},\nu_i\right)\right] - \Delta^*\nu_i\left[[U,p] + \gamma\frac{2}{R^2}pU_Z\right]$$

$$\delta W_{32}(\nu_i,\omega) = +\frac{\omega'}{R^2}\left(\psi,R^{-2}(\nu_i,\psi)\right) - \Delta^*\psi\frac{\omega'}{R^4}(\psi,\nu_i) + F\frac{\omega''}{R^4}[\nu_i,\psi] - F[\omega,\psi]\left[R^{-2}\Delta^*\nu_i + \left(R^{-2},\nu_i\right)\right] + \Delta^*\nu_i\frac{1}{R^2}\gamma p\omega'$$

$$\delta W_{33}(\nu_i,\chi) = +\frac{1}{R^2}\left(R^{-2}(\chi,\psi),R^{-2}(\nu_i,\psi)\right) - \frac{1}{R^4}\Delta^*\psi\left(R^{-2}(\chi,\psi),\nu_i\right) + \left[\Delta^*\nu_i\frac{1}{R^2} + \left(R^{-2},\nu_i\right)\right]F\nabla\bullet\frac{F}{R^4}\nabla_{\perp}\chi$$
$$+\frac{F}{R^6}\left[\nu_i,(\chi',\psi)\right] - \frac{F}{R^4}\left[\chi',R^{-2}(\nu_i,\psi)\right] + \frac{1}{R^6}\Delta^*\psi F[\chi',\nu_i] - \frac{F^2}{R^8}(\chi'',\nu_i) + \Delta^*\nu_i R^{-4}\left((\chi,p) + \gamma p\Delta^*\chi\right)$$

References

[1] W. Park, E. Belova, G. Y. Fu, et al., Phys Plasmas **6,** 1796 (1999)
[2] S. C. Jardin, J. Breslau, N. Ferraro, J. Comput. Phys, **226** (2007) 2146
[3] N. Ferraro and S. Jardin, Phys Plasmas **13** (092101) 2006
[4] C. Sovinec, A. H. Glasser, G.A. Gianakon, et al, , J. Comput. Phys. **195** (2004) 355
[5] E. Caramana, J. Comput. Phys. **96** (1991) 484
[6] I. Bernstein, E. Frieman, M. Kruskal, and R. Kulsrud, Proc. Royal Soc. 244 1740 (1958)
[7] S. Jardin, J. Comp. Phys. **200** (2004) 133
[8] H. Strauss, Phys. Fluids **19** (1976) 134
[9] R. Hazeltine et al, Phys Fluids **28** (1985) 2466, R. Fitzpatrick et al, Phys Plasmas **11** (2004) 4713
[10] R. Schmaltz, Phys. Lett. A, **82** 14 (1981)
[11] R. Grimm, J. Greene, J. Johnson, in Meth. Comput Phys., Academic Press, 1976
[12] J. W. Demmell, J. R. Gilbert, Y. S. Li, SUPERLU: Users Guide, U. C. Berkeley, 2003
[13] E. Seegyoung, M.S Shephard, Engineering with Computers **22**:197—213 (2006)
[14] X. Li, M.S. Shephard, M.W. Beall, Comp. Meth. Appl. Mech. Engng. **194**:4915--4950(2005)
[15] O. Sahni , J. Muller, K.E. Jansen, M.S. Shephard. and C.A. Taylor , Comput. Methods Appl. Mech. Engrg. **195**(41-43):5634—5655 (2006)

Journal of Physics: Conference Series **125** (2008) 012045 doi:10.1088/1742-6596/125/1/012045

Toward petascale shock/turbulence computations

Johan Larsson, Eric Johnsen, Sanjiva K. Lele, and Parviz Moin

Center for Turbulence Research, Stanford University, Stanford, CA 94305

E-mail: jola@stanford.edu (J. Larsson)

Abstract. Progress toward highly resolved simulations of shock/turbulence interaction is described, with a focus on the scientific reasons for studying this problem as well as the computational issues involved. We illustrate the importance of computing shock/turbulence interaction at Reynolds numbers of at least one order of magnitude larger than previously done. We also illustrate the need to study the effect of turbulent Mach number on the interaction process. Scaling data for the code is used to estimate the computational resources needed for these novel simulations on petascale systems.

1. Introduction

Most fluid flows of scientific and engineering interest involve and are strongly affected by turbulence. For example, turbulence has a great effect on the process of mixing, which in turn has a major effect on flow separation, combustion, and other phenomena. When interacting with a shock wave, the state of turbulence changes dramatically. Clearly, then, the interaction of turbulence with a shock wave is a key feature in many high-speed engineering systems. The most fundamental problem in this area is that of isotropic turbulence passing through a nominally normal shock. This problem has received considerable attention in past theoretical [1], experimental [2], and computational (e.g., [3, 4]) studies. The linear theory by Ribner [1] predicts that the velocity and vorticity variances are amplified during the shock interaction and that there is an inviscid adjustment region behind the shock where turbulence adjusts to its postshock state. This theory is formally valid in the limits of infinite Reynolds number and zero turbulent Mach number, but has thus far [3, 4] been tested only at Reynolds numbers of $Re_\lambda \approx 20$. At these low Reynolds numbers, the viscous decay behind the shock is comparable to or larger than the inviscid adjustment, which makes direct comparisons to the linear theory difficult. Furthermore, the spectrum of turbulence is not truly broadband at such Reynolds numbers. The ultimate objective of the present study is to use direct numerical simulation (DNS) to produce a set of highly resolved databases at much higher Reynolds number, to be used in examination of the validity of the linear theory as well as in *a priori* and *a posteriori* studies of how to model shock/turbulence interaction in the context of large eddy simulation (LES) and engineering turbulence modeling.

2. Numerical method

The *Hybrid* code (see [5, 6]) solves the compressible Navier-Stokes equations for a perfect gas using solution-adaptive finite-difference schemes on Cartesian (but stretched) grids. Near shock waves or other discontinuities, a fifth-order accurate weighted essentially non-oscillatory (WENO) scheme is used [7]. In these regions, the equations are discretized in divergence (or conservative) form, thereby ensuring convergence to the correct weak solution. Away from shock waves, a sixth-order accurate central difference scheme is used. This scheme has nominally zero numerical dissipation, which is an important attribute for the prediction of broadband turbulence (see [5]). To aid the nonlinear stability, the convective terms are discretized in the split form

$$\frac{\partial \rho f}{\partial x_j} \rightarrow \frac{1}{2}\left(\frac{\partial \rho f}{\partial x_j} + f\frac{\partial \rho}{\partial x_j} + \rho\frac{\partial f}{\partial x_j}\right)$$

for $f = u_j, \rho u_i, \rho h_0$ in the conservation equations for mass, momentum, and total energy, respectively. Here ρ is the density, u_i is the velocity vector, $h_0 = e_0 + p/\rho$ is the total specific enthalpy, and e_0 is the total specific energy. The coupling of different discretizations globally conserves mass, momentum, and energy and was shown in [8] to be linearly stable. The semi-discrete system is integrated in time by using classic fourth-order, fully explicit Runge-Kutta.

The method is implemented by using C++ and is parallelized through domain decomposition, with communication handled by MPI. Being fully explicit, the method is expected to scale well. However, one should note that the solution-adaptive "hybrid" nature of the algorithm has implications for the computational cost. The WENO scheme is 5–10 times more expensive than the central difference scheme but is used only in the vicinity of shocks. Since shocks are two-dimensional surfaces, the number of grid points treated by WENO scales as N_d^2, where N_d is the number of points in each direction. Thus the fraction of points treated by WENO becomes ever smaller under grid refinement, which raises the possibility of "superlinear" speedup under grid refinement. However, it also raises the possibility of load-balancing problems because (with domain decomposition) a small subset of processors is likely to handle the more expensive WENO regions.

The code has been run on several machines. Here we present scaling data from the Franklin machine at the National Energy Researcher Scientific Computing Center.

A first test of the scalability is the weak scaling of the nearly incompressible Taylor-Green vortex in a periodic box. This problem does not generate shocks, and thus the WENO scheme is never used, implying a (theoretically) perfect load balance. Each full time step is timed in the code, as well as the cumulative time spent in communication (including waiting). The results are shown in figure 1 for $n = 1, 2, 4, \ldots, 4096$, where n is the number of processors. We note that the code is algorithmically different for $n = 1$; for example, the periodic boundary data is copied directly rather than through temporary communication arrays. For this reason the scaling efficiency is defined with respect to $n = 2$ as

$$\eta_{\text{weak}}(n, N) = \frac{t(2, 2 * N)}{t(n, n * N)},$$

where $t(n, N)$ is the CPU time for one step on n processors on a grid with N points. The efficiency is above 86% and 90% up to 4,096 processors with $N = 32^3 = 32,768$ and $N = 64^3 = 262,144$ points per processor, respectively. Representing a factor of 2000 increase in n, these efficiencies must be considered satisfactory.

3. Canonical shock/turbulence interaction

The most fundamental shock/turbulence interaction problem is that of isotropic turbulence passing through a nominally normal shock wave (or, equivalently, a normal shock passing through isotropic turbulence). The essence of the problem is shown in figure 2. The incoming turbulence is isotropic, as evidenced by the random orientation of the vortex cores. The shock compresses the turbulence in the x direction, increasing the vorticity and making the postshock turbulence axisymmetric with vortex cores predominantly oriented in the $y - z$ plane.

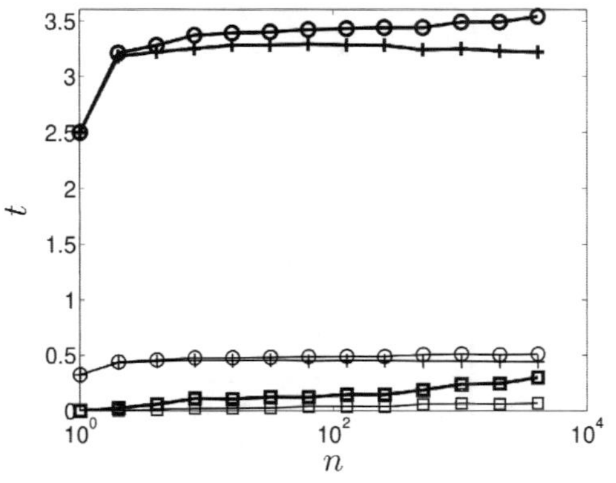

(a) Time in seconds for full step (circle), all communication (squares), and deduced compute time (plus).

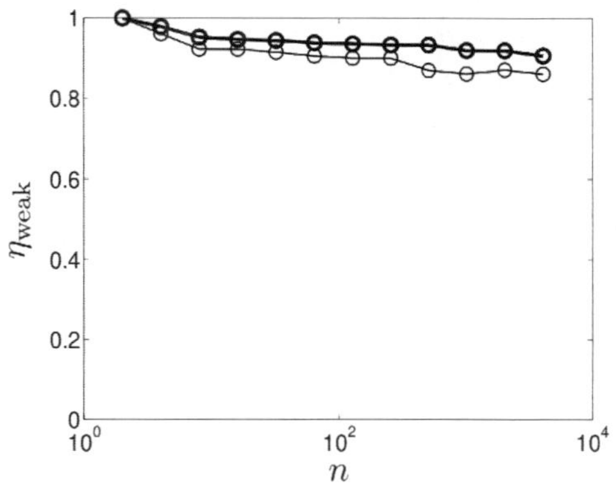

(b) Weak scaling efficiency η_{weak}.

Figure 1. Weak scaling with perfect load balancing, with base grid $N = 32^3$ (thin, lower curves in both plots) and $N = 64^3$ (thick, upper curves).

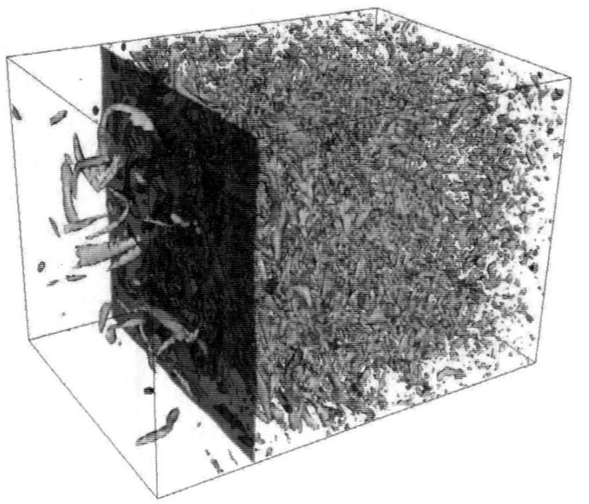

Figure 2. Snapshot of shock/turbulence interaction at $M = 2$, $M_t = 0.15$, and $Re_\lambda \approx 60$. The flow is from left to right, with the shock visualized by transparent isosurfaces of compression. Vortex cores are visualized by isosurfaces of the second invariant of the velocity gradient tensor, colored by the vorticity magnitude.

In order to minimize errors around the shock, a stretched grid is used in the streamwise direction with finer mesh spacing in the vicinity of the shock. The inflow turbulence is taken from a precomputed database of isotropic turbulence. At the outflow, a sponge region is used to gently drive the flow toward a laminar state. We note that the simulations presented here are underresolved and should be characterized as large eddy simulations rather than direct numerical simulations.

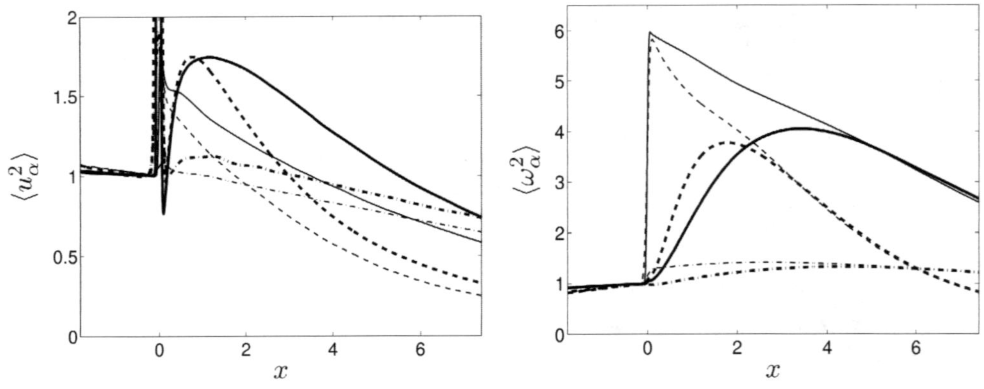

(a) Streamwise (thick) and transverse (thin) Reynolds stress.

(b) Streamwise (thick) and transverse (thin) vorticity.

Figure 3. Canonical shock/turbulence interaction on 590×128^2 grid. Variance of Reynolds stress and vorticity for $(M, M_t) = (2.0, 0.15)$ (solid), $(M, M_t) = (2.0, 0.30)$ (dashed), and $(M, M_t) = (1.1, 0.15)$ (dash-dotted).

The variances of velocity and vorticity fluctuations are shown in figure 3 for a sequence of cases. The transverse vorticity is directly amplified at the shock as a result of the compression, while the streamwise vorticity is initially unchanged. Behind the shock, the out-of-equilibrium turbulence adjusts toward an isotropic state, although the Reynolds stresses never reach isotropy in these runs. One outstanding question in shock/turbulence interaction is whether the turbulence truly returns to isotropy. The low Reynolds number of the present (and all previous) runs implies a quick viscous decay behind the shock, which overwhelms any return to isotropy. Large-scale calculations at higher Reynolds numbers are needed to illuminate this issue.

At high enough turbulent Mach number M_t, the turbulent pressure fluctuations become comparable to the pressure jump associated with the shock, which significantly alters the instantaneous shock profile. Some instantaneous profiles of the compression (negative dilatation) and density along the x axis are shown in figure 4 for such a case. The instantaneous structure of the shock varies wildly, from being twice as strong as average, to being replaced by a smooth compression wave, to being replaced by two weaker shocks. This intermittency is largely absent at lower values of M_t and represents a second outstanding fundamental question in shock/turbulence interaction.

3.1. *Scalability with imperfect load balancing*

The solution-adaptive algorithm uses the WENO scheme only around the shock, which, together with the local sponge region, implies a load imbalance. While the instantaneous extent of the WENO region is solution-dependent (and unknown), one can predict the average extent based on prior coarse-grid runs.

An example of the load imbalance is shown in figure 5(a). While the code timings should be averaged over longer times, clearly both the WENO scheme (processors 48-63) and the sponge region (processors 224-255) still induce higher costs. A simple *ad hoc* fix is to factor in the higher costs of the WENO scheme and the sponge region, along with the known locations of these regions, in the domain decomposition. As shown in figure 5(a), this leads to a better load balance and a small reduction in cost. We point out that the algorithm consists of several steps,

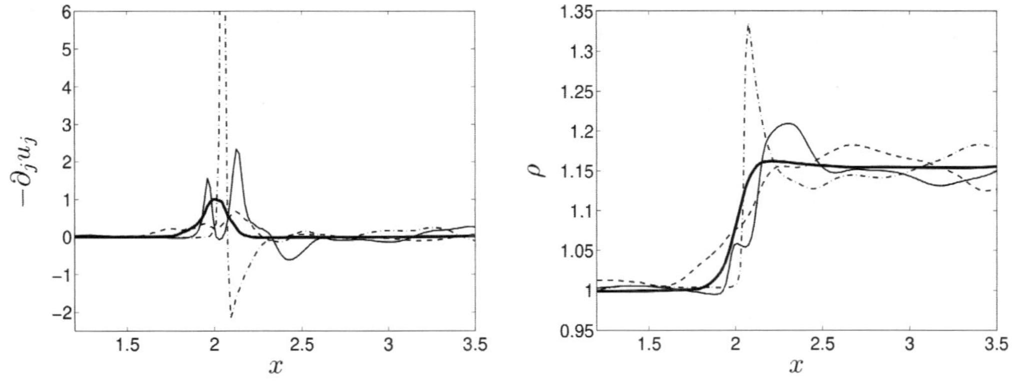

(a) Compression. (b) Density.

Figure 4. Canonical shock/turbulence interaction on 590×128^2 grid. Instantaneous (thin) and average (thick) profiles of compression (negative dilatation) and density for $(M, M_t) = (1.1, 0.15)$.

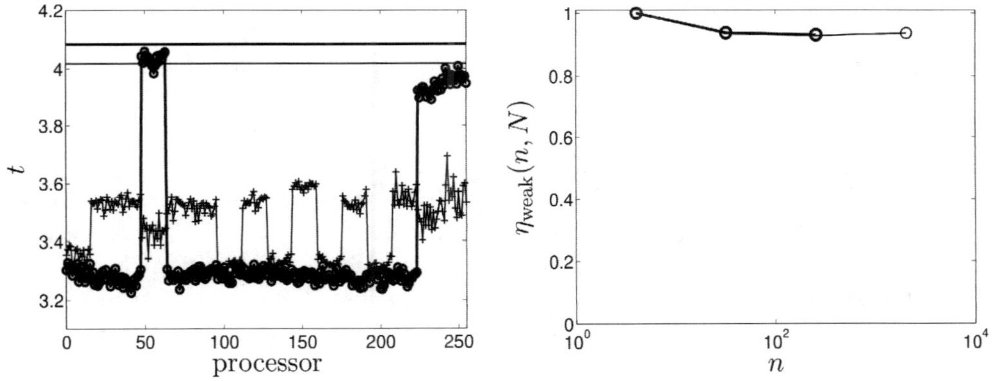

(a) Time in seconds for full step (solid) and deduced compute time (symbols) on $N = 1024 \times 256^2$ grid, with standard (thick, circle) and *ad hoc* (thin, plus) domain decomposition.

(b) Weak scaling efficiency $\eta_{\text{weak}}(n, N)$ with base grid $N = 32^3$ (thin) and $N = 64^3$ (thick). Efficiency defined with respect to $n = 4$.

Figure 5. Computational efficiency for shock/turbulence problem.

only one of which (the most expensive) incurs the higher cost of WENO and the sponge. Hence the *ad hoc* decomposition induces load imbalance in the remaining algorithmic steps, which explains why all processors spend significant time waiting for communication.

The weak scaling efficiency is above 92% for both 32^3 and 64^3 points per processor, going from $n = 4$ to $n = 2048$.

3.2. Cost estimates at higher Reynolds number

Well-resolved direct numerical simulation (DNS) requires adequate resolution of the viscous dissipation, which in isotropic turbulence amounts to having the maximum wavenumber $k_{\max} =$

Table 1. Estimates of total CPU time and Reynolds number upstream of the shock for a well-resolved DNS of $(M, M_t) = (2.0, 0.15)$ shock/turbulence interaction using $k_0 = 4$. $n_{1\text{Gb}}$ and $n_{24\text{h}}$ are the number of processors for 1 GB of memory and 24-hour turnaround, respectively.

N_y	Re_λ	CPU-hours	$n_{1\text{Gb}}$	$n_{24\text{h}}$
256	24	2.6×10^3	32	110
384	54	1.4×10^4	110	570
512	96	4.3×10^4	250	1800
768	220	2.2×10^5	850	9100
1024	390	6.9×10^5	2000	29000

$\pi N_y/L_y \gtrsim 1.5/\eta_k$, where N_y is the number of points in the transverse directions, L_y is the transverse domain size, and η_k is the Kolmogorov length scale. Consider turbulence at $M = 2$, $M_t = 0.15$, and some Re_λ immediately upstream of the shock. The shock compression modifies the Kolmogorov scale by about a factor of $(\rho_1/\rho_2)^{11/8}(p_2/p_1)^{3/8} \leq 1$, and hence the critical resolution is downstream of the shock. The upstream isotropic turbulence approximately satisfies $\eta_k \approx 0.5\lambda/\sqrt{Re_\lambda}$, where λ is the Taylor length scale. We need a relation between the wavenumber of peak energy k_0 and λ upstream of the shock. This is Re_λ-dependent, but a very crude estimate is to use the present preliminary data at $Re_\lambda \approx 60$, which gives $\lambda \sim L_y/(2\pi k_0)$. We then get, for a given transverse resolution N_y,

$$Re_\lambda \lesssim 0.006(N_y/k_0)^2$$

immediately upstream of the shock. This is a more stringent criterion than many previous studies but is necessary to adequately resolve the postshock dissipative motions.

Now consider the code performance. Using the data in figure 5, the code needs 1.6×10^{-5} seconds per grid point and step on 256 processors or more. Furthermore, the code uses about 1 GB of memory for 1.6×10^6 points per processor. Requirements on the domain size and the shock-normal grid spacing imply $3N_y$ points in the streamwise direction, and the preliminary runs needed about $47N_y$ time steps for statistical convergence. The final estimates, listed in table 3.2, make for sobering reading.

4. Summary and future work

The progress toward large-scale calculations of the canonical shock/turbulence interaction discussed here is part of a larger effort to developing improved computational methods for complex shock/turbulence interaction problems. The canonical shock/turbulence problem described here serves two purposes in this larger effort. First, it is a useful problem for computational efficiency assessment and improvement because it includes all the relevant algorithmic complexities. The *Hybrid* code has been shown to scale well up to 4096 processors, which is promising for future production runs. Work on profiling the code to identify single-processor bottlenecks is under way in collaboration with the SciDAC Performance Engineering Research Institute. Second, the shock/turbulence calculations will produce large databases at different combinations of turbulent and mean flow Mach numbers, at higher Reynolds numbers than previously possible. These databases will be of use both in theoretical developments, for example, for assessment of the validity of Ribner's linear theory at higher levels of compressibility, and in model development, for example, for *a priori* studies or *a posteriori* comparisons in subgrid scale modeling.

Acknowledgments
Financial support has been provided by the DOE Scientific Discovery through Advanced Computing (SciDAC) program, with additional support from the Natural Sciences and Engineering Research Council of Canada.

References

[1] Ribner H S 1953 Convection of a pattern of vorticity through a shock wave NACA Report 1164
[2] Agui J H, Briassulis G and Andreopoulos Y 2005 *J. Fluid Mech.* **524** 143–195
[3] Lee S, Lele S K and Moin P 1993 *J. Fluid Mech.* **251** 533–562
[4] Lee S, Lele S K and Moin P 1997 *J. Fluid Mech.* **340** 225–247
[5] Larsson J, Lele S K and Moin P 2007 *Annual Research Briefs* (Center for Turbulence Research) pp 47–57
[6] Larsson J, Cook A, Lele S K, Moin P, Cabot W, Sjögreen B, Yee H and Zhong X 2007 *J. Phys.: Conf. Ser.* **78** 012014
[7] Jiang G S and Shu C W 1996 *J. Comput. Phys.* **126** 202–228
[8] Larsson J and Gustafsson B 2008 *J. Comput. Phys.* **227** 2886–2898

Journal of Physics: Conference Series **125** (2008) 012046 doi:10.1088/1742-6596/125/1/012046

Particle-in-cell simulations for fast ignition

C Ren[1,2], J Tonge[3], G Li[1], F Fiuza[5], R A Fonseca[5], J May[3], W B Mori[3,4], L O Silva[5], T L Wang[4] and R Yan[1]

[1] Department of Mechanical Engineering and Laboratory for Laser Energetics, University of Rochester, Rochester, NY 14627, USA

[2] Department of Physics and Astronomy, University of Rochester, Rochester, NY 14627, USA

[3] Department of Physics and Astronomy, University of California, Los Angeles, CA 90095, USA

[4] Department of Electrical Engineering, University of California, Los Angeles, CA 90095,USA

[5] GoLP/Instituto de Plasmas e Fusao Nuclear, Instituto Superior Tecnico, 1049-001 Lisboa, Portugal

E-mail: cren@lle.rochester.edu

Abstract. The hole-boring scheme in fast ignition is studied via largee-scale, two-dimensional particle-in-cell simulations in two steps. First, laser channeling in millimeter-scale underdense plasmas is simulated. The results show a highly nonlinear and dynamic process involving longitudinal plasma buildup, laser hosing, channel bifurcation and self-correction, and electron heating to relativistic temperatures. The channeling speed is much less than the linear group velocity of the laser. Low-intensity channeling pulses are preferred to minimize the required laser energy. The channel is also shown to significantly increase the transmission of an ignition pulse. In the second step, the interactions of the ignition pulse and a hundred-critical-density plasma are simulated to study hot electron generation and transport. The results show that at ultra-high intensities, $I > 5 \times 10^{19}\mathrm{W/cm}^2$, most of the electrons transporting energy through 50μm of 100 times critical density plasma are in a relatively low energy range. The fraction of laser power that transits the dense plasma and is deposited into a dense core increases with laser intensity. Overall these results show the promise of using ultra-high-intensity ignition pulses in the hole-boring scheme.

1. Introduction

In the "hole-boring" scenario of fast ignition (FI) [1], an ignition pulse needs to first propagate through a millimeter-scale underdense plasma to reach a critical surface with a critical density $n_c = \omega_0^2 m_e/(4\pi e^2)$, where m_e and e are the electron mass and charge, respectively, and ω_0 is the pulse frequency. There the ignition pulse may continue to push forward into an overdense plasma through its ponderomotive pressure (hole-boring) and relativistic transparency. In the meantime its energy is absorbed by the plasma to generate energetic (so-called hot) electrons, which are to reach a dense core region and deposit their energy there to create a hot spot to start fusion reactions. In order to avoid energy loss of the ignition pulse in the millimeter-scale underdense plasma, a channeling pulse, which could be the prepulse of the ignition pulse or a separate pulse, has been proposed [1] to produce a low-density channel to reduce the nonlinear interactions of the ignition pulse in the underdense region.

Key information for the hole-boring scenario includes the energy needed for the channeling pulse, the ignition pulse to hot electron conversion efficiency, the spectrum of these electrons, and the transport of the hot electrons to the core. Many of the processes involved are highly nonlinear, and no adequate theory exists. In this paper we study the scenario in two simulations, both with the fully explicit particle-in-cell (PIC) code OSIRIS [2]. The first is on the laser channeling process [3]. The second is on the ignition pulse's interactions with the overdense plasma to study the hot electron generation and transport, assuming a low-density channel has been created through channeling and hole boring and the ignition pulse can reach the $n = 100n_c$ surface without much loss.

2. Laser channeling

Most previous experiments and simulations on laser channeling were done in 100 μ m-scale plasmas [4, 5, 6, 7, 8, 9, 10]. owever, the underdense region of an actual FI target is about 1000 μ m long. The residual plasma in the channel can continue to interact with the latter part of the channeling pulse and make the channeling process truly dynamic. To study the channeling process in more realistic settings, we have performed full-scale two dimensional (2D) PIC simulations with an underdense region consisting of a deuterium-tritium (DT) plasma as large as 987 μ m $\times 401\mu$ m and with a density profile exponentially rising from $0.1n_c$ to $1.02n_c$. The channeling pulse has a Gaussian transverse profile with a full-width-half-maximum (FWHM) intensity spot size of 16-47 μ m and a peak intensity between 10^{18} and 10^{20} W/cm^2. Both s and p polarization are used to infer 3D effects. The channeling process is simulated for up to 15 ps. We have also performed simulations with a preformed channel to study the transmission of the ignition pulse in the channel. The 2D simulations typically employ 79 million cells and 158 million particles and run 7×10^4 steps. In addition, we have also performed 3D simulations with a plasma volume of $(90\mu \text{ m})^3$. These runs employ 5.8 billion cells and 11.6 billion particles and run 10^4 steps. For details see [3].

Channeling in millimeter-scale plasmas indeed has many new phenomena that were not present in previous short-scale experiments and simulations, including plasma buildup to above critical density in front of the laser, laser hosing/refraction, and channel bifurcation and self-correction[3]. These phenomena are illustrated in figure 1.

As a result, the channeling speed oscillates and is much less than the laser linear group velocity (figure 2b). The simulations find that the effective channeling time T_c and the total energy E_c to reach the critical surface scale with the laser intensity I as $T_c \approx 2.9 \times 10^2 (I/10^{18}\text{W/cm}^2)^{-0.64}$ ps and $E_c \approx 1.7(I/10^{18}\text{W/cm}^2)^{0.36}$ kJ. The scaling for the first time shows that low-intensity channeling pulses are preferred to minimize the required energy but with an estimated lower bound on the intensity of $I = 5 \times 10^{18}$ W/cm^2 if the channel is to be established within 100 ps (figure 2a). To study the effect of the channel on the transmission of the ignition pulse, we have also performed simulations with a preformed channel. The preformed channel has a plasma density of $0.05n_c$, a width the same as that of the incoming pulse, and a length of 987μ m. The plasma temperature is the same as that outside the channel (1 keV). Figure 2c plots the forward Poynting flux within the channel received at $n_0 = n_c$, normalized to that entering the channel front, for an $I = 10^{19}$ W/cm^2 pulse. It shows that the front end of the pulse still suffers an energy loss but the main part can have an 80% transmittance. The increased transmission is mainly due to faster heating in a lower density plasma. In contrast, without the preformed channel, pulses of intensities of $I = 10^{19}$ W/cm^2 and $I = 10^{20}$ W/cm^2 have negligible transmission.

From a limited number of 3D runs, we have also found the channeling speed in 3D is greater than in 2D. This is because of stronger self-focusing in 3D, where the laser vector potential a scales with the spot size w as $a^2 \sim w^{-2}$, than in 2D, where $a^2 \sim w^{-1}$. This leads to a greater laser ponderomotive force and thus a greater channeling speed in 3D.

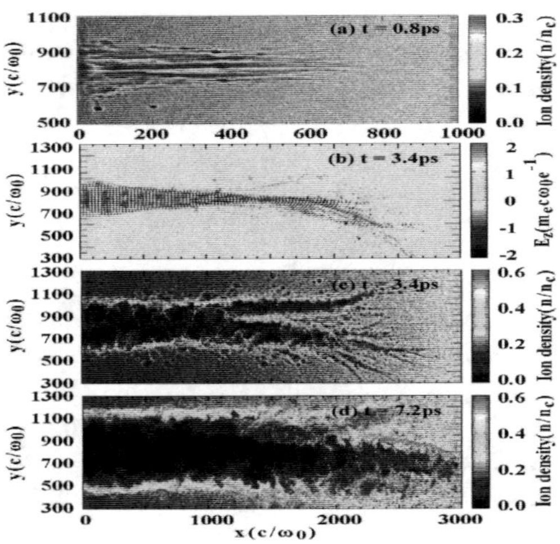

Figure 1. Results from a simulation with a 10 particle-per-cell, $I = 10^{19}\mathrm{W/cm}^2$, p-polarized laser and $n_0 = 0.1-0.3n_c$: (a) ion density at t=0.8 ps showing micro channels formed; (b) laser E-field showing laser hosing, (c) ion density showing channel bifurcation at t=3.4 ps, and (d) ion density at t=7.2 ps showing channel self-correction.

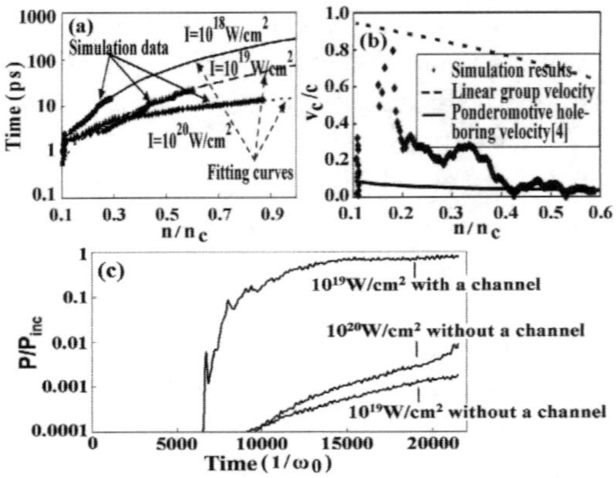

Figure 2. (a) Time for the channel to reach different initial densities under different pulse intensities and the corresponding fitting curves. Data was taken from the 1 particle-per-cell and $n_0 = 0.1 - 1.0n_c$ runs. (b) The channeling speed v_c in the $I = 10^{19}\mathrm{W/cm}^2$ case. (c) The forward Poynting flux within 1 laser spot size received at $n_0 = n_c$, normalized to that at the left entrance, for cases with and without a preformed channel.

3. Hot electron generation and transport

In the fast ignitor concept for inertial fusion a target is compressed to high density, up to 300g/cc, forming a collisional ($\nu_{ei} \approx \omega_p$ where ν_{ei} is the electron ion collision frequency and ω_p is the plasma frequency) plasma at the core of the target. The core is surrounded by collisionless plasma ($\nu_{ei} \ll \omega_p$) with a radial density profile dropping from above solid density to critical density (n_c) for a 1 μm laser in a few hundred microns. The energy flux generated by a high intensity laser and carried by hot electrons is used to ignite a small high density region of the core [1]. These hot electrons must couple through collisions to the high density core. This has significantly constrained the design of fast ignition experiments by limiting the intensity of the ignition laser to approximately $5 \times 10^{19} W/cm^2$ [11]. However, at this intensity with spot size of $20\mu m$ (the desired spot size of the spot at the core) the laser power would only be .3 PW while the desired power must be greater than .83 PW at 100% coupling efficiency [12]. If higher intensity can be used then this limit can be overcome. This question can be addressed using PIC simulations.

The focus of our 2D PIC simulations for laser absorption and electron transport is the transport of energetic electrons across the overdense collisionless plasma, the amount of energy delivered to the core, and the spectrum of the electrons that reach the core. In these simulations the initial conditions assume that the target has been compressed, and the channeling pulse has created a low-density channel in the path of the ignition laser. We have constructed a model that includes isolated boundaries, self-consistent generation of electrons at the laser plasma interface, self-consistent movement of the laser-plasma interface due to laser and plasma pressure, self consistent return currents, and an absorbing core. The simulation box is $19000\Delta \times 16400\Delta$ with the grid resolution Δ half a skin depth for $100n_c$, the maximum density within the target (figure 3). The isolated targets in the simulations are a 50 μm-radius overdense plasma with a 20μ m-radius dense core offset 20μm to the right of center. The purpose of the offset is to increase the distance that the fast electrons must traverse from the laser-plasma interaction region to the target core. The collisional core is modeled by a velocity drag on the electrons. The drag models collisions with a massive species at rest in the lab frame in the core. The drag force, $F_{drag} = A(v/c)^{-2}/\gamma$, is proportion to the inverse of the particle energy for relativistic particles with velocity near the speed of light and applies only for electrons with energy 20KeV< E <10MeV. The drag coefficient, A, is set so that a free streaming 1 MeV beam of electrons would penetrate 10μm into the core. This allows simulations to run for long times, at least 2.5 ps, without high-energy electrons refluxing through the target. The integrated line density from the target edge to the core is equal to the line density of a target with an exponentially rising density profile and with a 100μ m-scale length per decade from the target edge to a 10^{23}/cc density. This mass distribution is more compact than actual targets for simulation purposes. A 1μ m-wavelength ignition laser with a spot size of 20μ and intensities of 5×10^{19} W/cm^2, 2×10^{20} W/cm^2, and 8×10^{20} W/cm^2 are used.

The simulation can be broken into three regions: the laser-plasma interaction region where a flux of hot electrons is generated, the overdense layer of collisionless plasma through which the hot electrons are transported, and the collisional core. The electrons in the laser-plasma interaction region form a distribution such that the energy is transported to the core while allowing return currents to replenish electrons at the laser-plasma interface. In general the electron momentum distribution has characteristics of a relativistic thermal bulk with lower energy electrons moving toward the laser-plasma interface and energy transport from higher-energy electrons in the bulk directed toward the core.

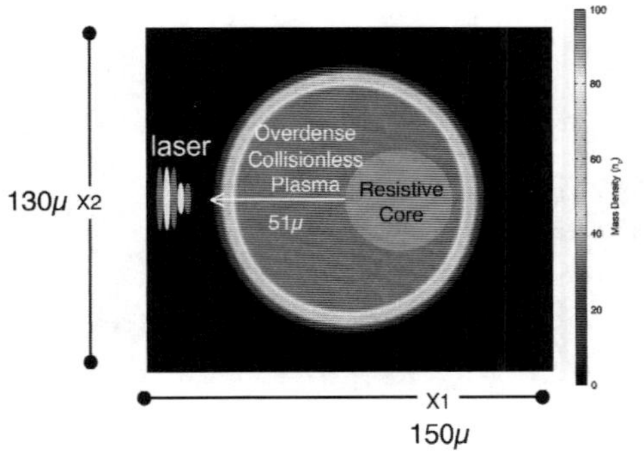

Figure 3. Schematic of target in simulation box. A laser strikes the target on the left generating hot electrons. The hot electrons traverse a large volume of overdense collisionless plasma and are absorbed into the core.

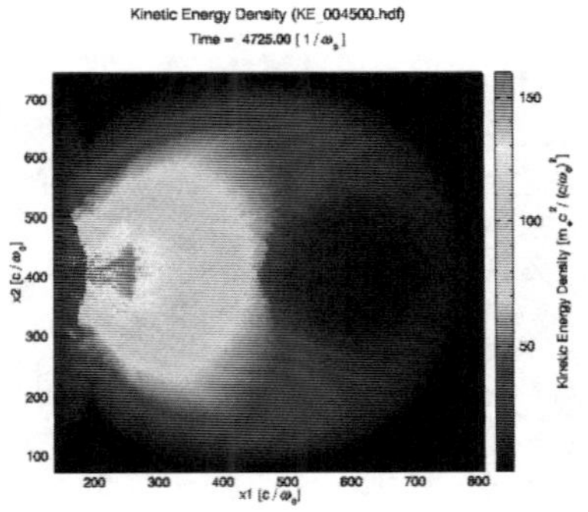

Figure 4. Spatial distribution of electron energy density in the target for I = $2 \times 10^{20} W/cm^2$ laser after 2ps. A sharp discontinuity exists at the laser-plasma interface (on the left) where the laser and hot plasma are in pressure balance. A hot spot builds up in front of the laser, and hot electrons are absorbed into the core.

The spatial distribution of electron kinetic energy shows a sharp interface where the energy is absorbed from the laser, as illustrated in figure 4. At this interface the electron energy density is in pressure balance with the energy density of the laser. Downstream a hot spot has developed in the collisionless plasma between the laser-plasma interface and the core. This hot spot is caused by wave turbulence generated as the hot electrons move through the bulk toward the core. The amount of energy that is transported into the core and the efficiency with which that energy is deposited determine the viability of any fast ignition scheme. The coupling efficiency, defined as power transmitted to the core normalized by the laser power, increases with laser intensity, while the bulk of this energy is still carried by electrons that are energetically able to interact with the target core. This is the advantage of higher-intensity lasers. Ultra-high-intensity ignition lasers deliver greater power because of both higher efficiency and higher energy density.

4. Summary

We have simulated the hole-boring scenario in FI in two PIC simulations. The laser channeling simulations have shown that, despite channel bifurcation from laser hosing, a low-density straight channel can be created with a speed approaching the hole-boring speed. Low-intensity channeling pulses are preferred to minimize the required laser energy. The channel is also shown to significantly increase the transmission of the ignition pulse to 80%. The simulations on hot electron generation and transport have shown that even an $I = 8 \times 10^{20}\mathrm{W/cm^2}$ pulse converts its energy mainly to hot electrons whose energy is suitable for FI. This is in contrast to previous predictions based on the ponderomotive scaling [13] that ultra-intense pulses would produce electrons too energetic to stop in the core. These results open up the prospect of using ultra-intense pulses in FI.

Acknowledgments

This work was supported by U.S. Department of Energy under Grants Nos. DE-FG02-06ER54879, DE-FC02-04ER54789, DE-FG52-06NA26195, and DE-FG02-03ER54721 and by FCT (Portugal). Simulations were carried out at the National Energy Research Scientific Computing Center through an INCITE award and on the UCLA DAWSON Cluster under grant no. NSF-Phy-0321345.

References

[1] Tabak M, *et al.* 1994 *Phys. Plasmas* **1** 1626
[2] Hemker R G 2000 UCLA Ph.D. thesis; see also Fonseca R A, *et al.* 2002 *Lecture Notes in Computer Science* **2331** 342 (Heidelberg: Springer-Verlag)
[3] Li G, Yan R, Ren C, Wang T L, Tonge J and Mori W B 2008 *Phys. Rev. Lett.* **100** 125002
[4] Young P E, *et al.* 1995 *Phys. Rev. Lett.* **75**, 1082; *Phys. Plasmas* **2** 2825
[5] Borghesi M, *et al.* 1997 *Phys. Rev. Lett.* **78** 879
[6] Fuchs J, *et al.* 1998 *Phys. Rev. Lett.* **80** 1658
[7] Mackinnon A J, *et al.* 1999 *Phys. Plasmas* **6** 2185
[8] Tanaka K A, Hashimoto H, Kodama R, Mima K, Sentoku Y and Takahashi K 3283 *Phys. Rev. E* **60** 3283
[9] Pukhov A and Meyer-ter-Vehn J 1996 *Phys. Rev. Lett.* **76** 3975
[10] Sentoku Y, Kruer W, Matsoka M and Pukhov A 2006 *Fusion Science and Technology* **49** 278
[11] Tabak M, Clark D S, Hatchett S P, Key M H, Lasinski B F, Snavely R A, Wilks S C, Town R P J, Stephens R, Campbell E M, Kodama R, Mima K, Tanaka K A, Atzeni S and Freeman R 2005 *Phys. Plasmas* **12** 057305
[12] Atzeni S 1999 *Phys. Plasmas* **6** 3316
[13] Wilks S C, Kruer W L, Tabak M and Langdon A B 1992 *Phys. Rev. Lett.* **69** 1383
[14] Ren C, Tzoufras M, Tsung F S, Mori W B, Amorini S, Fonseca R A, Silva L O, Adam J C and Heron A 2004 *Phys. Rev. Lett.* **93** 185004
[15] Tzoufras M, Ren C, Tsung F S, Tonge J W, Mori M, Fiori M, Ronseca F A and Silva L O 2006 *Phys. Rev. Lett.* **95** 105002

Scientific and computational challenges of the Fusion Simulation Project (FSP)

William M. Tang
Princeton University, Princeton Plasma Physics Laboratory, Princeton, NJ 08544, USA

E-mail: wtang@princeton.edu

Abstract. This paper highlights the scientific and computational challenges facing the Fusion Simulation Project (FSP). The primary objective is to develop advanced software designed to use leadership-class computers for carrying out multiscale physics simulations to provide information vital to delivering a realistic integrated fusion simulation model with unprecedented physics fidelity. This multiphysics capability will be unprecedented in that in the current FES applications domain, the largest-scale codes are used to carry out first-principles simulations of mostly individual phenomena in realistic 3D geometry while the integrated models are much smaller-scale, lower-dimensionality codes with significant empirical elements used for modeling and designing experiments. The FSP is expected to be the most up-to-date embodiment of the theoretical and experimental understanding of magnetically confined thermonuclear plasmas and to provide a living framework for the simulation of such plasmas as the associated physics understanding continues to advance over the next several decades. Substantive progress on answering the outstanding scientific questions in the field will drive the FSP toward its ultimate goal of developing a reliable ability to predict the behavior of plasma discharges in toroidal magnetic fusion devices on all relevant time and space scales. From a computational perspective, the fusion energy science application goal to produce high-fidelity, whole-device modeling capabilities will demand computing resources in the petascale range and beyond, together with the associated multicore algorithmic formulation needed to address burning plasma issues relevant to ITER – a multibillion dollar collaborative device involving seven international partners representing over half the world's population. Even more powerful exascale platforms will be needed to meet the future challenges of designing a demonstration fusion reactor (DEMO). Analogous to other major applied physics modeling projects (e.g., ASCI), the FSP will need to develop software in close collaboration with experimental researchers and validated against experimental data from tokamaks around the world. Specific examples of expected advances needed to enable a premier comprehensive integrated modeling capability are discussed.

1. Introduction

From the perspective of DOE Energy Undersecretary Raymond Orbach, the overarching objective of the Fusion Simulation Project (FSP) [1, 2] is to "produce a world-leading predictive simulation capability that will be of major benefit to the overall science and mission goals of the U.S. Fusion Energy Science Program." Such a capability must be (!) an important asset for optimizing U.S. participation in ITER – a multibillion dollar collaborative burning plasma fusion experiment involving seven international partners representing over half the world's population; (2) relevant to major current and planned toroidal fusion devices both nationally and internationally; and (3) strategically vital to U.S. interests in developing a future demonstration reactor (DEMO). This paper highlights the associated major scientific and computational challenges, which are truly formidable and will demand a strong alliance between DOE's Fusion Energy Science (FES) and Advanced Scientific Computing Research (ASCR) programs for developing advanced software designed to use leadership-class computers for carrying out multiscale physics simulations to provide information vital to delivering a realistic integrated fusion simulation model with unprecedented physics fidelity. Such a modern multiphysics capability will be unprecedented in that in the current FES applications domain, the largest-scale codes are used to carry out first-principles simulations of mostly individual phenomena in realistic 3D geometry while the integrated models are much smaller-scale, lower-dimensionality codes with significant empirical elements used for modeling and designing experiments. The integrated modeling capability of the FSP is expected to embody the theoretical and experimental understanding of magnetically confined thermonuclear plasmas and to provide a living framework for the simulation of such plasmas as the associated physics understanding continues to advance over the next several decades. Substantive progress on answering the outstanding scientific questions in the field will drive the FSP toward its ultimate goal of developing a reliable ability to predict the behavior of plasma discharges in toroidal magnetic fusion devices on all relevant time and space scales.

If successful, the FSP can be expected to enhance the return on U.S. investments in fusion experiments and help identify world leadership opportunities for the U.S. fusion program. In particular, the FSP will better enable the study of burning plasmas and greatly aid the U.S. role in the operation of the ITER experiment and in harvesting the associated scientific knowledge. This could enable discovery of new modes of operation, with possible performance enhancements and improvements needed for a demonstration fusion reactor. From a computational perspective, the fusion energy science application goal to produce high-fidelity, whole-device modeling capabilities will demand computing resources in the petascale range and beyond, together with the associated multi-core algorithmic formulation needed to address ITER burning plasma issues. Even more powerful exascale platforms will be needed to meet the future challenges of designing a demonstration fusion reactor. Analogous to other major applied physics modeling challenges (e.g., NNSA's Advanced Strategic Computing Initiative, ASCI), the FSP software will need to be developed in close collaboration with experimental researchers and validated against experimental data from tokamaks around the world. Examples of expected advances needed to enable a comprehensive integrated modeling capability are the following:

- Effective coupling of state-of-the-art codes for the plasma core and plasma edge regions
- Effective coupling of state-of-the-art codes for magnetohydrodyamics (MHD) and auxiliary heating of the plasma via RF waves
- Development of more realistic reduced models based on results obtained from the DNS-type (direct numerical simulation) major codes which use petascale capabilities
- Development of advanced frameworks and workflow management methods needed for code coupling
- Development of an appropriate verification and validation effort to ensure reliable predictive capability

1.1 General perspective

While many of the technologies used in ITER will be the same as those required in an actual demonstration power plant, further science and technology are needed to achieve the 2500 MW of continuous power with a gain of 25 in a device of similar size and field. Accordingly, strong R&D programs are needed to harvest the scientific knowledge from ITER and leverage its results. Advanced computations in tandem with experiment and theory [3] are essential in this mission – a point well-illustrated in the FSP Workshop Report's first figure [4], which depicts the imperative for the FSP to leverage ongoing investments in FES' base theory and experimental programs, in ASCR's computer science and applied math programs, and in the interdisciplinary SciDAC program. The associated research demands the accelerated development of computational tools and techniques that aid the acquisition of the scientific understanding needed to develop predictive models that can prove superior to extrapolations of experimental results. This is made possible by access to leadership-class computing resources that allow simulations of increasingly complex phenomena with greater physics fidelity. With ITER and leadership-class computing being two of the most prominent missions of the DOE Office of Science, whole device integrated modeling, which can achieve the highest possible physics fidelity, is a most worthy exascale-relevant project for producing a world-leading realistic predictive capability for fusion.

This should prove to be of major benefit to U.S. strategic considerations for energy, ecological sustainability, and global security.

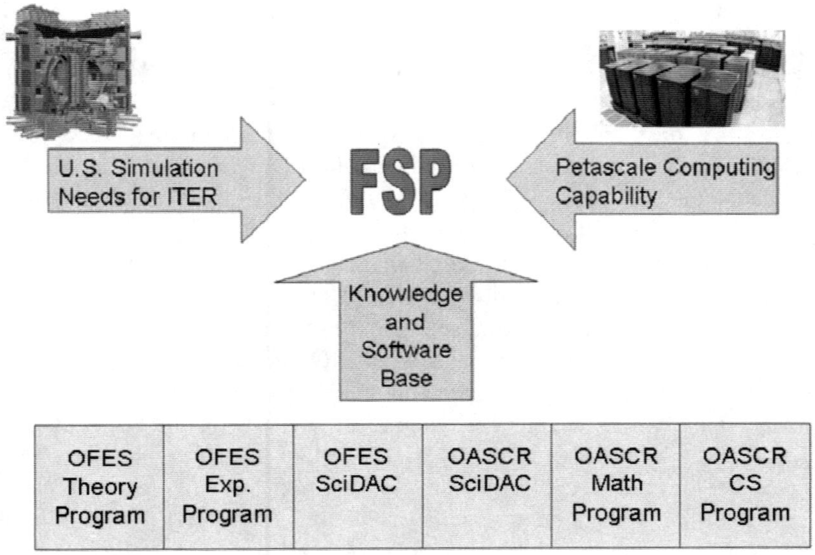

Figure 1. Illustration of expected strong connections to DOE's highest priority missions involving ITER and leadership computing as well as to FES theory and experiment, ASCR computer science and applied mathematics, and to the SciDAC program (DOE FSP Workshop Report, July 2007)

1.2 Situation analysis

Experiment and traditional theory are essential components of the fusion energy science research program with a key overarching goal being to reach a level of scientific understanding to enable an accurate predictive capability of a burning tokamak plasma. An experiment encompasses all the realistic physics but is limited in its scalability by the hardware. Traditional theory makes simplifications to first-principles equations to enable analytical solutions in special limits. Simulation ideally bridges the gap of experiment and traditional theory by taking advantage of state-of-the-art development in applied mathematics, computer science, and high-performance computers [3]. However, the associated challenge of closing the gap between idealized (largely empirically based) modeling and high-end first-principles-based simulations of mostly individual plasma dynamics is formidable. While the ideal goal is still far away, current progress has shortened the learning curve for developing more complete multiphysics models that are closer to the first-principles equations. Basically, the challenge involves capturing into more realistic reduced modules the most important physics trends produced by the aforementioned large-scale first-principles codes that successfully simulate important individual phenomena in realistic 3D geometry. An example is the ongoing development of physics-based transport models [5] that incorporate the predictions of advanced microturbulence simulations that are enabled by access to leadership-class computing platforms at the terascale and beyond.

A reasonable approach for the FSP to integrate the most complete physics models available is to begin with binary integration efforts (e.g., SciDAC proto-FSP projects SWIM [6] and CPES [7]) and then extend to higher-dimensional integrations to provide a progressively more realistic simulation of experiments. Since many of the physics models are still rapidly evolving, a complementary task involves providing a framework for incorporating new physics as it evolves. An example of this type of effort is the SciDAC proto-FSP FACETS project [8]. There are many other examples of smaller efforts aimed at integrating various physics components with a common set of challenges being: (1) keeping up with the constant evolution of validated physics models and (2) dealing with the incompatibility of various physics components (e.g., with respect to coding structure, dimensionality, and scales). Since the relevant scientific areas are not equally developed, it will be necessary to judiciously select an integration strategy working in concert with theory and experiment. An overarching scientific and computational challenge for the FSP is accordingly to provide a living framework for physics as it advances, making it as straightforward as possible to keep the models up to date, while enforcing standard interfaces to minimize the difficulty of different modules (models) communicating with each other and thereby easing the process for strongly coupled multiscale integration.

As evident from the DOE FSP Workshop Report (July 2008), it is fortunate that an excellent collaborative relationship between fusion energy scientists supported by FES and the computer science/applied math scientists supported by ASCR already exists. Much of the admirable depth of such alliances is due to DOE's truly interdisciplinary Scientific Discovery through Advanced Computing (SciDAC) program, which has now been in place for over six years. The FSP not only builds on and updates the original ISOFS/FSP J. Dahlburg Report in 2002 (strongly endorsed by FESAC at that time) [1] and the D. Post FSP Steering Committee Report in 2004 [2] but has also been encouraged by U.S. ITER leadership [4]. Consistent with the recommended levels in the Dahlburg Report, the targeted budget for the proposed FSP is around $25M per year with a 15-year timeline. As noted in the earlier reports, this is in line, for example, with the $25M per year allocated to just the University Alliances portion of the Accelerated Scientific Computing Initiative (ASCI) Program over the past decade. A lesson learned from major integrated modeling projects in areas such as ASCI is that in order for the FSP to succeed, a partnership of all the stakeholders including fusion physicists, applied mathematicians and computer scientists, with fully supported computing resources and stable funding is essential. The development of the advanced multiphysics codes targeted by the FSP is expected to take advantage of the ongoing FES investments in basic theory, the ongoing SciDAC program, and new developments involving joint experiment-theory-modeling efforts to predict and improve tokamak performance as time progresses. It will also be complemented by the expertise residing in ASCR's computer science and applied mathematics programs to help develop the needed algorithms and improved mathematical models capable of producing high-performance software compatible with the challenging multicore architectures at the petascale and beyond.

2. Key scientific challenges for the FSP

The general consensus from the U.S. fusion energy science research community is that the most prominent and urgent scientific issues impacting the burning plasma program and the successful operation of the ITER experiment include (1) disruption effects (large-scale macroscopic events producing rapid termination of plasma discharges), including avoidance and mitigation; (2) pedestal (steep-spatial gradient) formation and transient divertor heat loads in the plasma periphery; (3) tritium migration and impurity transport; (4) performance optimization and scenario modeling; and (5) plasma feedback control. While undoubtedly a longer list of important scientific questions could be generated (e.g., as found in table 2.1 of the DOE FSP Workshop Report [4]), these were identified to be the most important and compelling scientific challenges facing the FSP. Independent confirmation of their importance and relevance comes from the fact that the European fusion simulation effort is organized around precisely these five areas [9]. Each of the five questions is a computational grand challenge in its own right and requires an integrated simulation capability. This is made clear by table 2.1 of the DOE FSP Workshop Report [4], which is essentially a matrix of how the properties of tokamak plasmas depend on a variety of physical processes. The fact that the matrix is nearly full illustrates that these physics properties are mutually dependent to a large extent and powerfully makes the case for an integrated multiphysics approach. A more helpful picture of what is actually included in item (4) (the key topic of performance optimization and scenario modeling) is depicted in figure 2 (figure 2.2 in [4]). This figure captures a vast amount of physics knowledge, with each topic in its own right requiring detailed physics understanding.

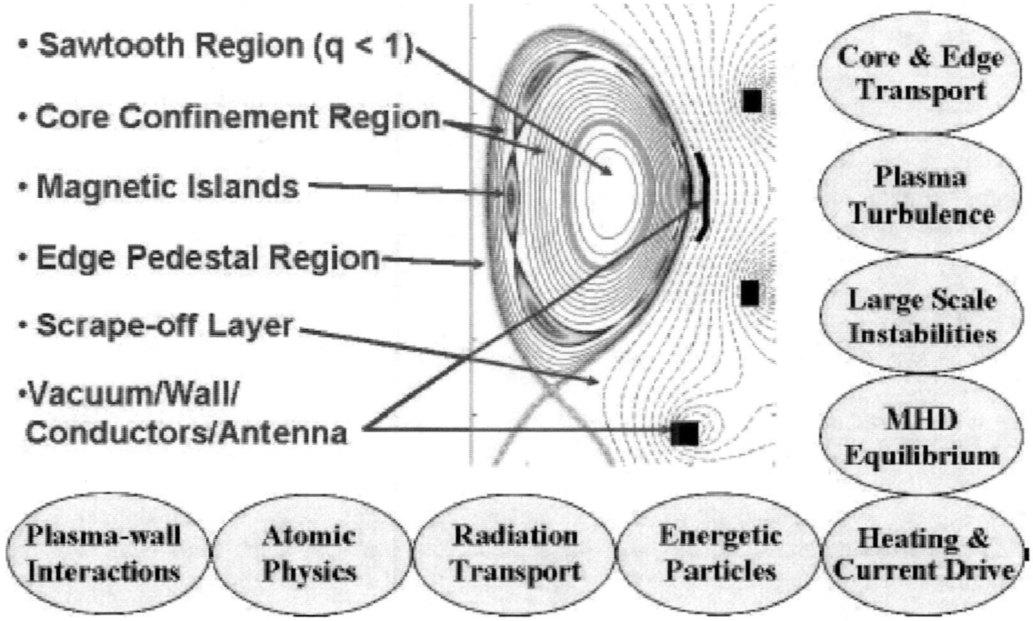

Figure 2. Illustration of the interacting physical processes within a tokamak discharge.

The five issues can be grouped into three topics (1–3) focusing on improved scientific understanding of physical processes, and two topics (4, 5) focusing on new tools for operational control of ITER experiments. In reality, item (4) overlaps scientific understanding and operational control. Actually, the science challenges (1–3) require integrated simulations of a different character than the operational challenges (4, 5), with the former requiring integration of a few "first-principles solvers" of high dimensionality and physics fidelity, while the latter require a larger number of reduced-dimensionality models. An associated future modeling challenge for the FSP is to decide whether it is feasible for these to all be part of a larger, integrated code.

The five integrated simulation capabilities listed will each also require improved physics modeling components to address (1) core and edge turbulence transport; (2) large-scale (MHD) instabilities; (3) sources and sinks of heat, momentum, current, and particles; and (4) energetic particle effects. Each of these components has its own set of major scientific and computational challenges. For example, since energetic particles are produced both by auxiliary heating and by fusion reactions in burning plasma experiments such as ITER, a prominent scientific challenge is to carry out self-consistent nonlinear simulations on the transport time-scales of energetic-/alpha-particle-driven Alfven instabilities in the presence of sources and sinks.

The physics modeling components listed here are, in fact, the traditional fusion topical science areas encompassing everything within the plasma boundary and are currently being addressed at various levels of depth and success in the FES SciDAC projects as well as within the ongoing base FES research program. They are in a sense too high level (and broad) to actually be specific components of the FSP but will, as noted, be separately pursued in a complementary manner in the base and SciDAC FES programs. The expected FSP approach is accordingly to focus on the five critical science issues to deliver reliable predictive tools that would be useful for fusion research in the not too distant future.

Of the five critical areas, the challenge of improved understanding of disruption effects, including avoidance and mitigation, is arguably the most important due to their potential to damage the actual plasma confinement device. When the plasma pressure exceeds linear MHD stability limits, a tokamak plasma is usually observed to disrupt. An important associated scientific challenge is to determine how close to such limits a disruption-free plasma discharge could be confidently expected to be sustainable. From the computational perspective, a formidable but worthy challenge is to effectively assemble the large (and diffuse) database of disruption-relevant events on tokamaks worldwide and then apply the most advanced analysis methodologies to search for key insights into disruption avoidance and mitigation. Other relevant issues include the identification of precursor signals for disruptions and the methods to mitigate the impact of disruption by minimizing the heat load in localized spots. In terms of ITER relevance, an outstanding scientific challenge is the experimental demonstration that a technique or combination of techniques will allow reliable mitigation of disruption effects via control of local heat loads, vertical/horizontal forces, and runaway electron currents. This will demand commencing an associated integrated modeling effort to provide a sound basis for extrapolation to ITER operation. In addition, producing a reliable predictive capability for important performance-limiting MHD instabilities such as neoclassical tearing modes is a major scientific challenge in its own right. This requires the inclusion of long mean-free-path dynamics (e.g., poloidal and toroidal ion flow damping in the presence of 3D magnetic perturbations), which will demand the development of formalisms beyond standard resistive MHD theory.

Pedestal formation and transient divertor heat load impact both the fusion core performance and requirements for materials facing the plasma. Unfortunately, the knowledge in this important area is mostly empirical, with very little scientific understanding to guide predictive modeling. Theoretically, it is very challenging because it is a multiscale regime, there is no obvious separation of scale for MHD and transport here, atomic physics is important, and the geometry is truly 3-dimensional. In contrast to the years of experience simulating core turbulence and MHD, the simulation of the edge is only beginning, albeit with some significant signs of progress in projects such as the SciDAC proto-FSP CPES [7]. The scientific challenge of better understanding of the edge must be met because information on the edge is needed in order to reliably simulate the performance of the whole device. A practical approach would be to first develop a reduced model based on current knowledge with the expectation that such a model will be improved as more knowledge becomes available.

Development of reliable predictive modeling of tritium migration is another important scientific challenge for the FSP because it can move through the plasma edge region to locations where it is difficult to remove. Effectively dealing with the key problem of tritium retention will require the development of improved capabilities for addressing associated atomic physics and material science issues. In addition, since fusion power production can be degraded by impurity dilution of the deuterium-tritium (DT) fuel, the transport and influx of impurities also needs to be properly predicted – especially with respect to ITER's planned use of carbon facing components for high heat flux regions of the divertor in advance of the DT phase, followed by the installation of tungsten targets before DT operation. The FSP will likely also need to deal with the broader challenge of improved understanding of plasma-surface interactions. In particular, extensive research and development will be needed to establish the physics basis for ITER reference scenarios which are expected to focus on environments with tungsten/beryllium plasma facing components. Associated plasma surface issues include sputtering erosion/redeposition (and sputtered impurity transport/plasma-contamination), erosion-dominated component lifetime, dust formation, flaking, tritium migration and trapping via co-deposition in beryllium, and tritium trapping in bulk tungsten.

Since each ITER discharge is expected to cost about $1M, improvement of performance optimization (including sustaining maximum fusion power production) and of scenario modeling capabilities to plan new experiments is important for the FSP. This will require effective tracking and incorporating significant advances in state-of-the-art physics understanding from the four topical science areas. Since some areas are more mature than others, first-principles codes will need to be complemented by reduced models – with both categories of codes needing to be rigorously validated against experiments. Tangible progress in identifying the most reliable models and including them in an integrated multiphysics simulation capability can be expected to provide important insights into the nonlinear coupling between the intertwined physics. Validating the resultant capability against experiments will be an essential and beneficial learning process for the FSP. The associated FSP modeling tools would be most welcome for ITER scenario development activities that include current ramp-up, current ramp-down, vertical stabilization, toroidal field ripple assessment, and various heating and current drive issues. In addition, moving from present-day experiments with tens of seconds duration to ITER discharges dominated by alpha-self-heating and lasting thousands of seconds will present a major challenge for controlling plasma current and pressure.

Plasma feedback control is particularly challenging for ITER because the associated burning plasma regime is a fundamentally new one where self-coupling dynamics are expected to be stronger. This involves discharge start-up, shape control, and disruption avoidance. These applications require mainly integrating MHD physics and have demonstrated progress via the use of modulated heating and current drive and the application of non-axisymmetric fields. Other capabilities, such as profile control to get to advanced performance regimes, will require knowledge of particle and momentum transport, about which there is currently little scientific understanding. Another key issue of serious concern to ITER involves controlling the edge localized modes (ELMs) since they can damage the divertor and impact the rapid erosion of plasma facing materials. This will require a good edge pedestal model, which does not currently exist. Several schemes are currently under study experimentally to mitigate or eliminate ELMs – including edge ergodization by introducing external resonant magnetic-field perturbations and modulating the frequency of pellet injection in line with ELM frequencies ("pacemaking"). Transforming what we learn from such experimental studies into a reliable, predictive model constitutes an important scientific and computational challenge for the FSP.

In addition to the multi-physics modeling challenges themselves, the FSP also faces the following key readiness questions:

(1) Which of the emerging or maturing simulation approaches appear most promising in the next 5 years to best make use of multicore petascale computational resources to accelerate their readiness for integration?
(2) What portions of the integrated model covering the multiple physics features depicted in figure 2 have been satisfactorily completed and when might the FSP expect to have an initial working version for experimental validation tests?

At the most generic level, whole-device integrated modeling encompasses the entire discharge duration. In particular, the plasma can be particularly delicate during the formation and shutdown periods as parameters change quickly and sudden changes in electromagnetic conditions may cause significant damage to the superconducting magnets. Indeed, desired operational scenarios for tokamaks are sensitive to control of the formation phase. It is also important to keep in mind that there are many nondisruptive instabilities that can cause the plasma to rapidly lose much of its stored energy, often coincident with a rearrangement of the current profile. Although there is little or no danger of machine damage, the plasma control system will likely not be able to return the plasma to the desired operating scenario following such an event. Figure 3 (as first shown in the DOE FSP Workshop Report [4]) is helpful in placing the integrated system simulation capability within the operational context of actual fusion experiments.

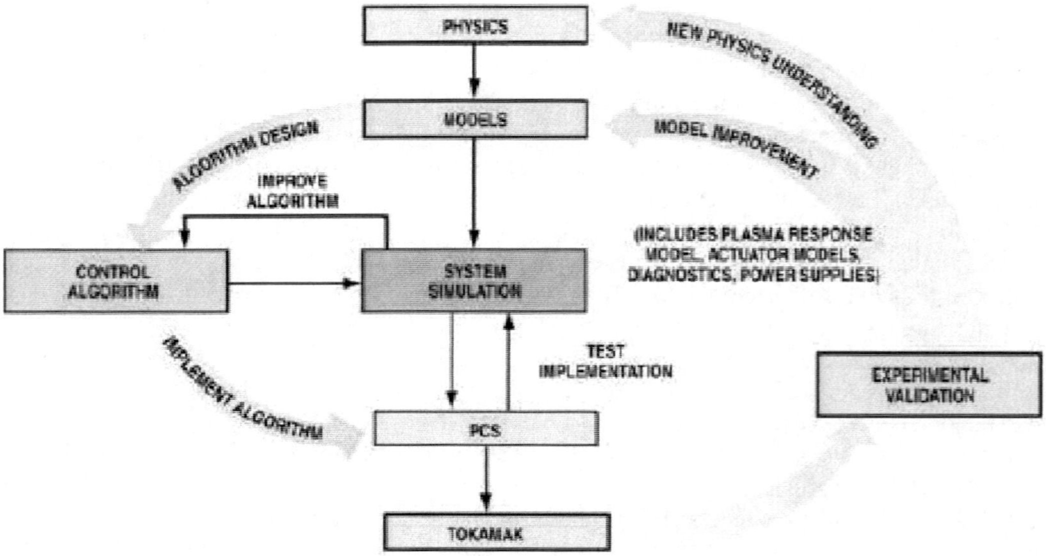

Figure 3. Schematic of integrated plasma control processes which is used by ITER (PCS here stands for a real-time plasma control system).

Given the real-time needs of the integrated plasma control process, the orange box labeled "system simulation" represents a reduced "control-level" model capable of rapid execution. It clearly cannot encompass the first-principles whole-device simulation model diagrammed in figure 2, which would take far too long to run. The plasma control community is already heavily engaged in developing a "Plasma Control System for ITER" with responsibility for this effort led by the Europeans, with significant U.S. participation. In order for the FSP to contribute effectively to this ongoing international activity, it must be capable of interfacing with what is already being developed and be able to add value rather than attempt to compete. In particular, an important scientific and computational challenge for the FSP in this context is to provide a modeling capability to enable effective exploration of ways to address how to control a burning plasma in regimes where the alpha power significantly exceeds external heating and the associated large scale instabilities may not be controllable by the use of modulated heating methods.

More generally (as also noted in the FSP Workshop Report [4]), the FSP will need to address the computational issue that the current state-of-the-art software tools for plasma feedback control are limited in their integration with other physical effects and their implementations are characterized as not extensible. Improvements provided by the FSP in this critical area of need would have a most beneficial effect on actual experimental operations. This raises the broader question of priorities when addressing the scientific and computational challenges of the FSP: Which are the most critical and which would most benefit from being addressed earliest? At least two integrated simulation software projects are required: (1) implementing some version of figure 2 using a combination of first principles kinetic and fluid solvers for attacking science issues 1–3 and (2) implementing figure 3 using reduced models. If such a strategy is adopted, an associated challenge is to efficiently connect the first-principles model to the control-level model – including proper validation activities.

3. Verification and validation challenges for the FSP

Establishing the physics fidelity of advanced physics modules and the associated integrated software is clearly recognized by the FES and ASCR communities as a major scientific challenge for the FSP [4]. As in the case of the ASCI project, a plan for verification and validation that would rely on ongoing theory and experimental research needs must be properly defined and executed.

Verification is basically the assessment of the degree to which a code correctly implements the chosen physical model. From a purely mathematical perspective, this involves applied math and computer science exercises in the execution of algorithms and touches on issues associated, for example, with numerical approximations, mesh discretization, temporal discretization, iterative solution of nonlinear equations, and statistical sampling error issues. However, it is important to keep in mind that verification in the context of the FSP must also include systematic comparisons against theoretical physics predictions. For example, codes are often developed to study highly nonlinear or even turbulent stages of plasma instabilities. The linear and weakly nonlinear phases of such instabilities are usually well described by analytic or semi-analytic theories that can be used to verify the code accuracy in the same regimes. Benchmarking codes with theoretical predictions is an essential tool to verify the code convergence in the limits where the theories are valid. When numerical difficulties prevent the solution of the most comprehensive mathematical models, alternate theoretical approximations may be necessary, or a reduced set of equations may have to be formulated to expedite numerical solutions. Other examples include replacement of kinetic equations with fluid equations or reductions of the range of finite gyro-radius corrections—with the consequence of limitations in the predictive capability of such codes when applied to actual experimental situations. Another fundamental issue of code verification concerns the limitations on physical parameters set by the numerical solvability of the mathematical model. For example, computational considerations often limit the magnitude of the magnetic Reynolds number (MRN) used in resistive magnetohydrodynamic simulations. Typical values of MRN used in the simulations are orders of magnitude less than the actual values in relevant fusion plasmas. Thus, it becomes crucially important to develop a strategy for assessing the effects of such limitations on the validity and relevance of the numerical results.

Though theoretical predictions can help in verifying code accuracy, analytic or semi-analytic solutions are often not available to verify codes in highly nonlinear and turbulent regimes. In such cases, a “cross-code” verification can be a satisfactory approach. Cross-code verification requires the systematic comparison of multiple codes which use different numerical algorithms (finite difference, finite elements, spectral methods, implicit methods, explicit methods, etc.) and/or different mathematical models (Particle-in-Cell, Vlasov or Continuum, Hybrid PIC-Fluid, etc.). In general, the FSP will need to complement conventional mathematical methods for verification of codes with systematic verification through comparisons with theoretical predictions and cross-code benchmarking.

Validation is basically the assessment of the degree to which a code "describes the real world" [4]. The challenge for the FSP is to develop specific strategies with regard to what needs to be validated and how it would be done. Clearly, simulations can never perfectly model physical reality, but can nevertheless be superior to empirical extrapolation if they can demonstrate a reasonable level of agreement with reliable results from systematic experimental measurements. In particular, documentation from such validation tests including sensitivity analysis should include data from both the simulation and experiment along with descriptions of data reduction techniques and error analysis. In particular, validation needs to be more than just carrying out experimental tests of the model. Writing code with today's version of the model is only the first step since the coding needs to be sufficiently flexible to accept periodic updating of the physics models. The improved physics models can be assimilated as significant advances in understanding occur. Examples include a better theoretical understanding of the edge pedestal physics from combining kinetic and fluid models; of wave-heating dynamics from combining Fokker Planck codes and full wave models with finite orbit physics of energetic particles; and of transport physics by including self consistent radial electric fields originating from neoclassical theory/simulation on transport time scales. Any one of these processes involves complex nonlinear dynamics, the improved understanding of which should be developed with a combination of experiment, theory, and modeling. Figures 4 and 5 show examples respectively of a possible approach to the development and validation of transport codes (figure 4) as well as to RF codes (figure 5) – both of which are currently being pursued under the SciDAC FES program. The benefit of such approaches is that the accuracy and value of both the models and experiments improve together over time and are integrated with ongoing experiments and theory. Once properly validated, the model would be ready to be assimilated into the FSP project, which then integrates and tests multiple physical packages into a "whole" system.

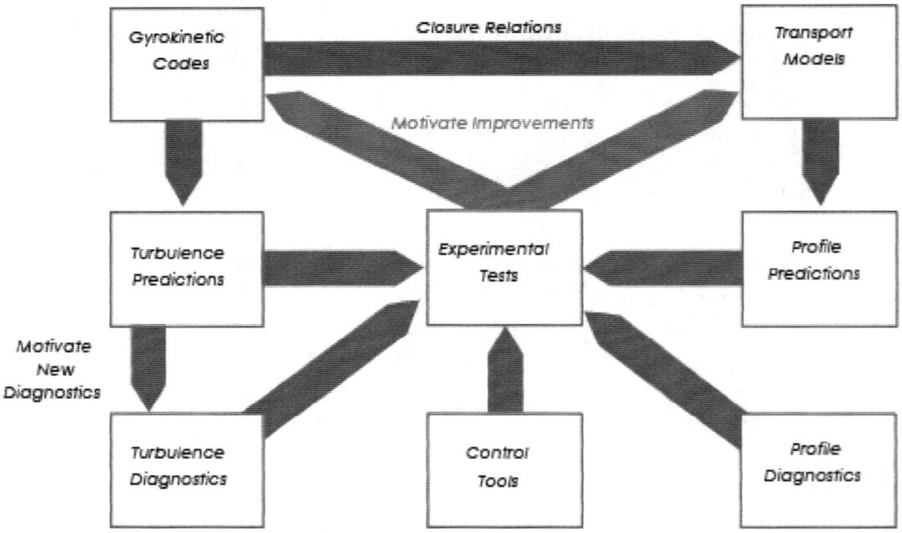

Figure 4. A fully predictive capability is best approached with the combined efforts of modeling and experiment. The above example refers to the development of a predictive understanding of transport in the plasma core.

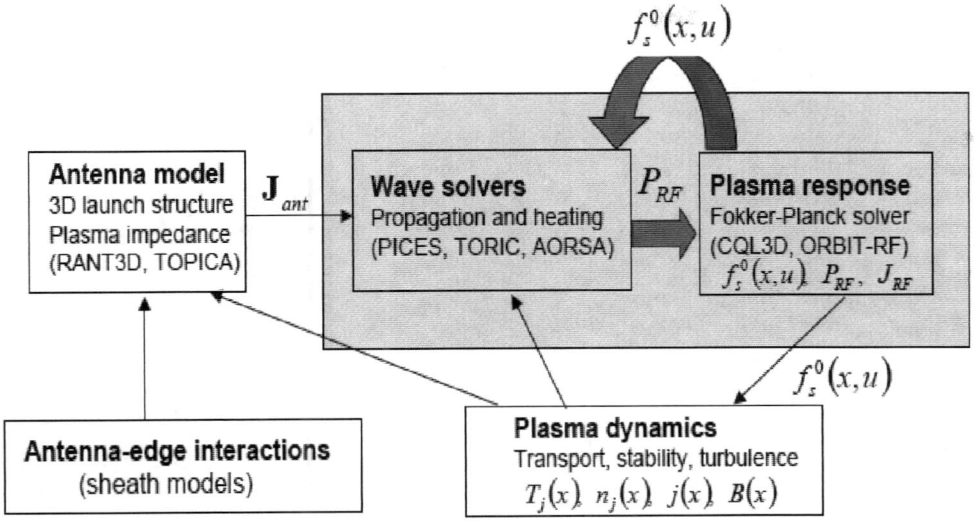

Figure 5. A fully predictive capability in the RF wave-particle interaction area is indicated in the above diagram.

A reliable predictive FSP code can be built only if the individual building blocks have a solid theoretical foundation and careful experimental validation. Basically, theoretical models that capture the targeted science should be implemented and used in the context of self-consistent simulations that can be compared with experimental data, which in turn should be analyzed and organized to facilitate comparisons with the simulation results. Since experiments rely on diagnostics, it is important to build up competence in 2-D and 3-D diagnostic capabilities. Furthermore, it is particularly beneficial to validation efforts if each diagnostic is implemented into the codes in a "synthetic" form. It is one of the most powerful tools for validating codes against experiments and has already produced impressive results in a number of cases involving RF physics modeling [10] and core turbulence validation studies [5, 11]. Development and application of synthetic diagnostics are necessary because almost none of the actual experimental diagnostics can currently provide local data accurate enough to make precise predictions for testing codes. Current devices are capable of performing experiments in all five of the critical scientific areas identified earlier. However, a properly designed validation exercise may require resources not already available, such as diagnostics and/or control actuators. FSP validation efforts will likely depend on the development of specific experimental tools needed to validate the models together with dedicated experiments for validation purposes. In addition, further progress is required in several areas of fusion theory since limitations in current fundamental theoretical understanding will continue to hinder progress in efficient code development. Thus, increased effort for novel development of both laboratory and synthetic diagnostics and for certain aspects of the base theory program will be needed to provide key complementary support to the FSP.

4. Computational Science Challenges for the FSP

The computer science, applied mathematics, and infrastructure (both hardware and its associated software) challenges to produce the needed high-performance software tools for achieving the fusion energy science goals of the FSP are formidable. In order to capture the needed multiscale (spatial and temporal) fidelity at both the individual physics and the integrated (multiphysics) level, the FSP mathematical models must be well posed as well as amenable to solutions via efficient numerical methods. Key areas such as simulations of resistive MHD effects, kinetic turbulence dynamics, high-frequency wave-heating, and energetic-particle-driven instabilities in realistic geometry involve mathematical issues encompassing numerical stability, adaptive mesh methodology, space-discretization approaches, efficiency of sparse matrix solvers, accuracy of splitting methods, and bifurcation analysis. In particular, the efficient scaling of MHD codes on multi-core platforms at the petascale will need to be aggressively pursued – perhaps via nonlinearly implicit preconditioned, Jacobian-free Newton-Krylov methods [12]. One of the most daunting challenges facing the FSP is to develop an integrated software product when the scientific basis for a number of key components continues to rapidly evolve. This will obviously require that some of the most advanced methods of computer science and applied mathematics be effectively integrated with advances in computational and theoretical plasma physics. In formulating the strategic approach to do so, an associated imperative will be to ensure that national and international best practices and "lessons learned" for large software development projects with strong verification and validation components (e.g., ASCI) are taken into account. Software projects are unique in the risks associated with them, to a degree that arguably exceeds even experimental device construction. A project of this magnitude and complexity needs to be able to quantify the risk associated with each key part of the software project and to have appropriate backup solutions and/or recovery methods identified. Accordingly, a key challenge is for this high-performance FSP software to be developed with an eye toward enabling it to be flexible enough to weather the likely evolution of the hardware architecture as well as that of the associated systems software.

With regard to the overarching technical challenge of dealing with the extreme multiscale nature of simulating tokamak plasmas for long integration times, a generic approach and vision need to be properly articulated, for example, for how the "macro/micro" coupling challenge can be achieved even if provided "infinitely powerful" computing power in the future. Although the computational platforms at the petascale and beyond will demand efficient "extreme" parallelization methods, it should nevertheless be kept in mind that the algorithmic challenges are not all parallel. Also, while the scale of the simulation output datasets suggests the likelihood of significant overlap in FSP requirements for data analysis, visualization, and data management with other disciplines, it is also likely that the FSP will need specific tools and new algorithms specifically devised for the characteristics of fusion problems. For example, there will be challenges associated with the development and application of needed tools for improved visualization and pattern recognition of edge plasma events such as edge localized modes (ELMs), turbulent transport structures (e.g., "blobs"), and plasma structures around the magnetic separatrix including magnetic islands.

A clearly dominant trend in the leadership computing hardware development area is the continued addition of more and more multiple CPU cores onto the same chip to deliver high aggregate computing performance. Current estimates are that the number of cores per chip is expected to increase by an order of magnitude in less than five years and two orders of magnitude in less than a decade. Together with the high-density, low-power packaging approach to construct large-scale massively parallel, distributed-memory, multicore computers, this suggests that a petascale or multi-petascale parallel machine in the next decade could reach 10 million CPU cores. The formidable challenge here is of course to develop new methods to effectively use such dramatically increased parallel computing power. Such methods will be necessary to achieve accelerated scientific discovery in the FSP as well as in general for fusion energy science and many other application domains. If the fusion energy science applications were only able to effectively use a small fraction of the cores on a CPU, major efforts would obviously be needed to accelerate development of innovative new methods for per processor performance.

Following are examples of outstanding computational challenges in the fusion energy science application area:

- Efficient scaling of MHD codes beyond terascale levels to enable higher-resolution simulations with associated greater physics fidelity

- Efficient extension of global PIC and continuum codes into fully electromagnetic regimes to capture the fine-scale dynamics relevant not only to transport but also to help verify the physics fidelity of MHD codes in the long-mean-free-path regimes appropriate for fusion reactors
- Development of framework and workflow capabilities for large-scale data management associated with advanced integrated codes
- Development of innovative data analysis and visualization to deal with increasingly huge amounts of data generated in simulations at the petascale and beyond

With regard to infrastructure challenges, FSP research enabled by leadership-class computing capabilities can be expected to demand much greater computer time and associated support than presently available. For example, a single run using the global particle-in-cell code developed within the SciDAC Gyrokinetic Particle Simulation Center (GPSC) [11] carried out at present to investigate the long-time evolution of turbulent transport requires around 100K cores * 240 hours = 24M CPU hours. Since the current version of a plasma edge code developed within the SciDAC Center for Plasma Edge Simulations (CPES) [7] requires roughly the same amount of time, the actual coupled simulations of the core and edge regions noted earlier could demand approximately 50M CPU hours. If additional dynamics (such as the modeling of the RF auxiliary heating [10]) were also included, then the need for computational resources at the exascale would be a reasonable expectation.

A practical operational challenge for the FSP is to deal with the need for simulations to be performed in a time critical fashion (for the interpretation of shot data or for experimental planning against a timeline). This will give rise to some new requirements for deadline-driven data assimilation methods and other quasi-real-time methods that will offer new challenges to both systems architecture and operational infrastructure. The precise nature of the deadline driven requirements are not yet fully articulated but are expected to be a key element in future requirements planning for ITER.

In general, a requirements and risk analysis for the FSP associated with its computational tools and infrastructure challenges will be needed to determine the appropriate level of direct investment and the expected major increase in capability due to normal developments in the field. Successful examples are evident in the SciDAC program of joint work funded by ASCR and FES in areas including mathematical techniques, computational libraries, collaboration technology, data analysis, and advanced visualization tools. These efforts will need to be enhanced and resourced with the support levels tightly coupled to the science and engineering goals of the FSP. The associated FSP computational and software infrastructural challenges and requirements will need to be communicated early and often to those organizations providing computational and data capabilities for the Office of Science, such as the leadership-computing centers, ESnet, and NERSC.

5. Summary

The integrated modeling capability developed through the FSP should be an embodiment of the theoretical and experimental understanding of confined thermonuclear plasmas. As such, substantive progress on answering the outstanding scientific questions in the field will drive the FSP toward its ultimate goal of developing a reliable ability to predict the behavior of plasma discharges in toroidal magnetic fusion devices on all relevant time and space scales. It faces the formidable challenge of closing the current gap between idealized (largely empirically based) modeling and high-end, first-principles-based simulations of mostly individual plasma dynamics. This involves capturing into more realistic reduced modules the most important physics trends produced by the high-performance first-principles codes that successfully simulate important individual phenomena in realistic 3D geometry. There are major computer science, applied mathematics, and infrastructure (both hardware and its associated software) challenges for the

FSP to produce the needed high-performance software tools for achieving its fusion energy science goals. In order to capture the needed multiscale (spatial and temporal) fidelity at both the individual physics and the integrated (multiphysics) level, the FSP mathematical models must be well-posed as well as amenable to solutions via efficient numerical methods, including the ability to deal with the increasingly complex multicore systems at the petascale and beyond.

Regarding the scope of the activity, the FSP challenge includes identifying what the deliverables are and their scientific merit. It is necessary, for example, to identify to what extent existing software components can be efficiently integrated now and to what extent "from-scratch" implementations would be required. More specifically, there is a need to better articulate how to selectively integrate into FSP the ongoing SciDAC program, the SciDAC proto-FSP integration projects, including SWIM, CPES, and FACETS, and new developments involving joint experiment-theory-modeling efforts as well as the expertise residing in OASCR's computer science and applied math programs. This effort will entail identifying the targeted scientific and computational challenges in the form of clear compelling problems with measurable goals. While the FSP Workshop Report [4] is rather generic and "all-inclusive" in many respects, it nevertheless contains valuable information for making the case that this project can succeed in answering questions in a timely way that experiment and traditional theory by themselves cannot.

The five critical scientific issues discussed in Section 2 of this paper have generally been identified to be the most important and compelling scientific challenges facing the FSP [4, 9]. Each of the five questions is a computational grand challenge in its own right, and requires an integrated simulation capability because the associated physics properties are mutually dependent to a large extent. However, an integration effort encompassing all five of these challenging issues from the beginning is too large a step to be practical. To be practically achievable, the FSP should begin with more modest integration efforts that exhibit a compelling level of verification and validation (V&V). Indeed, the FSP cannot succeed without a viable V&V effort, which will likely require enhancement of the existing fusion diagnostic program and an increased synthetic diagnostic development initiative involving personnel with skills in both diagnostic methods and code expertise. Basically, theoretical models that capture the targeted science should be implemented and used in the context of self-consistent simulations that can be compared with experimental data, which in turn should be analyzed and organized to facilitate comparisons with the simulation results.

The FSP should be a repository of the latest physics as it evolves. In this sense it cannot be a "stand-alone" project. It must be properly coordinated with theory, experiment and fundamental simulation. More specifically, a proper implementation of the FSP will demand an effective plan for developing "advanced scientific modules" via utilization of results from the FES base theory program, the SciDAC FES program, new insights from joint experiment-theory-modeling efforts, and the expertise residing in ASCR's computer science and applied mathematics programs. It should establish and maintain strong connections with relevant international projects and also draw on the large experience base from existing scientific software development projects from other fields.

In order to be successful, the FSP should not be "everything to everyone." It must be focused and project-driven with well-identified software deliverables that the stakeholders fully support. It is expected to provide great value-added for addressing critical FES programmatic needs, for enhancing return on U.S. fusion investments, and for helping to identify world leadership opportunities for the U.S. fusion program. A successful FSP will better enable the study of burning plasmas and greatly aid the U.S. role in the operation of the ITER experiment and in harvesting the associated scientific knowledge. This could enable discovery of new modes of operation, with possible extensions of performance enhancements and improvements needed for a demonstration fusion reactor. Increasingly reliable whole-device, multiphysics modeling capabilities in Fusion Energy Sciences will demand computing resources at the petascale range and beyond to address ITER burning plasma issues. Even more powerful exascale platforms will be needed to meet the future challenges of designing a demonstration fusion reactor.

Acknowledgments

A major portion of the source material in this paper came from the FSP Workshop Report [4] led by A. Kritz and D. Keyes and especially from the assessment of the Fusion Simulation Project carried out by the Fusion Energy Science Advisory Committee (FESAC) FSP Panel and summarized in the "FESAC Fusion Simulation Project (FSP) Panel Final Report" (October, 2007) led by the author. He is most grateful for the insightful contributions from his colleagues on the panel: Riccardo Betti (U. Rochester), Jeffrey Brooks (Purdue U.), Vincent Chan (General Atomics), Thom Dunning (NCSA, U. Illinois), Charles Greenfield (General Atomics), Brian Gross (GFDL), Michael Norman (UCSD), Miklos Porkolab (MIT), and Rick Stevens (ANL), and from valuable discussions with David Keyes (Columbia U.), Phil Colella (LBNL), Arnold Kritz (Lehigh U.), Doug Kothe (ORNL), and Ned Sauthoff (U.S. ITER Project Director). The helpful comments and suggestions from the referee of this manuscript are also much appreciated.

References

[1] Dahlburg, J, Corones J, Batchelor D, Bramley R, Greenwald M, Jardin S, Krasheninnikov S, Laub A, Leboeuf J, Lindl J, Lokke W, Rosenbluth M, Ross D and Schnack D 2001 *J. Fusion Energy* **20** 135

[2] Post D, Batchelor D, Bramley R, Cary J, Cohen R, Colella P, and Jardin S 2004 *J. Fusion Energy* **23** 1

[3] Tang W, Chan V 2005 *Plasma Phys. and Controlled Fusion Research* **47** (2): R1-34

[4] Kritz A and Keyes D eds. 2007 *DOE Fusion Simulation Project Workshop Report* http://www.science.doe.gov/ofes/programdocuments/reports/FSPWorkshopReport.pdf

[5] Holland C 2008 Validating simulations of core turbulence simulations: Current status and future directions *Proc. SciDAC 2008*

[6] Batchelor D 2008 Simulation of wave interactions with magnetohydrodynamics *Proc. SciDAC 2008*

[7] Chang C S 2008 Toward a first-principles integration simulation of tokamak edge plasmas *Proc. SciDAC 2008*

[8] Cary J, 2008 First results from core-edge parallel composition in FACETS project *Proc. SciDAC 2008*

[9] Bécoulet A, Strand P, Wilson H, Romanelli M and Eriksson L 2006 Integrated tokamak modelling: The way towards fusion simulators (International Atomic Energy Conference paper TH/P2-22)

[10] Bonoli P, Batchelor D, Berry L, Choi M, D'Ippolito D, Harvey R, Jaeger F, Myra J, Phillips C, Smithe D, Tang V, Valeo E, Wright J, Brambilla M, Bilato R, Lancellotti V and Maggiora R 2007 Evolution of nonthermal particle distributions in radio frequency heating of fusion plasmas *Proc. SciDAC 2007*

[11] Lin Z, 2008 Verification and validation of petascale simulation of turbulent transport in fusion plasmas *Proc. SciDAC 2008*

[12] Chacon L 2008 Scalable parallel implicit solvers for 3D magnetohydrodynamics *Proc. SciDAC 2008*

Journal of Physics: Conference Series **125** (2008) 012048 doi:10.1088/1742-6596/125/1/012048

Photons, photosynthesis, and high-performance computing: challenges, progress, and promise of modeling metabolism in green algae

Christopher H Chang[1], Peter Graf[1], David M Alber[1], Kwiseon Kim[1], Glenn Murray[2], Matthew Posewitz[2] and Michael Seibert[3]

[1] Materials and Computational Science Center, National Renewable Energy Laboratory, 1617 Cole Boulevard, Golden, CO 80401, USA

[2] Colorado School of Mines, Golden, CO 80401, USA

[3] Energy Sciences Directorate, National Renewable Energy Laboratory, 1617 Cole Blvd., Golden, CO 80401, USA

E-mail: christopher_chang@nrel.gov

Abstract. The complexity associated with biological metabolism considered at a kinetic level presents a challenge to quantitative modeling. In particular, the relatively sparse knowledge of parameters for enzymes with known kinetic responses is problematic. The possible space of these parameters is of high-dimension, and sampling of such a space typifies a combinatorial explosion of possible dynamic states. However, with sufficient quantitative transcriptomics, proteomics, and metabolomics data at hand, these challenges could be met by high-performance software with sampling, fitting, and optimization capabilities. With this in mind, we present the High-Performance Systems Biology Toolkit **HiPer SBTK**, an evolving software package to simulate, fit, and optimize metabolite concentrations and fluxes within the space of rate and binding parameters associated with detailed enzyme kinetic models. We present our chosen modeling paradigm for the formulation of metabolic pathway models, the means to address the challenge of representing such models in a precise and persistent fashion using the standardized Systems Biology Markup Language, and our second-generation model of H_2-associated *Chlamydomonas* metabolism. Processing of such models for hierarchically parallelized simulation and optimization, job specification by the user through a GUI interface, software capabilities and initial scaling data, and the mapping of the computation to biological questions is also discussed. Moreover, we present near-term future software and model development goals.

1. Introduction

1.1. Project goals and relationship to photobiological H_2 production

The project brings together (a) experimental high-throughput methods and large-scale systems biological modeling to address photobiological H_2 production through the development of reusable biological reaction models and (b) software tools to fit parameters in these models to transcriptomic, proteomic, and metabolomic data sets. With the *in silico* cell as the ultimate destination in mind, we are designing modular simulation software to permit multilayered biological simulations, with metabolic and regulatory simulations run on separate sets of compute nodes, in conceptually separate layers reflecting the underlying biological separation of timescales associated with metabolic flux and genetic regulation. We have begun with a metabolic simulation code, called **HiPer SBTK**, to optimize metabolite concentrations, fluxes, and network sensitivities within flexible, user-defined constraints, as well as a basic graphical interface for straightforward configuration of complex jobs. In this paper, we discuss our chosen modeling paradigm, the status of the metabolic model surrounding H_2 production in the green alga *Chlamydomonas reinhardtii*, developments in the simulation software including sensitivity optimization and the user interface, and the direction of project development for the near future.

2. Modeling metabolism

2.1. The need for model persistence

In moving from desktop-scale simulations of a small set of biochemical reactions to genome-scale simulations in the high-performance computing (HPC) arena, a paradigm shift must occur in the way we think of biological models. A complete representation of metabolism for a single organism implies model networks with thousands of nodes and edges. Detailed kinetic modeling with thermodynamic constraints requires a great deal of care and thought to be invested in the actual construction and encoding of the model itself. Furthermore, assumptions and decisions in the face of uncertain or missing parameter values and mechanistic information must be documented and retained if reproducibility is sought. Moreover, the massive resources expended in answering cutting-edge questions with HPC are incongruent with trivially constructed or maintained biological models.

Such large-scale modeling benefits from provenance tracking and workflow technologies, standard languages for the expression of arbitrarily complex models, and detailed annotation based on ontologies accepted by the biological community. In developing a model of metabolism in *Chlamydomonas reinhardtii* most closely related to H_2 production and photosynthesis, we have taken maximal advantage of community resources, such as the XML-based SBML, MathML, Gene Ontology, Systems Biology Ontology, and the Resource Description Framework for model annotation. Where tools are either lacking or require specialization for the modeling paradigm we have chosen, our development depends on freely available and widely used languages (Python, Java, C/C++), and libraries (libSBML, Swing).

2.2. Standard ontologies and languages in biology

As the term is employed in the field of computational biology, an ontology is a domain-specialized system for classification of the constituent concepts in that domain, and their interrelationships [1]. For a complex system such as the metabolome, proteome, or transcriptome, the establishment of an associated ontology is a critical step in the unambiguous assignment of meaning for the parts and the whole of a model. For example, the term "membrane" in a model can encompass a generalized entity (say, an abstract two-dimensional space of infinite or arbitrary extent in the model containing certain metabolic species defined by their concentrations), a specific entity (a cell outer membrane or any one of several distinct membranes defining separate compartments in an organelle), or a particular instance of a specific entity (e.g., the inner membrane of "mitochondrion #3"). Because we cannot query a model for clarification of meaning, clear and precise expression benefits greatly from one or more accessible ontologies, such as SBO, GO [2], and ChEBI [3].

The SBO specifically targets the constituent concepts of mathematical rate laws. Here, the primary goal is to identify the role of participants in the rate equations, for example, substrate, product, the type and nature of a constant or species in the equation, the form of an equation, and continuous vs. discrete frameworks. The Gene Ontology further covers genes and gene products, while ChEBI categorizes and makes explicit the small-molecule constituents of a modeled biological system. By combining terms from these ontologies with elements from the BioModels initiative through a standard annotation mechanism such as RDF [4], most or all of the ambiguity possible in a biochemical model may be removed. Additionally, model provenance metadata may be encoded through Dublin Core Metadata Initiative [5] terms within RDF/XML.

The remaining task is the actual expression or instantiation of a particular model. We have chosen the Systems Biology Markup Language [6] for this purpose. SBML has several direct and associated benefits, including

- use of the now-standard XML paradigm, allowing for unambiguous expression of model concepts and automated validation of model form and non-semantic content;
- use of the associated library libSBML [7], for input and output of models, model parsing and translation, and automated validation of semantic content;
- use of SOSlib [8], a collection of routines for the construction, analysis, and solution of ODE systems; and
- support for mathML [9] and RDF [4] markup.

The standard textual format of SBML allows direct editing of biological models with a simple text or more powerful XML editor or the use of more specialized tools for construction, modification, and visualization.

2.3. Modularity in metabolism

In the interest of present and future model curation, genome-scale metabolic models should ideally be expressed in as modular a fashion as possible. To a first approximation, metabolism is well suited to such model development, as it is often conceived of as highly modular. However, closer consideration presents several challenges. First, although biochemical pathways are usually represented as neat modules, these usually visually present only one or two compounds relevant to the chemical transformation being considered, and leave off to the side chemical species considered ubiquitous (e.g., protons) or of lesser interest to the primary "goal" of the pathway (e.g., redox cofactors NAD^+/NADH in the oxidation of glucose to pyruvate in glycolysis). When all chemicals are considered, metabolism loses much of its initially perceived modularity. Second, the modeler must decide how to separate a more or less modular set of reactions from the complete metabolic network. Intracellular compartments clearly separate reactions physically, for example, but the modeler must decide whether a corresponding conceptual separation exists and warrants distinct model coding. Although standard pathway definitions exist, these cannot serve all purposes: organisms are differentiated in part by their chemical diversity, and the conceptual boundaries in their metabolic networks will not necessarily map neatly onto templates. Thus, a certain amount of arbitrariness must be expected in complex model development, mitigated by the deployment of careful annotation and standard ontologies (see above).

2.4. SBML tools

Once the appropriate decisions have been made, a kinetic model will consist of several separate SBML files that must be merged prior to simulation and characterization. The SemanticSBML effort [10] addresses both the requirement to merge multiple SBML models and the invocation of semantic annotations for model components. However, the associated software dependencies and extraneous capabilities for our purposes have prompted us to develop a set of Python modules, which we call SBMLTools.

The most critical and painstaking issues in designing SBMLTools revolve around conflict detection between submodels. That is, in situations where two or more submodels contain elements referring to the same construct (i.e., elements of the same type sharing identical `id` attribute values), the merging tool determines if the elements conflict with each other. We have implemented a strict form of conflict detection that reports a conflict if any subelements or attributes differ (including situations where attributes are present in one model, but absent in another). If elements do conflict, the merging tool does not know which to place into the merged model, so the tool reports the conflict and halts. This approach differs from that of SemanticSBML, which prompts the user to decide which element is to go into the merged model. We plan to improve the reporting feature of the conflict detection routines in our merge tool to provide more information about the conflict and specifically which sets of submodels conflict with other sets of submodels.

In addition to the merging tool, we have developed several other tools in the course of building the *C. reinhardtii* metabolic model and have come to realize the importance of these tools to our work. Furthermore, we expect other researchers working with systems biology models have similar experiences. We are, therefore, developing a framework to serve as a plug-in manager for systems biology tools to enable the community to work on a common tools platform.

2.5. *Flux versus kinetic modeling*

Once the topology of a metabolic network has been defined in sufficient detail, a number of analyses and simulation methods can be brought to bear. If we consider the nonredundant list of m compounds and the non-redundant list of n chemical reactions that make up a metabolic network, we may define an $m \times n$ matrix $\mathbf{S}$, called the stoichiometric matrix, whose elements S_{ij} have magnitudes equal to the stoichiometry of compound i in reaction j. Compounds that are consumed in a reaction (reactants) as written are assigned negative values, those produced (products) positive values. Even in the absence of information about the reaction rate laws, $\mathbf{S}$ encodes a great deal of information about the associated metabolism, including the network topology and permissible combinations of fluxes in the steady state, when compound concentrations are constant with time despite fluxes producing and consuming the compound. With sufficient physical or empirical constraints on the permissible fluxes, flux balance analysis [11, 12] and metabolic control analysis [13] can elucidate much about the steady-state behaviors of metabolism. More detailed incorporation of thermodynamic concerns with respect to individual reactions leads to energy balance analysis [14, 15]. For the most part, applications of these methods have focused on fluxes characterized by what we here refer to as "noncatalytic" mass-action kinetic equations of the form

$$\frac{\mathrm{d}[P]}{\mathrm{d}t} = k_1[A]^2[B] - k_{-1}[P][Q]$$

for the reversible chemical reaction

$$2A + B \rightleftharpoons P + Q.$$

Application of this modeling paradigm has often reproduced experimental observations (e.g., [16, 17, 18]). Because the kinetic constants are treated as parameters, however, a good fit to an experimental dataset does not guarantee a satisfactory fit to other datasets with these parameters, or even with the same network structure [19]. Furthermore, accurate network dynamics cannot be captured by such a model, given that the flux responses will not mirror those known from the study of enzyme rate response to compound concentrations.

Biochemical reactions are catalyzed by enzymes that exhibit several important properties with respect to reaction dynamics:

- Reaction chemistry occurs at a specific "active" site on the enzyme, which can be saturated (bound by either reactants, intermediates, or products for a time interval and thus can be unavailable for reaction independent of compound concentrations).
- The actual reactive steps involving chemical bond rearrangements are preceded and followed by binding and nonreactive conformational rearrangements, which may or may not be kinetically limiting.
- ¡olecules not involved in the reaction may act as regulators, modulating the reaction rate through competitive binding with reactants or products or through indirect means by changing the energy landscape for the enzyme conformational ensemble.

Hence, the effective reaction rate will include catalyst concentrations, rate constants distinct from those characterizing the uncatalyzed reaction, and terms to account for the effects of saturation, binding events, and regulation.

Derivation of enzymological rate equations is a well-established procedure, as is the assignment of kinetic constants from experimental data [20]. Several studies have undertaken the arduous process of modeling metabolic pathways from known enzymological kinetics [19, 21, 22], with insightful results. Two key objections to this type of kinetic modeling are that (1) the transferability of equation forms among organisms is unknown, the measurement of the kinetic parameters can be a long and arduous process, and we can never hope to have complete characterization of all enzymatic mechanisms for every organism of interest; and, (2) the transferability of enzyme kinetic behavior between the test tube and the cellular interior is debatable [23].

The philosophy behind the modeling component of this SciDAC project is that implementation of the correct *form* of flux equations trumps issues of parameter transferability and complicated parameter content. Experimental data captures the behavior of a cellular ensemble, hence fine-grained models that capture the metabolic response of a single cell accurately may not reproduce high-throughput experimental data. However, the kinetic equations derived from *in vitro* studies of enzymes should be able to represent the qualitative features of the ensemble average response to metabolic perturbations. The kinetic constants derived from studies of dilute enzyme solutions may admittedly not be appropriate for catalysis in the cellular milieu. However, if the metabolic model takes on the correct form of the rate equations, our software together with sufficient high-throughput metabolomics data can produce a set of effective kinetic constants that fit the known network behavior. The deployment of HPC resources can mitigate the computational burden associated with large, complex models, and permit mining the richer dynamical content of such models. Indeed, accurate responses of flux to changes in compound or enzyme concentrations permits greater confidence in the results acquired from study of dynamic responses to metabolic perturbations. In addition, a model formulated from enzymological rate equations has a more direct connection with experimentally observable properties of the component enzymes. This has two important ramifications. First, the model may be improved by extraction of the enzyme from the organism (or cloning and heterologous expression) followed by enzymological characterization. Second, metabolic engineering is made more straightforward by direct molecular biological manipulation of an enzyme's measurable kinetic properties, rather than indirect manipulation of an effective parameter that may hide complexity spread locally and nonlocally over the enzyme's metabolic context.

We have generated a second-generation metabolic model of Chlamydomonal metabolism, built from a combination of published pathway models and postulated pathways specific to the alga. The detailed model of the tricarboxylic acid cycle, oxidative phosphorylation, and mitochondrial transport from Wu et al. [22] has been combined with the glycolysis model of Mulquiney and Kuchel [24], and a basic model of *Chlamydomonas* fermentative and H_2-producing pathways known to date [25]. We have adopted the modeling paradigm in the paper by Wu, et al., which by the authors' estimation represented an unprecedented level of complexity and data integration at the time of publication. This model is attractive for several reasons. Reaction thermodynamics are captured in the formulation of the reversible kinetic equations (Haldane constraints). Furthermore, the effects of compound ionization on ion pools (H^+, K^+, and Mg^{2+}) in each compartment, and the associated changes in the pool concentrations, are calculated explicitly. Perhaps most important, a systematic workflow can be used to construct new models, by virtue of the marvelously detailed Appendices in Wu et al.'s publication. The primary challenges to this approach are the (re)encoding in SBML, model merging, thorough model annotation, and validation through trial simulations. Our workflow, however, permits testing at several stages prior to achievement of the "final" model. Figure 1 illustrates the pathways currently completed or being encoded.

2.6. Relationships to high-throughput experimental methods

Ultimately, the project goal is to develop an *in silico* model of metabolism that captures all reliable experimental data available for *C. reinhardtii*. We emphasize that any modeling effort is an iterative one, constantly incorporating new data as time progresses. Thus, a means of fitting experimental datasets is essential and has been incorporated into our software (*vide infra*). At this point, it may be useful to describe more specifically the role of high-throughput datatypes in the modeling process.

Transcriptomics Changes in transcript levels are a first approximation to changes in expression levels in response to perturbations in environmental conditions, or to changes in the genetic background. Such changes are not always propagated to the protein level and are in general semi-quantitative at

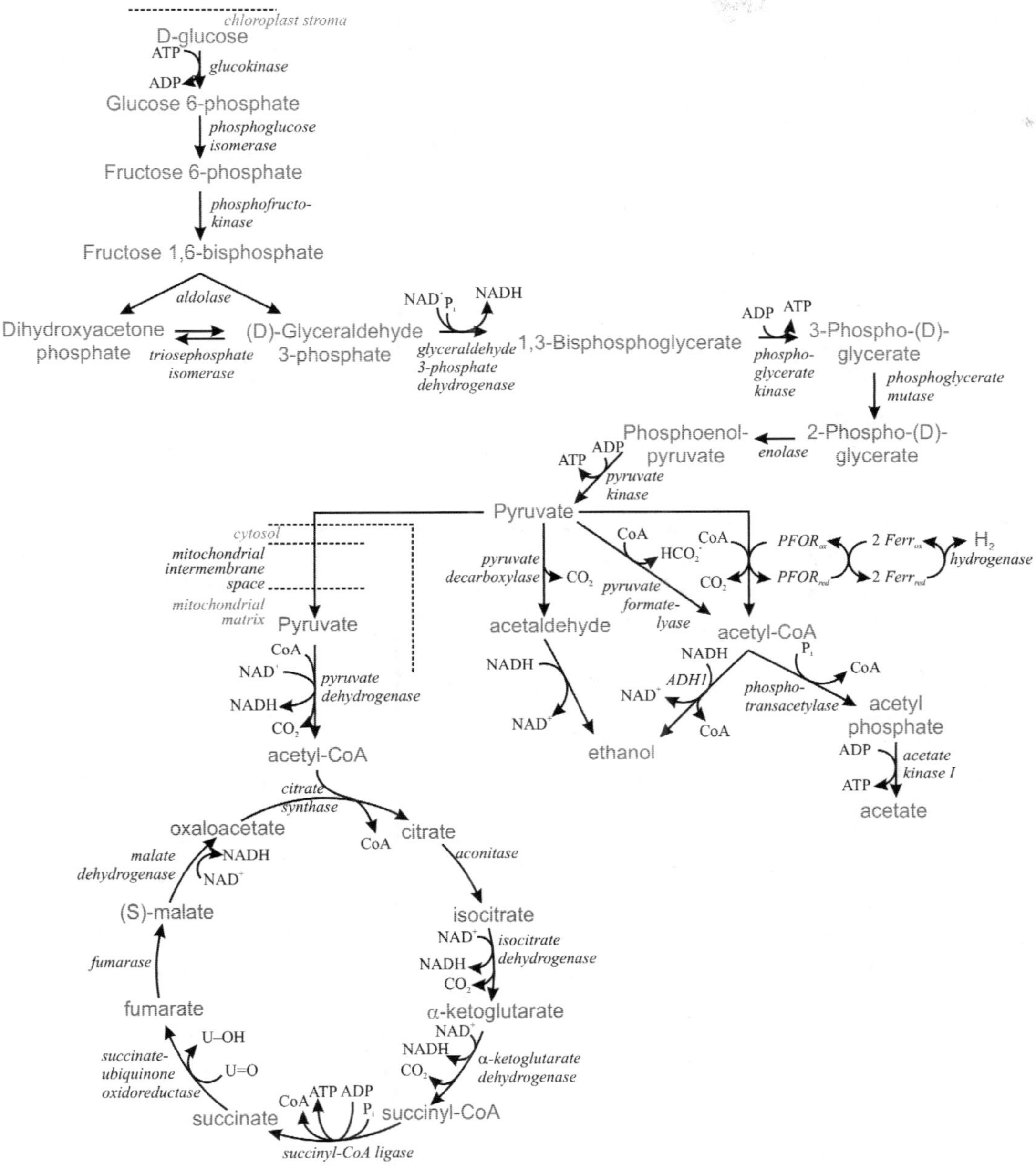

Figure 1. Metabolic pathways currently encoded as the critical biochemical interconversions surrounding H_2 metabolism in *C. reinhardtii*. Primary metabolites are in sans serif font, and color-coded to indicate their compartmentalization in the model: green for chloroplast, red for mitochondrial matrix. Currency metabolites are in serif typeface, enzymes and proteins are italicized. Details of cross-membrane transport are omitted for clarity.

best [26]. Nevertheless, transcriptomic data serves as a valuable input to test a model's behavior and predicted outputs. Transcriptomics is thus not useful to the first pass of model construction but permits coarse-grained validation.

Proteomics Like transcriptomics, the measurement of changes to protein expression levels allows model validation. However, proteomics approaches also permit absolute quantitation of protein levels in the cell and, as such, can provide approximate $[E]$ in the model. This is critical for moving the simulation from abstract to a concrete representation of a particular organism under particular circumstances. Thus, proteomics has a role both in model construction and model validation.

Metabolomics Direct measurement of compound concentrations and fluxes is of course the datatype *par excellence* for establishing the inputs and expected outputs of a metabolic model. While proteomic data can fit $[E]$, metabolomic data provides the actual targets for fitting metabolite levels.

2.7. Answering biological questions

Our modeling approach and software are suited to answer a variety of biological questions, including the following:

Kinetic parameter estimation For quantitative kinetic modeling, an accurate estimation of the parameters associated with kinetic laws is critical to assigning "reality" to a model. Assuming that an abundant amount of metabolomics data is available and that the enzyme levels in a model are known, one may fit the metabolomics data in the space of kinetic constants, keeping E fixed as "known" quantities. By fixing certain kinetic parameters to values either measured or extracted from the literature, the available space of possible values is further limited. With a large enough quantity of high-precision data to fit against, and sufficient parameters fixed, one may hope to find a unique solution.

Characterization of network capabilities Starting from either a well estimated kinetic network, or an assumed one, one can explore the capabilities of the network directly in the space of either kinetic parameters k or enzyme levels E by varying subsets of E or k, with the remainder either held fixed or left free to vary. In the former case, one would be scanning through a multidimensional space of k and/or E, characterizing the metabolic state achieved in a forward solution to the set of differential equations while considering certain $\{k,E\}$ as constant. This may be useful for characterization and visualization of a subspace of the metabolic space, perhaps around a set of optimal values. The latter case of leaving a set of $\{k,E\}_1$ free while scanning through another set $\{k,E\}_2$ would be typical of an optimization, where we are interested in characterizing optimum solutions. For example, referring to figure 1, we may wish to visualize with respect to the quantities (phosphotransacetylase $k_{cat,forward}$, [pyruvate decarboxylase], and pyruvate kinase $K_{M,ADP}$) the H_2 fluxes maximized with respect to the disparate quantities (PFOR $k_{cat,forward}$, [ferredoxin], and ADH1 K_{M,NAD^+}), leaving all other model parameters fixed at given values. Such a visualization may take the form of a cube of points, with the dimensions of the cube referring to phosphotransacetylase $k_{cat,forward}$, [pyruvate decarboxylase], and pyruvate kinase $K_{M,ADP}$ and with each point possessing a color encoding the maximal H_2 production possible by varying PFOR's $k_{cat,forward}$, [ferredoxin], and the ADH1 K_{M,NAD^+} within user-specified ranges, normalized over the actual maximum H_2 flux values.

Optimization of network behaviors Extending the example in the previous paragraph, perhaps the user is interested only in the single point in the cube of data points representing the overall maximal H_2 flux with respect to the six parameters phosphotransacetylase $k_{cat,forward}$, [pyruvate decarboxylase], and the pyruvate kinase $K_{M,ADP}$, PFOR's $k_{cat,forward}$, [ferredoxin], and the ADH1 K_{M,NAD^+}. Here, all six parameters can be allowed to vary freely, all others being held fixed as before. The calculation would yield a single optimal H_2 flux, rather than a cube of visualizable values.

Sensitivity characterization One may also consider how strongly a metabolite concentration or flux responds to a change in kinetic parameter or enzyme concentration in the immediate vicinity of an arbitrarily given or optimal value. This sensitivity is of clear utility in metabolic engineering efforts.

If, for example, H_2 flux is found to vary strongly with PFOR $k_{cat,forward}$, and not at all with ADH1 K_{M,NAD^+}, then the former would be the more valuable engineering target.

Sensitivity optimization If one were to desire a highly predictable response in the face of varying external conditions or mutagenic pressures, one may wish to minimize the sensitivity of a metabolite concentration or flux with respect to $\{k, E\}$. On the other hand, biosensor applications might benefit from the maximization of a sensitivity. In addition, one could speculate that present-day organisms represent evolutionary optima of maximum stability. Our software's ability to optimize sensitivity with respect to kinetic parameters permits quantitative study of these issues and opens up relatively uncharacterized areas for investigation.

3. Software

In this section we describe the HPC software we are developing to address the above issues.

3.1. Software structure

The high-performance simulation and optimization of metabolic models expressed in SBML are a complex task. For this we have developed a stack of software that includes translation of the SBML to a compiled language (C), simulation of the resulting ODEs using a state-of-the-art ODE integrator (SUNDIALS [27]), characterization via systematic parameter scanning applications, and optimization via both a gradient-based local optimizer based on TAO [28] and a scatter search-based global optimizer that incorporates the gradient-based optimizer into its local search phase. This stack of software has been described more thoroughly elsewhere [29] and is summarized in figure 2. From the HPC standpoint, the main features of our approach are (1) automated translation of SBML models into C, resulting in many orders of magnitude faster simulation over systems operating on merely interpreted models, and (2) exploitation of parallelism wherever possible. In particular, our global kinetic parameter estimation code for sensitivity optimization is now hierarchically parallel, with scaling to thousands of processors possible (see below).

3.2. Job setup from the user's perspective

As described above, the target user of HPC systems biology software is not necessarily a computer expert. Therefore, a component of our project is development of a graphical user interface to our high performance simulation and optimization tools. Figure 3 shows the main window of our controller tool. This tool is written in Java for cross-platform deployment. The paradigm we envision as an initial, easily implementable first step is that the GUI, running on the biologists' desktop machines, prepares all the necessary input files to run a large-scale simulation, characterization, or optimization on a compute cluster. Meanwhile, the main translation-simulation-optimization software stack is built only on the cluster. The biologist needs only to transfer the input files prepared via the GUI to the cluster and submit the associated parallel job. In this way the biologist is freed from the minutiae of the input files, required only to transfer a few files and submit a job. This simple mechanism will suffice for the initial deployment of our software.

3.3. Concentration/flux optimization

The SBML model of an organism results, mathematically, in a system of ODEs

$$\frac{dy}{dt} = f(y, k, E),$$

where y is a vector of metabolic species concentrations, k is a vector of kinetic parameters, and E is a vector of enzyme levels. From an optimization standpoint, our SciDAC project has two distinct phases.

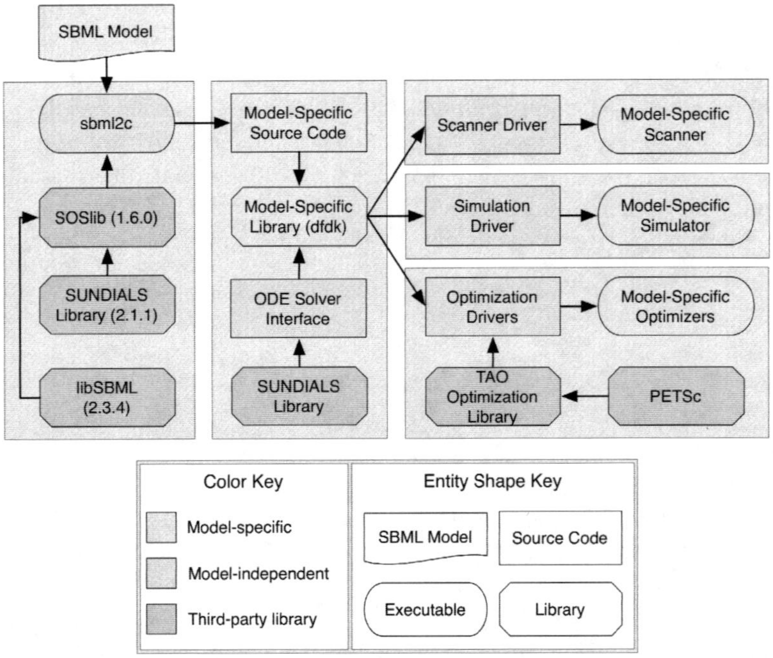

Figure 2. Software stack, dependencies, and workflow. The software features model translation with `sbml2c`, a model-specific library (`dfdk`) for evaluating various quantities, and a set of driver programs. Our set of tools consists of the elements in green and yellow.

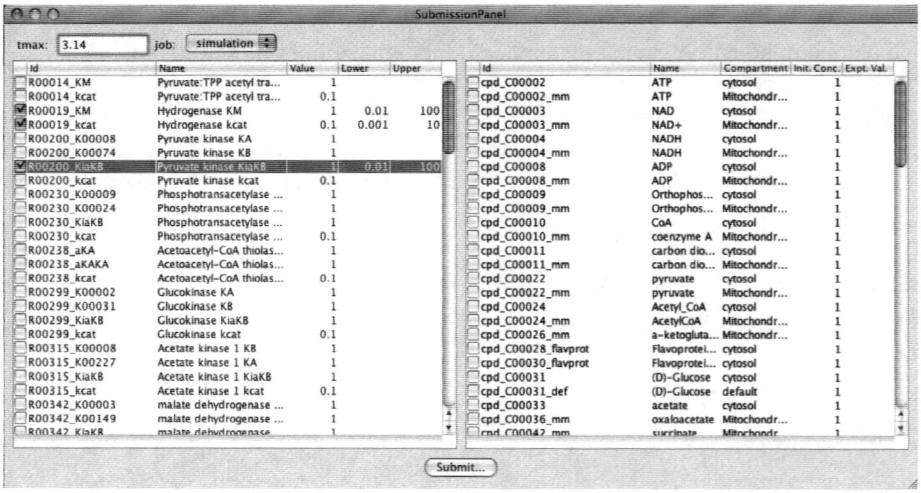

Figure 3. Prototype simulation and optimization user interface. The panel is built from a combination of the SBML file being simulated and existing optimization control files. Here, the user specifies which species and parameters will be "active" in the optimizations, how long to run the simulation, initial values of parameters and species, bounds of parameters, etc. Clicking the "submit" button results in the generation of all the necessary input files to launch the specified HPC scan or optimization on a compute cluster.

First, there is development of an accurate model. This involves solving, for fixed enzyme levels E, the data-fitting problem

$$\min_{k\in\Omega}\sum_{i=1}^{N_s}|y_i^c-y_i^e|^2, \tag{1}$$

where y_i^c and y_i^e are computed and experimental values, respectively, of the ith metabolic species, N_s is the number of metabolites, and Ω is the domain in which prior knowledge has determined the kinetic constants k must lie. Our parallel scatter search software solves such problems efficiently [30]. The second phase involves optimizing quantities with respect to E for fixed (at this stage known) k. We do not address optimizations with respect to E here.

3.4. Sensitivity optimization

It is posited that "robustness to perturbation" is a necessary quality of living organisms. To formulate and automate this notion, we have therefore developed software to solve the following:

$$\min_{k\in\Omega}\sum_{i,j}\left|\frac{\partial y_i}{\partial k_j}\right|^2, \tag{2}$$

that is, to minimize the squared Frobenius norm $f(k)\equiv\sum_{i,j}\left|\frac{\partial y_i}{\partial k_j}\right|^2$ of the so-called *sensitivity matrix* $\mathbf{dy}/\mathbf{dk}$ with respect to the kinetic parameters k. The objective function for this task is much more computationally expensive than in the data fitting task. It involves not just the computed values of species concentrations but their *derivatives* with respect to k. To apply gradient-based optimization to such problems requires, then, computing the *second derivatives* of species values with respect to k. This, however, allows us to introduce a further level of parallelism into our software, resulting in a hierarchically parallel, highly scalable, system. For this, we calculate finite-difference derivatives of the Frobenius norm f in parallel, at the same time as the scatter search-based global optimization is evaluating potential solutions in parallel. This two-level hierarchical parallelism is summarized in figure 4.

3.5. Performance: scaling studies

Initial scaling studies have been performed for a 64-parameter sensitivity minimization on our prototype model of *C. reinhardtii*. In less than 10 generations (approximately 1000 evaluations), the global optimizer achieves a several orders of magnitude reduction of sensitivity over the lowest sensitivity in the initial population (50 evaluations). Furthermore, the scaling studies suggest that the code will scale reasonably to thousands of processors, even on this relatively small problem (in the future, problem sizes will be larger by factors of hundreds or thousands). The reason is that the parallelism of finite difference calculations over components is extremely well balanced; for example, there is almost perfect scaling in sharing a single gradient calculation with respect to a vector of 64 components among 64 processors. This allows maximal use of processors at the lowest level, so even modest parallelism at the next level (e.g., 20 groups of processors evaluating the pool of new candidate solutions that scatter search generates at each iteration) will result in thousands of processors efficiently used. Figure 5 summarizes our findings to date.

4. Future

We are actively pursuing several areas:

Model expansion For maximal overlap with the metabolism currently undergoing scrutiny in the laboratory, we plan to add detailed models for starch degradation, light-driven photosynthetic electron transport, and the Calvin cycle. Template models for the latter two pathways have been

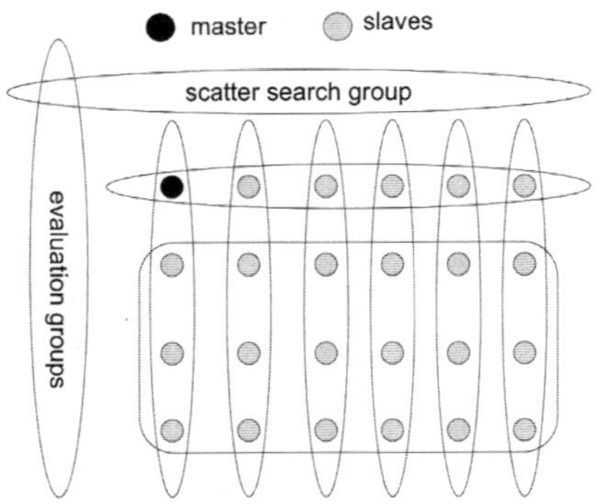

Figure 4. Schematic of the two-level master-slave hierarchical parallelism employed by our scatter search-based global sensitivity minimization code. At each iteration, there are some number (e.g., 70) candidate solutions to evaluate. The master distributes each of these dynamically to each member of the scatter search group. These, in turn, divide the gradient calculations up among the processors in the evaluation group in which they are rank 0. The figure shows the layout for a 24-processor run in which 4 processors work together to evaluate each candidate solution.

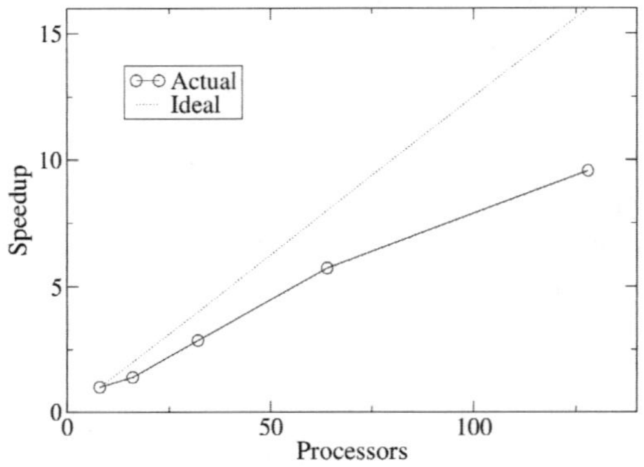

Figure 5. Initial scaling tests of hierarchically parallel scatter search code on global 64-parameter sensitivity minimization. The runs were compared after approximately 600 evaluations. Each evaluation consists of a gradient-based local search; the 600 evaluations require approximately 7,000 objective function and gradient calculations; each of these requires 65 (number of parameters plus one) simulations of the model, amounting to roughly 450,000 calculations of the full sensitivity matrix dy/dk. Speedup for n processors is defined here as the ratio of the wall time using n processors to that for 8 processors (the smallest number we used).

identified[31, 32], and the starch degradation pathways in *Chlamydomonas* are undergoing active investigation [33].

Model fitting As the sophistication of our *C. reinhardtii* model increases, our model-fitting tools will be further tested and expanded. We have prepared for treating the problem as a constrained optimization function by implementing constraints such as flux bounds, and we anticipate integrating a constrained nonlinear local optimizer into our global scatter search code.

Experimental data incorporation We have so far been developing and testing software with idealized, simulated data. We are now at a developmental stage in the model and simulation software to begin incorporation of real metabolic data from *C. reinhardtii*, the use of which will raise such inevitable limitations as underdetermination and data uncertainty. However, this next stage represents an exciting turning point where we move from development and testing, to answering real questions about biological systems.

Parallelism The hierarchical parallelism described above is in a prototype stage. Work on making the master-slave approach more scalable is ongoing. In particular, simple algorithms to decide when to use the master *only* as a manager of tasks, so there are never bottlenecks while the master is busy running a simulation, will be implemented. Furthermore, for truly large-scale parallelism, a peer-to-peer global optimization scheme will be considered, in which there is not the bottleneck of the one overall master orchestrating the process.

Relation of kinetic equations based modeling to flux balance analysis Ongoing collaboration with applied mathematicians involves bringing the flux balance analysis approach to bear on our metabolic modeling software. In particular, we are working toward a multiscale approach in which kinetic equations-based modeling (i.e., writing and using explicit forms of rate laws) is used where possible and appropriate, and gaps in such detailed knowledge are filled by recourse to a more stoichiometric matrix-based approach.

List of abbreviations

[*x*]: molar concentration of species *x*; *ADH1*: alcohol dehydrogenase 1; ATP/ADP: adenosine triphosphate/diphosphate; ChEBI: Chemical Entities of Biological Interest; CoA: coenzyme A; `dfdk`: subpackage of HiPer SBTK for integration of ODEs and evaluation of objective functions; Ferr: ferredoxin; GO: Gene Ontology; GUI: graphical user interface; HPC: high-peformance computing; mathML: Mathematical Markup Language; NAD^+/NADH: oxidized/reduced nicotinamide adenine dinucleotide; ODE: ordinary differential equation; PFOR: Pyruvate:ferredoxin oxidoreductase; P_i: inorganic phosphate; RDF: Resource Description Framework; RDF/XML: W3C specification of XML syntax for RDF; SBML: Systems Biology Markup Language; `sbml2c`: subpackage of HiPer SBTK for conversion of SBML models to c; SBO: Systems Biology Ontology; SOSlib: SBML ODE Solver; U=O/U–OH: oxidized/reduced ubiquinone; XML: Extensible Markup Language

References

[1] Schulze-Kremer S 2002 *In Silico Biol.* **2** 0017
[2] Consortium T G O 2000 *Nat. Genet.* **25** 25–9
[3] Degtyarenko K, de Matos P, Ennis M, Hastings J, Zbinden M, McNaught A, Alcántara R, Darsow M, Guedj M and Ashburner M 2008 *Nucl. Acids Res.* **36** D344–D350
[4] Beckett D (ed) 2004 *RDF/XML Syntax Specification (Revised)* Resource Description Framework (RDF) (W3C)
[5] Powell A and Johnston P 2003 Guidelines for implementing Dublin Core in XML Tech. rep. Dublin Core Metadata Initiative
[6] Hucka M, Finney A, Sauro H M, Bolouri H, Doyle J C, Kitano H, Arkin A P, Bornstein B J, Bray D, Cornish-Bowden A, Cuellar A A, Dronov S, Gilles E D, Ginkel M, Gor V, Goryanin I I, Hedley W J, Hodgman T C, Hofmeyr J H, Hunter P J, Juty N S, Kasberger J L, Kremling A, Kummer U, Le Novère N, Loew L M, Lucio D, Mendes P, Minch E, Mjolsness E D, Nakayama Y, Nelson M R, Nielsen P F, Sakurada T, Schaff J C, Shapiro B E, Shimizu T S, Spence H D, Stelling J, Takahashi K, Tomita M, Wagner J and Wang J 2003 *Bioinformatics* **19** 524–531
[7] Bornstein B J, Keating S M, Jouraku A and Hucka M 2008 *Bioinformatics* **24** 880–1
[8] Machné R, Finney A, Müller S, Lu J, Widder S and Flamm C 2006 *Bioinformatics* **22** 1406–1407
[9] Ausbrooks R, Buswell S, Carlisle D, Dalmas S, Devitt S, Diaz A, Froumentin M, Hunter R, Ion P, Kohlhase M, Miner R, Poppelier N, Smith B, Soiffer N, Sutor R and Watt S 2001 Mathematical markup language (MathML) Version 2.0 (Second Edition)
[10] Schulz M, Uhlendorf J, Klipp E and Liebermeister W 2006 *Genome Informatics Series* **17** 62–71
[11] Palsson B 2006 *Systems Biology: Properties of Reconstructed Networks* (New York: Cambridge Univ. Press)
[12] Lee J M, Gianchandani E P and Papin J A 2007 *Briefings in Bioinformatics* **prov**
[13] Klipp E, Herwig R, Kowald A, Wierling C and Lehrach H 2005 *Systems Biology in Practice: Concepts, Implementation and Application* (Weinheim: Wiley-VCH)
[14] Beard D A, Babson E, Curtis E and Qian H 2004 *J. Theor. Biol.* **228** 327–33
[15] Beard D A, Liang S d and Qian H 2002 *Biophys. J.* **83** 79–86
[16] Feist A M, Henry C S, Reed J L, Krummenacker M, Joyce A R, Karp P D, Broadbelt L J, Hatzimanikatis V and Palsson B 2007 *Mol. Sys. Biol.* **3** 121
[17] Jamshidi N and Palsson B Ø 2007 *BMC Sys. Biol.* **1** 26
[18] Oh Y K, Palsson B Ø, Park S M, Schilling C H and Mahadevan R 2007 *J. Biol. Chem.* **282** 28791–9
[19] Teusink B, Passarge J, Reijenga C A, Esgalhado E, van der Weijden C C, Schepper M, Walsh M C, Bakker B M, van Dam K, Westerhoff H V and Snoep J L 2000 *Eur. J. Biochem.* **267** 5313–29
[20] Segel I H 1975 *Enzyme Kinetics* Wiley Classics Library, 1993 ed (New York: John Wiley & Sons)
[21] Beard D A 2005 *PLoS Comp. Biol.* **1** e36
[22] Wu F, Yang F, Vinnakota K C and Beard D A 2007 *J. Biol. Chem.* **282** 24525–37
[23] Grima R and Schnell S 2006 *Biophysical Chemistry* **124** 1–10
[24] Mulquiney P J and Kuchel P W 1999 *Biochem. J.* **342** 581–96
[25] Mus F, Dubini A, Seibert M, Posewitz M C and Grossman A R 2007 *J. Biol. Chem.* **282** 25475–86
[26] Drăghici S 2003 *Data Analysis Tools for DNA Microarrays* (Boca Raton: Chapman & Hall/CRC) p 25

[27] Hindmarsh A C, Brown P N, Grant K E, Lee S L, Serban R, Shumaker D E and Woodward C S 2005 *ACM Trans. Math. Softw.* **31** 363–396 ISSN 0098-3500

[28] Benson S, McInnes L C, Moré J, Munson T and Sarich J 2007 TAO user manual (revision 1.9) Tech. Rep. ANL/MCS-TM-242 Mathematics and Computer Science Division, Argonne National Laboratory http://www.mcs.anl.gov/tao

[29] Chang C, Alber D, Graf P, Kim K and Seibert M 2007 *Journal of Physics: Conference Series* **78** 012011

[30] Alber D, Chang C, Graf P, Kim K and Seibert M 2008 *in preparation*

[31] Berry S and Rumberg B 2000 *Bioelectrochemistry* **53** 35–53

[32] Zhu X G, de Sturler E and Long S P 2007 *Plant Physiology* **145** 513–26 supplementary Information link included in Link to PDF field

[33] Posewitz M C, Smolinski S L, Kanakagiri S, Melis A, Seibert M and Ghirardi M L 2004 *Plant Cell* **16** 2151–2163

AUTHOR INDEX

AUTHOR INDEX

AUTHOR INDEX

AUTHOR INDEX

AUTHOR INDEX

AUTHOR INDEX

AUTHOR INDEX

AUTHOR INDEX

AUTHOR INDEX

Institute of Physics Publishing
Dirac House, Temple Back
Bristol BS1 6BE UK

ISBN: 978-1-60560-608-8

Institute of Physics Publishing

Scientific Discovery Through Advanced Computing Conference

SciDAC 2008

Journal of Physics: Conference Series Vol. 125

IOP Publishing

July 13-17 2008
Seattle, Washington

Volume 2 of 2

Institute of Physics Publishing

Scientific Discovery Through Advanced Computing Conference

SciDAC 2008

Journal of Physics: Conference Series Vol. 125

July 13-17 2008
Seattle, Washington

Volume 2 of 2

Printed from e-media with permission by:

Curran Associates, Inc.
57 Morehouse Lane
Red Hook, NY 12571
www.proceedings.com

ISBN: 978-1-60560-608-8

Some format issues inherent in the e-media version may also appear in this print version.

IOP Publishing
The Public Ledger Building
Suite 929
150 South Independence Mall West
Philadelphia, PA 19106
USA

Phone: (215) 627-0880
Fax: (215) 627-0879

info@ioppubusa.com

Scientific Discovery Through Advanced
Computing Conference
2008

TABLE OF CONTENTS

VOLUME 1

VOLUME 2

SciDAC 2008

Rick Stevens 2008 *J. Phys.: Conf. Ser.* **125** 011001 (1p) doi: 10.1088/1742-6596/125/1/011001 Help

Rick Stevens

The fourth annual Scientific Discovery through Advanced Computing (SciDAC) Conference was held June 13–18, 2008, in Seattle, Washington. The SciDAC conference series is the premier communitywide venue for presentation of results from the DOE Office of Science's interdisciplinary computational science program.

Started in 2001 and renewed in 2006, the DOE SciDAC program is the country's – and arguably the world's – most significant interdisciplinary research program supporting the development of advanced scientific computing methods and their application to fundamental and applied areas of science. SciDAC supports computational science across many disciplines, including astrophysics, biology, chemistry, fusion sciences, and nuclear physics. Moreover, the program actively encourages the creation of long-term partnerships among scientists focused on challenging problems and computer scientists and applied mathematicians developing the technology and tools needed to address those problems.

The SciDAC program has played an increasingly important role in scientific research by allowing scientists to create more accurate models of complex processes, simulate problems once thought to be impossible, and analyze the growing amount of data generated by experiments. To help further the research community's ability to tap into the capabilities of current and future supercomputers, Under Secretary for Science, Raymond Orbach, launched the Innovative and Novel Computational Impact on Theory and Experiment (INCITE) program in 2003. The INCITE program was conceived specifically to seek out computationally intensive, large-scale research projects with the potential to significantly advance key areas in science and engineering. The program encourages proposals from universities, other research institutions, and industry. During the first two years of the INCITE program, 10 percent of the resources at NERSC were allocated to INCITE awardees. However, demand for supercomputing resources far exceeded available systems; and in 2003, the Office of Science identified increasing computing capability by a factor of 100 as the second priority on its Facilities of the Future list. The goal was to establish leadership-class computing resources to support open science. As a result of a peer reviewed competition, the first leadership computing facility was established at Oak Ridge National Laboratory in 2004. A second leadership computing facility was established at Argonne National Laboratory in 2006.

This expansion of computational resources led to a corresponding expansion of the INCITE program. In 2008, Argonne, Lawrence Berkeley, Oak Ridge, and Pacific

Northwest national laboratories all provided resources for INCITE. By awarding large blocks of computer time on the DOE leadership computing facilities, the INCITE program enables the largest-scale computations to be pursued. In 2009, INCITE will award over half a billion node-hours of time.

The SciDAC conference celebrates progress in advancing science through large-scale modeling and simulation. Over 350 participants attended this year's talks, poster sessions, and tutorials, spanning the disciplines supported by DOE. While the principal focus was on SciDAC accomplishments, this year's conference also included invited presentations and posters from DOE INCITE awardees. Another new feature in the SciDAC conference series was an electronic theater and video poster session, which provided an opportunity for the community to see over 50 scientific visualizations in a venue equipped with many high-resolution large-format displays.

To highlight the growing international interest in petascale computing, this year's SciDAC conference included a keynote presentation by Herman Lederer from the Max Planck Institut, one of the leaders of DEISA (Distributed European Infrastructure for Supercomputing Applications) project and a member of the PRACE consortium, Europe's main petascale project. We also heard excellent talks from several European groups, including Laurent Gicquel of CERFACS, who spoke on `Large-Eddy Simulations of Turbulent Reacting Flows of Real Burners: Status and Challenges', and Jean-Francois Hamelin from EDF, who presented a talk on `Getting Ready for Petaflop Capacities and Beyond: A Utility Perspective'.

Two other compelling addresses gave attendees a glimpse into the future. Tomas Diaz de la Rubia of Lawrence Livermore National Laboratory spoke on a vision for a fusion/fission hybrid reactor known as the `LIFE Engine' and discussed some of the materials and modeling challenges that need to be overcome to realize the vision for a 1000-year greenhouse-gas-free power source. Dan Reed from Microsoft gave a capstone talk on the convergence of technology, architecture, and infrastructure for cloud computing, data-intensive computing, and exascale computing (1018 flops/sec).

High-performance computing is making rapid strides. The SciDAC community's computational resources are expanding dramatically. In the summer of 2008 the first general purpose petascale system (IBM Cell-based RoadRunner at Los Alamos National Laboratory) was recognized in the top 500 list of fastest machines heralding in the dawning of the petascale era. The DOE's leadership computing facility at Argonne reached number three on the Top 500 and is at the moment the most capable open science machine based on an IBM BG/P system with a peak performance of over 550 teraflops/sec. Later this year Oak Ridge is expected to deploy a 1 petaflops/sec Cray XT system.

And even before the scientific community has had an opportunity to make significant use of petascale systems, the computer science research community is forging ahead with ideas and strategies for development of systems that may by the end of the next decade sustain exascale performance. Several talks addressed barriers to, and strategies for, achieving exascale capabilities. The last day of the conference was devoted to tutorials hosted by Microsoft Research at a new conference facility in Redmond, Washington. Over 90 people attended the tutorials, which covered topics ranging from an introduction to BG/P programming to advanced numerical libraries.

The SciDAC and INCITE programs and the DOE Office of Advanced Scientific Computing Research core program investments in applied mathematics, computer science, and computational and networking facilities provide a nearly optimum framework for advancing computational science for DOE's Office of Science. At a broader level this framework also is benefiting the entire American scientific enterprise. As we look forward, it is clear that computational approaches will play an increasingly significant role in addressing challenging problems in basic science, energy, and environmental research.

It takes many people to organize and support the SciDAC conference, and I would like to thank as many of them as possible. The backbone of the conference is the technical program; and the task of selecting, vetting, and recruiting speakers is the job of the organizing committee. I thank the members of this committee for all the hard work and the many tens of conference calls that enabled a wonderful program to be assembled. This year the following people served on the organizing committee: Jim Ahrens, LANL; David Bader, LLNL; Bryan Barnett, Microsoft; Peter Beckman, ANL; Vincent Chan, GA; Jackie Chen, SNL; Lori Diachin, LLNL; Dan Fay, Microsoft; Ian Foster, ANL; Mark Gordon, Ames; Mohammad Khaleel, PNNL; David Keyes, Columbia University; Bob Lucas, University of Southern California; Tony Mezzacappa, ORNL; Jeff Nichols, ORNL; David Nowak, ANL; Michael Papka, ANL; Thomas Schultess, ORNL; Horst Simon, LBNL; David Skinner, LBNL; Panagiotis Spentzouris, Fermilab; Bob Sugar, UCSB; and Kathy Yelick, LBNL.

I owe a special thanks to Mike Papka and Jim Ahrens for handling the electronic theater. I also thank all those who submitted videos. It was a highly successful experiment.

Behind the scenes an enormous amount of work is required to make a large conference go smoothly. First I thank Cheryl Zidel for her tireless efforts as organizing committee liaison and posters chair and, in general, handling all of my end of the program and keeping me calm. I also thank Gail Pieper for her work in editing the proceedings, Beth Cerny Patino for her work on the Organizing Committee website and electronic theater, and Ken Raffenetti for his work in keeping that website working. Jon Bashor and John Hules did an excellent job in handling conference communications. I thank Caitlin Youngquist for the striking graphic design; Dan Fay for tutorials arrangements; and Lynn Dory, Suzanne Stevenson, Sarah Pebelske and Sarah Zidel for on-site registration and conference support. We all owe Yeen Mankin an extra-special thanks for choosing the hotel, handling contracts, arranging menus, securing venues, and reassuring the chair that everything was under control.

We are pleased to have obtained corporate sponsorship from Cray, IBM, Intel, HP, and SiCortex.

I thank all the speakers and panel presenters. I also thank the former conference chairs Tony Metzzacappa, Bill Tang, and David Keyes, who were never far away for advice and encouragement.

Finally, I offer my thanks to Michael Strayer, without whose leadership, vision, and persistence the SciDAC program would not have come into being and flourished. I am honored to be part of his program and his friend.

Rick Stevens
Seattle, Washington
July 18, 2008

SciDAC 2008

Michael Strayer 2008 *J. Phys.: Conf. Ser.* **125** 011002 (1p) doi: 10.1088/1742-6596/125/1/011002 (Help)

Michael Strayer

Welcome to Seattle and the 2008 SciDAC Conference. This conference, the fourth in the series, is a continuation of the PI meetings we first began under SciDAC-1. I would like to start by thanking the organizing committee, and Rick Stevens in particular, for organizing this year's meeting.

This morning I would like to look briefly at SciDAC, to give you a brief history of SciDAC and also look ahead to see where we plan to go over the next few years.

I think the best description of SciDAC, at least the simulation part, comes from a quote from Dr Ray Orbach, DOE's Under Secretary for Science and Director of the Office of Science. In an interview that appeared in the SciDAC Review magazine, Dr Orbach said, `SciDAC is unique in the world. There isn't any other program like it anywhere else, and it has the remarkable ability to do science by bringing together physical scientists, mathematicians, applied mathematicians, and computer scientists who recognize that computation is not something you do at the end, but rather it needs to be built into the solution of the very problem that one is addressing'.

Of course, that is extended not just to physical scientists, but also to biological scientists.

This is a theme of computational science, this partnership among disciplines, which goes all the way back to the early 1980s and Ken Wilson. It's a unique thread within the Department of Energy. SciDAC-1, launched around the turn of the millennium, created a new generation of scientific simulation codes. It advocated building out mathematical and computing system software in support of science and a new collaboratory software environment for data. The original concept for SciDAC-1 had topical centers for the execution of the various science codes, but several corrections and adjustments were needed. The ASCR scientific computing infrastructure was also upgraded, providing the hardware facilities for the program.

The computing facility that we had at that time was the big 3 teraflop/s center at NERSC and that had to be shared with the programmatic side supporting research across DOE. At the time, ESnet was just slightly over half a gig per sec of bandwidth; and the science being addressed was accelerator science, climate, chemistry, fusion, astrophysics, materials science, and QCD.

We built out the national collaboratories from the ASCR office, and in addition we built Integrated Software Infrastructure Centers (ISICs). Of these, three were in applied mathematics, four in computer science (including a performance evaluation research center), and four were collaboratories or Grid projects having to do with data management.

For science, there were remarkable breakthroughs in simulation, such as full 3D laboratory scale flame simulation. There were also significant improvements in application codes – from factors of almost 3 to more than 100 – and code improvement as people began to realize they had to integrate mathematics tools and computer science tools into their codes to take advantage of the parallelism of the day. The SciDAC data-mining tool, Sapphire, received a 2006 R&D 100 award. And the community as a whole worked well together and began building a publication record that was substantial.

In 2006, we recompeted the program with similar goals – SciDAC-1 was very successful, and we wanted to continue that success and extend what was happening under SciDAC to the broader science community. We opened up the partnership to all of the Offices of Science and the NSF and the NNSA. The goal was to create comprehensive scientific computing software and the infrastructure for the software to enable scientific discovery in the physical, biological, and environmental sciences and take the simulations to an extreme scale, in this case petascale. We would also build out a new generation of data management tools. What we observed during SciDAC-1 was that the data and the data communities – both experimental data from large experimental facilities and observational data, along with simulation data – were expanding at a rate significantly faster than Moore's law. In the past few weeks, the FastBit indexing technology software tool for data analyses and data mining developed under SciDAC's Scientific Data Management project was recognized with an R&D 100 Award, selected by an independent judging panel and editors of R&D Magazine as one of the 100 most technologically significant products introduced into the marketplace over the past year.

For SciDAC-2 we had nearly 250 proposals requesting a total of slightly over $1 billion in funding. Of course, we had nowhere near $1 billion. The facilities and the science we ended up with were not significantly different from what we had in SciDAC-1. But we had put in place substantially increased facilities for science. When SciDAC-1 was originally executed with the facilities at NERSC, there was significant impact on the resources at NERSC, because not only did we have an expanding portfolio of programmatic science, but we had the SciDAC projects that also needed to run at NERSC. Suddenly, NERSC was incredibly oversubscribed.

With SciDAC-2, we had in place leadership-class computing facilities at Argonne with slightly more than half a petaflop and at Oak Ridge with slightly more than a quarter petaflop with an upgrade planned at the end of this year for a petaflop. And we increased the production computing capacity at NERSC to 104 teraflop/s just so that we would not impact the programmatic research and so that we would have a startup facility for SciDAC. At the end of the summer, NERSC will be at 360 teraflop/s. Both the Oak Ridge system and the principal resource at NERSC are Cray systems; Argonne has a different architecture, an IBM Blue Gene/P. At the same time, ESnet has been built out, and we are on a path where we will have dual rings around the country, from 10 to 40 gigabits per second – a factor of 20 to 80 over what was available during SciDAC-1.

The science areas include accelerator science and simulation, astrophysics, climate modeling and simulation, computational biology, fusion science, high-energy physics, petabyte high-energy/ nuclear physics, materials science and chemistry, nuclear physics, QCD, radiation transport, turbulence, and groundwater reactive transport modeling and simulation. They were supported by new enabling technology centers and university-based institutes to develop an educational thread for the SciDAC program. There were four mathematics projects and four computer science projects; and under data management, we see a significant difference in that we are bringing up new visualization projects to support and sustain data-intensive science.

When we look at the budgets, we see growth in the budget from just under $60 million for SciDAC-1 to just over $80 for SciDAC-2. Part of the growth is due to bringing in NSF and NNSA as new partners, and some of the growth is due to some program offices increasing their investment in SciDAC, while other program offices are constant or have decreased their investment. This is not a reflection of their priorities *per se* but, rather, a reflection of the budget process and the difficult times in Washington during the past two years.

New activities are under way in SciDAC – the annual PI meeting has turned into what I would describe as the premier interdisciplinary computational science meeting, one of the best in the world. Doing interdisciplinary meetings is difficult because people tend to develop a focus for their particular subject area. But this is the fourth in the series; and since the first meeting in San Francisco, these conferences have been remarkably successful.

For SciDAC-2 we also created an outreach magazine, SciDAC Review, which highlights scientific discovery as well as high-performance computing. It's been very successful in telling the non-practitioners what SciDAC and computational science are all about.

The other new instrument in SciDAC-2 is an outreach center. As we go from computing at the terascale to computing at the petascale, we face the problem of narrowing our research community. The number of people who are `literate' enough to compute at the terascale is more than the number of those who can compute at the petascale. To address this problem, we established the SciDAC Outreach Center to bring people into the fold and educate them as to how we do SciDAC, how the teams are composed, and what it really means to compute at scale.

The resources I have mentioned don't come for free. As part of the HECRTF law of 2005, Congress mandated that the Secretary would ensure that leadership-class facilities would be open to everyone across all agencies. So we took Congress at its word, and INCITE is our instrument for making allocations at the leadership-class facilities at Argonne and Oak Ridge, as well as smaller allocations at NERSC. Therefore, the selected proposals are very large projects that are computationally intensive, that compute at scale, and that have a high science impact. An important feature is that INCITE is completely open to anyone – there is no requirement of DOE Office of Science funding, and proposals are rigorously reviewed for both the science and the computational readiness.

In 2008, more than 100 proposals were received, requesting about 600 million processor-hours. We allocated just over a quarter of a billion processor-hours. Astrophysics, materials science, lattice gauge theory, and high energy and nuclear physics were the major areas. These were the teams that were computationally ready for the big machines and that had significant science they could identify. In 2009, there will be a significant increase amount of time to be allocated, over half a billion processor-hours. The deadline is August 11 for new proposals and September 12 for renewals. We anticipate a significant increase in the number of requests this year.

We expect you – as successful SciDAC centers, institutes, or partnerships – to compete for and win INCITE program allocation awards. If you have a successful SciDAC proposal, we believe it will make you successful in the INCITE review. We have the expectation that you will among those most prepared and most ready to use the machines and to compute at scale.

Over the past 18 months, we have assembled a team to look across our computational science portfolio and to judge what are the 10 most significant science accomplishments. The ASCR office, as it goes forward with OMB, the new administration, and Congress, will be judged by the science we have accomplished. All of our proposals – such as for increasing SciDAC, increasing applied mathematics, and so on – are tied to what have we accomplished in science. And so these 10 big accomplishments are key to establishing credibility for new budget requests.

Tony Mezzacappa, who chaired the committee, will also give a presentation on the ranking of these top 10, how they got there, and what the science is all about. Here is the list – numbers 2, 5, 6, 7, 9, and 10 are all SciDAC projects.

Rank	Title
1	Modeling the Molecular Basis of Parkinson's Disease (Tsigelny)
2	Discovery of the Standing Accretion Shock Instability and Pulsar Birth Mechanism in a Core-Collapse Supernova Evolution and Explosion (Blondin)
3	Prediction and Design of Macromolecular Structures and Functions (Baker)
4	Understanding How Lifted Flame Stabilized in a Hot Coflow (Yoo)
5	New Insights from LCF-enabled Advanced Kinetic Simulations of Global Turbulence in

	Fusion Systems (Tang)
6	High Transition Temperature Superconductivity: A High-Temperature Superconductive State and a Pairing Mechanism in 2-D Hubbard Model (Scalapino)
7	PETsc: Providing the Solvers for DOE High-Performance Simulations (Smith)
8	Via Lactea II, A Billion Particle Simulation of the Dark Matter Halo of the Milky Way (Madau)
9	Probing the Properties of Water through Advanced Computing (Galli)
10	First Provably Scalable Maxwell Solver Enables Scalable Electromagnetic Simulations (Kovel)

So, what's the future going to look like for us? The office is putting together an initiative with the community, which we call the E3 Initiative. We're looking for a 10-year horizon for what's going to happen. Through the series of town hall meetings, which many of you participated in, we have produced a document on `Transforming Energy, the Environment and Science through simulations at the eXtreme Scale'; it can be found at http://www.science.doe.gov/ascr/ProgramDocuments/TownHall.pdf .

We sometimes call it the Exascale initiative. Exascale computing is the gold-ring level of computing that seems just out of reach; but if we work hard and stretch, we just might be able to reach it. We envision that there will be a SciDAC-X, working at the extreme scale, with SciDAC teams that will perform and carry out science in the areas that will have a great societal impact, such as alternative fuels and transportation, combustion, climate, fusion science, high-energy physics, advanced fuel cycles, carbon management, and groundwater.

We envision institutes for applied mathematics and computer science that probably will segue into algorithms because, at the extreme scale, we see the distinction between the applied math and the algorithm *per se* and its implementation in computer science as being inseparable. We envision an INCITE-X with multi-petaflop platforms, perhaps even exaflop computing resources. ESnet will be best in class – our 10-year plan calls for having 400 terabits per second capacity available in dual rings around the country, an enormously fast data communications network for moving large amounts of data.

In looking at where we've been and where we are going, we can see that the gigaflops and teraflops era was a regime where we were following Moore's law through advances in clock speed. In the current regime, we're introducing massive parallelism, which I think is exemplified by Intel's announcement of their teraflop chip, where they envision more than a thousand cores on a chip. But in order to reach exascale, extrapolations talk about machines that require 100 megawatts of power in terms of current architectures. It's clearly going to require novel architectures, things we have perhaps not yet envisioned. It is of course an era of challenge. There will be an unpredictable evolution of hardware if we are to reach the exascale; and there will clearly be multilevel heterogeneous parallelism, including multilevel memory hierarchies. We have no idea right now as to the programming models needed to execute at such an extreme scale. We have been incredibly successful at the petascale – we know that already. Managing data and just getting communications to scale is an enormous challenge. And it's not just the extreme scaling. It's the rapid increase in complexity that represents the challenge.

Let me end with a metaphor. In previous meetings we have talked about the road to petascale. Indeed, we have seen in hindsight that it was a road well traveled. But perhaps the road to exascale is not a road at all. Perhaps the metaphor will be akin to scaling the south face of K2. That's clearly not something all of us will be able to do, and probably computing at the exascale is not something all of us will do.

But if we achieve that goal, perhaps the words of Emily Dickinson will best summarize where we will be. Perhaps in her words, looking backward and down, you will say:

I climb the `Hill of Science'
I view the landscape o'er;
Such transcendental prospect
I ne'er beheld before!

Journal of Physics: Conference Series **125** (2008) 012049 doi:10.1088/1742-6596/125/1/012049

Developing improved MD codes for understanding processive cellulases

M F Crowley[1], E C Uberbacher[2], C L Brooks III[3], R C Walker[4], M R Nimlos[1] and M E Himmel[1]

[1]National Renewable Energy Laboratory, Golden, CO 80401,USA
[2]Oak Ridge National Laboratory, Oak Ridge, TN 37831, USA
[3]University of Michigan, Ann Arbor, MI 48109, USA
[4]San Diego Supercomputer Center, La Jolla, CA 92093, USA

Abstract. The mechanism of action of cellulose-degrading enzymes is illuminated through a multidisciplinary collaboration that uses molecular dynamics (MD) simulations and expands the capabilities of MD codes to allow simulations of enzymes and substrates on petascale computational facilities. There is a class of glycoside hydrolase enzymes called cellulases that are thought to decrystallize and processively depolymerize cellulose using biochemical processes that are largely not understood. Understanding the mechanisms involved and improving the efficiency of this hydrolysis process through computational models and protein engineering presents a compelling grand challenge. A detailed understanding of cellulose structure, dynamics and enzyme function at the molecular level is required to direct protein engineers to the right modifications or to understand if natural thermodynamic or kinetic limits are in play. Much can be learned about processivity by conducting carefully designed molecular dynamics (MD) simulations of the binding and catalytic domains of cellulases with various substrate configurations, solvation models and thermodynamic protocols. Most of these numerical experiments, however, will require significant modification of existing code and algorithms in order to efficiently use current (terascale) and future (petascale) hardware to the degree of parallelism necessary to simulate a system of the size proposed here. This work will develop MD codes that can efficiently use terascale and petascale systems, not just for simple classical MD simulations, but also for more advanced methods, including umbrella sampling with complex restraints and reaction coordinates, transition path sampling, steered molecular dynamics, and quantum mechanical/molecular mechanical simulations of systems the size of cellulose degrading enzymes acting on cellulose.

1. Introduction

Lignocellulose could potentially serve today as the dominant renewal biofuel source, capable of meeting our needs for a liquid transportation fuel, if it could be efficiently and economically depolymerized into its component glucose for microbial fermentations [1, 2]. One major obstacle to the efficient and effective use of cellulose is the high resistance of crystalline cellulose to chemical and biological hydrolysis [3]. Even the most efficient cellulase enzymes available are not effective enough to depolymerize cellulose at a rate or cost that can meet the needs of commercial scale biofuel production. Cellobiohydrolase I (Cel7A) from *Trichoderma reesei* is one of the most effective exoglucanases known and in many ways the dominant cellulase in the biosphere. Cel7A is a multidomain enzyme, consisting of a large catalytic domain containing an extensive active site tunnel and a small binding module joined to one another by a linker peptide. Through a series of mechanisms yet unknown, a single cellodextrin chain is removed from the crystal and fed into the active site tunnel of the catalytic domain for hydrolysis to cellobiose. Cel7A is hypothesized to proceed along the target cellodextrin in a processive manner, cleaving one cellobiose unit per catalytic event, see figure 1 below. However, experimental evidence for the processive motion hypothesis is limited. Although much is known about the structure of Cel7A, many questions remain about the functioning of this enzyme, which can be considered a "protein machine." Among them are the roles of the carbohydrate binding module (CBM) and the linker peptide connecting the CBM to the catalytic domain. We have shown in previous work that the CBM recognizes and binds to the cellulose surface [4]. Furthermore, we have postulated that the linker works in a "springlike" motion to enable the enzyme to move on the cellulose surface during catalysis of a cellulose chain; where the catalytic hydrolysis of a cellulose strand by the moving catalytic domain compresses the length of the linker, and once a short enough length is reached, the stored potential energy allows the linker to extend, freeing the binding module from its current position on the cellulose surface [5]. The mechanism by which the catalytic domain translates cellodextrin substrates toward the active site is also not understood. If improved Cel7A enzymes are to be developed using protein engineering, rational design methods require some baseline understanding for structure/function relationships.

Figure 1. Computational model of Cellobiohydrolase I (Cel7A) from *Trichoderma reesei* on a cellulose 100 surface displaying the large catalytic domain (left), linker (middle single strand), and cellulose binding module (right small domain). A cellodextrin strand is shown peeled out of the surface of the cellulose and threaded into the catalytic tunnel of Cel7A. The solvating water and the lower portion of the cellulose fiber are not shown in this visualization.

Given that there are no experimental methods available to probe this complex behavior at the molecular level, we have proposed using improved molecular dynamics simulations to study this problem from the perspective of the system ensemble, which includes cellulose, the enzyme, and water. To accomplish this work, which focuses on a cellulase-cellulose model of approximately 800,000 atoms, part of the model is shown in figure 1, one must develop improved MD codes capable of taking full advantage of new leadership class computers. Computational codes exist for performing simulations of up to millions of atoms on as many as thousands of processors with some degree of scaling efficiency, namely, Amber [6], LAMMPS [7], and NAMD [8], However, these codes have limited functionality beyond simple and steered MD and some simple sampling techniques. There is a great need for improving the scaling efficiency of MD programs and to migrate more complex simulation and sampling methods into highly scalable programs. Many of the necessary methods for studying the proposed mechanisms of cellulase enzymes and the cellulose substrate are available in CHARMM [9] but are not currently available to be used on systems the size of a cellulase/cellulose system for the length of simulation that would produce meaningful results within a reasonable timeframe. This SciDAC project will design highly parallel and highly efficient programs by both improving the parallel efficiency of existing programs and algorithms and creating new codes and new algorithms for the most important methods for use on current and future terascale and petascale computers.

2. Computational methods for studying cellulose and cellulase

In the most primitive approach to studying at the molecular level of detail the cellulase mechanism for degrading cellulose to cellobiose, simple molecular dynamics is used to produce many nanoseconds and hopefully microseconds of molecular motions. There is likely little to be gained from such simulations due to the low likelihood of important events occurring on that time scale since this enzyme acts on the seconds to minutes timescale. Some of the equilibrium structural and dynamical features of the system have been extracted from multi-nanosecond simulations of the enzyme and substrate including structuring of water around each of the three domains of the enzyme and how it restructures on the cellulose when the enzyme is present. For these simulations to continue to the microsecond time, the improvements in scaling and parallel efficiency are necessary.

The important parts of the mechanism are those responsible for the processive motion of the enzyme and the catalytic event. The processive motion is one or a series of rare events including decrystallization of the cellulose surface, movement of a free cellodextrin strand into the catalytic tunnel, movement of the cellodextrin through the tunnel, the catalytic event, leaving of the cellobiose product, and finally moving of the cellodextrin into reaction position again. Initial simulations using currently available MD programs show that these are both complex and seem to have several very high barriers and are thus rare events requiring much more sophisticated methods than simple MD. Our development wish list for highly parallel and scalable methods specifically targeted at solving the cellulase problem includes umbrella sampling using complex restraints, steered and targeted MD, QMMM, transition path searching and sampling methods, and thermodynamic integration and perturbation methods. Current and new scientific methods and models are also needed to improve the quality of the simulations and to enhance both the speed and sampling of the numerical experiments. These include using new solvent models such as the water models with off-center charges, implicit solvent models, and new polarizable force fields, none of which are available in highly parallel and scalable MD codes.

3. Program optimization - code development and enhancement

The primary focus of the code development work for this project is on computational methods that are the most useful to scientists studying large biological systems, and in particular on the methods that are initially needed for the biological application chosen, cellulase action. Our primary aim is to make these methods as efficient and parallel as possible without sacrificing versatility and functionality. Many of the most valuable molecular modeling methods such as thermodynamic integration [10], free energy perturbation, pulling or steered MD, umbrella sampling [11], and hybrid quantum mechanical/molecular mechanics (QMMM) [12, 13] have been restricted to smaller molecular systems, 30,000 atoms or less or are very limited in their implementations. These restrictions are a function of both their limited scaling and a lack of knowledge of how these algorithms behave when used for modeling larger systems. These approaches, which largely represent an extension of standard MD methods, are typically limited to around 64 CPUs or to very simple restraints. Not long ago, these CPU counts were considered reasonably large. The development of both the algorithms and the implementations of the computational methods have been largely guided by limitations in the resources available and the characteristics of the existing computers. Many of the existing methods have proven useful scientific tools for studying systems of moderate size. However, the current implementations are not designed for large systems or massively parallel computers by current or projected standards and are only marginally efficient for even the smaller subsystem simulations needed for the cellulase study. We will study the more important algorithms to uncover the limitations to extending to large systems and petascale machines with many thousands of processors in both the implementations and the methods themselves.

Journal of Physics: Conference Series **125** (2008) 012049 doi:10.1088/1742-6596/125/1/012049

Our efforts to improve parallel and scalar efficiency will be combined with a rational approach, to where the effort will be most effective by studying the theory and algorithmic aspects of several computational methods that go beyond simple classical MD and are essential to most studies that model biomolecular systems using MD for more than single trajectory studies. As we extend capabilities of existing tools and methods to run on larger and faster computers and to handle larger molecular models, it is important that we consider which problems are appropriate for the available tools and whether or not the modeling really answers the questions of interest. Our primary focus will be on the methods necessary to answer the questions posed in the Cel7A processive enzyme study, and if, even with petascale computers, these questions can be answered with the approaches that have been traditionally used. Simply running longer trajectories of all-atom models and extending to larger and larger systems for similar time-domain simulations may not yield the increase in information expected from linear-scaling arguments. For instance, larger molecules, because of motions on much larger spatial scales and exponentially larger phase space degrees of freedom, require significantly more time to undergo structural changes and concerted motions than do smaller molecules. For these reasons, we will focus heavily on evaluating the questions that are appropriate for the molecular systems and computational hardware we are targeting. As an essential result of this effort we will provide a gauge of the confidence that can be placed on the results of simulations using the algorithms, methods, theory, and software produced.

The methods we are interested in implementing into an improved code are essential tools of molecular modeling in general and are critical to the cellulase problem. These methods rely on the underlying MD engine of the program in which they are embedded, such as AMBER and CHARMM. The state-of-the-art molecular-modeling technology is well developed in some programs, such as LAMMPS and NAMD, that scale very well to thousands of processors but are limited to simple classical MD and a few simple sampling enhancing methods with limited functionality. The other more traditional simulation packages, such as CHARMM and AMBER, have a wealth of methodologies for detailed studies of the physical properties and behaviors of solvated macromolecular systems. Our primary aim is to enhance the massively parallel computations with the most useful and most widely used methods for studying the physical chemistry of biomolecular systems by implementing the already proven efficient parallel spatial-decomposition methods in the framework of existing less-efficient but highly functional and validated programs. The methods we wish to make highly efficient have not been applied to systems of millions of atoms that are currently considered as appropriate for computers with many thousands of processors. Little is known about whether the methods will produce meaningful results on large systems, or whether there are characteristics of the theory that will prevent either massively-parallel implementations or application to very large systems from being useful. We are implementing algorithms for several methods on massively parallel architectures and evaluating the validity, efficiency, and reliability on the cellulase system as a test bed, but with general use software as the target product. Our studies use existing MD packages, CHARMM]9], AMBER [6], and LAMMPS [7] as starting points both for the studies and code development and for validation of the scientific results produced by the implementations. The specific use and function of each program and the approach to the problem are described below.

The algorithms for parallel MD can be grouped into two general classes: replicated data and spatial decomposition. The traditional MD programs use the replicated data model for several reasons, not least of which is historical since they were originally written before there were parallel computers. The advantage of the replicated data model is that nearly all the important modeling methods can be made parallel when every processor has all the current data. However, there is a low ceiling to the amount of scaling that can be achieved with such an approach. On the other hand, the spatial decomposition method is highly scalable and current MD programs using this method can scale to thousands of processors with tolerable efficiency. The problem with spatial decomposition is that the more complex MD methods are not spatially localized in character. Many of the important methods break the scaling efficiency of the highly parallel codes and thus have not been implemented. Our aim is to study the scientific methods of standard MD studies and to develop new algorithms that will scale as efficiently as possible in the spatial decomposition framework. Although some of the methods will degrade parallel performance their usefulness outweighs the cost of using them. Some of the methods will not degrade parallel performance but will require the development of new approaches to their parallel implementation. In summary, the aim is to implement the scientific methods, study their scientific and performance behaviors, and develop the most effective and efficient implementations that preserve the most scientific benefit. The particular methods and features for our study are as follows:

- Spatial decomposition implementation in existing replicated-data program
- Spatially decomposed implicit solvent model algorithms
- New fast explicit solvent and non-bond list algorithms
- Polarizable Force Fields in spatial decomposition programs
- Lone pairs/Extra points (TIP4P and TIP5P water)
- General restraints and umbrella sampling
- Thermodynamic integration (TI)
- Nudged elastic band (NEB) and more refined path-finding bead methods
- Steered molecular dynamics (SMD)
- Semi-empirical QMMM methods with Ewald, PME, and implicit solvent models

The need for increased sampling is clear for larger molecular systems. As yet, there are only estimates for what the sampling needs are, and some guesses about what the issues are for both highly parallel approaches and larger systems. The methods of thermodynamic integration, umbrella sampling, and nudged elastic band are critically dependent on converged sampling. We will explore the issues for large systems and generate both analysis tools for the large-scale modeling and highly parallel codes. Many of the methods described above require numerous runs with different starting conditions or different parameters such as umbrella restraints, temperatures, or pulling forces and can be run together to use a large scale machine such as a 100 teraflop to petaflop computer even for the simulations of the subsystems.

4. Program validation and benchmarking

The National Leadership Computing Facility (NLCF) machines at ORNL are soon going to offer computing power on order of hundreds of teraflops. A one-petaflop machine is expected in the next few years. These systems will provide a hardware and software infrastructure that is substantially different from existing cluster supercomputers. One of the critical challenges of such a system is the utilization of extra computing resources. We have demonstrated that the MD calculations are capable of exploiting additional computing resources that are at the expense of reduced memory capacities and bandwidth [14]. Hence we anticipate that the existing application frameworks will exploit the multi-core resource of the next-generation MPP systems effectively. Based on our preliminary profiling studies, we expect to achieve 50-100 ns/day of MD simulation for biomolecular systems with 500K to 1M atoms. By using the portable performance tools such as the Tool and Analysis Utilities (TAU) from the University of Oregon [15] and vendor-specific performance tools like the Cray Performance Analysis Toolset (PAT), we will investigate and explore the factors limiting the achievable performance of petascale software simulations and the underlying systems. Analysis of the performance data reveals the performance and scaling bottlenecks in the code implementation as well as in the algorithms. Collaboration with algorithm designers, scientific library developers, application scientists, and performance engineers will enable the extreme-scale scaling of these bio-molecular application frameworks. In future, this task scope will also include the possibility of exploring alternative programming models for the application's key kernels in order exploit the underlying architecture efficiently. For example, the use of Co-Array Fortran on the Cray X1 has demonstrated significant performance benefits for Cray systems. The first goal of this strategy is to migrate LAMMPS, CHARMM, and Amber to existing high-performance hardware and verify them for correctness using the CHARMM force field against output from CHARMM on a trusted platform. A set of validation models has been created to compare the results of modified and new code with the trusted output for correctness of energies, forces, and trajectories. Once codes have been verified, optimization for both speed and parallel performance will be performed. A set of benchmarking simulations has been created to monitor the progress and performance of new work. A baseline of performance has been created for several platforms from which our improvements will be measured.

The simulations proposed here are expected to execute for weeks to months, a situation that requires additional support and capabilities from the target supercomputing systems as well as the applications. The mean time between failures (MTBF) for the latest generation of supercomputers with tens to hundreds of thousands of processors is projected to be a few hours. To benefit most from the computing power and to avoid loss of vital simulation data for long-running experiments, we are developing a new low-level optimized MD kernel that is able to self-monitor and is failure-tolerant. This work is being performed in collaboration with several hardware vendors including IBM, Cray, and SRC, as well as in collaboration with a SciDAC proposal "Center for Reliability, Availability and Serviceability for Petascale High-End Computing" (PI: Stephen Scott, ORNL). The novel capabilities of this MD kernel will include memory image check pointing for faster restart and progress migration to new processors. One of the novel features of this new MD kernel will be the ability to dynamically resize avoiding any lost productivity due to stopping and restarting. On the petascale machines failure of a single node will lead to the failure of the whole simulation. With this functionality the simulation will be able to continue on a decreased number of processors. In another situation, even on an oversubscribed machine there are idle processors when the scheduler is waiting for more processors to become available so that they can be assigned to the next job in queue. In such cases, as additional processors become available this novel MD kernel will expand on the additional processors; similarly shrink to fit a decreased number of processors and release processors to the system to be assigned other jobs. Such functionality will allow efficient utilization of the petascale computing resources. This powerful kernel will be designed to be modular and flexible. It will be able to be plugged with the functionalities of different programs such as CHARMM and AMBER, thus allowing the user to take advantage of these popular packages on the peta-scale systems. Moreover, we are working closely with hardware vendors to perform low-level optimization of each platform. thereby allowing science applications to benefit most from the available hardware.

Acknowledgments

This work is performed with support from the U.S. Department of Energy Office of Science through a SciDAC award titled "Understanding the Processivity of Cellobiohydrolase I."

[1] Schubert C 2006 Can biofuels finally take center stage *Nature Biotechnol.* **24** 777-784

[2] Lynd L R, Laser M S, Bransby D, Dale B E, Davison B, Hamilton R, Himmel M E, Keller M, McMillan J D, Sheehan J and Wyman C E 2008 How biotech can transform biofuels *Nature Biotechnol.* **26** 169-172

[3] Himmel M E, Ding S-Y, Johnson D K, Adney W S, Nimlos M R, Brady J W and Foust T D 2007 Biomass Recalcitrance: Engineering Plants and Enzymes for Biofuels Production *Science* **315** 804-807

[4] Nimlos M R, Matthews J F, Crowley M F, Walker R C, Chukkapalli G, Brady J W, Adney W S, Cleary J M, Zhong L and Himmel M E 2007 Molecular modeling suggests induced fit of Family I carbohydrate binding modules with a broken chain cellulose surface *Protein Engineering and Design* **20** 179-187

[5] Zhao X, Rignall T R, McCabe C, Adney W S and Himmel M E 2008 Molecular simulation evidence for processive motion of Trichoderma reesei Cel7A during cellulose depolymerization *Chem. Physics Lett.* (in press)

[6] Case D A, Cheatham III T E, Darden T, Gohlke H, Luo R, Merz Jr. K M, Onufriev A, Simmerling C, Wang B and Woods R J 2005 The Amber biomolecular simulation programs *J. Computational Chemistry* **26** 1668-1688

[7] Plimpton S 1995 Fast parallel algorithms for short-range molecular-dynamics *J. Computational Physics* **117** 1-19

[8] Phillips J C, Braun R, Wang W, Gumbart J, Tajkhorshid E, Villa E, Chipot C, Skeel R D, Kale L and Schulten K 2005 Scalable molecular dynamics with NAMD *J.f Computational Chemistry* **26** 1781-1802

[9] Brooks B R, Bruccoleri R E, Olafson B D, States D J, Swaminathan S and Karplus M 1983 CHARMM: A program for macromolecular energy, minimization, and dynamics calculations *J. Computational Chemistry* **4** 187-217

[10] Simonson T 2001 *Computational Biochemistry and Biophysics,* ed. O M Becker, *et al.* (New York: Marcel Dekker, Inc.)

[11] Torrie G M and Valleau J P 1977 Non-physical sampling distributions in Monte-Carlo free-energy estimation - umbrella sampling *J. Computational Physics* **23** 187-199

[12] Walker R C, Crowley M F and Case D A 2006 The implementation of a fast and efficient hybrid QM/MM potential method within Amber 9.0 Sander module *J. Comput. Chem.*

[13] Warshel A and Levitt M 1976 Theoretical studies of enzymic reactions - dielectric, electrostatic and steric stabilization of carbonium-ion in reaction of lysozyme *J. Molecular Biology* **103** 227-249

[14] Alam S, Vetter J S, Agarwal P K and Geist A 2006 Performance characterization of molecular dynamics techniques for biomolecular simulations. *Proc. 11th ACM SIGPLAN symposium on Principles and practice of parallel programming* (New York: ACM)

[15] Shende S and Malony A D 2006 The TAU parallel performance system *International J. High Performance Computing Applications* **20** 287-331

Journal of Physics: Conference Series **125** (2008) 012050 doi:10.1088/1742-6596/125/1/012050

The smallest cells pose the biggest problems: high-performance computing and the analysis of metagenome sequence data

Robert A Edwards

Mathematics and Computer Science Division, Argonne National Laboratory, Argonne, IL 60439, USA

Department of Computer Science, San Diego State University, San Diego, CA 92182, USA

E-mail: redwards@mcs.anl.gov

Abstract.

New high-throughput DNA sequencing technologies have revolutionized how scientists study the organisms around us. In particular, microbiology – the study of the smallest, unseen organisms that pervade our lives – has embraced these new techniques to characterize and analyze the cellular constituents and use this information to develop novel tools, techniques, and therapeutics. So-called next-generation DNA sequencing platforms have resulted in huge increases in the amount of raw data that can be rapidly generated. Argonne National Laboratory developed the premier platform for the analysis of this new data (mg-rast) that is used by microbiologists worldwide. This paper uses the accounting from the computational analysis of more than 10,000,000,000 bp of DNA sequence data, describes an analysis of the advanced computational requirements, and suggests the level of analysis that will be essential as microbiologists move to understand how these tiny organisms affect our every day lives. The results from this analysis indicate that data analysis is a linear problem, but that most analyses are held up in queues. With sufficient resources, computations could be completed in a few hours for a typical dataset. These data also suggest execution times that delimit timely completion of computational analyses, and provide bounds for problematic processes.

1. Introduction

Small organisms are influencing everything we do and how we do it. These organisms, called microbes because they are only visible under the microscope, are more abundant and much more diverse than the plants, trees, and other life we see all around us. Microbes affect us in both positive ways, providing cheese, bread, wine, and antibiotics for example; they also affect us in negative ways, causing disease, decay, rotting timber, and so forth.

Biologists have struggled for many years to understand the role of microbes in different environments. What particular microbes are in an environment, and what are they doing there? With this knowledge in hand, researchers want to begin manipulating the microbes. Like almost all known life, microbes use DNA (deoxyribonucleic acid) as their genetic material. This complex chemical encodes proteins that perform all cellular functions, from creating new cell walls, to propelling the cells toward food and away from noxious chemicals, to both making and using amino acids. A typical microbe is only about 10^{-6} meters long (1 micron) but has about 2,000,000 base pairs of DNA. Each base pair comprises one combination of the four chemical compounds adenine, guanine, cytosine, and thymidine (abbreviated A, G, C, and T, respectively). The particular combination of letters encodes for the genes that encode the proteins. By sequencing the complete complement of DNA in a microbe (its *genome*), and by comparing that to sequences that have been experimentally validated, microbiologists are able to characterize all of the genes in a microbe, and all of the functions that microbe is performing [1]. Rather than studying a single microbe in isolation, more recently a new approach has been taken to understand the role of microbes in complex communities in the environment. This new technique is called *metagenomics* to imply that it is more than just a single genome that is being sequenced. The first of these studies was most widely performed in the oceans [2]; subsequent studies have been done in most major microbial habitats, including the human body [3].

These transformative studies of microbes in their habitats have been assisted by new technologies that have revolutionized the speed and accuracy with which the DNA sequence can be determined. Until recently most sequencing was performed using a technique first described in 1977 [4], and only minor modifications and improvements to the basic approach had been made. High throughput was largely achieved by robotization and miniaturization of the existing technology. A typical sequencing machine would generate the DNA sequence of 96 fragments simultaneously. The first of the so-called next-generation sequencing technologies to market, massively parallel pyrosequencing developed by 454 Life Sciences, Inc. (now part of Roche, Inc.) enabled the generation of hundreds of thousands of DNA sequences at a time [5]. The increased volume of sequences came at a cost of decreased length of each single sequence. In the traditional approach a single DNA sequence could exceed 1,000 letters, but the new pyrosequencing approach could generate only about 100 letters from a single piece of DNA. Over time, however, the technology has matured, and pyrosequencing currently generates 400 letters from a single DNA fragment, and about 500,000 fragments in a single reaction. All of this is accomplished in less than eight hours. In addition to generating more data, pyrosequencing is about tenfold to one hundred-fold cheaper than Sanger sequencing (depending on economies of scale). This revolution in technology has enabled microbial researchers to take a sample from the environment, extract the DNA, and generate the sequence of a sample of that DNA in less than a week. The DNA sequence generated by this approach is the raw input for the advanced computing that determines the functions of those DNA fragments and is the focus of this paper.

2. Automated analysis pipeline for metagenomes

As researchers generated more and more sequence data, the need for an accurate, automated, high-throughput annotation and analysis platform became pressing. The SEED database was first developed to annotate complete microbial genomes, using a new approach that exploited the clustering in microbial genomes of genes that perform related functions (it is unknown why this happens: there are several hypotheses but none that is conclusively supported by all of the data). The SEED was used as the basis of the metagenomics annotation and analysis pipeline, called the *m*eta*g*enomics *r*apid *a*nnotation using *s*ubsystem *t*echnology (mg-rast). This pipeline is depicted in figure 1.

The pipeline consists of a series of discrete steps that are carried out in series on the whole dataset:

(i) Preprocessing
Incoming DNA sequence is checked to ensure that it is really DNA sequence (sometimes our users submit corrupted files, binary files from word-processing software, and so forth). There is also a known artifact of the high-throughput sequencing technology, that occasionally the exact same sequence is generated more than once. These sequences are removed, and each sequence is provided a unique identifier so that it can be tracked through subsequent steps.

(ii) Similarity computation

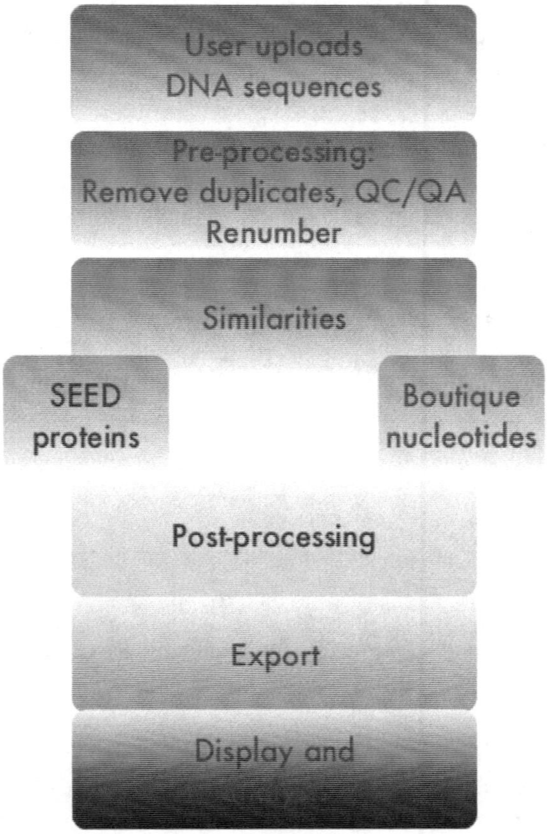

Figure 1. The metagenomics rapid annotation pipeline (mg-rast) consists of four steps: preprocessing, similarity computation, postprocessing, and export.

The sequences are compared to a number of databases of previously characterized and well understood sequences. These include a database that has all of the sequences from complete genomes from all three domains of life, and four different databases that contain sequences of certain genes of specific interest to microbiologists. (They are typically interested in these genes because they cannot change very quickly and thus can be used to describe the time-series relationships between organisms.)

(iii) Postprocessing
After the DNA sequences have been compared, the results are parsed to generate a description of each DNA sequence in the sample, and the entire sample itself.

(iv) Export
In the final step, reports are generated for the user either to view via an online interface or to download and peruse at will.

The mg-rast service was released in beta version in April 2007 and opened for unfettered public access in July of the same year. The service is freely available to all researchers, although registration with a valid email address is required so that we may contact the users in the case of trouble in the system. Since the initial release, over 800 sequence analysis jobs have been submitted by 99 different users, comprising in excess of 10,000,000,000 base pairs (10 Gbp) of DNA sequence; and the mean sample submitted by our users is currently 15,000,000 base pairs (15 Mbp).

Table 1. Computational resources used in the mg-rast pipeline.

Name in Figures	Processor	Memory	Number of units
bio-big	16 Intel Xeon CPU X7350 @ 2.93GHz	123822988 kB	1
CGAT	8 Intel Xeon CPU X5365 @ 3.00GHz	16436064 kB	1
DGRI	8 Intel Xeon CPU E5335 @ 2.00GHz	16440024 kB	1
Intel	1 Intel Pentium 4 CPU 3.00GHz	3636148 kB	2
MacPro	8 Intel Xeon CPU X5365 @ 3.00GHz	16387776 kB	2
mg-rast	8 Intel Xeon CPU E5450 @ 3.00GHz	16436064 kB	1
RAST	8 Intel Xeon CPU X5365 @ 3.00GHz	16436060 kB	1
PPC	2 PPC970FX, altivec supported	4042892 kB	45

1

3. Computation

The entire mg-rast computation is currently being performed on a hybrid cluster. Initially developed as an Apple PowerPC cluster running Debian Sarge, over time these machines have been supplemented with Intel Xeon processors also running linux (table 1). The cluster is a shared resource, and in addition to the mg-rast is also used to annotate complete microbial genomes through a related service, the RAST (*r*apid *a*nnotation using *s*ubsystems *t*echnology) [6]. Jobs on the cluster are scheduled using the Sun Grid Engine®, and the accounting system was used as the basis of all timing studies presented in this work. All accounting times were extracted using the PERL Schedule::SGE module written by the author and available from the Comprehensive Perl Archive Network (CPAN) or by contacting the author.

Throughout this paper the input data is measured in terms of base-pairs; individual letters of DNA uploaded by the end users. Of course, this equates to bytes of data with standard unicode encoding. Wall time is generally measured in seconds throughout the analysis.

4. The four stages of the computation

The four separate stages of the computation occur in series: each needs to be completed before the next is started. The only exception is the multiple different computations calculated during the similarity processing stage. Each different database can be compared in parallel, since each is an independent entity. At this stage, the input data is also routinely fragmented into smaller units to be processed. The goal is to achieve a single unit of computation that takes between one and two hours to complete, an empirically determined balance between fairness with other users of the clusters; limitations of IO capability and scaling; and sufficient throughput to complete the analysis in a timely fashion.

4.1. Preprocessing

As shown in figure 2, the preprocessing step is defined by two distinct populations. The first population displays linear complexity with input size, and represents those tasks that complete without problem. The preprocessing step is not memory limited, does not require complex calculations, and there does not appear to be a significant difference between the Intel or PPC architectures used in this approach. The total time taken to process the data is approximately 2 seconds per million base pairs of input sequence, or less than a minute for the average metagenome. The second population consists of those problematic jobs alluded to earlier. It is at this stage of the computation that user-derived errors are usually identified. Malformed files, bad sequence data, and so on result in larger processing times for smaller jobs. Those jobs likely represent the situations where endless cycles are performed before a submission is terminated. With this knowledge jobs can be automatically terminated if they take too long to process and the administrators notified of potential problems.

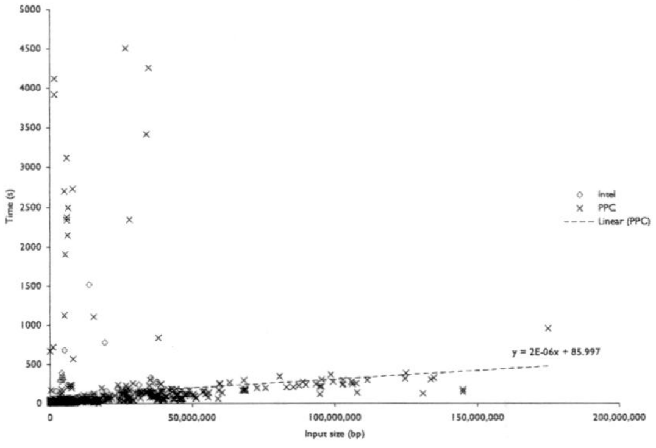

Figure 2. Preprocessing is characterized by two distinct populations. The first computes with linear complexity compared to input size, and represents those jobs that complete without problem. The second population is more scattered, takes longer to compute, and represents those tasks that are problematic.

4.2. Similarity

The similarity computation, as noted above, consists of several different computations run in parallel. The largest and most involved comparison is between the DNA sequence submitted by the user and the known sequences from all of the genomes that have ever been sequenced. (For efficiency reasons, we don't currently automatically compute a comparison between all the metagenomes that have been submitted.) In this case, additional processing is added to the computation because the DNA sequence from the user is compared to the known sequences in protein space. Recall that DNA encodes protein (albeit indirectly through an intermediate molecule called mRNA). Therefore, given a particular DNA sequence, the protein sequence can be generated computationally. However, because there are only four DNA letters (the base pairs G, A, T, and C) but twenty amino acids, nature has evolved to use a three-letter register for the translation of DNA sequence to protein sequence. The sixty-four possible combinations of DNA letters encode the twenty amino acids with some redundancy (more than one 3-tple encodes for a single amino acid). Therefore, for any given DNA sequence there are *six* possible protein sequences (three each in both the forward and reverse directions). Although this doesn't increase the complexity of the calculation, it does significantly increase the computation time required to complete the analyses.

As shown in figure 3, the similarity against the SEED protein database is significantly demanding, taking almost seven seconds per hundred base pairs of sequence on the PPC architecture and seven seconds per thousand base pairs of sequence on the 16-core Intel Xeon architecture.

In contrast to the large amount of data in the computation of the similarity to all known sequences, the comparison to more boutique databases is decidedly more rapid (figure 4), with between 3,000 and 12,500 sequences being processed per second depending on architecture.

4.3. Postprocessing and export

The final two steps of the procedure are to parse the results from the similarity computations and assert functions based on those results, and then to export the data into standard formats commonly used by members of the bioinformatics community. These two processes, as shown in figure 5, have linear complexity and are completed rapidly for small sequence datasets, and in less than two hours for the largest datasets.

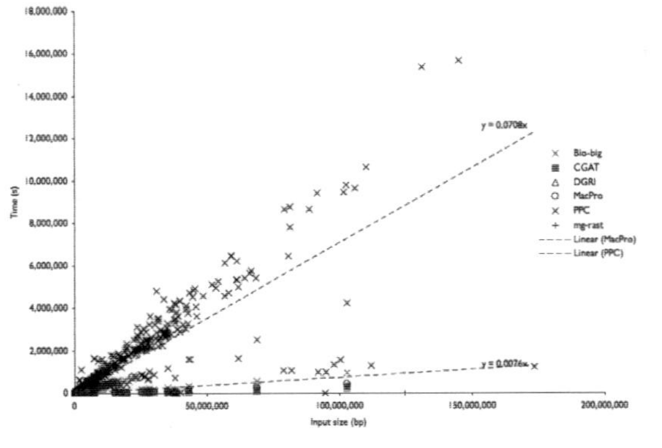

Figure 3. Calculating the similarity between the user-supplied sequences and the known sequences in the publicly available databases is the most time-consuming tasking. It is therefore the task that is most amenable to improvement in the whole pipeline. Using sixteen-core architecture it takes approximately seven seconds to process 1,000 base pairs of DNA sequence.

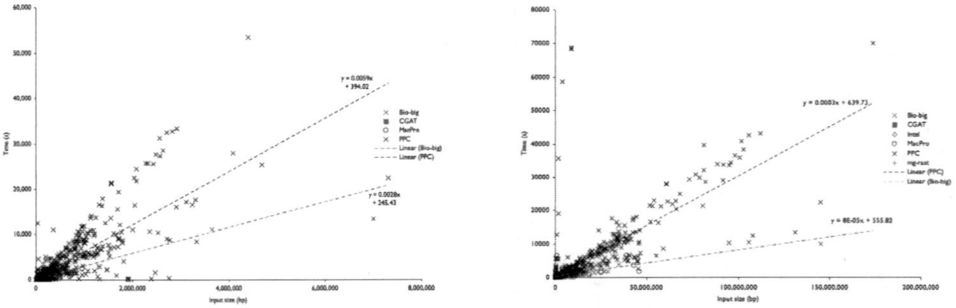

Figure 4. Comparison of all the sequences against the RDP [7] (a; left) and the GreenGenes database [8] (b; right) is much quicker than comparison to all known proteins.

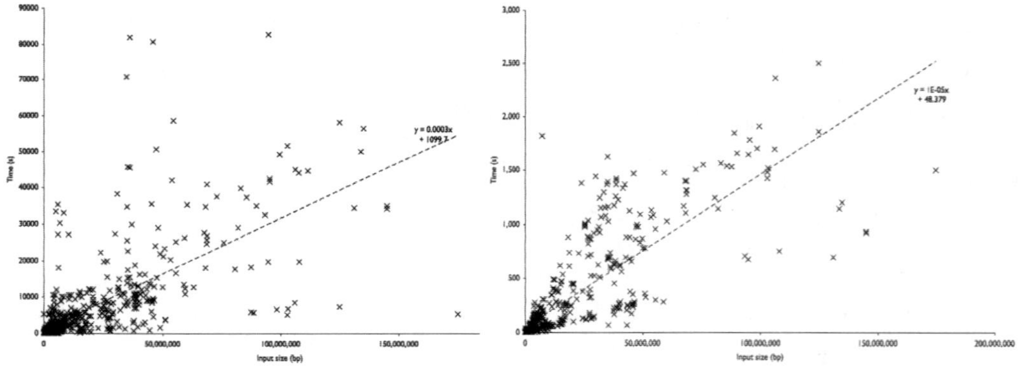

Figure 5. Postprocessing and data export are not computationally intensive steps, taking less than two hours for most large datasets. In these figures all comparisons are done using the PPC machines.

5. Total elapsed time

Given the linear nature of the processes involved – all calculations have a computational time complexity of O(n) – it was surprising to see that the total elapsed time for the calculation of all similarities between sequences was not linear (figure 6). As shown in figure 6 (right), the amount of data being processed by the metagenomics server is rapidly increasing, and about 1 Gbp of sequence data (1 Gbyte) is being processed per month. As noted above, this queue is not solely used for the metagenomics analyses, and other data being processed on this platform shows a similar trend (data not shown). The limiting factor, therefore, in processing the data

Journal of Physics: Conference Series **125** (2008) 012050 doi:10.1088/1742-6596/125/1/012050

Table 2. Time taken to compute each of the steps in the pipeline for a typical 100 Mbp DNA sample with no queuing.

Preprocessing	Similarity (SEED)	Postprocessing	Export
200 seconds	700,000 seconds (194 hours)	300 seconds	10 seconds

in the metagenomics platform is currently access to appropriate high-performance computing, and specifically the amount of time a job spends in the queue.

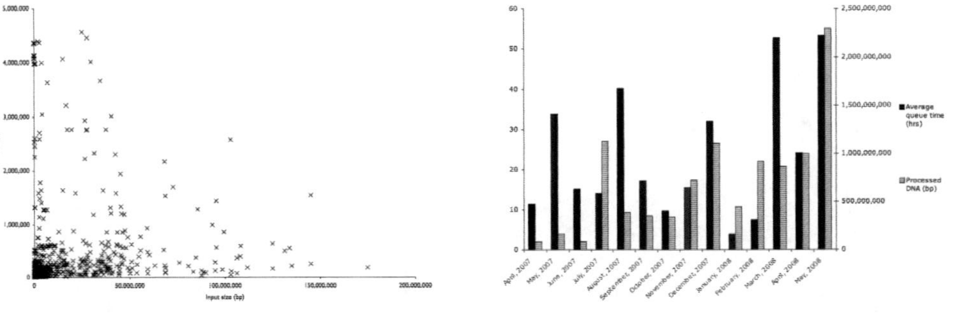

Figure 6. Total elapsed time is not linear overall (left). There is a correlation between the amount of data processed per month and the amount of time spent in the queue. (Data for June 2008 is incomplete and not shown.)

6. Conclusions

This analysis has shown that it is possible to compute the complete comparisons of metagenomes against pertinent databases in linear time, and the computation should be able to keep pace with the generation of new sequences. However, we are currently in a period of rapid deployment of sequencing machines around the country. It is rumored that the number of second-generation pyrosequencing units is fast approaching 300 in North America, with more coming online every day. These machines could generate an estimated 0.1 Tbp (100,000,000,000 bp) of sequence per day. We need to ensure that the increases in computational efficiency and computer speed maintain our ability to process data for the end users. Moreover, the bulk of the processing happens very quickly, even though they have $O(n)$ computational complexity. Of course, as n increases these computations will be challenging, but as long as increased processes speed tracks increased DNA sequencing capacity these computations can be completed in a timely manner. However, the single largest computational step, the comparison of the DNA sequences to the known protein sequences is not only increasing with $O(n)$ complexity, it is taking a significantly longer time to compute than any other step in the process. Better algorithms are required for comparing the sequences; algorithms to triage sequences that will not generate significant similarity may reduce overall computation. More computational horsepower also is needed to complete the analyses in an appropriate timeframe. Currently a single run of a pyrosequencing machine produces about 100 Mbp of data; the time taken for the individual steps in the pipeline is shown in table 2.

Acknowledgments
I thank Robert Olson for assistance with the SGE analysis and insightful discussions; all of the developers of the metagenomics RAST server, RAST server, the SEED, and the National Microbial Pathogen Resource (NMPDR) for assistance, dedication, and tutelage. Special thanks to Rick Stevens and Ross Overbeek for support and guidance. Part of this work was funded by the National Institute of Allergy and Infectious Diseases, National Institutes of Health, Department of Health and Human Services, under Contract HHSN266200400042C. This work was also funded by the U.S. Dept. of Energy under Contract DE-AC02-06CH11357.

References

[1] Overbeek R, Begley T, Butler R M, Choudhuri J V, Chuang H Y, Cohoon M, de Crecy-Lagard V, Diaz N, Disz T, Edwards R, Fonstein M, Frank E D, Gerdes S, Glass E M, Goesmann A, Hanson A, Iwata-Reuyl D, Jensen R, Jamshidi N, Krause L, Kubal M, Larsen N, Linke B, McHardy A C, Meyer F, Neuweger H, Olsen G, Olson R, Osterman A, Portnoy V, Pusch G D, Rodionov D A, Ruckert C, Steiner J, Stevens R, Thiele I, Vassieva O, Ye Y, Zagnitko O and Vonstein V 2005 *Nucleic Acids Res* **33** 5691–5702 ISSN 1362-4962 (Electronic)

[2] Dinsdale E A and Edwards R A 2007 *Oceanography* **20** 56

[3] Turnbaugh P J, Ley R E, Hamady M, Fraser-Liggett C M, Knight R and Gordon J I 2007 *Nature* **449** 804–810 ISSN 1476-4687 (Electronic)

[4] Sanger F, Nicklen S and Coulson A R 1977 *Proc Natl Acad Sci U S A* **74** 5463–5467 ISSN 0027-8424 (Print)

[5] Margulies M, Egholm M, Altman W E, Attiya S, Bader J S, Bemben L A, Berka J, Braverman M S, Chen Y J, Chen Z, Dewell S B, Du L, Fierro J M, Gomes X V, Godwin B C, He W, Helgesen S, Ho C H, Irzyk G P, Jando S C, Alenquer M L I, Jarvie T P, Jirage K B, Kim J B, Knight J R, Lanza J R, Leamon J H, Lefkowitz S M, Lei M, Li J, Lohman K L, Lu H, Makhijani V B, McDade K E, McKenna M P, Myers E W, Nickerson E, Nobile J R, Plant R, Puc B P, Ronan M T, Roth G T, Sarkis G J, Simons J F, Simpson J W, Srinivasan M, Tartaro K R, Tomasz A, Vogt K A, Volkmer G A, Wang S H, Wang Y, Weiner M P, Yu P, Begley R F and Rothberg J M 2005 *Nature* **437** 376–380 ISSN 1476-4687 (Electronic)

[6] Aziz R K, Bartels D, Best A A, DeJongh M, Disz T, Edwards R A, Formsma K, Gerdes S, Glass E M, Kubal M, Meyer F, Olsen G J, Olson R, Osterman A L, Overbeek R A, McNeil L K, Paarmann D, Paczian T, Parrello B, Pusch G D, Reich C, Stevens R, Vassieva O, Vonstein V, Wilke A and Zagnitko O 2008 *BMC Genomics* **9** 75 ISSN 1471-2164 (Electronic)

[7] Cole J R, Chai B, Farris R J, Wang Q, Kulam S A, McGarrell D M, Garrity G M and Tiedje J M 2005 *Nucleic Acids Res* **33** D294–6 ISSN 1362-4962 (Electronic)

[8] DeSantis T Z, Hugenholtz P, Larsen N, Rojas M, Brodie E L, Keller K, Huber T, Dalevi D, Hu P and Andersen G L 2006 *Appl Environ Microbiol* **72** 5069–5072 ISSN 0099-2240 (Print)

Toward petascale computing in geosciences: application to the Hanford 300 Area

Glenn E Hammond[1], Peter C Lichtner[2], Richard Tran Mills[3] and Chuan Lu[4]

[1]Hydrology Group, Environmental Technology Division, Pacific Northwest National Laboratory, Richland, WA 99352, USA

[2]Hydrology, Geochemistry, and Geology Group, Earth and Environmental Sciences Division, Los Alamos National Laboratory, Los Alamos, NM 87545, USA

[3]Computational Earth Sciences Group, Computer Science and Mathematics Division, Oak Ridge National Laboratory, Oak Ridge, TN 37831-6015, USA

[4]Energy & Geoscience Institute, University of Utah, Salt Lake City, UT 84108, USA

E-mail: glenn.hammond@pnl.gov, lichtner@lanl.gov, rmills@ornl.gov, clu@egi.utah.edu

Abstract. Modeling uranium transport at the Hanford 300 Area presents new challenges for high performance computing. A field-scale three-dimensional domain with an hourly fluctuating Columbia river stage coupled to flow in highly permeable sediments results in fast groundwater flow rates requiring small time steps. In this work, high-performance computing has been applied to simulate variably saturated groundwater flow and tracer transport at the 300 Area using PFLOTRAN. Simulation results are presented for discretizations up to 10.8 million degrees of freedom, while PFLOTRAN performance was assessed on up to one billion degrees of freedom and 12,000 processor cores on Jaguar, the Cray XT4 supercomputer at ORNL.

1. Introduction

As part of the SciDAC groundwater science application area, PFLOTRAN [1–3] is being used to simulate radionuclide transport at the U.S. Department of Energy's Hanford Site in southeastern Washington State. Significant simulation and modeling efforts were put forth in the early 1990s in an attempt to determine the rate at which uranium within the groundwater would migrate toward and discharge into the Columbia River. Unfortunately, these models lacked the sophistication necessary to accurately simulate uranium transport, predicting that the uranium would be removed or flushed out of the Hanford 300 Area subsurface by ambient groundwater flow within a decade. Today, nearly 15 years later, uranium concentrations at the site remain relatively unchanged.

Several possibilities exist attempting to explain the discrepancies between the predicted and observed transport of uranium. These include (1) fluctuations in the stage of the Columbia river and its impact on groundwater velocities, (2) the vadose zone contribution to the source term, (3) inadequacies in the uranium transport conceptual model due to overly simplistic reactions representing the sorption of uranium on sediment grains; and, (4) uncertainty of the uranium source term mass concentrations and leaching rates.

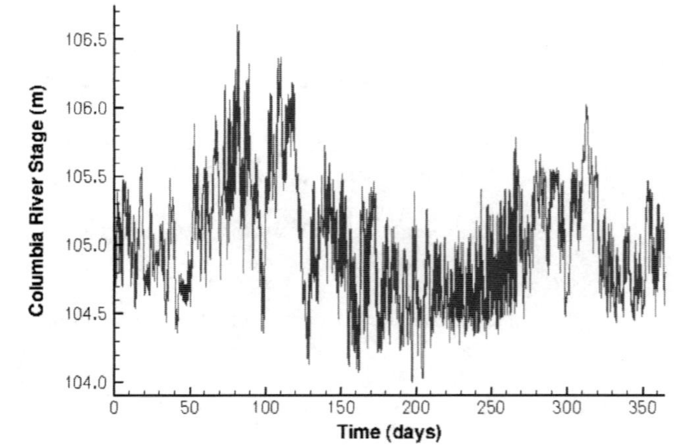

Figure 1. Fluctuation in the Columbia river stage.

The conceptual model for flow at the Hanford 300 Area is complicated by the highly permeable Hanford Unit and the rapidly fluctuating Columbia River shown in figure 1 producing changes in magnitude and direction of flow. Diurnal (daily) fluctuation in river stage can be up to 1.5 meters, while seasonal fluctuation can exceed 2.5 meters. The variable release of water from Priest Rapids Dam upriver is the cause of the diurnal fluctuation, while seasonal variations are due to snow melt, irrigation, and so forth. The river stage fluctuation along with the highly permeable Hanford Unit directly impacts the water table beneath the Hanford 300 Area, causing river water to flow into the area during high stage and retreat during low stage.

The soils beneath the Hanford 300 Area are composed of layers of high- and low-permeability soil, the top layer of which is the highly permeable Hanford Unit located near the water table. The Hanford Unit is composed of highly permeable cobbles, gravels, and sands. Hydraulic conductivities within the Hanford Unit are on the order of a thousand to tens of thousands of meters per day. In comparison, the Ringold Units below the Hanford Unit exhibit much lower hydraulic conductivities of 0.01 to 150 meters per day. Thus, within the Hanford Unit, where all original uranium sources are believed to have resided, groundwater has the potential of flowing rapidly with very small pressure gradients in the aquifer.

This work extends [1] to incorporate injection of a tracer into the highly variable flow field generated by the Columbia river stage fluctuations. It also incorporates the use of Richard's equation for variably saturated flow. Results of the flow simulations are compared with well head data from the 300 Area site, and tracer transport is compared on several levels of discretization.

2. PFLOTRAN capabilities

PFLOTRAN is a parallel multiphase-multicomponent reactive flow and transport code for modeling subsurface processes. The code is written in Fortran 90 using a modular, object-oriented approach (see figure 2) and is composed of several modules for flow and transport, with the option of running the modules in coupled or decoupled mode. PFLOTRAN is capable of simulating fluid flow though porous media with fluid phases air, water, and supercritical CO_2. PFLOTRAN-generated fluid flow velocities or fluxes and phase saturation states are used by the transport module to compute solute transport. Within PFLOTRAN, transport and reaction are fully coupled. PFLOTRAN's problem domain is discretized spatially using the integrated finite volume approach, with fully-implicit backward-Euler time differencing.

PFLOTRAN's parallel paradigm is based on domain decomposition where the computational problem domain is divided into subdomains, with one domain assigned to each processor. PFLOTRAN leverages PETSc [4] solvers and data structures to link these subdomains within a single parallel code. Since PFLOTRAN is founded these PETSc data structures, the code is capable of utilizing the full suite of algorithms available in the library. In addition, PETSc also provides linkage to a variety of external software packages (e.g., Hypre, Trilinos, Zoltan). Within PFLOTRAN, the distributed array (DA) is the key PETSc data structure that dictates the layout and decomposition of the three-dimensional parallel domain. From the DA, PETSc generates parallel matrices (Mat) and vectors (Vec) with consistently mapped local and global indexing for the parallel communication required to solve the problem. Therefore, although an in-depth understanding of parallel communication paradigms such as the message passing interface (MPI) is helpful, it is not required since PETSc hides the communication from the end user. For example, when employing the Newton-Raphson method to solve a nonlinear system of equations in parallel using the PETSc nonlinear solver or SNES, the programmer essentially provides functions to PETSc that evaluate the residual and compute the Jacobian for the nodes that reside locally on each processor. PETSc then uses these functions to compute a solution vector, which is then returned to PFLOTRAN for boundary condition updates, I/O, and so forth. Thus, the application scientist is able to focus more on the science (in this case, subsurface physics and chemistry) rather than solvers, preconditioners, and so forth. Furthermore, because PETSc supports almost any modern computer platform and because PFLOTRAN will run anywhere that PETSc is supported, our code is highly portable.

Recently, PFLOTRAN has demonstrated a relative parallel efficiency of 79% at 12,000 cores based on a strong-scaling study performed with a 500 million node problem (Hanford 300 Area) executed on 3,000-12,000 cores. Figure 3 shows the wall-clock time versus the number of processor cores run on Jaguar at ORNL using dual-core processors. Scaling is reasonable and the slight reduction in performance is attributed to load imbalance due to the use of inactive nodes in the calculation. As proof-of-concept for petascale computing, PFLOTRAN has been run on a similar one billion node ($4096 \times 2048 \times 128 = 1{,}073{,}741{,}824$ nodes) problem. Timings for a single step run on 1024 to 4096 cores are presented in figure 4.

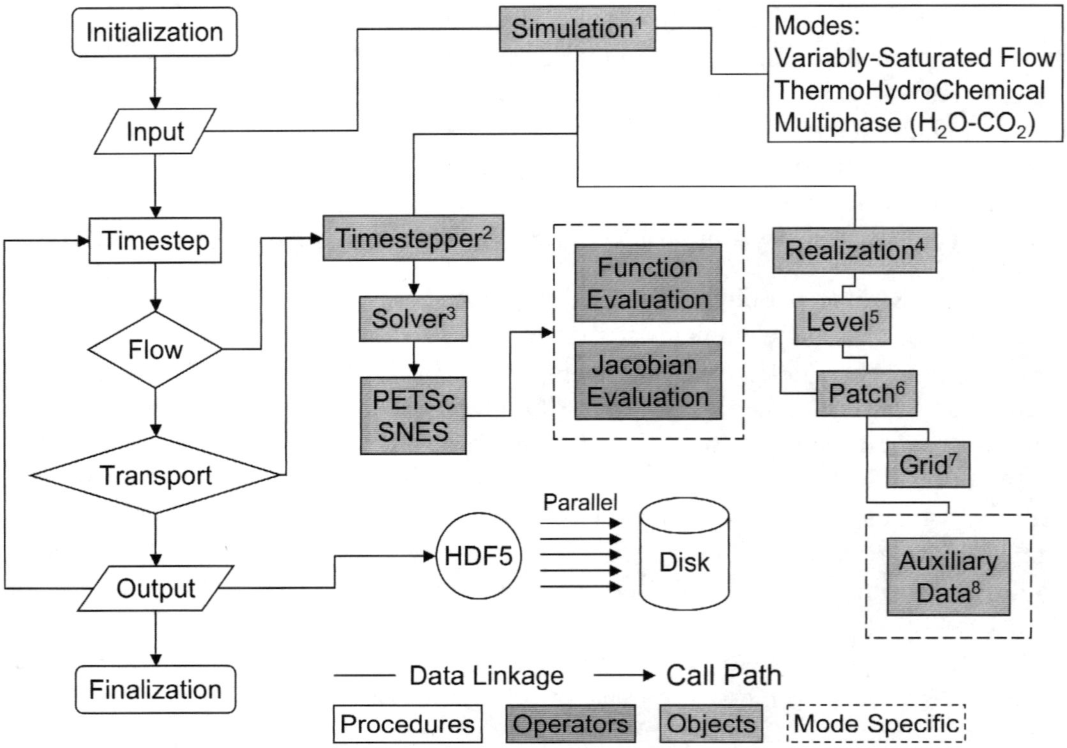

Figure 2. PFLOTRAN flow diagram illustrating use of procedures, operators, objects, and mode specific operators and objects. Flow chart definitions: (1) Simulation object: Highest level data structure providing all information for running a simulation; (2) Timestepper object: Pointer to Newton-Krylov solver and tolerances associated with time stepping; (3) Solver object: Pointer to nonlinear Newton and linear Krylov solvers (PETSc SNES/KSP/PC) along with associated convergence criteria; (4) Realization object: Pointer to all discretization and field variables associated with a single realization of a simulation; (5) Level object: Pointer to discretization and field variables associated with a single level of grid refinement within a realization; (6) Patch object: Pointer to discretization and field variables associated with a subset of grid cells within a level; (7) Grid object: Pointer to discretization within a patch; (8) Auxiliary Data object: Pointer field variables within a patch.

2.1. Governing equations

The governing equations that describe uranium transport at the Hanford 300 Area are based on Richards equation for fluid flow in a partially saturated medium. In this approximate formulation the gas phase is assumed to be inert. Richards equation is coupled to the advection-diffusion-reaction equation for solute transport. Richards equation may be represented in the form

$$\frac{\partial}{\partial t}\big(\varphi s \rho\big) + \boldsymbol{\nabla} \cdot \rho \boldsymbol{u} \;=\; \mathcal{S}, \tag{1}$$

where φ denotes the porosity of the porous with saturation s giving the fraction of pore volume filled with water, ρ denotes the fluid density, $\mathcal{S}$ refers to a source/sink term, and $\boldsymbol{u}$ denotes the

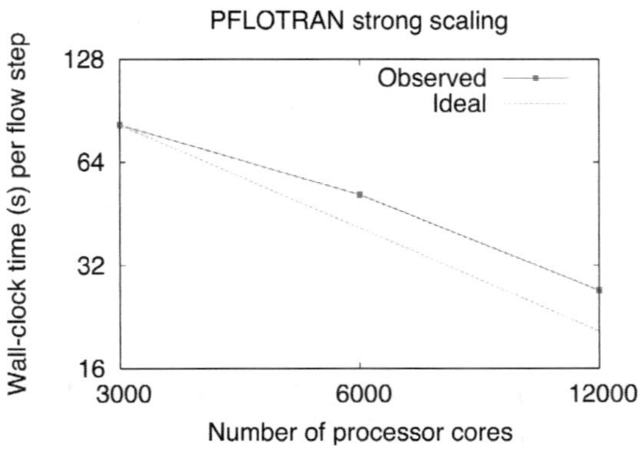

Figure 3. Strong scaling speedup curve for 500 million node problem run on Jaguar.

One-billion node performance on Jaguar

# procs	newton its	bcgs its	time(sec)
1024	16	1697	1143
2048	16	1626	697
4096	16	1720	325

Figure 4. Jaguar performance for one billion node problem.

Darcy velocity defined by

$$\boldsymbol{u} = -\frac{\kappa\kappa_r}{\mu}\boldsymbol{\nabla}\big(P - \rho g z\big), \quad (2)$$

with fluid pressure P, viscosity μ, vertical distance z, saturated permeability κ, relative permeability κ_r, and acceleration of gravity g. The relative permeability is a nonlinear function of saturation based on the van Genuchten equation.

For the purposes of this contribution, solute transport is described by the nonreactive scalar transport equation

$$\frac{\partial}{\partial t}\big(\varphi s C\big) + \boldsymbol{\nabla}\cdot\big(\boldsymbol{u}C - \varphi s D \boldsymbol{\nabla} C\big) = \mathcal{S}_C, \quad (3)$$

for concentration C, diffusion/dispersion coefficient D, and source/sink term $\mathcal{S}_C$. The transport equation is coupled to flow equation, equation(1), through the saturation s and Darcy flow velocity $\boldsymbol{u}$.

2.2. Numerical olution

PFLOTRAN uses a finite-volume spatial discretization combined with backward-Euler time differencing to solve the system of governing equations for subsurface flow and transport. Upwinding is used for the advective term in the transport equations. The system of nonlinear algebraic equations resulting from the discretization is solved using the highly scalable nonlinear solver framework (SNES) in PETSc. The framework allows a wide variety of solver algorithms; for the calculations described here, we used an inexact Newton-Krylov method with cubic line-search. The Jacobian systems are solved via a BiCGStab method with a block-Jacobi (domain-decomposition) preconditioner, with ILU(0) applied on each block.

3. Hanford 300 Area application

The PFLOTRAN model of the Hanford 300 Area consists of a computational domain measuring 1350×2500×20 meters (x, y, z) with orientation aligned with the Columbia River at 14° of west of north (figure 5). The base of the model lies at 90 meters elevation above sea level. Three grid resolutions are simulated in this work:

- 20 meter horizontal (x-y)$\times$1 meter vertical (z) = 170,000 degrees of freedom (170K dof)
- 10 meter horizontal $\times$ 0.5 meter vertical = 1,350,000 degrees of freedom (1.35M dof)
- 5 meter horizontal $\times$ 0.25 meter vertical = 10,800,000 degrees of freedom (10.8M dof)

Stratigraphy was mapped to each grid from the Hanford EarthVision database [5]. The two predominant geologic units simulated in the current conceptual model are the Hanford and Ringold units. These units are shown near the top of figure 6, a 60-meter deep representation of the problem domain viewed from the southeast. Hanford 300 Area hydraulic properties were assigned based on data provided by **(author?)** [6]. Transient hydrostatic and seepage face boundary conditions were assigned to the western inland and river boundaries, respectively, with time-varying datums and gradients. Specified surface recharges were assigned to the top of the model domain, while the north and south boundaries were no flow. Solute transport was simulated through an infinite rate-limited dissolution source term (normalized tracer concentration) specified near the center of the Hanford 300 Area Integrated Field-Scale Subsurface Research Challenge (IFC) site (see figure 5). Source zone region: area = 1600 m^2 (107K dof), 2500 m^2 (1.35M, 10.8M dof); elevation = 104–110 m.

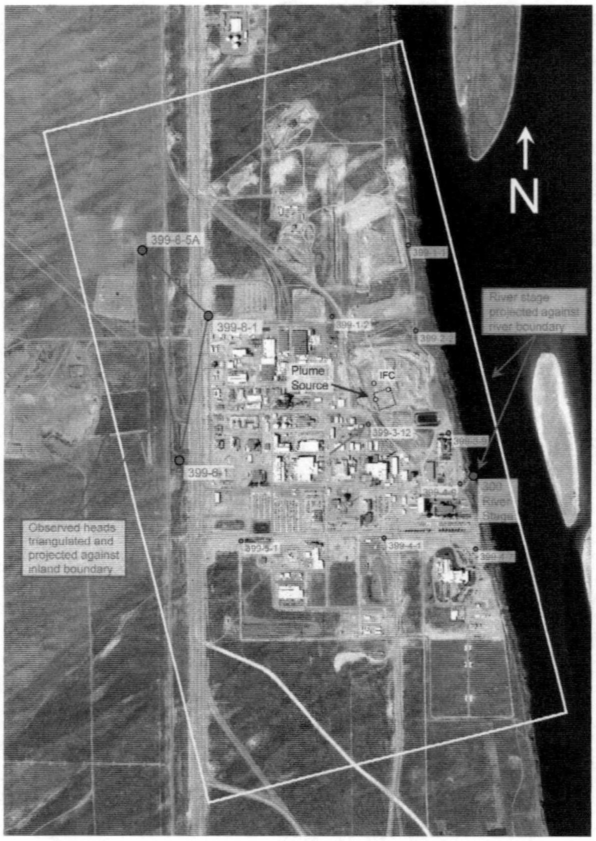

Figure 5. Layout of Hanford 300 Area showing the location of the IFC and monitoring wells.

PFLOTRAN simulations run to date focus on establishing and improving the accuracy of the Hanford 300 Area flow solution and the subsequent fluxes used for simulating transport. Simulations were initialized to steady state (based on 10 am May 1, 1992 conditions), restarted, and run in transient mode for 7500 hours simulation time (10 am May 1, 1992 to 10 am March 9, 1993).

Figure 7 illustrates x-y cross sections of the pressure (z = 90m) and tracer concentration (z = 105m) at 7500 hours of simulation time (10 am, Mar. 9, 1993) for the 170K, 1.35M, and 10.8M dof grids, respectively. For the 170K dof scenario, 128-processor cores on ORNL's Jaguar supercomputer required 50 minutes to complete the 7500-hour simulation. Simulations involving 1.35M and 10.8M dofs used 1024 and 600 processor cores and ran for 2.3 and 16 hours, respectively.

Journal of Physics: Conference Series **125** (2008) 012051 doi:10.1088/1742-6596/125/1/012051

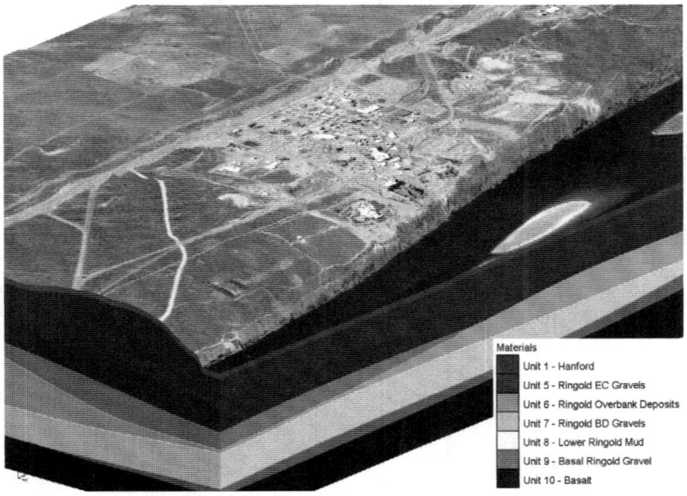

Figure 6. Hanford 300 Area stratigraphy (z scale = 20x, z axis ranges 70–130 meters).

Figure 8 compares the piezometric head observed at monitoring well 399-3-12 (see location in figure 5) to the PFLOTRAN heads predicted nearest to the well for each of the simulation scenarios. River stage is also provided as a reference, and is far more oscillatory than the observed head at the well. Figure 8(b) is a detailed view of figure 8(a) between 3250 and 3750 hours. All three PFLOTRAN grid resolutions produce nearly identical piezometric heads that slightly overestimate the observed head and are more oscillatory.

Predicted PFLOTRAN pore water flow velocities computed at the center of the IFC site are plotted in figure 9 for the various grid resolutions. The detailed plot (figure 9(b)) reveals that the coarse grid (170K dof) model underestimates peak velocities by as much as 60%, whereas the velocities are more consistent for the 1.35M and 10.8M dof simulations, although it is not clear that velocities have converged.

4. Conclusions

High-performance computation enables the solution of large, 3D high-resolution problem domains beyond what is possible on a single-processor workstation and with reasonable turnaround times, especially needed for calibration/optimization runs. Higher-resolution flow simulations may be necessary to improve the accuracy of flow velocities employed in radionuclide transport simulations. Specifically with regard to the Hanford 300 Area: PFLOTRAN variably saturated flow simulations using datasets provided by **(author?)** [6, 7, 8] produce piezometric heads slightly higher and more oscillatory than those observed in the field. They also result in a fairly consistent offset from the the observed wells over time (the same holds for other well observation data compared to the predicted heads). Calibration of hydraulic parameters (i.e., permeability) and the inland boundary condition is needed to improve agreement between observed and predicted piezometric heads. Convergence of flow velocities with higher-resolution grids needs to be investigated further to rule out numerical artifacts in the solution.

Acknowledgments

This research is supported by the U.S. Department of Energy's Office of Science under the Scientific Discovery through Advanced Computing (SciDAC) program. Computing time was provided by a 2008 DOE INCITE award for the project "Modeling Reactive Flows in Porous Media." This research used resources of the National Center for Computational Sciences at Oak Ridge National Laboratory, which is supported by the Office of Science of the U.S. Department of Energy under Contract No. DE-AC05-00OR22725.

References

[1] Hammond G E, Lichtner P C and Lu C 2007 *SciDAC 2007 Scientific Discovery through Advanced Computing* (*Journal of Physics: Conference Series* vol 78) ed Keyes D (Boston, Massachusetts: IOP Publishing) p 012025

[2] Lu C and Lichtner P C 2007 *SciDAC 2007 Scientific Discovery through Advanced Computing* (*Journal of Physics: Conference Series* vol 78) ed Keyes D (Boston, Massachusetts: IOP Publishing) p 012042

[3] Mills R T, Lu C and Lichtner P C 2005 *SC2005* SuperComputing 2005 (Seattle, WA: IOP Publishing)

[4] Balay S, Buschelman K, Eijkhout V, Gropp W D, Kaushik D, Knepley M G, McInnes L C, Smith B F and Zhang H 2004 PETSc users manual Tech. Rep. ANL-95/11 - Revision 2.1.5 Argonne National Laboratory

[5] Thorne P D, Bergeron M P, , Williams M D and Freedman V L 2006 Groundwater data package for Hanford assessments Tech. Rep. PNNL-14753 Rev. 1 Pacific Northwest National Laboratory, Richland, WA

[6] Williams M D, Rockhold M L, Thorne P D and Chen Y 2008 Three-dimensional groundwater models of the Hanford 300 Area Tech. Rep. PNNL-XXXX (in review) Pacific Northwest National Laboratory, Richland, WA

[7] Campbell M D 1994 Monitoring groundwater and river interactions along the Hanford Reach of the Columbia River Tech. Rep. PNL-9437 Pacific Northwest National Laboratory, Richland, WA

[8] Campbell M D and Newcomer D R 1992 Automatic measurement of water levels within the 300-FF-5 boundary Tech. Rep. PNL-7874 Pacific Northwest National Laboratory, Richland, WA

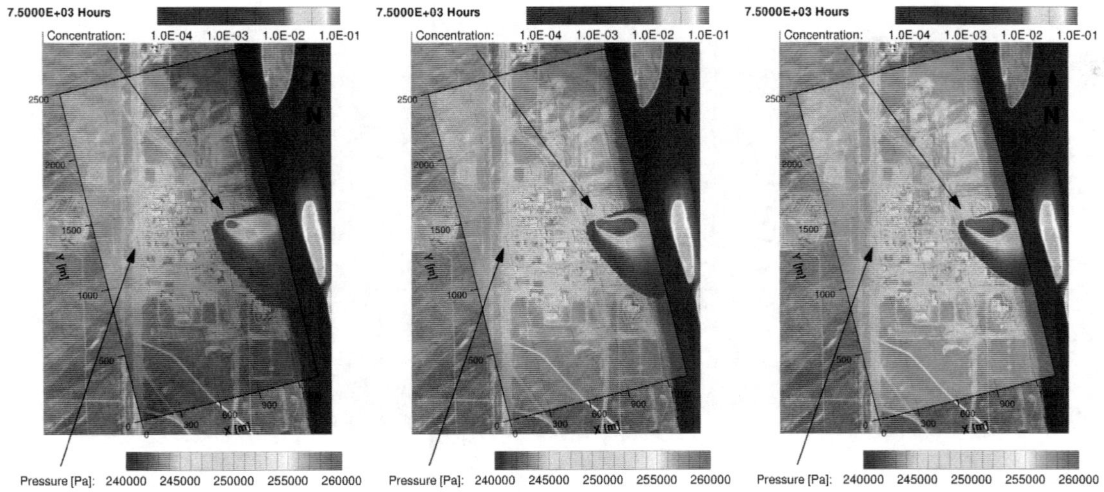

Figure 7. Comparison of tracer plume for different grid resolutions corresponding to 170K (left), 1.35M (center), and 10.8M (right) nodes at 7500 hours.

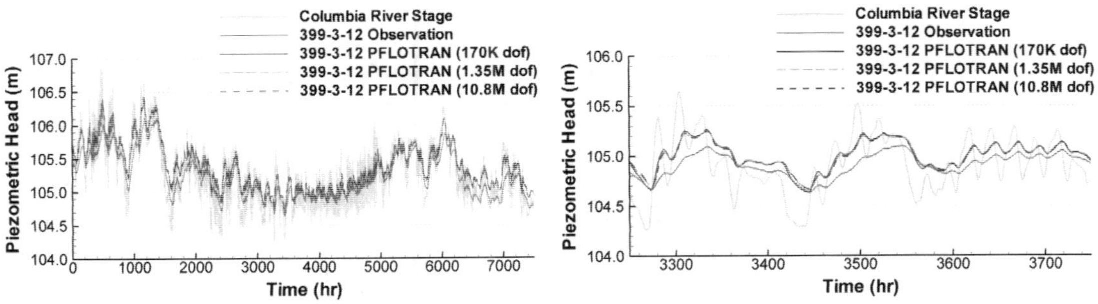

(a) Piezometric head over 7500 hour time interval.

(b) Detailed view of (a) over time interval 3250–3750 hours.

Figure 8. Comparison of observed versus predicted head at Well 399-3-12 and Columbia river stage.

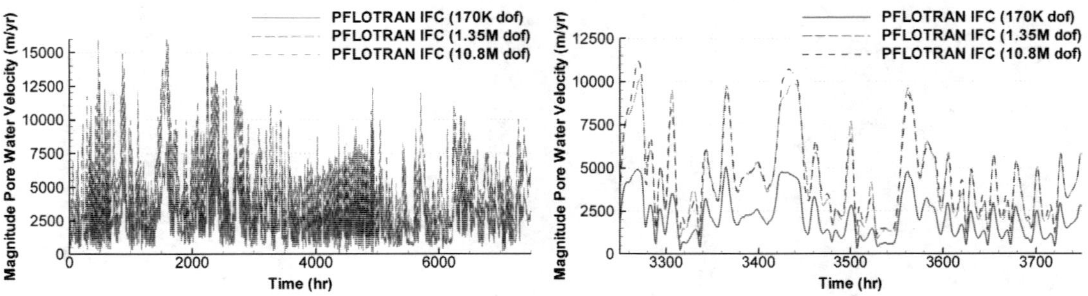

(a) Pore velocity over 7500 hr time interval.

(b) Pore velocity over time interval 3250–3750 hr.

Figure 9. Magnitude of predicted pore water velocities at IFC site for different grid resolutions.

Journal of Physics: Conference Series **125** (2008) 012052 doi:10.1088/1742-6596/125/1/012052

Bringing high-performance computing to the biologist's workbench: approaches, applications, and challenges

Christopher S Oehmen and William R Cannon
Pacific Northwest National Laboratory, Richland, WA 99354, USA

E-mail: Christopher.Oehmen@pnl.gov

Abstract. Data-intensive and high-performance computing are poised to significantly impact the future of biological research which is increasingly driven by the prevalence of high-throughput experimental methodologies for genome sequencing, transcriptomics, proteomics, and other areas. Large centers such as NIH's National Center for Biotechnology Information, The Institute for Genomic Research, and the DOE's Joint Genome Institute) have made extensive use of multiprocessor architectures to deal with some of the challenges of processing, storing and curating exponentially growing genomic and proteomic datasets, thus enabling users to rapidly access a growing public data source, as well as use analysis tools transparently on high-performance computing resources. Applying this computational power to single-investigator analysis, however, often relies on users to provide their own computational resources, forcing them to endure the learning curve of porting, building, and running software on multiprocessor architectures. Solving the next generation of large-scale biology challenges using multiprocessor machines—from small clusters to emerging petascale machines—can most practically be realized if this learning curve can be minimized through a combination of workflow management, data management and resource allocation as well as intuitive interfaces and compatibility with existing common data formats.

1. Introduction

The role of high-performance computing across the spectrum of science domains is as varied as the domains themselves. In astronomy, high energy physics, and climate modeling, just to name a few, one can hardly imagine the fields *without* advanced computing because of the sheer volume and multiscale nature of the challenges driving those fields, and the complexity of the algorithms that are utilized to understand at a mechanistic level the relationships between entities. In domains such as chemistry and physics, computational studies—often driven by high-performance computing—are more and more commonly recognized as the third pillar of investigation, alongside the conventional pillars of experimentation and theory. Using high-performance computing to advance the forefront of computational chemistry, for example by investigating larger and more complex systems and interactions using fundamental theory, is therefore extremely valuable to chemistry at large, in part because computational studies can be validated by and augment the knowledge gained from experimental studies.

In contrast, the application of high-performance computing to biology has been more conservative in scope. Although living systems obey the same natural laws that govern physics and chemistry, brute-force application of the equations describing these laws is an intractable strategy because of the size and complexity of the systems. Biologists have instead focused on solving subproblems that can be reasonably approximated using physics, chemistry, information theory, statistics, cellular automata, or other formulations. High-performance computing in biology has most prominently been used in three directed ways: (1) as a capacity resource to drive throughput for data analysis or data mining [1-6], (2) enabling large-scale simulation and modeling [7-11], and (3) as a mathematical tool driving algorithms aimed at revealing underlying

Journal of Physics: Conference Series **125** (2008) 012052 doi:10.1088/1742-6596/125/1/012052

properties of biological systems [12. 13]. These applications of high-performance computing are described in more detail in the following sections.

1. High-throughput data analysis and data mining includes bioinformatics and image analysis where the primary task is pattern matching or feature extractions. Commonly, the data of interest in this category is growing exponentially with time, so the immediate challenge is simply to provide the throughput required to perform the analysis. Examples of this rapidly growing data include genome sequence data, experimental process data (like transcriptomics or proteomics) and cell imaging data whose availability is driven by capture and analysis capacity of engineered components and increasing optical resolution.

2. High-performance Grid and cluster computing has also played a prominent role in large-scale biological simulations. Large-scale biophysical or biosystems models are increasing their demand for high performance computing as multiscale techniques are devised and implemented. Scalability is a challenge in this area primarily because of increasing complexity of underlying theoretical process descriptions and the computational challenges associated with integrating models across spatial and temporal scales—a challenge reminiscent of other scientific domains such as astrophysics where some processes occur at submicroscopic spatial scale and extremely short time scales, critically linked to processes that evolve over distance and time scales for observable celestial objects and beyond.

3. The underlying properties of biosystems are often inferred by using techniques such as network inference or graphical analysis. The goal is to map the behavior of biological systems onto a well-characterized mathematical representation (*e.g.* a network or graph) so that trends, relationships, and features might be inferred based on mathematical properties of the representation. Often, this reduced representation allows for more straightforward analysis of the system leveraging advances in mathematical theory, and can often reveal underlying structure that is not evident without the mathematical metaphor.

For these three general categories much work has been done with respect to high performance algorithm design and implementation. However, making use of these advances is often restricted to specialized users, or else it is implemented using a service-oriented model where computing infrastructure is not directly controlled by the end user. While this has the benefit of freeing the bench biologist from having to endure the learning curve of using high performance computing, a remote execution model also prevents potential users from taking full advantage of the increasing availability of *local* clusters and emerging architectures as well as emerging algorithms and software that can run on them. Historically, this has been acceptable to the biology community largely because the scope of computational task that a single investigator could realistically be expected to address did not often require advanced computing resources. However, as the cost of high-throughput technologies such as genome sequencing, transcriptomics, and proteomics put them within reach of a growing community of investigators, the need for increased computational capacity for individual investigators will likely grow as well.

The goal of this paper is to offer a perspective on how high-performance computing has and will continue to evolve to meet the ever-changing needs of biology from the standpoint of the changing needs of individual investigators. We also enumerate some future challenges that will need to be addressed to facilitate this evolution. Specific examples are included to illustrate how advanced architectures have traditionally been used in biology and how their increasing availability and demand are leading to new usage models where computing is more tightly integrated into analytical workflows. We focus on bioinformatics and proteomics applications in high performance computing because of their potential for immediate impact to the bench biologist and because of the availability of optimized computational implementations in these areas.

2. High-performance computing in genomic analysis

Large-scale genome sequencing centers are among the most visible users of high-performance computing in biology. The volume of published genomic sequence data continues to increase exponentially, driven by ongoing improvements in throughput of sequencing technologies. Commercialization by Illumina, 454 Life Sciences, and Applied Biosystems has led to relatively inexpensive access to genome sequencing technologies. Even for these high-throughput systems sequencing genomes is done by sequencing many fragments of a genome, then trying to assemble the final genome by correctly "connecting" the fragments. This is a challenging problem, not only because of the (albeit nominal) uncertainty inherent in the sequencing process, but by the presence of confounding sequences (such as highly repetitive regions of a genome). Often, high performance computing is needed to solve this "assembly problem" because of the combinatorial complexity associated with matching and ordering sequence fragments. Following assembly is the annotation process where encoding regions (genes) and their corresponding proteins are mapped onto the assembled genome. Assembly and annotation for a typical genome both require substantial computational power, as illustrated by the fact that most sequencing systems are shipped with multiprocessor clusters attached.

Making these sequences available to the general biology community is a driving priority after sequencing, assembly, and annotation are done. It is most often accomplished by providing datasets and analytical tool suites via websites or web services, as in the case of The Institute for Genomic Research (TIGR, http://www.tigr.org/), DOE's Joint Genome Institute (JGI, http://jgi.doe.gov/) or NIH's National Center for Biotechnology Information (NCBI, http://www.ncbi.nlm.nih.gov/). Many of these tools allow users to formulate a query on sequence or sequence-related data to find conserved patterns between some set of sequences and a reference set of sequences. Often, when a user submits a query to one of the large-scale genome centers, the task is routed via an Internet transaction to a supercomputing resource that queues the request and processes the data based on the query's priority and the current system load. This procedure results in a high degree of throughput for the sequencing center (i.e., a large number of queries that can be handled in a given time interval) and a reasonable turnaround for users who need only a handful of queries processed. In this way, users can interact with a preprocessed large-scale dataset of biosequence data, searching for conserved patterns between species of interest.

One of the most commonly used tools for finding these conserved patterns is via pairwise sequence alignment using an algorithm called BLAST [14, 15], which rapidly calculates an optimal alignment for a pair of gene or protein sequences (or translated versions thereof) producing a set of scores that describes the quality and statistical reliability of that alignment being due to common inheritance vs. the chance of a random alignment of the same quality. BLAST searches are at the heart of many areas of bioinformatics analysis, including genome annotation, phylogenic studies, ortholog identification, and others. Given BLAST's prevalence in bioinformatic studies, it is not surprising that accelerating this pattern matching process to drive multiple genome-scale analysis has been the focus of much high performance computing algorithm development. Parallel sequence analysis applications include ScalaBLAST [2], MPI BLAST [16], pioBLAST [17], TurboBLAST [18], beoBLAST [19], and others [20-25]. What these implementations have in common is that they (1) attempt to accelerate the rate of BLAST calculations in proportion to available compute resources, and most often (2) require users to obtain and install specialized software to use those resources. With the exception of applications like Soap-HT-BLAST [24], which allows users to schedule multiple BLAST queries using web services over distributed resources, the prevailing approach of these software tools is to assume that users will run applications on homogeneous clusters or on Grid architectures.

The two access models to high-performance resources mentioned so far represent polar opposites in terms of user interaction. At one extreme is the Web services model that is typified by launching remote tasks on dedicated, but remote resources (i.e., submitting a BLAST query through the NCBI BLAST webpage). The advantage of this model is that details of utilizing and interacting with the high performance resources are hidden from users. The primary disadvantage is that the Web services that drive these approaches are not always easily modified for use on compute resources when those resources are controlled directly by users. At the opposite end of the spectrum are the software packages that must be compiled and/or installed directly by end users on high-end systems. The advantage of this approach is that these solutions can be used to take advantage of whatever resources users have. However, users are often forced to learn to use high performance systems and schedulers and, in the worst case, must build the software themselves.

To address the evolving need for biology users to run large-scale calculations on their own dedicated resources, there is a growing trend toward solutions that bring local, dedicated high-performance computing into the analytical pipeline of individual researchers. One way to achieve this is to natively construct a web interface to dedicated resources. This allows multiple users at a single location to use their local resources but to dedicate those resources in a configuration that is more amenable to a small number of large-scale calculations than a large number of individual calculations.

An example of this approach is illustrated in figure 1. We have deployed a website [26] allowing internal Pacific Northwest National Laboratoryy users to run ScalaBLAST jobs on a dedicated cluster. The idea is to provide an intuitive interface that allows users to submit ScalaBLAST jobs that are run in parallel. The user model is not the traditional view in which users submit queries in small batches or one at a time. Instead, it is expected that whole or multiple-genome size queries are requested. The entire local cluster can be dedicated to servicing these requests effectively serving a local community of researchers who need larger-scale computational tasks performed without requiring any of them to endure the learning curve of high performance computing and without them having to compete with worldwide users on large-scale centralized resources. Additionally, the priority and scheduling of jobs can be tailored to the specific local community. This approach can be straightforwardly modified to accommodate different partitions of the ScalaBLAST job, enabling large scale all vs. all or collections of genome-specific calculations to be performed. In this way, certain predetermined ways of using BLAST can be encapsulated using a small number of run-time parameters and uploaded files to support a wide range of larger-scale users on a dedicated system. A similar interface can be created for other ScalaBLAST installations, making it possible for other users to more intuitively take advantage of their local computing resources. This same access model can be applied broadly to other computational tools, simplifying access for users and at the same time allowing them to take advantage of the increasing availability of multiprocessor computing resources at their native institutions.

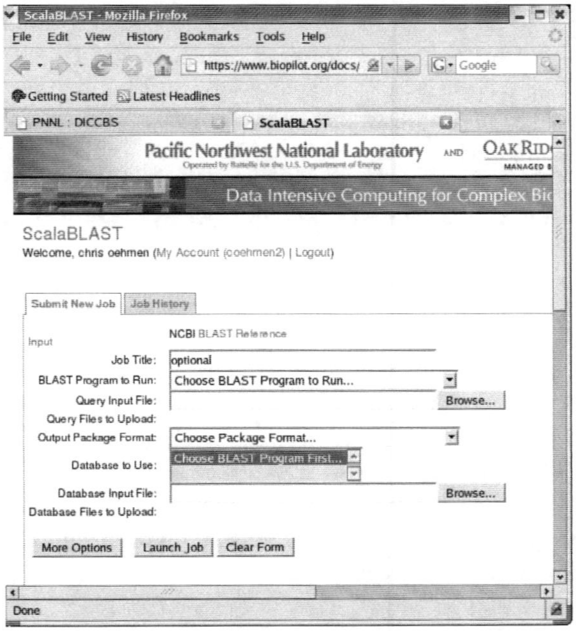

Figure 1. ScalaBLAST launching website for internal users. Users with access to the website can launch parallel ScalaBLAST calculations using a dedicated multiprocessor cluster. Scheduling of the parallel compute task is done automatically when the user clicks the "launch job" button. The system automatically interfaces the native scheduler to give status of pending, in process and completed jobs.

Taking this concept one step further to more closely integrate high-performance computing with the end users leads to scientific workflow systems that abstract away many of the details of complex tasks such as scheduling, data moving, and launching analysis tools on high-performance resources. In this approach, analytical data-driven pipelines can be constructed using a sequence of computational tasks, each viewed as a "component." An example of such a pipeline for genome sequencing would be to have high-performance components for genome assembly linked with gene prediction tools and automated annotation using sequence similarity. Each of these components can also be used in other workflows, but together they could form a powerful core capability for genome sequencing, yet be available and accessible to individual investigators.

3. High-performance computing in proteomics

A second area where high-performance computing has had a strong presence in biology is proteomics, defined as high-throughput analysis of expressed proteins from cell populations, in this example as measured by mass spectrometry (MS). A typical approach in MS-based proteomics is to analyze the *peptides* derived from cellular proteins rather than the proteins themselves because the smaller peptides are more amenable to mass measurement. The challenge, however, is that the thousands of proteins derived from a cell can result in millions of potential peptides, *each* of which must be identified before proteins can be reliably identified. As a result, the speed of computational analyses of MS proteomics data has always been at the forefront of priorities for analytical chemists. Because many of the early proteomics facilities used desktop servers for data analysis, farms of servers were the most straightforward approach to increasing throughput of MS data. For example, the PNNL proteomics center (ncrr.pnl.gov) now holds over 70 TB of MS data regarding hundreds to thousands of experiments.

Because of the sheer size of the data and the elaborate computational pipeline to analyze the data, interpretation of the data by a bioinformaticist may be limited to working with the processed and stored results contained in the center's database rather than real-time interaction or even re-analysis with minimally proccessed MS data. This limitation can restrict the type of analysis that can be performed by an analyst.

As discussed above, there is a growing trend toward solutions that bring local, dedicated high-performance computing into the analytical pipeline of individual analysts. As with

ScalaBLAST, PNNL has deployed a website [26] allowing internal users to run MSPolygraph [6], a high-performance peptide identification software tool, on a dedicated cluster. Our goal is to incorporate the high performance MSPolygraph interface into a desktop workflow used by bioinformatics analysts to interpret and mine the proteomics data. To that end, we are incorporating analysis tools into a computational environment for this purpose, the Biological Resource Manager (BRM) [27]. BRM is a software environment that provides the user with data retrieval and integration capabilities and user-directed access to a variety of analysis tools. Using BRM to access the supercomputing resource, analysts will be able to reanalyze and mine large sets of proteomics data quickly. In one example workflow, output from a large-scale MSPolygraph analysis run can be visualized by overlaying identified peptides onto a visual representation of the species' chromosome. Using the visual analysis tool Platform for Proteomics Peptide and Protein data exploration (PQUAD) [28], the mined peptides can be overlaid on any of six open reading frames located anywhere on the chromosome and color-coded by experimental condition to enable users to browse for peptides or proteins of interest. The rapid analysis of the large datasets and facile visualization of the results also allow analysts to identify miscalled genes and misidentified peptides. This analysis model enables bioinformatics analysts to use supercomputing resources that are integrated into a workflow, and as a result brings a new level of analysis capabilities that otherwise would be out of reach.

4. Bringing high-performance computing to the workbench: present and future challenges

If enabling scientists to harness the power of ever-growing resources like sequence data and high-throughput experimental methodologies is the goal, then a natural priority is helping bench biologists to incorporate high-performance computing into their normal routine. As we transition further into the metagenome era, typical investigations are more and more likely to focus not on single genes or proteins, but on complex interactions at the community level and beyond. Certainly software tools will need to be developed to address some of these emerging challenges, but it is likely that on some level these new solutions will be driven by large-scale analysis, based on some combination of sequence analysis, experimental processing, modeling and simulation, graph or network theory, and other areas. Implementing these solutions efficiently will necessitate evolving paradigms for access and interactions with high-performance architectures. The historical notion of web-based access to remotely controlled resources will need to be augmented by more tightly coupled interactions between compute resources and end users. We have presented a spectrum of alternatives to this conventional model, including access to local, dedicated resources through web interface or web services; application workflow environments that interface with high performance computing through a component-based model; and a tightly integrated application model where users can dynamically direct high performance computing resources for iterative hypothesis refinement.

These approaches are not meant to represent an exhaustive list but instead are meant to underscore the shifting user scenario in modern biology, as the landscape of questions that can be addressed by bench biologists continues to expand rapidly. The cost of obtaining large-scale data continues to decrease to well within the reach of individual investigators and the volume of public data continues to grow exponentially. At the same time, availability of multiprocessor architectures is increasing. As a consequence, high-performance computing is poised to play an increasingly prominent role in biological sciences. However, significant challenges remain for bringing high-performance computing all the way to the workbench. A few of these challenges are enumerated below.

Computational science challenges

Keeping up with the evolving data deluge in biology will require constantly evolving computational algorithms and data management techniques. Analysis that copes with current data sizes will quickly be overwhelmed in a matter of months. Emphasis should be placed on efficiently scaling algorithms so that continued advances in computational infrastructure have the greatest impact in biology. To support those algorithms advances in high-performance data management, including provenance capture, archiving, accessibility, metadata management and storage and retrieval, will become even more crucial for repeatability of results and dissemination of new information.

Mathematical challenges
Current analysis that is reliable at a given statistical confidence will quickly become unacceptable as datasets continue to grow, overwhelming search problems with false positives, for instance. In tandem with improvements in *performance* of computational algorithms, biology will need increasingly powerful *statistical models* just to maintain the current capacity to evaluate biological hypotheses.

Technical challenges
One concern facing computational biology is that many computational applications, including pattern matching, feature searching, graph analysis, and others, are often not limited by floating-point capacity of systems on which they are running. Rather, they are limited by availability, latency, or bandwidth to memory, disk, or interconnect. New computing paradigms and new approaches to hardware are needed that more naturally map to the core computational tasks associated with biological applications.

Policy challenges
The model of tightly integrating high-performance computing into an analytical pipeline presumes the availability of dedicated compute resources. The prevailing way in which clusters, high end systems and emerging architectures are deployed does not match this model. Rather, most systems are multi-user batch systems. We have presented workflows that interact with such systems, but real-time access will always be preferred when analysis is meant to be interactive, rather than offline. Unlike many other computational domains, evaluating a biological hypothesis is often approached in a highly iterative way, making the traditional batch mode of operation a less than desirable access model. In addition, conventional batch supercomputing is also often only granted through highly controlled channels to individuals trusted to interact intimately with the compute resource. Transparently integrating high-performance computing into biological workflow models presumes a lower degree of interaction between the end user and the resource (preferably no direct interaction), opening the possibility for a more lax user access policy. However, such user access is a challenging notion to most conventional security models.

Pushing scientific frontiers in biology
Increasing the utility of high-performance computing for the biology community has the potential to enable a revolution in the scope and breadth of biological investigation. Solving some of the scientific, technical, and procedural issues to enable this revolution will pave the way for practical and impactful advances in biology to address large-scale challenges in renewable energy, climate management, advancing human health, and many other areas.

Acknowledgments
Dr. Oehmen and Dr. Cannon are supported by the Data-intensive Computing for Complex Biological Systems project funded by the *Office of Advanced Scientific Computing Research,* and under the *LDRD Program* at the Pacific Northwest National Laboratory, a multiprogram national laboratory operated by Battelle for the U.S. Department of Energy under Contract DE-AC06-76RL01830; and by the Director, Office of Science, Office of Biological and Environmental Research, Life Sciences Division, U.S. Department of Energy under Contract No. DE-AC03-

76SF00098. The Molecular Science Computing Facility (MSCF) in the Wiliam R. Wiley Environmental Molecular Science Laboratory (EMSL) is a national scientific user facility sponsored by the U.S. DOE, OBER and located at PNNL.

The authors thank Lee Ann McCue and Terence Critchlow for helpful discussions and Darren Curtis, Daniel Crawl, Elena Peterson, Anuj Shah, and Douglas Baxter for technical work on web interfaces and workflow design for ScalaBLAST and NWPolygraph applications.

References

[1] Muin M and Fontelo P 2006 Technical development of PubMed interact: an impoved interface for MEDLINE/PubMed searches *BMC Med. Inform. Deci.s Mak.* **6**:36

[2] Oehmen C and Nieplocha J 2006 ScalaBLAST: A scalable implementation of BLAST for High Performance Data-Intensive Bioinformatics Analysis *IEEE Trans. Parallel. Dist. Sys.* **17**:740-9

[3] Singh D, Trehan R, Schmidt B and Bretschneider T 2008 Comparative phyloinformatics of virus genes at micro and macro levels in a distributed computing environment *BMC Bioinformatics* **9 suppl 1**:S23

[4] Thallinger G, Trajanoski S, Stocker G and Trajanoski Z 2002 Information management systems for pharmacogenomics *Pharmacogenomics* **3**:651-667

[5] Wang C and Lefkowitz E 2004 SS-Wrapper: a package of wrapper applications for similarity searches on Linux clusters *BMC Bioinformatics* **5**:171

[6] Cannon W, Jarman K, Webb-Robertson B-J M, Baxter D, Oehmen C, Jarman K, Heredia-Langer A, Auberry K and Anderson G 2005 Comparison of probability and likelihood models for peptide identification from tandem mass spectrometry data *J. Proteome Res.* **4**:1687-1698

[7] Gao J, Ma S, Major D, Nam K, Pu J and Truhlar D 2006 Mechanisms and free energies of enzymatic reactions *Chem Rev* **106**:3188-31209

[8] Hereld M, Stevens R, Lee H and van Drongelen W 2007 Framework for interactive million-neuron simulation *J. Clin. Neurophysiol.* **24**:189-196

[9] Lins R, Vorpagel E, Guglielmi M and Straatsma T 2008 Computer simulation of uranyl uptake by the rough lipopolysaccharide membrane of Pseudomonas aeruginosa *Biomacromolecules* **9**:29-35

[10] Shaikh S, Jain T, Sandhu G, Latha N and Jayaram B 2007 From drug target to leads--sketching a physiochemical pathway for lead molecule design in silico *Curr Pharm Des* **13**:3454-3470

[11] Sun Y, Shen B, Lu Z, Jin Z and Chi X 2008 GridMol: a grid application for molecular modeling and visualization *J. Comput Aided Mol. Des.* **22**:119-129

[12] Zhang B, Park B, Karpinets T and Samatova N 2008 From pull-down data to protein interaction networks and complexes with biological relevance *Bioinformatics* **24**:979-986

[13] Zhu M and Wu Q 2008 Transcription network construction for large-scale microarray datasets using a high-performance computing approach *BMC Genomics* **9 suppl 1**:S5

[14] Altschul S, Gish W, Miller W, Myers E and Lipman D 1990 Basic local alignment search tool *J. Mol. Biol.* **215**:403-10

[15] Altschul S, Madden T, Schaffer A, Zhang J, Zhang Z, Miller W and Lipman D 1997 Gapped BLAST and PSI-BLAST: a new generation of protein database search programs *Nucleic Acids Res.* **25**:3389-33402

[16] Darling A, Carey L and Feng W-C 2003 The design, implementation, and evaluation of mpiBLAST. In: *Proc. ClusterWorld,* (San Jose, CA)

[17] Lin H, Ma X, Chandramohan P, Geist A and Samatova N 2005 Efficient data access for parallel BLAST. In: *19th International Parallel and Distributed Processing Symposium (IPDPS),* (Denver, CO: IEEE CS Press)

[18] Bjornson R, Sherman A, Weston S, Willard N and Wing J 2002 TurboBLAST(r): A parallel implementation of BLAST built on the TurboHub. In: *16th International Parallel and Distributed Processing Symposium (IPDPS),* (Fort Lauderdale, FL: IEEE CS Press)

[19] Grant J, Dunbrack R J, Manion F and Ochs M 2002 BeoBLAST: distributed BLASTand PSI-BLAST on a Beowulf cluster *Bioinformatics* **18**:765-766

[20] Braun R, Pedretti K, Casavant T, Scheetz T, Birkett C and Roberts C 2001 Parallelization of local BLAST service on workstation clusters. *Future Generation Computer Systems* **17**

[21] Camp N, Cofer H and Gomperts R 1998 High-throughput BLAST.

[22] Hokamp K, Shields D, Wolfe K and Caffrey D 2003 Wrapping up BLAST and other applications for use on UNIX clusters *Bioinformatics* **19**:441-442

[23] Kim H-S, Kim H-J and Han D-S 2003 Hyper-BLAST: A parallelized BLAST on cluster system. In: *International Conference on Computational Science,* pp 213-222

[24] Wang J and Mu Q 2003 Soap-HT-BLAST: high-throughput BLAST based on web services *Bioinformatics* **19**:1863-1864

[25] Muriki K, Underwood K and Sass R 2005 RC-BLAST: Towards a portable, cost-effective open source hardware implementation. In: *HICOMB 2005, 4th IEEE international workshop on high performance computational biology* (Denver, CO)

[26] Curtis D, Peterson E and Oehmen C 2008 A secure web application providing public access to high-performance data intensive scientific resources. In: *4th annual international conference on web information systems and technologies,* (Funchal, Madeira - Portugal

[27] Shah A, Singhal M, Klicker K, Stephan E, Wiley H and Waters K 2007 Enabling high-throughput data management for systems biology: the Bioinformatics Resource Manager *Bioinformatics* **23**:906-909

[28] Webb-Robertson , Peterson E, Singhal M, Klicker K, Oehmen C, Adkins J and Havre S 2007 PQuad-- a visual analysis platform for proteomic data exploration of microbial organisms *Bioinformatics* **23**:1705-1707

Journal of Physics: Conference Series **125** (2008) 012053 doi:10.1088/1742-6596/125/1/012053

Coupling graph perturbation theory with scalable parallel algorithms for large-scale enumeration of maximal cliques in biological graphs

N. F. Samatova[1,2,+], M. C. Schmidt[1,2,*], W. Hendrix[1,2,*], P. Breimyer[1,2], K. Thomas[3] and B.-H. Park[2]

[1]Computer Science Department, North Carolina State University, Raleigh, NC 27695, USA
[2]Computer Science and Mathematics Division, Oak Ridge National Laboratory, Oak Ridge, TN 37831, USA [3]Cray, Inc. Seattle, WA 98104, USA
[*]Both authors contributed equally. [+]Corresponding author.

E-mail: samatovan@ornl.gov

Abstract. Data-driven construction of predictive models for biological systems faces challenges from data intensity, uncertainty, and computational complexity. Data-driven model inference is often considered a combinatorial graph problem where an enumeration of all feasible models is sought. The data-intensive and the NP-hard nature of such problems, however, challenges existing methods to meet the required scale of data size and uncertainty, even on modern supercomputers. Maximal clique enumeration (MCE) in a graph derived from such biological data is often a rate-limiting step in detecting protein complexes in protein interaction data, finding clusters of co-expressed genes in microarray data, or identifying clusters of orthologous genes in protein sequence data. We report two key advances that address this challenge. We designed and implemented the first (to the best of our knowledge) parallel MCE algorithm that scales *linearly* on thousands of processors running MCE on real-world biological networks with thousands and hundreds of thousands of vertices. In addition, we proposed and developed the *Graph Perturbation Theory* (GPT) that establishes a foundation for efficiently solving the MCE problem in *perturbed* graphs, which model the uncertainty in the data. GPT formulates necessary and sufficient conditions for detecting the differences between the sets of maximal cliques in the original and perturbed graphs and reduces the enumeration time by more than 80% compared to complete recomputation.

1. Introduction

Many biological problems often reduce to graph problems, with the maximal clique enumeration (MCE) problem being ubiquitous. The solutions of the MCE problem are used, for example, to align 3-dimensional protein structures [1], to integrate genome mapping data [2], to identify coexpressed genes [3], to identify common secondary structure elements of proteins [4], to detect protein-protein interaction complexes [5], to cluster similar mass spectrometry spectra [6], and to find clusters of orthologous genes [7]. The MCE problem is NP-hard [8], and its run time, in practice, scales exponentially with the problem size, unless $P = NP$ [9]. The challenge remains of how to scale an MCE algorithm to graphs with hundreds, thousands, or even millions of vertices, which are not unusual for the problems considered in these examples.

The problem is exacerbated when, in addition to the size of the graphs, the uncertainty, or noise, in the data from which these graphs are derived is taken into account. In this case, multiple solutions of the MCE problem are often sought for various "perturbed" graphs. Perturbations may be induced by filtering out some edges due to applied edge weight cutoffs or by adding edges based on additional orthogonal information sources. For example, two genes in a gene expression network can be viewed as coexpressed (i.e., connected by an edge) if their Pearson correlation derived from microarray data is above a certain threshold; various thresholds will correspond to different network perturbations. Likewise, two proteins in a protein-protein interaction network can be considered interacting if, in addition to genomic-context information (e.g., their neighborhood colocation on the genome), mass spectrometry pull-down experiments become available. The challenge is to support such "what-if" explorations of biological graphs that are not only large but also uncertain and changing.

In order to address both challenges, two major, interrelated advancements have been achieved. On the one hand, we developed a scalable, parallel MCE algorithm. It not only efficiently handles the data-intensive nature of the MCE problem but offers linear speedup, even using thousands of processors to run the algorithm on real-world biological graphs with thousands or hundreds of thousands of vertices. To the best of our knowledge, this is the first parallel MCE algorithm that scales to this number of processors on such large-scale, real-world problem sizes.

On the other hand, we proposed and advanced a new theory, which we call *Graph Perturbation Theory* (GPT), that establishes a foundation for solving graph problems in *perturbed* graphs. The intuition behind GPT is quite simple: if a solution to the "reference" or "unperturbed" graph is known, then it can be used to find the *exact* solution for the "perturbed" graph more efficiently than complete recomputation, especially when the perturbation is relatively small. Specifically, we formulated necessary and sufficient conditions for the maximal cliques that are induced or eliminated by the addition or removal of an arbitrary number of edges to a "reference" graph. Based on this theory, we produced a practical MCE algorithm implementation for perturbed graphs that enumerates the maximal cliques of a perturbed graph by using efficient indexing of features derived from the MCE solution for the "unperturbed" graph to detect the changes in the composition of maximal cliques induced by the target perturbations. We demonstrated more than 80% efficiency improvement compared to the traditional enumeration of maximal cliques in protein interaction networks for multiple organisms, even when the number of added edges (i.e., perturbations) ranged between 20% and 136%.

2. Parallel MCE algorithm

Our current parallel MCE algorithm is a parallelization of the widely used method of enumerating maximal cliques developed by Bron and Kerbosch (BK) [10]. Our previously developed *pClique* [11], the first parallel MCE algorithm, extends the algorithm of Kose et al [12] (dubbed as KOSE). In principle, KOSE is identical in spirit to the BK algorithm; it branches using alphanumeric ordering. However, whereas BK is a recursive algorithm with depth-first search (DFS) branching, KOSE is a serialized algorithm with breadth-first search (BFS) branching (see [13] for DFS and BFS definitions), which allows cliques of size k to be generated from cliques of size $k-1$. Consequently, all maximal cliques are produced in lexicographic order, which is an invaluable asset in certain applications. Nevertheless, the BFS branching strategy inevitably makes KOSE memory-intensive. Although *pClique* improves KOSE performance by using bit-vector manipulation of common neighbors, the huge memory requirements remain unchanged. This limitation affects both the size of the graphs that can be handled by *pClique* and the speedup achieved by using more processors. As a result, on an SGI Altix 3700 machine, *pClique* achieved a speedup factor of just 91 on 256 processors for a graph of 2,895 vertices and 10,914 edges. This nonideal scaling motivated the development of a parallel DFS-based BK algorithm.

To enable the parallelization of BK, we proposed an effective decomposition of the BK search tree, in which leaf nodes represent maximal cliques, into subtasks of generating the child nodes of an interior node. To allow this decomposition, we introduced a *candidate path* data structure containing the minimal information required for a *BK* search subtree exploration [7, 14].

The difficulty in applying this decomposition lies in the fact that the nature of the *BK* search tree makes it impossible to determine a well-balanced distribution of subtasks *a priori*. In particular, the size and number of cliques in the subtree beneath a particular tree node are unknown until that subtree has been fully generated. Thus, certain "unlucky" computing elements may generate subtrees with more and larger maximal cliques than others. Without allowing these overloaded computing nodes to transfer work to underloaded nodes, the execution time for different computing elements may differ greatly (see figure 1).

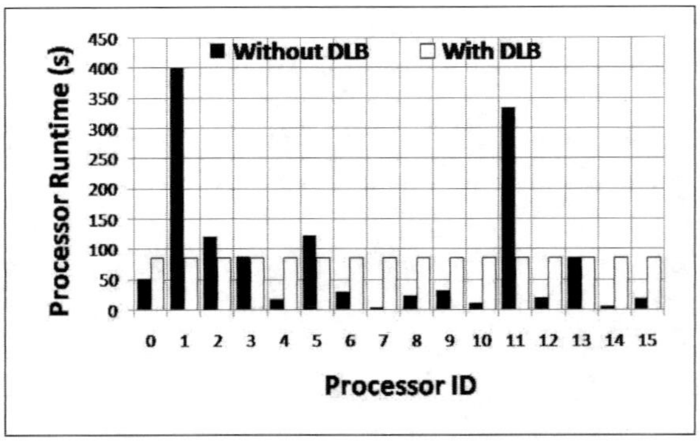

Figure 1. Clarifying example of the impact of dynamic load-balancing (DLB) on the execution times of the various processes. The black bars give the finishing times of the 16 processes used to run the parallel algorithm without dynamic load balancing. The white bars represent the finishing times with dynamic load balancing. The graphs were obtained by running the parallel algorithm on the *Shewanella oneidensis* gene expression graph.

The load balancing scheme we proposed intelligently couples a dynamic (runtime) work stealing process [15] with a stack splitting procedure [16] in order to minimize the idle (noncomputing) time over all processors. The amount of work for a computing element can be measured by the number of candidate path structures left in that computing element's stack. However, because the number of candidate paths remaining in a stack decreases gradually and increases rapidly over the course of the algorithm's execution, predicting exactly when a computing element will become idle is virtually impossible. To overcome this, we implemented a receiver-initiated scheme, allowing a computing element to become almost idle (stack size below some threshold) before requesting more work.

When this threshold is reached, the idle computing element requests more candidate paths from another randomly chosen computing element. (This process, called *random polling*, is one of the most efficient methods of requesting work when the underlying architecture of the computing system is unknown [15].) If the randomly chosen computing element has work available, it sends the candidate path structures most likely to represent large subtrees, a procedure motivated by the concept of *stack splitting* [16]. If the responding computing element has no work, then the requesting computing element selects another random computing element and repeats the request, with the program terminating when all computing elements are idle.

Figure 2 shows that the speedup of the algorithm is linear, and thus the initialization and idle time are small relative to the total execution time.

Parallel execution of the program is achieved by generating multiple *processes*, each capable of spawning multiple *threads* [17]. Interprocess communication is performed using MPI communication, and the threaded behavior of the application is enabled using POSIX threads

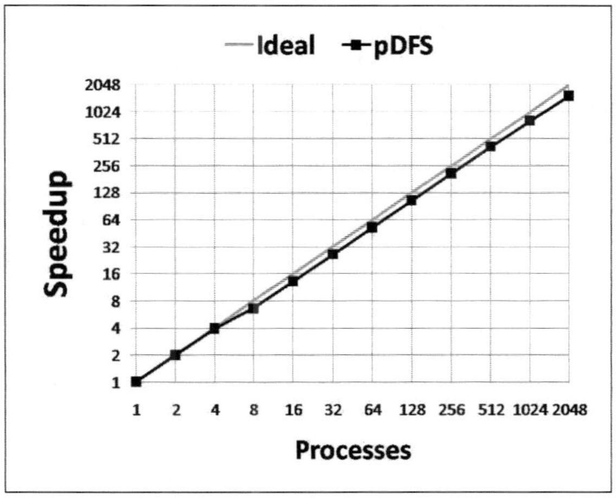

Figure 2. Speedup of the parallel algorithm on the *Saccharomyces cerevisiae* protein interaction network with between 1 and 2,048 processes on a Cray XT4.

(Pthreads). Each process is assumed to have its own memory that its associated threads share. This "hybrid" parallelism is motivated by the fact that many modern high-performance machines consist of clusters of symmetric multiprocessing (SMP) units. By combining both shared-memory and distributed memory parallelism techniques, better performance is achieved. In addition, the implementation is portable across different computer architectures.

3. Graph perturbation theory and algorithms

The basic idea behind graph perturbation theory is to examine the differences (added or removed edges, in our case) between two graphs—an "original" graph, for which the maximal cliques have already been enumerated, and a "perturbed" graph—and to list only the set of maximal cliques that are introduced and destroyed by the perturbation. By leveraging the enumeration of the original graph, the maximal clique enumeration of the perturbed graph may be calculated more quickly. Intuitively, if the perturbation between the two is relatively small, the two "difference sets" will be smaller than the full enumeration for the perturbed graph.

The following basic definitions are necessary before setting out our theory. Let G and G_{new} denote the original and perturbed graphs, respectively, and let $\mathcal{C}$ and $\mathcal{C}_{new}$ be the maximal clique enumeration of each. Define the difference sets $\mathcal{C}_+ = \mathcal{C}_{new} \setminus \mathcal{C}$ and $\mathcal{C}_- = \mathcal{C} \setminus \mathcal{C}_{new}$.

Theorems 3.1 and 3.2 establish simple necessary and sufficient conditions for containment in C_+ and C_-, respectively.

Theorem 3.1. *$C \in \mathcal{C}_+$ if and only if C is a maximal clique in G_{new} that contains some edge being added to G.*

Proof. Clearly, if a maximal clique $A \in \mathcal{C}_+$ contained some edge being added to G, C would not be a clique in G, so $A \in \mathcal{C}_+$.

Let $C \in \mathcal{C}_{new}$ be such that C contains no edge being added to G. As such, C is be a clique of G. Thus, either $C \in \mathcal{C}$, or there must be some $C' \in \mathcal{C}$ that strictly contains C. By our definition of $\mathcal{C}_+$, $C \notin \mathcal{C}_+$ if $C \in \mathcal{C}$. Also, since no edges are being removed from G, C' would have to be in $\mathcal{C}_{new}$, but this contradicts our assumption that C is a maximal clique in G. □

Theorem 3.2. *A maximal clique $C \in \mathcal{C}_-$ if and only if C is a clique in G and C is a subset of some $C' \in \mathcal{C}_+$.*

Proof. Let C be an arbitrary maximal clique of $\mathcal{C}_-$. Since no edges are being removed from G and C is a clique in G, C must be a clique in G_{new}. Thus, $C \notin \mathcal{C}_{new}$ if and only if C is not maximal, and C is not maximal if and only if there exists some clique $C' \in \mathcal{C}_{new}$ such that C is a proper subset of C'. Such a C' could not be a maximal clique of G as this would contradict the maximality of $\mathcal{C}_-$, so $C \in \mathcal{C}_+$. □

By Theorem 3.1, we know that all cliques of $\mathcal{C}_+$ are maximal cliques of G_{new} containing some edge being added to G. Thus, to calculate this set, we use a modified version of the BK algorithm.

On weighted protein-protein interaction networks for nine different organisms from [18], where the weight of each edge represents the probability two proteins interact, we generated graphs by applying thresholds at probabilities 0.75 and 0.70. The perturbations introduced by lowering the threshold to 0.70 accounted for 20–40% of the edges in the networks for 6 of the 9 organisms. The networks for *S. typhimurium* and *M. tuberculosis* saw healthy perturbations of 48% and 68%, respectively, but the *E. coli* network underwent a full 136% change in its number of edges.

After calculating the maximal clique enumeration for the network of each organism using the cutoff 0.75, we calculated the maximal clique enumeration for the graph induced by the cutoff 0.70 via both the perturbed graph algorithm as well as a single-threaded version of the original BK implementation. The percentage improvement of the perturbational algorithm over BK appears in figure 3.

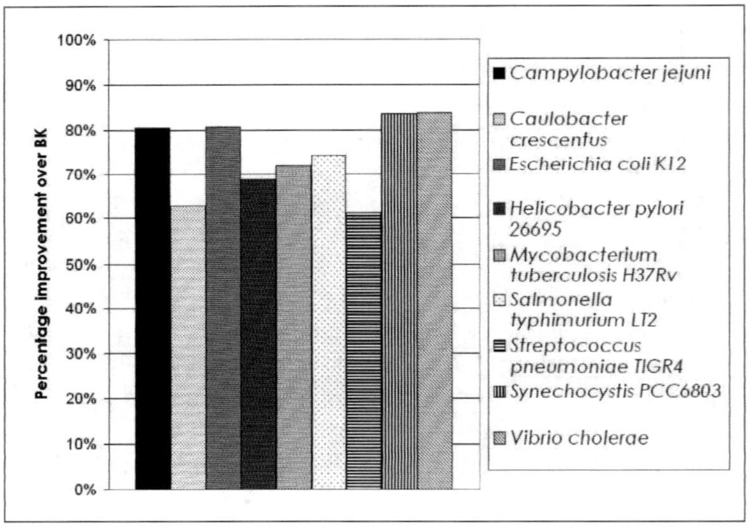

Figure 3. Percentage runtime improvement of perturbational algorithm over BK for the induced perturbations

As shown in the figure, all clique enumerations were produced by the perturbational algorithm in 50–85% less time than full recalculation by BK – even for the *E. coli* network, where more edges were added than existed in the original graph. While these results favor the perturbational algorithm, the algorithm performed better on this very large perturbation than would be suggested by intuition. Better performance is typically observed under smaller (less than 20%) perturbations applied to the "reference" graph (results are not reported).

4. Conclusion

We reported a novel capability for efficient enumeration of maximal cliques in biological graphs derived from large-scale, uncertain, and dynamically changing biological data. We demonstrated the first parallel MCE algorithm that scales *linearly* on thousands of processors for real-world biological networks with thousands and hundreds of thousands of vertices. We proposed the *Graph Perturbation Theory* (GPT) that takes advantage of the solution provided by parallel MCE on the "reference" graph to significantly reduce the time required to solve the MCE problem on the 'perturbed' graphs. We developed a practical implementation of the perturbed MCE algorithm that utilizes efficient database indices, constructed using the GPT theory, to achieve improved performance. The application of the MCE algorithms to real-world biological networks across multiple organisms has been demonstrated.

Acknowledgments

The authors are thankful to Cray Inc. for the access to large-scale Cray XT systems and the insights into the code optimization and benchmarks. This research has been supported by the "Exploratory Data Intensive Computing for Complex Biological Systems" project from U.S. Department of Energy (Office of Advanced Scientific Computing Research, Office of Science). The work of NFS was also sponsored by the Laboratory Directed Research and Development Program of Oak Ridge National Laboratory. Oak Ridge National Laboratory is managed by UT-Battelle for the LLC U.S. D.O.E. under contract no. DEAC05-00OR22725.

References

[1] Chen Y and Crippen G M 2005 *Protein Science* **14** 2935–46
[2] Harley E, Bonner A and Goodman N 2001 *Bioinformatics* **17** 487–494
[3] Rokhlenko O, Wexler Y and Yakhini Z 2007 *Bioinformatics* **23** e184–e190
[4] Grindley H M, Artymiuk P J, Rice D W and Willett P 1993 *Journal of Molecular Biology* **229** 707–721
[5] Zhang B, Park B H, Karpinets T and Samatova N F *Bioinformatics (Oxford, England)* **24** 979–86
[6] Tabb D L, Thompson M R, Khalsa-Moyers G, VerBerkmoes N C and McDonald W H 2005 *Journal of the American Society for Mass Spectrometry* **16** 1250–61
[7] Park B H, Samatova N F, Karpinets T, Jallouk A, Molony S, Horton S and Arcangeli S 2007 *SciDAC 2007* vol 78 (Boston, Massachusetts)
[8] Lawler E L, Lenstra J K and Kan A H G R 1980 *SIAM Journal on Computing* **9** 558–565
[9] Garey M R and Johnson D S 1979 *Computers and Intractability: A Guide to the Theory of NP-Completeness* (WH Freeman & Co. New York, NY, USA)
[10] Bron C and Kerbosch J 1973 *Communications of the ACM* **16** 575–577
[11] Zhang Y, Abu-Khzam F, Baldwin N, Chesler E, Langston M and Samatova N 2005 *Supercomputing, 2005. Proceedings of the ACM/IEEE SC 2005 Conference* p 12
[12] Kose F, Weckwerth W, Linke T and Fiehn O 2001 *Bioinformatics* **17** 1198–1208
[13] Cormen T, Leiserson C E, Rivest R L and Stein C 2001 *Introduction to Algorithms* 2nd ed (McGraw-Hill)
[14] Park B H, Schmidt M, Thomas K, Karpinets T and Samatova N F 2008 *Upcoming in Proceedings of IPDPS 2008*
[15] Kumar V, Grama A Y and Vempaty N R 1994 *Journal of Parallel and Distributed Computing* **22** 60–79
[16] Finkel R and Manber U 1987 *ACM Trans. Program. Lang. Syst.* **9** 235–256
[17] Thomas K, Samatova N F, Schmidt M and Park B H 2008 *Upcoming in Proceedings of CUG 2008*
[18] Flannick J, Novak A, Srinivasan B S, McAdams H H and Batzoglou S 2006 *Genome Research* **16** 1169–1181

Journal of Physics: Conference Series **125** (2008) 012054 doi:10.1088/1742-6596/125/1/012054

Hybrid numerical methods for multiscale simulations of subsurface biogeochemical processes

T D Scheibe[1], A M Tartakovsky[1], D M Tartakovsky[2], G D Redden[3], P Meakin[3,4], B J Palmer[1] and K L Schuchardt[1]

[1]Pacific Northwest National Laboratory, Richland, WA, 99352, USA
[2]University of California, San Diego, Department of Mechanical and Aerospace Engineering, La Jolla, CA, 92093, USA
[3]Idaho National Laboratory, Idaho Falls, ID, 83415, USA
[4]Physics of Geological Processes, University of Oslo, USA

E-mail: tim.scheibe@pnl.gov

Abstract. Many subsurface flow and transport problems of importance today involve coupled non-linear processes that occur in media exhibiting complex heterogeneity. Problems involving biological mediation of reactions fall into this class of problems. Recent experimental research has revealed important details about the physical, chemical, and biological mechanisms that control these processes from the molecular to laboratory scales. We are developing a hybrid multiscale modeling framework that combines discrete pore-scale models (which explicitly represent the pore space geometry at a local scale) with continuum field-scale models (which conceptualize flow and transport in a porous medium without a detailed representation of the pore space geometry). At the pore scale, we have implemented a parallel three-dimensional Lagrangian model of flow and transport using the smoothed particle hydrodynamics method and performed test simulations using millions of computational particles on the supercomputer at the Environmental Molecular Sciences Laboratory. We have also developed methods for gridding arbitrarily complex pore geometries and simulating pore-scale flow and transport using parallel implementations of grid-based computational fluid dynamics methods. Within the multiscale hybrid framework, we have coupled pore- and continuum-scale models to simulate coupled diffusive mixing, reaction, and mineral precipitation, and compared the results with conventional continuum-only simulations. The hybrid multiscale modeling framework is being developed using a number of SciDAC enabling technologies including the Common Component Architecture, advanced solvers, Grid technologies, scientific workflow tools, and visualization technologies.

1. Introduction

Hybrid multiscale numerical modeling methods combine multiple models defined at fundamentally different length and time scales within the same overall spatial and temporal domain. In many cases, the models also have fundamentally different ways of representing the physical, chemical, and biological processes. For example, several models in the materials science and physics literature couple molecular dynamics (MD) simulations at the molecular scale to continuum mechanics (typically finite element) simulations at larger scales [15, 2, 3].

Such simulations are motivated by classes of problems in which large-scale phenomena of interest (e.g., overall material failure and deformation) are strongly influenced by processes occurring at much smaller scales (e.g., microfracture propagation) that are not well represented by effective or averaged processes and/or properties. Executing an exhaustive simulation of processes at the smallest scales for a domain of engineering significance is currently impractical, and likely to remain so for a very long time. However, the hybrid multiscale approach, in which a small-scale model with high resolution is used in a small fraction of the overall domain and is linked to a large-scale model with coarse resolution over the remainder of the overall domain, can provide necessary efficiency of characterization and computation that may render solution of these problems practical.

This SciDAC science spplication is developing and applying hybrid numerical techniques to simulate complex subsurface flow and biogeochemical reactions of relevance to subsurface contaminants at U.S. Department of Energy sites. Two related science application partnerships (SAPs) support this science application through the implementation of the Common Component Architecture for parallel model integration and development of a scientific workflow and data management environment to facilitate execution of complex model processes and management of simulation input and output data. This paper introduces a subsurface problem that motivates the hybrid multiscale approach and describes progress to date.

2. Motivation

Important subsurface biogeochemical processes (e.g., microbial respiration or complexation of solutes with mineral surfaces) are currently best defined by events that take place at very small scales, typically ranging from molecular to cellular to small batch and column experimental scales (with time scales of minutes to days). However, the scales at which we are interested in predicting phenomena for engineering applications (e.g., fate and transport of contaminants in aquifers or geologic carbon sequestration) are very large, typically ranging from meters to many kilometers (with time scales of months to years or even a million years in the case of nuclear waster repositories). This disparity in scales creates computational challenges that motivate a hybrid modeling approach in which pore- and continuum-scale models execute concurrently within different portions of a model domain. In this paper, we focus specifically on porous media systems that exhibit strong coupling, with feedback, between flow, transport, and reaction processes, such as reactions involving biofilm development [12] and/or mineral precipitation [10]. In these systems, the biogeochemical reactions (controlled by flow and transport processes) lead to changes in pore structure, which in turn modify local flow and transport behavior. This coupling/feedback leads to strong non-linearity in apparent continuum-scale behavior, and this behavior is not well represented by averaged properties and state variables.

An example is provided by the laboratory experiment in which two solutes mix and react to precipitate calcium carbonate minerals within a very narrow zone in the center of the experimental cell. We have performed pore- and continuum-scale simulations of these mixing and reaction processes. The pore-scale simulation results indicate that strong mixing of the two solutes is limited to a very local region (only a few grain diameters wide) and the mixing is inhibited by mineral precipitation leading to a strong negative feedback or coupling between the reaction and transport across the narrow mixing zone in which precipitation occurs. The conditions strictly required for validity of a continuum model are violated in the mixing region. Nevertheless, we found that a continuum model was able to qualitatively reproduce the observed phenomena if adaptive grid resolution was used in the mixing zone at a scale comparable to that of the grains of the porous media. However, the continuum model did not quantitatively predict the amount of mineral precipitation because of discrepancies between the pore- and continuum-scale effective reaction rates. Details of the experiment and models are reported in [11]. The hybrid multiscale modeling approach is well suited to this problem, as it allows for explicit pore-scale simulation in the mixing region while using the less computationally demanding continuum methods in the majority of the model domain.

3. Hybrid multiscale modeling

Published reviews of hybrid multiscale modeling concepts are provided by [5, 6]. Hybrid multiscale modeling methods have been most widely applied in the fields of materials science and chemical engineering, in which atomic-scale models of molecular dynamics (MD) have been linked to continuum-scale models of material deposition, strength, deformation, and failure. A recent review of multiscale methods applied to materials science is given by [2]. Hybrid multiscale methods have also been applied in chemical engineering, catalysis and reactor processes (e.g., [14]), life sciences (e.g., [13]), and fluid mechanics (e.g., [4]). Previous applications in subsurface simulation are extremely limited [1].

We have implemented a two-scale hybrid model of diffusive mixing, reaction, and precipitation described in [11]. In this work, favorable comparisons were drawn between analytical solutions, a single-resolution model using high resolution throughout the model domain, and a two-scale hybrid model. Current efforts are focused on incorporating advection at both scales and on implementing a coupled SPH-finite element hybrid model.

4. Component-based pore-scale model

We have developed a parallel three-dimensional SPH code utilizing the Common Component Architecture (CCA) programming framework. SPH is a particle-based numerical method that has been adapted for simulation of pore-scale flow and reactive transport [8, 9]. In SPH simulations continuous fields are represented by a superposition of smooth bell-shaped functions centered on a set of Lagrangian particles, and this provides a meshless discretization of the continuum hydrodynamic fields (i.e., Navier-Stokes equations). The SPH particles carry attributes such as the solvent and solute masses that allow transport and reactions to be simulated. The value of the continuum fields (e.g., velocity, concentration, density) is computed by convolving the values associated with neighboring particles using a smooth weighting function. SPH simulations are relatively easy to implement, and they can handle complex internal boundaries, free surface and moving boundaries with relative ease. A typical SPH simulation consists of a loop to integrate the equations of motion coupled with a force evaluation at each time step. The force evaluations involve interactions between each particle and nearby particles within a finite neighbourhood. For parallel simulation, particles are distributed among processors in spatial blocks. This approach requires communication of information among processors for particles that are near subdomain boundaries and for particles that move across boundaries within a time step. SPH code components include a particle integration component, force evaluation component, communication layer component, and setup and I/O components. The parallel SPH code has been developed using the CCAFFEINE implementation of CCA and runs on the supercomputer "MPP2" at the Environmental Molecular Sciences Laboratory (EMSL). Test simulations have been executed using the pore geometry of a sand pack measured by x-ray microtomography (courtesy of Brian Wood, Oregon State University) and the results were visualized by members of the SciDAC Institute for Ultrascale Visualization at UC Davis (see figure 1). We are currently working on incorporating reactions into the parallel code and testing the code on scientifically relevant and challenging pore-scale simulation problems.

5. Scientific workflow

The application of high-performance computing to subsurface simulations creates the challenge of ever-increasing model complexity. Hybrid multiscale models as described above are one example of simulations of complex systems that generate large amounts of output data. Simulating uncertain processes often involves sensitivity analyses, Monte Carlo simulations, or other approaches that require the execution of multiple simulations with different input sets. Managing the resulting data (including metadata about the relationship among different simulations and recording of the process for a given simulation study) is a challenging task in this environment.

To address this issue, we have applied workflow concepts and an integrated data and provenance system to automate these processes while providing verifiable and repeatable records. Our application

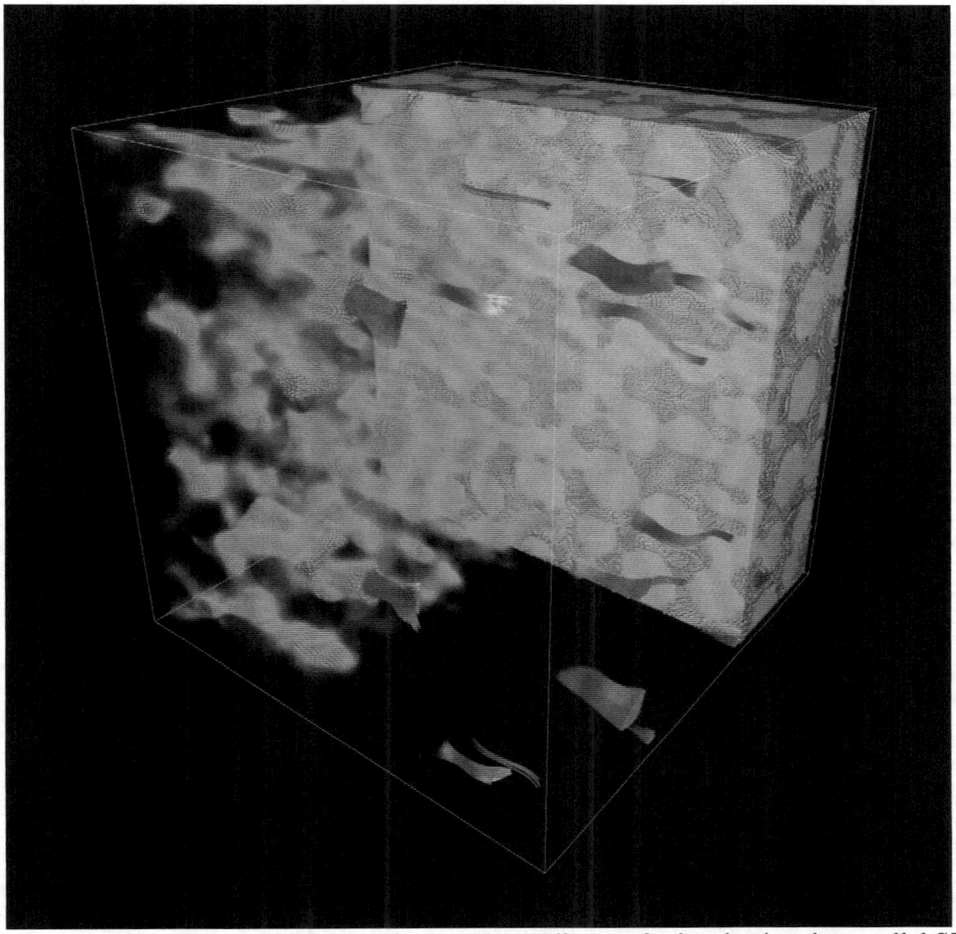

Figure 1. Visualization of water flow through a porous medium, calculated using the parallel SPH code developed by this project. Colored particles are SPH computational particles, where the color represents the local flow velocity (red is large, green is small). White regions represent solid grains of the porous medium. In the foreground portion of the image, the solid grains have been removed to clearly show the fluid particles and local streamlines. Visualization courtesy of Kwan-Liu Ma and Chad Jones, Institute for Ultrascale Visualization, University of California at Davis.

is based on the Kepler workflow tool, developed by the SciDAC Scientific Data Management (SDM) Center. Our application of workflow is unique in that it strives to seamlessly combine automated workflow with manual research steps to create a unified record of the overall research process. We have developed a prototype workflow environment that utilizes a graph-based representation of model relationships as described in [7]. We have applied this prototype to support experimental design of ongoing experiments on mixing-controlled reactions in porous media similar to those in [11].

6. Concluding remarks

Our research addresses the development of multiscale hybrid numerical methods for simulation of subsurface flow and biogeochemical reactions. We have developed and published results from an initial hybrid two-scale model of diffusion and reaction in a model porous medium. We have developed a 3D pore-scale parallel SPH code using CCA technology and are currently testing and applying that code. We have applied pore- and continuum-scale codes to a mixing-controlled precipitation reaction problem and published results demonstrating the need for hybrid methods. We have developed a prototype scientific workflow system and applied that system to parameter studies for design of the mixing-controlled reaction experiments.

Acknowledgments
This research is supported by the U. S. Department of Energy's Office of Science under the Scientific Discovery through Advanced Computing (SciDAC) program.

References

[1] Balhoff M T, Thompson K E and Hjortso M 2007 Coupling pore-scale networks to continuum-scale models of porous media *Computers & Geosciences* **33**:393-410
[2] Csanyi G, Albaret T, Moras G, Payne M C and De Vita D 2005 Multiscale hybrid simulation methods for material systems *J. Physics-Condensed Matter* **17**:R691-R703
[3] Curtin W A and Miller R E 2003 Atomistic/continuum coupling in computational materials science *Modelling and Simulation in Materials Science and Engineering* **11**:R33-R68
[4] Nie X B, Chen S Y, Robbins W N E and Robbins M O 2004 A continuum and molecular dynamics hybrid method for micro- and nano-fluid flow *J. Fluid Mechanics* **500**:55-64
[5] Oden J T, Prudhomme S, Romkes A and Bauman P T 2006 Multiscale modeling of physical phenomena: Adaptive control of models *SIAM J. Sci. Computing* **28**:2359-2389
[6] Rabczuk T, Xiao S P and SauerA 2006 Coupling of mesh-free methods with finite elements: basic concepts and test results *Communications in Numerical Methods in Engineering* **22**:1031-1065
[7] Schuchardt K, Chin G J, Chase J, Daily J and Scheibe T D 2008 Experiences in the design of scientific workflows for subsurface sciences **IEEE Trans. Automation Science and Engineering** (in review)
[8] Tartakovsky A M and Meakin P 2005 A smoothed particle hydrodynamics model for miscible flow in three-dimensional fractures and the two-dimensional Rayleigh-Taylor instability *J. Computational Physics* **207**:610-624
[9] Tartakovsky A M, Meakin P, Scheibe T D and Eichler West R M 2007 Simulations of reactive transport and precipitation with smoothed particle hydrodynamics *J. Computational Physics* **222**:654-672
[10] Tartakovsky A M, Meakin P, Scheibe T D and Wood B D 2007 A smoothed particle hydrodynamics model for reactive transport and mineral precipitation in porous and fractured porous media, Water Resources Research, 43 (2007).
[11] A. M. Tartakovsky, G. Redden, P. C. Lichtner, T. D. Scheibe and P. Meakin, Mixing-induced precipitation: Experimental study and multiscale numerical analysis *Water Resources Res.* **44**
[12] Tartakovsky A M, Scheibe T D and Meakin P 2008 Pore-scale model for reactive transport and biomass growth *J. Porous Media* (in press)
[13] Villa E, Balaeff A, Mahadevan L and Schulten K 2004 Multiscale method for simulating protein-DNA complexes *Multiscale Modeling & Simulation* **2**:527-553
[14] Vlachos D G, Mhadeshwar A B and Kaisare N S 2006 Hierarchical multiscale model-based design of experiments, catalysts, and reactors for fuel processing *Computers & Chemical Engineering* **30**:1712-1724
[15] Wang C Y and Zhang X 2006 Multiscale modeling and related hybrid approaches *Current Opinion in Solid State & Materials Science* **10**:2-14

Cellulosic ethanol: progress towards a simulation model of lignocellulosic biomass

Loukas Petridis and Jeremy C Smith

Center for Molecular Biophysics , Oak Ridge National Laboratory, Oak Ridge, TN, USA

E-mail: smithjc@ornl.gov

Abstract. A CHARMM molecular mechanics force field for lignin is derived. Parameterization is based on reproducing quantum mechanical data of model compounds. Partial atomic charges are derived by the examination of methoxybenzene:water interactions. Dihedral parameters are optimized by fitting to critical rotational potentials, and bonded parameters are obtained by optimizing vibrational frequencies and normal modes. The force field is validated by performing a molecular dynamics simulation of a crystal of a lignin fragment molecule and comparing simulation-derived structural features with experimental results. Together with the existing force field for polysaccharides, this work will enable full simulations of lignocellulose.

1. Introduction

The plant cell wall is made of cellulose microfibrils embedded in a matrix of polysaccharides (hemicellulose and pectins) and lignin [1]. Plant cell wall structure has come under renewed interest recently in the context of the production of bioethanol from the enzymatic hydrolysis of lignocellulosic biomass [2]. Cellulosic ethanol production is a two-stage process involving first the hydrolysis of cellulose by cellulases to smaller oligosaccharides and then fermentation of the sugars to ethanol. The hydrolysis step is the bottleneck of the process because of the natural resistance, or "recalcitrance," of plant cell walls to degradation [2].

Two main physical factors contribute to biomass recalcitrance [3]. First, cellulose is found in crystalline fibrils the compact structure of which impedes enzymatic access. In comparison, amorphous cellulose is readily digested by enzymes [4]. Second, matrix polysaccharides and lignin coat the cellulose fibril and act as a physical barrier preventing enzymes from reaching the cellulose. Confirmation of the contribution of lignin to biomass recalcitrance is provided by the finding that removing lignin from biomass increases the cellulose-hydrolysis yield from $\sim$ 20% to 98% [5].

Lignin is a heterogeneous biopolymer found in the plant cell wall; for a review see [6]. It is formed by radical polymerization of three alcohol monolignols: p-coumaryl, coniferyl, and sinapyl, shown in figure 1a. The chemical composition and structure of lignin is highly heterogeneous, varying significantly between plant species and even within different parts of the same plant cell wall. Although complex, lignin is composed primarily of three units p-hydroxyphenyl (H), guaiacyl (G), and syringyl (S), derived by oxidation of p-coumaryl, coniferyl and sinapyl respectively. An illustration of one G and one S units is shown in figure 1b.

Five types of linkages connect the units, leading to the formation of the long lignin biopolymer. The most common linkage (50–80% probability) is β-O-4$'$, connecting the oxygen of the hydroxyl

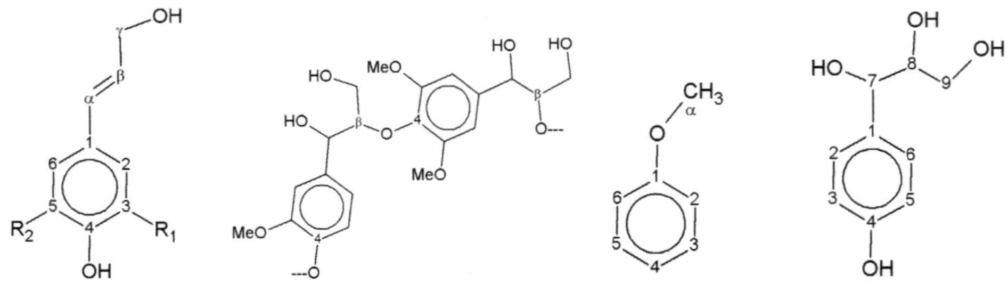

(a) monolignols (b) G and S units with β-O-4′ link (c) anisole (d) PHP

Figure 1: (a) The tree monolignols: p-coumaryl ($R_1 = R_2 = H$), coniferyl ($R_1 = H$, $R_2 = OMe$) and sinapyl ($R_1 = R_2 = OMe$). (b) A guaiacyl unit connected with a β-O-4′ linkage to a syringyl unit. (c) Model compound anisole. (d) Model compound PHP.

on the phenyl ring in one unit with the second tertiary carbon of the other; see figure 1b. Other common linkages are β-β', α-O-4′ and β-5′.

Computer simulation is a powerful tool for complementing experiment in obtaining an understanding of the molecular-level structure and dynamics of lignocellulose. The computational studies of lignin [7, 8] employed the CHARMM27 empirical force field, which was developed to model proteins rather than lignin. In this paper we present the first essential step toward the accurate computer simulation of lignin: the derivation of a molecular mechanics force field, complementing the CHARMM force field. Together with the existing force field for polysaccharides [9], this work will enable full simulations of lignocellulose.

2. Parameterization strategy

In this section we outline the general strategy employed to obtain the force field of lignin. The CHARMM potential energy function [10] of a molecule is approximated by a sum over bonded interactions (bond and angle vibrations and dihedral rotations) and nonbonded interactions (Coulomb electrostatic interactions and the Lennard-Jones potential for the van der Waals interactions). The strength of these interactions is determined by a set of parameters, such as bond, angle and dihedral force constants, partial atomic charges and Lennard-Jones parameters. These empirical parameters are molecule dependent and must be optimized to model the specific molecule prior to performing simulations. This optimization step is referred to as parameterization and its application to the case of lignin is the main task of this work.

This parameterization of lignin follows the main procedure of parameterization of proteins [11] and ethers [12] for the CHARMM force field. Lignin also has a linear ether bond, but different from those examined in [12] in that the oxygen is bonded to a phenyl ring and a tertiary carbon. For this reason it was found necessary to create a new atom type, OET, to represent the lignin's ether oxygen. Parameters were optimized by considering two factors. Firstly, the target data was reproduced as closely as possible. Second, compatibility with the existing CHARMM force field was ensured by restricting optimization to parameters that did not already exist.

Two model compounds were used, since extensive quantum mechanical (QM) calculations on a full lignin dimer are computationally expensive. The first model system, methoxybenzene, also known as anisole (see figure 1c), incorporates the basic features of a β-O-4′ link, an ether oxygen bonded to a tertiary and an aromatic carbon. Anisole was used to obtain all parameters involving the ether oxygen atom. The second compound (see figure 1d) is p-hydroxyphenyl (PHP), the simplest lignin unit. PHP was used to obtain all lignin parameters not involving the ether oxygen.

The optimization strategy for the new parameters is summarized below. Equilibrium values for bonds, angles and dihedrals were taken from MP2/6-31G* QM optimized geometries and were not further revised. The van der Waals parameters were taken unaltered from the existing protein [11] and ether [12] CHARMM force fields. Parameterization of the van der Waals parameters for the new atom type, OET, was not deemed necessary since all three ether oxygen types in reference [12] have the same Lennard-Jones parameters. This is a strong indication that the new atom-type in the present work, which is also an ether oxygen, will have the same van der Waals parameters.

Partial atomic charges were optimized by using a supramolecular approach with a model compound (anisole) interacting with one water molecule. The charges on the new atom type OET and its adjacent carbons were optimized by examining the minimum interaction energies and distances between anisole and a water molecule. The convergence criterion for charges was a $< 5\%$ difference between QM and molecular mechanics (MM) energies. The HF/6-31G* energy was computed varying only the inter-molecular distance so as to locate the minimum-energy separation and interaction energy. Anisole and water were fixed at their MP2 optimized and TIP3P [13] geometries, respectively. For comparison between the QM and MM data, the QM energies were scaled by a factor of 1.16 so as to correct for the absence of polarization and attractive Lennard-Jones contributions in the Hartree-Fock calculations, but no correction was made for basis set superposition error [12, 14].

After completing the nonbonded interactions, parameters for dihedral rotations were deduced from QM (MP2/6-31G*) potential energy surfaces. In cases where QM surfaces were well reproduced, the value of the force constant that gave the smallest error was chosen. In the cases where the shapes were not as well reproduced emphasis was placed on the low-energy region < 3 kcal/mol.

The remaining bond and angle parameters were derived by reproducing QM vibrational frequencies and normal modes with the automated frequency matching method (AFMM) [15]. The DFT/B3LYP/SBKJC/6-31G* level of theory was used and a frequency scaling factor of 0.9614 was used to compensate for the use of the harmonic approximation to the potential energy surface [16]. Good starting values of force constants were provided for AFMM by considering similar existing parameters in ethers, phenol and alcohols. Bond and angle optimization was repeated until the merit function (equation 1) reached a value of less than 60 cm^{-1}. The iterative procedure was repeated until all convergence criteria were satisfied.

3. Parameter optimization

3.1. Partial atomic charges

The partial charges were adjusted so as to reproduce minimum distances and interaction energies between anisole and a water molecule (see, for example, [17]). The use of this supramolecular approach ensures that the derived charges take into account effects of polarization. Two geometries were considered in this supramolecular approach, the first (d_0) with water lying on the phenyl plane and the second (d_{120}) with the water hydrogen pointing at the position of the lone pair of the ether oxygen. A list of all final atomic charges is shown in table 1.

In order to mimic the effect of electronic polarizability, which is not explicitly taken into account in additive force fields, atomic charges were purposely overestimated. This leads to an enhanced molecular dipole moment, with the QM gas-phase dipole moment being 1.42 Debyes whereas the MM value is 1.66 Debyes. Table 2 shows the comparison of the MM and QM interaction energies and distances, which were used to optimize the anisole charges. The empirical calculations reproduce well the scaled QM interaction energies, with the error being less than 3%. The empirical model gives distances about ~ 0.3Å shorter than the QM values, a result of intentionally overestimating the gas phase charges to obtain good condensed phase properties. In the previous general force field for ethers a similar behavior was observed with a

Journal of Physics: Conference Series **125** (2008) 012055 doi:10.1088/1742-6596/125/1/012055

Table 1: A list of the anisole atoms with their respective charges. Atom names refer to figure 1c and atoms types follow the CHARMM27 force field, with the new atom type labeled as OET.

Atom Name	Atom Type	Charge
C_α	CT3	-0.060
$H_{\alpha 1}$, $H_{\alpha 2}$, $H_{\alpha 3}$	HA	0.090
O	OET	-0.280
C_1	CA	0.070
C_2, C_3, C_4, C_5, C_6	CA	-0.115
H_2, H_3, H_4, H_5, H_6	HP	0.115

~ 0.3Å difference between QM and MM [12].

A further calculation was performed to ensure that the partial atomic charges of table 1, which were derived using a model compound, can be transferred to lignin. The minimum interaction energies and distances between a whole lignin dimer (G and S units connected with a β-O-4′ linkage), shown in figure 1b, and a TIP3P water molecule were obtained without further refinement of the parameters. This supramolecular complex is labeled as "dimer" in table 2. As in the case of anisole and water, the agreement between the QM and MM interaction energies was excellent, further justifying the use of the charges in Table 1 for the β-O-4′ lignin linkage. Furthermore, the lignin dimer had two methoxy groups at position 3 and 5 of the phenol ring, indicating that the presence of the methoxy group does not alter the charge of the ether oxygen.

3.2. Bonded parameters

3.2.1. Dihedral rotations Dihedral rotations around the β-O-4′ linkage play a significant role in determining the configuration of the lignin macromolecule, and therefore special care was taken to obtain good parameters for the equivalent dihedrals of the model system. These are ω_1: X-C_1-O-C_α, where X refers to wild atom types, and ω_2: C_1-O-C_α-H. The optimization was based on reproducing quantum-chemically obtained adiabatic energy surfaces, where the selected dihedral (ω_1 or ω_2) is held constant while the remaining degrees of freedom are allowed to relax to a constrained energy minimum. An example of such potential energy surfaces is shown in figure 2a, where it can be seen that the MM surfaces closely follow the target QM data.

The remaining dihedral parameters of lignin that do not involve the ether oxygen were deduced from the more complex rotational potential energy profiles of the second model compound, PHP. Four dihedrals were examined: ω_3=C_2-C_1-C_7-X, ω_4=C_1-C_7-O_7-H_{O7}, ω_5=C_1-C_7-C_8-X, and ω_6=X-C_8-C_9-X. All other dihedral parameters were obtained from the existing CHARMM force field. To accurately model a specific dihedral potential (e.g., ω_3), it was found

Table 2: Minimum scaled interaction energies (kcal/mol) and distances (Å) between water:anisole (d_0 and d_{120}) and water:lignin-dimer (dimer).

Geometry	Interaction Energy		Interaction Distance	
	QM	MM	QM	MM
d_0	-4.01	-3.96	2.15	1.82
d_{120}	-3.18	-3.09	2.16	1.87
dimer	-3.93	-4.02	2.10	1.81

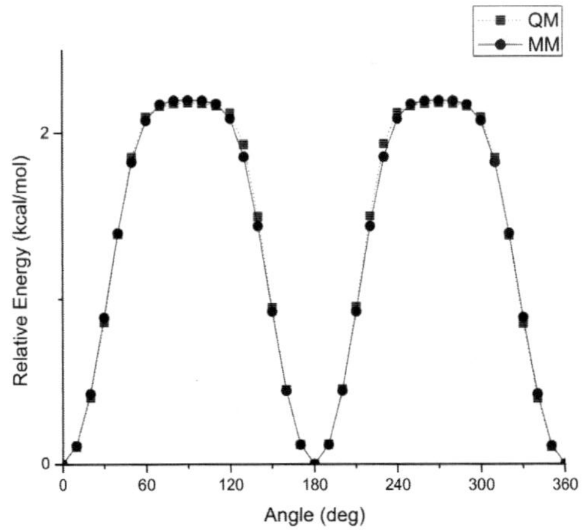

(a) ω_1

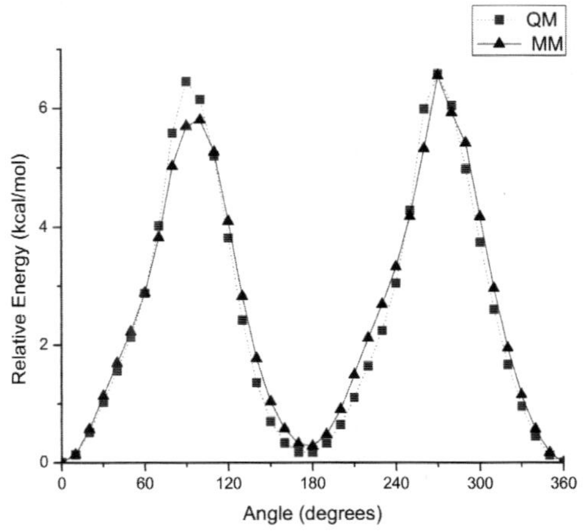

(b) ω_3

Figure 2: Potential energy for rotation around the *(a)* ω_1= C_2-C_1-O-C_α dihedral of anisole *(b)* ω_3=C_2-C_1-C_7-X dihedral of PHP.

to be necessary to constrain the remaining three (ω_4, ω_5 and ω_6) to their equilibrium values.

An example of the resulting energy surfaces is shown in figure 2b. Although the agreement between the QM and MM data is not perfect, the rather complex shapes are reproduced satisfactorily. During parameterization, special care was taken to reproduce as accurately as possible the low energy ($\leq$ 3 kcal/mol) regions since it is these regions that will be thermally most-frequently sampled.

3.2.2. Bond and angle vibrations The remaining bonded parameters (bonds and angles) were optimized to reproduce vibrational frequencies and eigenvector projections derived from QM calculations. For this the automated frequency matching method [15] (AFMM) was employed, which optimizes the MM parameter set until the best fit with the QM reference set is obtained. AFMM requires both the eigenfrequencies and eigenvectors of the MM set to match with the QM data. This is an important aspect of the method, since it avoids incorrect mode matching and thus misleading reproduction of vibrational frequencies. The merit function is given by

$$\sigma^2 = \frac{\sum_i^{3N-6} (\omega_i^{qm} - \omega_i^{mm})^2}{3N-6}, \tag{1}$$

where ω^{qm} and ω^{mm} refer to the vibrational frequencies obtained with QM and the MM methods respectively and N is the number of atoms of the molecule.

The resulting plots of ω^{mm} against ω^{qm} for ansisole and PHP are shown in figure 3 and indicate satisfactory matching between MM and QM modes. After optimization the root mean square deviation from the reference set was found to be $\sigma = 51.6\ cm^{-1}$ for anisole and $\sigma = 55.6\ cm^{-1}$ for PHP, similar in range to previous parameterizations involving AFMM, which quote values of $\sigma = 47 - 94\ cm^{-1}$ [19, 20]. A list of the bonded parameters of lignin can be found in [18].

3.3. Force field validation

In the final part of this work, the parameter set was tested without further adjustment against a condensed phase experimental property of lignin that was not used during the parameterization.

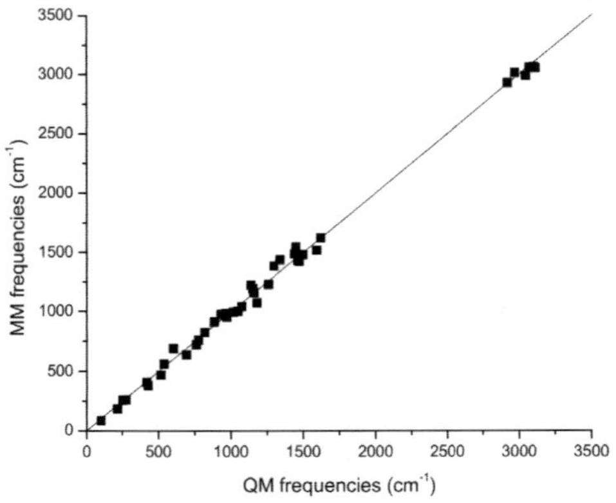

(a) anisole

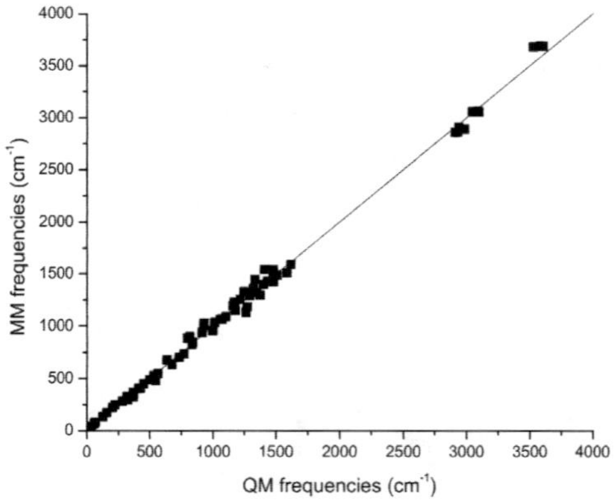

(b) PHP

Figure 3: Vibrational frequencies of model compounds anisole and PHP. The plotted line shows the ideal fit between QM and MM data

Because of the highly heterogeneous structure of lignin, the most appropriate experimental data to use is the crystal structure of a lignin-subunit dimer, erythro-2-(2,6-Dimethoxy-4-methylphenoxy)-1-(4-hydroxy-3,5-dimethoxyphenyl)propane-1,3-diol (EPD) [21]. The single crystal -ray diffraction study revealed a triclinic $P\bar{1}$ structure whose unit cell dimensions are listed in table 3. In order to mimic as closely as possible the conditions under which the experiment was run, the MD simulation was performed for 64 unit cells using periodic boundary conditions while keeping the temperature and pressure fixed at their experimental values.

The MD unit cell dimensions were allowed to vary during the simulation and their time averages are shown in table 3. The cell dimensions were found to be close to the experimental values and the system remained triclinic. The unit cell underwent a moderate expansion, with a 5% increase in volume. After aligning the MD coordinates with the experimental structure, the root mean square deviation between the experimental and calculated structure was found to be 0.173 ± 0.033 Å.

It is also of particular importance that the current force field models well the β-O-4′ linkage that plays an important role in the conformation of lignin. For this reason the time average of the two dihedrals, d_1 and d_2, that define the β-O-4′ linkage was compared with the experimental crystal values. The two dihedrals are (numbering scheme in Fig. 1d): d_1=C_5-

Table 3: Unit cell properties of small-molecule-dimer from crystal structure and MD simulation.

Cell Dimension	Experiment	MD
A (Å)	8.69	8.73 ± 0.02
B (Å)	8.90	8.93 ± 0.01
C (Å)	13.11	13.68 ± 0.03
α (deg)	73.85	74.48 ± 0.05
β (deg)	86.15	86.30 ± 0.01
γ (deg)	83.06	83.06 ± 0.02
cell volume (Å^3)	966	1020

C_4-O-C_8' = $77.9^o \pm 6.3$, compared to the experimental value of 80.0^o and d_2= C_4-O-C_8'-C_7' = $-148.5^o \pm 5.5$, compared to the experimental value of -152.8^o. As with previous results, the simulation results are in accord with experiment.

4. Discussion

This work presents a molecular mechanics force field for lignin that is compatible with the CHARMM potential energy function. The parameterization was based on reproducing quantum-mechanically derived target data. Special care was taken to correctly describe the most common lignin linkage: the β-O-4$'$ bond. The partial atomic charge of the oxygen and carbon atoms participating in the linkage were derived by examining interactions between a lignin fragment model compound and a water molecule. Dihedral parameters were obtained by reproducing QM potential energy profiles, with emphasis placed on reproducing accurately the thermally sampled low energy regions. The remaining bond and angle parameters were derived using the AFMM method. In order to test the validity of the force field a simulation of a lignin-dimer crystal was performed. The overall good agreement between the structural properties of the MD run and the experiment provide confidence that the force field can be used in simulation of biomass.

The accurate computer simulation of lignin in lignocellulose will present significant challenges. Unlike many biological macromolecules that have been studied with molecular simulation, both the chemical and three-dimensional structures of lignin are relatively poorly researched. However, the present force field provides a basis for constructing molecular models of lignin systems, and, in combination with a range of biophysical measurements, significant progress in determining structures of lignocellulosic biomass can be expected in the near future.

Acknowledgments

We acknowledge funding from the U.S.A. Department of Energy (DOE) Bioenergy Science Center and DOE Field Work Proposal ERKP704.

References

[1] Cosgrove D J 2005 *Nat. Rev. Mol. Cell Biol.* **6** 850–861
[2] Himmel M E, Ding S Y, Johnson D K, Adney W S, Nimlos M R, Brady J W and Foust T D 2007 *Science* **315** 804–807
[3] Zhang Y H P, Ding S Y, Mielenz J R, Cui J B, Elander R T, Laser M, Himmel M E, McMillan J R and Lynd L R 2007 *Biotechnol. Bioeng.* **97** 214–223
[4] Fan L T, Lee Y H and Beardmore D R 1981 *Biotechnol. Bioeng.* **23** 419–424
[5] Mooney C A, Mansfield S D, Touhy M G and Saddler J N 1998 *Bioresour. Technol.* **64** 113–119
[6] Davin L B and Lewis N G 2005 *Curr. Opin. Biotechnol.* **16** 407–415
[7] Besombes S and Mazeau K 2005 *Plant Physiol. Biochem* **43** 299–308
[8] Besombes S and Mazeau K 2005 *Plant Physiol. Biochem* **43** 277–286
[9] Kuttel M, Brady J W and Naidoo K J 2002 *J. Comp. Chem.* **23** 1236–1243
[10] Brooks B R, Bruccoleri R E, Olafson B D, States D J, Swaminathan S and Karplus M 1983 *J. Comp. Chem.* **4** 187–217
[11] MacKerell A D *et al* 1998 *J. Phys. Chem. B* **102** 3586–3616
[12] Vorobyov I, Anisimov V M, Greene S, Venable R M, Moser A, Pastor R W and MacKerell A D 2007 *J. Chem. Theory Comput.* **3** 1120–1133
[13] Jorgensen W L, Chandrasekhar J, Madura J D, Impey R W and Klein M L 1983 *J. Chem. Phys.* **79** 926–935
[14] Mackerell A D and Karplus M 1991 *J. Phys. Chem.* **95** 10559–10560
[15] Vaiana A C, Cournia Z, Costescu I B and Smith J C 2005 *Comput. Phys. Commun.* **167** 34–42
[16] Scott A P and Radom L 1996 *J. Phys. Chem.* **100** 16502–16513
[17] Chen I J, Yin D X and MacKerell A D 2002 *J. Comp. Chem.* **23** 199–213
[18] Petridis L and Smith J 2008 *J. Comp. Chem.* (to appear)
[19] Cournia Z, Smith J C and Ullmann G M 2005 *J. Comp. Chem.* **26** 1383–1399
[20] Vaiana A C, Schulz A, Wolfrum J, Sauer M and Smith J C 2003 *J. Comp. Chem.* **24** 632–639
[21] Langer V, Lundquist K and Miksche G E 2005 *Acta Crystallogr. Sect.* **E**: *Struct. Rep. Online* **61** O1001–O1003

Journal of Physics: Conference Series **125** (2008) 012056 doi:10.1088/1742-6596/125/1/012056

Simulation and modeling of synuclein-based "protofibril" structures: as a means of understanding the molecular basis of Parkinson's disease

I F Tsigelny, Y Sharikov, M A Miller and E Masliah

E-mail: itsigeln@ucsd.edu

Abstract. The protein α-synuclein (aS) is a 140 amino acid cytosolic protein of unknown function expressed abundantly in neurons and is associated with both environmentally induced and inherited forms of Parkinson's disease, the neurological disorder caused by progressive death of dopaminergic neurons in the substantia nigra. Here we demonstrate that aS aggregation leads to protofibril formation that accompanies aS binding to membranes. The specific surfaces of aS that bind to membranes are likely developed during the folding maturation process, where the 3D combinations of amino acids that are bound to these surfaces are very specific. Using molecular modeling and simulation technique, we elucidate the folding pathway of aS into protofibril structures and its configurations that bind to membranes. These membrane-binding surfaces can be targeted with pharmaceutical intervention for dissolution.

1. Introduction

The protein α-synuclein (aS) is a 140 amino acid cytosolic protein of unknown function expressed abundantly in neurons, and is associated with both environmentally induced and inherited forms of Parkinson's disease, the neurological disorder caused by progressive death of dopaminergic neurons in the *substantia nigra*. Evidence for the involvement of aS in disease progression comes from the association of Parkinsons with α -synuclein overexpression (caused by genomic triplication of the encoding genes) and with inherited forms that with three specific mutations in the aS gene (A53T, A30P, and E46K). Both cases are characterized by development of aggregated protein inclusions (Lewy bodies) in the cytosol and on the membrane of surviving neurons. The connection between development of Lewy bodies and dopaminergic neuron apoptosis is still unclear. For a number of years, Parkinson's disease progression was thought to reflect the impact of massive fibril accumulation, which eventually prevents normal cell functions. Recent evidence suggests that a small intermediate "protofibril" structure may actually be the toxic species (see [1, 2] and references in them). It is therefore important to examine the mechanism of aS aggregation and binding of these aggregates to the cell membrane, as a means of understanding the biochemical and biophysical mechanisms leading to onset of the disease.

Here we demonstrate that aS aggregation leads to protofibril formation that accompanies aS binding to membranes. The specific surfaces of aS that bind to membranes are likely developed during the folding maturation process, where the 3D combinations of amino acids that are bound to these surfaces are very specific. Using molecular modeling and simulation technique we elucidate the folding pathway of aS into protofibril structures and its configurations that bind to membranes. These membrane-binding surfaces can be targeted with pharmaceutical intervention for dissolution.

1.1. Unstructured proteins dynamics and membrane interactions

Well-known proteins and peptides such as aS, beta-amyloid, and prion that are responsible for a number of neurodegenerative diseases have one common feature – all of them are naturally unstructured proteins that do not have any stable conformation. To understand their pathology, one must develop tools to assess the behavior of possible conformations that molecule occupies in solution, as well as the potential for interaction of each conformation with the surface of the cell membrane.

The experiments we conducted take the view that these proteins exist in solution, while changing conformations. From time to time, the unstructured protein makes contact with the surface of the membrane, in whatever conformation it currently occupies. The protein will remain in contact with the membrane for a time that is defined by the energy of interaction between the membrane and the protein. If the energy of interaction is weak, the protein molecule will dispatch from the surface of the membrane back to the solution after a relatively short time. This cycle of binding and release will continue, and the protein will sample many conformations as the process goes on. If a protein has a favorable membrane binding conformation, the protein will remain on the surface of the membrane longer. Ultimately, if the time of residence on the membrane is long enough, the protein may start penetration into the outer layer of the membrane. In this period of time, the membrane may also stabilize the conformation of protein, by providing specific regions of hydrophobic and hydrophilic structure.

In the altered environment near the membrane surface, additional molecules of the same protein may come in contact with the first molecule located on the membrane and create dimers. Depending on the energetics of this interaction, these can propagate into higher-ordered multimers with time. The process is complex and does not have clearly defined steps, boundaries, or even initial molecular conformations to work with. Solving this problem is intractable (with current computational resources) if one wishes to conduct molecular dynamics simulations of all possible conformations and all possible interactions. Accordingly, we chose to address this problem by a sparse-sampling methodology that relies on representative "snapshots" of the system at various points in time and in various time regimes, rather than on complete end-to-end simulations of the entire process.

2. Modeling and Simulation

1.2. Computational approach to unstructured proteins

For study of protein aggregation (oligomerization) on the membrane with possible organization of pores, as well as its subsequent penetration into the membrane, we would require the following methods:

- Prediction of membrane-contacting surfaces of proteins
- Molecular dynamics simulation of protein molecules in water and on the membrane
- Molecular docking of the protein molecules on the membrane
- Molecular dynamics simulation of protein oligomers in the membrane

We have created a set of programs and workflows to address this challenge.

2.2. Membrane contacting regions recognition

Figure 1 shows the first block program, called MAPAS [3], to assess how strongly the protein interacts with the membrane (membrane-contacting potential). It contains scoring functions developed using methods of modeling of amino-acids disengagement and Poisson-Boltzmann free-energy calculations. The MAPAS program initially calculates all possible membrane-contacting surfaces of a given protein and then calculates the membrane philicity score and predicts the best membrane-contacting surface. This program was tested on a number of known membrane-contacting proteins from the PDB and showes good correspondence to the experimental data.

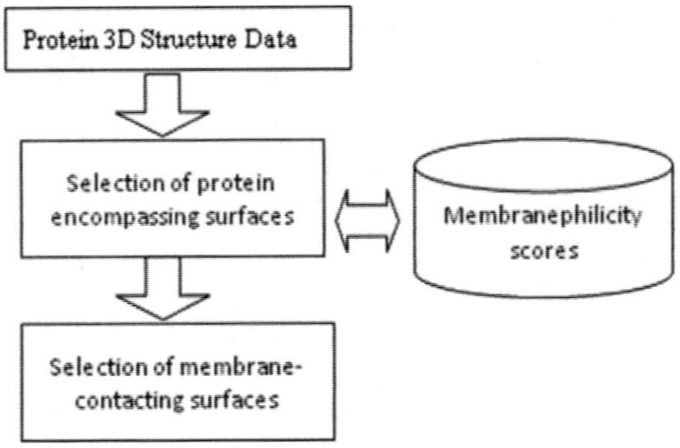

Figure 1. Algorithm of membrane-contacting surfaces selection.

2.3. Initial molecular dynamics in water environment and shape clustering

A set of molecular dynamics (MD) runs for unstructured proteins is a necessary step to define possible conformations that are more stable than others in solution. The MD simulations are conducted on the IBM Blue Gene computer and take quite a long time, but these are a necessary step in the entire modeling mentioned above. The calculations are described further in [4]. Because aS has no strongly favored low-energy conformers, we had to create a strategy for selection of possible "temporarily" stable conformers. We used a shape-clustering method to select a representative set of aS conformations obtained during MD simulations (we will refer to them as "conformers") during MD in water solution, and we evaluated their tendencies to associate with a phospholipid bilayer. To the input of the clustering program we submit all conformers from MD runs. The clustering program selects most populated clusters (figure 2). The core representatives of each cluster are used for farther docking on the membrane.

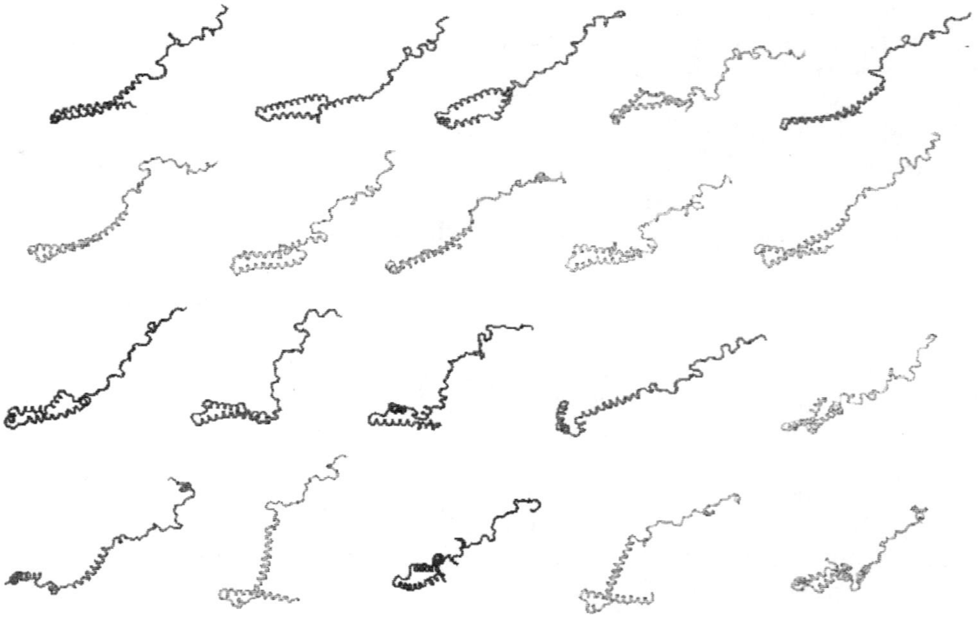

Figure 2. Most populated 20 clusters.

This way we select most common conformations of the protein (in this case aS). From a statistical point of view, these have the highest probability of existence through the time of the MD run. In the case of unstructured proteins we cannot use more stringent criteria because there are very small differences between the energies of these conformers.

2.4. Selection of conformers for further molecular docking on the membrane

We next selected from among the statistically most represented conformers to identify those conformers that are most likely to interact strongly with a phospholipid bilayer. These properties were calculated using the program MAPAS. We then selected the conformers with the highest membrane-contacting scores to examine for docking and possible formation of higher level aggregates at the membrane surface. Figure 3 shows the representative conformer cluster with the highest membrane contacting score.

Our simulation experiments show that aS dimers at the membrane can be divided into two groups. The first group adopts conformations at the membrane surface are energetically favorable for further propagation into ringlike structures (we call these "propagating dimers"). Propagating dimers support the oligomerization with subsequent protein molecules on the membrane. This process continues until a ringlike oligomer structure is organized on the membrane. We want to emphasize that at all steps of oligomerization the process can be stopped by some specific "nonpropagating" docking. In this situation no ringlike structure would be created. The second group does not adopt conformations that present a docking surface with sufficiently low energy of interaction to bind a third aS molecule. These forms do not proceed to form ringlike structures (we call these "nonpropagating dimers").

We note that docking of unstructured proteins on a lipid bilayer is completely different from docking simulations conducted in solution. First, the bilayer is not homogeneous but has regions of very high and very low polarity that are in close proximity. Second, contact between the bilayer and the protein excludes a significant part of the protein surface from the docking process. A key postulate of our work is that the outcome of a docking event between unstructured proteins is radically different when it occurs simultaneously with a membrane surface than when it occurs in solution. We believe that only through the membrane interaction can "propagating dimers" of aS form. We have already seen that docking to a bilayer excludes a significant part of protein surface from docking with other proteins.

Figure 3. Conformation of aS after 4.5 ns molecular dynamics (green) in the best membrane-contacting position selected by MAPAS (pink – membrane). This conformation had the best score of membranephilicity and has been selected for farther docking experiments on the membrane.

The entire surface facing the membrane on figure 3 is excluded from further docking. It is exclusion of this region that makes possible the formation of propagating dimers, by excluding a region that has a strong tendency to form nonpropagating dimers.

Journal of Physics: Conference Series **125** (2008) 012056 doi:10.1088/1742-6596/125/1/012056

Figure 4 illustrates how the membrane-contacting surfaces of aS are changed during molecular dynamics. Black ellipses enclose the parts of the molecule that contact the membrane. Evolution of these surfaces is conducted through MD runs. Our program assesses each of the snapshots to select the conformers for the largest membrane-contacting surface (that is scored not by the surface of contact but by the surface of contact with the membrane by the specified amino acids). The important point of the proposed mechanism is that the protein has to have sufficient "membranophilic" surface to exist on the membrane for enough time to contact other protein molecules to create a "core" propagating dimer that would have enough energy of interaction with the membrane to start a process of membrane penetration. Oligomerization will continue through the binding of subsequent molecules to this core dimer.

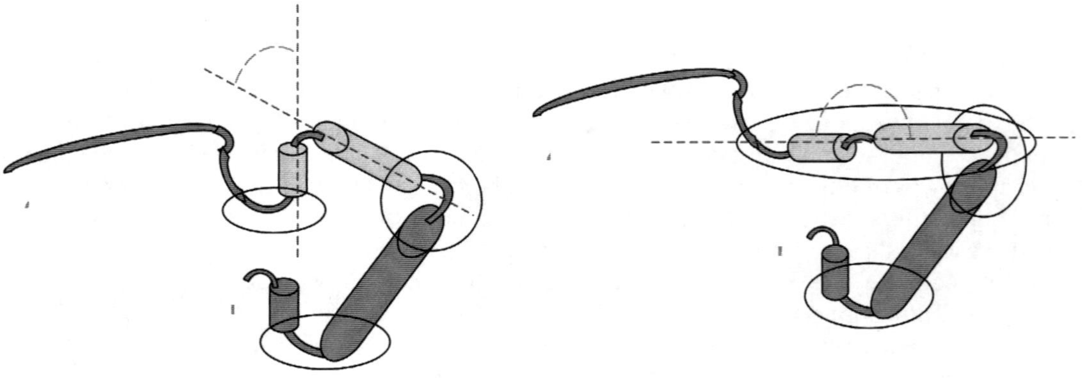

Figure 4. Evolution of aS membrane-contacting surfaces during molecular dynamics.

2.3. Molecular docking and protein oligomerization

A simple schematic representation of the stages of the organization of a pore in the membrane is shown in figure 5. In the initial stages (A) a protein molecule contacts the membrane and undergoes some conformational changes (B,C). These changes are simulated using MD of protein-membrane system. The resulting conformers are used for docking that can be done using various programs, for example, HEX, DOT, the Docking module of Insight (Accelrys), and (D, E). The final simulation is conducted with the ringlike oligomer in the membrane. This system is located in the water box, and unrestrained MD is conducted.

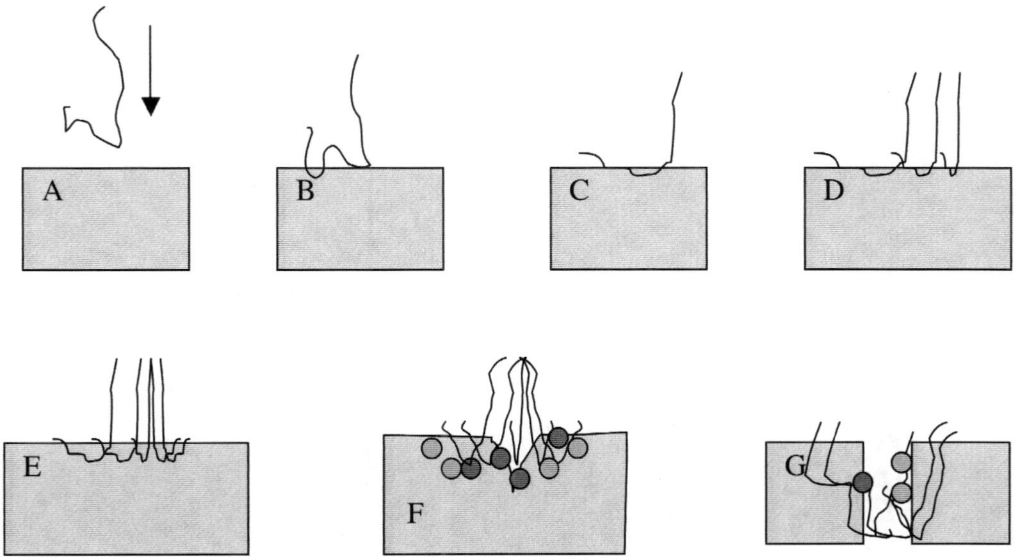

Figure 5. Scheme of pore organization in the membrane by aS molecules: A. Initial approach of aS, B. Contact of aS with the membrane, C. Conformational changes of aS due to this contact, D. Docking of the next aS molecules in the upper layer of the membrane, F. Interaction of aS molecules with the specific regions of phospholipids (red) and cholesterol (green), start of the oligomer penetration to the membrane with the damage of the phospholipids layer in the middle of the ringlike structure, G. Organization of the phospholipids elimination from this region with the pore organization.

After organization into a ringlike oligomer on the membrane, simulations suggest that the aggregated protein starts to embed more deeply into the membrane, eventually penetrating to the other side of the membrane. This provides a clear mechanism for the toxicity of aS in small aggregates (protofibrils) noted above. The ring structures created by aS self-assembly in simulations possess a central opening that would create a pore in the membrane. This could quickly lead to collapse of membrane ion gradients and cell death. For example, the concentration of calcium outside the cell is a much higher than inside the cell, so this concentration difference-driven influx will be significant. Calcium can start the caspases-based cell death cascade that leads to the neuron death (a main factor in Parkinson's disease). An important consideration is whether the process of oligomerization can be separated at very earlier stages. One process may go toward the formation of ring-like structures (protofibrils) and the other toward the large fibrils. The last process may actually be less harmful to the cell than the first one.

From the point of practical application of the abovementioned theoretical studies, it is very important to understand the mechanism of oligomerization and penetration of the oligomer through the membrane. At all stages (schematically shown on the figure 5) the molecular contacts that are responsible for continuation of the pore organization process can be elucidated.

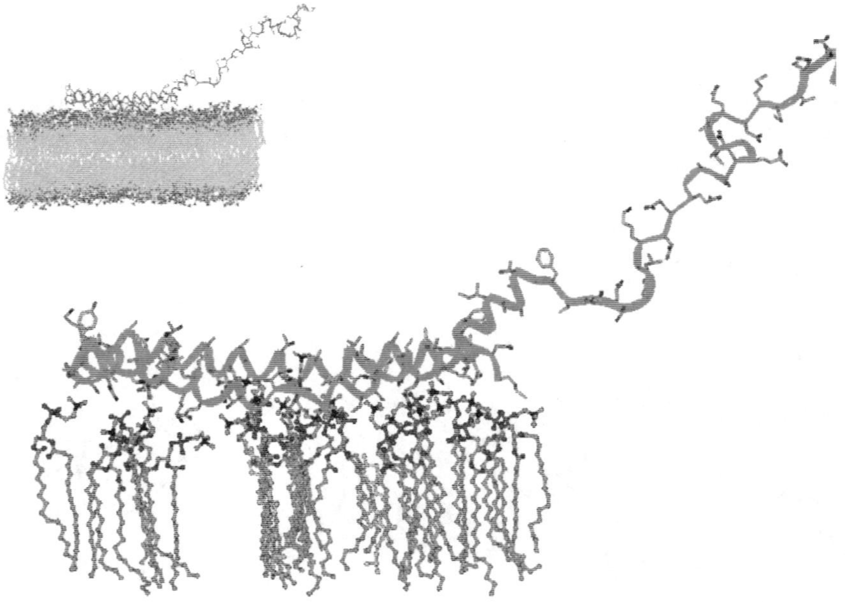

Figure 6. Phospholipids and cholesterol molecules that have direct contact with the aS molecule on the initial stages of penetration (selected from the entire membrane molecular set). The inset shows the entire picture of initial membrane-protein interactions.

Our studies of the aS interactions identified specific putative contacts for aS aggregation/self assembly at the membrane surface. Preliminary results show that interfering with these putative contacts can decrease the aggregation of aS. The results are already being used for drug design.

References

[1] Moore D J, West A B, Dawson W L and Dawson T M 2005 Molecular pathophysiology of Parkinson disease *Annu. Rev. Neurosci.* **28:**55-84

[2] Bertoncini C W, Jung Y S, Fernandez C O, Hoyer W, Griesinger C, Jovin T M and Zweckstetter M 2005 Release of long-range tertiary interactions potentiates aggregation of natively unstructured {alpha}-synuclein *Proc. Natl. Acad. Sci. USA* **102:**1430-1435

[3] Sharikov Y, Walker R C, Greenberg J, Kouznetsova V, Nigam S, Miller M A, Masliah E and Tsigelny I 2008 MAPAS: a tool for predicting membrane-contacting protein surfaces *Nature|Methods* **5:**119

[4] Tsigelny I F, *et al.* 2007 Dynamics of alpha-synuclein aggregation and inhibition of pore-like oligomer development by beta-synuclein *FEBS J.* **274:**1862-1877

Quantum Monte Carlo algorithms for electronic structure at the petascale; the endstation project

Kenneth P. Esler[1], Jeongnim Kim[1,2], David M. Ceperley[1,2,3]
National Center for Supercomputing Applications[1]
Materials Computation Center[2];Department of Physics[3]
University of Illinois at Urbana-Champaign, Urbana, IL, USA

Wirawan Purwanto, Eric J. Walter, Henry Krakauer, Shiwei Zhang
Department of Physics, College of William and Mary, Williamsburg, VA, USA

Paul R. C. Kent
Center for Nanophase Materials Sciences, Oak Ridge National Laboratory, Oak Ridge, Tennessee, USA

Richard G. Hennig[4], Cyrus Umrigar[5]
Department of Material Science[4]; Department of Physics[5]
Cornell University, Ithaca, NY, USA

Michal Bajdich, Jindřich Kolorenč, Lubos Mitas
Department of Physics, North Carolina State University, Raleigh, NC, USA

Ashok Srinivasan
Department of Computer Science, Florida State University, Tallahassee, FL, USA

Abstract. Over the past two decades, continuum quantum Monte Carlo (QMC) has proved to be an invaluable tool for predicting of the properties of matter from fundamental principles. By solving the Schrödinger equation through a stochastic projection, it achieves the greatest accuracy and reliability of methods available for physical systems containing more than a few quantum particles. QMC enjoys scaling favorable to quantum chemical methods, with a computational effort which grows with the second or third power of system size. This accuracy and scalability has enabled scientific discovery across a broad spectrum of disciplines. The current methods perform very efficiently at the terascale. The quantum Monte Carlo Endstation project is a collaborative effort among researchers in the field to develop a new generation of algorithms, and their efficient implementations, which will take advantage of the upcoming petaflop architectures. Some aspects of these developments are discussed here. These tools will expand the accuracy, efficiency and range of QMC applicability and enable us to tackle challenges which are currently out of reach. The methods will be applied to several important problems including electronic and structural properties of water, transition metal oxides, nanosystems and ultracold atoms.

1. Computational quantum mechanics and quantum Monte Carlo

"The underlying physical laws necessary for a large part of physics and the whole of chemistry are thus completely known, and the difficulty is only that the exact applications of these laws lead to equations much too complicated to be soluble." – Paul Dirac, 1929

This quotation, colloquially dubbed *Dirac's Challenge*, reflects the understanding that, at least in principle, the exact solution of the Dirac equation (or its non-relativistic counterpart, the Schrödinger equation) would yield all the information necessary to predict the behavior of matter at normal energy scales, but finding the solution by traditional pencil and paper methods is very difficult.

For a system containing N electrons, the Schrödinger equation is a partial differential equation in $3N$ dimensions. Because of the high dimensionality of the problem, exact solutions to these equations have thus far been found only simple cases such as one dimensional systems, so that Dirac's Challenge is still unmet. Nonetheless, much of theoretical physics and chemistry has been devoted to finding ever more accurate *approximate* solutions to these equations governing the behavior of matter at the quantum level. Through the decades, the coupling of more advanced methods with the exponential increase in computing power has enabled us to progress from qualitatively correct models based on empirically determined parameters to truly *ab initio* calculations, i.e. those starting *from the beginning*. These calculations, taking the Challenge head on, have as their input only the atomic numbers of system's constituent atoms.

Of these *ab initio* methods, the mostly widely known are based on *density functional theory* (DFT). The basis of these methods is an approximate mapping of the $3N$-dimensional equation onto a coupled set of N three-dimensional equations. In this mapping, the details of the correlated motion of individual electrons are approximated in an average "mean-field" sense. DFT has enjoyed a tremendous degree of success, and continues to do so, with new insights being gained every day.

Nevertheless, there are many outstanding problems for which the level of accuracy provided by DFT has proved insufficient. As an example, DFT predicts a number of known insulators, including important transition metal oxides, to be metallic. In these cases, methods which treat the many-body electron more directly are required. The most accurate method available, known as *full configuration interaction*, aims at a near-exact solution to the Schrödinger equation. Unfortunately, the required computation scales exponentially with N, the number of electrons in the system, limiting its application to atoms and small molecules. Other methods from quantum chemistry address larger systems by truncating the many-particle basis intelligently using chemical insight. These provide highly accurate approximate solutions, but are still very limited in system size since the cost scales with a high power of N.

Quantum Monte Carlo (QMC) methods achieve very high accuracy by treating the Schrödinger equation in the original $3N$-dimensional space using a stochastic sampling of the many-body wave function. Since the cost of Monte Carlo sampling is relatively insensitive to dimensionality, large systems can be simulated at reasonable computational expense. In current implementations, the cost scales as N^2 to N^3, depending on the quantities of interest, although order-N methods are currently under development. Since electron correlation is treated explicitly, QMC can achieve accuracy similar to the best quantum chemistry methods, but can be applied to much larger systems, including nanoclusters, macromolecules, and crystals. For these reasons, it is generally accorded the distinction of being the most accurate method available for systems of more than a few atoms. Furthermore, the methods are extremely general and have been successfully applied to molecules, fluids, perfect crystals, and material defects. Finally, the methods afford several avenues for parallelization which, when combined, will allow very efficient operation at the petascale.

Though still a rapidly evolving method, QMC has already contributed significant advancements across a range of disciplines. The Ceperley-Alder calculation of the correlation energy of the homogeneous electron gas [1] has become the basis of all modern functionals for DFT methods. Calculations of the phase diagram and properties of hydrogen under high temperature and compression have provided invaluable data necessary for models of Jovian planets [2, 3]. QMC was able to predict the correct lowest-energy structures for silicon nanoclusters, where DFT had failed [4]. QMC calculations of the pressure-volume curve and frequencies of lattice vibrations for materials under ultra-high compression have been shown to produce more accurate results than DFT [5].

Looking over the last 50 years, one of the remarkable developments has been the growth of the molecular dynamics (MD) and Monte Carlo simulation of atomic and molecular systems. This started in the 50's with the pioneering applications of Metropolis, Rosenbluth, Teller [6], Wood and Alder to the simplest possible hard sphere model of atoms. In the 60's, Verlet, Rahman, Berne and others developed the methodology to treat the more realistic Lennard-Jones systems, with further developments to treat more complicated systems using semi-empirical forces. A major leap forward occurred in 1985 with the discovery by Car and Parrinello [7], that a CRAY-1 class computer could use DFT calculated forces to perform a MD simulation of bulk liquid silicon. Clearly, improved computer capability can be used either to increase system size and duration *or* accuracy of the interatomic forces. The processing power is now sufficient to replace the DFT evaluation of energy and forces with more accurate QMC calculations. This will allow for simulation of systems undergoing chemical changes, such as during combustion, with a much higher degree of reliability, and validate chemical models that can be used for larger scale calculations.

The significant science advances through the use of QMC methods would not be possible without the phenomenal advance in computer power that has occurred over the past decades. The breakthrough QMC simulations have been enabled by the state-of-art, high-performance computers of every generation: hard-core bosons on a CDC 6600 [8]; ground state of solid hydrogen at high pressures on CRAY XMP and CYBER 205 [2]; electronic and structure properties of carbon/silicon clusters, Grossman-Mitas on HP 9000/715 cluster and Cray Y-MP; MLW, Williamson-Hood-Grossman [9] and Mitas-Grossman continuous DMC-MD [10] on the LLNL teraflop cluster; Coupled Electron-Ion Monte Carlo simulations of dense hydrogen, Ceperley et al. on Linux clusters and most recently on Cray XT3/4 at ORNL. Until recently, QMC methods have been able to exploit the steady increase in clock rates and enhancement of performance enabled by advances in CMOS VLSI technology for faster and bigger simulations without any significant modification in the algorithms and their implementations. However, over the last few years, clock rates have been relatively flat due to problems with heat dissipation and the structure of the computational components has undergone significant changes, including multi-core processors and powerful accelerators such as GPUs (Graphics Processing Units) for general-purpose scientific computing. These hardware trends have significant ramifications on an application performance.

The Endstation Project for quantum Monte Carlo aims to enable ground-breaking science with QMC methods at the petascale systems that have become available at the DOE leadership computing facilities. We are developing new algorithms and software solutions that can take advantage of the current hardware trends based multi-core processors. In this paper, we introduce the various methods grouped under the umbrella term of quantum Monte Carlo, and describe the QMC algorithms that have been effectively employed on current large-scale HPC systems. Then, we address the challenges posed by the recent hardware trends and propose solutions for the QMC to run efficiently at the petascale and beyond.

2. Quantum Monte Carlo (QMC) Methods

In quantum mechanics, all physically observable quantities for a system containing N electrons can be computed from the $3N$-dimensional *wave function*, $\psi(\mathbf{r}_1, \ldots, \mathbf{r}_N)$. The exact wave function for the system satisfies the time-independent Schrödinger equation,

$$\hat{\mathcal{H}}\psi = E_0\psi, \tag{1}$$

where E_0 is the ground-state energy for the system and $\hat{\mathcal{H}}$ is the Hamiltonian, or total energy operator, given by

$$\hat{\mathcal{H}} = \sum_{i=1}^{N}\left[-\frac{1}{2}\nabla_i^2 + V_{\text{ext}}(\mathbf{r}_i)\right] + \sum_{i<j}\frac{1}{|\mathbf{r}_i - \mathbf{r}_j|}, \tag{2}$$

Where $V_{\text{ext}}(\mathbf{r}_i)$ is the *external* potential generated by the nuclei, applied electric fields, etc. For any approximate, or *trial* wave function, $\psi_T(\mathbf{r}_i)$, we compute an energy as the expected value of $\hat{\mathcal{H}}$,

$$E_T = \frac{\int d^{3N}\mathbf{R}\ \psi_T^*(\mathbf{R})\hat{\mathcal{H}}\psi_T(\mathbf{R})}{\int d^{3N}\mathbf{R}\ |\psi_T(\mathbf{R})|^2}, \tag{3}$$

where $\mathbf{R}$ is a $3N$-dimensional vector representing the positions of the N electrons. It can be shown that E_T satisfies a *variational principle* which guarantees that $E_T \geq E_0$ for all ψ_T. This implies that the lower the energy a given ψ_T has, the closer it is to the exact solution of the Schrödinger equation. Several many-body methods, including QMC and quantum chemistry methods, seek to find these low energy solutions, but differ fundamentally in how they evaluate (3).

2.1. Variational Monte Carlo

In variational Monte Carlo (VMC), the integrand in (3) is rewritten as a product of a probability density, $\pi(\mathbf{R}) \equiv |\psi_T(\mathbf{R})|^2$, and the *local energy*, $E_L(\mathbf{R}) \equiv [\hat{\mathcal{H}}\psi_T(\mathbf{R})]/\psi_T(\mathbf{R})$. Metropolis Monte Carlo is then used to generate a large number,N_{MC}, of sample points, $\{\mathbf{R_i}\}$ distributed according to $\pi(\mathbf{R})$. The trial energy, E_T, may then be estimated by

$$E_T \approx \frac{\sum_i^{N_{\text{MC}}} E_L(\mathbf{R}_i)}{N_{\text{MC}}}. \tag{4}$$

This estimate becomes exact in the limit of $N_{\text{MC}} \to \infty$, with a statistical error proportional to $N_{\text{MC}}^{-\frac{1}{2}}$.

In VMC, the trial function is written in an analytic form with a number of optimizable parameters, $\{\alpha_n\}$. These parameters can then be adjusted to minimize E_T, yielding the best wave function possible given the analytic form. The direct evaluation of many-dimensional integrals by stochastic sampling enables us to employ highly accurate variational/trial wave functions which can capture crucial many-body effects in an efficient manner. In contrast, quantum chemistry methods represent ψ_T as a linear combination of many determinants, for which the integral in (3) can be evaluated analytically. This choice leads to the difficulties in scaling those methods to large systems mentioned above.

The trial wave functions most commonly used in quantum Monte Carlo applications to problems in electronic structure are written in the Slater-Jastrow form, $\Psi_T(\mathbf{R}) = \Psi_A(\mathbf{R})\exp\left[J(\{\alpha_n\}, \mathbf{R})\right]$. The antisymmetric part Ψ_A is usually given by a determinant of a matrix whose elements, A_{ij}, are given by evaluating orbital $\phi_j(\mathbf{r})$ at the position of the ith electron. The orbitals are real or complex functions of a three-dimensional variable, and are expanded in a linear basis. For systems up to about a thousand electrons, the evaluation of the orbitals incurs the dominant computation cost in a typical QMC simulation. For larger systems, the N^3 scaling of the determinant evaluation becomes dominant.

Electron correlation is included explicitly in the wave function through a parametrized Jastrow correlation factor, $J(\{\alpha_n\}, \mathbf{R})$. The factor includes several terms, the most important of which prevent electrons from coming close too each other. This is critical since the electrostatic repulsion between electrons would make such a configuration energetically unfavorable. In actual calculations these parameters are optimized to obtain the wave function which yields the lowest energy. The form used for the Jastrow factor allows the dominant correlation effects to be captured with relatively few parameters. For example, comparison of wave functions for C_{20} clusters showed that 20 variational parameters in Slater-Jastrow VMC wave function achieved the same accuracy as $\sim 10^6$ variational parameters in a traditional quantum chemistry approach.

2.2. *Diffusion Monte Carlo*

Once a trial wave function is optimized within VMC, it is further improved through a stochastic projection process known as diffusion Monte Carlo (DMC). It is straightforward to show that for $\tau \to \infty$ the operator $\exp(-\tau\hat{\mathcal{H}})$ projects out the ground state of a given symmetry from any trial function with nonzero overlap. Such a projection operator can be readily implemented in a Monte Carlo approach [11].

A central concept in the DMC algorithm is a single point in the $3N$-dimensional space known as a *walker*. In the algorithm, an ensemble of walkers, termed the *population*, is propagated stochastically from generation to generation. In each propagation step, the walkers are moved through position space. At the end of each step, each walker may reproduce itself, be killed, or remain unchanged, depending on the value of the local energy at that point in space, $E_L(\mathbf{R})$. This *branching* process causes fluctuations in the population size from generation to generation. Averaging the local energy over all walkers and generations yields an energy significantly lower than the VMC energy, but still guaranteed to be an upper bound of the true energy.

Formally, the DMC method is exact provided the *nodes* of the true wave function (i.e. the locus of points for which $\psi(\mathbf{R}) = 0$) are known. In practice, the nodes are assumed to be those of the trial wave function ψ_T. This gives rise to a positive-definite bias known as the *fixed-node error*, which is, in almost all cases, very small. The fixed-node DMC method has proven to be remarkably robust, accurate and effective. For the trial wave functions outlined above, the fixed-node approximation introduces only small errors which allow one to obtain $\sim$95% of the correlation energy and agreement with experiments within 1-2% for ground state properties such as the lattice constants, bulk modulus, and enthalpies of formation. High accuracies have been found in essentially all systems we have studied: atoms, molecules, clusters and solids. Furthermore, new forms of trial functions and optimization techniques promise to further improve the accuracy.

2.3. *The phaseless auxiliary-field QMC method*

The recently developed phaseless auxiliary-field (AF) method [12–14] is a complimentary method to DMC. The AF QMC method adds several new dimensions to the QMC methodology. It provides a different route to the sign problem [15, 16] from fixed-node DMC, because its random walks take place in a manifold of Slater determinants [12, 17]. Applications show that this often reduces the reliance of the solution on the quality of the trial wave function. The potential for making QMC even more accurate is of crucial significance to the proposed research, given the central goal of robust algorithms with high predictive power. The AF method also expands the application areas of QMC. It is formulated in a Hilbert space defined by any one-particle basis. In complex systems such as transition metal oxides, the use of an effective basis in which to carry out the random walks could lead to significant gains in efficiency. By truncating

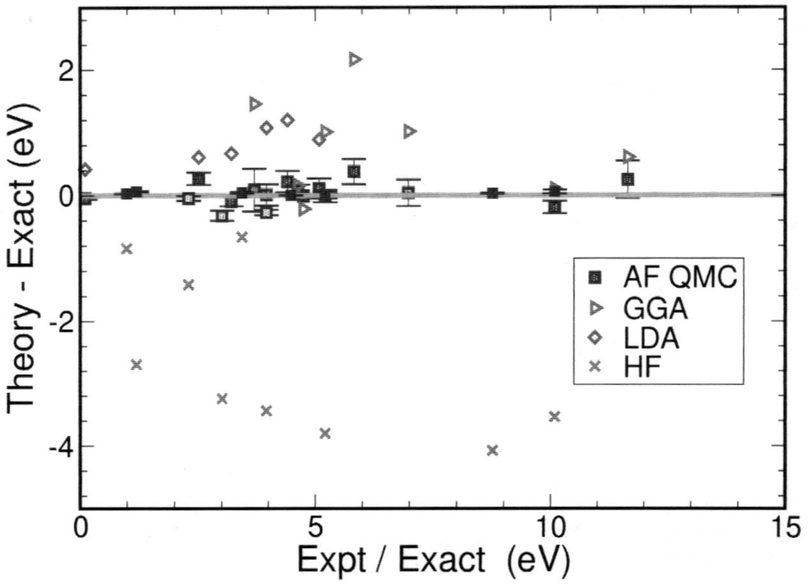

Figure 1. Calculated binding energies of molecules compared with experimental values. The discrepancy between theory and experiment is plotted versus the exact energy. The QMC is fed a trial wave function to start, which is taken directly from DFT [with either local-density approximation (LDA) or generalized-gradient approximation (GGA) functionals] or Hartree-Fock (HF). The corresponding DFT or HF results are also shown. As can be readily observed, the AF QMC results are in excellent agreement with experiment and are significantly more accurate than those from DFT and HF.

the basis (*e.g.*, through a down-folding approach [18]), we can do QMC calculations on *realistic model* Hamiltonians in which higher and irrelevant energy scales can be systematically removed.

To date, the new AF QMC method has been applied to close to 100 systems, including transition metal oxide molecules [19], simple solids [12, 20], post-*d* elements [21], van der Waals systems [22], and in molecules in which bonds are being stretched or broken [23, 24]. In these calculations we have operated largely in an automated mode, inputting only the DFT or Hartree-Fock solutions. The method demonstrated excellent accuracy, consistently able to correct errors in the mean-field trial wave function as summarized in Figure. 1.

3. Breakthrough petascale QMC simulations

We are in an era of exponential growth both in machine performance and in algorithm developments (as far as simulation is concerned). There have been many exciting advances in QMC methods in recent years, including new forms of wave functions (*e.g.* backflow, geminals and pfaffians) [25–28], methods for efficiently optimizing all the parameters in the wave functions [29–33], efficient methods to compute forces and energy differences [34–38], methods to extrapolate finite-size systems to the thermodynamic limit [20, 39–41], (*e.g.* twist average boundary conditions and scaling methods) [42], approaches for coupled electron Monte Carlo and molecular dynamics [3], more accurate methods for treating nonlocal pseudopotentials [43], new techniques to calculate observables [44, 45], algorithms which scale better with the number of electrons (linear or quadratic scaling) [9], and advances in auxiliary-field approaches to do QMC in different manifolds [12–14]. By building on these improvements, we will be able to demonstrate important science accomplishments with petascale facilities.

3.1. Coupled-Electron Monte Carlo (CEIMC) and Molecular Dynamics

During the last ten years we developed a method to perform a simultaneous simulation of electrons and ions, the Coupled Electron-Ion Monte Carlo (CEIMC) method [46], similar in spirit to the Car-Parrinello method [7], but having several advantages over that method. Of most importance is the use of explicit correlation using Reptation Quantum Monte Carlo (similar to DMC), leading to a more accurate Born-Oppenheimer surface, for all configurations of the ions, not just near equilibrium configurations. The CEIMC method can use path integrals for protons, which due to their small mass, are also quantum particles. A remarkable feature of the CEIMC method, is that adding ionic path integrals and twist averaging does not intrinsically slow down the simulation over the use of periodic boundary condition and classical protons. The CEIMC method has been used [3] to study the high-pressure phase diagram of hydrogen with computer time about the same as for corresponding *ab initio* Molecular Dynamics (AIMD) approaches. This opens opportunities for studies of light nuclei quantum effects such as proton tunneling in applications involving hydrogen at finite temperatures.

As an alternative to CEIMC, we have developed, in collaboration with J.C. Grossman from UC Berkeley, a new approach for obtaining QMC energies along the *ab initio* molecular dynamics paths generated in the DFT framework. This enables us to obtain QMC energy for an arbitrary point along the MD ionic trajectories with high efficiency. Our approach leverages the QMC propagation of walkers in the electronic position space by coupling it to the ionic MD steps so that the sampling ensemble follows the evolving wavefunction. By using this coupling and correlation between the QMC sampling points, we can speed up the calculations of energies by a factor 20 to 50 when compared to separate, point-by-point DMC calculations. We have tested this on small hydrogenated silicon clusters at high temperture for a few picoseconds with time step of about 0.5 fs. In addition to the ground state, we also carried out the calculation of excited states along the path, thus gaining insight into the temperature and impact of dynamics on the excitations. Finally, we applied this method to 32 water molecules with periodic boundary conditions, and we were able to calculate the heat of vaporization of water in an excellent agreement with experiment (unlike the DFT methods which were off by about 30%) with statistical error bar of $\sim$1 kcal/mol. This was a significant advance since computing thermodynamic quantities is often orders of magnitude more expensive than static calculations. Currently, we are working on expanding the study to larger systems.

3.2. Transition metal oxides

Transition metal oxides (TMOs) are a class of materials which exhibit a diverse set of unusual behaviors, with applications ranging from high temperature superconductivity to high-density magnetic storage, battery technology, catalysis, and spintronics. The electronic ground state of these systems is determined by several competing energetic contributions, including charge transfer between metallic ions and oxygens, large exchange and correlation energies in the d-electron subshells, crystal field effects, Jahn-Teller and other distortion related effects, etc. Because each of these contributions is of similar size, the application of pressure or chemical doping can drive transitions which change the fundamental properties of the system.

Unfortunately, the same competing energy scales that give rise to rich behaviors in materials have also made them a challenge for theory. Because each contribution to the total energy is small, even relatively small errors introduced by more approximate methods can result in qualitative mispredictions. Very basic characteristics such as equilibrium lattice constants, the energetic ordering of structural phases, electronic gaps, etc., require years of fine-tuning of mean-field theories and/or addition of *ad hoc* parameters. For example, in the LDA+U approach an adjustable parameter, U, is introduced to approximately account for the energy resulting from the double occupancy of a $3d$ orbital [47]. The complexity of many-body effects in these systems calls for new methods and strategies.

The Mott transition in TMOs such as MnO, CoO, and NiO is a prototypical problem. A half century ago, Mott [48] described how electron-electron interactions could induce an insulating state in these materials with a transition to a metallic state under pressure. Evidence of the Mott transition has recently been observed in the pressure-induced metalization of manganese oxide (MnO) [49, 50]. As such, it is an excellent testing case for expanding the applicability of QMC to transition metal solids. At high pressures (about one million atmospheres), the material undergoes a structural phase transition and very close to that also an electronic phase transition to a metal with essentially zero magnetic moment on the Mn ions. The DFT approaches, including extended DFT methods such as LDA+U, have significant difficulties to explain this transition. For example, with proper tuning of the correlation parameter, U, one can reproduce some, but not all, of the experimental observations, indicating that this requires much higher accuracy of the treatment of the electron correlation. Our preliminary QMC calculations provided very encouraging results for the estimation of the band gap and cohesive energy of the MnO solid at the ambient conditions. Therefore we are well poised to address the challenge of describing correctly the electronic structure of two different phases of TMO systems.

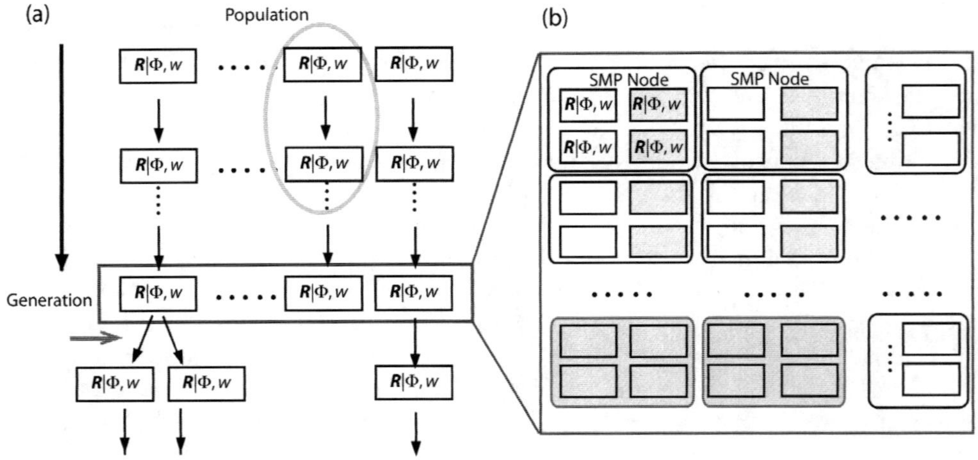

Figure 2. (a) Computational structure of DMC and AF QMC. Each column is a random walk stream. The number of columns gives QMC population size. The red arrow indicates a population control, where weights $\{w\}$ are adjusted and birth/death can occur. Shown in green is one QMC step for one walker. Each walker is a configuration of an N-particle system for DMC (shown as $\mathbf{R}$) or a Slater determinant which looks like a DFT solution for AF QMC (shown as $|\Phi\rangle$). (b) The parallel processor view of a generation after a population control. The population is distributed among SMP nodes to balance computational loads between the population control steps. Each SMP node can utilize multiple threads, e.g., two walkers for each thread on a dual-core node of this example. A group of SMP nodes as indicated by grey boxes can share single-particle orbitals whose storage needs exceed the physical memory of a SMP node.

4. Quantum Monte Carlo: Algorithms, Scalability, and Performance on current HPC systems

The vast majority of the computational time in an accurate QMC simulation is spent in diffusion Monte Carlo (DMC) and phaseless auxiliary-field QMC (AF AMC) methods. A variational optimization of the wave function is required to make the DMC run efficiently, but this step is almost trivially parallelizable and can be treated as a special case of DMC. For this reason, we will focus our discussion on the algorithm and parallelization of DMC and AF QMC and analyze their performances as implemented in current QMC codes.

Both DMC and AF QMC sample a multi-dimensional space by propagating an ensemble of walkers (population) stochastically from generation to generation. This common algorithm is illustrated in Figure 2(a). The key difference of the two methods lies in what a walker represents. In DMC, a walker is a $3N$-dimensional configuration of an N-particle system, while a walker in AF DMC is a Slater determinant in a Hilbert space. A QMC step is carried out independently over the walkers. This allows QMC methods to be efficiently parallelized over many processors by distributing the walkers to each processor. At the end of a generation, the branching process results in a fluctuating population and the walkers are redistributed so that the next QMC step can be executed with a good load-balance. A typical number of walkers is $10^3 - 10^4$ for each population. A feedback mechanism is used to keep the average number of walkers at the target population size.

Figure 2(b) describes the current parallelization scheme of DMC/AF QMC algorithms for a given generation. Each virtual processor, typically a MPI node on a Symmetric Memory Processor (SMP) node, holds in its memory a small number of walkers and proceeds with a propagation of local walkers. Current QMC codes, which employ simple load balancing schemes and a lock-step propagation of each generation, scale well to a few thousand processors. As long as each processing element, *e.g.*, a SMP node of Fig. 2(b), is responsible for several walkers, transient load imbalances and communication time result in a few percent loss in efficiency. Figure 3 shows a pilot AF QMC calculation on the Cray XT3 at ORNL. The parallel speedup is shown for a model system of a P_2 molecule in a supercell. The calculation has a population size of $\mathcal{M}$ =10,000. We see that a parallel efficiency of over $\simeq$85% is achieved on up to 2000 MPI nodes, when the average number of walkers per node is 5. The deviation from perfect scaling is due mostly to load imbalance at the limit of few walkers per node. A similar parallel efficiency is achieved for DMC calculations of a wide range of systems. While DMC and AF QMC share the common parallelization and load-balance schemes in propagating a population over the generations, the operations for each QMC step are fundamentally different.

The computational effort of DMC is dominated by that of the wave functions. Accurate functional forms are essential to achieve high quality results with small statistical error. The dominant computation for a large system is set by the antisymmetric part Ψ_A of the trial wave function: i) N^3 scaling to update Slater determinants; and ii) $N - N^3$ scaling to fill in the Slater determinants with one-particle orbitals. An example is given in Fig. 4 (a).

An analysis of computational kernels [53] is presented in Fig. 4 (b). Of three alternative

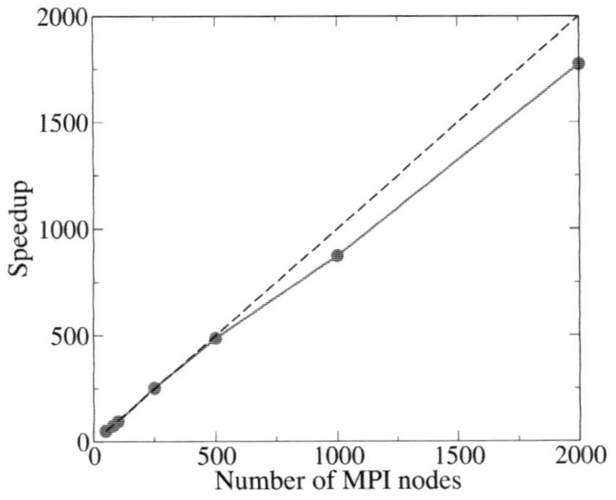

Figure 3. The parallel efficiency of plane-wave AF QMC for a model system, the P_2 molecule in a supercell with a single-particle basis of 9939 plane-waves. The speedup on a Cray XT3 (dual-core AMD Opteron Opteron 2.6 GHz CPUs) is plotted as a function of the number of MPI nodes. "Hard scaling" is shown, with the number of walkers fixed at $\sim$10,000 (the total fluctuates due to population control). The dashed line represents perfect scaling. Efficiencies of $\simeq$85% are achieved up to 2000 nodes, where there are only $\sim$5 walkers per node on average.

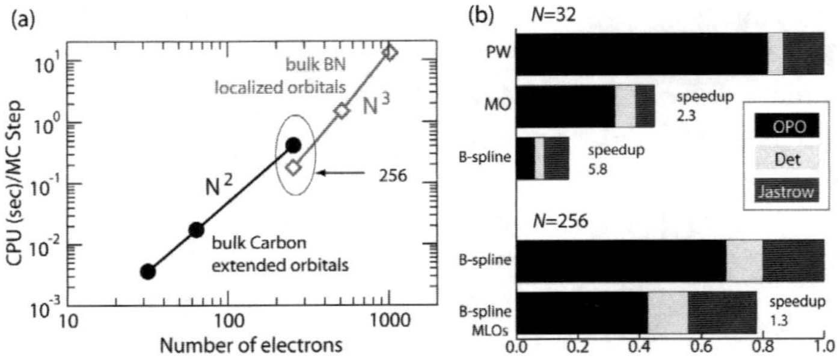

Figure 4. (a) CPU time per Monte Carlo (MC) step for bulk carbon (diamond) and BN (zincblend) [51]. Each MC step involves random moves of N electrons. (b) Breakup of many-body wave function computations for 32- and 256-electron systems using one-particle orbitals in a plane-wave (PW), molecular-orbital (MO) basis and on a real-space grid with B-spline methods [52].

representations of the one-particle orbitals, the use of B-spline interpolation of orbitals on a grid speeds up the computation sixfold over use of a plane-wave (PW) basis set, while the cost with molecular-orbital (MO) basis set falls between B-spline and PW. The gain in computation time of real-space methods become increasingly large as the system size grows. The time to evaluate one-particle orbitals can be further reduced by using maximally localized orbitals (MLOs) [54]. The speedup at $N = 256$ is a direct consequence of using the localized orbitals. For large systems, the performance of determinant updates and Jastrow function evaluations, which typically scale as N^2, becomes as important as one-particle orbital evaluations.

In AF QMC, each walker is a Slater determinant: $|\phi\rangle = |\varphi_1, \varphi_2, \cdots, \varphi_N\rangle$ [55], where N is the number of electrons, and each electron orbital is given by $\varphi_i = \sum_{j=1}^{M} c_{j,i}|\chi_j\rangle$, where $|\chi_j\rangle$ is a one-electron basis state, and M is the size of the basis. The main operations include (i) propagating a walker and (ii) calculating the so-called local energy. A walker propagates by

$$e^{\hat{h}(\sigma)}|\phi\rangle \rightarrow |\phi'\rangle, \tag{5}$$

where $\hat{h}(\sigma)$ is an independent-electron Hamiltonian which depends on the external auxiliary fields, σ. The fields are sampled by the random walk and fluctuate both spatially and *temporally*. The local energy is given by

$$E_L(\phi) = \frac{\langle\Psi_T|\hat{H}|\phi\rangle}{\langle\Psi_T|\phi\rangle}, \tag{6}$$

where $|\Psi_T\rangle$ is a Slater determinant (i.e., of the same form as $|\phi\rangle$).

There are presently two main flavors of the phaseless AF QMC method, from two different choices of the one-electron basis: (i) plane-wave and norm-conserving pseudopotential, and (ii) Gaussian basis sets. In (i), a simulation supercell or twist-average boundary conditions. A basis function is specified by $|\chi_j\rangle \propto \exp(i\mathbf{G}_j \cdot \mathbf{r})$, where $\mathbf{G}_j$ is a reciprocal lattice vector. As in plane-wave DFT calculations, our basis set consists of plane waves with $|\mathbf{G}_j|^2/2 \leq E_{\text{cut}}$, where the cutoff energy, E_{cut}, controls the size of the basis, $M \propto E_{\text{cut}}^{3/2}$. The dimension of the AF σ is $\mathcal{O}(8\,M)$. The computational cost of each MC step (advancing all components of σ) scales roughly as $N^2\,M \ln M$. In (ii), the size of the basis is much smaller, but the cost of each step currently scales as M^3. The calculation of the local energy is similar to the calculation of the variational energy in a Hartree-Fock calculation.

5. Transition from the Terascale to the Petascale QMC.

As demonstrated in Figure 3, current QMC algorithms scale well to a few thousand processors provided that each processing element has a sufficient number of local walkers. They also highlight the limitation of the current QMC implementation for the petascale systems of $10^4 - 10^5$

compute cores. There are several critical impediments in changing from a terascale performance to a petascale one. Consider two types of computations of interest, neither possible on today's computers:

First, consider the computation of a system of ten thousand electrons versus the current limitation of about one thousand electrons. Given a computation time scaling of between the second and third power of the number of electrons, each iteration would take between a hundred and a thousand times as long, assuming minimal change in processor speed. However, the number of needed walkers would not necessarily increase, so the strategy of simply increasing the number of walkers would vastly increase the needed wall clock time to reach convergence. The memory requirements for storing the trial functions would increase in the same way [56].

Second, consider our grand goal, namely the ability to perform classical MD using quantum forces. The utility of these MD calculations will be determined by the amount of wall clock time needed to advance a single time step; *i.e.* to calculate the forces to sufficient accuracy. This will require walkers to advance enough steps so that they become decorrelated with their starting values. Adding additional walkers beyond what is needed would not be effective because this critical number of steps is fixed by the physics of the system, not the number of processors.

Both of these examples argue that we must exploit the shared memory capability of a single node. First, having the processors on a given node share the wave function reduces tje memory requirement by the number of cores per node. Second, by partitioning the wave function calculation of a single walker across the cores on a node, the wall clock time of a single iteration will be reduced by the same amount.

5.1. Multithreading

The architecture of current and proposed petascale systems is based on multicore Symmetric Memory Processor (SMP) nodes. While the memory per core is likely to remain at the current size of few gigabytes, the number of cores per SMP node is expected to grow steadily over the years and hence also the shared memory per SMP node. The current QMC codes can reach a petascale performance without major redesign, simply by utilizing the increased shared memory capacity and improved performance of multi-threaded software. We have implemented the simplest loop-parallelization over the walkers on a node using an MPI/OpenMP hybrid scheme as depicted by the white and yellow boxes in Figure 2(b). The parallel efficiency remains high for QMC as we increase the number of OpenMP threads while fixing the average number of walkers on a SMP node. In fact, DMC scales nearly perfectly with respect to the number of threads: the additional parallelism from multithreading allows more walkers per node and improves the overall parallel efficiency and load balance among SMP nodes within the hierarchical partitioning. Evaluations of different terms of a trial wave function, single particle orbitals and dense/sparse updates of determinants can be readily parallelized by OpenMP threads, although care must be taken to schedule multiple tasks with varying computation loads to maximize the parallelism while minimizing data dependency and overhead of thread management.

5.2. Multi-level algorithmic partitioning

In many applications, components of the existing QMC algorithms can be cast into averages over a set of variables or parameters, providing opportunities for efficient subdivision. Such possibilities enable grouping of processors and assigning each group an instance of the variable or parameter subset. For example, in calculations of systems with periodic boundary conditions, especially for metals, carrying out calculations with different twist angles (*i.e.*, Bloch **k**-vectors) [42] reduces the statistical errors and the finite system-size errors. A group of processors operate on a single twist-angle but averages are taken over all the groups. In the CEIMC method which uses QMC to compute electronic energy differences and Path Integral Monte Carlo to perform simulation of the protonic (or light ionic) degrees of freedom, every protonic time slice can be assigned to a group of processors, and the overall average of the energy difference used to decide on movement of the entire path, thus allowing much lower temperatures than one can do on current platforms.

6. Challenges for QMC at the petascale and beyond

6.1. Load balancing in DMC and AF QMC

Current QMC codes, which employ simple load balancing schemes and a lock-step propagation of each generation, scale well to a few thousand processors. As long as each processing element is responsible for several walkers, transient load imbalances and communication time, result in a few percent loss in efficiency. A finite number of generations is required for the population to reach equilibrium, prior to which no useful data may be collected. Since the total effort required to reach equilibrium scales with the population size, there is an advantage to keeping the population as near to the minimum required population as possible. Thus, it would be highly desirable to have an algorithm which would allow the simulation to approach the *dilute limit* of one to two walkers per node without sacrificing efficiency due to load imbalance.

Development of just such a method is underway. It aims to achieve near perfect efficiency despite load imbalance by relaxing the requirement that all walkers be at the same generation number at all time. This removes the need for barrier synchronizations and blocking communication and allows the communication required for walker redistribution to be overlaid with computation. A simple, but effective, implementation of such a scheme could be done with a manager/worker model. At set wall-clock intervals, each node would send a packed message containing the number of walkers on each node and the physical properties of those walkers which are the output of the simulation. They would then continue propagating their local population of walkers. Meanwhile, the manager would decide on an optimal walker redistribution to be effected by pointwise transfers of walkers between pairs of nodes. It would send messages to each node that needs to participate in communications. Workers would occasionally check for such posted messages, and comply by sending or receiving walkers as appropriate. If the scale requires, this scheme can be efficiently generalized to a two-level hierarchy.

To allow scaling to simulations containing $\mathcal{O}(10^4)$ electrons, the work required to propagate a single walker may be distributed with threads across cores sharing a memory address space. Present indications suggest that a typical petascale machine will have on the order 32 cores in an SMP node. If care is taken to avoid threading at too fine a scale, it should be possible to achieve efficient distribution of single-walker propagation over this number of cores.

6.2. Fault tolerance and reproducibility

A significant concern for the petascale drive in general is the issue of fault tolerance. The sheer number of components involved in a single calculation on the order of 10^5 to 10^6 processing cores may reduce the mean time between failures (MTBF) to below the average run time for a job. To avoid wasting large amounts of CPU time, it will thus be necessary to have codes which are robust to failure of individual processors. The manager/worker architecture introduced for efficient load balancing can easily be designed to accommodate the failure of individual nodes. In the simplest implementation, a node which has failed to communicate for some time beyond the designated wall-clock interval will be invalidated and its walkers removed from the pool. Since this removal is triggered by a random event (i.e. hardware failure), it should not introduce bias into the walker population. Furthermore, the proposed load balancing scheme should allow the remainder of the nodes to proceed without loss of efficiency.

6.3. Very large systems and orbital storage

The memory usage of quantum Monte Carlo for complex materials is dominated by the storage of single-particle orbitals. In the simplest implementation, this usage scales quadratically with system size. Just as in density functional calculations, this scaling stems from the delocalized character of the orbitals and can be reduced to linear scaling for large systems using localized orbitals [9, 54]. Furthermore, we have developed several new methods to reduce the memory required for orbital storage through the use of nonuniform meshes and by exploiting symmetry. We anticipate that the memory per SMP node on upcoming petascale machines will be sufficient for a wide range of problems, since the complete *read-only* orbital data can be shared by multiple threads on a SMP node.

For a sufficiently large, disordered system, the memory required for orbital storage may exceed what is available on a single SMP node. In this case, it should be possible to distribute the orbitals among groups of tightly coupled nodes. Within each group of a few nodes, electron positions would be broadcast in an all-to-all fashion. Each node would then evaluate the orbitals stored in its local, shared memory at the required coordinates. The nodes would then send the data back to the necessary nodes in an operation akin to a small, distributed matrix transpose. This communication may be overlaid with the computation required for other parts of wave function evaluation, and should thus incur very little performance penalty.

Using an efficient B-spline basis, complex orbital evaluations can be performed at a rate of approximately 4 million per second per core on current processors, thus generating data at a rate of 64 MB/s per core. For this scheme to work well, the point-to-point bandwidth should exceed 64 $N_{\rm core}$ MB/s. This requirement is significantly exceeded on most present terascale and sub-petascale machines, and will continue to be if the present ratio of interconnect bandwidth to FLOPS is maintained in future machines.

In the limit of a large system, AF QMC will parallelize the calculation by splitting each walker, a Slater determinant, across multiple processors. If we have $\mathcal{M}$ walkers, the calculation will look like $\mathcal{M}$ loosely coupled DFT calculations. Each walker, *i.e.*, each "DFT stream", will reside on n processors. Typically $\mathcal{M}$ is larger than, say, 300, to ensure that population control bias is small. This means that $n < 350$ for the expected petascale machine with more than 100,000 processors. Current implementation of DFT is well into this regime. For example, a plane-wave DFT calculation on an 807-atom FePt nanoparticle has been performed on the Cray XT3 by team member Kent and co-workers, achieving a parallel efficiency of about 75% on 512 processors, and scaling to at least 2048 processors. This scaling is already sufficient for the proposed AF QMC petascale calculations. Because population controls occur infrequently in the QMC (every $\mathcal{O}(10)$ steps), the grain size will be exceedingly large relative to load-balancing cost. In other words, the QMC efficiency should closely follow that of the DFT with respect to the number of processors, and be very weakly dependent on the total number of processors. In developing the plane-wave AF QMC kernel, we will need to address several additional technical issues, such as the calculation of the local energy $E_L(\Phi)$ and the force bias [12, 57], but the road-map is clear.

6.4. Pseudorandom number generation

At the petascale, Monte Carlo simulations will require one to two orders of magnitude more pseudorandom numbers than have ever been used to date. Since the statistical quality of the random streams has never been tested at this scale, it is critical to verify the integrity of this fundamental input. Work is currently underway to test the Scalable Parallel Random Number Generator (SPRNG) suite of generators, utilizing Jaguar, the Cray XT4 at ORNL with over 30,000 cores.

We tested the Multiplicative Lagged Fibonacci Generator in SPRNG. This generator passed all the statistical tests in the SPRNG test suite, both tests of sequential random number generators and tests of parallel random number generators. Around 28,000 random number streams were used in these tests, with a total of around 10^{13} random numbers per test. We also tested this generator with a 2D Ising model using the Metropolis algorithm. In this test,

256 random number streams are used on a 16×16 lattice. Such computation is repeated with 512 different sets of streams. We verified the accuracy of each set of streams, and then verified the cumulative results of the entire sets of streams. We believe that this is the largest test of random number generators, both in terms of the number of streams used and the total number of random numbers consumed. These tests give us confidence that even at the petascale, our simulations will be free of bias from correlated random numbers.

7. Summary

We have outlined some of the most relevant ideas which will enable to fulfill the potential of petaflop machines for solving breakthrough many-body quantum problems. We plan to test these devlopments on some of the current problems which are at the forefront of attention of computational practitioners in this field. The role of flagship supercomputers in science is that of *time machines*: whatever the computational scientists do on supercomputers today will be available to scientists at large in 10-15 years. The goal of our scientific thrust is to prepare and develop the underlying methodologies, algorithms and implementations for the benefit of broad scientific communities in physics, chemistry, materials science and other fields.

Acknowledgments

The QMC Endstation project is supported by the U.S. Department of Energy (DOE) under Contract No. DOE-DE-FG05-08OR23336. This research used resources of the National Center for Computational Sciences and the Center for Nanophase Materials Sciences, which are sponsored by the respective facilities divisions of the offices of Advanced Scientific Computing Research and Basic Energy Sciences of DOE under Contract No. DE-AC05-00OR22725.

References

[1] D M Ceperley and B J Alder 1980 *Phys. Rev. Lett.* **45** 566
[2] D M Ceperley and B J Alder 1987 *Phys. Rev.* B **36** 2092–2106
[3] Pierleoni C, Ceperley D M and Holzmann M 2004 *Phys. Rev. Lett.* **93** 146402
[4] L Mitas, J C Grossman, I Stich, J Tobik 2000 *Phys. Rev. Lett.* **84** 1479
[5] Maezono R, Ma A, Towler M and Needs R 2007 *Phys. Rev. Lett.* **98** 025701
[6] N Metropolis, A W Rosenbluth, M N Rosenbluth, A H Teller and E Teller 1953 *J. Chem. Phys.* **21** 1087–1092
[7] Car R and Parrinello M 1985 *Phys. Rev. Lett.* **55** 2471–2474
[8] Kalos M H, Levesque D and Verlet L 1974 *Phys. Rev.* A **9** 2178–2195
[9] A J Williamson, R Q Hood and J C Grossman 2001 *Phys. Rev. Lett.* **87** 246406
[10] J C Grossman and L Mitas 2005 *Phys. Rev. Lett.* **94** 056403
[11] W M C Foulkes, L Mitas, R J Needs and G Rajagopal 2001 *Rev. Mod. Phys.* **73** 33
[12] Zhang S and Krakauer H 2003 *Phys. Rev. Lett.* **90** 136401
[13] Al-Saidi W A, Zhang S and Krakauer H 2006 *J. Chem. Phys.* **124** 224101
[14] Suewattana M, Purwanto W, Zhang S, Krakauer H and Walter E J 2007 *Phys. Rev.* B **75** 245123 (pages 12) URL `http://link.aps.org/abstract/PRB/v75/e245123`
[15] D M Ceperley and B J Alder, J Chem Phys **81**, 5833 (1984); Shiwei Zhang and M H Kalos, Phys Rev Lett **67**, 3074 (1991); J. B. Anderson in *Quantum Monte Carlo: Atoms, Molecules, Clusters, Liquids and Solids*, Reviews in Computational Chemistry, Vol. 13, ed. by Kenny B. Lipkowitz and Donald B. Boyd (1999)
[16] Zhang S 1999 *Quantum Monte Carlo Methods in Physics and Chemistry* ed Nightingale M P and Umrigar C J (Kluwer Academic Publishers) cond-mat/9909090
[17] Zhang S, Carlson J and Gubernatis J E 1997 *Phys. Rev.* B **55** 7464
[18] Andersen O K and Saha-Dasgupta T 2000 *Phys. Rev.* B **62** R16219
[19] Al-Saidi W A, Krakauer H and Zhang S 2006 *Phys. Rev. B* **73** 075103
[20] Kwee H, Zhang S and Krakauer H 2008 *Phys. Rev. Lett.* In press.
[21] Al-Saidi W A, Krakauer H and Zhang S 2006 *J. Chem. Phys.* **in press** physics/0608298
[22] Al-Saidi W A, Krakauer H and Zhang S 2007 *The Journal of Chemical Physics* **126** 194105 (pages 8) URL `http://link.aip.org/link/?JCP/126/194105/1`
[23] Al-Saidi W A, Zhang S and Krakauer H 2007 *The Journal of Chemical Physics* **127** 144101 (pages 8) URL `http://link.aip.org/link/?JCP/127/144101/1`

[24] Purwanto W, Al-Saidi W A, Krakauer H and Zhang S 2008 *The Journal of Chemical Physics* **128** 114309 (pages 7) URL `http://link.aip.org/link/?JCP/128/114309/1`
[25] P Lopez Rios, A Ma, N D Drummond, M D Towler and R J Needs 2006 *Phys. Rev.* E **74** 066701
[26] Casula M and Sorella S 2003 *The Journal of Chemical Physics* **119** 6500–6511 URL `http://link.aip.org/link/?JCP/119/6500/1`
[27] Casula M, Attaccalite C and Sorella S 2004 *J. Chem. Phys.* **121** 7110
[28] Bajdich M, Mitas L, Drobny G, Wagner L and Schmidt K E 2006 *Phys. Rev. Lett.* **96** 130201
[29] Umrigar C J and Filippi C 2005 *Phys. Rev. Lett.* **94** 150201
[30] Umrigar C J, Toulouse J, Filippi C, Sorella S and Hennig R G 2007 *Phys. Rev. Lett.* **98** 110201
[31] Toulouse J and Umrigar C J 2007 *J. Chem. Phys.* **126** 084102
[32] Sorella S, Casula M and Rocca D 2007 *J. Chem. Phys.* **127** 014105
[33] Toulouse J and Umrigar C J 2008 *J. Chem. Phys.* **128** 174101
[34] Filippi C and Umrigar C J 2000 *Phys. Rev.* B **61** R16291
[35] Filippi C and Umrigar C J 2002 *Recent Advances in Quantum Monte Carlo Methods, Vol. III* (World Publishing: ed. by W. A. Lester and S. M. Rothstein)
[36] R Assaraf and M Caffarel 2003 *J. Chem. Phys.* **119** 10536
[37] Mosé Casalegno M M and Rappe A M 2003 *J. Chem. Phys.* **118** 7193
[38] S Chiesa, D M Ceperley, and S Zhang 2005 *Phys. Rev. Lett.* **94** 036404
[39] L M Fraser, W M C Foulkes, G Rajagopal, R J Needs, S D Kenny and A J Williamson 1996 *Phys. Rev.* B **53** 1814
[40] Williamson A J, Rajagopal G, Needs R J, Fraser L M, Foulkes W M C, Wang Y and Chou M Y 1997 *Phys. Rev.* B **55** 4851
[41] Chiesa S, Ceperley D M, Martin R M and Holzmann M 2006 *Phys. Rev. Lett.* **97** 076404
[42] C Lin F H Z and Ceperley D M 2001 *Phys. Rev.* E **64** 016702
[43] Casula M 2006 *Phys. Rev.* B **74** 161102
[44] Assaraf R and Caffarel M 1999 *Phys. Rev. Lett.* **83** 4682
[45] Toulouse J, Assaraf R and Umrigar C J 2007 *J. Chem. Phys.* **126** 244112
[46] Ceperley D M and Dewing M 1999 *J. Chem. Phys.* **110** 9812
[47] Kasinathan D, Kunes J, Koepernik K, Diaconu C V, Martin R L, Prodan I D, Scuseria G E, Spaldin N, Petit L, Schulthess T C and Pickett W E 2006 *Phys. Rev.* B **74** 195110 URL `http://link.aps.org/abstract/PRB/v74/e195110`
[48] Mott N F 1949 *Proceedings of the Physical Society. Section A* **62** 416
[49] Patterson J R, Aracne C M, Jackson D D, Malba V, Weir S T, Baker P A and Vohra Y K 2004 *Phys. Rev.* B **69** 220101 (pages 4)
[50] Yoo C S, Maddox B, Klepeis J H P, Iota V, Evans W, McMahan A, Hu M Y, Chow P, Somayazulu M, Hausermann D, Scalettar R T and Pickett W E 2005 *Phys. Rev. Lett.* **94** 115502 (pages 4)
[51] Dell PowerEdge 1955 Blade Server Cluster, Intel64 (Clovertown, 2.33GHz, quad core) and InfiniBand (IB) interface.
[52] Performance analysis by PerfSuite on 1.6 GHz Itanium 2 (6.4 GFLOPS peak).
[53] Kufrin R perfSuite, http://perfsuite.ncsa.uiuc.edu/
[54] Reboredo F A and Williamson A J 2005 *Phys. Rev.* B **71** 121105(R)
[55] Shiwei Zhang 2003 *Theoretical Methods for Strongly Correlated Electron Systems* (Springer-Verlag) chap 2, pp 39–74
[56] For the one-particle orbitals in a plane-wave and molecular-orbital basis, the memory usage scales N^2 for the system size N. The use of localized orbitals can lead to a linear scaling
[57] Purwanto W and Zhang S 2004 *Phys. Rev.* E **70** 056702

Computational Challenges of Large-Scale Long-Time First-Principles Molecular Dynamics

P. R. C. Kent

Center for Nanophase Materials Sciences, Oak Ridge National Laboratory, Oak Ridge, Tennessee, USA

E-mail: kentpr@ornl.gov

Abstract. Plane wave density functional calculations have traditionally been able to utilize the largest available supercomputing resources. We analyse the scalability of modern projector augmented wave implementations to identify the challenges in performing molecular dynamics calculations of large systems containing many thousands of electrons. Benchmark calculations on the Cray XT4 demonstrate that global linear algebra operations are the primary reason for limited parallel scalability. Plane-wave related operations can be made sufficiently scalable. Improving parallel linear algebra performance is an essential step to reaching longer timescales in future large scale molecular dynamics calculations.

1. Introduction

Over the last two decades, density functional theory (DFT)[1, 2] has developed into a routinely applied electronic structure methodology used in diverse areas of materials science, chemistry, nanoscience, and condensed matter physics. The accurate prediction of the properties of materials at the atomistic level requires systems be treated at the quantum mechanical level. By mapping the many-body electronic structure problem of Schrödinger's equation onto a effective mean field theory, DFT reduces the otherwise intractable equation into a set of coupled equations that can be solved to useful accuracy on today's computers, even for relatively large systems. Nevertheless, this problem remains a challenging numerical task.

DFT is in principle exact but in practice an approximate functional must be utilized, effectively making all results for non-trivial systems approximate. However, the development of so-called gradient corrected functionals (GGAs) in the 1990s, such as PW91[3], enabled significantly improved predictions of the properties of molecules, and by extension significantly improved predictions of e.g. chemical reactions, liquids, and interface physics. While many failure cases of the functionals exist (e.g. many transition metal oxides are incorrectly predicted as metals; dispersion forces/van der Waals interactions are barely accounted for), many systems of scientific interest can be modeled *in silico* with sufficient accuracy to provide understanding or guide engineering decisions and research directions. Of particular note, DFT is routinely able to make accurate predictions of chemical reactions and systems where substantive changes in the underlying chemical bonding occurs. These are systems where classical and empirical force fields cannot be applied and where the only option is to use quantum mechanical methods. The utility of DFT has consequently resulted in demands to simulate ever larger systems and to simulate the dynamical properties of systems over increasing timescales.

In this paper we concentrate on the plane wave pseudopotential implementations[4, 5] of DFT. These methods are readily parallelizable and have traditionally be able to make effective use of the largest available supercomputers. Although there are no limitations of the size of system that can be studied with DFT, given sufficient computational resources, in practice there are limitations on the achievable system size due to the time taken by the calculations. For molecular dynamics calculations this limitation is particularly relevant: although the computational costs of the methods scale as the second to third power of the system size, the methods only have limited computational scalability. Consequently the amount of simulated time that can be performed in one week or one month of computation tends to reduce with increased system size. The imminent availability of supercomputers containing $O(10^5)$ processing cores provides a severe challenge to traditional DFT methodologies, if they are to continue to run at the largest scale. We focus on Born-Oppenheimer molecular dynamics calculations where classical molecular dynamics is performed on the atomic positions with quantum mechanical forces determined by DFT. Recently proposed improved Born-Oppenheimer integrators (e.g [6]) make these methods particularly attractive where the objective is the longest possible simulation time.

In the following sections we first give an outline of DFT (Sec. 2) and then describe the key elements for parallelization in Sec. 3. In Section 4 we show experimental timings and scaling for several molecular dynamics calculations. The presented data has been obtained on Cray XT4 machines with dual core 2.6 Ghz Opteron processors, franklin at NERSC and an older version of the current jaguar machine at NCCS. In Section 5 we analyze the generic underlying reasons for the observed timings and scaling. Conclusions, identifying several areas requiring improved algorithms and/or numerical implementations are presented in Section 6.

2. Density functional theory

Density functional theory, as applied in electronic structure calculations, is well established and there exists many reviews on the subject, e.g. [4, 5]. Here we give a brief overview of the method to identify the key quantities and challenges for parallelization.

To solve for the quantum mechanical properties of a system containing N electrons at zero temperature requires - in principle - the solution of the time-independent Schrödinger equation,

$$\hat{\mathcal{H}}\psi = E_0\psi, \tag{1}$$

where E_0 is the ground-state energy for the system and $\hat{\mathcal{H}}$ is given by

$$\hat{\mathcal{H}} = \sum_{i=1}^{N}\left[-\frac{1}{2}\nabla_i^2 + V(\mathbf{r}_i)\right] + \sum_{i<j}\frac{1}{|\mathbf{r}_i - \mathbf{r}_j|}, \tag{2}$$

Here $V(\mathbf{r}_i)$ is the potential generated by the nuclei, but could in principle contain the potential due to applied fields. In Kohn-Sham DFT this highly coupled N-body interacting problem is replaced by a set of N single particle Schrödinger equations interacting only through a common external potential, V_{ext}:

$$\hat{h}\phi_i = \epsilon_i\phi_i \tag{3}$$

where $\hat{h}$ is given by

$$\hat{h} = -\frac{1}{2}\nabla^2 + V + V_{\text{ext}}, \tag{4}$$

and there are N single-particle wavefunctiona ϕ_i.

The external potential, $V_{\text{ext}}(\mathbf{r})$, is computed from a functional of the electron density $\rho(\mathbf{r})$, where

$$\rho(\mathbf{r}) = \sum_{i=1}^{N} |\phi(\mathbf{r})|^2 \tag{5}$$

Kohn-Sham DFT is exact provided the exact functional relationship between the ground state electron density and the external potential is known ($V_{\text{ext}} = F[\rho]$). Unfortunately the exact functional is not known for real materials, so approximations such as the local density approximation (LDA)[7] and generalized gradient approximation (GGA) are applied. These functionals incorporate the subtleties of the electron-electron interactions in an approximate way. Hence, provided a sufficiently accurate approximation exists for a given system of interest, the electronic structure problem is reduced to solving N coupled equations self-consistently.

Here we concentrate on periodic boundary conditions plane wave pseudopotential calculations[5]. In these calculations the core electrons around each ion are replaced by an effective (pseudo) potential. This allows a relatively small plane wave Fourier expansion to be used to represent the single particle wavefunctions, ϕ_i, since the remaining valence electrons typically have smooth wavefunctions. In traditional applications the individual wavefunctions are strictly orthogonal to each other,

$$< \phi_i | \phi_j >= \delta_{ij}. \tag{6}$$

Determining the ground state electron density ρ and corresponding set of single particle wavefunctions $\{\psi_i\}$ is a challenging task due to the coupled nature of the equations. Typically iterative methods are utilized whereby an estimated initial set of wavefunctions is optimized to approximately satisfy Eq.4 for an initial density. The density is then recomputed for the new wavefunctions and "mixed" with the old density. This process is repeated until the wavefunction and density reach self-consistently. Common methods for optimizing the wavefunctions include conjugate gradient techniques[5], Davidson diagonalization[8], and a family of functional minimization techniques such as residual minimization [9, 10]. From a programming perspective, these methods can be cast in a common framework. Fig. 1 illustrates a typical self-consistent loop. How to arrange the data to perform the numerical operations described by this loop efficiently is the subject of the next section.

During the 1990s two substantial advances were made to traditional plane-wave pseudopotential methods: the development of the so-called "ultrasoft" pseudopotential [11] and the projector-augmented wave (PAW)[12] formalisms. The former allows both for more accurate calculations through the use of improved pseudopotentials and for the use of greatly shortened Fourier expansions. The latter formalism additionally makes rigorous connections to all-electron calculations[13], can be yet more accurate, and allows for the straightforward reconstruction of the non-pseudised wavefunctions. These methods result in considerable extra numerical and programming complexity but do not change the overall self-consistent scheme. For brevity details of these methods are not discussed further in this article, but we note that due to the increased accuracy and significantly shortened Fourier expansions these methods have become the pseudopotential methods of first choice for accurate materials simulation. All of the calculations presented in Section 4 utilize the PAW method. The scalability of calculations using the ultrasoft method is essentially the same.

3. Implementation of plane wave density functional methods on massively parallel supercomputers

In this section we discuss the implementation of plane wave calculations in parallel. As outlined in Section 2, the numerical implementation of the methods reduces to solving a coupled set of single particle Schrödinger equations. There are many methods for solving these equations, each

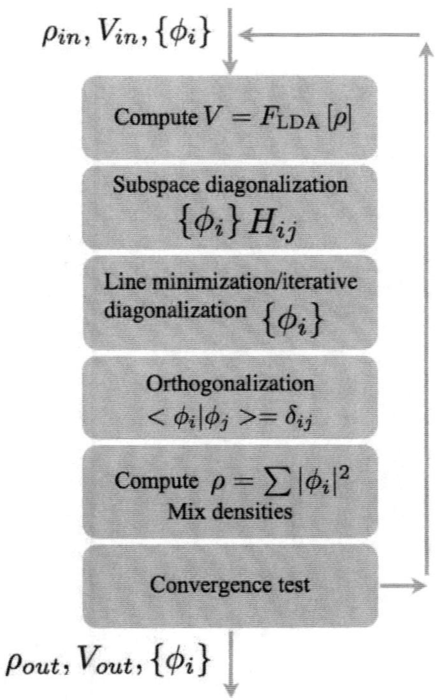

Figure 1. Flow chart illustrating a typical DFT self-consistent loop. The density ρ is calculated self-consistently from the set of single-particle wavefunctons $\{\phi_i\}$. The exact ordering of the subspace diagonalization, iterative diagonalization and orthogonalization steps is implementation dependent. In some methods, e.g. Davidson diagonalizaton, certain operations are combined. Adapted from Refs. [10] and [4].

with areas of optimum performance, but the general requirements and numerical operations remains similar between them. Figure 1 outlines a typical self-consistent loop in a modern DFT code.

To solve for the self-consistent set of wavefunctions $\{\phi_i\}$ and density ρ, the wavefunctions are repeatedly optimised (e.g. via a conjugate gradients approach) subject to the orthogonality requirement. Optimisation is frequently performed with the orthogonality constraints being imposed only periodically. So-called subspace diagonalization, which can be considered as a form of preconditioning, may be performed either implicitly or explicitly (Fig. 2). At this step a matrix $H_{ij} =< \phi_i|\hat{h}|\phi_j >$ is computed, diagonalized, and the resultant eigenvectors used to unitarily transform the set of wavefunctions $\{\phi_i\}$. This step includes wavefunctions that are unoccupied in the ground state and hence influences convergence of the algorithm. In practice, this step is highly desirable to accelerate convergence and is usually applied every iteration of the self-consistency loop. Orthogonalization of the wavefunctions requires a similar flow of operations to subspace diagonalization, except that the computed matrix is the overlap matrix $S_{ij} =< \phi_i|\phi_j >$ and the orthogonality constraint is imposed via LU^{-1} decomposition of S.

Scalable parallization of plane-wave DFT calculations requires efficient parallelization of all steps of the self-consistency loop to avoid an Amdahl's law limitation on overall scalability. This requires (1) efficient parallelization of all plane-wave operations involving the wavefunctions ϕ_i, and (2) efficient parallelization of all linear algebra operations in orthogonalization and subspace diagonalization. Traditionally the focus for scalability has been on the plane wave operations, but we shall show in Section 5 that, using current algorithms and numerical techniques, the

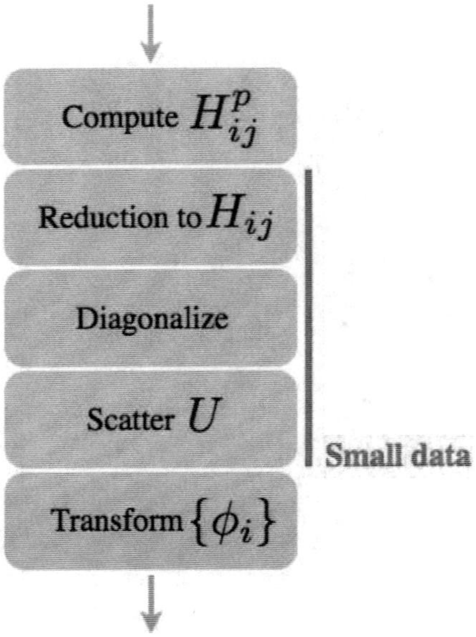

Figure 2. Flow chart illustrating subspace diagonalization operations. Contributions to each element $\{i, j\}$ of the subspace matrix H are calculated on each processor p. These contributions are combined globally and the matrix diagonalized. The eigenvectors are used to unitary transform the set of single particle wavefunctions $\{\phi_i\}$. The first and last operation involve all plane wave coefficients, but the remaining operations involve a matrix of size equal to the number of states (bands/orbitals) in the calculation, which is small in the context of massively parallel simulations.

linear algebra represents a considerable challenge to scalability. The underlying technologies of most codes are currently message passing using MPI[14], Fourier transforms using optimized serial libraries such as FFTW[15], and parallel linear algebra libraries such a SCALAPACK[16].

3.1. Parallelization of plane wave operations

In the plane-wave pseudopotential method the wavefunctions ϕ_i are expanded in a Fourier series

$$\phi_i = \sum_G c_{i,G} e^{i\mathbf{G}.\mathbf{r}}, \tag{7}$$

where the many plane waves G can be distributed over a number of parallel processors. The plane wave expansion is chosen such that the plane waves fit within a 3D sphere defined by an energy cutoff $|G|^2 < E_{\text{cut}}$.

In a periodic condensed matter system, wavefunctions are sampled from multiple $\mathbf{k}$ points across the Brillouin zone, which adds an additional $\mathbf{k}$ index to the wavefunctions. This adds a trivially parallelizable $\mathbf{k}$ point loop between evaluations of ρ in Fig. 1. However, in molecular dynamics calculations a single $\mathbf{k}$ point is typically sufficient, so we drop the additional index.

To compute $\hat{h}\psi_i$ it is necessary to compute the kinetic energy operator $-\frac{1}{2}\nabla^2$. This operator is diagonal in reciprocal or Fourier space. Other operators, such as the addition of the potential V are most straighforwardly computed in real space. Computation of $\hat{h}\psi_i$ consequently proceeds via a fast Fourier transform (FFT) to reciprocal space, computation of the kinetic energy operator, then Fourier transforming back to compute the potential terms. Efficient parallelization consequently depends on the availability of an efficient distributed 3D parallel

FFT. In practice this is implemented via a series on 1D FFTs and a series of MPI[14] all-to-all communications (or equivalent). Computation of operations such as $< \phi_i|\hat{h}|\phi_j >$ proceeds similarly, except that after computation of $\hat{h}\phi_j$ the plane-wave coefficients of ϕ_i must be available.

In early plane-wave DFT codes, plane-wave parallelization was achieved by distributing the plane-wave coefficients over all processors. However, all-to-all operations represent a significant impendiment to scaling to thousands of processors with the typical data sizes of plane wave calculations. Modern codes running on large machines instead divide the available processors into several groups, each containing a subset of wavefunctions. Groups are chosen so that FFTs only require communication between processors in a single group. Computation of quantities involving inter-group properties, such as H_{ij} proceeds via a redistribution (transpose) of the data. Optimization of the group size is required for each system, based on the number of plane waves in each wavefunction and the performance of the underlying processors and parallel interconnect. Several wavefunctions may be transformed simultaneously, reducing the effect of communications latency. Despite the simplicity of this approach, we will show in Section 5 that this approach is already sufficiently scalable for current supercomputers. We note that more sophisticated methods involving the overlap of multiple FFTs to hide communications latency and bandwidth limitations [17, 18] are effective for small systems where FFTs performance is particularly important.

3.2. Parallelization of linear algegra operations

To parallelize linear algebra operations, direct use is made of well-established parallel inear algebra libraries such as SCALAPACK[16]. Scability is then determined by the scalability of the underlying algorithms employed by the libraries for operations such as (1) computing all eigenvectors of a Hermitian matrix and (2) computing an LU^{-1} decomposition of a matrix.

In the data presented in Sec. 4, we utilize SCALAPACK routines such as PDSYEVZ/PZHEEVX or PDSYEVD/PZHEEVD which implement methods based on bisecection and inverse iteraition (PDSYEVX,PZHEEVX) and the divide and conquer approaches (PDSYEVD/PZHEEVD). We benchmark the algorithms and run them at or close to optimum scaling. For typical molecular dynamics problems where substantial eigenvalue clustering does not occur, we do not find substantial ($\geq 1.4\times$) performance differences between these algorithms.

A secondary but still relevant operation necessary for efficient parallelization is the transformation of data layout necessary to use the parallel linear algebra libraries. Fig. 2 illustrates the major step of subspace diagonalization. Each processor p typically computes a contribution to every element of the matrix H_{ij}. These contributions must be summed and redistributed into the appropriate layout for the linear algebra library. In the case of SCALAPACK this is a block-cyclic distribution. Many applications implement a global summation of H_{ij}, but this operation includes unnecessary redundant work, since each processor does not need the full summed matrix, only the contributions to the local block-cyclic data distribution. For large simulations we avoid redundant communications and benchmark various algorithms based on explicit reduction to the block-cyclic layout. The optimum strategy is highly dependent on the data sizes, underlying communications hardware, and even MPI library versions and settings, due to cross-overs between various communications algorithms.

4. Example large-scale calculations

To illustrate the challenges of large scale calculations for long time molecular dynamics calculations, in this section we present several examples including relevant timing and data sizes. For these calculations the key challenge is to run the simulations at good parallel efficiency while completing the simulations on an acceptable *human* timescale. All of the calculations presented could be completed using fewer processors, but in most cases the increase in turnaround time would not be acceptable. These calculations have all been performed as part of ongoing scientific research for user projects for the Center for Nanophase Materials Sciences at ORNL. Although we consider these calculations to run efficiently, they have been chosen to represent typical large scale calculations and not to represent the largest possible or best optimized calculations. In Section 5 we present more detailed timing information and identify the key limitations of the method affecting these calculations.

All calculations were performed using a version of the VASP code[19, 20, 21, 10, 13] optimized for the Cray XT series, as described in Section 3. This code implements the plane-wave projector augmented wave method. Pseudopotentials were evaluated in real-space. Timings were obtained using simple calls to MPI_WTIME, and have been confirmed by profiling measurements of selected runs as well as simple wall clock timings of actual simulations. With appropriate optimization work we expect other projector augmented wave codes would deliver similar scalability and performance for equivalent calculations. Due to the use of the projected augmented wave method relatively small plane wave cutoffs of $\sim 400eV$ can be utilized for these calculations, greatly reducing the overall number of plane wave coefficients to be distributed compared to norm-conserving calculations.

4.1. Water interacting with oxide surfaces

In Figure 3 we show a simulation cell consisting of a slab of titanium dioxide and several monolayers of water molecules. The aim of the simulation was to study the dynamical processes occurring at the interface between water and this prototype oxide surface, necessitating relatively long (14ps) molecular dynamics calculation with comparatively small timesteps (0.5fs) for Born-Oppenheimer simulations.

At the chosen plane wave cutoff of 400eV, each of the 1632 orbitals ϕ in the calculation contained $\sim$ 75K plane wave coefficients. Distributing these orbitals over up to 64 processors was sufficient to give good scalability while still maintaining good performance for serial linear algebra operations. Using 1024-2048 processors we obtained a "real-world" average time per Born Oppenheimer step of 1 minute, including all input/output overhead, over the entire 14ps simulation.

The measured time of 1 minute per 0.5 fs step in this simulation is sufficient to obtain 10ps of simulation time for each wall-clock week of simulation time. At most supercomputer sites, $\geq$ 1024 processors cannot be reserved for periods of more than one day due to their non-dedicated nature. Consequently the overall real-world simulation time for these simulations was closer to one month. Hence for simulations at this scale, run on shared facilities, simulation times of 40ps might be realized in a period of time acceptable to most researchers, but longer timescales are much less practical.

4.2. Nanomagnetism

In Figure 4 we show the timings and scaling achieved for inedividual electronic iterations of simulations of ferromagnetic FePt nanoparticles. These nanoparticles are of interest for high-density magnetic storage owing to the large magnetic anisotropy (and hence thermal stability) of FePt. One electronic iteration corresponds to a single loop around Figure 1. The simulation cell consists of a single FePt nanoparticle surrounded by vacuum.

The aim of these simulations is to identify the closely coupled interrelation between the physical, electronic, and magnetic structures of the nanoparticles. Molecular dynamics based optimization[22] was used to explore different possible ground state geometries. The calculations were spin polarized.

For $Fe_{399}Pt_{408}$ atom nanoparticles we observed a near-perfect speedup to around 512 processors. Around 2048 processors significant reductions in performance occurred with an estimated extrapolated parallel scaling limit of 4096 processes at $\sim$ 40 seconds/iteration. For

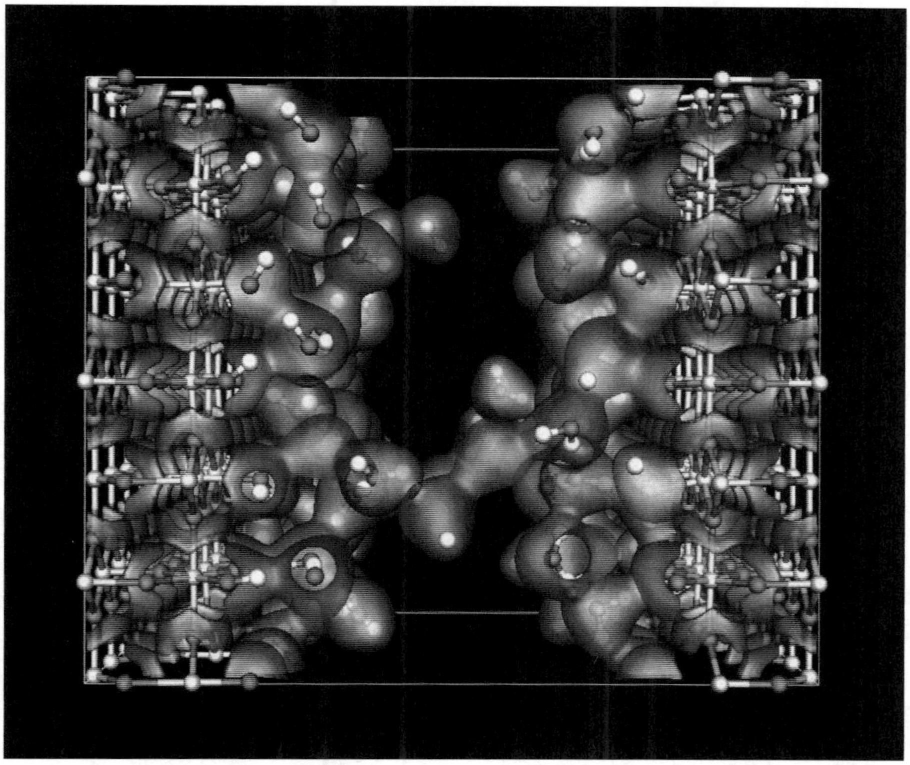

Figure 3. Simulation cell containing 108 TiO_2 units and 108 H_2O monomers used for long time molecular dynamics calculation. Allowing for unoccupied states 1632 bands/orbitals are used in the calculations. At a typical plane wave cutoff of 400eV/29.4Ry the finest FFT grids are 180x192x240. An isosurface of the electron density is shown in blue. Courtesy J. Sofo, Penn State University.

the larger $Fe_{657}Pt_{632}$ atom nanoparticle we measured a very similar performance curve, except shifted to higher processor counts. Good speedups are obtained up to 4096 processors, with a parallel scaling limit of around 10000 processors. This last limit corresponds to 8 processors per atom. Note also that due to the use of the projector augmented wave methods and 2 Gbyte/processor memory, these systems could be run on as few as 64 and 512 processors, albeit with less attractive run times.

The curves obtained for the FePt nanoparticles are prototypical scaling curves that we have observed over many calculations of many different electronic systems. Although exceptions exist, for the class of materials considered here, using common algorithms we typically find good scaling up to around one atom per processor, with continued scalability for up to ~ 5 times processors per atom. Scaling is increased to greater processor counts if higher plane wave cutoffs are selected for increased accuracy.

4.3. Nanocatalysis

In Figure 5 we illustrate a system for which we would like to perform long time dynamics with rapid turnaround. The supercell contains a platinum nanoparticle surrounded by water molecules. For this simulation we would aim to study how solvation influences the surface structure and catalytic properties of the metal nanoparticle.

The supercell contains 201 Pt atoms and 426 water monomers. At typical plane wave cutoffs

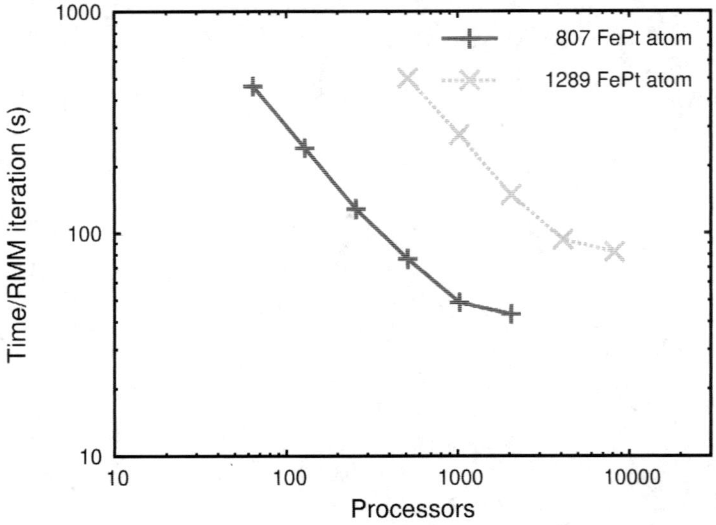

Figure 4. Overall scaling achieved for electronic steps in ferromagnetic FePt nanoparticles. The times shown are for updates to both spins; half for paramagnetic calculations.

we allow 160K plane wave coefficients per orbital. Timings as per Fig. 4 find 14 seconds are sufficient for each electronic iteration on 2048 processors. However, profiling reveals that 25 percent of this time is spent in subspace diagonalization as opposed to iterative diagonalization of the wavefunctions.

5. Limitations of current software, systems, and algorithms

In this section we analyse the timings obtained for a 1289 atom nanoparticle calculation to determine the reason for limited parallel scalability of current simulations. The limited scaling is intrinsic to the current algorithms that are widely used in plane wave DFT calculations. Although these scaling limitations can be concealed by soft-scaling calculations (using systems of increasing atom count), practical calculations only require the incorporation of a certain length scale to obtain converged results. Similarly, increased plane wave count is not required in general, but results in apparent increased scalability. In the context of long time molecular dynamics calculations, we wish to accelerate the simulation of a system of fixed size at fixed accuracy, so that longer simulation times can be reached in an acceptable real-world timescale.

In Figure 6 we show a breakdown of the time taken to complete key stages of a subspace diagonalization operation (Fig. 2). Similar trends are obtained for orthogonalization, although the orthogonalization steps are quicker overall due to reduced complexity. The graph shows that the time taken to build the Hamiltonian H_{ij} on each processor decreases in proportion to processor count. The time taken to transform the set of orbitals $\{\phi_i\}$ improves similarly. However, the time taken so setup and perform the parallel diagonalization is approximately constant above 512 processors. For each processor count we measured the best combination of SCALAPACK eigenvector solver and matrix redistribution algorithms (Section 3.2).

Figure 6 shows that limited scalability of subspace diagonalization operations occurs due to the limited scalability of the linear algebra operations. The FFT and related operations used for plane-wave operation on the single particle wavefunctions are, at present, sufficiently scalable: improvements to these algorithm using, e.g., a scheduled overlap of computation and communication[17, 18], would not significantly improve the scalability of the methods. (These algorithms would be welcome however for small scale calculations with fewer orbitals, where the

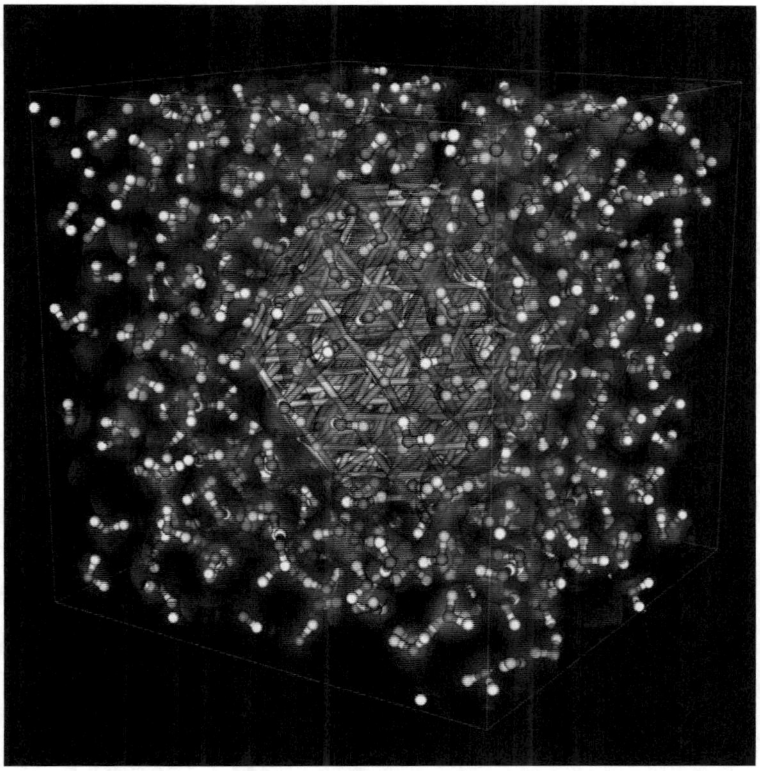

Figure 5. Simulation cell containing a 201 atom Pt nanoparticle and 427 H2O monomers (1482 total atoms). Allowing for unoccupied states, 3456 bands/orbitals are used in the calculations. At a typical plane wave cutoff of 400eV/29.4Ry the finest FFT grids are 256x256x256. An isosurface of the electron density is shown in blue.

linear algebra operations are less costly.).

The timings shown in Figure 6 show that when characterizing electronic structure calculations it is important to refer to not only the atom count, but also the plane wave count per orbital (which determines the amount of FFT related work) and the total number of bands in the calculation (which determines the size of global linear algebra operations). The FFT related work can be made to scale better than the global linear algebra related work. Computation of all eigenvectors of the Hamiltonian matrix requires N^3 work, where N is the number of bands in the calculation. However, this work cannot be parallelized in proportion to its cost: we find on the Cray XT4 that optimum performance is obtained from SCALAPACK for $p \approx N/10$. Hence for the datapoint at 4096 processors, only 1024 processors are used in the diagonalization, a 25% efficiency. This imposes a scaling limit on the calculation via the parallel Amdahl's law ("scaling is limited by the least scalable part of the calculation"). We find similar limitations on other parallel systems with fast processors and fast interconnects.

We wish to note that straightforward changes to the well-established algorithms can be used to improve scalability, but at the loss of some generality: less frequent subspace diagonalization and orthogonalization can be used in some systems, particularly those with a large electronic gap. For some systems we have also determined that use of single or mixed-precision computation can deliver significant speedups, but for a sufficiently large system double precision accuracy is eventually required. Speedups occur with the use of single precision not only because of the faster single precision floating point operations on many processors, but also the reduction in

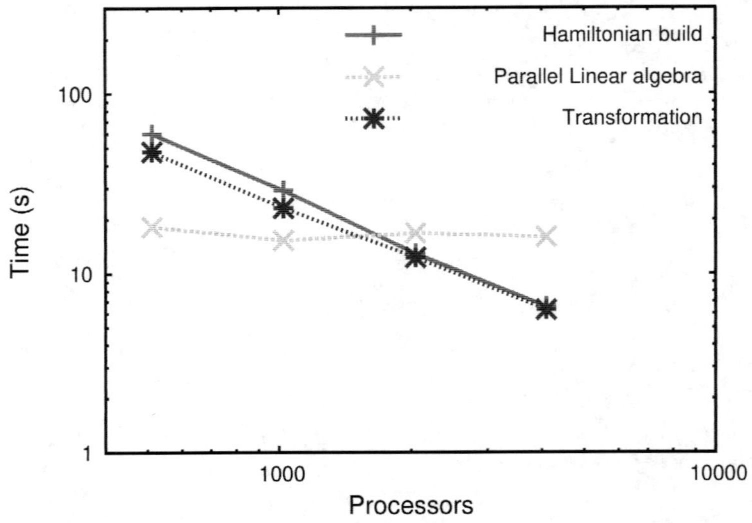

Figure 6. Scaling achieved for subspace diagonalization operations for calculation of spin-polarized 1289 atom nanoparticle with $\sim$ 7600 bands.

bandwidth demands for the parallel operations.

Our benchmarks identify a demand for improved parallel numerical libraries that have been optimized to achieve best time to solution given the availability of a comparative excess of processors. Many benchmarks of diagonalization algorithms and implementations benchmark only the time to solution for a fixed and relatively small number of processors. In the electronic structure context we have shown that there is an excess of computational power available and that, potentially, a numerical algorithm that is optimized for parallel communications might do better than those currently available. Due to the importance of these algorithms, we expect that considerable workspace would be made available in applications if a faster time to solution could be achieved. Alternatively, improved self-consistency algorithms that reduce the number of electronic iterations to reach self-consistency and overall demands on linear algebra are desired.

Finally we note that in the absence of numerical methods and implementation with improved scalability, the present limitation provide further motivation for the development and implementation of linear scaling algorithms. By exploiting the finite length scale decay of the density matrix in insulators, these methods exploit locality in the wavefunctions to avoid global dense linear algebra and instead utilize sparse methodologies.[23, 24, 25, 26]

6. Conclusions

We have shown that current implementations of plane wave density functional theory to large systems are limited in scalability primarily by the limited scalability of global linear algebra operations, particularly in modern implementations of the projector augmented wave and ultrasoft pseudopotential methods when the number of plane waves is limited. Parallel fast fourier and related plane wave operations are sufficiently scalable and little affect the overall parallel scalability of large calculations. Improving parallel scalability is critical to accessing longer time scales for systems of current interest: we identify a demand for improved parallel linear algebra libraries optimized to achieve best time to solution on an excess of processors.

7. Acknowledgments
PRCK thanks T. C. Schulthess for support and J. Larkin at the Cray Center of Excellence at ORNL for assistance with the calculations. This research used resources of the National Energy Research Scientific Computing Center and National Center for Computational Sciences, as well as the Center for Nanophase Materials Sciences, which are sponsored by the respective facilities divisions of the offices of Advanced Scientific Computing Research and Basic Energy Sciences of the U.S. Department of Energy.

References
[1] Hohenberg P and Kohn W 1964 *Phys. Rev.* **136** 864
[2] Kohn W and Sham L 1965 *Physical Review* **140** 1133
[3] Perdew J P and Wang Y 1992 *Phys. Rev. B* **45** 13244
[4] Martin R M 2004 *Electronic Structure: Basic Theory and Practical Methods* (Cambridge University Press)
[5] Payne M, Teter M, Allan D, Arias T and Joannopoulos J 1992 *Reviews of Modern Physics* **64** 1045–1097
[6] Niklasson A M N 2008 *Physical Review Letters* **100**
[7] Perdew J P and Zunger A 1981 *Phys. Rev. B* **23** 5048
[8] Davidson E R 1983 *NATO Advanced Study Institute* (*C* vol 113) (Plenum) chap Methods in Computational Molecular Physics, p 95
[9] Wood D and Zunger A 1985 *Journal of Physics A-Mathematical and General* **18** 1343–1359
[10] Kresse G and Furthmuller J 1996 *Phys. Rev. B* **54** 11169
[11] Vanderbilt D 1990 *Physical Review B* **41** 7892–7895
[12] Blöchl P 1994 *Phys. Rev. B* **50** 17953
[13] Kresse G and Joubert J 1999 *Phys. Rev. B* **59** 1758
[14] Snir M and Gropp W 1998 *MPI: The Complete Reference* 2nd ed (The MIT Press)
[15] Frigo M and Johnson S 2005 *Proceedings of the IEEE* **93** 216–231
[16] Blackford L S, Choi J, Cleary A, D'Azevedo E, Demmel J, Dhillon I, Dongarra J, Hammarling S, Henry G, Petitet A, Stanley K, Walker D and Whaley R C 1987 *ScaLAPACK Users' Guide (Software, Environments, Tools)* (Society for Industrial Mathemetics)
[17] Goedecker S, Boulet M and Deutsch T 2003 *Computer Physics Communications* **154** 105–110
[18] Vadali R, Shi Y, Kumar S, Kale L, Tuckerman M and Martyna G 2004 *Journal of Computational Chemistry* **25** 2006–2022
[19] Kresse G and Hafner J 1993 *Phys. Rev. B* **47** 558
[20] Kresse G and Hafner J 1994 *Phys. Rev. B* **49** 14251–14269
[21] Kresse G and Furthmüller J 1996 *Comput. Mat. Scie.* **6** 15
[22] Bitzek E, Koskinen P, Gaehler F, Moseler M and Gumbsch P 2006 *Physical Review Letters* **97** 170201
[23] Brazdova V and Bowler D R 2008 *J. Phys.: Condens. Matter* **20** 275223
[24] Bowler D, Miyazaki T and Gillan M 2002 *Journal of Physics-Condensed Matter* **14** 2781–2798
[25] Skylaris C K, Haynes P D, Mostofi A A and Payne M C 2008 *Journal of Physics-Condensed Matter* **20** 064209
[26] Fattebert J and Gygi F 2006 *Physical Review B* **73** 115124

Journal of Physics: Conference Series **125** (2008) 012059 doi:10.1088/1742-6596/125/1/012059

Large-scale ab initio nanostructure electronic structure calculations for energy applications

Lin-Wang Wang
Computational Research Division, Lawrence Berkeley National Laboratory, Berkeley, CA 94720, USA

E-mail: lwwang@lbl.gov

Abstract. Nanostructures have often been used to construct third-generation solar cells and for solid state lighting. To fully explore the potential for such energy applications, we need to understand the electronic and optical properties of the corresponding nanosystems. These include the quantum confinement effects, the electron hole separations, the exciton binding energy, and optical absorption spectrums. In the past ten years, we have developed a systematic approach to study such properties of thousand atom nanostructures based on ab initio calculations and large -cale comutations. In this paper, we present a few examples using such an approach to study the nanostructure properties related to energy applications.

1. Core/shell nanowire for solar cells

One of the challenges for massive solar cell deployment to solve the global energy crisis is to find cheap and abundant material. Unfortunately, the most abundant and environmentally benign semiconductors (e.g., the metallic oxides) all have band gaps too large for optimal solar cell application. One possibility is to stack two such large band gap materials together. If they have a type-II band alignment – which means one material's valence band maximum (VBM) and conduction band minimum (CBM) are both higher than the other material's VBM and CBM, respectively – then the overall band gap of the whole system will be reduced, thereby making it suitable for solar cell applications. Since the small band gap optical absorption happens only near the interface of these two materials, it is necessary to make such a device in nanoscale, so the interface area occupies a big proportion of the whole system.

We have studied ZnO/ZnS nanowires [1] using the charge patching method (CPM) [2] and direct density functional theory (DFT) calculations. In the nanowire, ZnO is the core, which is surrounded by a shell of ZnS. Both ZnO and ZnS are abundant and environmentally friendly materials. Moreover, these semiconductors are very stable. Although the band gaps of ZnO and ZnS are both larger than 3 eV, the combined system has a overall band gap of about 2 eV, which makes a theoretically possible solar cell efficiency of 23%. The absorption amplitude of the system close to the band gap is similar to that of bulk Si. Although this is not as strong as the direct band gap material, it nevertheless makes such systems feasible for solar cell applications. In figure 1, we show the geometry of the nanowire and its optical absorption spectrum. We see that there are strong absorption below the bulk ZnO and ZnS band gaps. In figure 2, we show the hole (VBM) and electron (CBM) wave functions. The electron wave function is localized in the core, while the hole wave function is localized in the shell. There is an advantage for such separation. In many solar cell devices using both inorganic nanocrystallines and organic polymers, the hole is conducted by the conjugated polymers. Thus, the hole localized at the outer shell can be readily transferred to polymers attached to the periphery of the wire.

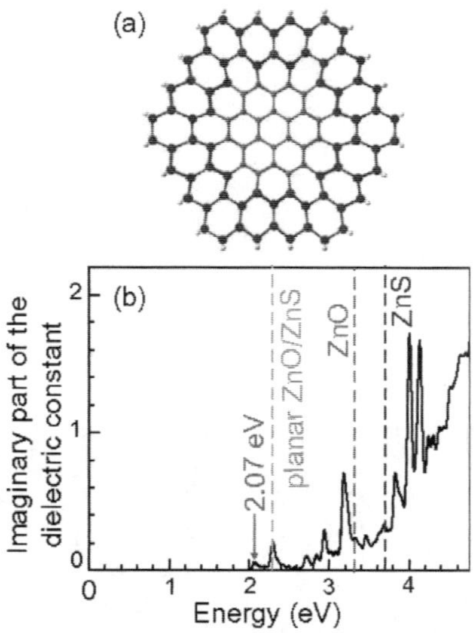

Figure 1. ZnO/ZnS core/shell nanowire (a), and the optical absorption spectrum of the nanowire (b). The core is ZnO while the shell is ZnS. The vertical dashed lines in (b) indicate the band gap of the corresponding bulk, while the planar ZnO/ZnS indicates the band gap of a ZnO/ZnS planar superlattice.

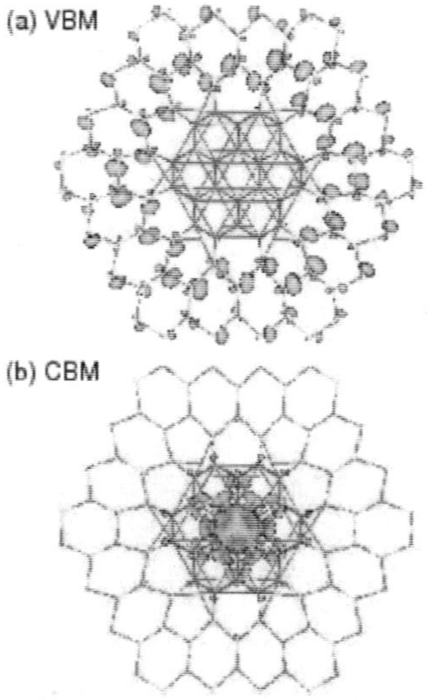

Figure 2. Valence band maximum (VBM) wave function (a) and conduction band minimum (b) states of the ZnO/ZnS nanowire.

2. A spherical core in a nanorod

Besides core/shell nanowires, we have studied other core shell structures. While spherical core/shell quantum dots have been synthesized for many semiconductor materials, recently, a nonsymmetric rod has been synthesized with one spherical core embedded in one end of the rod [3,4]. Such structure is interesting because it provides the opportunity to manipulate the electronic states through band alignment, elastic strain, surface dipole moment, and quantum confinement effect. Many of these factors are absent from spherical core/shell structures and infinitely long core/shell nanowires.

In the system we studied, a CdSe spherical core is embedded in a CdS nanorod. We first used the valence force field (VFF) method to relax the atomic positions. The charge patching method is then used to construct the ab initio DFT quality charge density of the nanosystem. This charge density is then used to calculate the local density approximation (LDA) potential in the single particle Schrodinger's equation. To overcome the well-known band gap problem of the LDA, we have modified the nonlocal parts of the pseudopotential so that the band gaps and effective masses of the resulting bulks system are correct [5]. We call the final Hamiltonian the LDA+C Hamiltonian. The band edge states of this LDA+C Hamiltonian is solved using our folded spectrum method (FSM)[6].

We have studied two nanorods with different surface passivations. Although both passivations are not realistic, they do provide idealistic models for different type of passivations. Realistic passivation modeling is not possible at this time since the atomic detail of the real passivation is not known. In our first model passivation, both Cd atoms and S atoms exist at the surface, and any atom with only one bond connected to the rest of the nanorod is removed; we will call this model Cd+S termination model hereafter. The surface atoms are then passivated with pseudo hydrogen atoms. For Cd, it is passivated by hydrogen atoms with 1.5 electron, while for S, it is passivated by hydrogen atoms with 0.5 electron. These hydrogen atoms are placed at the centers of the cut-off bonds at the surface. In our second passivation model, only Cd atoms are at the surface, including Cd atoms with only one bond connecting to the rest of the nanorod; we will call this the Cd termination model. The main difference of these two models is that they provide very different overall dipole moment of the system. In the Cd+S model, since the surface Cd-H and S-H bonds (especially at the periphery) are very different, each of these bonds provides a small dipole moment, and they sum up to give a large net surface dipole moment. This surface dipole moment will induce an internal electric field, which will strongly affect the internal electronic structure of the system. On the other hand, the bond dipoles of the Cd termination model cancel each other out among different bond directions. As a result, its net surface dipole moment is negligibly small.

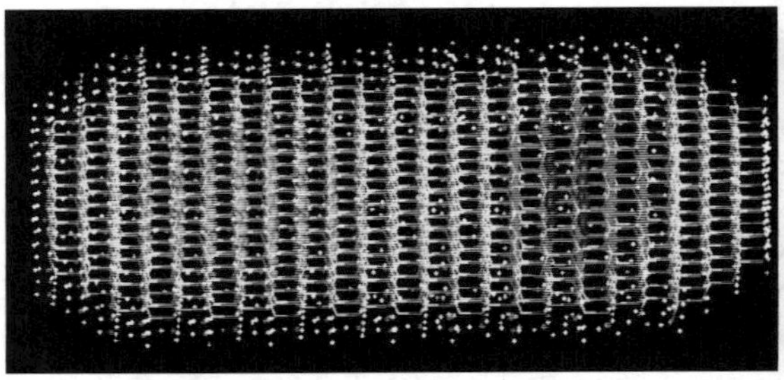

Figure 3. Spherical CdSe core (the pink color) embedded in a CdS nanorod (yellow color). The surface is terminated with Cd atoms, and the small white dots are the artificial hydrogen atoms. The blue and green are isosurfaces of hole and electron wave function squares, respectively.

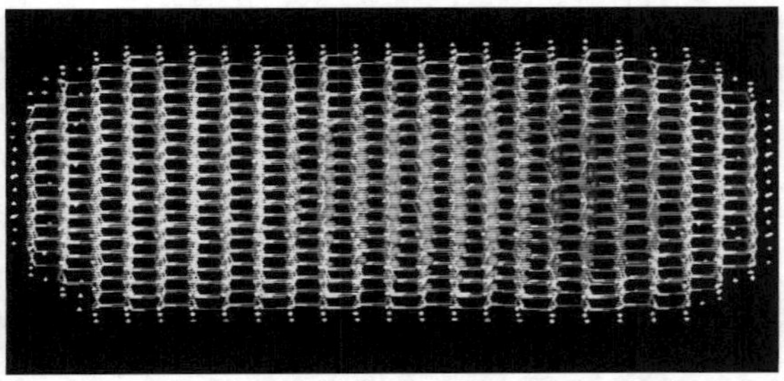

Figure 4. Spherical CdSe core (the pink color) embedded in a CdS nanorod (yellow color). The surface is terminated by Cd and S atoms. The blue and green are isosurfaces of hole and electron wave function squares, respectively.

In figure 3, we show the VBM and CBM states of the Cd terminated core/rod. As we can see, the hole state is completely localized inside the CdSe core, while the electron has been pushed outside the core, far away in the CdS rod. Note that, in the natural band alignment (e.g., if both CdSe and CdS have their fully relaxed natural lattice constants), the bulk CdS CBM state is above the bulk CdSe CBM state. Thus the electron should be inside the CdSe material. In the core/rod structure, however, the CdSe is compressed from its bulk lattice constant, as a result, its bulk CBM level has been pushed up, above the bulk CdS level. As a result, the electron state is located in the CdS. It is away from the CdSe core due to the quantum confinement effect. In figure 4, we show the VBM and CBM states of the Cd+Se terminated core/rod structure. The hole wave function is still inside the CdSe core, but now the electron wave function is much closer to the CdSe core. The reason is that the surface dipole moment of this system has provided an internal electric field, which pushes the electron to the right hand side, toward the CdSe core. Through this example, we see that how one can use band alignment, strain, dipole moment, and quantum confinement to manipulate the electron and hole localization, which is critical for many device applications.

3. Internal electric field and dipole moment of nanorods

As demonstrated in the example of the CdSe core inside a CdS nanorod, the dipole moment of the nanosystem can play an important role in the electronic structure of the system. We therefore carried out a detailed study of the dipole moment in nanocrystalline. To do this, we have used our newly developed linear scaling three dimensional fragment method (LS3DF) [7]. This method divides a nanosystem into small fragments with each fragment containing less than 100 atoms. The fragments are mutually overlapped, and their surfaces are artificially passivated. After the charge densities of all the fragments are calculated, they are patched together with both positive and negative prefactors. The negative fragments are used so that the effects of the artificial boundaries due to the subdivision can be cancelled out among different fragments. The global potential of the whole system is solved selfconsistently with the patched global charge density via a Poisson equation. Since the computationally expensive Schrodinger's equations are solved only for each individual fragments, the whole calculation scales linearly to the number of fragments, thus to the size of the system. The calculation can also be massively parallelized due to the independence of the Schrodinger's equation for different fragments. Our LS3DF code has been run on the near full Cray XT4 Franklin machine with 17,280 processors at NERSC and has reached 35 Tflops, which is 40% of its theoretical peak [8].

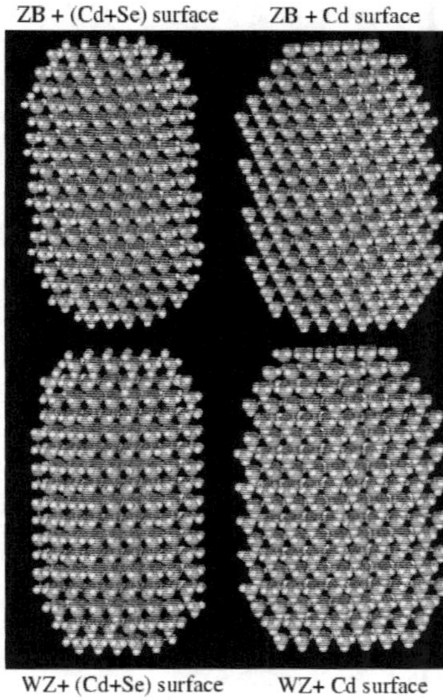

Figure 5. Four CdSe quantum rods calculated for their total dipole moments. ZB and WZ denote their crystal structures, and Cd+Se and Cd are for their surface terminations.

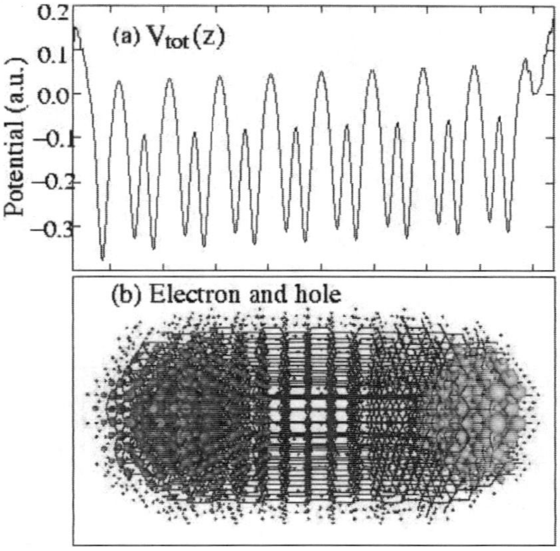

Figure 6. Internal electric potential (a) and electron and hole isosurface plots for the WZ+Cd nanorod. Its total dipole moment is 71 a.u. The red and green isosurfaces are for electron and hole, respectively.

We used the LS3DF code to calculate the self-consistent charge density and the corresponding dipole moments of the CdSe nanorods. To study the effects of different surface passivations, we used the Cd+Se termination, and Cd only termination as described in the previous section. Again, pseudo-hydrogen atoms are used to saturate the dangling bonds. To study the effect of permanent bulk dipole contribution, we tested zinc blend (ZB) bulk crystal structure and wurtzite (WZ) crystal structure. These four systems are depicted in figure 5. They are system-1, $Cd_{954}Se_{718}$, ZB structure, Cd-terminated; system-2, $Cd_{961}Se_{724}$, WZ structure, Cd-terminated; system-3, $Cd_{715}Se_{718}$, ZB structure,

Cd+Se terminated, $Cd_{714}Se_{724}$; system-4, WZ structure, Cd+Se terminated. Their calculated total dipole moments are -13.1, 71.0, -21.5, 51.8 a.u., respectively. To show the effects of these dipole moments, we plotted the electron and hole wavefunctions of the system-2 in figure 6. As one can see, the electron and hole are localized at completely the opposite ends of the nanorod due to the internal electric field induced by the total dipole moment of the system.

4. Electron and hole states in P3HT

There are many nanostructure electronic and optical applications that involve both inorganic nanocrystalline and organic polymers. Thus, it will be interesting to calculate the electronic structures of the organic polymers in the same footing as we calculate the inorganic semiconductor nanocrystallines. Recently, we have applied our charge patching method to organic polymers [9]. The accuracy we obtained is similar to that of inorganic nanocrystals, which is 20-30 meV in the absolute eigen energy error. The error for the eigen energy differences among adjacent states is much smaller, typically in a few meV.

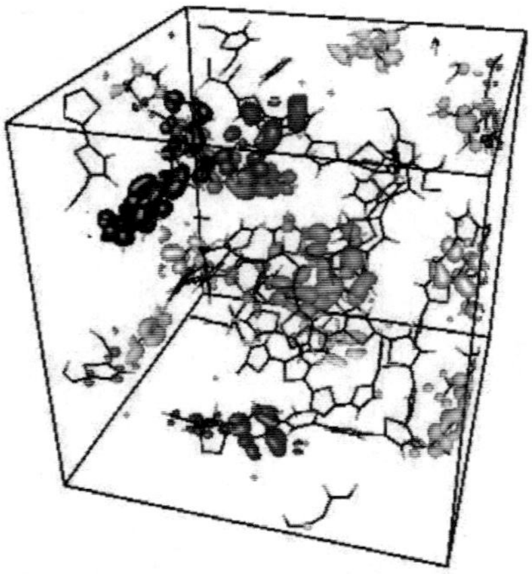

Figure 7. The first seven conduction band states in an energy ascending order: dark blue, green, light blue, red, pink, black, and gray. The side chains of the P3HT are not shown for clarity.

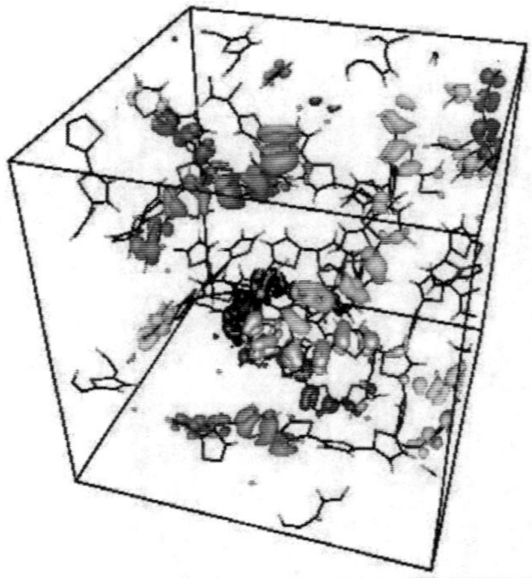

Figure 8. The first seven valence band states in an energy descending order: dark blue, green, light blue, red, pink, black, and gray.

Poly(3-hexylthiophene) P3HT is an often-used organic polymer for electronic and optical applications. After annealing, P3HT likes to form local plate like crystal domains, and between the domains they are also amorphous like structure. While the electronic structure of the perfect crystal can be studied easily by small system direct ab initio calculations, the electronic structures in the amorphouslike regions are more difficult to be calculated. Here we have applied our CPM to study the electronic structures in such amorphous regions. The empirical force field method, which is accurate in describing the nonchemical interactions between the polymers, is used in molecular dynamics simulation and simulated annealing to obtain the morphologies and atomic structures of the polymers. Once the atomic positions are obtained, CPM is applied to get the charge density of the system. As in the inorganic nanocrystalline, the charge density is then used to calculate the potential of the system, and the folded spectrum method [6] is used to calculate the band edge states.

Figure 7 shows the atomic structures of the P3HT molecules and the corresponding conduction band states. In the calculation, there are 5 P3HT chains, each with 20 pentagon units. In total, there are 2510 atoms. They are placed in a periodic supercell with the average density equal to that of the usual P3HT polymer blend. Simulated annealing is performed based on the empirical force field, and the amorphous state atomic structures are obtained. In figures 7 and 8, for clarity, only the main P3HT chains are shown, not the side chains. The seven conduction band states shown in figure 7 have an average energy separation of about 30 meV. Figure 8 shows the first seven valence band states for the same molecule configuration as in figure 7. The average energy separations between these seven valence state is about 43 meV, which is larger than the corresponding conduction band value. In figures 7 and 8, the isosurface for each state encloses 50% of its electron charge. From figures 7 and 8, we can see that the wave functions are localized to 2 or 3 pentagon units in most cases. They are not extended to all the units in a given chain. Besides, there are strong correlations between electron and hole states. That means if a segment is good for a band edge electron state, it is probably also good for a band edge hole state. The localizations of the electron and hole states will have significant impacts on their mobilities. Our calculations shed light on the electronic properties of this widely used, yet poorly understood complicated polymer system.

5. Conclusion

Energy applications of nanostructures require the understanding of their electronic structures and optical properties. We have developed a comprehensive approach to study such properties, especially for electrons, hole localizations, quantum confinement effects, and optical properties. In the current paper, we have demonstrated how to use our approach to study these properties for different systems with ab initio accuracy. Our approach includes a non-self-consistent charge patching method and self-consistent linear scaling three-dimensional fragment method. With massively parallel computers, it is now possible to calculate thousand-atom nanosystems on a routine basis. This is a major achievement compared to ten years ago, when thousand-atom systems could be calculated only with empirical pseudopotentials or with tight-binding models.

Acknowledgments
This work was supported by the US Department of Energy, Office of Science, Basic Energy Science under LAB03-17 initiative (Theory, Modeling and Simulation in Nanoscience), contract No. DE-AC02-05CH11231. We used the computational resources of the National Energy Research Scientific Computing Center (NERSC) and National Center for Computational Science (NCCS), both are supported by the Office of Science of the U.S. Department of Energy.

References

[1] Schrier J, Demchenko D O, Wang L W and Alivisatos A P 2007 Optical properties of ZnO/ZnS and ZnO/ZnTe heterostructures for solar cell applications *Nanolett.* **7:**2377
[2] Wang L W 2002 Charge-density patching method for unconventional semiconductor binary systems *Physical Rev. Letters* **88**:256, 402
[3] Carbone L, *et al.* 2007 Synthesis and micrometer-scale assembly of colloidal CdSe/CdS nanorods prepared by a seeded growth approach *Nano Lett.* **7:**2942
[4] Talapin D V, Nelson J H, Shevchenko E V, Aloni S, Sadtler B and Alivisatos A P 2007 Seeded growth of highly luminescent CdSe/CdS nanoheterostructurees with rod and tetrapod morphologies *Nano. Lett.* **7:**2951
[5] Li J and Wang L W 2005 Band-structure-corrected local density approximation study of semiconductor quantum dots and wires *Phys. Rev. B* **72:**125325
[6] Wang L W and Zunger A 1994 Solving Schrodinger's equation around a desired energy: Application to silicon quantum dots *J. Chem. Phys.* **100:**2394
[7] Wang L W, Zhao Z and Meza J 2008 A linear scaling three dimensional fragment method for large scale electronic structure calculations *Phys. Rev. B* **77:**165113
[8] Wang L W, Lee B, Shan H, Zhao Z, Meza J, Strohmaier E and Bailey D Linear scaling 3D fragment method for large scale electronic structure calculations (unpublished).

[9] Vulmirovic N and Wang L W 2008 Charge patching method for electronic structure of organic systems *J. Chem. Phys.* **128:**121102

Hierarchical petascale simulation framework for stress corrosion cracking

Priya Vashishta[1], Rajiv K Kalia[1], Aiichiro Nakano[1], Efthimios Kaxiras[2], Ananth Grama[3], Gang Lu[4], Stephan Eidenbenz[5], Arthur F Voter[5], Randy Q Hood[6], John A Moriarty[6] and Lin H Yang[6]

[1]University of Southern California, Los Angeles, CA 90089-0373, USA
[2]Harvard University, Cambridge MA 02138, USA
[3]Purdue University, West Lafayette, IN 47907, USA
[4]California State University Northridge, Northridge, CA 91330, USA
[5]Los Alamos National Laboratory, Los Alamos, NM 87545, USA
[6]Lawrence Livermore National Laboratory, Livermore, CA 94550, USA

E-mail: priyav@usc.edu

Abstract. We are developing a scalable parallel and distributed computational framework consisting of methods, algorithms, and integrated software tools for multi-terascle-to-petascale simulations of stress corrosion cracking (SCC) with quantum-level accuracy. We have performed multimillion- to billion-atom molecular dynamics (MD) simulations of deformation, flow, and fracture in amorphous silica with interatomic potentials and forces validated by density functional theory (DFT) calculations. Optimized potentials have been developed to study sulfur embrittlement of nickel with multimillion-to-multibillion atom MD simulations based on DFT and temperature dependent model generalized pseudopotential theory. We have also developed a quasi-continuum method embedded with quantum simulations based on DFT to reach macroscopic length scales and an accelerated molecular dynamics scheme to reach macroscopic time scales in simulations of solid-fluid interfaces that are relevant to SCC. A hybrid MD and mesoscale lattice Boltzmann simulation algorithm is being designed to study fluid flow through cracks.

1. Introduction

Corrosion is an enormously complex technological and economic problem with an annual cost of about 3% of the U.S. gross domestic product. The basic requirements for the operation of structural systems exposed to corroding conditions under stress loads are safety and reliability. Such safe and reliable operation is endangered by the uncertainties in stress corrosion cracking (SCC). To prevent SCC and to predict the lifetime beyond which SCC may cause failure requires that we understand the atomistic mechanisms underlying SCC; that is the conditions influencing initiation, dynamics, and growth rates of SCC.

The science and engineering of SCC provide a rich and challenging set of applications involving simultaneous chemical, physical, and material processes. To achieve the sound atomistic/mechanistic basis that would allow designers to optimize the performance of structures and systems in chemically aggressive environments, we must develop first-principles models for all aspects of SCC, including the role of pH, electrochemical potential, material microstructure, local temperature, and local stress.

2. Scalable and adaptive algorithms for hierarchical simulations

We are developing scalable parallel algorithms for quantum molecular dynamics (QMD), large-scale atomistic simulations – namely, molecular dynamics (MD) based on quantum-based model generalized pseudopotential theory (MGPT) and reactive force fields (F-ReaxFF) – and for accelerated MD simulations. The algorithms are based on our embedded divide-and-conquer (EDC) algorithmic framework combined with adaptive multigrids. Scalability has been achieved on 10^5 processors through linear-scaling algorithms and performance-optimization techniques. For example, we have developed a highly efficient, modular, and effective reactive MD simulation code employing state-of-the-art data structures, algorithms, and numerical methods. This code reduces the memory footprint by a factor of up to 3 and improves speed by a factor of up to 4, compared to other reactive codes. The code employs dynamic compressed octrees that minimize storage and time for fast neighborhood computations. The terms are themselves computed through composite lookup (the entire potential function is indexed into a dynamic table). Since the arguments to these potential functions tend to be highly clustered (bond lengths, angles, and torsion terms), the use of dynamic lookups results in significant speedups, while being memory efficient. Charge equilibration is achieved using an efficient predictor-corrector generalized minimal residual (GMRES) iterative scheme. This scheme is shown to achieve excellent solution times and convergence rates, while being memory efficient (small Krylov subspace). We have demonstrated 201 billion-atom MD, 1.59 billion-atom reactive MD, and 17.7 million-atom (1.56 trillion grid points) quantum-mechanical MD in the framework of divide-and-conquer density functional theory on adaptive multigrids with the parallel efficiency as high as 0.99 on 196,608 Blue Gene/L processors.

3. Damage, flow, and fracture in amorphous silica

We have performed large-scale MD simulations to investigate shear-induced void deformation and nanoductility and defect nucleation and migration in nanoindentation of a-SiO_2. The simulations are based on an interatomic potential, which incorporates ionic and covalent effects through a combination of two-body and three-body terms. The interatomic potential was validated by comparing the MD simulation results for structural and mechanical properties of a-SiO_2 with experimental measurements and quantum mechanical calculations based on density functional theory performed by our SciDAC team member Lin Yang at LLNL.

3.1 Damage and cracking in shearing silica glass

One set of SCC simulations is focused on shear-induced deformation and breakup of nanovoids with and without water in silica glass. The simulations involve up to 633 million atoms. The initial diameter of the void ranges between 3 and 50 nm, which covers nearly the entire range of void sizes observed in the damage zone in dynamic fracture simulations [1] and quasi-static stress corrosion cracking experiments [2] on silica glass.

Figure 1 shows the degree of deformation for a void in silica glass at room temperature. The initial diameter of the void is 10nm. As time evolves, the void shape changes to ellipsoidal and damage in the form of nanocracks nucleates on the void surface at the two ends of the short axis of the ellipsoid; see the snapshot in figure 1(a). The same kind of deformation and damage are also observed for voids of initial diameters 3 nm and 50 nm. Stress calculations reveal that nanocracks form along the direction of maximum tensile stress.

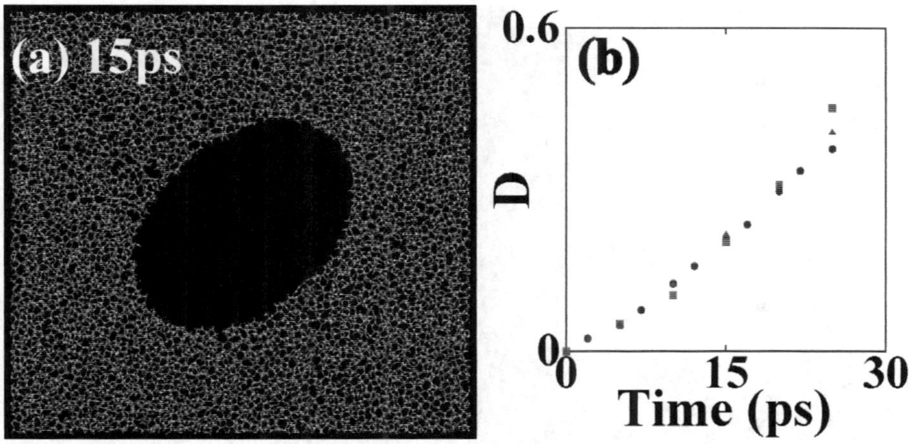

Figure 1. (a) Shear deformation of a spherical void of diameter 10nm to an ellipsoid after 15 ps; (b) time variation of the deformation parameter, *D*, for voids of initial diameters 3 nm (blue circles), 10 nm (red squares), and 50 nm (green triangles). Shear rate $=10^{12}\ s^{-1}$.

Figure 1(b) shows the time variation of the deformation parameter, *D*, for initially spherical voids of diameters 3 nm (blue), 10 nm (red), and 50 nm (green). The deformation parameter is defined as $D = (L-B)/(L+B)$, where *L* and *B* are the major and minor axes of the ellipsoid, respectively. In all three cases *D* increases linearly with time up to 15 ps, which is an indication that the void deformation is elastic up to a shear strain of 15%. To distinguish between elastic and plastic void deformations, we switch off the shear strain rate at a certain value of the strain and let the system relax without shear. We find that if the shear strain rate is dropped to zero before the strain reaches 15%, voids recover their initial spherical shapes and sizes. This situation confirms that the void deformation is elastic so long as the strain is below 15%.

Plastic deformation in shearing silica glass appears after 15 ps (strain > 15%) as the deformation parameter begins to increase non-linearly with time. Figures 2(a) and 3(b) are snapshots showing long, thin voids with nanocracks on the surfaces near the middle of the long axis. Initially, these were spherical voids of diameters 3 nm and 50 nm. In both cases, the voids become threadlike before breakup, which occurs above a strain of 40%. Figure 2(c) shows plastic deformation of a void, which was initially a sphere of diameter 10 nm. The snapshot taken at 35ps shows a long, thin void with ends deformed roughly like an *S* shape. There are nanocracks on the surface and near the ends of the void. As time evolves, nanocracks grow and the void becomes more elongated, and at 40 ps the void fragments; see figure 2(d).

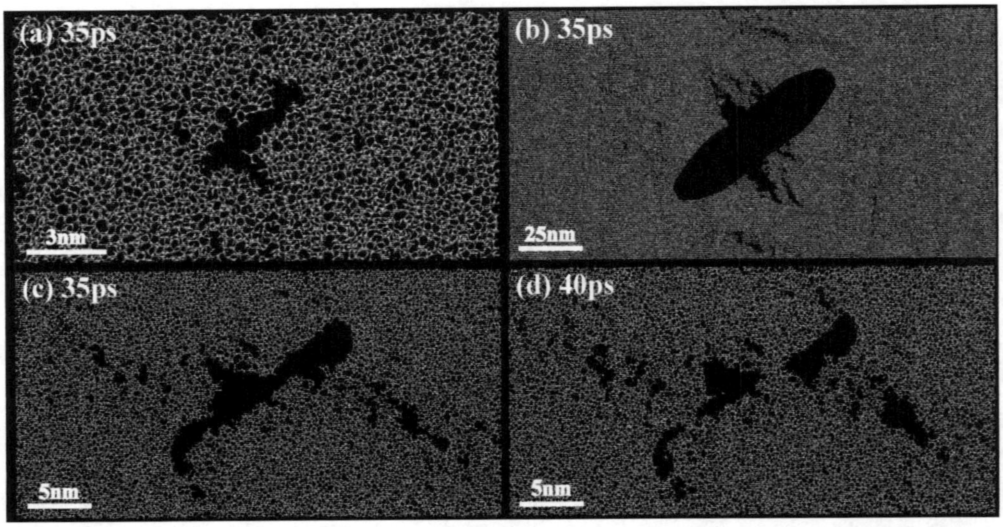

Figure 2. Panels (a) and (b) show plastic deformations of voids of initial diameters 3 nm and 50 nm, respectively. Panel (c) shows a plastically deformed void that was initially a sphere of diameter 10 nm; and panel (d) shows its breakup at 40 ps.

Analyses of elastic-to-plastic void deformation and crack initiation and growth reveal a novel mechanism involving strain-enhanced defect transport. In the unstrained a-SiO_2, each Si atom (yellow) is connected to four O atoms (red) in the form of a SiO_4 tetrahedron, and these tetrahedra are linked into nanometer size –Si-O-Si-O– rings through corner-sharing O atoms. In figure 3(a), green and blue regions represent 6- and 9-member rings, respectively, at a strain of 5%. The magenta atom is a bridging O, and the blue atom is a threefold coordinated Si atom in the 9-member ring. Figure 3(b) shows that at a strain of 8% the magenta O atom becomes nonbridging while the blue Si atom remains undercoordinated. The grey ring with these Si and O defects is a 13-member ring. At a strain of 10%, the green Si and magenta O atoms become fully coordinated by bonding with each other. The blue Si is still undercoordinated but now belongs to a 14-member ring (yellow). Nonbridging oxygen and threefold coordinated silicon atoms play pivotal roles in the nucleation of cracks through the enlargement of –Si-O-Si-O– rings and in void deformation and shear flow in silica glass.

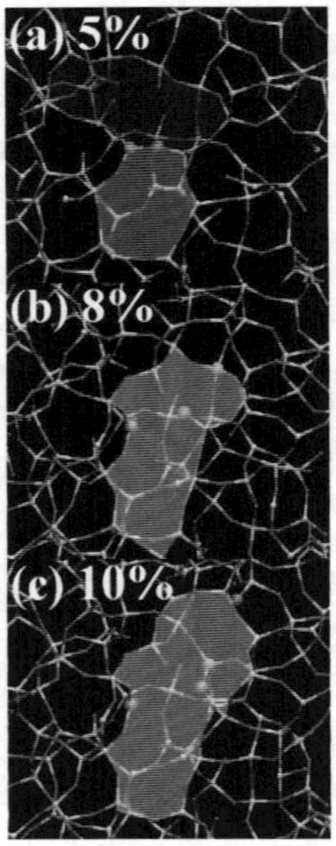

Figure 3. Defects responsible for shear deformation and flow in silica glass.

We have also investigated changes in the void shape and elastic-to-plastic transition in silica glass at 1200K. In contrast to the case at room temperature, the voids are partially filled with atoms released from the void surfaces and the density of these atoms inside the voids increases with time. The voids still undergo spherical-to-ellipsoidal shape transformation, but the time variation of the deformation parameter is slightly less than that at room temperature. Nanocracks appear on void surfaces with the onset of plasticity. Voids deform into long ellipsoidal shapes with pointed ends, and their breakup is again preceded by threadlike structures just as at room temperature.

Taken together, MD simulations reveal that shear-induced void deformation, damage, and flow in silica glass have the same underlying mechanism involving Si-O bond breaking and the migration of threefold coordinated silicon and non-bridging oxygen defects on –Si-O-Si-O– rings. The density of these defects remains small and nearly constant in the elastic deformation regime. With the onset of plasticity at a shear strain of 15%, the defect density increases and reaches a plateau soon after a void breaks up. Despite the enormous differences in spatio-temporal scales, the observed shape changes and fragmentation of voids in silica glass are remarkably similar to the deformation and breakup of macroscopic inviscid drops at high shear rates.

3.2 Defect migration and recombination in nanoindentation of silica glass

Multimillion-atom MD simulations are performed to investigate plasticity and flow mechanisms in nanoindentation of amorphous silica [3]. The simulations reveal migration of nonbridging oxygens and their recombination with undercoordinated silicons in the densified plastic region under and the material pileup region around the indenter. The migration involves chains of defects switching bonds in string-like processes. These defects were postulated by Mott to explain shear flow in silica glass.

Figure 4(a) shows a snapshot of the system at the maximum indentation depth. The amorphous film contains 40 million atoms and has dimensions 100×100×60 nm^3. The indenter, a square-based rigid pyramid with an apical angle of 90°, interacts with the substrate through a repulsive potential. We have used two kinds of indenters to study the effect of the indenter curvature: an atomistically sharp indenter and an indenter whose end is spherical with a diameter of 10 nm. Figure 4(b) displays the applied load in the MD simulation as a function of the penetration depth of the indenter. From the load and residual indentation impression, we obtain the hardness, H, of a-SiO_2. In the case of an atomistically sharp indenter, H = 7.2 GPa, and for the slightly blunt indenter H = 8.0 GPa. The AFM indentation experiment on fused silica yields a hardness value of 10 GPa.

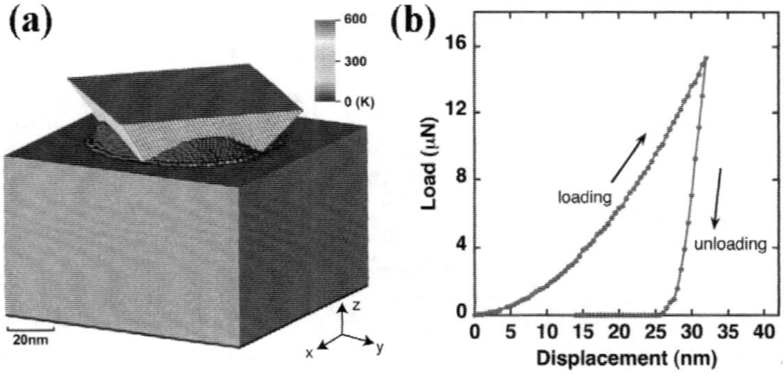

Figure 4. (a) Pileup around the indenter at the maximum indentation depth. Indentation causes localized damage creating high-temperature area underneath the indenter, which facilitates pile-up through plastic flow. The colors represent temperature. (b) Load versus displacement during loading and unloading phases of indentation.

Figure 5 shows snapshots of the indented silica film at indentations depths (a) h = 11 nm and (b) 32 nm . For clarity, only a slice of the system without the indenter is shown here. At low indentation depths (e.g., h = 11nm), the material close to the bottom of the indenter densifies but there is no material pile-up around the indenter. At the maximum indentation depth of 32 nm, we observe significant pileup and a 20% increase in the mass density of a-SiO_2 beneath the indenter relative to the mass density of bulk silica glass under ambient conditions (2.2 g/cc). In the pileup region shown in panel (b), the density is 18% lower (1.8 g/cc) than in the bulk system. During the unloading phase of the simulation, the silica substrate exhibits slight elastic recovery and a decrease in the density under the indenter. Panel (c) is a snapshot of a slice of the system after the indenter is pulled out of the substrate. Here the density just beneath the indenter goes back to the normal bulk a-SiO_2 density (2.2 g/cc), but there is still a considerable pileup around the indenter.

Detailed analyses of atomistic configurations reveal how plastic deformation and pileup occur in the nanoindentation of a-SiO_2. In the normal SiO_2 glass, each Si atom is bonded to 4 O in the form of a tetrahedron and each O bonds with two Si atoms to connect neighboring SiO_4 tetrahedra. These corner-sharing tetrahedra form closed –O–Si–O–Si–O– rings of various sizes, with a peak in the distribution around 6-member rings, that is, 6 Si–O connected pairs. According to Mott, undercoordinated Si and nonbridging O point defects diffusing on –Si-O-Si-O– rings are responsible for plastic flow in silica glass. These point defects in the silica network glass play a central role in the plastic deformation and pileup we observe in both nanoindentation simulations. The defects are involved in bond-switching events, where either a silicon or an oxygen atom severs its bond with one of its neighbors and shortly thereafter (~ 10 ps in simulation 1) forms a new bond with a different neighbor atom. Figures 6 (a)–(c) show how a bond-switching event occurs in the case of a nonbridging oxygen atom. The three atoms involved in this event are labeled as O1 (initial oxygen defect), Si2 (bond switching silicon) and O3 (the final oxygen defect). Initially, O1 is a nonbridging oxygen, whereas Si2 and O3 have normal coordinations (4 O neighbors for Si2 and 2 Si neighbors for O3). At the transition state, the Si2-O3 bond breaks, and these two atoms become undercoordinated. Subsequently, Si2 forms a new bond with O1, and they both become fully coordinated, whereas the oxygen labeled O3 remains a defect.

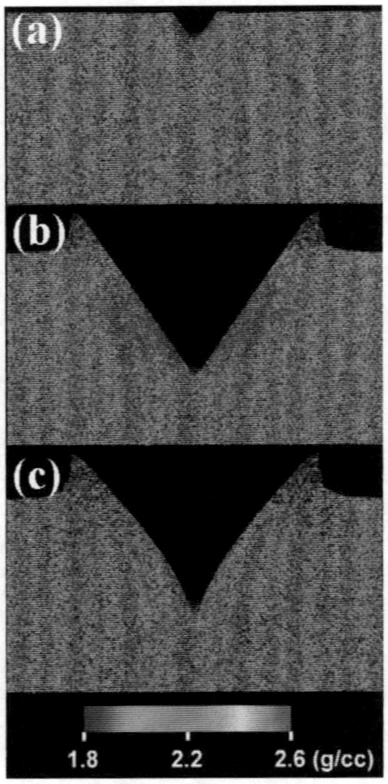

Figure 5. Density distribution before pileup (a), at the maximum indentation-depth (b), and after unloading (c).

Another bond-switching event we observe involves the annihilation of a nonbonded pair of undercoordinated silicon and oxygen atoms. Figures 6 (d)-(f) show this pair-annihilation event. The four atoms involved in this event are labeled as Si1 (an undercoordinated silicon), O2 (bond-switching oxygen), Si3 (bond-switching silicon), and O4 (a non-bridging oxygen). Initially, O2 is bonded with Si3 and they both have normal coordinations of 2 and 4, respectively. In the transition state, O2-Si3 bond breaks, and they both become undercoordinated. In the final state, Si1 forms a bond with O2 and Si3 with O4 and now these four atoms have normal coordinations.

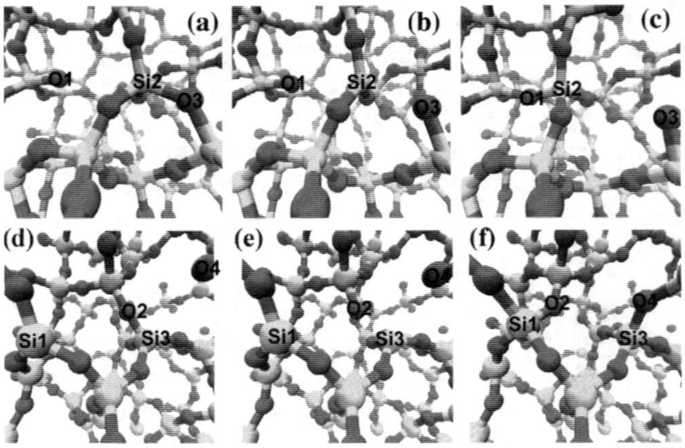

Figure 6. (a)-(b): Atomic configurations showing transport of oxygen defects; (a) the initial state, (b) the transition state, and (c) the final state. The oxygen defect migrates by switching covalent bond from Si2-O3 to O1-Si2. (d)-(f): Atomic configurations showing an annihilation event for a pair of defect atoms; (d) the initial state, (e) the transition state, and (f) the final state.

Our simulations also reveal a defect transport mechanism in which a defect migrates a considerable distance via a chain of bond-switching events. This is an important mechanism in the plastic flow of silica glass. Panels (a) and (b) in figure 7 and show two examples of a bond-switching chain event. In panel (a), an undercoordinated silicon migrates via a bond-switching chain involving seven atoms and in panel (b) the annihilation of a defect pair is mediated by a bond-switching chain of six atoms. Here yellow and red spheres represent positions of silicon and oxygen atoms, respectively, after each multiatom bond-switching event. In the transport event for a silicon defect, the defect effectively migrates about 1nm in 10 ps.

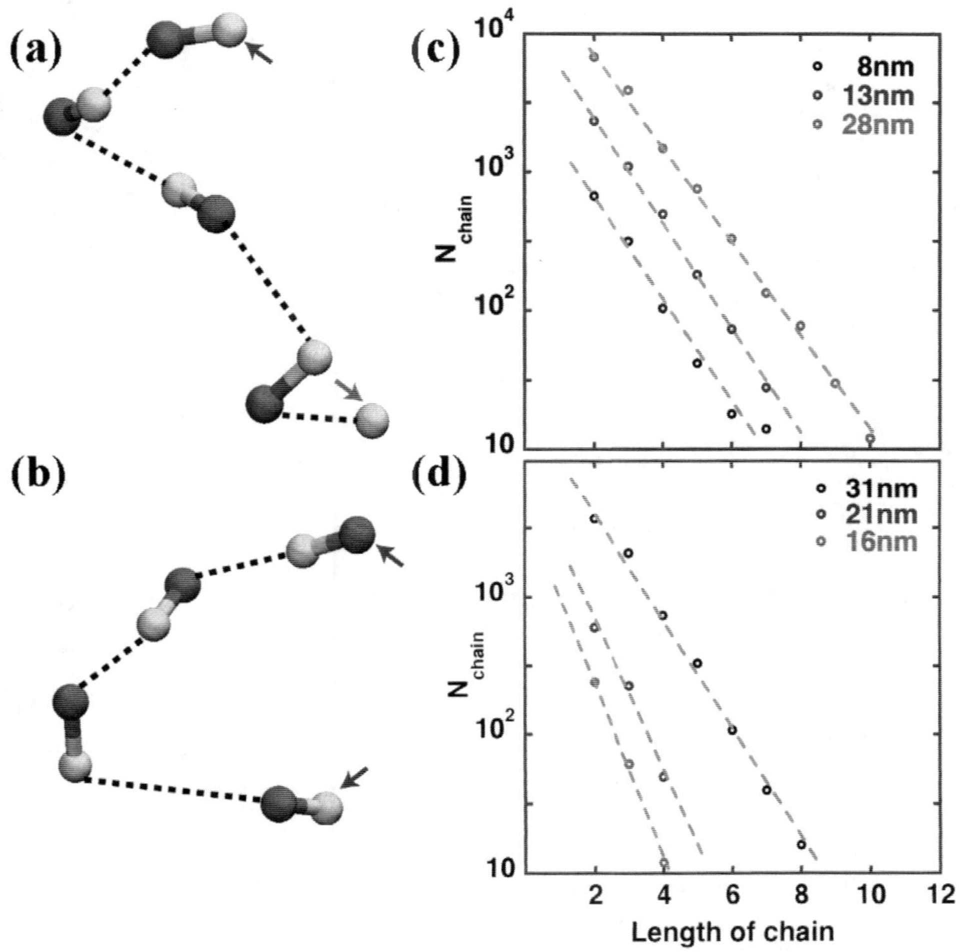

Figure 7. Snapshots of defect migration pathways ((a) and (b)) in a ball-and-stick representation. Yellow and red spheres represent silicon and oxygen atoms. Black dotted lines indicate covalent bonds before defect migration. Blue and red arrows represent initial and final positions of defect atoms, respectively. (a) Migration of a silicon defect. (b) Annihilation of a pair of silicon and oxygen defects. Panels (c) and (d) show the length of the string-like defect motion as a function of the number of migration events during loading and unloading, respectively. Mechanical loading by the indenter induces longer migration pathways.

The length of the chain of bond-switching events varies with mechanical loading. Panels (c) and (d) in figure 7 show the dependence of the number of bond-switching events N_{chain} on the chain length l at several indentation depths. In the loading phase, longer bond-switching chains appear as the indentation depth increases. Note that the slopes at the three indentation depths are almost constant in the semi-log plot, which implies exponential distributions $N_{chain} \propto \exp(-l/l_0)$ with nearly the same characteristic length l_0. The slopes differ significantly before (h = 31nm) and after (h = 16 and 21nm) the indenter is pulled out of the substrate, which shortens the chain length.

4. Hybrid atomistic/mesoscale simulations of fluid flow through cracks

We are designing an algorithm to interface MD with a mesoscale approach to fluid flow simulation, namely, the lattice Boltzmann (LB) method, which solves a minimal form of the Boltzmann kinetic equation ignoring all details of molecular motion except those needed to recover hydrodynamic behavior at the macroscopic scale. For hierarchical fluid flow simulations, we have developed an extensible LB computational framework for terascale-to-petascale parallel/distributed platforms. The LB framework includes scalable algorithms and tools as well as capabilities for handling, analyzing, and visualizing petabyte simulation datasets. It allows collaborative construction and execution of complex, multi-component, computationally demanding hybrid numerical simulations of fluid flow. We have achieved scalability beyond 10^5 processors through linear-scaling algorithms and performance-optimization techniques. Recently, we have performed LB simulations of fluid flow through a network of cracks in amorphous silica on a cluster of PlayStation3 consoles (figure 8).

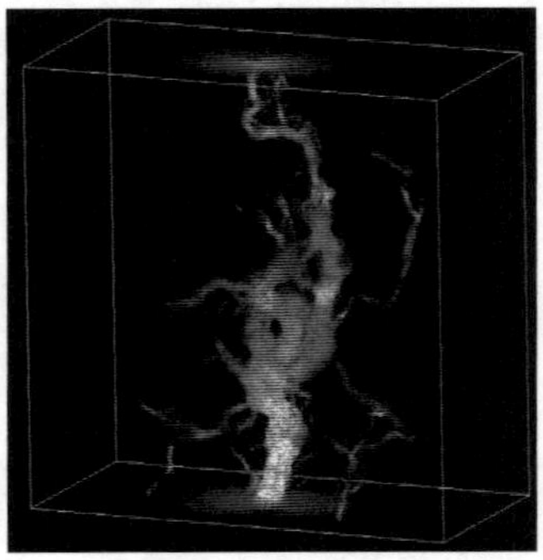

Figure 8. Lattice Boltzmann flow simulation through nanopores and cracks in silica glass. The silica simulations are performed with MD for systems between 10^6 and 10^9 atoms.

Currently, we are developing an automated model transitioning to embed higher fidelity MD simulations concurrently inside LB simulations only when and where they are required (figure 10). We are using a domain decomposition algorithm based on Schwartz alternating method to implement hybrid MD-LB simulations. In this scheme, the system is decomposed into (1) a purely atomistic region described by MD, (2) a continuum region described solely by LB, and (3) an overlapping MD-LB region. Since the MD region is open and particles move out of the MD into the LB region, it is necessary to design an efficient algorithm to insert particles back into the MD region to keep the average particle density constant. This is a challenging task, especially in the case of a dense fluid.

We are using a highly efficient algorithm called USHER to introduce particles effectively even in a dense fluid. The attractive features of USHER are as follows:

1) If the initial random position of the inserted particle is too close to an existing particle, USHER solves this problem in one update.
2) The displacement in each step is bounded by an upper limit.
3) The final energy of the inserted particle is close to the average energy per particle in the system.

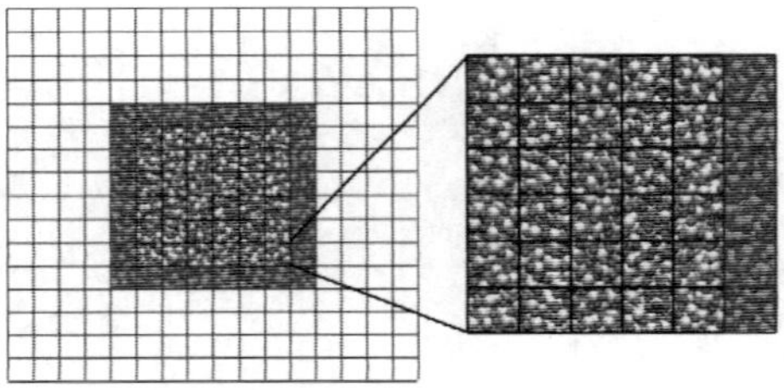

Figure 9. Domain decomposition for the hybrid MD/LB simulation, where squares are LB cells, yellow and red spheres are atoms in MD, and the shaded region represents the Schwartz alternating MD/LB coupling.

5. Optimized potentials for MD simulations based on DFT and temperature-dependent model generalized pseudopotential theory

The LLNL group is tasked with providing the capability to perform ultrascale atomistic simulations to study SCC in materials with quantum-mechanical accuracy. To this end, they are developing a new hybrid QMD-MD/MGPT simulation algorithm for materials properties to be used in SCC applications. This algorithm directly links first-principles DFT quantum simulations, namely, quantum molecular dynamics (QMD) [4], with large-scale atomistic simulations, namely, molecular dynamics with quantum-based model generalized pseudopotential theory (MGPT) interatomic potentials [5-7]. By developing this capability for small systems in the solid and in the liquid, one can dramatically extend the effective QMD time scale and harvest optimized *temperature-dependent* MGPT potentials in the long time limit, as shown schematically in figure 10. In this process, there will be a real-time information feedback loop between the alternate QMD and MD/MGPT segments of the simulation, with both snapshot and accumulated QMD information being used to achieve optimization of the potentials. The optimized potentials can then be reused in ultra-large *linear-scaling* MD/MGPT atomistic simulations of materials properties. The initial goal is to simulate one billion atoms at high temperature in prototype systems (e.g., Mo, Ni) via MD/MGPT with the accuracy to a direct QMD simulation on one hundred atoms.

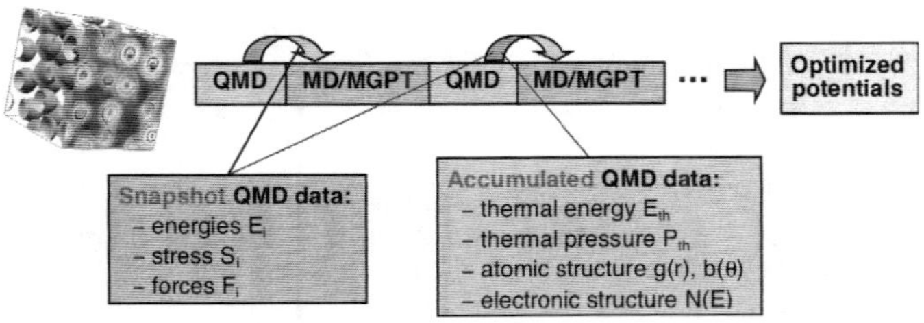

Figure 10. Schematic representation of the hybrid QMD-MD/MGPT simulation algorithm. A real-time information feedback loop between the alternate QMD and MD/MGPT segments of the simulation will allow optimized and reusable MGPT potentials.

The research accomplishments for the past year have been primarily in three areas: (1) progress in developing the hybrid QMD-MD/MGPT simulation algorithm and optimized temperature-dependent MGPT potentials for a Mo prototype; (2) calculation of DFT data bases of fundamental materials properties for Ni and Ni_3S_4; and (3) development of preliminary $T = 0$ MGPT potentials for Ni using the DFT data base. We have been greatly aided in this work with a large *LLNL Institutional Computing Grand Challenge* allocation of dedicated high-performance computer time (140,000 CPU hours per week) on the new 44 Tflops Atlas parallel machine at LLNL. This past year we successfully competed a renewal proposal to continue this allocation for a second year (through May 2009) [8].

5.1 Hybrid QMD-MD/MGPT algorithm for optimized potentials

There are four main technical tasks in the development hybrid QMD-MD/MGPT simulation capability.

5.1.1 Development and testing of an operational QMD-MD/MGPT simulation code.

During the past year we have fully adapted our fast portable quantum simulation code *FEQMD* [9] to perform efficient hybrid QMD-MD/MGPT simulations on Atlas. The resulting parallel QMD-MD/MGPT code is now called *P^3MD*. In QMD mode, this code uses the well-established plane-wave pseudopotential method to solve the finite-temperature DFT quantum-mechanical equations for the valence and outer core electrons in the system at each time step in the simulation, thereby providing accurate instantaneous forces on the ions to advance the molecular dynamics. Efficient parallelization is obtained by distributing the plane-wave basis functions, electronic energy bands and reciprocal-space k points across a group of nodes in an optimum way. A custom parallel 3D fast Fourier transform (FFT) routine has been developed and used to transfer information between reciprocal space, where the DFT equations are solved, and real space, where the forces are needed. In MD/MGPT mode, *P^3MD* evaluates the real-space, finite-temperature, many-body MGPT potentials at each time step, including two-, three- and four-ion interactions, to provide corresponding quantum-based energies and forces. This is done using the fast matrix version of MGPT, which also provides the most general functional form of the potentials currently available based on noncanonical *d* bands. In the QMD and MD/MGPT modes, the same molecular dynamics equations are solved, and only the input forces are different. In the hybrid QMD-MD/MGPT simulation, *P^3MD* performs constant-volume, constant-temperature molecular dynamics in a chosen fixed-shape computational cell. Ion positions and velocities are passed forward from one segment of the simulation to the next to provide initial conditions. QMD snapshot and accumulated information is passed forward to test and refine the MGPT potentials at the beginning of each MD/MGPT simulation segment.

5.1.2 Development and application of a robust simulation algorithm to produce optimized MGPT potentials.

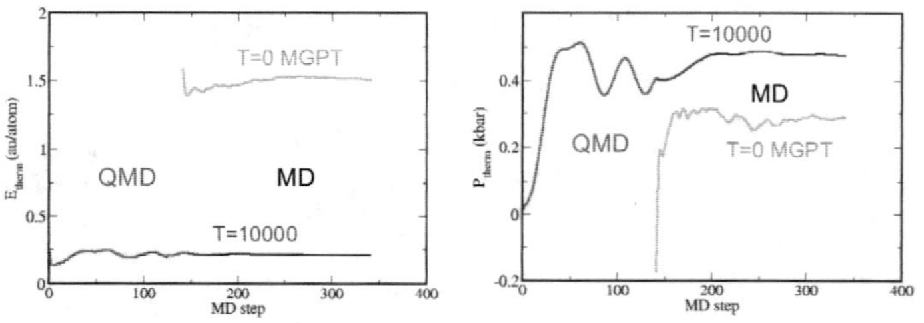

Figure 11. Thermal energy and pressure in Mo at T = 10,000 K and an atomic volume of 105.1 au, as obtained in 54-atom hybrid QMD-MD/MGPT simulations. Use of temperature-dependent 10,000-K MGPT potentials allows a smooth join of the QMD and MD/MGPT simulation segments and rapid convergence of the derived potentials.

Developing an algorithm to produce optimized MGPT potentials involves two main considerations: (1) the functional form of the potentials and what quantities are allowed to be optimized; and (2) the specific QMD data to be used to optimize the potentials and how that data is introduced. At present we have adopted a provisional functional form based on first-principles theory and focused on developing an effective strategy for (2) in the context of a Mo prototype system. The potential form currently used is a finite-temperature generalization of the well-established DFT-derived $T = 0$ form we successfully developed in previous research and includes the introduction of noncanonical d bands. The main parameters to be refined are a small set of potential coefficients plus two d-band variables. The available QMD data base of information for that refinement is extremely rich and includes snapshot energies and forces for thousands of ion configurations and accumulated information on thermal energies and pressures, atomic structure (pair and bond-angle distribution functions), electronic densities of states, and electron density. In our first-generation algorithm, we have used as potential input data a blend of fixed basic properties in the bcc phase (calculated at the temperature of interest) and 20-30 random snapshot energies from the QMD simulations of the liquid. This scheme has produced an excellent set of derived Mo potentials over a range of volumes at temperatures of 5000 and 10,000 K. Figure 11 shows the calculated thermal energy and pressure obtained in the first two segments of a representative hybrid QMD-MD/MGPT simulation, demonstrating the smooth join across the interface and the rapid convergence of the derived potentials. Figure 12 shows the corresponding excellent predicted atomic structure of the liquid obtained from these potentials. In both figures we have contrasted the results obtained with the present temperature-dependent MGPT potentials with those obtained using existing $T = 0$ potentials. The contrast is indeed striking and clearly demonstrates both the validity of our present algorithm and the importance of electron temperature to the calculated high-temperature physical properties.

5.1.3 Establishment of a QMD test database and corresponding test simulations on derived MGPT potentials.

An important additional test of the derived MGPT potentials is their transferability to different structural environments at the same volume and temperature. A relevant and very sensitive test of this transferability is the calculation of the high-pressure melt curve, which requires the potentials to accurately describe both the high-temperature solid and the liquid simultaneously. This application is particularly timely because high–pressure melting in transition metals is currently highly controversial with static and dynamic high-pressure experiments yielding conflicting results. We believe that an accurate theoretical melt curve in the case of Mo can be obtained from two-phase co-existence MD simulations using a minimum cell size of

256 atoms. We require, however, both a baseline QMD result and test MD/MGPT results. The former is quite challenging computationally and, in fact, has never been directly calculated for a high-Z metal like Mo. Our goal is to calculate 4-5 high-quality QMD melt points that are sufficient to define a smooth melt curve. To date we are about halfway to that goal. Preliminary two-phase MD/MGPT melt points have also been obtained from our derived potentials at 5000 and 10,000 K. In 2009, we plan to complete the QMD melt curve for Mo and fully test our derived potentials over a range of pressures and temperatures.

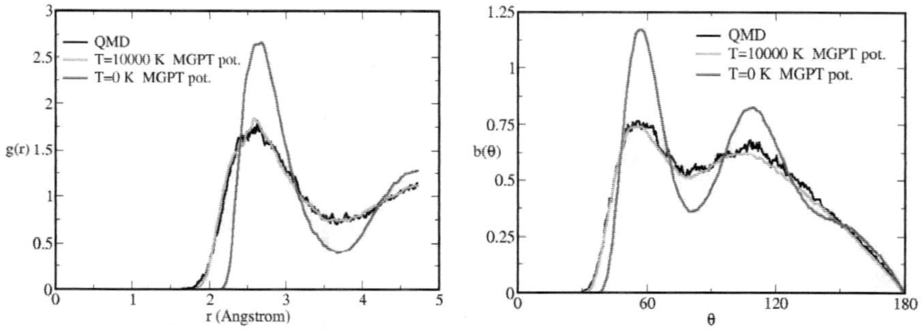

Figure 12. Atomic structure of liquid Mo at $T = 10{,}000$ K and an atomic volume of 105.1 au, as obtained from direct QMD simulations, from MD/MGPT simulations with present converged 10,000 K potentials, and from MD/MGPT simulations with existing $T = 0$ potentials. On the left: pair correlation function, $g(r)$; on the right: bond-angle distribution function, $b(\theta)$. All results are for 54-atom simulations.

5.1.4 Billion-atom MD/MGPT demonstration simulations with final optimized potentials.
These simulations will be performed once final optimized MGPT potentials for Mo are established.

5.2 First-principles materials property data bases for Ni and Ni_3S_4
In addition to on-the-fly high-temperature QMD data, it is useful to have static $T = 0$ DFT databases of fundamental properties for interatomic-potential development on SCC relevant materials. This year such databases have been calculated for pure nonmagnetic Ni over an extended pressure range and for polymidymite Ni_3S_4 at ambient pressure. These calculations have been done with the same DFT plane-wave pseudopotential method used in the QMD simulations.

5.2.1 Nonmagnetic Ni properties at high pressure.
The pressure-dependent structural and mechanical properties of non-magnetic Ni have been calculated over a wide pressure range extending to above 100 GPa. The calculations use modern generalized gradient approximation (GGA) corrections to the exchange-correlation potential and energy within density-functional theory. To perform these calculations, we have constructed a plane-wave-based pseudopotential for Ni by solving for the all-electron Ni atom in the reference state $4s^{1.98}$, $4p^{0.02}$, and $3d^8$. The energies of the core electrons were obtained by solving the scalar relativistic equations, while the valence electrons were treated nonrelativistically. The pseudopotential was modeled in the non-local norm-conserving Troullier-Martins form, with a total of 10 valence electrons generated using a plane-wave cutoff of 80 Ry. There is one projector for each component of angular momentum. The cutoff radii for the pseudized orbitals are 1.51, 1.51 and 1.32 Å for the 4*s*, 4*p*, and 3*d* valence orbitals, respectively. The *d*-component was chosen to be local. To improve the efficiency of our simulations, we transformed the pseudopotential into the Kleinman-Bylander form. The calculated zero-temperature structural and mechanical properties include the cold equation of state; fcc elastic moduli and zone-boundary phonons; fcc, bcc and hcp

structural energies; and the unrelaxed fcc vacancy formation energy. The observed ambient-pressure fcc phase is found to be thermodynamically stable throughout the pressure range studied. The calculated bulk and shear moduli and the *L* - and *X*-point zone-boundary phonons are displayed in figure 13. The elastic moduli and phonons are in reasonably good agreement with experiment [10] at ambient pressure and remain real and positive at high pressure, demonstrating that the fcc phase is mechanically stable in this regime. In the calculation of the vacancy formation energy for fcc Ni, we compared the total energy for a supercell with N = 108 atoms and no vacancy, [E_{tot} (N, W_N)], and the total energy for a supercell with N-1 atoms and one vacancy, placed at the center, [E_{tot} (N-1, W_{N-1})]. This procedure maximizes the cancellation of numerical errors in the calculation of E^f_{vac}, which is computed as $E^f_{vac} = E_{tot}$ (N-1, W_{N-1}) – (N-1)/N E_{tot} (N, W_N). Here W_{N-1} and W_N are the respective supercell volumes, such that W_N (P) = NW_0 (P), where W_0 (P) is the bulk equilibrium atomic volume at pressure, P. For comparison, we also calculated the fully relaxed vacancy formation energy for fcc Ni at zero pressure. This was found to be 1.92 eV, which is in reasonable agreement with the experimental value of 1.79 eV [11].

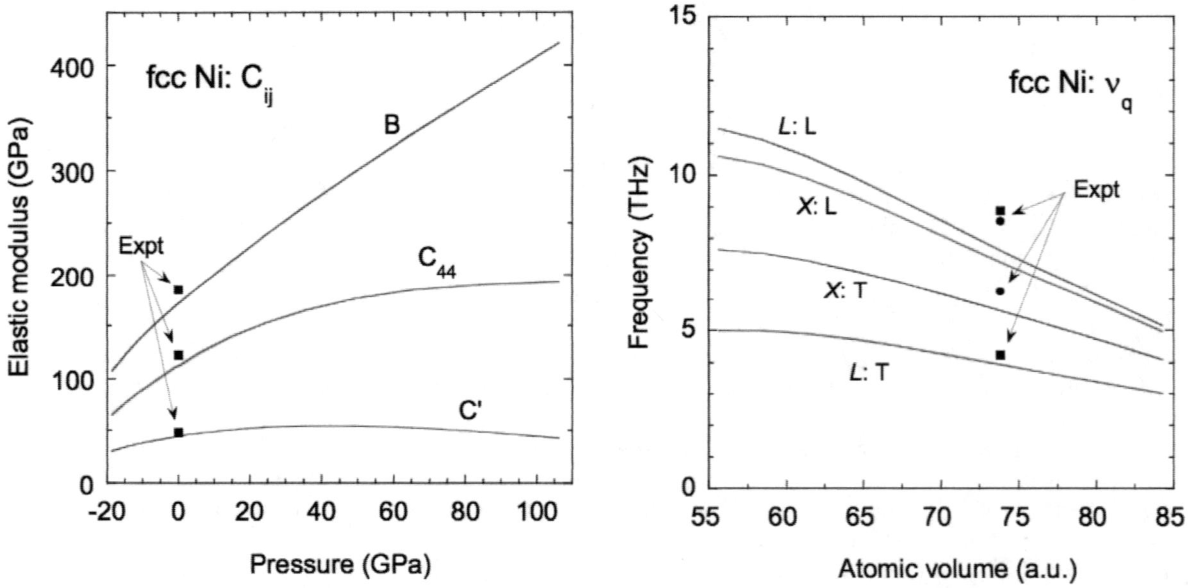

Figure 13. Bulk and shear elastic moduli (left panel) and *L*- and *X*-point zone-boundary phonons (right panel) for nonmagnetic fcc Ni, as calculated from the present first-principles DFT electronic-structure calculations.

5.2.2 Materials properties of polymidymite Ni_3S_4.

Polymidymite Ni_3S_4 is a sulfur-rich system and a good prototype system for the fundamental understanding of the electronic and structural properties associated with the SCC problem of sulfur concentrated at or near Ni grain boundaries. Polydymite Ni_3S_4 has two types of Ni atoms and one type of S atoms, $Ni^{I}Ni_2^{II}S_4$, in a cubic structure. The Ni^{I} atom occupies 1/8 of the tetrahedral site, the two Ni^{II} atoms occupy half of the octahedral sites, which produces a trigonal distortion along the [111] direction, and the four S atoms are on an approximate close-packed fcc lattice (see figure 14). We have performed first-principles DFT electronic-structure calculations on the Ni_3S_4 solid at ambient pressure. The bulk properties of this system are important material parameters that will be used for interatomic-potential development on S-rich Ni-S systems. The calculated lattice constant for the ideal structure is 9.489 Å and three cubic elastic constants are C_{11} = 201 GPa, C_{12} = 81 GPa and C_{44} = 115 GPa. In addition, we have calculated the electronic properties of this system. In particular, the density of the states at the Fermi level $N(E_F)$ = 31.1 states/eV-cell, with 56 atoms per cell. This leads to a band-structure electronic-heat-capacity constant of g = 6.542 mJ/mol-K^2, where 1 mol refers to 1 mol of Ni_3S_4.

5.3 Quantum-based interatomic potentials for Ni

Using the above DFT database for Ni, we have begun to develop T = 0 quantum-based MGPT potentials for this metal over the -20 to 100 GPa pressure range. In this regard, Ni requires certain different considerations than in the case of Mo, our prototype metal for the hybrid QMD-MD/MGPT simulations. In Mo the direct multi-ion *d*-state interactions are of primary importance and are retained through four-ion interactions, while *sp-d* hybridization is of secondary importance and is retained only in the leading volume term of the total energy. This is the standard approach for MGPT [5-7]. In Ni, on the other hand, the *sp-d* hybridization is of primary importance and the three- and four-ion interactions are of secondary importance. We have, therefore, now developed an advanced form of MGPT that restores the *sp-d* hybridization from the full first-principles generalized pseudopotential theory in the two-ion pair potential, in a suitably screened form, and we have applied the new approach to Ni. In both the standard and advanced forms on MGPT, the potentials are volume dependent in the bulk, although the inclusion of the hybridization necessarily imparts a longer range to the two-ion potentials. Preliminary advanced MGPT potentials for Ni at three atomic volumes, including equilibrium, are displayed in figure 15. Optimization of these $T = 0$ potentials is currently in progress. The advanced MGPT methodology can be readily imported into the QMD-MD/MGPT algorithm to develop corresponding temperature-dependent potentials for Ni and Ni-based materials, and this is now planned as part of the SCC project.

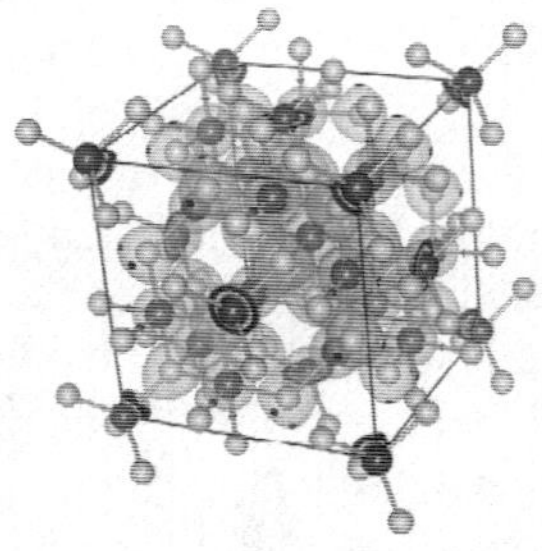

Figure 14. Crystal structure and electronic charge density of polymidymite Ni_3S_4. There are two types of Ni atoms in this system: the NiI atom occupies 1/8 of the tetrahedral site and the two NiII atoms occupy half of the octahedral sites. There are a total of 56 atoms in the supercell, including 24 Ni atoms and 32 S atoms.

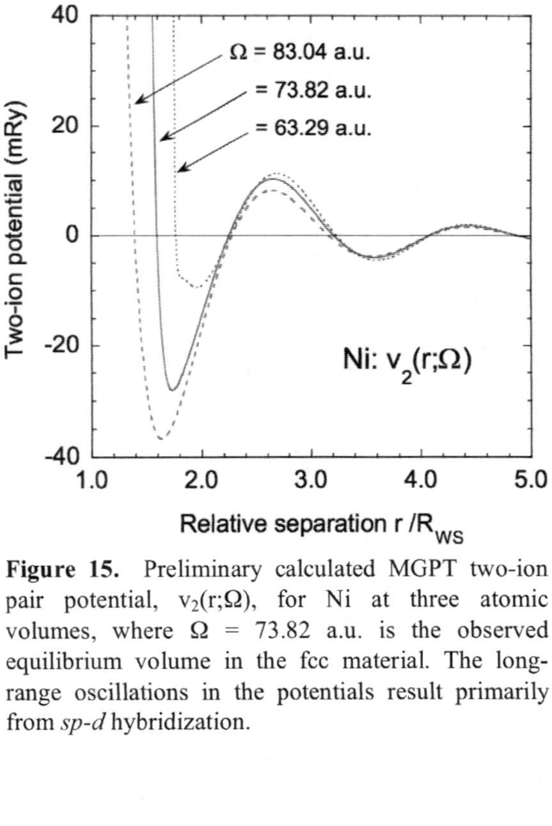

Figure 15. Preliminary calculated MGPT two-ion pair potential, $v_2(r;\Omega)$, for Ni at three atomic volumes, where Ω = 73.82 a.u. is the observed equilibrium volume in the fcc material. The long-range oscillations in the potentials result primarily from *sp-d* hybridization.

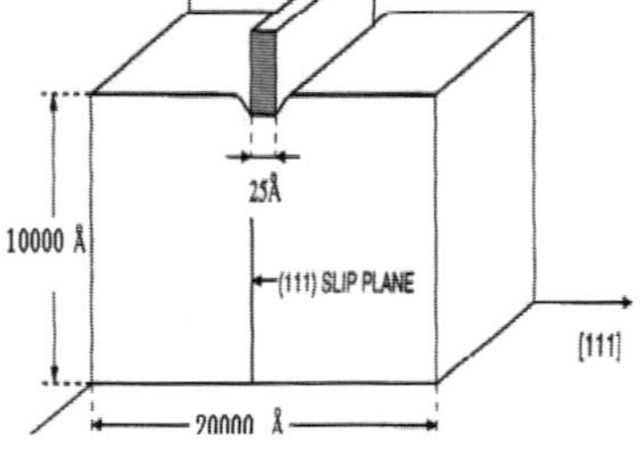

Figure 16. Schematic representation of nanoindentation of Al thin film.

6. Quasi-continuum modeling from electrons to finite elements

6.1 Development of QCDFT method

Lu at California State University Northridge has developed a concurrent multiscale method that makes it possible to simulate multi-million to billions atoms effectively based on the density functional theory (DFT).The method, termed QCDFT, is formulated within the framework of quasicontinuum method, and with DFT as its sole energetic input; that is, there is no empiricism in the formulation. The only underlying energy functional in QCDFT is either Kohn-Sham DFT or orbital-free DFT. The unimportant degrees of freedom are systematically coarse-grained out by finite-element interpolation scheme. The development of QCDFT represents a major progress in quantum simulation of materials properties and opens door for many applications that are beyond the reach of existing quantum simulation methods. The code has been parallelized with MPI. The QCDFT method has been applied to nanoindentation of an Al thin film. The results suggest that QCDFT is an excellent method for quantum simulation of materials properties at length-scales that are relevant to experiments [12]. Figure 16 shows a schematic representation of the sample and the indenter. Figure 17 presents an overview of the entire system and domain partition in the nanoindentation simulation.

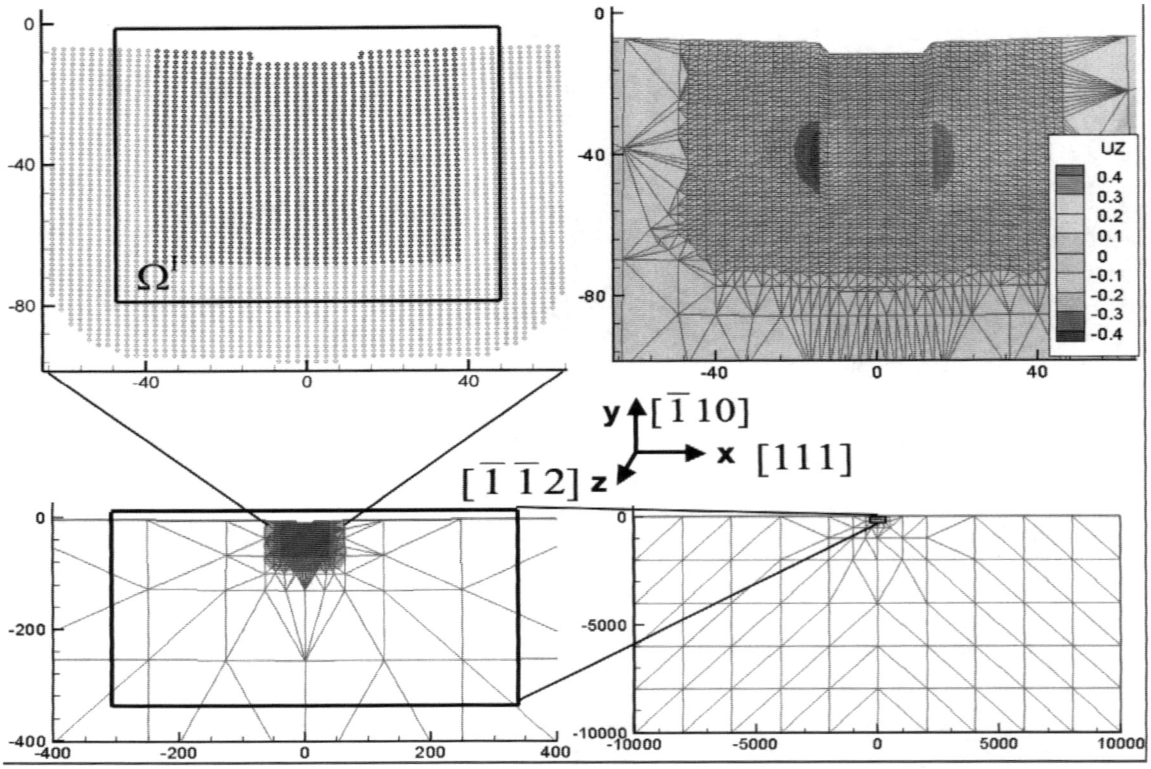

Figure 17. Domain partition in QCDFT with nanoindentation. DFT atoms are blue and buffer atoms are green. The energy of the finite-elements and the interaction energy between the buffer atoms and DFT atoms are calculated by DFT as well.

6.2 Validation of EAM potentials using sparse grid and first-principles calculations

We have demonstrated that the sparse grid method can be used to greatly reduce the necessary first-principles DFT calculations in order to validate the accuracy of empirical potentials. We have examined three different EAM potentials of Al by computing energy and stress as a function of uniform hydrostatic deformations. The EAM results deviate significantly from the corresponding DFT values for large compressions, while the results agree reasonably well to the DFT values for expansions. The failure of the EAM potential is attributed to the uniform electron density approximation adopted in the EAM theory. In compressed configurations, the local electron density gradient increases considerably, and the uniform electron density approximation is no longer valid, hence EAM fails. We have identified a critical value of the electron density gradient beyond which EAM energetics deviate significantly from the DFT values. We have tested this critical value for the case of the generalized stacking fault energy surface calculations and found that for the run-on stacking fault where the maximum local density gradient exceeds the critical value, EAM indeed fails in giving the correct energy (see figure 18). For all other stacking faults where the critical value is not exceeded, EAM gives reasonable results. We have proposed a solution to improve the transferability of EAM potentials by considering the gradient of electron density in the analytical expression of EAM energy—the embedding function should depend not only on electron density but also on its gradient.

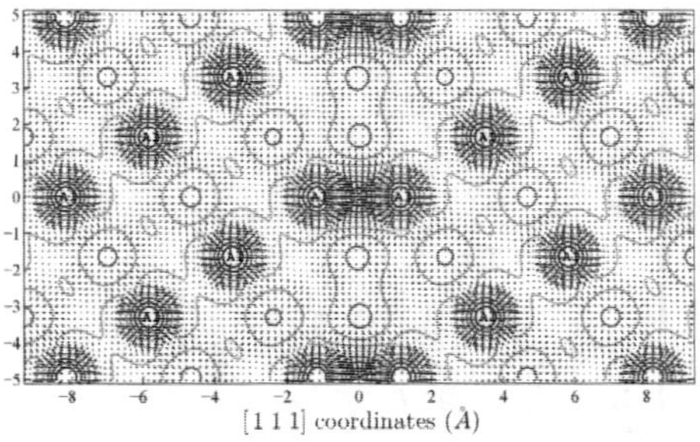

Figure 18. The valence charge density and its gradient for a run-on stacking fault of Al. The length and direction of the arrows represent the magnitude and direction of the charge density gradient. Two layers of Al atoms stack right on top of each other in the [111] direction (in the middle of the figure) constitute a run-on stacking fault.

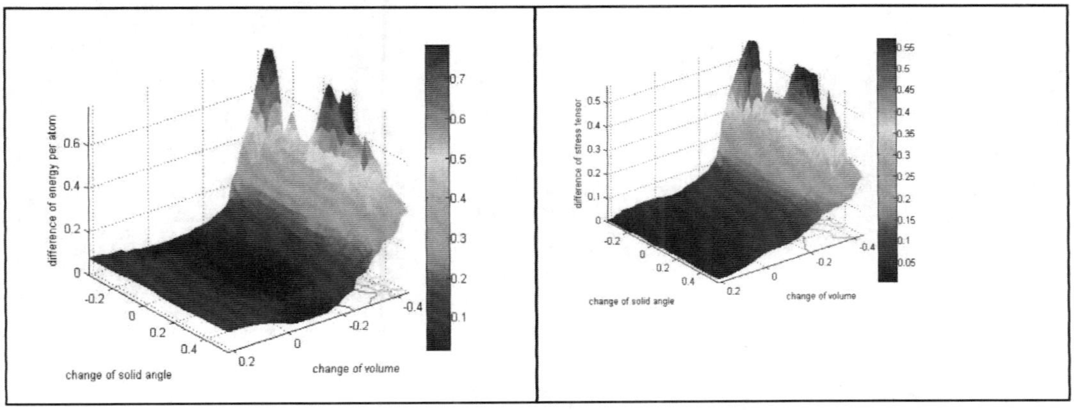

Figure 19. Difference between Ercolessi-Adams embedded atom method (EAM) potential and first-principles (VASP) calculations for the energy ($dE = E_{\mathrm{EAM}} - E_{\mathrm{VASP}}$) versus change of solid angle of the deformed cell and the change of cell volume in the left panel. The difference of stress tensor $dS = ||S_{\mathrm{EAM}} - S_{\mathrm{VASP}}||$ is shown in the right panel. The results are represented in a sparse grid with only 8433 points to represent 1,291,467,969 points in a uniform grid with the same precision. The sparse grid is being incorporated in the QCDFT method to perform DFT calculations for local region.

7. Accelerated molecular dynamics for stress corrosion cracking

As discussed above, accurate simulation of stress-corrosion cracking depends on building a hierarchical multiscale framework. At the largest scale, the externally applied stress is imposed on the system, requiring a continuum-level (e.g., finite-element or quasi-continuum) treatment. This information is passed to successively finer, more accurately described levels. Ultimately, at the finest level, a small number of the most crucial atoms at the crack tip (region A, order 100 atoms) are modeled with a full electronic structure calculation (e.g., DFT). To reach appropriate time scales, accelerated molecular dynamics (AMD) simulations are employed on a small region (region B, ~1000 atoms) that includes this DFT region A as a subset. The atoms in region B other than the DFT atoms are described with a highly accurate semiempirical potential. After the entire system has been brought into thermal equilibrium using MD coupled to FEM, a shell of atoms outside region B is held fixed at their time-averaged positions to provide proper boundary conditions for region B while an AMD simulation is performed within that region. This AMD simulation proceeds until a thermally activated transition takes this subsystem to a new state. For example, one bond at the crack tip may break, allowing the crack to extend by a few angstroms. At this point, communication with the rest of the system is require, to allow the strain field of the overall system to re-equilibrate with the modified region B. The newly relaxed shell of atoms around region B is then fixed in place again and new AMD is performed until the next reactive event. This procedure is repeated to advance the infrequent-event dynamics of the entire system.

In technologically relevant systems, the crack tip may be in contact with a liquid phase, and this liquid may play a critical role in promoting (or perhaps even inhibiting) the fracture. In the procedure above, if standard MD is employed to evolve the active region, the liquid poses no particular problem, but in the context of accelerated molecular dynamics, the liquid is problematic because the barriers separating one potential energy basin from another are extremely low. While AMD methods are powerful for compressing the time scale of transitions from long time scales (e.g., microseconds or milliseconds) down to the picosecond time scale, direct application of any of the AMD methods in this solid-liquid case will not give any computational boost because basin-to-basin transitions in the liquid are already occurring on the picosecond time scale.

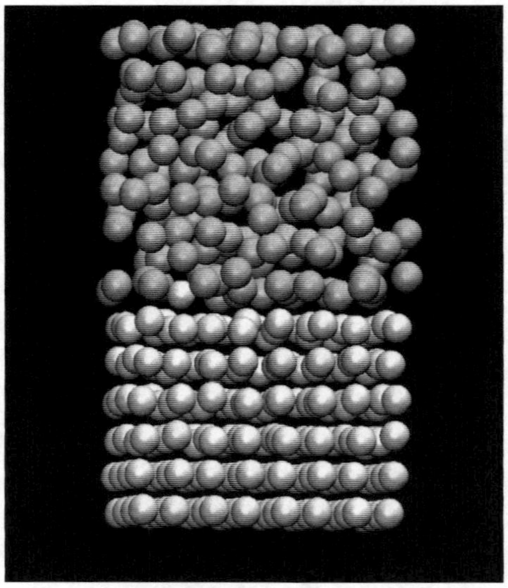

Figure 20. Silver adatom on Ag(100) surface in contact with Lennard-Jones liquid at $T = 600$K.

As a step toward solving this problem, we are developing a variation on one of the AMD methods, parallel-replica dynamics [2], to treat the dynamical evolution of a solid surface that is in contact with a liquid phase. The key is to define the concept of a super-state, so that the separation of time scales between intra-super-state and inter-super-states dynamics is restored. We then apply parallel-replica dynamics at the super-state level. A simple strategy to reach this goal is to define super-states simply based on the configuration of the slow degrees of freedom (DOF) in the system, i.e., every state in which the slow DOF are in the same configuration belongs to the same super-state, notwithstanding the state of the fast DOF. If the typical time scale over which the fast DOF evolve is sufficiently faster than that of the slow DOF, the validity of the super-state parallel replica approach should be assured and it should give a computational boost.

For typical solid/liquid interfaces—for example, a metal in contact with a liquid like water at room temperature—this segregation of DOF into fast and slow subsets should be easy to define: coordinates of the liquid atoms form the fast subset while coordinates of the solid atoms form the slow subset, since transitions in such a liquid will occur thousands of time faster than rearrangements of the solid atoms.

To test this approach, we have carried out super-state parallel replica simulations of the diffusion of an adatom on a silver (100) surface (modeled using an embedded atom method potential) in contact with a film of a prototypical fluid (modeled using a Lennard-Jones potential), as shown in figure 20. The distribution of 500 adatom hopping times at 600K is shown in figure 21 for a direct simulation using conventional molecular dynamics and for a super-state parallel replica simulation. The results clearly show that transition statistics are equivalent for the two approaches, implying that our super-state definition restores the time scale separation essential for the validity of the AMD methods. In this case the parallel replica simulation was quite efficient, with a parallel efficiency of around 0.8 on 8 processors, despite the presence of very fast transitions in the liquid. We expect that at lower temperatures, much greater boosts will be possible.

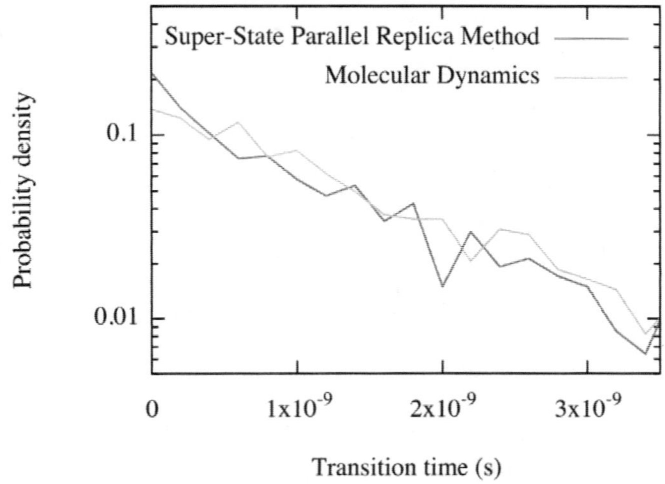

Figure 21. Distribution of hopping times for an adatom at a solid-liquid interface at T=600 K.

Note that the agreement between MD and AMD is not due to a negligible effect of the liquid on the adatom dynamics. Indeed, as shown in figure 22, the transition rate of the adatom on the wet surface is around 1.6 times slower than on the dry surface at 600K, increasing to a factor 3.5 at 500K. The rate corrections due to the liquid are thus significant and have a nontrivial temperature dependence.

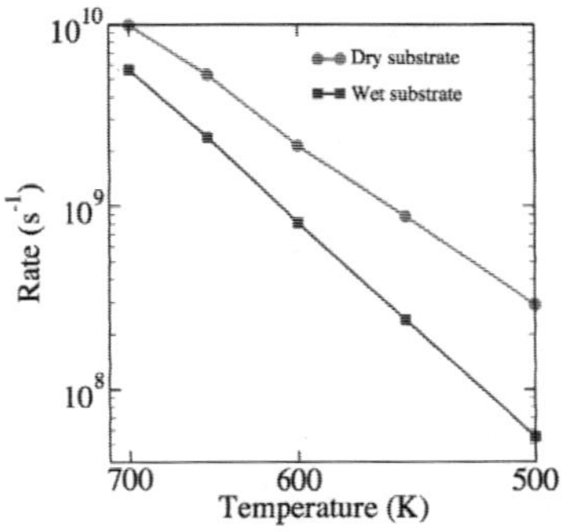

Figure 22. Temperature dependence of the hopping rate for an adatom at a dry interface and at a wet interface.

While these results are encouraging, challenges remain. For example, correctly treating the dynamics of *dissolved* solid atoms is problematic since their diffusion rate in the liquid will be effectively *decelerated.* Also, we must recognize when a solid atom is still part of the solid and when to declare it liquidlike, to prevent dramatic loss of boost. Similarly, it may be necessary to begin treating liquid atoms as solid if they react with the solid atoms, or if they form persistent ordered structures. We are developing methods to determine these characteristics on the fly.

References

[1] Chen Y, Lu Z, Nomura K,Wang W, Kalia R K, Nakano A and Vashishta P 2007 Interaction of voids and nanoductility in silica glass *Phys. Rev. Lett.* **99**:155506

[2] Célarié F, Prades S, Bonamy D, Ferrero L, Bouchaud E, Guillot C and Marlière C 2003 Glass breaks like metals, but at the nanometer scale *Phys. Rev. Lett.* **90**:075504-1
[3] Nomura K, Chen Y-C, Kalia R K, Nakano A and Vashishta P 2008 Defect Migration and Recombination in Nanoindentation of Silica Glass", *Phys. Rev. Lett.* (submitted)
[4] Hood R Q, Yang L H and Moriarty J A 2008 Quantum molecular dynamics of uranium at high pressure and temperature *Phys. Rev. B* (in press)
[5] Moriarty J A, Belak J F, Rudd R E, Söderlind P, Streitz F H and Yang L H 2002 Quantum-based atomistic simulation of materials properties in transition metals *J. Phys.: Condens. Matter* **14**:2825
[6] Moriarty J A, Benedict L X, Glosli J N, Hood R Q, Orlikowski D A, Patel M V, Söderlind P, Streitz F H, Tang M and Yang L H 2006 Robust quantum-based interatomic potentials for multiscale modeling in transition metals *J. Mater. Research* **21**:563
[7] Moriarty J A, Glosli J N, Hood R Q, Klepeis J E, Orlikowski D A, Söderlind P and Yang L H 2008 "Quantum-Based Atomistic Simulation of Metals at Extreme Conditions," in *TMS 2008 Annual Meeting Supplemental Proc. Vol. I: Materials Processing and Properties* (TMS, Warrendale, PA) 313
[8] Moriarty J A, Yang L H, Hood R Q, Vashishta P, Kaxiras E and Voter A F 2008 Toward petascale atomistic simulations with quantum-level accuracy 3rd Institutional Computing Grand Challenge Proposal (LLNL)
[9] Yang L H 2000 in *Industrial Strength Parallel Computing* ed. A Koniges (Morgan Kaufmann Publishers: San Francisco) 297
[10] Birgeneau R J, Cordes J, Dolling G and Woods A D B 1964 Normal modes of vibration in nickel *Phys. Rev.* **136**:A1359
[11] Schultz H and Ehrhart P 1991 in *Atomic Defects in Metals* ed. H Ullmaier, Landolt-Börnstein, New Series, Group III (Springer, Berlin)
[12] Zhang X and Lu G 2007 Quantum mechanics/molecular mechanics methodology for metals based on orbital-free density functional theory *Phys. Rev. B* **76**:245111

Journal of Physics: Conference Series **125** (2008) 012061 doi:10.1088/1742-6596/125/1/012061

Atomistic Structure and Nucleation of Nanoprecipitates in Thermoelectric PbTe-$AgSbTe_2$ Composite

Jihui Yang
Materials and Processes Laboratory, GM R&D Center, Warren, MI 48090, USA

E-mail: jihui.yang@gm.com

Abstract. Increasing awareness and concern for energy resources and the environment has prompted the recent worldwide effort on thermoelectric (TE) materials research and technology development. Advances in TE technology can have a significant impact on the automotive industry in terms of fuel economy improvements by generating electricity from waste heat and augmenting air conditioning efficiency.

1. Introduction

A critical component of developing automotive TE technologies is the discovery and development of highly efficient TE materials. From the fundamental materials research perspective, this development is motivated by the more than half-century-long challenge of simultaneously enhancing the electrical conductivity and thermopower and lowering the thermal conductivity of materials and of understanding the underlying mechanisms.

2. Approach

Many recent advances in thermoelectric materials are attributed to nanoscale phenomena. A prerequisite for understanding properties of complex nanocomposites is an accurate description of the atomistic structure of their nanoscale constituents. Our combined large-scale density functional calculations and HRTEM measurements resolve the nanostructural features and provide a realistic view of the $AgSbTe_2$ nanoprecipitate nucleation and interaction in PbTe matrix at atomic scale.

3. Results and impact

These findings provide crucial insights for exploration of superior nanocomposite thermoelectric materials.

Journal of Physics: Conference Series **125** (2008) 012062 doi:10.1088/1742-6596/125/1/012062

Computational nuclear structure: Exploring nuclei through INCITE

D J Dean
Physics Division, Oak Ridge National Laboratory, Oak Ridge, TN 37831-6251, USA

E-mail: deandj@ornl.gov

Abstract. Computing provided by the INCITE program is essential to efforts to disentangle the nuclear puzzle. The goal of the computational nuclear structure INCITE project is to provide a comprehensive description of all nuclei and their reactions, by using a variety of computationally intensive techniques.

1. Introduction

Nuclei comprise 99.9% of all baryonic matter in the universe and are the fuel that burns in stars. The rather complex nature of the nuclear forces among protons and neutrons generates a broad range and diversity in the nuclear phenomena that we observe. As we have seen during the past decade, developing a comprehensive description of all nuclei and their reactions requires theoretical, computational, and experimental investigations of rare isotopes with unusual neutron-to-proton ratios. We call these nuclei exotic, or rare, because they are not typically found on Earth. They are difficult to produce experimentally because they typically have extremely short lifetimes. The goal of a comprehensive description and reliable modeling of all nuclei—light, heavy, and superheavy—represents one of the great intellectual opportunities for physics in the twenty-first century.

2. Computational approach

Although the interactions of nuclear physics differ from the electromagnetic interactions that dominate chemistry, materials, and biological molecules, the theoretical methods and many of the computational techniques to solve the quantum many-body problems are shared (figure 1). This INCITE project includes four primary computational paradigms for investigations of nuclei. We use Green's function Monte Carlo to explore nuclear forces in light nuclei and neutron drops. We utilize basis expansion methods (configuration interaction in chemistry, interacting shell model in nuclear physics) that use exactly the same techniques—that of diagonalizing the Hamiltonian matrix. We have developed nuclear coupled cluster techniques for ab initio calculations in medium mass nuclei. For heavier nuclei, we employ density functional theory, which is built on theorems showing the existence of universal energy functionals for many-body systems. These methods overlap in certain regions of the table of nuclei, as shown in figure 1.

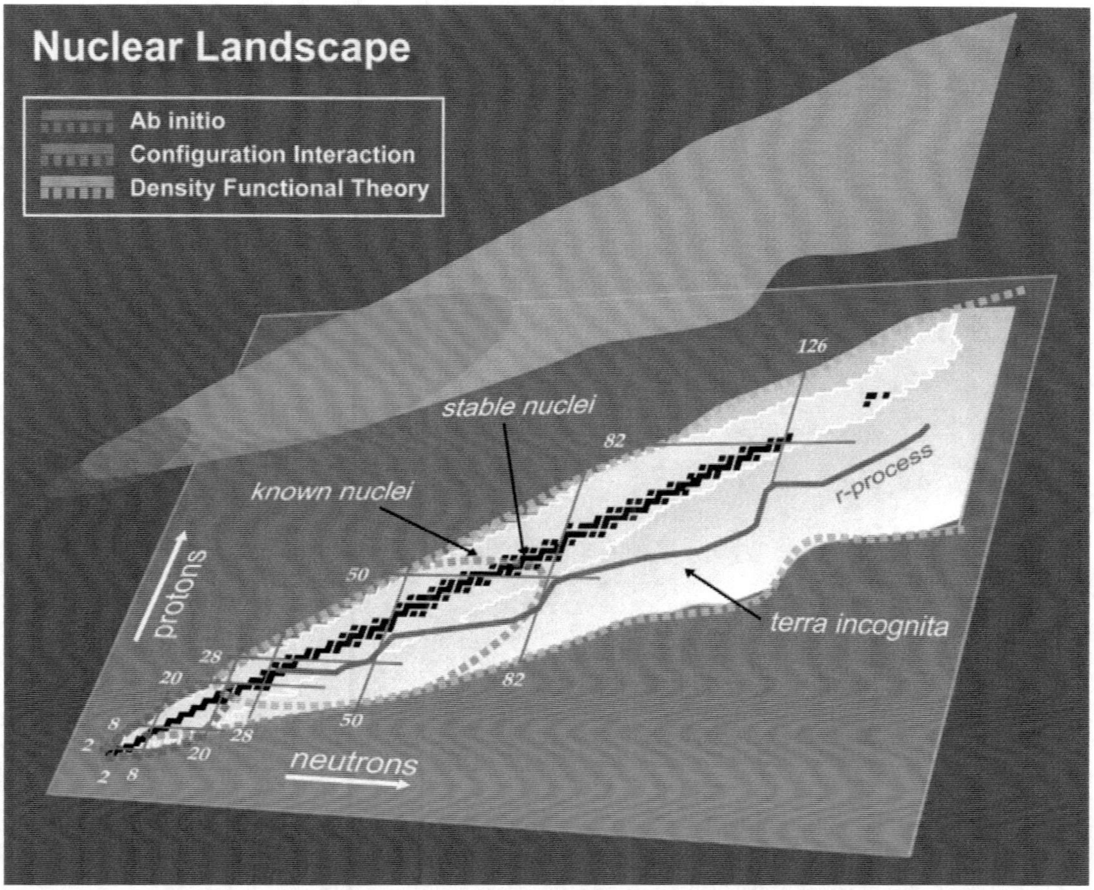

Figure 1. Theoretical methods and computational techniques used to solve the nuclear many-body problem. On this chart of the nuclides in the (N,Z)-plane, the black squares represent stable nuclei and the yellow squares indicate unstable nuclei that have been produced and studied in the laboratory. The many thousands of these unstable nuclei yet to be explored are indicated in green (terra incognita). Except for the lightest nuclei where it has been reached experimentally, the neutron drip line (the rightmost border of the nuclear landscape) has to be estimated on the basis of nuclear models—hence it is very uncertain because of the dramatic extrapolations involved. The red vertical and horizontal lines show the magic numbers, reflecting regions where nuclei are expected to be more tightly bound and have longer half-lives. The anticipated path of the astrophysical r-process responsible for nucleosynthesis of heavy elements is also shown (purple line). The thick dotted lines indicate domains of major theoretical approaches to the nuclear many-body problem. For the lightest nuclei, ab initio calculations (Green's function Monte Carlo, no-core shell model, coupled cluster method), based on the bare nucleon-nucleon interaction, are possible (red). Medium-mass nuclei can be treated by configuration interaction techniques (interacting shell model, green). For heavy nuclei, we employ density functional theory based on self-consistent/mean field theory (blue). By investigating the intersections between these theoretical strategies, one aims at nothing less than developing the unified description of the nucleus.

Acknowledgments
This work was supported by the U.S. Department of Energy under Contract No. DE- AC05-00OR22725 with UT-Battelle, LLC (Oak Ridge National Laboratory).

Journal of Physics: Conference Series **125** (2008) 012063 doi:10.1088/1742-6596/125/1/012063

Getting ready for Petaflop capacities and beyond: a utility perspective

J F Hamelin and J Y Berthou
EDF R&D, 1, Av du Général de Gaulle, BP 408, 92141 Clamart Cedex, France

E-mail: jean-francois.hamelin@edf.fr, jy.berthou@edf.fr

Abstract. Why should EDF, the leading producer and marketer of electricity in Europe, start adding teraflops to its terawatt-hours and become involved in high-performance computing (HPC)? In this paper we answer this question through examples of major opportunities that HPC brings to our business today and, we hope well into the future of petaflop and exaflop computing. Five cases are presented dealing with nondestructive testing, nuclear fuel management, mechanical behavior ofnuclear fuel assemblies, water management, and energy management. For each case we show the benefits brought by HPC, describe the current level of numerical simulation performance, and discuss the perspectives for future steps. We also present the general background that explains why EDF is moving to this technology and briefly comment on the development of user-oriented simulation platforms.

1 Simulation background at EDF: a set of competencies and codes backing the operation of a standardized nuclear fleet

EDF background in numerical simulation is deeply rooted in its nuclear PWR program launched in the early 1970s and in its particular responsibilities as an architect and owner-operator. (EDF does not buy turnkey plants but takes responsibility for the global design: it then specifies and assembles different parts provided by various vendors like Areva for the “nuclear island.”) With installed capacity of 63 GWe and 58 standardized nuclear units, EDF boasts Europe’s largest nuclear fleet and 17% of the world nuclear capacity, enabling production of 428 TWh without CO_2. It is fully responsible for nuclear safety and for meeting all the stringent rules and controls imposed by external independent bodies to the nuclear industry. From the onset, EDF has chosen to develop a high level of competence in order to successfully manage this program and make the most of an operationnal feedback currently exceeding 1200 reactor-years. This was one of the reasons for establishing strong in-house Engineering and R&D teams and has proved useful to constantly stick to the highest degrees of safety while keeping costs under control and operational performance well within the best world standards.

In this context, numerical simulation has been recognized from the beginning as an indispensable tool. It has been used for a long time in such important operational issues as optimizing day-to-day production or choosing the safest and most effective configurations for nuclear refuelling; but most of its advance toward higher levels of performance has been driven by the constant need to better explain complex physical phenomena behind maintenance issues, assess the impact of potential modifications or new vendor technology, and anticipate changes in operating or regulatory conditions.

his permanent progress toward finer and broader physics simulation has been capitalized over the years in a number of codes that make the increase of knowledge quickly transferred from research to operation. Figure 1 shows how Code Aster, a nonlinear structure mechanics code, has been constantly enriched over the past 20 years, with 100k lines of code being added annually over the past ten years.

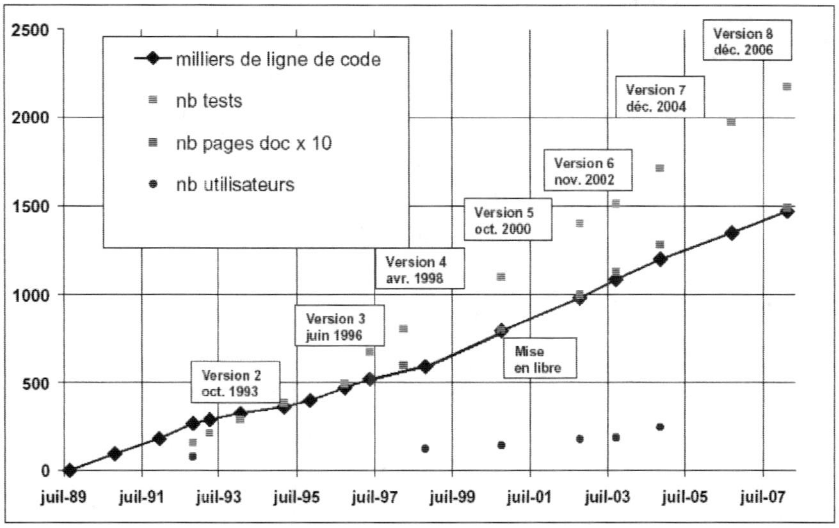

Figure 1. Structure mechanics *Code_Aster* continuous improvement capitalized over 20 years.

All those codes, together with the precise knowledge of their capacities and limits, represent precious assets as they not only contain the right physics modelling for the problems that we have to meet but also integrate a high degree of qualification and quality assurance that make them suitable for safety studies. As a consequence the overall pricetag of some of them can reach the 100 M€ line.

Table 1 shows the main disciplinary codes currently in use at EDF in the nuclear domain, most of them being developed by EDF or co-developped with academic or industrial partners. About 150 EDF R&D engineers are directly involved into the development of those codes and associated simulation platforms.

Table 1. Main codes used for nuclear applications at EDF (Ab initio and dynamic molecular material codes have not been included in this list.) CEA: Commissariat à l'Energie Atomique. LAMEL is a joint laboratory with Lille University and EDF.

Name	ASTER	CYRANO	SATURNE/ SYRTHES	NEPTUNE	THYC	TELEMAC	TRIPOLI	COCCIN ELLE	CARMEL
Domain	Non-linear Structure thermo-mechani cs	Fuel rod Thermo-mechanics	Mono-phasic fluid thermo-dynamics	Di-phasic fluid dynamics (CFD)	Di-Phasic fluid dynamics (porous media)	Free surface fluid dynamics	Stochastic neutronics (reference studies)	Diffusion neutronics (industrial studies)	Electro-magnet-ism
Owner	EDF	EDF	EDF	EDF/CEA	EDF	EDF	CEA	EDF	EDF/ LAMEL
Size (k lines)	1500	150	450	120	100	190	250	200	20

2 Petaflop computing: opening simulation to quantitative and statistical margin assesment in tenser operating conditions

Most of the above-mentioned codes have been launched 10 or 20 years ago, at times when available computing power was scarce and massively parallel computing unknown in industry. Those obvious limitations made possible only simplified physics modelling, geometrical approximations, mean value envelope calculations, involving a number of parameters derived from expertise and matched with reference experimentations. All of these proved effective in carrying out studies that would generally stay far from design margins or require moderate optimization. However, those needs are rapidly evolving into a situation that is bringing a number of structural changes to our company:

- As nuclear assets are ageing, quantitative and credible estimation of operation margins is a key factor to curb maintenance costs, allow for performance optimization (through renewed fuel management for instance), or help extend a plant's lifetime.
- More competitive and complex markets, the industrial development of renewable and less predictable energy sources like wind power, combined to the urge of minimizing energy consumption and peak carbon-intensive production call for renewed and much finer management of physical and financial production portfolio and associated risks.
- Climate change and subsequent environmental modifications, floods, droughts, and storms together with increasing environmental awareness are bringing up new problems where accurate quantitative prediction at both global and local scales is much needed, some of them with very high stakes such as increased protection of nuclear power plants and dams against floods and quakes or nuclear waste repository design.

Against this tense operating backdrop, we have been convinced from the start that the advent of high-performance computing (HPC) in the petaflop range would bring new opportunities, not only for nuclear, but also for a wider range of business applications. As shown in figure 2, we chose to increase dramatically the computing power made available for our teams in order to boost HPC-based simulation and demonstrate that thinking petaflop was no R&D utopia but could provide quick return for business.

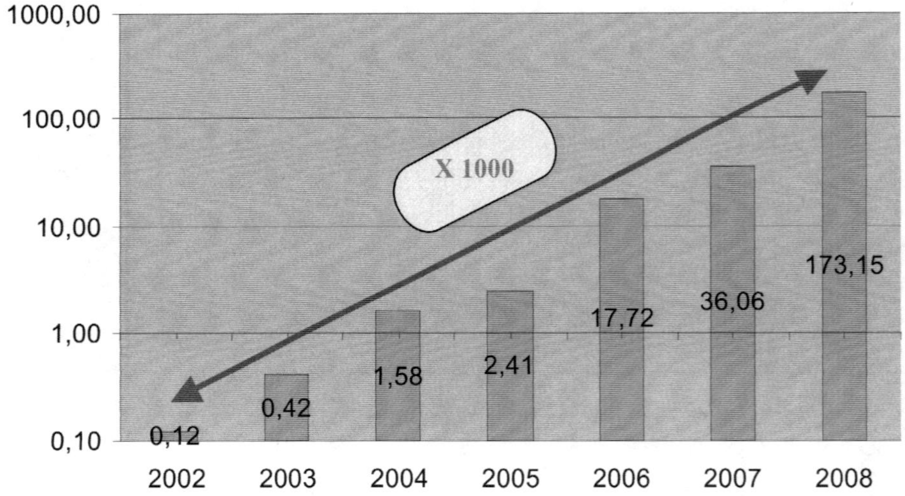

Figure 2. Evolution in computing power at EDF (Tflops).

Two years after we installed our first 23 Tflops IBM BG/L machine, and only months after we added another 100 Tflops BG/P, this vision has been confirmed by facts. The following paragraphs will report on some of the practical advances that have been already attained and the perspectives that they open for future, and even more ambitious applications. As shown in figure 3, we have started work along three technically distinctive "HPC paths", shortly described below, which roughly correspond to increasing complexity and costs.

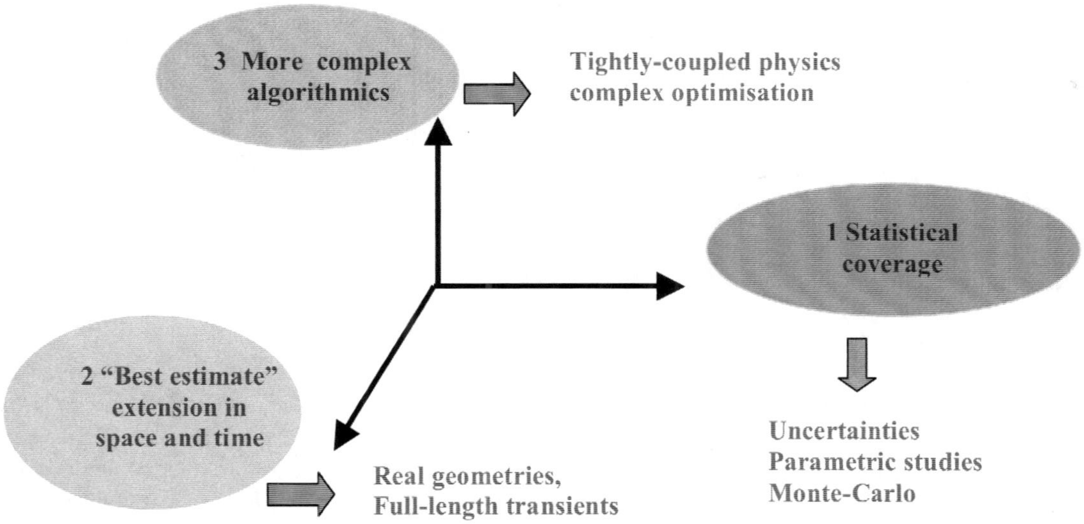

Figure 3. Three paths to HPC benefits.

1. The first and easiest move that we made consisted in using HPC parallel brute-force calculations to **provide statistical coverage with unchanged industrial-grade codes**, through thousands of parallel single simulations (representing as many combinations between parameters values, events, operating conditions), thus allowing for uncertainties management and parametric studies. These kinds of high-throughput approaches represent a broad spectrum of opportunities as they enable to switch from single, often worst-case calculations to stochastic methods that give more insight on rare events distribution. However, even if this category of completely uncoupled calculations has the smallest cost tag, first experiments show that it can challenge both the code itself (which is tried against a much wider domain) and the parallel architecture where data flows management have to be carefully designed.

2. At the same time we started to advance on the more classical path of tightly coupled calculations in order to **extend the use of existing "best estimate" simulation technology** to systemwide analysis, with the relevant physical space and time coverage. The intention was to move from the calculation of average operating values (the kind of calculations that were made at the design stage with sometimes heavy margins to cover uncertainties and approximations) to the evaluation of much finer phenomena and punctual extrema in space and time that can yield precious information on operating margins. A typical example, described later, is extending the use of 3D-CFD to simulate the flow of water within a complete nuclear vessel with submillimetric prediction of temperatures and pressures across time. The main cost for this second type of HPC benefits is bringing the best estimate codes to the strong-scaling capacities that are necessary to keep calculation times within acceptable limits. On top of that, production of the fine meshing required and visualization of the large datasets produced become nontrivial tasks, calling for specific R&D efforts.

3. The third approach, which we are only beginning to touch on, involves not only scaling existing codes but **developing new HPC simulation technology**, currently unavailable because meaningless in the absence of large computing power. Closely coupled multiphysics simulations as well as complex optimization fall in this category. This type of development is still at initial R&D stage but will obviously yield the biggest medium-term returns. Of course, categories 2 and 3 can, and often must, be combined with stochastic simulation: initial petaflop-grade requirements can then quickly turn into near-exaflops needs.

3 Nondestructive testing

Nondestructive testing is a major issue for nuclear operation. Annual costs range well over 100 M€/year , as numerous tests are systematically conducted during maintenance operations to ensure and demonstrate the safety of critical parts of plants like the nuclear vessel, steam generators, and associated piping. It is therefore essential to do the following:

- Demonstrate the ability of the various procedures (radiographic, ultrasonic or eddy-current)
- Detect potential defects
- Optimize preparation in order to guarantee safe, reliable, and timely operation during the very constrained outage schedules (any problem occurring at this time can generate high-costs delays)

This is usually achieved through preparatory experiments carried out on physical models by highly skilled engineers. However, this approach has met with a number of difficulties:

- Some models can be costly to design because of difficulties to engineer faithful geometrical shapes and defects.
- Selecting a methodology for a particular control requires a choice between several options and testing robustness against a number of uncertain parameters that are difficult and costly to reproduce extensively during the experimental phase; radiographic testing, for instance, means selecting type and position of source (selenium, iridium, cobalt, etc.) and type and position of film, taking into account possible misalignments or incomplete knowledge of the precise state of installation under control (water level in the pipes, etc.).
- Physical models do not easily allow for parametric study on the size and location of defects.

A first attempt at using high-performance simulation for radiographic testing has proved successful in overcoming these difficulties and improving performance while reducing costs and delays [1]. Simulation here means simulating the thousands of billions of photons emitted by the radioactive source used in a real test, calculating direct and scattered paths, and imaging the impacts at the simulated film level as, represented in figure 5.

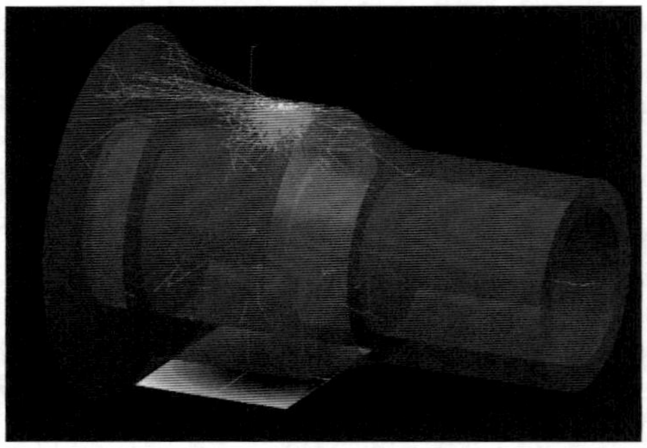

Figure 5. Simulated radiographic testing showing paths of gamma photons in a pipe.

This is a comparatively easy job as it requires only a simple Monte Carlo algorithm (particle paths are independent, thus allowing "natural" parallelization of the code) with a boxed-in device recording photons as they reach the film. (The main result of the calculation is the ratio between direct and scattered radiation, which is a characteristic data for a given test configuration, equivalent to a signal to noise ratio; most photons don't reach the film; scattered photons that reach the film blur the crisp image from direct radiation and limit the capacity to detect defects. This ratio is extrapolated by smoothing from a partial calculation.) However, a complete calculation would have to deal with up to 5×10^{16} photons, which is currently not feasible for lack of a petaflop computer fully used during several months! In practice, we validated an extrapolation method from a semi-reference calculation carried out with our Moderato code, using 1200 billion photons and 145,000 CPU-hours on 1000 cores of an IBM BG/L. This enabled the simulation of the most challenging cases, such as preparing for the inspection of bimetallic connections in the coolant systems. The difficulty here stems from the panoramic exposure and the extreme thickness of the part, 100 mm steel piping. A 150-billion photon calculation was run in less than one day on 1000 BG/L cores and included source sensitivity studies such as cone opening and type.

But this is far from being the end of the story. Similar advances are expected for ultrasonic testing [2], which is used to control many critical parts of the primary circuit (welds, core internal surface, etc.). The challenge is to move from a mostly 2D analysis, enough to deal with fatigue cracks, to full 3D simulations, able to represent the interaction with complex ramified and disoriented stress corrosion cracks. We used the 3D code ATHENA, developed with the help of INRIA (the French national institute for research in computer science and control). It uses an explicit finite-difference scheme to solve the 3D elastodynamic equations, a fictitious domains method to solve the crack-wave interaction problem, and PML (Perfectly Matched Layers) to model infinite domains. This leads to problems with up to 10 billions meshes (at least 10 meshes are required per wavelength – 0.5 mm at 2 MHz), or several tens of billions of degrees of freedom, and will need executions on more than 20,000 cores to keep response time under two days. Athena parallelization is under way with INRIA, and the appropriate level of performance should be achieved during 2009.

4 Nuclear fuel management

EDF reloads about 50 of its nuclear reactors every year. Standardized reload assemblies are used for each plant type (900 MWe, 1300 MWe, and 1450 MWe) called "reference fuel management," defining the uranium enrichment to be used (enrichment varies from 3 % to 4.5 % ^{235}U), leading to 12 or 18 months cycle length.and the number of fresh fuel assemblies to be loaded. An industrial calculation chain enables to perform reload design and safety studies [3]. Gaining flexibility in loading patterns would allow for better adjustment of cycle lengths to market needs and yield significant performance improvement of the whole nuclear fleet. However, assessing the safety of those new patterns requires an increased number of full core transient calculations, often complex, difficult, and expensive.

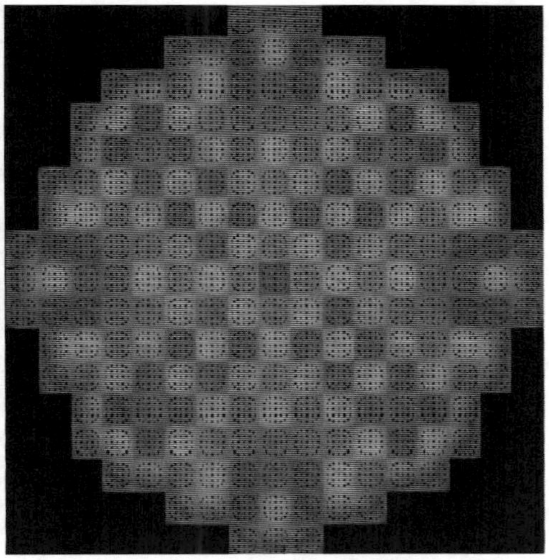

Figure 6. Refuelling a nuclear power plant (left) ; simulation result showing the most solicited assemblies within the core (right).

We currently are studying a different option that replaces transient calculations by a bounding approach, resulting in much simpler, static core simulations but repeated a very large number of times in order to cover the full combination of parameters that define possible states of the nuclear core at any given moment. The main values to be explored are the following:

- burn-up levels (at least 3 or 4),
- xenon (which results from iodine radioactive decay and is a neutron absorber) three-dimensional distributions (several tens),
- power levels (typically 10), and
- control rod positions for each group of rods (several hundreds).

Altogether, up to 200,000 full-core static computations are necessary, using the neutronics code Coccinelle, each one calculating the 3D power distribution in the core (1-minute calculations on a standard core). Preliminary simulations have been run on 400 cores, lasting roughly 20 hours. So far, these studies have been conducted for R&D purposes only but, in the short term, might join the set of industrial tools EDF uses for preliminary safety analysis. In the long term, they might even be used for demonstrating the safety of individual reloads: computation time will have to be brought down to less than one hour, a task that should be comparatively easy with about 10k cores. But the most difficult challenge will be to demonstrate a very high level of hardware and software reliability, as required for field operation.

These kinds of studies are typical of a large set of high-stakes applications where the massive parametric calculations capacities brought by HPC must be carried out "against the clock": in a matter of days for engineering-type processes, and in a matter of hours or minutes for field operation processes.

5 Mechanical behavior of nuclear fuel assemblies and nuclear vessel interns

Precise estimation of thermo-mechanical effects in the nuclear reactor is of paramount importance for such high-stakes safety and maintenance issues as nuclear vessel lifespan, axial stability of nuclear assemblies, or premature wearing of fuel rods induced by fretting (this does not count the major thermo-mechanical effects induced by neutronics reaction within the fuel rods, which constitute a separate study field). This kind of problem requires more and more accurate calculations of hydraulic flows within the core and make it highly desirable to move to 3D CFD codes applied to the full geometrical complexity of the internal structure of the vessel. HPC is a key enabler to apply those "best estimate" codes at such a complex level and reach a better management of conservative assumptions associated with more simple flow modeling (like porous media codes).

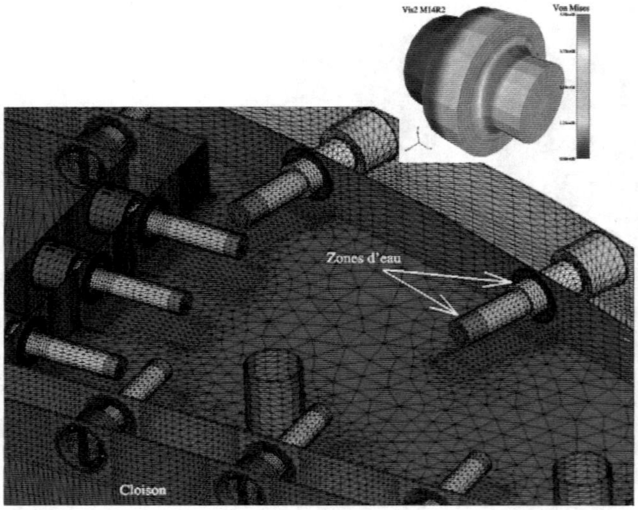

Figure 7. Flow simulation in complex geometries within a nuclear vessel allows fine temperature evaluation at screw level (top right).

One of the first thermal hydraulics applications of HPC at EDF was the precise evaluation of mechanical properties of hundreds of screws used to hold the peripheral shielding part in the nuclear core. Fine calculations of temperature-induced mechanical constraints under the screw head implied thermo-hydraulic simulations spanning not only the full multimetric height of the core but also the submillimetric scales at the screw level, taking into account complex geometries as can be seen on figure 7. This first calculation was run in 2006 on 200 processors over 11 days.

However, much more refined calculations are necessary in order to solve even more complex issues such as the precise fuel assembly and fuel rods mechanical load induced by water flow within the core. In 2007, a better understanding of the effect of mixing grids (millimetric details shown in figure 8) on the mechanical loading in fuel rods was made possible thanks to a CFD stationary flow calculation around part of a fuel assembly. This involved 100 million cells meshing and 1 month of calculation with the code Saturne on 8000 BG/L processors.

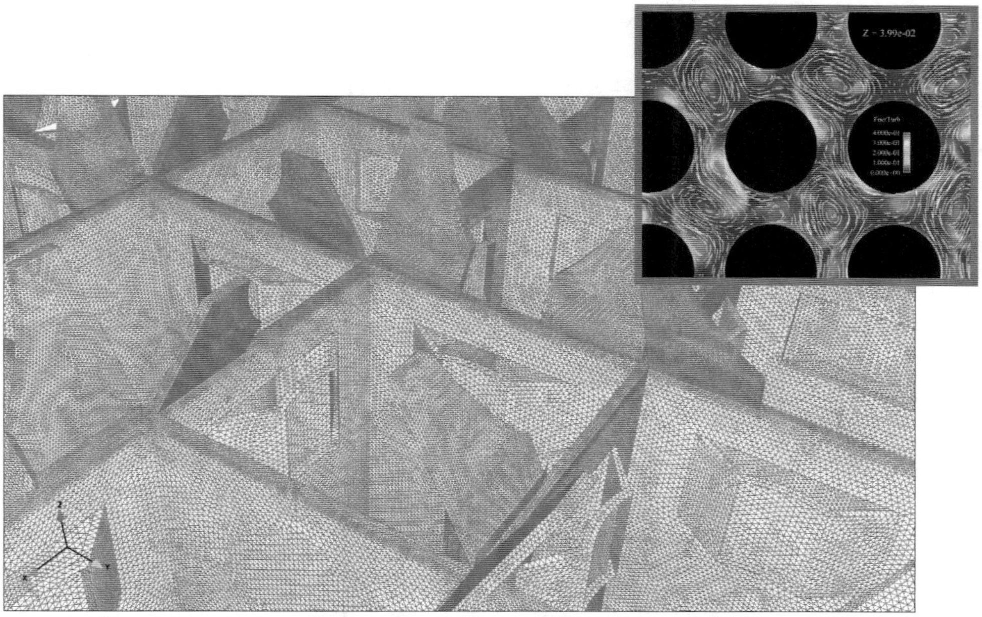

Figure 8. Submillimetric meshing allows for fine evaluation of mixing grids effects on the water flow in fuel assemblies.

But this level of detail should also be reached at the full core level: this would allow for instance to account for intrinsic dissymetries brought by different geometries of different vendors assemblies used at the same time in a single nuclear vessel. We anticipate that this full-scale simulation, which may require as much as a 10-billion cell meshing and several weeks computing on a 10-petaflop computer, will become possible in about five years. Combined with similar changes in neutronics simulation, this would yield significant qualitative changes in the way safety and performance studies can be carried out for nuclear vessel environments.

6 Water management

In France, 70 percent of surface waters flow through an EDF dam or by an EDF plant. Water management is a strategic issue for both protecting powerplants against environmental phenomena (dimensionning dykes to resist exceptionnal events like the December 1999 storms in Europe) and for managing such effects as the dilution of warm water from nuclear plants in the rivers (excessive temperature rise can force plant shutdown in hot summers). In addition, complex underground water flow is a key aspect to nuclear waste repository design, a very high-stakes project currently under study in France.

The multidimensional simulation set of tools TELEMAC has been designed and used for more than 20 years to conduct such studies. It allows the modeling of the main hydraulic and environmental phenomena such as tides, floods, waves, transfer of pollutants, and sediments in surface and underground waters. However the growing importance of environmental issues is calling for much more complex studies that may involve both long-distance and local three-dimensional phenomena (pollutant discharges, thermal plumes, sediments transportation, etc.), turbulence and stratification (soft water discharges of the Durance river dams into the salty water Berre Lake), or very complex surface flows (waves breaking against dykes or offshore windmills). Telemac is currently being adapted to HPC in order to provide the dramatic performance increase that is needed to tackle those new challenges, as the following example will show.

Following reassessment of extreme meteorological events, a number of dam spillways must be adapted or added to accommodate worst-case floods that can be twice as severe as in the initial design. Very complex flows occur at the spillway level and at the bottom of the dam. Accurate prediction is necessary to ensure that the particular design can indeed provide the expected extra throughput and that the dam bottom structure will not be affected by extreme turbulence. Those studies have been started with physical-scale models like the one in figure 9 that was until recently the only way to represent such highly turbulent free-surface flows (experimental studies take as much as 12 months). However, HPC is rapidly changing that by enabling an alternative simulation method, called SPH, or smooth particles hydraulics; SPH is a Lagrangian approach that models complex water flows by a set of macroscopic particles interacting with each other through repulsive forces and displaced according to the impulse received).The appproach until recently had remained confined to much smaller test cases for lack of computing power.

In less than six months, the corresponding code Spartacus 3D has been ported to BG/P through a partnership with IBM and its scalability improved to a level where it can now accommodate the number of particles required for this kind of industrial-grade study. First runs with 6 million particles have just been achieved on 8000 cores, simulating a reference dam break. The encouraging results should lead to a full-size simulation of the Goulours dam complex spillway at the end of 2008, for validation purposes, opening the way to fully numerical studies (including parametric design) in the near future [4, 5, 6, 7].

Figure 9. Physical model of Goulours dam showing complex flow from spillway.

7 Energy management

One of the most challenging problem EDF has to solve, and also one with the most immediate and heavy societal and financial stakes, is constantly balancing demand and production with minimum risk of not meeting demand, minimum environmental impact, and minimum costs. At a given moment, decisions have to be made regarding which production units to use, with particular emphasis on the proportion coming from hydraulic plants. Always choosing to use near-zero-cost water stocks from the dams (when available) is far from being the best decision when considering the consequences over one or two years. It is therefore necessary to formulate a more elaborate strategy that optimizes current decisions against many possible scenarios for the future, up to three years ahead. This is a tremendously complex stochastic dynamic programming problem, where computing needs grow exponentially with the number of energy stocks and with uncertainty factors.

As a consequence, current models and algorithms (e.g., the Bellman algorithm) can use only four energy stocks (three aggregated water stocks and one stock of demand-side options that can be activated to off-load consumption peaks) in order to deliver results in the limited time frame expected by field operators (overnight is a maximum). Taking into account many more stocks is necessary to reach a better optimization, for instance: more water stocks (corresponding to a finer decomposition of existing ones), fuel stocks, futures bought on the market, SO_2 maximum emissions, all those parameters being currently managed by experts outside the main optimization loop. In addition there is also a need to better represent the many types of constraints that limit operation of the various units such as start-up and shutdown ramp constraints, minimum operating and stopping delays, dam dependencies in a hydraulic valley.

Massive HPC power, combined with algorithmic efforts to reduce the "curse of dimensionnality" of the Bellman algorithm, is currently being tested to evaluate the benefits of a 10-stocks optimization allowing for dynamically hedging with electricity market available futures contracts. This optimization, over a of 1.5 years with 1-day time steps, will use 32,000 BG/P cores to keep processing time under 3 hours. Depending on the improvement desired over current methodologies, this kind of application will clearly require petaflops-class computing in the near future [8,9].

8 End-user simulation environment

Advancing individual solvers performance is not enough to bring high-performance simulation to the end-user. Each community needs a much broader set of tools in order to conduct industrial studies: CAD, mesh generation, data setting tools, computational scheme editing aids, visualization, and so forth.

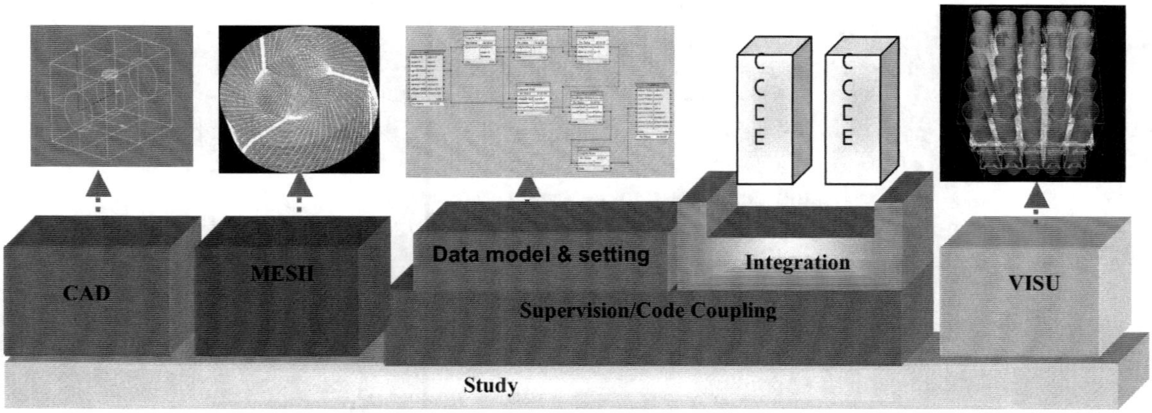

Figure 10. The Salome platform, www.platform-salome.org.

In early 2000 EDF, together with CEA and other industrial and academic partners, started the development of an integrated toolbox Salome [10,11], with the following aims:

- Reduce the cost of complex simulation platforms by mutualizing a set of common tools: pre- and postprocessing, calculation distribution and supervision, and so on.
- Boost performance through easy integration of multiple solvers for multiphysics studies (via a common data model).

Salome contains about 500,000 lines of code (85 % C++, 10 % Python) and has been used since 2007 for major applications such as the structural mechanics platform SimuMeca, which serves a community of about 200 in EDF Engineering and R&D Divisions.

Salome's technical prerequisites rely on widely used open source solutions. Computational results are stored in HDF5 format (or XML for small pieces of data) in order to ensure portability and efficiency. Communication between user interfaces, computational engines, and codes is performed with CORBA, which makes them interoperable whatever the implementation programming language or their localization. A common data model and associated libraries make possible structured data exchange, allowing better modularity between components. Graphical representation of input data (geometry, meshes) and results is made with the VTK visualization library, which is easy to use and offers a high level of abstraction. In the near future, Paraview will also be included in Salome. Data manipulation, computation supervision, and exploitation of simulation results may be done through a graphical interface or via a Python script. These two ways of performing simulations, thanks to the Salome communication bus, enable the Salome user to interactively define its study via the graphical interface and then to quickly automate the simulation process by reloading the Python script generated by Salome. Python also offers advanced users numerous services for complex studies or parametric computations.

If Salome has been proved to be well adapted for sequential and moderately parallel simulations it is evolving in order to support massively parallel computing. Work is under way on what we consider three critical components:

Journal of Physics: Conference Series **125** (2008) 012063 doi:10.1088/1742-6596/125/1/012063

Mesh generation. In 2007 it took several months to produce the 10^8 cells mesh for the simulation of part of a fuel assembly with the CFD code Saturne, compared to "only" 1 month of calculation needed on 8000 BG/L processors. Generating a 10^{10} cells mesh as targeted in 2015 requires future meshing tools to provide parallel meshing, automatic hexahedral meshing, mesh healing, CAD healing for meshing, and dynamic mesh refinement. A collaboration with Distene (Spin-off initiated by INRIA) is engaged on these issues.

Visualization. We are currently putting together a 6 by 3 meter, 23 Mpixels Power Wall, together with parallel visualization tools and clusters, in order to boost our visualization capacities of the most advanced simulations. However, the sheer volume of data produced by petaflop calculations, storage and network limitations, and multisite teams make it necessary to further advance R&D in a number of areas [12]. Under the CARRIOCAS project and the SCOS project we have started work on remote and collaborative multiuser visualization, parallel visualization, and parallel and distributed file systems.

Supervising and code coupling tool. In 2006, EDF and CEA started development of YACS, a new generation of supervisor, intended to handle a parallel multiphysics coupling scheme named PACO++ [13], developed by INRIA through a portable parallel extension to CORBA. PACO++ defines a parallel object as a collection of identical CORBA objects whose execution model is SPMD (single program multiple data). Special attention has been given by EDF team to the support of distributed arguments. The arguments can be distributed on both the client and the server sides. PACO++ provides mechanisms (static and dynamic) to specify the data distribution on both sides. SInce the two data distributions may differ, data redistribution may be required during the communication between the client and the server.

9 Concluding remarks

As we have seen from the preceding examples, high-performance simulation for our industry is not just a possibility for a hazy and distant future: as of mid-2008 work is under way on 10 major cases where this kind of technology opens clear and unique opportunities. One of them is already in operationnal use; three more should yield industrial-grade results this year. HPC has been raising opportunities across a large scale of business activities. Far from being confined to nuclear operation, it is now being seriously considered for applications as different as revewable energy development (advancing photovoltaic performance), marketing (pricing new services), or financial risk management.

This process started with a decision to bring HPC machines to R&D teams, even though no detailled needs for such supercomputers was available. Building from this initial move, a few computer science engineers and applied mathematicians, together with some advanced numerical physicists, succeeded in "spreading the HPC word" within the company by showing first results meaningful for business decision makers. These efforts in turn made it obvious that this technology would not only improve existing processes but could change them altogether for more radical optimizations. Innovative thinking is now well on the way, fueled by a better perception of the new HPC capabilities and perspectives.

At the same time HPC raised new concerns:

- Are underlying models sound enough to make sense when stretched to new performance levels? How to extend qualification domain when experimental data is scarce or very costly to access?
- What level of model refinement is coherent with unavoidable remaining uncertainties?
- How can we build decision-makers' confidence when models and calculations complexity make it more and more difficult to check against one's own intuition? What kind of new expertise should be developped to secure a broader use of those new capabilities?
- At which time horizons and costs can we expect the computing reliability and service level necessary for a reliable use in the field?

Most of those questions remain open today and we try to handle them step by step through practical feedback in real operational situations. On the technology side, besides pre and post processing issues raised by the huge increase of data , we are facing the immediate problem of scaling existing codes on a very large number of cores. Figure 11 shows that three applications are already using 10k processors or more.

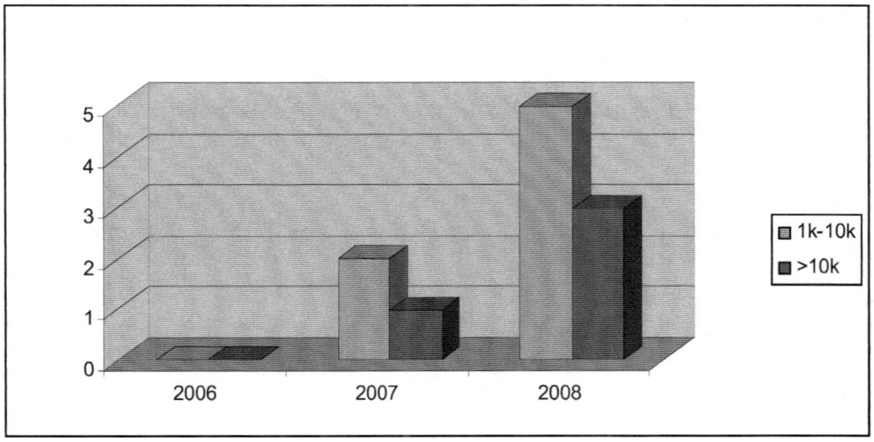

Figure 11. Number of applications using more than 1k processors.

This has been achieved only two years after we bought our first BG/L machine, through both in-house efforts and partnerships with academics and IBM to port and scale our codes. However successful we have been in this initial period, we know that future advances may be at a higher cost:

- ASTER, for instance, our structure-mechanics code, hardly scales today on more than a few tens of processors, because we have not yet found a general enough solution that would match the very wide array of its modeling capacities, and the direct solving methods that give it its industrial-grade robustness. (Three "nonintrusive" ways of scaling ASTER are currently being looked at with academic partners: at the solver level, testing parallel direct solvers for large, sparse matrixes; at the modeling level, implementing a domain decomposition method called FETI (Finite Element Tearing and Interconnecting); and at the simulation control level using a code-coupling library adapted to parallel computing to coordinate many ASTER runs within a domain decomposition approach.)
- On the other end, our fluid dynamics code Saturne that currently scales over 10 k cores, will have to support runs on 100 K or possibly one million cores in order to yield expected performance on already identified challenges. At this level of parallelism we know that we will have to deal with hardware fault tolerance, potentially heterogeneous/hybrid computing nodes, hierarchical multilevel memory, and critical load balancing.
- Dealing with large domains (billions of cells, tens millions of particles), also raises numerical stability and precision computation problems that could induce algorithmic and numerical solver adaptations, as we have already discovered.

Solutions to this kind of generic HPC obstacle lie very much in the hands of academic research, universities and computer vendors R&D. EDF is willing to take part in this effort through direct R&D inputs and partnerships with the relevant initiatives and communities in Europe and worldwide, as well as by publicizing our industrial cases and focusing attention on the scientific challenges that they pose, some of which we believe can be nearly as daunting as the most difficult problems in more fundamental energy research.

Acknowledgments
We express our gratitude to the following EDF R&D colleagues who contributed significant input to this paper: Christian Caremoli, Ange Caruso, Olivier Dubois, Estelle Desroches, Yvan Fournier, Vincent Lefebvre, Jean-Daniel Mattei, and Xavier Warin.

References

[1] Bremnes O and Schumm A 2006 Modeling the influence of the water level on radiographic inspection of pipes, *ECNDT*

[2] Diaz J, Fouquet T, Duwig V, Schumm A and Chassignole B 2006 Structural noise in modelling welds *ECNDT* 2006, September 25–29

[3] Blanchon F, Ha-Duong T and Planchard J 1998 Numerical methods for solving the reactor kinetics equations, *Progress in Nuclear Energy* **22**(2):173–180

[4] Violeau D, Issa R, Benhamadouche S, Saleh K, Chorda J and Maubourguet M 2008 Modelling a fish passage with SPH and Eulerian codes: The influence of turbulent closure *3rd SPHERIC International Workshop* (Lausanne) June 3–6

[5] Moulinec C, Issa R, Emerson D and Gu X 2008 High-Performance Computing SPH: Towards a hundred million particle simulation *3rd SPHERIC International Workshop* (Lausanne) June3–6

[6] Issa R, Decung F, Razafindrakoto E, Moulinec C, Latino D, Boiteau O and Violeau D 2008 HPC for Spartacus-3D SPH code and applications to industrial environmental flows *Parallel CFD*

[7] Issa R, Violeau D, Thibault G, Moulinec C and Latino D 2008 HPC for Spartacus-3D SPH code and applications to real environmental flows *3rd SPHERIC International Workshop* (Lausanne) June 3–6

[8] Vialle S, Warin X, Makassikis C and Mercier P 2008 Stochastic control optimization and simulation applied to energy management: From 1-dimensional to N-dimensional problem distributions, on clusters and Blue Gene supercomputers *Parallel Computing* (to be published)

[9] Vialle S, Warin X and Mercier P 2008 An N-dimensional stochastic control algorithm for electricity asset management on PC cluster and Blue Gene supercomputer *Para 2008* (to be published)

[10] Ribes A and Caremoli C 2007 SALOME platform component model for numerical simulation *COMPSAC* (Beijing, China)

[11] Bergeaud V and Tajchman M 2007 Application of the SALOME software architecture to nuclear reactor research *SCS Spring Simulation Multiconference on High Performance Computing Symposium* (Norfolk, VA)

[12] Jourdain S, Forest J, Mouton C, Nouailhas B, Moniot G, Kolb F, Chabridon S, Simatic M and Mallet L 2008 ShareX3D, a scientific collaborative 3D viewer over HTTP *Web3D Symposium* (Los Angeles) August 9–10

[13] Pérez C, Priol T and Ribes A 2004 PaCO++: A parallel object model for high performance distributed systems *HICSS-37*, p 274a (IEEE Computer Society Press, Hawaii) January

Journal of Physics: Conference Series **125** (2008) 012064 doi:10.1088/1742-6596/125/1/012064

Time-dependent density functional theory applied to superfluid nuclei

Aurel Bulgac[1] **and Kenneth J Roche**[2]
[1]Department of Physics, University of Washington, Seattle, WA 98195-1560, USA
[2]Computer Science and Mathematics Division, Oak Ridge National Laboratory, Oak Ridge, TN 37831-6173 , USA

E-mail: bulgac@phys.washington.edu and rochekj@ornl.gov

Abstract. We describe the response of superfluid nuclei to any external time-dependent probe within an extension of the density functional theory, the time-dependent superfluid local density approximation (TD-SLDA). All quasi-particle wave functions (qpwfs) are represented on a three-dimensional spatial grid to allow for the breaking of all possible symmetries, and subsequently these functions are evolved in time using a high order multi-step algorithm. Mathematically this problem is a system of coupled, time-dependent, nonlinear partial differential equations, which when discretized becomes equivalent to a nonlinear classical mechanics problem with $\propto N_x^6 \approx 10^{12}$ degrees of freedom, where N_x is number of spatial mesh points in one direction. The lattice representation of the qpwfs allows for fast and accurate evaluation of various spatial derivatives required for the construction of various physical quantities that appear in the theory. The total number of independent qpwfs is of the order of spatial mesh points, and the system is evolved for tens to hundreds of thousands of time steps. We describe how we implemented the TD-SLDA method and report some of the challenges encountered.

1. Introduction

In condensed matter physics and quantum chemistry density functional theory (DFT) and time-dependent density function theory (TD-DFT) are commonly used approaches to describe the ground state properties of electronic systems [1], otherwise known as normal fermionic systems. However, there is an extended class of physical systems for which these are inappropriate tools, namely, the superfluid systems [2], the most common electronic incarnation of which are the superconductors. While for normal systems the number of single-particle wave functions needed in a (TD-)DFT description is equal to the number of particles, in the case of superfluid systems this number could exceed the number of particles by a very large factor. The larger the critical temperature of the system, the larger this factor is and the more demanding the computational complexity becomes. Superfluids, for which the pairing gap is sufficiently large when compared to the Fermi energy of the system, can be encountered in a wide range of situations, such as fermionic cold atoms in traps, high T_c-superconductors, nucleons in nuclei and neutron stars, and quarks in high-density QCD matter.

2. TD-SLDA

Mathematically the TD-SLDA requires the solution of a large set of coupled time-dependent nonlinear partial differential equations:

$$i\frac{\partial\psi_n(\vec{r},t)}{\partial t} = [h(\vec{r},t) + V_{ext}(\vec{r},t)]\psi_n(\vec{r},t), \tag{1}$$

where i is the imaginary unit, $\psi_n(\vec{r},t)$ are complex (wave) functions of time and 3D space coordinates, and each of them is labeled by an index n that takes usually both discrete and continuous values. $h(\vec{r},t)$ is a 3D partial differential operator depending in a nonlinear manner on the entire set of complex functions $\psi_n(\vec{r},t)$ (a feature routinely referred to as self-consistency), and $V_{ext}(\vec{r},t)$ is an arbitrary function of time and 3D space coordinate that models a specific external probe of the system. Spin and isospin coordinates are not displayed for simplicity of presentation. Typically $h(\vec{r},t)+V_{ext}(\vec{r},t)$ is a Hermitian operator, and for that reason the above system of nonlinear partial differential equations has an infinite number of conserved quantities, but that, however, does not make it an integrable system. After time-dependent solutions have been computed, one can extract various densities and through a time-dependent Fourier transform extract physical observables of interest. The number of various densities needed is of order of ten or so in most cases.

3. Implementation of TD-SLDA

For the studies we envision, in principle 3,000 to 10,000 nuclei are ultimately of interest. Simple estimates show that to extract the observables with a reasonable accuracy and over a frequency interval of interest for a particular nucleus, of the order 10^4 to 10^5 time steps must be performed for a given nucleus.

Taking a time step means solving equationuations (1). In order to solve these equations the system is placed on a 3D spatial lattice with a total $N_{3D} = N_x \times N_y \times N_z$ mesh points. The size of the lattice and the prescribed lattice spacing are calibrated according to the total number of nucleons of the nucleus to be studied. The number of complex functions $\psi_n(\vec{r},t)$ at a given time is $O(N_{3D})$, that is, the number of labels n of the qpwfs. Each of these functions has an internal complexity that requires additional indices to label spin, isospin, and particle-hole degrees of freedom.

Here we explicitly compute the variation of the energy of the system on the lattice. The first action our code takes after some system diagnostics is that the lattices in both coordinate and momentum space are initialized. Next, the kinetic energy is evaluated. This process is simplified by the fact that often this term is diagonal in k-space in our chosen representation. Also, the potentials have to be initialized at each point on the coordinate lattice. In the nuclear problem this includes central, Coulomb, spin-orbit, and pairing potentials. From the qpwfs we construct various densities, which are bilinear combinations of the qpwfs, (e.g., the proton and neutron number densities, anomalous, kinetic energy, spin-orbit densities) on the lattice at each coordinate. With the densities initialized and the potential initialized, the system energy and the particle numbers of the system are evaluated.

The densities are bilinear combinations of the qpwfs, the complex conjugates of the qpwfs, and of the various spatial partial derivatives of the qpwfs. The plane wave lattice representation allows for fast and accurate computation of partial derivatives. The operation is simply two discrete Fourier transforms and an element-by-element array multiply per qpwf. For 3D lattices of up to 128^3 it was shown on various architectures that an FFTW [3] evaluation of partial derivatives is about a factor of seven at most the time required for a simple multiplication of two 3D arrays element-by-element [4]. If the lattice spacing is chosen appropriately for each physical system to be studied, the accuracy of this representation of the wave functions can easily attain almost machine precision [5].

To perform the time evolution we have chosen a multistep predictor-modifier-corrector method due to Adams-Bashforth-Milne (ABM) [6]. The predictor and modifier are calculated for each qpwfs at each lattice site. This computation is a function of the qpwfs and their time derivatives at multiple time steps and includes operations such as discrete Fourier transforms. The modified densities are computed, and these terms are subsequently used to modify the potentials. From the modified densities and potentials, the corrector for each qpwf is calculated at each site. The corrected densities and potentials are constructed, total energy and total particle number are evaluated, and time is officially iterated a time step. Other methods can be easily implemented, but this choice has some advantages, not necessarily unique to this method. The ABM algorithm requires only two rhs evaluations per time step, it is a fifth-order method and is apparently also very stable numerically for long time intervals for the class of problems we envision studying. Take a 40^3 lattice as example, the system energy and number density are numerically conserved, showing a variance of $O(1/10^6)$ over a thousand time steps during the evolution of 66796 qpwfs. The algorithm has also been tested on a simpler problem, a time-dependent homogeneous unitary Fermi gas, where a system of about 30,000 differential equations was integrated for about 300,000 time steps [7].

A major drawback of multi-step algorithms is the heavy memory requirements. This problem can be managed a number of ways but we are pursuing two particular approaches: out-of-core and aggregated distributed memory.

In the out-of-core technique, it is assumed that not enough resources are available to store in memory all the data required to compute system observables within a time step. Thus, the information required to describe each quasi-particle state is either stored in memory or on disk in files. To evolve the system requires read and write operations on files stored on disk. This approach suffers from the poor performance of swapping data via disk accesses during each phase of computation. It has the advantage that larger lattices can be used to study the system. For instance, $N_x = N_y = N_z = 100$ lattice yields 1046604 qpwfs and would require $\sim 800 \times 2^{40}$ bytes of memory to run the problem without access to the filesystem. With swapping, this system can be computed on a single node allocation where there are 4×2^{30} bytes of RAM. For lattices greater than 130^3, one is forced to reconsider the out-of-core approach since even swapping with the current data structures is unwieldy to allocating memory for a base unit of work. However, for realistic problems (40^3 or greater) the operational complexity is $O(2^{40}$ flop) or greater per time step, and so the return time of the calculation would be unacceptable not just because of time spent swapping, but also because of time spent computing the work that goes into computing a physical time step.

The aggregated distributed-memory approach essentially requests the number of compute nodes sufficient to fit the entire problem in physical memory and then relies on network communication to organize the computation. There is some redundancy not present in the sequential code implementation or an out-of-core scheme since each process in the distributed memory approach needs a copy of the lattices as well as global system observables. The mapping that we use to reduce both data and instruction complexity relates MPI [8] process ranks to disjoint ranges of canonical labels of the qpwfs. Each process manages the same number of functions and all other data derived from those functions within a time step (differing by at most a single qpwf between MPI processes). Thus, both the storage and instruction complexity are balanced per process. In order to compute the various densities of the system, each process simultaneously evaluates the partial contribution of the qpwfs it manages locally on its copy of the lattices. The complete densities are then constructed by communication over the distributed network via a global reduction of these partial values at each point of the lattice followed by a broadcast of the total densities. The updated densities of the system are used to modify the system potentials and construct the values of the qpwfs at the next time step. This approach is still memory bound as the lattice dimensions grow. For instance, as of today and without any modifications, the current code cannot study lattices larger than $\sim 60^3$ on the 7832 Quad Core AMD Opteron [9] compute nodes of the DOE's Cray XT [10] system in the National Center for Computational Sciences at Oak Ridge National Laboratory [11]. This equates to evolving 226156 qpwfs, a number that may already be more than sufficient to study most nuclei of interest.

The discretization described here also has the advantage that it exhibits outstanding strong scaling trends; see table 1 for an example. The idea of weak scaling for an algorithm is that we

Table 1. Strong scaling on dual-core AMD Opteron Cray XT. The problem complexity is fixed at 28288 qpwfs on a 30^3 lattice for a single time step. MPI processes (PEs) are added thus reducing the storage and work load per process. The time reported is the wall-time in seconds expired to compute the single time step for a given number of processes.

PEs	576	1152	1728	2304
NWF/PE	48	24	16	12
[s]/TS	56.2	28.8	19.3	14.92

increase the problem complexity by some factor, increase the PE count by the same factor, and the runtime remains constant. We are studying [cost ($, time, instructions, etc.) / lattice site / qpwf / PE / time step] as a weak scaling metric. Tables 2 and 3 provide reference material for the weak scaling behavior of the code on `jaguar.ccs.ornl.gov` .

Table 2. Hardware event profile data from Quad Core AMD Opteron(tm) Cray XT for 10 time steps of TD-SLDA evolution.

Nx^3	30^3	40^3	50^3
QPWF	28288	66796	130528
PEs	168	942	3626
[s]	297.31	296.15	319.03
INS	141538488856847	787635078990076	3133849257252106
FLOP	33786997017008	184227473630636	822772291175810

The current software implementation of the TD-SLDA discretization exists in both the Fortran and C programming languages. Distributed memory parallel computing systems designed with both Intel Xeon(tm) and AMD Opteron(tm) multicore chipsets have been successfully tested to conserve significant system observables as a function of time. The GNU Compiler Collection (GCC) [12], Portland Group (PGI) [13], and Intel [14] compilers have been used during testing on these systems. The FFTW and MPI [8] libraries have been used and the code employs the PAPI api [15] to provide chip-level diagnostics of the hardware during code execution. We plan to pursue a hybrid version of the code capable of exploiting the benefits of both the out-of-core and aggregated distributed memory techniques.

Acknowledgments

We thank Y. Yu for major help at the initial stages of coding the theory and the UNEDF SciDAC collaboration for support through the US Department of Energy under grant No. DE-FC02-07ER41457. This research is sponsored by the Office of Advanced Scientific Computing Research; U.S. Department of Energy. The work was performed at the Oak Ridge National Laboratory, which is managed by UT-Battelle, LLC under Contract No. De-AC05-00OR22725.

Table 3. Weak scaling assertion of TD-SLDA on Quad Core AMD Opteron(tm) Cray XT (reference Table [2]). The ratios of the runtimes are reported in column 1. These should be 1 in the ideal case. Column two is the weak scaling factor as it is the ratio of the more complex problem hardware allocation to the less complex problem hardware allocation. Column 3 is the ratio of the total executed instructions for the more complex problem to the total executed instructions of the less complex problem. Ideally, this number would be greater than or equal to the number in column 2. Since this is not ideal, remember to skew the result by the value in column 1. Column 4 is the ratio of the total floating point instructions executed for the more complex problem to the total floating point instructions executed by the less complex problem. Again, we expect this number to be consistent with the weak scaling factor. Thus, for these runs, the weak scaling of the TD-SLDA code is not perfect but is basically excellent.

$\frac{T[50^3]}{T[40^3]} = 1.07(7)$	$\frac{PE[50^3]}{PE[40^3]} = 3.84(9)$	$\frac{INS[50^3]}{INS[40^3]} = 3.97$	$\frac{FLOP[50^3]}{FLOP[40^3]} = 4.46$
$\frac{T[50^3]}{T[30^3]} = 1.07$	$\frac{PE[50^3]}{PE[30^3]} = 21.58$	$\frac{INS[50^3]}{INS[30^3]} = 22.14$	$\frac{FLOP[50^3]}{FLOP[30^3]} = 24.35$
$\frac{T[40^3]}{T[30^3]} = .996$	$\frac{PE[40^3]}{PE[30^3]} = 5.60(7)$	$\frac{INS[40^3]}{INS[30^3]} = 5.56$	$\frac{FLOP[40^3]}{FLOP[30^3]} = 5.45$

References

[1] Dreizler R M and Gross E K U 1990 *Density Functional Theory: An Approach to Quantum Many-Body Problem* (Berlin: Springer).
[2] Bulgac A 2007, Phys. Rev. A **76** 040502(R).
[3] The Fastest Fourier Transform in the West, `www.fftw.org`
[4] Aichinger A and Krotscheck E 2005, Comp. Mat. Science **34** 188.
[5] Littlejoh R G, Cargo M, Tucker Carrington J, Mitchell K A, and Poirier B, 2002, J. Chem. Phys. **116** 8691.
[6] Hamming R W 1962 *Numerical Methods for Scientists and Engineers*, (New York: McGraw-Hill).
[7] Bulgac A and Yoon S 2008, in preparation.
[8] Message Passing Interface, `www.mpi-forum.org`
[9] Advanced Micro Devices (AMD), Software Optimization Guide for AMD Family 10h Processors, rev. 3.0.6 (2008); Compiler Usage Guidelines for AMD64 Platforms, rev. 3.22 (2007).
[10] Cray Inc. , `www.cray.com`
[11] National Center for Computational Sciences, `www.nccs.gov`
[12] The GNU Project, `www.gnu.org`
[13] The Portland Group Compilers, `www.pgroup.com`
[14] Intel, `developer.intel.com`
[15] Performance Application Programming Interface, `icl.cs.utk.edu/papi`

Journal of Physics: Conference Series **125** (2008) 012065 doi:10.1088/1742-6596/125/1/012065

Performance engineering challenges: the view from RENCI

RJ Fowler[1], T Gamblin[1], AK Porterfield[1], P Dreher[1], S Huang[1] and B Joó[2]

[1] Renaissance Computing Institute, UNC, Chapel Hill, NC 27516, USA
[2] Thomas Jefferson National Laboratory, Newport News, VA 23606, USA

E-mail: rjf@renci.org, tgamblin@cs.unc.edu, akp@renci.org

Abstract. Trends in chip technology and system design are causing a revolution in high-performance computing. The emergence of multicore processor chips, the construction of very large computing systems, and the increasing need to deal with power and energy issues in these systems are three of the most significant changes. We focus on the way that these trends have created a new set of challenges in the area of performance engineering, the measurement, analysis, and tuning of computing systems and applications. We discuss these changes and outline recent work at the Renaissance Computing Institute to meet these challenges.

1. Trends in high-performance computing

As it has over its entire brief history, computing technology continues to evolve at a rapid pace. Since the emergence of microprocessor-based clusters, this has generally been felt as incremental improvements in single-thread performance. Recently, however, several trends are introducing qualitative changes, especially in high-performance computing.

Instead of continuing to exploit increased clock rates, and an increased degree of implicit, instruction level parallelism (ILP), the improved performance of the current generation of chips comes from increased explicit, multi-core, and multi-thread parallelism.

Technology improvements make it feasible to build high-end systems with many tens- to hundreds- of thousands of chips. Combined with new generations of processor chips, these systems will have a degree of thread-level parallelism in the millions. The sheer scale of these systems forces everything, including performance analysis, to be done in parallel.

Power and energy considerations impinge on every decision. Power density, efficiency, and cooling issues are driving the trend towards explicit parallelism. Power and cooling infrastructure are dominating facility costs. Energy costs are a large, and increasing, component of the total cost of system ownership.

Each of these trends is having a substantial impact on the general area of "performance engineering" in high-performance computing. We describe ongoing work at RENCI to address the first two of these issues. Power and energy are overarching problems that are driving the other trends, but specifics of dealing with them are beyond the scope of this paper. Within the present context they are a matter of running parallel systems as efficiently as possible.

2. Resource-centric performance tools and analysis

2.1. Processors in an era of single-thread performance gains

The impressive increase in microprocessor performance has been a consequence of several several related trends. The linear feature size on chips commodity chips has shrunk from tens of microns to a range of 45 to 65 nanometers; this has led to a quadratic increase in the number of devices per unit area. The size of commodity dies has also increased. The increase in devices per chip, usually known as Moore's law, has permitted designers to improve performance by adding instruction level parallelism (ILP) by building deeply pipelined, out-of-order, superscalar processors. The decreasing feature size has also improved transistor switching time and dramatically higher clock rates. Memory density has increased at a similar pace; and large, complex caches have mitigated the much slower rate of increase of memory speed.

We are emerging from a period in which "performance engineering" of applications was relatively simple and effective. For many applications, it was sufficient to "ride the Moore's law wave"; why waste effort optimizing code if computers will double in speed in the next eighteen months anyway? For other classes of application, this was a "golden age of optimization" with program transformations, applied manually by "performance specialists" or automatically by optimizing compilers [1]. Even applications that were bound by memory bandwidth could achieve some ILP by exploiting memory interface protocols to overlap concurrent memory operations. Average memory concurrency of only 6.25 can yield bandwidths of over 4 GB/sec on systems that would otherwise only achieve 640 MB/sec (one 64-byte block per 100ns).

On high-ILP systems, performance issues were a matter of how all of the instructions in flight interacted with one another in the pipelines and the memory system. For example, whether a particular memory reference is a cache miss is not as important as how much of the miss latency can be hidden by concurrent instructions or by a hardware prefetcher.

2.2. The era of explicit parallelism

The overhead incurred to achieve ILP increased much faster than the performance gains realized. At the same time, the amount of circuit-level parallelism and power consumed has increased commensurately. Hence, the processor industry has embraced a more efficient explicitly-parallel design strategy. Clock rates temporarily decreased, and now grow much more slowly.

With the advent of explicitly parallel, multicore, multithread processor chips, the execution context of each instruction now includes all the concurrent instructions in all the other threads and cores. In such environments, the interactions of threads (cores) competing for shared resources will dominate performance.

In a single application, some interthread interactions will be explicit scheduling and synchronization. Implicit contention for shared resources is as much of a problem. To illustrate this, we ran independent copies of the NAS parallel benchmarks on quad-core Intel and AMD processors and measured the throughput [2]. While throughput increased from one to four threads, total run time increased to as much as 80% for MG and SP, yielding a parallel efficiency of $1/1.8 = 0.55$ for nominally independent executions. On Intel Clovertown, parallel efficiency was as low as 0.43 for four copies of MG. Other measurements on AMD chips show dramatic increases in the number of DDR memory operations (especially for DDR page conflicts). The impact of inter-thread interactions is clearly large, and this effect can have a larger impact on performance than local code tuning within a single thread.

Since costs are increasingly less attributable to a single thread, tools that focus on "first-person" performance measurement [3, 4] are less able to measure and diagnose crucial performance problems in this domain. At RENCI, we are pursuing the approach of developing "resource centric" measurement and analysis tools. Our focus is on the use of shared resources and interactions among active threads during periods of high utilization.

Information generated by tools needs to be available for off-node analysis and reporting tools.

We are pursuing on-node, on-line performance introspection. The current state of our prototype will be presented with the poster accompanying this paper at the SciDAC 2008 conference.

3. Challenges of system scale

The degree of parallelism in top systems is increasing dramatically as a result of simultaneous trends toward much larger node counts, multisocket node architectures, and multicore/multithread chips. At such scales, Amdahl's law will be a harsh master for applications, system software, and even for performance tools. Sequential algorithms for measurement, data collection, analysis, and presentation of performance data on these systems are already untenable.

Current popular performance tools [3, 4, 5] focus primarily on measurement, storing large volumes of data to disk and doing *post-mortem* analyses on other systems. Profiling tools [3, 4] reduce data volume by compiling summary statistics of metrics over time, but they lose valuable temporal information that could be used for root cause analysis. Tracing tools [3, 5] preserve this information, but the output is too large to store at petascale without significant perturbation. Further, sequential analysis of such data on workstations or small clusters would be too slow to be usable, even if petascale data volumes could be practically stored. Developers could measure, analyze, and tune on small-scale systems and problem instances. This situation can certainly help in some cases, but small-scale performance is no guarantee of large-scale behavior.

Almost by definition, driving problems for very large computer systems use the entire system and run for days or weeks. Otherwise, it would be preferable to run longer but more efficient jobs on smaller systems. Furthermore, the codes to solve these problems use the most advanced algorithms, which are often adaptive. Addressing the problem of performance measurement and analysis, if only to confirm that the code is running as well it can, requires operating at very large scale for long times. Cost prohibits conducting such runs only to measure performance, so tool overheads must be acceptable in a production environment.

3.1. Approach

To address these issues, we are conducting research and development on scalable tools that filter and analyze data *in-situ*. At runtime, the tools have access to the same computing power and parallelism as the application. Using low-overhead parallel filtering algorithms such as multiscale wavelet compression, we can reduce and analyze the data before it is stored to disk or even moved out of the computing fabric. These techniques also allow hierarchical, interprocessor analyses in real time, facilitating adaptive measurement and even steering.

3.2. Work to Date

AMPL [6] is a tracing tool that uses population sampling techniques to reduce trace data volumes in large applications. AMPL's output size scales sublinearly with the number of nodes in the parallel system, and sample overhead can be further reduced by monitoring homogeneous groups of processes separately.

Load balance is an important problem for petascale performance but a difficult challenge for tool designers, as it requires system-wide, inter-process measurements over the duration of program execution. The reason is that adaptive solvers and irregular data partitioning methods can dynamically introduce imbalances. It is thus important to not record and store data that scales linearly with problem size and run length. Furthermore, to diagnose the causes of imbalance, it is important to attribute observed problems to the region of application code that caused them. Without knowledge of application semantics, this task is difficult, so we have developed a model for characterizing the sources of computational load in SPMD codes, and a tool that uses the model to dynamically extract and store multidimensional load data efficiently [7]. This tool uses parallel wavelet transforms to reduce data volume by 2-3 orders of magnitude with very low error. Compression time is fast enough for online, production use.

We have proposed, with our collaborators, a scalable component framework for rapid development of application-specific petascale performance tools [8]. Our framework enables existing scalable components (measurement, data aggregation, analysis) to be used modularly in conjunction with each other and to be reconfigured dynamically to fit the needs of applications.

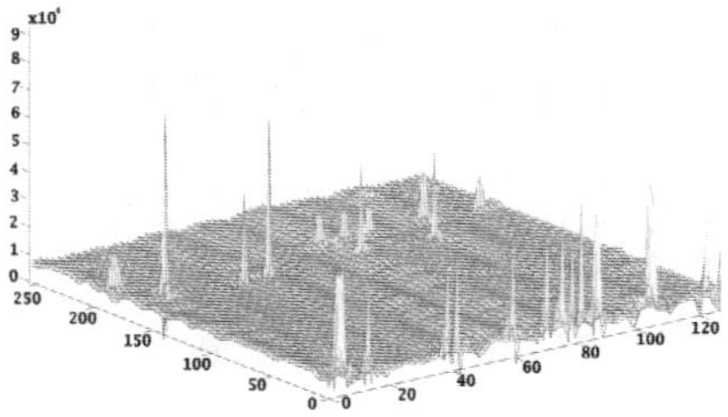

Figure 1. Transient load in ParaDiS collision computation for 128 processes and 256-timesteps.

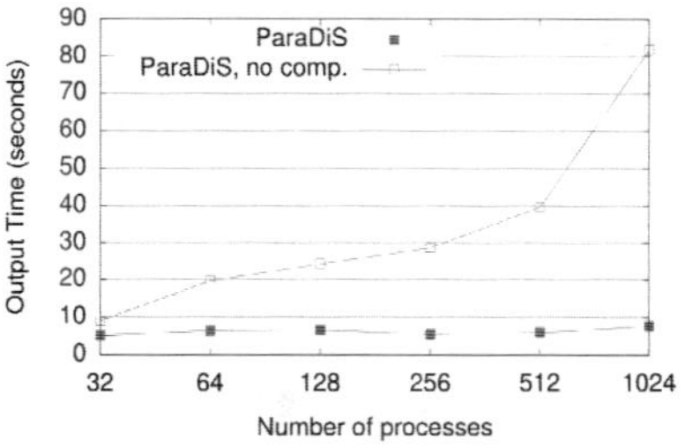

Figure 2. Scaling behavior of load data output for ParaDiS, with and without compression.

3.3. Dynamic load-balance case study

We used our load-balance tool to monitor ParaDiS [9], a crystal dislocation dynamics code. ParaDiS includes a data-dependent computation phase in which dislocations are split or merged when they collide. Figure 1 shows transient load spikes in this phase for a 128-processor run. Figure 2 shows elapsed time to gather over 100 such plots for 256 time steps of ParaDiS on Blue Gene/L for increasing processor counts, both with and without our tool. Without our tool, the time to collect and compress the data scales linearly with the number of processors in the system. With it, data compresses to a volume handled well by Blue Gene/L's I/O backbone, and we achieve constant scaling.

4. Case study: LQCD codes on multicore

Quantum chromodynamics (QCD) is the theory of the strong force in the Standard Model of subatomic physics. Since planned Lattice QCD (LQCD) calculations are expected to consume hundreds of teraflop-years it is vital that LQCD code be as efficient as possible. As part of the SciDAC LQCD project we are engaged in the evaluation and tuning of LQCD codes.

In common LQCD simulations, one phase involves a global matrix inversion. For a fixed size problem, the cost of the inversion phase increases super-linearly with the number of computational nodes used. Restated, for a fixed global problem size, inverter efficiency improves as each node's local problem size is increased.

The other phases are dominated by local computation and nearest neighbor communication. Implementations are extremely sensitive to memory traffic. Hence, highly tuned implementations of these phases perform very efficiently when they can be done in cache. Said differently, local kernel efficiency decreases when local problem size becomes larger than the cache size. We discussed the opposing efficiency trends in a previous article [10] in this series.

A key computational kernel for LQCD is the Wilson Dslash operator. Researchers at JLab in collaboration with ECCC at Edinburgh, UK have been adding multithread support to Dslash to better utilize multicore hardware. A comparison of several multicore implementations is presented in a related paper in this proceedings [11]. Prototypes compared include

"MPI everywhere" and hybrid implementations using MPI and either OpenMP, or QMT, a multithreading library specifically for QCD applications.

In conjunction with Jlab, RENCI researchers are endeavoring to explain the details of the performance of these multithreaded implementations [11]. The performance differences between the hybrid versions are due to differences in thread scheduling. Each iteration of the Dslash test involves multiple parallel regions separated by MPI operations in sequential regions. In the OpenMP version we observed that CPU utilization for each core averaged only about 60% and this is confirmed by CPU cycle measurements within each thread. Hence, the computational cores block and go to a halt state between parallel regions, requiring relatively heavyweight operations to restart. The amount of work done in the parallel regions, particularly for smaller local problem sizes, is insufficient to amortize this overhead. The OMP_WAIT_POLICY control in OpenMP version 3 will mitigate this problem.

Of particular concern are issues of data and thread affinity, sharing data vs. competing for shared cache space, and general issues of cache locality. In the QMT version, we have diagnosed two areas for improvement. The version used in [11] uses explicit AMD "non-temporal" prefetches and stores. Eliminating extra prefetch instructions and using the native caching policies results in a 4-core, 1-chip speedup of about 10 percent. This is reflected as a greatly reduced L3 cache miss rate. About one quarter of the memory operations are now initiated by the hardware prefetcher.

QMT uses spin locks for synchronization and task-dispatching. Hardware performance counter measurements show that four QMT threads running an 8^4 problem on a single 2.1 GHz AMD Barcelona chip generate approximately an average of 2.1 GBytes/second of memory traffic attributed mostly to the spin-lock. This memory traffic is uniform across threads indicating that there's no indication of imbalance. The memory traffic is, however, all generated during the sequential phases in which MIP operations occur and may thus be a source of interference. We are continuing our efforts to diagnose and improve the performance of this code.

5. Concluding remarks

In this brief paper, we have endeavored to present an overview of some the challenges of todays high performance computing environments and to discuss recent activities at the Renaissance Computing Institute to meet those challenges.

Acknowledgments

This work was supported by the DOE SciDAC program under award numbers DE-FC02-06ER25764 (PERI) and DE-FC02-06ER41445.

References

[1] Allen R and Kennedy K 2001 *Optimizing Compilers for Modern Architectures* (San Francisco, CA: Morgan Kaufmann)
[2] Porterfield A, Fowler R and Neyer M 2008 *Workshop on Managed Multicore systems MMCS* (Boston, MA)
[3] Shende S and Maloney A 2006 *International Journal of High Performance Computing Applications* **20** 287–331
[4] Mellor-Crummey J, Fowler R, Marin G and Tallent N 2002 *The Journal of Supercomputing* **23** 81–104
[5] Brunst H, Hoppe H C, Nagel W E and Winkler M 2001 *Computational Science - ICCS 2001* pp 751–760
[6] Gamblin T, Fowler R J and Reed D A 2008 *Proceedings of the International Parallel and Distributed Processing Symposium (IPDPS)*
[7] Gamblin T, de Supinski B R, Schulz M, Fowler R J and Reed D A 2007 *In Submission*
[8] de Supinski B R, Fowler R, Gamblin T, Mueller F, Schulz M and Ratn P 2008 *International Workshop on Scalable Tools for High-End Computing (STHEC)* (Kos, Greece)
[9] Bulatov V, Cai W, Hiratani M, Hommes G, Pierce T, Tang M, Rhee M, Yates K and Arsenlis T 2004 *Supercomputing 2004 (SC'04)*

[10] Zhang Y, Fowler R, Huck K, Malony A, Porterfield A, Reed D, Shende S, Taylor V and Wu X 2007 *J. Phys: Conf. Ser* **78** 5pp
[11] B Joo and U C 2008 *these proceedings* (Preprint arXiv:0806.2312v1 [hep-lat])

Continuing progress on a lattice QCD software infrastructure

Bálint Joó on behalf of the USQCD Collaboration

Thomas Jefferson National Laboratory, Newport News, VA 23606, USA

E-mail: bjoo@jlab.org

Abstract. We report on the progress of the software effort in the QCD application area of SciDAC. In particular, we discuss how the software developed under SciDAC enabled the aggressive exploitation of leadership computers, and we report on progress in the area of QCD software for multi-core architectures.

1. Introduction

Large-scale numerical lattice quantum chromodynamics (QCD) simulation programs require extensive software infrastructure [1]. In this contribution we report on some advances in the SciDAC-supported software work in the USQCD national program [2] during the past year. Details about the scientific results from numerical QCD simulations can be found in [3].

Considerable effort was invested this past year in preparation for using the leadership computing facility at Argonne and Oak Ridge national laboratories. We have also carried out investigations into the efficient use of multicore architectures for QCD. Our data sharing efforts have progressed, and we continue to extend and refine standards within our collaboration and worldwide. This article is organized as follows. In Section 2 we present some highlights of our exploitation of leadership computing resources. We report on our threading research in section 3. We discuss data sharing in section 4 and briefly consider other activities in section 5.

2. Exploiting leadership facilities for QCD

During the last year, all of our large community codes have been successfully ported to leadership systems of interest. In particular, the Chroma [4] and MIMD Lattice Collaboration [5] (MILC) codes have been deployed on the Cray XT, Blue Gene/P, and recent cluster hardware, while the Columbia Physics System (CPS) [6] has been ported to the Blue Gene/P. Considerable effort has been invested in the optimization of high-performance components for these architectures.

We show in the left half of figure 1 the performance of the MDWF solver on the BlueGene/P at Argonne. MDWF is an optimized domain wall fermion (DWF) conjugate gradients inverter package, developed at MIT [7], which incorporates several DWF operator variants. One can see that the single-precision solver achieves 800 Mflops/core (about 25% of peak). The solver performance is roughly constant over a variety of volumes, indicating strong scaling, and the recursive data ordering on node results in effective use of cache for the larger problems. MDWF has been interfaced with Chroma for production use by Jefferson Lab (JLab) staff.

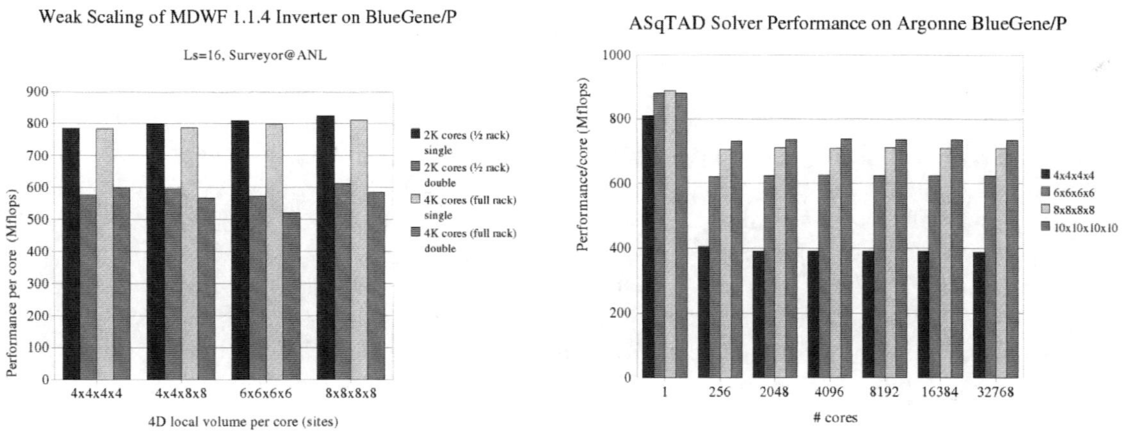

Figure 1. Performance on the Blue Gene/P for MDWF [7] (left) and AsqTAD (right) inverters. The MDWF measurements were made with Chroma interfaced to MDWF 1.1.4 on Surveyor. The AsqTAD numbers are from the MILC Code on Intrepid. Both machines are at the Argonne Leadership Computing Facility.

The MILC researchers have optimized their AsqTAD conjugate gradients solver for the Blue Gene/P and the Cray XT series and their gauge force for a variety of platforms. We show the performance of their AsqTAD inverter on the Argonne Blue Gene/P in the right-hand plot of figure 1. One can see that the weak scaling is excellent all the way out to 32K cores.

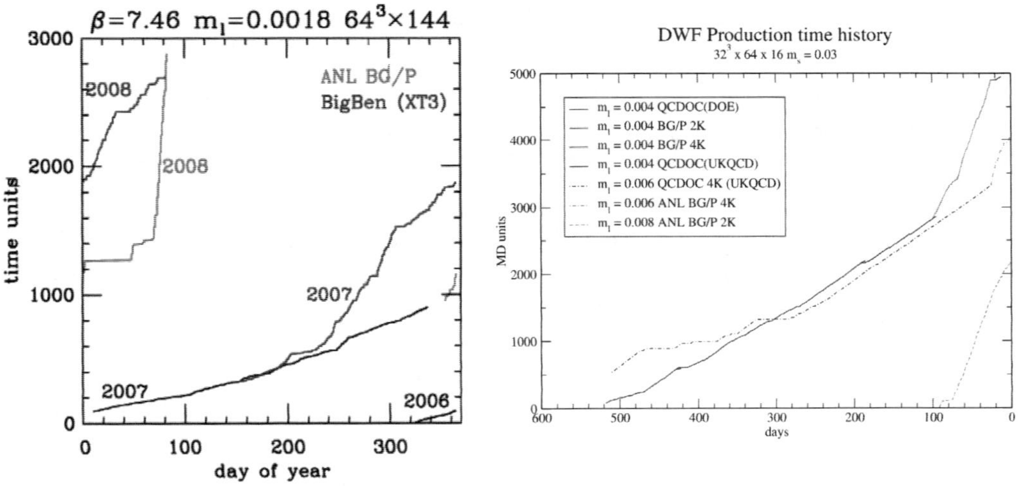

Figure 2. Production history of an AsqTAD staggered fermion run through late 2006, 2007, and early 2008 (left) and for some domain wall fermion production runs (right).

To emphasize the impact the leadership machines have had on our data production, we show in figure 2 the production history of both an AsqTAD run and some DWF runs over the past year. The AsqTAD production was performed by using the MILC code, while the domain wall fermion production was carried out by colleagues based at Columbia University and Brookhaven National Laboratory using the CPS code. The dramatic increases in production brought about by the Blue Gene/P for both sets of data production (red lines) are clearly evident.

In summary, USQCD has completed successful ports of its applications to leadership hardware and is now reaping the benefits in terms of science production.

3. Threading investigations

In preparation for the arrival of multicore hardware, JLab staff, in collaboration with EPCC in Edinburgh, UK, have added threading support to one of our key computational kernels, called *Wilson Dslash* [8]. This kernel is a four-dimensional nearest-neighbor operator. We show a 2D schematic picture of the communications patterns in this operator in figure 3 for a pure MPI implementation on the left, versus a hybrid-threaded one on the right. Two potential efficiency gains are immediate. First, one can see that threading on-node eliminates on-node messaging (green arrows in the figure). Second, the threading effectively coalesces multiple messages that would have been sent by individual cores into fewer, larger messages sent by the node (red arrows, ellipses in figure), strategy that may be advantageous in some networks.

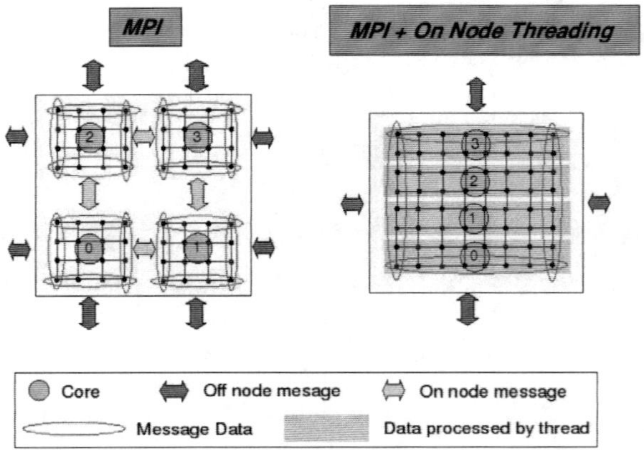

Figure 3. Communications patterns of the Wilson Dslash kernel in 2D on a quad core node for a full MPI application (left) and a hybrid threaded-MPI application (right).

Multithreading on a node is typically realized by using OpenMP or some custom thread library. We have developed a lightweight thread library called QCD Multi-Thread (QMT) [9], which enables a data-parallel programming technique similar in spirit to OpenMP: Work is supplied to QMT by calling the `qmt_call()` library function, with a callback procedure that can perform part of the desired work. QMT then invokes this function, with different parts of the problem from different threads. When `qmt_call()` completes, it calls a barrier among the threads to synchronize them. In the work described here, we used QMT with a queue-based barrier, optimized for the MOESI cache coherence protocol on AMD Barcelona cores.

Our numerical experiments consisted of running the Wilson Dslash operator on several nodes of the Jaguar Cray XT4 system at ORNL, either as a pure MPI or as a hybrid MPI-threaded application, using alternately both OpenMP and QMT for the threading. We performed our tests on a single node and then repeated them on 16 nodes that could be mapped as a 2^4 processor grid communicating in four directions. The tests were repeated using several local volumes: 2^4, 4^4, 6^4, and 8^4. In particular, the 2^4 volume is our hard scaling limit with all sites on the surface, and the 8^4 volume is typically too large to be cache resident. We have found the 6^4 volume most efficient, with 4^4 volume case less so because of a worse surface-to-volume ratio.

Our results are shown in figure 4. Figure 4(a) shows that on a single node the QMT and MPI performances are essentially identical. Looking at results for 16 nodes, we see that the threaded performances are much improved for the 4^4 and 6^4 volumes over the pure MPI case. In the cases of the smallest and largest volumes, there seems to be no difference between the pure MPI and

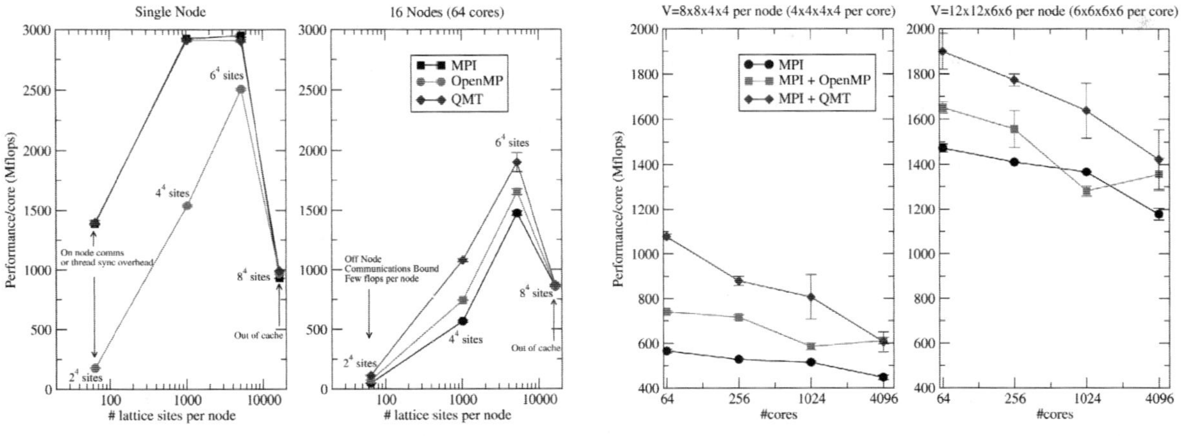

a) b)

Figure 4. (a) Effects of threading the Wilson Dslash operator on a single node (left) and a partition of 16 nodes (right) of Jaguar. (b) Weak scaling of the performance of the Wilson Dslash operator on Jaguar for two fixed local volumes: 4^4 sites (left) and 6^4 sites (right), to 4096 cores.

hybrid threaded versions. In the situations where threading results in a gain, using QMT for on-node threading results in a higher performance than when using OpenMP.

We then performed a weak scaling benchmark for the 4^4 and 6^4 local volumes in an attempt to scale the gains from the hybrid-threaded approach up to a large partition. Our results are shown in figure 4(b). One can see that for both local volumes, a performance advantage is maintained over pure MPI for as far out as 4096 cores when using QMT. When using OpenMP, the weak scaling appears quite erratic, but typically performance is less than the QMT case except for the the 4096 core partition size.

Our interpretation of these results is as follows. Since single node tests suggest no gain from eliminating on-node messages, we surmise that in our application, threading gains performance over pure MPI as a result of the collation of off-node messages. This gain is likely to be network and OpenMP implementation dependent. Our limiting 2^4 volume is completely communications bound with very few flops to overlap with communication, hence the low performance in that case for all the approaches tried. In the cases of the 4^4 and 6^4 volumes, message collation reduces the number of messages and increases their size, thus taking better advantage of the Cray network. In the 8^4 volume case we fall out of cache, and all approaches appear to perform equally.

In this investigation, we have neglected issues that arise in multisocket NUMA architectures such as thread, process, and memory affinity. We have also not explored general multicore aspects such as the abundance of floating point power versus the comparative lack of memory bandwidth. Partitioned Global Address Space languages such as UPC provide a natural programming model for NUMA based architectures, and aggressive prefetching and double buffering may alleviate the memory bandwidth issue to some degree. Some of these concerns are investigated in [10], and we intend to explore these issues more fully in future work.

4. Data sharing and Grid-related efforts

Our data-sharing efforts have continued in the past year by publishing many of our gauge configurations on the International Lattice Data Grid (ILDG) [11] and some through other channels [12, 13]. Our ILDG infrastructure is based jointly at Fermi National Accelerator Laboratory, where the storage element is hosted and managed, and at JLab, where one can find the Metadata and File Catalog Web Services. Some 11,000 configurations are now published in 16 ensembles through the ILDG. Conversely, in the past year several U.S. researchers have joined the ILDG virtual organization, in order to use published data shared through ILDG.

Within USQCD data sharing has moved forward through the definition of file formats for quark propagators and the implementation of software to read and write the standard within the QIO library. Application codes have also been modified to read and write these files. European collaborators have defined propagator formats that are compatible with the USQCD format, and there is some hope that worldwide propagator sharing will eventually be formalized in the ILDG.

5. Other activities

The USQCD software program continues its work and collaboration in many other areas not discussed here for lack of space, including the application of workflows to QCD, algorithmic developments, improvements to data analysis, code optimization, visualization and the use of Grid technologies.

6. Summary

In this contribution, we presented an overview of software progress in lattice QCD in the U.S. over the past year, with emphasis on performance achieved on leadership computers, and our work on multicore architectures. Much of this work was carried out in international collaboration, in particular with colleagues in the UK. We intend to continue progress in the software area in the future in order to carry on with our highly successful exploitation of available resources and to continue producing high quality scientific results and discoveries from lattice QCD.

Acknowledgments

We thank Steve Gottlieb (Indiana University), James Osborn (ALCF, Argonne), and Chulwoo Jung (BNL) for providing the data and plots for figures 1 and 2. The MDWF Inverter was written by Andrew Pochinsky at MIT and can be downloaded from [7].

This research used resources of the National Center for Computational Sciences at Oak Ridge National Laboratory, which is supported by the Office of Science of the Department of Energy under Contract DE-AC05-00OR22725. This research used resources of the Argonne Leadership Computing Facility at Argonne National Laboratory, which is supported by the Office of Science of the U.S. Department of Energy under contract DE-AC02-06CH11357.

The USQCD software effort is funded by the SciDAC program of the U.S. Department of Energy.

References

[1] Joó J 2007 USQCD Collaboration *J.Physics, Conference Series* **78** 012034
[2] USQCD web page: `http://www.usqcd.org`
[3] Kronfeld A 2008 Quantum chromodynamics with advanced computing *Proc. SciDAC Conf. 2008*
[4] Edwards R G and Joó 2005 *J, Nucl. Phys B1* **40** (Proc. Suppl) 832 `arXiv:hep-lat/0409003v1`, Chroma Code Web Page: `http://usqcd.jlab.org/usqcd-docs/chroma`
[5] MILC code web page: `http://www.physics.utah.edu/~detar/milc/index.html`
[6] CPS code web page: `http://qcdoc.phys.columbia.edu/cps.html`
[7] MDWF web page: `http://www.mit.edu/~avp/mdwf`
[8] McClendon C *Jlab preprint, JLAB-THY-01-29* `http://www.jlab.org/~edwards/qcdapi/reports/dslash_p4.pdf`
[9] Chen J, Watson III W and Mao W 2007 *Proc. HPC-Asia*
[10] Fowler R J, *et al.* Performance Engineering Challenges: The view from RENCI *Proc. SciDAC Conf. 2008*
[11] USQCD ILDG web page: `http://www.usqcd.org/ildg`

[12] BNL Lattice Archive: `https://lattices.qcdoc.bnl.gov`
[13] Gauge Connection Lattice Archive: `http://qcd.nersc.gov`

Quantum chromodynamics with advanced computing

Andreas S Kronfeld
Fermi National Accelerator Laboratory, Batavia, IL 60510, USA
(for the USQCD Collaboration)
E-mail: ask@fnal.gov

Abstract. We survey results in lattice quantum chromodynamics from groups in the USQCD Collaboration. The main focus is on physics, but many aspects of the discussion are aimed at an audience of computational physicists.

1. Introduction and background

Quantum chromodynamics (QCD) is the modern theory of the strong nuclear force. The physical degrees of freedom are quarks and gluons. The latter are the quanta of gauge fields, in some ways analogous to photons. The crucial difference is that gluons couple directly to each other, whereas photons do not. A consequence of the self-coupling is that the force between quarks does not vanish at large distances (as the Coulomb force $e^2/(4\pi r^2)$ does) but becomes a constant $F_{\mathrm{QCD}} \approx 800$ MeV fm$^{-1} \approx 15,000$ N. It would, thus, require a vast amount energy to separate quarks out to a macroscopic distance. Indeed, long before the separation becomes large enough to measure, the energy stored in the gluon field "sparks" into quark-antiquark pairs. This phenomenon of QCD explains why freely propagating quarks are not observed in nature, and it is called confinement.

QCD is a quantum field theory, which means that from the outset one must deal with mathematical objects that are unfamiliar even to many physicists. The most widely used theoretical tool for quantum field theories is relativistic perturbation theory, as developed for quantum electrodynamics (QED) by Feynman, Schwinger, Tomonaga and others sixty years ago (1). Perturbation theory provides the key connection between the mathematical theory with experiments in QED (2) and the Glashow-Weinberg-Salam theory merging QED with the weak nuclear force (3).

A textbook example of perturbative quantum field theory is to calculate how virtual pairs induce a distance dependence on the coupling in quantum gauge theories. In QED, one considers the fine structure constant,

$$\alpha = \frac{e^2}{4\pi\hbar c}, \tag{1}$$

where $-e$ is the charge of the electron. In QCD, one has the strong coupling α_s, related to the gauge coupling g by analogy with equation (1). When virtual pairs are taken into account, α and α_s are not constant but depend on distance; they are said to "run." The running, depicted in figure 1, is completely different in the two theories. In QED α becomes constant at distances $r > \hbar/m_e c$ (the electron's Compton wavelength), but at short distances it grows. In QCD α_s grows at long distance; perturbative running no longer makes sense once r is in the confining regime. At short distances—or, equivalently, high energies—the QCD coupling is small enough that perturbation theory can again be used. For example, the cross section for an electron-positron pair to annihilate and produce a quark-antiquark pair can be reliably computed as a power series in α_s. The quark and antiquark each fragment into a jet of hadrons—mostly pions, but some protons and neutrons, too—and general properties such as the energy flow and angular distributions of these jets can be traced back to the quark and antiquark. In this way, the experiments can measure quark properties, and these are found to agree with theoretical calculations (4).

QCD and the electroweak theory form the foundation of the Standard Model of elementary particles. They are so well established that many particle physicists consider the chromodynamic and electroweak gauge symmetries, and the associated quantum-number assignments of the quarks and leptons, to be laws of nature. We cannot conceive of a more comprehensive theory that does not encompass them. But there is more to the Standard Model. We know that something must spontaneously break the electroweak symmetry and that something (perhaps the same thing) must generate masses for the quarks and leptons. The Standard Model contains interactions that *model* these phenomena. The model interactions are consistent with observations yet also incomplete. It is hoped that insights obtained with the Large Hadron Collider (LHC), to commence operations at CERN later this year, will enable particle physicists to solve many of the puzzles raised by the Standard Model.

The purpose of this paper is to cover QCD, particularly in the strong-coupling regime where perturbation theory is insufficient. It is worth bearing in mind, however, that there is a close connection between many of the results discussed below and some of the big questions in particle physics. A simple example is the masses of the quarks. To determine the quark masses (which because of confinement cannot be measured), we have to calculate the relation between quark masses and measurable quantities, such as hadron masses. Quark masses are interesting for many reasons, the most unsettling of which is as follows. The mass of the top quark is larger than that of the bottom quark; see figure 2. Similarly, the mass of the charmed quark is larger than that of the strange quark. If this simple pattern held with the up and down quarks—the ones that make up proton and neutrons—then the up quark would be more massive than the down. But then protons would decay to neutrons, positrons, and neutrinos ($p \to ne^+\nu$). The positrons would find electrons and annihilate to photons.

This universe would consist of neutron stars surrounded by a swarm photons and neutrinos, and nothing else. Our universe is not at all like this, because neutrons are more massive than

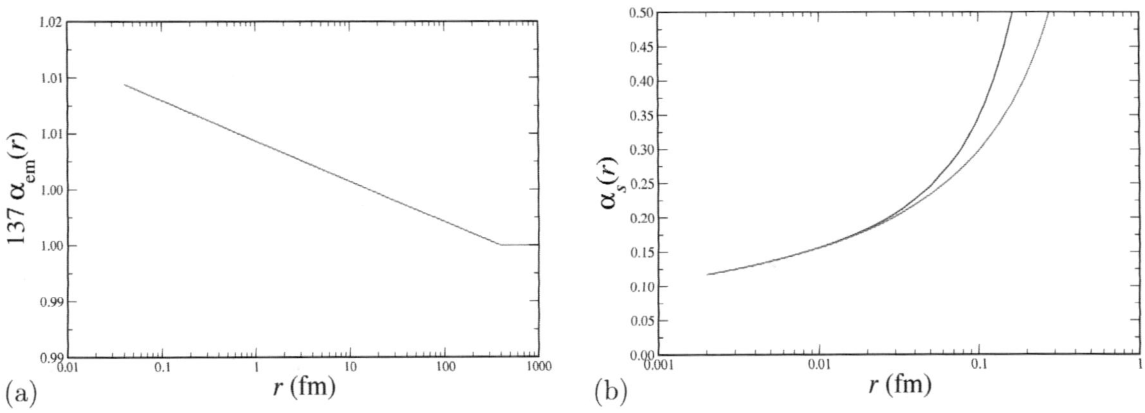

Figure 1. Running of (a) the QED and (b) the QCD coupling with distance. For QCD the red curve shows the perturbative result, and the (higher) blue show the effect of confining forces.

protons. The allowed decay reaction is $n \to pe^- \bar{\nu}$, leading to an abundance of protons and electrons, and making possible chemistry and biochemistry. With QCD we trace the neutron-proton mass difference back to the down-up quark masses and then to the deeper origin of quark masses. The pattern of quark masses is necessarily not simple, and the search for simple explanation of a messy pattern occupies many particle theorists. Below we give further examples where QCD is essential for working out the mysteries of particle physics.

QCD is also the cornerstone of modern nuclear physics. The simplest nucleus is nothing but the proton, a bound state of two up quarks and one down quark. The neutron is a bound state of two down quarks and one up quark. The confining force is very, very strong. By comparison, the force traditionally called the strong nuclear force is a residue of the fundamental chromodynamic force. This is similar to van der Waals forces among molecules, which are electromagnetic forces between neutral objects with structure leading to a distribution of electric charge. Some basic problems are, therefore, to understand nucleon structure directly from QCD, to study the excitation spectrum of nucleons (and their cousins with strangeness and other flavors), and to see how few-nucleon (or, more generally, few-hadron) systems interact.

Another active area of research in nuclear QCD is the properties of quark matter at high temperature and density. This regime is important to cosmology, because an epoch of the early universe consisted of hot quark matter, and to astrophysics, because, for example, neutron stars are composed of dense quark matter. In the laboratory, nuclear physicists create hot, dense quark matter by colliding heavy ions, at present with Brookhaven's RHIC, soon with (the heavy-ion mode of) CERN's LHC, and someday with GSI's FAIR.

At strong coupling it is necessary to go beyond perturbation theory. In a relativistic quantum field theory, such as QCD, this is not easy. An important obstacle is that a quantum field associates one (or perhaps a few) degrees of freedom with every point in space-time. Hence, the number of degrees of freedom is uncountably infinite. Furthermore, the fluctuations at short distances, if added up naively, contribute an ultraviolet divergence to anything of interest. These fluctuations are treated by a set of ideas known as *renormalization* (5). In short, one must find a tool that (correctly) handles both strongly coupled fields and the renormalization of short-distance fluctuations.

This tool is lattice field theory or, for gauge theories like QCD, lattice gauge theory. The idea, which goes back at least as far as Heisenberg, is to replace continuous space (or, in practice, space-time) with a discrete grid, or lattice. Then the number of degrees of freedom is countable. If the lattice is of finite extent, it occupies a finite physical volume, and, moreover, the number of degrees of freedom is finite. In 1974, Kenneth Wilson showed how to maintain gauge symmetry with a discrete lattice (6). In this and subsequent work, he and others elaborated how the lattice provides a mathematically rigorous definition of quantum field theory via functional integrals (7).

The functional integral for quantum fields on a finite lattice is a familiar object, a multi-dimensional integral, that can be evaluated on a computer. Before downloading some publicly available code (8) and trying to run it on a laptop, let us pause to contemplate how many variables of integration are reasonable (for physics). One wants to have the spacing between

Quark flavors in the Standard Model:

$$\begin{pmatrix} u \\ d \end{pmatrix} \quad \begin{pmatrix} c \\ s \end{pmatrix} \quad \begin{pmatrix} t \\ b \end{pmatrix}$$

$m_u < m_d$ $m_c > m_s$ $m_t > m_b$
!!!

What explains the pattern of masses?

Figure 2. The six *flavors* of quarks come the three *generations* with a doublet in each generation. The doublet structure plays a role in the electroweak interactions. For example, the upper (lower) entries have electric charge $+2/3$ $(-1/3)$.

sites on the lattice small enough to resolve hadron structure, and one wants the volume large enough that the boundary conditions do not modify the structure. If one supposes that the ratio of these two lengths is 32, then the dimensions of the integrals of interest are

$$\texttt{gluon d.o.f.} = 8 \times 4 \times 32^3 \times 128 > 10^8, \tag{2}$$

because gluons come in 8 colors with 4 polarization states each. The factor 128 presumes a space-time lattice with a time extent long (in relativistic units) compared to the spatial extent; this is common practice.

To cope with integrals of such large dimension, the only practical technique is Monte Carlo integration with importance sampling. The weight guiding the importance sampling is e^{-S}, where S is the classical action for the random sample of gluon variables. In functional integral formalisms for quantum mechanics, the expression for the action is the defining equation of the physical system. The development of algorithms to generate these samples is a vibrant subject, covered in part by Bálint Joó's poster here at the SciDAC conference. This paper will not dwell on algorithms, but it is important to mention one more complication.

Quarks are fermions and, as such, must satisfy the Pauli exclusion principle. In the functional integral formalism, this is handled by introducing anticommuting Grassmann variables for fermions. The integration is no longer Riemann (or Lebesgue) integration but a formal procedure called Berezin integration. To make a long story short, in lattice QCD we always can and do carry out the Berezin integration by hand. The outcome is that the weight for the importance sampling becomes $\det M \exp(-S_{\text{gluons}})$, where M is a sparse matrix with space-time indices. The matrix M is $N \times N$, where

$$N = \texttt{quark d.o.f.} = n_f \times 3 \times 2 \times 2 \times 32^3 \times 128 > 10^8. \tag{3}$$

The quark field represents n_f flavors with 3 colors and 2 spin states of quark and antiquark for every flavor and color. The matrix M is sparse because it is a lattice version of a differential operator, the Dirac operator (generalized to QCD). Incorporating $\det M$ into the weight is computationally demanding, increasing by two or three orders of magnitude the amount of floating-point operations needed to generate a new configuration of gluons.

The lattice is a helpful device mathematically and computationally, but it is not physical. To obtain a real result of the chromodynamics of continuum spacetime, one must work out the integrals for a sequence of lattices, and take the limit $a \to 0$ in a way that respects renormalization. Taking a smaller and smaller increases the computing burden as onerously as $a^{-(4+z)}$. The exponent z depends on the algorithm and is typically around 1 or 2. The 4 in the exponent is the key reason why numerical lattice QCD requires the most advanced computing facilities available. No algorithm can reduce it, because it is the dimension of space-time.

Most lattice-QCD calculations are carried out in two steps. The first is to generate an ensemble of gluon fields, with a certain lattice spacing and quark masses. This requires the computationally super-demanding $\det M$. To generate an ensemble of useful size, one needs several hundred, perhaps even a few thousand, samples of the gluon field. Computers of the highest available capability are needed for this step. In recent years, lattice gauge theorists have carried out this step with the special-purpose computer QCDOC, large clusters of PCs, and, more recently, leadership-class machines constructed under the auspices of the DOE's Office of Advanced Scientific Computing Research (ASCR). These ensembles of gluon fields are valuable: within the U.S. they are generated by and for particle and nuclear physicists and shared under various agreements. File-sharing is facilitated by formats and software developed by the International Lattice Data Grid (9).

The second step is to mine these ensembles for physics. A simple example is to compute the quantum mechanical amplitude for a proton to propagate from one point to another, and study the behavior of the amplitude as the separation varies: this is what one does to compute the proton mass, in fact. The key ingredients are a few rows of the inverse of the matrix M, introduced above. With the same ensemble one can study protons, pions, and more uncommon hadrons such as the charmed strange pseudoscalar meson (or D_s for short). Different rows of M^{-1} (i.e., different flavors, colors, spins, and space-time separations) are combined in various ways to obtain the quantum numbers of the hadrons in the problem at hand. Because each problem is different, and there are so many of them, we must attack this problem with the highest capacity computers. The job mix here is heterogeneous, with many users whose job streams range from many large jobs to extremely many medium-sized jobs. The challenge is to develop systems that scale well (for the large jobs), while maintaining the flexibility needed to handle the heterogeneity. The USQCD Collaboration has designed clusters of PCs that deliver this capacity in a cost-effective way (10).

The remainder of this paper focuses on physics. It is organized according to four principal themes of lattice gauge theory:

- Determination of fundamental parameters of the Standard Model of particle physics
- Study of nucleon structure, the hadron spectrum, and hadron interactions
- Simulation of the thermodynamic properties of QCD at nonzero temperature and density
- Lattice gauge theories beyond QCD, such as those appearing in models of electroweak symmetry breaking

Roughly speaking, the first and last are particle physics, and the second and third are nuclear physics. But the lines are blurry: for example, some of the simplest calculations of proton structure (the second item) may play a role more accurate calculations of pp scattering cross sections at the LHC. The way that scientific goals drive the computing needs of lattice gauge theory are discussed further, along the lines of these themes, in public whitepapers of the USQCD Collaboration (8).

2. Standard Model particle physics

In quantum electrodynamics, the fundamental parameters are the masses of electrically charged particles and the fine structure constant α. Similarly, in quantum *chromo*dynamics, the fundamental parameters are the masses of colored particles (the quarks) and the strong coupling α_s. Just as α can be determined in many ways (2), α_s can be determined at high energies from jet cross sections with perturbative QCD, and at low energies from the hadron spectrum with lattice QCD. Two recent results are (evaluated at distance $r = \hbar/m_{Z^0}c$)

$$\alpha_s = \begin{cases} 0.1172 \pm 0.0022 & \text{perturbative QCD} \quad \text{jet shapes} \quad \text{Ref. (11)} \\ 0.1170 \pm 0.0012 & \text{lattice QCD} \quad \text{hadrons} \quad \text{Ref. (12)} \end{cases}, \tag{4}$$

where both error bars reflect experimental and theoretical uncertainties. The first result is determined at high energies, around 100 GeV, in the regime where quarks can be treated as nearly free. The second is determined at low energies, of a few GeV and lower, where quarks are confined. Both are linked soundly to the defining equations of QCD. The agreement demonstrates the richness and broad validity of QCD and should make any enthusiastic scientist say “Wow!”

A straightforward application of lattice QCD is to work out how hadron masses depend on quark masses. Quark masses are interesting for several reasons. As fundamental constants of nature, they are interesting in their own right. As discussed in the introduction, the pattern of quark masses is puzzling and, hence, a motivation to search for extensions of the Standard Model that would contain a simple explanation.

Three flavors of quarks—top, bottom, and charm—have masses large enough to influence some high-energy scattering and decay processes. Their masses can be determined with perturbative QCD as well as lattice QCD. (The results agree (13).) The other three quarks—up, down, and strange—have masses so small that the full bound-state problem must be considered. Lattice QCD, therefore, provides the best information. Some recent results are the following (14; 15):

$$m_u = 1.9 \pm 0.2 \text{ MeV}/c^2, \quad (5)$$
$$m_d = 4.6 \pm 0.3 \text{ MeV}/c^2, \quad (6)$$
$$m_s = 88 \pm 5 \text{ MeV}/c^2. \quad (7)$$

One remarkable aspect of these results is that the strange quark's mass is somewhat smaller than had been thought on the basis of less reliable techniques for attacking strongly-coupled QCD. This point is, perhaps, of interest mostly to particle physicists, but a further one is of broader interest. The up and down masses—the constituents of protons and neutrons and, thus, everyday matter—are very small, within an order of magnitude of the electron mass (0.511 MeV/c^2). Only about 1–2% of the proton mass (938.3 MeV/c^2) or neutron mass (939.6 MeV/c^2) is accounted for by the quark masses. Consequently, most of their masses—and the mass of everyday matter—arises from confinement, namely, the binding energy of the gluons and the (relativistic) kinetic energy of bound quarks. This is a stunning insight, to be revisited in Section 5.

The caption of figure 2 states that the pairings of quarks in each doublet stem from the electroweak interactions. That is not the full story. There is no reason for the eigenstates of mass to be the eigenstates of the interaction term with W gauge bosons. The two are related by a unitary transformation, $V_{\rm CKM}$, namely,

$$\begin{pmatrix} d \\ s \\ b \end{pmatrix}_W = V_{\rm CKM} \begin{pmatrix} d \\ s \\ b \end{pmatrix}_{\rm mass}, \quad V_{\rm CKM} = \begin{pmatrix} V_{ud} & V_{us} & V_{ub} \\ V_{cd} & V_{cs} & V_{cb} \\ V_{td} & V_{ts} & V_{tb} \end{pmatrix}. \quad (8)$$

Here the initials CKM stand for Cabibbo-Kobayashi-Maskawa; Cabibbo introduced equations (8) for two generations, and Kobayashi and Maskawa were the first to study the ramifications for three generations.

The left-hand side of the first equation denotes the electroweak basis. The labels on the elements of $V_{\rm CKM}$ are the weak isopartner and the mass eigenstate, because the transition amplitude for $W \to \bar{b}u$, for example, is proportional to V_{ub}. The study of physical processes related to $V_{\rm CKM}$ is called flavor physics, because the central issues are to understand why there are several flavors of quarks and what lends them their separate identities.

As a unitary matrix, $V_{\rm CKM}$ contains complex entries. Some of the phases can be absorbed into unobservable phases of the quark wave functions. For only two generations, one physical parameter remains, and it is real. For three generations, as in nature, three physical parameters are real and one phase remains. The complex couplings imparted by the phase lead to physical processes that proceed at different rates for particles and the corresponding antiparticles. Such reactions are needed to explain the abundance of matter and the dearth of antimatter in the universe. Among particle physicists, this phenomenon is called CP violation. It is clearly intriguing to know whether the CP violation of the CKM matrix is enough to account for the matter-antimatter asymmetry of the universe. According to present measurements and theoretical understanding, it is insufficient, so one would like to know what other interactions violate CP.

In the Standard Model, there is a scalar boson called the Higgs boson. Section 5 will explain more about it. For now let us note that both the quark masses and the CKM matrix have their

Table 1. Decay constants f_P of pseudoscalar mesons (here $P \in \{\pi, K, D, D_s\}$). Values in MeV. Experimental determinations assume a suitable element of the CKM matrix element consistent with all flavor data.

Meson	MILC (15)	FNAL (23)	HPQCD (24)	Experiment (Ref.)	Deviation
π	128 ± 3	–	132 ± 2	130.7 ± 0.4 (13)	0.4σ
K	154 ± 3	–	157 ± 2	159.8 ± 1.5 (13)	1.7σ
D^+	–	201 ± 19	207 ± 4	206 ± 9 (33)	0.1σ
D_s	–	249 ± 17	241 ± 3	277 ± 9 (34)	3.8σ

origin in matrices of couplings between quark fields and the Higgs field. Therefore, the puzzles of CP violation and the pattern of the quark masses are tied together by the Standard Model.

To make progress in flavor physics, lattice-QCD calculations are needed to interpret the experimental measurements. In a schematic form the rate Γ is given by

$$\Gamma = \begin{pmatrix} \text{known} \\ \text{factor} \end{pmatrix} \begin{pmatrix} \text{CKM} \\ \text{factor} \end{pmatrix} \begin{pmatrix} \text{QCD} \\ \text{factor} \end{pmatrix}, \tag{9}$$

where the "known factor" consists of well-measured physical constants and numerical factors like 4π. The paradigm is to measure as many flavor-changing processes as possible, and use the Standard-Model formulae to over-determine the CKM matrix. If all the determinations are consistent, then one can set limits on non-Standard sources of flavor and CP violation. To do so, the QCD factor must be computed, and often the only way to do so is with lattice QCD.

There are dozens of relevant measurements, but because the CKM matrix is unitary, the elements of the CKM matrix must satisfy constraints. A particular vivid one is the unitarity triangle that stems from the orthogonality of columns of a unitary matrix:

$$V^*_{ud}V_{ub} + V^*_{cd}V_{cb} + V^*_{td}V_{tb} = 0, \tag{10}$$

which traces out a triangle in the complex plane. The triangle and the present uncertainties from various constraints is shown in figure 3, dividing each side by $V^*_{cd}V_{cb}$. Except for the wedge labelled $\sin\beta$ the uncertainties can be reduced with lattice calculations of the QCD factor.

Lattice calculations for a wide range of observables relevant to the CKM matrix have appeared over the past few years, and further improvements are under way. These include calculations of properties of strange kaons (15; 17; 18; 19; 20; 21), charmed D and D_s mesons (22; 23; 24), and beautiful B and B_s mesons (25; 26; 27; 28; 29; 30). For recent reviews, including references to work in Asia, Australia, and Europe, see (31; 32).

The past several months witnessed an unexpected development in flavor physics, in the field of leptonic decays of pseudoscalar mesons. (Pseudoscalar bosons have spin 0 and negative parity; the most common examples are the pion and kaon.) One can consider the measured decay rates to determine a QCD matrix element called the decay constant, and denoted f_π (for the pion), f_K (for the kaon), and so on. Table 1 shows the results of some SciDAC-supported calculations of decay constants and the corresponding measurements. The calculations agree very well with the measurements, except in the case of the charmed strange pseudoscalar meson D_s.

This set of circumstances is odd, because the computation is easiest for the D_s. Algorithms for the propagators of light quarks slow down as the quark mass m_q is decreased. As a consequence, properties of hadrons containing up or down quarks are reached via a (controlled) extrapolation in m_q. This introduces a source of uncertainty that is absent for the D_s, which, as the lightest bound state of a charmed quark and a strange antiquark, can be computed by working directly at the physical quark masses.

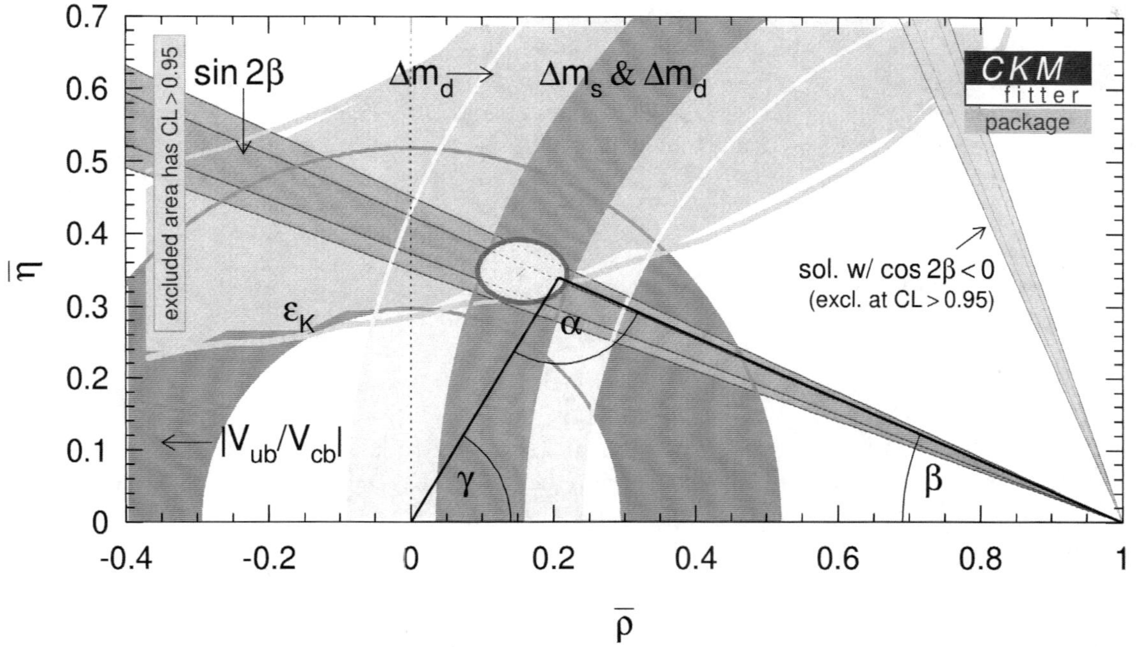

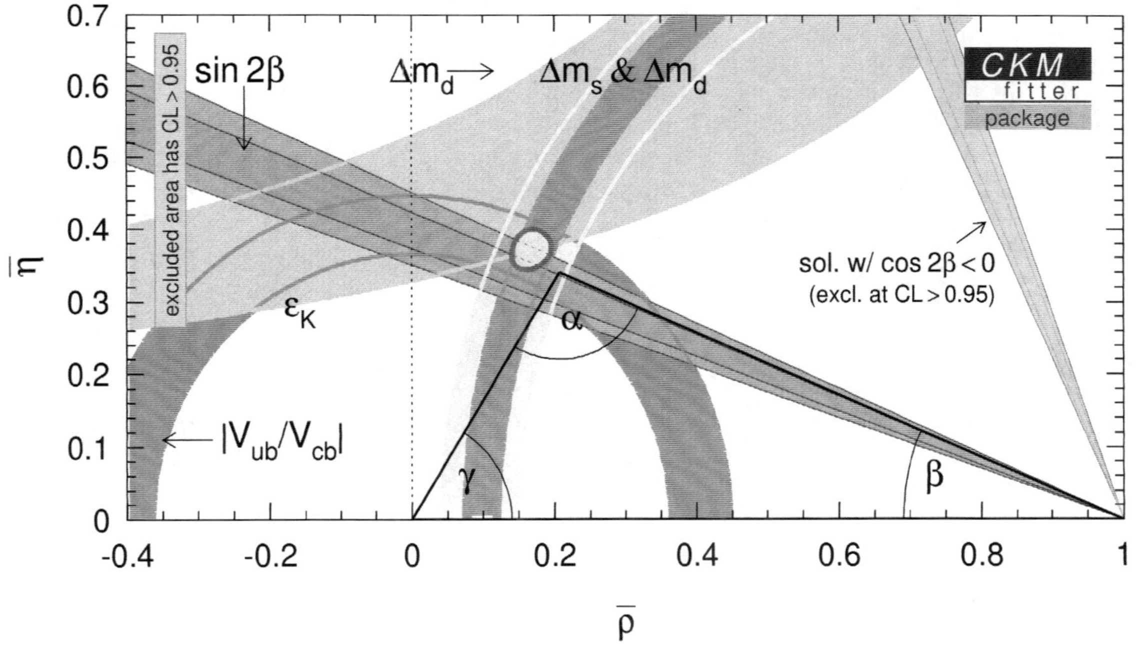

Figure 3. Constraints on the CKM unitarity triangle from experiments and QCD calculations, highlighting those where lattice QCD plays a role. The upper panel shows the current status, and the lower a forecast of how the uncertainties can be reduced with lattice QCD (16). In the lower panel, one can envision tension between the lime-green hyperbola and the blue wedge, depending on how the central values evolve and measurements and lattice-QCD computations improve. The current status of the angles α and γ is omitted, because better measurements are needed; they are expected from LHCb.

There is no obvious explanation for this discrepancy within the Standard Model. Unless there is a real blunder somewhere, the discrepancy points to a non-Standard particle mediating the decays $D_s \to \mu\nu$ and $D_s \to \tau\nu$, thereby changing the interpretation of the "measured" f_{D_s}. It turns out to be rather easy to devise models of such interactions (34). Given the agreement of CLEO's new measurement of f_{D^+} (33), new W' bosons and charged Higgs bosons seem unlikely, making leptoquarks—particles with baryon and lepton number—are the most plausible agents of decay.

3. Nucleon structure and nuclear physics

Let us turn now to nuclear physics, where the importance of lattice QCD is difficult to overstate. Many nuclear laboratories, such as Jefferson Laboratory in the U.S., have turned their attention to the structure of the proton (the simplest nucleus!) (35), the excitation spectrum of baryons (36), and processes that may produce glueballs. On the theoretical side, the previous decade witnessed dramatic growth in the application of effective field theories to understand nuclear interactions (37). These techniques connect a wealth of nuclear and hadronic data to the QCD Lagrangian. Lattice QCD is an irreplaceable tool to interpret the new and upcoming experiments at JLab and to compute the so-called low-energy constants of the effective theory (38).

A topic at the interface between nuclear and particle physics concerns the distributions of partons (a generic term for quarks and gluons) inside the proton. In high-energy collisions of $p\bar{p}$ (at the Tevatron) or pp (at the LHC), the transverse motion of partons can be neglected. To compute cross sections (to produce, say, a Higgs boson), the key property needed is the distribution of the longitudinal momentum. From a nuclear perspective, this is also one of the most basic aspects of proton structure. Usually one studies parton distributions q as a function of the fraction x of longitudinal momentum carried by a parton, $0 < x < 1$. Fits to scattering data can determine $q(x)$ only over a range, $x_{\text{min}} < x < x_{\text{max}}$, limited by kinematics and statistics. Lattice QCD can compute moments of q: $\langle x^n \rangle_q = \int_0^1 dx\, x^n q(x)$. Clearly the two kinds of information are complementary. The experimental range can be extended with *ad hoc* functions possessing the right asymptotic forms at the endpoints, and in this way an Ansatz-guided experimental determination of $\langle x^n \rangle_q$ is possible. One finds good agreement with lattice calculations (39). To explore nucleon structure further, the next step is to probe the transverse structure and compute the so-called generalized parton distributions (39).

The axial charge of the nucleon is another topic of keen interest, because it appears in the description of β decay. It is also related to the relative spin polarization of up and down quarks in the proton. Some recent results with lattice QCD are

$$g_A = \begin{cases} 1.21 \pm 0.08 & \text{Ref. (40)} \\ 1.20 \pm 0.07 & \text{Ref. (41)} \end{cases}, \tag{11}$$

showing good agreement with experiment ($g_A = 1.269 \pm 0.003$). Work is under way to reduce the theoretical uncertainty.

In all fields of physics, spectroscopy is a time-honored approach to learn about the dynamics of the underlying phenomena of interest. In nuclear and hadronic physics, therefore, an important area of investigation is the excitation spectrum of hadrons, particularly baryons. Excited states present additional challenges for lattice QCD, but the tools needed are available and have been demonstrated to work (42; 43). This subject extends to include topics such as the intriguing Roper resonance (44) and electromagnetic transitions of the form $\gamma N \to N^*$, where N^* is an excited nucleon (45).

Because the gluons couple to each other, it is conceivable that there are hadrons consisting essentially of gluons, with no valence quarks. (Virtual quark-antiquark pairs appear in all hadrons.) The most persuasive evidence that glueballs do indeed exist come from lattice-QCD calculations. In an approximation omitting virtual quark pairs, the evidence is clear that stable gluon-only bound states arise (46; 47). There are two challenges with glueballs. One, obviously, is to extend the lattice calculations to full QCD with virtual quark-antiquark pairs. The other is to identify them unambiguously in the laboratory (48). Experiments with γp collisions are expected to produce such states and are planned at the 12 GeV upgrade of CEBAF, the accelerator at Jefferson Lab. To plan this program it is essential to know the photocouplings of the mesons (exotic or not). It is possible to compute these photocouplings in lattice QCD, as has been demonstrated for charmonium ($\bar{c}c$ mesons) (49). An extension of this work to light mesons, including exotics and glueballs, is under way.

A long-term goal of nuclear physics is to compute many-body interactions from QCD. In addition to providing a better understanding of the nucleus, these calculations are relevant to mesonic atoms, namely, those containing a positively and negatively charged meson, and to strangeness in neutron stars. Although in some cases lattice QCD results can be compared to measurements, the more interesting situation is where this cannot be done. The aim here is to gain confidence in the reliability of these difficult calculations, so that one can use them (with robust uncertainty estimates) in applications where it is otherwise impossible to acquire the needed information any other way.

An interesting example is the lattice calculation of the πK scattering lengths, of which there are two, $a_{1/2}$ and $a_{3/2}$, depending on whether the πK system has isospin $I = \frac{1}{2}$ or $\frac{3}{2}$. A recent lattice-QCD calculation finds (50)

$$a_{1/2} m_\pi c/\hbar = +0.1725 \pm 0.0017^{+0.0023}_{-0.0156}, \quad (12)$$

$$a_{3/2} m_\pi c/\hbar = -0.0574 \pm 0.0016^{+0.0024}_{-0.0058}, \quad (13)$$

in convenient dimensionless units. These are in rough, but not spectacular, agreement with other determinations using chiral perturbation theory or the Roy-Steiner equations. For example, an analysis based on the Roy-Steiner method (51) finds

$$a_{1/2} m_\pi c/\hbar = +0.224 \pm 0.022, \quad (14)$$

$$a_{3/2} m_\pi c/\hbar = -0.0448 \pm 0.0077, \quad (15)$$

employing an extrapolation from high energies down to the threshold (where the scattering length is defined); see (38) for more discussion. A more direct experimental determination is proposed by the DiRAC Collaboration, which would form $\pi^- K^+$ atoms (52). In this sense, Eqs. (12) and (13) represent something exciting: a theoretical prediction awaiting definitive experimental confirmation.

Multimeson systems are proving to be a fruitful area of research, with further work on $\pi\pi$ (53), KK (54), and even states with as many as twelve pions (55). Once methods have been vetted in mesonic systems, the next step is to consider nucleon-nucleon (56) and nucleon-hyperon (57) systems. This is more difficult, because the signal-to-noise ratio in the Monte Carlo estimate of baryonic correlation functions is worse. Here the methods of effective field theories can be used to extend feasible lattice-QCD calculations to a wider range of nuclear phenomena. The hyperon-nucleon interaction is of interest, because there is speculation that the large Fermi energy may make it energetically favorable for some nucleons to transmute to hyperons (58).

4. QCD thermodynamics

The discussion in the introduction of the computational scope of numerical lattice QCD mentions that, but does not explain why, the spacetime lattice usually has a time extent much longer than

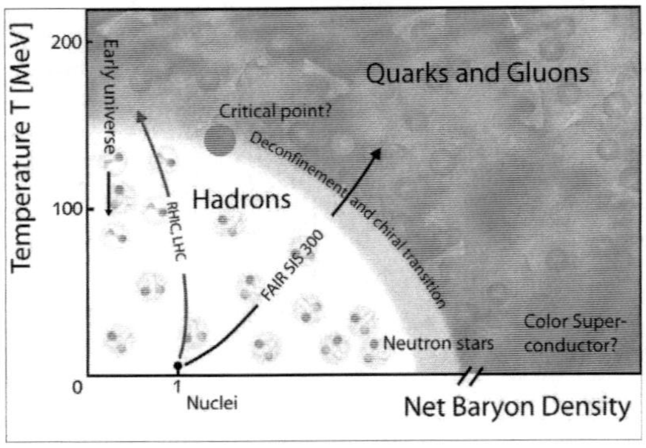

Figure 4. Phase diagram of QCD, showing conjectured phases and phenomena of physical interest. Accelerator experiments at RHIC and LHC probe small chemical potential, around the crossover regime. At the planned Facility for Antiproton and Ion Research (FAIR) experiments can reach larger chemical potential. Image from the CBM Collaboration (60).

the spatial dimension, for example, $32^3 \times 128$. The reason is that the functional integral actually describes a grand canonical ensemble (in the sense of thermodynamics) at temperature T,

$$k_B T = \hbar c / N_t a, \tag{16}$$

where N_t is the number of sites in the time direction, a is the lattice spacing, and k_B is Boltzmann's constant. In the applications to particle and nuclear physics discussed in Sections 2 and 3, one wants $T = 0$ and, hence, takes N_t large enough for the Boltzmann suppression $e^{-mc^2/k_BT} = e^{-N_t amc/\hbar}$ to eliminate thermal effects, where m is a hadron mass.

There are, however, many areas of research where a nonzero temperature and a nonzero chemical potential arise. For nonzero temperature, one simply makes N_t smaller, typically $4 \leq N_t \leq 12$ in current work. Such lattices take somewhat less memory, but not less computing because thermodynamics obviously requires calculations at several temperatures. It is conceptually easy to introduce a nonzero chemical potential into lattice gauge theory, but the conceptually clean approach is computationally beyond present resources. For small chemical potential, several approaches are available and efficient (59).

The physical phenomena of interest are summarized by the phase diagram in figure 4. The key feature is a phase transition from hadronic matter to a substance called the quark-gluon plasma. The transition is expected to be first order and end at a critical point at chemical potential $\mu \neq 0$. Between the critical point and the $\mu = 0$ axis the transition is a rapid crossover. Lattice-QCD calculations, as well as measurements of heavy-ion collisions at RHIC and LHC, probe the crossover, which is relevant to the cooling of the early universe. Neutron stars appear at high baryonic density, namely, high μ. At even higher densities, several phases with color superconductivity are thought to exist. Apart from heavy-ion collisions (so far at small μ) much of this phase diagram is unexplored, and lattice gauge theory offers the only *ab initio*, quantitative, theoretical tool to see if it really reflects QCD.

The central concept in QCD thermodynamics is the equation of state (EoS), namely, the relationship between the pressure p and energy density ϵ as the temperature T and chemical potential μ are varied. The EoS reveals the relevant degrees of freedom. In QCD, are they hadrons? Or quarks and gluons? At low temperatures, hadrons are clearly present, and phenomenological models of hadron gases can describe QCD thermodynamics. At very high temperatures, perturbation theory in $\sqrt{\alpha_s(\hbar c/k_BT)}$ can be used. But near the transition, nonperturbative methods are needed.

The first step is to find the transition temperature, which has now been done with 2+1 flavors of realistic sea quarks (61; 62). The different physics below, near, and above the critical temperature T_c also has ramifications for controling systematic errors. Discretization effects from the lattice, for example, depend on the basic degrees of freedom. (For a crossover it is, strictly speaking, incorrect to speak of a critical temperature. But the critical point in the phase diagram is close enough to the $\mu = 0$ axis that thermodynamic variables change rapidly at T_c.)

The next step is to map out the thermal contribution to the trace of the energy-momentum tensor, $\operatorname{tr}\Theta = \epsilon - 3p$. From the temperature dependence of $\epsilon - 3p$ other thermodynamic variables, such as the entropy density, can be worked out. This also has now been done with 2+1 flavors of realistic sea quarks (63; 64), including at small μ (65). As the system is heated, even to $2T_c$, the behavior of $\epsilon - 3p$ shows that the system does not turn rapidly into weakly interacting quarks and gluons.

To make contact with heavy-ion collisions, the last step would be to take the output of lattice QCD as input to hydrodynamic models. With these models one would then compute experimental observables, such as elliptic flow and quarkonium suppression. With current computing resources, the EoS has been pinned down with 20–25% accuracy. To make a big impact on hydrodynamic modeling, however, it will be necessary to reduce the error to 5%. This is conceptually straightforward to achieve, but it will be computationally challenging. Part of the strategy is to reduce discretization effects, which scale as a^2 (a is the lattice spacing), while the net cost scales as a^{-11}.

5. Non-Standard particle physics

As mentioned above, particle physics awaits a new era when the LHC starts producing physics results. Commencing later this year, the LHC will collide two beams of 7 TeV (TeV = teraelectronvolt) protons. (Amusingly, some particle physicists now call this energy scale the terascale, deliberately alluding to terascale computing. It would be unkind to tell them that in computing we are moving on to the petascale.) In particular, the LHC will address the central problem of particle physics, which is the origin of the W and Z boson masses. The fact that these masses do not vanish demonstrates that some mechanism breaks the electroweak gauge symmetry spontaneously. "Spontaneous symmetry breaking" means simply that, although the equations of motion are symmetric, the solution chosen is not symmetric. For example, in the Standard Model a scalar field breaks electroweak symmetry, via a technique invented by Higgs. Interactions between the Higgs field and the fermions also generate (in the Standard Model) the quark and charged lepton masses and the CKM matrix. Therefore, flavor physics will remain interesting in the LHC era, and the physics program discussed in Section 2 will need to be continued and extended.

In this section, however, I would like to consider applications of lattice gauge theory to understanding the mechanism of electroweak symmetry breaking. Since we suspect that the Standard Higgs sector is far from the full story, it is possible (some would say likely) that strongly coupled fields are involved. For example, instead of being an elementary field, the Higgs boson (and its siblings, the longitudinal polarization states of the W and Z bosons) could be composite. The constituents would then form a whole spectrum of states, just as quarks bind into many different hadrons. Even if the Higgs is elementary at the terascale, other agents of electroweak symmetry breaking could be strongly coupled.

Numerical work on theoretical bounds on the Higgs mass are a long-standing topic in lattice field theory (66; 67). The issue is that the Higgs mass is proportional to its self-coupling, which runs at short distances in a way similar to α in QED; see figure 1. The perturbative running becomes invalid once the coupling is too large, and nonperturbative methods are required. A subtle relation arises between the Higgs mass and a scale where new phenomena imply that the Standard Model breaks down. To make a long story short, one can in this way bound the Higgs mass from above. A similar analysis applied to the Yukawa coupling between the Higgs field and the top quark leads to a lower bound on the Higgs mass, as discussed in a recent comprehensive analysis (68).

A popular class of non-Standard models is to replace the Higgs sector with a strongly interacting sector that is, in some respects at least, analogous to QCD. In fact, absent a stronger source of symmetry breaking, the pions of QCD would break the electroweak symmetry, leading to W and Z masses around 100 MeV. What one needs, then, is a new quantum number, called technicolor, and a gauge force to bind the corresponding particles, called techniquarks, together. These dynamics are posited to take place at terascale energies and, by analogy with QCD, generate vector boson masses around 100 GeV, as observed. The earliest models of technicolor followed the example of QCD too closely and have been ruled out by experimental measurements. One way around the experimental constraints is to assume that the coupling evolves more slowly than in QCD. This is called "walking" (i.e., slower than running) technicolor. For a wide range of scales below the confining scale, the coupling is too strong for perturbation theory to be relied on, so nonperturbative work with lattice gauge theory is called for. In the past year, the walking hypothesis has indeed been tested, by examining the scaling behavior of the technipion mass, the vector meson mass, and the technipion decay constant (69). Even more recently, the evolution of α_s in (a TeV-scale version of) QCD with several flavors has been computed (70), to see how many flavors are needed to having a walking coupling, and how a walking coupling changes the physics.

The most popular way to relieve the theoretical problems of the Standard Model is to introduce supersymmetry, a symmetry that transforms fermions into bosons, and vice versa. In these models, every known particle (quarks, leptons, gauge bosons) comes with another state known as a superpartner. In the minimal supersymmetric model, the Higgs sector contains three neutral bosons, a charged pair, and their superpartners. No superpartner has been observed, so at the terascale and below supersymmetry must be broken somehow. Usually supersymmetric models are conceived of as weakly coupled, but, since they are gauge theories, strong coupling is not to be dismissed out of hand. It is therefore of great interest to formulate and simulate supersymmetric gauge theories on a spacetime lattice. This has been an extremely fruitful theoretical field during the past several years (71; 72), and now first numerical simulations are being carried out (73; 74). This rapid development yields hope that nonperturbative studies of supersymmetric models will be possible, including studies of supersymmetry breaking.

6. Perspective

Every now and then one hears the question, "How much computing would it really take to solve QCD on a computer?" The questioner is (usually) well meaning, but most experts would agree that the question is not well conceived. The calculations for some of the simplest hadronic quantities are finally becoming mature, with analyses that forthrightly address all sources of uncertainty. This class of observable, sometimes called "gold-plated" (75), consists of masses of particles that are stable (with respect to the strong interactions), not too close to thresholds, and of hadronic transition matrix elements with one or zero of these stable hadrons in the final and initial states. Many, but not all, aspects flavor physics (Section 2), and some aspects of hadronic structure (Section 3), lie in this class, and the first solid round of serious computation is nearing its end.

But this is the end of the beginning. In these nearly mature fields, it turns out that precision is important, at least for the foreseeable future. For example, to gain persuasive evidence of new phenomena in flavor physics, 5–25% accuracy is not enough. A better target (given the stakes and the quality of the corresponding experiments) is 1–3%. If the history of kaon physics is a guide, it will eventually be important to go further. Happily, the need for and clear path to precision has persuaded the particle-physics community that the resources needed for lattice flavor physics are worth the cost.

Even a quick glance at Section 3 reveals that the large majority of interesting problems in nuclear physics require much more computing resources than gold-plated observables. By definition the excited hadrons are not stable under the strong interaction. They will be harder to pin down, and it is possible that we will not know how hard until the work currently under way reaches some of its milestones. The calculations of scattering lengths, while impressive, must be repeated with several values of the physical volume to ensure that the mathematical formalism used to extract them works as it should. Given the importance of nuclear lattice QCD in guiding nuclear experimentation, one would hope for full support from experimenters for the needed computing resources.

The "how much?" question should be rephrased. It is really a class of questions, "How much computing would it take to compute my favorite observable?" Of course, the answer depends on your observable and its wider context. It should be clear, however, that physics with either precision and complexity leads to questions whose answers will need one or two orders of magnitude more computing than what is available now.

Another reason the "how much?" question is ill-conceived is that it often overlooks the heterogeneous job mix needed to attack many problems in lattice QCD. It is not enough to generate huge ensembles of lattice gauge fields, with ever smaller lattice spacing and ever smaller quark masses, on the world's highest-capability computer. This step is necessary, creating a gold mine of information. It is not enough, however, merely to open a mine; the gold must be extracted and, here, we need a large capacity of computing that is flexible in every way (while having significant capability of its own).

We do not know how broad the future of lattice gauge theory will be. If the LHC discovers a strongly interacting sector beyond the Standard Model, it is likely that chiral fermions play a role. This is simply because, at the terascale, we already know that left-handed fermions and right-handed fermions are different fields. But chiral fermions possess some conceptual and computational challenges not found in QCD. A strongly coupled terascale will bring many more particle theorists to lattice gauge theory, and this larger community will clamor for computing that is up to the task.

Acknowledgments

This work has been supported in part by the United States National Science Foundation and the Office of Science of the U.S. Department of Energy (DOE). Software development and hardware prototyping within the USQCD Collaboration are supported by SciDAC. Fermilab is operated by Fermi Research Alliance, LLC, under Contract DE-AC02-07CH11359 with the US DOE.

References

[1] Schwinger J (ed) 1958 *Selected Papers in Quantum Electrodynamics* (New York: Dover) ISBN 0-486-60444-6

[2] Kinoshita T 1996 *Rept. Prog. Phys.* **59** 1459–1492

[3] Alcaraz J *et al.* (LEP) 2007 Precision electroweak measurements and constraints on the Standard Model (*Preprint* `arXiv:0712.0929 [hep-ex]`)

[4] Dixon L J 2007 Hard QCD processes at colliders, in *23rd International Symposium on Lepton-Photon Interactions at High Energy* (*Preprint* `arXiv:0712.3064 [hep-ph]`)

[5] Lepage G P 1989 What is renormalization? in *From Actions to Answers* ed DeGrand T and Toussaint D (Singapore: World Scientific) (*Preprint* `arXiv:hep-ph/0506330`)

[6] Wilson K G 1974 *Phys. Rev.* D **10** 2445–2459

[7] Glimm J and Jaffe A M 1987 *Quantum Physics: A Functional Integral Point of View* (New York: Springer) ISBN 0-387-96476-2

[8] USQCD Collaboration URL `http://www.usqcd.org/`

[9] Detar C E (ILDG) 2006 *PoS* **LATTICE2007** 009 (*Preprint* `arXiv:0710.1660 [hep-lat]`)

[10] Holmgren D J 2005 *Nucl. Phys.* B Proc. Suppl. **140** 183–189 (*Preprint* `arXiv:hep-lat/0410049`)

[11] Becher T and Schwartz M D 2008 A precise determination of α_s from LEP thrust data using effective field theory (*Preprint* arXiv:0803.0342 [hep-ph])

[12] Mason Q *et al.* (HPQCD) 2005 *Phys. Rev. Lett.* **95** 052002 (*Preprint* arXiv:hep-lat/0503005)

[13] Yao W M *et al.* (Particle Data Group) 2006 *J. Phys.* G **33** 1–1232

[14] Mason Q, Trottier H D, Horgan R, Davies C T H and Lepage G P (HPQCD) 2006 *Phys. Rev.* D **73** 114501 (*Preprint* arXiv:hep-ph/0511160)

[15] Bernard C *et al.* (MILC) 2006 *PoS* **LATTICE2007** 090 (*Preprint* arXiv:0710.1118 [hep-lat])

[16] Van de Water R S (USQCD) 2007 *J. Phys.* Conf. Ser. **78** 012079

[17] Aubin C *et al.* (MILC) 2004 *Phys. Rev.* D **70** 114501 (*Preprint* arXiv:hep-lat/0407028)

[18] Beane S R, Bedaque P F, Orginos K and Savage M J (NPLQCD) 2007 *Phys. Rev.* D **75** 094501 (*Preprint* arXiv:hep-lat/0606023)

[19] Gamiz E *et al.* (HPQCD) 2006 *Phys. Rev.* D **73** 114502 (*Preprint* arXiv:hep-lat/0603023)

[20] Antonio D J *et al.* (RBC and UKQCD) 2008 *Phys. Rev. Lett.* **100** 032001 (*Preprint* arXiv:hep-ph/0702042)

[21] Boyle P A *et al.* (RBC and UKQCD) 2008 *Phys. Rev. Lett.* **100** 141601 (*Preprint* arXiv:0710.5136 [hep-lat])

[22] Aubin C *et al.* (Fermilab Lattice and MILC) 2005 *Phys. Rev. Lett.* **94** 011601 (*Preprint* arXiv:hep-ph/0408306)

[23] Aubin C *et al.* (Fermilab Lattice and MILC) 2005 *Phys. Rev. Lett.* **95** 122002 (*Preprint* arXiv:hep-lat/0506030)

[24] Follana E, Davies C T H, Lepage G P and Shigemitsu J (HPQCD) 2008 *Phys. Rev. Lett.* **100** 062002 (*Preprint* arXiv:0706.1726 [hep-lat])

[25] Okamoto M *et al.* (Fermilab Lattice and MILC) 2005 *Nucl. Phys.* B Proc. Suppl. **140** 461–463 (*Preprint* arXiv:hep-lat/0409116)

[26] Dalgic E *et al.* (HPQCD) 2006 *Phys. Rev.* D **73** 074502 (*Preprint* arXiv:hep-lat/0601021)

[27] Gray A *et al.* (HPQCD) 2005 *Phys. Rev. Lett.* **95** 212001 (*Preprint* arXiv:hep-lat/0507015)

[28] Bernard C *et al.* (Fermilab Lattice and MILC) 2006 *PoS* **LAT2006** 094

[29] Dalgic E *et al.* (HPQCD) 2007 *Phys. Rev.* D **76** 011501 (*Preprint* arXiv:hep-lat/0610104)

[30] Laiho J (Fermilab Lattice and MILC) 2006 *PoS* **LATTICE2007** 358 (*Preprint* arXiv:0710.1111 [hep-lat])

[31] Della Morte M 2007 *PoS* **LATTICE2007** 008 (*Preprint* arXiv:0711.3160 [hep-lat])

[32] Jüttner A 2007 *PoS* **LATTICE2007** 014 (*Preprint* arXiv:0711.1239 [hep-lat])

[33] Eisenstein B I *et al.* (CLEO) 2008 Precision measurement of $B(D^+ \to \mu^+\nu)$ and the pseudoscalar decay constant f_{D^+} (*Preprint* arXiv:0806.2112 [hep-ex])

[34] Dobrescu B A and Kronfeld A S 2008 *Phys. Rev. Lett.* **100** 241802 (*Preprint* arXiv:0803.0512 [hep-ph])

[35] Hägler P 2007 *PoS* **LATTICE2007** 013 (*Preprint* arXiv:0711.0819 [hep-lat])

[36] McNeile C 2007 *PoS* **LATTICE2007** 019 (*Preprint* arXiv:0710.0985 [hep-lat])

[37] Kaplan D B 2007 Effective field theory as the bridge between lattice QCD and nuclear physics, in *Quark Confinement And The Hadron Spectrum 7* ed Ribeiro J *et al.* (New York: AIP) (*Preprint* arXiv:nucl-th/0611025)

[38] Beane S R, Orginos K and Savage M J (NPLQCD) 2008 Hadronic interactions from lattice QCD (*Preprint* arXiv:0805.4629 [hep-lat])

[39] Hägler P *et al.* (LHPC) 2008 *Phys. Rev.* D **77** 094502 (*Preprint* arXiv:0705.4295 [hep-lat])

[40] Edwards R G *et al.* (LHPC) 2006 *Phys. Rev. Lett.* **96** 052001 (*Preprint* arXiv:hep-lat/0510062)

[41] Yamazaki T *et al.* (RBC and UKQCD) 2008 *Phys. Rev. Lett.* **100** 171602 (*Preprint* arXiv:0801.4016 [hep-lat])

[42] Sasaki S, Blum T and Ohta S 2002 *Phys. Rev.* D **65** 074503 (*Preprint* arXiv:hep-lat/0102010)

[43] Basak S *et al.* (LHPC) 2007 *Phys. Rev.* D **76** 074504 (*Preprint* arXiv:0709.0008 [hep-lat])

[44] Mathur N *et al.* 2005 *Phys. Lett.* B **605** 137–143 (*Preprint* arXiv:hep-ph/0306199)

[45] Lin H W, Cohen S D, Edwards R G and Richards D G 2008 First lattice study of the N-P_{11}(1440) transition form factors (*Preprint* arXiv:0803.3020 [hep-lat])

[46] Vaccarino A and Weingarten D 1999 *Phys. Rev.* D **60** 114501 (*Preprint* arXiv:hep-lat/9910007)

[47] Morningstar C J and Peardon M J 1999 *Phys. Rev.* D **60** 034509 (*Preprint* arXiv:hep-lat/9901004)

[48] Sexton J, Vaccarino A and Weingarten D 1995 *Phys. Rev. Lett.* **75** 4563–4566 (*Preprint* arXiv:hep-lat/9510022)

[49] Dudek J J, Edwards R G and Richards D G 2006 *Phys. Rev.* D **73** 074507 (*Preprint* arXiv:hep-ph/0601137)

[50] Beane S R *et al.* (NPLQCD) 2006 *Phys. Rev.* D **74** 114503 (*Preprint* arXiv:hep-lat/0607036)

[51] Buettiker P, Descotes-Genon S and Moussallam B 2004 *Eur. Phys. J.* C **33** 409–432 (*Preprint* arXiv:hep-ph/0310283)

[52] Dimeson Relativistic Atom Complex Collaboration (DiRAC) URL http://dirac.web.cern.ch/DIRAC/

[53] Beane S R *et al.* (NPLQCD) 2008 *Phys. Rev.* D **77** 014505 (*Preprint* arXiv:0706.3026 [hep-lat])

[54] Beane S R *et al.* (NPLQCD) 2008 *Phys. Rev.* D **77** 094507 (*Preprint* arXiv:0709.1169 [hep-lat])

[55] Detmold W *et al.* (NPLQCD) 2008 Multi-Pion States in Lattice QCD and the Charged-Pion Condensate (*Preprint* arXiv:0803.2728 [hep-lat])

[56] Beane S R, Bedaque P F, Orginos K and Savage M J (NPLQCD) 2006 *Phys. Rev. Lett.* **97** 012001 (*Preprint* arXiv:hep-lat/0602010)

[57] Beane S R *et al.* (NPLQCD) 2007 *Nucl. Phys.* **A794** 62–72 (*Preprint* arXiv:hep-lat/0612026)

[58] Kaplan D B and Nelson A E 1986 *Phys. Lett.* B **175** 57–63

[59] Schmidt C 2006 *PoS* **LAT2006** 021 (*Preprint* arXiv:hep-lat/0610116)

[60] Compressed Baryonic Matter Collaboration (CBM) URL http://www.gsi.de/fair/experiments/CBM/

[61] Bernard C *et al.* (MILC) 2005 *Phys. Rev.* **D71** 034504 (*Preprint* arXiv:hep-lat/0405029)

[62] Cheng M *et al.* (RBC and Bielefeld) 2006 *Phys. Rev.* **D74** 054507 (*Preprint* arXiv:hep-lat/0608013)

[63] Bernard C *et al.* (MILC) 2007 *Phys. Rev.* D **75** 094505 (*Preprint* arXiv:hep-lat/0611031)

[64] Cheng M *et al.* (RBC and Bielefeld) 2008 *Phys. Rev.* D **77** 014511 (*Preprint* arXiv:0710.0354 [hep-lat])

[65] Bernard C *et al.* (MILC) 2008 *Phys. Rev.* D **77** 014503 (*Preprint* arXiv:0710.1330 [hep-lat])

[66] Dashen R F and Neuberger H 1983 *Phys. Rev. Lett.* **50** 1897

[67] Heller U M, Klomfass M, Neuberger H and Vranas P M 1993 *Nucl. Phys.* B **405** 555–573 (*Preprint* arXiv:hep-ph/9303215)

[68] Fodor Z, Holland K, Kuti J, Nogradi D and Schroeder C 2007 *PoS* **LATTICE2007** 056 (*Preprint* arXiv:0710.3151 [hep-lat])

[69] Catterall S and Sannino F 2007 *Phys. Rev.* D **76** 034504 (*Preprint* arXiv:0705.1664 [hep-lat])

[70] Appelquist T, Fleming G T and Neil E T 2008 *Phys. Rev. Lett.* **100** 171607 (*Preprint* arXiv:0712.0609 [hep-ph])

[71] Kaplan D B 2004 *Nucl. Phys.* B Proc. Suppl. **129** 109–120 (*Preprint* arXiv:hep-lat/0309099)

[72] Giedt J 2006 *PoS* **LAT2006** 008 (*Preprint* arXiv:hep-lat/0701006)

[73] Catterall S and Wiseman T 2007 *JHEP* **12** 104 (*Preprint* arXiv:0706.3518 [hep-lat])

[74] Catterall S and Wiseman T 2008 Black hole thermodynamics from simulations of lattice Yang-Mills theory (*Preprint* arXiv:0803.4273 [hep-th])

[75] Davies C T H *et al.* (HPQCD, MILC, and Fermilab Lattice) 2004 *Phys. Rev. Lett.* **92** 022001 (*Preprint* arXiv:hep-lat/0304004)

Enabling petascale science: data management, troubleshooting, and scalable science services

A Baranovski[1], K Beattie[6], S Bharathi[2], J Boverhof[6], J Bresnahan[3,4], A. Chervenak[2], I Foster[3,4,5], T Freeman[3,4], D Gunter[6], K Keahey[3,4], C Kesselman[2], R Kettimuthu[3,4], N Leroy[7], M Link[3,4], M Livny[7], R Madduri[3,4], G Oleynik[1], L. Pearlman[2], R Schuler[2] and B Tierney[6]

[1]Fermi National Accelerator Laboratory, Batavia, IL 60510, USA
[2]Information Sciences Institute, University of Southern California, Marina del Rey, CA 90292, USA
[3]Computation Institute, University of Chicago and Argonne National Laboratory, Chicago, IL 60637, USA
[4]Mathematics and Computer Science Division, Argonne National Laboratory, Argonne, IL 60439, USA
[5]Department of Computer Science, University of Chicago, Chicago, IL 60637, USA
[6]Lawrence Berkeley National Laboratory, Berkeley, CA 94720, USA
[7]Department of Computer Science, University of Wisconsin, Madison Madison, WI 53719, USA

E-mail: annc@isi.edu

Abstract. The Center for Enabling Distributed Petascale Science is developing serviced to enable researchers to manage large, distributed datasets. The center projects focus on three areas: tools for reliable placement of data, issues involving failure detection and failure diagnosis in distributed systems, and scalable services that process requests to access data

1. Introduction

DOE science applications in such diverse areas as astrophysics, biology, chemistry, combustion, fusion, high energy physics, nanoscience, and nuclear physics are generating and analyzing up to petabytes of data per year. The Center for Enabling Distributed Petascale Science (CEDPS) project [1, 2] works with DOE application science communities to provide services required to move datasets where they are needed; to enable analysis near the data; and to detect and recover from failures in the distributed environment. Each of these tasks is challenging in a petascale environment because of the need to coordinate numerous shared resources, including CPUs, storage, and networks.

The three main components of the CEDPS project are as follows:

- **Data Placement:** Developing tools and techniques for reliable, high-performance, secure, and policy-driven placement of data within a distributed science environment
- **Troubleshooting:** Addressing cross-cutting issues relating, in particular, to *failure detection and diagnosis* in distributed systems
- **Scalable Services:** Supporting programs that are packaged as scalable science services that process many requests to access and analyze data

In this paper, we describe the ongoing work of the CEDPS project in each of these areas.

2. Data management

We are developing tools and techniques for reliable, high-performance, secure, and policy-driven placement of data in a distributed science environment. This work includes end-to-end data distribution services that implement different data placement behaviors. In addition, we are significantly enhancing the GridFTP data transfer service to provide management of the space, bandwidth, connections, and other resources needed to transfer data to and from a storage system. We are working with a range of DOE application communities to deploy, apply, and evaluate these new capabilities.

2.1. GridFTP enhancements

In this section, we discuss new functionality and recent developments that we have incorporated into the Globus GridFTP framework [1]. Specifically, we describe the following four initiatives.

1. GridFTP over SSH, which enables GridFTP client "globus-url-copy" [2] to use SSH security for authentication and for exchanging commands and responses on the GridFTP control channel.
2. GridFTP multicasting, which allows efficient transfer of a single data set to many locations, as well as a GridFTP overlay network with many GridFTP servers act as routers forwarding packets.
3. A plugin for GridFTP that enforces usage policies on storage systems managed by Lotman.
4. Concurrency, a client-side optimization to improve the performance of lots of small files transfers

2.1.1. GridFTP over SSH. GridFTP traditionally uses Grid Security Infrastructure (GSI) [4] for establishing secure connections as a strong, robust, and flexible means of securing messages. In some situations, however, it is preferable to use the SSH security mechanism. We enhanced the GridFTP server and the globus-url-copy client to use SSH for authentication and command/response exchange. GridFTP-over-SSH leverages the fact that an SSH client can remotely execute programs by forming a secure connection with SSHD. In this case, all of the standard I/O from the remote program is routed back to the client through the secure SSH channel. The standard I/O is used as the secure channel to exchange control channel messages.

2.1.2. *GridFTP multicasting*. The multicasting capability allows for efficient transfer of a single data set to many locations. The goal is to use available network cycles effectively to move the total amount of data (destinations * file size) at network speeds. The Globus GridFTP server uses Globus XIO [5] for its I/O. We added a multicasting capability by creating a new XIO drive. Clients may add the new XIO multicast driver to their GridFTP server's I/O stack. Globus XIO provides a modular driver abstraction. As the GridFTP server writes data blocks to its file system, the data blocks first pass through the multicast driver, which passes the data block to other GridFTP servers. The multicast driver can be configured to forward data to the next server and not to write that data to disk. This configuration allows creation of a network overlay where many GridFTP servers act as routers, forwarding packets until they reach their final destination, where they are written to disk.

2.1.3. *Lotman plug-in for GridFTP*. Lotman is a storage management tool from University of Wisconsin. We created a plug-in for GridFTP to communicate with Lotman to enforce space usage for data transferred to storage systems that are managed by Lotman. Lotman supports disk quotas for users on a storage system in the form of lots. Before writing data onto the disk, GridFTP server communicates with the Lotman and ensures that the user quota limit is not violated. One advantage of this plug-in is that transfer failures due to a destination storage system running out of space during a transfer can be avoided if the file size is known in advance; the GridFTP server can check whether there is enough space at the destination before starting the transfer.

2.1.4. *Concurrency*. In [1, 2] , we discussed the pipelining optimization in GridFTP to improve the performance of many small files transfers. In modern high-speed networks, datasets less than 100 MB are too small for the underlying protocols like TCP to utilize the maximum capacity of the network. To further enhance the performance of small files transfers, we added a feature called "concurrency" to the

GridFTP client. This feature helped DOE's Advanced Photon Source user facility at Argonne to transfer more than a terabyte of data to Australia at a rate 30 times faster than traditional data transfer mechanisms such as SCP.

2.2. *Data placement services*

The CEDPS team's staged development of data placement service functionality began with the implementation of a priority-based bulk transfer service for managing many concurrent data transfer operations. This functionality is needed to provide reliable distribution of data according to the plan decided by the higher-level placement service. In addition to the bulk transfer service, we have implemented a web service interface for submitting data placement requests. A paper describing the initial data placement service implementation was published at the Petascale Data Storage Workshop in November 2007 [3].

2.2.1. *Integration of cata placement and workflow management.* In addition to development of the functionality of the data placement services, a significant focus of our work has been to understand the interaction between data placement services and workflow management systems, which are increasingly being used to manage sophisticated scientific analyses consisting of thousands of interdependent tasks. Through measurements and simulations, we have done substantial work in understanding how data placement can improve the performance of data-intensive computational workflows running on distributed computing resources, clusters and supercomputers. The CEDPS team did an initial study of placement service options that evaluated the performance of a data placement service in combination with the Pegasus workflow management system. We performed initial measurements that demonstrated the performance advantages of prestaging data sets onto computational resources before execution of workflows; this work was presented at the Grid2007 Conference [4]. In addition, we modified an existing Grid simulator to experiment with different data placement algorithms operating in conjunction with a workflow manager. A paper describing these simulation results for a variety of data placement algorithms was submitted to the SC2008 conference [5]. Moreover, in a paper presented at the 3rd International Workshop on Workflow Systems in e-Science (WSES 08), we take a broad view of the challenges of data placement and data management at all stages of the workflow lifecycle [6].

2.2.2. *Simplified data replication tools.* To make data replication services more accessible to CEDPS users, we have designed and developed a set of tools to simplify key data replication usage scenarios. Existing tools for data replication include the Globus RLS replica catalog and the Globus GridFTP Server. Our new tools integrate the capabilities in the replica catalogs and GridFTP server for common data operations such as getting, putting and copying replicas.

3. CEDPS troubleshooting

3.1. *Introduction*

Many of today's Grids have ongoing performance and reliability problems that have yet to be addressed. As far back as 2003, the Grid2003 project, now Open Science Grid (OSG) [7], saw a 30% job submission failure rate [8], with 90% of the failures caused by problems such as disk filling errors, gatekeeper overloading, and network interruptions. Although many improvements have been made, the Grid continues to grow in size and complexity. The net improvement is often small: for example, the LCG Grid [9] is still reporting about a 25% error rate [10],.

Troubleshooting Grid middleware can be difficult because of the large number of interconnected components. For example, a single action, such as reliably transferring a directory of files, can result in the coordination of a wide suite of loosely coupled software tools. Each of these systems typically creates its own log files using their own log format, semantics, and identifiers. To troubleshoot a problem as it cascades from one component into the next, this information must be combined to form a logically consistent trail of activity. The immediate goal of the CEDPS troubleshooting area is to replace the current project-specific ad hoc techniques used in distributed computing with a more powerful unified toolkit. The long-term goal is that any software that navigates multiple layers of interfaces can leverage the CEDPS troubleshooting tools to more simply and definitively analyze failure and performance issues.

3.2. Logs

We have written a recommendations guide for Logging Best Practices [11] that combines good instrumentation practices with log format guidelines. The practices and log format go hand in hand so that there are standard representations for all the essential elements: timestamps, *start* and *end* events to wrap operations, unique identifiers, and status codes. Arbitrary additional named attributes can be added by the application. The log event format is simply a line of ASCII "name=value" pairs: this is highly portable, human-readable, and works well with line-oriented tools and syslog-ng.

3.3. Logging framework

A logging framework is required to bring all the various types of log files together and make sense of them. There are four main components to our logging framework: log collector, log parser, database loader, and database. Loading a subset of logs into a relational database enables sophisticated data mining. Historical queries can provide baseline performance information, and this database can be used for post-mortem anomaly detection.

For the log collector we use *syslog-ng* [12]. *syslog-ng* allows us to aggregate the data and filter logs based on program name, log level, and even a regular expression on message contents. Syslog-ng is very scalable: if a particular collector gets over-loaded, one can just bring up another collector on another machine, and send half the logs to each collector.

To convert logs into a common format we have written a log parser component. This log parser is implemented with a plug-in architecture to make it easy to add new parsers. Parsed logs are then fed to our database loader, which loads the records into a MySQL database.

3.4. OSG deployment

The Open Science Grid is a distributed computing infrastructure for large-scale scientific research. It is built and operated by a consortium of universities, national laboratories, scientific collaborations and software developers. The CEDPS project has been working closely with OSG to design and deploy a centralized log collection framework based on syslog-ng. We have written a pacman [13] package for the Virtual Data Toolkit (VDT) [14], which is deployed on OSG.

After much discussion with OSG site administrators, it was determined that sending a large of amount of logs to the Grid Operations Centers (or GOC) would not work due to security and privacy concerns at several sites. If sites choose, they can forward their logs to the GOC directly to aid in troubleshooting. The OSG logging architecture is shown in figure 1.

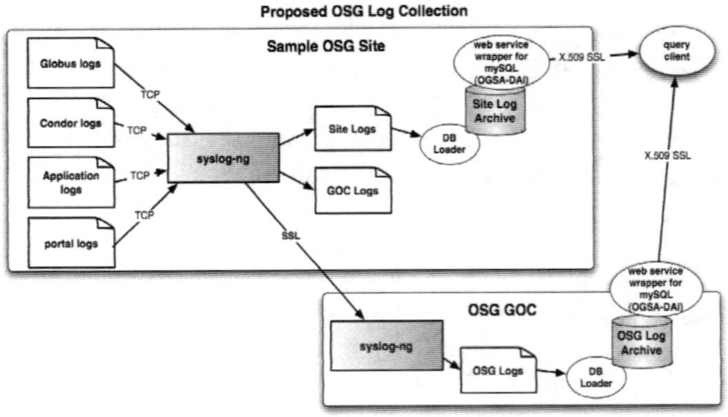

Figure 1. Sample OSG syslog-ng deployment.

3.5. Sample results

We have reworked the logging for Globus v4.2 to use the new "best practices" format for most Globus services. We have found a number of examples where this has made troubleshooting much easier. For example, the GridFTP server [15] now wraps authentication, authorization, and transfers with *start* and *end* events. If we see the following logs events: authn.start, auth.end, authz.start, authz.end, transfer,start; but are missing the transfer.end log and there are no other error messages, then we now can tell this is likely a firewall issue on the data port. Before this additional logging, it was very difficult to determine why the GridFTP transfer was hung.

While our initial motivation was to build this framework for troubleshooting issues, we have found it to be very useful for more detailed accounting as well. For example, the Southern California Earthquake Center (http://www.scec.org/) uses the Grid to run several seismological modeling codes that simulate earthquake processes. They need to run many thousands of small independent jobs (800K in a typical run) over a period of one to two days. They wanted to better understand their load balancing across the available resources, and used our logging framework to load their "kickstart" logs [16] into our database. We were then able to easily extract information such as the number of jobs and job run-time broken down by hour for a typical day long run.

3.6. Availability and next steps
All the components of our logging framework are available as part of our NetLogger Toolkit, available at http://acs.lbl.gov/NetLogger/. Our next focus will be on the integration of OSGA-DAI [17]. We hope that OSGA-DAI will solve the authorization problem of allowing users to see their own logs, but not the logs of others. Performance issues with OSGA-DAI will also be explored. We also plan to focus on further improvements to our Logging Best Practices recommendations and implementation. Currently, BP-formatted logs simplify the process of diagnosing problems that occur within a single service, or among multiple services in a single hosting environment (such as a Globus container). We plan to expand our logging efforts to include information exposing the relationships between subtasks of workflows that span multiple sites.

4. CEDPS scalable services
We are developing tools and techniques for the construction, operation, and provisioning of scalable science services. We are creating service construction tools that take application code (simulation or data analysis, program or library) and wrap it as a remotely accessible service, with appropriate interfaces, authorization, persistence, provisioning, and other capabilities. We are constructing provisioning tools for dynamic management of the computing, storage, and networking resources required to execute a service, and the configuration of those resources to meet application requirements.

4.1. Service construction tools
The CEDPS team has been working on developing tools to easily wrap applications. This year marked some changes in the technologies to be used for wrapping legacy applications into Grid services. We started working on a Java-based tool called Grid Remote Application Virtualization Interfaces (gRAVI). This tool is built on top of a Grid service-building tool called Introduce. The following are the current capabilities of the gRAVI:

- Autogeneration of services using standard Grid security and flexible authorization models
- Integration of gRAVI with centralized registries to enable automatic publishing of services enabling discovery
- Development of strongly typed services

4.2. Application outreach
- **APS:** The Advanced Photon Source scientists used gRAVI to wrap applications and created a workflow that orchestrates two gRAVI-created application services that run analytics.
- **caBIG:** Using gRAVI, the cancer Biomedical Informatics Grid project was able to quickly wrap an application called hierarchical clustering into a Grid service and successfully demonstrate this at the caBIG conference. The tool will let the users create a strongly typed Grid service and would enable the execution of the application as a Grid job on a cluster.

4.3. Future work
We are working on generating code automatically that would enable the transfer of input files needed for the execution of the application from the user's machine and also transfer the generated output files back

to the user's machine. We are also developing provisioning services using virtual workspaces and WS-GRAM services.

Acknowledgments
This work was supported in part by the Department of Energy's Scientific Discovery through Advanced Computing 2 program under grant DE-FC02-06ER25757 and by the Mathematical, Information, and Computational Sciences Division subprogram of the Office of Advanced Scientific Computing Research, Office of Science, U.S. Department of Energy, under contracts DE-AC03-76SF00098 and DE-AC02-06CH11357.

References

[1] Baranovski A, *et al.* 2007 Enabling distributed petascale science *J. Physics: Conference Series* **78**
[2] Schopf J M, Chervenak A, Foster I, Fraser D, Gunter D and Tierney B 2007 End-to-end data solutions for distributed petascale science *CTWatch Quarterly* **3**(4)
[3] Chervenak A L and Schuler R 2007 A data Ppacement service for petascale applications Presentation at Petascale Data Storage Workshop (Supercomputing '07, Reno, NV)
[4] Chervenak A, Deelman E, Livny M, Su M, Schuler R, Bharathi S, Mehta G and Vahi K 2007 Data placement for scientific applications in distributed environments Presentation at 8th IEEE/ACM Int'l Conference on Grid Computing (Austin, Texas)
[5] Bharathi S and Chervenak A 2008 Data staging strategies and their impact on workflow performance *Proc. SC2008 Conference*
[6] Deelman E and Chervenak A 2008 Data management challenges of data-intensive scientific workflows Presentation at 3rd International Workshop on Workflow Systems in e-Science (WSES 08) in conjunction with CCGrid 2008 Conference (Lyon, France)
[7] Open Science Grid Consortium 2006 http://www.opensciencegrid.org/
[8] Foster I, Gieraltowski J, Gose S, Maltsev N, May E, Rodriguez A, Sulakhe D, Vaniachine A, Shank J and Youssef S 2004 The Grid 2003 production Grid: principles and practice Presentation at 13th IEEE International Symposium on High Performance Distributed Computing
[9] LCG Grid 2008 http://www.gridpp.ac.uk
[10] Aggarwal M, Colling D, McEvoy B, Moont G and van der Aa O 2006 A statistical analysis of job performance within LCG Grid Presentation at CHEP06 (Mumbai, India)
[11] CEDPS Project 2007 Logging best practices guide http://www.cedps.net/wiki/index.php/LoggingBestPractices 2007
[12] syslog-ng 2008 http://www.balabit.com/products/syslog-ng/
[13] Pacman 2008 http://physics.bu.edu/~youssef/pacman/
[14] VDT 2008 http://vdt.cs.wisc.edu/
[15] Allcock W, Bresnahan J, Kettimuthu R, Link M, Dumitrescu C, Raicu I and Foster I 2005 The Globus striped GridFTP framework and server Presentation at IEEE Supercomputing Conference (Seattle, WA)
[16] Vöckler J S, Mehta G, Zhao Y, Deelman E and Wilde M 2006 Kickstarting remote applications Presentation at 2nd International Workshop on Grid Computing Environments (Tampa, FL)
[17] University of Edinburgh 2008 Open Grid Services Architecture Data Access and Integration (OGSA-DAI) http://www.ogsadai.org.uk/

High-performance metadata indexing and search in petascale data storage systems

A W Leung*, **M Shao**†, **T Bisson**†, **S Pasupathy**†**and E L Miller***
*Storage Systems Research Center, University of California, Santa Cruz, CA 95064, USA
†NetApp, Inc., Sunnyvale, CA 94089, USA

E-mail: {aleung,elm}@cs.ucsc.edu, {minglong,tbisson,shankarp}@netapp.com

Abstract. Large-scale storage systems used for scientific applications can store petabytes of data and billions of files, making the organization and management of data in these systems a difficult, time-consuming task. The ability to search file metadata in a storage system can address this problem by allowing scientists to quickly navigate experiment data and code while allowing storage administrators to gather the information they need to properly manage the system. In this paper, we present Spyglass, a file metadata search system that achieves scalability by exploiting storage system properties, providing the scalability that existing file metadata search tools lack. In doing so, Spyglass can achieve search performance up to several thousand times faster than existing database solutions. We show that Spyglass enables important functionality that can aid data management for scientists and storage administrators.

1. Introduction

Recent years have seen an enormous increase in the amount of data placed on storage systems. This is particularly true in the scientific community where research, development, and experimentation have become largely digital processes. Scientific applications demand both scale and performance from their data storage systems. These applications produce petabytes of data spread across billions of files, and often have running times tied directly to how quickly they can access data. Building large scale, high performance storage systems required by scientific applications has been a major area of research for over two decades [1, 2, 3, 4, 5]; however, while these systems can provide the necessary performance, the scale of these systems has given rise to a new challenge: managing and organizing petabytes of data and billions of files.

The mere presence of billions of files and petabytes of data makes it difficult for users to find and organize their data and complicates management decisions for storage administrators, particularly in scientific environments. An experiment can produce millions of files that must be analyzed to produce useful results, and both users and administrators must manage the files from many experiments. For scientists, this situation can waste time and can cause important results to be missed. Administrators face more difficult decisions, potentially jeopardizing data on the storage system.

A significant portion of the data management problem can be addressed through the ability to rapidly search file metadata in a storage system; the importance of search in a storage system is comparable to its importance on the Web. Large-scale systems consist of a sea of loosely organized data identified by a location (path name in a storage system and URL on the web) that must be known in order to access it. Search allows users to use metadata about the file to locate it without requiring the path to be known, a technique far superior to the manual browsing and organizing done today. In a storage system, metadata search allows a scientist to quickly search for relevant experimental results, and it provides a means for administrators to gain important information about the data they manage.

Recently, tools such as Apple's Spotlight and Google Desktop have become available to users of desktop file systems. While these tools are effective in small-scale file systems, they lack the scalability to address file metadata search in large-scale storage systems. because they do not make any significant optimizations or designs for file search. Instead, they rely upon traditional indexing techniques, such as relational database systems, that are designed for very different workloads, resulting in systems that are often too slow to search and too slow to update and require too many system resources, such as memory and disk space.

In this paper, we discuss the design and implementation of scalable systems for searching and indexing file metadata that are capable of addressing the search needs of scientists and their storage administrators. We first present background information for search in large-scale scientific computing. We then discuss Spyglass [6], a system we have designed for searching large-scale storage systems that is up to several thousand times faster than existing database solutions. Next, we provide examples of how Spyglass can aid scientific research. Finally, we discuss related and future work and then conclude.

2. Background

File metadata search works by building an index of fields in a file's inode, such as owner and size, and its extended attributes, which are manually defined key-value pairs. An example of an extended attribute key-value pair may be a key (e.g., `experiment_runtime`) and value (e.g., `2 hours`). An example metadata query might ask, "Which experiments did I run last week that took less than 20 minutes to run?" An index is a data structure that allows the rapid identification of files that contain specific metadata values. User queries can quickly be satisfied by simply looking up the information in the index. Without an index, the file system must be manually traversed, reading each file's metadata, in order to satisfy a query, a process far too slow to be effective.

Many existing systems, such as Apple's Spotlight and Linux's Tracker, index metadata using relational database systems (DBMS). DBMSs build relational tables in which each column in a table defines a different metadata field and each row defines a different file's metadata values. In general, queries are satisfied by searching the table for files that match a query's predicates. In large-scale storage systems, however, scanning a table can be very slow because the table is so large. Thus, B+-tree indexes are often built on the columns that need fast search. A B+-tree is a data structure that allows the files with specific values for a key to be quickly located through logarithmic search of the key's sorted value space, eliminating much of the needless searching required by a table scan.

However, a DBMS does not make any optimizations for storage system workloads; instead, the design is based on data warehousing workloads from the 1970s. As a result, a generic DBMS is not a "one size fits all solution" for indexing data, a sentiment echoed by the DBMS community itself [7]. File systems have common access patterns [8, 9] that can be exploited to improve performance. For example, requests are not evenly distributed across storage namespace locations, meaning some subtrees and directories are more popular than others. Also, requests have locality of reference: if a subtree of a directory was recently accessed, it is likely to be accessed again soon. Additionally, metadata values tend to be clustered in small parts of the file system namespace. For example, most files owned by a user are located in their home directory, with others in specific project directories in which the user is actively working. By ignoring many such properties of file systems, DBMSs have a number of inefficiencies that limit their scalability.

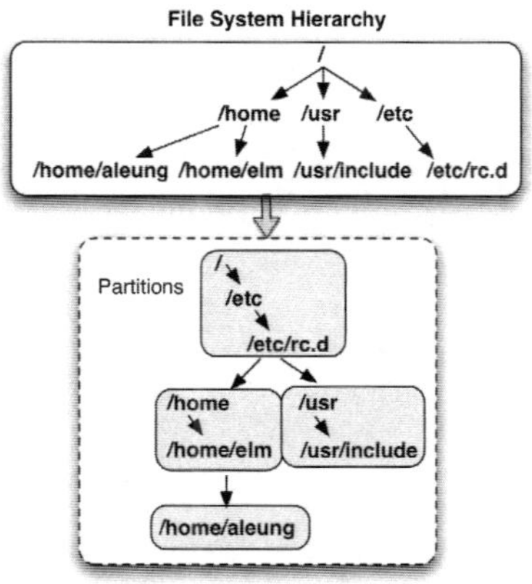

Figure 1. Hierarchical Partitioning. Spyglass partitions the index based on the storage system's namespace. This allows search and indexing to be controlled at the finer granularity of subtrees. We can exploit the namespace's locality and reference properties to reduce a query's search space and improve performance.

3. Spyglass design

Our metadata indexing system, Spyglass, introduces two index design methods that are optimized for large-scale storage systems: *hierarchical partitioning* and *partition versioning.* Hierarchical partitioning takes advantage of the storage system's locality properties by partitioning the index into a number of smaller partitions, based on the file system namespace, with each partition responsible for indexing one or more subtrees in the namespace. An example of this partitioning is shown in figure 1, in which each partition indexes files from different subtrees in the namespace. The key idea is that hierarchical partitioning allows search and indexing to be controlled at the granularity of subtrees, which cannot be done when all data is stored in a single index. Using this, we can take advantage of many storage system workload properties. For example, metadata attributes tend to be clustered in the storage namespace, meaning that, in most cases, few partitions are required to satisfy a query, greatly reducing a query's search space. Similarly, some parts of the namespace are frequently accessed, so caching the corresponding index partitions in memory can greatly improve performance for common-case queries.

Partition versioning simplifies operation and improves performance of index updates and allows users to search over past file metadata. Spyglass does this by batching changes made to file metadata over an interval (the length of the interval is a file system parameter) and then applying each batch to the index as a new index version. Thus, when the index is updated, a new version does not overwrite old an old version but instead allows old metadata to be maintained and searched by the user. This feature is important for management and trend queries where metadata information overtime is needed. To save space, each version stores only metadata changes since the previous version, rather than storing a full copy of all file metadata in the system; thus, partition versions are incremental. Also, each partition manages its own versions, allowing partition versioning to take advantage of storage workload properties. For example, because only a few popular namespace locations frequently change, it is usually the case that only a few partitions must be versioned. Likewise, versions for popular partitions can be kept in cache with the partition itself, meaning most queries over multiple versions are very fast.

These optimizations to file system indexes both reduce resource usage and improve performance. Because each index partition covers only a small part of the file system namespace, each partition can be individually "customized" to the metadata it contains. For example, while there may be thousands of users who have created files in the file system, an individual index partition may only contain files owned by fewer than ten users, allowing the index to dramatically compress the representation of the users' ID. The index can perform similar compression on other attributes, including timestamps, permissions, file types, and other extended attributes.

Spyglass further improves performance by using *signatures* to quickly identify partitions that may be relevant for a particular query. For example, each index partition lists the users that own files it indexes. If a query requests files owned by user A, only those files that contain user A's files are even considered by the query program. For the localized queries common in file system usage, this technique quickly rules out most of the index partitions, leaving only the partitions that might contain positive results for the query.

To demonstrate the advantages of using the partitioned, versioned indexes in Spyglass, we conducted a number of experiments that compare Spyglass performance with existing DBMS solutions using real-world large-scale storage system metadata traces. We found that Spyglass requires between 5 to 8 times less disk space and updates between 8 to 44 times faster than DBMS systems. The reason is that DBMSs use a B+-tree for each metadata attribute that needs fast query performance, each of which adds space and update overhead. Spyglass indexes each metadata value just once by using multi-dimensional data structures. We also found that Spyglass query performance was at least 10 times faster than DBMSs and up to several thousand times faster in some cases, particularly those in which users were able to localize their search to only a part of the storage namespace. Because hierarchical partitioning allows control over the granularity of the search, a query's search space can be significantly reduced, reducing the total amount of index data that must be processed for the query.

4. Example applications

Fast metadata search enables scientists and large-scale storage administrators to improve their efficiency and better do their jobs. Suppose a scientist runs a long experiment that produces a million files. Using existing file system interfaces, such as `xattr_set()`, extended attributes about the experiment can be set and indexed in Spyglass. A scientist may want to create extended attributes that describe whether the experiment was successful, how long it took, or the results contained in the files. Without any method to search these files, the scientist must manually navigate the files, a slow and imprecise process, or use an external index to manage the files, an approach that cannot leverage tight integration with the file system. Using Spyglass, however, the scientist can simply search for the desired files using queries such as "Which experiment files that I created yesterday were successful and finished in less than 1 hour?" Additionally, by using the versioning abilities of Spyglass, the scientist can ask trend queries about experiments. Using a chemical experiment as an example, a scientist could ask how the pH levels in particular acidity experiments been changing over the past week, a question that is difficult, if not impossible, to answer with existing tools.

Storage administrator who manage a storage system for scientific data can use Spyglass to aid their management decisions. For example, the administrator can easily find which old, unsuccessful experiments to delete by asking, "Which unsuccessful experiment files have not been accessed in over a month?" Additionally, the administrator can improve data security by searching for and encrypting all files that have "nuclear bomb test" attributes.

5. Future work

Scalable file search in large-scale storage systems still presents a number of unsolved challenges. We plan to extend our existing work on metadata indexing in Spyglass to several of these other areas. We are planning on exploring techniques that apply hierarchical partitioning to improve the scalability of content-based search. Content-based search allows keywords to be automatically extracted from files for indexing and differs from metadata search in the number of possible keywords that can be searched. Extracting keywords efficiently is also a challenge, especially for large scientific files, which we also plan to explore in the future. Given that many high-performance storage systems are distributed, indexing solutions for these system must also be distributed. Thus, we also plan to look at scalable communication protocols for index consistency and query execution.

6. Conclusions

The scientific communities need for scalable storage, high-performance storage has yielded storage systems capable of storing petabytes of data across billions of files. However, the scale of these systems has made managing data on them a difficult task. The ability to search file metadata in a storage system can greatly ease the burden on scientists and storage administrators. However, existing file search tools lack the scalability effectively search large-scale storage systems, largely because they are optimized for storage systems.

In this paper, we presented Spyglass, a scalable metadata search and indexing system that is designed specifically for storage systems. By designing using an index that exploit storage properties, Spyglass is able to achieve significantly better query performance than existing DBMS solutions, while using less disk space. We showed through example how its functionality can be used to aid the data management in the scientific community.

Acknowledgments

This work was supported in part by the Department of Energy under the Petascale Data Storage Institute (award DE-FC02-06ER25768) and industrial sponsors of the Storage Systems Research Center at UC Santa Cruz, including Agami Systems, Data Domain, Hewlett Packard, Hitachi, LSI, NetApp, Seagate, and Symantec. We also thank our colleagues in the SSRC and NetApp's Advanced Technology Group for their insightful feedback, which greatly improved the quality of the research.

References

[1] Corbett P F and Feitelson D G 1996 The Vesta parallel file system *ACM Trans on Computer Systems* **14** 225–264

[2] Gibson G A, Nagle D F, Amiri K, Butler J, Chang F W, Gobioff H, Hardin C, Riedel E, Rochberg D and Zelenka J 1998 A cost-effective, high-bandwidth storage architecture *Proc. 8th Conf. on Architectural Support for Programming Languages and Operating Systems (ASPLOS)* (San Jose, CA) 92–103

[3] Miller E L and Katz R H 1997 RAMA: an easy-to-use, high-performance parallel file system *Parallel Computing* **23** 419–446

[4] Schmuck F and Haskin R 2002 GPFS: a shared-disk file system for large computing clusters *Proc. 1st Conference on File and Storage Technologies (FAST)* 231–244

[5] Weil S A, Brandt S A, Miller E L, Long D D E and Maltzahn C 2006 Ceph: a scalable, high-performance distributed file system *Proc. Symp. Operating Systems Design and Implementation (OSDI)*

[6] Leung A W, Shao M, Bisson T, Pasupathy S and Miller E L 2008 Spyglass: fast, scalable metadata search for large-scale storage systems Tech. Rep. UCSC-SSRC-08-01 Storage Systems Research Center, University of California Santa Cruz

[7] Stonebraker M and Cetintemel U 2005 "One size fits all": an idea whose time has come and gone *Proc. 21st Intl. Conference on Data Engineering* (Tokyo, Japan) 2–11

[8] Leung A W, Pasupathy S, Goodson G and Miller E L 2008 ' Measurement and analysis of large-scale network file system workloads *Proc. USENIX Annual Technical Conference*

[9] Roselli D, Lorch J and Anderson T 2000 A comparison of file system workloads *Proc. USENIX Annual Technical Conference* 41–54

Journal of Physics: Conference Series **125** (2008) 012070 doi:10.1088/1742-6596/125/1/012070

New science on the Open Science Grid

The Open Science Grid Executive Board on behalf of the OSG Consortium:

Ruth Pordes[1], Mine Altunay[1], Paul Avery[2], Alina Bejan[3], Kent Blackburn[4], Alan Blatecky[5], Rob Gardner[3], Bill Kramer[6], Miron Livny[7], John McGee[5], Maxim Potekhin[8], Rob Quick[8], Doug Olson[6], Alain Roy[6], Chander Sehgal[1], Torre Wenaus[8], Mike Wilde[3] and Frank Würthwein[10]

[1]Fermi National Accelerator Laboratory, Batavia, IL 60510, USA

[2]University of Florida, Gainesville, FL 32611, USA

[3]University of Chicago, Chicago, IL 60607, USA

[4]California Institute of Technology, Pasadena, CA 91125, USA

[5]Renaissance Computing Institute· Chapel Hill, NC 27517, USA

[6]Lawrence Berkeley National Laboratory, Berkeley, CA 94720, USA

[7]University of Wisconsin, Madison, Madison, WI 53706,USA

[8]Indiana University, Bloomington, IN 47405, USA

[9]Brookhaven National Laboratory, Upton, NY 11973, USA

[10]University of California, San Diego, La Jolla, CA 92093, USA

E-mail: ruth@fnal.gov

Abstract. The Open Science Grid (OSG) includes work to enable new science, new scientists, and new modalities in support of computationally based research. There are frequently significant sociological and organizational changes required in transformation from the existing to the new. OSG leverages its deliverables to the large-scale physics experiment member communities to benefit new communities at all scales through activities in education, engagement, and the distributed facility. This paper gives both a brief general description and specific examples of new science enabled on the OSG. More information is available at the OSG web site: www.opensciencegrid.org.

1. Science on the OSG

The Open Science Grid (OSG) provides a nationally distributed computational infrastructure for science research where computing and storage owned by members of an open consortium are made accessible to and shared with other members through installation of OSG software and subscription to OSG services. The majority of the

Figure 1. OSG Sites, June 2008

resources are from and current use is by the large physics collaborations: ATLAS, CDF, CMS, D0 and LIGO. This new way of doing their science has already paved the way for a substantially expanded infrastructure, which now includes access to the computing and storage facilities at four DOE laboratories, more than forty universities from twenty states in the US, as well as six sites offshore, and diversified use of the facility to include more than ten other scientific domains (see figure 1).

The mission of the OSG includes extending the shared computational environment to include new participants – new scientists, new campuses, new communities, and new scales – and to ultimately transform the landscape for future research and discovery. Such a transformation is neither light nor fast. It takes months or years of continual communication and support to achieve the cultural as well as the technical adaptation to move an individual and/or a community group from an application environment that assumes dedicated local processing and data repositories to a distributed system of a dynamic set of heterogeneous remote computing and data resources. This implies moving from a local environment characterized by deterministic response times and interactive access to the job and data file execution nodes to a distributed system where performance is achieved via the aggregate overall throughput, where some of the resources are unavailable (inaccessible, preoccupied) for use at any particular time, where the data must be transferred from and to remote processing units, and the resources are shared subject to autonomous local policies of prioritization and access.

The OSG facility provides support, operational security, common software, and other Gridwide facility services such as accounting, monitoring, and resource information on which all OSG members rely [2]. For the LHC experiments these OSG services form part of the "critical service" baseline of the World Wide LHC Computing Grid [3]. OSG is thus part of the federated global infrastructure enabling science across more than thirty-five countries on four continents. The facility provides monitoring, management, response and problem-solving frameworks that leverage available expertise, promote common best practices and approaches, and provide natural dissemination and growth of the knowledge base.

Another important aspect of the OSG is the provision and support of a common, integrated, supported set of software components, the Virtual Data Toolkit (VDT) [4], available on many platforms for installation and use by resources providers and users of the infrastructure. By being available for many variants of Linux (and in client mode for MacOSX and AIX), the VDT provides an anchor to define common interfaces to the services (job execution, data movement and storage, security, management and monitoring, etc.) on and through the OSG. The components of the VDT are determined by the needs of the communities; and the processes to build, test, and release the software ensure managed evolution and extension of the capabilities offered.

It has been eighteen months since the start of the SciDAC-2/NSF-funded OSG project. We describe here the types and examples of the new science enabled through the work of the OSG community.

2. New scientists

OSG education and engagement activities work to help and teach new scientists – and, in the case of education, specifically young new scientists – to adapt and run data and compute intensive applications on the existing distributed infrastructure.

OSG grid schools, lasting one to several days, have combined lectures and hands-on laboratories to teach the fundamentals of Grid technologies. To date, they have enabled more than 300 students to actually run jobs across and transport data between specific OSG sites that advertise support for the "education community." Selected students also attend the longer, residential school, the International Summer School on Grid Computing '08, now cosponsored by the OSG. Students receive a grounding in computer science fundamentals of distributed computing and then work in teams as prototypical communities in a competition to develop working integrated scientific applications, running across a set of (locally) distributed computers. In additional, several faculty now rely on OSG materials for their Grid computing courses.

OSG engagement has pioneered the method of "immersive, embedded support" [5]. OSG staff, together with those from the companion NSF-sponsored EIE-4CI project, spend from several days to several months focusing on a particular scientific application. They first help an individual or small subgroup run jobs on the distributed infrastructure. They soon realize faster aggregate response and benefit from parallel execution of computations in order to produce the output to test their ideas and hypotheses. The greatest benefit is seen by those applications that can easily be parallelized, with little interprocess communication, and typically include parameter sweeps, simulation, or other high-throughput computing components. After initial demonstrations, viral communication and demonstration of the benefits then bring more individuals, and ultimately the new community to the table.

Two concrete examples on the OSG in the past year are discussed below.

Dr. Catherine Blake, an assistant professor in the School of Information and Library Science at the University of North Carolina, has developed a method for retrieving, analyzing, and finding relationships in published research texts from multiple disciplines. "Claim Jumping through Scientific Literature" enables researchers in a given field to collect relevant published information from other fields of study. The sentence structure of millions of sentences is now generated in less than 30 hours on OSG resources, which the researchers had previously attempted on their local desktops and which would have taken several years. Analysis of individual sentences is a good application for a high-throughput facility such as OSG; the end-to-end job-run naturally carves up into smaller processing, pieces. As Dr. Blake told members of the OSG "Reducing the processing time has enabled us to investigate otherwise prohibitively complex research questions."

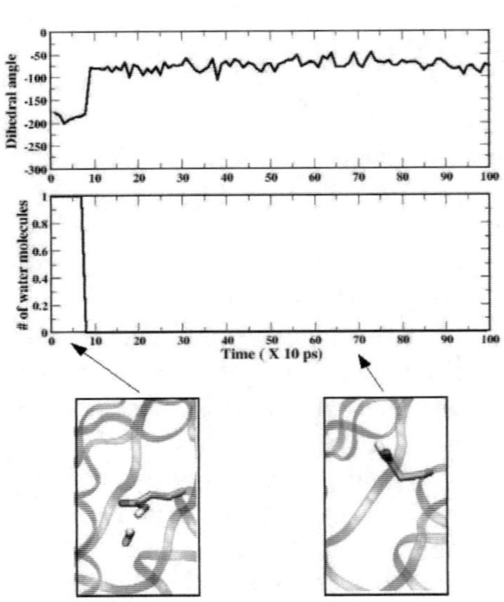

Figure 2. MD simulations reveal that the staphylococcal nuclease protein residue can twist into another conformation, usually forcing water molecules to leave. But which conformation is adopted and for what proportion of the time? And what happens to the water structure around these conformations? (Images courtesy of Johns Hopkins University)

Dr Ana Damjanovic of the National Institutes of Health and Johns Hopkins University uses CHARMM (Chemistry at HARvard Macromolecular Mechanics), a popular application developed at Harvard University for modeling the structure and behavior of molecular systems to investigate the interactions between proteins and water (figure 2). The OSG applications support staff adapted the PANDA workflow tools, developed for the LHC ATLAS experiment, to enable short CHARMM jobs that do not need MPI, to run across OSG sites. The tools provide for automated job resubmission on failure and the submission of hundreds of jobs in a sequence. The JHU group adapted their habits and models to submit large numbers of jobs simultaneously. Simulations on OSG were used to study hydration of the side chain of the internal Glu-66 in staphylococcal nuclease. Results of the study have been submitted [6].

3. New science institutions

The OSG implements an architecture of federated infrastructures where the local-area and the wide-area are connected through locally managed interfaces. This architectural foundation allows new collaborating institutions and regional groups to locally and independently adopt OSG software and services in a self-organized and self-managed framework. It allows the understanding of and trust in the technologies and methods to develop locally before introducing the added challenges of accessing and being accessed by the remote "anonymous" ensemble.

The large physics communities, the main scientific drivers of the OSG, have data and computing systems that span hundreds of universities worldwide. Through OSG and its peer EGEE infrastructure in Europe (and other national grid infrastructures such as NorduGrid in Scandinavia), the LHC experiments have a goal of a providing continuously operating, robust (>99% availability of) critical services and resources. Data is distributed at sustained rates of hundreds of megabytes per second [7]. More than 300,000 jobs are executed daily. A scientist at any site expects to retrieve and access data from any other site in the global mesh and to seamlessly run their data analysis and simulation codes at any collaborating institution that has the data available. New institutions dynamically join the facility when they are accepted into the physics collaboration and as they are ready to provide local resources and/or access and analyze the data. This is bringing about a cultural advance in the ability of university physics departments, far from the accelerator site and central computing farms and tape stores in Switzerland, to sustain active local research groups while at the same time contributing to common scientific discovery worldwide. Examples are the more than fifteen Tier-2's (experiment-funded) and ten Tier-3 (locally funded) universities in the U.S. that are part of the OSG infrastructure, as well as the emerging Tier-2 in South Africa, and six sites in Central and South America.

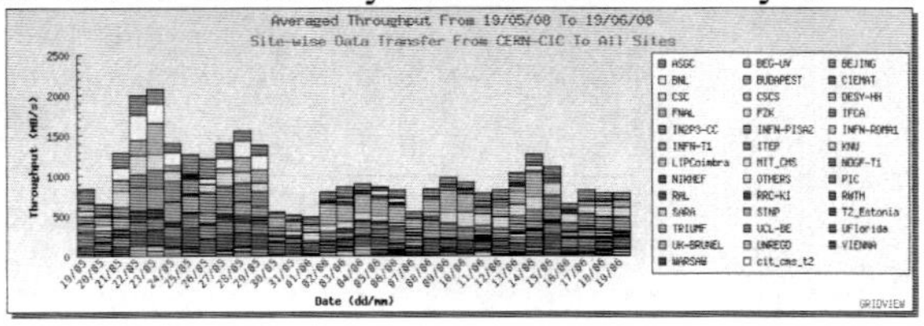

Figure 3: Data distribution throughput for the LHC experiments from CERN to centers worldwide. Courtesy of Ian Bird, CERN.

OSG actively promotes this model through collaboration and partnerships with campus and regional infrastructures in the US and elsewhere. Each organization may adopt some, or all, of the OSG software and services as needed and provide OSG gateways for the bidirectional transport of jobs and data. The ability and reach of institutions is expanded in new ways by fostering local expertise and support as well as leading to broader participation in national and international cyber-infrastructure.

The New York State Grid (NYSGrid) consortium and the Clemson University Campus Infrastructure are distinct examples that have been built up over the past twelve months. NYSGrid formed from a local consortium of universities in New York State that partners with the OSG to allow use of and support for the OSG VDT. About half of the sites directly interface to the OSG as well as participate in the local NYSGrid services and policies. The system administrators of one or two sites participate in OSG activities and then interface with, educate, and inform the rest. Dr Andrew Schultz, a computational chemist in the laboratory of Professor David Kofke, was engaged in running his applications on OSG via this model. Within two weeks of being introduced to OSG in spring 2008, he was systematically using resources and within two months was running across more than ten sites on the OSG and NYSGrid itself and with throughput of more than 250 CP-days per day [8].

The Clemson University Grid is a cross-campus Condor pool of Windows computers with an aggregate of more than 2,000 cores. Researchers, faculty and students have agreed to share their resources locally. This local infrastructure is interfaced to the OSG via a Linux head-node that allows jobs and data to be controlled as they flow between the OSG and the Clemson grid. Resources on this ensemble of clusters were almost saturated within six months of building the environment. Researchers in industrial engineering have been among the first communities to adapt their codes and benefit from this campuswide shared facility.

4. New scientific modalities

The OSG aims to support science at all scales – from the single PI to the largest, multithousand member scientific collaborations. Technologies and processes are built to support groups of individuals from the small to the large, cascading of subgroups within these groups and, eventually, dynamic grouping and regrouping to support ad hoc, short-term collaborations stimulated through collaborative and interdisciplinary thought and work.

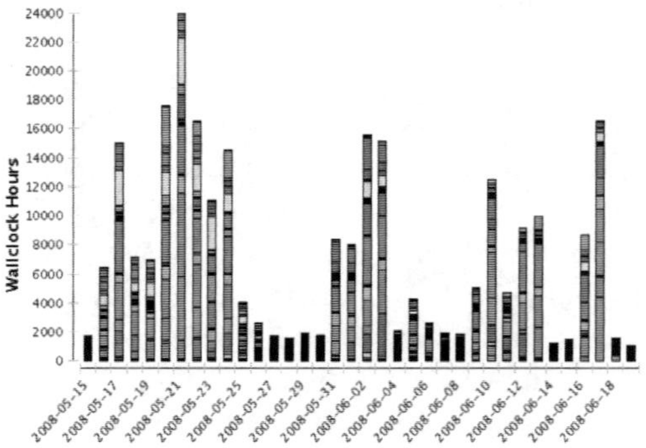

Figure 4. Cyclic use of OSG by the Engagement community.

The OSG model of sharing resources across multiple communities leads to a smoothing out of the periodicity in use by each individual collaboration. This periodicity follows the natural cycle of proposing the methods and algorithms to do the research, coding and debugging of the software to implement them, production runs to obtain the outputs, and then the offline analysis, interpretation, and publication of the results. The timelines of these periodicities can vary greatly. Use of the OSG is typically significant only while the production runs are being done. Figure 4 shows four such cycles of use of the OSG for the engagement community over the past six weeks. The periodicity for the large physics experiments is typically greater before the accelerator data arrives and is expected to be non-existent in general afterwards due to the overlapping of the four phases of data scientific analysis described above.

The OSG methodology of working toward an "application-ready" infrastructure and "distributed-infrastructure-ready" applications allows for (1) sustained aggregate throughput of the whole facility to smooth out the cycle in usage due to these periodicities and (2) throughput for each community to be as large and responsive as possible when needed. The OSG model of up-front agreement on the policies for sharing, in concert with technologies to support easily changing the policies as needed, allows for dynamic use of the system without the per-event overhead of negotiation and allocation at the time of the actual need.

5. Summary

Through its collaborative community-based activities using existing methodologies and tools, the OSG immediately benefits and stimulates new science. In addition, the activities of the OSG help evolve the organization and technical frameworks of scientific collaborations to broadly transform how collaborative data and compute intensive science will be done in the future.

Acknowledgments

This work was supported by the Office of Science, U.S. Department of Energy, SciDAC program under Contract DE-FC02-06ER41436 and the National Science Foundation Cooperative Agreement, PHY-0621704. Fermilab is supported by the U.S. Department of Energy under contract No. DE-AC02-76CH03000.

References

[1] Pordes R, *et al.* 2007 The Open Science Grid *J. Phys.: Conf. Ser.* **78:**012057

[2] Pordes R, *et al.* 2007 The Open Science Grid - Its Status and Implementation Architecture", to appear in *JPCS Proc. International Conference on Computing in High Energy and Nuclear Physics (CHEP 07)*

[3] Bird I, *et al.* 2005 Deploying the LHC computing grid - the LCG service challenges *Local to Global Data Interoperability - Challenges and Technologies*

[4] The Virtual Data Toolkit 2008 http://vdt.cs.wisc.edu

[5] National Science Foundations Award CI-Team Implementation Project 2008 Embedded

Immersive Engagement for Cyberinfrastructure

[6] Damjanovi Ana, *et al.* 2008 An Open Science Grid study of the coupling between conformation and water content in the interior of a protein *J. Physical Chemistry B* (submitted)

[7] Schulz M, *et al.* 2007 Building the WLCG file transfer service *JPCS Proc. International Conference on Computing in High Energy and Nuclear Physics (CHEP 07)*

[8] Benjamin K M, Schultz A J and Kofke D A 2007 Virial coefficients of polarizable water: Applications to thermodynamic properties and molecular clustering *J. Phys. Chem. C* **111:**16021

Combinatorial algorithms for computational science and engineering

Erik G Boman[1], Doruk Bozdag[2], Umit V Catalyurek[2], Karen D Devine[1], Assefaw H Gebremedhin[3], Paul D Hovland[4] and Alex Pothen[3]

[1] Discrete Algorithms and Math Department, Sandia National Laboratories, Albuquerque, NM 87185, USA
[2] Biomedical Informatics, and Electrical and Computer Engineering, Ohio State University, Columbus, OH 43210, USA
[3] Department of Computer Sciences and Computing Research Institute, Purdue University, West Lafayetter, IN 47907, USA
[4] Mathematics and Computer Science Division, Argonne National Laboratory, Argonne, IL 60439, USA

Abstract. We discuss recent activities in the SciDAC Applied Mathematics Institute for Combinatorial Scientific Computing and Petascale Simulations (CSCAPES).

1. Introduction

The Institute for Combinatorial Scientific Computing and Petascale Simulations (CSCAPES) was established in 2006 to help contribute to the SciDAC mission by developing infrastructural algorithmic and software technologies for supporting high-performance computing. Its scope encompasses four areas: load-balancing and related tasks in parallel processing (such as data migration and task scheduling); automatic differentiation (AD), a technology for transforming computer source code for computing a function into code for computing its derivatives; basic ingredients in linear solver technology; and runtime data and iteration reordering to improve performance in irregular computation.

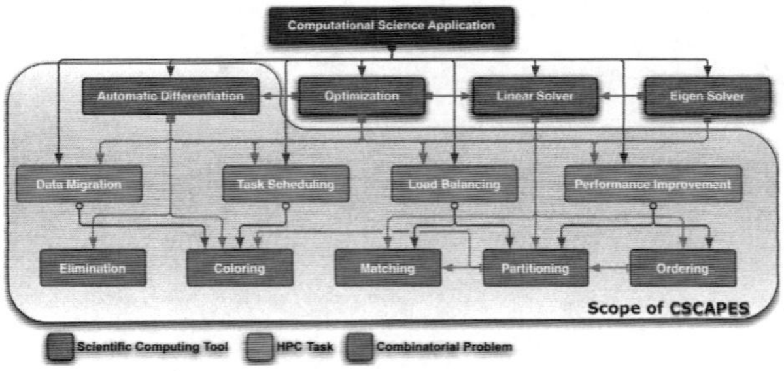

Figure 1. Key research areas in CSCAPES and their relationship to a typical SciDAC application. An arrow from A to B indicates that A in some sense uses B.

The focus in CSCAPES in each of these areas is on the formulation (modeling) and solution (algorithms) of the underlying fundamental combinatorial problems, which are often phrased by using graphs or hypergraphs. Load balancing, the task of distributing data and work in a large-scale computation among the processors of a machine to minimize total execution time, can be phrased as a partitioning problem on a graph or, with more accuracy, on a hypergraph. When computational dependency among subtasks is modeled by using a graph, a distance-1 coloring of the graph can be used to determine a scheduling of the tasks for concurrent computation. Distance-2 coloring and several other specialized variants are used to model matrix partitioning problems that arise in the efficient computation of sparse Jacobians and Hessians using AD. In AD, exploitation of associativity and commutativity in the chain rule of differential calculus leads to a variety of vertex or edge elimination problems in the directed acyclic graph used to represent the computation of a function and its derivative. Various types of matchings in graphs are used in the solution of sparse linear systems (numerical preprocessing and block triangular decomposition) and in the coarsening phase of multilevel methods for graph partitioning. Work and storage reduction in direct solvers for large sparse systems calls for vertex ordering problems in graphs. Graph and hypergraph based models are used to capture the needs in reordering the nodes and elements within an unstructured mesh to improve runtime performance.

CSCAPES researchers are developing sequential and parallel algorithms and software for the combinatorial problems outlined in the previous paragraph. The load-balancing and parallelization capabilities are being deployed through the Sandia-housed Zoltan software toolkit. Similarly, the AD capabilities are being deployed through the tools being developed at Argonne in the OpenAD framework. In addition, serial as well as parallel codes (for coloring, matching, and reordering) are being separately developed at the academic institutions in CSCAPES, with the ultimate goal of being integrated with the major software outlets, Zoltan and OpenAD. Both Zoltan and the tools in OpenAD predate CSCAPES and have been under continuous development for several years with sustained support from DOE; similarly, the coloring work in CSCAPES was previously supported by the National Science Foundation. Zoltan's internal interface is currently being redesigned to improve support for integration of third-party libraries. For example, it has recently acquired an interface to the parallel ordering functionality available in PT-Scotch.

The load-balancing and partitioning facilities in Zoltan have been used to enable a wide variety of applications in structural mechanics, chemical engineering, groundwater flow, biological systems, electronic circuit simulations, and molecular dynamics. Likewise, the tools in OpenAD have been applied to a broad range of applications: modeling breast cancer, climate, weather, semiconductor devices, power networks, and groundwater; atmospheric chemistry; computational fluid dynamics; the network-enabled optimization server (NEOS); water reservoir simulation; and chemical kinetics.

In the remainder of this short paper we provide a few nuggets of our recent works, to give a flavor of the ongoing research and development activities within CSCAPES. For more information, consult the references listed in this paper and the website http://www.cscapes.org.

2. Dynamic hypergraph repartitioning and 2D matrix partitioning

In parallel adaptive computations, processor work loads may become unbalanced over time because of changes in the computations. In adaptive mesh refinement, for example, the refinement and coalescing of elements may cause significant load imbalance. This kind of application requires *dynamic* load balancing. Dynamic load balancing has several goals with complicated trade-offs: load balance in the new data distribution; low communication cost within the application, low cost to migrate data from the old distribution to a new one, and short repartitioning time.

CSCAPES researchers have recently developed a hypergraph-based repartitioning method [1]. In this method, new vertices are added to represent each part. Then, each of these vertices is connected to the objects in the same part with new (weighted) edges to model the data migration cost. The resulting hypergraph accurately accounts for both the communication volume and the migration cost to move the data and can be partitioned using existing algorithms and software.

In figure 2, results from applying this hypergraph repartitioning method (HG-Repart) to an

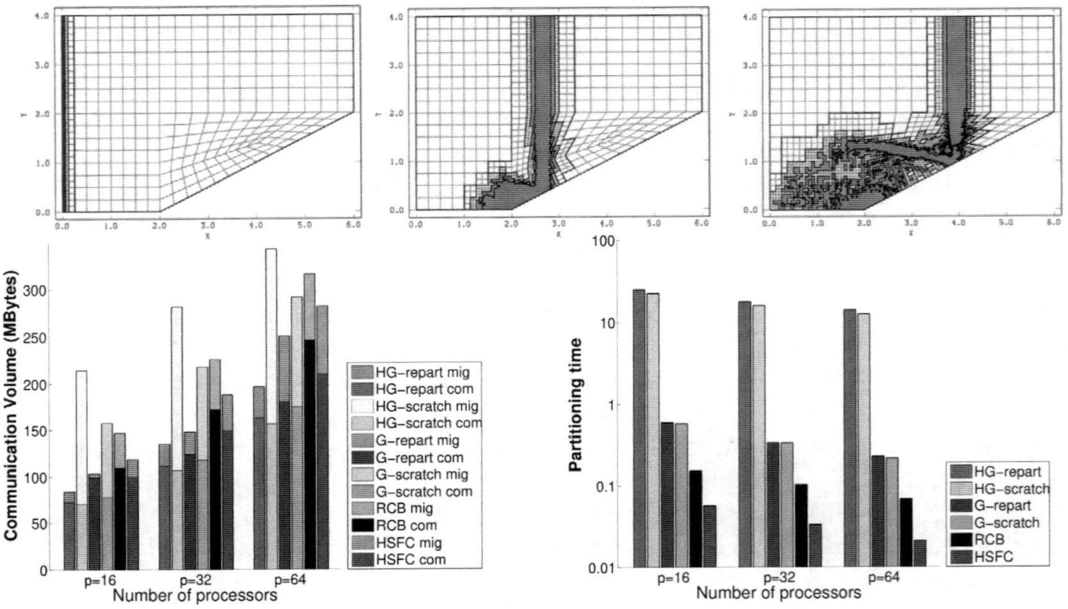

Figure 2. (Top) Sample adaptive finite element meshes at time-steps 0, 54, and 108; the smallest mesh has 132,209 nodes, and the largest has 1,380,266 nodes. (Bottom) A comparsion of communication volume and execution time over 109 time-steps of an adaptive finite element simulation on 16, 32, and 64 processors of a Linux cluster (dual 2.2 GHz AMD Opterons with 4 GB RAM and a Myrinet network).

adaptive mesh refinement problem from the ALEGRA shock physics code [2] is shown. HG-Repart is compared with other connectivity-based approaches—"scratch-remap" hypergraph partitioning (HG-Scratch) and graph-based partitioning (G-Scratch) and repartitioning (G-Repart)—and the coordinate-based methods recursive coordinate bisection (RCB) and Hilbert space-filling curve partitioning (HSFC). Scratch-remap methods compute partitions without accounting for existing part assignments and then remap parts to try to minimize migration costs. While the coordinate-based methods do not model communication and migration costs explicitly, they are implicitly incremental, making them viable for repartitioning mesh-based problems. The repartitioning hypergraph method HG-repart produced lower total cost (application communication volume plus data migration communication volume) than all other methods. The coordinate-based methods were much faster than graph- and hypergraph-based methods, but their total cost was high. Execution time for HG-repart was greater than G-repart, indicating the need for faster heuristics in the hypergraph implementation for applications with relatively low and homogeneous connectivity.

The multiplication of a sparse matrix by a vector is an important kernel in scientific computing. CSCAPES researchers have long studied how to optimize the performance of this operation in parallel by reducing communication volume [3]. Recently, new partitioning algorithms have been developed for novel two-dimensional data distributions that can be computed efficiently using current partitioning software [4, 5].

3. Parallel coloring

CSCAPES researchers have recently developed a framework for parallelizing greedy distance-1 coloring algorithms on distributed-memory computers [6]. Formulated in a generic manner, the techniques employed in the framework include: careful exploitation of features of the initial data distribution; speculation—maximizing concurrency by tentatively tolerating inconsistencies and

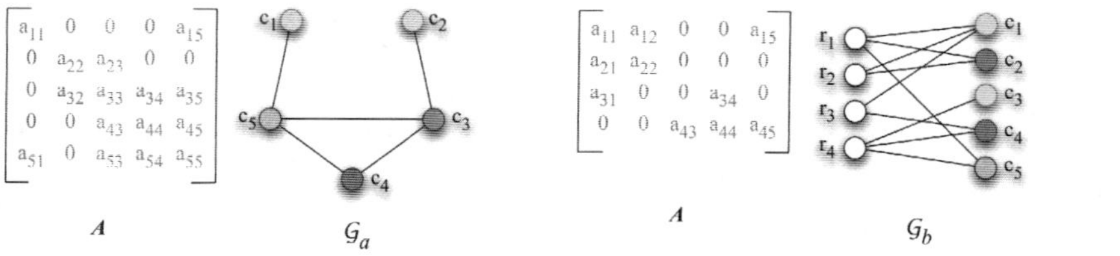

Figure 3. Structurally orthogonal column partition of a derivative matrix A and its representation as a distance-2 coloring in the adjacency graph $G_a(A)$ when A is symmetric, and bipartite graph $G_b(A)$ when A is nonsymmetric.

graph name	$\|V\|$	$\|E\|$	Degree max	colors (seq)
pkustk13	94,893	3,260,967	299	303
shipsec5	179,860	4,966,618	125	140
pwtk	217,918	5,708,253	179	180
inline_1	503,712	18,156,315	842	843
ldoor	952,203	22,785,136	76	112

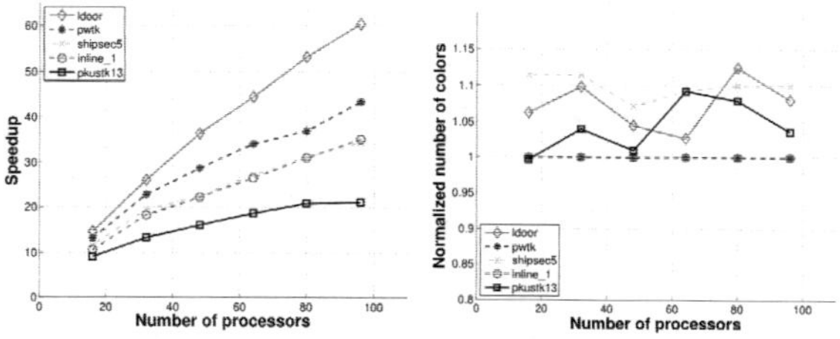

Figure 4. Performance of the parallel distance-2 coloring algorithm on select application graphs. The maximum degree in a graph is a lower bound on the optimal number of colors needed for distance-2 coloring. A greedy sequential algorithm gave near optimal (or perhaps optimal) solution for these test graphs. The parallel algorithm used in the worst case only 12% more colors.

then detecting and resolving them; randomization; and infrequent, coarse-grain communication among processors as opposed to frequent, fine-grained communication. In [6], several specialized algorithms designed using the framework have been presented and evaluated; experiments carried out on moderate size machines demonstrated good scalability.

In a work that is currently being finalized [7], the distance-1 coloring framework has been extended for the distance-2 coloring problem, an archetypal model for partitioning problems that arise in the efficient computation of *sparse* Jacobian and Hessian matrices using AD (see figure 3). To address the challange of accessing color information of distance-2 neighbors, efficient means of information exchange between processors hosting a pair of vertices that are two edges away from each other are introduced. Experiments run on up to a hundred processors have shown that the resulting parallel distance-2 coloring algorithm scales well and gives a solution very close to the sequential variant, which in turn often gives a near-optimal solution (see Figure 4). MPI implementations of the distance-2 coloring algorithm as well as the distance-1 coloring algorithm discussed earlier have been incorporated in Zoltan.

OInHessian computation, the distance-2 coloring model does not exploit the available symmetry. Models that exploit symmetry to various degrees exist. The most accurate model for a direct method is *star coloring* and that for a substitution method is *acyclic coloring*. CSCAPES investigators have in a recent work exploited the structures of two-colored induced subgraphs in star and acyclic coloring to design novel sequential algorithms that have been shown to be superior to previously known approaches [8]. The serial software they have developed has been incorporated into the operator-overloading-based AD tool ADOL-C. Experiments using ADOL-C on Jacobian computation in a simulated moving bed process, a method used in chromatographic separation in chemical engineering, and Hessian computation on an electric power grid problem have shown that the use of the coloring techniques in large-scale problems reduces overall run time by several orders of magnitude [9, 10].

4. Automatic differentiation

Suppose the derivatives of $y = \prod_{i=1}^{n} x_i$ with respect to all x_i need to be computed. An algorithm to compute this function is $y_1 = x_1; y_i = x_i * y_{i-1}, i = 2, \ldots, n$. One can then compute the derivatives using $\overline{y}_n = 1; \overline{y}_i = x_{i+1} * \overline{y}_{i+1}, \overline{x}_{i+1} = y_i * \overline{y}_{i+1}, i = n-1, \ldots, 1$. Then, $\partial y/\partial x_i = \overline{x}_i$. This algorithm to compute the derivatives corresponds to the reverse mode of automatic differentiation. The reverse mode is optimal in this case and within a factor two of optimal for all scalar functions. Now, consider the case of $y = w^T \prod_{i=1}^{n} M_i v$, where w and v are vectors and the M_i are dense, rectangular matrices (y remains a scalar). One might compute y using $b_0 = w^T; b_i = b_{i-1}M_i, i = 1, \ldots, n; y = b_n v$. Then, the derivatives with respect to the elements of the M_i can be computed optimally (or nearly so) using $\overline{b}_n = v; \overline{b}_{i-1} = M_i \overline{b}_i, \overline{M}_i = \overline{b}_i b_{i-1}, = n, \ldots, 1$. However, if one computes the more general sequence $a_1 = w^T M_1 v_1$, $a_2 = w^T M_1 M_2 v_2$, $\ldots$, $a_n = w^T M_1 M_2 \ldots M_n v_n$, where $y = \sum_{i=1}^{n} a_i$, and then computes the derivatives of each a_0 using the algorithm above (summing to produce the derivatives of y), the result is far from optimal. The problem is that although the chained vector-matrix products from the left can be reused from one a_i to the next, the chained matrix-vector products from the right cannot be reused because the v_i vary. To illustrate the problem, suppose all of the M_i are square matrices of size $n \times n$. Then, the cost of computing the function is $O(n^3)$ while the cost of computing the derivative matrices is $O(n^4)$. If, however, one recasts the computation as $b_0 = w^T; b_i = b_{i-1}M_i, a_i = b_i v_i, i = 1, \ldots, n$; then, one can simply recur backward on the b_k sequence, with an extra term thrown in to account for the inner product with x_k. This apprroach brings the cost back down to $O(n^3)$. Thus, in developing methods to compute the derivatives of a function, it is important to consider the full algorithm used to compute the function, not just subcomponents. We identified a way to use a similar technique to reduce the cost of computing the gradient of a function arising in computational chemistry from $O(n^5)$ to $O(n^4)$ [11]. This reduction will have a major impact on the size of problems that can be solved.

Acknowledgments

Sandia is a multiprogram laboratory operated by Sandia Corporation, a Lockheed Martin Company, for the U.S. Department of Energy's National Nuclear Security Administration under contract DE-AC04-94AL85000. The work at Argonne was supported by the Mathematical, Information, and Computational Sciences Division subprogram of the Office of Advanced Scientific Computing Research, U.S. Department of Energy, under Contract DE-AC02-06CH11357. The CSCAPES Institute is supported by the U.S. Department of Energy's Office of Science through grant DE-FC-0206-ER-25774, as part of its SciDAC program.

References

[1] Catalyurek U V, Boman E G, Devine K D, Bozdag D, Heaphy R T and Riesen L A 2007 Hypergraph-based dynamic load balancing for adaptive scientific computations *Proc. 21th Intl. Parallel and Distributed Processing Symposium (IPDPS'07)*

[2] Boucheron E A, Brown K H, Budge K G, Burns S P, Carroll D E, Carroll S K, Christon M A, Drake R R, Garasi, C G, Haill T A, *et al.* 2002 ALEGRA: User Input and Physics Descriptions Version 4.2 *Sandia National Laboratories Tech. Report SAND2002-2775*

[3] Çatalyürek U V and Aykanat C 999 Hypergraph-partitioning based decomposition for parallel sparse-matrix vector multiplication *IEEE Trans. Parallel and Dist. Systems* **10(7)** 673–693

[4] Boman E G 2008 A nested dissection approach to sparse matrix partitioning *Proc. Appl Math. Mech.* to appear

[5] Wolf M M, Boman E G and Hendrickson B 2008 Optimizing sparse matrix-vector multiplication by corner partitioning *Proc. PARA08* to appear

Journal of Physics: Conference Series **125** (2008) 012071 doi:10.1088/1742-6596/125/1/012071

[6] Bozdag D, Gebremedhin A H, Manne F, Boman E G and Catalyurek U 2008 A framework for scalable greedy coloring on distributed memory parallel computers *J. Parallel and Distributed Computing* **68(4)** 515–535
[7] Bozdag D, Catalyurek U, Gebremedhin A H, Manne F, Boman E G and Ozguner F 2008 Distributed-memory parallel graph coloring algorithms for Jacobian and Hessian computation (*Preprint*)
[8] Gebremedhin A H, Tarafdar A, Manne F and Pothen A 2007 New acyclic and star coloring algorithms with application to computing Hessians *SIAM J. Sci. Comp.* **29(3)** 1042–1072
[9] Gebremedhin A H, Pothen A, Tarafdar A and Walther A 2008 Efficient computation of sparse Hessians using coloring and automatic differentiation *INFORMS J. Computing* to appear
[10] Gebremedhin A H, Pothen A and Walther A 2008 Exploiting sparsity in Jacobian computation via coloring and automatic differentiation: a case study in a Simulated Moving Bed process *The Fifth International Conference on Automatic Differentiation* to appear
[11] Shepherd R 2008 (Personal communication)

SciDAC 2008 IOP Publishing
Journal of Physics: Conference Series **125** (2008) 012072 doi:10.1088/1742-6596/125/1/012072

Data management and analysis for the Earth System Grid

The Earth System Grid Center for Enabling Technologies Team:

D N Williams[3], R Ananthakrishnan[1], D E Bernholdt[7,9], S Bharathi[8], D Brown[5], M Chen[7], A L Chervenak[8], L Cinquini[5], R Drach[3], I T Foster[1,9], P Fox[5] , S Hankin[6], V E Henson[3], P Jones[4], D E Middleton[5,9], J Schwidder[7], R Schweitzer[6], R. Schuler[8], A Shoshani[2], F Siebenlist[1], A Sim[2], W G Strand[5], N Wilhelmi[5] and M Su[8]

[1] Argonne National Laboratory, Argonne, IL 60439, USA
[2] Lawrence Berkeley National Laboratory, Berkeley, CA 94720, USA
[3] Lawrence Livermore National Laboratory, Livermore, CA 94550, USA
[4] Los Alamos National Laboratory, Los Alamos, NM 87545, USA
[5] National Center for Atmospheric Research, Boulder, CO 80307, USA
[6] National Oceanic and Atmospheric Administration (PMEL), Seattle, WA 98115, USA
[7] Oak Ridge National Laboratory, Oak Ridge, TN 37831, USA
[8] University of Southern California, Information Sciences Institute, Marina del Ray, CA 90292, USA

E-mail: williams13@llnl.gov, don@ucar.edu, itf@mcs.anl.gov, bernholdtde@ornl.gov

Abstract. The international climate community is expected to generate hundreds of petabytes of simulation data within the next five to seven years. This data must be accessed and analyzed by thousands of analysts worldwide in order to provide accurate and timely estimates of the likely impact of climate change on physical, biological, and human systems. Climate change is thus not only a scientific challenge of the first order but also a major technological challenge. In order to address this technological challenge, the Earth System Grid Center for Enabling Technologies (ESG-CET) has been established within the U.S. Department of Energy's Scientific Discovery through Advanced Computing (SciDAC)-2 program, with support from the offices of Advanced Scientific Computing Research and Biological and Environmental Research. ESG-CET's mission is to provide climate researchers worldwide with access to the data, information, models, analysis tools, and computational capabilities required to make sense of enormous climate simulation datasets. Its specific goals are to (1) make data more useful to climate researchers by developing Grid technology that enhances data usability; (2) meet specific distributed database, data access, and data movement needs of national and international climate projects; (3) provide a universal and secure web-based data access portal for broad multi-model data collections; and (4) provide a wide-range of Grid-enabled climate data analysis tools and diagnostic methods to international climate centers and U.S. government agencies. Building on the successes of the previous Earth System Grid (ESG) project, which has enabled thousands of researchers to access tens of terabytes of data from a small number of ESG sites, ESG-CET is working to integrate a far larger number of distributed data providers, high-bandwidth wide-area networks, and remote computers in a highly collaborative problem-solving environment.

1. Context: climate change and Earth system modeling

For several decades, scientists have been working to understand the Earth's delicate ecosystem and intricate climate balance. Notably, they have sought an understanding of potential changes in these systems induced by naturally and human-generated greenhouse gases (see figure 1). Naturally occurring greenhouse gases—such as carbon dioxide—help absorb the sun's rays, keeping the planet warm enough to support life. Over the past century, however, humans have released vast quantities of greenhouse gases into the atmosphere and have simultaneously disrupted Earth's natural carbon sequestration process by cutting down forests. Although calculating the CO_2 sequestration of various types of trees is complicated, it is estimated that one mature tree consumes between 40 and 50 pounds of CO_2 per year and releases enough oxygen into the atmosphere to support two human beings [1].

The burning of fossil fuels and deforestation will undoubtedly continue well into the 21st century. Moreover, the effects of these activities are already being felt worldwide. The ten warmest years ever-recorded have all occurred since 1990, and rapid reductions in the extent of mountain glaciers have been observed on five continents. Atmospheric CO_2 has increased over the past century from 280 parts per million (ppm) to over 360 ppm— easily the largest concentration of CO_2 in the atmosphere since humans evolved. Oceans are also on the rise as ocean temperature increases, causing thermal expansion of the oceans as well as the melting of both land ice and polar icecaps. Rising sea levels threaten disastrous inundation of low-lying coastal countries, many of which are densely populated.

Together, the atmosphere, oceans, land, ice, and biosphere constitute a complex system that is heavily influenced by greenhouse gas emissions. The importance of understanding greenhouse gas emissions and their potentially devastating impact on this earth system cannot be overstated. Indeed, the quality of human life on earth, and perhaps even human survival itself, may depend on reaching that understanding.

To help understand these climatic imbalances and potential impacts, scientists rely on powerful computers and complicated computer codes to simulate the Earth's climate. Numerical climate models allow scientists to improve their understanding of the earth system by running exhaustive computational experiments aimed at studying the behaviors of individual chemical, biological, and geophysical processes, as well as relationships among these individual processes. Computational experiments performed with these models allow scientists to model and study all aspects of climate change to help understand future climate. Simulations can also be used to study the effectiveness of proposed approaches to mitigating the impact of climate change.

Climate models are also used to study regional climate changes as well as high-impact characteristics of climate such as typhoons, droughts, and floods. Accurately resolving these and other processes and discovering long-term patterns such as temperature distribution or the melting of land and sea ice, requires the application of intricate mathematical formulas to data at points on ever finer grid resolutions. Hence, as computing power increases, so does the opportunity to improve the computer models (see figure 2).

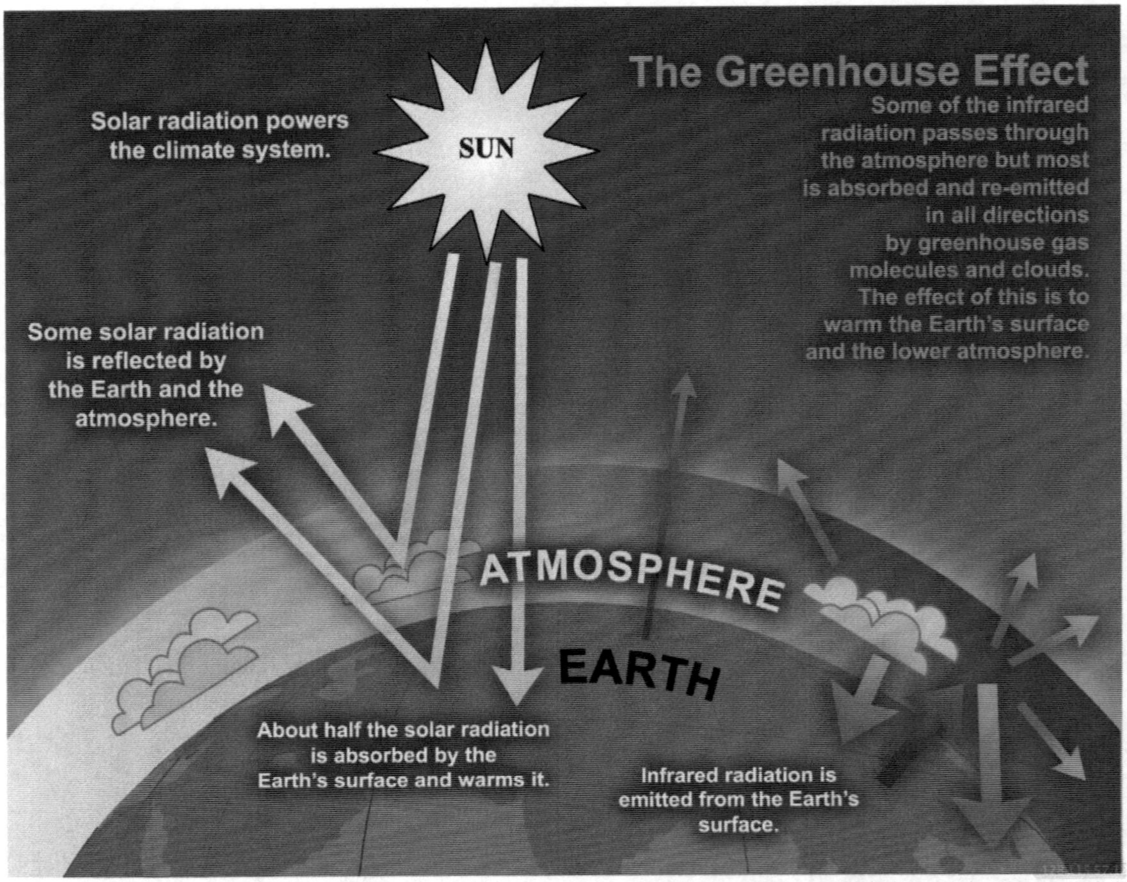

Figure 1. Depiction of the natural greenhouse effect. (Image courtesy of the Intergovernmental Panel on climate Change (IPCC) Fourth Assessment Report [2].)

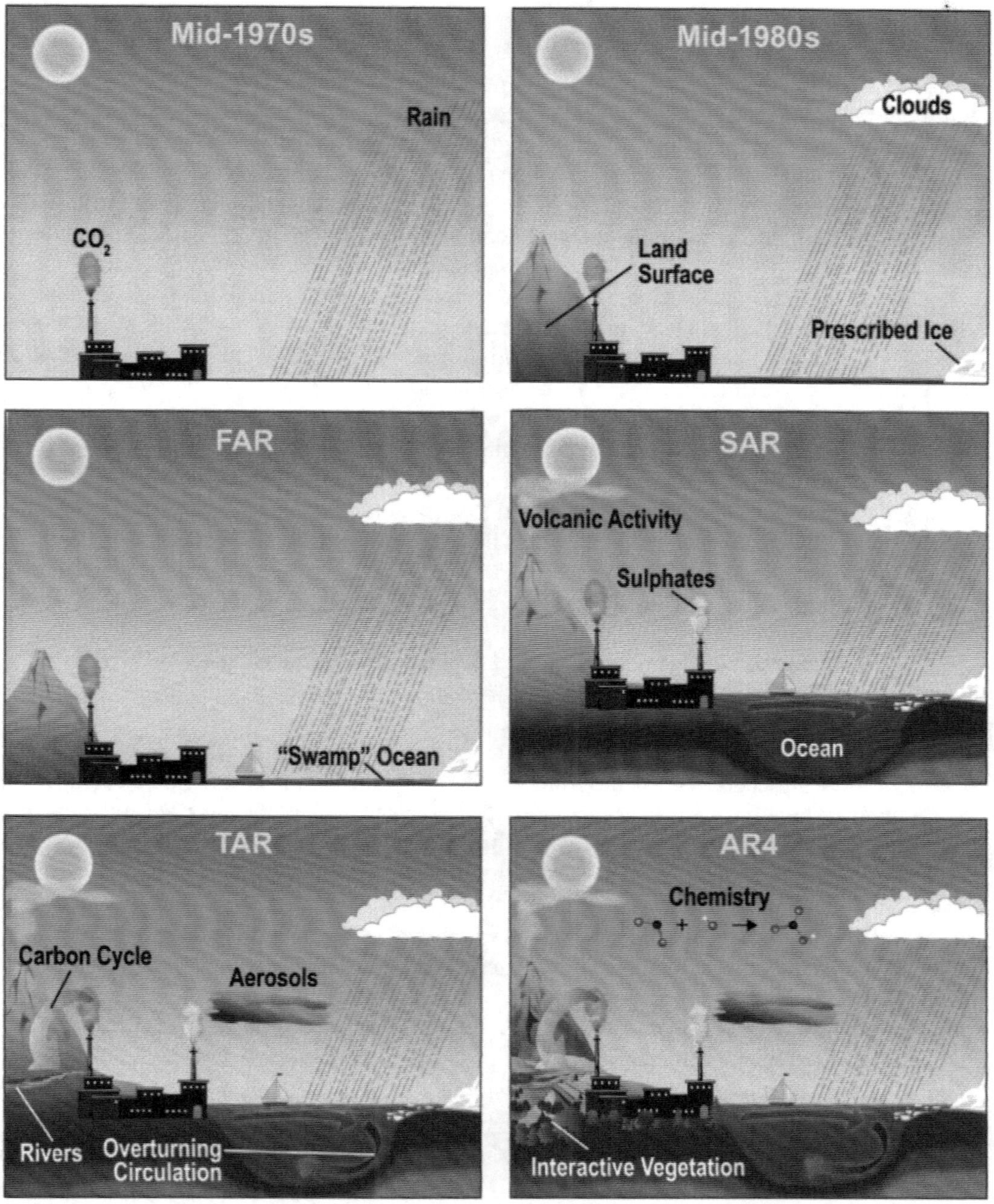

Figure 2. Progression of the complexity in the coupled models over the past few decades, where FAR represents the Intergovernmental Panel on Climate Change (IPCC) First Assessment Report, SAR represents the IPCC Second Assessment Report, TAR represents the IPCC Third Assessment Report, and AR4 represents the IPCC Fourth Assessment Report. (Image courtesy of the Intergovernmental Panel on climate Change (IPCC) Fourth Assessment Report [2].)

2. The climate data challenge and the Earth System Grid

A consequence of increased computing power and more comprehensive climate models is a dramatic increase in data output describing the more complex spatial and temporal Earth system model. Over the past decade, model output has increased exponentially—moving from megabytes, to gigabytes, to terabytes, and now petabytes. The worldwide climate community is expected to generate hundreds of petabytes of simulation data within the next five to seven years.

Using these data will not be easy. Indeed, merely examining or sorting data will be problematic, partly due to the enormous quantity of data involved, and partly due to the fact that the storage for all the data will be distributed worldwide. Far more than just climate model output data will be involved. For accurate study of the climate, in situ, satellite, biogeochemistry, and ecosystems data will also be required. Because of the high cost to generate the data (in both computer and human time), they must be carefully managed to maximize and preserve their value to the community.

Data providers want to make their data available to others but will do so only if they have confidence in distribution processes they understand and (to the extent necessary) control. Data users need to be able to find and access data of interest, and must have confidence in the origins and integrity of the data and the associated metadata. Because most users will have considerably less storage (and network connectivity) than the major centers where simulation data are generated and archived, data users will need assistance from the archive sites in performing common data reduction, analysis, and visualizations, all of which must be done on the "server side" to reduce the volume of data sent over the network and stored at the user's location.

The Earth System Grid (ESG) Center for Enabling Technologies (ESG-CET) [3, 4] was established to address these technical challenges. Specifically, it was charged with addressing the management and use of extremely large and diverse sets of climate data, aiming to build a "science gateway" to climate resources. Funded by the SciDAC program through the offices of Advanced Scientific Computing Research (ASCR) and Biological and Environmental Research (BER), ESG-CET intends to develop approaches and technologies that also benefit other scientific application areas that entail management of petascale data.

The remainder of this paper examines some of the issues associated with data management at this scale, and how the ESG-CET team is addressing those issues.

3. The ESG-CET

The ESG-CET project did not begin from scratch: it is built on the considerable existing infrastructure of the first-generation Earth System Grid [5], developed over the past several years, and on the associated suite of open-source software that ESG provides to access and manage climate data, information, models, analysis tools, and visualization tools. ESG-CET will also benefit from the considerable experience and user-feedback obtained through use of the first-generation system. ESG users now number nearly 10,000, and ESG data holdings total more than 235 terabytes.

The climate modeling community have become enthusiastic supporters and users of ESG, which has led in turn to requests and expectations that it will play an expanded role in data dissemination in the future. Hence, the ESG-CET project is the "next phase" in ESG development. The ESG-CET team is now designing new products and services that will enable users to find and obtain data from spinning disk or tertiary storage, migrate data from one location to another, and enable a number of workflow capabilities across a worldwide network of climate data archives. Further, this will occur at a scale several orders of magnitude larger than the community has seen to date.

Working closely with key customers and major collaborators, the ESG-CET team has established requirements for the next phase of the project. These customers and collaborators include the following:

- The Intergovernmental Panel on Climate Change (IPCC) Fifth Assessment Report (AR5)
- The Computational Climate End Station (CCES) at the DOE Leadership Computing Facility at Oak Ridge National Laboratory (ORNL)
- Groups addressing the data management and dissemination needs of the Community Climate System Model (CCSM) and related model development efforts
- A variety of regional and other related efforts within the climate community requiring similar data management and dissemination capabilities

This broad set of target users represents a significant change from the earlier project, for which CCSM development was the primary focus, and supporting the IPCC AR4 activity was added mid-project.

3.1. Distributed partners and virtual organizations

The requirements that have had the most significant impact on the architecture of the next-generation ESG environment have to do with the breadth and number of sites in the distributed system.

In the current production ESG environment, two primary portals service data archives at five sites. The two portals are located at the National Center for Atmospheric Research (NCAR) and the Program for Climate Model Diagnosis and Intercomparison (PCMDI) at Lawrence Livermore National Laboratory. Users connecting to these portals can access data archives at Los Alamos National Laboratory (LANL), Lawrence Berkeley National Laboratory (LBNL), LLNL, NCAR, and ORNL. Historically, NCAR and ORNL have served as the primary computational sites for development and science runs of CCSM and its precursors and collectively hold more than 198 TB of simulation data. Over 35 TB of data from the IPCC AR4 was collected from modeling centers in the 13 participating countries and consolidated at the PCMDI (at LLNL) for publication through ESG.

Planning for AR5, which is the major driver of requirements for ESG-CET, is still very much in progress, but the expectation is that the core simulation data will exceed 2 PB, with perhaps an order of magnitude more supplementary data. Rather than trying to consolidate all of this data at a single site, AR5 planners expect data to be hosted at the participating modeling centers, or perhaps consolidated within the participating nations. Assuming that AR5 participation will be similar to AR4, ESG-CET will need to support between one and two-dozen data archives in the next-generation ESG. Discussions have been held with several other groups interested in publishing specialized data collections through ESG; this has led to a plan for 20–30 data nodes, with the possibility of more. These sites will be located throughout the world, and only a handful will be at institutions where the ESG-CET team has a presence.

Each of these sites and special collections may be a significant research and data management operation in its own right, much more than a mere partner in ESG. Therefore, extending the ESG "virtual organization" (VO) will often be a matter of *federating* the ESG VO with other, established VOs. Indeed, these other groups may already have in place their own data services, user bases, and other organizational aspects. Though federation of VOs and data systems is an area of active research, a goal of ESG-CET is that ESG should be able to use identities and data archives managed by other VOs to the maximum extent possible, rather than requiring that users maintain multiple accounts, replicate data archives, etc. At the same time, federating external VOs must provide the "owners" of the data published in ESG with adequate information about (and control over) who has access to their data.

3.2. Data management

Recent trends in climate simulation are to use higher resolution grids (see figure 3), to include additional components in the simulation (for example, incorporating biogeochemical effects), and to rely more on ensembles of simulations rather than individual simulation runs. All these trends will increase both the number and sizes of the files comprising individual datasets and will increase the number of datasets to be published in ESG.

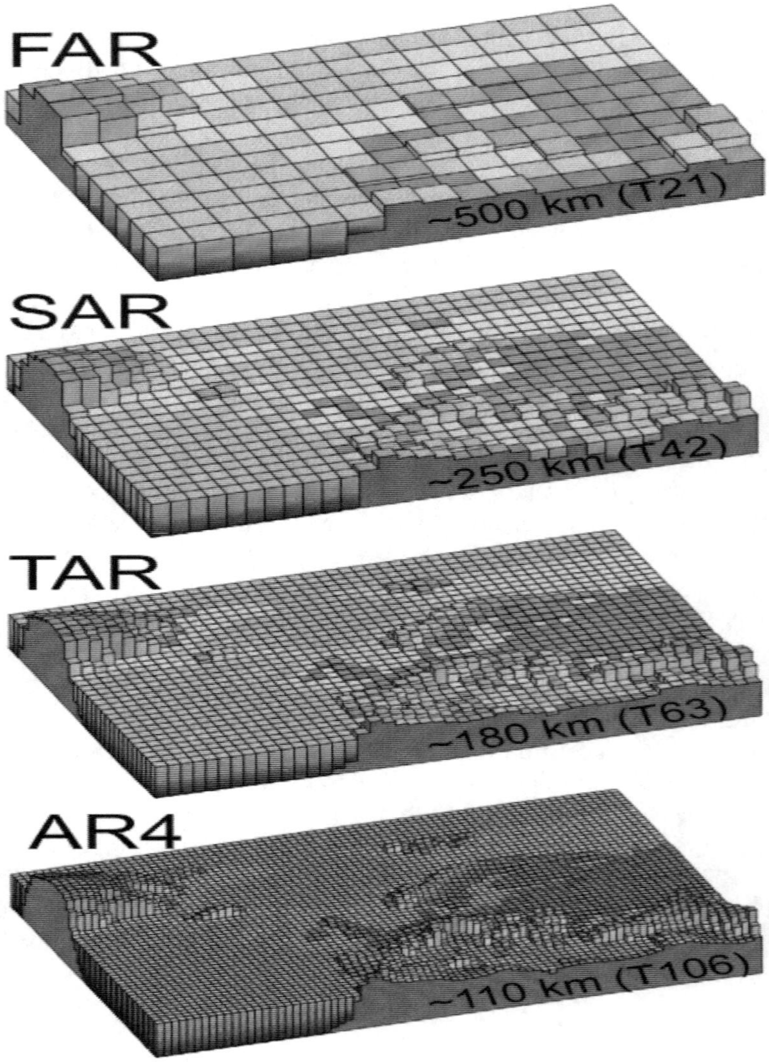

Figure 3. Progression of climate model grid resolution used in the IPCC Assessment Reports: FAR (1990), SAR (1996), TAR (2001), and AR4 (2007). (Image courtesy of the Intergovernmental Panel on Climate Change Fourth Assessment Report [2].)

The current ESG holdings, which include the entire IPCC AR4 archive and the past six years of joint DOE/NSF climate modeling simulation data, total roughly 235 TB. A reasonable estimate anticipates that the AR5 modeling campaign alone will involve several petabytes of data. In addition, holders of other collections of a similar magnitude have expressed interest in joining ESG. Thus, ESG-CET planning for the next generation system includes the capability of handling one to two orders of magnitude more data than the current system, easily reaching into the tens of petabytes. Planning for this scale has implications throughout ESG, from search and discovery through processing, analysis, and delivery of the results, and provides an important motivation for ESG-CET's emphasis on server-side analysis and subsetting capabilities.

The technology that will be used to store all of this data is also an important factor in the design of the next-generation ESG. While some sites will be able to field dedicated disk farms or use large-scale filesystems, (such as are being deployed at many HPC centers) many sites can accommodate these quantities of data only on hierarchical mass storage systems. The ESG design must allow for the fact that such data may take time to retrieve from the mass storage system before it can be processed and delivered to the user. Also, the organization of the data may make some user requests extremely expensive to carry out when that data comes from deep storage. One such example might be the extraction of the complete time history of a single variable from a dataset that is organized as one time step per file. These factors will have affects across the ESG system; planning to deal with them has led to discussions with model development teams about the organization of the output produced directly from the simulations (which also has a significant impact on I/O performance during the simulations).

3.3. New user communities

As ESG has become established in the climate modeling community as a widely used resource, the project has sought to broaden the range of users served. ESG-CET is focusing particularly on the needs of two types of users.

At one end of the spectrum are relatively small numbers of "power" users. These are researchers working with cutting-edge models that might produce huge volumes of data, use specialized grids, or have other novel characteristics. Such users may require especially detailed metadata for their models. They also may be particularly interested in high-end analyses and visualizations of their data, using sophisticated "thick" clients on their desktop computers, or perhaps more powerful computer resources for those tasks.

At the other end of the spectrum are the analysts who examine the impacts of climate change on the local and regional environments, populations, and infrastructures. This so-called impacts community is estimated to be an order of magnitude (or more) larger than the developers and immediate users of global climate models. Users in the impacts community tend employ digested or simplified information derived from one or many global climate simulations. They generally do not have a deep understanding of the science behind the global climate data, which primarily serves as inputs or boundary conditions for their models. They need the ability to discover and access model data in ways specific to their applications, including mapping to relevant coordinate systems, delivering data compatible with widely used geographic information systems, and creation of an appropriate and carefully documented suite of analyses and visualizations.

Both of these communities, as well as users who fall somewhere between the power user and the local analyst, have historically been part of ESG's primary user base and have strong needs for powerful data management and analysis capabilities, which are key areas of development in the ESG-CET project.

4. ESG-CET architecture overview

Given the ambitious nature of ESG-CET's future work, it is evident that the design and architecture of the ESG must be modified to enable the manipulation, management, and access of petabyte-scale datasets. A useful step is to review the current ESG architecture and then introduce the principal features of the new ESG-CET architecture.

4.1. Current ESG architecture

The current ESG architecture (see figure 4) went into production in late 2004. It consists of four major components designed to allow the climate community to share model data sets and enhances the scientist's ability to discover and access data sets of interest. The design also affords straightforward access to the tools required to analyze and interpret the data. The four major components of the architecture are as follows:

1. *Client applications*. The most important client application is the ESG portal, which provides a convenient, browser-based interface to ESG services. Other applications include user-level tools for data publication and assorted tools for data analysis and visualization.
2. *High-level and ESG-specific services*. These services span ESG resources and provide capability such as site-to-site data movement, distributed, reliable metadata access, and data aggregation and filtering.

3. *Globus/Grid infrastructure*. This provides remote, authenticated access to the shared ESG resources, enabling access to credentials, data and metadata, replica location, and submission and management of computational jobs.
4. *Database and application servers*. These servers include the basic resources on which ESG is constructed and includes the computational resources used for creation of virtual data, disk caches and mass storage systems used to hold data assets, database servers used to store metadata, and application servers for hosting ESG portal services.

The current ESG production data services span multiple domains including metadata, security, data transport, data aggregation and subsetting, usage metrics, and services monitoring. The cohesive integrated architecture is built in Java upon the Tomcat servlet engine and Struts web framework technologies. The production web portals allow data users and data providers to access the full spectrum of ESG high-level data services for publication, search and discovery, description, download, aggregation, and subsetting.

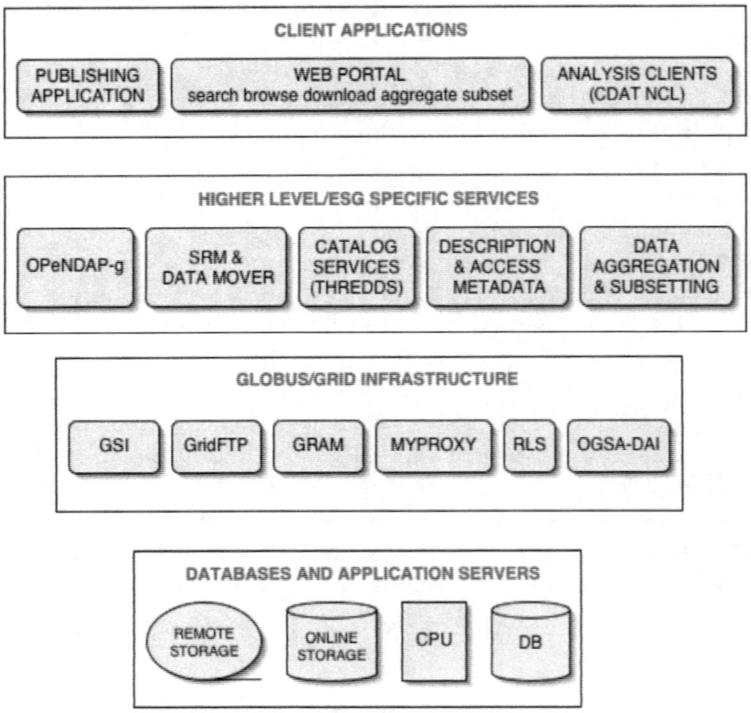

Figure 4. The four major components in the current ESG architecture.

4.2. New ESG-CET architecture

The current ESG architecture was designed to support a fixed (and small) set of well-known U.S. archive sites operating ESG nodes with essentially the same capabilities as a federated system, fronted by a single web-based portal as the primary interface to the user. In order to meet coming needs, this architecture must be greatly generalized. The next generation of the ESG must provide for a larger number of distributed sites located throughout the world. Further, the sites will have widely differing capabilities, and ESG must allow them to selectively federate, cooperate, or operate in standalone fashion, accommodating a variety of means for user access, including multiple portals and service- or API-based access, and data delivery mechanisms. The architecture must also be robust in the face of system and network failures at the participating sites.

In order to meet these expanded scientific and technical requirements, the new software framework (see figure 5) is designed to support a rich combination of interacting system components, while at the same time remaining fairly simple, lightweight, highly customizable, and easily deployable.

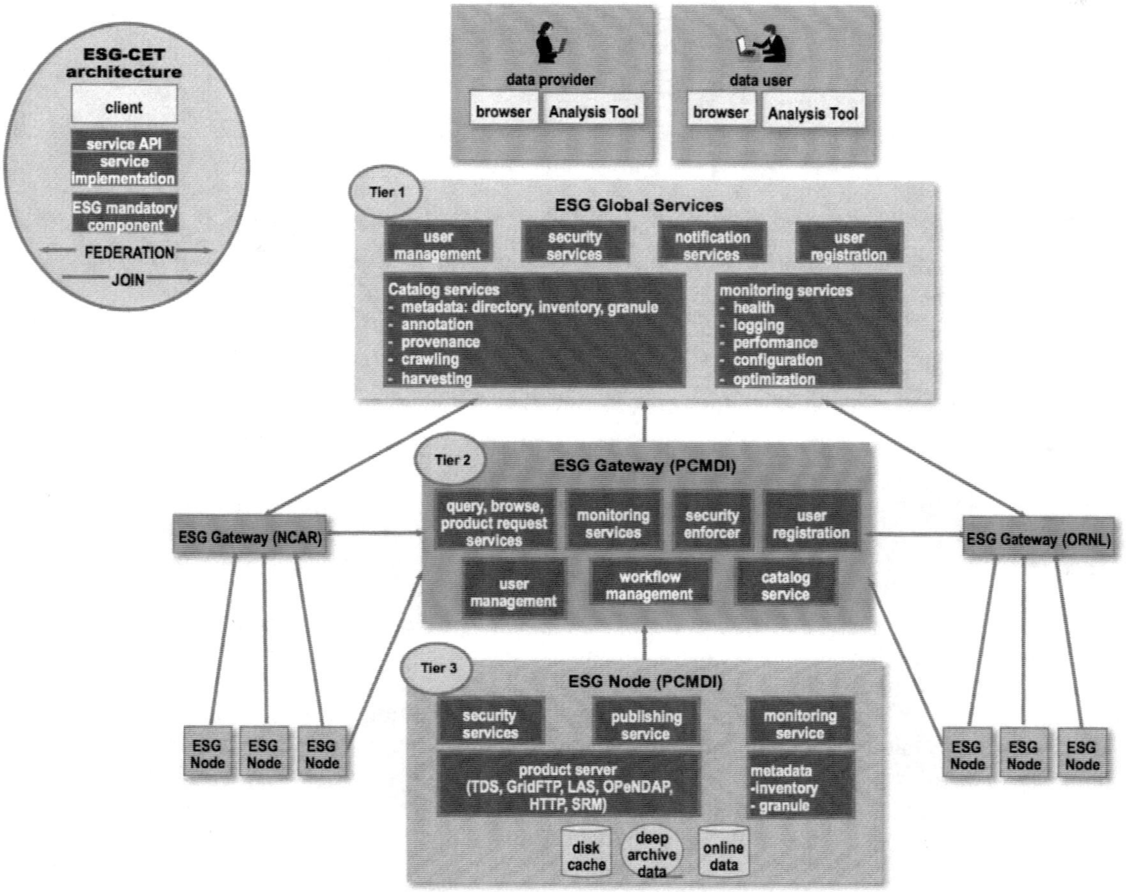

Figure 5. The new ESG-CET architecture, showing the three-tier data services.

The new ESG-CET framework will support both individual "ESG nodes" and collaborating, thematic ESG Gateways. ESG nodes are configured to offer a core set of ESG functionality for local data, metadata and computational resources, ESG Gateways federate multiple ESG nodes and provide access to distributed resources. In order to prepare for the ever-larger climate archives—expected to be in the petabytes—the new infrastructure is based on a three tier data service model:

1. *Tier 1 (Global metadata services for search and discover):* comprises a set of services providing shared functionality across the world-wide ESG-CET federation. An overall single sign-on authentication and authorization scheme will allow a registered user to access resources across the whole system, and to find data throughout the federation, independent of the site at which a search is launched. Tier 1 services will be managed by the ESG-CET staff.
2. *Tier 2 (Data gateways as data-request brokers)* comprises a limited number of ESG Data Gateways which act as brokers handling data requests to serve specific user communities. Services deployed on a Gateway include the user interface for searching and browsing metadata, for requesting data (including analysis and visualization) products, and for orchestrating complex workflows. At least initially, gateways will be managed directly by ESG-CET engineering staff.
3. *Tier 3 (ESG nodes with actual data holdings and metadata accessing services)* includes the actual data holdings and reside on a (potentially large) number of federated ESG Nodes, which host those data and metadata services needed to publish data onto ESG and execute data-product requests through an ESG Gateway. A single ESG Gateway serves data requests to many associated ESG nodes: for example, more than 20 institutions are expected to set

up ESG nodes hosting data as part of the IPCC Fifth Assessment. Personnel at local institutions will operate ESG nodes.

4.3. ESG-CET rollout plans

To reach the project goals, a distributed testbed based on the new architecture is scheduled to be in place by late 2008 to early 2009. The minimum testbed requirements include secure and reliable access to federated metadata, federated portals, a minimal set of server-side analysis functions, the ability to distribute data aggregation, publish data, and deliver data. Five U.S. institutions will participate as part of the initial ESG testbed: LLNL (ESG gateway, ESG node); NCAR (ESG gateway, ESG node); ORNL (ESG gateway, ESG node); LANL (ESG node); and LBNL (ESG node). In preparation for a truly global network of climate data services, the testbed will be expanded rapidly to include ESG nodes at several international sites operated by ESG collaborators.

5. Meeting user needs for data discovery and access

In the early days of global climate modeling, data management was *ad hoc* and required a great deal of manual intervention. Datasets were stored wherever the group that generated them could find space, in formats specific to the group or modeling code that generated them. Outsiders who wished to access this data needed to obtain special access to the storage resources and needed to acquire knowledge of the data location, metadata information, and appropriate data translators. Often users became discouraged by either not finding data or finding data that was unintelligible to them. In many instances, data providers would not provide data to outsiders, fearing that the requesting user would not know how to interpret their data correctly. Consequently, a majority of the data sets never received proper scientific scrutiny.

If data was acquired from the data providers, accessing and using it were often painful. Users generally had to develop special code to put the data in a useable format; often the formats were only good for that specific user's needs. Hence many scientists ended up writing their own programs to ingest, manipulate and display their data; and the software produced was generally not user-friendly, reusable, or portable. Often, metadata was kept in files separate from the data, which led to data loss and/or replicated data. Just locating metadata and associated data could be a tedious manual process. With no formal process and far too much done by hand or documented only in the heads of data providers, data management was virtually nonexistent. The communitywide inefficiency and redundancy in these efforts were a significant diversion of time away from the actual science.

As the community recognized the need to facilitate the exchange of data, they began to standardize on storage formats and metadata conventions. The Climate and Forecast (CF) [9] metadata conventions data standards have been well accepted in the community, as has the netCDF file format. The CF conventions define metadata that provide a definitive description of what each variable represents, and the spatial and temporal properties of the data. It enables extraction and comparison of different data sources from different facilities, and facilitates quality assurance.

While these developments helped with the interpretation of data, they did not address the problems of access to data. This was the primary motivation for the initial development of ESG. Taking advantage of the metadata conventions, the first generation of ESG was able to provide search, discovery, and access to a wide range of datasets across key U.S. climate data archives without requiring users to have individual accounts at all of the participating sites.

Arguably, the biggest problem facing climate scientists in the future is not the absence of data but rather its abundance. Current ESG-CET development is making improvements to every aspect of the data management process for both data providers and users. Rather than relying on separate prepublication quality control procedures (which were the responsibility of the providers), the new data publishing system provides built-in quality control on aspects of the metadata, which have been found essential to its effective use within the community. Through appropriate hooks additional project-specific automated quality assurance procedures can be incorporated. ESG-CET and several major data providers (i.e., the CCSM modelling team and the IPCC AR5) are collaborating closely to provide more useful metadata. For data users, the search and discovery capabilities are being expanded greatly. Users must be informed when published data they have downloaded has been removed or changed, and may also register interest in the publication of new datasets meeting certain criteria.

Other important aspects of data management are less visible to the user. In order to provide server-side processing capabilities that allow reduced download volumes, much more detailed metadata is required than for user-level search and discovery. These metadata, related to the structure and contents of individual files within the datasets (in some cases thousands or tens of thousands of files), must be harvested and catalogued by ESG. This is essential if ESG is to offer a reasonable user experience for server-side processing requests (regardless of where the data is actually stored). The increased emphasis on server-side processing also brings out the need to manage provenance information for any derived data.

6. Importance of server-side analysis

To address looming petascale data scenarios effectively, ESG must greatly enhance server-side capabilities for data analysis and visualization and provide for integration with existing community tools. By shifting processing away from the user workstation and into the ESG environment, it will not be necessary for every user to have petascale resources locally to process the data. Delivering the data needed by the user more precisely makes better use of limited wide-area network bandwidth. Both of these factors will contribute to a better overall experience for the user and facilitate more effective community access to large-scale climate data archives. Investigations are under way that will enable the ESG to cache or even preprocess and publish as separate datasets, some of the most popular requests.

6.1. Server-side analysis

The initial focus for server-side analysis is on integrating the Live Access Server (LAS) into the ESG environment. LAS provides a highly flexible web interface for a wide range of visualization, subsetting, analysis, and comparison operations. By default, it acts as a front-end for the Ferret analysis tool, though it can be configured for additional back-ends as well· Fourth-generation scripting languages are extremely popular in the climate community for analysis and visualization of model and observational data. The ESG-CET leadership includes parties responsible for three popular packages: the Climate Data Analysis Tools (CDAT) [10], the NCAR Command Language (NCL) [11] and its Python companion PyNGL, and Ferret [12], which enjoys a particular following in the ocean sciences community.

6.2. Enabling client-side interactions with ESG

The real power of the ESG system will become apparent when ESG is extended with APIs to support client-side applications, based on scripting languages and other tools. This will provide remote access to catalogs, data, and higher-level services for client applications. These applications will serve as drivers for ESG service interface design; the applications will be rewritten so that they can call on ESG services for metadata, catalog information, data, and analysis.

Like ESG itself, these applications have user communities in the thousands, so the combined potential for impact is believed to be quite large. However, such scenarios pose many challenges in terms of workflow definition and management, security, distributed services, and policy issues. The delivery of powerful access and analysis capabilities is a nontrivial task for several reasons:

- Many analysis procedures are written in powerful scripting language systems such as CDAT, NCL, and Ferret. These represent a considerable investment in software development that should be retained; however, also it is necessary to be able to apply the analysis procedures to large aggregated datasets that may be many terabytes in size. Thus it is essential to develop the ability to apply scripting language procedures to large aggregated datasets.

- These aggregated datasets may be located in archival storage, and indeed may be too large to move onto disk in their entirety at one time. Thus it is necessary to develop techniques and tools for managing the movement of data between archival storage and disk.
- As analysis services become more powerful, it is likely that the number of requests for these services will grow. The ESG must be capable of managing such workloads, and must coordinate the use of potentially many computing and storage resources.
- While it is relatively easy to provide a set of commonly used analysis procedures (e.g., "zonal mean"), it is probable that different communities and indeed different individual users will want to define their own procedures. It is important to enable users (at least on some computers) to define and upload their own analysis procedures, in a way that does not compromise the security or reliability of the server on which those procedures run.
- As more analysis procedures become available, it becomes increasingly important that users be able to discover their availability and determine their properties.

The basic analysis service framework, which will be developed in collaboration with the CEDPS SciDAC project, will provide for the wrapping of the analysis function code; the receipt, authorization, and dispatch of client requests; monitoring and control of execution, for purposes of troubleshooting, accounting, audit, management and so forth; and interactions with the back-end resources used to execute analysis functions.

A management interface will allow ESG administrators to monitor service delivery and control quality of service parameters such as the priorities given to different users and tasks. The backend provisioning component allows for the set of resources used for computational tasks to be varied over time, in response to changing load and priorities. The work here will focus on integration with climate data analysis frameworks (CDAT, NCL, and Ferret), the rest of the ESG software suite, and the mechanisms used to manage interactions with archival storage.

6.3. Example future analysis scenario

One example of how these capabilities will be used is illustrated in figure 6. A user searches the ESG-CET portal from a remote site. Required data are found at the distributed data node sites (e.g., NCAR deep storage archive and ORNL fast access disk). Using popular climate analysis tools (such as CDAT, NCL, Ferret), the user regrids the data where they physically reside before transferring the reduced data subset to the PCMDI gateway at LLNL, where further intercomparison diagnostics (such as computing a zonal mean) can be performed on the disparate data sets. Once the diagnostics are complete, the desired end products are delivered to the user's platform.

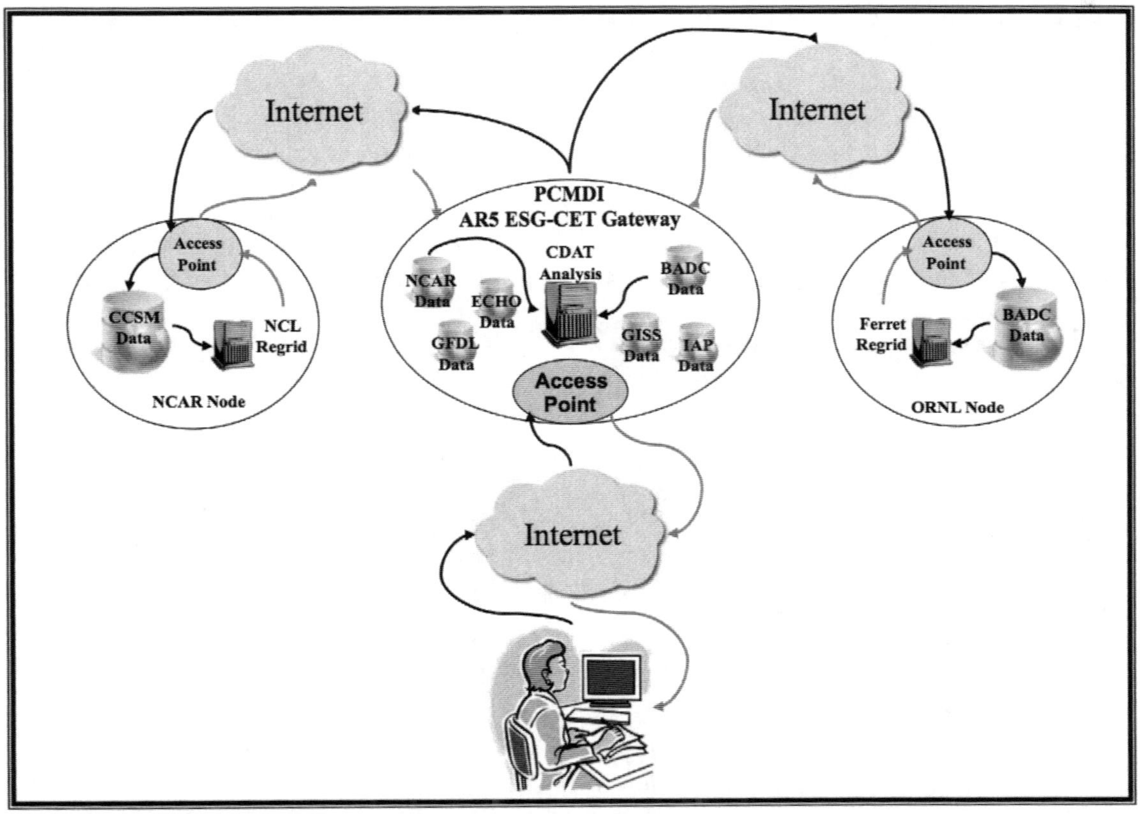

Figure 6. The user accessing and analyzing data via the new ESG-CET architecture.

7. Conclusions

ESG-CET's goal is ambitious—to meet, in a timely manner, the needs of a worldwide community for rapid and reliable access to, and analysis of, many petabytes of climate data. This goal can be attained, but only through a process of careful design and requiring considerable effort. In the process ESG users and designers will make important contributions to understanding how and why climate is changing and how best humanity can mitigate and/or adapt to that change. In order to be useful, the data and software that underpin climate change research must be made freely available to global change researchers worldwide, in a manner that allows convenient access, analysis, evaluation, discussion, intercomparison, and application. This requires a plan for an infrastructure and collaborative environment that links data centers, high-performance computing groups, users, models, data, and resources on a global scale. The creation of such an infrastructure and environment is vital to the success of climate change research. It demands continued investment in data management, analysis software, networking, and collaboration technologies. The methods and software developed in this work are likely to have some relevance to other projects in which large quantities of data must be shared and analyzed by distributed research teams, and the ESG can prove a valuable tool to scientific communities well outside the climate community.

Acknowledgments

This work is supported through the U.S. Department of Energy Office of Science, Offices of Advanced Scientific Computing Research and Biological and Environmental Research, through the SciDAC program. Argonne National Laboratory is managed by Argonne Chicago LLC under Contract DE-AC02-06CH11357. Information Sciences Institute is a research institute of the Viterbi School of Engineering at the University of Southern California. Lawrence Berkeley National Laboratory is managed by the University of California for the U.S. Department of Energy under contract No. DE-AC02-05CH11231. Lawrence Livermore National Laboratory is managed by the University of California for the U.S. Department of Energy under contract No. W-7405-Eng-48. Los Alamos National Security is managed by LLC (LANS) for the U.S. Department of Energy under the contract No. DE-AC52-06NA25396. National Center for Atmospheric Research is managed by the University Corporation for Atmospheric Research under the sponsorship of the National Science Foundation. Oak Ridge National Laboratory is managed by UT-Battelle, LLC for the U.S. Dept. of Energy under contract DE-AC-05-00OR22725. Pacific Marine Environmental Laboratory is under the National Oceanic and Atmospheric Administration's line office of Ocean and Atmosphere Research, lies within the U.S. Department of Commerce.

The ESG-CET executive committee consists of David Bernholdt, ORNL; Ian Foster, ANL; Don Middleton, NCAR; and Dean Williams, LLNL.

References

[1] McAliney M 1993 *Arguments for Land Conservation: Documentation and Information Sources for Land Resources Protection* (Trust for Public Land: Sacramento, CA)

[2] Intergovernmental Panel on Climate Change (IPCC) 4th Assessment Report 2007 http://www.ipcc.ch/ipccreports/ar4-wg1.htm

[3] Williams D N, Bernholdt D E, Foster I T and Middleton D E 2007 The Earth System Grid Center for Enabling Technologies: Enabling community access to petascale climate datasets *CTWatch) Quarterly* **3**(4)

[4] Bernholdt D, Bharathi S, Brown D, Chanchio K, Chen M, Chervenak A, Cinquini L, Drach B, Foster I, Fox P, Garcia J, Kesselman C, Markel R, Middleton D, Nefedova V, Pouchard L, Shoshani A, Sim A, Strand G and Williams D 2005 The Earth System Grid: supporting the next generation of climate modeling research *Proc. IEEE* **93**(3):485-495

[5] ESG. The Earth System Grid - turning climate model datasets into community resources 2004-2008 http://www.earthsystemgrid.org

[6] CCSM Carbon LAnd Model intercomparison Project (C-LAMP) 2007 http://www.climatemodeling.org/c-lamp/

[7] The Parallel Climate Model 2007 http://www.cgd.ucar.edu/pcm/

[8] The Parallel Ocean Program (POP) ocean circulation model 2007 http://climate.lanl.gov/Models/POP/

[9] CF. NetCDF climate and forecast metadata convention 2005-2008 http://cf-pcmdi.llnl.gov

[10] CDAT 2008 The Climate Data Analysis Tools (CDAT) http://www-pcmid.llnl.gov/software-portal/cdat

[11] NCL 2008 The NCAR Command Language (NCL) http://ngwww.ucar.edu/ncl/index.html

[12] Ferret: an analysis tool for Gridded and non-Gridded data 2008 http://ferret.wrc.noaa.gov/Ferret

[13] Fox P, Garcia J and West P 2007 OPeNDAP for the Earth System Grid *Data Science J.* (in preparation)

[14] Foster I, Nefefobva V, Ahsant M, Ananthakrishnan R, Liming L, Madduri R, Mulmo O, Pearlman L and Siebenlist F 2006 Streamlining Grid operations: definition and deployment of a portal-based user registration service *J. Grid Computing* **4**(2):135-144

[16] Shoshani A, Sim A and Gu J 2003 Storage resource managers: essential components for the Grid In Nabrzyski J, Schopf J M and Weglarz J eds. *Grid Resource Management: State of the Art and Future Trends* (Kluwer Academic Publishers)

Journal of Physics: Conference Series **125** (2008) 012073 doi:10.1088/1742-6596/125/1/012073

Science on the Grid with CMS at the LHC

Frank Würthwein

Univeresity of California, San Diego, La Jolla, CA 92093, USA

E-mail: fkw@ucsd.edu

Abstract. In late 2008 a new experimental facility, the Large Hadron Collider is expected to start data taking. Two large experiments, Atlas and CMS, are getting ready to explore the high-energy frontier of elementary particle physics at this facility. We briefly explain why this new facility is a once-in-a-lifetime opportunity for the present generation of particle physicists. Using the example of the CMS experiment, we explain the computing challenge of these large experiments, and how this challenge is addressed. In this context, we explain why Grid computing is an essential enabling technology to fully exploit the science potential of these experiments. We use data transfer and data analysis as examples to give a flavor of the present state of readiness of the globally distributed computing system for data taking, and what we had to do to get there.

1. A once-in-a-lifetime opportunity for discovery

Elementary particle physics strives to understand the fundamental building blocks of matter and its interactions, as well as the nature of space and time. To probe the smallest distances and thus the most fundamental particles and their interactions requires the largest energies. We thus refer to the limits of what can be probed experimentally in the laboratory often as the "high energy frontier."

One of the two crucial ingredients that determine the scientific reach of an experimental facility is thus simply the energy scale that it can probe. We understand nature at a given energy or distance scale by virtue of the kinds of fundamental particles that get produced at that energy, their mass and lifetime, the rates and processes by which they get produced and decay. To create a given energy scale in the laboratory, we collide beams of ordinary low mass particles such as protons head on with each other after accelerating them. The kinetic energy provided by the acceleration allows us to reach the desired energy scale. The second crucial ingredient that determines the scientific reach is thus the brightness of the beams, because it determines the rate of the collisions. Larger rate means access to collisions that are less likely and thus may yield radically new insights.

In late 2008, the Large Hadron Collider (LHC) is expected to start colliding protons on protons at a center of mass energy about an order of magnitude above anything ever reached, and with an eventual design luminosity two orders of magnitude beyond the previous generation of such colliders. The increase in energy alone results in 4–5 orders of magnitude increase in expected production rates for collisions at the high energy frontier. To understand this strong nonlinearity between beam energy increase and physics reach, one must realize that protons are compound objects made out of quarks and gluons. These quarks and gluons inside the protons usually collide to create the largest energies. The probability for two of these so-called partons to carry a significant fraction of a proton's momentum each is small. For example, the probability for a gluon-gluon collision at an energy of 1TeV is roughly 5 orders of magnitude larger at the 14 TeV center of mass energies of the LHC than at the 1.96 TeV proton anti-proton collision energies of the Tevatron, the present-generation hadron collider.

To increase the excitement further, we have reason to believe that the increase in energy provided by the LHC propels us into a regime that is qualitatively different, because it pushes us above the energy scale of electroweak symmetry breaking. Crudely speaking, one might think of this as being able to explore the other side of a phase transition for the first time. What we will find on the other side is object of much theoretical speculation, and very little experimental fact at this point.

Some sort of symmetry-breaking mechanism must be discovered, the most likely of which is the so-called Higgs mechanism. Somewhat more speculative, but still well motivated by cosmic observation of dark matter, is the search for "supersymmetry," Completely speculative scenarios include the possible observation of phenomena indicating additional spacial dimensions, as well as a host of other new phenomena. Many experimentalists are hoping to see something entirely unexpected and are preparing accordingly.

The first five years of exploration after the LHC starts colliding beam is thus an era with enormous discovery potential. An opportunity that is difficult to replicate, given the cost of the experiments and collider, and the myriad technical challenges that had to be overcome in its multiple decades long design and construction phase. To put things in historic perspective, the last energy increase like this happened in 1987, when the Tevatron surpassed SPS energies by going from 270 GeV to 1 TeV beam energies. That was 20 years ago and was only half as big a jump in energy, and without the additional motivation of jumping across a phase transition.

In short, we are at the start of a once-in-a-lifetime scientific opportunity!

2. The basic computing problem

The computing problem in experimental particle physics is fundamentally trivially parallel because of the nature of the data that we take with our experiments.

While the LHC beams collide at a rather rapid rate of once every 25 ns, we record only those collisions that pass a set of realtime trigger decisions. The rate of events we plan to record with CMS is thus a mere 200–300 Hz, as compared to the 40MHz rate at which the proton beams collide. Each of the recorded events is independent of all other recorded events. An event is thus an atomic unit of data that is processed by itself without any reference to any other event. The average size of an event is 1–2 MBytes when it comes out of the detector.

A nominal years worth of data taking is thus something like 200 -300 Hz times 10^7seconds times 1–2 MByte, or a total of 2-3 billion events per year that occupy 2-6 PByte of storage prior to any processing. To be fair, it is unlikely that operations of the LHC and CMS will be sufficiently smooth during the first year or two to generate this volume of "raw" data. Table 1 shows the storage and processing requirements according to the official budget planning process. (The current computing capacity requirements from the four LHC experiments are available via the WLCG Management Board web pages at http://lcg.web.cern.ch/LCG/MB/. The quoted numbers here are from August 24, 2007. It is likely that these requirements will continue to be revised, especially after experience is gained from the initial data taking.)

To appreciate the complexity of the processing of raw data before it can be meaningfully used to derive science results, the scale of the CMS experiment is worth noting. CMS is cylindrical, with the beamline at its center. It is approximately 21 meters long and 16 meters in diameter, weighs roughly 12,500 tons, and includes ~80 million electronic sensors to record information about each and every event: 80 million sensors and a collision every 25 ns is equivalent to 10 petabyte/s of raw information. It is immediately obvious that the bulk of the data compression must be happening in realtime on the detector itself. This is done via a combination of "zero-suppression" and "triggering". Zero-suppression means that channels below a threshold are not written out. The individual event sizes vary significantly with the physics that causes it, while the average event size depends on our ability to suppress noise signals in the 80 million sensors. Both of these are expected to change over time as accelerator and detector reach their design performance.

In addition to processing the data taken with the detector, we need very detailed simulations of both the physics as well as detector response. Crudely speaking, these simulations have three purposes: (1) to understand the detector response and develop physics object reconstruction algorithms; (2) to design analysis strategies to derive scientific results from the data, and (3) to determine the detection efficiencies that ultimately allow us to extrapolate from the yields we see to the rates at which nature produces the physics we see. In the asymptotic limit, once the accelerator and detector have reached design goals, we expect to simulate roughly as many events per year as we accumulate with the detector.

The average software performance ranges from 0.2 kSi2k seconds to 200 kSi2k seconds [1] [2]. The units here are the number of seconds it takes to process an event on a CPU that has the performance of 1000 SpecInt2000. The large range reflects the difference in processing requirements for simulation of events versus high-level filtering as is done in the final stages of an analysis. Primary reconstruction requires roughly 30 kSi2k seconds per event. Actual purchasing decisions are no longer based on SpecInt2000 but include appropriate scale factors to reflect changes in technology. The official RAM requirements by CMS are 1 GByte per batch slot. The actual consumption of real applications range from 0.5 GByte to 2 GByte, depending on the type of processing.

Given these requirements, a typical hardware procurement today is a rack full with 1U systems of Dual quad-core CPUs, 16 GByte of RAM, and a single gigabit NIC. The cluster is then built based on switched gigabit with a 10 gbps WAN uplink. Disk space is either provided as part of the 1U system, and virtualized via storage systems like dCache [3], or provided separately via commodity fileservers and then virtualized in the same way. Tier-2 centers generally do not have the budget to provide a global filesystem but, rather, use remote file I/O mechanisms. This apporach does not limit us in any way as files once committed to storage are never modified. Cache coherence is a luxory that is completely unnecessary for what we do.

In summary, the total CPU/disk/tape requirements are 30Million SpecInt2000 / 14.5 PB / 15.1 PB in 2008 and reach 221.1 million SpecInt2000 / 44PB / 67PB in 2012. This is a sufficiently large computing resource requirement that we decided to adopt a distributed computing model, with archival as well as processing centers distributed truly globally, involving (almost) all continents on Earth.

That decision, made primarily because of the total volume of resources needed, has significant sociological benefits as well. The CMS collaboration includes more than 2000 physicists from 180 institutions across almost 40 countries. Despite its size, this is fundamentally a research enterprise and thus depends on the creativity in addition to the hard work of its collaborators. Hosting data and processing as widely distributed as the collaboration itself provides for an opportunity to maximally engage the entirety of the distributed community. To be successful means to strike a reasonable balance between central control, and distributed self-organization. Grid technology is crucial to achieve this goal because it is from the ground up designed to enable virtual organizations and distributed control, in addition to providing the basic technologies for data distribution, job placement, and so forth.

CMS relies on three primary Grid infrastructures: EGEE, Nordugrid [4], and the Open Science Grid (OSG) [5] at a rough split of 60 percent , a few percent , and 40 percent, respectively. We expect to continue to rely on these Grids through the lifetime of the LHC physics program.

Taking the 20-year lifetime of the Tevatron (since 1987) as an estimate for the lifetime of the LHC physics program, it is immediately clear that this implies continued management of technology evolutions, ensuring smooth ongoing use of the global infrastructure. This by itself is a significant challenge.

Table 1. CMS computing resource requirements.

Requirement	2008	2010	2012
tape [PB]	15	38	67
disk [PB]	15	27	44
CPU [MSi2k]	30	140	220

To strike a balance between centrally organized and distributed and self-organizing, the CMS computing model is based on a multitiered hierarchy. Tier-0 and Tier-1 are fully controlled centrally. Tier-2 is a mix of central and local control, and Tier-3 is controlled completely by the local community that owns the site.

The 300–600 MByte/sec raw data that is coming out of the detector in Cessy, France, is sent to CERN, Geneva, for archival storage as well as "prompt reconstruction." We refer to CERN as the Tier-0 of the data-processing chain. From here, the data is immediately shipped to seven Tier-1 centers worldwide for "custodial storage." Custodial storage means that the set of seven Tier-1s together store a live copy of all the data on tape. CERN itself keeps an archival copy as a backup but does not have sufficient computing resources to process all the data to the point where it would be scientifically meaningful. This processing is the responsibility of the Tier-1 centers.

Each Tier-1 center is responsible for processing and most likely reprocessing many times over, all the data in its custodial storage. The purpose is to turn the digitized readings from the 80 million sensors into meaningful physics objects, applying calibration constants in the process. The derivation of these calibrations is in itself a complex set of processing steps of subsamples of the data, some of which were specifically recorded for this purpose. The processed data is now categorized into largely mutually exclusive datasets of well-defined provenance and physics content.

Once data is processed at a Tier-1, all Tier-1's synch up a reduced format copy of this data via the WAN. In parallel, the CMS community of a couple thousand physicists worldwide pull the processed data from the Tier-1's to 54 Tier-2's for scientific analysis. A given team of physicists working on a scientific analysis generally requires a handful of datasets and will process these many times over at a Tier-2 center. However, as no one Tier-1 center has all the data, each Tier-2 center will want to be able to pull data from each Tier-1 center in order to be scientifically agile, that is, to have access in principle to any dataset the physicists may need to study.

All the simulated data is produced at Tier-2 centers and from there transferred to custodial storage to one of the Tier-1's. Each Tier-1 thus serves a fixed set of Tier-2's as their regional archive for simulation data uploads.

All of this processing and simulating are very much an iterative process. Different scientific pursuits will need different levels of understanding of the detector, and some activities are feasible only after several iterations of calibrating the detector. At the same time, the physics object reconstruction tends to become more sophisticated as the understanding of the physical detector improves. All of this is a continued R&D effort and depends to a very large extent on the creativity of a very large number of individuals among the couple of thousand physicists from 180 institutions worldwide that collaborate on CMS.

3. Brief overview of technical solutions

It is impossible to even come close to doing justice to the technical solutions of our computing problem in this paper. Instead, we focus on providing a conceptual overview and refer to the literature where possible for further information.

Fundamentally, we look at the Grids as Grids of sites. Sites within a Grid provide a set of APIs for storing data [6][7][8][9], submitting jobs[10], as well as a well-defined execution environment. A given Grid guarantees uniformity of these APIs and execution environment, as well as some level of well understood interoperability across Grids. In addition, the Grids provide information and accounting systems, as well as a common security context, in terms of software as well as operations [11]. The execution environment includes a set of client tools, local disk space for the duration of the batch slot lease, and a set of environment variables that allow the job to find all of the above.

The Grids also provide a variety of services, all geared toward decreasing the total cost of operations of the globally distributed infrastructure for achieving the scientific goals of its user communities. Those services include middleware integration, packaging, testing, and certification; trouble shooting and trouble ticketing; support and training of user, developer, and provider communities, and cross-Grid coordination. In a nutshell, OSG, Nordugrid, and EGEE operate in a triangle of large user communities, computing resource providers, and middleware developers, each of whom brings his own specific needs and desires to the table. The Grids facilitate negotiations about procedures, schedules, development, and deployment roadmaps between these stakeholders and provide an operational underpinning, all in the interest of enabling scientific discoveries.

CMS uses this environment to distribute application software installations, place data, and submit jobs to all sites we care to use. For this purpose, we built our own data placement system (PhEDEx) [12], Dataset Bookkeeping System (DBS) [13], Calibration Replication (Frontier) [14], workload management systems for large scale data processing operations (ProdAgent) [15] and user analysis (CRAB) [16]. The entirety of the CMS specific application layer depends crucially on the Grid middleware and services.

The basic ideas are as follows. The experiment decides centrally which application software releases are "active." All of them then get installed uniformly at all sites via a centralized installation procedure. These installations are done via the job submission interface at the Grid sites. These releases are then available read-only at all batch slots on a Grid site. Users can not modify an application software installation.

Similarly, primary data processing is organized centrally, and again, all processing jobs are Grid job submissions. The produced datasets are then replicated on demand to the Tier-2's. Each user can request such a replication. However, a site data manager needs to sign off on the request. At the same time, some data replication is managed centrally. The sites thus have partial control over what data they are hosting. Replication is kept track of by the DBS.

The analysis job submission system takes as input a set of user-developed libraries, a release version this was developed for, a dataset name, and some guidance as to how many events to process per job. The system then packages the user code, determines the set of sites that have both the required release and the required data, and creates and submits jobs to those sites. The input data for each job is thus fixed at job creation time, and data must be placed prior to that. The output data may be delivered back to the user via the jobs output sandbox, or moved out at the end of the job to storage that is generally not local. This is configured by the user depending on the expected size of the output files.

If the output data volume is large, then it is desirable to be able to use it as input to another set of jobs later. To make this possible, a user may register the output data with a "local scope" DBS. Local scope here refers to the fact that this is a DBS that is privately owned by a user or group of users. This implies that users can configure job submissions such as to point to a local scope DBS for dataset resolution, instead of the official experiment wide DBS that records all the primary data. If privately produced data becomes of interest to a larger community, it can be published to the whole experiment by uploading from local DBS to global DBS. Only datasets in the global DBS can be transferred across the CMS data Grid. Users are thus confined to their "local" Tier-2 for any processing of data they produced and stored privately. All CMS physicists are assigned to at least one such local Tier-2. This assignment is generally based on the type of physics they are interested in. A Tier-2 thus hosts data and services for a "virtual community" of physicists that have some shared interest.

All job submission activities in CMS are tracked by the "CMS Dashboard" [17]. To do so, we have instrumented all submission systems to send messages via MonALISA [18] to a central repository at CERN. These messages are sent at submit time from the submit host, as well as before and after running the executable from the batch slot on the Grid where the job is scheduled to run.

4. Commissioning the globally distributed system

To commission the Grids for use by CMS, we have run extended "computing service challenges (CSA)" once a year, typically for 4–6 weeks. These challenges have had increasing scope in terms of both computing resources utilized and complexity of the activities. Since CSA06 in fall 2006, we have also had continued production operations, in addition to "site availability monitoring," which runs short representative test jobs several times a day at all sites. Describing all of these commissioning exercises within a 10 page paper is impossible. Instead, we will pick a few representative activities to give an impression of this global undertaking.

4.1. Data transfer commissioning

Within the CMS computing model, data has to flow across close to 500 WAN links, connecting 62 sites globally. The vast majority of the sites involved had little or no prior experience with large-scale data storage. Even the experienced sites had to go through a significant learning curve because the API, as well as the software artifacts that implemented them, was brand new. To deal with these issues head-on, CMS created a task force to work with all participating sites from June to December 2007 to achieve and maintain end-to-end connectivity between sites. A link was considered "commissioned" if it sustained $\geq$ 300GB data transfers for 4 out of 5 days. In order to remain commissioned, a link needs to have one day a week with $\geq$ 300 GB transferred per day. Roughly half the total number of links were commissioned during this period. The remaining links not yet commissioned are partly due to sites not being ready and partly due to one of the seven Tier-1s having problems for more than a week at a time and thus causing more than 50 links at once to be not commissioned. The problems encountered related predominantly to the storage and data transfer layers. Only 6% of the issues arising had their origin in problems with networking. During the last three months of 2007, CMS typically transfered 100 TBytes per day across its data Grid.

In early 2008, the commissioning metric was changed to requiring 20 MB/sec sustained for 24 hours. In addition, we instituted a round robin pick of 21 links every day. These links had to exceed 20 MB/sec for 12 hours. The intention of this change was to reflect more the expected data flow pattern of each Tier-1 to have a few high volume flows each day. With this new metric, a total of 350 links got commissioned for CSA08/CCRC08 in May 2008. Figure 1 shows the average I/O per day across all sites during the most recent challenge in May 2008. Daily totals ranged from 70 TB to 200 TB. In total, 3.6 petabytes of data were transfered across the CMS data Grid during that month.

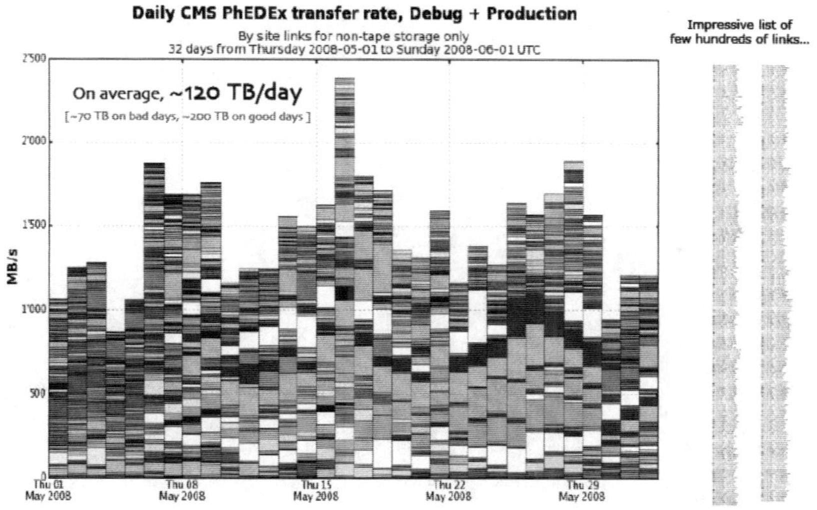

Figure 1. Average aggregate I/O rate per day during CCRC08. Different colours correspond to different links.

4.2. Data analysis commissioning

One of the most difficult problems in CMS computing is the support of data analysis by the entirety of the CMS physics community. It is difficult because it is the most diverse in many dimensions. It is distributed across the most sites globally and requires the most complete set of functionality at all of these sites. The user community is the most distributed, and its skillset is the most diverse, thus requiring easy-to-use tools. The applications are the most diverse because they are written by the users themselves to a significant degree. Accordingly, the performance characteristics in terms of I/O, CPU, and RAM requirements are as diverse as the community that writes the applications.

During May 2008, CMS went through a set of exercises referred to as CSA08/CCRC08. This included a variety of calibration exercises that started with simulations that included deliberate miscalibrations of the detector, as expected in the early data-taking periods. CSA08/CCRC08 thus included all the various processing, reprocessing, and analysis activities that are expected when data taking starts.

Figure 2 shows the analysis activities at the 25 most heavily used sites during this period. The "unknown" category refers to jobs for which the dashboard was unable to determine the location that they were running at. In total, the CMS dashboard reported $\sim$ 450 users submitting 1.5 million jobs across 86 sites during that period. The number of sites is significantly larger than the official 54 Tier-2 plus 7 Tier-1 plus CERN for two reasons. First, some sites have multiple entry points. For example, Fermilab (UCSD) shows up three (two) times among the 25 most active sites in figure 2. Sites do this for redundancy and scalability. Second, some of the sites active in analysis have no official commitment to CMS. We refer to these sites as "Tier-3". In the U.S. alone, 13 sites are part of the CMS data Grid but have no official service commitment to the collaboration, as compared to only 7 Tier-2 and 1 Tier-1. The Tier-3 are used primarily by their owners. CMS as a whole may use these sites on an opportunistic basis. While Tier-1 and Tier-2 receive significant support from the CMS project, Tier-3 operations depend much more on support from the Grids.

We note that 11 of the top 24 sites listed in figure 2 are OSG sites in the U.S. OSG in general and the U.S. in particular clearly hold a leadership position in CMS analysis computing. Among the remaining 13 sites in this list, only seven are Tier-2's, two in Italy, two in Germany, two in France, and one in Spain. Six of the seven Tier-1s are in the top 24, and so is CERN.

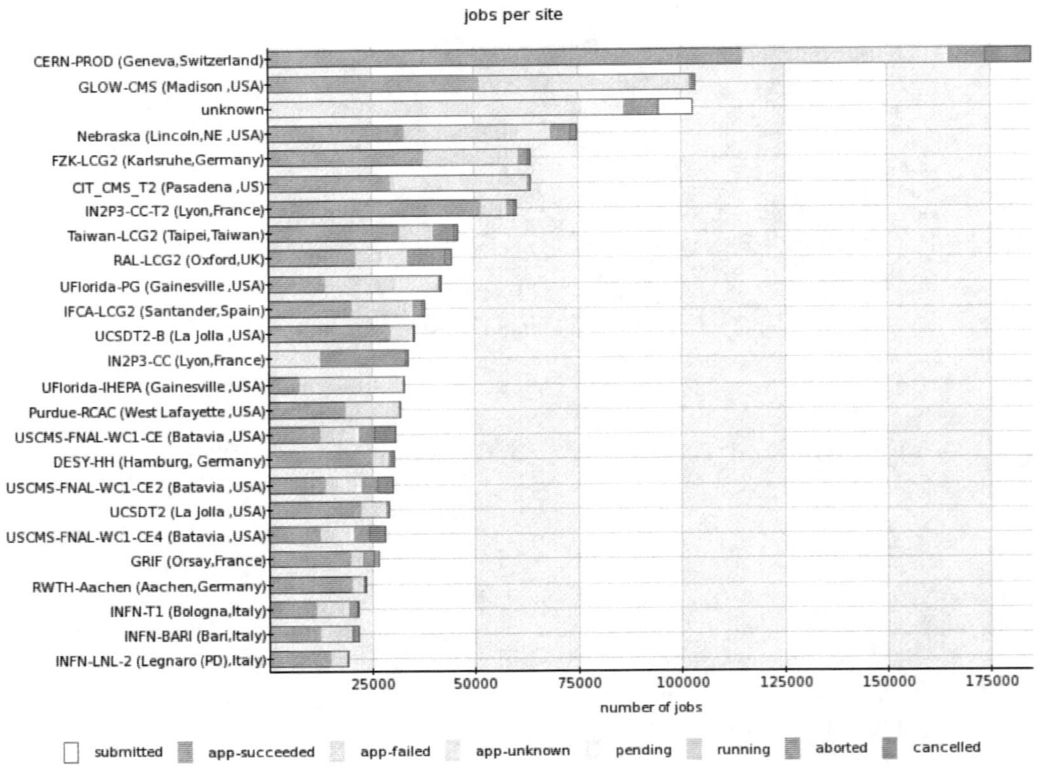

Figure 2. Number of analysis jobs per site during May 2008 for the top 25 most actively used sites in CMS. Fermilab, and UCSD show up more than once because the accounting is done by Grid entry point to the compute cluster. IN2P3 and Florida show up twice because there are two independent compute clusters at these institutions. Both clusters access the same storage in these two cases.

In grand total, slightly more than 50% of the 1.5 million jobs submitted were recorded as successfully completed. To understand the origin of the failures, we did a controlled experiment during the month of May 2008. We identified eight datasets, one from each Tier-1 plus CERN, and asked all Tier-2's to download at least one of them. We then verified write access to storage at each Tier-2 for output storing. A standard, well tested application was then run at the 40 participating Tier-2. A total of 230,000 jobs succeeded throughout the month, as compared to 44,000 failures. When analyzing the logfiles of all jobs, we found that the failures are predominantly due to storage problems at sites. In most but not all cases, those problems are detected and fixed by the site admins within 24 hours. However, during those 24 hours, many more jobs fail than could possibly succeed during the same timeframe because a job that fails in its storage access is very short. When ignoring storage failures, the success rate in this exercise is better than 99%.

The lesson we learned from this exercise is that we need to build into the CRAB system the ability to detect storage failures and stop submissions to that site until the storage issue is fixed. The details of how to accomplish this technically, as well as what operational procedures to adopt, are being worked out.

4.3. Site availability monitoring

After a site is installed properly once, the next challenge is to keep it that way. CMS has found it extraordinarily beneficial to develop a set of simple tests that cover all the functionality a site needs to provide. These tests are of varying complexity, and the set is run such as to reflect that. Success or failure is recorded for each test, together with detailed error logs. These tests are then run a few times a day at all sites, and the results are available via a web interface.

We expect from our site admins at Tier-0, 1 and 2 to check this web interface for their site at least daily and offer automated email upon failure. Once a month, we rank the sites according to their average availability according to these tests. For Tier-0 and Tier-1, we in principle strive for 24x7 support, while the Tier-2's are expected to provide support only during normal business hours. In fact, we find that support in terms of response time to failures at our most active Tier-2's is typically better than at some of the Tier-1s. Several of the Tier-1 sites cannot provide 24x7 operations for a variety of reasons, and many of the Tier-2 provide better than business hours support.

Typically, more than 50% of the sites achieve an availability of 80% or better in any given week, while over a 6 month period, only 20% of the sites achieve that. Availability here is calculated absolute; that is, scheduled downtimes count the same as "unscheduled" ones. In general, availability is better during service challenges, and worse right after. We interpret this to mean that operating a site still requires significant attention to details and is thus labor intensive.

5. Summary

The Large Hadron Collider facility at CERN, Geneva, Switzerland, is expected to start colliding protons on protons at unprecedented energies in late 2008. The start of data taking with the two high-energy frontier experiments Atlas and CMS will be the beginning of a once-in-a-lifetime opportunity for discovery in elementary particle physics.

The complexity and scale of these experiments, as well as their data volumes, requires worldwide distributed computing Grids. Grid technology is used in these experiments, not only to provide the computing resources required, but also to map onto the sociology in these experiments.

As an example, the CMS experiment has more than 2,000 collaborators across 180 institutions in 40 countries worldwide. To understand the detector and derive meaningful physics results will require the creativity and hard work of the entire collaboration. Grid technologies are allowing the collaboration to form virtual communities at so-called Tier-2 sites where most of the final analysis is expected to be done. In addition, it allows the seamless integration of resources at CMS institutions that don't have a computing service agreement with CMS. These so-called Tier-3 sites serve primarily the interests of their local owners, but can be used by the rest of the collaboration on an opportunistic basis. In the US alone, there are 13 such Tier-3 sites, as compared to 7 Tier-2.

Commissioning the Grids for CMS is a significant challenge because of its technical complexity, scale, and distributed nature, as well as distributed control. Excellent progress has been made, especially in the U.S. However, a lot of work remains to be done in order to maximally benefit scientifically from the financial investment into both the detector and the global computing infrastructure.

Acknowledgments

This research was supported in part by the U.S. Department of Energy under Grants No. DE-FC02-06ER41436 subcontract No. 647F290 (OSG), and DE-FG03-30ER40546 (UCSD Particle

Journal of Physics: Conference Series **125** (2008) 012073 doi:10.1088/1742-6596/125/1/012073

Physics), and the National Science Foundation under Grants No. PHY-0533280 (DISUN) and PHY-0612805 (CMS Maintenance & Operations).

References

[1] LCG web page 2008 http://lcg.web.cern.ch/LCG/MB/
[2] Della Negra M, *et al.* 2005 *CMS Computing Technical Design Report CERN-LHCC-2005-023*
[3] Furhmann P, *et al.* 2008 The dCache book http://www.dcache.org/manuals/Book/
[4] Ellert M, *et al.* 2007 Advanced resource connector middleware for lightweight computational Grids *Future Generation Computer Systems* **23**:219–240
[5] Pordes R, *et a.l* 2007 The Open Science Grid *J. Phys.: Conf. Ser.* **78**:012057
[6] Sim A, *et al.* 2008 *The Storage Resource Manager Specification Version 2.2* (GFD-R-P.129 Open Grid Forum)
[7] Abadie L, *et al.* 2007 Storage resource manager version 2.2: design, implementation, and testing experience *Proc. International Conference on Computing in High Energy and Nuclear Physics (CHEP 07)* 2007 *J. Phys.: Conf. Ser.*
[8] Abadie L, *et al.* 2002 Storage resource managers: recent international experience on requirements and multiple co-operating implementations *24th IEEE Conference on Mass Storage Systems and Technologies (MSST 2007)* 47–59
[9] Schulz M, *et al.* 2007 Building the WLCG file transfer service *JPCS Proc. International Conference on Computing in High Energy and Nuclear Physics (CHEP 07)* (to appear)
[10] Frey J, Tannenbaum T, Foster I, Livny M and Tuecke S 2002 Condor-G: a computation management agent for multi-institutional grids *Cluster Computing* **5**:237–246; Czajkowski K, Foster I, Karonis N, Kesselman C, Martin S, Smith W and Tuecke S 1998 A resource management architecture for metacomputing systems *Proc. IPPS/SPDP '98 Workshop on Job Scheduling Strategies for Parallel Processing* 62–82; Cecchi M, *et al.* 2007 The gLite workload management system *JPCS Proc. International Conference on Computing in High Energy and Nuclear Physics (CHEP 07)*
[11] Altunay M and Olson D 2008 *Open Science Grid Security Activities* (OSG document 749-v1) http://osg-docdb.opensciencegrid.org/cgi-bin/ShowDocument?docid=749; Welch V, Siebenlist F, Foster I, Bresnahan J, Czajkowski K, Gawor J, Kesselman C, Meder S, Pearlman L and Tuecke S 2003 Security for Grid services *Proc. 12th IEEE International Symposium on High Performance Distributed Computing (HPDC-12)* 48–57; Groep D, *et al.* 2007 glExec - gluing grid computing jobs to the Unix world to appear in *JPCS Proc. International Conference on Computing in High Energy and Nuclear Physics (CHEP 07)*
[12] Rehn J, *et al.* 2006 PhEDEx high-throughput data transfer management system 2006 *Proc. International Conference on Computing in High Energy and Nuclear Physics (CHEP 2006)*
[13] Afaq A, *et al.* 2007 The CMS dataset bookkeeping service *Proc. International Conference on Computing in High Energy and Nuclear Physics (CHEP 07) J. Phys.: Conf. Ser.* (in preparation)
[14] Blumenfeld B, Dykstra D, Lueking L and Wicklund E 2007 "CMS conditions data access using FroNTier", *Proc. International Conference on Computing in High Energy and Nuclear Physics (CHEP 07) J. Phys.: Conf. Ser.* (in preparation)
[15] Spiga D (for the CMS Collaboration) CMS workload management 2007 *Nucl.Phys.Proc.Suppl.* **172:**141–144
[16] Spiga D, *et al.* 2007 he CMS Remote Analysis Builder (CRAB) **LNCS 4873** 580–586
[17] ndreeva J, *et al.* 2007 Dashoard for the LHC Experiments *Proc. International Conference on Computing in High Energy and Nuclear Physics (CHEP 07) J. Phys.: Conf. Ser.* (in preparation)
[18] Legrand I C, *et al.* 2004 MonALISA: An agent based, dynamic service system to monitor, control, and optimize grid based applications *Proc. International Conference on Computing in High Energy and Nuclear Physics (CHEP 2004)* http://indico.cern.ch/getFile.py/access?contribId=89&sessionId=7&resId=0&materialId=paper&confId=0

New eigensolvers for large-scale nanoscience simulations

A Canning[1], O Marques[1], C Voemel[1], Lin-Wang Wang[1], J Dongarra[2], J Langou[2] and S Tomov[2]
[1]Computational Research Division, Lawrence Berkeley National Laboratory, Berkeley, CA 94720, USA
[2]Dept. of Computer Science, University of Tennessee, Knoxville, TN 37996, USA

E-mail: acanning@lbl.gov

Abstract. We present results for applications to nanosystems of state-of-the-art iterative eigensolvers based on conjugate gradients and variants of Davidson in the context of semi-empirical plane wave electronic structure calculations. We are concerned with the computation of electronic and optical properties of nanosystems using the Energy SCAN method to compute interior eigenstates around the band gap that determine their properties. Numerically, this interior Hermitian eigenvalue problem poses several challenges, with respect to both accuracy and efficiency. All the iterative eigensolvers are seeking the minimal eigenvalues of the folded operator with reference shift in the band-gap. The tested methods include standard conjugate-gradient (CG)-based Rayleigh quotient minimization, Locally optimal block-preconditioned CG (LOBPCG) and two variants of the (Jacobi-)Davidson method: JDQMR and GD+1. Our experimental results indicated that Davidson-type methods are often the fastest.

1. Introduction

The computation of the electronic properties of large nanostructures such as quantum dots is an important field of current research. Its significance is highlighted by a number of activities such as the DOE-funded SciDAC project Predicting the Electronic Properties of 3D Million-atom Semiconductor Nanostructure Architectures. In this paper we focus on the mathematical aspects of this project. Some of the physics applications of this project are discussed in another publication in this volume [1].

We are interested in electronic structure calculations based on the solution of an effective single-particle Schrödinger equation,

$$H\psi_i(r) = \{-\frac{1}{2}\nabla^2 + V(r)\}\psi_i(r) = E_i\psi_i(r),$$

where H denotes the approximate Hamiltonian. The atomic system is described by the potential $V(r)$, which we assume as an externally given empirical pseudopotential [2]. The set $\psi_i(r)$ denotes the orthogonal wave-functions (eigenstates) and E_i their corresponding energies. In the context of the self-consistent field iteration [3], a large number of eigenstates corresponding to all the electron states of the system need to be computed. On the other hand, only a small number of eigenstates are relevant for determining certain optical and electronic properties. Our task is to compute these states at the top of the valence and bottom of the conduction band. These states typically lie in the middle of the spectrum with too large a distance from the lowest (ground) state to efficiently compute all of them from the lowest up.

In a discrete, finite plane-wave basis, the single-particle Schrödinger equation directly translates into a Hermitian eigenvalue problem, and we use the same notation H for the matrix in question. The discrete Laplacian is the commonly computed in the plane-wave basis because it is diagonal. On the other hand, the discretized potential is diagonal in real space. Thus the resulting H becomes implicitly available only through matrix-vector products and via the fast Fourier transform (FFT). This together with the large system size requires the use of efficient iterative eigenvalue methods.

Our parallel ESCAN code [4] uses a folded spectrum approach [2] to find the interior eigenstates that we are looking for. Based on physical knowledge about the system, a reference energy E_{ref} was chosen, and then the smallest eigenvalues of the system $(H - E_{\mathrm{ref}})^2$ were computed via the PCG (parallel conjugate gradient) Rayleigh quotient. In our study, we compare this method with blocked CG solvers and variants of the (Jacobi-)Davidson method.

2. Overview of eigenvalue methods

The eigenvalues of H near $\square_{ref}$ can be ideally tackled by a shift-and-invert strategy, which conveniently maps those eigenvalues into the dominant and well-separated eigenvalues of $(H-\square_{ref} I)^{-1}$, while the eigenvectors remain unchanged. (In practice, operations involving $(H-\square_{ref}I)^{-1}$ are replaced by solutions of systems of linear equations through direct methods.) However, such a strategy is infeasible in the present context because the size of our systems makes the number of operations required for shift-and-invert too costly. In addition, in our codes the matrix H is never explicitly available: we can only afford to obtain the action of H on a vector. One may also consider the use of polynomial spectral transformations on $(H-\square_{ref} I)$ to enhance spectral properties of interest. The folded spectrum method (FSM) [2] is a typical illustration of this approach. However, it remains to be verified whether polynomials of low degree are the best approach to capture information about interior eigenstates.

A nonlinear CG algorithm can be used to successively minimize the Rayleigh quotient function $f(x_i)=(x_i^* Hx_i)/(x_i^* x_i)$, where the gradient is given by $\nabla f(x_i)= Hx_i - x_i(x_i^* Hx_i)/(x_i^* x_i)$. (The minimum of the Rayleigh quotient corresponds to the minimum eigenvalue of H). The method can also be implemented in a blocked form, whereby the minimization is simultaneously applied to a set of orthogonal vectors X_i. Together with a suitable preconditioner and FSM, this approach (PCG) has been used to solve a number of problems of practical interest. A potential improvement over the PCG algorithm is the locally optimal preconditioned conjugate gradient (LOPCG) method [5], which extends PCG by applying Rayleigh Ritz on $span(w_i, x_i, x_{i-1})$, where $w_i = P\nabla f(x_i)$ and P is the preconditioner. As before, the method is amenable to an implementation by blocks (LOBPCG).

A class of algorithms based on subspace or projection-based techniques can also be used to seek eigenstates of interest. This class includes the algorithms of Lanczos and Arnoldi. Starting with a vector q, these algorithms generate an appropriate basis for a Krylov subspace. Often the trade-off of these methods lies between the sizes of the basis. While the dimension of the subspace is usually much smaller than the dimension of the original problem, if we can afford to set it larger, then better approximations for the eigenvalue and eigenvectors are obtained as the computation progresses. However, finite-precision arithmetic requires explicit orthogonalization of the vectors generated, which implies significant cost for large subspaces. We note the Krylov-based techniques can also be implemented in blocked form. Often, subspaces of small dimensions do not produce good approximations for the, say k, eigenstates of interest. Then, some kind of restarting strategy with a proper set of vectors has to be employed, possibly a number of times until convergence for the wanted solutions is obtained. The ARPACK implementation of the Arnoldi algorithm, for example, first generates a subspace S of dimension $k+p$. Then, information obtained from the associated reduced problem T (i.e., the projection of H into S, which also has dimension $k+p$) is used to project S into another subspace of dimension k. In order to do this, the eigenvalue approximations obtained from T that are not relevant, θ_u, are used as shifts for a QR factorization of T, i.e. $T- \theta_u I=QR$, with Q unitary and R upper triangular. The resulting leading k columns of Q are used for reducing the dimensions of T and S, from which point the algorithm is restarted. The goal is to make the (restarting) subspace of dimension k poor in the directions of the irrelevant solutions, therefore promoting convergence to the sought solutions. This type of restart is referred to as *implicit* in the sense that spectral information is embedded into the restarting subspace. It has already enabled the computation of eigenstates of large systems. A variant of this restarting scheme is implemented in the thick-restart Lanczos method, which employs a subspace of dimension k for restart (as above), but which is explicitly built with Ritz vectors (eigenvectors approximations) obtained from Q.

More recently, variants of Davidson's method have been shown to be effective for solving a number of applications. Davidson's original algorithm targeted mainly the smallest eigenvalues of (strongly diagonal dominant) matrices arising from applications in Chemistry. It relies on the fact that in the case of a strongly diagonally dominant matrix, the inverse of its diagonal entries is a good approximation for the inverse of the matrix (and thus a good preconditioner). In this case, given an approximate eigenpair $(\underline{x}, \underline{\theta})$ and the residual $r=(\underline{\theta}I-A)\underline{x}$, the vector obtained from $t=(\underline{\theta}I-diag(A))^{-1}r$ is used to generate a new vector to be added to the subspace S. Depending on A and $\underline{\theta}$, however, the preconditioner $(\underline{\theta}I-diag(A))^{-1}$ may be ill conditioned. To remedy this situation, an orthogonal component correction has been proposed, which solves $(I-\underline{xx}^*)(A-\underline{\theta}\, I)(I-\underline{xx}^*)t=-r$ approximately (by means of the QMR algorithm for example). This is the essence of the Jacobi-Davidson algorithm. It allows for the generation of more general subspaces than the original algorithm and takes great advantage of the preconditioner.

3. Test results

We perform experiments on a 1,061-atom cadmium-selenium quantum dot and a 66,624 atom InAs quantum wire. The size of the Hamiltonian matrix for the quantum dot is 141,625 and 2.7 million for the quantum wire. The goal of the experiments is to compare the different iterative methods in terms of time to solution and robustness. The methods compared are a set of two conjugate gradient-like methods Banded-PCG and LOBPCG, PARPACK, and two (Jacobi-)Davidson-like methods GD+k (Olsen's algorithm) and JDQMR as implemented in the PRIMME [6] package. All of these methods are extremely robust. All the runs are performed on 32 processors of an IBM SP5 at the NERSC computing center. Various sets of parameters were tested for GD and JDQMR in all the cases, and we present the results for the best parameters.

Table 1: Matrix vector multiplications and time for calculation of 10 conduction band minimum (CBM) states for a 1061-atom CdSe quantum dot using a nonlocal empirical pseudopotential. The * indicates that the algorithm converged but missed relevant eigenvalues.

Algorithm (Folded)	Matrix-Vector Multiplications	Time (s)
Band-by-Band CG	7066	63
LOBPCG	6880	92
PARPACK*	9362	86
GD+k	2176	21
JDQMR	6212	52
Unfolded		
PARPACK	2434	24
GD+k	12813	144
JDQMR	6383	59

Table 2: Matrix vector multiplications and time for calculation of 5 valence band maximum (VBM) states for a 66,624-atom InAs quantum wire using a local empirical pseudopotential. The * indicates no convergence with the maximum number of allowed restarts.

Algorithm (Folded)	Matrix-Vector Multiplications	Time (s)
Band-by-Band CG	149726	7278
LOBPCG	56207	3690
PARPACK	*	*
GD+k	26326	1424
JDQMR	81364	3877
Unfolded		
PARPACK	*	*
GD+k	39828	2452
JDQMR	56441	2746

As can be seen from tables 1 and 2, GD+k gives the best results for the folded spectrum method and for the unfolded spectrum for the larger quantum wire. It can be five times faster than previously used methods. Our experiments have also indicated that although PARPACK may exhibit fast convergence in the unfolded spectrum for some cases (see table 1), it may miss eigenvalues of interest. A more detailed discussion of some of the results in this paper, along with more results on CdSe quantum dots of different sizes, can be found in [7].

As part of this project we have also developed preconditioners based on using a modified form of the solution of the bulk system of the nanostructure we wish to study. The theoretical basis for the success of this method is that, in the interior of the nanostructure, the properties of the system are bulk-like so the bulk solutions are a good approximation to the true solution in this region. The bulk solutions are easily derived in a negligible amount of time. We have implemented the bulk-band preconditioner in the PCG version of ESCAN (see [8] for details) and are in the process of implementing this preconditioner for the other solvers.

The parallel ESCAN code is available for download from http://icl.cs.utk.edu/doe-nano/software/pescan. This website gives more information about the results discussed here, and also a set of benchmark systems.

Acknowledgments

This work was supported by the U.S. Department of Energy, Office of Science, Office of Advanced Scientific Computing (MICS) and Basic Energy Science under LAB03-17 initiative (Theory, Modeling and Simulation in Nanoscience), contract Nos. DE-FG02-03ER25584 and DE-AC03-76SF00098. We also used computational resources of the National Energy Research Scientific Computing Center, which is supported by the Office of Science of the U.S. Department of Energy. We thank A. Knyazev and A. Stathopoulos for their advice regarding the use of the LOBPCG and the PRIMME packages.

References

[1] Wang L W 2008 Large scale ab initio nanostructure electronic structure calculations for energy applications *Proc. SciDAC 2008*

[2] Wang L W and Zunger Q 1994 Solving Schrödinger's equation around a desired energy: Application to silicon quantum dots *J. Chem. Phys.* **100**: 2394

[3] Payne MC, Teter M P, Allan D C, Arias T A and Joannopoulos J D 1992 Iterative minimization techniques for ab initio total-energy calculations: Molecular dynamics and conjugate

gradients *Rev. Mod. Phys.* **64(4):**1045

[4] Canning A, Wang L W, Williamson A, and Zunger A 2000 Parallel empirical pseudopotential electronic structure calculations for million atom systems. *J. Comp. Phys.* **160:**29

[5] Knyazev A V 2001 Toward the optimal preconditioned eigensolver: Locally optimal block preconditioned conjugate gradient method *SIAM J. Sci. Comput.* **23(2):**517

[6] Combs J R and Stathopoulos A 2006 PRIMME: Preconditioned iterative multiMethod eigensolver: Methods and software description (*Technical Report WM-CS-2006-08* Computer Science Dept. The College of William and Mary)

[7] Voemel C, Tomov S, Marques O, Canning A, Wang L-W and Dongarra J 2008 State-of-the-art eigensolvers for electronic structure calculations of large scale nano-systems *J. Comp. Phys.* (to appear)

[8] Voemel C, Tomov S, Wang L-W, Marques O and Dongarra J 2007 The use of bulk states to accelerate the band edge state calculation of a semiconductor quantum dot *J. Comp. Phys.* **223:**774

A posteriori error analysis of multiscale operator decomposition methods for multiphysics models

D. Estep[1], V. Carey[2], V. Ginting[3], S. Tavener[2], T. Wildey[4]

[1]Department of Mathematics and Department of Statistics, Colorado State University, Fort Collins, CO 80523
[2]Department of Mathematics, Colorado State University, Fort Collins, CO 80523
[3]Department of Mathematics, University of Wyoming, Laramie, WY 82071
[4]The Institute for Computational Engineering and Sciences, The University of Texas, Austin, TX 78712

E-mail: `estep@math.colostate.edu`

Abstract. Multiphysics, multiscale models present significant challenges in computing accurate solutions and for estimating the error in information computed from numerical solutions. In this paper, we describe recent advances in extending the techniques of *a posteriori* error analysis to multiscale operator decomposition solution methods. While the particulars of the analysis vary considerably with the problem, several key ideas underlie a general approach being developed to treat operator decomposition multiscale methods. We explain these ideas in the context of three specific examples.

1. Multiscale, multiphysics problems

Multiphysics, multiscale models that couple different physical processes interacting across a wide range of scales abound in the application areas of the SCIDAC program. Such models present significant challenges for computing accurate numerical solutions, which makes it very important to obtain accurate estimates of error in computed information. But, multiphysics, multiscale problems also raise significant challenges for error estimation. In this paper, we describe recent advances in computing *a posteriori* error estimates for multiscale operator decomposition, which is a powerful solution method for multiphysics, multiscale problems that is in widespread use in high performance computing.

1.1. Three examples of multiphysics models

We briefly describe three examples of multiphysics phenomena that arise in SCIDAC application domains. These examples illustrate the myriad ways in which physical processes are coupled.

Example 1.1. A thermal actuator. This is a MEMS (microelectronic mechanical switch) device modeled by a system of three coupled equations, each representing a distinct physical process acting on its own scale.

$$\begin{cases} \nabla \cdot (\sigma \nabla u_1) = 0, & x \in \Omega, \\ \nabla \cdot (\kappa(u_2) \nabla u_2) = \sigma(\nabla u_1 \cdot \nabla u_1), & x \in \Omega, \\ \nabla \cdot \big(\lambda \operatorname{tr}(E) I + 2\mu E - \beta(u_2 - u_{2,ref}) I\big) = 0, \quad E = \big(\nabla u_3 + \nabla u_3^\top\big)/2, & x \in \Omega \end{cases} \tag{1}$$

The first is an electrostatic current equation governing potential u_1 (current $J = -\sigma\nabla u_1$), the second is a steady-state energy equation for the governing temperature u_2, and the third equation is linear elasticity describing the steady-state displacement u_3.

This is an example of "parameter passing," in which the solution of one component is used to compute the parameters and/or data for another component. Note that u_1 can be calculated independently, while computing u_2 requires u_1. Computing u_3 requires u_2 and therefore u_1.

Example 1.2. The Brusselator problem. This is a model of chemical dynamics [1],

$$\begin{cases} \frac{\partial u_1}{\partial t} - k_1 \frac{\partial^2 u_1}{\partial x^2} = \alpha - (\beta+1)u_1 + u_1^2 u_2, & x \in (0,1), t > 0, \\ \frac{\partial u_2}{\partial t} - k_2 \frac{\partial^2 u_2}{\partial x^2} = \beta u_1 - u_1^2 u_2, & x \in (0,1), t > 0, \\ u_1(0,t) = u_1(1,t) = \alpha,\ u_2(0,t) = u_2(1,t) = \beta/\alpha, & t > 0, \\ u_1(x,0) = u_{1,0}(x),\ u_2(x,0) = u_{2,0}(x), & x \in (0,1), \end{cases} \tag{2}$$

where u_1 and u_2 are the concentrations of species 1 and 2, respectively.

This problem combines different physics - in this case, reaction and diffusion - in one equation. The generic picture for a reaction-diffusion equation is a relatively fast, destabilizing reaction component interacting with a relatively slow, stabilizing diffusion component. Thus, the physical components have both different scales and different stability properties.

Example 1.3. Conjugate heat transfer between a fluid and solid object. We consider the flow of a heat-conducting Newtonian fluid past a solid cylinder. The model consists of the heat equation in the solid and the equations governing the conservation of momentum, mass and energy in the fluid, where we apply the Boussinesq approximation to the momentum equations in the fluid. The temperature field is advected by the fluid and couples back to the momentum equations through the buoyancy term.

Let Ω_F and Ω_S be polygonal domains in $\mathbb{R}^2$ with boundaries $\partial\Omega_F$ and $\partial\Omega_S$ intersecting along an interface $\Gamma_I = \partial\Omega_S \cap \partial\Omega_F$. The complete coupled problem is

$$\begin{cases} -\mu\Delta u + \rho_0 (u\cdot\nabla) u + \nabla p + \rho_0 \beta T_F g = \rho_0 (1+\beta T_0)\, \mathbf{g}, & x \in \Omega_F, \\ -\nabla\cdot u = 0, & x \in \Omega_F, \\ -k_F \Delta T_F + \rho_0 c_p (u\cdot\nabla T_F) = Q_F, & x \in \Omega_F, \\ \text{interface} \begin{cases} T_S = T_F, \\ k_F(n\cdot\nabla T_F) = k_S(n\cdot\nabla T_S), \end{cases} & x \in \Gamma_I, \\ -k_S \Delta T_S = Q_S, & x \in \Omega_S, \end{cases} \tag{3}$$

where we supplement these equations by imposing Dirichlet and Neumann conditions for the velocity field respectively on boundaries $\Gamma_{u,D}$ and $\Gamma_{u,N}$ and Dirichlet and Neumann conditions for the temperature fields in the fluid and the solid respectively on boundaries $\Gamma_{T_F,D}$, $\Gamma_{T_F,N}$, $\Gamma_{T_S,D}$, and $\Gamma_{T_S,N}$. Here, ρ_0 and T_0 are reference values for the density and temperature, respectively, μ is the molecular viscosity, β is the coefficient of thermal expansion, c_p is the specific heat, k_F and k_S are the thermal conductivities of the fluid and solid, respectively, Q_F and Q_S are source terms and n is the unit normal vector directed into the fluid. Note that u is a vector.

This presents a class of problems where physics in different physical domains are coupled across a common boundary.

1.2. Challenges and goals for multiscale, multiphysics models

Challenges arising in the numerical solution of multiscale, multiphysics problems include producing accurate numerical solutions efficiently, dealing with complex stability properties, and transforming information across scales. Another complication is the range of applications of multiphysics models, including model prediction, sensitivity analysis, and parameter optimization, that require computation of solutions corresponding to wide range of data and parameters. We expect the solution behavior to vary significantly and the ability to obtain accurate numerical solutions therefore to vary as well.

On the other hand, predictive science and engineering have come to rely on high performance simulation of complex physical phenomena. In this context, it is critically important to accurately quantify the numerical error in any computed information. Another important goal is determining efficient ways to compute specific information accurately.

2. Multiscale operator decomposition

Multiscale operator decomposition is a widely used technique for solving multiphysics, multiscale models in SCIDAC simulations. The general approach is to decompose the multiphysics and/or multiscale problem into components involving simpler physics over a relatively limited range of scales, and then to seek the solution of the entire system through some sort of iterative procedure involving numerical solutions of the individual components (see figure 1). In general,

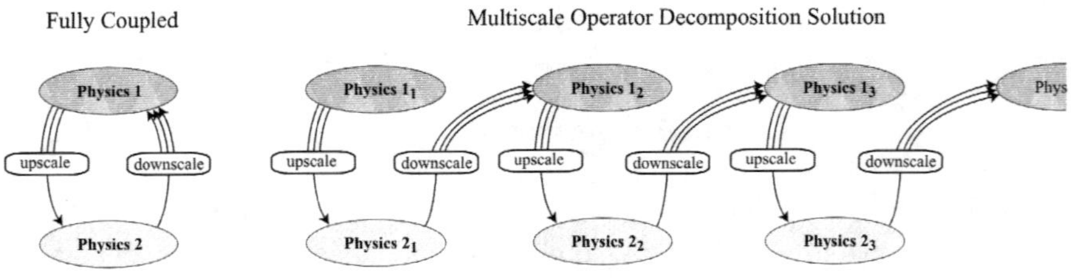

Figure 1. Left: Illustration of a multiscale, multiphysics model. Right: Illustration of a multiscale operator decomposition solution.

different components are solved with different numerical methods as well as with different scale discretizations.

Reasons that this approach is appealing include;

- There is generally a good understanding of how to solve a broad spectrum of single physics problems accurately and efficiently.
- It provides an alternative to accommodating multiple scales in one discretization.
- It allows for efficient high performance simulations.

However, multiscale operator decomposition generally has significant impact on the accuracy and stability of numerical solutions.

3. Analysis of multiscale operator decomposition

In general, multiscale, multiphysics problems have complex stability properties that defy accurate description via classic *a priori* analysis. We use duality and adjoint operators to quantify stability *a posteriori*. We focus on computing a particular quantity of interest determined by a linear functional. We combine these tools with variational analysis to produce accurate estimates of the effects of perturbation and error. By numerically solving the adjoint problem, we can compute very accurate error estimates.

Duality and adjoint operators have a long history of application in model sensitivity analysis and optimization, dating back to Lagrange; see [2, 3, 4, 5] for example. The application of these tools to *a posteriori* error estimation has a more recent history, see [6, 7, 8, 9, 10, 11, 12, 13].

Our main purpose is to describe recent advances in extending the techniques of *a posteriori* error analysis to multiscale operator decomposition solutions of multiphysics, multiscale problems. While the particulars of the analysis vary considerably with the problem, there are several key ideas underlaying a general approach being developed to treat operator decomposition multiscale methods, including the following:

- We identify auxiliary quantities of interest associated with information passed between physical components and solve auxiliary adjoint problems to estimate the error in those quantities.
- We deal with scale differences by introducing projections between discrete spaces used for component solutions and estimate the effects of those projections.
- The standard linearization argument used to define an adjoint operator associated with error analysis for a nonlinear problem may fail, requiring a new approach. The adjoint operator associated with a multiscale operator decomposition solution method is often significantly different than the adjoint associated with the original problem, which affects the stability of the method.
- In practice, solving the adjoint associated with the original fully-coupled problem may present the same kinds of computational challenges posed by the original problem, so attention must be paid to the solution of the adjoint problem.

We explain these ideas in the context of three examples.

3.1. Multiscale decomposition of triangular systems of elliptic problems

We capture the essential features of the thermal actuator model in Example 1.1 using a "one-way" coupled system

$$\begin{cases} -\nabla \cdot a_1 \nabla u_1 + \mathbf{b}_1 \cdot \nabla u_1 + c_1 u_1 = f_1(x), & x \in \Omega, \\ -\nabla \cdot a_2 \nabla u_2 + \mathbf{b}_2 \cdot \nabla u_2 + c_2 u_2 = f_2(x, u_1, Du_1), & x \in \Omega, \\ u_1 = u_2 = 0, & x \in \partial\Omega, \end{cases} \tag{4}$$

where a_i, b_i, c_i, f_i are smooth functions, with $a_1, a_2 \geq \alpha > 0$ on a bounded domain Ω in $\mathbb{R}^N$ with boundary $\partial\Omega$, and α is a constant.

We introduce the finite-element space $\mathcal{S}_{h,1}(\Omega) \subset H_0^1(\Omega)$, corresponding to a discretization $\mathcal{T}_{h,1}$ of Ω for the first component, and another finite element space $\mathcal{S}_{h,2}(\Omega)$, on a different mesh $\mathcal{T}_{h,2}$, for the second component. To evaluate integrals involving functions defined on different meshes, we introduce projections $\Pi_{i \to j}$ from $\mathcal{S}_{h,i}$ to $\mathcal{S}_{h,j}$, for example, interpolants or an L^2 orthogonal projection. We also define the bilinear forms, $\mathcal{A}_i(u_i, v_i) \equiv (a_i \nabla u_i, \nabla v_i) + (b_i(x) \cdot \nabla u_i, v_i) + (c_i u_i, v_i)$, $i = 1, 2$. We present the multiscale operator decomposition algorithm in figure 1.

Algorithm 1 Multiscale Operator Decomposition for Triangular Systems of Elliptic Equations

Construct discretizations $\mathcal{T}_{h,1}, \mathcal{T}_{h,2}$ and corresponding spaces $\mathcal{S}_{h,1}, \mathcal{S}_{h,2}$
Compute $U_1 \in \mathcal{S}_{h,1}(\Omega)$ satisfying $\mathcal{A}_1(U_1, v_1) = (f_1, v_1)$, for all $v_1 \in \mathcal{S}_{h,1}(\Omega)$.
Compute $U_2 \in \mathcal{S}_{h,2}(\Omega)$ satisfying $\mathcal{A}_2(U_2, v_2) = (f_2(x, \Pi_{1\to 2}U_1, \Pi_{1\to 2}DU_1), v_2)$, for all $v_2 \in \mathcal{S}_{h,2}(\Omega)$.

We observe that any errors made in the solution of the first component affect the solution of the second component. This turns out to be a crucial fact for *a posteriori* error analysis.

Example 3.1. We solve a system

$$\begin{cases} -\Delta u_1 = \sin(4\pi x)\sin(\pi y), & x \in \Omega \\ -\Delta u_2 = b \cdot \nabla u_1 = 0, & x \in \Omega, \\ u_1 = u_2 = 0, & x \in \partial\Omega, \end{cases} \qquad b = \frac{2}{\pi}\begin{pmatrix} 25\sin(4\pi x) \\ \sin(\pi x) \end{pmatrix}, \ \Omega = [0,1] \times [0,1], \tag{5}$$

using a standard piecewise linear, continuous finite element method to compute the quantity of interest $u_2(.25, .25)$. We solve for u_1 first and then solve for u_2 using independent meshes.

Using uniform meshes, evaluating the standard *a posteriori* error estimate for the second component problem while ignoring any effect arising from error in the solution of the first component, we estimate the error in the quantity of interest to be $\approx .0042$. However, the true error is $\approx .0048$, so there is discrepancy of $\approx .0006$ ($\approx 13\%$) in the estimate. This is a consequence of ignoring the error "transferred" from the solution of the first component.

If we adapt the mesh for the solution of the second component based on the *a posteriori* error estimate of the error in that component while neglecting the effects of the decomposition, the discrepancy becomes alarmingly worse. For example, we can refine the mesh until the estimate of the error in the second component is $\approx .0001$. But, we find that the true error is $\approx .2244$! See figure 2.

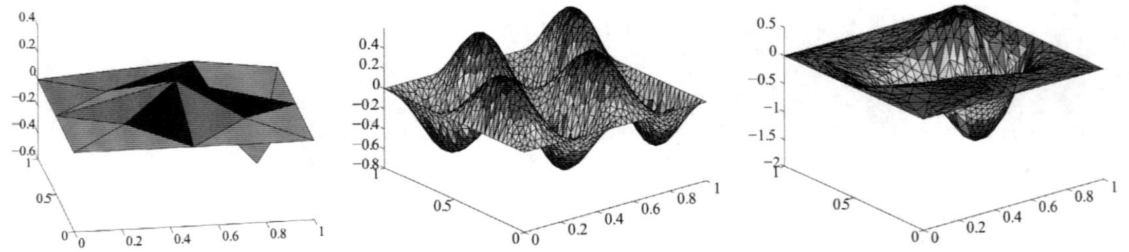

Figure 2. Multiscale operator decomposition solutions (u_1 on the left, u_2 in the middle) of the component problems of (5). We solve for u_2 using a mesh adapted to control the pointwise error of the numerical solution of u_2 while ignoring the effect of errors in the approximation of u_1. On the right, we plot the pointwise error of the computed u_2.

3.1.1. Description of the a posteriori analysis We seek the error in a quantity of interest given by a functional of $e_2 = U_2 - u_2$. Since we introduce some additional, auxiliary quantities of interest, we denote the primary quantity of interest by $(\psi, e) = (\psi_2^{(1)}, e_2)$. We use the adjoint operators

$$\mathcal{A}_i^*(\phi_i, v_i) = (a_i \nabla\phi_i, \nabla v_i) - (\mathrm{div}(b_i\phi_i), v_i) + (c_i\phi_i, v_i), \quad i = 1, 2.$$

We also use the linearization $Lf_2(u_1)(u_1 - U_1) = \int_0^1 \frac{\partial f_2}{\partial u_1}(u_1 s + U_1(1-s))\,ds$.

Noting that the solution of the first adjoint component is not needed to compute the quantity of interest, we define the primary adjoint problem to be

$$\mathcal{A}_2^*(\phi_2^{(1)}, v_2) = (\psi_2^{(1)}, v_2), \ \text{for all } v_2 \in \tilde{W}_2^1(\Omega).$$

The error representation formula for the transfer error is

$$\big(Df_2(U_1) \times e_1, \Pi_{2\to1}\phi_2^{(1)}\big) + \big(Df_2(U_1) \times e_1, (I - \Pi_{2\to1})\phi_2^{(1)}\big), \tag{6}$$

where $\Pi_{i\to j}$ is a projection from discretization i to discretization j. The first term in (6) is the error contribution arising from the transfer while the second term is significant when the approximation spaces are different. The transfer error is a (nominally nonlinear) functional of the error in u_1, defining an auxiliary quantity of interest. We approximate it by a linear functional and define the auxiliary quantity of interest,

$$\big(f_2(u_1) - f_2(U_1), \phi_2^{(1)}\big) \approx \big(Df_2(U_1) \times e_1, \Pi_{2\to 1}\phi_2^{(1)}\big) = \big(\psi_1^{(2)}, e_1\big).$$

We define the corresponding transfer error adjoint problem

$$\mathcal{A}_1^*(\phi_1^{(2)}, v_1) = \big(\psi_1^{(2)}, v_1\big) \text{ for all } v_1 \in \tilde{W}_2^1(\Omega). \tag{7}$$

The additional term $\big(Df_2(U_1) \times e_1, (I - \Pi_{2\to 1})\phi_2^{(1)}\big)$ is a linear functional, so we define another auxiliary quantity of interest

$$\big(\psi_1^{(3)}, e_1\big) = \big(Df_2(U_1) \times e_1, (I - \Pi_{2\to 1})\phi_2^{(1)}\big)$$

and the corresponding adjoint problem

$$\mathcal{A}_1^*(\phi_1^{(3)}, v_1) = \big(\psi_1^{(3)}, v_1\big) \text{ for all } v_1 \in \tilde{W}_2^1(\Omega). \tag{8}$$

We then define the weak residuals

$$\mathcal{R}_1(U_1, \chi; \nu) = (f_1(\nu), \chi) - \mathcal{A}_1(U_1, \chi), \quad \mathcal{R}_2(U_2, \chi; \nu) = (f_2(\Pi_{1\to 2}\nu), \chi) - \mathcal{A}_2(U_2, \chi).$$

The final error representation is therefore [14]

Theorem 3.2

$$\begin{aligned}(\psi, e) = \mathcal{R}_2(U_2, (I - \Pi_2)\phi_2^{(1)}; U_1) + \mathcal{R}_1(U_1, (I - \Pi_1)(\phi_1^{(2)} + \phi_1^{(3)})) \\ + \big(\Pi_{1\to 2} f_2(U_1) - f_2(\Pi_{1\to 2}U_1), \phi_2^{(1)}\big) + \big((I - \Pi_{1\to 2}) f_2(U_1), \phi_2^{(1)}\big).\end{aligned} \tag{9}$$

We emphasize that evaluating the integrals in (9) is far from trivial. We have used Monte-Carlo techniques with good results; see [14].

Example 3.3. In Example 3.1, we estimate the contributions to the error reported in that computation using the relevant portions of (9). To produce the adaptive mesh results shown in figure 2, we construct the adapted mesh using equidistribution based on a bound derived from the first term in (9), that is, neglecting the terms that estimate the transfer error.

Instead, we consider the system (5) for the quantity of interest equal to the average value of U_2. We begin with the same initial coarse meshes as in figure 2, but add the transfer error expression to the mesh refinement criterion. Adapting the mesh so that the total error in the quantity of interest for U_2 has error estimates less than 10^{-4} yields the meshes shown in figure 3. We see that the first component solve requires significantly more refinement than does the second component.

3.2. Multiscale decomposition of reaction-diffusion problems

Operator splitting for reaction-diffusion problems is the classic example of multiscale operator decomposition. We consider

$$\begin{cases} \dfrac{\partial u}{\partial t} = \epsilon\, \Delta u + f(u), & x \in \Omega, 0 < t, \\ \text{suitable boundary conditions}, & x \in \partial\Omega, 0 < t, \\ u(\cdot, 0) = u_0(\cdot) \end{cases} \tag{10}$$

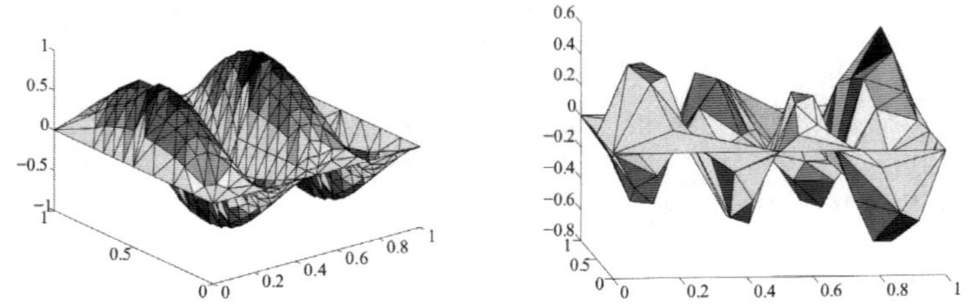

Figure 3. The adapted meshes resulting from the full estimate that accounts for "primary" and "transfer" errors. The transfer error dominates and the mesh for the first component is refined while the mesh for the second component is not. Nonetheless, the simulation yields an accurate value for the quantity of interest. It is interesting to compare this result with that shown in Fig. 2. Since the quantity of interest is computed from the second component, intuition might suggest that the mesh for the second component needs refinement. But, it is the mesh for the first component that is critical for determining accuracy.

where $\Omega \subset \mathbb{R}^d$ is a spatial domain and f is a smooth function. Employing the method of lines, we discretize in space using a continuous, piecewise linear finite element method with M elements and obtain the initial value problem: Find $y \in \mathbb{R}^M$ such that

$$\begin{cases} \dot{y} = Ay(t) + F(y(t)), & 0 < t \leq T, \\ y(0) = y_0, \end{cases} \tag{11}$$

where A is an $l \times l$ constant matrix representing a "diffusion component" and $F(y) = (F_1(y),$ $F_2(y), \cdots, F_l(y))^\top$ is a vector of nonlinear functions representing a "reaction component".

We next discretize in time using a discontinuous (dG) Galerkin method [15, 8]. With appropriate choices of quadrature, we can treat standard integration methods. We first discretize $[0, T]$ into $0 = t_0 < t_1 < t_2 < \cdots < t_N = T$ with diffusion time steps $\{\Delta t_n\}_{n=1}^N$, $\Delta t_n = t_n - t_{n-1}$. For each diffusion step, we choose a (small) time step $\Delta s_n = \Delta t_n / M_n$ with $\Delta s = \max_{1 \leq n \leq N} \Delta s_n$, and the nodes $t_{n-1} = s_{0,n} < s_{1,n} < \cdots < s_{M_n,n} = t_n$. We define the time intervals $I_n = [t_{n-1}, t_n]$ and $I_{m,n} = [s_{m-1,n}, s_{m,n}]$. See Fig. 4. The finite-element approximate solutions are sought in

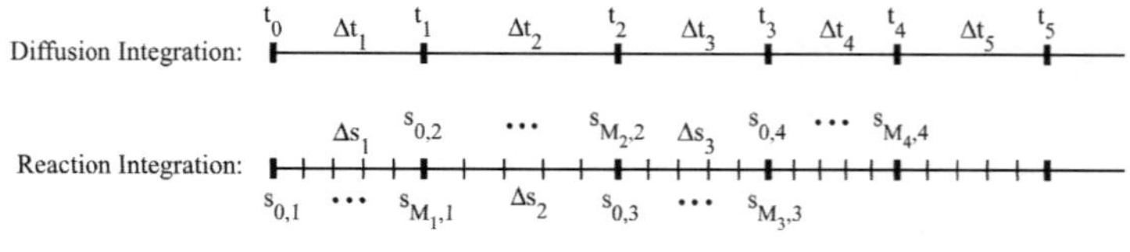

Figure 4. Discretization of time used for multiscale operator splitting.

piecewise polynomial spaces for the diffusion and reaction components, respectively:

$$\mathcal{V}^{(q_d)} = \{U : U|_{I_n} \in \mathcal{P}^{(q_d)}(I_n)\}, \quad \mathcal{V}^{(q_r)}(I_n) = \{U : U|_{I_{m,n}} \in \mathcal{P}^{(q_r)}(I_{m,n})\},$$

for $n = 1, \cdots, N$ and, for each n, $1 \leq m \leq M_n$. $\mathcal{P}^{(q)}(I)$ denotes the space of polynomials in $\mathbb{R}^l$ of degree q on interval I. We let $U_n^{+,-}$ denote the left- and right-hand limits of U at t_n and $[U]_n = U_n^+ - U_n^-$ the jump value of U at t_n.

Let $\tilde{Y}(t)$ be the piecewise continuous finite element approximation of the operator splitting with

$$\tilde{Y}(t) = \frac{t_n - t}{\Delta t_n}\tilde{Y}_{n-1} + \frac{t - t_{n-1}}{\Delta t_n}\tilde{Y}_n, \quad t_{n-1} \le t \le t_n.$$

The nodal values $\tilde{Y}_n$ are obtained from the following procedure:

Algorithm 2 Multiscale Operator Splitting for Reaction-Diffusion Equations

Set $\tilde{Y}_0 = y_0$
for $n = 1, \cdots, N$ **do**
  Set $Y^r{}^-_{0,n} = \tilde{Y}_{n-1}$
  for $m = 1, \cdots, M_n$ **do**
    Compute $Y^r|_{I_{m,n}} \in \mathcal{P}^{(q_r)}(I_{m,n})$ satisfying

$$\int_{I_{m,n}} \left(\dot{Y^r}, W\right) dt + \left([Y^r]_{m-1,n}, W^+_{m-1}\right) = \int_{I_{m,n}} (F(Y^r), W)\, dt \quad \forall\ W \in \mathcal{P}^{(q_r)}(I_{m,n}) \tag{12}$$

  end for
  Set $Y^d{}^-_{n-1} = Y^r{}^-_{M_n,n}$, compute $Y^d|_{I_n} \in \mathcal{P}^{(q_d)}(I_n)$ satisfying

$$\int_{I_n} (\dot{Y^d}, V)\, dt + \left([Y^d]_{n-1}, V^+_{n-1}\right) = \int_{I_n} (AY^d, V)\, dt \quad \forall\ V \in \mathcal{P}^{(q_d)}(I_n) \tag{13}$$

  Set $\tilde{y}_n = y^d(t_n^-)$
end for

Example 3.4. We illustrate the instability of operator splitting applied to the Brusselator problem (2). We apply a standard first order splitting scheme to a space discretization of the

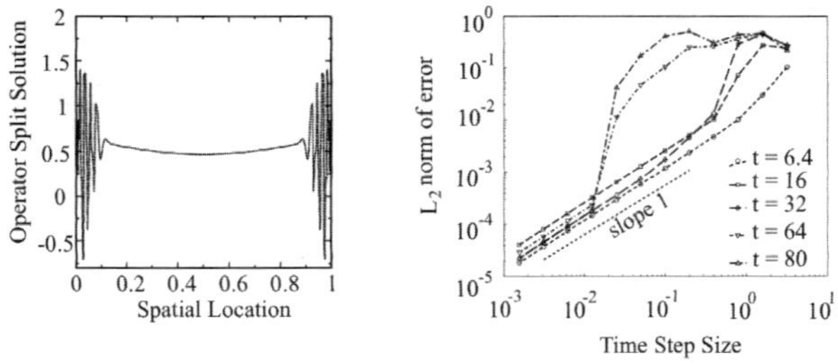

Figure 5. Left: Typical instability arising from multiscale operator splitting applied to Brusselator problem. Solution is shown at time 80. Right: Plots of the error in the L_2 norm versus time step size at different times.

Brusselator model with 500 discrete points with $\alpha = .6$, $\beta = 2$, $k_1 = k_2 = .025$ consisting of the trapezoidal scheme for the diffusion with time step of .2 and backward Euler scheme for the reaction with time step of .004. On the left of figure 5, we show a numerical solution that exhibits nonphysical oscillations that developed after some time. On the right, we show plots of the error versus time steps at different times. There is a critical time step above which the instability develops. Moreover, changing the space discretization does not improve the accuracy. In [16], it is demonstrated that a finer spatial discretization for a constant time step size leads to significantly more error in the long time solution.

An accurate *a posteriori* estimate of the error must account for the stability effects arising from operator splitting. The standard approach used to define an adjoint operator for analyzing nonlinear problems based on linearization of a perturbation equation fails. It turns out that the adjoint operators associated with the original problem and the multiscale operator decomposition method are fundamentally different. In the estimate below, this difference takes the form of "residuals" between certain adjoint operators. A practical difficulty with such a result is that solving the adjoint for the fully coupled problem poses the same multiphysics challenges as solving the original forward problem. We therefore develop a new hybrid *a priori* - *a posteriori* estimate that combines a computable leading order expression obtained using *a posteriori* arguments with a provably higher order bound obtained using *a priori* convergence result.

Note that $\tilde{Y}$ is a consistent numerical solution for an analytic version of the operator splitting and the expression for its error can be estimated using the standard *a posteriori* error analysis. We let ϑ^d define the adjoint solution associated with the diffusion component (13) and ϑ^r define the adjoint solution associated with the reaction component (12) satisfying respectively

$$\begin{cases} -\dot{\vartheta}^d = A^\top \vartheta^d(t), & t_n > t \geq t_{n-1}, \\ \vartheta^d(t_n^-) = \psi_n, & n = N, \cdots, 1, \end{cases} \quad \begin{cases} -\dot{\vartheta}^r = (\hat{F}'(y^r, Y^r))^\top \vartheta^r(t), & s_{m,n} > t \geq s_{m-1,n}, \\ \vartheta^r(s_{m,n}) = \psi^r_{m,n}, & n = N, \cdots, 1, m = M_n, \cdots, 1, \end{cases}$$

with ${\psi^r}_{M_n,n} = {\vartheta^d}^+_{n-1}$ and ${\psi^r}_{m,n} = \vartheta^r_{m,n}$ for $m < M_n$. Thus ϑ^r is continuous across the internal reaction time nodes $s_{m,n}$, $m = 1, \cdots, M_n - 1$. Here, $\hat{F}'(y^r, Y^r) = \int_0^1 F'(sy^r + (1-s)Y^r)\, ds$.

The nonlinearity complicates the analysis because the choice of trajectory around which to linearize is not clear. We cannot use the standard approach of linearizing the error representation because of the operator splitting. Instead, we assume that both the original problem and the operator split version have a common solution and we linearize each problem in a neighborhood of this common solution. For example, we assume that $y = 0$ is a steady state solution of both problems, which can be achieved by assuming that

$$\text{Homogeneity Assumption:} \quad F(0) = 0,$$

and we linearize in a region around 0. In terms of applications to reaction-diffusion problems, there are mathematical reasons for this assumption and it is satisfied in a great many applications. We can modify the analysis to allow for linearization around any known common solution, see [17].

For $n = 1, \cdots, N$, we define the three adjoint problems. The diffusion problem is simplest because it is linear:

$$\begin{cases} -\dot{\varphi}^d = A^\top \varphi^d(t), & t_n > t \geq t_{n-1}, \\ \varphi^d(t_n^-) = \psi_n^d. \end{cases} \tag{14}$$

It is convenient to let Φ_n^d denote the solution operator, so $\varphi^d(t_{n-1}) = \Phi_n^d \psi_n^d$. We require two adjoint problems to treat the reaction component. The difference between the problems is the function around which they are linearized,

$$\begin{cases} -\dot{\varphi}_1^r = \overline{F'(\tilde{Y})}^\top \varphi_1^r(t), & t_n > t \geq t_{n-1}, \\ \varphi_1^r(t_n^-) = \psi_n^r. \end{cases} \qquad \begin{cases} -\dot{\varphi}_2^r = \overline{F'(Y^r)}^\top \varphi_2^r(t), & t_n > t \geq t_{n-1}, \\ \varphi_2^r(t_n^-) = \psi_n^r. \end{cases} \tag{15}$$

If $\Phi_n^r(z)$ denotes the solution operator for the problem linearized around a function z, then we have $\varphi_1^r(t_{n-1}) = \Phi_n^r(\tilde{Y})\psi_n^r$ and $\varphi_2^r(t_{n-1}) = \Phi_n^r(Y^r)\psi_n^r$. We can now prove [17].

Theorem 3.5 *A hybrid* a posteriori - a priori *error estimate*

$$(\tilde{Y}_N - y_N, \psi_N) = \sum_{n=1}^{N} \left(\int_{I_n} (\dot{Y}^d - AY^d, \vartheta^d - \Pi\vartheta^d)\, dt + ([Y^d]_{n-1}, \vartheta^{d+}_{n-1} - \Pi\vartheta^{d+}_{n-1}) \right)$$
$$+ \sum_{n=1}^{N}\sum_{m=1}^{M_n} \left(\int_{I_{m,n}} (\dot{Y}^r - F(Y^r), \vartheta^r - \Pi\vartheta^r)\, dt + ([Y^r]_{m-1,n}, \vartheta^{r+}_{m-1,n} - \Pi\vartheta^{r+}_{m-1,n}) \right)$$
$$+ \sum_{n=1}^{N} (\tilde{Y}_{n-1}, (E_1 + E_2)\psi_n) + \mathbf{O}(\Delta t^{q_d+2}) + \mathbf{O}(\Delta t\, \Delta s^{q_r+1}),$$

where

$$E_1 = \frac{1}{2}\Delta t_n \left(A^\top \mathcal{F}(\tilde{Y}) - \mathcal{F}(\tilde{Y})A^\top \right), \; \mathcal{F}(\tilde{Y}) = \int_{I_n} \overline{F'(\tilde{Y})}\, dt, \quad E_2 = \left(\Phi^r_n(\tilde{Y}) - \Phi^r_n(Y^r) \right) \Phi^d_n.$$

The first expression on the right is the error introduced by the numerical solution of the diffusion component while the second expression is the error introduced by the numerical solution of the reaction component. The third expression on the right measures the effects of operator splitting, where E_1 is a leading order estimate for the effects of operator splitting while E_2 accounts for issues arising from the differences in linearizing around the global computed solution as opposed to the solution of the reaction component. Both of these quantities are scaled by the solution itself, so that these effects become negligible when the solution approaches zero. Finally, the remaining terms represent bounds on terms that are not computable but are higher order. In practice, we neglect those terms when computing an estimate.

Example 3.6.

We consider the Brusselator problem (2) with $\alpha = 2$, $\beta = 5.45$, $k_1 = 0.008$, $k_2 = 0.004$ and initial conditions $u_1(x, 0) = \alpha + 0.1\sin(\pi x)$ and $u_2(x, 0) = \beta/\alpha + 0.1\sin(\pi x)$, which yields an oscillatory solution. In this case, the reaction is very mildly unstable, with at most polynomial rate accumulation of perturbations as time passes. We use a 32 node spatial finite element discretization, resulting in an differential equation system with dimension 62. We note that in original form, the reaction terms do not satisfy the requirement $F(0) = 0$ so we linearize around the steady state solution c with with $c_i = \alpha$ for $i = 1, \cdots, N_e - 1$ and $c_i = \beta/\alpha$ for $i = N_e, \cdots, 2N_e - 2$, so that $F(c) = 0$.

Figure 6 compares the errors computed using $\Delta t = 0.01$ and $M = 10$ reaction time steps to the hybrid *a posteriori* error estimates. We show results for $[0, 2]$, when the solution is still in a transient stage, and at $T = 40$ when the solution has become periodic. All the results show that the exact and estimated errors are in remarkable agreement.

3.3. Multiscale decomposition of a fluid-solid conjugate heat transfer problem

We next consider the multiscale decomposition solution of the heat transfer problem described in Example 1.1. The weak formulation of (3) consists of computing $u \in V_F$, $p \in L^2_0(\Omega_F)$, $T_F \in W_F$ and $T_S \in W_S$ such that

$$\begin{cases} a_1(u, v) + c_1(u, u, v) + b(v, p) + d(T_F, v) = (f, v), \\ b(u, q) = 0, \\ a_2(T_F, w_F) + c_2(u, T_F, w_F) + a_3(T_S, w_S) = (Q_F, w_F) + (Q_S, w_S), \end{cases} \tag{16}$$

for all $v \in V_{F,0}$, $q \in L^2_0(\Omega_F)$, $w_F \in W_{F,0}$ and $w_S \in W_{S,0}$, where

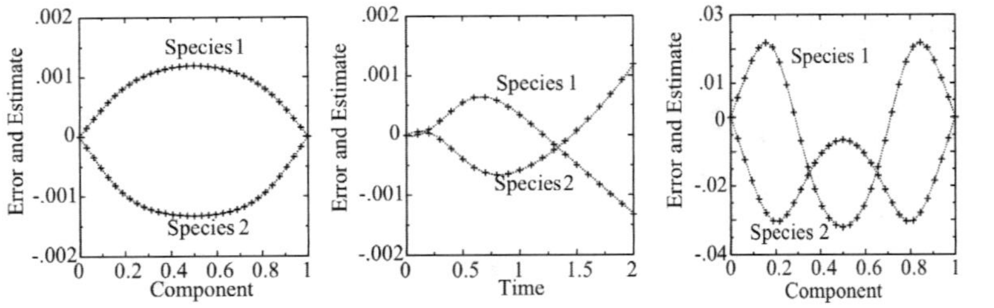

Figure 6. Brusselator results. Left: Comparison of errors against the spatial location at $T = 2$. Middle: Time history of errors at the midpoint location on $[0, 2]$. Right: Comparison of errors against the spatial location at $T = 40$. The dotted line is the exact error, and the (+) is the estimated error

$$\begin{aligned}
&a_1(u,v) = \int_{\Omega_F} \mu(\nabla u : \nabla v)\, dx, && a_2(T_F, w_F) = \int_{\Omega_F} k_F(\nabla T_F \cdot \nabla w_F)\, dx,\\
&a_3(T_S, w_S) = \int_{\Omega_S} k_S(\nabla T_S \cdot \nabla w_S)\, dx, && b(v,q) = -\int_{\Omega_F} (\nabla \cdot v) q\, dx,\\
&c_1(u,v,z) = \int_{\Omega_F} \rho_0\, (u \cdot \nabla)\, v \cdot z\, dx, && c_2(u,T,w) = \int_{\Omega_F} \rho_0 c_p (u \cdot \nabla T)\, w\, dx,\\
&d(T,v) = \int_{\Omega_F} \rho_0 \beta T g \cdot v\, dx, && f = \rho_0\, (1 + \beta T_0)\, g,
\end{aligned}$$

and

$$\begin{aligned}
V_F &= \left\{v \in H^1(\Omega_F) \,|\, v = g_{u,D} \text{ on } \Gamma_{u,D}\right\}, & V_{F,0} &= \{v \in V_F \,|\, v = 0 \text{ on } \Gamma_{u,D}\},\\
W_F &= \left\{w \in H^1(\Omega_F) \,|\, w = g_{T_F,D} \text{ on } \Gamma_{T_F,D}\right\}, & W_{F,0} &= \{w \in W_F \,|\, w = 0 \text{ on } \Gamma_{T_F,D}\},\\
W_S &= \left\{w \in H^1(\Omega_S) \,|\, w = g_{T_S,D} \text{ on } \Gamma_{T_S,D}\right\}, & W_{S,0} &= \{w \in W_S \,|\, w = 0 \text{ on } \Gamma_{T_S,D}\}.
\end{aligned}$$

To discretize, we construct independent locally-quasi-uniform triangulations $T_{F,h}$ and $T_{S,h}$ of Ω_F and Ω_S, respectively. We use the piecewise polynomial spaces

$$\begin{aligned}
V_F^h &= \left\{v \in V_F \mid v \text{ continuous on } \Omega_F,\ v_i \in P^2(K) \text{ for all } K \in \tau_{F,h}\right\},\\
Z^h &= \left\{z \in Z \mid z \text{ continuous on } \Omega_F,\ z \in P^1(K) \text{ for all } K \in \tau_{F,h}\right\},\\
W_F^h &= \left\{w \in W_F \mid w \text{ continuous on } \Omega_F,\ w \in P^2(K) \text{ for all } K \in \tau_{F,h}\right\},\\
W_S^h &= \left\{w \in W_S \mid w \text{ continuous on } \Omega_S,\ w \in P^2(K) \text{ for all } K \in \tau_{S,h}\right\},
\end{aligned}$$

and the associated subspaces

$$V_{F,0}^h = \left\{v \in V^h \mid v = 0 \text{ on } \Gamma_{u,D}\right\}, \quad W_{F,0}^h = \left\{w \in W_F^h \mid w = 0 \text{ on } \Gamma_{T_F,D}\right\},$$
$$W_{S,0}^h = \left\{w \in W_S^h \mid w = 0 \text{ on } \Gamma_{T_S,D} \text{ and } w = 0 \text{ on } \Gamma_I\right\},$$

where $P^q(K)$ denotes the space of polynomials of degree q on an element K. Note that $W_{S,0}^h$ is different from $W_{F,0}^h$ in an important way since $\Gamma_{T_S,D}$ does not include Γ_I. We let π_V, π_{W_F}, π_{W_S}, and π_Z be projections into V_F^h, W_F^h, W_S^h and Z^h respectively. We also use π_{W_F} and π_{W_S} to denote projections into W_F^h and W_S^h respectively along the interface Γ_I.

To compute a stable solution of the fluid equations, we choose V_F^h and Z^h to be the Taylor-Hood finite element pair satisfying the discrete inf-sup condition. In general, the convergence of the iteration defined by this algorithm depends on the values of k_S and k_F along the interface and the geometry of each region. Often, a "relaxation" is used to help improve convergence properties, which affects the analysis [18, 19].

Algorithm 3 Multiscale Decomposition Method for Conjugate Heat Transfer

k = 0
while $(\|T_S^{\{k\}} - \pi_S T_F^{\{k\}}\|_{\Gamma_I} > TOL)$ **do**
  k = k+1
  Compute $T_{S,h}^{\{k\}} \in W_S^h$ such that $T_{S,h}^{\{k\}} = \pi_{W_S} T_{F,h}^{\{k-1\}}$ along the interface Γ_I and

$$a_3(T_{S,h}^{\{k\}}, w) = (Q_S, w), \quad \forall w \in W_{S,0}^h, \tag{17}$$

  Compute $u_h^{\{k\}} \in V_F^h$, $p_h^{\{k\}} \in Z^h$ and $T_{F,h}^{\{k\}} \in W_F^h$ such that

$$\begin{cases} a_1(u_h^{\{k\}}, v) + c_1(u_h^{\{k\}}, u_h^{\{k\}}, v) + b(v, p_h^{\{k\}}) + d(T_{F,h}^{\{k\}}, v) = (f, v), & \forall v \in V_{F,0}^h, \\ b(u_h^{\{k\}}, q) = 0, & \forall q \in Z^h, \\ a_2(T_{F,h}^{\{k\}}, w) + c_2(u_h^{\{k\}}, T_{F,h}^{\{k\}}, w) = (Q_F, w) - (k_S(n \cdot \nabla T_{S,h}^{\{k\}}), w)_{\Gamma_I}, & \forall w \in W_{F,0}^h. \end{cases} \tag{18}$$

end while

3.3.1. Description of an a posteriori error analysis We define the errors $e_u = u - u_h^{\{k\}}$, $e_p = p - p_h^{\{k\}}$, $e_{T_F} = T_F - T_{F,h}^{\{k\}}$, and $e_{T_S} = T_S - T_{S,h}^{\{k\}}$. The adjoint problem for the quantity of interest $(\psi, e) = (\psi_u, e_u) + (\psi_p, e_p) + (\psi_{T_F}, e_{T_F}) + (\psi_{T_S}, e_{T_S})$ is

$$\begin{cases} -\mu\Delta\phi + \overline{c}_1^*(\phi) + \nabla z + \overline{c}_{2u}^*(\theta_F) = \psi_u, & x \in \Omega_F, \\ -\nabla \cdot \phi = \psi_p, & x \in \Omega_F, \\ -k_F\Delta\theta_F + \overline{c}_{2T}^*(\theta_F) + \rho_0\beta(\mathbf{g}\cdot\phi) = \psi_{T_F}, & x \in \Omega_F, \\ \begin{cases} \theta_F = \theta_S, \\ k_F(n\cdot\nabla\theta_F) = k_S(n\cdot\nabla\theta_S), \end{cases} & x \in \Gamma_I, \\ -k_S\Delta\theta_S = \psi_{T_S}, & x \in \Omega_S, \end{cases} \quad \begin{cases} \phi = \mathbf{0}, & x \in \Gamma_{u,D}, \\ \mu\dfrac{\partial\phi}{\partial n} = \mathbf{0}, & x \in \Gamma_{u,N}, \\ \theta_F = 0, & x \in \Gamma_{T_F,D}, \\ k_F(n\cdot\nabla\theta_F) = 0, & x \in \Gamma_{T_F,N}, \\ \theta_S = 0, & x \in \Gamma_{T_S,D}, \\ k_S(n\cdot\nabla\theta_S) = 0, & x \in \Gamma_{T_S,N}. \end{cases} \tag{19}$$

Here, we have used the linearizations

$$\overline{c}_1^*(\phi) = \frac{1}{2}\rho_0\nabla(u+u_h)\cdot\phi - \frac{1}{2}\rho_0(u+u_h)\cdot\nabla\phi - \frac{1}{2}\rho_0(\nabla\cdot(u+u_h))\phi,$$
$$\overline{c}_{2u}^*(\theta) = \frac{1}{2}\rho_0 c_p\nabla(T+T_h)\theta, \quad \overline{c}_{2T}^*(\theta) = -\frac{1}{2}\rho_0 c_p(u+u_h)\cdot\nabla\theta - \frac{1}{2}\rho_0 c_p(\nabla\cdot(u+u_h))\theta.$$

We solve (19) numerically using an iterative operator decomposition approach as for the forward problem independently of the forward iterations. In [18, 19], we derive estimates that require only adjoint solutions of the two component problems.

To write out the *a posteriori* error representation, we introduce an additional projection $\pi_{W_S}^0 : H^2 \to W_{S,0}^h$ defined such that for any node x_i. $\pi_{W_S}^0\theta_S(x_i) = \pi_{W_S}\theta_S(x_i)$ for $x_i \notin \Gamma_I$, and 0 otherwise. We can now prove the following theorem [19].

Theorem 3.7

$$\begin{aligned}(\psi, e) = (f, \phi - \pi_V\phi) - a_1(u_h^{\{k\}}, \phi - \pi_V\phi_1) - c_1(u_h^{\{k\}}, u_h^{\{k\}}, \phi - \pi_V\phi)\\ - b(\phi - \pi_V\phi, p_h) - d(T_{F,h}^{\{k\}}, \phi - \pi_V\phi) - b(u_h^{\{k\}}, z - \pi_Z z)\end{aligned} \quad (20)$$

$$\begin{aligned}+ (Q_F, \theta_F - \pi_{W_F}\theta_F) - a_2(T_{F,h}^{\{k\}}, \theta_F - \pi_{W_F}\theta_F)\\ - c_2(u_h^{\{k\}}, T_{F,h}^{\{k\}}, \theta_F - \pi_{W_F}\theta_F) + (Q_S, \theta_S - \pi_{W_S}\theta_S) - a_3(T_{S,h}^{\{k\}}, \theta_S - \pi_{W_S}\theta_S)\end{aligned} \quad (21)$$

$$+ \left(T_{S,h}^{\{k\}} - \pi_S T_{F,h}^{\{k\}}, k_S(n\cdot\nabla\theta_S)\right)_{\Gamma_I} + \left(\pi_S T_{F,h}^{\{k\}} - T_{F,h}^{\{k\}}, k_S(n\cdot\nabla\theta_S)\right)_{\Gamma_I} \quad (22)$$

$$+ \left(k_S(n\cdot\nabla T_{S,h}^{\{k\}}), \pi_{W_F}\theta_F\right)_{\Gamma_I} + \left(Q_S, \pi_{W_S}\theta_S\right) - a_3\left(T_{S,h}^{\{k\}}, \pi_{W_S}\theta_S\right). \quad (23)$$

The contributions to the error are threefold.

- Equations (20)–(21) represents the contribution of the discretization error arising from each component solve.
- Equation (22) represents the contribution from the iteration.
- The first term in (23) represents contribution of the transfer error, while the remaining terms represent the contribution arising from projections between two different discretizations.

Example 3.8.

For the flow past a cylinder, we solve the steady nondimensionalized Boussinesq equations in the fluid domain and the nondimensional heat equation in the solid domain. To simulate the flow of water past a cylinder made from stainless steel, we set the dimensionless constants $Pr = 6.6$ and $k_R = 30$, and choose the inflow velocity and the temperature gradient so that $Re = 75$, $Pe = 495$, $Fr = 0.001$, $Ra = 50$. The temperature gradient is imposed by setting different temperatures along the top and bottom boundaries, with a linear temperature gradient on the inflow boundary, and an adiabatic condition on the outflow boundary.

We show results for two quantities of interest. The first is the temperature in a small region in the wake, located approximately one channel width downstream of the center of the cylinder and 1/4 of a channel width below the upper wall. The second is temperature at a small region in the center of the cylinder. In each case, we base adaptivity on an element tolerance of 1×10^{-8}.

We show the final adaptive meshes for the flow and solid in figure 7. For the first quantity of interest, the flow mesh is most refined near the region of interest and upstream of the region of interest, locating more elements between the cylinder and the top wall than the cylinder and the bottom wall since the flow advecting heat to the region of interest passes above rather than below the cylinder. For the solid, the mesh is highly refined along the top in order to increase the accuracy of the normal derivative that is computed in the solid and used as a boundary condition in the fluid computation. Evidently, the normal derivatives elsewhere on the interface have less of an influence on the first quantity of interest. For the second quantity of interest, the mesh is highly refined upstream of the cylinder. We note that the refinement downstream of the cylinder corresponds closely to the recirculation region, and the mesh refinement is slightly asymmetric about the midplane of the channel due to the asymmetric initial mesh. The mesh in the solid is refined uniformly near the boundary, reflecting the fact that the error in the finite element flux makes a significant contribution to the error in the quantity of interest.

3.3.2. Loss of order and flux correction The meshes shown in figure 7 are highly refined near the interface. This result reflects the fact that there is significant error in the numerical flux passed between the components. It turns out that this pollutes the entire computation, so that overall the method loses an entire order of accuracy.

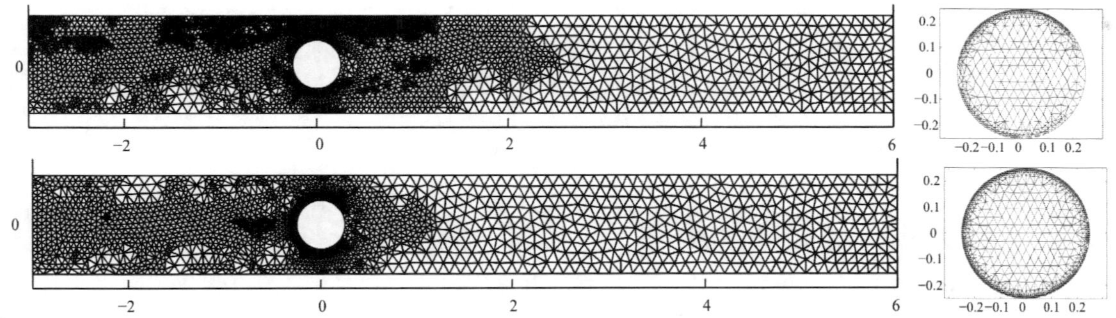

Figure 7. Upper: Final adaptive meshes in the fluid and solid when the quantity of interest is the temperature in a small region in the wake above the cylinder. Lower: Final adaptive meshes in the fluid and solid when the quantity of interest is the temperature in a small region in the center of the solid.

One way to compensate for the loss of order is by refining the mesh locally near the interface. Another way is to compute the particular information, in this case the flux on the interface, more accurately. It turns out that we can adapt a post-processing technique called flux correction developed originally by Wheeler [20] and Carey [21, 22] to recover boundary flux values with increased accuracy.

We denote the set of elements in $\tau_{S,h}$ that intersect the interface boundary by $\tau_{S,h}^{\Gamma_I} = \{K \in \tau_{S,h} \mid \overline{K} \cap \Gamma_I \neq \emptyset\}$, and we consider the corresponding finite element space $\Sigma_h = \{v \in P^2(K)$ with $K \in \tau_{S,h}^{\Gamma_I}$, $v(\eta_i) = 0$ if $\eta_i \notin \Gamma_I\}$, where $\{\eta_i\}$ denotes the nodes of element K. The degrees of freedom correspond to the nodes on the boundary. We compute $\sigma^{\{k\}} \in \Sigma_h$ satisfying

$$-\big(\sigma^{\{k\}}, v\big)_{\Gamma_I} = (Q_S, v) - a_3(T_{S,h}^{\{k\}}, v), \quad \text{for all } v \in \Sigma_h,$$

where $T_{S,h}^{\{k\}}$ solves (17). Since the dimension of the problem scales with the number of nodes on a boundary, it is relatively inexpensive to solve.

In the algorithm, (18) is replaced by the following: Compute $u_h^{\{k\}} \in V_F^h$, $p_h^{\{k\}} \in Z^h$ and $T_{F,h}^{\{k\}} \in W_F^h$ such that

$$\begin{cases} a_1(u_h^{\{k\}}, v) + c_1(u_h^{\{k\}}, u_h^{\{k\}}, v) + b(v, p_h^{\{k\}}) + d(T_{F,h}^{\{k\}}, v) = (f, v), \\ b(u_h^{\{k\}}, q) = 0, \\ a_2(T_{F,h}^{\{k\}}, w) + c_2(u_h^{\{k\}}, T_{F,h}^{\{k\}}, w) = (Q_F, w) - \big(\sigma^{\{k\}}, w\big)_{\Gamma_I}, \forall v \in V_{F,0}^h, q \in Z^h, w \in W_{F,0}^h. \end{cases} \tag{24}$$

Using the recovered boundary flux leads to a cancelation of the "transfer error" term in the error representation formula (23), which is the source of the loss of order. We can prove that using the recovered flux recovers the expected cubic order of convergence; see [19].

Example 3.9. The recovered accuracy is easily demonstrated by considering the adapted meshes produced by the modified algorithm. We show the final adaptive meshes for the solid in figure 8. There is no mesh refinement near the boundaries, indicating that the flux error is no longer dominant.

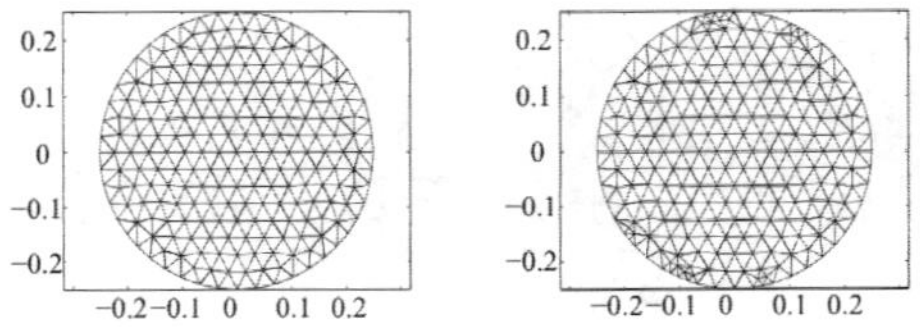

Figure 8. Left: final adaptive mesh in the solid when the quantity of interest is the temperature in a small region in the wake above the cylinder. Right: final adaptive mesh in the solid when the quantity of interest is the temperature in a small region in the center of the solid.

4. Conclusion

Multiphysics, multiscale models present significant challenges in terms of computing accurate solutions and for estimating the error in information computed from numerical solutions. In this paper, we describe recent advances in extending the techniques of *a posteriori* error analysis to multiscale operator decomposition solution methods. While the particulars of the analysis vary considerably with the problem, several key ideas underlie a general approach to treat operator decomposition multiscale methods that is under development. We explain these ideas in the context of three specific examples.

In this paper, we have minimized the effects arising from the solution of nonlinear and/or fully coupled systems by carefully choosing the models and results that are discussed. Referring back to figure 1, we generally expect multiscale operator decomposition to require a number of iterations between the physical components. This raises additional issues that are discussed in [18, 19, 23, 17, 24].

Acknowledgments

This work was supported in part by DOE (DE-FG02-04ER25620, DE-FG02-05ER25699, DE-FC02-07ER54909), NASA (NNG04GH63G), NSF (DMS-0107832, DMS-0715135, DGE-0221595003, MSPA-CSE-0434354, ECCS-0700559), INL (00069249), LLNL (B573139) and SNL (PO299784)

References

[1] Prigogine I and Lefever R 1968 *J. Chem. Phys* **48** 1695–1700
[2] Marchuk G I, Agoshkov V I and Shutyaev V P 1996 *Adjoint equations and perturbation algorithms in nonlinear problems* (Boca Raton, FL: CRC Press) ISBN 0-8493-2871-3
[3] Lanczos C 1997 *Linear Differential Operators* (Dover Publications)
[4] Cacuci D 1997 *Sensitivity and Uncertainty Analysis: Theory* vol I (Chapman & Hall/CRC)
[5] Marchuk G I 1995 *Adjoint equations and analysis of complex systems* (Kluwer)
[6] Eriksson K, Estep D, Hansbo P and Johnson C 1996 *Computational differential equations* (Cambridge: Cambridge University Press) ISBN 0-521-56312-7; 0-521-56738-6
[7] Eriksson K, Estep D, Hansbo P and Johnson C 1995 *Acta numerica, 1995* Acta Numer. (Cambridge: Cambridge Univ. Press) pp 105–158
[8] Estep D, Larson M G and Williams R D 2000 *Mem. Amer. Math. Soc.* **146** viii+109 ISSN 0065-9266
[9] Becker R and Rannacher R 2001 *Acta Numerica* 1–102
[10] Giles M and Süli E 2002 *Acta Numerica* 145–236
[11] Bangerth W and Rannacher R 2003 *Adaptive Finite Element Methods for Differential Equations* (Birkhauser Verlag)
[12] Paraschivoiu M, Peraire J and Patera A 1997 *Comput. Methods Appl. Mech. Engrg.* **150** 289–312
[13] Barth T J 2004 *A-Posteriori Error Estimation and Mesh Adaptivity for Finite Volume and Finite Element Methods* (*Lecture Notes in Computational Science and Engineering* vol 41) (New York: Springer)
[14] Carey V, Estep D and Tavener S 2006 *SINUM* To appear
[15] Estep D and Stuart A M 2002 *Math. Comp.* **71** 1075–1103 (electronic) ISSN 0025-5718
[16] Ropp D L, Shadid J N and Ober C C 2004 *J. Comput. Phys.* **194** 544–574 ISSN 0021-9991

[17] Ginting V, Estep D, Shadid J and Tavener S 2006 *SINUM* To appear
[18] Estep D, Tavener S and Wildey T 2006 *SINUM* To appear
[19] Estep D, Tavener S and Wildey T 2008 *JCP* Submitted
[20] Wheeler M 1974 *Numer. Math* **2** 99–109
[21] GF Carey SS Chow M S 1985 *Comp. Meth. in Applied Mech. and Engr.* **50** 107–120
[22] Carey G 1982 *Comp. Meth. in Applied Mech. and Engr.* **35** 1–14
[23] Carey V, Estep D and Tavener S 2008 *In preparation*
[24] Ginting V, Estep D and Tavener S 2008 *IMA J. Numer. Analysis* Submitted

Petascale algorithms for reactor hydrodynamics

Paul Fischer, James Lottes, David Pointer and Andrew Siegel

Argonne National Laboratory, Argonne, IL 60439, USA

E-mail: `fischer@mcs.anl.gov`

Abstract. We describe recent algorithmic developments that have enabled large eddy simulations of reactor flows on up to $P = 65,000$ processors on the IBM BG/P at the Argonne Leadership Computing Facility.

1. Introduction

Petascale computing is expected to play a pivotal role in the design and analysis of next-generation nuclear reactors. Argonne's SHARP project is focused on advanced reactor simulation, with a current emphasis on modeling coupled neutronics and thermal-hydraulics (TH). The TH modeling comprises a hierarchy of computational fluid dynamics approaches ranging from detailed turbulence computations, using DNS (direct numerical simulation) and LES (large eddy simulation), to full core analysis based on RANS (Reynolds-averaged Navier-Stokes) and subchannel models. Our initial study is focused on LES of sodium-cooled fast reactor cores. The aim is to leverage petascale platforms at DOE's Leadership Computing

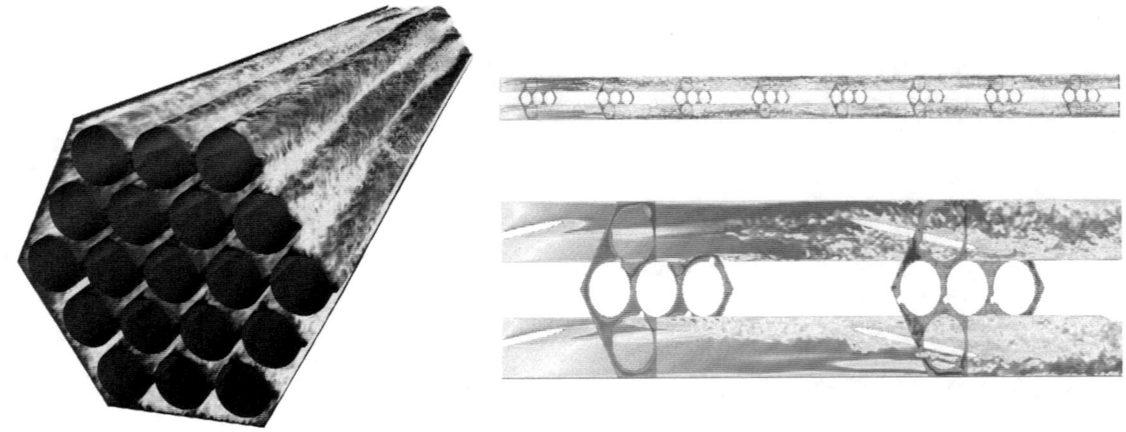

Figure 1. Turbulence in wire-wrapped subassemblies visualized by axial velocity distributions. Periodic boundary conditions applied over a single pitch, $z \in [0, H]$, are used for the 19-pin case (left) while inflow/outflow conditions at $z = 0/3H$ are used for the 7-pin case (right). Transition to turbulence from a uniform inlet flow occurs at $z < H/2$ in the 7-pin case (lower right).

Facilities (LCFs) to provide detailed information about heat transfer within the core and to provide baseline data for less expensive RANS and subchannel models [5].

Figure 1 shows recent simulations of turbulent coolant flow in 19- and 7-pin assemblies. Coolant passes through interior flow channels between the pins and through exterior edge and corner channels between the pins and the walls. The pins are separated by helically wrapped spacer wires that also divert flow between channels and enhance mixing. Questions to be answered by these initial studies include the degree of cross-assembly mixing induced by the spacer wires, the amount of bypass flow in the edge channels, and the magnitude of the pressure drop through the assembly. These fine-scale simulations are providing data and physical insight previously accessible only through experiment.

LES of full or even partial subassemblies is challenging. For pin diameter D, the hydraulic diameter for an interior channel is $D_h \approx 0.4D$. The wire pitch is $H \approx 60D_h$, which is much longer than the correlation length at typical Reynolds numbers, $Re_h := UD_h/\nu \approx 50,000$. The number of channels is roughly twice the number of pins, and the domain length is $\approx 10H$. Thus, for a 217-pin subassembly one is faced with LES in a channel of length equivalent to $\approx 400\times10\times60D_h = 240000D_h$. Fortunately, this figure can be reduced by exploiting symmetries. In particular, the relatively short entrance length ($< 60D_h$; see figure 1) for these flows permits the use of axial periodicity such that one only needs to simulate a single wire pitch. Even with periodicity, the range of scales encountered in a single subassembly has driven the computational requirements for our LES code, Nek5000, to unprecedented levels.

In the remainder of this paper, we describe recent algorithmic advances that are enabling scalable simulations on the DOE's new petascale architectures.

2. Petascale flow solvers

Our large eddy simulations are based on the spectral element code, Nek5000, which has been designed from the outset for distributed-memory platforms. The spectral element method (SEM) combines the geometric flexibility of finite elements with the minimal numerical dispersion and dissipation of spectral methods, making it ideally suited to LES in large complex domains, where one is interested in accurate propagation of the principal energy carrying eddies over long times. In the SEM, the solution within each of E elements is represented as a tensor-product of Nth-order Lagrange polynomials based on the Gauss-Lobatto-Legendre nodal points, resulting in $n \approx EN^3$ degrees of freedom per scalar field. Our current LES formulation relies on slight (resp., 1.25 and 5%) filtering of modes k=N-1 and N, which provides an energy drain at the unresolved grid scale. In well-resolved regions, the action of the filter is void and spectral accuracy is retained.

The principal computational bottleneck in simulating unsteady incompressible and low-Mach number flows is the elliptic problem governing the pressure, which must be computed implicitly at each timestep. For large unstructured problems, the resultant discrete Poisson problem is most efficiently solved using multilevel iterative methods embedded within a Krylov subspace projection scheme such as conjugate gradients or GMRES. For the SEM, we employ variational multigrid using *local* overlapping Schwarz methods for element-based smoothing at resolution N and $\approx N/2$, coupled with a *global* coarse-grid problem based on linear elements. (A detailed description of our solver can be found in [2, 3].)

The coarse-grid problem is a well-known impediment to scalability. In the past, we have used a projection-based scheme to solve the coarse problem $A\underline{x} = \underline{b}$. If A is sparse symmetric positive definite, then there exists a quasi-sparse matrix X with $O(n^\alpha)$ nonzeros, $1 < \alpha < 2$ such that $\underline{x} = XX^T\underline{b}$. For linear finite element discretizations of the 3D Poisson problem, one can compute $\underline{x}$ with a P-processor work complexity of $T_a \approx 10n^{5/3}/P$ and a communication complexity involving only $2\log_2 P$ contention-free messages of length $\leq 5n^{2/3}$ [6]. While this highly parallel and communication-minimal scheme has been quite effective for $(P,n) < (10^4, 10^5)$, its superlinear costs are ultimately limiting, and we have recently turned to algebraic multigrid (AMG) as an alternative.

Our AMG approach is guided by the two-grid asymptotic convergence rate, equal to the spectral radius of the error propagation matrix $\rho(\ (I-BA)(I-PA_c^{-1}P^TA)\ )$, where the (AMG) coarse operator is $A_c = P^TAP$. Here we see that the iteration is determined by the choice of *smoother*, B, and by the choice of the $n \times n_c$ *prolongation* matrix, P. P is further constrained by specifying that the coarse "C-variables" be a subset of the original variables. Ordering these variables last gives rise to the block forms

$$A = \begin{pmatrix} A_{ff} & A_{fc} \\ A_{cf} & A_{cc} \end{pmatrix}, \qquad P = \begin{pmatrix} W \\ I \end{pmatrix}, \qquad B = \begin{pmatrix} \hat{B}_{ff} & 0 \\ 0 & 0 \end{pmatrix}.$$

The key components of the AMG scheme are the choice of the C-variables, the prolongation weights (W), and the smoother. (We discuss the form of B below.)

We take $\kappa(D_{ff}^{-1/2}A_{ff}D_{ff}^{-1/2})$ to be a measure of coarsening quality, as it determines both the support required for the prolongation weights and the number of smoothing steps required. Our coarsening procedure assigns coarse variables in order to eliminate the maximal Gershgorin disc radii of $D_{ff}^{-1/2}A_{ff}D_{ff}^{-1/2} - I$, where D_{ff} is the diagonal of A_{ff}. The procedure stops once the a priori bound on the quality measure is below some tolerance.

Our prolongation weights are based on the energy-minimizing interpolation of Wan, Chan, and Smith [7], which is the solution to a constrained minimization problem that may be viewed as a tractable proxy to the problem of minimizing the departure of W from the minimal energy weights, $-A_{ff}^{-1}A_{fc}$, given a sparsity pattern on W. We determine the supports in a spirit similar to sparse approximate inverse (SAI) preconditioners by locally approximating the columns of $A_{ff}^{-1}A_{fc}$ and widening the support until the approximation is accurate to a given tolerance. For the smoother, $\hat{B}_{ff}$, we choose diagonal SAI accelerated in a Chebyshev polynomial (typically of degree 2 or 3). The choice of using "F-relaxation" was motivated by the fact that no additional benefit may come from using a full smoother in a two-level asymptotic convergence rate analysis and that, by approximately solving equations governed by A_{ff}, one may legitimately employ Krylov subspace (e.g., Chebyshev) acceleration.

The success of our AMG solver derives from its robust convergence properties (overall convergence is the same as with the direct XX^T solver) and the rapid reduction in the work with each level of the V-cycle. For the 19-pin case of figure 1, which starts with 120 million points, the AMG grid sizes are 417600, 267100, 83346, 37097, 19339, 7912, 2237, 459, 92, 15, 2, and 1. The challenge for AMG is to minimize communication overhead, as we discuss below.

Significant improvements in scalability derived from rewriting our central communication kernel, *gs()*, which implements the interface exchange (gather-scatter) between spectral elements. The *gs()* kernel is a general-purpose utility for implementing parallel matrix-vector products with a particularly lightweight interface. In a setup phase, one invokes *gs_handle* = *gs_setup(glo_num,*n_p*)*, where *glo_num()* is a local list of n_p integers containing a global identifier (e.g., column number) for each entry. For any repeated entries in *glo_num()* (locally or on other processors) the call *gs(gs_handle,* $\underline{u}$*,'+')*, will return the sum of the corresponding elements of $\underline{u}$. In short, if Q is an arbitrary Boolean matrix, *gs()* implements the symmetric operation QQ^T [1]. The code supports as its last argument any commutative/associative operator as well as a mode for vector fields of arbitrary dimension. The new setup routine discovers interprocessor connections using binning, which is implemented using a crystal-router- (CR-) based all-to-all communication [4]. The CR strategy is simply stated. Any processor $p < P/2$ having data needed by any processor $q \geq P/2$ sends its data to $p + P/2$. Processors in the upper half reciprocate. One then bisects the processor set and recurs, and the exchange terminates after

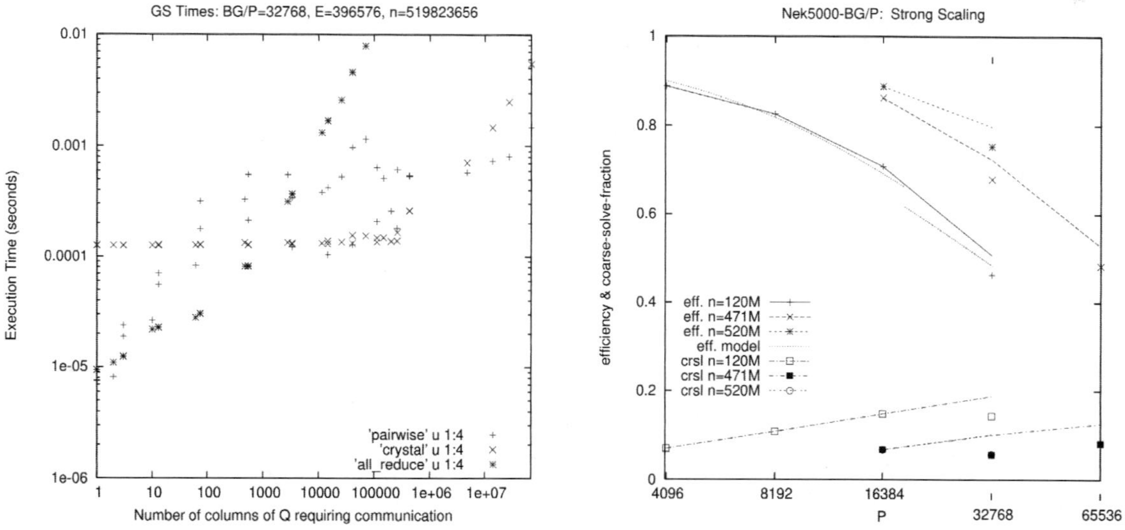

Figure 2. Nek5000 performance on BGP: (left) per-call communication costs for several implementations of the *gs()* kernel versus number of nontrivial columns for the case $P = 32768$; (right) efficiency and coarse solve fraction for problems with n=120M, 470M, and 520M gridpoints.

$\log_2 P$ phases. With this strategy, *gs_setup()* requires $\approx$ .09 seconds for n=150 million on 32768 processors of BG/L. (We remark that the stand-alone *gs()* code is available from the authors.)

While robust, general, and easy to use, *gs()* was originally designed for *nearest-neighbor* communications—effectively, Q sparse—and relied exclusively on pairwise processor exchanges to swap edge and face data. AMG, which also requires general unstructured communication, can have stencil widths of several hundred at the intermediate levels. A pairwise exchange strategy in this case would require hundreds of messages and suffer significant latency. To this end, we have restructured *gs()* to support several exchange strategies within the same interface and with the chosen strategy determined by a test that is invoked automatically at setup. In addition to pairwise exchanges, *gs()* now supports *all_reduce()*- and CR-based exchanges. The former approach simply puts the entries from each processor onto a null-vector of length n and then condenses the result using the native MPI *all_reduce()*. This is very fast on BG/P, which has a dedicated tree network for this task. The latter approach employs the CR scheme described above, with condensation taking place whenever shared values collide at intermediate stages. Figure 2, left, shows the performance envelope for the three strategies as a function of the number of columns of Q having more than one entry, taken over all calls to *gs setup()* for the 7-pin configuration of Fig. 1 with N=11. While the plot does not indicate the (important) multiplicity of each column of Q, it nonetheless gives some insight to the communication patterns. In particular, *pairwise* is generally fastest for the large sparse Qs, while *all_reduce()* is superior for the smallest systems and *CR* is optimal in the middle range.

3. Parallel scaling on BG/P

Strong scaling studies were performed for models shown in figure 1. Case 1 is the 19-pin model with E=359424 and N=7, corresponding to n=120 million points. Case 2 is the same model, but with N=11 (n=471 million) and Case 3 is the 7-pin $\times$ 3-pitch inflow/outflow configuration, with $E = 396576$ and N=11 (n=520 million). Each case is run for 100 timesteps without I/O.

Table 1 gives a detailed breakdown of the parallel overhead for Case 1. Reported times are:

Table 1. BG/P solution times (seconds) for 19-pin case, $N = 7$.

Case	Total	QQ^T	Coarse	all_reduce()
x4096	1994	125	1180	1.2
a4096	1112	125	192	1.4
b4096	846	126	25	1.
8192	460	88	22	1.
16384	266	64	20	1.

total; nearest-neighbor (QQ^T) exchanges for the SEM, including the $N/2$ multigrid exchanges; coarse-grid solve; and all vector reductions. The entry *x4096* uses the XX^T solver, while *a4096* uses a variant of the AMG solver which selects between the pairwise *gs()* exchange or *all_reduce()* to optimize the AMG communication. The entries *b4096* and below are for the current scheme, which enters a third communication option in the mix, namely, the *crystal_router()* routine.

Parallel efficiencies (figure 2, right) are derived from from the solution-time model, $T(P) = W_s + W_p/P$. For any value of P, given $T(P)$ and $T(P/2)$, one can estimate the serial-to-parallel work ratio $W_{sp} := W_s/W_p$, from which the efficiency is $\eta_p = (1+W_{sp})/(1+PW_{sp})$. For Case 1, $W_{sp} \approx$2.72e-5, based on an average over runs on 1, 2, 4, and 8 racks, with 4096 cores/rack. For Case 2, $W_{sp} \approx$1.17e-5, based on 4, 8, and 16 racks. For Case 3, $W_{sp} \approx$7.74e-6, based on 4 and 8 racks. At P=32768, Case 1 achieves roughly 50% efficiency, which is in fact reasonable given that n/P is only $\approx$ 3700 in this case. Case 2 sustains roughly 50% efficiency at P=65536 with $n/P \approx$ 7300 and Case 3 realizes 80% efficiency for P=32768 with $n/P \approx$ 15870. The fraction of time spent in the coarse grid is $<$ 20% for all cases.

4. Conclusion

We have presented algorithmic and performance considerations for petascale simulations of incompressible flows. Computations on the ALCF BG/P are already yielding important results in the analysis of reactor core flows, including establishment of core entry lengths.

Acknowledgments

This work was supported by the U.S. Dept. of Energy under Contract DE-AC02-06CH11357. Computer time on the Argonne Leadership Computing Facility was provided through a 2008 DOE Office of Science INCITE Award. We thank Hank Childs and David Bremer from the Lawrence Livermore National Laboratory VisIt group for assistance with data analysis.

References

[1] Deville M, Fischer P and Mund E 2002 High-order methods for incompressible fluid flow (Cambridge University Press)
[2] Fischer P and Lottes J 2004 In R Kornhuber, R Hoppe, J Priaux, O Pironneau, O Widlund and J Xu eds *Domain Decomposition Methods in Science and Engneering* (Springer: Berlin)
[3] Fischer P, Loth F, Lee S, Smith D and Bassiouny H 2007 *Comput. Methods Appl. Mech. Engrg.* **196** 3049–3060
[4] Fox G C, Johnson M A, Lyzenga G A, Otto S W, Salmon J K and Walker D W 1988 *Solving Problems on Concurrent Processors* (Prentice-Hall: Englewood Cliffs, NJ)
[5] Pointer W, Fischer P, Siegel A and Smith J 2008 *Proc. Int. Congress on Advances in Nuclear Power Plants* (Anaheim, CA) Paper 8252
[6] Tufo H and Fischer P 2001 *J. Parallel Distrib. Comput.* **61**:151–177
[7] Wan W L, Chan T F and Smith B 1999 *SIAM J. on Sci. Comp.* **21(4)**:1632–1649

Computational science research in support of petascale electromagnetic modeling

L Lee, V Akcelik, L Ge, S Chen, G Schussman, A Candel, Z Li, L Xiao, A Kabel, R Uplenchwar, C Ng and K Ko

Stanford Linear Accelerator Center, Menlo Park, CA 94025, USA

E-mail: liequan@stanford.edu

SciDAC CET/Institute Collaborators: E Ng, X Li, C Yang (LBNL), L Dianchin (LLNL), K Devine, E Boman (SNL), B Osting, D Keyes (Columbia), X Luo, M Shepard (RPI), R Barrett, S Hodson, R Kendall (ORNL), W Gropp (UIUC), Z Bai, K Ma(UCDavis) and O Ghattas (UTAustin)

Abstract. Computational science research components were vital parts of the SciDAC-1 accelerator project and are continuing to play a critical role in the SciDAC-2 accelerator project, the Community Petascale Project for Accelerator Science and Simulation (ComPASS). Presented here are recent advances and achievements in the area of computational science research in support of petascale electromagnetic modeling for accelerator design analysis, which include shape determination of superconducting RF cavities, mesh-based multilevel preconditioner for solving highly indefinite linear systems, moving window using h- or p-refinement for time-domain short-range wakefield calculations, and improved scalable application I/O.

1. Introduction

Particle accelerators, such as the Spallation Neutron Source (SNS), Linac Coherent Light Source (LCLS), Rare Isotope Accelerator, Large Hadron Collider, and proposed International Linear Collider, are pivotal experimental facilities for discoveries in physical sciences. Electromagnetic modeling of those existing or proposed billion-dollar-class accelerator facilities often requires petascale computing and drives the computational science research in the related areas. Computational science research components were vital parts of the SciDAC-1 accelerator project [1, 2, 3, 4] and continue to play a critical role in the SciDAC-2 accelerator project, the Community Petascale Project for Accelerator Science and Simulation (ComPASS). Presented here are recent advances and achievements in the area of computational science research in support of petascale electromagnetic modeling for accelerator design and optimization.

2. Shape determination of superconducting RF cavities

Shape deviations of the real cavity from the design may significantly impact cavity response. A shape determination tool has been developed at the Stanford Linear Accelerator Center (SLAC) to infer the unknown shape deviations using measurable quantities such as cavity frequencies, field distributions and external quality factor values. A nonlinear least square optimization problem constrained by a complex Maxwell eigenvalue problem is solved in the shape determination process:

$$\underset{\mathbf{e}_j,k_j,\mathbf{d}}{\text{minimize}} \;\; \mathcal{J}(\mathbf{e}_j,\mathbf{k}_j,\mathbf{d}) \;\; \text{subject to} \;\; \mathbf{K}\mathbf{e}_j + ik_j\mathbf{W}\mathbf{e}_j - k_j^2\mathbf{M}\mathbf{e}_j = 0 \tag{1}$$

$$\mathbf{e}_j^H\mathbf{M}\mathbf{e}_j = 1 \;\; and \;\; \Re(\mathbf{e}_j)^T\mathbf{M}\Im(\mathbf{e}_j) = 0, \tag{2}$$

where $(\mathbf{e}_j, k_j)$ is the jth eigenpair and $\mathbf{d}$ contains the shape deviation parameters. $\Re$ denotes the real part and $\Im$ the imaginary part. More discussion about the complex quadratic eigenvalue problem equation(1) can be found in [1].

The objective function $\mathcal{J}$ includes a weighted least-squares misfit of the computed and measured cavity response. A Gauss-Newton algorithm with a discrete adjoint method is used to solve the above nonlinear least square minimization problem. This problem is ill-posed, and a truncated singular value decomposition method is used to overcome the ill-posedness. Each nonlinear iteration of the algorithm requires solution of the forward problem (complex Maxwell eigenvalue problem), solution of adjoint problems (linear system of equations), and evaluation of the eigenvector and eigenvalue sensitivities. The algorithm is tested with examples, and the solution of the shape deviation converges within a handful of nonlinear iterations.

The shape determination tool has been used to find the causes of high external quality factor in a high gradient prototype cryomodule at Jefferson Lab [5]. The algorithm predicted that the manufactured cavity was deformed significantly from the original shape, and its length was 6 to 8 mm shorter than that of the original design depending on the choice of inversion parameter set. The results agree well with those from the measurements.

3. Mesh-based multilevel preconditioner

In analyzing eigen-modes of accelerator cavities, the governing Maxwell's equations can be simplified to the following harmonic vector wave equation.

$$\nabla \times \left(\frac{1}{\mu}\nabla \times \overrightarrow{\mathbf{E}}\right) - \varepsilon k^2 \overrightarrow{\mathbf{E}} = 0 \tag{3}$$

With finite-element discretization using tangentially continuous Nedelec basis functions [6], $\overrightarrow{\mathbf{E}} = \sum_i x_i \overrightarrow{\mathbf{N}}_i$, the above equation becomes an eigenvalue problem:

$$\mathbf{K}\mathbf{x} = k^2\mathbf{M}\mathbf{x}. \tag{4}$$

Here $\mathbf{K} = \int_\Omega \frac{1}{\mu}\left(\nabla \times \overrightarrow{\mathbf{N}}_i\right) \cdot \left(\nabla \times \overrightarrow{\mathbf{N}}_j\right) \, d\Omega.$

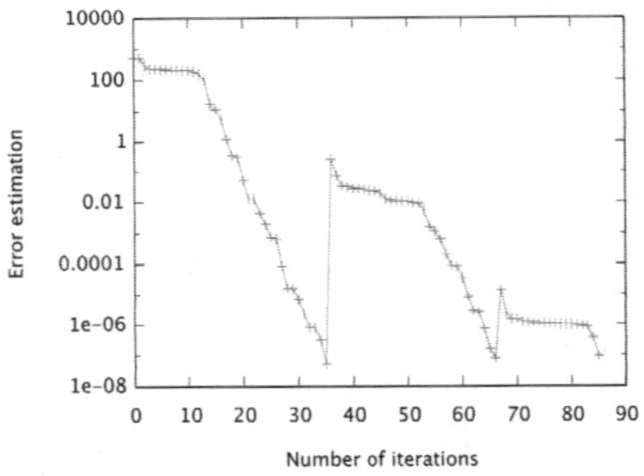

Figure 1. Convergence history using GMRES with multilevel preconditioner on solving a shifted linear system. The error estimation of inner iterations of GMRES is also plotted.

Moreover, $\mathbf{M} = \int_\Omega \varepsilon \overrightarrow{\mathbf{N}}_i \cdot \overrightarrow{\mathbf{N}}_j \, d\Omega$. Note that matrix $\mathbf{M}$ is symmetric positive definite and matrix $\mathbf{K}$ is symmetric positive semi-definite with a large null space. Since the interior eigenvalues are of the interest in the accelerator cavity modeling, a shift-and-invert transform is often applied in the process of solving the above eigenvalue problem. This requires the solution of a highly indefinite linear system in every eigenvalue iteration, which is notoriously difficult to solve with iterative methods:

$$(\mathbf{K} - \sigma\mathbf{M})\,\mathbf{x} = \mathbf{b}, \tag{5}$$

where σ is a prescribed shift close to the eigenvalues of the interest.

A mesh-based multilevel preconditioner has recently been developed for solving equation (5). Given the preconditioner system $\mathbf{Ax} = \mathbf{b}$, the vector $\mathbf{b}$, which is defined on a dense mesh, is first restricted onto a coarse mesh and becomes $\mathbf{b^c}$ (from now on, the variables with superscript c denote those on the coarse mesh). Then $(\mathbf{K}^c - \sigma\mathbf{M}^c)\,\mathbf{x}^c = \mathbf{b}^c$ is solved with a sparse direct solver since its size can be very small. At the last step, the vector $\mathbf{x}^c$ is prolongated back onto the dense matrix and is regarded as the solution of the preconditioner system $\mathbf{x}$. For the restriction operation, a linear system is solved as follows:

$$(\mathbf{M}^c + \alpha\mathbf{K}^c)\mathbf{b}^c = \int_\Omega \left(\overrightarrow{\mathbf{F}}(\mathbf{b}) \cdot \overrightarrow{\mathbf{N}}_i^c + \alpha(\nabla \times \overrightarrow{\mathbf{F}}(\mathbf{b})) \cdot (\nabla \times \overrightarrow{\mathbf{N}}_i^c)\right) d\Omega, \tag{6}$$

where α is a positive parameter to be specified and $\overrightarrow{\mathbf{F}}(\mathbf{b}) = \sum_\mathbf{i} \mathbf{b_i}\overrightarrow{\mathbf{N}}_\mathbf{i}$ on the dense mesh. Note that the matrix $\mathbf{M}^c + \alpha\mathbf{K}^c$ is symmetric positive definite when α is positive and the above linear system can be solved using conjugate gradient with incomplete Cholesky preconditioner. A similar linear system is solved for the prolongation operation. It is shown in [7] that the above operation balanced both the error of the interpolating field and that of the curl of the interpolating field, which is important in electromagnetic modeling. Therefore, it is more effective to use the $\mathbf{M}^c + \alpha\mathbf{K}^c$ than to use the mass matrix $\mathbf{M}^c$ alone.

Figure 1 shows the convergence history of the GMRES with this mesh-based multilevel preconditioner for solving a shifted linear system. With 85 iterations, the converged solution with residual of less than 10^{-7} is achieved using this method. As a comparison, it took about 518 iterations to achieve the same converged solution using conjugate gradient (CG) with SSOR preconditioner and it failed to converge using CG with incomplete factorization-based preconditioner. Work on the parallel implementation of this multilevel preconditioner is in progress.

4. Moving window with h- or p-refinement

In calculating short-range wakefield inside an accelerator structure, only the small region in the vicinity of the particle beam is required in the simulation. A moving window technique, in which the domain of the simulation is limited to a small region of interest near the beam and moves with it, can greatly save the computational efforts. This technique has been widely used in the finite-difference simulation but not in the finite-element simulation with unstructured grids. In this section, we give a brief introduction of finite-element time-domain (FETD) method used in T3P. Moving window techniques with curvilinear tetrahedral meshes using either h-refinement or p-refinement are then presented.

Figure 2. A couper region of an accelerating cavity.

In T3P, the following inhomogeneous vector wave equation is solved numerically for the electric field $\overrightarrow{\mathbf{E}}$:

$$\nabla \times \left(\frac{1}{\mu}\nabla \times \int_{-\infty}^{t} \overrightarrow{\mathbf{E}}\,\mathrm{d}\tau\right) + \varepsilon\frac{\partial^2}{\partial t^2}\int_{-\infty}^{t} \overrightarrow{\mathbf{E}}\,\mathrm{d}\tau = -\overrightarrow{\mathbf{J}}, \tag{7}$$

where $\overrightarrow{\mathbf{J}}$ is the electric current density and ε and μ are the electric permittivity and magnetic permeability, respectively.

With finite-element spatial discretization [6], $\int_{-\infty}^{t} \vec{\mathbf{E}}(\mathbf{x},\tau)\,\mathrm{d}\tau = \sum_i e_i(t) \cdot \vec{\mathbf{N}}_i(\mathbf{x})$, and the implicit Newmark-beta scheme [8] for temporal discretization, the following linear system is solved for each time step:

$$\begin{aligned}\left(\mathbf{M} + \beta(\Delta t)^2\mathbf{K}\right)\mathbf{e}^{n+1} = &\left(2\mathbf{M} - (1-2\beta)(\Delta t)^2\mathbf{K}\right)\mathbf{e}^{n} - \left(\mathbf{M} + \beta(\Delta t)^2\mathbf{K}\right)\mathbf{e}^{n-1} \\ &- (\Delta t)^2\left(\beta\mathbf{f}^{n+1} + (1-2\beta)\mathbf{f}^{n} + \beta\mathbf{f}^{n-1}\right),\end{aligned} \tag{8}$$

where matrices $\mathbf{K}$ and $\mathbf{M}$ are the same as in equation (4) and vector $\mathbf{f} = \int_\Omega \vec{\mathbf{N}}_i \cdot \vec{\mathbf{J}}\, d\Omega$ is the discretized representation of the current density. Note that the four discretized vectors $\mathbf{e}^n$, $\mathbf{e}^{n-1}$, $\mathbf{f}^n$, and $\mathbf{f}^{n-1}$, need to be transferred from the current window configuration to the next window configuration when the particle beam moves out of the window.

4.1. P-refinement

A window is defined with its front and back boundaries perpendicular to the velocity of the particle beam. The order of the basis functions for the tetrahedral elements inside the window, p, is set to be nonzero, while outside the window, p is zero. This effectively makes the number of degrees of freedom (DOF) zero outside the window, thereby reducing the computational efforts. A padding zone is put between the front of the beam and the front boundary of the window so that the particle beam will stay in the window for a while. By changing the size of the padding zone, the window size is adjusted. When the particle beam moves out the padding zone, the window will move forward a given distance. The vectors $\mathbf{e}^n$, $\mathbf{e}^{n-1}$, $\mathbf{f}^n$, and $\mathbf{f}^{n-1}$ as shown in Eq (8) are transferred elementwise according to the changes of the finite-element order p's. If the order, p, increases, zeroes are filled for those additional coefficients. On the other hand, if p decreases, the corresponding coefficients are dropped.

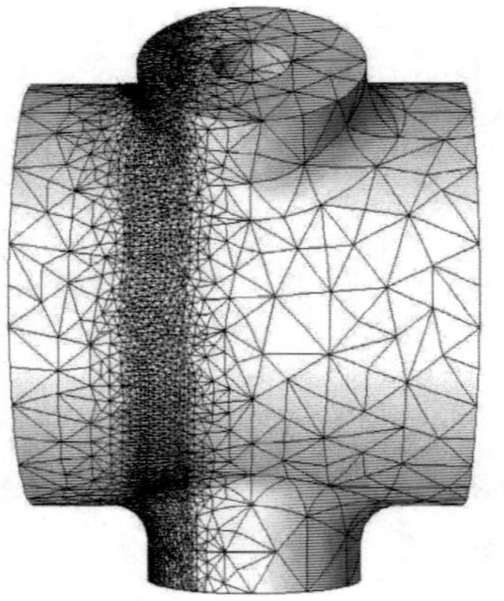

Figure 3. Picture of a mesh for the coupler region.

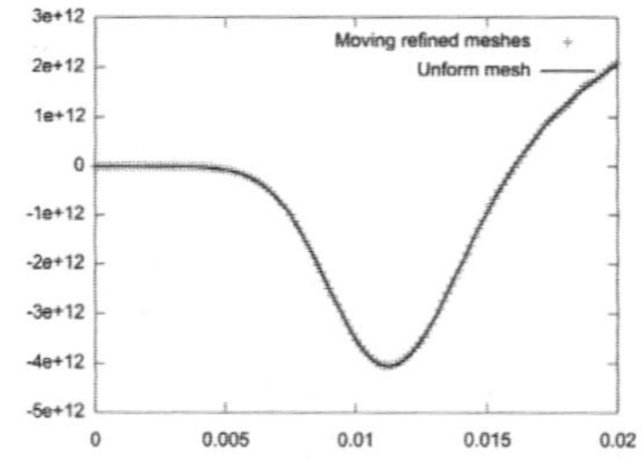

Figure 4. Wakefield comparison with moving h-refined meshes and with a uniform mesh.

The short-range wakefield of a coupler shown in Figure 2 is calculated with the moving window through p-refinement on a curvilinear tetrahedral mesh with 13 million elements. During the simulation, the window moved five times with each window having 2.37 million, 1.08 million, 1.02 million, 1.02 million, 1.50 million, and 1.78 million elements, respectively. The run time of the simulation using the moving window technique is only *one-tenth* of that using all 13 million elements.

4.2. H-refinement

A series of windows with padding zones is defined in the vicinity of the moving particle beam. The corresponding series of meshes is generated by scientists from the SciDAC ITAPS Center [9, 10] with a dense mesh region inside each window and a coarse mesh region outside the window. Figure 3 shows one such mesh for the coupler region shown in figure 2. When the particle beam moves out of one window and into the next window, the vectors $\mathbf{e}^n$ and $\mathbf{e}^{n-1}$

are transferred onto the next mesh using the projection as in equation (6) while the vectors $\mathbf{f}^n$ and $\mathbf{f}^{n-1}$ are recalculated with the new mesh. The wakefield monitors in a pillbox cavity with moving h-refined meshes and with a uniform mesh are plotted in figure 4.

The results are in remarkable agreement, which indicates the validity of our h-refined moving window technique. Further work on moving window through combined h-refinement and p-refinement is in progress.

5. Scalable application I/O

In this section, we present the improvement made on the parallel application I/O in the SLAC's simulation tool Omega3P and T3P.

Because of the lack of the parallel I/O specification in the netcdf standard, a "rank 0 write all" approach for writing out the results in a parallel simulation was used in the SLAC's code. This approach was nonscalable, and the time for writing out data exceeded the time for the rest of the simulation when the number of executing processors was more than 4,000. With scientists at Oak Ridge and Argonne, the application I/O has been overhauled with parallel netcdf API [11]. With parallel netcdf API the execution time for writing out data in the Omega3P simulations was reduced by two orders of magnitude.

Curvilinear tetrahedral elements are used in SLAC's finite-element-based application suite for high-fidelity modeling. To save disk space, only the mid-points of the curved edges are stored in the mesh file. During the simulation, all the mid-points of the curved edges are read in from the file, and only those belonging to the local processor are kept in the memory. A more optimized read algorithm has been developed by taking advantage of fast interprocessor communication. First the mid-points are partitioned with the number of processors. Each processor read in its partitioned portion of the mid-points in the second step. In the third step, an `MPI_Allgatherv()` is invoked to get all the mid-points for the selection of the mid-points in the local mesh. This scheme provides a scalable read algorithm and has a huge time-saving on NCCS Jaguar computer when number of processors in the simulation is large.

With the above-mentioned application I/O improvement and various communication pattern enhancements, SLAC's FETD code T3P has been used efficiently with a *half-billion* degrees of freedom to simulate an 8-cavity cryomodule with the INCITE allocation [12].

Acknowledgments

This work is supported by the U.S. Department of Energy under contract number DE-AC02-76SF00515. The work used resources of the National Center for Computational Sciences at Oak Ridge National Laboratory and resources of the National Energy Research Scientific Computing Center, which are supported by the Office of Science of the Department of Energy under Contract DE-AC05-00OR22725 and DE-AC02-05CH11231, respectively.

References

[1] Lee L Q, and *et al.* 2005 *J. Phys.: Conf. Ser.* **16** 205–209
[2] Lee L Q, Akcelik V, Chen S, Ge L, Prudencio E, Li Z, Ng C, Xiao L and Ko K 2006 *Proc. of SciDAC 2006 Conference* (Denver, Colorado, USA)
[3] Lee L Q, and it et al. 2007 *J. Phys.: Conf. Ser.* **78** 012040
[4] Akelik V, Ko K, Lee L Q, Li Z, Ng C K and Xiao L 2008 . *Computational Physics* **227** 1722–1738
[5] Rimmer R and Li Z 2008 Private communication
[6] Sun D K, Lee J F and Cendes Z 2001 *SIAM J. Sci. Comput.* **23** 1053–1076
[7] Lee L Q, Candel A and Kabel A 2008 On projecting discretized electromagnetic fields (In preparation)
[8] Newmark N M 1959 *J. Engineering Mechanics Division* **85** 67–94
[9] Diachin L 2008 The SciDAC ITAPS center http://www.itaps-scidac.org/
[10] Shephard M and *et al.* 2008 *Proc. SciDAC 2008 Conference* (Seattle, WA)
[11] Li J, Liao W K, Choudhary A, Ross R, Thakur R, Gropp W, Latham R, Siegel A, Gallagher B and Zingale M 2003 *Proc. SC'03* (Phoenix, AZ)

[12] Lee L Q, Ng C, Ko K, Li Z and Kabel A 2008 *2008 INCITE Program*

Peridynamics for multiscale materials modeling

E Askari[1], F Bobaru[2], R B Lehoucq[3], M L Parks[4], S A Silling[5], O Weckner[6]

[1,6] Applied Math, The Boeing Company, Seattle, WA 98124, USA
[2] Department of Engineering Mechanics, University of Nebraska-Lincoln, Lincoln, NE, 68588-0526, USA
[3,4] Applied Mathematics and Applications, Sandia National Laboratories Albuquerque, NM 87185, USA
[5] Multiscale Dynamic Materials Modeling, Sandia National Laboratories, Albuquerque, NM 87185, USA

E-mail:
abe.askari@boeing.com,fbobaru2@unl.edu,rblehou@sandia.gov,mlparks@sandia.gov, sasilli@sandia.gov,olaf.weckner@boeing.com

Abstract. The paper presents an overview of peridynamics, a continuum theory that employs a nonlocal model of force interaction. Specifically, the stress/strain relationship of classical elasticity is replaced by an integral operator that sums internal forces separated by a finite distance. This integral operator is not a function of the deformation gradient, allowing for a more general notion of deformation than in classical elasticity that is well aligned with the kinematic assumptions of molecular dynamics. Peridynamics effectiveness has been demonstrated in several applications, including fracture and failure of composites, nanofiber networks, and polycrystal fracture. These suggest that peridynamics is a viable multiscale material model for length scales ranging from molecular dynamics to those of classical elasticity.

1. Introduction

The response of materials to the environments and loads occurring in practice all require an understanding of mechanics at disparate spatial and temporal scales. Such "multiscale" understanding is a fundamental challenge for next generation materials modeling; see [1] for information and references to the existing literature. A currently popular multiscale approach couples two or more well-known models, for example, molecular dynamics and classical elasticity [2], each of which is useful at different scales. An alternative approach is to develop a single multiscale material model that remains valid and useful over a wide range of temporal and spatial scales.

The onerous practical limitations of molecular dynamics and the limited validity of classical elasticity has led to generalized continuum theories purporting to supply a single multiscale material model. Such theories are motivated by introducing a length-scale absent in classical elasticity. These generalized continua are certainly not new; see [3, 4, 5] for overviews and citations to the abundant literature. However, these theories develop material models with a length-scale dependence by augmenting the displacement field with supplementary fields (e.g., rotations) that provide information about fine-scale kinematics, or by using higher-order gradients of the displacement field, or by averaging local strains and/or stresses.

This paper presents an overview of peridynamics [6, 7], a generalized continuum theory that employs a nonlocal model of force interaction. Specifically, the stress/strain relationship of classical elasticity is replaced by an integral operator that sums internal forces separated by a finite distance. This integral operator is not a function of the deformation gradient, allowing for a more general notion of deformation than classical elasticity that is well aligned with the kinematic assumptions of molecular dynamics.

2. Peridynamic primer

This section reviews peridynamic fundamentals including kinematics, balance of angular and linear momentum, length scale, a relationship with Newtonian mechanics, and discretization. We refer the reader to [8] demonstrating that linearized peridynamics can reproduce nonlinear dispersion relationships for short wavelengths, in contrast to linear elasticity.

2.1. Kinematics and internal force density

The spatial point $\mathbf{y}$ (in the deformed configuration) is related to the material point $\mathbf{x}$ (in the reference configuration) via $\mathbf{y}(\mathbf{x},t) = \mathbf{x}+\mathbf{u}(\mathbf{x},t)$ where $\mathbf{u}(\mathbf{x},t)$ is the displacement of $\mathbf{x} \in \mathcal{B} \subset \mathbb{R}^3$. We say the body $\mathcal{B}$ undergoes deformation if the displacement $\mathbf{u}(\mathbf{x},t)$ is nonzero for some $\mathbf{x} \in \mathcal{B}$. Deformation represents the mechanism through which continuum mechanics relates deformation to the internal forces of a body. Classical elasticity assumes that the displacement field is continuously differentiable at every $\mathbf{x} \in \mathcal{B}$ so that

$$\mathbf{u}(\mathbf{x}',t) - \mathbf{u}(\mathbf{x},t) = \nabla_{\mathbf{x}}\mathbf{u}(\mathbf{x},t)(\mathbf{x}'-\mathbf{x}) + O(\|\mathbf{x}'-\mathbf{x}\|^2),$$

where $\nabla_{\mathbf{x}}$ denotes the gradient operator with respect to the material point $\mathbf{x}$. If $\mathbf{F}(\mathbf{x},t) := \mathbf{I} + \nabla_{\mathbf{x}}\mathbf{u}(\mathbf{x},t)$ denotes the deformation gradient, then

$$\mathbf{y}(\mathbf{x}',t) - \mathbf{y}(\mathbf{x},t) \approx \mathbf{F}(\mathbf{x},t)(\mathbf{x}'-\mathbf{x}).$$

In words, the body deformation $\mathbf{y}(\mathbf{x}',t) - \mathbf{y}(\mathbf{x},t)$ can be *approximated* by the deformation gradient acting on the *bond* $\mathbf{x}'-\mathbf{x}$. By relying on the true deformation $\mathbf{y}(\mathbf{x}',t) - \mathbf{y}(\mathbf{x},t)$, peridynamics avoids assumptions on the smoothness of the displacement field, in contrast to classical elasticity. The mapping from a bond $\mathbf{x}'-\mathbf{x}$ to $\mathbf{y}(\mathbf{x}',t) - \mathbf{y}(\mathbf{x},t)$, or equivalently $\mathbf{u}(\mathbf{x}',t) - \mathbf{u}(\mathbf{x},t)$, is in general nonlinear. The use of the deformation gradient is tantamount to approximating this nonlinear map with a local linear approximation.

Denote by $\mathbf{y}', \mathbf{y}$ the values $\mathbf{y}(\mathbf{x}',t), \mathbf{y}(\mathbf{x},t)$, respectively, and consider the integral operator

$$\int_{\mathcal{B}} \mathbf{k}\left(\mathbf{y}'-\mathbf{y},\mathbf{x}',\mathbf{x}\right)\, d\mathbf{x}' \tag{1}$$

representing the (internal) force density. The integral operator is nonlocal because material points $\mathbf{x}' \neq \mathbf{x}$ are involved. The constitutive equation relating deformation to internal force density is contained in the kernel $\mathbf{k}$ that has units of force per volume squared. As an example, consider a proportional microelastic material [9] with the peridynamics kernel (derivable from a stored energy function)

$$\mathbf{k}\left(\mathbf{y}'-\mathbf{y},\mathbf{x}',\mathbf{x}\right) = cs\frac{\mathbf{y}'-\mathbf{y}}{\|\mathbf{y}'-\mathbf{y}\|}, \qquad s = \frac{\|\mathbf{y}'-\mathbf{y}\| - \|\mathbf{x}'-\mathbf{x}\|}{\|\mathbf{x}'-\mathbf{x}\|}, \tag{2}$$

where $c/\|\mathbf{x}'-\mathbf{x}\| > 0$ is the stiffness per unit volume squared. The corresponding force density operator of classical elasticity is given by

$$\mathbf{C}(\mathbf{x},t) := \nabla_{\mathbf{x}} \cdot \mathbf{P}\left(\mathbf{F}(\mathbf{x},t)\right), \tag{3}$$

where $\mathbf{P}$ denotes the Piola-Kirchoff stress tensor and $\nabla_{\mathbf{x}}\cdot$ the divergence with respect to $\mathbf{x}$. The operator $\mathbf{C}$ is local because its value at a point $\mathbf{x}$ depends, via the deformation gradient, only on the value of the stress tensor at that same point. The Piola-Kirchoff stress tensor is a constitutive equation, for example, a mapping from strain to stress tensors incorporating the material behavior.

The peridynamics force density operator (1) obviates the need for the stress/strain relationship implicit in (3). Such relationships are based on the idea of contact forces (interactions between material in direct contact) and follow from Cauchy's postulate [10, pp. 60–61]. Peridynamics is fundamentally different in that material separated by a finite distance may exert forces on each other. The paper [11] establishes conditions under which a peridynamic stress tensor $\boldsymbol{\nu}(\mathbf{x},t)$ exists such that

$$\nabla_{\mathbf{x}} \cdot \boldsymbol{\nu}(\mathbf{x},t) = \int_{\mathcal{B}} \mathbf{k}\left(\mathbf{y}' - \mathbf{y}, \mathbf{x}', \mathbf{x}\right)\, d\mathbf{x}'.$$

But even in this case, peridynamics is fundamentally different from classical elasticity in that the peridynamics stress tensor represents a sum of forces per unit area *through* $\mathbf{x}$ instead of *at* $\mathbf{x}$, as is the case for the Piola-Kirchoff stress tensor $\mathbf{P}$.

The bond-based peridynamics model originally introduced in [6] assumed that $\mathbf{k}(\mathbf{y}'-\mathbf{y},\mathbf{x}',\mathbf{x})$ is collinear with $\mathbf{y}'-\mathbf{y}$. This assumption results in an effective Poisson ratio $= 1/4$ for isotropic, linear, microelastic materials. The paper [6, § 15] also briefly described how the bond-based peridynamics kernel may be augmented so that a Poisson ratio $\neq 1/4$ can be achieved. Recently, in [7], the bond-based peridynamics model of [6] was generalized to that of a *state*-based peridynamics model for which $\mathbf{k}(\mathbf{y}'-\mathbf{y},\mathbf{x}',\mathbf{x})$ may be a force state; for example, it may depend on the collective behavior at $\mathbf{x}$ and $\mathbf{x}'$. Peridynamics states is the continuum equivalent of the multibody potentials of classical particle mechanics that go beyond central force interactions (or the extension discussed in [6, § 15].)

2.2. Balance of linear and angular momentum

The peridynamic global balance of linear momentum is

$$\int_{\Omega} \rho(\mathbf{x})\ddot{\mathbf{u}}(\mathbf{x},t)\, d\mathbf{x} = \int_{\Omega}\int_{\mathcal{B}/\Omega} \mathbf{k}(\mathbf{y}'-\mathbf{y},\mathbf{x}',\mathbf{x})\, d\mathbf{x}'\, d\mathbf{x} + \int_{\Omega} \mathbf{b}(\mathbf{x},t)\, d\mathbf{x}, \tag{4}$$

where $\rho(\mathbf{x})$ and $\mathbf{b}(\mathbf{x},t)$ are the mass and body force densities, respectively. This balance holds over any $\Omega \subset \mathcal{B}$ if and only if the peridynamic kernel is antisymmetric, for example, $\mathbf{k}(\mathbf{y}'-\mathbf{y},\mathbf{x}',\mathbf{x}) = -\mathbf{k}(\mathbf{y}-\mathbf{y}',\mathbf{x},\mathbf{x}')$. The first term on the right-hand side of (4) is the analogue of the contact force in classical elasticity; and when $\Omega = \mathcal{B}$, this force is zero implying that external (to $\mathcal{B}$) contact forces play no role in peridynamics. We may define this first term to represent the force that the material in $\mathcal{B}/\Omega$ exerts on the material in Ω.

The local/nonlocal distinctions drawn can be given a mechanical interpretation. This force is nonlocal precisely because material inside and outside of Ω interacts. In contrast, the classical theory of elasticity restricts the force interaction to the surface of Ω. The peridynamic global balance of angular momentum is

$$\int_{\Omega} \mathbf{y} \times \rho(\mathbf{x})\ddot{\mathbf{u}}(\mathbf{x},t)\, d\mathbf{x} = \int_{\Omega}\int_{\mathcal{B}} \mathbf{y} \times \mathbf{k}(\mathbf{y}'-\mathbf{y},\mathbf{x}',\mathbf{x})\, d\mathbf{x}'\, d\mathbf{x} + \int_{\Omega} \mathbf{y} \times \mathbf{b}(\mathbf{x},t)\, d\mathbf{x} \tag{5}$$

for $\Omega \subset \mathcal{B}$. The first term on the right-hand side of (5) represents the torque of Ω and may be replaced by $\int_{\Omega}\int_{\mathcal{B}/\Omega} \mathbf{y} \times \mathbf{k}(\mathbf{y}'-\mathbf{y},\mathbf{x}',\mathbf{x})\, d\mathbf{x}'\, d\mathbf{x}$ if and only if $\mathbf{k}(\mathbf{y}'-\mathbf{y},\mathbf{x}',\mathbf{x})$ is collinear with $\mathbf{y}'-\mathbf{y}$ as in bond-based peridynamics.

In classical elasticity, the locality of force interactions also implies that there is no length scale independent of specific material behavior (e.g., outside of material inhomogeneities.) On the other hand, the nonlocal force density operator (1) does contain a length-scale, the peridynamic *horizon*—a nonzero volume over which material exerts forces. For instance, let the horizon be given by

$$\mathcal{H}(\mathbf{x}) = \{\mathbf{x}' \in \mathcal{B} \,|\, 0 \leq \|\mathbf{x}' - \mathbf{x}\| < \delta_{\mathbf{x}}\} \subset \mathcal{B}$$

so that the first terms on the right-hand side of (5) and (4) are replaced with $\int_\Omega \int_{\mathcal{H}(\mathbf{x})} \mathbf{y} \times \mathbf{k}(\mathbf{y}' - \mathbf{y}, \mathbf{x}', \mathbf{x})\, d\mathbf{x}'\, d\mathbf{x}$ and $\int_\Omega \int_{\mathcal{H}(\mathbf{x})/\Omega} \mathbf{k}(\mathbf{y}' - \mathbf{y}, \mathbf{x}', \mathbf{x})\, d\mathbf{x}'\, d\mathbf{x}$, respectively, for $\Omega \subset \mathcal{H}(\mathbf{x})$. The horizon may vary with $\mathbf{x}$ and/or the forces about $\mathbf{x}$ so that the largest horizon over the body defines a maximum length scale.

The recently accepted paper [12] explains how the general state-based peridynamic material model converges to the classical elastic material model as the length scale, or horizon, vanishes, (assuming that the underlying deformation is sufficiently smooth). Building on the analysis of [12] and [11], a collapsed peridynamics stress tensor $\boldsymbol{\nu}^0(\mathbf{x}, t)$ is derived that is a function of the deformation gradient at $\mathbf{x}$ and so is a Piola-Kirchoff stress tensor. This limiting Piola-Kirchhoff stress tensor is differentiable and its divergence represents the force density due to internal forces. The limiting, or collapsed, stress-strain model satisfies the conditions in the classical theory for angular momentum balance, isotropy, objectivity, and hyperelasticity, provided the original peridynamic constitutive model satisfies appropriate conditions.

A body composed of discrete particles (e.g., atoms) can be represented as a peridynamic body. For example, suppose a set of discrete particles is given with reference positions $\mathbf{x}_i$ and masses m_i, $i = 1, 2, \ldots, n$. Let the force exerted by particle j on particle i after deformation of the system be denoted by $\mathbf{F}_{j,i}(t)$. Define a peridynamic body by $\rho(\mathbf{x}) = \sum_i m_i \delta(\mathbf{x} - \mathbf{x}_i)$ and $\mathbf{k}(\mathbf{y}' - \mathbf{y}, \mathbf{x}', \mathbf{x}) = \sum_i \sum_{j \neq i} \mathbf{F}_{j,i}(t) \delta(\mathbf{x}' - \mathbf{x}_j) \delta(\mathbf{x} - \mathbf{x}_i)$ for all $\mathbf{x}$, $\mathbf{x}'$ in $\mathbb{R}^3$, where $\delta(\mathbf{x})$ denotes the Dirac delta function or density (with units of per volume). Global conservation of linear momentum (4) where Ω is a volume enclosing only $\mathbf{x}_i$ reduces to

$$m_i \ddot{\mathbf{u}}(\mathbf{x}_i, t) = \sum_{j \neq i}^{n} \mathbf{F}_{j,i}(t), \qquad i = 1, 2, \ldots, n,$$

which is the familiar statement of Newton's second law in the particle mechanics setting. Replacing the Dirac delta densities with more general probability densities provides a mechanism for converting the set of discrete particles into a (continuum) peridynamics material model.

2.3. Discretization

Let $\mathbf{x}_1, \ldots, \mathbf{x}_n$ be distinct points in $\mathcal{B}$, where $\mathcal{B} = \bigcup_{i=1}^n \Omega_i$ and $\mathbf{x}_i \in \Omega_i \subset \mathcal{H}(\mathbf{x}_i)$. The spatial discretization of (4) results in

$$\rho(\mathbf{x}_i) \ddot{\mathbf{u}}(\mathbf{x}_i, t) = \sum_{j \neq i} \mathbf{k}(\mathbf{y}_j - \mathbf{y}_i, \mathbf{x}_j, \mathbf{x}_i) V_j + \mathbf{b}(\mathbf{x}_i, t), \qquad i = 1, \ldots, n, \tag{6}$$

where V_j is the volume of Ω_j. Note that the sum is not over $n - 1$ but rather over the number of $\mathbf{x}_j \in \mathcal{H}(\mathbf{x}_i)$. An advantage of quadrature-based approaches is that they lead to three-dimensional, meshless Lagrangian implementations. A time integration scheme is then applied to (6); see [9, 13] for details. For a uniformly spaced $\mathbf{x}_i$, the midpoint quadrature rule gives a quadratic spatial rate of convergence. See [14, 15] for extensive verification studies in one and two dimensions (including adaptivity and comparisons with a commercial finite-element code), respectively. The papers [16, 17] discuss the use of more sophisticated quadrature for the internal force calculation, and [18] discusses an implementation within an engineering finite element analysis code.

3. Applications

We review three successful multiscale applications of peridynamics within engineering analyses. All of the results discussed here were obtained with the Emu [9] or the molecular dynamics code LAMMPS [19, 13]. Both implementations are meshless three-dimensional Lagrangian codes—a mesh generator is not required. The examples (including validation where available) provide concrete evidence that peridynamics has been successfully used for three-dimensional engineering analyses that strain, perhaps go beyond, the limits of classical elasticity, by exploiting the nonlocality of force central to peridynamics. This nonlocality allows peridynamics to model complex material behavior possessing an intrinsic length-scale. The three applications presented deal with fracture problems—a quintessential multiscale phenomena. A compelling advantage of peridynamics is the ability to deal with fracture in a straightforward fashion. This is accomplished by allowing bonds (e.g., between $\mathbf{x}'$ and $\mathbf{x}$) to break when stretched beyond some predetermined limit (see [9, 13] for a discussion) so enabling cracks initiate and propagate when and where it is energetically favorable—unguided crack growth and complex interaction among cracks is the rule. This is in contrast to standard fracture mechanics approaches; see [20] for a review. The reader is referred to [21, 22] for applications of peridynamics not discussed in our paper.

3.1. Polycrystal fracture

Brittle fracture is a catalyst for failure in ceramic materials subject to appropriate mechanical loading. Predicting the performance and structural integrity of ceramics requires a fundamental understanding of brittle fracture in polycrystalline ceramics at the grain-size level. Such an understanding can then be used to design advanced structural ceramic systems with increased impact resistance, higher thermal shock resilience, and so forth. Here we present a novel approach for simulating fracture in polycrystalline ceramics, which has been difficult to do in the past, in no small measure because of the complexity of the phenomenon.

Combined transgranular (cracks pass through the grains) and intergranular (cracks propagate between the grains) fracture can take place in brittle fracture of ceramics. A transition between the trans- and intergranular fracture is observed under certain loading. Experiments reported in [23] show that, while the main fracture mode is intergranular, local transgranular fracture also appears because of the existence of crystalline phase at grain junctions. In addition, crack deflection and crack bridging mechanisms were observed because of the presence of rodlike grains. Further evidence of the importance of accurate modeling of inter- and transgranular fracture in polycrystalline ceramics is presented in [24].

Models for simulating brittle fracture in polycrystalline ceramics that can include combined trans- and intergranular crack propagation have severe limitations, including the inability of modeling propagating cracks that naturally coalesce and/or branch; limitation to modeling only a single or a small number of cracks; and complicated algorithms that cannot extend to three dimensions.

The finite-element method applied to equations of classical elasticity has been the preferred tool used in numerical models of polycrystalline materials (see [25, 26, 27, 28]). Fracture in polycrystalline materials has been modeled by using cohesive-zone models initiated in [29, 30, 31]. Recently, some difficulties regarding time discontinuities in cohesive zone models have been pointed out in [32]. Bias of the crack propagation path and possible remedies are discussed in [33]. Mathematical and physical limitations, and constraints on cohesive laws are discussed in [34]. The most common approach for simulating brittle fracture of polycrystalline ceramics is to consider *only* intergranular fracture: cracks are restricted to grain boundaries (see, e.g., [35, 36, 37], and finite elements with a "soft-kill method" are used to propagate the crack along grain boundaries instead of using a cohesive zone model. Lattice spring models have been used to study brittle fracture and damage in polycrystalline materials by, for example, [38, 39]. A transition from intergranular to transgranular fracture with increasing grain boundary toughness is observed in [38]; however, these studies are limited to 2D quasi-static analysis.

Compared to the methods discussed above, peridynamics for analyzing crack initiation, propagation, and fragmentation in a rate-dependent mechanically loaded Voronoi polycrystalline ceramics has important advantages. These include inter- and transgranular fracture are direct consequences of the computations and they do not have to be postulated via ad hoc assumptions as is the case for the classical approach; mode-transition and mode-mixing of crack propagation is naturally captured by the peridynamic formulation; fracture at triple-junction points is not controlled by ad hoc assumptions but by the actual loading conditions in a region surrounding the triple-junctions; and the meshfree discretization of peridynamics eliminates the need for complicated meshing algorithms for Voronoi polycrystals that represents an impediment to three dimensional computations.

We now summarize some results for simulating polycrystalline fracture in silicon ([40] will contain more details). Single-crystal silicon has a cubic structure, with the elastic moduli c_{11} = 166 GPa, c_{44} = 80 GPa, and c_{12} = 64.5 GPa. Extensions to other types of crystals is immediate and do not require special considerations. We generate a Voronoi polycrystal over a square with dimensions of 1 cm by 1 cm. Using 120 randomly distributed x and y coordinates for the Voronoi cells seeds, we obtain a grain-size distribution that is close to a Weibull distribution. This number of cells in the Voronoi structure gives the same grain-size distribution for different realizations (see [40]). For each Voronoi cell we assign a certain, random, orientation angle (from a uniform distribution). We define the micro-stiffness for the peridynamic bonds to vary with their orientation inside the grain and match the effective stiffness. Thus, the micro-stiffness depends on the orientation of each grain and the orientation of the bond inside the grain.

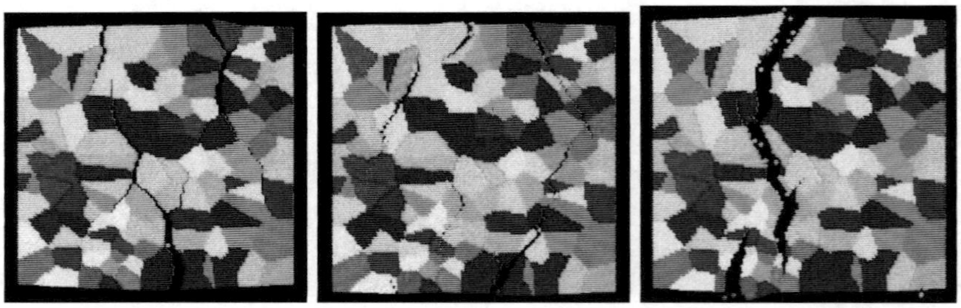

Figure 1. Fracture dependence on parameter β: β=0.25 (left), β=1 (center), and β=4(right).

The peridynamic bond relative elongation is given by s defined in (2), and the critical relative elongation at which a bond breaks is associated with the fracture energy and the peridynamic horizon via (see [9]) $s_0 = \sqrt{(5G_0)/(9\kappa\delta)}$, where G_0 is the material's (single crystal silicon) fracture energy and κ is the bulk modulus. Bonds that connect different grains have special properties that may be related to the properties of the grain boundaries. We select the micro-stiffness for such bonds to be the average of the corresponding directions in the two grains. For the critical relative elongation of the bonds that have ends in different grains we define an "interface strength coefficient" $\beta = s_0^{\mathrm{GB}}/s_0$, where s_0^{GB} is the critical relative elongation of the bonds that pass over a grain boundary, and s_0 is the value defined above for the single-crystal material. As we vary β from subunitary to superunitary numbers, we change the strength of the grain boundaries from lower than the single crystal to higher than the single crystal.

The tests are performed using EMU [9] on a two-dimensional grid of 100 nodes in each direction. Extensions to three-dimensional simulations are immediate and does not require special considerations (except for the increased computer storage and floating point computation). We apply velocity boundary conditions on two sides for stretching along the horizontal direction. The top and the bottom sides are free. The applied strain rate is constant equal to 50/s and the initial velocity distribution is consistent with this strain rate. We simulate 32 μs to reach an effective strain of 0.2%. We do not introduce initial cracks.

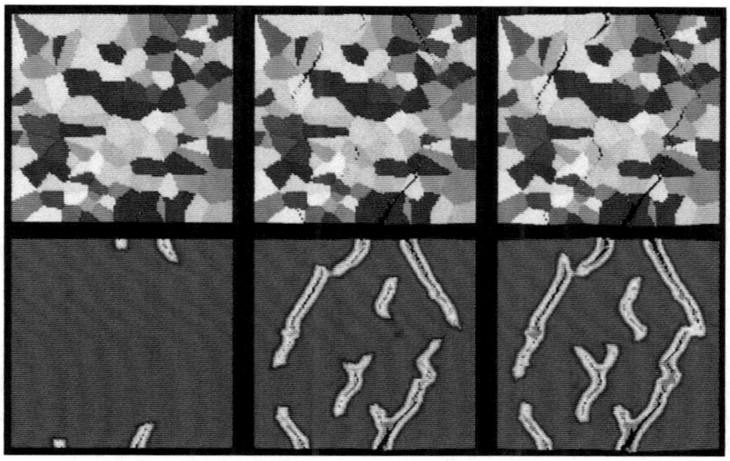

Figure 2. Evolution of damage and cracks in time for the $\beta = 1$ case. The top and bottom rows show the polycrystals boundaries and damage indices, respectively.

The plots in figure 1 show the polycrystal after 22 μs of the simulation for different values of the interface strength coefficient β. As β increases from 0.25 (left figure) to 1 (center) to 4 (right figure), we transition between purely intergranular fracture to pure transgranular fracture, passing through combined trans- and inter-granular fracture observed for average values of β. To better observe the ability of the peridynamic method to simulate crack propagation and crack interaction in polycrystalline ceramics we monitor the damage time-evolution for the case when $\beta = 1$ (see figure 2). We note that there are cracks that begin from the top and bottom edges, but also cracks that start in the interior of the sample, grow, branch, and finally join together.

3.2. Composite fracture and failure

The prediction of the spontaneous formation of a crack and its subsequent growth in load-carrying structures such as the wing or fuselage of an airplane presents a long-standing problem in computational solid mechanics. In a complex loading situation such as a bird strike, multiple cracks may occur simultaneously and interact. The path along which a crack propagates in a three-dimensional structure is generally not known in advance and must be determined as part of the solution of a computer simulation and ultimately depends on the underlying material. Recently anisotropic composite materials (such as CFRP) are replacing more traditional isotropic materials (such as high-strength aluminum alloys) in part because of their larger specific strength promises significant weight savings. A typical composite material used within the aircraft industry is called "prepreg" and consists of individual plies stacked on top of each other and bonded together in an autoclave oven. Each ply consists of extremely stiff fibers that are embedded in a relatively soft matrix. Such a material exhibits many different failure mechanisms that can be active at the same time: delamination of adjacent plies or microbuckling of individual fibers are examples of common failure mechanisms in composites that have no counterpart in metals. Resolving the myriad failure mechanisms is essential in a simulation so that the correct path along which a crack propagates is computed. Peridynamics is ideally suited for describing damage propagation in composite materials because of the ability to handle crack propagation when the crack path is not known a priori, the natural correspondence between fibers and peridynamic bonds, and the (infinitely) larger number of degrees of freedom in a peridynamic deformation state (in comparison to the classical local deformation gradient) that can capture a greater range of failure phenomena.

Anisotropy is introduced by allowing bonds in different directions to have different stiffness and failure properties. In the peridynamic model of a unidirectional fiber-reinforced lamina, the bonds in the direction of the fibers represent the fiber properties, while the bonds in all other directions represent the matrix properties. The fiber and matrix bonds have properties that are independent of each other, including both elastic response and failure properties.

The peridynamic model of a lamina is illustrated in figure 3. The stiffness of matrix and fiber bonds, c^m, c^f, are chosen to match the effective laminate stiffness E_{11} and E_{22}. The critical strains for bond breakage in tension, s_0^{mt} and s_0^{ft}, are used to match the critical failure strains in a 0 and 90 degree lamina. In compression, the matrix and fiber bonds are assumed to fail at the same critical strain, denoted s_0^c. This assumption is made because the microbuckling mechanism for lamina failure in compression involves simultaneous localization in both the resin and fibers; hence it is not possible to identify separate failure criteria for each in compression. A model of a laminate is constructed from a stack of lamina models, each with a particular fiber direction. The laminas interact with each other through special peridynamic bonds that capture the through thickness stiffness E_{33} as well as the interlayer shear stiffness G_{13} and the corresponding energy release rates for mode I and II that can be measured in an DCB and ENF tests. For more detailed descriptions see [41, 42, 43, 44].

Using this simple model, a wide variety of problems involving damage propagation in CFRP panels were studied. Figure 4 shows a stiffened notched composite panel under compressive loading. Only half of the panel is shown because of symmetry. Note that in addition to the notch in the skin, the middle stringer has also been cut. The aim of such as simulation is to predict the damage tolerance capabilities in an lower wing structure of a composite airplane.

3.3. Nanofiber networks

There is much interest in polymer nanofibers because of their desirable structural mechanical properties. In particular, they possess a large surface area to volume ratio, as well as high axial strength and extreme flexibility. The use of nanofibers is being explored in such diverse areas as composites, protective clothing, catalysis, electronics, biomedicine (including tissue engineering, implants, membranes, and drug delivery), filtration, and agriculture [45]. Recently, polymer nanofibers have also been proposed as a filters for chemical warfare defense [46]. The ability to tailor the properties of nanofiber networks and membranes for specific applications requires that an understanding of the mechanical properties of these networks and membranes through modeling and simulation. A peridynamic model for simulating the mechanical behavior and strength and toughness properties of three-dimensional nanofiber networks under imposed stretch deformation was proposed in [47].

In this numerical model, the peridynamic prototype microelastic brittle (PMB) [9] was augmented with a Lennard-Jones potential to mimic the effects of van der Waals forces. This augmented potential was applied to nanofiber networks similar to that shown in figure 5(a). The model domain is 400 nm wide by 400 nm high by 10 nm thick. The top, bottom, left, and right ends have a constant velocity boundary condition applied to induce a biaxial strain that eventually leads to the failure of the nanofiber network. Simulations performed with the PDLAMMPS code [13] are shown in figure 5. Detailed numerical simulation in [47] led to the observations that van der Waals interactions are important in increasing the strength and toughness of the nanofiber network, and that heterogeneity in the strength of bonds in the nanofiber network can increase its toughness and ductility. Additionally, reorientation due to breakage and deformation and a Poisson effect due to van der Waals induced accretion are the two main mechanisms that control the dynamic deformation and damage of nanofiber networks [47].

Acknowledgment

Sandia is a multiprogram laboratory operated by Sandia Corporation, a Lockheed Martin Company, for the United States Department of Energy under contract DE-AC04-94AL85000. The authors thanks Gail Pieper and Lori Diachin for their helpful comments.

References

[1] de Pablo J J and Curtin W A (eds) 2007 *Multiscale Modeling in Advanced Materials Research* (*MRS Bulletin* vol 32) (Material Research Society)
[2] Miller R E and Tadmor E B 2007 *Multiscale Modeling in Advanced Materials Research* (*MRS Bulletin* vol 32) ed de Pablo J J and Curtin W A pp 920–926
[3] Bažant Z P and Jirásek M 2002 *J. of Eng. Mech.* **128** 1119–1149
[4] Chen Y, Lee J D and Eskandarian A 2004 *Int. J. of Solids and Structures* **41** 2085–2097
[5] Eringen A C 2002 *Nonlocal continuum field theories* (Springer-Verlag NewYork, Inc)
[6] Silling S A 2000 *J. Mech. Phys. Solids* **48** 175–209
[7] Silling S A, Epton M, Weckner O, Xu J and Askari E 2007 *J. Elasticity* **88** 151–184
[8] Weckner O and Abeyaratne R 2005 *J. Mech. Phys. Solids* **53** 705–728
[9] Silling S A and Askari E 2005 *Comp. Struct.* **83** 1526–1535 EMU available at `www.sandia.gov/emu/emu.htm`
[10] Ciarlet P 1988 *Finite Element Method for Elliptic Problems* (North-Holland, Amsterdam)
[11] Lehoucq R B and Silling S A 2008 *J. Mech. Phys. Solids* **56** 1566–1577
[12] Silling S A and Lehoucq R B 2008 *J. Elasticity* Article in Press, available online
[13] Parks M L, Lehoucq R B, Plimpton S J and Silling S A 2008 *Comp. Phys. Comm.* Article in Press, available online
[14] Bobaru F, Yang M, Alves L F, Silling S A, Askari A and Xu J 2008 *Int. J. Num. Meth. Eng.* Submitted
[15] Bobaru F, Yang M, Alves L F, Silling S A and Askari A 2008 *Int. J. Num. Meth. Eng.* In preparation
[16] Emmrich E and Weckner O 2007 *Math Mod Analysis* **12** 17–27
[17] Emmrich E and Weckner O 2005 *Math and Mech of Solids* **12** 363–384
[18] Macek R and Silling S 2007 *Fin. Elem. Anal. Design* **43** 1169–1178
[19] Plimtpon S J 1995 *J. Comp. Phys.* **117** 1–19 LAMMPS available at `http://lammps.sandia.gov`
[20] Ingraffea A R 2007 *Encyclopedia of Computational Mechanics* (*Solids and Structures* vol 2) ed Stein E, de Borst R and Hughes T J R (Wiley) chap 11
[21] Gerstle W, Sau N and Silling S 2007 *Nuc. Eng. Des* **237** 1250–1258
[22] Silling S A and Bobaru F 2005 *Int. J. Non-Linear Mech.* **40** 395–409
[23] Lee B T, Han B D and Kim H D 2003 *Materials Letters* **58** 74– 79
[24] Sun X, Li G J, Guo S, Xiu Z, Duan K and Hu X 2005 *J of the American Ceramic Society* **88** 1536–1543
[25] Ghosh S and Mukhopadhyay S 1993 *Comp Meth Appl Mech Eng* **104** 211–247
[26] Ghosh S and Mallett R 1994 *Comp. Struct.* **50** 33–46
[27] Zavattieri P, Raghuram P and Espinosa H 2001 *J. Mech. Phys. Solids* **49** 27–68
[28] Zhang D, Wu M and Feng F 2001 *International Journal of Plasticity* **21** 801–834
[29] Barenblatt G 1959 *Applied Mathematics and Mechanics (PMM)* **23** 622636
[30] Dudgale D 1960 *J. Mech. Phys. Solids* **8** 100–104
[31] Barenblatt G 1962 *Advances in Applied Mechanics* **7** 55–129
[32] Papoulia K, Sam C and Vavasis S 2003 *Int. J. Num. Meth. Eng.* **58** 679–701
[33] de Borst R 2003 *Eng. Frac. Mech.* **70** 1743–1757
[34] Jin Z and Sun C 2005 *Int. J. Frac.* **134**
[35] Grah M, Alzebdeh K, Sheng P, Vaudin M, Bowman K and Ostoja-Starzewski M 1996 *Acta Materialia* **44** 4003–4018
[36] Holm E 1998 *Journal of the American Ceramic Society* **81** 455–459
[37] Maiti S, Rangaswamy K and Geubelle P 2005 *Acta Materialia* **53** 823 – 834
[38] Yang W, Srolovitz D, Hassold G and Anderson M 1990 *In Simulation and Theory of Evolving Microstructures* (The Metallurgical Society: Warrendale, PA) chap Microstructural effects in the fracture of brittle materials, pp 277–284
[39] Rinaldi A, Krajcinovic D, Peralta P and Lai Y C 2008 *Mechanics of Materials* **40** 17–36
[40] Bobaru F and Silling S 2008 *in preparation*
[41] Askari A, Xu J and Silling S A 2006 *Proc. 44th AIAA Aerospace Sciences Meeting and Exhibit* AIAA 2006-88 (Reno, NV)
[42] Silling S A, Askari A, Nelson K, Weckner O and Xu J 2008 *Proc. SAMPE Fall Technical Conference 2008*
[43] Xu J, Askari A, Weckner O and Silling S A 2008 *J. Aero. Eng.* Article in Press

[44] Xu J, Askari A, Weckner O, Razi H and Silling S A 2007 *Proc. 48th AIAA/ASME/ASCE/AHS/ASC Structures, Structural Dynamics, and Materials Conference* (Honolulu, HI)
[45] Dzenis Y 2004 *Science* **304** 1917–19
[46] Ramaseshan R, Sundarrajan S, Liu Y, Barhate R, Lala N L and Ramakrishna S 2006 *Nanotechnology* **17** 2947–2953
[47] Bobaru F 2007 *Modelling Simul. Mater. Sci. Eng.* **15** 397–417

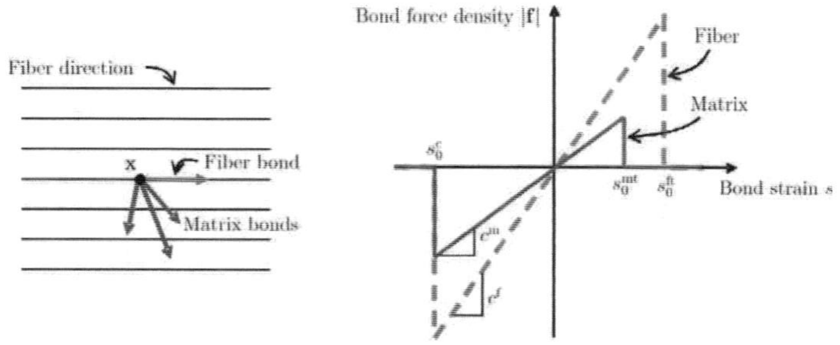

Figure 3. Peridynamic lamina model

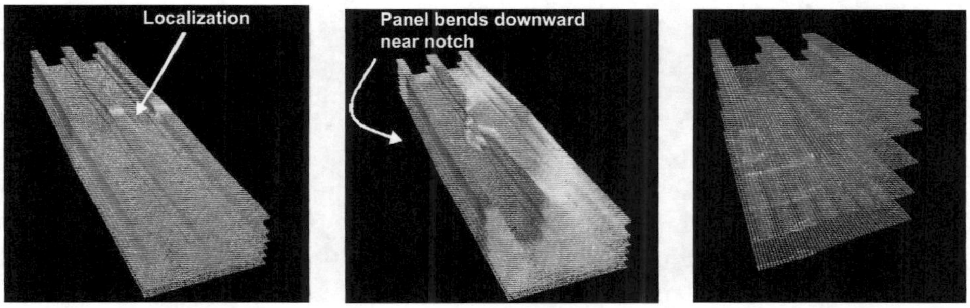

Figure 4. (Left to right) Axial displacement during damage growth; out of plane displacement growth; damage at a later time shows disbonding.

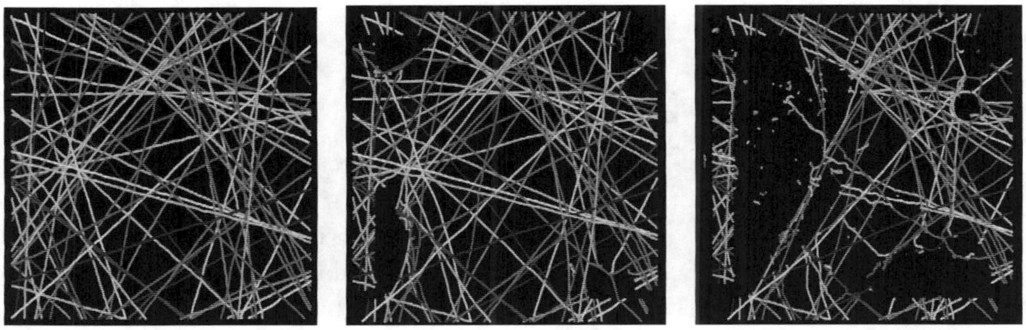

Figure 5. Peridynamic simulation of polymer nanofiber network under biaxial strain. (Left to right) Undeformed configuration; Biaxial strain of 17.6% (30 ns); Biaxial strain of 29.4% (50 ns). Colors indicate individual fibers.

Journal of Physics: Conference Series **125** (2008) 012079 doi:10.1088/1742-6596/125/1/012079

Evaluation of SuperLU on multicore architectures

Xiaoye S. Li

Lawrence Berkeley National Laboratory, Berkeley, CA 94720, USA

E-mail: xsli@lbl.gov

Abstract. The Chip Multiprocessor (CMP) will be the basic building block for computer systems ranging from laptops to supercomputers. New software developments at all levels are needed to fully utilize these systems. In this work, we evaluate performance of different high-performance sparse LU factorization and triangular solution algorithms on several representative multicore machines. We included both Pthreads and MPI implementations in this study and found that the Pthreads implementation consistently delivers good performance and that a left-looking algorithm is usually superior.

1. Introduction

The Chip Multiprocessor (CMP) systems will be the basic building blocks for computers ranging from laptops to supercomputers. Compared to the superscalar microprocessors exploiting high degree of instruction level parallelism, the CMP designs represent a paradigm shift that strikes better trade-offs between performance (parallelism) and energy efficiency. In theory, the CMPs can often be programmed the same way as the conventional SMPs, but the CMPs have lower memory bandwidth and abundance of fine-grained parallelism. Given the diversity of CMP designs, it is necessary, albeit difficult, to develop new software strategies at the system level as well as the application level in order to fully utilize the hardware resources.

In this paper, we study the factorization and triangular solution kernels in the sparse direct solver SuperLU [1] on two leading CMP systems. Our goal of this study is twofold. First, we would like to evaluate performance of the existing implementations on the new CMP architectures, and secondly, we would like to identify the inefficiencies in the algorithms and/or implementations and the ways to improve them for the new architectures.

2. Experimental machines

Our testing systems include an Intel Colvertown, a Sun VictoriaFalls, and an IBM Power5. The last one contains a conventional SMP node. Table 1 summarizes the key architectural features of the three systems used in this study. The sources come from [2, 3, 4].

The Intel Colvertown consists of two sockets, each with two pairs of dual-core Xeon chips (Core2Duo), with total eight processors (Dell PowerEdge 1950 dual-socket). Each core runs at 2.33 GHz with a peak performance of 9.3 Gflops (4 flops per cycle) and has a private 32 KB L1 cache. Each chip (two cores) share a 4 MB L2 cache. Each socket has access to a Front Side Bus (FSB) delivering 10.6 GB/s. The two independent FSBs are connected to the memory controller which interfaces to the DRAM channels, delivering 21.3 GB/s read memory bandwidth and 10.6 GB/s write bandwidth.

The dual-chip Sun VictoriaFalls contains 16 SPARCv9 cores, in which each CMP is a Niagara2 chip with 8 cores. Each core runs at 1.16 GHz with a peak performance of 1.16 Gflops, and has a private 8 KB L1 cache. All eight cores share a 4 MB L2 cache. In addition, each core

Table 1. Summary of the experimental machines.

	Intel Colvertown	Sun VictoriaFalls	IBM Power5 (575)
Core type	superscalar (4)	multithreaded (8)	superscalar (4)
Clock (GHz)	2.3	1.16	1.9
L1 Dcache	32 KB	8 KB	32 KB
DP Gflops	9.3	1.16	7.6
# Sockets	2	2	8
# cores/socket	4	8	1
L2 cache	4 MB/2-cores (16 MB)	4 MB/socket (8 MB)	1.92 MB /core (32 MB L3$/node)
DP Gflops	74.7	18.7	60.8
DRAM GB/s read	21.3	42.6	200
write	10.6	21.3	–
Byte/flop ratio	0.29	0.44	3.29
Power/socket (Watts)	160 (max)	84 (max)	500 (measured [3])

supports eight hardware threads, and the entire dual-chip system provides a total of 128 threads. The two sockets are interconnected via External Coherence Hubs (ECH). There are altogether 8 FBDIMM memory channels, delivering the aggregate DRAM bandwidth of 42.6 GB/s for read and 21.3 GB/s for write.

The IBM p575 Power5 is a scalable distributed-memory high-performance computing system consisting of conventional SMP nodes. The entire system (bassi at NERSC) has 111 compute nodes, each of which has 8 Power5 processors running at 1.9 GHz and has a shared-memory pool of 32 GBytes. Each processor has a peak performance of 7.6 GFlops (4 flops per cycle) and has a private 32 KB L1 cache. We use only one SMP node in this study.

3. Overview of the algorithms and implementations

3.1. Factorization in `SuperLU_MT` *using Pthreads or OpenMP*

`SuperLU_MT` [8] was first developed with Pthreads, targeted for the SMPs of modest size (e.g., 32 processors). Recently, we have added OpenMP support. The factorization uses a panel-based *left-looking* algorithm, with partial pivoting and possibly with diagonal preference to better preserve sparsity. The kernel is based on supernode-panel update, which invokes multiple calls to BLAS 2, effectively achieving BLAS 2.5 speed. Parallelization uses an asychronous and barrier-free dynamic algorithm to schedule both coarse-grained and fine-grained parallel tasks and achieve a high level of concurrency. A globally shared task queue is used to store the ready panels in the column elimination tree; and whenever a thread becomes free, it obtains a ready panel from the task queue. The coarse-grained task is to factorize the independent panels in the disjoint subtrees, while the fine-grained task is to update panels by previously computed supernodes. The scheduler facilitates the smooth transition between the two types of tasks, and maintains load balance dynamically. Figure 1(a) illustrates the left-looking factorization scheme and the dynamic scheduling method using the elimination tree.

3.2. Factorization in `SuperLU_DIST` *using MPI*

In order to address the scalability issues, the parallel algorithm in `SuperLU_DIST` [5] is significantly different from that in `SuperLU_MT`. Using the supernode partition, we perform a two-dimensional (nonuniform) block-cyclic matrix-to-processor mapping. The factorization uses a block-based *right-looking* algorithm which comprises abundance of parallelism during the block

Table 2. Properties of the test matrices. Minimum degree algorithm was applied to the structure of $|A| + |A|^T$. "fill-ratio" denotes the ratio of number of nonzeros in $L + U$ over that in A; "Mean S-node" refers to an average number of columns in a supernode.

	Application	Dimension	Nonzeros in A	Fill-ratio	Mean S-node
g7jac200	economic model	59,310	837,936	40.2	1.9
stomach	duodenum model	213,360	3,021,648	45.4	4.0
torso1	2D model of torso	116,158	8,516,500	3.1	4.0
twotone	nonlinear anal. circuit	120,750	1,224,224	9.3	2.3

outer-product updates to the Schur complements. We use the elimination DAGs to identify block dependencies, and a look-ahead scheme to overlap communication with computation on the critical path. Figure 1(b) illustrates the 2D block-cyclic partition and distribution for a sparse matrix.

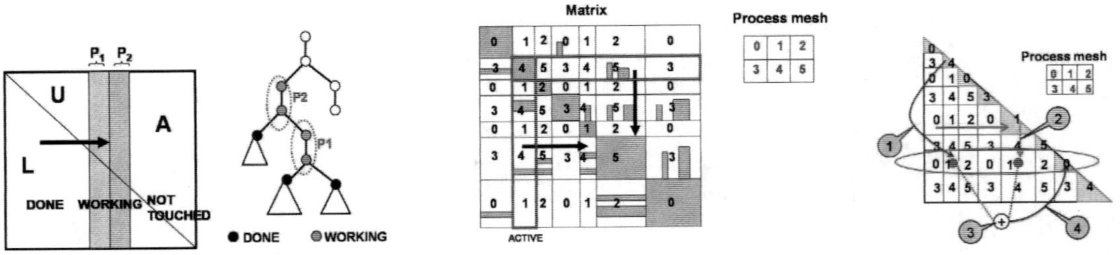

(a) 1D left-looking factorization. (b) 2D right-looking factorization. (c) 2D triangular solve.

Figure 1. Illustration of the parallel algorithms in `SuperLU_MT` (a) and `SuperLU_DIST` (b, c).

3.3. Triangular solution in `SuperLU_DIST`

The triangular solution phase in `SuperLU_MT` is not yet parallel; therefore we evaluate only the parallel algorithm in `SuperLU_DIST`. When solving $Lx = b$, where L is a lower triangular matrix, the ith solution component is computed as $x_i = (b_i - \sum_{j=1}^{i-1} L_{ij} \cdot x_j)/L_{ii}$. Therefore, computation of x_i needs some of the previous solution components $x_j, j < i$, depending on the sparsity structure of the i-th row of L. This sequentiality often poses scaling hurdle for a parallel algorithm. Another hurdle to achieve good performance is the much lower arithmetic density as measured by flops per byte of DRAM access or communication, compared to factorization. In `SuperLU_DIST`, the parallel triangular algorithm uses the same 2D block-cyclic distribution as used in the factorization phase. Figure 1(c) illustrates such a distribution and the solution procedure. The processes owning the diagonal blocks (called *diagonal processes*) are responsible for computing the corresponding blocks of the x components. Consider one block row of the L matrix, as circled in figure 1(b), computation of the corresponding x entry needs to proceed following the steps ①, ②, ③ , and ④. Communication is necessary when the data reside on different processes, as in steps ① and ④.

4. Experimental results

Table 2 presents the characteristics of our benchmarking matrices, which are available from the University of Florida Sparse Matrix Collection [6]. We benchmarked the Pthreads version of `SuperLU_MT`, and `SuperLU_DIST` using MPICH [7].

Table 3 shows the parallel factorization times of the two solvers on the Clovertown. The time includes both symbolic and numerical factorization. First, when we used MPICH configuration with `ch_p4` device for communication through sockets, the code slowed down significantly beyond two or four cores. After we switched to `ch_shmem` setup, we obtained respectable speedup. So

Table 3. Factorization time in seconds on Intel Clovertown.

Matrix	Threads or Tasks		1	2	4	8	Speedup
g7jac200	SuperLU_MT		32.78	17.91	12.41	10.60	3.1
	SuperLU_DIST	ch_shmem	28.10	15.95	11.06	7.57	3.9
		ch_p4	28.62	22.98	56.31	62.39	
	speedup ratio (_MT/_DIST)		1.00	1.03	1.01	0.80	
stomach	SuperLU_MT		64.38	37.15	20.39	17.24	3.7
	SuperLU_DIST	ch_shmem	43.45	25.91	15.81	13.64	3.4
		ch_p4	44.28	27.84	210.99	264.58	
	speedup ratio (_MT/_DIST)		1.00	0.99	1.10	1.10	
torso1	SuperLU_MT		9.43	4.92	2.87	2.20	4.3
	SuperLU_DIST	ch_shmem	9.43	5.83	4.55	4.76	2.2
		ch_p4	9.62	7.23	54.77	76.32	
	speedup ratio (_MT/_DIST)		1.00	1.12	1.49	1.99	
twotone	SuperLU_MT		6.80	4.05	2.32	1.83	3.9
	SuperLU_DIST	ch_shmem	18.08	10.17	7.55	7.21	2.1
		ch_p4	18.34	12.19	47.30	60.99	
	speedup ratio (_MT/_DIST)		1.00	0.95	2.26	1.86	

for a large distributed system comprising manycore chips, it is imperative to be able to configure MPICH in a *hybrid* device mode — `ch_shmem` within socket and `ch_p4` across sockets. Currently, this hybrid mode is not avaialbe. Second, we examine the single core performance. We would expect that `SuperLU_DIST` outperforms `SuperLU_MT`, because the former uses BLAS 3, whereas the latter uses only BLAS 2.5. This is true only with two matrices, g7jac200 and stomach, which have relatively denser L and U factors (the fill ratios are over 40; see table 2), and hence BLAS 3 plays a larger role. For sparser problems, the algorithms are memory-bound. We believe the worse performance of `SuperLU_DIST` is mainly due to more memory traffic of the right-looking algorithm, especially more memory write operations. Third, we examine the speedups of the two codes. The last column of table 3 shows the speedup obtained when creating eight threads or MPI tasks. The best speedup is 4.3 and is less than what we observed on conventional SMP processors [8]. After performing code profiling, we found that the overhead of the scheduling algorithm using the shared task queue and the synchronization cost using mutexes (locks) are quite small. Further study is needed to understand where the time goes. Lastly, `SuperLU_MT` usually achieves more speedup than `SuperLU_DIST`. This can be seen in the row "speedup ratio (`_MT/_DIST`)" associated with each matrix. In some cases, `SuperLU_MT` achieves a factor of two more speedup than `SuperLU_DIST`.

Table 4 shows the parallel factorization times on the Sun VictoriaFalls. The single-thread performance of `SuperLU_DIST` is usually better than that of `SuperLU_MT`. This is probably because the machine has a higher byte-to-flop ratio (see table 1) compared to Clovertown, hence it does not penalize an algorithm that is memory-bandwidth demanding, such as the right-looking algorithm in `SuperLU_DIST`. However, the coarse-grained task parallelism supported by MPI programming does not match the fine-grained multithreading architecture; MPICH uften crashes when more than 16 tasks are generated. The Pthreads program is much more robust, and `SuperLU_MT` can effectively use 64 threads. Similar to the Clovertown, `SuperLU_MT` usually achieves more speedup than `SuperLU_DIST`. In some case, `SuperLU_MT` achieves a factor of 2 more speedup than `SuperLU_DIST`.

For the parallel triangular solution, we compare the eight-core Clovertown with the eight-processor Power5 SMP node. The parallel runt imes are tabulated in table 5. The columns

Table 4. Factorization time in seconds on Sun VictoriaFalls ("f" indicates an MPI failure).

Matrix	Threads or Tasks	1	2	4	8	16	32	64	128
g7jac200	SuperLU_MT	480.84	244.24	126.16	68.93	40.22	28.47	**23.95**	24.80
	SuperLU_DIST	283.44	153.18	83.09	49.20	31.70	f	f	f
	speedup ratio (_MT/_DIST)	1.00	1.06	1.09	1.15	1.24			
stomach	SuperLU_MT	1212.97	620.58	319.85	168.04	90.01	56.51	**53.54**	62.37
	SuperLU_DIST	598.49	329.28	183.90	116.22	85.56	f	f	f
	speedup ratio (_MT/_DIST)	1.00	1.06	1.13	1.33	1.79			
torso1	SuperLU_MT	201.05	102.09	52.51	27.41	15.16	11.56	**10.23**	11.34
	SuperLU_DIST	101.68	58.25	32.53	21.83	17.06	f	f	f
	speedup ratio (_MT/_DIST)	1.00	1.12	1.18	1.46	2.01			
twotone	SuperLU_MT	113.12	60.09	31.50	17.18	11.17	8.17	**7.26**	7.90
	SuperLU_DIST	135.43	78.44	46.64	30.01	18.49	f	f	f
	speedup ratio (_MT/_DIST)	1.00	1.08	1.19	1.38	1.26			

Table 5. SuperLU_DIST triangular solution time in seconds on Clovertown and Power5.

		Current				Improved			
Matrix	Tasks	1	2	4	8	1	2	4	8
g7jac200	Clovertown	0.39	0.79	0.76	2.94	0.30	0.28	0.29	0.44
	Power5	0.61	0.68	0.46	0.39	0.43	0.39	0.28	0.22
stomach	Clovertown	0.93	1.21	3.79	6.74	0.77	0.74	0.53	0.90
	Power5	1.24	1.29	0.86	0.75	0.92	0.77	0.59	0.46
torso1	Clovertown	0.28	0.52	1.98	3.22	0.21	0.29	0.32	0.45
	Power5	0.31	0.41	0.27	0.24	0.22	0.24	0.18	0.13
twotone	Clovertown	0.46	1.51	4.42	7.52	0.32	0.44	0.47	0.80
	Power5	0.71	0.97	0.69	0.58	0.44	0.52	0.44	0.34

labeled "Current" correspond to the current released code, and the columns labeled "Improved" refer to the new implementation as a result of this study. As seen in the table, the current code runs much more slower with more cores involved on the Clovertown. A similar trend was also observed on the VictoriaFalls. After profiling various parts of the code, we found that the slowdown is due to many calls of `MPI_Reduce`, which can take over 75% of the time using eight cores. In figure 1(b), each diagonal process needs to know which off-diagonal processes will have sum contributions to be sent to the diagonal process. To compute this count, every process holds a 0/1 flag indicating whether this process has nonzero blocks. Then all the processes in each row communicator perform an `MPI_Reduce` (by SUM) over the flags, with root being the diagonal process. Overall, each block row corresponds to one such reduction operation.

The improvement we have made is the following. Instead of performing many reductions with one integer, we allocate a flag array of integers, the size of which is the number of block rows owned by each process. Each entry is the flag associated with one block row. Then all the processes in the respective process row perform *only one* reduction operation on this flag array. This has greatly reduced the memory and communication latency cost. On eight-core Clovertown, the improvement is significant, ranging from 6- to 9-fold. Even on the conventional SMP node, such as eight-CPU Power5, we also obtained 63% to 84% improvement.

Acknowledgments
This research was supported in part by the Director, Office of Advanced Scientific Computing Research of the U.S. Department of Energy undercontract number DE-AC03-76SF00098. We used the multicore clusters with the PSI project and the RADlab at UC Berkeley, and the resources at the National Energy Research Scientific Computing Center.

References

[1] Li X S 2005 overview of SuperLU: Algorithms, implementation, and user interface *ACM Trans. Mathematical Software* **31(3)**:302–325

[2] Phillips S 2007 Victoriafalls: Scaling highly-threaded processor cores *HOT CHIPS 19: A Symposium on High Performance Chips* (Stanford, CA)

[3] Shalf J 2008 (Private communications)

[4] Williams S 2008 (Private communications)

[5] Li X S and Demmel J W 2003 SuperLU_DIST: A scalable distributed-memory sparse direct solver for unsymmetric linear systems *ACM Trans. Math. Soft.* **29(2)**:110–140

[6] Davis T A 1997 *University of Florida Sparse Matrix Collection* http://www.cise.ufl.edu/research/sparse/matrices

[7] *MPICH – A portable implementation of MPI* http://www-unix.mcs.anl.gov/mpi/mpich1/

[8] Demmel J W, Gilbert J R, and Li X S 1999 An asynchronous parallel supernodal algorithm for sparse Gaussian elimination. *SIAM J. Matrix Analysis and Applications* **20(4)**:915–952

Embedded boundary grid generation using the divergence theorem, implicit functions, and constructive solid geometry

Terry J Ligocki, Peter O Schwartz, Julie Percelay and Phillip Colella
Computational Research Division, Lawrence Berkeley National Laboratory, Berkeley, CA 94720, USA
E-mail: TJLigocki@lbl.gov

Abstract.
To construct finite-volume methods for PDEs in arbitrary dimension to arbitrary accuracy in the presence of irregular boundaries, we show that estimates of moments, integrals of monomials, over various regions are all that are needed. If implicit functions are used to represent the irregular boundary, the needed moments can be computed straightforwardly and robustly by using the divergence theorem, Taylor expansions, least squares, recursion, and 1D root finding. Neither a geometric representation of the irregular boundary nor its interior is ever needed or computed. The implicit function representation is general and robust. Implicit functions can be combined via constructive solid geometry to form complex boundaries from a rich set of primitives including interpolants of sampled data, for example, 2D/3D image data and digital elevation maps.

1. Finite-volume methods and flux computations

The overall goal of this work is to numerically compute solutions to PDE on domains with irregular boundaries in an arbitrary dimension Cartesian space, $\mathbb{R}^D$, to an arbitrary accuracy by using finite-volume methods [1]. The PDE are written as conservation laws that contain a divergence of a "flux", $\nabla \cdot \boldsymbol{F}$, for example,

$$\nabla \cdot \beta(\nabla u) = \rho, \quad \boldsymbol{F}(u) = \beta(\nabla u) \qquad \text{(Elliptic PDE)}$$

$$\frac{\partial u}{\partial t} = \nabla \cdot \rho(\nabla u), \quad \boldsymbol{F}(u) = \rho(\nabla u) \qquad \text{(Parabolic PDE)}$$

$$\frac{\partial u}{\partial t} + \nabla \cdot \boldsymbol{F}(u) = S(u), \quad \boldsymbol{F}(u) \text{ is given.} \qquad \text{(Hyperbolic PDE)}$$

The PDE are solved on a discretization of an irregular domain, Ω, defined implicitly using an "implicit" function, $\phi : \mathbb{R}^D \to \mathbb{R}$, namely,

$$\Omega = \{\boldsymbol{x} : \boldsymbol{x} \in \mathbb{R}^D, \ \phi(\boldsymbol{x}) < 0\}. \qquad (1)$$

Implicit functions are chosen to represent Ω for several reasons:

- The implicit function, $\phi(\boldsymbol{x})$, is defined everywhere in $\mathbb{R}^D$, and thus Taylor expansions of ϕ and functions of ϕ (e.g., $\boldsymbol{n}(\boldsymbol{x}) = \nabla\phi(\boldsymbol{x})/|\nabla\phi(\boldsymbol{x})|$) can be computed anywhere they exist if ϕ is smooth enough.

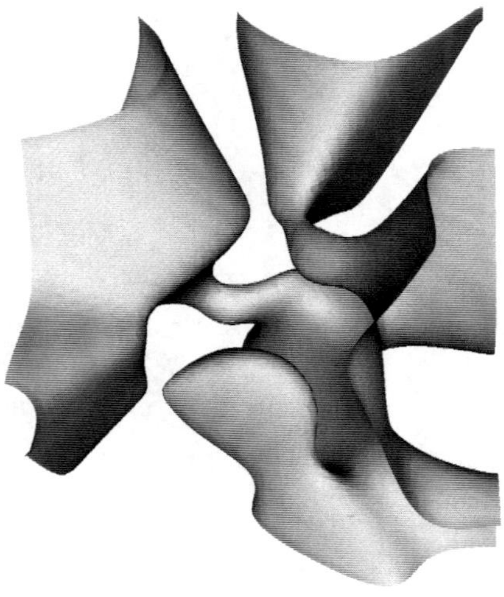

(a) Polynomial function of three variables

(b) SF Bay digital elevation map data

Figure 1. Domains generated via an implicit functions.

- Implicit functions can be used to represent a rich set of geometric shapes (see figure 1) either in functional form or as interpolants of discrete, sampled data (e.g., 2D/3D image data [3, 4, 5], digital elevation maps).
- Implicit functions can be easily restricted to lower dimensions. This is exploited in the algorithm that computes moments (see Section 2).
- Implicit functions can be extended to arbitrary dimensions allowing computations to be done in phase spaces or space-time.
- Ω can implicitly evolve in time if ϕ is allowed to change with time, possibly as a function of other PDE variables or time directly.

In this embedded boundary approach, Ω will be discretized on a set of control volumes, $\mathcal{V}$, formed by intersecting rectangular cells with Ω:

$$\mathcal{V} = \{[\boldsymbol{i}h, (\boldsymbol{i}+\boldsymbol{e})h] \cap \Omega : \boldsymbol{i} \in \mathbb{Z}^D\}, \tag{2}$$

where $\boldsymbol{e} \in \mathbb{Z}^D$ and all its components are one and h is the size of the rectangular cells that are, in this case, square.

Given this formulation, the integral of the divergence of the vector field, $\nabla \cdot \boldsymbol{F}$, over the control volume, $V \in \mathcal{V}$, can be transformed by using the divergence theorem into an integral over the boundary of the volume, $A = \partial V$:

$$\int_V \nabla \cdot \boldsymbol{F} \, dV = \int_A \boldsymbol{F} \cdot \boldsymbol{n} \, dA \tag{3}$$

where $\boldsymbol{n}$ is the normal to A pointing out of V. In control volumes that do not intersect the boundary of Ω, this can be rewritten as A_-^d: $\boldsymbol{F}_- \cdot \boldsymbol{n} = -F_-^d$ on the low faces and as A_+^d: $\boldsymbol{F}_+ \cdot \boldsymbol{n} = F_+^d$ on the high faces by noting that on the faces the normal points in a coordinate direction, d. The result is the canonical "(sum of) difference in fluxes (integrated over faces)" formulation of finite volume methods. These methods are as accurate as the approximations to integrals of $F_\pm^d$ over the faces.

With these definitions, a finite-volume method for control volumes containing embedded boundaries can be defined by using the divergence theorem:

$$\int_V \nabla \cdot \boldsymbol{F}\, dV = \sum_{\pm=+,-} \sum_{d=1}^{D} \pm \int_{A^d_\pm} F^d_\pm\, dA + \int_{A_{EB}} \boldsymbol{F} \cdot \boldsymbol{n}_{EB}\, dA, \tag{4}$$

where $A^d_\pm = \{x : x \in V, x_d = (i_d + \frac{1}{2} \pm \frac{1}{2})h\}$, $A_{EB} = V \cap \partial\Omega$, and $\boldsymbol{n}_{EB} = \nabla\phi/|\nabla\phi|$. Note: $\boldsymbol{n}_{EB}$ is defined *everywhere* in Ω and not simply on $\partial\Omega$. This equation has still not been discretized; that is, it is exact. The integrals over the boundary are discretized by using an Pth order Taylor expansion of $\boldsymbol{F}$ and $F^d_\pm$ about points $\boldsymbol{x}_{EB} \in A_{EB}$ and $\boldsymbol{x}^d_\pm \in A^d_\pm$ that approximate the centroids of the domains of integration. In addition, we can expand $\boldsymbol{n}_{EB}$ about some point in the cell (which we take to be the origin of our coordinates), since it is a known smooth function. Using these Taylor expansions, we can rewrite equation (4) as follows:

$$\int_V \nabla \cdot \boldsymbol{F}\, dV = \sum_{0\le|\boldsymbol{p}|\le P} \frac{1}{\boldsymbol{p}!}\Bigg(\sum_{\pm=+,-} \sum_{d=1}^{D} \pm (\nabla^{\boldsymbol{p}} F^d_\pm) \int_{A^d_\pm} (\boldsymbol{x} - \boldsymbol{x}^d_\pm)^{\boldsymbol{r}}\, dA$$
$$+ \nabla^{\boldsymbol{p}}\boldsymbol{F} \cdot \sum_{0\le|\boldsymbol{r}|\le R} \frac{\nabla^{\boldsymbol{r}}\boldsymbol{n}_{EB}}{\boldsymbol{r}!} \int_{A_{EB}} (\boldsymbol{x} - \boldsymbol{x}_{EB})^{\boldsymbol{p}} \boldsymbol{x}^{\boldsymbol{r}}\, dA \Bigg) + O(h^{D+P+R}.) \tag{5}$$

Assuming $\boldsymbol{F}$ is discretized on the rectangular grid covered by Ω, one can approximate $\nabla^{\boldsymbol{p}}\boldsymbol{F}^d_\pm$ and $\nabla^{\boldsymbol{r}}\boldsymbol{F}$ to the appropriate order by finite differences. The remaining information required is in the form of moments, namely, integrals of polynomials computed over the V, $A^d_\pm$, and A_{EB}. Note that $\boldsymbol{x}^d_\pm$ and $\boldsymbol{x}_{EB}$ can also be computed from moments. Computing these is the analogue of grid generation for the embedded boundary method. No explicit representation of the irregular boundary is ever needed.

2. Moment computations

Approximating the needed moments to the necessary order can be accomplished by taking all monomials, $\boldsymbol{x}^{\boldsymbol{p}}$, of a given degree, P, where $\boldsymbol{p}$ is a multi-index and $|\boldsymbol{p}| = P$, and substituting $\boldsymbol{F}(\boldsymbol{x}) = \boldsymbol{x}^{\boldsymbol{p}}\boldsymbol{e}_d$ into equation (4), where $\boldsymbol{e}_d$ is the unit vector in direction d. Note that $\partial\boldsymbol{x}^{\boldsymbol{p}}/\partial x_d = p_d\boldsymbol{x}^{\boldsymbol{p}-\boldsymbol{e}_d}$ and replacing $\boldsymbol{n}_{EB}$ by its Taylor expansion; then each monomial yields the set of D equations defined by

$$p_d \int_V \boldsymbol{x}^{\boldsymbol{p}-\boldsymbol{e}_d}\, dV - n_d \int_{A_{EB}} \boldsymbol{x}^{\boldsymbol{p}}\, dA = \sum_{\pm=+,-} \pm \int_{A^d_\pm} \boldsymbol{x}^{\boldsymbol{p}}\, dA$$
$$+ \sum_{1\le|\boldsymbol{r}|\le R} \left(\frac{\nabla^{\boldsymbol{r}} n_d}{\boldsymbol{r}!} \int_{A_{EB}} \boldsymbol{x}^{(\boldsymbol{p}+\boldsymbol{r})}\, dA \right) + O(h^{D+P+R}) \tag{6}$$

If the integrals on the LHS are treated as unknowns and the RHS is treated as known, then the set of equations formed in this way for all monomials $\boldsymbol{x}^{\boldsymbol{p}}$ where $|\boldsymbol{p}| = P$ forms a system of linear equations. The number of unknowns is $N_{P-1} + N_P$ (where N_P is the number of monomials of degree P) and number of equations is DN_P. Since $N_{P-1} < N_P$, if $D \ge 2$, this is an overdetermined system and can be solved by least squares. The one-dimensional case can be solved explicitly by using 1D root-finding techniques and integration of monomials (see below).

Examining the RHS shows that the first integral is of the same form as the volume integrals on the LHS except the integral has been restricted to one less dimension and the monomial degree

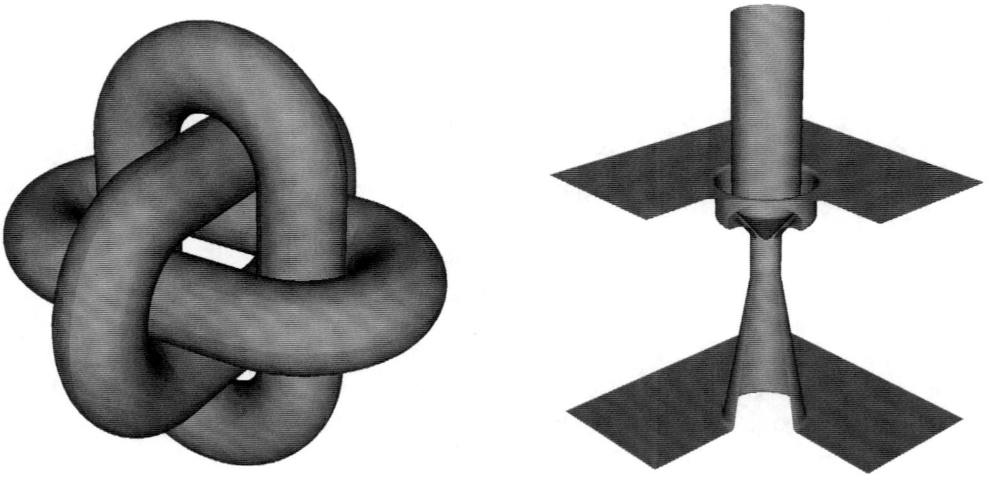

(a) Borromean (interlocked) rings (b) Gas jet nozzle

Figure 2. Domains generated via constructive solid geometry and multiple implicit functions.

is one more. Hence, the order of accuracy is the same, $((P+1)+(D-1)+R = P+D+R)$. Using recursion, we can reduce the dimension until $D=1$. At this point all the moments can be computed over A_d because A_d is the intersection of V with a line parallel to the x_d axis and the monomial only has one variable that isn't constant. The intersection can be found by using robust 1D root finding algorithms to get the end points of the 1D integral, which is then computed explicitly.

The second integral on the RHS involves moments over the A_{EB} in the current dimension but the degree has increased to $P+|r|$. Thus, they can be computed recursively. The recursion terminates because as the degree of the monomial moments increases, the dimension and order of accuracy stays the same. Thus, the value of R (the number of terms in the Taylor series) in the recursion decreases by $|r|$. Eventually, there are no terms of the Taylor series on the RHS so this recursion terminates.

Overall, the algorithm proceeds by setting up the equations to compute the needed moments over the needed regions as a least squares problem. Then the RHS is recursively computed recursing in both dimension and monomial degree. Finally, the original least squares problems are solved for the unknown moments, and these are used in the flux computations. Since the least squares problems are always well conditioned and the only computations involving the Ω are the intersections of the Ω with lines parallel to the coordinate axes, this algorithm is very robust and straightforward to implement.

3. Complex irregular boundaries

In order to construct more complex irregular boundaries, implicit functions can be composed into more complex implicit functions using constructive solid geometry, CSG. To do so, one must define the complement, intersection, and union of irregular domains, Ω_i, defined by implicit functions, ϕ_i. To this end we use the following correspondences.

$$\begin{aligned}
\Omega_i^{complement} &\Leftrightarrow -\phi_i \\
\Omega^{intersection} &\Leftrightarrow \max_i \phi_i \\
\Omega^{union} &\Leftrightarrow \min_i \phi_i
\end{aligned}$$

Further, coordinate transformations, ψ, of Ω can be implemented as

$$\Omega_\psi = \{\boldsymbol{x} : \boldsymbol{x} \in \mathbb{R}^D,\ \phi(\psi^{-1}(\boldsymbol{x})) < 0\}.$$

Examples of ψ include rotations, translations, and scaling. In addition to being straightforward to implement by using implicit functions, CSG is a very robust and intuitive method for building complex geometry objects and easily manipulating them (see figure 2).

Once constructed, only the value of the implicit function representing the irregular domain and its derivatives need to be evaluated in order to compute the necessary moments. This allows the embedded boundary method to perform efficiently even on workstations.

If good estimates of the bounds of the local Taylor series expansion of the implicit functions are available, these can be used to dramatically improve performance by subdividing space recursively and only evaluating the implicit function where an irregular boundary could exist, i.e., regions where $\phi(\boldsymbol{x})$ could be zero. Thus, the amount of work done is proportional to the number of cells containing the irregular boundary and not the entire space. This also allows cells with under resolved geometry to be detected and corrected by adaptively refining these cells. This has been found to occur routinely when the higher-dimension irregular domains are restricted to lower dimensions; for example, slices of a well-resolved unit sphere can be circles with arbitrarily small radii. Once adequate refinement has been done, either the computation can use the refined computational cells, or they can be coarsened to provide the moments at the original resolution.

4. Conclusions and future work

It has been shown that finite volume methods for PDE can be done in arbitrary dimension to arbitrary accuracy in the presence of irregular boundaries by computing moments, integrals of monomials, over various regions. If the irregular boundary is defined by implicit functions, computing these moments can be computed using the divergence theorem, Taylor expansions, least squares, recursion, and 1D root finding. As a result, an explicit representation of the irregular domain and boundary is never needed nor computed. The resulting computations are robust and efficient even for complex domains which can be easily defined using implicit functions and constructive solid geometry.

These methods can be extended to mapped grids by incorporating coordinate mappings in the flux and moment computations. In addition, these methods can be used in any computational context where the algorithm can be reduced to the computation of moments over irregular domains or boundaries, for example, the integration of functions over these regions.

Acknowledgment

This research was supported by the Office of Advanced Scientific Computing Research of the US Department of Energy under contract number DE-AC02-05CH11231.

References

[1] Colella P 2001 Volume-of-fluid methods for partial differential equations *Godunov Methods: Theory and Applications* E F Toro, 161–177

[2] Aftosmis M, Berger M J and Melton J 1998 Robust and efficient Cartesian mesh generation for component-based geometry *AIAA J.* **36(6)** 952–960

[3] Deschamps T, Schwartz P, Trebotich D, Colella P, Malladi R and Saloner D 2004 'Vessel segmentation and blood flow simulation using level sets and embedded boundary methods (Elsevier International Congress Series, 1268) 75–80

[4] Schwartz P, Adalsteinsson D, Colella P, Arkin, A P and Onsum M 2005 Numerical computation of diffusion on a surface *Proc. Nat. Acad. Sci.* **102** 11151–11156

[5] Malladi R, Sethian J A and Vemuri B C, Shape modeling with front propagation: A level set approach *IEEE Trans. Pattern Anal. Machine Intell.* **17**:158–175.

Simulation of pellet ablation for tokamak fueling with ITAPS front tracking

Roman Samulyak[1,2], Tianshi Lu[2], Paul Parks[3], James Glimm[1,2] and Xiaolin Li[1]

[1] Department of Applied Mathematics and Statistics, Stony Brook University, Stony Brook, NY 11794, USA
[2] Computational Science Center, Brookhaven National Laboratory, Upton, NY 11973, USA
[3] General Atomics, P.O. Box 85608, San Diego, CA 92186-5608

E-mail: rosamu@ams.sunysb.edu

Abstract. A magnetohydrodynamic numerical model and parallel software for the ablation of cryogenic deuterium pellets in the process of tokamak fueling has been developed based on the method of front tracking of ITAPS Center. The main features of the model are the explicit tracking of material interfaces, a surface ablation model, a kinetic model for the electron heat flux, a cloud charging and rotation model, and an equation of state accounting for atomic processes in the ablation cloud. The software was used for the first systematic studies of the pellet ablation rate and properties of the ablation channel in magnetic fields. Simulations revealed new features of the pellet ablation such as strong dependence of the radius of the ablation channel and ablation rate on the "warm-up" time and supersonic spinning of the ablation channel.

1. Introduction

The injection of small frozen deuterium-tritium pellets is an experimentally proven method of tokamak refueling [1]. Pellet injection is currently seen as the most likely refueling technique for the International Thermonuclear Experimental Reactor (ITER). In order to evaluate the efficiency of this fueling method, it is necessary to determine the pellet ablation rate.

The ablation of tokamak pellets and its influence on the tokamak plasma have been studied by using several analytical and numerical approaches (see [2] and references therein). The inherent limitation of the previous ablation models has been the the absence of a rigorous inclusion of important details of physics processes in the vicinity of the pellet and in the ablation channel and insufficient accuracy of numerical models due to extreme change of thermodynamics states on short length scales. The motivation of the present work is to improve both the accuracy of computational models by using the front tracking technology of the Interoperable Technologies for Advanced Petascale Simulations (ITAPS) Center [3] and physics modeling of the interaction of the pellet ablation channel with the tokamak magnetic field. The present work is a continuation of [4], which introduced sharp interface numerical MHD model for the pellet ablation but omitted effects of charging and rotation of the pellet ablation channel. Both sharp interface numerical techniques and the resolution of complex physics processes in the pellet ablation channel are important, not only for calculating pellet ablation rates and the fueling efficiency, but also for understanding striation instabilities in tokamaks [5].

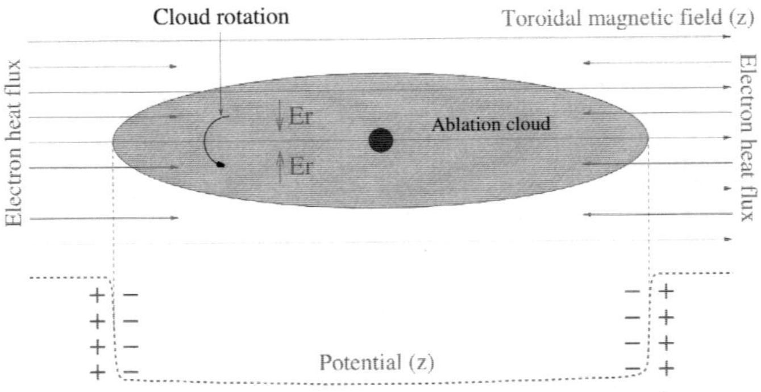

Figure 1. Schematic of physics processes associated with the ablation of deuterium pellet in a tokamak.

2. Main physics processes

In this section, we briefly summarize the main physics processes associated with the ablation of a cryogenic deuterium pellet in a tokamak magnetic field (see Fig. 1). Hot electrons traveling along the magnetic field lines hit the pellet surface causing a rapid ablation. A cold, dense, neutral gas cloud forms around the pellet and shields it from the incoming hot electrons. After the initial stage of ablation, the most important processes determining the ablation rate occur in the cloud. The cloud away from the pellet heats up above the dissociation and then the ionization levels and partially ionized plasma channels along the magnetic field lines. As shown in [4], this process can be described by the system of magnetohydrodynamics (MHD) equations in the low magnetic Reynolds number approximation. The plasma cloud stops the incident plasma ions at the cloud / plasma interface, while the faster incident electrons penetrate the cloud where their flux is partially attenuated depending on the cloud opacity. We employ the kinetic model for the hot electron-plasma interaction proposed in [6]. The tendency of the background plasma to remain neutral confines the main potential drop to a thin sheath adjacent to both end-faces of the cloud. Inside the cloud, the potential slowly changes along each field line. Since the cloud density and opacity vary radially, the potential inside the cloud varies from field line to field line, causing $E \times B$ cloud rotation about the symmetry axis. The potential can be explicitly found by using kinetic models for hot currents inside the cloud [5]. The fast cloud rotation widens the ablation channel, redistributes the ablated gas, and changes the ablation rate. We believe that it also causes striation instabilities observed in pellet injection experiments [5].

3. Numerical method

In this section, we describe numerical ideas implemented in the FronTier-MHD code. The system of MHD equations in the low magnetic Reynolds number approximation is a coupled hyperbolic-elliptic system in a geometrically complex moving domain. We have developed numerical algorithms and parallel software for 2D and 3D simulations of such a system [7] based on the ITAPS front-tracking technology [8].

The numerical method uses the operator splitting. We decouple the hyperbolic and elliptic parts of the MHD system for every time step. The mass, momentum, and energy conservation equations are solved first without the electromagnetic terms (Lorentz force). We use the front tracking hydro code FronTier with free interface support for solving the hyperbolic subsystem. In general case, the electromagnetic terms are found from the solution of the Poisson equation for the electric potential in the conducting medium using the embedded boundary method, as described in [7]. In the 2.5D axisymmetric model for the pellet ablation, the elliptic step is eliminated as the current density in each point of the ablation cloud is a function of the hydrodynamic state and some integrated quantities along the corresponding magnetic field line. Solving of a modified 3D Poisson problem is required in our current work on 3D pellet ablation. At the end of the time step, the fluid states are integrated along every grid line in the longitudinal direction in order to obtain the electron heat deposition and internal hot currents. The heat deposition changes the internal energy and temperature of fluid states and therefore the electrical conductivity. The Lorentz force and the centrifugal force are then added to the momentum equation.

FronTier represents interfaces as lower-dimensional meshes moving through a volume-filling grid (Fig. 2). The traditional volume-filling finite-difference grid supports smooth solutions located in the region between interfaces. The location of the discontinuity and the jump in the solution variables are defined on the interface. A computational stencil is constructed at every interface point in the normal and tangential direction, and stencil states are obtained through interpolation. Then Euler equations, projected on the normal and tangential directions, are solved. The normal propagation of an interface point employs a predictor-corrector technique. We solve the Riemann problem for left and right interface states to predict the location and states of the interface at the next time step. Then a corrector technique is employed that accounts for fluid gradients on both sides of the interface. Specifically, we trace back characteristics from the predicted new interface location and solve Euler equations along characteristics. After the propagation of the interface points, the new interface is checked for consistency of intersections. The untangling of the interface at this stage consists in removing unphysical intersections and rebuilding a topologically correct interface. The update of interior states using a second-order conservative scheme is performed in the next step. The tracked interface allows us to avoid the integration across large discontinuities of fluid states and thus eliminate the numerical diffusion. The FronTier code has been used for large-scale simulations on various platforms including the IBM BlueGene supercomputer New York Blue at Brookhaven National Laboratory.

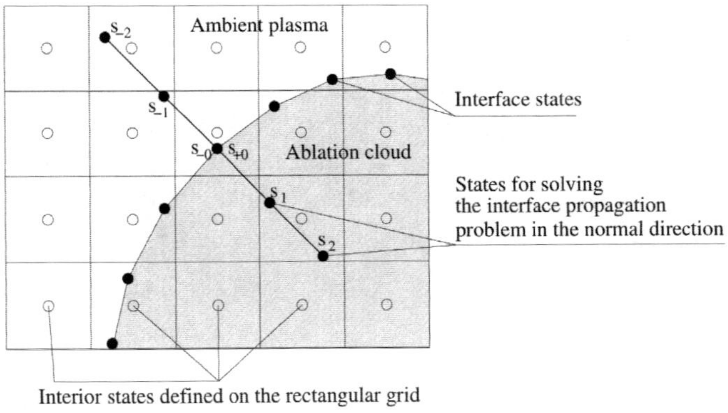

Figure 2. Rectangular grid, interface, and states for the method of front tracking. States contain density, momentum, and energy density of the fluid (plasma) and references to the EOS model and other parameters.

4. Simulation results

The numerical pellet ablation model was benchmarked in [4] with the theoretical Neutral Gas Shielding model [9] and previous pure-hydrodynamic numerical simulations [6]. We reproduced the double transonic layer in the ablation cloud induced by atomic processes of dissociation and ionization (Fig. 3) and obtained an excellent agreement of cloud properties and ablation rates. Performing 2D pure hydrodynamic simulations, we explained the factor of 2.2 reduction of the pellet ablation rate of the axially symmetric model compared to the spherically symmetric one. In the literature, it was attributed to the directional heating [6]. We showed that the directional heating is responsible for only 18% of the ablation rate reduction, and the effect is caused mainly by the difference of ablation rates induced by the Maxwellian and monoenergetic electron heat fluxes [4].

Then we performed the first systematic studies of pellet ablation rates and channel properties in magnetic fields. The ionization of the pellet ablation cloud by the electron heat flux leads

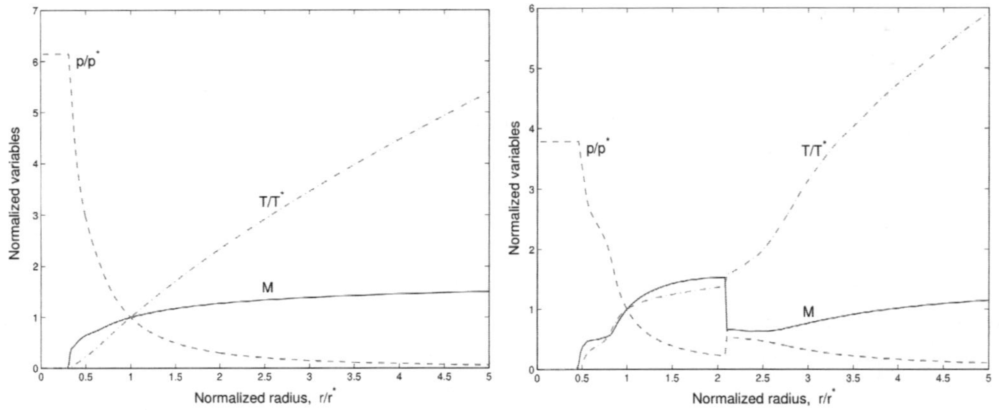

(a) Polytropic EOS (b) Plasma EOS

Figure 3. Normalized ablated gas profiles at $10\,\mu s$ in 1D spherically symmetric model of (a) ablation without atomic processes (polytropic EOS), and (b) with atomic processes (plasma EOS). The solid, dashed, and dash-dotted lines are M, p/p^*, and T/T^* as functions of r/r^*, respectively.

to the channeling of the ablation flow along magnetic field lines. We found that this effect is sensitive to the parameter "warm-up time" and the cloud rotation. Longer warm-up time leads to a slower increase of temperature and wider ablation flow channels. The rotation of the cloud also widens the ablation channel. The channel radius increased from 2.3 cm without cloud rotation to 2.8 cm with rotation. The cloud pressure was lowered by 7% and the pressure peak in the cloud was shifted to $r = 1.5$ cm. Along the z-axis, the density was lower with cloud rotation due to the centrifugal force, which corresponded to less shielding. Fig. 4 shows isosurfaces of pressure in the steady-state ablation cloud.

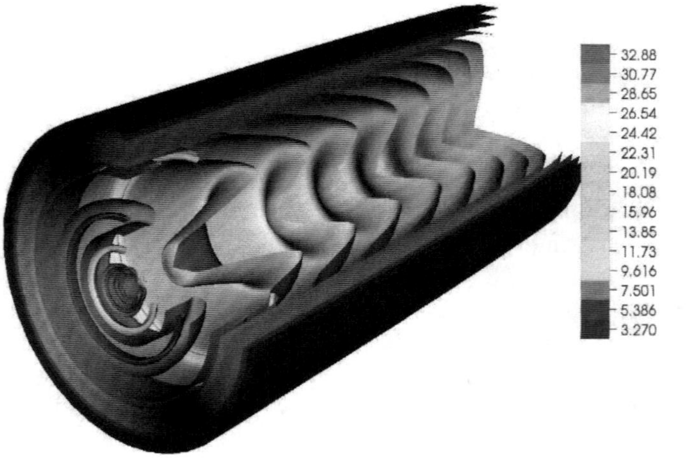

Figure 4. Isosurfaces of pressure in the steady-state ablation cloud.

Figure 5 plots the Mach number of the rotational velocity, $M_\theta = u_\theta/c$, for the steady-state flow with cloud rotation. It has transonic distribution, and the sonic points are located in the ablation channel close to the channel boundary. With cloud rotation, the steady-state ablation rate increased from 195 g/s to 260 g/s. We conclude that the cloud rotation increases the ablation rate because of to the widening of the ablation channel.

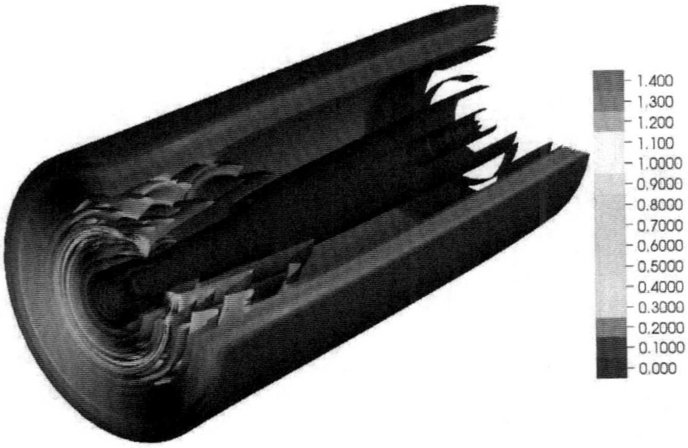

Figure 5. Isosurfaces of rotational Mach number in the steady-state ablation cloud.

5. Conclusions and future plans

Using the front tracking technology of the ITAPS Center, we have developed novel mathematical models, numerical algorithms, and computational software for the numerical simulation of free surface magnetohydrodynamic (MHD) flows of conducting liquids and flows of partially ionized plasmas in the presence of phase transitions and high power particle beams. The software was applied to the numerical simulation of the pellet ablation for tokamak fueling. In our recent work, the pellet ablation rate and lifetime in magnetic fields were systematically studied for the first time and compared with theory and experimental databases. Simulations revealed several new features of the pellet ablation. In the MHD simulations, the Lorentz force funnels the ablation flow into an extended plasma shield, which intercepts the incident plasma heat flux and reduces the ablation rate, depending on the rise time of the heat flux seen by the pellet. Shorter warm-up times lead to narrower ablation channels and reduced ablation rate. This new feature implied that pellets transversing strong plasma gradients, as in the edge pedestal region of the ITER plasma, could have significantly lower ablation rates (higher fueling efficiency) if injected at higher velocity. Using a new model for the potential distribution in the ablation channel, we also demonstrated the supersonic rotation of the channel, a phenomenon most likely causing the striation instabilities. The study of striation instabilities will be the focus of our future research.

Acknowledgments

This research utilized resources at the New York Center for Computational Sciences at Stony Brook University/Brookhaven National Laboratory which is supported by the U.S. Department of Energy under Contract No. DE-AC02-98CH10886 and by the State of New York.

References

[1] Baylor L, *et al.* 2000 *Phys. Plasmas* **7** 1878
[2] Pegourie B 2007 *Plasma Phys. Control. Fusion* **49** R87
[3] http://www.itaps-scidac.org/
[4] Samulyak R, Lu T and Parks P 2007 *Nuclear Fusion* **47** 103
[5] Parks P 1996 *Plasma Phys. Control. Fusion* **38** 571
[6] Ishizaki R, Parks P, Nakajima N and Okamoto M 2004 *Phys. Plasmas* **11** 4064
[7] Samulyak R, Du J, Glimm J and Xu Z 2007 *J. Comput. Phys.* **226** 1532
[8] Du J, Fix B, Glimm J, Li X, Li Y and Wu L 2006 *J. Comput. Phys.* **213** 613
[9] Parks P and Turnbull R 1978 *Phys. Fluids* **21** 1735

Curved mesh correction and adaptation tool to improve COMPASS electromagnetic analyses

Xiaojuan Luo[1], Mark Shephard[1], Lie-Quan Lee[2], Cho Ng[2] and Lixin Ge[2]

[1]Rensselaer Polytechnic Institute, Troy, NY 12180, USA
[2]Stanford Linear Accelerator Center (SLAC), Menlo Park, CA 94025, USA

E-mail: xluo@scorec.rpi.edu, shephard@scorec.rpi.edu, liequan@slac.stanford.edu, cho@slac.stanford.edu and lge@slac.stanford.edu

Abstract. SLAC performs large-scale simulations for the next-generation accelerator design using higher-order finite elements. This method requires using valid curved meshes and adaptive mesh refinement in complex 3D curved domains to achieve its fast rate of convergence. ITAPS has developed a procedure to address those mesh requirements to enable petascale electromagnetic accelerator simulations by SLAC. The results demonstrate that those correct valid curvilinear meshes can not only make the simulation more reliable but also improve computational efficiency up to 30%.

1. Introduction
SLAC has been successfully taking advantage of higher-order finite elements [1] to perform analyses for the design of next-generation accelerators which are regarded as critical to basic energy research [2, 3, 4]. The short-range wakefield calculations in electromagnetic analysis using the higher-order elements requires the meshes must be properly curved to the 3D complex geometric domains and adaptively control refinement around the particles beams that need sufficiently smaller mesh size than the rest of the domain. The common straight-sided mesh generation procedures [5, 6] can not automatically generate valid curvilinear meshes to meet those requirements. The invalid curved meshes or overrefined meshes lead to infeasibly large problem sizes, inaccurate results, or possible failure of the simulations. The DOE SciDAC center ITAPS has been working with SLAC to develop a procedure that applies Bezier mesh curving and size-driven technologies to address these mesh requirements. SLAC has successfully applied this procedure to generate meshes used in accelerator simulations. The results yield stable and reliable time-domain simulations and improve computational efficiency up to 30%.

2. Curved mesh correction and mesh adaptation dontrol tool
This section discusses the two key technical components – curved mesh correction and adaptive mesh refinement control – to generate valid curvilinear meshes that improve COMPASS electromagnetic analyses.

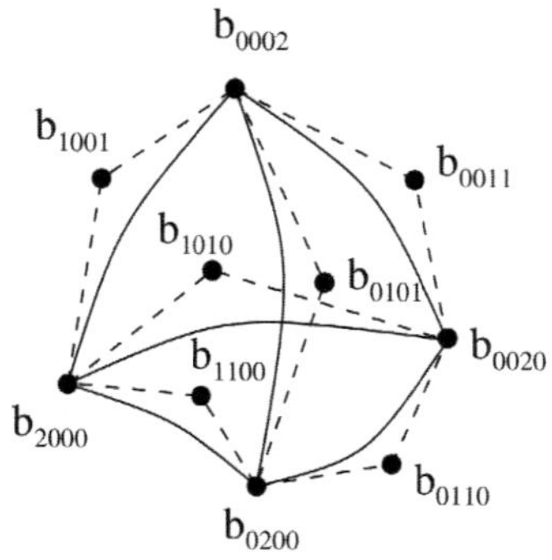

Figure 1. Bezier control points for a quadratic tetrahedral region.

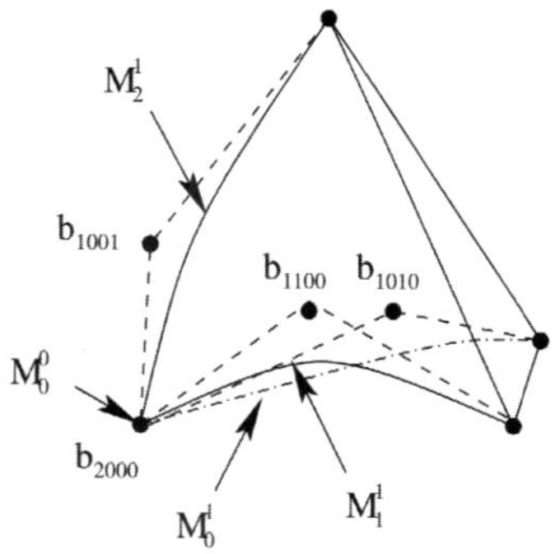

Figure 2. Computation of det(J) indicates that the mesh entities M_0^0, M_0^1, M_1^1 and M_2^1 are key mesh entities.

2.1. Curved mesh correction

The common approach to the construction of curved meshes is to apply a straight-sided mesh generation procedure [5, 6] and then curve the mesh edges and faces on the curved domain boundaries to the proper orders. This approach takes advantage of the conventional unstructured mesh generators to deal with the complexity of model geometry. However, the resulting meshes may become invalid because the curving of the mesh entities to model boundaries can lead to negative determinants of the Jacobian in the closures of curved elements. The curved mesh correction tool we developed applies Bezier polynomial representations [9] to define hierarchic higher-order shapes for topological mesh entities in their parametric coordinates. Figure 1 shows the Bezier control points for a quadratic tetrahedral region; $b^q_{|\boldsymbol{i}|}$ are the control points used to define the shapes of the Bezier mesh edges, faces, and regions, and $|\boldsymbol{i}| = i + j + k + l$ is the control point net index of a higher-order tetrahedral region in its parametric coordinates $\boldsymbol{\xi} = (\xi_i, \xi_j, \xi_k, \xi_l)$, $\xi_i + \xi_j + \xi_k + \xi_l = 1$.

The Bezier higher-order shapes provide an effective means to form a general validity check algorithm for curved elements. The algorithm takes advantage of the convex hull property to ensure that a valid curved element always has positive determinants of the Jacobian in its closures [7]. Given a qth-order Bezier tetrahedron mesh region, the determinant of the Jacobian J can be represented as

$$det(J) = \sum_{|\boldsymbol{i}|=r} C^r_{|\boldsymbol{i}|} c^r_{|\boldsymbol{i}|} \boldsymbol{\xi}^{|\boldsymbol{i}|}, \tag{1}$$

where $r = 3(q-1)$. $C^r_{|\boldsymbol{i}|}$ and $c^r_{|\boldsymbol{i}|}$ are the coefficients computed by the control points $b^q_{|\boldsymbol{i}|}$.

The convex hull property of Bezier polynomial indicates [9],

$$min(c^r_{|\boldsymbol{i}|}) \leq det(J) \leq max(c^r_{|\boldsymbol{i}|}). \tag{2}$$

Therefore, a curved tetrahedral region is valid in its closure as long as $min(c^r_{|\boldsymbol{i}|}) > 0$.

The Bezier curved mesh correction tool processes invalid curved elements one at a time by applying a set of local mesh modification operations on the key mesh entities. The computation of the determinants of the Jacobian can provide useful information to determine the key mesh entities and appropriate operations to correct the invalidity. As an example, figure 2 shows an invalid quadratic tetrahedral region, which has a negative determine of Jacobian at control point b_{2000}. Since the control points $b_{2000}, b_{1100}, b_{1010}$, and b_{1001} affect the computation of $det(J)$, the mesh entities M_0^0, M_0^1, M_1^1, and M_2^1 associated with those control points are key mesh entities,

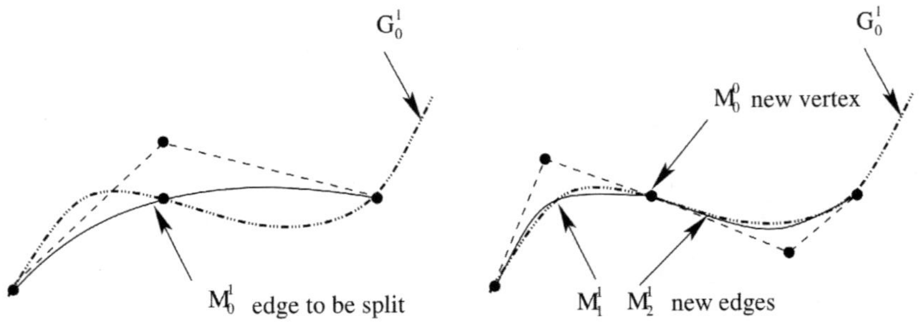

Figure 3. Before (left) and after (right) refinement of a quadratic curved mesh edge M_0^1 on model edge G_0^1. New mesh edges M_1^1, M_2^1 have been appropriately curved to the model boundaries.

and applying local mesh modifications on any of them can effectively make the curved element valid. Curved meshes for more complex domains used by SLAC for electromagnetic linear accelerator analysis are shown in Section 3.

2.2. Moving mesh adaptation control in curved domains

The size=driven mesh adaptation procedure [8] has been successfully applied in cardiovascular blood flow simulations [10], metal forming process [11], wave propagation simulations [12], and the other studies, and the results have demonstrated that computational efficiency can be substantially improved b using the isotropic or anisotropic adapted meshes to effectively resolve solution fields. The procedure has been extended to deal with curved meshes for higher-order finite elements to track the needed refinement around the particle beams for short-range wakefield time-domain electromagnetic simulations. The extended procedure maintains the existing functionalities developed for straight-sided meshes such as vertex-based size field specifications and selective local mesh modification applications [8]. In addition, the following two steps have been added in when the mesh is curved.

- The validity check algorithm described in equation 2 must be applied when the affecting cavities for a local mesh modification operation have curved mesh entities. This step ensures that resulting curved meshes are valid after applying the selected local mesh operation.
- Any newly created mesh entities on the curved domain boundaries must be properly curved to the model boundaries to ensure that the geometric approximation of the resulting adapted meshes is maintained. As an example, figure 3 shows the results of the procedure to split a quadratic curved mesh edge M_0^1 that is classified on the curved model edge G_0^1. The two new created mesh edges M_1^1 and M_2^1 are also curved to the model edge G_0^1.

Moving adaptively refined meshes for SLAC to perform short-range wakefield electromagnetic simulations is shown in Section 3.

3. Analysis results

3.1. Curvilinear meshes for FETD electromagnetic simulation

The wakefield effects of an 8-cavity cryomodule for the proposed International Linear Collider (ILC) are studied by using the FETD method. Figure 4 shows a snapshot of the electric field distribution excited by a beam in the ILC cryomodule. A curved mesh with 2.97 million quadratic isoparametric tetrahedral elements is used in this FETD simulation, resulting in about 20 million degrees of freedom. The simulation used 256 multistream processors on the Cray-X1E, a leadership-class facility at Oak Ridge National Laboratory. It took a total runtime of

300 wall-hours through multiple jobs with checkpointing for a complete run. Half a terabtye of data was generated.

From the initial given curvilinear mesh 1,583 invalid curved elements have been corrected by using the procedure discussed in Section 2.1. Figure 5 shows the curved mesh for one cavity of the model and the closeup mesh before and after curving.

The corrected curvilinear mesh not only leads to a stable time-domain simulation but also reduced computational cost by 30%.

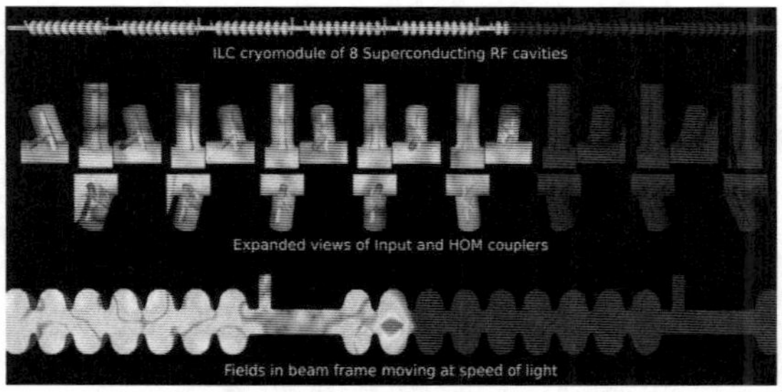

Figure 4. Snapshot of the electric field distribution excited by a beam in an 8-cavity cryomodule for the proposed International Linear Collider.

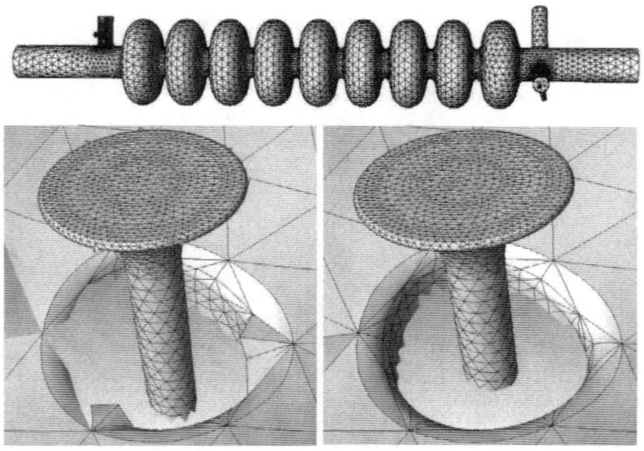

Figure 5. Curved mesh for one cavity, close-up mesh before and after curving, and local mesh cavity before and after applying edge swap to correct the invalid element.

3.2. Moving adaptive refined meshes for short-range wakefield calculations

A series of moving adapted meshes in a curved domain was generated by using the procedure described in Section 2.2 for short-range wakefield calculations by SLAC. Figure 6(a) shows the geometric model, which has some complex components in the middle of the domain. The initial location of the beam is at the left end of the domain, the desired mesh size inside the particle dense mesh is 1 and the size for the rest of the domains is 10. Figure 6 shows the moving adapted meshes up to step 5 to track the moving particle beams. The adaptively refined meshes have around $1 \sim 1.15$ million elements comparing to the uniform refined mesh with 6.5 million elements if the mesh size inside the particle beam domains is applied in the entire domain. The increase of the number of elements in the middle of domain is due to the complex geometries as shown in Figure 6(a). The computation effort of short-range wakefield calculations using the moving adaptive refined meshes can reduce by one order of magnitude compared to the uniformly refined mesh.

4. Conclusion

This paper has presented a procedure to track moving adaptive mesh refinement in curved domains. The procedure is capable of generating suitable curvilinear meshes to enable large-scale accelerator simulations. The procedure can generate valid curved meshes with substantially fewer elements to improve the computational efficiency and reliability of the COMPASS electromagnetic analyses. Future work will focus on the scalable parallelization of all steps for petascale simulations.

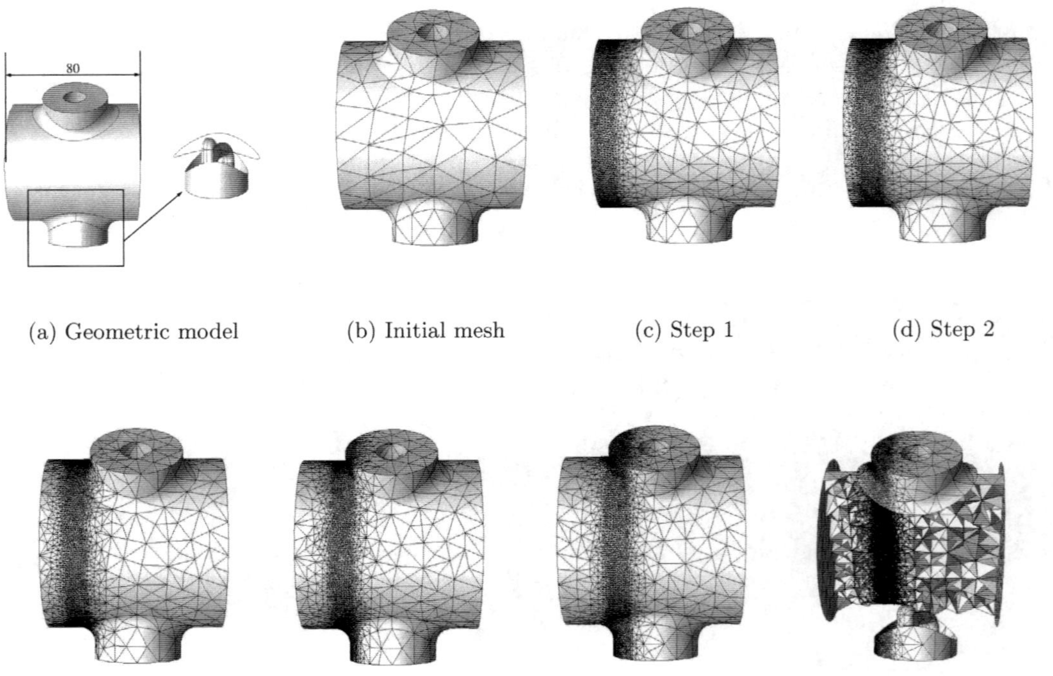

(a) Geometric model (b) Initial mesh (c) Step 1 (d) Step 2

(e) Step 3 (f) Step 4 (g) Step 5 (h) Interior

Figure 6. Moving adapted meshes in curved domain for short-range wakefield simulation.

Acknowledgments
This work is supported by U.S. Department of Energy under DOE grant number DE-FC02-06ER25769 and DE-AC02-76SF00515.

References

[1] Babuska I, Szabo BA and Katz IN 1981 *Int. J. Numer. Meth. Engng.* **18(3)**:515–545
[2] Akcelik V, Ko K, Lee LQ, Li ZH, Ng CK and Xiao LL 2008 *J. Comput. Physics* **227(3)**:1722–1738
[3] Xiao L, Adolphsen C, Akcelik V, Kabel A, Ko K, Lee LQ, Li Z and Ng CK 2007 in Proc. of 2007 Particle Accelerator Conference
[4] Lee LQ, Akcelik V, Chen S, G LX, Prudencio E, Schussman G, Uplenchwar R, Ng C, Ko K, Luo XJ and Shephard MS 2007 *J. of Physcis: Conference Series* **78**:012040
[5] CUBIT geometry and mesh generation toolkit. http://cubit.sandia.gov
[6] Simmetrix Inc. Enabling simulation-based design. http://www.simmetrix.com
[7] Luo XJ, Shephard MS, Obara RM, Nastasia R, Beall MW 2004 *Engineering with Computers* **20**:265–285
[8] Li XR, Shephard MS and Beall MW 2005 *Comp. Meth. Appl. Mech. Engng.* **194**:4915–4950
[9] Farin G 1992 Curves and Surfaces for Computer Aided Geometric Design. Academic Press
[10] Sahni O, Muller J, Jansen KE, Shephard MS. and Taylor CA (2006) *Comput. Methods Appl. Mech. Engrg.* **195(41-43)**:5634–5655
[11] Wan J 2006 An Automated Adaptive Procedure for 3D Metal Forming Simulations. PhD Thesis, Rensselaer Polytechnic Institute, New York
[12] Chevaugeon N, Hillewaert K, Gallez X, Ploumhans P and Remacle JF 2007 *J. Comput. Physics* **223(1)**:188–207

Performance of Embedded Boundary Methods for CFD with Complex Geometry

David Trebotich†, Brian Van Straalen, Dan Graves and Phil Colella

Applied Numerical Algorithms Group, Lawrence Berkeley National Laboratory, Berkeley, CA 94720, USA

†Center for Applied Scientific Computing, Lawrence Livermore National Laboratory, Livermore, CA 94550, USA

E-mail: trebotich1@llnl.gov

Abstract. In this paper, we discuss some of the issues in obtaining high performance for block-structured adaptive mesh refinement software for partial differential equations in complex geometry using embedded boundary/volume-of-fluid methods. We present the design of an adaptive embedded boundary multigrid algorithm for elliptic problems. We show examples in which this new elliptic solver scales to 1000 processors. We also apply this technology to more complex mathematical and physical algorithms for incompressible fluid dynamics and demonstrate similar scaling.

1. Introduction

We use a Cartesian grid embedded boundary (EB) method to discretize partial differential equations in the presence of complex geometry in 3D. In this approach, the irregular domain is discretized as a collection of control volumes formed by the intersection of the problem domain with the cubic Cartesian grid cells. The various operators are approximated by using finite volume differences on the irregular control volumes based on the divergence theorem. This approach has been used as the basis for second-order accurate methods for elliptic, parabolic and hyperbolic PDEs in two and three dimensions [1, 2], and have been combined using the predictor-corrector approach in [3] to solve the equations of incompressible fluid dynamics in irregular geometries [4]. One of the principal advantages of an embedded boundary volume-of-fluid method is that it is compatible with a robust and accurate level-set technique for grid generation of geometries obtained from image data [5, 6]. Furthermore, the embedded boundary approach is amenable to performance-boosting strategies such as adaptive mesh refinement (AMR) and parallelism. Here, we present the algorithm design for high performance computations of elliptic problems using adaptive embedded boundary methods and extend this computational approach to the multiple elliptic operators needed to solve the PDEs that describe incompressible fluid dynamics.

2. AMR multigrid algorithm

We begin with the design of a non-EB adaptive multigrid algorithm for elliptic problems. *AMR Multigrid* is a natural extension of a multigrid iteration algorithm, with this particular version being based on [7]. A pseudo-code description of the algorithm is given in figure 1. The operator

```
procedure Solve(l, l^{base})
R := ρ − L(φ)
while (||R|| > ε||ρ||)
   AMRVCycleMG(l − 1, l^{base})
   R := ρ − L(φ)
end while

procedure AMRVCycleMG(l, l^{base})
   if (l = l^{max})
      R^l := ρ^l − L^{nf}(φ^l, φ^{l−1})
   end if
   if (l > l^{base})
      φ^{l,save} := φ^l on Ω^l
      e^l := 0 on Ω^l
      RelaxLevel(L^{nf}, e^l, R^l, n^{l−1}_{ref}, false)
      φ^l := φ^l + e^l
      e^{l−1} := 0 on Ω^{l−1}
      R^{l−1} := { Average(R^l − L^{nf}(e^l, e^{l−1}))   on C_{n^{l−1}_{ref}}(Ω^l)
                 { ρ^{l−1} − L^{comp,l−1}(φ)             elsewhere in Ω^{l−1}
      AMRVCycleMG(l − 1, l^{base})
      e^l := e^l + I_{pwc}(e^{l−1})
      R^l := R^l − L^{nf}(e^l, e^{l−1})
      δe^l := 0 on Ω^l
      RelaxLevel(L^{nf}, δe^l, R^l, n^{l−1}_{ref}, false)
      e^l := e^l + δe^l
      φ^l := φ^{l,save} + e^l
   else
      SolveLevel(L^{nf}, e^0, R^0) (solve L^{nf}(e^0) = R^0 on Ω^0)
      φ^0 := φ^0 + e^0
   end if
```

Figure 1. Pseudo-code description of the AMR multigrid algorithm.

L^{nf} is a two-level discretization of the Laplacian:

$$L^{nf}(\psi^l, \psi^{l-1,valid}) = D(\vec{G}^f(\psi^l, \psi^{l-1,valid})). \tag{1}$$

The smoothing operator `RelaxLevel`$(\varphi^f, R^f, r, final)$ performs a multigrid V-cycle iteration on φ^f for the operator L^{nf}. In this V-cycle, level f is only coarsened until it is a factor of 2 finer than the next coarser AMR level. The coarse-grid values required for the boundary conditions are assumed to be identically zero.

Code reuse is facilitated by using a *Template Method* design pattern [8]. The purpose of the *Template Method* design pattern is to define an algorithm as a fixed sequence of steps where one or more of the steps is over-ridden. In our case we have a hierarchy of algorithms that we wish to re-use across a family of applications, such as the various elliptic operators involved in solving the equations of incompressible fluid dynamics. The hierarchy of algorithms is defined by the specific *variable steps* that an application must provide to complete the algorithm.

3. EB AMR elliptic operator

The `EBAMRElliptic` operator implements the `AMRMultigrid` template to solve elliptic problems with complex geometry. Several issues must be addressed to obtain solutions to elliptic problems in complex geometry where EB is employed: flexible data structures for fast computations where special discretization stencils are employed, elimination of temporary data, minimization of communication, and operator-dependent load-balancing.

3.1. Serial performance

First, we optimized for performance in serial with the most significant EB-specific improvement – a factor of 100 improvement over our previous initial implementation – resulting from use of aggregate stencil operations where only pointers to memory and integer offsets are stored and recalled for calculations in irregular cells. Flux calculations in irregular cells involve specialized stencils. A caching technique was developed, based on pointers and integer offsets, in order to eliminate storage of temporary data structures, when, for example, updating data in place occurs. This turned previously high-order indirection access to single-order indirection and much shorter instruction depth, which is more suitable for current microprocessors.

The original relaxation strategy involved 2^D-color relaxation, requiring a data striding of 2^D. This was robust and stable but incurred a very high serial performance cost because of its poor reuse of cache-line data on cache-based processors. Careful analysis allowed us to replace this scheme with a 2-color red-black relaxation which both reduced the intralevel communication costs, and allowed greater re-use of data in cache lines on relaxation sweeps. The elimination of redundant application of boundary conditions also helped regularize the data access pattern.

3.2. Parallel optimizations

In addition to serial optimization, distributed-memory parallel performance requires algorithm modifications. Unlike the non-EB AMR elliptic solver, it is not adequate to perform a static load balance of boxes to processors based on the volume of a box. Prior to executing the `AMRMultigrid` algorithm, each AMR level is rebalanced based on our `EBEllipticLoadBalance` routine. `EBEllipticLoadBalance` computes the Laplacian operator on each box and measures the actual time taken. This data is then passed to the load balance routine that subdivides the space-filling curve. Space-filling curve algorithms are a means of ordering boxes within a level to minimize intra-level communication. We use a Morton ordering approach. If D is the spatial dimension of the problem, Morton ordering [9] is a 1-1 mapping of Z^D onto Z with good locality: the fraction of nearest neighbors in Z^D of the inverse image of an interval $\mathcal{I} \subset Z$ of length M whose Morton indices are not in $\mathcal{I}$ is $O(M^{-1/D})$. See [10] for details.

Some additional optimizations in parallel result from minimizing communication in the relaxation scheme. A "lazy" relaxation scheme needs only to exchange data with other boxes once per iteration, not the number of colors per iteration, because of the checker board operations. While this optimization is not quite the optimal strategy in serial, it is a significant improvement in parallel. Similarly, we can eliminate multiple applications of boundary conditions in this "lazy" routine.

Box edge and corner communication can also be minimized by trimming the number of ghost cells needed to communicate boundary conditions (e.g., Neumann requires one ghost cell per cell, Dirichlet, three). This optimization does not have a significant impact on the amount of data needing to be communicated, but it does have a significant impact on the number of processors a given processor must communicate with, thus reducing the number of messages. This improves latency, and intralevel relaxation communication is latency bound, not bandwidth limited.

Inconsistent stencils, that is, stencils where fluxes on embedded boundaries are simply determined by the cell-centered value of the irregular cell and not by proper second-order extrapolations, led to faster multigrid iterations; however, multigrid convergence was a factor of 5 worse and, thus, not an overall win.

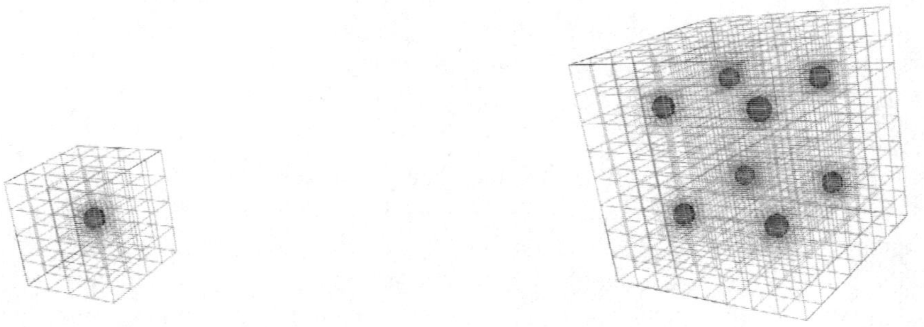

Figure 2. (L) One replication unit for weak scaling test. Base grid resolution is $32\times32\times32$, with 3 levels of adaptivity, factor of 2 refinement, and maximum grid resolution for and individual box in the AMR layout of $16 \times 16 \times 16$. (R) Replicated spheres and AMR meshes for 64 processors, a factor 8 replication.

3.3. Results for Poisson's equation

To demonstrate scaling and performance of our `EBAMRElliptic` operator, we constructed a weak scaling test. In weak scaling, the size of the problem is increased with the number of processors, as opposed to holding the size of the problem fixed as in strong scaling (see [10] for details). For example, we solve Poisson's equation around a sphere inside of a box on 8 processors at a base grid resolution of $32 \times 32 \times 32$ with 3 levels of adaptivity, factor of 2 grid refinement. This problem represents one replication unit. For 16 processors the number of cells and length is doubled in one dimension, adding another sphere such that the base grid resolution is $64\times32\times32$ and the domain is a rectangular prism. The replication is repeated for 32 processors in another dimension. Figure 2 shows a factor of 8 replication for the 64 processor test problem in the suite.

In figure 3, we show plots of wall clock time for the elliptic solver part of the calculation on the Cray XT4 for 8 to 1024 processors. We normalize the times by the time to solve the smallest problem (one replication unit, 8 processors, or 0.074 seconds). We observed 54% efficient scaled speedup over range of 8-256 processors, corresponding to a wall clock time of 0.074-0.135 seconds. For higher concurrencies, we observe 44% efficient scaled speedup over range of 8-1024 processors.

3.4. Implementation and results for incompressible Navier-Stokes

We extended the EB elliptic solver performance optimizations to a predictor-corrector algorithm for incompressible, viscous flow (see [3], for example). The elliptic solver is re-used across multiple operators in one advance timestep: two Poisson equations for the pressure projections that enforce incompressibility (advection and update steps); three Helmholtz equations for the components of velocity; and an additional Poisson equation for the projection filter in the update. Performance is also optimized by making certain diagnostics and derived quantities optional such as norm calculations for checking incompressibility of the velocity field, as well as I/O.

We performed a weak scaling test for the `EBAMRINS` solver that is based on replication similar to that for `EBAMRPoisson`, but with a coarser base grid by a factor of 2. In figure 3 we demonstrate a scaling pattern for the `EBAMRINS` solver that is very similar to that of `EBAMRPoisson`. We note that this study was focused primarily on elliptic scaling, but the pattern indicates that the hyperbolics scale very well.

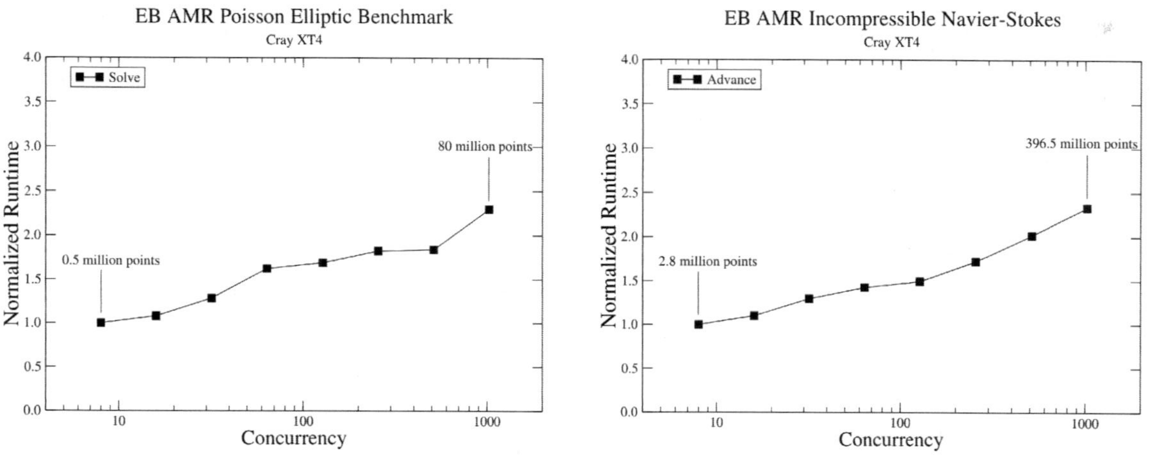

Figure 3. (L) Weak-scaling results for EB AMR Poisson solver from 8 to 1024 processors. Replication of grids for scaling test is shown in figure 2. (R) Weak-scaling results for EB AMR Incompressible Navier-Stokes solver from 8 to 1024 processors. Replication of grids is same as for EB AMR Poisson except grid resolution is a factor of 2 coarser in order to fit the larger EB AMR INS application problem onto same numbers of processors.

4. Conclusion

We have demonstrated the scalability of adaptive embedded boundary methods to 1000 processors for elliptic problems with complex geometry and have extended this technology to an incompressible Navier-Stokes solver. Aggregate stencil operations have all but eliminated the time needed to perform calculations for the special stencils in irregular cut cells resulting from the embedded boundary approach. Operator-dependent load balancing with Morton ordering helps minimize intra-level communication. An efficient relaxation scheme has further improved performance of elliptic solvers by better caching and cache re-use. In conclusion, embedded boundary methods are a fast, efficient and simple technology for handling grid generation for complex geometry and, when combined with adaptive mesh refinement, provide a powerful high-resolution tool for modeling multi-scale fluid dynamics in complex geometry.

References

[1] Colella P, Graves D T, Keen B and Modiano D 2006 A Cartesian grid embedded boundary method for hyperbolic conservation laws *J. Comput. Phys.* **211**:347–366

[2] Schwartz P O, Barad M, Colella P and Ligocki T J 2006 A Cartesian grid embedded boundary method for the heat equation and Poisson's equation in three dimensions *J. Comput. Phys.* **211**:531–550

[3] Bell J B, Colella P and Glaz H M 1989 A second-order projection method for the incompressible Navier-Stokes equations *J. Comput. Phys.* **85**:257–283

[4] Trebotich D, Miller G H, Colella P, Graves D T, Martin D F and Schwartz P O 2005 A tightly couple particle-fluid model for DNA-laden flows in complex microscale geometries *Computational Fluid and Solid Mechanics* 1018–1022

[5] Deschamps T, Schwartz P, Trebotich D, Colella P, Saloner D and Malladi R 2004 Vessel segmentation and blood flow simulation using level-sets and embedded boundary methods *International Congress Series* vol. 1268, pp 75–80

[6] Schwartz P, Adalsteinsson D, Colella P, Arkin A P and Onsum M 2005 Numerical computation of diffusion on a surface *Proc. Natl. Acad. Sci. USA* **102**:11151–11156

[7] Martin D and Cartwright K 1996 Solving Poisson's equation using adaptive mesh refinement (Technical Report M96/66, University of California, Berkeley Electronic Research Laboratory) Code and documentation are available at `http://barkley.me.berkeley.edu/~martin/public_html/AMRPoisson.html`

[8] Gamma E, Helm R, Johnson R and Vlissides J M 1994 *Design Patterns: Elements of Reusable Object-Oriented Software* (Addison-Wesley)

[9] Morton G M 1996 A computer oriented geodetic data base and a new technique in file sequencing (Technical report, IBM Ltd., Ottawa, Ontario)
[10] Colella P, Bell J, Keen N, Ligocki T, Lijewski M and Van Straalen B 2007 Performance and scaling of locally-structured grid methods for partial differential equations *J. Physics: Conference Series* **78**:012013

Embedded boundary algorithms and software for partial differential equations

Phillip Colella, Dan Graves, Terry Ligocki, David Trebotich[1] and Brian Van Straalen

Applied Numerical Algorithms Group, Lawrence Berkeley National Laboratory, Berkeley, CA 94720, USA [1]Center for Applied Scientific Computing, Lawrence Livermore National Laboratory, Livermore, CA 94550, USA

E-mail: bvstraalen@lbl.gov

Abstract. In this paper, we give an overview of a set of methods being developed for solving classical PDEs in irregular geometries, or in the presence of free boundaries. In this approach, the irregular geometry is represented on a rectangular grid by specifying the intersection of each grid cell with the region on one or the other side of the boundary. This leads to a natural conservative discretization of the solution to the PDE on either side of the boundary. Stable and robust hyperbolic and linear elliptic/parabolic solvers have been designed and implemented. Example applications of this approach are shown for compressible and incompressible gas dynamics problems in complex geometries, and for surface diffusion in a cell membrane.

1. Introduction

Structured grid numerical methods based on the finite-volume approach have a number of significant computational and numerical advantages: regular predictable memory access, implicit structure (thus saving the memory and computation required for data interpretation), higher accuracy for less computation due to stencil error term cancellation, fast and robust geometric multigrid solvers, and efficient extensions to block-structured locally refined grids to provide multiresolution capabilities. To provide comparable simulation capabilities in the presence of complex geometry, we use the embedded boundary (EB) approach. Domains with irregular boundaries are represented by the intersection of the domain with Cartesian cells, thus forming irregular control volume(s). Partial differential operators in conservation form are then approximated by using a finite-volume discretization of the divergence operator, i.e. as a sum of fluxes through faces. For hyperbolic PDEs the fluxes are functions of the conserved variables, while for elliptic/parabolic PDEs they are proportional to the gradient of the solution. Both take as the basis of their discretization

$$\frac{1}{V}\int_V \nabla \cdot \vec{F} = \frac{1}{\kappa h}(\sum_i \pm\alpha_i F^i) + \alpha_B \vec{F}^B \cdot \vec{n} \qquad (1)$$

(figure 1, left). Fluxes on faces that have area fractions α that are not equal to unity are constructed by interpolating fluxes centered at the centers of the Cartesian faces, shown in 2D in figure 1, right. The figures in 1 illustrates the kinds of geometric data we need to compute accurate simulations: boundary centroid, boundary normal $\vec{n}$, face centroids: $[\bar{x}, \bar{y}, \bar{z}]$, aperture

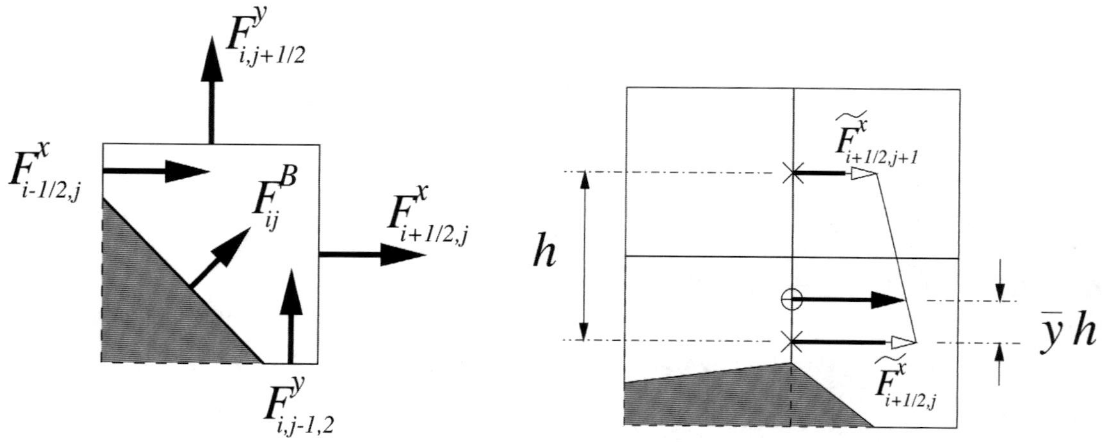

Figure 1. Left: fluxes that make up a conservative update of an irregular cell. Right: constructing a non-unit area fraction aperture flux by interpolation.

area fractions: α, volume fractions: κ. The computation of these quantities is the analogue of mesh generation for body-fitted grid methods.

2. Mesh generation

The volume centroids, volume fractions, area centroids, area fractions, and boundary normal required in (1) can be extracted from surface triangulations through careful application of computational geometry in a very general fashion (as done in [5]). We have developed an alternative approach, based on representing the irregular domain Ω in terms of an implicit function $\phi : \mathbb{R}^D \to \mathbb{R}$

$$\Omega = \{\boldsymbol{x} : \phi(\boldsymbol{x}) < 0\} \ , \ \partial\Omega = \{\boldsymbol{x} : \phi(\boldsymbol{x}) = 0\}. \qquad (2)$$

Implicit functions are standard representations for image data, where they are constructed using level-set methods for segmentation of discrete representations of images; and in geophysical data, in the form of digital elevation maps. It is also straghtforward to build complex domains out of simpler ones using composition. Given two domains represented by implicit functions ϕ_1 and ϕ_2, then the union of these domains is represented by the implicit function $\phi = min(\phi_1, \phi_2)$, and the intersection by $\phi = max(\phi_1, \phi_2)$. In addition, the EB mesh generation problem for domains defined by implicit functions can be reduced, by means of repeated applications of the divergence theorem, to finding the intersection of coordinate lines with the zero set of ϕ, together with a set of well-conditioned linear least-squares problems for the moments for all higher dimensions. The details of this algorithm are described elsewhere in these proceedings [2].

In addition to the real-valued quantities associated with each control volume and cell face, we also need to compute and store the topological information describing the entire collection of control volumes and their connectivity. Every $[i, j, k]$ Cartesian cell in the index space is classified as either *Inside*, *Outside* or *Irregular*. Cells *Outside* have have a positive value for the the implicit function everywhere and are outside the computational domain. The opposite holds for all *Inside* cells. *Inside* cells have unit volume fraction and unit aperture area fractions and centroids on the zero axis. In *Irregular* cells the implicit function change sign. In a similar fashion, boxes in the Cartesian index space $[i, j, k]_{low}, [i, j, k]_{high}$ can also be classified as *Outside* if every cell in the region is outside, *Inside* if every cell is inside, or *Irregular* if any cell in the region is irregular.

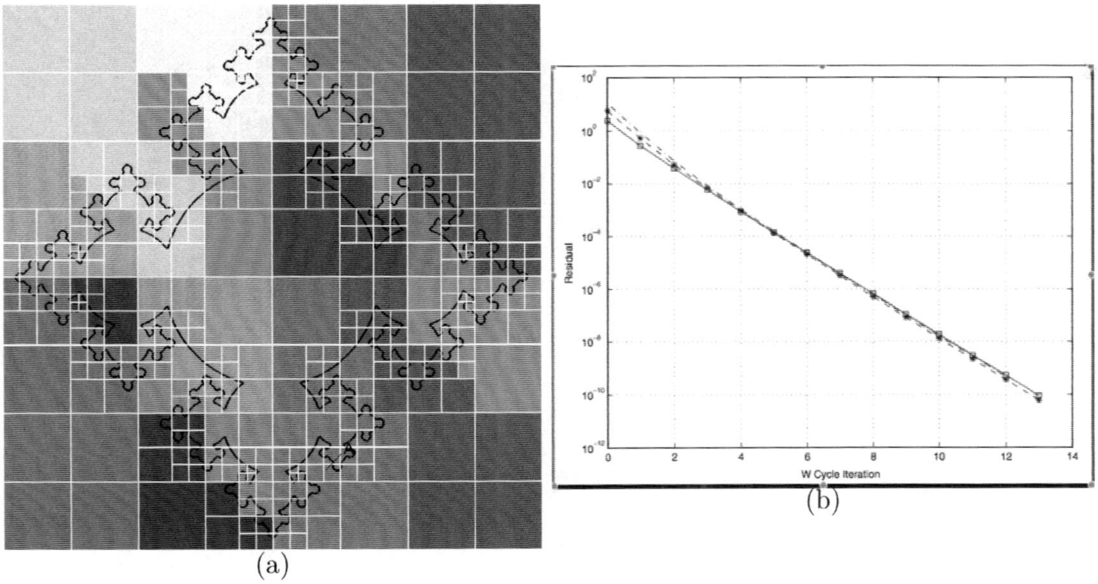

Figure 2. (a) `EBIndexSpace` database object. Colors represent processor assignments. White represents box boundaries. Black represents the zero of the implicit function. The implicit function in this case is a composite solid geometry fractal design created from recursive `unions` of spheres. (b) Plot of the max norm of the partial volume-weighted residual versus W-cycle iteration. The graphs are for meshes of size 64 (□), 128 (*), and 256(-)

3. Parallel Computation

Embedded boundary methods admit a straightforward extension of the domain-decomposition approach used in other structured-grid and block-structured AMR methods [6]. The computational domain at a given level of refinement is covered by a collection of boxes in the Cartesian index space, each of which is assigned to a processor. For each box, one can define the corresponding EB data. Then the organization of parallel computation is the same as for the non-EB case, for example, applying operators by iterating over the boxes on each processor, exchanging ghost cell values.

Chombo does not compute the geometric moments on-the-fly, but builds a distributed database based on a recursive bisection parallel decomposition of the Cartesian index space. Starting with the overall problem domain we query the implicit function if the box covering the domain is *Inside/Outside/Irregular.* In a typical complex geometry problem this will be *Irregular.* The box is divided into into two equal subboxes, and the algorithm recurses into each subbox. The recursion stops when a region is either *Inside* or *Outside* or a minimum box size is reached.

This algorithm creates a covering set of the domain. A space-filling curve is constructed for this collection. These boxes are then assigned to processors with a heuristic load-balancing algorithm. Each processor then visits each of its *Irregular* boxes and proceeds to query each $[i, j, k]$ cell in the box and make the *Inside/Outside/Irregular* query. *Irregular* cells have their geometry moments computed and stored. A sample distribution to processors, shown by colors, is given in figure 2a. *Inside* and *Outside* boxes are represented sparsely, with just an integer flag. *Irregular* boxes are stored as a dense mask of cell flags. *Irregular* cell data is stored in auxiliary sparse data structures. This then forms the `EBIndexSpace` object.

The data distribution for the EB grid generation is not same as for the simulation unknowns. After each regridding operation in an AMR simulation there is a geometry data localization operation. Each level of grids has an associated `EBISLayout` that represents a locally cached

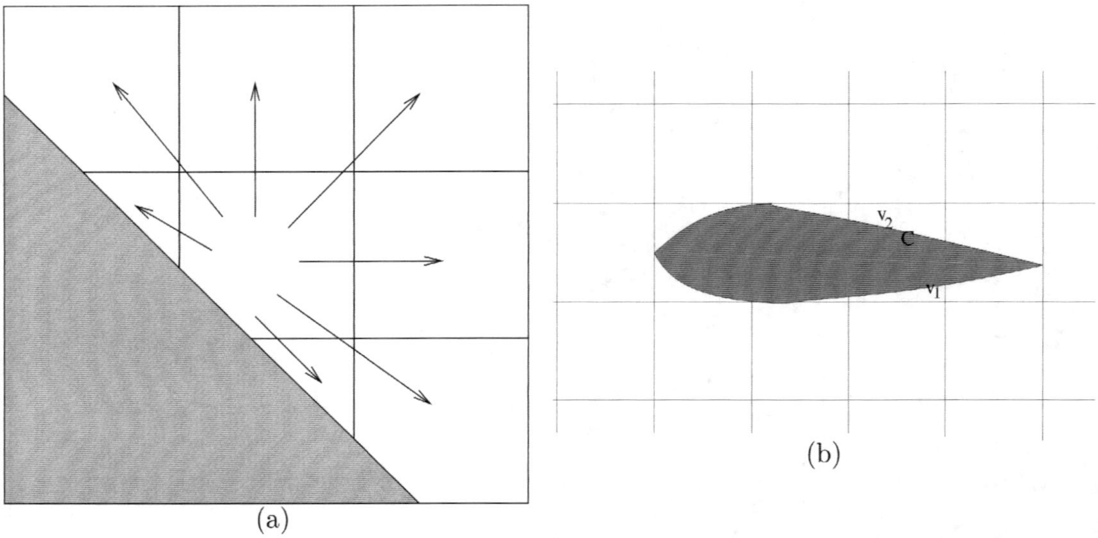

Figure 3. (a) Redistribution of nonconservative update to neighboring cells (b) Coarsened `EBIndexSpace` showing a multivalued cell

duplication of the geometric data for the grids that are on-processor.

4. Solvers

In the applications being developed using the EB algorithms and software, we have typically used semi-implicit methods. These are based on expressing the governing partial differential equations are as a decomposition in terms of operators of classical type, i.e. *hyperbolic*, *elliptic* and *parabolic*. Explicit methods are used for discretizing hyperbolic operators, while fast linear solvers are required for elliptic PDEs and for implicit discretizations of parabolic terms. These are combined with implicit methods for advancing parabolic systems and implicit solvers for elliptic constraints. The overall algorithms are robust and efficient with stability properties that are insensitive to the resolution of the geometric details.

4.1. Hyperbolic Solvers

For hyperbolic operators, we are solving

$$\frac{\partial U}{\partial t} + \nabla \cdot \vec{F} = 0. \tag{3}$$

We would like to discretize (3) in time using an explicit method, with (1) for the spatial discretization. As written there is a severe restriction on the explicit time step because the volume fraction κ can be arbitrarily small. This is the "small cell problem" that appears in most cut/embedded cell methods. The approach we have taken [9] is to compute two different approximations of $\nabla \cdot F$: a nonconservative approximation that ignores the embedded boundary and is known to be stable for the regular grid spacing CFL condition$(\nabla \cdot F)^{NC}$, and a conservative but possibly unstable approximation $(\nabla \cdot F)^{C}$. A lineaqr combination of these fluxes is a stable discretization, but adding a conservative and nonconservative update results in a nonconservative update. To recover conservation there is a redistribution of the nonconservative update to neighboring cells (see figure 3a).

To compute the stable nonconservative update can require the computation of a flux on an aperture that is entirely outside the simulation domain. The details are complicated but involve the existence of smooth extensions of solution derivatives normal to the boundary face.

4.2. Elliptic Solvers
One of the principal advantages of Cartesian grid discretizations of elliptic PDEs is the efficacy of geometric multigrid in providing algorithmically efficient, high-performance solvers [4]. This advantage persists in the case of EB discretizations(fig. 2b), but leads to a more complicated implementation [7].

A key step in multigrid algorithms is *coarsening*. To restrict the residual to the next multigrid level, one performs geometric averaging of the grid by a small interger factor, usually two. In the non-EB case, computing the relationship between the locations of the coarse and fine data involves simple integer arithmetic. In the EB case, both the data access and the averaging operations are more complicated. Coarser geometry moments are generated by sums and averages of the finest geometry values. The volume fraction in a coarse cell must be the sum of the volume fractions of the finer covering cells, similar for area apertures, and so on. This prevents spurious residual sources during restriction and prolongation. It is also essential that coarsening a geometry preserves the topology of the finer EB representation. The consequence of this requirement is that geometric features at coarser levels of the `EBIndexSpace` will have a length scale that is smaller the grid spacing, which leads to the "thin body problem". The field needs to maintain separate smooth extensions for each boundary face. Thus the cell shown in figure 3b has more than one value. These multicells must be preserved and not merged during coarsening. Merging these cells results in multigrid that does not converge because the coarser level generates corrections that do not respect the finer level's boundary conditions. `EBIndexSpace` must maintain these cells in auxiliary sparse data structures at each level of multigrid refinement. All this means the `EBIndexSpace` needs to maintain successively coarsened representations of the geometry as part of the database.

5. Results
In figure 4a, we show an inviscid gas dynamics calculation solved using the explicit EB method based on a second-order unsplit Godunov scheme [9], in r-z coordinates. This problem is motivated by the problem of gas jet formation in novel laser wakefield accelerator designs.

In figure 4b, we show a calculation of a viscous incompressible flow problem using a semi-implicit projection method. The hyperbolic advective terms are computed using a version of the method in [9]. These are then used as a right-hand side for a second-order implicit Runge-Kutta scheme [10] for the parabolic viscous terms to create a second-order accurate stable approximation for the viscous forces, and requiring the solution of a Helmholtz equation. Finally, an approximate update for the velocity field obtained from the incrementing the velocity at the old time by the advective and viscous terms is projected to extract its divergence-free part, requiring the solution of a Poisson equation. There are additional operations that appear in the non-EB algorithms for supporting the use of AMR, similar to those described in [12]. This algorithm permits us to take time steps that are only limited by the CFL condition $\Delta t = O(h/v_{\mathrm{max}})$.

Finally, we show an example that illustrates the value of an implicit function representation of the geometry. In this case, microscopy data is turned into signed distance function of a cell membrane. This allows us to generate an `EBIndexSpace` corresponding to an annular region of fixed thickness around the membrane. If the annulus has a thickness on the order of the grid spacing h, then applying homogeneous Neumann boundary conditions on the surfaces of this region and solving the 3D parabolic diffusion equation results in an $O(h^2)$ approximation of diffusion within the membrane [13].

6. Future work
The embedded boundary approach can be extended to all regimes of existing APDEC simulation capabilities. The first major new technology will be the use of the *Allspeed* approach

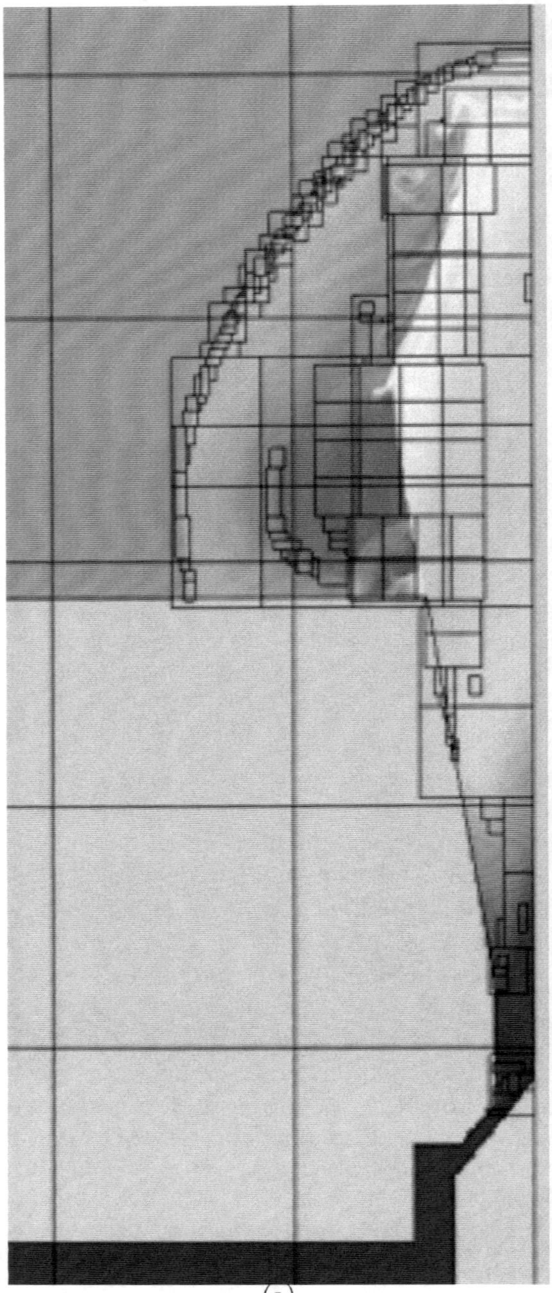

(a)

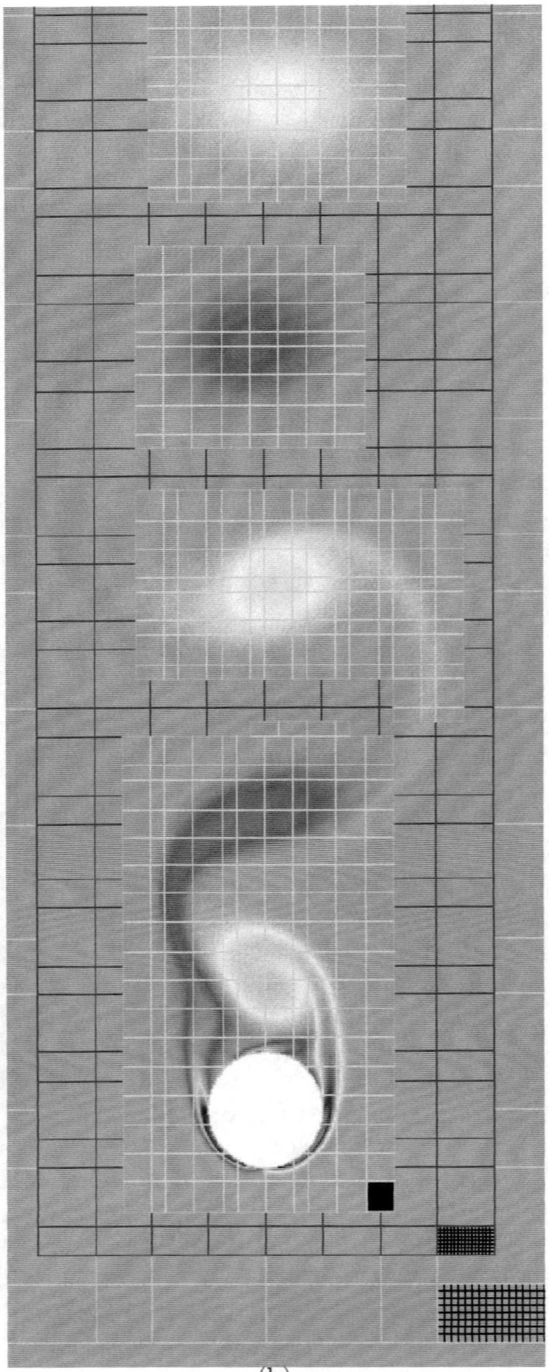

(b)

Figure 4. (a) Compressible gas dynamics flow from high pressure reservoir into a vacuum through a nozzle. (b) Kármán vortex street behind cylinder at Re=300. Effective fine grid resolution 2048x4096. Color scaled for magnitude of vorticity [-175, 175]. Images made with the VisIt visualization package.

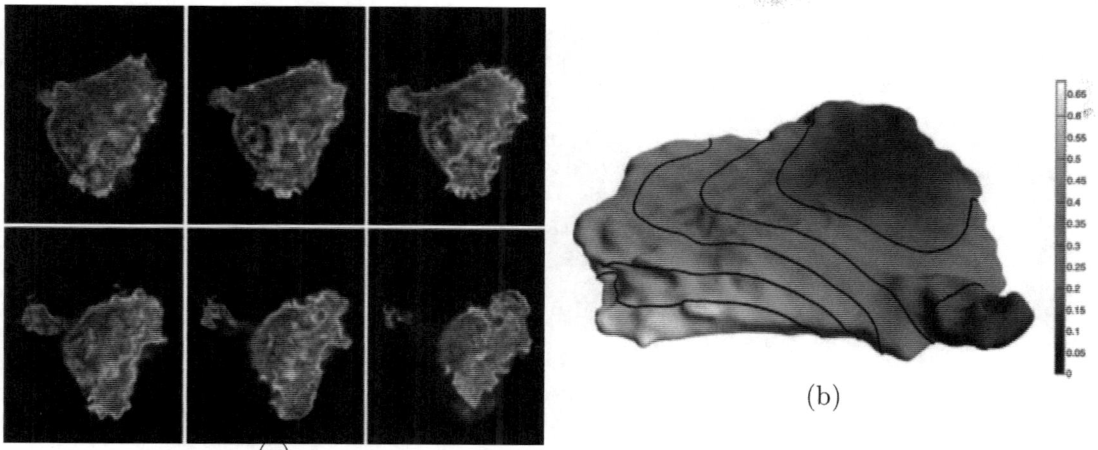

Figure 5. (a) 2D slices through a gray-scale image of an hl-60 cell obtained by using deconvolution microscopy. Image is ≈2,000 times actual size. (b) Solution for surface diffusion at time = 50 s. The time step is 5.0 s.

to compressible flows with significant acoustic energy modes [14]. Originally developed to handle transonic flows, the APDEC center is now using this approach to handle a range of multiwavescale physics.

With *Allspeed* in place we will be able to bring together the combustion modeling effort in APDEC with the embedded boundary approach and start performing full engineering device combustion modeling with full chemistry and at resolutions that do not utilize sub-grid models.

The EB boundary can also represent a defined or tracked interface. Development of a multifluid modeling capability is underway, as well as basic research into using EB for moving interfaces A movinf interface creates higher dimension space-time apertures, which are easily handled with the divergence theorem concepts.

Embedded Boundary Chombo, an open-source software release containing all the features and examples shown in this paper, is expected by the end of FY08.

Acknowledgment

Research supported by the Office of Advanced Scientific Computing Research of the US Department of Energy at the Lawrence Berkeley National Laboratory under contract number DE-AC02-05CH1123, and at the Lawrence Livermore National Laboratory under contract DE-AC52-07NA27344.

References

[1] Colella P 2001 Volume-of-fluid methods for partial differential equations *Godunov Methods: Theory and Applications* (E F Toro) 161–177

[2] Ligocki T, Schwartz P, Percelay J and Colella P 2008 Embedded boundary grid generation using the divergence theorem, implicit functions, and constructive solid geometry *Proc. SciDAC 2008*

[3] Martin D F, Colella P and Graves D T 2006 A cell-centered adaptive projection method for the incompressible Navier-Stokes equations in three dimensions (LBNL report LBNL-62025)

[4] Martin D F and Colella P 1996 Solving Poisson's equation using adaptive mesh refinement (UC Berkeley Electronics Research Laboratory report No. UCB/ERL M96/66)

[5] Aftosmis M, Berger M J and Melton J 1998 Robust and efficient Cartesian mesh generation for component-based geometry *AIAA J.* **36**(6):952–960

[6] Colella P, Bell J, Keen N, Ligocki T, Lijewski M and Van Straalen B 2007 Performance and scaling of locally-structured grid methods for partial differential equations *Proc. SciDAC 2007*

[7] Schwartz P, Barad M, Colella P and Ligocki T J 2006 A Cartesian grid embedded boundary method for the heat equation and Poisson's equation in three dimensions (LBNL-56607) *J. Comput. Phys.* **211**:531–550
[8] Colella P, Graves D, Van Straalen B and Trebotich D 2008 Performance of embedded boundary methods for CFD with complex geometry *Proc. SciDAC 2008*
[9] Colella P, Graves D T, Keen B J and Modiano D 2006 A Cartesian grid embedded boundary method for hyperbolic conservation laws (LBNL-56420) *J. Comput. Phys.* **211**:347–366
[10] Twizell E H, Gumel A B and Arigu M A 1996 Second-order, l0-stable methods for the heat equation with time-dependent boundary conditions *Advances in Computational Mathmatics* **6**:333–352
[11] Colella P and Woodward P 1984 The piecewise parabolic method (PPM) for gas-dynamical simulations *J. Comp. Phys.* **54**(1):174–201
[12] Martin D and Colella P 2000 A cell-centered adaptive projection method for the incompressible Euler equations *J. Comput. Phys.* **163**:271–312
[13] Schwartz P, Adalsteinsson D, Colella P, Arkin A and Onsum M 2005 Numerical computation of diffusion on a surface *Proc. Nat. Acad. Sci.* **102**:11151–11156 1999
[14] Colella P and Pao K 1999 A projection method for low speed flows *J. Comput. Phys.* **149**:245–269

A Newton-Krylov solver for implicit solution of hydrodynamics in core collapse supernovae

Daniel R Reynolds[1], F Douglas Swesty[2] and Carol S Woodward[3,]

[1] Department of Mathematics, University of California at San Diego, La Jolla, CA 92093-0112, USA

[2] Department of Physics and Astronomy, SUNY at Stony Brook, Stony Brook, NY 11794-3800, USA

[3] Center for Applied Scientific Computing, Lawrence Livermore National Laboratory, Livermore, CA 94551, USA

E-mail: drreynolds@ucsd.edu, dswesty@mail.astro.sunysb.edu, cswoodward@llnl.gov

Abstract. This paper describes an implicit approach and nonlinear solver for solution of radiation-hydrodynamic problems in the context of supernovae and proto-neutron star cooling. The robust approach applies Newton-Krylov methods and overcomes the difficulties of discontinuous limiters in the discretized equations and scaling of the equations over wide ranges of physical behavior. We discuss these difficulties, our approach for overcoming them, and numerical results demonstrating accuracy and efficiency of the method.

1. Introduction

Because of a large disparity in time scales, a multitude of relevant physical processes, and solution components spanning a wide range of scales, modeling the collapse of a massive stellar core to produce either a hot proto-neutron star or a supernova explosion proves to be an inherently challenging numerical problem. The model must describe the hydrodynamic behavior of hot, dense matter with complex composition and must describe the flow of radiation, in the form of neutrinos, through the matter. During the cooling phase of these processes, neutrinos carry off most (approximately 99%) of the gravitational potential energy released by the collapse of the star's core, with the remaining 1% or so of the energy contributing to the observed supernova explosion. This general scenario was confirmed through detection of neutrinos by two independent experiments from supernova SN1987A [1, 2]. Simulations of such phenomena must account for both the explosion and cooling phases, covering a time period of tens of seconds [3]. Thus algorithms for these simulations must allow high accuracy over long periods of time.

In such models a set of combined fluid dynamic and radiation transport equations must be solved to describe the evolution of the system from its initial state. If explicit time integration methods are applied for the hydrodynamic processes, time steps will be limited by the Courant-Friedrichs-Levy (CFL) stability constraint, typically requiring steps less than 1 μs [4]. Furthermore, coupling between matter and radiation occurs on a time scale even faster than the CFL limit, resulting in significant stiffness in the problem. To date, no radiation-hydrodynamic simulation for evolution of a collapsed core over the period of tens of seconds has been possible using explicit methods. Explicit calculations [5, 6] have been limited to short ($< 1s$) periods of time. Long timescale cooling calculations [3, 7] have had to assume the collapsed core is in hydrostatic equilibrium. Since implicit methods allow high-order-accurate solutions, can be robust for stiff problems and are not limited by CFL stability constraints, their use is thus highly desirable for these models.

The use of implicit methods introduces its own set of challenges beyond those strictly pertaining to the relevant physics. First, the governing equations are nonlinear and sparse, suggesting the use of Newton-Krylov (hereafter NK) iterative methods [?]. Second, discretizations employed to capture the behavior of hydrodynamic shocks present in the system introduce limiters that result in discontinuities in the nonlinear system. These discontinuities are problematic from the standpoint of the convergence theory of NK methods. A third difficulty with using implicit methods arises from the fact that the model equations describe the evolution of matter and radiation over a range of densities and physical length scales spanning many orders of magnitude from the center of the collapsing core to the outer edges of the computational domain. This wide range of values means that the discretized equations involve scaling issues that can present difficulties for NK methods.

Another element of complexity arises because ofthe size of the discrete systems. Radiation transport models of core collapse supernovae must describe the evolution of the distribution of neutrinos across a large number of frequencies, thus adding a spectral dimension to the space-time discretization and increasing the size of the resulting nonlinear and linear systems that must be solved. In multiple spatial dimensions, this problem then requires the use of massively parallel computing resources in order for the memory and computation requirements to even be tractable. Thus, the approach required for solution of the discrete systems must be scalable to large parallel computing platforms.

In this work, our goal has been to develop a nonlinear solver framework employing Newton-Krylov methods robustly on a variety of radiation-hydrodynamic discretization schemes. NK methods are typically used for large, sparse problems through combining Newton's method with a scalable preconditioned Krylov method for solving the linear systems within each nonlinear iteration [?]. We targeted NK methods for these problems because of the fast, quadratic convergence properties of Newton's method. NK methods have been very effective in solving large, nonlinear systems in a number of application areas; for details we refer the reader to the review article [9]. In order to effectively test our NK methodology on the applications of interest, we consider the problem of cooling the collapsed core in spherical symmetry (thus creating a 2-D nonlinear system in space and neutrino frequency).

This paper is organized as follows. In the next section, we present the implicit fluid dynamics models we use. The third section describes our NK method, and the fourth section discusses some software packages that provide efficient and robust implementations of NK methods. The fifth section presents numerical results, and the last section gives some conclusions.

2. Implicit hydrodynamics models

While our ultimate goal is to use NK methods to solve the full set of radiation hydrodynamic equations in an implicitly coupled fashion, in this paper we consider only the most difficult portion of the system from a nonlinear solver viewpoint, the nonlinear hydrodynamic equations. We handle the nonlinear radiation transport equations through an operator-split approximation. The use of NK methods to treat the neutrino radiation has been described elsewhere [10], and future work will investigate a full nonlinear coupling of these two implicit systems. Our focus in this work is to establish a robust methodology to effectively deal with the challenges of implicit hydrodynamics in the context of proto-neutron star and supernova simulations. The implicit approach we consider addresses both Lagrangean and Eulerian formulations of compressible hydrodynamics with differing spatial meshes and discretization schemes.

We first consider the Langragean formulation that is applicable in the initial stellar collapse phase of a supernova. In this phase, a comoving mass coordinate m follows the collapse over

Journal of Physics: Conference Series **125** (2008) 012085 doi:10.1088/1742-6596/125/1/012085

many orders of magnitude in the radial coordinate r (which evolves in time). The Lagrangean formulation for compressible hydrodynamics is given by [11]

$$\frac{Dv}{Dt} = \frac{Gm}{r^2} - 4\pi r^2 \left(\frac{\partial(P_m + P_R)}{\partial m}\right) \equiv F^{lv}, \quad (1)$$

$$\frac{Dr}{Dt} = v \equiv F^{lr}, \quad (2)$$

$$\frac{D}{Dt}\left(\frac{E_m}{\rho}\right) = -P_m \frac{D}{Dt}\left(\frac{1}{\rho}\right) \equiv F^{le}, \quad \text{and} \quad (3)$$

$$0 = \frac{\partial m}{\partial r} - 4\pi r^2 \rho \equiv G^{lm}, \quad (4)$$

corresponding to conservation of momentum, the definition of velocity, conservation of energy, and self-gravity. In (1)-(4), v is the matter velocity, ρ is the matter density, E_m is the matter energy density, and P_m is the matter pressure. The radiation pressure P_R is considered an input from the radiation solver in a standard operator-split approach. The operator D/Dt is used to denote the Lagrangean (comoving frame) time derivative. The system of equations is closed through an equation of state of the form

$$E_m = E_m(T, \rho) \quad \text{and} \quad P_m = P_m(T, \rho), \quad (5)$$

in terms of the matter temperature T. The equations are spatially discretized using the Richtmyer-Morton finite-difference method using a staggered grid and an artificial viscosity scheme to treat shocks (see [11], p. 485). The discrete analogs of (1)-(4) for each cell form the coupled set of sparse, nonlinear equations that must be solved to advance each time step, as described in Section 3.1. The system is defined by the vector of unknowns and the vector of nonlinear equations. In this Lagrangean discretization the vector of unknowns is given by $x^T = \left(\rho_{i+1/2}, T_{i+1/2}, r_i, v_i\right), i = 1, \ldots, N$, where integer indices denote quantities defined at cell edges, half-integer indices denote quantities defined at cell centers, and N denotes the total number of active cells. The vector of nonlinear equations corresponds to the discretized versions of equations (1)-(4) within each cell. Let us denote this system of unknowns and equations as System I and note that the total size of the system is $4N$.

In the initial phase of stellar collapse it is desirable to employ a Lagrangean hydrodynamic scheme as the core of the star contracts from a radius of several thousand kilometers to just a few hundred kilometers [4]. The Lagrangean scheme allows coordinates to comove with the collapsing core. After collapse it is typically more desirable to consider an Eulerian formulation where the coordinates are fixed in space. We consider two common Eulerian formulations, differing in the choice of energy equation. One formulation employs a gas energy equation and the other a total energy equation. Each of these approaches has merits and limitations. The former approach ensures that the First Law of Thermodynamics is satisfied, while the latter ensures the conservation of total energy. In spherical symmetry the former approach is given by

$$\frac{\partial \rho}{\partial t} = -\frac{1}{r^2}\frac{\partial}{\partial r}\left(r^2 \rho v\right) \equiv F^{ec}, \quad (6)$$

$$\frac{\partial E_m}{\partial t} = -\frac{1}{r^2}\frac{\partial}{\partial r}\left(r^2 E_m v\right) - P_m \frac{1}{r^2}\frac{\partial}{\partial r}\left(r^2 v\right) \equiv F^{eg}, \quad \text{and} \quad (7)$$

$$\frac{\partial(\rho v)}{\partial t} = -\frac{1}{r^2}\frac{\partial}{\partial r}\left(r^2 \rho v^2\right) - \left(\frac{\partial(P_m + P_r)}{\partial r}\right) - \frac{Gm\rho}{r^2} \equiv F^{ev}, \quad (8)$$

corresponding to conservation of mass, evolution of gas energy, and conservation of momentum. These are augmented with the self-gravity equation (4), which remains unchanged from the Lagrangean formulation.

Such a formulation has been used in the explicit ZEUS-2D code [?]. We consider the implicit generalization of this scheme using a staggered mesh, an artificial viscosity, and a van Leer monotonic limiter to treat shocks (see [?] for details). In this discretization the vector of unknowns is given by $x^T = \left(\rho_{i+1/2}, T_{i+1/2}, v_i, m_i\right), i = 1, \ldots, N$, where the mass coordinate m evolves over time and the spatial coordinate r is held fixed. The vector of nonlinear equations defining this system corresponds to the discretized versions of (6)-(8) and (4) within each cell. We denote this as System II, and note that the total size of the system is again $4N$.

An alternative formulation replaces (7) with an equation for conservation of total energy,

$$\frac{\partial(E_m + \frac{1}{2}\rho v^2)}{\partial t} = -\frac{1}{r^2}\frac{\partial}{\partial r}\left(r^2\left((E_m + \frac{1}{2}v^2 + P_m)v\right)\right) \equiv F^{et}. \tag{9}$$

This approach is used in many finite-volume fluid dynamics methods. We consider the implicit extension of the widely used LCPFCT discretization scheme [13], which is an extension of the pioneering FCT scheme [14]. This approach uses a cell-centered finite-volume mesh with limiters to maintain monotonicity and capture shocks. In this discretization the vector of unknowns is given by $x^T = \left(\rho_{i+1/2}, T_{i+1/2}, v_{i+1/2}, m_{i+1/2}\right), i = 1, \ldots, N$, and the vector of equations is given by the discretized versions of (6), (8), (9) and (4) within each cell. We denote this as System III, and note the size of the system is also $4N$.

3. Integration and solution approach

This section overviews the implicit time integration approach, along with the fundamental components of our NK method.

3.1. Temporal wntegration with θ-scheme

We employ a two-level θ-scheme for time integration. This formulation allows switching between explicit and both first- and second-order implicit integration methods based on the continuous parameter θ. For the relevant spatially systems I, II, or III, this scheme is

$$\frac{V(x^n) - V(x^{n-1})}{\Delta t} = \theta F(x^n) + (1-\theta)F(x^{n-1}), \tag{10}$$

$$G(x^n) = 0, \tag{11}$$

where $F(x^n)$ represents the set of discretized time-dependent equations $\{(1)\text{-}(3)\}$, $\{(6)\text{-}(8)\}$, or $\{(6), (8), (9)\}$, evaluated at time t^n, and $G(x^n)$ represents the discretized constraint (4). We write $V(x)$ to denote that the time derivatives include functional forms of the unknowns; the form of V is readily apparent through inspection. When $\theta = 0$, this scheme reduces to the forward Euler explicit method. When $\theta = 1$ the scheme is the first-order backward Euler implicit method. When $\theta = 0.5$ the scheme is the second-order Crank-Nicholson implicit method.

3.2. Newton-Krylov nonlinear system solver

When advancing (10)-(11) with $\theta > 0$, a nonlinear system of equations given by

$$H\left(x(t^n)\right) = 0 \tag{12}$$

must be solved. Here $x(t^n) \in I\!R^{4N}$ is the vector of unknowns approximated at time t^n, and H is the vector comprising (10) and (11).

The Newton-Krylov algorithm was first developed for use in solving discrete nonlinear systems arising from PDEs [15]. This method leverages the quadratic convergence of Newton's method with the robustness and scalability of preconditioned Krylov methods. Details of the algorithm can be found in [9] and [?]. The basic algorithm for solution of the system $H(x) = 0$ is as followd.

Start with an initial condition x^0 at time t^0:
For each time step $n = 1, \ldots, N$, find $x^n \approx x(t^n)$ by Newton's method:

1. *Start with an initial Newton iterate $x^{n(0)}$ (typically $x^{n(0)} = x^{n-1}$)*
2. *For each Newton iteration $j = 1, 2, \ldots$*
   a. *Using a Krylov method, approximately solve for s^j,*
   $J_H\left(x^{n(j)}\right) s^j = -H\left(x^{n(j)}\right)$ *such that* $\left\| J_H\left(x^{n(j)}\right) s^j + H\left(x^{n(j)}\right)\right\| \leq ltol.$
   Within each Krylov iteration perform:
      i. *One matrix-vector multiply with $J_H\left(x^{n(j)}\right)$*
      ii. *One preconditioner solve*
   b. *Update the Newton iterate, $x^{n(j+1)} = x^{n(j)} + \lambda s^j$*
   c. *Test for convergence to stop the iteration, $\left\| H\left(x^{n(j+1)}\right)\right\| < htol.$*

Here, $J_H\left(x^{n(j)}\right) \in I\!R^{4N\times 4N}$ is the Jacobian of the nonlinear function $H(x)$ evaluated at the previous Newton iterate $x^{n(j)}$, and $\lambda \in (0, 1]$ is a line search parameter chosen to help globalize the method [?]. Information on Krylov methods and their use with Newton's method can be found in [?, 16]. The Krylov methods most commonly used within an NK framework are GMRES [17] and BiCGStab [18], as these are general to nonsymmetric systems as is typical in multi-physics applications.

As noted in the algorithm description above, Krylov methods require a matrix-vector multiply at each iteration. In fact, there is no other requirement or use of the linear system matrix. Since this matrix arises from the nonlinear system, instead of constructing $J_H(x)$ directly we can approximate the action of the Jacobian on a vector with a directional difference,

$$J_H(x)w \approx H(x + \sigma w) - H(x)/\sigma, \tag{13}$$

while still preserving convergence of the overall Newton method [19]. Thus, as long as the nonlinear function can be evaluated at each linear iteration, the Newton-Krylov solve can proceed without analytically calculating Jacobian matrices.

Although very robust, Krylov methods can often slow or stall in their convergence. As a result, we apply a right preconditioner, P, within Step 2a. via the transformation

$$J_H s = -H \quad \longrightarrow \quad (J_H P^{-1})(Ps) = -H. \tag{14}$$

3.3. Solver customizations for astrophysical hydrodynamics

While standard NK algorithms work in theory for sufficiently-differentiable nonlinear problems, astrophysical hydrodynamics presents a number of challenges to these methods. We describe the key ingredients required for application of NK algorithms to our systems.

As described in Section 2, flux limiters and artificial viscosity terms are included in the models with the goal of increasing the stability of numerical methods in the presence of shocks. Unfortunately, these terms are typically enabled through nondifferentiable "if" statements in a code. As such, the nonlinear functional $H(x)$ is not, in general, Lipschitz continuous, as is required for Newton convergence theory, and the equations themselves may even change between one evaluation of H and the next. In practice, such effects may result in nonconvergence of a matrix-free Krylov linear solver, since the multiple evaluations of H in (13) may be based on different equations, with possibly differing artificial viscosities or flux terms. To ameliorate this difficulty, we "freeze" the choice of these algorithmic nonlinearities throughout each Newton iteration. The resulting Krylov iteration converges without difficulty, leaving any differentiability issues to the more forgiving globalized Newton solver, which may require at most a few additional iterations to resolve the solution. We accomplish this "freeze" through approximating the Jacobian matrix $J_H(x)$ directly, again using finite differences, but now computing the specific Jacobian entries,

$$[J_H(x)]_{i,j} \approx H_i(x + \sigma e_j) - H_i(x)/\sigma, \quad i,j = 1,\ldots,4N, \tag{15}$$

where e_j corresponds to the unit vector formed with zeros in all but the jth entry, and $H_i(x)$ corresponds to the ith entry in the nonlinear residual vector. In addition, since the hydrodynamics equations involve a fixed stencil size, the large majority of these entries are zero. We exploit this feature through the use of *colored finite-difference* methods to perform the Jacobian approximation (15) using only a small fixed number of evaluations of H. Through approximating these entries, we effectively "smooth" out discontinuities present in the true Jacobian enabling nonlinear convergence. In addition, with these precomputed Jacobian entries, we easily construct a block-diagonal preconditioning matrix P through extracting only the spatially-local entries that couple the variables together within a given cell.

Typically, Newton methods are declared "converged" when a norm of the nonlinear residual is below a given tolerance, that is, $\|H(x)\| < htol$. For coupled systems, however, each equation may be expressed in a different scale (or units). As a result, even when a Newton iteration x is far away from the solution, some equations may have very "small" residuals compared with others. Hence, the choice of convergence criteria becomes very important, in that it must measure convergence of *all* equations, irrespective of whether the units result in large or small residuals. We employ a scaled root-mean-squared norm in measuring convergence,

$$\|H(x)\| = \left(\frac{1}{4N}\|DH(x)\|_2^2\right)^{1/2}, \tag{16}$$

where $D = \text{diag}\left(d_1^{-1}, d_2^{-1}, \ldots, d_{4N}^{-1}\right)$, with d_i corresponding to the typical magnitude of the given equation at that spatial location, effectively renormalizing all of the equations to remove unit discrepancies. To this end, we choose these scaling values as follows.

$$d_i = \begin{cases} 1, & \text{Equations (1), (6), and (9)} \\ r^{n-1}, & \text{Equation (2)} \\ E_m^{n-1}\rho^{n-1}, & \text{Equation (3)} \\ m, & \text{Equation (4), System I} \\ \rho^{n-1}, & \text{Equation (4), Systems II and III} \\ E_m^{n-1}, & \text{Equation (7)} \\ p^{n-1}/\rho^{n-1}, & \text{Equation (8)(sound speed)} \end{cases} . \tag{17}$$

Unlike standard nonlinear systems of equations, Systems I, II, and III involve variables with implicit positivity constraints; that is, density and energy must be strictly positive. For astrophysical problems, the values of these variables can vary by orders of magnitude throughout the spatial domain. As a result, small errors due to inexact solver tolerances or even floating-point precision can result in violations of constraints when solution values get small. However, the standard Newton-Krylov algorithm does not account for constraints on the variables. We employ a novel transformation to logarithmically rescale these sensitive variables for use within the Newton solver. For example, instead of using values for the density ρ, we use $\tilde{\rho} = \log(\rho)$ in the Newton solver and unknown vector x. This transformation adjusts the nonlinear functional so that it depends on the nicely scaled $\tilde{\rho}$ instead of ρ. The Newton updates are performed on the new unconstrained variable, $\tilde{\rho}^{n(j+1)} = \tilde{\rho}^{n(j)} + \lambda s^{j+1}$, which cannot result in constraint violations in the density $\rho = \exp(\tilde{\rho})$. We use such scalings in System I on the variables $\{\rho, T, r\}$, and in Systems II and III on the variables $\{\rho, T, m\}$. This type of transformation increases the nonlinearity of the functional $H(x)$ on the solver arguments (e.g., $\tilde{\rho}$), which in practice results in approximately one additional Newton iteration per step.

4. Software

Our implementation of the nonlinear solver portion of the simulation code uses the SciDAC TOPS solver package KINSOL, from the SUNDIALS suite [20, 21]. SUNDIALS is a C language package written with the goal of easily being included in existing physics codes. The implementation of Newton-Krylov methods within KINSOL is data structure neutral, in that the implementation does not depend on exactly what types of data structures the user code has for variables and equations. Robust implementations of NK methods are also available in the TOPS C++ package Trilinos [22] and the TOPS C package PETSc [23]. The physics routines in our code, RH1D, are written in F95 to achieve good performance. The core of the physics is contained in routines to evaluate the nonlinear equations. The code to accomplish the colored finite-difference approximation of the Jacobian is also written in F95 to obtain peak efficiency, as are all vector operations needed to implement the vector operations needed by SUNDIALS.

5. Numerical results

In this section we show results of applying the implicit solution method to two analytic solution problems and a proto-neutron star cooling problem.

5.1. Analytic solution testing

The analytic tests we performed consisted of shock tube problems, blast wave problems, and hydrostatic equilibrium problems. We lack space to recount these tests in full detail here, but we present highlights of the shock tube and blast wave. Our first reference problem is the Sod shock tube [24]. The numerical performance of the three hydrodynamic discretizations, described in Section 2, on this problem are shown in figures 1–3, along with the analytical reference solutions (solid lines). In these tests the time step was set by choosing a CFL factor of 0.5 and using

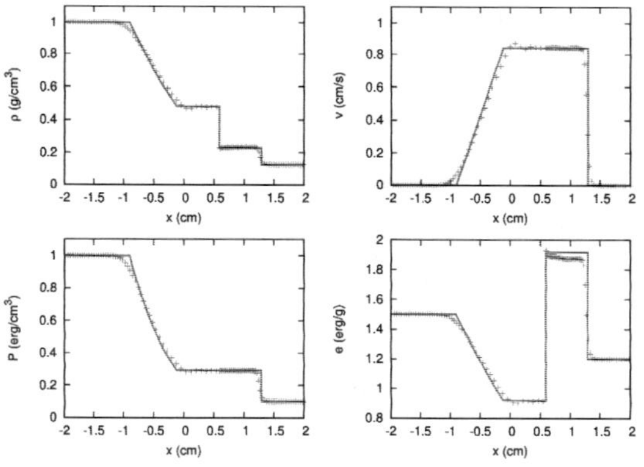

Figure 1. Sod shock tube for the implicit Lagrangean scheme, I.

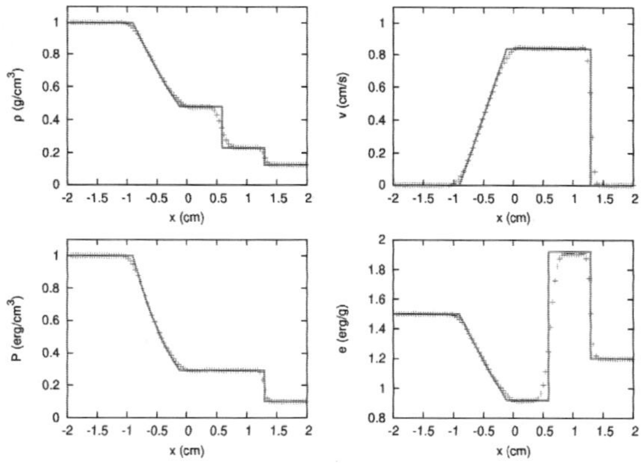

Figure 2. Sod shock tube test problem for the implicit staggered mesh scheme, II.

$\theta = 1$ for the Lagrangean scheme and $\theta = 0.55$ for the Eulerian schemes. The computational domain was discretized into $N = 100$ zones. The full details of the problem setup are described in [25, 10]. Our tests indicate that the implicit algorithms yield comparable numerical results to each other, and to comparable explicit algorithms (see [13, ?, 25] for comparison). Increasing resolution in these tests (figure 4) shows that the methods are converging as expected. We also present results from tests on the Sedov-Taylor blast wave problem. Results for the implicit Lagrangean algorithm are shown in figure 5. This problem is particularly challenging to the nonlinear solver because the blast-wave test creates an extremely dynamic set of conditions. The initial conditions for this test are described in [10] and are chosen to create a problem where the

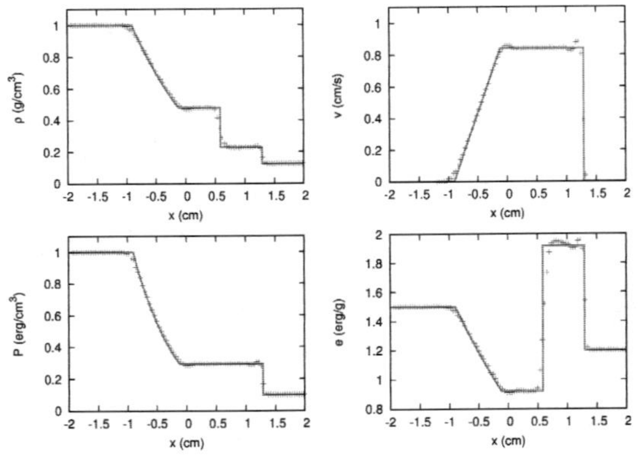

Figure 3. Sod shock tube test problem for the implicit FCT scheme, III.

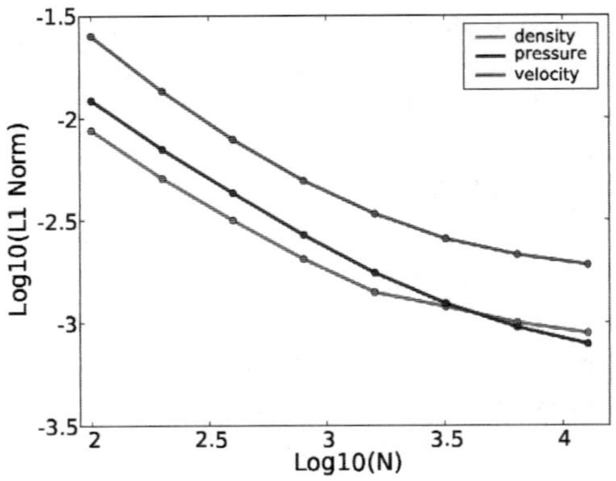

Figure 4. Convergence, of the implicit Lagrangean scheme for the Sod shock tube.

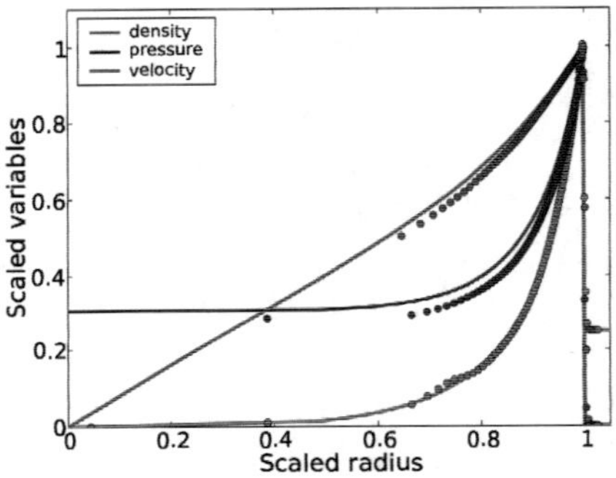

Figure 5. Analytic and implicit Lagrangean numerical solutions of the Sedov-Taylor problem. Radius is scaled by the analytic shock radius, and variables are scaled relative to the values at the shock front.

range of variables across the computational domain varies by eight orders of magnitude. The ensuing blast wave sweeps up material into a thin shell with a very strong shock discontinuity in all hydrodynamic variables. These physical discontinuities trigger limiters in the hydrodynamic algorithm, which in turn introduce mathematical discontinuities in the Jacobian of the nonlinear system. When a Lagrangean scheme is used to model this problem, the zones that follow the fluid are swept up by the blast wave resulting in few zones in the evacuated region near the origin and many zones near the shock front where physical quantities vary most rapidly. The results of this test are comparable to those of other schemes [25].

5.2. Proto-neutron star cooling test

To demonstrate the applicability of this algorithm to core collapse supernovae and proto-neutron star cooling problems, we have applied the code to a simplified problem that captures the hydrodynamic and radiative behavior of a collapsed stellar core while employing simplified microphysics. The proto-neutron star is constructed to be hot, initially static, and of approximately 1.4 solar masses that is heavily supported by radiation pressure. For the purposes of constructing initial data, the neutrino-radiation distribution is assumed to be static. The multigroup opacity of the material is set to mimic the opacity of that in a more realistic model, while the equation of state is taken to be an ideal gas for the sake of simplicity. The central density of the star is taken to be $\rho_c = 5 \times 10^{14} g/cm^3$. The initial distribution of radiation is chosen so that it contributes approximately one half of the pressure support in the center of the star, while diminishing radially outward from the center. This configuration yields a central temperature of approximately 10 MeV, in agreement with more realistic models of a hot proto-neutron star core taken from supernova studies. The cooling time scale for this problem is set by the opacity and density of the stellar material, which for these problem parameters results in a time scale of several seconds.

We evolve this model in an operator-split fashion by first applying our implicit hydrodynamics algorithm, followed by an implicit multi-group radiation transport algorithm. For comparison, an explicit hydrodynamic CFL restriction would yield a time step of approximately $\Delta t \approx 2.5 \times 10^{-8} s$, hence requiring $O(10^9)$ time steps and rendering even this spherically symmetric problem intractable. In contrast, the implicit algorithm can take a much larger time step since there are no fast-moving shocks or other short time-scale phenomena of interest involved in the problem. We used the implicit Lagrangean algorithm I to evolve the problem with a CFL factor of 10^3 (i.e., $\Delta t \approx 2.5 \times 10^{-5} s$); the results are shown in figurs 6–7. After initial transient behavior, the initially static configuration should gradually contract because the neutrino-radiation contribution to the pressure is lost as the neutrinos diffuse outward from the core. This is indeed seen in figure 6. The neutrino luminosity as a function of time is shown

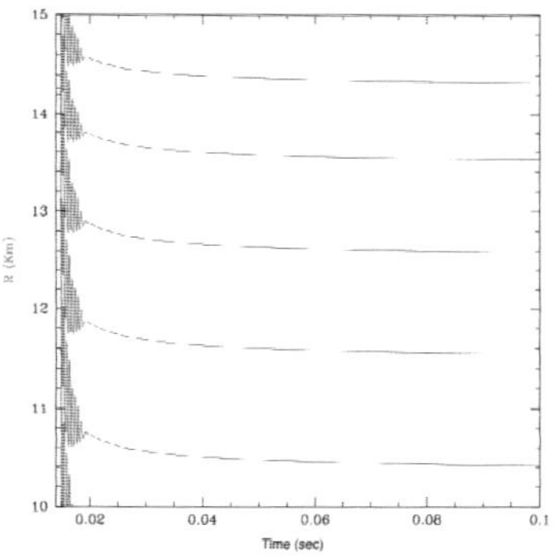

Figure 6. Trajectories of constant-mass contours showing gradual contraction.

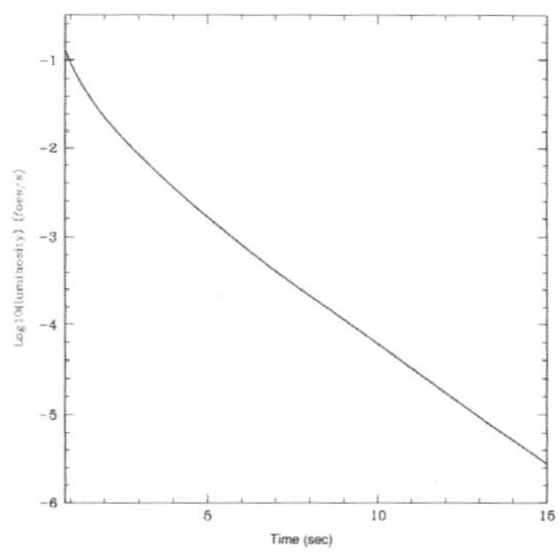

Figure 7. Luminosity of neutrinos in units of foes/s (1 foe = 10^{51} ergs).

in figure 7, which correctly decays over the long cooling time scale as the neutrinos gradually diffuse from the core.

6. Conclusions

We have developed a strategy that allows us to use NK methods to achieve an implicit solution for various Lagrangean and Eulerian hydrodynamic simulations. The solution uses a finite-difference scheme to achieve a quasi-smooth approximation of the Jacobian and employs a novel transformation to account for the large dynamic range of physical variables. The use of these methods allows us for the first time to visit multiple timescale problems that could not be attacked with explicit schemes.

Acknowledgments

DRR and CSW gratefully acknowledge support by the DOE Office of Science of the U.S. Dept. of Energy, through the TOPS SciDAC project. FDS gratefully acknowledges support by the

Office of Science of the U.S. Dept. of Energy, through SciDAC Award DE-FC02-01ER41185 and supercomputing support for this work at NERSC and NCCS. We also thank Eric Myra for many helpful discussions. The work of C S Woodward was performed under the auspices of the U.S. Department of Energy by Lawrence Livermore National Laboratory under Contract DE-AC52-07NA27344.

References

[1] Bionta R M, Blewitt G, Bratton C B, Caspere D and Ciocio A 1987 Observation of a neutrino burst in coincidence with supernova 1987A in the Large Magellanic Cloud *Phys. Rev. Lett.* **58**:1494–1496
[2] Hirata K, Kajita T, Koshiba M, Nakahata M and Oyama Y 1987 Observation of a neutrino burst from the supernova SN1987A *Phys. Rev. Lett.* **58**:1490–1493
[3] Burrows A and Lattimer J M 1986 The birth of neutron stars *Astrophys. J.* **307**:178–196
[4] Swesty F D, Lattimer J M and Myra E S *Astrophys. J.* The role of the equation of state in the "prompt" phase of type II supernovae **425**:195–204
[5] Keil W, Janka H-T and Mueller E 1996 Ledoux convection in protoneutron stars—a clue to supernova nucleosynthesis? *Astrophys. J.* **473**:L111+
[6] Mezzacappa A, Calder A C, Bruenn S W, Blondin J M, Guidry M W, Strayer M R and Umar A S 1998 The interplay between proto-neutron star convection and neutrino transport in core-dollapse supernovae *Astrophys. J.* **493**:848+
[7] Pons, *et al.* 1999 Evolution of proto-neutron stars *Astrophys. J.* **513**:780–804
[8] Kelley C T 1995 *Iterative Methods for Linear and Nonlinear Equations* vol. 16 in *Frontiers in Applied Mathematics* (Philadelphia, PA:SIAM)
[9] Knoll D A and Keyes D E 2004 Jacobian-free Newton-Krylov methods: a survey of approaches and applications *J. Comp. Phys.* **193**:357–397
[10] Swesty F D and Myra E S 2008 A numerical algorithm for modeling multigroup neutrino-radiation hydrodynamics in two spatial dimensions *Astrophys. J.* (submitted)
[11] Mihalas D and Weibel-Mihalas B 1999 *Foundations of Radiation Hydrodynamics* (Dover)
[12] Stone J M and Norman M L 1992 ZEUS-2D: a radiation magnetohydrodynamics code for astrophysical flows in two space dimensions, I: the hydrodynamic algorithms and tests *Ap. J. Suppl.* **80**:753–790
[13] LCPFCT 2008 http://www.lcp.nrl.navy.mil/lcpfct/
[14] Boris J P and Book D L 1973 Flux-corrected transport, I: SHASTA, a fluid transport algorithm that works *J. Comp. Phys.* **11**:38–69
[15] Brown P N and Saad Y 1990 Hybrid Krylov methods for nonlinear systems of equations *SIAM J. Sci. Statist. Comput.* **11**:450–481
[16] Freund R W, Golub G H and Nachtigal N M 1992 Iterative solution of kinear systems *Acta Numerica* A Iserles ed. (Cambridge Univ. Press: New York) 57–100
[17] Saad Y and Schultz M H 1986 GMRES: a generalized minimal residual algorithm for solving nonsymmetric linear systems *SIAM J. Sci. Stat. Comput.* **7(3)**:856–869
[18] van der Vorst HA 1992 Bi-CGSTAB: a fast and smoothly converging variant of BI-CG for the solution of nonsymmetric linear systems *SIAM J. Sci. Stat. Comput.* **13(2)**:631–644
[19] Brown P N 1987 A local convergence theory for combined inexact-Newton/finite-difference projection methods *SIAM J. Numer. Anal.* **24**:407–434
[20] SUNDIALS (SUite of Nonlinear and DIfferential ALgebraic Solvers) http://www.llnl.gov/casc/sundials
[21] Hindmarsh, *et al.* 2005 SUNDIALS: Suite of nonlinear and differential/algebraic equation solvers *ACM Trans. Math. Softw.* **31(3)**:363–396
[22] Trilinos http://software.sandia.gov/trilinos
[23] PETSC (Portable, Extensible Toolkit for Scientific Computation) http://www-unix.mcs.anl.gov/petsc
[24] Sod G A 1978 A survey of several finite difference methods for systems of nonlinear hyperbolic conservation laws *J. Comp. Phys.* **27**:1–31
[25] Swesty F D 1995 Implicit general relativistic hydrodynamic methods for modeling the late-time explosion mechanism in core collapse supernovae *Astrophys. J.* **445**:811–827

Gaining confidence in scientific applications through executable interface contracts

Tamara L. Dahlgren,[1] David E. Bernholdt,[2] Lois Curfman McInnes[3]

[1] Lawrence Livermore National Laboratory, Livermore, CA 94720, USA

[2] Computer Science and Mathematics, Oak Ridge National Laboratory, Oak Ridge, TN 37831, USA

[3] Mathematics and Computer Science, Argonne National Laboratory, Argonne, IL 60439, USA

E-mail: `dahlgren1@llnl.gov`, `bernholdtde@ornl.gov`, `mcinnes@mcs.anl.gov`

Abstract. Interface contract enforcement is intended to help scientists gain confidence in software built from third-party components. Unfamiliar components present increased risk of incorrect or unanticipated usage patterns and unexpected component behavior. Executable interface contracts can address these issues but may incur unacceptable overhead. Research into techniques for performance-driven contract enforcement pursues practical solutions to adapting the level of contract enforcement to performance constraints.

1. Introduction

Many aspects of modern high-performance scientific computing challenge the levels of confidence scientists and engineers have in software. Scientific applications continue to grow ever larger and more complex because ofthe desire to simulate more complex phenomena through higher fidelity models, evolving high-end computer architectures, and new algorithms to effectively utilize these systems. These goals are of particular concern to applications under development by SciDAC teams working in combustion [1], quantum chemistry [2], core-edge fusion modeling [3], and accelerator modeling [4]. Executable interface contracts – in the form of concise, human-readable, machine-processable specifications of behavioral constraints – can be used to gain confidence in applications built from components.

Verification and *validation* (V&V) are terms for the general processes used to gain confidence in software. Developers *verify* that their software conforms to its specifications – in terms of performing its computations. Researchers *validate* results from their computational models against appropriate external results through analytical models, experiments, and/or observations. Scientific computing software environments traditionally provide few facilities for aiding the V&V process, leading developers to build and rely on sets of tests. Thorough test suites can be challenging and time consuming to develop. In practice, errors often arise because developers are not able to anticipate every possible usage scenario and input data set.

Interface contracts offer a complementary approach to software verification. Current practice in programming defines only the *syntax* of the interface through specifications of methods and their parameters. Interface contracts provide a means of specifying some of the semantics associated with component interfaces – in terms of the behaviors associated with the methods. Contracts are formed by precondition, postcondition, and/or class-invariant clauses belonging to the interface specification, not the underlying implementation(s). *Precondition* clauses consist of assertions on properties that must hold prior to method execution. *Postcondition* clauses contain assertions that must hold upon method completion. *Class invariants* apply before and after execution of all defined methods. Because this approach specifies properties of the interface itself in a form amenable to automated enforcement, it allows *every* use of the interface to be treated as a test case, thereby enabling dynamic validation of program behavior under actual operating conditions. The result is improved testing, debugging, and runtime monitoring of software quality, thereby providing scientific software developers with a powerful tool for catching errors early and ensuring correct software usage during deployment.

Several techniques developed in the community focus on verifying computational results [5, 6, 7]. The most prominent approaches are *algorithm-based fault tolerance* (ABFT) and *result checking* (RC). ABFT can take the form of any one of a number of techniques taking advantage of the underlying mathematical computations to accumulate information about the correctness of the results. One example is the use of a check sum. RC, on the other hand, is akin to postconditions. It can involve a similar *a posteriori* computation used to compare with results. Although both ABFT and RC are useful for gaining confidence in computational results, they can be expensive and do not address the more general issue of using software correctly.

The enforcement of interface contracts also results in computational overhead. Verifying that a scalar parameter is nonzero is a fast operation, but confirming a vector or matrix has a non-negative norm scales with the size of the data. If the overhead of contract checking is too high, enforcement will be limited to selected test cases. However, the greatest chance of exposing errors and failures comes when the code is being used in "extreme" situations, such as in the largest-scale simulation runs. These are also the conditions under which users tend to be most sensitive to performance overheads. Standard practice in communities adopting contracts is to eliminate all checks or manually disable at least the more complex or expensive ones. Consequences of completely disabling contract enforcement range from spending days to weeks reproducing and debugging errors to making decisions or reporting findings based on erroneous information.

This research was undertaken as part of the SciDAC Center for Technology for Advanced Scientific Component Software (TASCS) [8] as part of the Common Component Architecture (CCA) [9] Forum's pursuit of scientific software component technologies. Components are encapsulated units of software interacting only through well-defined interfaces. The specification of contracts is a first step in providing automated tools using semantic information to increase confidence in the software and resulting scientific data.

This paper highlights some of the work and findings to date. Results are presented from experiments primarily using mesh components developed by the Interoperable Technologies for Advanced Petascale Simulations (ITAPS) program [10] SciDAC project, which is providing meshing and discretization tools for a variety of SciDAC applications. Mesh management is a critical part of many scientific applications so the benefits of interface contracts on mesh components can help improve the quality of many applications.

2. Interface contracts in Babel

One of the key reasons executable interface contracts are not ubiquitous in this domain is the inability to define them in common scientific programming languages. This research aims to make contracts broadly available through the Babel language interoperability toolkit [11]. Babel provides middleware tailored for high-performance scientific programming for the following languages: C, C++, Fortran, Java, and Python. Support for the specification and enforcement of interface contracts has been added to the Babel toolkit. Contracts are specified through the addition of clauses to the interfaces defined in the Scientific Interface Definition Language (SIDL). Statements to enforce contracts are automatically generated by the Babel compiler

```
double norm(in double tol)
   throws      /* Exceptions */
      sidl.PreViolation, NegativeValueException, sidl.PostViolation;

   require     /* Precondition clause */
      non_negative_tolerance: tol >= 0.0;

   ensure      /* Postcondition clause */
      non_negative_result: result >= 0.0;
      nearEqual(result, 0.0, tol) iff isZero(tol);
```

Figure 1. Vector norm contracts example.

along with the language interoperability middleware. This approach allows interface contracts to be defined once in the specification and applied to all associated implementations in any of the supported languages.

Figure 1 shows the SIDL specification of a *norm* method on a vector of doubles. In this case, `norm` is a function taking `tol` as an argument and returning a double `result`. The precondition clause consists of an assertion requiring the caller to pass a non-negative tolerance, `tol`, value. If the preconditions are satisfied, all implementations of the method must then ensure the return of a non-negative `result` within the specified tolerance of zero if and only if the values of all elements in the vector are zero (i.e., it is the *zero* vector).

Contract enforcement is automatically managed by the Babel runtime library. Built-in, contract clause-specific exceptions are raised whenever a violation is detected. Since SIDL and Babel are central to the widely used reference implementation of the CCA, interface contracts implemented in this manner can easily be used with CCA components.

3. Experiments

Research thus far focuses on identifying and developing interface contract enforcement sampling techniques driven by execution time overhead limits [12, 13, 14, 15]. Experiments are conducted to investigate the associated impacts. Data are collected to establish metrics for comparing enforcement strategies. This section summarizes basic strategies and metrics used in this research. Selected results and an overview of findings are also presented. More information and details can be found in [14, 15].

3.1. Enforcement policies

Interface contract enforcement policies reflect different strategies for checking contract clauses. Although this research involves the implementation of many policies, this paper focuses on three: *never*, *always*, and *adaptive fit*. The first two policies – *vever* and *always* – represent "all-or-nothing" strategies providing baseline data. The *Never* policy disables interface contract enforcement. The resulting execution times are used to establish enforcement overhead metrics. The *Always* policy enables full contract enforcement. That is, all interface contracts encountered during execution are checked. Consequently, the policy provides baselines for the total number of checked (or enforced) contracts and detected violations for each trial. *Adaptive fit* is one of three performance-driven sampling approaches. The policy checks contract clauses only if the accumulated enforcement execution time estimate is within the user-specified limit of the cumulative estimates outside the contracts.

Table 1. Summary of data collected in the latest study [14]. Experiments did not detect any interface contract violations for the program labeled **MA**.

Program	Software Component (Source)	Policy	Mean (over all trials) Enforcement Overhead	Contracts Enforced	Violations Detected
A	Simplicial Mesh (ITAPS)	*Always*	5%	100%	100%
		Adaptive fit	8%	65%	67%
AA	Simplicial Mesh (ITAPS)	*Always*	26%	100%	100%
		Adaptive fit	9%	58%	67%
MA	Simplicial Mesh (ITAPS)	*Always*	25%	100%	0%
		Adaptive fit	11%	0.27%	0%
MT	Volume Mesh (ITAPS)	*Always*	3%	100%	100%
		Adaptive fit	2%	0.35%	6%
VT	Vector (Babel)	*Always*	81%	100%	100%
		Adaptive fit	67%	59%	64%

3.2. Metrics

Three metrics are used to compare the effects of policies under study: enforcement overhead, contracts enforced, and violations detected. **Enforcement overhead** is the percentage difference in the average execution time of a policy above the cost of conducting the experiment without contract checking. That is, the overhead of a given policy is relative to the average execution time for experiments using the *Never* policy. **Contracts enforced** is computed based on the number checked for a policy relative to the total number checked using the *Always* policy. **Violations detected** for a policy is also relative to the total number of violations detected with the *Always* policy. Hence, the effects of sampling contracts are considered in terms of execution time, contract checks, and contract clause violations.

3.3. Findings

Table 1 summarizes data obtained from experiments using the policies described in Section 3.1. Experiments involve two mesh components from the ITAPS [10] SciDAC project and a vector component from Babel [11]. Details, based on results by trial, can be found in [14]. Performance-driven enforcement automatically adjusts the level of contract clause checking and tends to have better overall control of enforcement overhead than alternatives strategies considered in the study. As illustrated by programs **MT** and **A**, performance-driven enforcement is inappropriate for programs exercising contracts composed predominantly of scalar checks. However, performance-driven enforcement appears to be most suited to programs enforcing contract clauses whose checking is fast-to-moderately expensive relative to the amount of time spent in component methods.

Analysis of data across three studies [15] further indicates the quality of execution time estimates impacts checking and overhead. While coarse estimates enable performance-driven policies to keep the overhead closer to the target level, they preclude the detection of violations in trials of the first study. The most refined estimates in the final study enable performance-driven policies to detect significant numbers of violations in roughly half of the trials.

4. Summary

The goal of this research is to help scientists gain confidence in software built from third-party components using specialized interface contract enforcement strategies. The vision of scientists developing insights into and predictions about physical phenomena through component technologies is based on the idea of a repository containing multiple components conforming to the same interface specification. Behavioral specifications in the form of executable interface contracts are a well-known mechanism for ensuring compliance at run time. However, performance overhead concerns can be roadblocks to their adoption.

This research proposes the use and investigates the effects of performance-driven sampling as a means of controlling the impact of interface contract enforcement on program execution time. Target applications make numerous calls to methods with contracts and are expected to incur unacceptable overhead while enforcing all contracts. Experiments leveraging two ITAPS mesh components and a Babel vector component are described. Findings indicate performance-driven interface contract enforcement is a viable alternative to the traditional strategy of disabling enforcement for applications exhibiting target characteristics. Programs executing methods whose contracts contain assertions which are fast-to-moderate to check relative to the time spent executing their methods appear to benefit most from performance-driven strategies.

Future work includes further testing and enhancement of these capabilities in SciDAC applications, as well as incorporating these techniques into tools under development by the TASCS project for computational quality of service [16], or the automatic selection and configuration of components to suit a particular computational purpose and environment.

Acknowledgments
We thank those who contributed software and/or interface contracts to the studies: Lori Freitag Diachin (simplicial mesh), Carl Ollivier-Gooch (volume mesh, program **MT**), and Kyle Chand. This work was performed under the auspices of the U. S. Department of Energy's Scientific Discovery through Advanced Computing (SciDAC) program's Center for Technology for Advanced Scientific Component Software (TASCS). Lawrence Livermore National Laboratory is managed by Lawrence Livermore National Security, LLC under Contract DE-AC52-07NA27344. Oak Ridge National Laboratory is managed by UT-Battelle, LLC under Contract DE-AC-05-00OR22725. Argonne National Laboratory operates under Contract DE-AC02-06CH11357.

References

[1] Najm H (PI) Computational Facility for Reacting Flow Science (CFRFS) `http://cfrfs.ca.sandia.gov`
[2] Gordon M (PI) Chemistry Framework using the CCA `http://www.scidac.gov/matchem/better.html`
[3] Cary J (PI) Framework application for core-edge transport simulations `http://www.facetsproject.org`
[4] Spentzouris P (PI) Community Petascale Project for Accelerator Science and Simulation (COMPASS) DOE SciDAC project, 2007
[5] Hull T E, Cohen M S, Sawchuk J T M and Wortman D B 1988 *ACM Trans. Mathematical Software* **14** 201–217
[6] Prata P and Silva J G 1999 *Proc. Intl. Symp. on Fault-Tolerant Computing* pp 4–11
[7] Turmon M, Granat R, Katz D S and Lou J Z 2003 *IEEE Trans. on Computers* **52** 579–591
[8] Center for technology for advanced scientific component software (tascs) `http://tascs-scidac.org`
[9] Allan B A, Armstrong R, Bernholdt D E, Bertrand F, Chiu K, Dahlgren T L, Damevski K, Elwasif W R, Epperly T G W, Govindaraju M, Katz D S, Kohl J A, Krishnan M, Kumfert G, Larson J W, Lefantzi S, Lewis M J, Malony A D, McInnes L C, Nieplocha J, Norris B, Parker S G, Ray J, Shende S, Windus T L and Zhou S 2006 *Intl. J. High-Perf. Computing Appl.* **20** 163–202 URL `http://hpc.sagepub.com/cgi/reprint/20/2/163`
[10] Diachin L (PI) Center for Interoperable Technologies for Advanced Petascale Simulations (ITAPS) `http://www.scidac.gov/math/ITAPS.html`
[11] Kumfert G (PI) Babel `http://www.llnl.gov/CASC/components/babel.html`
[12] Dahlgren T L and Devanbu P T 2004 *Proc. Intl. Workshop on Software Engineering for High Perf. Computing System Appl.* (Edinburgh, Scotland) pp 64–69
[13] Dahlgren T L and Devanbu P T 2005 *Proc. 2nd Intl. Workshop on Software Engineering for High Perf. Computing System Appl.* (St. Louis, Missouri) pp 73–77
[14] Dahlgren T L 2007 *Proc. Intl. Symp. on Component-Based Software Engineering* (Medford, MA USA)

[15] Dahlgren T L 2008 *Performance-Driven Interface Contract Enforcement for Scientific Components* Ph.D. thesis University of California, Davis One Shields Avenue, Davis, CA, 95616
[16] Norris B, Ray J, Armstrong R, McInnes L C, Bernholdt D E, Elwasif W R, Malony A D and Shende S 2004 *Proc. Intl. Symp. on Component-Based Software Engineering* (Edinburgh, Scotland)

SciDAC 2008 IOP Publishing
Journal of Physics: Conference Series **125** (2008) 012087 doi:10.1088/1742-6596/125/1/012087

Managing locality in grand challenge applications: a case study of the gyrokinetic toroidal code

Gabriel Marin, Guohua Jin and John Mellor-Crummey

Department of Computer Science, Rice University, Houston, TX 77005, USA

E-mail: {mgabi,jin,johnmc}@cs.rice.edu

Abstract. Achieving high performance with grand challenge applications on today's large-scale parallel systems requires tailoring applications for the characteristics of the modern microprocessor architectures. As part of the US Department of Energy's Scientific Discovery through Advanced Computing (SciDAC) program, we studied and tuned the Gyrokinetic Toroidal Code (GTC), a particle-in-cell code for simulating turbulent transport of particles and energy in burning plasma, developed at Princeton Plasma Physics Laboratory. In this paper, we present a performance study of the application that revealed several opportunities for improving performance by enhancing its data locality. We tuned GTC by performing three kinds of transformations: static data structure reorganization to improve spatial locality, loop nest restructuring for better temporal locality, and dynamic data reordering at run-time to enhance both spatial and temporal reuse. Experimental results show that these changes improve execution time by more than 20% on large parallel systems, including a Cray XT4.

1. Introduction

As part of the application engagement efforts in the SciDAC Center for Scalable Application Development Software and the Performance Engineering Research Institute, at Rice University we have undertaken analysis and tuning of the Gyrokinetic Toroidal Code (GTC). GTC [0] is a particle-in-cell (PIC) code for simulations of turbulent transport in fusion plasmas such as the International Thermonuclear Experimental Reactor (ITER). GTC is the production code of two DOE SciDAC projects: GPS-TTBP and GSEP. An allocation of 8M processor-hours has been awarded for 2008 to explore GTC applications on the Cray XT4 computer at Oak Ridge National Laboratory.

During each time step, the PIC algorithm repeatedly executes `charge`, `solve`, and `push` operations. In the `charge` step, it deposits the charge from each particle onto grid points nearby. The `solve` step computes the electrostatic potential and field at each grid point by solving the Poisson equation on the grid. In the `push` step, the force on each particle is computed from the potential at nearby grid points. Particles move according to the forces on them. The version of GTC that we studied uses a domain decomposition along the toroidal direction. Each toroidal domain including one or more poloidal planes is assigned to a separate MPI process. Particles move across the boundary poloidal planes through MPI communications. Figure 1(a) shows one-eighth of the donut-shaped torus arrows representing ion movement. Figure 1(b) shows a poloidal plane consisting of rings equally distant from each other (in the radial direction) and uniformly distributed grid cells along each of the rings (in the θ direction).

(a) One-eighth of the toroid (b) a poloidal plane perpendicular to field lines.

Figure 1. Static view of the torus in the GTC simulation.

In our investigation, we used both modeling and measurement to identify and understand opportunities for improving GTC's performance. We used modeling [0] to understand the principal data reuse patterns in GTC and to identify opportunities for enhancing reuse in cache. We observed that performance degrades as the position of ions in the plasma evolves during the simulation. To alleviate this problem, we periodically reorder particles so that particles nearby in the simulated tokamak are close in memory. We present a mathematical model for computing the optimal sorting interval and an adaptive algorithm that uses this model to control sorting at run-time. We also modified GTC's communication code to reduce the rate at which locality degrades. This transformation has two benefits: it reduces the need for particle reordering, and it diminishes the performance loss due to locality degradation between each pair of sorting steps.

2. Single-node performance improvements

Tuning the performance of parallel applications requires not only achieving good scalability as the number of processors increases but also maximizing the application's performance on each individual node. Elsewhere [0], we present in greater detail the process of analyzing and tuning the single-node performance of GTC. For brevity, in this paper we present only a summary of our code transformations for improving data locality in GTC.

- *Static data structure reorganization.* We transposed the ion arrays; this transformation improves spatial reuse and eliminates about half of the cache misses for these arrays.
- *Loop fusion for charge deposition.* Our analysis identified that about 11% of all L3 cache misses are due to reuse of data between two loops that iterate over all particles in the charge deposition phase. By fusing these loops, we eliminated these cache misses.
- *Loop tiling and fusion for particle pushing.* During each time step, the velocities of ions are updated based on the electrical field. This calculation was performed in several loops. Our analysis identified that reuse between these loops accounts for about 20% of all L3 cache misses in the program. Fusing these loops and iterating over subsets of ions improved temporal reuse of the ion data in this phase.
- *Loop interchange for field smoothing.* In the calculation of the electrical field, a single loop nest was contributing a high fraction of TLB misses. Applying loop interchange eliminated all extraneous TLB misses in this phase.

3. Particle reordering

During each time step, each particle deposits charge onto cells related to the particle's position in space, a potential field is computed using the charge distribution among the cells, and each particle's position is updated after it is pushed by the resulting electric field. While the potential field computation involves only the cell data structures, the charge deposition and particle position updates involve interactions between particles and cells. For these phases, the best

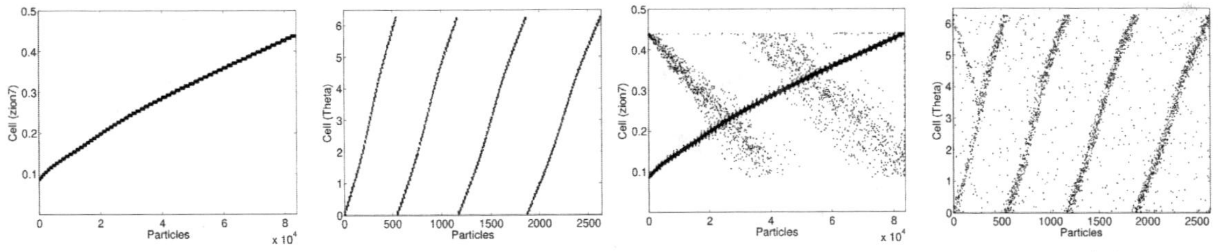

(a) (b) (c) (d)

Figure 2. Particle positions change over time: (a,b) time step 1; (c,d) time step 20.

locality is achieved when all particles in a cell are processed together. At program start, particles are sorted in cell order. However, as the simulation progresses there is a gradual loss of locality as particles migrate from one cell to another, and even to different processors.

Figure 2 shows scatter plots of the particle positions along the radial and θ directions at the time steps 0 and 20. The data was collected from a single processor in a four-processor simulation. Figures 2(a) and (b) show the initial radial positions of all particles and θ positions for the first four poloidal rings. Figures 2(c) and (d) show the particle positions at time step 20.

By storing particles in cell order, we can improve data locality in GTC beyond what is achievable using only the data structure and loop transformations described in Section 2. Because performance degrades only gradually over time and since sorting the particles after every time step is expensive, it is preferable to sort the particles only once in a while. The next section describes a mathematical model for computing the optimal sorting interval, τ, based on the cost of sorting and the rate at which performance degrades.

3.1. Understanding the optimal sorting interval

Let $f(x)$ be the function that describes how performance degrades over time. Thus, $f(x)$ represents the execution cost of the xth time step since the latest sorting step. We assume that performance per time step returns to the same base level after each sorting step, and that afterwards it degrades in accordance with function $f(x)$. We want to define $G(t)$, the benefit function for sorting at every t time steps and then determine $t = \tau$ for which $G(t)$ is maximized.

Let C be the cost of one sorting step. The benefit function over N time steps when we perform sorting every t time steps can be written as follows.

$$G(t) = \sum_{k=1}^{\frac{N}{t}-1} \left(\int_{kt}^{(k+1)t} f(x)\,dx - \int_0^t f(x)\,dx - C \right) = \int_0^N f(x)\,dx - \frac{N}{t}\int_0^t f(x)\,dx - \frac{N}{t}C + C \quad (1)$$

Note that $\int_0^N f(x)\,dx$ and C are independent of t, and thus we can remove them. Function $G(t)$ is maximized when the equation: $\frac{N}{t}\int_0^t f(x)\,dx + \frac{N}{t}C$ is minimized. The total number of time steps, N, is a constant and we can divide the equation by N. The problem reduces again to finding the value of t for which the following equation is minimized:

$$h(t) = \frac{1}{t}\left(\int_0^t f(x)\,dx + C \right) \quad (2)$$

This result is not surprising. It tells us that the optimal sorting interval is found when the average execution cost per time step of the program with sorting every t time steps is minimized.

3.2. An adaptive particle sorting algorithm

We could compute the optimal sorting interval analytically if we knew the function $f(x)$ that describes the performance degradation. We also could approximate $f(x)$ by applying regression to data collected from previous runs. However, we would have to repeat this process for every target architecture, each time we modify the code, and potentially for each program input.

Alternatively, we can compute the optimal sorting interval adaptively at run time. Based on equation 2, we need to know C – the cost of a sorting step, and the cost of the first t time steps – $e(t) = \int_0^t f(x)\,dx$, which we can compute iteratively at run-time. After each time step we update $e(t)$ and evaluate $h(t) = \frac{e(t)+C}{t}$. We stop when the value of $h(t)$ starts to grow, which gives us the value of t for which $h(t)$ is minimized. We added an adaptive sorting algorithm to GTC based on the above observations. The initialization phase of the sorting algorithm executes the following four steps algorithm along the normal execution of the program:

Step 0 *Compute C, the execution cost of a sorting step.* We let the program run for a fixed number of time steps, after which we invoke our particle reordering routine measuring its execution cost. In our implementation we benchmark the cost of sorting on the tenth time step of the program. The actual initialization phase starts on Step 1.

Step 1 *Evaluate $h(t)$.* After each time step we compute the value of $e(t)$ incrementally, evaluate $h(t)$, and update its minimum encountered value, h_{min}, as needed.

Step 2 *Compute a local optimum.* If the most recent value of $h(t) > h_{min}$, we have a candidate value for the local optimal sorting interval: $\tau_{local} = t - 1$.

Step 3 *Compute a global optimum.* For SPMD programs, such as GTC, all processes execute in interlocked steps. In order to achieve good performance, it is essential for all processes to perform particle reordering at the same time. The largest optimal interval across all processes is selected using a reduction operation. We use the maximum value over all processes because from our experience it is more costly to sort excessively than not to sort often enough.

Step 4 *Provide confidence in the global optimum.* In practice, the time per time step may fluctuate slightly from one time step to another, and we may find several local minima for $h(t)$ only a few time steps apart. For this reason, we do not stop the initialization algorithm when we find the first minima. Instead, if τ is our global sorting interval, we continue executing steps **1–3** for τ more time steps. If any process finds a new local optimum, the new optimal interval will be agreed upon by all processes in Step 3 of the algorithm, and the value of τ itself will increase in this way. However, if we do not find another minimum during the next τ steps, we are confident that the interval we found is optimal.

3.3. A new particle migration algorithm

In our study of GTC, we found that the strategy for migrating particles between poloidal planes contributed to the rate of locality loss observed for parallel runs. During execution, gaps in the particle array are created as particles migrate between processors. These gaps were being filled with particles from the end of the array, before newly received particles were appended at the end. Using particles from the end of the array that initially reside on the outermost rings of a plane to fill gaps in other rings accelerates the rate with which particles become unsorted.

We expect the number of particles sent and received by any local domain to be approximately equal, since the particle density should be rather uniform along the toroidal dimension. Moreover, we expect the number of particles sent and received through each subsection of a poloidal plane to be approximately equal. With these observations, and since particles are initially sorted in cell order, we modified the particle migration algorithm to fill the gaps in the particle array with newly received particles. Any leftover particles are appended at the end of the array, and any unfilled gaps are filled with particles from the end of the array.

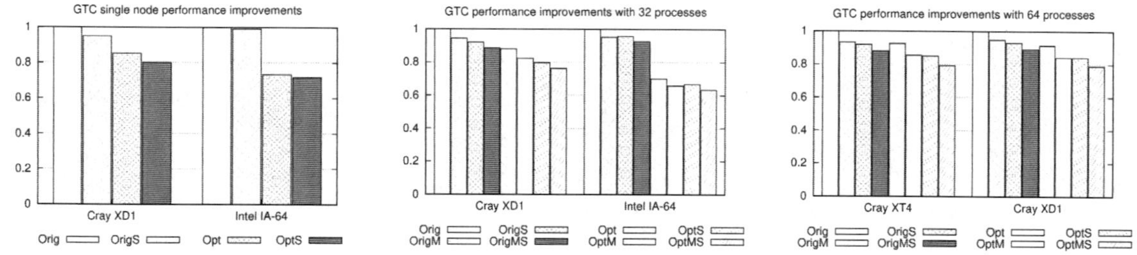

(a) single-node runs (b) 32-process runs (c) 64-process runs

Figure 3. Performance of GTC code with different code optimizations.

4. Results

We evaluated our optimizations on three parallel systems: a Cray XT4 system at the Lawrence Berkeley National Laboratory with one 2.6 GHz dual-core AMD Opteron processor and 4 GBytes of memory per node; a Cray XD-1 machine with two dual-core 2.2 GHz AMD Opteron CPUs and 8 GBytes of memory per node; and an Intel Itanium 2 cluster with two 900 MHz processors and 4 GBytes of memory per node. The last two clusters are located at Rice University. We compared up to eight versions of the GTC codes on the three parallel systems, with three types of optimizations turned on and off. In naming the versions, "Orig" represents the original version of the code, "Opt" includes the loop and data restructuring optimization, "M" includes the better migration approach, and "S" represents the adaptive sorting. Execution times were collected. The programs are compiled with the Fortran and C compilers from the Portland group with optimizations "-fast -fastsse" on the Cray machines, and with the Intel Fortran and C compilers with optimization "-fast" on the Intel Itanium cluster.

Figure 3(a) presents the single-node performance results of three optimized versions of the GTC codes on the Cray XD1 and Intel Itanium 2. All timing results are normalized to the execution time of the original version. Overall, the version with both optimizations applied achieves the best performance with a 27% reduction of execution time on the Itanium2 cluster.

Figures 3(b) and (c) compare the parallel execution time of all eight versions of the code. In figure 3(b), the optimized version "OptMS" with all three optimizations applied achieved the best performance with about 37% reduction of overall execution time on the Itanium2 cluster. The loop and data restructuring optimization contributed significantly to the overall improvement on the Itanium 2 cluster. In figure 3(c), the best version of the code with all three optimizations applied achieves approximately a 21% execution time reduction against the original version on both Cray machines.

5. Conclusions

Using today's microprocessor-based leadership computing platforms most effectively requires careful structuring of codes to make the most effective use of their multi-level memory hierarchies by exploiting both spatial and temporal reuse of data. For complex data-intensive scientific codes, exploiting locality is easier said than done. Significantly improving the performance of the Gyrokinetic Toroidal Code required comprehensive changes, including transposition of particle-centric data structures to improve spatial reuse, restructuring of procedures and loop nests to improve temporal reuse, adjustment of data movement associated with communication to avoid unnecessarily introducing disorder in the plasma, and periodically sorting particles by their coordinates to enhance spatial and temporal reuse. By paying careful attention to all of these factors affecting locality, we were able to reduce GTC's execution time on Opteron-based parallel systems by just over 20% and on Itanium2-based systems by approximately 37%.

References

[1] Lin Z, Hahm T S, Lee W W, Tang W M and White R B 1998 Turbulent transport reduction by zonal flows: massively parallel simulations *Science* **281(5384)**:1835–1837

[2] Marin G and Mellor-Crummey J 2008 Pinpointing and exploiting opportunities for enhancing data reuse *Proc. 2008 IEEE Intl. Symposium on Performance Analysis of Systems and Software*

HPCToolkit: performance tools for scientific computing

Nathan Tallent, John Mellor-Crummey, Laksono Adhianto, Michael Fagan and Mark Krentel

Department of Computer Science, Rice University, Houston, TX 77005, USA

E-mail: {`tallent, johnmc, laksono, mfagan, krentel`}`@cs.rice.edu`

Abstract. As part of the U.S. Department of Energy's Scientific Discovery through Advanced Computing (SciDAC) program, science teams are tackling problems that require simulation and modeling on petascale computers. As part of activities associated with the SciDAC Center for Scalable Application Development Software (CScADS) and the Performance Engineering Research Institute (PERI), Rice University is building software tools for performance analysis of scientific applications on the leadership-class platforms. In this poster abstract, we briefly describe the HPCToolkit performance tools and how they can be used to pinpoint bottlenecks in SPMD and multi-threaded parallel codes. We demonstrate HPCToolkit's utility by applying it to two SciDAC applications: the S3D code for simulation of turbulent combustion and the MFDn code for *ab initio* calculations of microscopic structure of nuclei.

1. Introduction

Harnessing the power of emerging petascale computational platforms to aid scientific discovery poses a grand challenge for computer science. Programming petascale systems requires both exploiting coarse-grained parallelism across nodes and making effective use of individual microprocessor-based nodes. Performance tools that help pinpoint bottlenecks and quantify opportunities for improvement are essential for tuning codes to use these platforms efficiently.

As part of activities associated with the SciDAC Center for Scalable Application Development Software (CScADS) and the Performance Engineering Research Institute (PERI), Rice University is developing HPCToolkit [1], a performance toolkit for accurately measuring and pinpointing performance bottlenecks. HPCToolkit uses novel techniques for measurement and analysis of parallel programs. In particular, it uses statistical sampling of hardware performance counters and attributes metrics to both the calling context in which they occur and program structure, including loops and inlined procedures. We describe here how we use HPCToolkit to pinpoint scalability and multicore performance bottlenecks. We demonstrate HPCToolkit's utility by applying it to two SciDAC applications: the S3D code for simulation of turbulent combustion [2] and the MFDn code for *ab initio* calculations of microscopic structure of nuclei [3]. Although we focus here on measurement and analysis to support manual tuning, the precise characterization of performance bottlenecks that we obtain is a prerequisite for more automatic approaches based on feedback-directed optimization.

2. Overview of HPCToolkit

HPCToolkit [1] consists of components for measuring the performance of fully optimized executables of parallel programs, analyzing application binaries to correlate measurements with program structure, and novel analysis techniques for pinpointing performance bottlenecks. We designed HPCToolkit to perform the following:

- *Work at binary level for language independence.* This enables HPCToolkit to support measurement and analysis of multi-lingual codes with external binary-only libraries.
- *Profile rather than adding code instrumentation.* Sample-based profiling is less intrusive than code instrumentation and requires only very modest data volume.

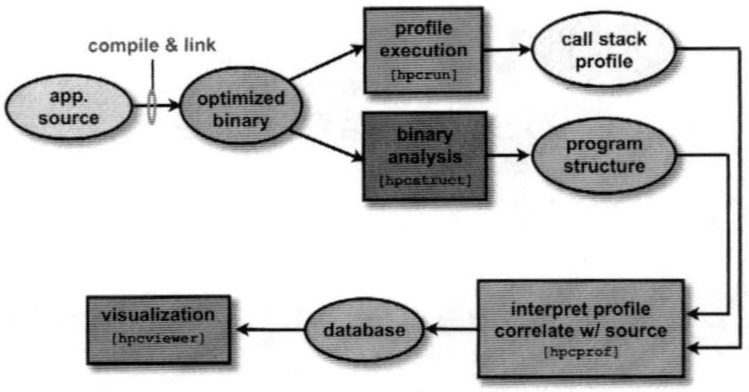

HPCToolkit workflow.

- *Collect and correlate multiple performance metrics.* Performance problems typically cannot be diagnosed with only one species of event.
- *Compute derived metrics to aid analysis.* Synthetic metrics, such as memory bandwidth consumed, often provide insight for optimization.
- *Attribute costs very precisely.* HPCToolkit is unique in its ability to associate measurements with dynamic calling context, loops, and inlined code.

2.1. Pinpointing and quantifying performance bottlenecks

We have developed several novel performance analysis techniques that provide deep insights into application performance on parallel platforms.

2.1.1. Pinpointing scalability bottlenecks in SPMD codes. HPCToolkit pinpoints scalability bottlenecks with an exquisite level of detail – all the way down to invidividual source lines in the program and the calling context in which the bottlenecks arose. Using our technique, one can do this for production runs of fully-optimized applications with run-time measurement overhead of only a few percent for arbitrary SPMD parallel programs including applications written using MPI [4]. Here, we briefly sketch how this process works.

Consider two executions of an application using weak scaling, one executed on p processors and the second executed on $q > p$ processors. In a weak scaling scenario, processors in each execution compute on the same amount of data. If the application exhibits perfect weak scaling, then the execution times should be identical on both q and p processors. In fact, if every part of the application scales uniformly, then we would expect this to hold in *each part* of the application. Using HPCToolkit, we measure the cost associated with each calling context in a program execution on each of p and q processors. We use a data structure called a *calling context tree* (CCT) to record our measurements. We expect that the execution time attributed to all pairs of *corresponding nodes*in the CCT to be identical. (Corresponding nodes in a pair of CCTs represent the same calling context, *e.g.*, `main` calls `solve`, which calls `sparse matrix-vector multiply`.) We use this expectation as the basis for identifying where performance scalability falls short. A similar strategy can also be applied to analyze strong scalability [4].

2.1.2. Identifying performance bottlenecks in multithreaded codes. We are extending HPCToolkit with two novel techniques for pinpointing, quantifying, and explaining performance bottlenecks in multithreaded programs. Our approach will support diagnosis of two types of inefficiencies: insufficient parallelism (which might be due to load imbalance or serialization) and parallelization overhead.

To pinpoint and quantify insufficient parallelism in executions of multithreaded programs, we will measure two quantities at sample points: the number of threads performing useful work ($\mathcal{W}$) and those that are idle ($\mathcal{I}$). If a sample event occurs in a thread that is idle, we ignore it. When a sample event occurs in a thread that is actively working, the thread records one sample in a metric representing the thread's work, and in a second metric the thread records a fractional sample $\mathcal{I}/\mathcal{W}$ to charge it a proportional share of its responsibility for not keeping the idle processors busy at that moment at that point in the program. From this, we can compute a metric representing the loss of parallelism and inefficiency in each program context.

To pinpoint parallel overhead, we observe that a compiler for a multi-threaded programming model, such as OpenMP, can tag statements in its generated code to indicate which are associated with parallelization overhead. Using binary analysis, we can recover information recorded by the compiler and identify instructions associated with these source lines and attribute any samples associated with these lines to parallelization overhead.

Values for insufficient parallelism and parallel overhead can be attributed to the loops, procedures, and calling contexts of a program. When they are combined with derived metrics measuring instruction mix, memory bandwidth, memory latency, and pipeline stalls, we can directly assess the effectiveness of a parallelization and provide guidance for how to improve it.

3. Case studies

Here we study the performance of two SciDAC codes to illustrate two uses of HPCToolkit: analysis of multi-core scaling and analysis of overall performance on a single core.

3.1. Multicore scaling of S3D

S3D uses direct numerical simulation (DNS) to study the micro-physics of turbulent reacting flows. The S3D code decomposes a 3D data volume among a collection of processes organized as a 3D logical mesh. Processes have same number of grid points and communicate using MPI. Most communication occurs between processes that are nearest neighbors in the 3D logical mesh.

To investigate the node performance of S3D, we studied two executions of an ethylene combustion simulation: one execution on a single core of a 1.9 GHz AMD quad-core Opteron 8347 processor (known as a Barcelona), and a second execution on all eight cores of a shared-memory node consisting of a pair of Barcelonas. Each of the cores has 512 MB of private L2 cache, and 2 MB of shared L3 victim cache. The nodes of the system serve as a proxy for the quad-core Barcelona nodes of the NCCS Cray XT4. For the model problem, on a single core, the code achieved .56 FLOPs/cycle; on the eight-core configuration, the model problem achieved 0.20 Flops/cycle for each core, yielding a total performance of 0.80 Flops/cycle per quad-core processor. Thus, in the eight core configuration, each core achieved only 36% of the performance of a single core in isolation. However, four cores of a quad-core processor achieved a 45% higher FLOP rate than a single core in isolation.

To understand the difference between the single core and eight core executions, we used HPCToolkit to collect a call path profile for each core for both the single-core and eight-core executions. Figure 1 compares the performance of the execution on a single core and the performance of one of the cores in the eight-core execution. This figure shows a screen capture from HPCToolkit's `hpcviewer` user interface, showing a *flat view* that correlates performance metrics with each loop nest in S3D. The `hpcviewer` display is in three parts: a source code window, a pane full of performance metrics of various types, and a navigation pane that relates values in the metric pane to the program source. The first three columns in the metric pane show cycles, L2 cache misses, and FLOPs for a single-core execution. The next three columns show the same metrics for one core of an eight-core execution. Each core is computing on a 30^3

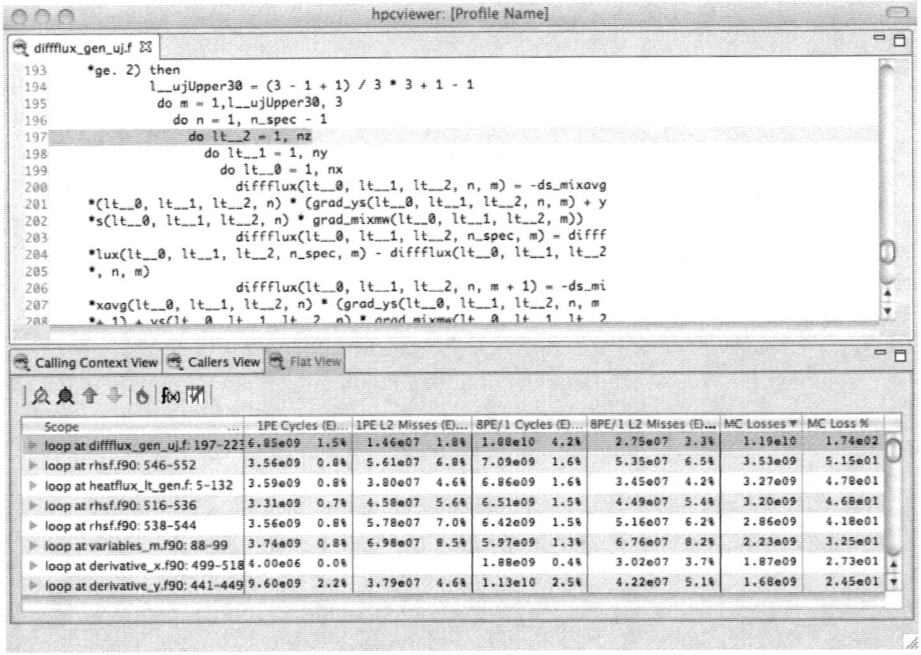

Figure 1. `hpcviewer` displaying a flat view of a timer-based call path profile for S3D.

volume of data. By comparing the cycles metric for one of the cores in eight-core execution with that of the single-core execution, we see that overall, the execution time on each core of the eight-core run is slower by a factor of 2.74 when all eight cores of the system are used. The last two columns of figure 1 demonstrate `hpcviewer`'s capability for computing arbitrary derived metrics on demand. Column seven of the metric pane shows a derived metric representing the multicore loss of performance, computed as the difference between the cycles on one of the eight cores and the cycles of a single-core execution. The final column of the display shows the percentage loss of performance in each loop nest, written in scientific notation.

In figure 1, the loops of the application are sorted according to their multicore loss, in descending order. The source pane shows part of the loop nest for which the loss was greatest. In the rightmost column of the metric pane, we see that its multicore loss for this loop was 174%. By comparing the L2 misses for one of the cores in an eight-core execution with those of a single core running in isolation, we see that the number of L2 misses increases by 88%. This loop, a diffusive flux calculation, is the most memory-intensive loop in the program. Inspection of the other top loops in the display shows multicore losses ranging from 24.5% to 51.5%. By sorting and ranking loop nests according to their multicore losses, `hpcviewer` focuses attention on the loop nests that deserve the most attention for performance tuning.

3.2. Single-core performance of MFDn

The "Many Fermion Dynamics — nuclear" code (MFDn) evaluates the many-body Hamiltonian and obtains the low-lying eigenvalues and eigenvectors using the Lanczos algorithm. The code shows good scaling and load balance on 15,000 cores, but is lacking in per-process efficiency.

To understand how much opportunity exists for improving the performance of the loop nests in MFDn, we used HPCToolkit to measure and analyze the code on one core of a 2.2 GHz dual-core Opteron. Figure 2 shows both measured metrics (cycles and Flops) and a derived metric FPWASTE. The FPWASTE column represents the difference between the theoretical peak

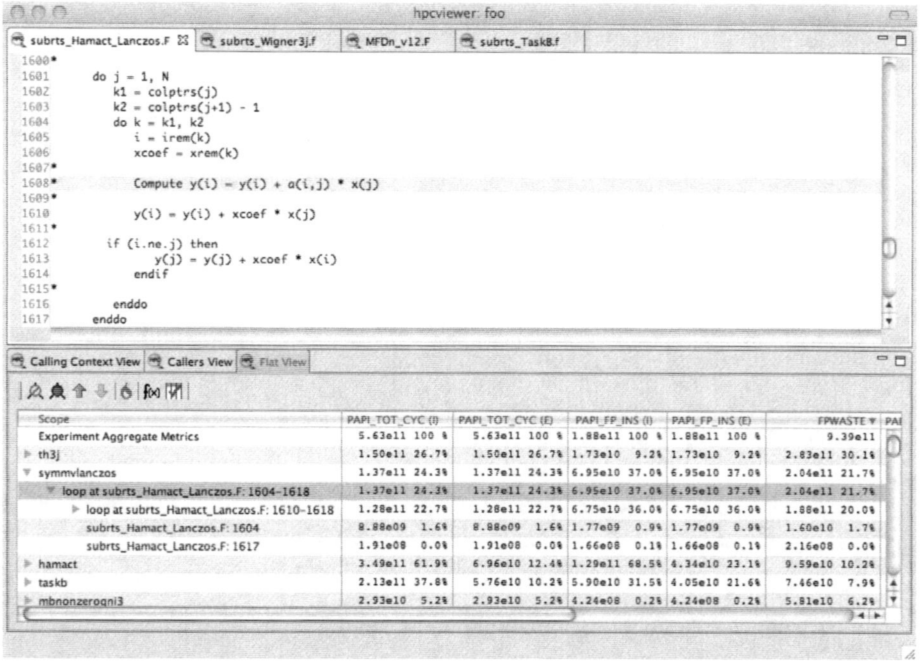

Figure 2. `hpcviewer` displaying floating point waste (FPWASTE) for the loop nest within MFDn's `symmvlanczos`.

Flops possible on the Opteron core and the actual Flops executed, that is, $(2 \times \mathrm{CYC}) - \mathrm{FP\_INS}$; this indicates how much unrealized opportunity for floating-point computation is associated with each loop. The highlighted loop computes the quantity $\mathbf{y} = \mathbf{y} + \mathbf{A}\mathbf{x}$, where $\mathbf{A}$ is represented in the *compressed sparse column* format. This loop accounts for roughly 20% of the FPWASTE in the execution. The matrix-vector multiply in the highlighted loop requires a load and store of a result value for each non-zero in the sparse matrix $\mathbf{A}$. If the matrix could be stored in *compressed sparse row format*, the calculation would perform one write and no loads of intermediate results per matrix row; this would improve the overall efficiency of this calculation.

4. Current status

Currently HPCToolkit is being used to analyze application performance on Opteron-based Linux clusters. As soon as kernel bugs are resolved on the Cray XT and Blue Gene/P, these tools will become available on the DOE's leadership class machines.

References

[1] Rice University 2008 HPCToolkit performance tools. `http://www.hipersoft.rice.edu/hpctoolkit`
[2] Monroe D 2002 *SciDAC Review* URL `http://www.scidacreview.org/0602/html/combustion.html`
[3] Vary J P, Shirokov A M and Maris P 2008 *ArXiv e-prints* **804** (*Preprint* 0804.0836)
[4] Coarfa C, Mellor-Crummey J, Froyd N and Dotsenko Y 2007 *ICS '07: Proceedings of the 21st Intl. Conference on Supercomputing* (New York, NY, USA: ACM) pp 13–22 ISBN 978-1-59593-768-1

SciDAC 2008 IOP Publishing
Journal of Physics: Conference Series **125** (2008) 012089 doi:10.1088/1742-6596/125/1/012089

PERI Auto-Tuning

David H Bailey[2], Jacqueline Chame[4], Chun Chen[4], Jack Dongarra[5], Mary Hall[4], Jeffrey K Hollingsworth[3], Paul Hovland[1], Shirley Moore[5], Keith Seymour[5], Jaewook Shin[1], Ananta Tiwari[3], Sam Williams[2] and Haihang You[5]

[1]Argonne National Laboratory, Argonne, IL 60439, USA
[2]Lawrence Berkeley National Laboratory, Berkeley, CA 94720, USA
[3]University of Maryland, College Park, MD 20742, USA
[4]USC/ISI, Marina del Rey, CA 90292, USA
[5]University of Tennessee, Knoxville, TN 37996, USA

E-mail: mhall@isi.edu

Abstract. The enormous and growing complexity of today's high-end systems has increased the already significant challenges of obtaining high performance on equally complex scientific applications. Application scientists are faced with a daunting challenge in tuning their codes to exploit performance-enhancing architectural features. The Performance Engineering Research Institute (PERI) is working toward the goal of automating portions of the performance tuning process. This paper describes PERI's overall strategy for auto-tuning tools and recent progress in both building auto-tuning tools and demonstrating their success on kernels, some taken from large-scale applications.

1. Introduction

As we are enter the era of petascale systems, as with earlier high-end systems, there is likely to be a significant gap between peak and sustained performance. Historically, the burden of achieving high performance on new platforms has largely fallen on the application scientists. To relieve application scientists of this burden, we would like to provide performance tools that are (largely) automatic, a long-term goal commonly called *auto-tuning*. This goal encompasses tools that analyze a scientific application, both as source code and during execution, generate a space of tuning options, and search for a near-optimal performance solution. There are numerous challenges to fully realizing this vision, including enhancement of automatic code manipulation tools, automatic run-time parameter selection, automatic communication optimization, and intelligent heuristics to control the combinatorial explosion of tuning possibilities. On the other hand, we are encouraged by recent successful results such as ATLAS, which has automatically tuned components of the LAPACK linear algebra library[1]. We are also studying techniques used in the highly successful FFTW library[2] and several other related projects [3-6].

The Performance Engineering Research Institute (PERI) is formalizing a performance tuning methodology used by application developers and automating portions of this process. Auto-tuning is one of the three aspects of performance tuning that are the focus of PERI, in addition to performance modeling and application engagement. In the context of these three aspects, the goal of PERI is to migrate automatic and semi-automatic prototypes into practice for a set of important applications. This document focuses on the PERI auto-tuning strategy (Section 2), recent progress in developing common interfaces to auto-tuning tools (Section 3), and tool infrastructure and experimental results (Section 4), followed by a conclusion.

2. PERI auto-tuning conceptual diagram

Figure 1 provides a conceptual diagram of auto-tuning in PERI. Several phases are shown, as described here, but this document focuses on transformation, code generation, and off-line search.

1. *Triage.* This step involves performance measurement, analysis and modeling to determine whether an application has opportunities for optimization.
2. *Semantic analysis.* This step involves analysis of program semantics to support safe transformation of the source code. The analyses include traditional compiler analyses to determine data and control dependences and can exploit semantic information provided by the user through annotations or information about domain-specific abstractions.
3. *Transformation.* Transformations modify source code to coax native compilers to generate better-performing implementations. Transformations such as loop unrolling may be parameterized to allow for input size and machine characteristic tuning.
4. *Code generation.* This phase produces a set of possible implementations to be considered.
5. *Offline search.* Offline search entails running multiple variants of the generated code to identify the best-performing implementation, with guidance from performance models or user input.
6. *Application assembly.* At this point, the optimized code components are integrated to produce an executable code, possible including instrumentation and support for dynamic tuning.
7. *Training runs.* Training runs involve a separate execution step designed mainly to produce performance data for feedback into the optimization process.

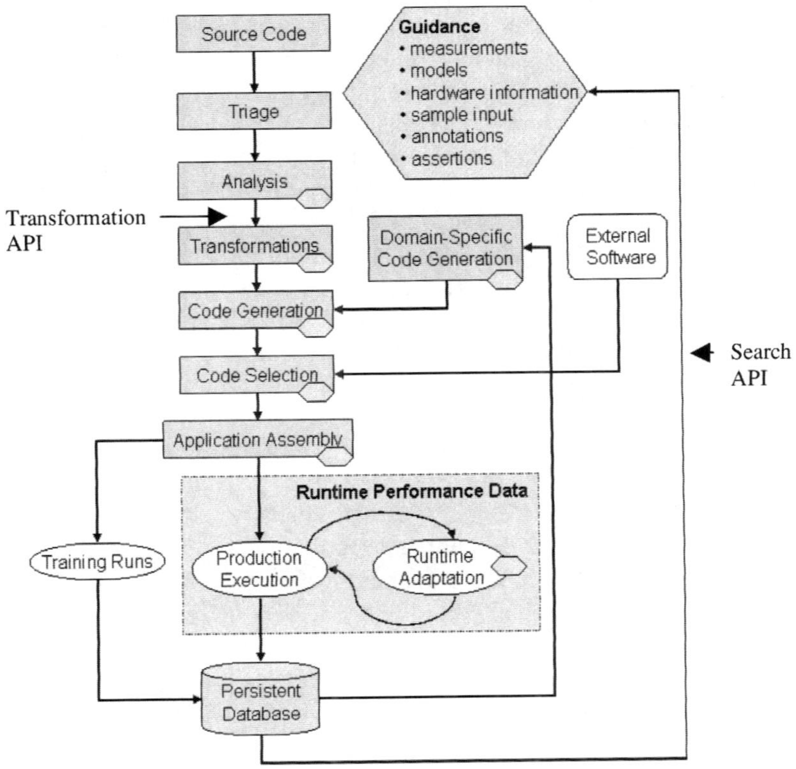

Figure 1. Flow diagram of the auto-tuning process.

8. *Online adaptation.* Optimizations may occur during production runs, especially for problems or machines whose optimal configuration changes during execution.

3. Evolving an auto-tuning system through common interfaces

The challenges in automating the performance tuning process require that the following three issues be addressed. First, the number of code variants for a complete application can be enormous. Strategies to avoid code explosion and to judiciously select what transformation techniques to apply to different sections of the application code are needed to keep the tuning time at manageable levels. Second, as the number of tuning parameters increases, the search space becomes high dimensional and exponential in size. Search algorithms that can cope with exponential spaces and deliver results within a few search iterations are needed. Finally, metrics that measure and isolate the performance of a section of code being optimized within an application is needed to accurately guide search algorithms.

Within PERI, five different research groups are developing auto-tuning *tools* to address these issues. These projects have complementary strengths and can, therefore, be brought together to develop an integrated auto-tuning *system*. Toward that end, we are working to develop a common framework to allow auto-tuning tools to share information and search strategies. Through common APIs, we can evolve an auto-tuning system that brings together the best capabilities of each of these tools, and also engage the broader community of tool developers beyond PERI researchers.

We have focused development of the interfaces on two portions of the auto-tuning process. Any compiler-based approach will apply code transformations to rewrite application code from its original form to one that more effectively exploits architectural features such as registers, caches, SIMD compute engines, and multiple cores. Commonly used code transformations include loop unrolling, blocking for cache, and software pipelining. Thus, we are designing a *transformation API* that will be input to the Transformation box in figure 1. This API provides a *transformation recipe* that describes how to transform original source into an optimized source representation. By accepting a common transformation recipe, the auto-tuning system permits code transformation strategies derived by PERI compilers and tools (or users) to be implemented using any transformation and code generation tools, such as Chill (USC/ISI), LoopProcessor (LLNL) and POET (UTSA). The API supports the specification of unbound transformation parameters that are then tuned using search algorithms. The initial draft of the API includes a naming convention for specifying language constructs in source code and code transformations available in Chill and POET.

A *search API* provides input into the empirical optimization process of running a series of experiments on actual hardware to determine the best optimized implementation. The search API allows the auto-tuning tools to exchange information about their available tuning options and constraints on the search space and to plug in different search algorithms. The common framework will support both auto-tuning using training runs (and recompilation) along with continuous optimization during production runs. For the search API, we are working on developing a simple and extensible language that standardizes the parameter space representation. Using the language, developers and researchers can expose tunable parameters to tuning frameworks. Relationships (ordering, dependencies, constraints, and ranking) between tunable parameters can also be expressed.

To understand the requirements of integrating auto-tuning tools developed independently by PERI researchers, we have been engaging in the integration of several PERI tools. Researchers at the University of Tennessee have integrated auto-tuning search (described below) with the ROSE LoopProcessor (LLNL) and the POET code generator (UTSA). Similarly, an initial integration of the Active Harmony system (UMD) and the Chill transformation framework (USC/ISI) is providing experience in how to effectively integrate these separate tools into an auto-tuning system.

4. Auto-tuning infrastructure and experimental results

We now describe recent performance results from applying auto-tuning to dense and sparse linear algebra kernels. The dense linear algebra experiments use our prototype compiler and code-generation tools, while the more difficult-to-analyze sparse linear algebra experiments use a library-based auto-tuning approach.

4.1 *Code generation and empirical search*

In conjunction with the previously described APIs, a goal in PERI is to easily substitute different code generators and search engines in the auto-tuning process. To that end, researchers at the University of Tennessee have developed a simple syntax for specifying the way to generate code to perform the evaluation of the code variants. In order to permit switching search techniques, the application-specific aspects of the evaluation process are described by a simple specification of the search bounds and constraints. This system allows evaluation of various search techniques to determine which work best in an auto-tuning context. We have evaluated some classic techniques such as genetic algorithms, simulated annealing, and particle swarm optimization, as well as some more ad hoc techniques such as a modified orthogonal search.

As a simple example of an optimization search space, figure 2 shows optimization of square matrix-matrix multiplication (N=400), using an exhaustive search over two dimensions: block sizes up to 128 and unrolling up to 128. The x and y axes represent block size and unrolling amount, respectively, while the z axis represents the performance in Mflop/s of the generated code. In general, we see the best results along the blocking axis with a low unrolling amount as well as along the diagonal where blocking and unrolling are equal, but there are also peaks along areas where the block size is evenly divisible by the unrolling amount. The best performance was found with block size 80 and unrolling amount 2. This code variant ran at 1459 Mflop/s compared to 778 Mflop/s for the naive version compiled with gcc.

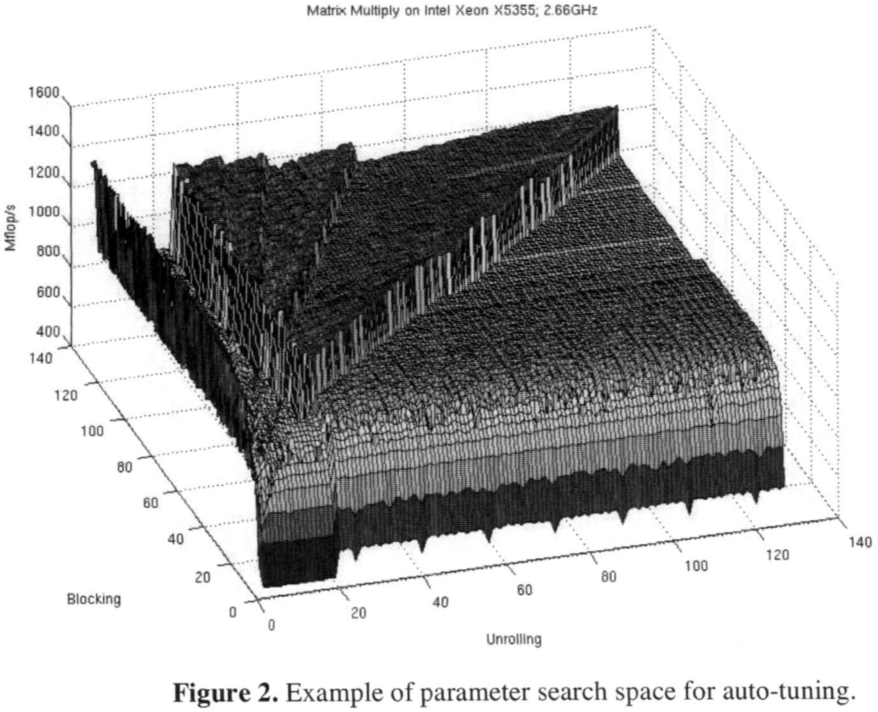

Figure 2. Example of parameter search space for auto-tuning.

4.2 Compiler-based infrastructure

Researchers at USC/ISI are developing the TUNE auto-tuning compiler and its constituent Chill transformation framework, which provide a compiler-based infrastructure for general application code. Chill takes source code and a transformation recipe with bound parameter values as input. Using a polyhedral framework, Chill extracts a mathematical representation of the iteration space and array accesses and composes the transformations in the recipe through rewriting rules, to generate optimized code. A parameter sweep interface takes a transformation recipe with unbound parameter values and ranges and constraints on parameter values to derive a series of alternative implementations.

TUNE has been used to achieve hand-tuned levels of performance on dense linear algebra kernels such as matrix-matrix multiply, matrix-vector multiply, and LU factorization for older architectures such as the SGI R10K, Pentium M, and Pentium D. Recent work to optimize matrix-matrix multiply for newer architectures is promising, but show gaps between compiler-optimized and hand-tuned performance. Performance results for the Jacquard system at LBNL are 37% below hand-tuned performance, and we attribute the gaps to code generation for SSE-3, instruction scheduling, and control of prefetch. Collaborating researchers at Argonne have been experimenting with manually scheduled inner-loop bodies, have achieved performance within 83% of peak, and are developing automatic approaches to close this gap. On the Intel Core2Duo, we have achieved performance that is 28% below near-peak hand-tuned results. The primary performance gap appears to be control of software prefetch at the source code level.

4.3 Auto-tuning sparse-matrix vector multiply for multi-core architectures

In a study of auto-tuning a sparse matrix-vector multiply (SpMV) kernel, PERI researchers at Lawrence Berkeley National Laboratory compared performance on a dual-socket Intel Core2 Xeon, single- and dual-core AMD processors, a quad-core Opteron, a Sun Niagara2, and an IBM Cell Broadband Engine. This study found that these processors differed markedly in the effectiveness of various autotuning schemes on the SpMV kernel [6]. In two subsequent studies [7,8], these researchers analyzed similar but more sophisticated automatic optimizations for SpMV, the explicit heat equation PDE on a regular grid (Stencil), and a lattice Boltzmann application (LBMHD). This auto-tuning approach employs a code generator that produces multiple versions of the computational kernels using a set of optimizations with varying parameter settings. The optimizations include TLB and cache blocking, loop unrolling, code reordering, software prefetching, streaming stores, and use of SIMD instructions.

The impact of each optimization varied significantly across architecture and kernel, necessitating a machine-dependent approach to auto-tuning in this study. In addition, detailed analysis revealed performance bottlenecks for each computation on the various systems mentioned above. The overall performance results showed that the Cell processor offers the highest raw performance and power efficiency for these computations, despite having peak double-precision performance and memory bandwidth that is lower than many other platforms in our study. The key architectural advantage of Cell is explicit software control of data movement between the local store (cache) and main memory, which is a major departure from conventional programming. These studies clarify that there is still considerable room for improvement in automatic tuning methods for all the candidate architectures.

Acknowledgment

This work was supported in part by the U.S. Department of Energy, under Contract DE-AC02-06CH11357.

References

[1] Whaley C, Petitet A and Dongarra J 2001 Automated empirical optimizations of software and the ATLAS project *Parallel Computing* **27** (1): 3-25

[2] Frigo M and Johnson S 1998 FFTW: an adaptive software architecture for the FFT *Proc. International Conference on Acoustics, Speech, and Signal Processing* (Seattle, Washington)

[3] Bilmes J, Asanovic K, Chin C W and Demmel J 1997 Optimizing matrix multiply using PHi-PAC: a Portable, High-Performance, ANSI C Coding methodology *Proc. International Conference on Supercomputing* (Vienna, Austria: ACM SIGARCH)

[4] Vuduc R, Demmel J and Yelick K 2005 OSKI: a library of automatically tuned sparse matrix kernels *Proc. SciDAC 2005, Journal of Physics: Conference Series*

[5] Chen C, Chame J and Hall M 2005 Combining models and guided empirical search to optimize for multiple levels of the memory hierarchy *Proc. Conference on Code Generation and Optimization*

[6] Williams S, Carter J, Oliker L, Shalf J and Yelick K 2008 Lattice Boltzmann simulation optimization on leading multicore platforms *Proc. International Parallel & Distributed Processing Symposium (IPDPS)* (Best paper, applications track)

[7] Williams S, Oliker L, Vuduc R, Shalf J, Yelick K and Demmel J 2007 Optimization of sparse matrix-vector multiplication on emerging multicore platforms *Proc. SC07* (ACM/IEEE)

[8] Williams S, Datta K, Carter J, Oliker L, Shalf J, Yelick K and Bailey D 2008 PERI – auto-tuning memory intensive kernels for multicore (available at http://crd.lbl.gov/~dhbailey/dhbpapers/scidac08_peri.pdf)

Journal of Physics: Conference Series **125** (2008) 012090 doi:10.1088/1742-6596/125/1/012090

A new DoD initiative: the Computational Research and Engineering Acquisition Tools and Environments (CREATE) program

D E Post for the CREATE Team: S. Arevalo, C. Atwood, P. Bell, T. D. Blacker, S. Dey, D. Fisher, D. A. Fisher, P. Genalis, J. Gorski, A. Harris, K. Hill, M. Hurwitz, R. P. Kendall, R. L. Meakin, S. Morton, E. T. Moyer, D. E. Post, R. Strawn, D. van Veldhuizen, L. G. Votta, S. Wynn, and G. Zelinski.

DoD High Performance Computing Modernization Program, Lorton, VA 22079, USA

E-mail: post@hpcmo.hpc.mil

Abstract. In FY2008, the U.S. Department of Defense (DoD) initiated the Computational Research and Engineering Acquisition Tools and Environments (CREATE) program, a $360M program with a two-year planning phase and a ten-year execution phase. CREATE will develop and deploy three computational engineering tool sets for DoD acquisition programs to use to design aircraft, ships and radio-frequency antennas. The planning and execution of CREATE are based on the "lessons learned" from case studies of large-scale computational science and engineering projects. The case studies stress the importance of a stable, close-knit development team; a focus on customer needs and requirements; verification and validation; flexible and agile planning, management, and development processes; risk management; realistic schedules and resource levels; balanced short- and long-term goals and deliverables; and stable, long-term support by the program sponsor. Since it began in FY2008, the CREATE program has built a team and project structure, developed requirements and begun validating them, identified candidate products, established initial connections with the acquisition programs, begun detailed project planning and development, and generated the initial collaboration infrastructure necessary for success by its multi-institutional, multidisciplinary teams.

1. A new paradigm for solving many of the major problems facing society

Since the invention of the electronic digital computer in the early 1940s, computing power has been increasing exponentially by a factor of ~ 1.7 each year from ~ 1 floating-point operations per second (Flops) to ~ 10^{15} Flops today (figure 1). This has led to enormous gains in our ability to develop, assess, and use information. Computers have revolutionized our ability to conduct national and international business and trade and to coordinate the diverse activities of society, both private and public. This revolution is also transforming the fundamental processes of science and engineering.

While computational science and engineering (CSE) has been making significant contributions to society, the future holds the promise of a major paradigm shift in problem solving from traditional methods toward CSE. Within 10 years, supercomputers will be able to perform 10^{18} calculations per second. With this computing power, we will be able to employ sufficient resolution to accurately model complex systems, deploy highly accurate computational mathematical algorithms, include all of the effects we know to be important in a system, model a complete system (e.g., an entire airplane or ship) in multiple dimensions, and be able to run enough problems to carry out useful parameter surveys.

While high-speed, powerful computers are the enabling, transformational technology, more than computers are needed (table 1). Sponsors use the information for decisions and designs and provide the resources and oversight. Code users apply software to solve problems. Code developers write and support the use of the software. Verification and validation ensure the accuracy of the calculations. Computers are needed to perform the calculations. Successful CSE requires all five elements.

Table 1. Essential elements of Computational Science and Engineering

Sponsors
Code users and designers
Codes (development and user support)
Verification and Validation
Computers

2. Challenges facing computational science and engineering

The establishment of a new technology usually takes several generations and many mistakes. The design methodology for steel and concrete suspension bridges took almost 200 years to mature While computational science and engineering has been making significant contributions to society, the future holds the promise of a major paradigm shift in problem solving from traditional methods toward CSE. Within 10 years, supercomputers will be able to perform 10^{18} calculations per second. With this computing power, we will be able to employ sufficient resolution to accurately model complex systems, deploy highly accurate computational mathematical algorithms, include all of the effects we know to be important in a system, model a complete system (e.g., an entire airplane or ship) in multiple dimensions, and be able to run enough problems to carry out useful parameter surveys [1].

Case studies of bridge-design errors led to the identification of "lessons learned." Their incorporation into the design of future bridges was essential for this technology to mature. CSE will be similar. Unfortunately, there have been only a few case studies of CSE projects [2, 3]. However, these have provided an important number of "lessons learned," and it is crucial to incorporate as many of them into future CSE projects as possible. Petroski points out that one learns the most from unsuccessful projects [1]. While there have been many failures in CSE software development [4], there have been almost no case studies of unsuccessful CSE projects, so special focus is needed on what little data exist for such projects.

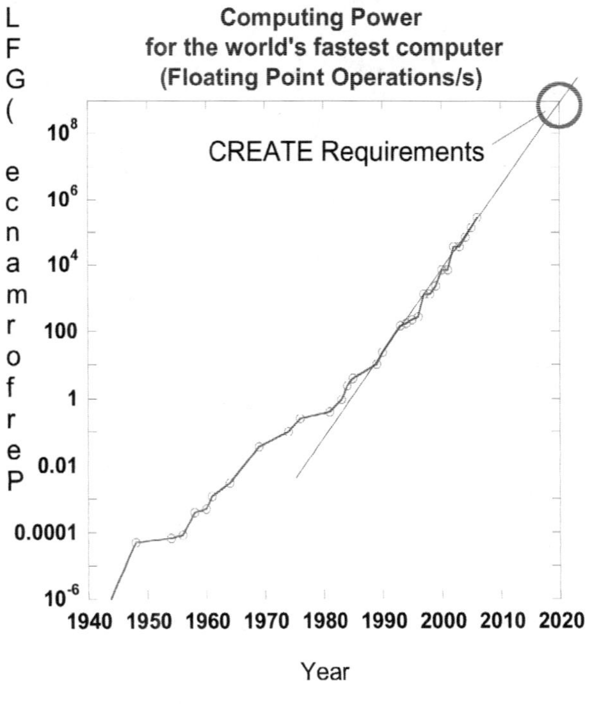

Figure 1. Growth in computer power from 1945 projected through 2020. From 1990 through 2020, the number of processors (cores) per computer (~ order of magnitude) is also plotted.

Based on our case studies, we present many "lessons learned" in this paper [4, 5]. These lessons may seem obvious – and they are. However, our case studies indicate that implementing them in real projects is not common. Few projects follow most of the lessons, and many follow only a few. Our case studies trace the failure of many projects to achieve their goals within schedule and budget, or management decisions to abandon the projects to violations of the practices and procedures captured in the "lessons learned." This has resulted in a tragic waste of the creative energy and efforts of many talented people and financial resources, and the loss of the potential benefits of those projects. Based on the case studies (e.g., [5]), we group the challenges for CSE into three categories:

1. Scientific and Engineering - How to develop software that captures the scientific and engineering aspects that must be calculated and interpreted accurately
2. Project Management - Organizing and executing the development of the complex CSE software that meets the needs of the users
3. Programming and Computing - Developing and supporting software that can execute on tomorrow's complex and rapidly evolving computer hardware

2.1 Scientific and engineering challenges

The scientific and engineering challenges of developing CSE codes reflect the complexity of the scientific or engineering subject matter. The codes must be able to accurately calculate the trade-off of many distinct physical phenomena over many orders of magnitude of time and distance scales. Since CSE codes represent models of nature, we must verify that the code has no important defects (verification) and establish that the model correctly includes all of the important effects (validation) [6, 7]. Incorrect and inaccurate calculations lead to wrong decisions and bad designs. Verification includes intensive software testing. Validation involves comparison of the code's predictions with experimental data. Existing verification and validation (V&V) methods are inadequate for the present generation of CSE codes and research is needed to develop better V&V methods. Accurate numerical solution algorithms that can efficiently use many processors (up to 10^9) will be necessary, but few such algorithms exist. Calculations of complex, multimaterial systems require multidimensional numerical representations of the objects. Methods for rapidly generating the required accurate geometries and numerical grids and meshes generally do not exist and need to be developed. Generating an accurate mesh for a large ship can take a team of analysts a year, far too long to benefit design decisions. Improved techniques for rapid analysis and visualization of the huge datasets generated by large-scale codes are required to ensure that computational results will provide useful information for making decisions or developing designs.

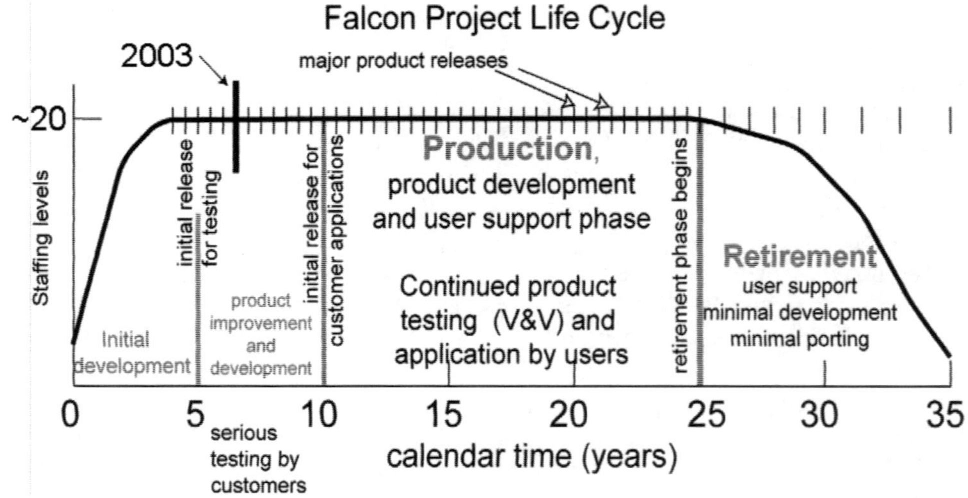

Figure 2. Life cycle of the Falcon code project [2].

2.2 Project management challenges

The development and support of complex scientific and engineering software are a complicated process. The software must meet the needs of the customers, namely, the designer or user community. It must provide accurate solutions for the problems that the users need to solve, and it must be convenient to use. The users' requirements are often difficult to determine and constantly change over the life cycle of the code. These issues are especially challenging if the users or customers are not the code developers as is the case for codes for design engineers. Many "scientific" codes are developed and run by the same groups who then publish the results of their investigations. These groups understand the limitations and range of validity of their codes. In contrast, many distinct groups are involved in the steps in Figure 3. Following development and V&V, the code will be used to develop a design for a system that will be built, tested and operated by distinct groups. Few of these disparate groups have a detailed knowledge of the strengths and weaknesses of the code. The quality of the code must therefore be very high since the quality of the final product depends on the quality of the code and the skill of the design engineer.

Development and initial deployment of large-scale CSE codes generally take large, multidisciplinary teams (~20–30 professional staff) on the order of 10 years (figure 2) or more [5]. The required funding is around $100M over a ten-year period. The development of large-scale code requires more organization and structure than the development of smaller codes. The team leaders must possess many skills, including subject matter, computational, managerial, financial, leadership, and team-building skills. Multi-institutional teams require close collaboration among the team members from the participating institutions. Each institution must provide adequate support for their portion of a distributed project. The development of large-scale codes is a complex, iterative process (figure 3). Large-scale code development is risky. As many as one-half or more large-scale CSE code projects fail to meet their initial goals in terms of performance, cost and schedule. Up to one-third (or more) are eventually abandoned after significant investment [5, 7].

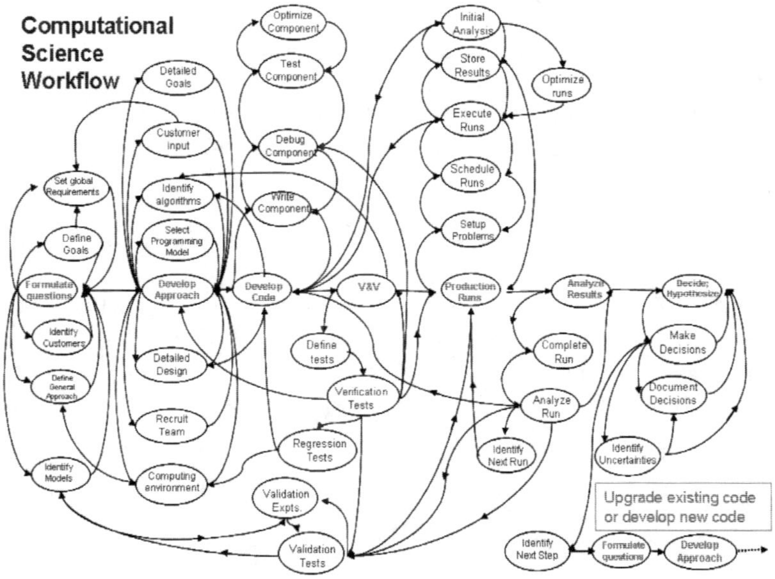

Figure 3. CSE code development and application workflow [8].

2.2 Programming challenges

Computers are becoming more complex as a result of massive parallelism and the use of heterogeneous processors. Computers in 2020 will likely have as many as 10^9 processors and execution speeds of 10^{18} Flops. Exploiting even a small portion of this power will require overcoming the architectural complexity of 10^9 or more processors connected through complicated networks. Codes will need to be able to achieve reasonably efficient performance (~ 5% or more of peak speed). It will be necessary to port codes to each generation of computers every three to four years, and different computer architectures every 10 years or so, as well as port codes to a variety of existing platforms. Code development tools for tomorrow's computers are inadequate and immature, when they exist at all.

3. Lessons learned

Case studies of many large-scale CSE code projects have identified the requirements for successful development and deployment of such codes (table 2).

Table 2. Essential elements of successful CSE code development projects [5]

1. Sponsor support for the life of the project
2. Requirements based on the evolving needs of the customers over the life of the project, i.e. strong customer focus
3. Verification and validation
4. Software project management designed for large-scale CSE codes
    - Realistic goals and schedules with adequate resources
    - Balance of long-term and short-term goals
    - Sufficient project orientation to organize and coordinate the work, but enough flexibility to solve unanticipated problems and meet new requirements
    - Multidisciplinary team with strong technical skills
    - Good team dynamics: trust, respect, cooperation and commitment
    - Collaboration methods and tools for multi-institutional code development teams
5. Software engineering designed for large-scale CSE codes
    - Risk identification and mitigation
    - Effective software engineering methods and tools that balance the need for structured development with the required degree of flexibility and agility
    - A flexible code architecture and infrastructure

The evidence indicates that it takes on the order of $100M for 10 years to build a large-scale CSE code [5]. To get this level of support, one has to first find someone who can and will pay $100M to solve their problem and then make a convincing case that you can develop a CSE tool that can solve their problem (table 3). It is not sufficient to have a good proposal to solve a problem one thinks is important. The idea must solve a problem that someone with $100M thinks is important. This is not easy. Not many groups have resources at that level, nor are they willing to wait up to 10 years for a code. In addition, CSE is not yet a widely accepted problem solving paradigm.

Table 3. Four elements for marketing a code development program

1. Identify a problem that needs to be solved, that CSE can solve, and that a sponsor will pay to have solved.
2. Establish that solving the problem can provide enough benefit to the sponsor that he will pay for the solution.
3. Establish that your group has credibility for solving the problem.
4. Present a compelling vision for solving the problem.

4. An Example: A new Department of Defense Initiative, The Computational Research and Engineering Acquisition Tools and Environments (CREATE) program

In 2008, the Department of Defense (DoD) initiated the CREATE program, a new $360M program to develop and deploy three computational engineering tool sets for the design of airplanes, ships and radio-frequency (RF) antennas. CREATE is a multiservice DoD program with support from the Office of the Secretary of Defense (~70%) and 30% matching "in-kind" resources from the armed services (Navy, Air Force and Army). CREATE is a 12-year program with a 2-year planning phase and a 10-year development phase. The program was conceived and planned based on the "lessons learned" described in the prior sections, andthe concept for CREATE was developed in 2005-2006. The proposal was developed and submitted during 2006-2007 as part of the DoD budget process (Program Objective Memorandum, POM) for the 2008 federal fiscal year. CREATE was approved in FY07 and started in FY08.

The CREATE proposal was built around four points: problem, benefit, credibility, and vision (table 3). We began by identifying DoD acquisition programs as organizations with severe problems that computational tools could help solve. Many major DoD acquisition programs are years behind schedule, billions of dollars over budget, and fail to meet performance goals [8]. The acquisition process is rigid and slow, and it is not able to respond rapidly to changing threats [9]. The benefit of an improved, flexible, and agile acquisition process would be timely delivery of weapons systems that met performance requirements at the projected cost and the ability of acquisition programs to respond to rapidly changing requirements. The U.S. would save billions of dollars, weapons systems would be available for operations at the scheduled time, and the systems would work as specified.

Part of acquisition program shortcomings are due to the steady decline in the size of the DoD engineering workforce (a factor of 5 to 10 over the past 35 years) [9]. Another challenge is the increased need to accelerate the introduction of new design concepts and technologies into new weapons systems. Historical experience is not a good approach for new designs based on concepts and technologies outside of the existing design experience. The present acquisition and design paradigm is based on an iterative, empirically based "design, build, test, fail, redesign, etc." approach. The CREATE proposal was based on the development and deployment of physics-based computational engineering tools to replace design methods based on historical experience. The CREATE tools would thus enhance the productivity of the acquisition workforce by enabling acquisition engineering to produce better designs more quickly. Using evidence and lessons-learned from our case studies, the credentials of our team and successful examples of DoD CSE applications, we established our credibility to build the proposed computational engineering tools within our estimated schedule and budget. We illustrated how use of CSE tools had helped other government agencies and industry solve problems similar to those facing DoD acquisition programs and achieve the desired benefits.

The physics-based CREATE tools will be able to produce optimized designs early in the design process. Design engineers could use the tools to identify design defects and correct them early in the design process before major schedule and budget commitments were made. Testing would be reduced by getting the design right the first time. Tests would be used to validate a mature design rather than identify the problems in an immature design. Experimental prototypes can be replaced with computational prototypes that can be constructed much more quickly and cheaply. Industry is beginning to employ this approach. One business used CSE to reduce their time to market from three years to less than 9 months, and credits their company's survival to the adoption of CSE.

4.1 Key elements of the CREATE program

Once the CREATE proposal was funded, we then began to plan the program in detail. We had embedded the lessons learned listed in table 2 into the initial program proposal. We were able to build on them so that our plans were consistent with what we had promised in the proposal. In the proposal process, we first ensured that our DoD sponsors understood that CREATE was a long-term program and that it would take $10M/year for at least 10 years to fully develop each suite of codes. Second, we had established connections to the prospective CREATE customers and users, the DoD acquisition programs. The proposal included endorsements of CREATE and promises of resource commitments of about 30% of the total CREATE budget from the acquisition programs. Third, we emphasized the importance of the elements in table 2 to our DoD sponsors so that they would understand and support CREATE during the early phases of the program prior to delivery of our major products. It was especially important for our DoD sponsors to understand our challenges and provide the institutional support as well as financial support to meet those challenges. These include resolving infrastructure issues and providing access to the personnel required to establish and execute a successful multi-institutional program.

We began to carefully build a strong leadership team for the CREATE program and the three major projects—aircraft, ships, and RF antennas. We solicited and received nominations for the leadership position from the offices of assistant secretaries of the Navy, Army, and Air Force as part of an open, advertised national search process. The position descriptions and selection criteria emphasized the need for a balance of subject matter expertise in ship, aircraft, or RF antenna design issues with expertise and experience in project management and leadership experience in large-scale computational science projects. Once the leadership team was in place, the team began to build the individual project teams and to develop plans emphasizing principles based on the "lessons learned" in table 2.

4.2 Requirements development and validation

To ensure sponsor support and customer focus, we examined the acquisition programs to identify gaps in their capabilities to design and deliver weapons systems involving ships, airplanes, and RF antennas. We then identified the capability gaps that would benefit from the application of computational engineering tools. Then we developed concepts and ultimately requirements for candidate tools to fill those gaps. Finally, based on the potential impact on high-priority programs, the maturity of the computational technologies and physics-based algorithms, and the budgeted resources and schedule, we selected a set of products to develop from the list of the candidate tools. We documented the gaps and requirements in an "Initial Capability Document" (ICD) patterned after the DoD Acquisition Process (Joint Capabilities Integration Development System, JCIDS) [10]. Mapping our process into the general DoD process helped us meet the expectations of the DoD community. We are using "Boards of Directors" composed of senior staff (generally at the flag rank, Senior Executive Service level or equivalent) for the targeted acquisition programs to validate the gaps and requirements described in the ICD. This approach ensures that the tools we plan to develop will meet the needs of the acquisition programs. The board members will vet our ICDs through their organizations before endorsing them, thus ensuring that we have correctly identified the gaps and requirements. Since the gaps and requirements evolve and change, we intend to revalidate them at least once a year to keep them current.

Studies of Systems of Systems in the DoD, of which large, complex codes are an example, show that a leading cause of failure of large-scale systems is that many do not update their requirements rapidly enough to fulfill the evolving needs of their customers [11]. The resultant system is thus not useful to the customers when it is finally delivered. Revalidating our requirements annually will help keep the projects on track. CREATE has formed small teams who are charged with "shadowing" the acquisition programs. These "Shadow-Operations" (Shadow-Ops) teams will identify opportunities for computational engineering tools to address problems in acquisition programs. The Shadow-Ops teams will then apply computational tools to solve acquisition engineering problems and share the results with the programs. This approach enables the Shadow-Ops teams to track the evolving requirements of the acquisition programs, helps establish the value of computational engineering analyses to the acquisition programs and helps facilitate the transition of computational engineering tools to the programs. The initial results of this approach indicate that it will be highly successful.

Verification and validation (V&V) are essential. The CREATE program is establishing strong connections to the DoD Test and Evaluation (T&E) community to ensure access to validation data. This community has been a strong supporter of the CREATE program, because CREATE tools will help the T&E community fulfill its testing role with fewer tests and greater efficiency and effectiveness. The goal is to establish a design paradigm in which designs will be based not on historical experience requiring a long series of exploratory tests, but on one in which designs will be developed using physics-based computational tools. Physics-based tools are much more effective for new concepts and technologies than reliance on historical knowledge. Such designs can be validated with a much smaller set of targeted tests. This has been a highly successful strategy for several businesses. A strategy that emphasizes the specification of tests as part of the development process is also being developed.

4.3 Code development strategy

As part of our planning process, we assessed the requirements of the acquisition programs and prioritized them to help identify the most useful tools that could be developed for ~$10M/year for 10 years. To minimize risks, we emphasized building on existing expertise and knowledge, and focused on codes that integrated four to five different physical effects. We started developing detailed project plans based on our initial gaps and requirements analysis. Those plans will evolve as the requirements change.

We have developed a strategy to achieve a good balance between long- and short-term deliverables. Without establishing value in the short-term, we will not be able to defend our long-term budget. However, if we do not keep a long-term focus, we probably will not be able to develop codes that can exploit 10^9 cores and 10^{18} Flops to achieve the goal of a paradigm shift from reliance on historical data and experience for design to the use of physics-based tools for new concepts. Our strategy is to begin by increasing the capability, parallel scalability, robustness and usability of existing computational engineering tools, and support their transition to the acquisition programs (figure 4). We will then use the "lessons learned" from this experience to design and build new tools designed to exploit the next generation of supercomputers in 2020 and transition them to the users when they are ready. The Shadow-Ops program described above will us help realize this strategy. It will also help keep the code developers closely tied to the user community and the user community's current requirements.

We are a developing a flexible and agile project management process that is designed to ensure the achievement of project goals and tasks while retaining the flexibility and agility to respond to events that occur during the development process (e.g., figure 3).

We have emphasized the formation of a good team with the range of experience and skills necessary to carry out the code development projects and transition the products to the customers. The DoD High Performance Computing Modernization Program (HPCMP) Institutes and Portfolios have helped to create a community of trained and skilled computational scientists who have formed the core of much of the CREATE leadership and code development teams. We have also emphasized subject matter expertise, and technical leadership and team building skills, rather than strong project management skills and background.

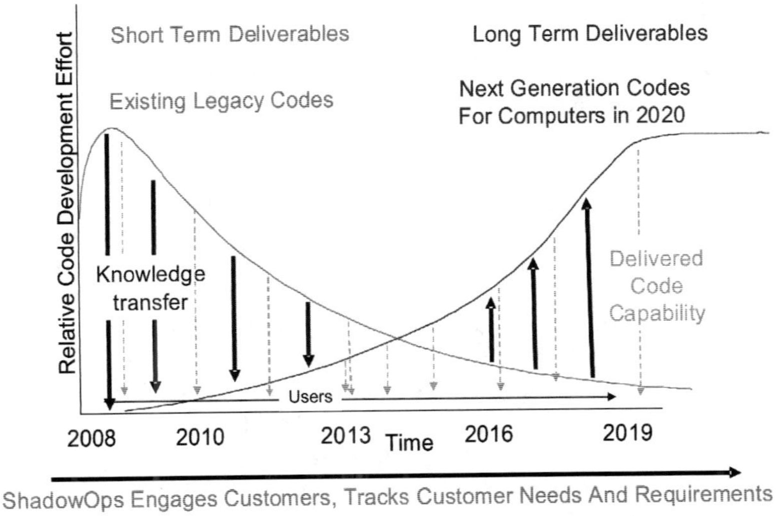

Figure 4. CREATE code development strategy.

Code development requires strong teams that have good team dynamics. Software is developed by teams. We have found that the level of sponsor support and the quality and cohesiveness of the team are three of the most reliable predictors of project success [12].

No single DoD site has sufficient staff to execute a CREATE project. We have therefore emphasized the development of effective methods for multi-institutional teams to work together effectively and efficiently. We have established a central server repository for code development files and documents. We are procuring and supporting collaboration tools, including several different types of high-quality video conferencing desktop tools that are essential for good communication and building trust and a sense of shared commitment and community within a highly distributed project. Computer security issues that inhibit communication among DoD facilities have made this challenging.

Large-scale code development has large risks [7]. Historically, half or more of such projects fail to meet their initial goals within the estimated budget and schedule. We have assessed the general risks for CSE [13]. The most critical risk is the potential loss of sponsor support [12]. For CREATE, we have therefore focused on providing short-term and long-term value to our customers and ensuring that our customers keep our sponsors informed of this value. Another major risk is the stability and competence of the development team. We concentrated on building a solid, competent leadership team because it is the first step in building a solid project team. Risk assessment and the development of mitigation plans will be a priority for FY09.

Developers of engineering and scientific codes have found that imposition of rigid processes, for example the Capability Maturity Model (CMM, CMMI) [14, 15], is not useful [5] Instead, we have concentrated on identifying the software engineering practices and tools that have proven useful for large-scale engineering projects [16, 17]. We are providing training in these practices and are establishing a central set of servers to provide those tools to the teams. This approach allows us to capture the best aspects of the process approaches like CMM but retain the flexibility and agility we need.

The CREATE computational engineering tool suite must be usable, modular, extensible and maintainable. We are emphasizing ease of use through extensive use of Graphical User Interfaces (GUI) for the Application Program Interface. Easy and convenient generation of geometries and meshes is a priority. Extensibility and maintainability are necessary to provide the new functionality that the users will require as their needs change and computer platforms evolve. Modularity and interoperability are essential to achieve these aspects. The products developed by each sub-team must be usable by the other subteams. For instance, many of the RF antenna modules must be usable by the ship and aircraft design tools. We have been developing a strategy based on a lightweight infrastructure that provides wrappers for each major set of modules (computational fluid mechanics, structural mechanics, control algorithm, etc.) to achieve a high degree of modularity. Each module has well-defined interfaces so that it can be used for multiple purposes. Lightweight means that the interfaces and wrappers are designed to handle our specific classes of problems, rather than broad classes of problems. Our "lightweight" infrastructures are "shallow" in the sense that the interfaces do not intrude too deeply into the code modules. The emphasis is on code-interoperability rather than tight integration. If a new need arises, the relevant wrappers are extended to accommodate the new interface and connectivity requirements. If necessary, enhanced functionality can be added to the module core. We have found that heavyweight frameworks that emphasize generic "plug and play" aspects are not as useful. Such frameworks attempt to address many different types of problems. This forces them to incorporate choices and assumptions that often result in too much rigidity to find technical solutions for different problems. Rigid frameworks have proven to be a major cause of failure of some large-scale CSE projects [4]. The lightweight infrastructure approach also allows us to deliver products quickly while retaining a long-term focus. Wrappers and GUIs to link existing major code modules (e.g., computational fluid dynamics or structural mechanics modules) can be built and delivered quickly to the user community within a year or two. We will build on this experience to design and develop new major modules and solvers (e.g., fluid dynamics, structural mechanics, etc.) that can take advantage of the next generation of computer with 10^5 to 10^9 processors, but will take 5 to 10 years to design and build.

We have been able to exploit these lightweight infrastructures to simultaneously provide short-term deliverables with immediate value while keeping a focus on meeting our long-term objective to provide codes that can exploit the next generation of computers (figure 4). The political reality is that if we cannot provide near-term value to the DoD, our long-term funding will be at risk in the highly competitive resource environment in the DoD. We are beginning to link several existing legacy codes with a user-friendly, light-weight framework to provide the ability to study integrated, multi-physics effects such as the interaction of aerodynamic air flow and the structural mechanics of an airframe. This is being accomplished in one to two years, much shorter than the five to ten years that experience indicates is required to develop a new computational fluid dynamics or structural mechanics module. This allows us to provide design engineers the capability to study the design implications of these interactions early in our project, and us the opportunity to learn about the strengths and weaknesses of existing codes and the real needs of the design community. We can then use this knowledge to design and develop new codes that can exploit the power of the next generation of computers.

5. Summary

We built the CREATE program on the "lessons learned" from the past 50 years of experience with large-scale CSE software. While some of these lessons are undocumented, many have been described in papers and reports (e.g., [18]). This experience has been essential for us to get the necessary resources and to plan and execute the CREATE program. CSE is a new, emerging paradigm. As noted earlier, many of these lessons seem obvious but we have found that very few projects incorporate them, resulting in the waste of people's time, the country's resources, and loss of the potential benefits of the project. It is crucial that everyone launching CSE projects examine and learn from the past experiences of others who have tried and succeeded, and especially those who have not succeeded. "Those who cannot learn from history are doomed to repeat it" [19]. Every new technological paradigm builds on the mistakes of prior efforts. We urge our CSE colleagues to share and document their experiences, both the good and especially the bad, so that our field can move forward, and society can have the full benefit of CSE tools.

Acknowledgments

The CREATE team is grateful for the support and advice from our colleagues including John Blair, Luise Couchman, Larry Davis, John Grosh, Cray Henry, Jeremy Kepner, Ed Kraft, Scott Littlefield, Robert Lucas, Andrew Mark, Paul Nielsen, Patrick Purtell, Cliff Rhoades, Al Shaffer, André van Tilborg, and John Young.

References

[1] Petroski H 1994 *Design Paradigms: Case Histories of Error and Judgment in Engineering* (New York: Cambridge University Press)
[2] Post D E, Kendall R and Whitney E M 2005 Case study of the Falcon code project In Johnson P (Ed.) *2nd Workshop on HPC applications, ACM/IEEE International Conference on Software Engineering* (St. Louis, MO: Association of Computing Machinery)
[3] Hatton L 1997 The T experiments: errors in scientific software *IEEE Computational Science and Engineering* **4:**27-38
[4] Post D E and Kendall R P 2004 Lessons learned from ASCI *International J. High Performance Computing Applications* **18:**399-416
[5] Post D E, KendallL R P and Lucas R F 2006 The opportunities, challenges, and risks of high performance computing in computational science and engineering In Zelkowitz M V (Ed.) *Advances in Computers* (New York: Elsevier)
[6] Oberkampf W and Trucano T 2002 Verification and validation in computational fluid mechanics *Progress in Aerospace Studies* **38**:209-272
[7] Post D E and Votta L G 2005 Computational science demands a new paradigm *Physics Today* **58:**35-41
[8] GAO 2007 General Accounting Office: Defense Acquisitions, Assessments of Selected Major Weapon Programs **167,** GAO-07-406SP.
[9] Kaminski P G, *et al.* 2008 *Pre-Milestone A and Early-Phase Systems Engineering* (National Academy of Sciences)
[10] Garrett G A and Rendon R G 2006 *U.S. Military Program Management, Lessons Learned and Best Practices* (West Sussex: Kogan Page Publishers) CJCSI 3170.01C
[11] Fisher D A 2006 An emergent perspective on interoperation in systems of systems **66** CMU/SEI-2006-TR-003
[12] Demarco T and Lister T 2003 *Waltzing with Bears, Managing Risk on Software Projects* (New York: Dorset House Publishing)
[13] Kendall R P, *et al.* 2007 A proposed taxonomy for software development risks for high-performance computing (HPC) scientific/engineering applications **39** CMU/SEI-2006-TN-039
[14] Chrissis M B, Konrad M abd Shrum S 2003 *CMMI: Guidelines for Process Integration and Product Improvement* (Addison-Wesley Professional)

[15] Humphrey W 1989 *Managing the Software Process* (Addison-Wesley Professional)
16] Wilson G 2006 Software carpentry, getting scientists to write better code by making them more productive *Computing in Science and Engrg.* **8:**66-69
[17] Post D E and Kendall R P 2007 Large-scale computational and engineering code development and production workflows *Use of High Performance Computing in Meteorology* Mozdzynski G (ed.) (London: World Scientific)
[18] Post D E and Kendall R P 2006 Large-scale computational scientific and engineering project development and production workflows *CTWatch Quarterly* **2**
[19] Santayana G 1905 *Reason in Common Sense* (New York: Charles Scribner's Sons) 504

Journal of Physics: Conference Series **125** (2008) 012091 doi:10.1088/1742-6596/125/1/012091

Sunfall: a collaborative visual analytics system for astrophysics

Cecilia R Aragon, Stephen J Bailey, Sarah Poon, Karl Runge and Rollin C. Thomas
Lawrence Berkeley National Laboratory, Berkeley, CA 94720, USA

E-mail: CRAragon@lbl.gov

Abstract. Computational and experimental sciences produce and collect ever-larger and complex datasets, often in large-scale, multi-institution projects. The inability to gain insight into complex scientific phenomena using current software tools is a bottleneck facing virtually all endeavors of science. In this paper, we introduce Sunfall, a collaborative visual analytics system developed for the Nearby Supernova Factory, an international astrophysics experiment and the largest data volume supernova search currently in operation. Sunfall utilizes novel interactive visualization and analysis techniques to facilitate deeper scientific insight into complex, noisy, high-dimensional, high-volume, time-critical data. The system combines novel image processing algorithms, statistical analysis, and machine learning with highly interactive visual interfaces to enable collaborative, user-driven scientific exploration of supernova image and spectral data. Sunfall is currently in operation at the Nearby Supernova Factory; it is the first visual analytics system in production use at a major astrophysics project.

1. Introduction

Many of today's important scientific breakthroughs are being made by large, interdisciplinary collaborations of scientists working in geographically widely distributed locations, producing and collecting vast and complex datasets. These large-scale science projects require software tools that support not only insight into complex data but collaborative science discovery. The technique of "visual analytics," defined by Thomas in *Illuminating the Path* [1] as "analytical reasoning facilitated by interactive visual interfaces," combines statistical algorithms and advanced analysis techniques with highly interactive visual interfaces that support collaborative work. Such approaches offer scientists the opportunity for in-depth understanding of massive, noisy, and high-dimensional data. Astrophysics in particular lends itself to a visual analytics approach due to the inherently visual nature of much astronomical data (including images and spectra).

One of the grand challenges in astrophysics today is the effort to comprehend the mysterious "dark energy," which accounts for three-quarters of the matter/energy budget of the universe. The existence of dark energy may well require the development of new theories of physics and cosmology. Dark energy acts to accelerate the expansion of the universe (as opposed to gravity, which acts to decelerate the expansion). Our current understanding of dark energy comes primarily from the study of supernovae.

The Nearby Supernova Factory (SNfactory) [2] is an international astrophysics experiment designed to discover and measure Type Ia supernovae in greater number and detail than has ever been done before. These supernovae are stellar explosions that have a consistent maximum brightness, allowing them to be used as "standard candles" to measure distances to other galaxies and to trace the rate of expansion of the universe and how dark energy affects the structure of the cosmos. The SNfactory receives 50–80 GB of image data per night, which must be processed and examined by teams of domain experts within 12–24 hours to obtain maximum scientific benefit from the study of these rare and short-lived stellar events.

To facilitate the supernova search and data analysis process and enable scientific discovery for project astrophysicists, we developed Sunfall (SuperNova Factory AssembLy Line), a collaborative visual analytics system for the Nearby Supernova Factory that has been in production use for over 18 months. Sunfall incorporates sophisticated astrophysics image processing algorithms, machine learning capabilities including boosted trees and support vector machines, and astronomical data analysis with a usable, highly interactive visual interface designed to facilitate collaborative decision making. An interdisciplinary group of physicists, astronomers, and computer scientists (with specialties in machine learning, visualization, and user interface design) were involved in all aspects of Sunfall design and implementation.

The major contribution of this paper is the description of a novel visual analytics framework for supernova search and followup that is currently in production use and has significantly improved all aspects of the scientific data flow. Additional contributions include (1) an evaluation of multiple machine learning algorithms and their effectiveness in supernova search, (2) intelligent and highly interactive visual interfaces for facilitating scientific insight into supernova data, and (3) novel image-processing algorithms for supernova search, including Fourier contour analysis for image processing. Lessons learned from the development of this framework will have value for the design of future automated transient search projects such as the SuperNova Acceleration Probe (SNAP) [3] or Large Synoptic Survey Telescope (LSST) [4].

This paper is organized as follows. Section 2 describes the astrophysics background for supernova detection and spectral analysis, including the SNfactory project data flow. Section 3 contains information on previous systems built for other supernova experiments. Section 4 discusses the Sunfall design approach, and Section 5 describes the Sunfall architecture, including its four major components: Search, Workflow Status Monitor, Data Forklift, and Supernova Warehouse. Section 6 discusses our conclusions and lessons learned, and Section 7 presents ideas and plans for future work.

2. Science background

The discovery of dark energy is primarily due to observations of Type Ia supernovae at high redshift (up to $z = 1$, or a lookback time of about 8 billion years) [5,6]. This supernova cosmology technique hinges on the ability to reliably compare luminosities of high-redshift events to those at low redshift, necessitating detailed study of low-redshift events. Large-scale digital sky surveys are now being planned in order to constrain the properties of dark energy. These studies typically involve high-volume, time-constrained processing of wide-field charge-coupled device (CCD) images, in order to detect Type Ia supernovae and capture detailed images and spectra on multiple nights over their lifespans.

A stellar explosion resulting in a Type Ia supernova will occur on average only once or twice per millennium in a typical galaxy of 400 billion stars and, then, will remain visible for only a few weeks to a few months. Additionally, the maximum scientific benefit is obtained if the supernova is discovered in the first two weeks before it attains peak brightness. To further add to the challenge, the imaging data from which supernovae must be detected are extremely noisy, corrupted with spurious objects such as cosmic rays, satellite tracks, asteroids, and CCD artifacts such as diffraction spikes, saturated pixels, and ghosts.

The Nearby Supernova Factory is an international project designed to tackle these ambitious goals and to accumulate the largest homogeneously calibrated spectrophotometric (containing both images and spectra) dataset of Type Ia supernovae in the "nearby" redshift range ($0.03 < z < 0.08$, or about 0.4 to 1.1 billion light years distant) ever studied [2]. On a typical night, 50–80 GB of data (approximately 30,000 images containing 600,000 potential supernovae) are received by SNfactory image-processing software. These data are processed overnight; the best candidates are selected each morning by humans for further follow-up measurements. Likely candidate supernovae are sent to a dedicated custom-built spectrograph for follow-up imaging and spectrography. The resulting spectra typically each contain 2000–3000 data points of flux as a function of wavelength. The SNfactory database is expected to be released to astronomers worldwide and become a definitive resource for measurements of dark energy. This program is the largest data volume supernova search currently in operation.

The SNfactory obtains wide-field imaging data from the Near Earth Asteroid Tracking program (NEAT) [7] using the 112 CCD QUEST-II camera of the Mt. Palomar Oschin 1.2 m telescope [8], covering 8.5 square degrees per exposure. (For comparison, the full moon covers 0.2 square degrees.)

Images are transferred from Mt. Palomar via the High Performance Wireless Research and Education Network (HPWREN) to the High Performance Storage System (HPSS) at the National Energy Research Scientific Computing Center (NERSC) in Oakland, California. Each morning, SNfactory search software running on NERSC's 700-node computing cluster, the Parallel Distributed Systems Facility (PDSF), matches images of the same area of the sky, processes them to remove noise and CCD artifacts, then performs an image subtraction from previously observed reference images on each set of matched images (figure 1) [9-11].

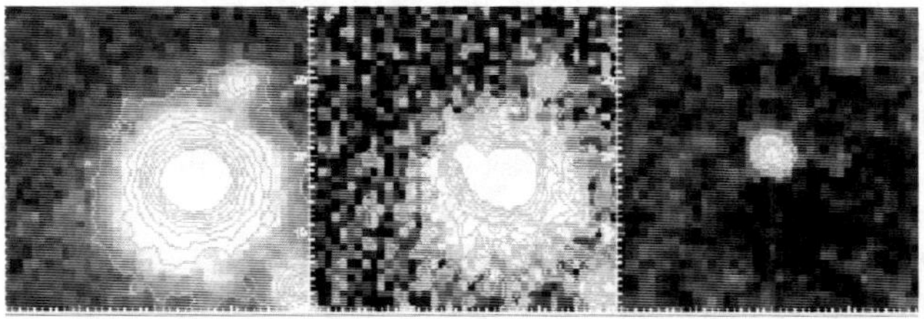

Figure 1. Example supernova images, from left to right: reference image of galaxy, new image of galaxy plus new bright spot, subtracted image reveals new supernova.

Supernova candidates are identified from among the over 600,000 objects processed per night by a set of image features including object shape (roundness and contour irregularity computed from Fourier contour descriptors) [11,12], position, distance from nearest object, and motion. These features are used as input to machine learning algorithms that select candidates to be sent to humans for scanning and vetting [9,13]. Promising supernova candidates that pass the human scanning and vetting procedure are sent for confirmation and spectrophotometric follow-up by SNIFS (the SuperNova Integral Field Spectrograph) [2] on the University of Hawaii 2.2 m telescope on Mauna Kea.

Candidates are imaged through a 15 x 15 microlens array on SNIFS and the spectral data is saved to a database in France. Currently the SNfactory has collected spectrophotometric data on over 150 Type Ia supernovae, and more data is being gathered each night.

Spectra are the best key to understanding the actual physics of exploding stars. Understanding Type Ia supernovae through their spectra is important because it places constraints on their presupernova evolution and the immediate stellar environment of the event and provides clues that can be turned into correlations between peak brightness and the physics of the explosion. With a better understanding of the physics of Type Ia supernovae through their spectra, it is believed that their utility as standardized candles for cosmological distance measurements can be refined for their use in future precision cosmology experiments such as SNAP, DESTINY, or ADEPT [3, 14, 15].

3. Related work

Since the discovery of dark energy via studies of Type Ia supernovae nearly a decade ago [5,6], astrophysicists have developed several large-scale supernova searches to collect as much data as possible on these rare events. These searches typically rely on custom image subtraction software containing highly complex, hand-tuned heuristics and "cuts" to extract supernova candidates. They are often plagued by large numbers of false positives, which must then be screened out by humans in a labor-intensive process. For example, the 2005 Sloan Digital Sky Survey II (SDSS-II) supernova program generated approximately 4,000 objects per night, which needed to be visually checked by humans for verification [16]. The ESSENCE and SNLS supernova searches both resulted in 100–200 objects to scan per night, with a significantly smaller input data load than the SNfactory [17]. Techniques used in these surveys will not scale to future, much larger datasets.

After supernova data has been collected, they are typically stored in a database accessible via a web-based interface. These web sites often present information in tabular form, listing the supernova coordinates and providing links to images. They are designed to provide all the necessary information for astronomers to observe supernovae with their own telescopes. Examples are the Supernova Legacy Survey (SNLS) and Sky Survey (SDSS) websites [18,19]. In [16], Sako et al. describe the SDSS software tools for supernova search and follow-up. They include sets of scripts that perform functions similar to parts of Sunfall, but require human intervention at numerous stages during the supernova search and follow-up process. None of these groups has built a complete, automated visual analytics system that encompasses the entire data search, analysis, and scientific discovery process.

4. Sunfall design process

To design an effective collaborative visual analytics system for the SNfactory, we first conducted, in mid-2005, an extensive, two-month evaluation of the existing SNfactory procedures and environment. Data sources used for evaluation included individual interviews, observation of team members performing typical project tasks, review of existing source code, literature reviews, examination of other supernova search projects, and consultation with physicists and computer scientists with relevant experience building similar scientific software systems.

We conducted over 100 hours of interviews with LBNL scientists, postdoctoral researchers, and students and with SNfactory collaboration members outside LBNL. This included 19 current and former team and collaboration members, and 12 scientists with relevant experience outside the SNfactory collaboration. We also performed a detailed software review of over 150,000 lines of existing SNfactory legacy code in C++, IDL, Perl, and shell scripts.

We established the following requirements for the Sunfall software framework: It must encompass the entire scientific data capture, processing, storage, and analysis process, enable collaborative, time-critical scientific discovery, and incorporate SNfactory legacy code (custom astronomical image-processing algorithms). It must reduce the number of false positives sent to humans and improve the quality of the subtraction image processing. It must automate repetitive data transfers and other manual tasks to leverage domain experts' unique processing and image recognition skills to the maximum extent.

The Sunfall user interface was designed and implemented using participatory and iterative design techniques; for example, interactive prototypes were used to evaluate areas where existing interfaces did not support scientists' workflow. Scientists' feedback sometimes led to major redesigns of the interface. The system was implemented from the beginning with this possibility in mind, so that changes were easily made and accepted as an appropriate part of the development process [16]. Sunfall has been in operation at the SNfactory since 2006.

5. Sunfall architecture and components

Sunfall consists of four major components: Search, Workflow Status Monitor, Data Forklift, and Supernova Warehouse (figure 2). This section describes each component's structure in detail.

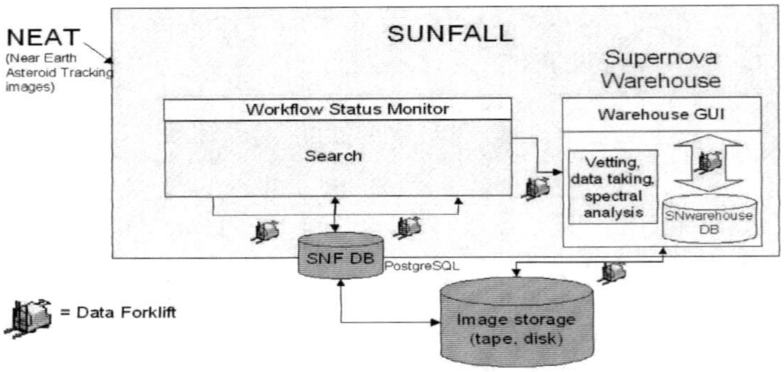

Figure 2. Sunfall architecture diagram, depicting the four components (Search, Workflow Status Monitor, Data Forklift, and Supernova Warehouse) and data flow between the components.

Search handles image processing and subtraction and includes machine learning algorithms and novel Fourier contour descriptor algorithms to reduce the number of false positive SN candidates. The Workflow Status Monitor is a web-based monitor that facilitates collaboration and improves project scientists' situational awareness of the data flow by displaying all relevant workflow (search pipeline) status data on a single site.

The Data Forklift is a middleware mechanism consisting of a coordinator and a suite of services to automate astronomical data transfers in a secure, reliable, extensible, and fault-tolerant manner. The Data Forklift also provides the middleware for the other three components, transferring data between heterogeneous systems, databases and formats securely and reliably.

The Supernova Warehouse (SNwarehouse) is a comprehensive supernova data management, workflow visualization, and collaborative scientific analysis tool. The SNwarehouse contains a PostgreSQL database, Forklift middleware, and a graphical user interface implemented in Java.

5.1. Search

The supernova search component of Sunfall is an example of *analytic discourse*, defined in *Illuminating the Path* as "the interactive, computer-mediated process of applying human judgment to assess an issue," [1] applied to the realm of astrophysics. Sunfall Search has increased efficiency of supernova detection and enabled more effective human intervention by reducing the number of false positive candidates by nearly an order of magnitude [6,13] and, through the use of intelligent interfaces, by increasing human efficiency in evaluating candidates by a factor of four.

The legacy search software computed approximately 19 photometric and geometric features on each of over 600,000 supernova candidate subimages per night and applied threshold "cuts" to each of these features. Subimages that satisfied all thresholds (1,092 on a typical night in April 2006, just before the new software was put into production) were sent to human scanners who selected potential supernova candidates for follow-up study. The majority of subimages that passed the thresholds were false positives that humans had to manually reject. These manually tuned thresholds were brittle and often resulted in both missed SNe and too many false positives for human scanners to evaluate daily. The process of scanning on the order of 1,000 subimages took 6-8 people several hours per day.

Machine learning algorithms, including support vector machines and boosted trees, were incorporated into the search software. By replacing simple threshold "cuts" on the image features with classifiers of the high-dimensional data in the feature space, we reduced the number of false positives by an order of magnitude (from 1,094 to 113 on typical nights in 2005 and 2006). This resulted in labor savings of nearly 90%, where the process of scanning now takes one person a couple of hours each day. Additionally, system tests determined that the new algorithms had a true positive ratio equal to or better than the original algorithm (in other words, they did not miss any more actual supernovae). The data is depicted in figure 3.

Currently, of the approximately 100 visually scanned subimages per night, about 10-30 are flagged as containing potential supernova candidates. These candidates (about .001% -.01% of the objects originally imaged) are sent to the SNIFS spectrograph [2] for spectrophotometric screening and follow-up. Typically from two to five of the objects examined by SNIFS will finally be determined to be supernovae, of which (on average) about one per night is the sought-after Type Ia supernova. Thus of the objects originally imaged, somewhere near .0002% prove to be Type Ia supernovae. This process is described in more detail in [9,13].

For confirmation and selection of promising supernova candidates, the scanners use a visual interface to view and analyze the candidates. This visual scanning interface was redesigned to increase interactivity and allow humans to focus exclusively on the imagery itself. Previously, operators of the interface spent much of their time cutting and pasting long filenames, pausing the scanning process to look up data on several different external web sites, and waiting for image data to be retrieved from tape storage. Previously, this process took two minutes per candidate on average; with Sunfall, a decision could be made within 30 seconds. By applying intelligent algorithms to the search data itself, we were able to leverage the unique human abilities that enable the final detection of true supernova candidates and not only minimize the number of false positives but also increase the efficiency of the spectroscopic follow-up.

5.2. Workflow Status Monitor

The Workflow Status Monitor is a web-based dashboard application, designed to improve project scientists' situational awareness of the data flow by synthesizing diverse information flows from various systems and displaying critical workflow status data on a single site (figure 4).

The supernova search software is highly parallelized as 30,000 images are queued for processing in a multistage pipeline that runs on NERSC's 700-node computing cluster, the Parallel Distributed Systems Facility (PDSF). Nodes frequently go down or jobs fail, and failures must be detected promptly and jobs resubmitted quickly due to the time-critical nature of the search. Detection of such failures was challenging and time-consuming before the deployment of the Workflow Status Monitor. The monitor displays graphs and visual displays of job completion times on PDSF, job queues, PDSF node uptimes, and disk vault loads. The interface was prototyped, evaluated for efficacy by project scientists, and implemented via PHP on a web site hosted on an SNfactory server running SuSE Linux.

The dashboard also displays date and time information at project locations in several formats relevant to astronomers. One author's previous experience with a status monitor for the Mars Exploration Rovers [21] demonstrated that apparently minor features, such as a clock that displayed the time on Mars and at various project locations on Earth, yielded tremendous improvements in scientists' situational awareness and may very well have prevented critical errors. Additional information displayed includes telescope scheduling, supernova candidates found, and current operational status of telescopes, spectrographs, and cameras used in processing SNfactory data.

5.3. Data Forklift

Each night, over 50 GB of heterogeneous, distributed supernova data arrives at the SNfactory and needs to be processed, managed, analyzed, queried, and displayed – all within a time period of 24 hours or less (ideally 12 hours). The data is distributed over wide geographic distances; some of it is hosted on systems outside the SNfactory's control, and some on unreliable systems with frequent downtime.

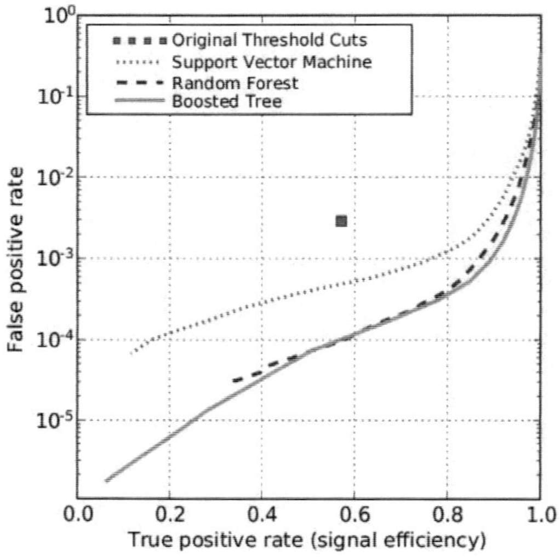

Figure 3. Comparison of Boosted Trees (solid line), Random Forest (dashed line), and Support Vector Machine (dotted line) for false positive identification fraction vs. true positive identification fraction. The lower right corner of the plot represents ideal performance.(from S. Bailey et al. [13]).

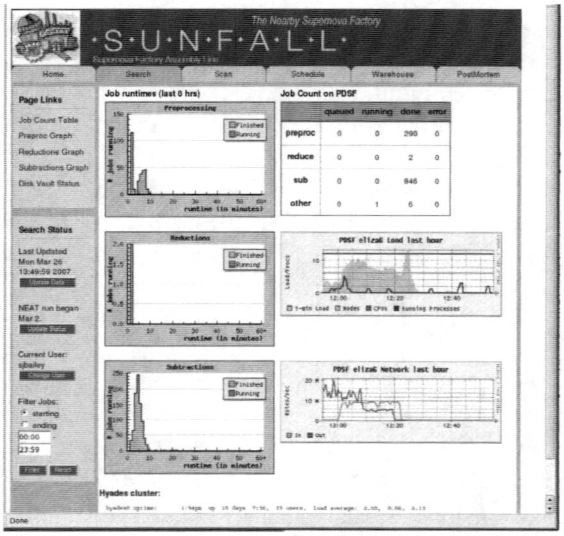

Figure 4. Sunfall Workflow Status Monitor. This web-based dashboard presents a single-page overview of many relevant information flows from multiple systems and, according to project scientists, increases operator efficiency significantly.

To solve these data management problems, we designed and implemented the Sunfall Data Forklift, a suite of services for automated data retrieval, storage, movement, querying, and staging for display. The Data Forklift consists of a coordinator and a set of individual services, each customized for its particular task, but having key attributes in common.

The Data Forklift supports "different place, asynchronous" collaborative scientific work by facilitating data and information transfer amongst a geographically separated team. Because of the time-critical nature of the data collection (telescopes must be operated at night and are located in different time zones), tasks must take place at distinct, specified times.

All Data Forklift services share the following properties:

- The mechanism resides at a central location (an SNfactory server), but processes data remotely from widely separate geographic locations, including making connections to external machines not under the SNfactory's control.
- All data transfers are secure, utilizing encrypted channels and authentication.
- The Forklift services and coordinator are reliable and self-restarting. The Forklift coordinator restarts itself if its process dies or if the server is rebooted.

- The mechanism is fault-tolerant; services detect unreliable connections and recover from errors in data transfer. Partially transferred files are detected and retransferred, and the process restarts the transfer at the point of failure.
- The mechanism is extensible (new tasks can easily be added). The Forklift coordinator was designed so that new services can easily be added to a central table.
- The Data Forklift enables data retrieval from external sources. Many astronomical databases are web-based, and designed for interactive rather than automated retrieval. Forklift services handle such interactions in a fault-tolerant manner. SNfactory spectral data is stored at a central processing database in France and must be retrieved according to external requirements.
- User actions can trigger Data Forklift requests. For example, whenever a supernova candidate is saved to the Supernova Warehouse (SNwarehouse) database, several Forklift services are automatically started, retrieving related image data from tape storage, accessing external web-based asteroid, galaxy, and other astronomical databases, converting image files into display format, and staging data for display.
- Forklift services act as middleware for the Supernova Warehouse, retrieving data from internal and external databases, performing program logic, and staging data for display by the SNwarehouse GUI.
- The Data Forklift is written in Perl and runs under SuSE Linux 9.3 on an SNfactory server.

5.4. Supernova Warehouse

The Supernova Warehouse (SNwarehouse) is a comprehensive supernova data management, workflow visualization, and collaborative scientific analysis application. It consists of a PostgreSQL database hosted on a dedicated SNfactory database server running SuSE Linux 9.3, Data Forklift services as middleware, and a graphical user interface (GUI) written in Java. SNwarehouse supports collaborative remote asynchronous work in several different ways.

Collaboration members can access the GUI from any networked computer worldwide via a Forklift remote deployment mechanism. Security is provided via password authentication and encrypted communication channels. SNwarehouse furnishes project scientists with a shared workspace that enables easy distribution, analysis, and access of data. Collaboration members can view, modify, and annotate supernova data, add comments, change a candidate's state, and schedule follow-up observations from work, home, while observing at the telescope, or when attending conferences. This access is critical due to the 24/7 nature of SNfactory operations. All transactions are recorded in the SNwarehouse database, and the change history and provenance of the data re permanently stored (records cannot be deleted in order to maintain the change history) and continuously visible to all authenticated users.

SNwarehouse centralizes data from multiple sources and supports task-oriented workflow. Project members perform well-defined tasks, such as vetting, scheduling, and analyzing targets, which collectively accomplish the goal of finding and following Type Ia supernovae. Typically, an individual or small group performs a given task, and the results of the task provide inputs for the next task in the workflow, often performed by another set of group members. Thus, the inputs and outputs of any task must be well defined and easily recognizable.

5.4.1. SNwarehouse overview. SNwarehouse's interaction design takes the approach of overview, filter, and drill down to details. The main overview page (figure 5) displays two tightly coupled representations of the list of targets registered in the database. The top visualization plots the targets in the sky; below is a sortable, tabular representation of the same data. From here one can drill down as needed based on the desired task.

5.4.2. Supernova candidate vetting. "Vetting" involves a process of scientific analytic discourse where the scientist must quickly decide, based on limited data, how to allocate scarce and expensive telescope time to the night's supernova candidates. At this point, before spectra are taken, it is often unknown whether a target candidate is a supernova, so the scientist must gather as much relevant data as rapidly as possible to make an informed prediction. This task involves retrieving any available images of the target's location in the sky prior to discovery and querying several external astronomical databases.

Once a target is saved as a candidate supernova, the target name will appear in the SNwarehouse overview table color-coded in red, indicating that the target is newly discovered. On a rolling basis, a background process checks if previous data products exist on disk. If found, these data products are registered with the database, and a "!" appears next to the target name, a flag signaling newly found images. The combination of these two visual clues allows the scientist to quickly see which targets need evaluation. Once this determination is made, the vetter will then open the Details View for the target (figure 6).

At this stage, the vetter is trying to determine whether this target is potentially a Type Ia supernova, based on how consistently the magnitude is rising in comparison to standard Ia's. The Details View helps the vetter make this determination in two ways. First, a visualization of the magnitudes of all target observations against a standard Ia lightcurve (figure 6) allows the vetter to quickly determine whether the magnitudes are rising like those of a typical Ia. In addition, the vetter can compare the discovery image with prior images taken at this particular set of coordinates in the sky (figure 6).

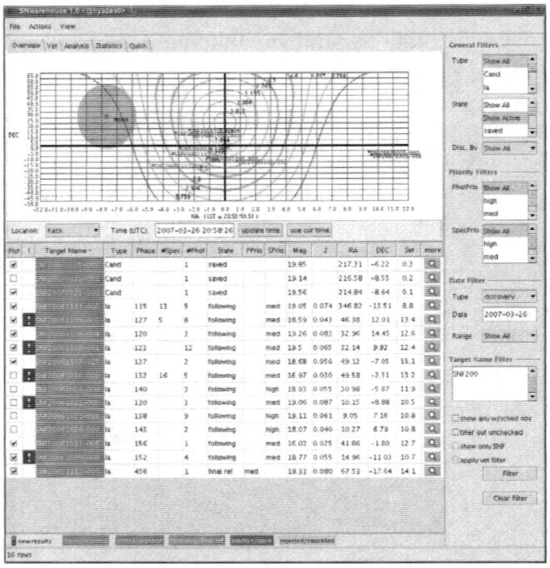

Figure 5. SNwarehouse overview tab. The Sky visualization at the top plots declination vs. hour angle (astronomical sky coordinates) in a view preferred by domain experts. The green lines represent airmass (the thickness of the atmosphere) at particular coordinates at the specified time.

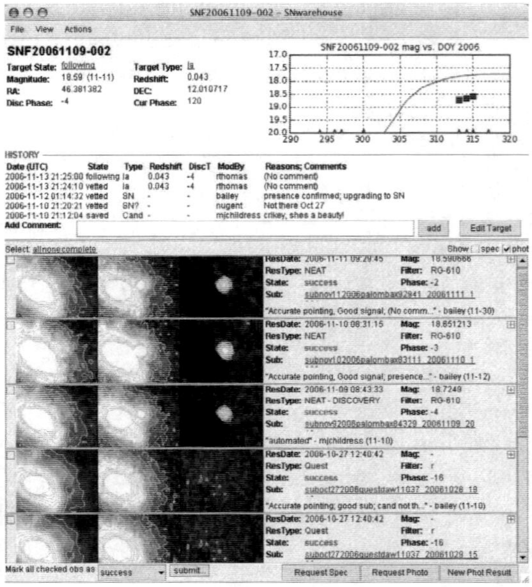

Figure 6. SNwarehouse details view. Comparison to standard lightcurve is in upper right-hand corner. The five rows at the bottom were imaged on different dates; note the target supernova is visible at the time of discovery but not on prior dates.

5.4.3. Observation scheduling. Prior to followup observation at the telescope, a schedule is made based on the vetter's assessments to order and allocate telescope time. In SNwarehouse, the scheduler starts by filtering for only those targets with observation requests. With this list, the scheduler must order and assign exposure times for each target using a variety of techniques. Exposure times are automatically determined using a lookup table based on the phase and redshift of the target. Considering these exposure times, the scheduler must also order the target list according to when the target will be visible in the sky by the telescope. A custom visualization offers insight for this task.

The Sky visualization, described in [22] and also shown at the top of figure 5, depicts the positions of targets in the sky at a given time and ground location. The green lines represent airmass (atmosphere thickness) for target coordinates at the specified time. The blue line is the horizon, and the red halo around the moon is the "lunar exclusion zone," where light cast by the moon makes it difficult to view faint targets. The yellow circle depicts the sun. Major telescope names and corresponding latitudes and longitudes are displayed on a drop-down menu so the visualization can be used world-wide. The time can be changed so the viewer can plan observations for the remainder of the night. If the target appears within the blue horizon lines, it is visible to the telescope, assuming good weather. The scheduler must note the rise and set times of each target when creating the schedule for the night.

5.4.4. Data taking. During each night of spectral observations with SNIFS at the University of Hawaii 2.2 m telescope, an observer needs to point the telescope at each scheduled target in order. In the event of weather or mechanical problems, a rescheduling decision must be made quickly and efficiently. A visual data-taking interface facilitates this process. Much of the observation procedure is automated, allowing the observer to concentrate on troubleshooting, rescheduling, and determining the success of each target observation.

5.4.5. Analysis and post-mortem. Once observation at the telescope is complete, the data products (images and spectra) from the telescope are registered with the database and marked with a "!" in the SNwarehouse overview page. The next task is spectral processing and analysis, known as the "post-mortem" process. The scientist performing post-mortem filters for targets with new data products that are being followed (highlighted in green and blue, depending on the type of followup) and opens up the details view to evaluate the telescope results (figure 7).

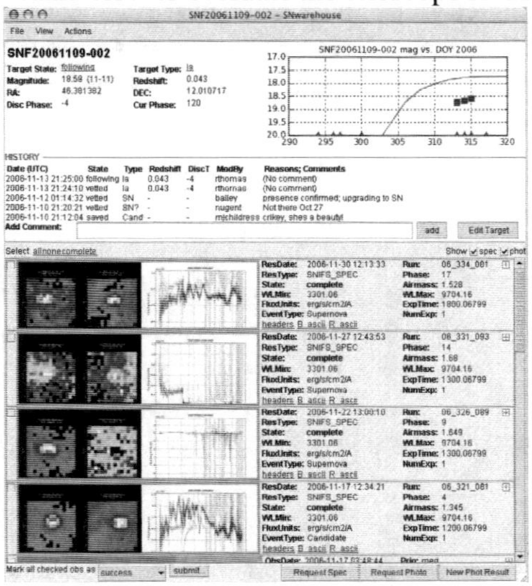

Figure 7. SNwarehouse details view depicting spectral data observations. The small images on the left-hand side of each observation subwindow indicate the accuracy and signal strength of the received data.

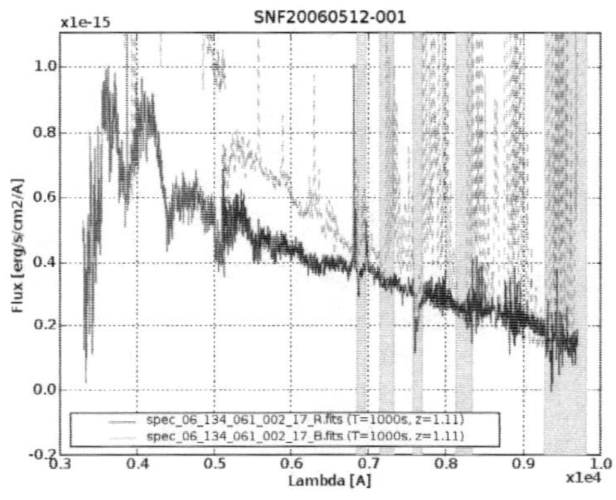

Figure 8. A spectrum from a very early Type Ia supernova, 15 days before peak brightness. The green and blue plots indicate the spectral data. The spiky grey lines in the background depict the background sky spectrum. This supernova is just starting to show the features that define a Type Ia.

The raw data from the SNIFS spectrograph is complex and requires a significant amount of processing in order to yield meaning to the scientist. A visual depiction of the accuracy of the pointing and signal strength provides much more information more quickly than tables of numeric data. The two small images on the left-hand side of each observation subwindow are a custom visualization designed to indicate whether the telescope accurately pointed at the target and if a good signal was received by the spectrograph (figure 7). Color-coding and position indicate the accuracy and signal strength of the received data. If there is a bright dot in the center of both squares, that is a clear indicator of a "good pointing."

An early Type Ia supernova (captured well before peak brightness) will display a certain pattern in its spectral plot (figure 8). A later supernova (imaged a few days past peak brightness) will show other characteristic spectral lines and features. Spectral data are plotted in green and blue. The spiky grey lines depict the spectrum of the background sky. The broad grey bands represent areas of atmospheric absorption. Because of the complexity of the data (thousands of points of flux vs. wavelength for each observation) and the necessity to make rapid, accurate decisions in order to maximize the use of limited, expensive telescope time, visualization provides the most efficient solution to the problem. The scientist can also access the raw numeric data by clicking on any of the images or spectra.

6. Conclusion

Sunfall, a collaborative visual analytics system in operation at a large-scale astrophysics project, has demonstrated that such systems that facilitate scientific analytic discourse and computer-supported collaborative work can have a positive impact on data-intensive science. In the process of design and implementation, we learned that an interdisciplinary team incorporating specialists from several fields, including scientific domain experts, will be most effective in designing an effective visual analytics system for science. Additionally, the framework architecture, efficient visual interfaces, and algorithms developed for Sunfall will be transferable to future large-scale automated transient alert pipelines such as SNAP or LSST.

7. Future Work

Current plans for improvements to Sunfall include the development of a multidimensional visual analytics tool for supernova spectra classification. We are currently applying several different clustering algorithms to the normalized spectral data. Tools are under development to apply various transformations to the calibrated and normalized spectra. The end goal is feature detection and similarity detection across supernova spectra. The optimal presentation of these clusters is still an open problem. We hope these visual analytic tools will facilitate the search for new correlations between classes of supernovae.

Acknowledgments

We thank the scientists of the SNfactory collaboration for their time and detailed feedback. This work was supported by the Director, Office of Advanced Scientific Computing Research, Office of Science, of the U.S. Department of Energy under Contract No. DE-AC02-05CH11231 through the Scientific Discovery through Advanced Computing (SciDAC) program's Visualization and Analytics Center for Enabling Technologies (VACET), and by the Director, Office of Science, Office of High Energy Physics, of the U.S. Department of Energy under Contract No. DE-AC02-05CH11231, and by a grant from the Gordon & Betty Moore Foundation. This research used resources of the National Energy Research Scientific Computing Center, which is supported by the Office of Science of the U.S. Department of Energy under Contract No. DE-AC02-05CH11231.

References

[1] Cook Thomas J and Cook K 2005 *Illuminating the Path: The Research and Development Agenda for Visual Analytics* (National Visualization and Analytics Center, IEEE)

[2] Aldering G, *et al.* 2002 Overview of the Nearby Supernova Factory *Proc. SPIE* **4836**:61.

[3] SNAP 2007 Supernova Acceleration Probe, http://snap.lbl.gov

[4] Izevic Z, et al. (LSST) 2008
arXiv:0805.2366v1, http://www.lsst.org/overview/overview_v1.0.pdf

[5] Perlmutter S, *et al.*1999 Measurements of Omega and Lambda from 42 High-Redshift Supernovae *Astrophys. J.* **519:**565–586

[6] Riess A, *et al.* 1998 Observational evidence from supernovae for an accelerating universe and a cosmological constant *Astron. J.* **116**:1009–1038

[7] NEAT 2007 Near Earth Asteroid Tracking, http://neat.jpl.nasa.gov

[8] QUEST 2008 Palomar-QUEST Survey 2002
http://hepwww.physics.yale.edu/quest/palomar.html

[9] Romano R, Aragon C, and Ding C 2006 Supernova Recognition Using Support Vector Machines *Proc. 5th Intl .Conf. of Machine Learning Applications.*

[10] Wood-Vasey W 2004 Rates and progenitors of Type Ia Supernovae (Ph.D. dissertation, University of California, Berkeley)

[11] Aragon C and Aragon D 2007 A Fast Contour Descriptor Algorithm for Supernova Image Classification *Proc. SPIE Symposium on Electronic Imaging: Real-Time Image Processing.*

[12] Zahn C and Roskies R 1972 Fourier descriptors for plane closed curves *IEEE Trans. Computers*, **21**:269–281

[13] Bailey S, Aragon C, Romano R, Thomas R, Weaver B and Wong D 2007 How to find more supernovae with less work: object classification techniques for difference imaging *Astrophys. J.* **665**:1246B
[14] Benford D and Lauer T 2006 Destiny: a candidate architecture for the Joint Dark Energy Mission *Space Telescopes and Instrumentation I: Optical, Infrared, and Millimeter, Proc. SPIE*
[15] ADEPT, Advanced Dark Energy Physics Telescope, 2006 http://www.physorg.com/news73758591.html
[16] Sako M, *et al.* 2008 The Sloan Digital Sky Survey-II supernova survey: search algorithm and follow-up observations *Astronomical J.* **135**:348-373
[17] Astier P, *et al.* (SNLS) 2006 *A&A* **447**:31-48
[18] SNLS, SuperNova Legacy Survey 2007 http://www.cfht.hawaii.edu/SNLS/
[19] SDSS, Sloan Digital Sky Survey, 2007 http://www.sdss.org
[20] Aragon C and Poon S 2007 The impact of usability on supernova discovery *Workshop on Increasing the Impact of Usability Work in Software Development, CHI 2007: ACM Conference on Human Factors in Computing Systems*
[21] Mak R, Walton J, Keely L, Heher D, and Chan L 2005 Reliable Service-Oriented Architecture for NASA's Mars Exploration Rover Mission *IEEE Conf. Aerospace*
[22] Aragon C, Poon S, Aldering G, Thomas R, and Quimby R 2008 Using Visual Analytics to Maintain Situational Awareness in Astrophysics *Proc. IEEE Visual Analytics Science Tech.*

COVE: a visual environment for ocean observatory design

Keith Grochow[1], Mark Stoermer[2], Deb Kelley[3], John Delaney[3] and Ed Lazowska[1]

[1]Department of Computer Science, University of Washington, Seattle, WA 98195, USA [2]Center for Environmental Visualization, University of Washington, Seattle, WA 98195, USA [3]School of Oceanography, University of Washington, Seattle, WA 98195, USA

E-mail: `keithg@cs.washington.edu`, `mstorm@u.washington.edu`, `dskelley@u.washington.edu`, `jdelaney@u.washington.edu`, `edl@cs.washington.edu`

Abstract. Physical, chemical, and biological ocean processes play a crucial role in determining Earth's environment. Unfortunately, our knowledge of these processes is limited because oceanography is carried out today largely the way it was a century ago: as expeditionary science, going to sea in ships and measuring a relatively small number of parameters (e.g., temperature, salinity, and pressure) as time and budget allow. The NSF Ocean Observatories Initiative is a US$330 million project that will help transform oceanography from a data-poor to a data-rich science. A cornerstone of this project is the deep water Regional Scale Nodes (RSN) that will be installed off the coasts of Washington and Oregon. The RSN will include 1500 km of fiber optic cable providing power and bandwidth to the seafloor and throughout the water column. Thousands of sensors will be deployed to stream data and imagery to shore, where they will be available in real time for ocean scientists and the public at large. The design of the RSN is a complex undertaking, requiring a combination of many different interactive tools and areas of visualization: geographic visualization to see the available seafloor bathymetry, scientific visualization to examine existing geospatially located datasets, layout tools to place the sensors, and collaborative tools to communicate across the team during the design. COVE, the Common Observatory Visualization Environment, is a visualization environment designed to meet all these needs. COVE has been built by computer scientists working closely with the engineering and scientific teams who will build and use the RSN. This paper discusses the data and activities of cabled observatory design, the design of COVE, and results from its use across the team.

1. Introduction

The oceans have always been an important focus of scientific study and exploration. They cover over 70% of the Earth's surface, moderate its climate by being a stabilizing force for the environment, and are a significant source of food for a large part of the planet. Because of the increasing impact of global warming, understanding ocean processes is increasingly central to predicting how our climate will evolve during the next century, and the impact on ocean and human health. To date, our ability to collect data about the oceans and simulate important interlinked processes has been extremely limited compared to the need.

A relatively new solution to this problem is ocean observatories that continually collect and analyze diverse data from the oceans in real time. Over the next few years, the National Science Foundation (NSF) Ocean Observatory Initiative (OOI) will create an ocean observatory of unprecedented scale. The Regional Scale Nodes (RSN) portion of the OOI, installed off the Washington and Oregon coasts, will support sensors from the ocean surface to deep in the seafloor, connected to cables delivering power and bandwidth. The goal of this endeavor is to build a flexible platform allowing hundreds of scientists from various fields to conduct experiments, provide real-time sensor and data access through the Internet, and create a vast archive of data. The diverse range of users and unprecedented collection of data require a system that makes design and exploration of the observatory layout and experiments easy and intuitive.

Designing the observatory requires several tasks. Observed data for the sites must be collected and simulation models generated to understand the processes and possible issues at each experiment location. Different options must be explored and analyzed to determine optimal layouts for data collection within budgetary and technical constraints. The reasoning for alternative layouts must be presented to the science users to determine the effectiveness of various designs. These tasks must be carried out, not only to create the infrastructure, but regularly over the life of the system as science teams extend the observatory with new sensors.

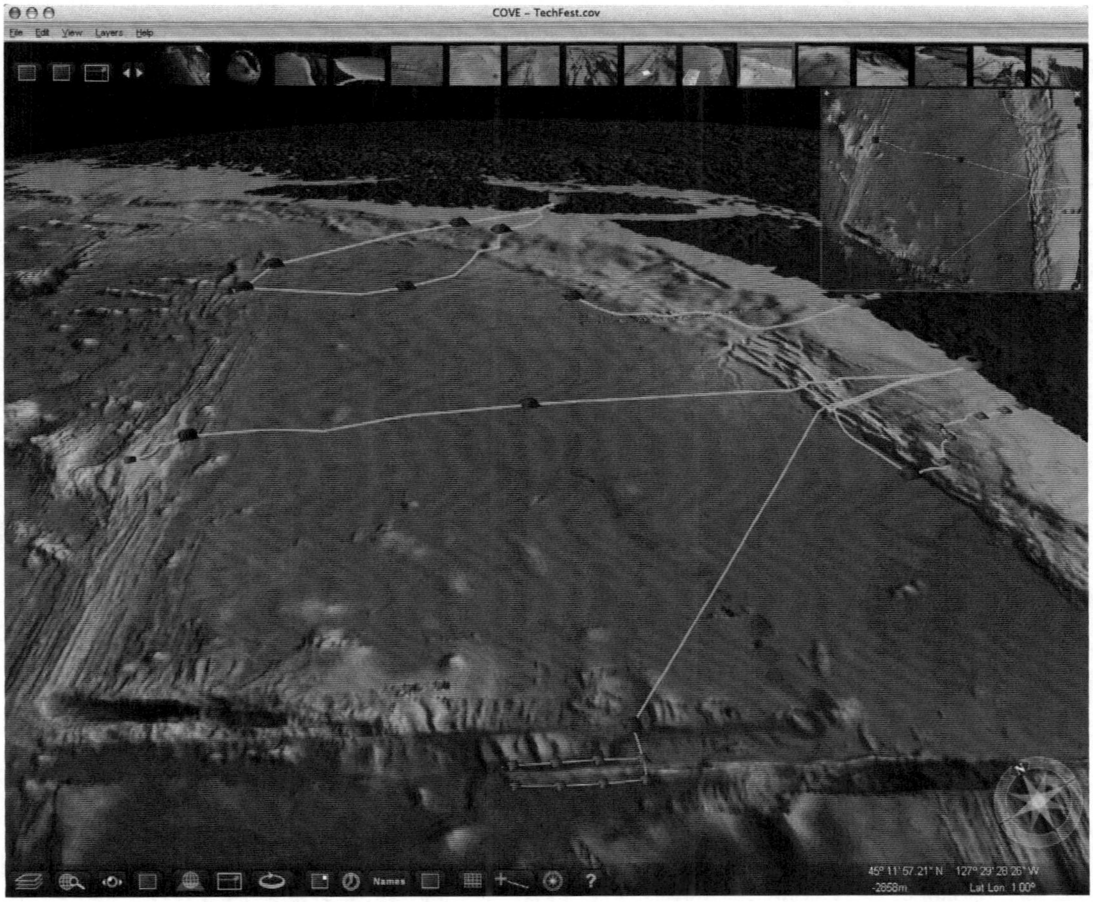

Figure 1. COVE: Common Observatory Visualization Environment.

Visualization is particularly important to these tasks for several reasons. Sparse observed datasets require human expertise to fill in gaps and identify important trends. Oceanographic models require domain experts to assess their validity. At a proposed experimental site, scientists from diverse fields will be dealing with myriad new data types, planning 10–15 year experiments with hundreds of sensors in the context of existing seafloor bathymetry (seafloor terrain), sensor layouts, and environmental parameters (e.g., lava morphology, areas of hot spring venting, and gas hydrate deposition). Effective visual tools are crucial to allow human cognition to rapidly determine patterns, note anomalies, and determine the validity of data based on environmental context.

Although several tools provide solutions to some of these problems, none effectively solves them all. Georeferencing browsers such as Google Earth and Microsoft Live Earth (hereafter referred to as geobrowsers) provide a familiar physical metaphor for visualizing such large geographic layouts. Their multi-resolution interface affords immediate context for data, allows easy layering of datasets, and makes georeferencing data intuitive. But to date they do not support changing 3D bathymetry and deliver little visual analysis capability, limiting their usefulness for oceanographic investigation. They also have limited ability for direct manipulation of sophisticated cabled layouts and sensor deployment planning. Tools such as Fledermaus™or GeoZui3D allow importing and manipulation of bathymetry but do not have tools for rich interactive data exploration in scientific experiments or sensor layout.

Observatory designers must therefore currently move between multiple systems to create, analyze, and visualize their designs. This situation leads to inefficiencies in exploration and planning, makes sophisticated analysis pipelines seldom reusable, and leaves little opportunity to share processes or data. This scenario will be exacerbated when the RSN moves beyond the initial design phase and enables large-scale participation by the global science community. A system that integrates support for the above areas enhances planning of complex seafloor and water column experiments, science-engineering integration, and knowledge sharing by avoiding the need to use several different applications and interfaces.

COVE, the Common Observatory Visualization Environment, is a visualization system that provides a solution to these problems (figure 1) Based on investigation into the visualization needs for large-scale ocean observatory design, COVE was developed with scientists and engineers to meet those needs. Building from the geobrowser model, historical datasets for the sites are provided on layers to the user for inclusion in their current view. Behind these layers is a flexible web service-based architecture and an XML-based data handling and tagging scheme to make shared storage and access of geolocated historical data easy. New instrument layouts and extensions of core infrastructure can be easily applied with direct manipulation techniques and shared among the design team. Visualizations explaining these layouts can be shared for collaborative design discussions across the team or for presentation to diverse groups to explain decisions.

COVE has been used extensively by the team to create the core observatory design for the RSN. It has provided a common ground to investigate and present those designs to experts from across various fields of oceanography and earth science. Plans are being made to extend its capabilities for continuing work on the observatory in areas of operations and education and ultimately visualization of real-time datasets as they stream from the observatory.

2. Related work

Our general approach is closely related to the work by Wood *et al* for interactive visualization of a large spatio-temporal dataset [1], where Google Earth was used as a front end to allow users to quickly mash up geovisualizations. Geobrowsers such Google Earth [2], World Wind [3], and Virtual Earth [4] provide a very accessible interface to geographic data. Their focus on the earth globe paradigm and zooming interface have made them popular for viewing simple geospatial data. They facilitate labeling of locations, georeferencing maps and images, and sharing through XML-based documents. The Google Ocean project [5] by Magic Instinct Software provides an example of using geobrowsers for oceanographic data and shows the limitations of their use: bathymetry is limited to 2D images overlaid on the ocean surface, there are no data display or analysis tools, and there are no interactive layout capabilities.

Geobrowsers have been used to display Earth observatory data, as demonstrated by the EarthScope project [6]. They display seismometer locations with sensor detail available by clicking on a geolocated object. World Wind from NASA was used to provide a more extensive observatory solution through its plug-in interface [7]. Spyglass [8] is an observatory-specific system that supports viewing of instruments, their connections, and detail about specific nodes. It allows visualization features such as interpolation of real-time data values through a plugin interface. Geospatial context for the nodes is limited to a 2D map layer.

Journal of Physics: Conference Series **125** (2008) 012092 doi:10.1088/1742-6596/125/1/012092

Two desktop tools for visualizing general oceanographic data currently available are the Interactive Data Viewer (IDV) [9] and GeoZUI3D [10]. IDV was built for the atmospheric community on top of VisAD. It supports sophisticated 3D model visualization (isosurfaces, contours, volumetric), terrain maps, and data animation but with no inherent support for oceanographic data. GeoZUI3D [11] (and its commercial version Fleidermaus) integrates multiple 3D bathymetry sets and geolocated 2D and 3D objects for oceanographic visualization and recently added the ability to play back time-varying 3D point data. Both these tools have sophisticated user interfaces that are often difficult for scientists to use and do not provide interactive layout tools.

3. COVE

To create a visualization system that would meet the unique needs of this user and data environment, we worked closely with the ocean observatory design team for several months to understand their requirements and develop possible visualization solutions. Our approach involved collecting artifacts, recording events, and carrying out interviews with the primary participants on the team. Current software and nonsoftware solutions were analyzed to determine possible improvements to the systems in use. Based on this investigation, prototypes were created to test hypotheses. We found that constant prototyping was an extremely useful tool to bridge the gap between various views of the best solution.

With a project of this scope, a team of diverse users was involved in the design decisions at various stages, which we grouped into three primary user categories:

- **Core Design Team**: a small group of scientists that creates a conceptual layout of the system to define infrastructure for the major science questions explored by the observatory
- **Engineering Team**: a group that integrates with the Core Design Team to generate technical analysis, itemized costing and implementation schedules
- **Principal Investigators**: individual scientists who extend the infrastructure by adding sensors to collect data for their specific experiments.

Driven by the design schedule for the RSN, our work to date has focused primarily on systems to aid the Core Design Team. The design process in place before COVE included use of multiple software applications and digital mapping tools that were not integrated. Proposed designs were created by the project team and captured using word-processing or spreadsheet documents and static visualizations that were manually created. These visualizations were used for community discussions and reviews. When necessary for outside communication, an onsite graphics group was employed to design high-resolution visuals.

Several limitations were identified in this process. The design solutions were hard to share among team members since they were often using different tools. Changes were time-consuming and expensive since there was no automated way to update costing or cabling. Comparison of different models required reviewing spreadsheets or paper visuals. It was difficult to view experimental sites from different angles or at different resolutions. Moreover, it was not possible to overlay key information on a new layout, such as bedrock characteristics or pre-existing cable route information, to easily determine design flaws.

Working with the science team, several requirements were identified for a more usable system to address these needs. The goal was an integrated architecture that would address all three of primary requirements areas:

- **Collecting and Registering Data**: All data needed to be in a common environment, rapidly accessible, and viewable in different ways so that it could easily be used across the team and visually checked for accuracy.
- **Creating Observatory Designs**: All infrastructure (nodes, cabling, connectors) and sensors, associated metadata, and costing needed to be integrated into the design so that technical limitations and cost-design benefits were readily visible.

- **Communicating and Collaborating**: Layouts and views of experimental/sensor sites needed to be easily sharable with mechanisms for comments by other users and presentation to diverse audiences.

To meet these requirements we created COVE, shown in figure 1. COVE is an immersive, direct manipulation 3D environment based on the geobrowser model for easy layering of georeferenced bathymetry, images, and datasets. It provides easy mechanisms to create, edit, and share sensor layouts with the team and science user community. And it allows histories of changes and decisions to be captured for later availability. Each of these areas is discussed in more detail below.

3.1. Architecture

The primary goals of our architecture were to provide (1) rich cross-platform, interactive 3D graphics, flexible internal design to quickly prototype new solutions and (2) the ability to use the web to transparently scale our data-handling needs.

We had hoped to use an existing geobrowser as the basis for our work on COVE, but we found that because of limitations in bathymetry and programming accessibility (Google Earth) or lack of cross-platform support at the time (World Wind), existing clients were not an option. Web browser-based solutions were also not viable since they did not provide the level of 3D interaction required. COVE was therefore implemented as a C++ open source project on top of the OpenGL graphics interface. A cross-platform interface was built using the Fast Light Tool Kit (FLTK). COVE is designed in a modular manner to allow porting to geobrowser platforms as it becomes feasible.

To provide flexibility, we use XML for both file formats and internal record handling. As a file format, this was an obvious choice based on existing geobrowser formats. For record handling, we expose the capability to read and write the XML text for a record at any time. This allows the state of a record to be viewed at any point as text output and edited as text in the UI without needing to continually add new dialogs. Another advantage is that undo/redo is easily handled by saving text versions of records at each modification. Further, users can add arbitrary tags to any record to handle unanticipated data needs without an update to the system.

3.2. Collecting and registering data

A first step in observatory design is to gather datasets providing information about the proposed infrastructure and experimental locations. These include variable resolution bathymetry datasets for underwater terrain, geological maps, images and video of the seafloor, observed and simulated datasets, and site features such as existing telecommunication cables, navigation hazards (e.g., shipwrecks) and legally protected zones.

A zoomable geobrowser interface allows the user to interact with datasets across the entire observatory while retaining context. It provides a viewable environment from the thousands of square miles covered by the RSN, down to a few meters around a specific sensor in an experiment. It also provides a simple layering metaphor for the data to help organize different data types. Figure 2 displays an example of different layered datasets in the same area of the RSN. The first image shows existing buoy locations and surface geology. In the second image, these datasets are turned off or made semi-transparent to examine recorded earthquakes more closely.

Bathymetry for the seafloor comes in a large variety of resolutions: from low resolution 100 km sampled sets based on satellite data, down to 1 meter samples. The terrain sets are commonly in flux as new versions become available, and certain terrain sets are shared only among a small group of scientists, so arbitrary file-based terrain handling is necessary. Our terrain engine uses a quad-tree based tiling system and a greedy algorithm to find the highest resolution point that can be interpolated from the currently loaded datasets. Where bathymetry for an area is undersampled, Catmull Rom splines provide a smoothing function for the surface to avoid faceting of the final terrain mesh. Maps and images can be overlaid on the terrain, as well as user-selected gradient sets with contours or binned color gradient values to highlight depth ranges.

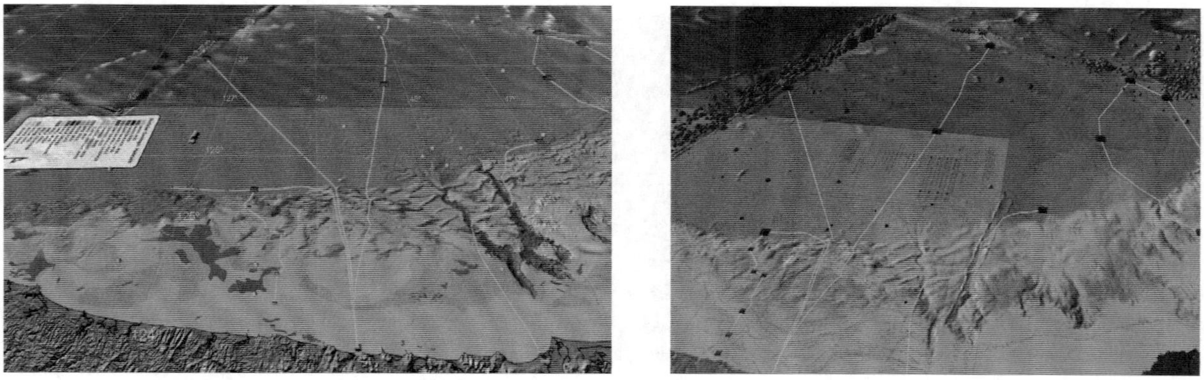

Figure 2. COVE provides layering to quickly review relevant datasets. Here surface geology and buoy locations are easily replaced by earthquake fault maps and recorded seismic events.

COVE provides simple geolocated visualization of observed and simulated data over time. The goal is not to deliver the exhaustive capabilities of scientific visualization packages but rather to provide a set of interactive visual techniques most commonly used by the scientists. Data can be viewed as points, lines, vectors, planes, or isosurfaces using gradients to define the data color. The available visualization methods are expected to be expanded in the future as new scientific needs arise.

3.3. Creating observatory designs

The next step is to facilitate easy generation of sensor layout and cabling. COVE provides a direct manipulation drag-and-drop interface to add and position instruments. The instruments can be represented by icons, common 3D shapes, or instrument specific 3D models. Information summaries for an instrument can quickly be viewed in a popup window (make, model, data format, power consumption, etc.), or the user can be redirected to a detailed web page for the instrument. Based on metadata entered for the sensors, the system has the ability to detect intersensor interference and visualize the interference range.

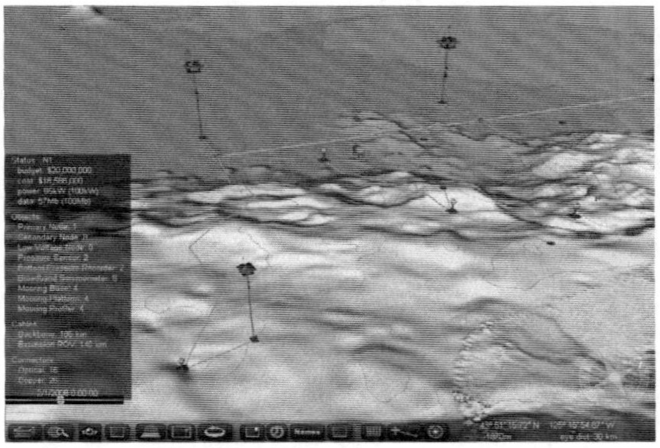

Figure 3. An example of a detailed layout off the Hydrate Ridge site.

Figure 4. Here we have an annotation popup for communicating with a colleague about a site.

Default cabling is provided between new instruments and is automatically updated as instruments are moved. The cables properly follow the bathymetry, and problems such as extreme depth change can be detected and examined in a cabling window, presenting a cross-section view of the cable layout. Cable handles can be arbitrarily added and positioned for maneuvering the cable around obstacles in the terrain or changes in the surface geology of the seafloor. For areas with insufficient bathymetry to completely determine cable positioning, seafloor surveys can be planned by directly dragging a survey field over the area in question.

A set of instruments and cabling can be removed or inserted using common cut, copy, and paste procedures, as well saving them under an instrument template. Templates allow easy insertion of complex instrument packages, such as mooring-based sensors or geodetic arrays. A subset of the layout can also be moved to another layer, making it easy to build up the observatory a section at a time. This also enables different designs to be viewed on top of each other to quickly determine differences. Session-based undo/redo is always available, making it safe and easy to explore different layout options.

Site metrics such as cost, power needs, and bandwidth on user-selected sections of the observatory, are automatically updated on the screen for the user during these editing sessions in a heads-up display. By having budgets readily visible and costs automatically updated with changes, layouts can be quickly refined to meet specific goals (figure 3). Moreover, detailed reports can be viewed in a table format or saved to a file for further analysis in tools such as spreadsheets.

3.4. Communicating and collaborating

Since the observatory is built by a team of scientists and engineers from many different disciplines and locations, sharing information across the team is crucial. COVE offers a set of tools to facilitate effective communication and collaboration. For simple file-based communication, users can always transfer their current work file to another user. Necessary datasets supporting the file layers are pulled from the data repository as allowed by access control. For communication outside the team, COVE allows output to high-resolution images (limited only by graphics memory), movies, and KML files that can be displayed in readily available programs such as Google Earth.

When scientists are satisfied with their changes locally, they can share the design to the data repository, where it is subject to simple check-in and check-out procedures to ensure integrity. In order to allow multiuser editing of the observatory, it can easily be split into multiple layers with different users working on each one. To be kept apace of changes in the repository, users can register to receive messages in COVE (much like an RSS feed), should someone make a change to layouts or data that they care about. They can also register interest in a specific location and be informed of any new occurrences or posts.

New views can easily be created and saved by users to invoke a specific set of camera, layer, and visual settings. These can be used locally for examining work from various points of view or posted to the data repository for sharing with the team. Once a view is posted, another user can investigate and click on the view to activate it in COVE. Views can also be automated, like a slide show, to provide a guided tour through different aspects of the observatory. These views are available as thumbnails in the interface for easy selection and navigation through the observatory.

Users also can place annotations at any location or time in COVE. These annotations are represented visually by a familiar thumbtack icon and include the user, time, and urgency of the comment. Users can click on these to see what has been shared by their colleagues and then add comments if they wish. Because these annotations are stored in the repository with other team data, they are available on the web as well as through COVE and can be filtered and sorted to quickly find relevant information.

4. Results

The results for COVE were positive on many fronts and demonstrate the potential for this type of observatory design tool.

We observed many interactions that highlighted the value of integrating visualization tasks. Scientists would create new layouts and then use saved views to evaluate the results from various angles, particularly between 2D top-down and 3D perspective views. Maps of fault lines and surface geology were displayed to help with interactive cable layout, while using the heads-up display to ensure that

Journal of Physics: Conference Series **125** (2008) 012092 doi:10.1088/1742-6596/125/1/012092

budgets were met. Since the terrain engine allowed collected bathymetry sets to be easily added, the team quickly built up a multigigabyte bathymetry repository for use when creating layouts. The scientists found the layering model for datasets intuitive and integration with Google Earth data formats a real benefit.

COVE was used to help create the primary infrastructure for the RSN in preparation for a recent NSF design review. It allowed different core cabling alternatives to be explored, and helped convince the team of the necessity for a significant change from a ring to a star based core configuration. It was also the key tool for creating visuals to explain the design at the review and to the other observatory teams. The review was deemed a big success for this team, and our users have been vocal in praising COVE as a key contributor to their achievements.

A very encouraging sign is the interest from user groups not part of our original study, who have asked that COVE be extended to handle their observatory visualization needs in areas such as daily operations and inventory tracking. We have also been approached by Earth sensor observatory teams to determine how to extend our solution to accommodate their needs.

Future plans include working aggressively to build in workflow and scripting capabilities. As noted above, these offer a scalable means to load and transform more datasets, while allowing COVE to be optimized for visual interaction. Workflow solutions are also well suited to real-time data flows inherent in the upcoming operations phase of ocean observatories. Another potential area of research is stronger integration with web browsers to determine the effectiveness of web-based collaborative solutions with COVE.

More Information about COVE

For more information about the COVE project, including available downloads of the most recent version, one can visit our website at `http://www.cs.washington.edu/homes/keithg/oceans.html`. More information about the Ocean Observatory Initiative and RSN is available at `http://www.neptune.washington.edu/`.

Acknowledgments

The authors thank Microsoft's Technical Computing Initiative (particularly Roger Barga and Jim Gray), the University of Washington School of Oceanography, and the Monterey Bay Aquarium Research Institute for data and technical advice. This work was supported in part by grants from Microsoft Corporation and the National Science Foundation.

References

[1] Wood J, Dykes J, Slingsby A and Clarke K 2007 *IEEE Transactions on Visualization and Computer Graphics* **13** 1176–1183 ISSN 1077-2626

[2] Google 2008 Google Earth `http://earth.google.com`

[3] NASA 2008 World Wind `http://worldwind.arc.nasa.gov`

[4] Microsoft 2008 Virtual Earth `http://www.microsoft.com/virtualearth`

[5] Magic Instinct Software 2008 Google Oceans `http://www.justmagic.com/GM-GE.html`

[6] Newman H B, Ellisman M H and Orcutt J A 2003 *Commun. ACM* **46** 68–77 ISSN 0001-0782

[7] Coughlan J C and Hogan P 2006 *AGU Fall Meeting Abstracts* A811+

[8] Buschmann C, Pfisterer D, Fischer S, Fekete S P and Kröller A 2005 *SIGBED Review* **2** 1–6 ISSN 1551-3688

[9] Unidata 2008 Integrated Data Viewer (IDV) `http://www.unidata.ucar.edu/software/idv/`

[10] Ware C, Plumlee M, Arsenault R, Mayer L and Smith S 2001 *Proc. MTS/IEEE Conf.* vol 3 pp 1960–1964

[11] Arsenault R, Ware C, Plumlee M, Martin S, Whitcomb L, Wiley D, Gross T and Bilgili A 2004 *Proc. IEEE OCEANS Conf.* vol 22 pp 743–747

Time-varying multivariate visualization for understanding terrestrial biogeochemistry

Robert Sisneros[1], Markus Glatter [1], Brandon Langley[1], Jian Huang[1], Forrest Hoffman[2] and David J.Erickson III[2]

[1] University of Tennessee, Knoxville, TN 37996, USA
[2] Oak Ridge National Laboratory, Oak Ridge, TN 37831, USA

E-mail: {sisneros, glatter, langley, huangj}@cs.utk.edu, {hoffmanfm, ericksondj}@ornl.gov

Abstract. Petascale computing has brought forth a transformational way of doing science. To the global effort on studying climate change, this shift has enabled not only tools more functional and more powerful than before but also a scientific exploration more comprehensive than before. In this work, we report our efforts to employ recent ultrascale visualization technologies (SciDAC Ultravis) to study model comparison in terrestrial biogeochemistry datasets produced by computation (SciDAC C-LAMP). While many of the current efforts are specific to climate modeling research, our method of location-specific summarizing visualization of extreme and normal relative distribution patterns is generally applicable to other fields of computational sciences.

1. Introduction

Our driving application is to visualize SciDAC-caliber [1] datasets from global coupled climate-carbon cycle model simulations produced by different models. We draw our research motivation from the current thrust to generate accurate simulations of the global carbon cycle that model the interactions and feedbacks between the terrestrial biosphere and the climate system.

The Carbon-Land Model Intercomparison Project (C-LAMP), www.climatemodeling.org/c-lamp [2], was initiated to allow the international scientific community to thoroughly test and compare such terrestrial biogeochemistry models through a set of carefully crafted experiments. Well-defined metrics have been established for comparison of model results against best-available observational datasets, and models are graded on their scientific performance with respect to these metrics. Visualization tools and diagnostics are particularly helpful in uncovering model differences and discovering ways for improving individual models.

The visualization aspects of this task are very demanding for several reasons. First, there are a large number of variables involved in each simulation. Exacerbated by the need to study multiple runs in a cohesive manner, the combinatorial space that needs to be explored is overwhelming, even just the task of studying two variables from two simulation runs. For instance, scientists already have some empiric understanding of how net ecosystem exchange relates to net primary productivity. Does the relationship exist as expected in a petascale simulation, including across different time spans of different runs? This model of investigative study and the need of high interactive rates currently present a challenge for large data visualization.

2. C-LAMP

The purpose of this model-measurement intercomparison is to allow the international scientific community to evaluate the performance of biogeochemical models normally coupled to general circulation models (GCMs) [2]. Terrestrial models are scored based on their performance as compared to best-available site, field, and satellite observations through a rigorous set of metrics. To this end, we encourage others to provide feedback on the experimental protocol, the metrics used to evaluate model performance, and the observational datasets available for use in the intercomparison. The C-LAMP project bridges the climate modeling community and the measurement community. This role of C-LAMP enables significant potential of model improvement and more comprehensive measurement campaigns.

The C-LAMP project conducts two types of experiments. In the first type of experiments, biogeochemical land surface models are forced with an improved NCEP/NCAR reanalysis climate dataset. In these offline runs, the objective is to examine the ability of the models to reproduce surface carbon and energy fluxes at multiple sites and to examine the influence of climate variability, prescribed atmospheric CO_2 concentrations, and land cover change on terrestrial carbon fluxes during the 20th century, and specifically during the period for which the reanalysis data are available (1948-2004).

In the second type of experiment, an active atmosphere model is used to couple energy flows between the atmosphere and the terrestrial biosphere. However, atmospheric CO_2 follows a prescribed trajectory for both steady-state and transient components of the experiment. The prescribed CO_2 is radioactively active and sea surface temperatures (SSTs) and ocean carbon fluxes are prescribed. The objective of these simulations is to examine the effect of a coupled biosphere-atmosphere on carbon fluxes and climate during the 20th century.

3. Parallel query-driven visualization

Decadal to century time scale climate simulations typically output a large number of two- and three-dimensional variables at regular intervals, usually monthly. While identifying features is difficult, it is actually intuitive and practical to select only a subset of the data from within that high-dimensional variable space, to obtain a qualitative understanding of the overall results. This approach can be easily implemented if the resulting dataset can be stored entirely in-core. However, to handle larger datasets, a more sophisticated solution is necessary.

Our solution [3] involves designing specialized scalable visualization data servers with large-scale parallelism. Our index system indexes general data items, including vertices, voxels, or particles, and is independent of grid type. The core data structure for indexing is an optimized M-ary search tree. The tree structure amounts only to ~1% the size of the whole dataset. The dataset can be stored externally on hard drives in a compressed manner. Only parts of the data that are used by the scientist are decompressed (and cached). The compression rates vary from dataset to dataset. For some typical datasets, we obtained a 20x compression rates, while we could obtain as low as a 4x compression on highly turbulent or noisy datasets. Using these rates, we conservatively estimate that a mid-sized cluster could already support parallel visualization of a dynamically queried dataset of 1 terabyte.

The M-ary tree uses a large branching factor and serves the role of metadata to guide a search process. The branching factor is one of the primary differences between this search data structure and previous data structures, such as interval tree, k-d tree, quadtree and octree. Because of the large branching factor, M, the M-ary search tree requires little storage space. The data is not stored in the tree, but in a linear list sorted by a key function. The leaf nodes of the tree store only pointers to the respective data records in a sorted linear list. We have discovered that conventional methods to access records by traversing the tree are too expensive, in terms of both caching performance and the large number of addressing operations. We use a novel method to accelerate range searches in an M-ary tree to address this, optimized specifically for multivariate datasets [3].

Data items are partitioned into groups by round-robin assignment according to high-dimensional space filling curve [8] order in attribute space. We use this type of data partition to distribute data amongst all visualization data servers equally. Thereby, we are able to achieve a nearly optimal load-balance for almost all kinds of queries. The M-ary search tree is then used to manage the data on each server. Our approach is relatively easy to deploy on networked commodity computers, whether clustered or not. The necessary number of parallel data servers depends on the size of the dataset.

4. Concurrent visualization of multiple patterns

It is common for scientific visualization production tools to provide side-by-side images showing various results of significance. This is particularly true when handling time-varying datasets with a large number of variables. However, it is often desirable to have a general scientific visualization method to summarize the data into the fewest possible images. As shown in [5] this is a hard problem. It is often impossible without reducing the data of each variable. One way to do so is by feature detection and extraction. Previous authors [6-7] have presented a number of successful example methods based on feature extraction. A common limitation of these methods, however, is the requirement to have features accurately defined before a visualization can be created. This is particularly hard when the purpose is to provide initial concepts of unknown features involving unprecedented numbers of concurrent variables.

For comparative visualization of unknown model differences in C-LAMP, we use a location-specific summarizing method to visualize extreme and normal relative patterns among multiple concurrent attributes [4]. With each attribute representing a different physical variable, a different model or simulation run, we provide a novel comparative perspective of the data. Our motivation is to guide a user's attention to a much-reduced subset of a large and otherwise incomprehensible multivariate dataset. In order to do such guidance without requiring too much user input, we base our methods on relative distribution patterns of multiple variables. In the overall problem space formed by all concurrent attributes, we term each subset of interest in our framework an attribute-specific subspace, because each subspace can eventually be represented by one tag that is in direct correspondence with an attribute.

The input to our method [4] is a general dataset that has, at each location, an associated attribute set. Each attribute must have a value at each location (point). We normalize all attributes to a canonical range. Also, the user supplies attribute target values that allow us to determine if the value at any location is "good," that is, close to its target value. Common examples are: highest possible temperature around the globe, heaviest seasonal rainfall around the globe or longest period of drought around the globe. At each location, classification is based on which attribute at this location is the closest to its target value. Each location is classified and assigned to at most one attribute-specific subspace. When all values on a location are sufficiently far away from the target values, this location is thresholded and assigned to no subspace. If it is a dataset with 20 concurrent variables, after classification we always have 20 attribute-specific subspaces. To render, each subspace gets a different color. In renderings, thresholded locations appear as empty spaces.

Let us illustrate how we intend the technique to be used in the following. A typical time-varying dataset contains a set of variables associated with each point. We can simply treat the temporal values of a single variable associated with each spatial location as different attributes. Please note in our method, extreme vs. normal are equivalent because it's just a matter of picking an extreme value or an average value as the target. In this case, we show, in a location specific manner, for this one variable the timesteps in which interesting features arise. This could be used to discern which timesteps would be most valuable in that variable's evolution over time.

5. Results and discussion

To application scientists, parallel query-driven visualization serves as a data reduction, which we have accelerated to interactive rates [3]. The concurrent visualization method based on attribute subspaces is a subsequent step to summarize multiple patterns according to their relative strengths so that overall trends can be visually compared, analyzed and comprehended [4].

Figure 1. A location-specific summarizing visualization of extreme and normal relative distribution patterns among CASA and CN models in C-LAMP 1.2 and 1.4 runs. The variables considered are NEE (red color), NPP (orange color), and TLAI (yellow color) in January and July 1990.

As typical with any climate modeling efforts, the same models would be run under different conditions. To demonstrate our approach and typical use cases, we show three different types of runs: (i) control run (C-LAMP 1.2), (ii) varying climate transient run (C-LAMP 1.3) and (iii) varying climate, CO_2 and N deposition transient run (C-LAMP 1.3). Under each type of run, there are two models from NCAR Community Climate System Model Version 3 (CCSM3), specifically CCSM3-CASA and CCSM3-CN. Hence we have in total 6 different simulation scenarios to visualize. The time span considered is the decade from 1990 till 1999. In this section, we show major patterns of three variables: net ecosystem exchange of carbon (NEE), net primary production (NPP) and total leaf area index (TLAI). While we find that, in reference to the general climate trends of these three variables, the differences among models are visually discernible in several cases, within both models the results from 1.2 and 1.3 runs seem visually indistinguishable. For reason of space, we do not show results from the 1.3 run.

When viewing more than one variable in one setting, differences in models become apparent. In figure 1, "high," "low" and "mid" mean that the target values for all three variables are chosen to be global maximum, global minimum and mid-way between global extremes, respectively. In terms of relative strength (or relative variation patterns) among NEE, NPP and TLAI, CASA and CN largely agree with high and low distributions in both January and July in 1990 (peak summer and peak winter months). The mid range distributions do reveal differences in high latitude areas, large deserts and mountainous (high elevation) regions.

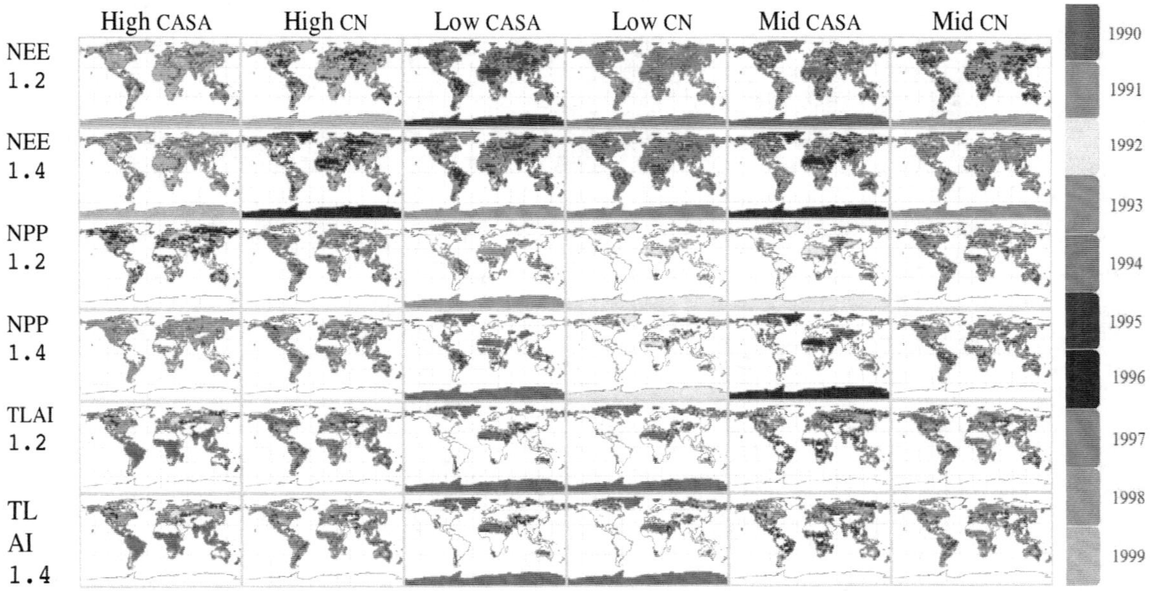

Figure 2. Year-long average computed over 1990-1999. For each of NEE, NPP and TLAI and in 1.2 and 1.4 runs, which year is the closest to the decadal maximum (High), mid-point of the possible range (Mid), and the decadal minimum value (Low)?

In figure 2 with TLAI, 1.2 and 1.4 are visually similar. The heavier coupling in 1.4 does not seem to change relative strength in temporal distribution of yearly averages in the decade. Both CASA and CN seem to agree that 1990 has the lowest TLAI in the decade and in the geographic regions of barren lands (Arctic, Antarctic, Sahara and Tibet). As to relatively high TLAI in the decade, CASA still favors 1990 (closer to decadal-global extremes) in South America and sub-Sahara Africa. Also the earlier three years have high TLAI in all other areas of the globe. This pattern of earlier years in 90's showing high relative TLAI is not in CN runs. With both NEE and NPP, CASA and CN largely agree, but are, however, off by a year or two (both for high and lows) in some cases. For the NPP modeling results (both High-CASA and Low-CASA), the Amazon mostly shows the first half of the 90's to have lower values. In CN, the Amazon shows high NPP more uniformly distributed throughout the decade. No year during the 90s showed "abnormally" low NPP in CN models.

As another example, we calculate Pearson's correlation coefficient between pairs of variables for each simulation scenario over 1990-1999 (figure 3). By viewing all models concurrently, some model trends are clearly visible. Since 1.2 and 1.3 are so similar, only 1.2 is represented in this image, that is, no visible green or yellow. C-LAMP 1.4 is represented as blue and light blue (CASA and CN), so blue in an image represents the more coupled the run. NEE-NPP shows stronger positive correlation in more coupled runs in Eurasia, whereas NEE-TLAI shows stronger positive correlation more in less coupled runs in the same area. With respect to correlation we would expect this viewing to appear completely random, but the presence of contiguous areas of identical color highlight geographical regions of interest with a model.

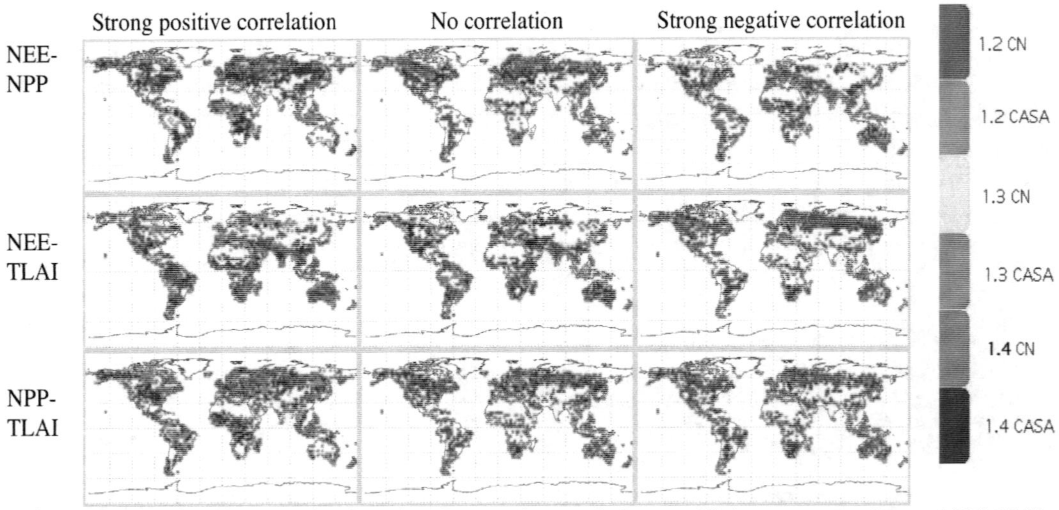

Figure 3. Model differences in bi-variate Pearson's correlation coefficient, computed over 1990-1999.

In the future, we plan to engage in even closer collaboration with C-LAMP researchers because few of the model differences we have identified so far have clear explanations from the modeling community. We also plan more research for evaluating model differences in the context of model biases from real observational data.

Acknowledgment

The primary source of funding for this work is provided through Institute of Ultra-Scale Visualization (www.ultravis.org) under the auspices of the SciDAC program within the U.S. Department of Energy (DOE). This work is also supported in part by NSF grant CNS-0437508 and a DOE Early Career PI grant awarded to Jian Huang (No. DE-FG02-04ER25610). The Carbon-Land Model Intercomparison Project (C-LAMP) are partially sponsored by the Climate Change Research Division (CCRD) of the Office of Biological and Environmental Research (OBER) and the Computational Science Research and Partnerships (SciDAC) Division of the Office of Advanced Scientific Computing Research (OASCR) within DOE Office of Science. This research used resources of the National Center for Computational Science (NCCS) at Oak Ridge National Laboratory (ORNL), which is managed by UT-Battelle, LLC, for DOE under Contract No. DE-AC05-00OR22725.

References

[1] Scientific Discovery through Advanced Scientific Computing 2000 Office of Science (U.S. Department of Energy, Washington, D.C.)

[2] Hoffman F, Covey C, Fung I, Randerson J, Thornton P, Lee Y-H, Rosenbloom N, Stockli R, Running S, Bernholdt D E and Williams D 2007 Results from the Carbon-Land Model Intercomparison Project (C-LAMP) and availability of the data on the Earth System Grid (ESG) *J. Physics: Conference Series* **78:**012026

[3] Glatter M, Mollenhour C, Huang J and Gao J 2006 Scalable data servers for large multivariate volume visualization *IEEE Trans. Visualization and Computer Graphics* **12**(5):1291-1299
[4] Sisneros R, Johnson C R and Huang J 2008 Concurrent viewing of multiple attribute-specific subspaces *Proc. EuroVis'08* (Eurographics/IEEE VGTC Symposium on Visualization, Eindhoven, NL)
[5] Taylor R 2002 Visualizing multiple fields on the same surface *IEEE Computer Graphics and Applications* **22**(3):6–10
[6] Weiler M, Botchen R P, Stegmaier S, Ertl T, Huang J, Jang Y, Ebert D S and Gaither K P 2005 Hardware-assisted feature analysis and visualization of procedurally encoded multifield volumetric data *IEEE Computer Graphics and Applications* **25**(5):72–81
[7] Walter J D and Healey C G 2001 Attribute preserving dataset simplification *Proc. IEEE Visualization 2001* 113–120
[8] Pascucci V and Frank R 2001 Global static indexing for real-time exploration of very large regular grids *ACM/IEEE Conf. Supercomputing (SC'01)* 2–2
[9] Hoffman F, Hargrove W, Erickson D J and Oglesby R J, Using clustered climate regimes to analyze and compare predictions from fully coupled General Circulation Models *Earth Interactions* **9**(10):1-27

Sapphire: experiences in scientific data mining

Chandrika Kamath

Lawrence Livermore National Laboratory, Livermore, CA 94551, USA

E-mail: `kamath2@llnl.gov`

Abstract.

The size and the complexity of the data from scientific simulations, observations, and experiments are becoming a major impediment to their analysis. To enable scientists to address this problem of data overload and benefit from their improved data collecting abilities, the Sapphire project team has been involved in the research, development, and application of scientific data mining techniques for nearly a decade. In this paper, I first describe the Sapphire system architecture that was motivated by the needs of a diverse set of applications. Then, using examples from different domains, I discuss our experiences in mining science data and some of the challenges we faced in analyzing data ranging from a megabyte to several terabytes.

1. Introduction

Over the past ten years, the Sapphire project team (https://computation.llnl.gov/casc/sapphire) at Lawrence Livermore National Laboratory has been involved in the research, development, and application of scientific data-mining techniques to various problems involving data from observations, experiments, and simulations. This paper summarizes our approach to addressing the diverse needs of applications and describes some of the lessons learned in the analysis of scientific datasets.

At the start of the project, discussions with domain scientists indicated that we needed to support a multitude of requirements. The raw data available for analysis was in the form of images, structured or unstructured mesh data with physical variables at each mesh point, or time-series data collected by different sensors. The type of analysis required was problem dependent and ranged from identifying and characterizing structures of interest to classifying structures belonging to a certain category or class. Image data was often noisy, with the noise statistics varying with the sensors used to collect the data. There was also interest in building predictive models, where each experiment or simulation was characterized by several input and output variables and the goal was to predict the output variables for a given set of input variables or to identify key input variables to predict the values of the output variables.

To support these diverse requirements, we designed and built a software toolkit (figure 1) [1] with separate modules for different tasks such as denoising, background subtraction to identify moving objects in video, dimension reduction to identify key characteristics of objects, pattern recognition for clustering, and classification. Recognizing that for each task, different methods were likely to be appropriate based on the data, we used object-oriented techniques to provide a uniform interface to these algorithms. The focus on modularity also carried over to the packaging of the software, with different libraries supporting different tasks, allowing a user to link to only what is necessary for the solution of their problem.

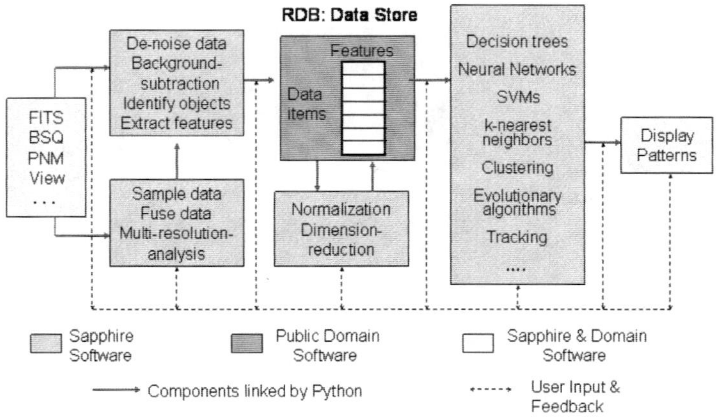

Figure 1. Sapphire system architecture. The compute-intensive parts are highlighted in blue, the domain specific parts in yellow, and the public-domain data store in pink.

Sapphire software incorporates several algorithm advances resulting from our research. It is written in C++; and its modularity, as well as the use of object-oriented design concepts, allows the easy addition of new algorithms as needed. A dataset is usually analyzed by linking several modules together, either through a driver code in C++ or via Python. We have applied Sapphire software to a broad range of problems in different domains, including the following:

- Classification of bent-double galaxies in observational astronomy data [2]
- Separating signals, such as El Niño and volcano signals, from surface temperature data obtained from climate simulations [3]
- Similarity-based object retrieval in two- and three-dimensional simulation data [4]
- Detection of human settlements in satellite images [5]
- Identification of key features associated with edge-harmonic oscillations (EHOs) in sensor data from a tokamak [6]
- Detection and tracking of moving objects in video sequences [7, 8]
- Estimating missing features in multimedia information retrieval [9]
- Comparing simulation and experimental data for code validation [10]
- Characterizing and tracking bubbles and spikes in three-dimensional simulations of the Rayleigh-Taylor instability [11, 12]
- Classification of orbits in Poincaré plots [13]
- Detection of blobs in experimental images from fusion plasma [14]

Using examples from three problems – one each in observational, experimental, and simulation data – I next describe in more detail some of the issues and challenges in mining scientific data.

2. Classification of bent-double galaxies

Our goal in the analysis of the data from the FIRST survey (http://sundog.stsci.edu) was to develop a classifier to discriminate between bent-double and non-bent double galaxies [2]. Bent-double galaxies are of interest to astronomers as they indicate the presence of clusters of galaxies. The astronomers first identified bent-doubles visually in the images and then cross-validated potential bent-doubles with other astronomical surveys. This not only was subjective but also became infeasible as the survey grew to nearly a million galaxies. Figure 2 shows

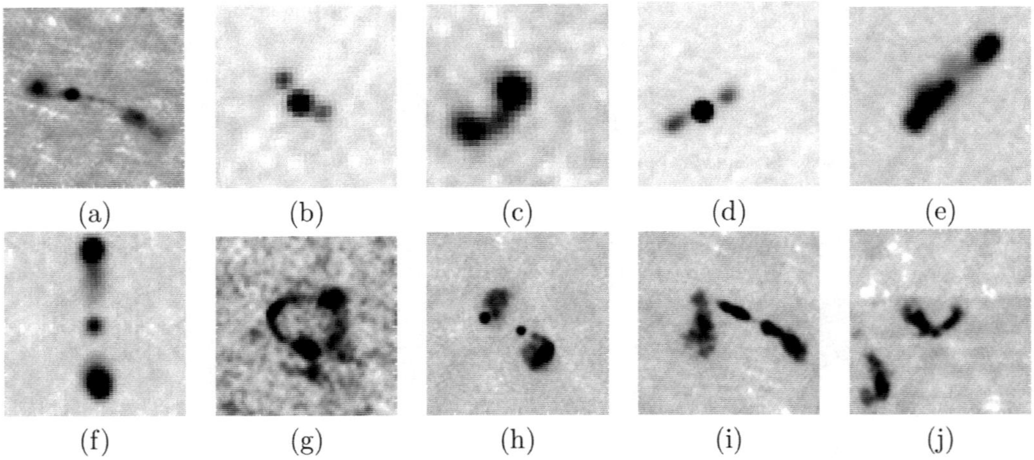

Figure 2. Example radio sources from FIRST (a)-(c) bent-doubles. (d)-(f) non-bent doubles (g)-(j) complex galaxies. Note the similarity between (b) and (d).

some examples of bent-double and non-bent-double galaxies, as well as some galaxies whose morphology is complex enough to be of interest to the astronomers.

Data from the FIRST survey are available in two forms — image maps and a catalog. The catalog is obtained by processing an image map and fitting two-dimensional elliptic Gaussians to each galaxy. Each entry in the catalog describes a single Gaussian and includes features such as the RA and Dec (i.e. coordinates) for the center of the Gaussian, the lengths of the major and minor axes, the peak flux, and the position angle of the major axis (degrees counterclockwise from North). Because of an upper limit on the number of Gaussians used, galaxies with a complex morphology are not approximated well by the catalog. For simpler galaxies, however, the astronomers believed that it made sense to focus on the catalog in our analysis.

Our first task was to identify the catalog entries that formed a single galaxy. Any Gaussians within 0.96 arc minutes of each other were considered to belong to one galaxy. If there were four or more such Gaussians, the galaxy was considered complex enough to be of interest and therefore flagged as such. Galaxies with only a single Gaussian were ignored because they were unlikely to be "doubles," bent or otherwise. We then focused on the two- and three-entry galaxies separately as they are described by different features, resulting in feature vectors of different length.

Once the catalog entries forming a galaxy were identified, they were used to extract features representing the galaxy. Our work on the extraction of features is covered in detail elsewhere [2, 15, 16]. In this paper, I focus on the challenges encountered when we use the features to classify the galaxies as bent-doubles or non-bent-doubles. As an example, I consider three-entry galaxies and decision tree classifiers.

We faced two major issues in classification: the size and quality of the training data and the accuracy of the results. The training data generated manually by the astronomers was relatively small and unbalanced, with 167 bent-doubles and 28 non-bent doubles. Our goal was to use these 195 galaxies to build a decision tree model (with at most 10% error rate) which could automatically classify 15,000 three-entry galaxies without a class label.

We first used our original set of 195 training examples to refine the set of features until the error rate on 10 runs of 10-fold cross validation dropped below 10%. In the process, we removed features that were not very discriminating and identified more robust techniques to extract the features so that they were insensitive to small changes in the data.

We next used the tree to classify galaxies which had not been assigned a label. Our

Method	Gini		Gain Ratio		Info Gain	
	No Pruning	Pruning	No Pruning	Pruning	No Pruning	Pruning
Single tree	22.79 (0.31)	19.77 (0.18)	22.62 (0.27)	19.83 (0.15)	22.77 (0.39)	19.71 (0.41)
Histogram-based single tree	21.73 (0.34)	20.46 (0.29)	20.85 (0.39)	18.81 (0.17)	22.56 (0.32)	20.96 (0.39)
Histogram-based 10 trees	18.69 (0.28)	18.27 (0.30)	18.10 (0.16)	18.02 (0.22)	18.22 (0.18)	18.42 (0.34)
Sampling-based 10 trees (50%)	18.21 (0.23)	17.31 (0.17)	17.83 (0.20)	17.48 (0.15)	17.79 (0.19)	17.58 (0.18)
Adaboost 10 trees	21.87 (0.42)	20.40 (0.45)	22.37 (0.53)	22.56 (0.47)	20.50 (0.45)	20.75 (0.43)
Bagging 10 trees	19.40 (0.28)	18.35 (0.34)	18.98 (0.34)	18.12 (0.26)	18.98 (0.36)	18.52 (0.35)
ArcX4 10 trees	20.48 (0.39)	20.12 (0.20)	21.67 (0.35)	22.48 (0.41)	21.06 (0.22)	19.77 (0.37)

Table 1. Cross-validation error (std error) using different classification methods, using 485 instances.

original plan was to have the astronomers validate these class labels. Though we built a simple interactive tool to make the process easy, we found that it was still rather tedious, subjective, and inconsistent as the labels assigned by an astronomer were subject to the drift common to human labelers. Therefore, we were able to validate only 290 galaxies, of which 92 were bents and 198 non-bents, to compensate for the original unbalanced data.

We evaluated the performance of various tree-based classification algorithms on this new training set of 485 galaxies. These algorithms include the commonly used techniques of Adaboost, bagging, and ArcX4, as well as two methods we proposed, called ASPEN, for Approximate Splitting for Ensembles. One of these methods uses a histogram instead of sorting all the feature values at each node, and the other uses a sample of the instances at each node to make the decision. We consider three splitting criteria: Gini, Gain ratio, and Info Gain [17]. Table 1 presents the cross-validation error (and standard error) of 10 runs of 10-fold cross-validation for this data both with and without pruning. We observe that pruning reduces the error rate by avoiding overfitting, which is to be expected. Also, the use of ensemble methods, where more than one tree is created from the same training set using randomization and the results combined using voting, work better. This is again to be expected as decision trees have high variance, which is reduced through the randomization and voting. The best results are obtained with the two ASPEN approaches, followed closely by bagging.

We also observe that the error rate is now closer to 20% though the tree built with the original training data had an error rate of less than 10%. This, too, is expected because the refinement of features to reduce the error rate was done for the original training set which has a disproportionate number of bent-doubles. It is unlikely that the same set of features would be the best for the enhanced training set. In addition, the 290 galaxies added after validation constituted mainly galaxies that we (non-astronomers) found it difficult to classify. Unfortunately, the astronomers too had problems with these galaxies, either disagreeing on a label or being inconsistent with their labeling when shown the same galaxy on different occasions.

Ideally, we should have again refined the features for the new training set to see if we could reduce its error. However, as we looked more closely at the galaxies that were being misclassified, we found that these were the difficult-to-classify galaxies, where even the astronomers disagreed on the class label. This suggested that there was a limit to the accuracy we could expect given the quality and size of the training data. Fortunately, the astronomers were interested in a ranked list of bent-double galaxies as they wanted to schedule telescope time to further study these galaxies. Therefore, we used the different algorithms to rank order the bent-doubles, with a galaxy identified as bent-double by n algorithms being ranked higher than one classified as bent-double by m algorithms, if $m < n$- This approach allowed the astronomers to prioritize the galaxies which they wanted to investigate further via a telescope.

This classification step in the identification of bent-double galaxies illustrates several challenges. First, the training data may be small, unbalanced, and of poor quality, with mislabeled instances. We must therefore be careful in interpreting the accuracy results. Further, achieving high accuracy may not necessarily be the goal of the scientists; given the subjectiveness and inconsistency in labeling, it is not clear whether high accuracy can always be achieved or is desired.

3. Identification of key features for EHOs

Sometimes the goal in a scientific application is not to build a predictive model but to discover a set of features that may provide insights into the phenomena of interest. In our work in the analysis of sensor data from fusion experiments at the DIII-D tokamak, we were interested in determining which sensor variables were key to the presence of edge-harmonic oscillations (EHOs) in the tokamak. The preferred mode of operation of a tokamak is the high-confinement mode. This comes at a significant cost due to effects of edge localized modes (ELMs) which can cause rapid erosion or even destruction of some components. Recently, a "quiescent H-mode" of operation, without ELMs, has been observed in the DIII-D tokamak. Associated with this mode is a phenomena known as the edge harmonic oscillation (EHO), which can be identified both visually and automatically using simple rules derived from the visual analysis.

Our goal in this analysis was to identify which of the variables being measured by different sensors were relevant to the presence of EHOs. First, with input from the physicists, we extracted the values of 37 candidate variables that describe approximately 700 experiments, each lasting about 6 seconds. Each 50 ms time window of each experiment received a binary label (high/low EHO-ness) using the program that detects EHOs. Next, we preprocessed the data to discard the time windows that contain at least one missing value caused by either an inactive sensor or data from all sensors not being sampled at the same rate. A visual examination of box-plots and histograms of the data revealed many outliers. Using the median value of each variable in each time window eliminated some outliers, but since many still remained, we decided to eliminate the time windows that contained at least one variable in the top or bottom percentile of its range. After the preprocessing, our training set consisted of 41818 instances. Note that unlike the problem of classification of bent-double galaxies, our training data for this problem is not small as it was generated using automated techniques.

The key challenge in this problem was the validation of the results from feature selection algorithms. The scientists did not have any preconceived notion of what features were likely to be important; therefore, we used several approaches to gain confidence in the results. First, as described earlier, we improved the quality of the data so our results were not skewed by the presence of outliers. Second, we tried several different algorithms for feature selection. While we did not expect all techniques to give identical results, we would trust the results more if several techniques selected the same set of features as important. We did not consider one of the most commonly used techniques, namely, principal component analysis (PCA), as it linearly transforms the data into a lower dimensional space, whereas the scientists were interested in a subset of the original 37 features. Third, we introduced a "sentinel" random feature that is uniformly distributed in the interval [0,1]. The rank of this feature was another measure of how much we could trust the results from an algorithm.

We considered the following feature selection techniques:

- The *PCA filter* [18] is a technique derived from the PCA. It first performs a PCA on the data. Then, starting with the eigenvector corresponding to the smallest eigenvalue, it discards the variable with the largest coefficient (in absolute value) in that vector. It then proceeds to the eigenvector corresponding to the next smallest eigenvalue and discards the variable with the largest coefficient, among the variables not discarded earlier. The process is continued until all variables are ranked.
- The *distance filter* uses the Kullback-Leibler (KL) distance between histograms of feature values to estimate how well a feature can separate the data into the two classes, EHOs and non-EHOs. We first discretize the numeric features using $b = \sqrt{|D|/2}$ equally-spaced bins, where $|D|$ is the size of the training data. The histograms are normalized to estimate the probability, $p_j(d = i|n)$, that the j-th feature takes a value in the i-th bin of the histogram, given a class n. For each feature j, we calculate the class separability as

$$\Delta_j = \sum_{m=1}^{c} \sum_{n=1}^{c} \delta_j(m, n),$$

where c is the number of classes (2, in our case) and $\delta_j(m,n)$ is the KL distance between the distributions corresponding to classes m and n for feature j:

$$\delta_j(m, n) = \sum_{i=1}^{b} p_j(d = i|m) \log \left(\frac{p_j(d = i|m)}{p_j(d = i|n)} \right).$$

The features are ranked by sorting them in descending order of the distances Δ_j as a good feature will have a large value of the KL distance.

- The *Chi-square filter* ranks features by sorting them in descending order of Chi-square statistics computed from their contingency tables. The contingency tables have one row for every class and the columns correspond to possible values of the feature. Numeric features are represented by histograms, so the columns of the contingency table are the histogram bins. The Chi-square statistic for feature j is

$$\chi_j^2 = \sum_i \frac{(o_i - e_i)^2}{e_i},$$

where the sum is over all the cells in the contingency table, o_i stands for the observed value, and e_i is the expected frequency of items.

- The *stump filter* is based on the approach used in decision trees to determine the split at each node. The filter considers only the split at the root node of the tree that is based on all the data. Each feature is examined in turn and a split found that minimizes (maximizes) an impurity (purity) measure. We rank the features using the Gini index [17],

$$\sum_b \frac{n_b}{n} \left(1 - \sum_c \left(\frac{n_{bc}}{n_b} \right)^2 \right)$$

, where n is the total number of instances, n_b is the number of instances in branch b, and n_{bc} is the number of instances of class c in branch b.

- The *boosting approach* is an iterative method where in each iteration, the algorithm ranks the features that have not been selected so far and adds the highest-ranking feature to the feature subset. Then, a classifier (naive Bayes, in our case) is trained, and weights are assigned to the examples so that misclassified ones have higher weight. These weights are used in the KL distance method to rank the features unselected thus far.

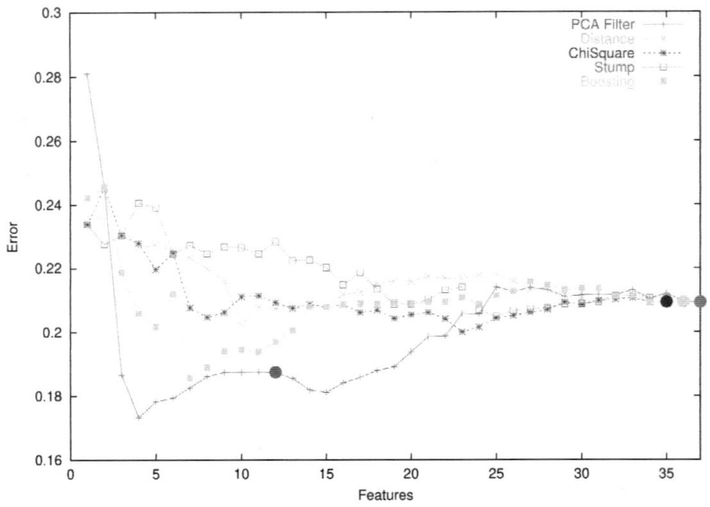

Figure 3. Error rates varying with the number of features for the fusion data. The large dots represent the rankings of the random noise feature.

Figure 3 presents the error rates of a naive Bayes classifier trained on increasingly large feature subsets [6]. The PCA filter produced a compact feature subset that results in the lowest classification error of 17.3% compared to 20.9% obtained with all the features. Note, however, that the PCA filter ranks the noise feature (indicated by the red dot) highly. This would lead us to doubt the features selected by this method, especially as the features ranked high by the PCA filter are ranked much lower by other methods. The remaining four methods are remarkably consistent in the features they select as important: six features were ranked in the top ten by four methods, and an additional three were ranked in the top ten by three methods.

This example illustrates that the quality of the data must be checked before we start the analysis and we must use caution in our interpretation of the results. Further, the use of more than one analysis method is often necessary when the phenomena being analyzed is poorly understood.

4. Counting bubbles and spikes in Rayleigh-Taylor simulations

I next describe a problem where not only are the phenomena being simulated poorly understood but the dataset is so large that one has to be ingenious in solving the problem. Our goal was to identify, count, and track coherent structures known as bubbles and spikes in three-dimensional simulations of the Rayleigh-Taylor instability. This occurs when an initially perturbed interface between a heavier fluid on top of a lighter fluid is allowed to grow under the influence of gravity. The fingers of lighter fluid penetrate the heavier fluid in what are referred to as 'bubbles', while 'spikes' of heavier fluid move into the lighter fluid. With time, these structures, which are initially distinct, continue to evolve; in the process, they may grow, split, merge with surrounding structures, or shrink in size relative to other structures that grow and overtake them.

Our work in using image analysis to count and track the bubbles and spikes is documented elsewhere [11, 12]. Here, I focus on two aspects of the work, namely, the exploitation of domain-specific information and the sensitivity of the results to parameters in the analysis algorithms.

As an example, I will use the larger of two datasets we analyzed, which was obtained using a direct numerical simulation on a three-dimensional computational domain in the form of a cube with uniformly spaced grid points in the x, y, and z directions, and $\Delta x = \Delta y = \Delta z$. This allowed us to consider the domain as a three-dimensional image, with a grid-point corresponding

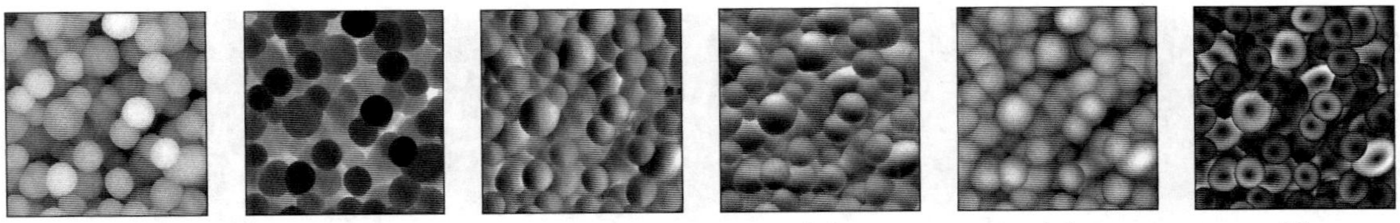

(a) (b) (c) (d) (e) (f)

Figure 4. A 300 × 300 sub-image of the top view of the bubble surface illustrating the (a) height of the surface from original fluid interface; (b) pressure; (c) x-velocity;(d) y-velocity; (e) z-velocity; (f) magnitude of the x-y velocity.

to a pixel. There are 3072 grid points along each dimension; the simulation is run for 248 time steps, and five variables are output in single precision at each time step: the pressure, the density, and the x, y, and z velocities at each grid point, resulting in an 80-terabyte dataset. Note that unlike an image, the data at each grid point are floating point values, not integers. The simulation was run on a parallel machine using 16,384 processors at early time and 65,536 processors at late time. It took 17 days to complete, accounting for a total of 2303 single-CPU years. The 3072^3 grid points were partitioned among the processors in vertical columns.

The large size and the distributed nature of the data were not the only challenges faced in the analysis of the data. A key issue was the lack of a precise definition for a bubble, especially one that holds over all time. Our approach was to start with a small subset of the data at every n-th time step, experiment with various algorithms and parameter settings, and then try the best ones on ever larger subsets of the data.

We used a two-phase approach. In the first phase, we first converted the 3-D data into 2-D by using a simple region-growing technique to identify the boundary of the bubbles (spikes) and then considering the image formed by the top (bottom) view of this surface. In the second phase, we counted the bubbles and spikes in these 2-D images. Figure 4 shows a 300 × 300 subset of the full 3072 × 3072 images of the top view of various variables at the bubble boundary. These include the height-map, which is the height of a point on the bubble surface relative to the original interface; the pressure; the $x-$, $y-$ and $z-$ velocities; and the magnitude of the x-y velocity. These images are at time step 50 in the simulation, where the initial perturbation has grown to form clearly identifiable bubbles. As the bubbles grow, merge, split, and die off, their shape changes considerably and it is no longer easy to identify the extent of each bubble.

We next applied image segmentation techniques to the data in figure 4(a) to identify the extent of each bubble, and thus obtain their count. We used a simple region-growing technique which considers neighboring pixels with nearly the same height as forming a single region as shown outlined in red in figure 5 panel (a). Panel (b) shows the regions after cleanup to remove small regions, merge regions completely within each other, and remove odd shaped regions.

We had also observed that the x-velocity at the bubble surface was negative on the left half and positive on the right; similarly, the y-velocity was negative on the bottom and positive on the top. So, combining the two images could identify the center region of a bubble. Figure 4(e) shows the magnitude of the x-y velocity which is small (indicated by the dark pixels) at both the center of a bubble as well as its perimeter. We could isolate the center pixels as the height near the center did not change as rapidly as near the perimeter (see Figure 5(a)). The centroid of these center pixels formed the tip of a bubble, as highlighted in green in figure 5(c).

This exploitation of the domain information led to a much faster algorithm for counting bubble tips, taking only 8 seconds to process a 3072 × 3072 image in contrast with 2800 seconds for the region-growing approach. As shown in [11], the results from several variants of both approaches are very similar, building our confidence in the results.

For further verification, we considered the sensitivity of the results to the choice of parameters

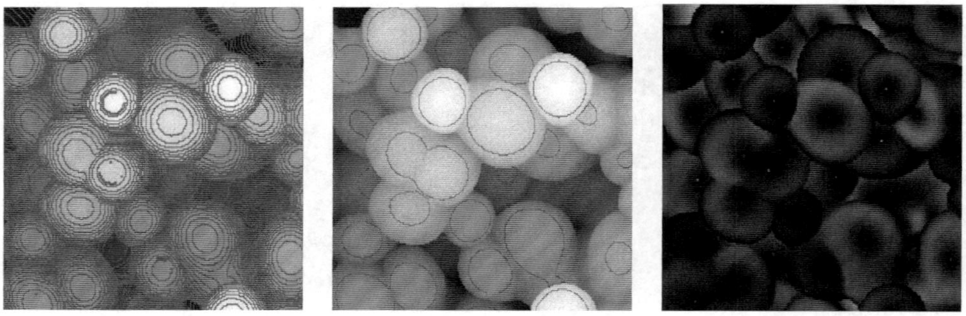

(a) (b) (c)

Figure 5. The 300 × 300 height map image from Figure 4 showing the result of segmentation using region growing (a) before and (b) after cleanup. (c) The magnitude of the x-y image with the bubble tip indicated in green.

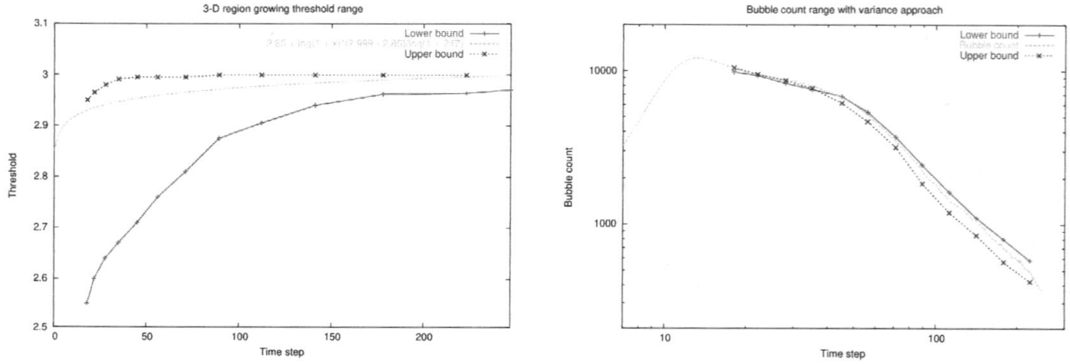

(a) (b)

Figure 6. (a) The values of the threshold used in phase I of bubble counting. The threshold used in the analysis is shown in green, while the blue and red curves illustrate the range over which we can vary the threshold without a large variation in the bubble counts, as shown in (b).

in our algorithms. In the second phase of the analysis, namely, the identification of bubbles in the two-dimensional images, we could visually check that we were counting almost all the bubbles. However, we also needed to confirm that the results of the first phase were not sensitive to the choice of parameters. Our simple region growing algorithm used a single threshold on the density variable. We varied the value of this threshold and observed how the bubble counts changed when we used a fixed algorithm and set of parameters for phase 2. Since were working with three-dimensional data, we conducted this study at select time steps. Figure 6 shows in green the threshold we used in our analysis and in blue and red the range over which we could change this threshold without a large change in the bubble counts. These plots show that small changes in the threshold parameter do not result in large changes in the bubble counts.

This problem illustrates the challenges faced in analysis of massive data sets, especially when the structures of interest in the data are poorly defined. It then becomes important to plan the analysis carefully, exploiting domain information where possible and conducting sensitivity studies to gain confidence in the results.

5. Summary

In this paper, I described the work done by the Sapphire team in the analysis of scientific data. Using three examples, I illustrated some of the challenges in scientific data mining and our solution approaches. These include improving the quality of data to handle outliers and unbalanced training data; using multiple algorithms and sensitivity analysis to gain confidence in the results; and exploiting domain knowledge where possible to simplify the analysis, improve the quality of the data, and increase the size of the training set.

Acknowledgments

I gratefully acknowledge the contributions of the Sapphire project team to both the implementation of the software and the analysis of data from various applications. The domain scientists graciously shared their data and expertise. The work was partially funded by the DOE Office of Science SciDAC program, the DOE NNSA ASC program, and the LDRD program at Lawrence Livermore National Laboratory.

This work performed under the auspices of the U.S. Department of Energy by Lawrence Livermore National Laboratory under Contract DE-AC52-07NA27344.

References

[1] Kamath C 2006 *Journal of Physics Conference Series, Volume 46* pp 500–504
[2] Kamath C, Cantu-Paz E, Fodor I and Tang N 2002 *IEEE Computing in Science and Engineering* **4** 52–60
[3] Fodor I and CKamath 2003 *Proceedings, Independent Component Analyses, Wavelets, and Neural Networks, SPIE Proceedings, Volume 5102* (SPIE) pp 25–36
[4] Kamath C, Cantú-Paz E, Cheung S C, Fodor I and Tang N 2005 *Next Generation of Data Mining Applications* ed Kantardzic M and Zurada J (New York, NY: John Wiley) pp 211–232
[5] Kamath C, Sengupta S K, Poland D N and Futterman J A H 2003 *Image Processing: Algorithms and Systems II (Proceedings of SPIE/IS&T Volume 5014)* pp 270–280
[6] Cantú-Paz E, Newsam S and Kamath C 2004 *Proceedings of the SIGKDD International Conference on Knowledge Discovery and Data Mining* pp 788–793
[7] Cheung S C and Kamath C 2005 *Eurasip Journal on applied signal processing* **14** 2330–2340
[8] Kamath C, Gezahegne A, Newsam S and Roberts G M 2005 *Image and Video Communications and Processing, Proceedings of SPIE Volume 5685* pp 442–453
[9] Bagherjeiran A, Love N S and Kamath C 2007 *Proceedings, IEEE International Conference on Image Processing, Volume II* pp 233–236
[10] Kamath C and Miller P L 2007 *IEEE International Conference on Image Processing, Volume III* pp 525–528
[11] Kamath C, Gezahegne A and Miller P L 2006 Analysis of Rayleigh-Taylor instability, Part I: Bubble and spike count Tech. Rep. UCRL-TR-223676 Lawrence Livermore National Laboratory
[12] Gezahegne A and Kamath C 2008 *Proceedings, IEEE International Conference on Image Processing* to appear
[13] Bagherjeiran A and Kamath C 2006 *Proceedings of the SIAM International Conference on Data Mining* pp 574–579
[14] Love N S and Kamath C 2007 *Proceedings, Applications of Digital Image Processing, XXX, SPIE Conference 6696*
[15] Kamath C, Cantú-Paz E, Fodor I and Tang N 2001 *Data Mining for Scientific and Engineering Applications* ed Grossman R, Kamath C, Kegelmeyer W, Kumar V and Namburu R (Boston, MA: Kluwer) pp 95–114
[16] Fodor I K and Kamath C 2001 *Computational Statistics and Data Analysis* **41** 91–122
[17] Breiman L, Friedman J, Olshen R A and Stone C 1984 *Classification and Regression Trees* (Boca Raton, Florida: CRC Press)
[18] Joliffe I T 1973 *Applied Statistics* **22** 21–31

Journal of Physics: Conference Series **125** (2008) 012095 doi:10.1088/1742-6596/125/1/012095

Quantitatively driven visualization and analysis on emerging architectures

Patrick McCormick, Erik Anderson, Steven Martin, Carson Brownlee, Jeff Inman, Mathew Maltrud, Mark Kim, James Ahrens and Lee Nau

Los Alamos National Laboratory, Los Alamos, NM 87545, USA

E-mail: pat@lanl.gov

Abstract. We live in a world of ever-increasing amounts of information that is not only dynamically changing but also dramatically changing in complexity. This trend of "information overload" has quickly overwhelmed our capabilities to explore, hypothesize, and thus fully interpret the underlying details in these data. To further complicate matters, the computer architectures that have traditionally provided improved performance are undergoing a revolutionary change as manufacturers transition to building multi- and many-core processors. While these trends have the potential to lead to new scientific breakthroughs via simulation and modeling, they will do so in a disruptive manner, potentially placing a significant strain on software development activities including the overall data analysis process. In this paper we explore an approach that exploits these emerging architectures to provide an integrated environment for high-performance data analysis and visualization.

1. Introduction

Simulation and modeling provide a cornerstone for many efforts within the broad scientific community. The hardware requirements for these applications range from a single desktop computer to machines with thousands of processors. Over the past ten years the increasing clock speeds of microprocessors has provided the performance improvements required to meet the ever-increasing demands of computational science. The fundamental processor design supporting these tasks is undergoing a revolutionary change that will impact systems at all scales. Furthermore, datasets are continuing to increase in both resolution and complexity. These factors place a significant challenge to successfully leveraging the power of future large-scale computing resources. This section provides an overview of the current trends in microprocessor design, the growth of data sizes and complexity, and discusses the impact they will have on the overall data analysis and visualization process.

While improvements in semiconductor manufacturing continue to allow for increasing transistor densities, the physical limits of these designs (heat dissipation and power consumption) are placing a significant strain on supporting faster clock speeds. In response, the computer industry has switched to building chips assembled from several simplified, more power efficient processor cores. With this change, future performance improvements will be achieved not by increasing clock frequencies but by increasing the amount of parallelism within processors. Over the next several years this trend will lead to fundamental changes in the way we design algorithms and develop software.

Coupled with the changes in the design of central processing units (CPUs), graphics hardware has rapidly evolved over the past decade from a fixed-function hardware pipeline, into a programmable parallel processor with hundreds of cores. These highly parallel processors have computational and memory bandwidth rates that significantly exceed the capabilities of the most recent CPU designs. For example, AMD and NVIDIA have recently announced new graphics processors (GPUs) with approximately one teraflop of (theoretical) single-precision peak performance [1, 2]. (Note that while both processors support double-precision operations, they are expected to perform at significantly lower rates on the order of 100 gigaflops.) Furthermore, several efforts have shown that GPUs can provide substantial performance improvements and can be successfully used in cluster-based computing environments [3, 4, 5]. Although there are distinct disadvantages and limitations of using GPUs for general-purpose computation, manufacturers are slowly introducing hardware features and programming tools that are making them more suitable for the scientific community. Finally, the improvements in graphics processors are also supporting the development of new interactive rendering and visualization techniques that were difficult, or impossible, on previous generations of the hardware. Much like the changes to CPU architectures, the substantial levels of parallelism, and the nuances of programming models and hardware designs used in GPUs represent a significant challenge to developing efficient, high-performance software.

Further trends within the microprocessor industry suggest that it is likely that many features of GPUs will be incorporated as additional cores within the traditional homogeneous core designs used today. IBM's Cell Broadband Engine provides an early glimpse into a chip that leverages this heterogeneous core design [6]. While such designs have the potential to improve performance, it is likely to further complicate programming models. The extent to which software can fully leverage increasing levels of parallelism within a single chip, in the form of homogeneous or heterogeneous cores, is likely to be the limiting factor on overall performance in the future. This revolution in processor design will have a broad impact across all of modeling and simulation – including data analysis and visualization operations. This will become critical as we confront the continuing challenge of ever-increasing data set size and complexity.

Over the past decade we have seen the sizes of datasets increase along with the growth in processing power. This situation has quickly overwhelmed our capabilities to explore, hypothesize, and thus fully interpret the underlying details within these datasets. In addition to increasing spatial resolution, we are now producing data with ever more complexity. This complexity is due to increasing numbers of variables that are required to faithfully represent the behavior of complex and dynamic systems and advanced techniques such as adaptive grid structures. As computational resources begin to reach peta- and exa-scale sizes, we will face a significant challenge in addressing the needs of analyzing and visualizing these data. There are two fundamental issues that arise when processing datasets of this magnitude. First, efficiently and effectively processing enormous datasets will require computational capabilities and capacities similar to those that were used to produce the data. It is simply the case that we are well on the way to producing more data than we can possibly store. Second, the set of algorithms and supporting code used to comprehend the myriads of information contained within massive datasets will also increase in complexity. Each of these aspects will be directly influenced by the architectural issues discussed above. The challenge of developing software that scales with the number of processor cores will directly impact our ability to deal with increasing amounts of information and thus limit our ability to make new scientific discoveries.

2. Impact on the scientific workflow

The performance growth of computers has allowed them to become a fundamental part of scientific research. This has allowed us to replace or augment the traditional approaches used in the pursuit of understanding complex systems ranging from the interactions of subatomic particles, to the large-scale phenomena responsible for the formation of the universe. In this paper we explore a subset of the data analysis challenges faced by applications as they confront rapidly changing computer architectures and the continuing onslaught of data. Of particular importance to our discussion is the relationship between the overall data analysis process and the scientific method. Figure 1 presents a simplified version of a workflow for the associated set of tasks. The stages of this process consist of an iterative series

Journal of Physics: Conference Series **125** (2008) 012095 doi:10.1088/1742-6596/125/1/012095

of both qualitative and quantitative studies of data. The qualitative presentations assist in the overall understanding of the data. In comparison, a combination of numerical studies and quantitatively driven visualization operations contribute to the interpretation and assessment of the results. Together these two sets of tasks provide support by gathering observable, empirical and measurable data to assist in the formulation and testing of hypotheses. Throughout this process visualization plays a critical role in helping scientists understand large amounts of information. As the size and complexity of datasets grow, it becomes increasingly difficult to display information at interactive rates. This situation affects the ability to leverage the full power of our visual system; thus directly impacting the efficiency of the workflow.

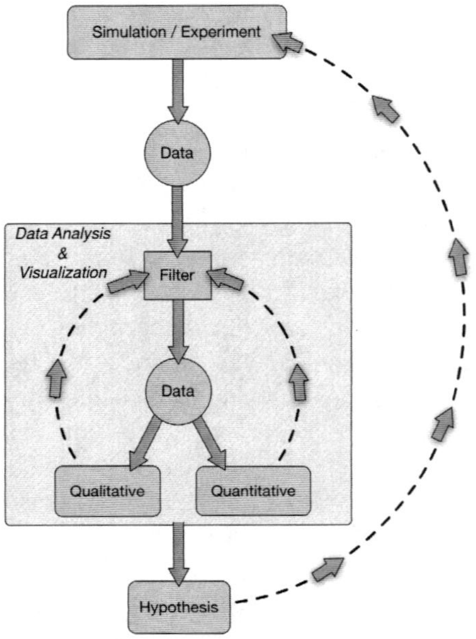

Figure 1. Workflow for data analysis and visualization relative to the overall scientific method. The results from a simulation, or experiment, are run through a series of filters that are used to produce both qualitative and quantitative results. This process is repeated until a hypothesis is tested and/or found to be true with further experimentation.

A wide variety of tools is available to scientists to use during this workflow. These tools range from commercial and open-source products to custom application-centric software. It is not uncommon to find scientists within the same fields using different software to accomplish the same tasks. This set of tools is often very disjoint in terms of both the features they provide as well as their abilities to leverage high-performance computing resources and efficiently process large datasets. For example, while many simulations require hundreds to thousands of processors, the resulting data is often processed using tools that are capable of running only on a single desktop computer. In the worst case, this lack of cohesion can significantly impact productivity and often leaves vast portions of data unexplored. The difficulty of developing analysis software that can process extremely large datasets, leverage the power of large-scale computing resources, and meet the specific needs of individual application domains presents an significant roadblock to the scientific community. We believe a portion of these challenges can be addressed by providing users with direct programmatic control over fundamental numerical and visualization operations.

3. Direct programming

Many of our past efforts have focused on providing interactive visualization and rendering algorithms for dealing with large-scale data sets. Not surprising, such approaches have solved only a very small portion of the overall challenge faced by today's scientific applications. Although they represent possible building blocks, these algorithms fail to fully address the higher-level issues related to providing users with a set of flexible tools for exploring and evaluating data in both a qualitative and quantitative fashion. As the popularity of several software packages has shown, it is beneficial to have the ability to directly perform mathematical operations when exploring and analyzing data. For example, the success of the Python programming language combined with libraries like `matplotlib` speaks for itself [7, 8]. However, these tools rarely address the full set of challenges associated with scaling to process large datasets and only a few are beginning to leverage the full capabilities of emerging hardware. Addressing this challenge has been the fundamental focus of the Scout project, which provides a parallel programming language for data analysis and visualization [9, 10]. The use of an explicit parallel language helps to hide the nuances of the programming models of the underlying architectures – including multicore CPUs, GPUs, and cluster-based environments. Furthermore, analysis, visualization, and rendering computations can all be carried out within the language (and supporting runtime environment) and potentially improve the productivity of the workflow discussed in Section 2. As we have previously presented, we see this approach as playing a fundamental role in an integrated data analysis and visualization framework aimed at assisting in the generation of new scientific insights and results [11].

Before presenting an advanced set of operations using Scout, we begin with two simple examples that present the basic functionality and provide an introduction to the programming model. Scout is based on the ideas and structures of the C* and CM FORTRAN programming languages [12, 13]. Both of these languages, initially designed for Thinking Machines supercomputers, support a data-parallel programming paradigm that fits well with other languages used within the community and can be matched to the capabilities of the underlying hardware. Scout extends these languages by adding structures that support visualization and rendering operations that support controlling the color generated for each resulting pixel. All of the examples presented in this paper present language keywords in red and built-in language functions as blue.

```
// Simple 2D color-mapped slice example.
// pt = potential ocean temperature.

mask(pt,-999); // mask out land values

render with(shapeof(pt[:,:,0:0]) {
  // use hue-saturation-value space
  // to color blue to red (cool to warm)
  image = hsv(240*(1.0-norm(pt)),1,1);
}
```

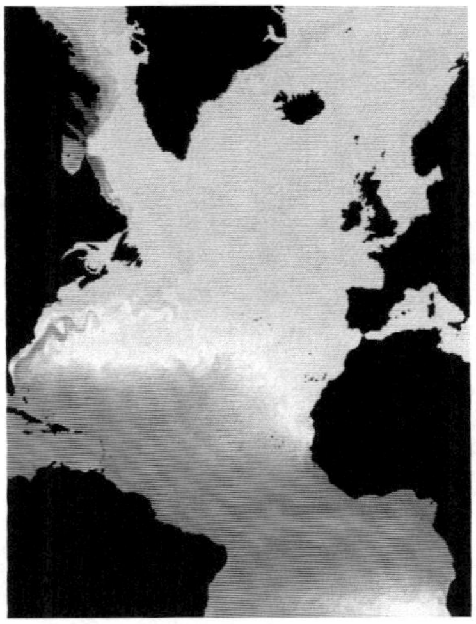

Figure 2. Sample Scout code showing a simple two-dimensional visualization operation. The `mask` operator removes locations with a value of −999 from any computations.

The first example shown in figure 2 shows how a two-dimensional slice can be extracted from a three-dimensional field, and then rendered according to the minimum and maximum data values within the slice. The `with` construct can be best thought of as enabling a set of processors that execute the operations contained within the enclosing braces in parallel. In this example a range operator (`[:,:,0]`) is used to select all the values over the x and y dimensions and only the first slice of the z dimension of the supplied dataset – producing a two-dimensional array of data values. The body of the supporting `with` statement then simply normalizes the input values into the range $[0.0, \ldots, 1.0]$ and colors each resulting pixel in the output `image` by mapping them into the $[0, \ldots, 240°]$ hue range (blue to red) of the HSV color space [14].

In a similar fashion, we can compute an isosurface and color it by another field. Figure 3 shows the code for and results of executing this operation on the same dataset presented in figure 2. Like the previous example, this code also enables a set of parallel processors but these are only active over the generated locations within the isosurface; therefore, the enclosed `range` operation is only for the salinity values over the surface and not a global data range. During this process, Scout will leverage a combination of both CPU and GPU resources to complete the task; this is done regardless of the computational platform being used, be it a laptop computer or cluster of systems. The next section provides further details on Scout's ability to leverage the power of multicore and cluster-based computing resources.

```
// Isosurface at 20.0 degrees.
render isosurface with(temp, 20.0) {
  // Find the range of salt values over
  // the generated surface.
  float smin, smax, srange;
  range(salt, smin, smax);
  srange = smax - smin;
  // Color output pixels by salt range
  // (blue to red).
  image = hsv(240 * ((salt - smin)/srange),
              1, 1);
}
```

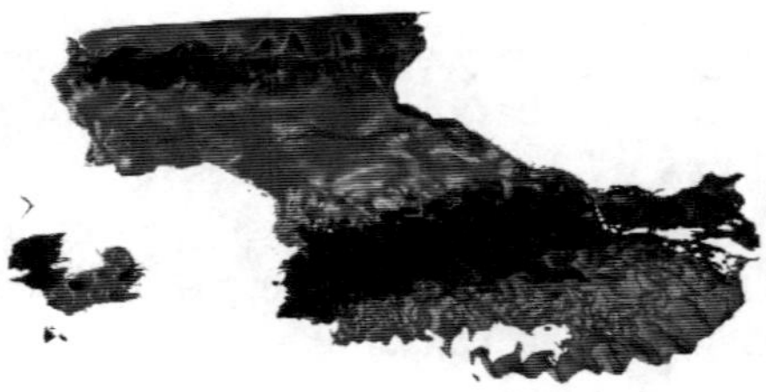

Figure 3. Sample Scout code showing a simple isosurface of temperature at 20 degrees Celsius colored by corresponding salinity values. Note that the inclosing `with` block forces computations to occur only over the points within the generated surface.

4. Examples

The examples presented in the previous sections used fundamental parts of Scout's basic operations that were originally designed to leverage the power of the GPU within a single desktop computer. In this section we present two detailed examples exploiting new features that are currently being developed. The first example presents our findings for leveraging the power of a cluster system built from multicore CPUs and GPU processors.

4.1. Leveraging Multi-core and Cluster Resources

Although they have received very little attention within the research community, conventional two-dimensional visualization operations comprise a significant majority of the techniques used by the scientific community to reveal structure and the relationships between multiple variables. These techniques serve as an important link between computational results and traditional methods of understanding observed values. Unfortunately, many of the software packages used to produce these plots do not scale to support large data or high-performance computing resources. Thus many users are left running batch processes, or significantly down sampling data, to produce these simple results. In this section we present the results of using Scout's underlying runtime system to support an interactive, programmable scatter plot that can run on systems ranging from laptops to visualization-centric clusters.

As we have previously discussed, the ability to leverage the parallelism of a multi- or many-core architecture will play a key role in guaranteeing improved performance and interactive data analysis and visualization. Scout's runtime software layers attempt to achieve this goal by providing a low-level set of fundamental operations that can be combined to provide more advanced capabilities. These routines are written to exploit parallelism at both the node and cluster levels and are based upon the instructions introduced by Blelloch, and other parallel programming languages, for data parallel computing [15]. For example, consider the following set of steps required to prepare for the rendering of a traditional scatter plot comparing two variables and coloring the resulting points by a third field.

- The classic `min` and `max` reductions must be used to find the range of values within all three input variables. As was presented in figure 3, Scout adds support for a `range` operation that combines both the minimum and maximum value reductions into a single operation.
- In order to color the points by the third variable, each point must also maintain the knowledge of its location in the computational grid. This generated set of coordinate values closely mimics Blelloch's single-dimensional `index` operation (Scout extend this to support the generation of multidimensional vectors of data). As an implementation detail, these coordinates serve as a lookup table within the graphics hardware allowing us to leverage the GPU's high-bandwidth memory system to provide improved rendering rates.
- The two variables that serve as data points along the x and y axes must be combined to produce a set of points that can be sent to the graphics hardware for display. Once again using Blelloch's terminology, this operation is a *merge* and can be can be implemented using a modified version of the `flag-merge` instruction (we will refer to this operation as `merge`). (The use of Blelloch's terminology is helpful for discussion, but it does not necessarily provide the details of an efficient implementation on today's CPU or GPU hardware.)

We are actively working on developing efficient versions of these operations on emerging architectures. Scout's runtime layer supports execution on multicore systems ranging from a single desktop to clusters. Figure 4 shows the performance results for running on a single node with eight processor cores and 1 core on each of 16 nodes (top images). These results were run on a cluster containing dual quad-core 2.0 GHz Opteron processors and NVIDIA Quadro FX 5600 graphics cards with 1.5 GB of video memory. As expected, the simplicity of these operations allows them to scale well and we are able to achieve an interactive scatter plot functionality for moderately sized datasets (see the bottom right image of figure 4). This is built directly on the foundation that will eventually enable the scatter plots to be programmatically controlled in the same manner as the examples presented above and the rendering of point data previously presented [11]. The fundamental issue with achieving reasonable multicore performance is avoiding the impact of memory contention and various systems-level issues such as affinity and system noise. This is clearly shown by the difference in scaling behavior between the eight cores within a single node and the same effect across 16 nodes with multiple sets of active cores (bottom left image of figure 4).

In the current implementation the GPU's primary role is rendering, which is the current bottleneck in overall performance. The key challenge in shifting the GPU to a role of a computational coprocessor is to amortize out the costs of data movement between the CPU and GPU memories. Current benchmarks show that the bandwidth for these transfers are approximately 2.0 to 2.5 GB/sec depending upon the data sizes and recent additions to the CUDA API now allow for asynchronous communication [16]. We are actively working on overlapping transfers between both CPU and GPU computations to increase the level of concurrency during the operation of these operations – we believe this is one of the key issues to achieving a significant percentage of peak performance for multi- and many-core processors. Finally, we are actively working on incorporating the results of the CUDA Data Parallel Primitives Library that directly relate to GPUs and Blelloch's fundamental set of operations [17, 18]. This concept of a

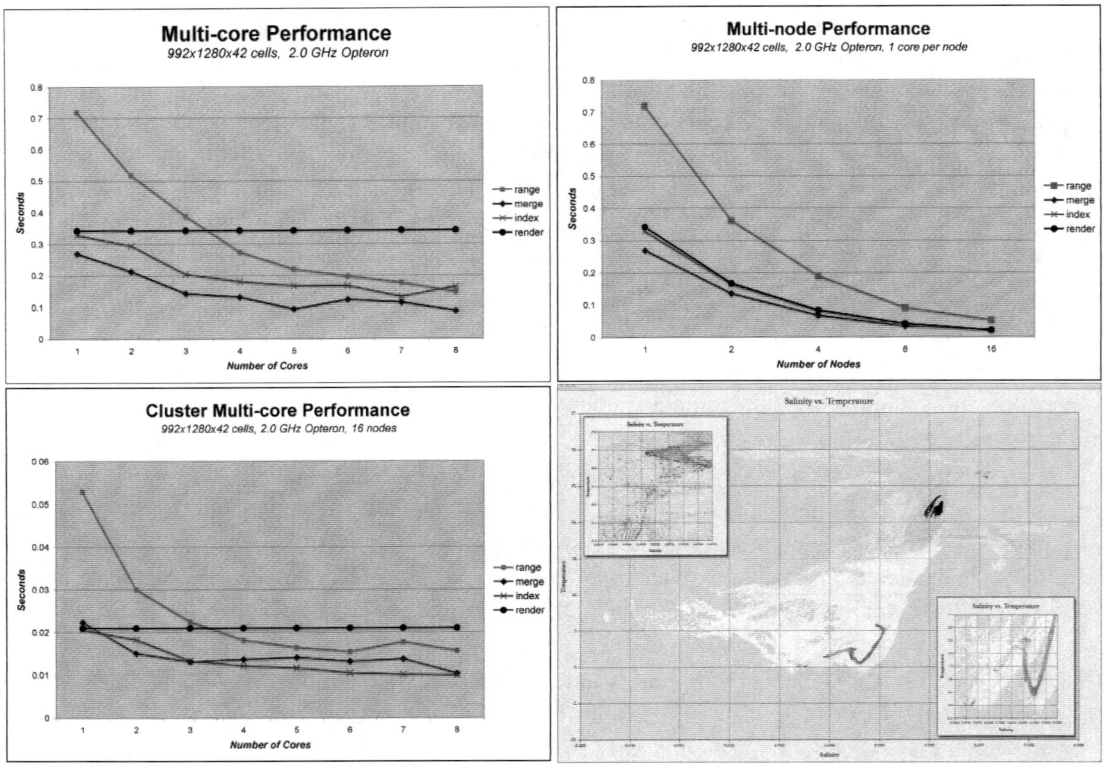

Figure 4. Performance scaling comparison for the computational portions of the scatter plot operations on a single node with an eight core CPU (top left), 16 cluster nodes using only a single core (top right), and 16 nodes with a varying number of cores in use. Although the benchmarked operations are extremely simple, the results show the clear disadvantage of memory contention for the multi-core, single-node, timings. In addition to the basic `range`, `merge`, and `index` operations the performance graphs include rendering time for comparison. The scatter plot tool built on these fundamental operations is shown in the lower right image. The scatter plot shows ocean temperature vs. salinity.

fundamental set of operators for programming complex, parallel systems will serves as a building block for further, higher-level operations related to the analysis of large-scale data.

4.2. Multivariate Example for Climate Modeling

Scientific modeling of the world's oceans, specifically the North Atlantic and Arctic Oceans, is a topic of current research. Complex systems of differential equations govern the computations performed in the model, resulting in large, high-dimensional spatio-temporal data sets. These data contain information regarding the physical aspects of the ocean circulation (such as temperature, salinity, and current velocity), as well as variables describing abundances of nutrients (such as nitrates and silicates), dissolved gasses (such as oxygen and carbon dioxide), and both phytoplankton and zooplankton. This makes size and number of dimensions to consider confounding factors in both validating the operation of the model and analyzing the data it produces.

In order to verify the model behaves properly, scientists must manually inspect each dimension of the data. This examination must take into account not only the underlying mathematical behavior, but the individual data being used as input to the computation. Even for small and relatively simple models, this process is daunting and time-consuming. Existing tools do little to mitigate the problems associated with scientific model verification. With the size and complexity of simulations increasing as more powerful computing equipment is made available, a new suite of tools and verification methods is required.

While powerful analysis tools can be leveraged to form meaningful hypotheses from a simulation, equally powerful tools must be developed to both analyze and ensure the data produced by the code accurately portrays the underlying mathematical model. Tools commonly used to visualize and otherwise process simulation results are often ineffective when verifying the accuracy of a model. However, an environment equipped with a programmable interface, such as Scout, provides a flexible platform on which both visualization and verification can operate with a minimal amount of programming by the scientist.

One method to verify the proper behavior of a simulation is to examine the correlations between variables as the simulation progresses in time. These correlations can then be related back to the original mathematical description of the simulation in order to verify relationships between the variables in the model. In order to quickly find correlations between the many dimensions of such a large dataset, a fast and robust method must be used. While image-based correlations can be computed using each dimension of the data, this process is computationally expensive [19]. However, as the goal of this work is to reduce the number of dimensions of interest to a manageable subset rather than to perform pattern matching and recognition, a faster approximation can be formulated.

The examination and manipulation of an image's spectral properties is common in the field of image processing [20] . Since the frequency content of an image or time series is both unique and easy to compute using various Time-Frequency Representations (TFRs),it is possible to compute correlations between dimensions of a datum taking only their representations in the frequency (or time-frequency) domain into consideration. This approach has several advantages over the more straightforward image-space correlations common to pattern matching and recognition. Firstly, TFRs are agnostic to the range of the input data - no normalization of the input or removal of any data masks is required to adequately correlate any two dimensions. Secondly, the domain over which correlations must be computed is smaller than the image domain, offering an increase in processing speed. Finally, correlation computations are composed of several atomic, parallel operations. This natural parallelization is well-used by the Scout data-parallel programming language, making it accessible to the scientists analyzing the data.

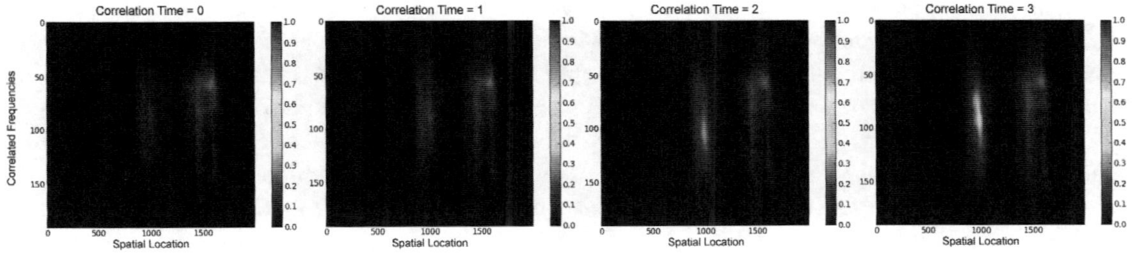

Figure 5. Correlation study performed between the presence of Diatom-based Carbon and SiO_3 dissolved in the water. As can be seen, the two data are correlated throughout time. By holding one dimension's time step constant, a relationship in time can be seen as the degree of correlation changes as time steps are processed. Artifacts due to land masses have been removed. The x-axis represents the spatial location of the correlation while the y-axis describes the correlated frequencies.

Figure 5 examines a small subset of correlation results that describe the relationships between two variables in a North Atlantic Oceanographic simulation. It should be noted here that, while the computation of image-space correlates still takes place, the smaller domain under consideration substantially accelerates the computation when compared to attempting to correlate the original input volumes. Additionally, the data-parallel nature of Scout further enhances the speed in which the results are computed while the direct programming of the system provides scientists a means to ensure all calculations performed are meaningful. This figure depicts a correlation over time between two variables in the input data set across a single spatial dimension. As the vertical dimension here is frequency, it is inferred that the correlation is not unique to a single frequency, but rather a frequency band indicating that the spatial data is not changing at a uniform rate. While, in this case, only the horizontal axis of the resulting image represents a spatial dimension, Fourier-based transforms are often easily applied to any number of dimensions, allowing correlations to be found between any set of spatial or temporal dimensions. However, using image-based correlates, we are able to describe the *lag* of the correlations. This feature is useful in correlating variables changing together yet isolated from each other in space.

As in Section 4.1, the functional building-blocks of this example tie closely to the data-parallel operators outlined by Blelloch. With the exception of the time-frequency and correlation calculations, which are easily parallelized, this method can be expressed in terms data-parallel operations as follows:

- Extract the variable to which all others should be correlated against using Scout's range selection operator (see the example code in figure 2).
- For each vector (in parallel) in the selected dimension (time or space), compute a time-frequency representation of it.
- For each corresponding vector (again, in parallel) compute the time-frequency representation.
- Compute the correlation between each corresponding vector.
- Find the maximum correlation value and its location using `max` and `max-index` reduction operators.
- Sort the resulting correlations based on their value and select the *N* most highly correlated variables to further explore.

These full set of operations, and any necessary support functionality (such as mean and standard deviation computations), are currently being added to Scout and will be exposed to the language for direct programmability by end users.

5. Conclusion

As the amount of information being generated and acquired is increasing in both size and complexity, new systems must be developed to handle the challenges presented by these growing datasets. Recently, CPU architectures are reaching physical design limits in terms of power consumption and heat dissipation. To solve this, manufacturers are producing CPUs with multiple cores, or processing units, on a single die. This new generation of architectures comes with increased performance but increased programmatic complexity. The Scout programming language and environment provide scientists with a variety of scalable, atomic, functions that can be used to scale their explorations to take advantage of the computational resources available to them, from a single desktop workstation, to a cluster consisting of hundreds of processors. In addition, this system combines the operations needed for both qualitative and quantitative visualization and analysis. Our goal is to enable scientists to more effectively explore their data in its entirety, without the need for sub-sampling or cropping large portions of their data. We hope this will eventually improve the overall scientific workflow by improving the process of gathering observable, empirical and measurable data to assist in the formulation and testing of hypotheses.

References

[1] AMD 2008 Amd stream computing http://ati.amd.com/technology/streamcomputing/index.html
[2] NVIDIA 2008 Nvidia tesla c1060 computing processor http://www.nvidia.com/object/tesla_c1060.html
[3] Owens J D, Houston M, Luebke D, Green S, Stone J E and Phillips J C 2008 *Proceedings of the IEEE* **96** 879–899
[4] Goddeke D, Wobker H, Strzodka R, Mohd-Yusof J, McCormick P and Turek S 2008 *International Journal of Computational Science and Engineering* to appear
[5] GPGPU 2008 General-purpose computation using graphics hardware http://gpgpu.org
[6] Kahle J A, Day M N, Hofstee H P, Johns C R, Maeurer T R and Shippy D 2005 *IBM Journal of Research and Development* **49** 589–604
[7] Python 2008 Python programming language http://www.python.org
[8] Python 2008 Matplotlib / pylab – matlab style python plotting http://matplotlib.sourceforge.net/

[9] McCormick P S, Inman J, Ahrens J P, Hansen C and Roth G 2004 *VIS '04: Proceedings of the conference on Visualization '04* (Washington, DC, USA: IEEE Computer Society) pp 171–178 ISBN 0-7803-8788-0
[10] McCormick P S, Inman J T, Ahrens J P, Mohd-Yusof J, Roth G and Cummins S 2007 *Parallel Computing* **33** 648–662
[11] Ahrens J, Heitmann K, Habib S, Ankeny L, McCormick P, Inman J, Armstrong R and Ma K L 2006 *Journal of Physics Conference Series* **46** 526–534
[12] Thinking Machines Corporation 1991 *C* User's Guide*
[13] Thinking Machines Corporation 1991 *CM Fortran User's Guide*
[14] Wikipedia 2008 Hsl and hsv http://en.wikipedia.org/wiki/HSL_color_space
[15] Blelloch G E 1990 *Vector models for data-parallel computing* (Cambridge, MA, USA: MIT Press) ISBN 0-262-02313-X
[16] NVIDIA 2008 Learn more about cuda http://www.nvidia.com/object/cuda_learn.html
[17] Sengupta S, Harris M, Zhang Y and Owens J D 2007 *Graphics Hardware 2007* pp 97–106
[18] Harris M, Sengupta S and Owens J D 2007 *GPU Gems 3* ed Nguyen H (Addison Wesley) chap 39, pp 851–876
[19] Castleman K R 1996 *Digital image processing* (Upper Saddle River, NJ, USA: Prentice Hall Press) ISBN 0-13-211467-4
[20] Gonzales R and Woods R 1992 *Digital Image Processing* (Addison-Wesley)

Remote rendering for ultrascale data

Kenneth Moreland[1], Daniel Lepage[2], David Koller[2] and Greg Humphreys[2]

[1]Sandia National Laboratories, P.O. Box 5800 MS 1323, Albuquerque, NM 87185-1323
[2]University of Virginia, 151 Engineer's Way, P.O. Box 400740, Charlottesville, VA 22904-4740

E-mail: kmorel@sandia.gov

Abstract. The mission of the SciDAC Institute for Ultrascale Visualization is to address the upcoming petascale visualization challenges. As we move to petascale computation, we are seeing a trend not only in the growth but also in the consolidation of computing resources. As the distances between user and petascale visualization resources grow, the expected performance of the network degrades, especially with respect to latency. In this paper we will explore a technique for remote visualization that leverages unstructured lumigraph rendering. This technique will provide an interactive rendering experience regardless of the network performance to the remote visualization resource. The unstructured lumigraph rendering can also replace many of the other level-of-detail techniques currently used that have problems that are exasperated by petascale data.

1. Introduction

As visualization and simulation platforms continue to grow and change, so to do the challenges and bottlenecks. Although image rendering was once the limiting factor in visualization, rendering speed is now often a secondary concern. In fact, it is now common to perform interactive visualization on clusters without specialized rendering hardware. The new focus on visualization is in loading, managing, and processing data. One consequence of this is the practice of co-locating a visualization resource with the supercomputer for faster access to the simulation output [1, 2].

Most of us are not fortunate enough to have a supercomputer or associated visualization resource located in our office, yet we still require access to these resources. A remote visualization capability allows us to provide the widest access to our resources; anyone with network capabilities can perform large-scale visualization from his desktop using remote visualization.

The distance between scientist and computing/visualization resources, as a trend, is growing. The U.S. Office of Science, through programs such as SciDAC, is encouraging scientists from different disciplines and different organizations to collaborate; the U.S. Department of Defense and Department of Energy are consolidating resources such as supercomputers; universities are reaching out to researchers around the world. As the distance between user and visualization resource becomes larger, remote visualization becomes more important and more challenging. One of the goals of the SciDAC Institute for Ultrascale Visualization is to ensure that remote visualization remains an interactive experience, even when accessing petascale data on the far side of the world.

Journal of Physics: Conference Series **125** (2008) 012096 doi:10.1088/1742-6596/125/1/012096

2. Remote rendering

Unlike most large-scale parallel processing, visualization tends to be an interactive process by nature. The parallel visualization must be controlled remotely to make its use convenient and, in some cases, possible. Thus, most large-scale turn-key visualization systems, such as ParaView, VisIt, and EnSight, provide some way to access the parallel visualization remotely. The architecture of these systems varies, but all comprise a client, run on a local desktop or laptop, that connects, via a socket, to one or more servers running on a remote parallel machine. For large-scale problems, it is critical that all processing, including rendering, occurs on the server. The parallel rendering usually employ one of the "sort-last" algorithms for their scalability with respect to the geometry size [3, **?**, 4].

Given enough computational resources, a sort-last parallel rendering algorithm should be able to render ultrascale data in a fraction of a second. At times, however, this is not possible either because the physical resources are strained or because the constant overheads of the parallel rendering algorithm are too high. For these circumstances, some scalable visualization tools provide a special "interactive" render mode in which quality can be sacrificed for speed (although a high-quality "still" render mode is always available). For example, ParaView [5] provides geometric decimation [6], image decimation, and lossy image compression to achieve interactive rates.

When the client and server are connected over a wide area network, the latency of the network may make interative rendering from the server impossible. When rendering locally on the client, the visualization tool has a delicate balance between quality of the rendering and how much geometry it can store. Too often, the local client renderings either underrepresent the data or require too much data to be pulled from the server.

We are investigating a new technique for approximate interactive rendering in remote visualization using image-based techniques. As shown in figure 1, high-quality still images generated by the server are cached on the client. This cache of images is used to synthesize images from new viewpoints during interactive rendering.

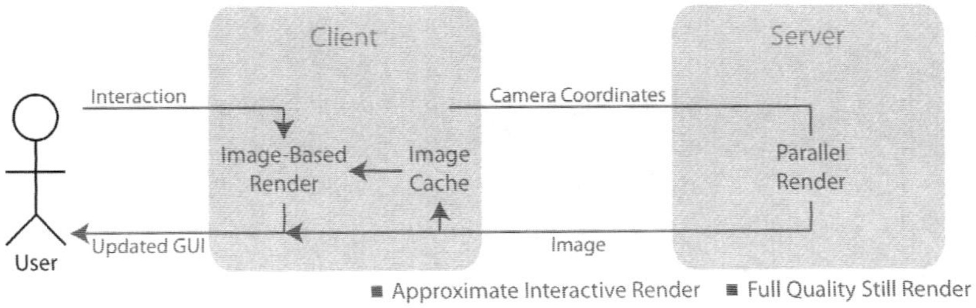

Figure 1. Update cycles for remote rendering with a local image-based approximate interactive rendering in the client.

Using images rendered from the server to synthesize other approximate images has two distinct advantages. First, it is scalable with respect to the size of the geometry being represented. Like the sort-last parallel rendering employed on the server, the overhead of the technique is independent of geometry size. Second, image-based rendering can leverage the rendering already being done on the server. Lengthy preprocessing of the data is not necessary, as is the case with geometric decimation.

3. Unstructured lumigraph rendering

We perform the client-side image-based rendering by extending the unstructured lumigraph rendering technique [7]. This method constructs a blending field across the image plane representing how the set of previously rendered input images will be combined. The color at each pixel of the new image is determined by projecting the input images onto a geometric proxy and merging the resulting images according to the blending weights.

3.1. The proxy

We use a 3D triangle mesh proxy to map pixels from prior rendered images to pixels in the image plane of an interpolated view. The quality of the synthesized image increases when the proxy closely approximates the shape of the actual dataset. However, high-quality interpolated views can be generated even when the proxy detail is low, as a coarse hull of the 3D model [8], or even a simple bounding volume such as a rectangular prism or a sphere will still produce high-quality renderings given a sufficiently dense set of known images (see figure 2).

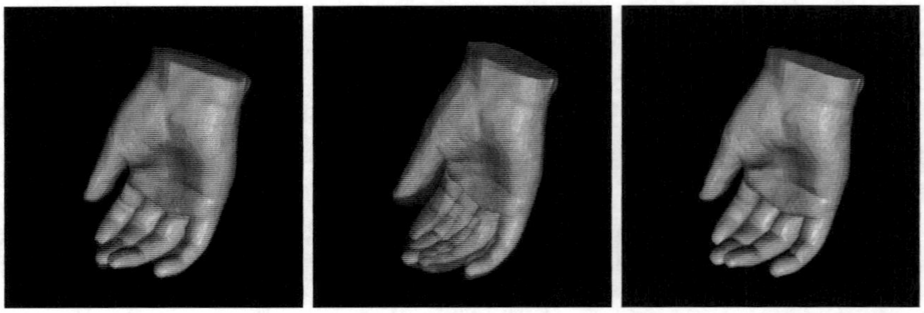

Figure 2. Left: An interpolated view of a 3D mesh model, projected onto a high-quality proxy. Center: The same view, using an axis-aligned bounding box as the proxy; the image quality is reduced. Right: Using the bounding box proxy again, but interpolating a greater density of sample images. The result is comparable with the high-quality proxy image but does not require knowledge of the actual geometry.

3.2. Sampling the blending field

Rather than recomputing the blending field at every point in the image plane, we choose a discrete set of sample points and triangulate them, interpolating the blending weights linearly across each triangle as in [7]. We place a sample point at every vertex of the visible faces of the proxy geometry, at the center of projection for every known view, and in a regular grid across the image plane. We add the proxy edges and the square edges of the regular grid as constraints, and then perform a conforming Delaunay triangulation of the resulting 2D mesh in the image plane (figure 3).

3.3. Projecting back to the proxy

Each triangle of this triangulation either is outside the bounds of the proxy geometry or is a visible subsection of a face of the proxy. We project these triangles back onto the proxy by raytracing from the camera through the center of each triangle. If the ray does not intersect the proxy, then the triangle is discarded; otherwise, the ray identifies a unique face of the proxy that contains the triangle. In the latter case, we intersect rays through each of the triangle's vertices with the plane of the face to get the world-space coordinates of the sample triangle on the proxy.

One advantage of this method over that of [7] is that it allows concave proxies without repeating triangles. A sample triangle located in front of several proxy faces is processed only once, for the nearest face.

3.4. Blending weights

We assign each prerendered input image a weight at each sample point. The client obtains an exact image from the server whenever the user stops moving. Hence, it is important to ensure that when our technique is used to synthesize a view that coincides with a portion of a known prerendered view, this known view has weight 1 within that portion, with all other views having weight 0 (the epipolar consistency property).

We view each input image as a collection of known rays in the lightfield and compute the weights by comparing the rays out of the known viewpoints with those from the new view. This approach allows us to synthesize new views from a mix of perspective and orthographic cameras, as both are common in scientific visualization.

The system of penalties used by [7] does not provide true epipolar consistency, so we have modified the blending weights. We compute four weights for each known image at each sample point, ranging from zero to infinity. One weight is based on whether the sample point is occluded in the view of the known image, one on how close the sample point is to the edges of the known image's view frustum, one on the angle between the ray from the known image to the point and the ray from the new view to the point, and one based on the difference between the distances from the known and new views to the sample point. The total weight per camera per sample point is the product of these four weights.

For each sample point, we then find the k cameras with highest weights. We subtract the kth highest weight from the weights of the $k-1$ cameras with higher weights, then divide each of them by the sum of these $k-1$ nonzero weights so that they sum to one. This gives us a uniform blending field that sums to one at every visible point on the proxy, and that will vary smoothly when the synthesized viewpoint moves.

3.5. Rendering

We render each sample triangle m times, where m is the number of cameras that have nonzero weights for at least one of the three vertices. Each time, we enable the corresponding camera's texture and perform projective texturing using the blending weight as the alpha value. This results in an image that interpolates the k nearest known views to produce a new view. For the example images shown in figures 2 and 3, $k = 4$.

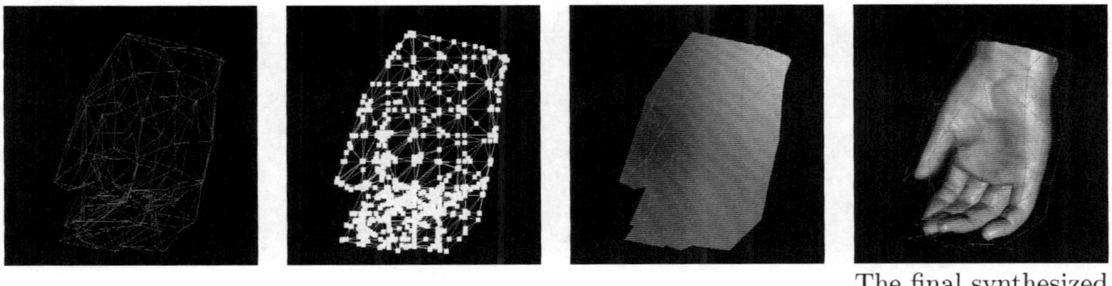

A coarse geometric proxy for mesh | The sample points of the blending field | The resulting blending field | The final synthesized view, with the proxy overlaid

Figure 3. Stages of the unstructured lumigraph rendering.

3.6. Prefetching images

The unstructured lumigraph rendering technique is most effective when the input images available for interpolation densely sample the nearby visualization space for the new view being rendered on the client. We attempt to roughly model and predict the user's navigation path through the 3D environment using a number of heuristics, and then accurately prefetch corresponding rendered images from the server. Our prefetcher adapts the volume of its requests to the state of the network traffic and server load.

4. Conclusions

The basic algorithms for the unstructured lumigraph rendering as well as the heuristics for prefetching images are implemented. Our current work involves applying these techniques to large-scale scientific data. Much of this involves integrating lumigraph rendering into the ParaView application, as demonstrated in figure 4. Our current challenges include adding multithreading to the rendering routines to allow prefetching to happen during idle times and to generate proxy geometries of general datasets.

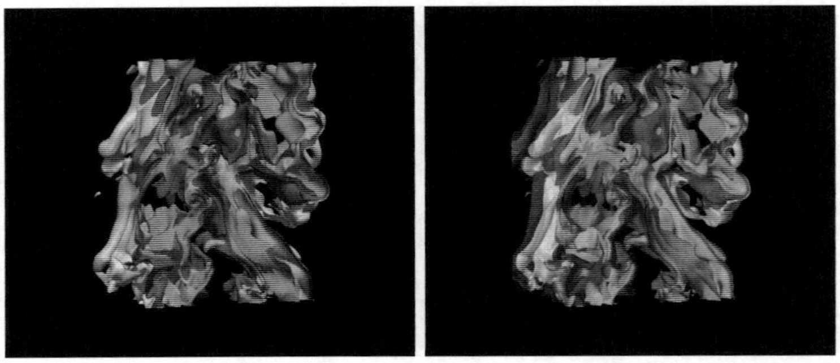

Figure 4. Turbulent combustion data rendered within ParaView (data courtesy of Jacqueline Chen, Sandia National Laboratories). Left: Full geometry render. Right: Lumigraph rendering with simple bounding box proxy.

Acknowledgments

This work was done in part at Sandia National Laboratories. Sandia is a multiprogram laboratory operated by Sandia Corporation, a Lockheed Martin Company, for the United States Department of Energy's National Nuclear Security Administration under contract DE-AC04-94AL85000.

References

[1] Ahern S 2007 Petascale visual data analysis in a production computing environment *Journal of Physics: Conference Series (Proceedings of SciDAC 2007)* vol 78

[2] Cedilnik A, Geveci B, Moreland K, Ahrens J and Farve J 2006 Remote large data visualization in the ParaView framework *Eurographics Parallel Graphics and Visualization 2006* pp 163–170

[3] Ma K L 1994 Parallel volume rendering using binary-swap image composition *IEEE Computer Graphics and Applications* **14** 59–68

[4] Wylie B, Pavlakos C, Lewis V and Moreland K 2001 Scalable rendering on PC clusters *IEEE Computer Graphics and Applications* **21** 62–70

[5] Squillacote A H 2007 *The ParaView Guide* ParaView 3 ed (Kitware, Inc.) ISBN-13: 978-1-930934-21-4

[6] Lindstrom P 2000 Out-of-core simplification of large polygonal models *Computer Graphics (Proceedings of SIGGRAPH 2000)* 259–262

[7] Buehler C, Bosse M, McMillan L, Gortler S and Cohen M 2001 Unstructured lumigraph rendering *Computer Graphics (Proceedings of SIGGRAPH 2001)* 425–432

[8] Buehler C, Bosse M, McMillan L, Gortler S J and Cohen M F 2001 Unstructured lumigraph rendering *Computer Graphics (Proceedings of SIGGRAPH 2000)* 425–432

Streaming visualization for collaborative environments

M Hereld[1], E Olson[2], M E Papka[1] and T D Uram[1]
[1] Argonne National Laboratory, Argonne, IL 60439, USA
[2] The University of Chicago, Chicago, IL 60637, USA

E-mail: papka@anl.gov

Abstract. Connecting expensive and scarce visual data analysis resources to end-users is a major challenge today. We describe a flexible mechanism for meeting this challenge based on commodity compression technologies for streaming video. The advantages of this approach include simplified application development, access to generic client components for viewing, and simplified incorporation of improved codecs as they become available. In this paper we report newly acquired experimental results for two different applications being developed to exploit this approach and test its merits. One is based on a new plugin for ParaView that adds video streaming cleanly and transparently to existing applications. The other is a custom volume rendering application with new remote capabilities. Using typical datasets under realistic conditions, we find the performance for both is satisfactory.

1. Introduction

The problem of sharing visualization resources in the form of computational power and data storage is one of the important challenges that all users of HPC face. These resources are scarce and geographically distant from many users and from other resources. Analysis of the results of large simulations must use these resources, and yet the user may not be conveniently near to them. Many approaches for solving this problem center on connecting remote users to the analysis engines and raw data through tools that present relatively condensed representations locally. Bandwidth and latency prove to be the key limiting parameters that govern the success of the solution. Bandwidth limits the amount of data that can be viewed, while latency limits the responsivity of the control system that enables interaction with the data and analysis.

The need for a new paradigm in handling pixels is being expressed in several areas of ongoing work. At the more traditional end of the scale are the solutions that consider end-point displays as containing graphics horsepower according to the usual workstation standards. Multi-display systems using WireGL [1], Chromium [2], and similar solutions are of this ilk. A central problem for architectures based on this paradigm is that operating system integration with the display adaptor has evolved to provide powerful graphics pipelines connecting the local application to the local pixels: breaking into this pipeline while retaining high performance is difficult. At the other end of the scale, solutions that put pixels in the prime position (as opposed to graphics primitives) are providing perhaps the most flexible and neutral paradigm, typically at the expense of performance. SAGE [3], VNC [4], and the many progeny of VNC are among the approaches that fit into this category. The venerable and ancient X Window System lies between these extremes.

2. Our approach

In devising a solution to this problem we are motivated by several characteristics of distributed scientific collaborations:

- The size of scientific datasets has been growing exponentially, and the rate of growth will continue to increase.
- Storage and compute resources are often centralized.
- Collaborations are increasingly multi-institutional, requiring communication among groups of people distributed geographically.

Video codecs designed for low-bandwidth, high-frame-rate applications by using inter- and intra-frame compression and motion compensation appear a practical intermediate format for transmitting visualization data. Streaming video over the network links provides near-immediate views of the data. Streaming over multicast brings the further benefit of one-to-many distribution while accruing no incremental network or processing burden to the sender.

Based on these considerations, our approach is to build interfaces between visualization tools and streaming video tools to allow remote collaborators to see views of the data generated by compute resources at the sender site. We also aim to allow the remote sites to interact with the visualization from a distance. The sections below detail how we have applied this approach to two applications, vl3 [5] and ParaView [6].

2.1. vl3: client-side stream aggregation

The vl3 volume visualization application was developed at Argonne National Laboratory. It exploits the latest features of graphics hardware (for example, programmable shaders) to optimize performance and includes support for parallel rendering on clusters. The vl3 head node composites segments rendered in the data domain and ships the results on streams targeting segments in the image domain (figure 1).

The receiving client, a modified version of vic, the Open Mash (http://www.openmash.org) streaming video tool, assembles the streams into a single video for display. The receiving client has also been modified to accept mouse and keyboard interaction and to relay those interactions back to vl3 running on the cluster, to effect changes in orientation, color map, and display. This model allows any receiver to control the rendering, subject to network security constraints between the client and the cluster nodes. By streaming the output frames as video, vl3 enables remote viewing of and interaction with the data, avoiding the common hindrances of transfer of large datasets, access to remote compute resources and familiarity with complex visualization software.

2.2. ParaView streamer: a remote visualization plugin

Because ParaView is a mature and feature rich application, and is widely used for visualizing data from many disciplines, it was a natural target for extension into collaborative scientific visualization. The ParaView plugin architecture provides a natural way to add streaming video for remote visualization, producing data streams in a generally accessible format and at lower bandwidth. When loaded, the streamer plugin adds controls to the ParaView toolbar to allow the user to specify the

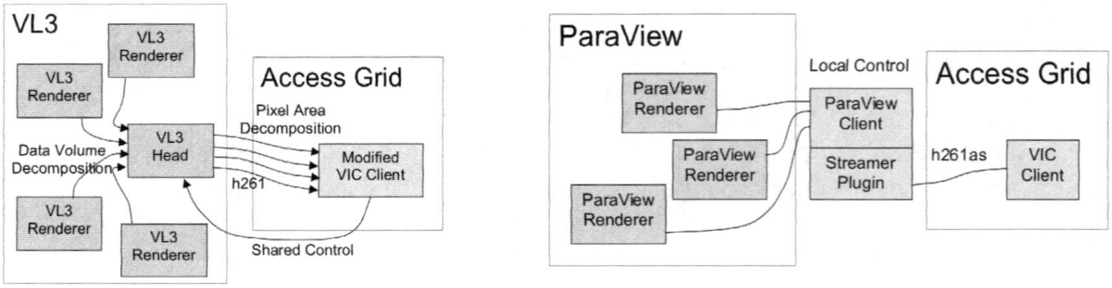

Figure 1. Comparison of the vl3 and streaming ParaView application architectures.

destination address to which frames should be streamed, and to start and stop streaming. Additionally, ParaView can be launched directly from within the Access Grid [7], where it will automatically set the destination to the current venue.

The streaming plugin subscribes to render events in the main ParaView render area, updating an internal buffer. A separate thread streams frames from the buffer at native resolution at 30 frames per second, independent of ParaView's rendering back-end. The format used for the video stream is h.261 with modifications to support arbitrary size frames, known informally as h261as. Support for this codec is included in recent releases of vic and Access Grid distributions. Because the h.261as codec includes support for motion vectors and interframe compression, the stream bandwidth drops appreciably when the rendering is unchanged. Lossy codecs give generally acceptable results in interactive settings where the user controls the focus of the visualization and exploration resolves ambiguities introduced by quantization errors.

3. Measurements

To study the performance of these two applications, we devised an experiment to capture data rates for all streams as a function of time and activity. In both cases, we used a dataset that typified our use of the system. While one member of the team manipulated the interactive controls to rotate, translate, and zoom the visualization, the other annotated the activities with single letter codes added to a running log with automatic time stamps at 1-second intervals. The client code reported the bandwidth data at 1-second intervals as well. The experimental design emphasizes measurement of a realistic use case over a significant time period. In settling on this methodology, we were driven by the observation that the instantaneous performance depends on the complexity of interactions between the data, the streaming codec, the POV, and the rate of change of the POV. Significant variation in the measured performance is to be expected from moment to moment. Quantitative modeling would be extremely difficult and of questionable value.

The results from both experiments are presented in figures 2 and 3. Each synopsis includes an image taken of the data in a typical pose to give an indication of both the visual complexity of each and of the variety of data in common use. The annotated performance data is reported on the right-hand panel. Annotations show when the user was idle at the controls (perhaps studying the image) or was manipulating the view with rotate, translate, or zoom. Each created a log of at least 10 minutes of interactive browsing. In the vl3 experiment, four separate streams from rendering processes were

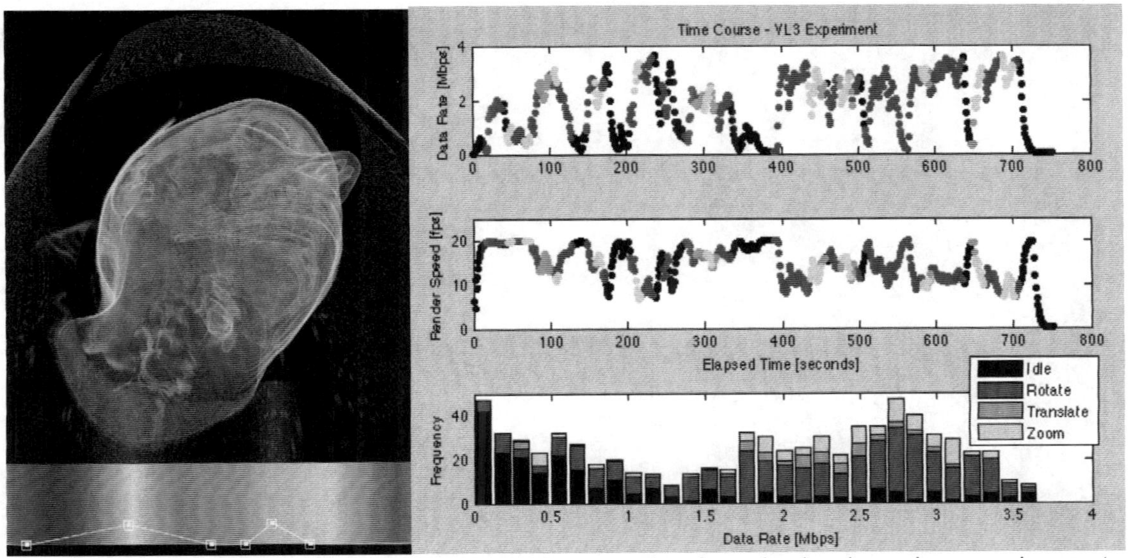

Figure 2. The vl3 streaming experiment. At left is a snapshot from the data browsing experiment. At right is a summary of the performance data from a timed exploration. (CT data courtesy of the University of Chicago)

bonded at the client to form the final image. The data rate reported in the figures is the sum of the individual stream data rates. In the ParaView experiment, the visualization was generated by one render process and the values reported in the figures are for the h.261as stream as received by the vic client.

Despite the complexity of interactions between dataset, rendering speed, viewpoint, changing viewpoint, data-domain compositing speed, and video codec, it is possible to make a few observations about the results shown in figures 2 and 3.

- The frame rate varied by a factor of two or so over the course of both experiments. The bandwidth varies more, but its impact is capped by the maximum frame rate.
- When interaction becomes idle, it is clear that the codec responds rapidly to the stable image content by approaching zero load on the network – even though the back-end may be laboriously re-rendering the static image at a very low frame-rate.
- Except for a large peak from the low bandwidth required during idle time, the distribution in both cases is not particularly remarkable. Judging from the stacked distributions in the histogram, different activities do not present significantly different demands: for example, rotation seems no more or less taxing than zoom.
- Frame rates can dip below 10 fps in the vl3 test, presumably depending on the relative complexity of the image structure and fraction of the image area it fills.

On the other hand, because of the complex interactions enumerated above, it doesn't seem possible to pin performance on any one factor. Measurements of the vl3 system (not included in the above plots) show that the frame rate at the vic client can be limited by effects anywhere along the pipeline depending on circumstances: by a single h.261 channel (codec + network bandwidth) when the complex moving visualization is placed in one quadrant of the pixel area; or by the rendering back-end when the viewpoint is changing (to keep the codec busy) and the data fills the field of view.

5. Conclusions

We have described our approach to providing interactive remote visualization to collaborative environments. Based on video streams and widely used nonproprietary codecs, this approach can provide high quality visualization at low development cost to a range of standard clients. We have described two applications based on the idea and presented newly acquired experimental results from realistic performance tests.

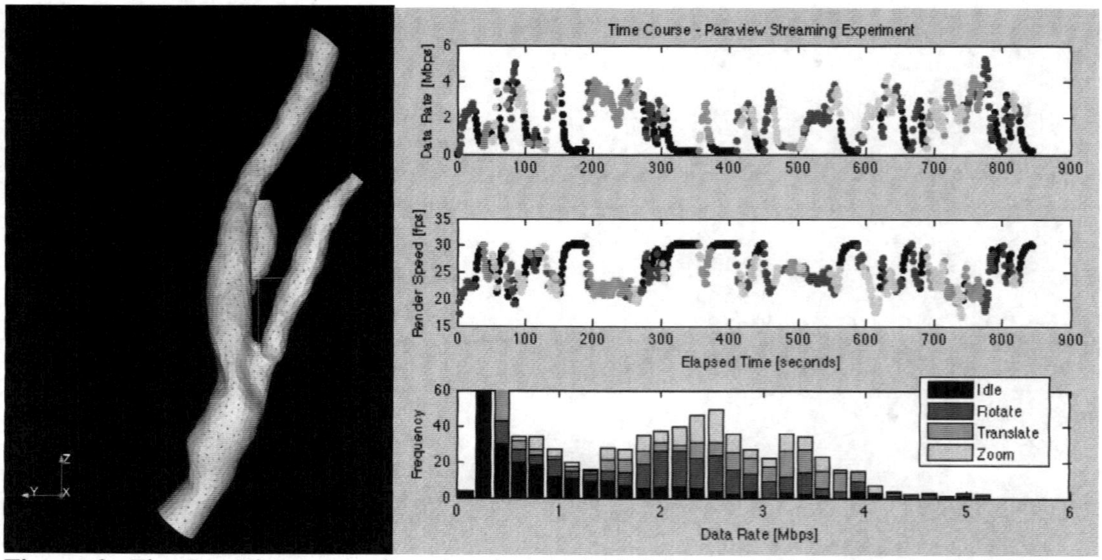

Figure 3. The ParaView streaming experiment. At left is a snapshot from the data browsing experiment. At right is a summary of the performance data from a timed exploration. (Arterial tree data courtesy of Brown University)

The experimental results are promising: in both applications we verified acceptable performance; not surprising, we note significant variation in the bandwidth required – with dependence on data, orientation, scale, and motion; the required bandwidth in both cases was no more than several megabits-per-second, a value that is already commonplace in typical research settings. We plan future work in this area to include incorporation of improved codecs, notably h264, and adding shared control to the ParaView application.

Acknowledgments
We acknowledge Joseph Insley for his help in this work; the University of Chicago TeraGrid RP support team; and the Australian National University for its modified h261 codec. This work has been supported in part by the Mathematical, Information, and Computational Sciences Division subprogram of the Office of Advanced Scientific Computing Research, Office of Science, U.S. Department of Energy, under Contract DE-AC02-06CH11357. Additional supported provided by the National Institutes of Health/National Library of Medicine, under Contract N01-LM-3-3508 and the National Science Foundation under Grant No. ANI-01-22296.

References

[1] Humphreys G, Eldridge M, Buck I, Stoll G, Everett M and Hanrahan P 2001 *Proc. 28th Annual Conference on Computer Graphics and Interactive Techniques SIGGRAPH '01* (ACM,: New York) 129–140
[2] Humphreys G, Houston M, Ng R, Frank R, Ahern S, Kirchner P D and Klosowski J. T. 2002 *ACM Trans. Graph.* **21**(3):693–702
[3] Jeong B, Renambot L, Jagodic R, Singh R, Aguilera J, Johnson A and Leigh L 2006 *ACM/IEEE Supercomputing 2006 Conf. (SC'06)*
[4] Richardson T, Stafford-Fraser Q, Wood K R and Hopper A 1998 *IEEE Internet Computing*, **2**(1):33–38
[5] Binns J, Dech F, Papka M E, Silverstein J C and Stevens R 2005 *From Grid to HealthGrid* pp. 70–79
[6] Ahrens J, Geveci B and Law C 2005 *The Visualization Handbook*, C D Hansen and C R. Johnson eds. (Elsevier Inc.) pp. 689–716
[7] Stevens R 2003 *The Grid: Blueprint for a New Computing Infrastructure* 2nd ed I Foster and C Kesselman, eds. (Morgan Kaufmann)

Journal of Physics: Conference Series **125** (2008) 012098 doi:10.1088/1742-6596/125/1/012098

Multiscale Morse theory for science ciscovery

Valerio Pascucci[1] **and Ajith Mascarenhas**[2]

[1] Scientific Computing and Imaging Institute, School of Computing, University of Utah, Salt Lake City, UT 84112, USA

[2] Center for Applied Scientific Computing, Lawrence Livermore National Laboratory, Livermore, CA 94551, USA

E-mail: pascucci@acm.org

Abstract. Computational scientists employ increasingly powerful parallel supercomputers to model and simulate fundamental physical phenomena. These simulations typically produce massive amounts of data easily running into terabytes and petabytes in the near future. The future ability of scientists to analyze such data, validate their models, and understand the physics depends on the development of new mathematical frameworks and software tools that can tackle this unprecedented complexity in feature characterization and extraction problems. We present recent advances in Morse theory and its use in the development of robust data analysis tools. We demonstrate its practical use in the analysis of two large scale scientific simulations: (i) a direct numerical simulation and a large eddy simulation of the mixing layer in a hydrodynamic instability and (ii) an atomistic simulation of a porous medium under impact. Our ability to perform these two fundamentally different analyses using the same mathematical framework of Morse theory demonstrates the flexibility of our approach and its robustness in managing massive models.

1. Introduction

Accurate analysis of massive space-time scientific simulation data introduces unique algorithmic challenges due to the impossibility of fitting all data into main memory and the multiscale nature and complexity of the features in such data. In this papers we focus on the latter challenge. In particular, we study the development of a sound mathematical framework for feature definition and design robust efficient algorithms for discovering and quantifying trends in the data. We also aim at algorithms that provide the user with the flexibility to control the scale of interest and to produce visualizations and collect statistics on the data.

In this paper, we describe ideas from Morse theory that enable effective feature characterization using the powerful language of topology. In addition, our approach is based on a set of combinatorial constructs that are implemented exactly (no numerical approximations) yielding a highly robust data analysis framework that can segment and quantify features at several scales for massive datasets.

We illustrate the effectiveness of our analysis techniques with two applications. In the first application, we enable the discovery and quantification of four major mixing trends in a fundamental hydrodynamic instability simulation. Our techniques also allowed for the first time a feature-based comparison and validation of simulations developed with two different techniques. The hydrodynamic instability (a Rayleigh-Taylor simulation) was developed at Lawrence Livermore National Laboratory [8, 9]. This analysis is described in greater detail in

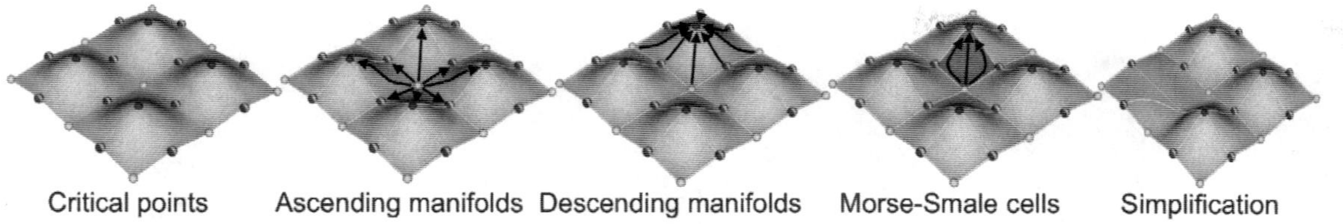

Figure 1. MS-complex construction, simplification and topologically valid approximation: (a) Morse function with critical points shown; (b) stable manifolds; (c) unstable manifolds; (d) MS-complex; (e) MS-complex and manifold after simplification. Maxima are solid blue, minima are solid white, and saddles are mixed.

[15, 17]. In the second application, we enable the detection of core structures in the atomistic simulation of a porous medium and their tracking in time as a projectile impacts the medium causing structural failures. This analysis is described in detail in [12]. For a more detailed description of Morse-theory based analysis see [7, 10, 11, 3, 4, 20].

2. Morse theory for robust multi-scale feature analysis

To begin, we present some background from Morse theory [16, 18] and from combinatorial and algebraic topology [1, 19].

Smooth maps on manifolds. Let $f\colon \mathbb{M} \to \mathbb{R}$ be a smooth map. A point $x \in \mathbb{M}$ is a *critical point* of f if the gradient of f vanishes at x, and the value $f(x)$ is a *critical value.* Noncritical points and noncritical values are called *regular points* and *regular values*, respectively. A critical point x is *nondegenerate* if the Hessian (matrix of second-order partial derivatives) at x is nonsingular. The *index* of a critical point x, denoted by index x, is the number of negative eigenvalues of the Hessian. For $d = 3$ there are four types of nondegenerate critical points: the *minima* (index 0), the 1-*saddles* (index 1), the 2-*saddles* (index 2), and the *maxima* (index 3). A function f is *Morse* if all critical points are nondegenerate with distinct values.

Morse-Smale complex. An *integral line* is a maximal path on $\mathbb{M}$ whose tangent vectors agree with the gradient of f. The *stable manifold* of a critical point x is the union of x and all integral lines that *end* at x. The *unstable manifold* of x is defined symmetrically as the union of a critical point x and all integral lines that *start* at x. One can superimpose the stable and unstable manifolds of all critical points to create the *Morse-Smale complex* (or MS complex) of f [10, 4], see figure 1(a)–(d). The *nodes* of this complex are the critical points of f, its *arcs* are integral lines starting or ending at saddles, and its *regions* are the nonempty intersections of stable and unstable 2-manifolds. More details on the definition of the MS-complex on 2-manifold triangle meshes and algorithms to compute it are given by Bremer et al. [4].

We briefly describe recent work on 3D Morse-Smale complex computation; See [13] for details. We use a single sweep to construct both the combinatorial structure of the complex and the geometric structure of the ascending manifolds. A second sweep uses this segmentation to guide construction of the geometry of the descending manifolds in a manner consistent with the topology identified in the first step. We view the construction of the ascending and descending manifolds as an iteration through dimensions, where the interior of each n-manifold is computed via region growing, and its $n-1$-manifold boundaries are recursively computed. Figure 2 shows how we iterate through dimensions while constructing the ascending manifolds.

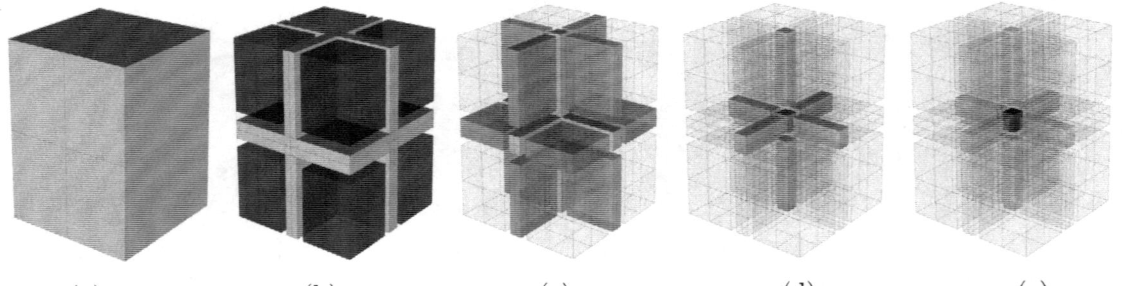

(a) (b) (c) (d) (e)

Figure 2. (a) The algorithm at each iteration identifies interior and boundary vertices of an ascending n-manifold. (b) Using minima as seed points, we grow ascending 3-manifolds, finding vertices that are interior (blue) to a single 3-manifold. The grey region depicts the boundary between ascending 3-manifolds. (c) We next compute ascending 2-manifolds by first identifying interior points. An ascending 2-manifold (green) separates exactly two ascending 3-manifolds. (d) Ascending 1-manifolds (yellow) separate a unique set of ascending 2-manifolds. (e) Finally, the maxima (red) are found as the separating sets between ascending 1-manifolds.

Simplification. It is often useful to simplify an MS-complex to remove noise and to perform multi-scale function analysis. Following [4] we perform *cancellations* of arc-connected maximum-saddle and minimum-saddle critical point pairs to simplify an MS-complex. Cancellations are ranked by their *persistence* —the absolute difference in function value between the canceled critical point pair. Figure 1(e) and (f) shows an example of a topological simplification and a corresponding approximation of f.

Computation. In practice, one usually deals with piecewise linear (PL)-functions given at the vertices of a triangulation. See [4, 5] for a detailed discussion on how to translate concepts from the generic smooth functions discussed above to PL-functions.

Starting from saddles, we compute the arcs of the MS-complex as steepest non-crossing monotone lines. Because we avoid mesh refinement to handle degeneracies, and instead directly handle merged lines and multi-saddles we can use efficient static data structures to store the triangulation and can compute MS-complexes of large data sets common in simulation.

3. Analysis of turbulent mixing in hydrodynamic instabilities

Understanding the turbulent mixing of fluids is one of the fundamental research problems in the area of fluid dynamics. Turbulent mixing occurs in a broad spectrum of phenomena ranging from boiling water to astrophysics and nuclear fusion. Rayleigh-Taylor instability occurs when two fluids of different density are accelerated opposite the mean density gradient. That is, a heavier fluid is accelerated against a lighter fluid by a force such as gravity. Figure 3 shows the mixing layers and the progression of the mixing process. The heavy fluid accelerates downward, forming "spikes," while the light fluid moves upward forming "bubbles." The bubbles and spikes are thought to be one way to characterize the large-scale behavior of the mixing process. Scientists analyzing these simulations are particularly interested in the number of bubbles (and spikes) and their respective evolution. Large-scale models have been proposed based on bubble dynamics in which bubble growth, movement, and interaction are modeled [2]. Our analysis is performed on the envelope surfaces describing the boundary between undisturbed and "mixed" fluids.

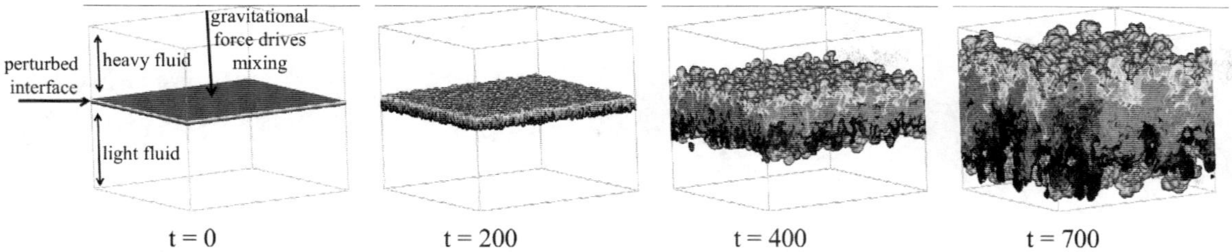

Figure 3. An overview of the 1152^3 simulation (periodic in x and y) of the Rayleigh-Taylor instability at start ($t = 0$), early ($t = 200$), middle ($t = 400$), and late ($t = 700$) time. The light fluid has a density of 1.0, the heavy fluid has a density of 3.0. Two envelope surfaces (at densities 1.02 and 2.98) capture the mixing region. The boundaries of the box show the density field in pseudocolor. heavy fluid is red and the light fluid is blue. Totally mixed fluid and the midplanes to study mixing trends.

Segmentation of bubbles. One of the challenges in analyzing the mixing behavior is that there exists no prevalent mathematical definition for what constitutes a bubble/spike. In general, a bubble can be understood as a three dimensional feature composed of lighter density fluid moving upwards (in the Z-direction) into a heavier density fluid. We use the topological concepts introduced in Section 2 to define bubbles, spikes and other features of interest. Consider the images of the segmented mixing envelope surface at different times shown to the left, bottom, and right of the plot in figure 4. During early time steps (figure 4 upper left and middle left) it is natural to consider the mixing envelope as a time-varying functional surface defined over the XY-plane and associate local maxima with bubbles. This analogy fails at later time steps because the surface becomes non-functional. However, we can generalize this approach by treating the envelope surface as the domain of a function whose value at a point x is the Z-coordinate of x. It is natural to connect the maxima of this function to bubbles and compute the stable manifold of each maximum as a segmentation of the surface into bubbles. As can be seen in figure 4, this segmentation corresponds very well to the human notion of a bubble. Symmetrically, we use the unstable manifolds of minima to define spikes. Potentially, other functions could be defined on the envelope surface that would result in a robust segmentation as well. For example, the Z-velocity at all of the points on the envelope surface could be incorporated to capture the fact that bubbles should be moving upwards into the heavy fluid. Given the complexity of the problem, it has been determined to use the simplest most intuitive segmentation.

In general, topological-based segmentations are often linked to important features: maximal and minimal Z-velocities on the midplanes correspond to cores of rising and falling sections of fluids; density extrema correspond to pockets of unmixed fluids. Topological methods are flexible and enable analysis of a variety of phenomena using the same methodology. Furthermore, the MS-complex can be computed combinatorially [10, 4] which translates into provably correct and stable algorithms which are crucial when dealing with large and complex data.

Multiscale analysis and persistence selection. The MS-complex, just as any other segmentation, captures noise as well as features. A simplification scheme can be used to remove noise and construct a series of approximations at decreasing resolution. Unlike many other techniques, topological segmentations allow a simplification scheme that is optimal in the L_∞-norm. One can formulate the problem of coarsening a segmentation in the following manner: given a function f and a segmentation S of the domain of f, what is the minimal change on f such that S is coarsened? If the segmentation one considers is the MS-complex of f, then it can be shown [4] that canceling a critical point pair with persistence p in f requires an approximation $\hat{f}$ with $||f - \hat{f}||_\infty \geq p/2$. Therefore, canceling critical points in order of increasing persistence corresponds to an L_∞-optimal simplification.

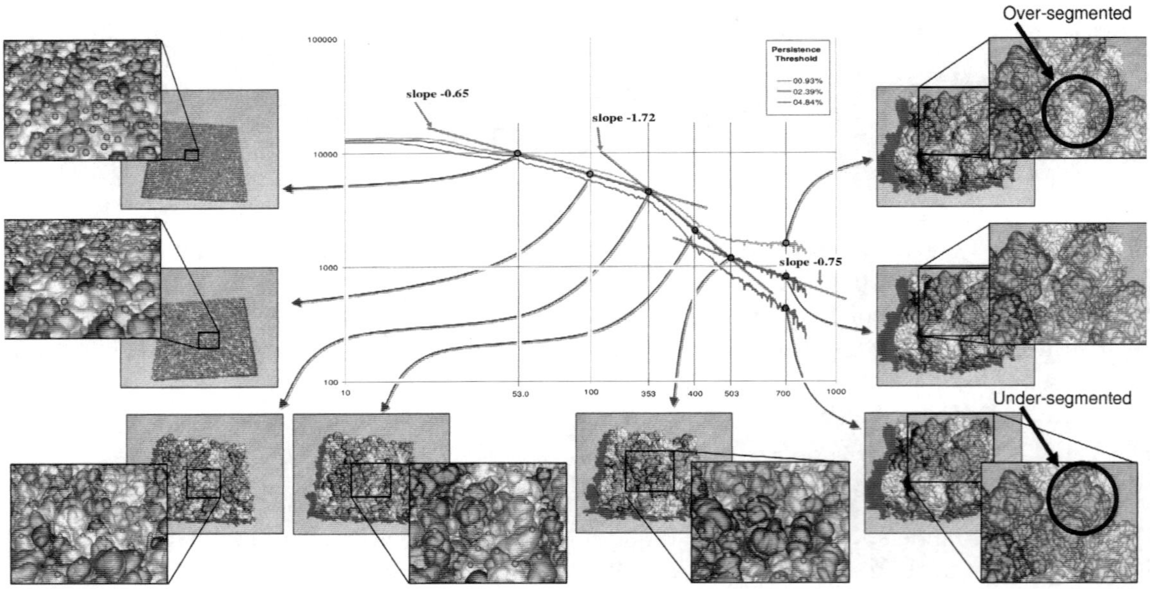

Figure 4. Depiction of bubble counts of the envelope surface for three persistence values. Three regions of power-law behavior are shown by the curve fits in gray to the 2.39% persistence curve. To the right of the plot, the MS-segmentation at three persistence values for time 700. To the left and below the plot, the bubble segmentation along the 2.39% persistence curve at various times. Each maximum along with the Morse cells of its child-maxima are colored the same.

For each MS-complex, we compute a sequence of cancellations that optimally simplify the complex down to its minimal configuration. We can thus define a *family of segmentations* of the envelope surface ranging from persistence $p = 0.0$, where both signal and noise features are segmented, to persistence $p = 1.0$ (full function range), where the entire surface is collapsed into a single component. We can then create statistics showing the number of bubbles over time using a range of persistence thresholds. As shown in Section 3.1, the mixing behavior can differ significantly across scales. Using the simplification sequences, we can capture the behavior on all scales without recomputing the MS-complex. Domain scientists interact with a visualization of the segmented surface and select an appropriate persistence value based on their physical intuition of a correct segmentation of bubbles.

3.1. Results: Bubble counts and quantification of mixing phases

The input data consists of 758 time steps, each time step containing about 5.8 GB of density data defined on a 1152^3 grid. Our analysis was performed on 68 dual-processor nodes of a cluster running Linux. For each time step, we extract an isosurface at density value 2.98 and compute the Morse-Smale complex with function f set to the z-coordinate of each point. Isosurface extraction takes about 15 seconds at early time steps and about 30 seconds at the late time steps where the surface is more convoluted. Morse-Smale complex computation takes about 2 seconds at early time-steps and about 25 seconds at late time-steps. We compute a digest of all required data from the Morse-Smale complex that allows us to calculate bubble counts for any choice of persistence efficiently.

Figure 4 describes the count for three choices of persistence values. The inset figures show the segmentation of the isosurface into bubbles. It was not clear what value of persistence was a good choice or if the persistence value woudl have to change over time. Because our analysis pipeline is flexible, and because we can quickly compute the Morse-Smale complex and bubble count at

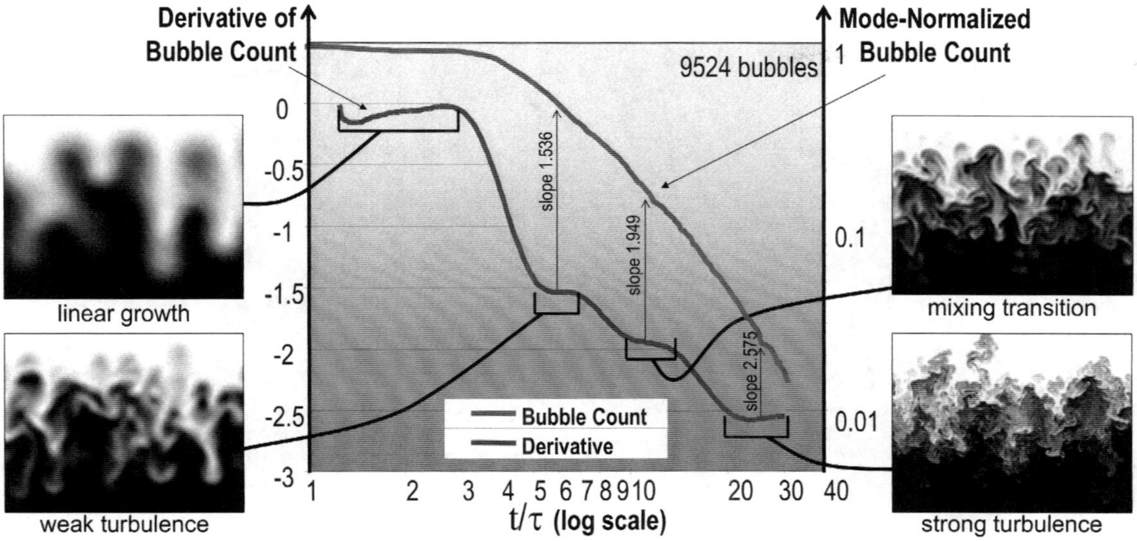

Figure 5. Bubble count and its derivative. The derivative curve shows four regions where the slope is virtually constant, which correspond to distinct stages of the turbulent mixng process.

any choice of persistence from the digest, we can provide analysts with a variety of visualizations and statistics that they can study and choose an appropriate persistence that produces results that make sense with respect to the application domain. Based on feedback from simulation scientists, we chose a persistence value of 2.39% for the late stages of the mixing. Notice, though, that each of the bubble-count curves show three linear (power-law) regimes. Although the slopes of the three curves are different, the times at which thier slopes change are the same, showing that a temporal analysis based on variation of derivatives is scale independent (robust to the chioce of input parameters).

Figure 5 shows the results of a more detailed analysis of the bubble count made by the application scientitists. Here we have a linear variation of persistence from an early value of nearly zero to the late value of 2.39%. The plot reveals four phases of the mixing process, and provides for the first time quantification of the mixing rates and transition times between the different stages.

3.2. Feature-based comparison of large eddy simulations and direct numerical simulations

Our analysis techniques provide a method to compare two different data sets of the same phenomenon simulated using different classes of simulations. Large eddy simulations (LES) require smaller resolution meshes and less computational resources than direct numerical simulations (DNS). Grid resolutions for LES capture large-scale effects but must model the sub-grid scale effects explicitly. On the other hand, DNS grid resolutions must be fine enough to capture all scales.

Validating the results of these two types of simulations applied to the same phenomenon is of great importance to scientists. We have conducted a feature-based comparison that is shown in figure 6. The LES grid dimension was 1152^3 and the DNS grid dimension was 3072^3. We perform the analysis as described previously and compute bubble counts for both types of simulations. After normalizing the counts, we can plot them together and compare them. Figure 6 shows that the simulations are in close agreement and produce similar trends in the mixing layer.

Journal of Physics: Conference Series **125** (2008) 012098 doi:10.1088/1742-6596/125/1/012098

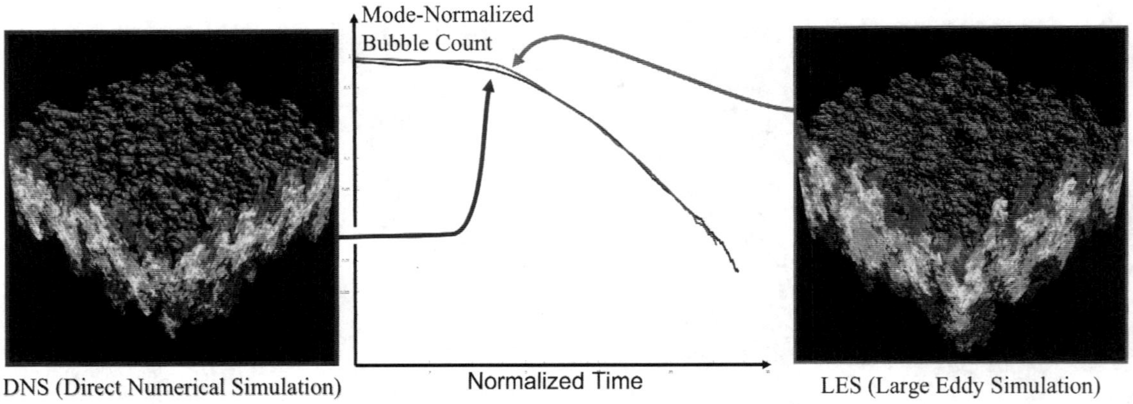

Figure 6. Feature-based comparison and validation of large eddy simulation (LES) and direct numerical simulation (DNS) of Rayleigh-Taylor instabilities.

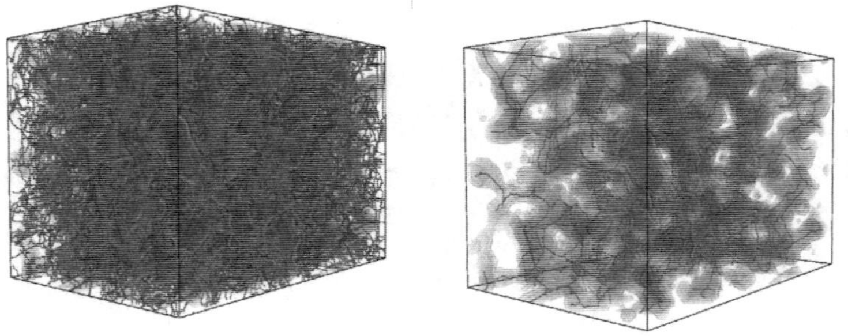

Figure 7. The initial computation of the MS complex for the full dataset (left) is simplified revealing the graph structure (right) of the porous solid.

4. Core structure of a porous solid

Porous materials, such as the aerogel used in the Stardust mission [14], are employed in many scientific and industrial applications. Atomistic simulations [6] have been developed to understand their structural properties under a variety of conditions including under impact by a projectile. Detecting the core structure of the porous solid and tracking this structure under impact aids analysis of the simulation and provides important insight into the behaviour of the solid.

We compute the core structure of a porous solid using two methods. First, we compute the Morse-Smale complex using the incremental algorithm presented in [13], with the finger removing and arc smoothing modifications. Then, we analyze the critical point pairs, and filter the arcs to extract the core structure. The full complex shown in Figure 7, is used to plot the distribution of 2-saddle-maximum pairs, shown in figure 8. The features we are interested in are the arcs that connect a low 2-saddle to a high maximum; these critical point pairs are in the top left of the distribution. Flat regions in the integral along each axis reveal that the stable threshold for maxima is 1.5, and the stable threshold for 2-saddles is -0.8. By cancelling all arcs that do not entirely cross the range $[-0.8, 1.5]$, we remove the artifacts and noise. The core structure is extracted as the arcs connecting 2-saddles to maxima that remain after simplification and are entirely contained in the isosurface for isovalue zero. For this particular application, we are interested in the connectivity of the porous solid, therefore we omit arcs that are connected to the structure at only one endpoint from the final core structure.

The results were generated by using an off-the-shelf personal computer, a 3.4 GHz Pentium 4, with 2 GB of memory. The porous solid was represented as real-valued samples on a $230 \times 230 \times 375$ regular grid. The total time required for computation of the initial MS complex was 6 hours 32 minutes and 45 seconds. Additional processing to attain the graph structure took 32 seconds. The total time required to create the clean distance field using the propagation

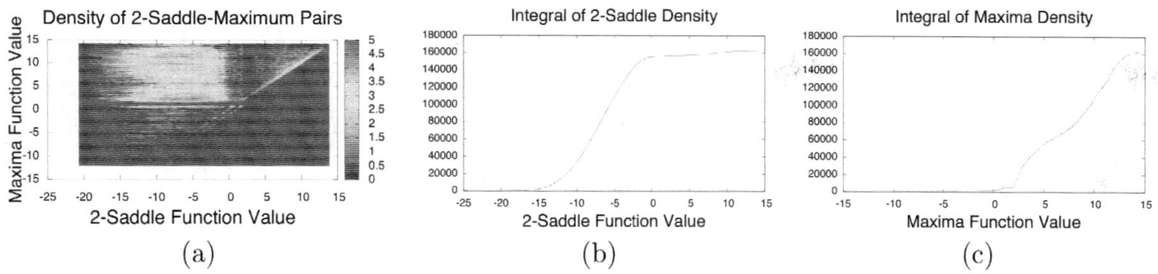

Figure 8. Distribution of the 2-saddle-maximum pairs (a). The color scale shows the number of occurrences of 2-saddle-maximum pairs where the 2-saddle has function value along the x-axis, and the maximum has function value along the y-axis. We integrate along the x-axis (b) and along the y-axis (c) to find stable threshold values.

method starting with the same input dataset was 91 seconds. After the initial computation, exploration and further simplification can be done interactively. For large datasets, computation of the full MS complex is not possible; however, we can still perform analysis on a subset of the data to attain the distribution of critical point pairs, assuming features are distributed nearly uniformly in the dataset. The clean distance field method is an efficient implementation that can be applied to large data and benefits from such an analysis.

5. Time-dependent impact data

We have used our methods to explore a simulated dataset of a particle impact on the porous material at several time steps. By computing the clean distance field, we can obtain the density of the porous solid as the ratio of the number of sample points interior to the zero isosurface to the number of sample points outside. The clean distance field ensures that all sample points identified in this way contribute to the structure of the porous material. The structural analysis and comparison between time steps allow us to obtain an important result: there is significant densification of the foam below the crater wall. Such analysis provides a simple, quantitative answer to the amount of densification, as shown in figure 10. In this figure we show the density profile at different times as a function of depth for slices of (a) the whole sample and (b) the center of the sample, including the crater. It can be seen in (b) that the density increases by a factor of two below the crater.

We also have computed the core structure using the MS complex for each time step. Using the distance measures described in [12], we can compare qualitatively how the structure of the material changes between time steps. Figure 9 is a visual depiction of the displacement of filament segments at different times after impact. Note the large displacements near the crater, and the nearly zero displacement well below the crater.

In both figures we can see that density barely changes in the bottom one-third of our sample. This is partly due to the fact that the foam is extremely efficient at absorbing the impact shock wave. Key statistics of the core structure for each time steps are summarized in Table 1.

The ratio of cycle counts before and after the impact supports this observation, as approximately two-thirds of the cycles are destroyed. The ratio of the total length of the filaments before and after the particle impact implies that volume of material displaced by the crater is approximately one-half the volume of the rest of the material. Since this ratio is fairly close to the ratio of the cycle counts, we can say that the majority of the filaments that were

Metric	t=500	t=12750	t=25500	t=51000
# Cycles	762	340	372	256
Total Length	34756	24316	23798	18912

Table 1. Statistics for each time step.

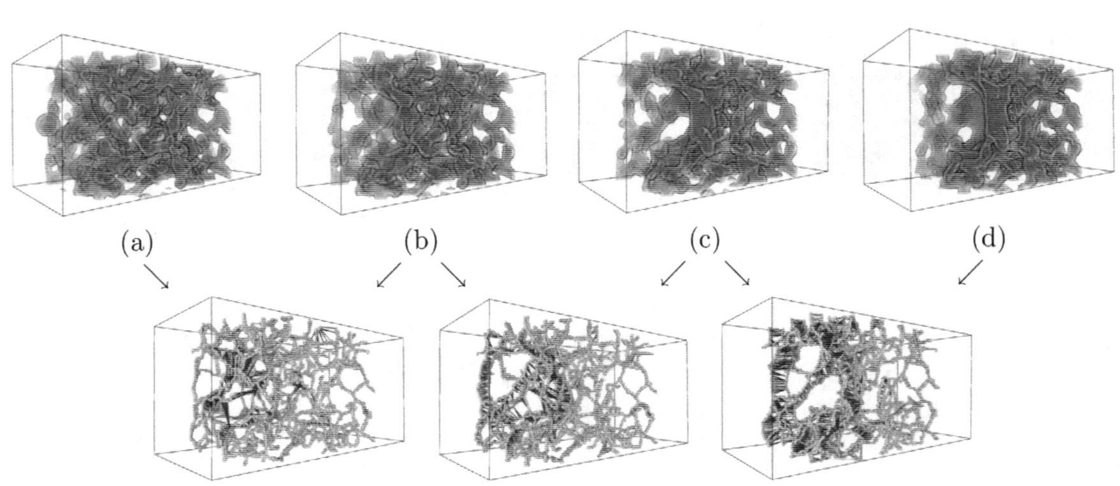

Figure 9. *Top:* Volume rendering of the impact of the ball entering the porous solid from the left at time step 500 (a), time step 12750 (b), time step 25500 (c), and time step 51000 (d). *Bottom:* We compare the core structures of consecutive time steps. The yellow dots represent the core structure of the initial time step, and the teal dots represent the core structure of the next time. The closest arcs between in the core structures at the different time steps are connected via blue and red line segments. The length of these segments corresponds to the displacement of the arc.

broken happened to be in the interior of the crater. The sum of the Hausdorff distances between the time steps is 98.6, giving the maximum distance that any element of the material travelled during the impact. This number is surprisingly high, corresponding to the entire depth of the crater; it indicates that the material of the filaments first hit by the particle was displaced along the trajectory of the particle. The average distance between closest pairs in the graphs of the consecutive timemsteps was less than 5.0, indicating that the displacement did not propagate into the material, outside the direct path of the particle.

Densification of the foam will vary as a function of impact velocity, and a quantitative characterization of such a function might help to narrow down such velocities when they are not known. In addition, the fact that the foam is getting denser would change, for large particle fluxes, the foam's mechanical and thermal properties.

6. Conclusion

Morse theory provides a sound mathematical framework to study functions on manifolds. This framework can be extended to analyze piecewise linear functions created from real-world scientific data. Using this foundation we develop robust and flexible software that can be fine-tuned to the analysis of a variety of datasets. We present the analysis of two datasets: Rayleigh-Taylor instability simulations and porous solid atomistic simulations.

As current trends continue, computing power will increase rapidly and scientists will develop more sophisticated and larger scale simulations. We believe techniques, such as those described in this paper, will become crucial to address the challenge of analyzing these scientific datasets, and accelerate scientific discoveries.

References

[1] P. S. Alexandrov. *Combinatorial Topology.* Dover, Mineola, NY, USA, 1998.

[2] U. Alon and D. Shvarts. Two-phase flow model for rayleigh-taylor and richtmeyer-meshkov mixing. In *Proc. Fifth International Workshop on Compressible Turbulent Mixing.* World Scientific, 1996.

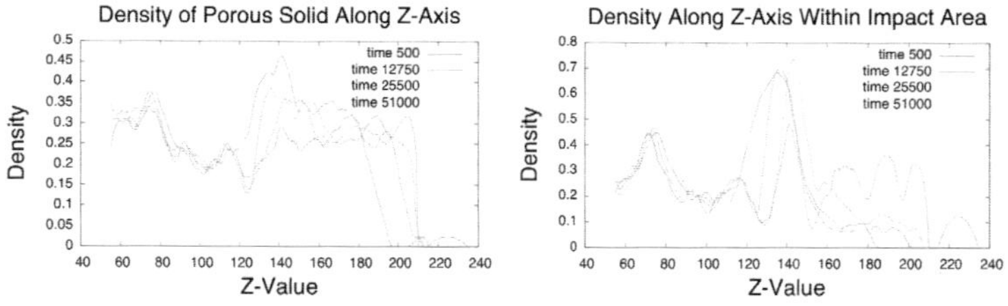

Figure 10. Computation, at each time step, of the filament density of the material (ratio of samples inside the interface surface over those outside the interface). We compute this density profile for the entire data (left) and a cylinder through the impact crater (right).

[3] P.-T. Bremer, H. Edelsbrunner, B. Hamann, and V. Pascucci. A multi-resolution data structure for two-dimensional Morse-Smale functions. In G. Turk, J. J. van Wijk, and R. Moorhead, editors, *Proc. IEEE Visualization '03*, pages 139–146, Los Alamitos California, 2003. IEEE, IEEE Computer Society Press.

[4] P.-T. Bremer, H. Edelsbrunner, B. Hamann, and V. Pascucci. A topological hierarchy for functions on triangulated surfaces. *IEEE Trans. on Visualization and Computer Graphics*, 10(4):385–396, 2004.

[5] P.-T. Bremer and V. Pascucci. A practical approach to two-dimensional scalar topology. page to appear. 2006.

[6] E. M. Bringa, J. U. Cazamias, P. Erhart, J. Stolken, N. Tanushev, B. D. Wirth, R. E. Rudd, and M. J. Caturla. Atomistic shock hugoniot simulation of single-crystal copper. *Journal of Applied Physics*, 96(7):3793–3799, 2004.

[7] K. Cole-McLaughlin, H. Edelsbrunner, J. Harer, V. Natarajan, and V. Pascucci. Loops in reeb graphs of 2-manifolds. In *Proceedings of the 19th Annual Symposium on Computational Geometry*, pages 344–350. ACM Press, 2003.

[8] A.W. Cook, W. Cabot, and P.L. Miller. The mixing transition in rayleigh-taylor instability. *J. Fluid Mech.*, to appear, 2004.

[9] A.W. Cook and P. E. Dimotakis. Transition stages of rayleigh-taylor instability between miscible fluids. *J. Fluid Mech.*, 443:66–99, 2001.

[10] H. Edelsbrunner, J. Harer, and A. Zomorodian. Hierarchical Morse-Smale complexes for piecewise linear 2-manifolds. *Discrete Comput. Geom.*, 30:87–107, 2003.

[11] H. Edelsbrunner, D. Letscher, and A. Zomorodian. Topological persistence and simplification. In *FOCS '00: Proceedings of the 41st Annual Symposium on Foundations of Computer Science*, page 454, Washington, DC, USA, 2000. IEEE Computer Society.

[12] A. Gyulassy, M. Duchaineau, V. Natarajan, V. Pascucci, E. Bringa, A. Higginbotham, and B. Hamann. Topologically clean distance fields. *IEEE Transactions on Visualization and Computer Graphics (to appear)*, pages xxx—xxx, 2007.

[13] A. Gyulassy, V. Natarajan, V. Pascucci, and B. Hamann. Efficient computation of morse-smale complexes for three-dimensional scalar functions. *IEEE Transactions on Visualization and Computer Graphics (to appear)*, pages xxx—xxx, 2007.

[14] Friedrich Horz, Ron Bastien, Janet Borg, John P. Bradley, John C. Bridges, Donald E. Brownlee, Mark J. Burchell, Miaofang Chi, Mark J. Cintala, Zu Rong Dai, Zahia Djouadi, Gerardo Dominguez, Thanasis E. Economou, Sam A. J. Fairey, Christine Floss, Ian A. Franchi, Giles A. Graham, Simon F. Green, Philipp Heck, Peter Hoppe, Joachim Huth, Hope Ishii, Anton T. Kearsley, Jochen Kissel, Jan Leitner, Hugues Leroux, Kuljeet Marhas, Keiko Messenger, Craig S. Schwandt, Thomas H. See, Christopher Snead, I Stadermann, Frank J., Thomas Stephan, Rhonda Stroud, Nick Teslich, Josep M. Trigo-Rodriguez, A. J. Tuzzolino, David Troadec, Peter Tsou, Jack Warren, Andrew Westphal, Penelope Wozniakiewicz, Ian Wright, and Ernst Zinner. Impact Features on Stardust: Implications for Comet 81P/Wild 2 Dust. *Science*, 314(5806):1716–1719, 2006.

[15] D. Laney, P. T. Bremer, A. Mascarenhas, P. Miller, and V. Pascucci. Understanding the structure of the turbulent mixing layer in hydrodynamic instabilities. *IEEE Transactions on Visualization and Computer Graphics*, 12(5):1053–1060, 2006.

[16] Y. Matsumoto. *An Introduction to Morse Theory. Translated from Japanese by K. Hudson and M. Saito.* American Mathematical Society, 2002.

[17] P. Miller, P.-T. Bremer, W. Cabot, A. Cook, D. Laney, A. Mascarenhas, and V. Pascucci. Application of morse theory to analysis of rayleigh-taylor topology. In *Proocedings of the 10th International Workshop on the Physics of Compressible Turbulent Mixing*, 2006.

[18] J. W. Milnor. *Morse Theory.* Princeton Univ. Press, New Jersey, USA, 1963.

[19] J. R. Munkres. *Elements of Algebraic Topology.* Addison-Wesley, Redwood City, CA, USA, 1984.

[20] V. Natarajan, Y. Wang, P.-T. Bremer, V. Pascucci, and B. Hamann. Segmenting molecular surfaces. *Comput. Aided Geom. Des.*, 23(6):495–509, 2006.

Visualization and parallel I/O at extreme scale

R B Ross[1], T Peterka[1], H-W Shen[2], Y Hong[2], K-L Ma[3], H Yu[3], and K Moreland[4]
[1]Mathematics and Computer Science Division, Argonne National Laboratory, Argonne, IL 60439, USA
[2]Department of Computer Science and Engineering, The Ohio State University, Columbus, OH 43210, USA
[3]Department of Computer Science, University of California at Davis, Davis, CA 95616, USA
[4]Sandia National Laboratories, Albuquerque, NM 87185, USA

E-mail: rross@mcs.anl.gov

Abstract. In our efforts to solve ever more challenging problems through computational techniques, the scale of our compute systems continues to grow. As we approach petascale, it becomes increasingly important that all the resources in the system be used as efficiently as possible, not just the floating-point units. Because of hardware, software, and usability challenges, storage resources are often one of the most poorly used and performing components of today's compute systems. This situation can be especially true in the case of the analysis phases of scientific workflows. In this paper we discuss the impact of large-scale data on visual analysis operations and examine a collection of approaches to I/O in the visual analysis process. First we examine the performance of volume rendering on a leadership-computing platform and assess the relative cost of I/O, rendering, and compositing operations. Next we analyze the performance implications of eliminating preprocessing from this example workflow. Then we describe a technique that uses data reorganization to improve access times for data-intensive volume rendering.

1. Introduction

As with theory and experimentation, computation has become a critical component of scientific discovery. In the quest to find answers to ever more challenging questions through computational techniques, we find ourselves constructing ever larger and more capable systems. Today we are pushing the scale of compute systems to reach a petaflop and beyond, and it has become increasingly important that all resources in the system are used as efficiently as possible. While we commonly use the floating-point operation rate as a measure of the performance of large-scale compute systems, most members of the scientific computing community realize that other characteristics of the system, such as the rate of memory access, more often determine the resulting application performance.

Because advances in storage have trailed behind those of processors and networks, data access has become a serious challenge for many applications running at large scale. Storage systems are being hampered by a number of issues. First, storage hardware bandwidths are not increasing at nearly the rate of processing hardware, meaning that more devices (disks) must be used in each successive generation of storage systems in order to keep pace with compute capabilities. Second, while significant algorithmic advances have taken place in scientific applications, much storage software – namely commercial parallel file systems – are using algorithms, interfaces, and architectures designed over a decade ago for much smaller scale systems. Working around the deficiencies of these old approaches can be very difficult. Third, scientists are often not using storage systems in the most effective manner. Tuning of data access is typically one of the last steps taken when improving an application code to operate more efficiently at scale, so many applications still perform I/O the same way they did many years ago, on very different systems. This might mean that all processes in a parallel application push data through a single node, or it might mean that every process creates its own independent data file when checkpointing, neither of which is likely to result in high performance. High-level libraries such as HDF5 [1] and Parallel netCDF [2] have been developed to provide more natural and efficient interfaces when used properly, but adopting these interfaces is a time-consuming task, and even with these libraries significant tuning is often necessary.

The analysis phase brings an additional set of I/O challenges. The conventional computational workflow has the simulation application deposit results on storage, and those results are processed by independent analysis tools as a separate step, after the simulation is complete. The separation of these two phases results in significant traffic to and from the storage system, because any data needed for analysis purposes must be stored for later access. The end result is that applications spend significant time waiting on I/O to complete, or they store less data, or store data less often, than they might otherwise. Moreover, the data written by the simulation code may not be in a form amenable to analysis, resulting in I/O patterns that make poor use of the storage hardware, even if the analysis code itself has been written using best practice access methods. These challenges require their own specific solutions, distinct from those developed to help with checkpointing and other output during the simulation phase, and developing these solutions typically involves both analysis and I/O expertise.

The SciDAC Institute for Ultra-Scale Visualization is working to enable extreme-scale knowledge discovery by advancing the state of visualization technologies, fostering awareness of and communication about new visualization technologies, and by putting these technologies into the hands of application scientists, especially SciDAC application teams. Institute research falls into one of three areas (figure 1). Work in architectures for visualization explores the viability of various types of compute resources with respect to scientific visualization. In improving and developing new fundamental algorithms, we seek to enable new visualization capabilities or apply existing capabilities to new types of data. Efforts in interactivity, interfaces, and tools seek to provide new ways to view and interact with data that help scientists more quickly and easily gain insight from their results. Overall, these three areas fit together to cover visualization challenges from end-to-end: from the underlying computational resources up to the user.

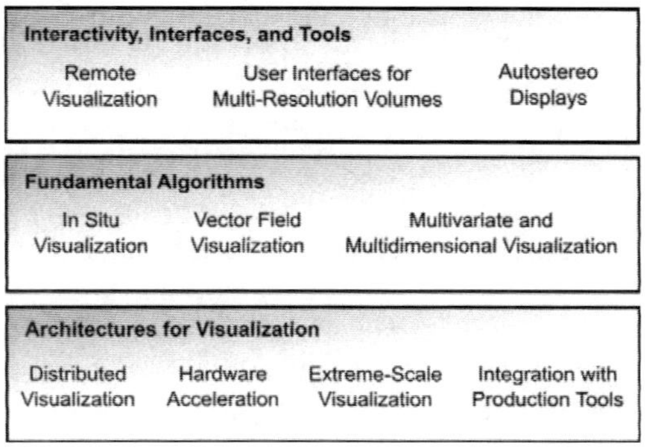

Figure 1. Research topics in the Institute.

Because of the technical challenges and the importance of effective visual analysis tools for extreme scale data, the Institute has a number of ongoing efforts in this area. One such effort is the addition of a parallel I/O layer to Paraview/VTK. This layer allows for the use of MPI-2 [3] I/O routines to improve performance on large-scale systems by eliminating bottlenecks in the data path, while serial I/O options are provided for portability. A small set of readers using this new parallel I/O layer are under development, and preliminary benchmarking is under way. By enhancing the widely used Paraview tool, we hope to enable more effective I/O for a broad range of applications.

Additional research is necessary to develop solutions for the analysis of data at extreme scales. In the following two sections we discuss Institute efforts in visualization for extreme scale data. In Section 2 we cover our work in volume rendering on the IBM Blue Gene/P architecture, and in Section 3 we describe our work in data organization for efficient I/O during visualization. In Section 4 we conclude and point to possible future approaches for reducing the I/O demands of analysis.

2. Volume rendering of large-scale data

One example of the Institute's research efforts is parallel volume rendering on leadership-class resources such as the Blue Gene/P at Argonne National Laboratory. Using the same compute architecture for visualization as for simulation can provide three benefits for large-scale data:

- Data movement from the simulation architecture to a visualization architecture can be avoided.
- The massive resources of a leadership machine, including high-speed interconnection network and parallel file system, can be brought to bear on the visualization problem.
- This is an intermediate step toward in situ visualization.

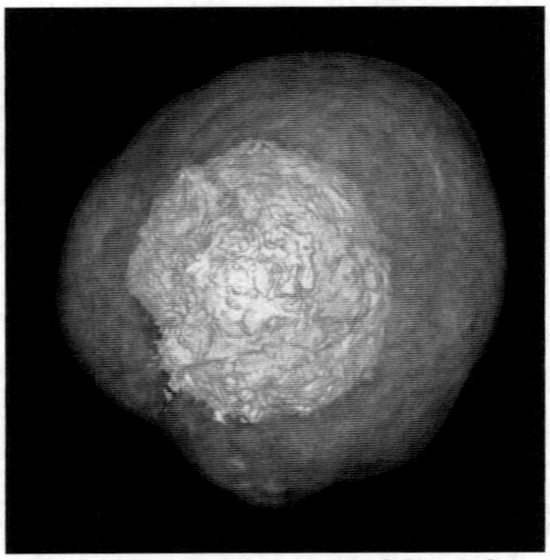

Figure 2. Image resulting from volume rendering of astrophysics dataset on IBM Blue Gene/P. Original image is 1600x1600 pixels.

At first glance, using a leadership computing resource for analysis might not make sense. The vector parallelism provided by modern GPUs enables peak rendering rates that are on average 100 to 1000 times faster than the software rendering rate of a low-power general-purpose processor such as BG/P's PowerPC. However, modern datasets that are gigabytes or larger per time step demand parallel processing for visualization, including parallel I/O. In this scenario, the bottleneck is not rendering time, but data movement between storage and processors. Thus, we want a visualization system that excels at these tasks, not solely by rendering. Fortunately, such visualization architectures, with robust parallel I/O and communication infrastructures already exist in the BG/P and Cray XT systems.

This work is in collaboration with John Blondin of North Carolina State University and Anthony Mezzacappa of Oak Ridge National Laboratory, who are working to understand the origin of spherical accretion shock instability (SASI) in supernova core collapse [4]. The simulation seeks to understand how the SASI is driven by global acoustic modes trapped in the supernova interior. By manipulating the transparency of rendered data (i.e., the transfer function) in our volume rendering code, the scientists can quickly see different combinations of variables and isolate features of interest. In their dataset, five scalar variables are stored in regular grid locations up to 1120^3, in netCDF format. Up to 200 time steps are stored in a simulation run, each in a separate file.

2.1. Phases of a visual analysis algorithm

Most parallel visual analysis algorithms can be divided into four main tasks: input, render, composite, and output. In classical parallel visualization with smaller datasets and systems, it was customary to ignore input costs and focus on rendering and compositing. However, the expanding size and complexity of data shift the burden onto the I/O system, meaning this cost must be recognized and included in our analyses as an integral component of parallel visualization [5].

Journal of Physics: Conference Series **125** (2008) 012099 doi:10.1088/1742-6596/125/1/012099

Parallel rendering algorithms are usually grouped into one of three types: sort-first, sort-middle, and sort-last [6]. We use the sort-last algorithm, which distributes the dataset among processors first and then gathers the partial results last. This is appropriate when the size of data dominates the problem size, as it does in most modern scientific visualizations that are data-intensive.

During the input phase, cores read a time step in parallel. By using MPI-2 features such as collective I/O as an interface to the parallel file system, all cores simultaneously read only their designated portion of the dataset. This approach increases I/O bandwidth due to parallel data paths between compute cores and storage, compared to a single data pathway from storage to a root node. Moreover, the entire dataset need not fit into the memory of a single core; the memory requirement is distributed among all of the cores.

Parallel input of a dataset tacitly assumes that a predefined domain decomposition of the data has already been established. Distribution can be static or dynamic; while dynamic distribution has been shown to produce good load balance at small data and system scales [7], it is costly at large scales and we do not believe it appropriate for tens of thousands of cores. Rather, we implement a static scheme that decomposes a time step into block-shaped subvolumes. By dividing into a multiple of blocks per processor core (usually 4–16) and allocating cores to blocks in a round robin manner, we have balanced load to greater than 50% rendering efficiency across 16K cores. As the next paragraph explains, some decrease in efficiency is unavoidable even with perfect load balance, because optimizations act locally, not globally across the data volume.

The rendering algorithm within a single core is classic software ray casting [8] with some optimizations and quality improvements, with the goal of maximizing rendering quality and minimizing per-frame preprocessing costs. Once the core has its subvolume of data and has applied the transfer function to map a physical quantity into a chromatic one, it casts rays from all pixels through its data, sampling the data at evenly spaced intervals along each ray. The spacing along a ray is controllable and proportional to the data spacing. Lighting is computed and includes diffuse, specular, and ambient components [9]. Next, the sample points along each ray are blended front-to-back using the "over" operator [10] to compute a single (R, G, B, A) pixel value. This process terminates prematurely if a maximum opacity is reached, called early ray termination. Other optimizations such as hierarchical levels of detail, octree data structuring, or empty space skipping are not used in our algorithm.

Compositing, like other classical parallel visualization tasks, has been extensively published but not at the system scales that we address in our research. We currently use the direct-send algorithm, mainly for its simplicity [11]. Direct-send is a two-step process. First, each process receives subimages from those processes that contribute to the portion of the image that the process is responsible for compositing. The process performs the same "over" blending operation that is performed for an individual ray during the rendering phase, but this time blending completed pixels from the subimages. Second, all processes send their completed subimages to the root. Unlike rendering, which is only approximately load balanced, the work in the compositing stage is evenly divided among processes. We have demonstrated effective capability of direct-send compositing up to 16K cores. Our results show nearly constant compositing time through 8K cores, and then a gradual increase beyond that. Network contention accounts for the majority of the compositing time at this point. In the first step a large number of small messages are exchanged, and each process has only a few pixels to composite. In the second step the finished subimages are gathered at the root process using MPI_Gather. Both communication steps are expensive at this scale. Because binary swap compositing [12] has lower complexity in the number of messages, we plan to analyze its performance at these scales in the future.

Thus far, image sizes have been limited to 2K x 2K pixels, or 4 megapixels, so output bandwidth to storage has not been a bottleneck. This will change as images become larger, and we will experiment with collective writes of the output image to eliminate the need to send all subimages to one root process, which can be a point of contention in the communication system.

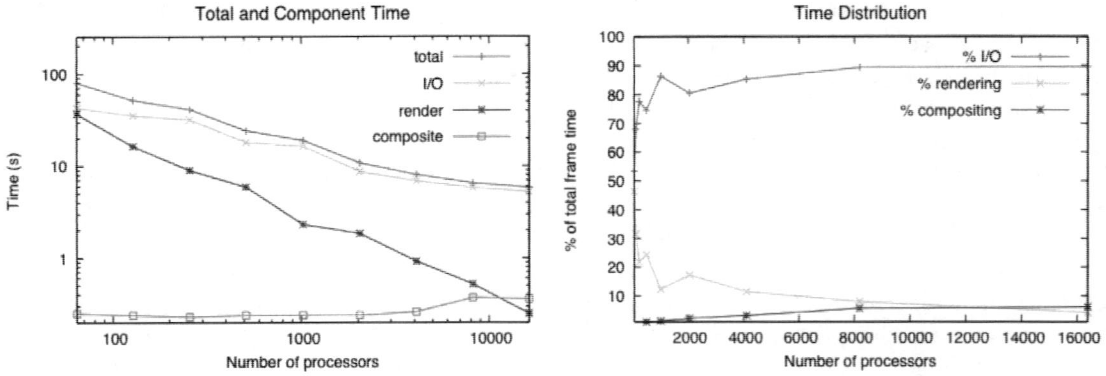

Figure 3. The time distribution between three stages of the volume-rendering algorithm for a 1600^2-pixel image, in seconds (left) and as a percentage of run time (right). Time is dominated by I/O.

2.2. Experimental results

Figure 3 shows how the duration that one time step requires is distributed among I/O, rendering, and compositing, from 64 to 16K cores. The jagged steps at smaller numbers of cores are caused by imperfect load balancing at certain number of processes. Overall, the time is dominated by I/O, consuming nearly 90% at 16K cores. Beyond 16K, compositing is more expensive than rendering, which is one indication that a new compositing algorithm is needed if we expect to scale further.

The dependence of large-scale visualization on I/O performance, made clear by figure 3, demonstrates the need to (1) continue to improve the I/O bandwidth and latency of the parallel file systems of architectures such as BG/P, and (2) devise ways to hide or mitigate the effects of I/O latency within visualization applications. While storage researchers attempt to improve storage performance overall through techniques such as advanced I/O forwarding [13], in Section 3 we discuss one effort in the Institute to reduce the I/O demands of visual analysis algorithms.

2.3. To preprocess or not to preprocess

Simulation output files are often preprocessed in order to reduce data sizes prior to moving data across the wide area, to extract variables of interest during analysis, or to transform the data into a format more amenable to processing by general-purpose analysis tools. In this context we consider this operation "preprocessing" because it must be performed before the analysis phase (in other contexts it might be termed "postprocessing" because it occurs after the simulation). This preprocessing step doesn't generate any new results, and as such it can be thought of as overhead in the scientific workflow. Reducing or completely eliminating preprocessing, especially for multivariate time varying data, presents an opportunity to improve the I/O performance of an application.

In our example astrophysics dataset, each time step is a netCDF file consisting of five record variables: pressure, density, and velocity in three directions. Each of these five quantities is a scalar value on a regular 3D Cartesian grid. For a grid size of 1120^3, one time step is approximately 25 GB. In the past, we would preprocess each time step to extract a single variable of interest, for example pressure, and write this into a separate, 32 bit raw data format. We would then proceed to volume render the sequence of pressure time steps, as described above.

This approach is expensive and needs to be streamlined if we expect it to scale further. First, there is the cost of preprocessing time. The extraction of a single variable from a single time step of the above

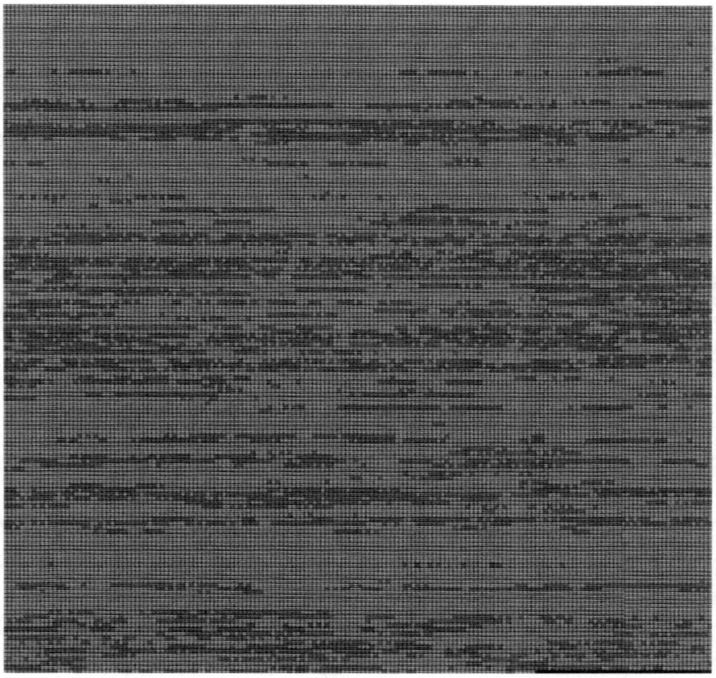

Figure 4. Depiction of a 2-dimensional representation of data blocks from the Visible Woman dataset stored on a single I/O server. Green (light) boxes represent data blocks that were not accessed during the sample operation. Red (dark) blocks show regions that were accessed. This view of data access highlights the sparse nature of data accesses on each server during a single visualization operation. This type of access will result in many disk seeks, reducing I/O throughput, and significantly reduces the effectiveness of prefetching.

dataset takes 2.5 minutes, using a serial netCDF reader on a single process of the same BG/P machine (and storage system). While this may not seem expensive because it is done offline, a dataset consisting of multiple variables and hundreds of time steps can require hours or days of preprocessing time. Additional storage space is also required: typically the original netCDF files are kept in addition to the extracted files.

To avoid these overheads, we have implemented direct parallel reading of netCDF data files within our volume rendering code. Using the parallel netCDF library [2], we can perform collective I/O from multiple cores as we did previously from a file holding a single variable. Our initial tests indicate that this parallel read takes 10.8 seconds for the 1120^3 volume on 16K cores, which is approximately twice the time taken to read the preprocessed data file. We are currently investigating ways to improve this performance; but given how quickly analysis occurs for this dataset and the simplicity from the workflow perspective, this approach of reading directly from the simulation output seems appropriate in this case.

3. Organizing data for visualization

In the previous section we discussed the benefits of eliminating one type of preprocessing from the scientist's workflow. When we explored directly reading data from the astrophysics dataset in the above example, the performance degradation seen was due to the fact that we were accessing only 20% of the data in the time step and that data was spread out in the file, causing noncontiguous I/O access, where accesses for a single process are spread out across the file [14]. In datasets with more variables, in unstructured datasets where data might be spread less regularly throughout the file, or in cases where some data might not be needed for algorithmic reasons, the performance gap could be much wider between simulation data and preprocessed data access rates. While storage researchers are investigating methods to improve performance for analysis workloads with noncontiguous access patterns [15], we are examining methods to reorganize data within preprocessed data files to improve locality of access and, in doing so, improve I/O access rates.

Our algorithm is based on knowledge of data distribution in parallel file systems such as the Parallel Virtual File System (PVFS) [16], Lustre [17], and GPFS [18]. In these systems, data are striped across I/O servers in a round-robin manner. Figure 4 shows the I/O access patterns from a sample I/O server for a typical rendering run on a dataset of 512x512x1728 voxels stored on a PVFS file system with 16 I/O servers and 1 Kbyte strip size. This figure shows that the data I/O access patterns can be very sparse. When the data blocks needed for a visualization query are scattered across a large file, two problems can occur. First, these blocks may be striped onto only one or a few I/O servers, which can result in I/O congestion. Second, the data blocks on individual I/O servers may be stored noncontiguously, which will increase the number of disk seeks on that server during access, reducing effective throughput.

Generally, when volume rendering is applied to a large dataset, for a given user-defined transfer function and a particular view direction, there can be up to 70% of data blocks that are outside the value range of interest, and hence meaningless to load. Among those meaningful blocks, half of them can be occluded by other data blocks and thus become invisible. These blocks are also unnecessary to load. Our approach is to deploy a more efficient data distribution strategy to group together data blocks that are more likely to be accessed together to avoid noncontiguous I/O requests when visualizing large scale datasets. To group data blocks, we analyze the visibility pattern of the data blocks using *visibility feature vectors* for data blocks as signatures to identify data blocks that have similar access patterns. By reorganizing these data blocks based on the calculated access patterns, the number of necessary noncontiguous I/O can be greatly reduced.

3.1. Visibility feature vectors

When volume rendering is implemented by using the standard ray casting algorithm, a block's visibility can be estimated if we are given a list of sample values collected by the viewing rays starting from the image plane and before they reach the volume block. This is possible because we can evaluate the discrete volume rendering equation using the sample values and estimate the accumulated opacity. To represent the sample values for each ray, we can use a *per-ray sample histogram*. For a volume block under a particular view direction, we can combine the per-ray sample histograms from all the pixels within the projection footprint of the block. This combined sample histogram can characterize the block's visibility in the sense that if two blocks share a very similar sample histogram, they are likely to have similar visibility. This is the basic idea behind our visibility feature vector.

A visibility feature vector is a *n*-dimensional tuple in the format of $[h_1, h_2, ..., h_n]$, where each dimension represents a sample view direction. The entry h_i is the corresponding histogram with respect to the *i*th view direction. The purpose of the visibility feature vector is to keep an overall description of a block's visibility status for all possible view directions. Each block has a visibility feature vector, based on which it is possible to compare the visibility characteristics among multiple blocks. The blockwise visibility feature vector is constructed based on the data only and does not depend on run time transfer functions. By clustering data blocks based on the visibility feature vectors, it is possible to identify those blocks with approximately the same visibility status, and hence the same usage pattern. In addition to volume rendering, visibility feature vectors can be applied to other scalar data visualization techniques such as isosurfaces by configuring an appropriate transfer function. Isosurfacing can be implemented by setting the desired isovalue completely opaque while leaving other isovalues transparent. Invisible isosurfaces can be quickly estimated as in volume rendering using the per-ray sample histograms and the run-time transfer function.

To improve the I/O performance and minimize the number of noncontiguous I/O requests, we group the data blocks that share similar visibility characteristics and place them together in a file. When copying the file onto the parallel file system, data blocks are striped along various I/O servers in terms of pre-calculated clusters: the blocks within the same cluster are distributed in a round-robin manner into a

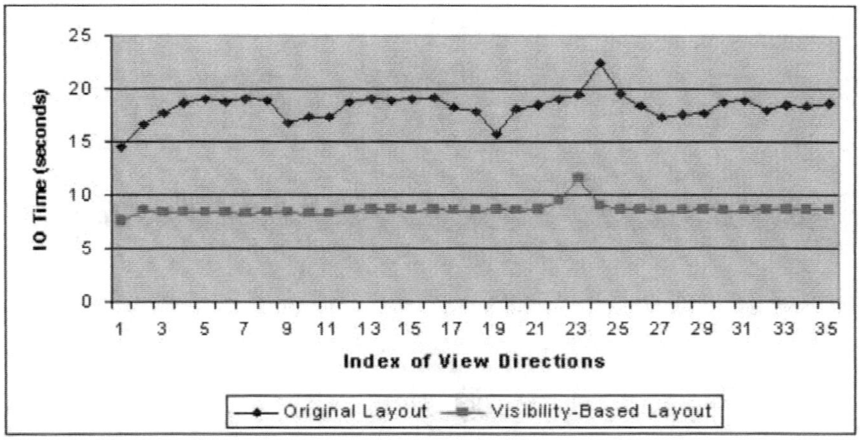

Figure 5. Comparison of I/O time between original file layout and visibility-based file layout using the Visible Woman dataset and a 1 Kbyte PVFS strip size.

sequence of I/O servers. This new layout strategy has the advantage that since the data blocks in the same cluster are likely to be accessed together, distributing the I/O load to all servers can avoid placing the load among only a few I/O servers. It can also avoid issuing multiple noncontiguous small IO requests in the individual I/O servers. Our experimental tests showed that the new file layout greatly reduces the I/O cost.

3.2. Experimental results

The test dataset for our I/O performance experiment is a 512x512x1728 dataset called the "Visible Woman." All our tests were run on an IBM Blue Gene/L supercomputer at Argonne National Laboratory, which has up to 2048 700 MHz PowerPC 440 cores, each with a double-pipeline-double-precision floating-point unit (FPU), a cache subsystem with built-in DRAM controller and the logic to support multiple communication sub-systems. The dual FPUs give each Blue Gene/L node a theoretical peak performance of 5.6 GFLOPS. In our experiments, the Visible Woman volume dataset is partitioned into 110,592 16^3 volume blocks and stored on a PVFS file system across all 16 I/O servers. All of our experiments were conducted using 64 processors.

We computed the visibility feature vectors described above. In the case of this data set, the alignment of the data set is such that view direction variations along the z-axis produce only small variations in resulting images. For this reason the sampled view directions were rotated along the x-axis and y-axis with a 20-degree increment in each step, with a fixed z-axis position, which produced a total of 64 sampled view directions. It took approximately 2.3 hours using 64 nodes of the Blue Gene/L to sample the view directions and build the corresponding visibility feature vectors. Based on the visibility feature vectors, a clustering process was conducted to group those blocks with similar feature vectors. Data blocks in the same cluster are placed contiguously in a new file; generating this new file took an addition 13 minutes. The PVFS file system will stripe the data in a round-robin manner to several I/O servers, placing these blocks near one another on individual servers. Assuming that we have correctly clustered data, I/O requests in this format should be serviced significantly more quickly than they would have been serviced in the original configuration.

We conducted a series of experiments to test the I/O performance of our algorithm. Two file layouts were tested on the Visible Woman dataset: one was the original layout where the data blocks were stored sequentially according to their spatial locations, and the other was the new layout using our visibility-based clustering. In order to ensure that the I/O performance improvement was stable under different transfer functions, four transfer functions were used for each layout, and 35 view directions were sampled to test with each transfer function. During the run-time rendering process, after visibility culling, each processor reads the needed data blocks using MPI-IO. In our implementation we first use MPI_File_set_view() to adjust individual process's view of the data in the file. Then we call MPI_File_read_all() to utilize the noncontiguous I/O optimizations provided in MPI-IO. For each view direction, the I/O time was collected for each processor and the average time was calculated as the metric to compare between the two layouts. We tested using 16 I/O servers with three strip sizes – 1 Kbyte, 4 Kbyte, and 8 Kbyte – to explore the relationship between strip size and our layout algorithm.

Figure 5 shows a comparison of I/O performance between the two file layouts with the 1 Kbyte PVFS strip size for a sample transfer function, clearly demonstrating the improvements achieved by the visibility-based file layout. For each view direction the same number of data blocks were read in both file layouts. PVFS performed more efficiently when using the visibility-based layout. Results show up to 50% reduction of I/O time by using the visibility-based file layout under different view directions. Similar results were found for the other three transfer functions and with the 4 Kbyte and 8 Kbyte strip sizes, although access times improved slightly for both layouts as the strip size became more closely matched to the data block size. Moreover, the variances of I/O time under different view directions with the visibility-based layout were far less than those with the original layout. This is because in the visibility-based layout, there existed fewer numbers of noncontiguous I/O requests compared with the original file layout.

4. Conclusions

Optimizing access to storage is becoming increasingly important, not just during simulation run-time but also during analysis. Numerous approaches have been pursued to date, each with its own advantages and disadvantages. Directly reading simulation output makes sense when the simulation output is relatively well suited to analysis, as was the case in our astrophysics volume rendering example. In other cases aggressive preprocessing prior to analysis can provide substantial performance improvement, as seen in the Visible Woman example. Another option is to avoid this postsimulation analysis altogether, by performing analysis at simulation run-time.

In situ processing is the reduction, transformation, analysis, or viewing of data as it is being computed, using the same architecture as the computation [19]. This technique has two distinct advantages. First, this technique allows analysis routines to operate on data while it is still in memory, potentially significantly reducing the I/O demands of the overall workflow. Second, the technique naturally enables the application scientist to see the results of the simulation as it proceeds, because these results are generated as the simulation progresses. The Institute has explored in situ processing previously [19], including a study of in situ visualization for steering of a massively parallel earthquake simulation [20].

Solutions exist to enable in situ analysis in applications, but current portable solutions require that applications be developed in specific problem solving environments [21–25]. This requirement prevents in situ analysis from being adopted by a wider community of users, despite the technique's substantial benefits in time to solution and ability to track progress. A more accessible approach to in situ processing is desperately needed to address the performance limitations of storage resources in supercomputing systems. Because applications are increasingly using more descriptive I/O interfaces to interact with storage, from database interfaces such as SQL to multidimensional data interfaces such as the Hierarchical Data Format (HDF5) and Network Common Data Format (netCDF) libraries, one option might be to incorporate analysis techniques into the I/O path, hiding analysis routines under interfaces already in place in the applications and transparently enabling in situ processing.

Acknowledgments

This work was supported in part by the Mathematical, Information, and Computational Sciences Division subprogram of the Office of Advanced Scientific Computing Research, Office of Science, U.S. Department of Energy, under Contract DE-AC02-06CH11357. Work is also supported in part by NSF through grants CNS-0551727 and CCF-0325934, and DOE with agreement No. DE-FC02-06ER25777.

References

[1] *The Hierarchical Data Format*, http:// hdf.ncsa.uiuc.edu/HDF5/
[2] Li J, Liao W-K, Choudhary A, Ross R, Thakur R, Gropp W, Latham R and Siegel A, Gallagher B, and Zingale M 2003 Parallel netCDF: A high-performance scientific I/O interface *Proc. Supercomputing 2003*
[3] The MPI Forum 1997 MPI-2: extensions to the message-passing interface
[4] Blondin J M, Mezzacappa A, and Demarino C 2003 Stability of standing accretion shocks, with an eye toward core collapse supernovae *The Astrophysics J.* **584**(2):971
[5] Peterka T, Yu H, Ross R and Ma K-L 2008 Parallel volume rendering on the IBM Blue Gene/P *Proc. Eurographics Parallel Graphics and Visualization Symposium 2008*
[6] Molnar S, Cox M, Ellsworth D and Fuchs H 1994 A sorting classification of parallel rendering *IEEE Computer Graphics and Applications* **14**(4):23–32
[7] Marchesin S, Mongenet C and Dischler J-M 2006 Dynamic load balancing for parallel volume rendering *Proc. Eurographics Symp. Parallel Graphics and Visualization 2006*
[8] Levoy M 1988 Display of surfaces from volume data *IEEE Computer Graphics and Applications* **8**(3):29–37
[9] Max N L 1995 Optical models for direct volume rendering *IEEE Trans. Visualization and Computer Graphics* **1**(2):99–108
[10] Porter T and Duff T 1984 Compositing digital images *Proc. 11th Annual Conf. on Computer Graphics and Interactive Techniques* 253–259
[11] Hsu W M 1993 Segmented ray casting for data parallel volume rendering *Proc. 1993 Parallel Rendering Symposium* 7–14
[12] Ma K-L, Painter J S, Hansen C D and Krogh M F 1994 Parallel volume rendering using binary-swap compositing *IEEE Computer Graphics and Applications* **14**(4):59–68
[13] Iskra K, Romein J W, Yoshii K and Beckman P 2008 ZOID: I/O-forwarding infrastructure for petascale architectures *Proc. 13th ACM SIGPLAN Symp. on Principles and Practice of Parallel Programming* 153–162
[14] Ching A, Choudhary A, Liao A, Ross R and Gropp W 2004 Evaluating structured I/O methods for parallel file systems *International J. High Performance Computing and Networking* **2**:133–145
[15] Gu P, Wang J and Ross R 2008 Bridging the gap between parallel file systems and local file systems: A case study with PVFS *Proc. International Conf. on Parallel Processing 2008* (to appear)
[16] Carns P, Ligon W B, Ross R and Thakur R 2000 PVFS: A parallel file system for Linux clusters. *Proc. 4th Annual Linux Showcase & Conference*
[17] Cluster File Systems 2002 Lustre: A scalable, high performance file system *White paper*
[18] Schmuck F and Haskin R 2002 GPFS: A shared-disk file system for large computing clusters *Proc. File and Storage Technologies Conference* 231–244
[19] Ma K-L, Wang C, Yu H and Tikhonova A 2007 In-situ processing and visualization for ultrascale simulations *J. Physics* 78
[20] Tu T, Yu H, Ramirez-Guzman L, Bielak J, Ghattas O, Ma K-L and O'Hallaron D R 2006 From mesh generation to scientific visualization: An end-to-end approach to parallel supercomputing *Proc. ACM/IEEE Supercomputing 2006 Conference*
[21] Johnson C, Parker S, Hansen C D, Kindlmann G L and Livnat Y 1999 Interactive simulation and visualization *IEEE Computer* **32**(12):59–65
[22] Parker S and Johnson C 1995 SCIRun: A scientific programming environment for computational steering *Proc. Supercomputing 1995*
[23] Tuchman A, Jablonowski D and Cybenko G 1991 Run-time visualization of program data *Proc. IEEE Visualization 1991* 255–261
[24] Jablonowski D, Bruner J D, Bliss B and Haber R B 1993 VASE: The Visualization and Application Steering Environment *Proc. Supercomputing 1993*
[25] Geist G A, Kohl J A and Papadopoulos P M 1997 CUMULVS: Providing fault-tolerance, visualization, and steering of parallel applications *International J. High Performance Computing Applications* **11**:224–236

Software Infrastructure for exploratory visualization and data analysis: Past, present, and future

Cláudio T. Silva[1,2] **and Juliana Freire**[2]

[1]Scientific Computing and Imaging Institute, University of Utah, Salt Lake City, UT 84112, USA

[2]School of Computing, University of Utah, Salt Lake City, UT 84112, USA

E-mail: `csilva@sci.utah.edu`, `juliana@cs.utah.edu`

Abstract. Future advances in science depend on our ability to comprehend the vast amounts of data being produced and acquired, and scientific visualization is a key enabling technology in this endeavor. We posit that visualization should be better integrated with the data exploration process instead of being done after the fact – when all the science is done – simply to generate presentations of the findings. An important barrier to a wider adoption of visualization is complexity: the design of effective visualizations is a complex, multistage process that requires deep understanding of existing techniques, and how they relate to human cognition. We envision visualization software tools evolving into "scientific discovery" environments that support the creative tasks in the discovery pipeline, from data acquisition and simulation to hypothesis testing and evaluation, and that enable the publication of results that can be reproduced and verified.

1. Introduction

As scientists need to deal with large volumes of heterogeneous data, such as medical MRI and CT scans, water salinity measurements, engine temperature measurements, and results of computational experiments, there is an increasing need for techniques and usable tools to help them analyze and make sense out of these data. Although the term "visualization" is somewhat recent, generally accepted to having been coined for the 1987 NSF report on scientific visualization, visualization has been used as a means of "data understanding by visual representation or other visual means" for hundreds of years. What separates the old from the new is the availability of advanced computing capabilities, including modern computer graphics techniques, which form the backbone of modern visualization research.

While computer graphics techniques are an integral and important part of visualization, it is important to contrast the two fields. In the more traditional computer graphics, used to make movies or games, the goal is to produce visually engaging (beautiful) imagery that "appears plausible." In visualization, *accuracy* is more important than aesthetics. As such, users must be aware of errors and uncertainty present in the visualizations so that they are not misled by the resulting images [1]. Furthermore, the process of extracting insight from data goes beyond the generation of imagery and needs to encompass the complete scientific discovery pipeline [2–4].

Several visualization techniques have been developed that greatly improve our ability to comprehend large amounts of complex data. Figure 1 shows two images of a CT tooth dataset that clearly exemplify the the capabilities of advanced visualization techniques.. On the left, we

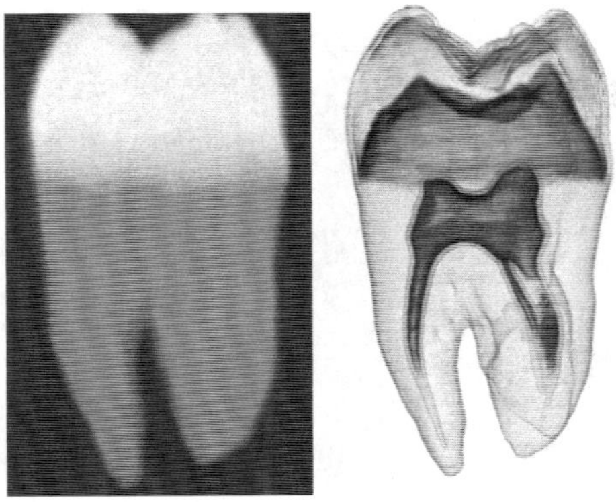

Figure 1. Maximum intensity projection (MIP) volume rendering (left) versus full direct volume rendering (right). Image courtesy of the Scientific Computing and Imaging (SCI) Institute.

show a maximum intensity projection (MIP) image, of the type we could compute in the early 1980s. On the right, we display a full volume rendering that uses multidimensional transfer functions [5]. While in the MIP visualization, we are limited to seeing overall density changes, using state-of-the art volume rendering, we are able to precisely extract the different tooth structures, and study them in detail.

In spite of the benefits of visualization, currently visualization is not well integrated with the scientific process. In this paper, we discuss some of the barriers to a wider adoption of visualization techniques in science and new research directions that aim to close this gap. One issue is the lack of communication between the visualization community and the potential users in different scientific areas. Scientists are often unaware of the state of the art and often use older techniques that may suffer from deficiencies (e.g., in performance or visualization quality). We need to make visualization techniques more widely available. We also need tools that integrate these techniques and that guide users through the process of creating insightful visualizations. To paraphrase Don Norman: we should design tools that make us smarter [6].

Presentation versus exploratory visualization. Most contemporary scientists often see the best visualizations as finished products, for instance, in the form of short videos that present detailed explanations of some complex phenomena. To this end, web sites such as `www.scivee.tv`, have been created as a forum for scientists to present their research by contributing videos. One of the earliest, and still best, examples of a visual presentation using advanced visualization techniques is the National Center for Supercomputing Application (NCSA) classic "Study of a Numerically Modeled Severe Storm." This short video presents the physics of a severe storm in a clear and concise manner. It is divided into several segments, each of which builds up intuition for the underlying characteristics of the simulation. After introducing many of the important aspects, the movie ends by present a complex visual representation of the overall phenomena by using many visualization techniques at the same time. This movie provides a very good example of the use of visualization for summarizing and presenting scientific results. It was highly influential, and it is routinely used as a case study. For instance, Tufte [7] writes an excellent critic of this classic and proposes a number of improvements to the presentation.

While this use of visualization techniques for the presentation of complex technical results is extremely important, in our view the area can play an even more important role in the sciences if it is integrated into the scientific progress. Many scientists take the view that visualization is what is done "after the fact"—when all the science is done—to generate presentations of the findings. However, we believe the real benefits of visualization become more pronounced when used as a part of the data exploration process.

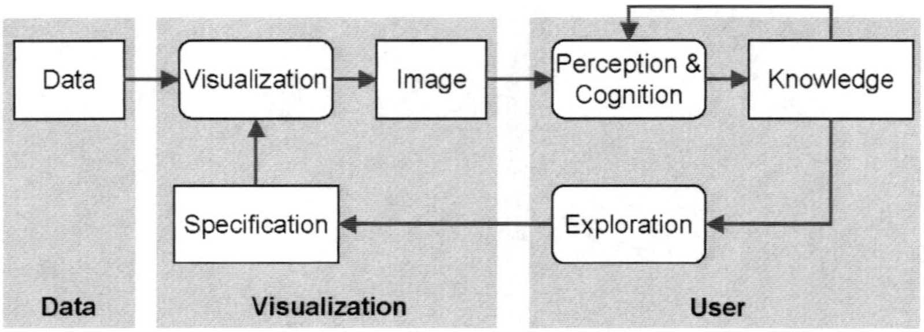

Figure 2. A model for exploratory visualization. Adapted from J. van Wijk [2].

We take the view that future advances in science depend on the ability to comprehend the vast amounts of data being produced and acquired. Visualization is a key enabling technology in this endeavor: it helps people explore and explain data through software systems that provide a static or interactive visual representation. A basic premise of visualization is that visual information can be processed at a much higher rate than raw numbers and texr – as the cliché goes: "A picture is worth a thousand words."

Despite the promise that visualization can serve as an effective enabler of advances in other disciplines, the application of visualization technology is non-trivial. The design of effective visualizations is a complex process that requires deep understanding of existing techniques, and how they relate to human cognition. Although there have been enormous advances in the area, the use of advanced visualization techniques is still limited.

Challenges in exploring data through visualization. Data exploration through visualization requires scientists to go through several steps. To successfully analyze and validate various hypotheses, one must pose several queries, correlate disparate data, and create insightful visualizations of both the simulated processes and observed phenomena.

As illustrated in figure 2, a scientist needs to assemble and execute complex workflows that consist of data set selection, specification of series of operations that need to be applied to the data, and the creation of appropriate visual representations, before finally viewing and analyzing the results. Often, insight comes from comparing the results of multiple visualizations created during the exploration process, for example, by applying a given visualization process to multiple datasets generated in different simulations, by varying the values of certain visualization parameters, or by applying different variations of a given process (e.g., which use different visualization algorithms) to a dataset.

The challenge of this exploration process is that for a visualization to be insightful, it needs to be both effective and efficient. This requires a combination of art, technology, and science to reveal information that is otherwise obscured. Assessing the value of an insight or measuring how much insight is required in the analysis of the data is a difficult task because the scientist may not even know what aspects or features the data contains [8]. Therefore, visualization can be regarded as an open-ended problem with no single answer.

The process is further complicated when viewed in the larger context of reproducible science. In particular, scientists and engineers need to expend substantial effort managing data (e.g., scripts that encode computational tasks, raw data, data products, images, and notes) to ensure that the visualizations are reproducible by themselves as well as others in the scientific community.

Despite the enormous progress in scientific visualization research, existing tools do not support the analytical reasoning process that scientists use in their work. There is little or no support for linking data, visualizations, and knowledge. For instance, as insight is generated over time, the findings are not linked to supporting evidence, and scientific publications to a large extend stand on their own with little hard evidence of the scientific facts.

2. Incomplete history of scientific visualization tools

To see where we stand today in terms of visualization and data analysis tools for scientific discovery, it is useful to take a look at some of the tools that have been developed by the visualization community over the past two decades.

The Application Visualization System (AVS) [9] was one of the earliest and most influential visualization environments developed in the 1980s. It was based on a dataflow model and aimed to provide an easy-to-use, powerful system for supporting the filter/map/render pipeline. The IBM Data Explorer (DX) [10] and the IRIS Explorer are two other systems from the same period. Each system had its own innovations; e.g., IBM DX had a very flexible data model. Testimony to their effectiveness, these dataflow-based visualizations systems are still widely used today, over 20 years since they were originally developed; and they can be seen as precursors of current scientific workflow systems.

A dataflow is a directed graph where nodes represent computations and edges represent streams of data: each node or module corresponds to a procedure that is applied on the input data and generates some output as a result. The flow of data in the graph determines the order in which the processing nodes are executed. In the dataflow model, a change in one variable forces the recalculation of other connected variables (similar to inter-dependent cells in a spreadsheet). Furthermore, computation is stateless: the output of a module is functionally determined by the module and its inputs. In visualization, it is common to refer to a dataflow network as a "visualization pipeline." The actual computation is performed by the modules by operating on the input data. It is sometimes useful to think of the modules as performing different parts of the computation. Common types include modules that generate data (sources), consume data (sinks), and operate on data by applying algorithms to the data (processes). In their most general form, each module has a set of inputs, a set of outputs, a set of parameters, and, in certain systems, some form of GUI that can be used for interacting with the modules. Tthese inputs and outputs are often referred to as "ports." It is useful to enforce "types" in these systems, allowing connections only between ports of compatible types.

A "visualization pipeline" provides a means to create visual representations of raw data. It is fairly common to break up (at least, conceptually) the visualization pipeline into three main phases that each data item needs to pass through to be visualized: *filter*, *map*, and *render*. In practice, there might be steps that need to happen before (e.g., reading the data from disk and uncompressing it) or after (e.g., image delivery). Also, depending on the exact visualization algorithm being applied, one can think of "skipping" a step. This "3-stage pipeline" is a bit too simplistic but, despite its potential faults, serves as the basis of a general framework for setting up visualization pipelines.

The goal of the filtering phase is to prepare data for visualization. It takes the raw simulated or sensored data and transforms it into another form that is more informative or less voluminous. It might also generate auxiliary information that is not readily available but is necessary for later processing steps. For instance, it might do data resampling, interpolation, or gradient calculation. The mapping phase will turn the filtered data into geometric primitives that can be rendered. For instance, it can generate OpenGL display lists from triangulated models or create polygonalized versions of glyphs. The rendering phase takes the outputs of the map phase and creates images, taking advantage of graphics hardware when possible.

The early 1990s brought us Kitware's Visualization Toolkit (VTK) [11], which is an open source, object-oriented toolkit based on the dataflow programming model. For efficiency, the core components of the system are written in a compiled language (C++). For flexibility and extensibility, an interpreted language (e.g., Python, Tcl, or Java) can be used for higher-level applications. This scripting capability and a large core set of algorithms have promoted VTK to its current status as one of the most popular visualization packages for researchers. VTK is widely used around the world, and many tools have been developed on top of it. Its open-source license has enabled the development of a number of influential end-user visualization tools. Two such tools developed under DOE funding are ParaView and VisIt. The ParaView project started by James Ahrens at Los Alamos National Laboratory [12] was aimed not only at extending VTK into a parallel framework but also at developing a turnkey end-user tool that did not require users to explicitly build dataflow graphs. VisIt [13] had a similar goal, but its design was dictated by the visualization needs of Lawrence Livermore National Laboratory. In particular, the development team led by Hank Childs decided to keep the look and feel of LLNL's MeshTV visualization tool.

In the 1990s, we also saw the start of the movement toward integrating scientific computing, parallelism, and visualization. SCIRun [14] and GRASPARC (GRAphical Support for PARallel Computing) [15] are two great examples of systems developed during this time. For the most part, other than the ability to generate complex visualizations, these systems lacked other features that were explicitly aimed at supporting the scientific process. One notable exception is that as part of their pioneering work on the GRASPARC system, Brodlie and colleagues proposed the use of a history mechanism that allowed scientists to steer an ongoing simulation by backtracking a few steps, changing parameters, and resuming execution.

3. Building tools to support scientific discovery

While visualizations in the form of plots, pictures, animations, and interactive 3-D applications are very important to generate insight, recently there has been a new research trend that is based on the premise that it is possible to do more to help humans be more effective at scientific discovery [3, 4, 16]. Pirolli and Card [17, 18] study the sensemaking process (i.e., the process of looking for meaning in a body of data with the intent of solving particular problems). They propose categories for the overall process and explore opportunities where software can help. The goal is to "enhance" the discovery process by providing software tools that support existing tasks, for example, better organization tools that improve the utilization of working memory. Social data analysis, a process by which multiple people collectively analyze data, has recently been proposed as a means to uncover new paths of discovery, and tools have been proposed that are aimed at facilitating collaboration and knowledge reuse.

Very recently, Schrinivasan and van Wijk [16] proposed Aruvi, a system for supporting the analytical reasoning process. Their system is based on three main components: a *data view*, where a user can interact with visual representations of the data (this is effectively what is available in traditional visualization systems, such as the ones that were discussed in the previous section and the main visualization supported in their system is the scatterplot), a *navigation view* which provides a visual representation of the exploration process itself, basically it records the history of exploration; and a *knowledge view*, where users can maintain analysis artifacts (e.g., hypothesis, conclusions) with direct links to the plots that support those conclusions. While the type of analysis that can be done with Aruvi is somewhat limited, the system clearly shows the potential for building tools that can be used for improving the support for discovery.

Social Web sites and web-based communities (e.g., Flickr, Facebook, Yahoo! Pipes), which facilitate collaboration and sharing between users, are becoming increasingly popular. An important benefit of these sites is that they enable users to leverage the wisdom of the crowds. For example, in Flickr, users, in a mass collaboration approach, tag large volumes of pictures. These tags, in turn, help them to more easily find pictures they are looking for. A notable social data analysis system is IBM's ManyEyes. This system is based on the blog paradigm and enables a large number of users to collaboratively analyze data. Users can upload datasets into the site and create visualizations from a dozen or so predefined types. Anyone can modify the existing visualizations (always in a nondestructive manner) and comment on each other's work, just as in a normal blog. The discussions are persistent, and they provide a log that can be used for understanding any conclusions reached by the participants. It has been reported that ManyEyes gets over 500,000 visits a month, clearly highlighting the appeal of open discussion.

While it would be interesting to use the same technology for supporting scientific discovery, there are many challenges [19]. The heterogeneity of the data, its size, and its location greatly complicate the data analysis pipelines. Whereas existing "social data analysis" (SDA) systems require that data be uploaded to a central location for analysis, in many scientific applications that manipulate large volumes of data, this is not feasible. Furthermore, the visualization process is likely to require staged processing and a larger variety of underlying visualization techniques than what is currently supported by systems such as ManyEyes. Data analysis generates more data (e.g., graphs, visualizations) that add to the overflow of information scientists need to deal with. Thus, ad hoc approaches to data exploration, which are widely used in the scientific community, have serious limitations. In particular, scientists and engineers need to expend substantial effort managing data (e.g., scripts that encode computational tasks, raw data, data products, images, and notes) and to record provenance information so that basic questions can be answered, such as: Who created a data product and when? When was it modified and by whom? What was the process used to create the data product? Were two data products derived from the same raw data? Not only is the process time-consuming but also error-prone. The absence of provenance makes it hard (and sometimes impossible) to reproduce and share results, to solve problems collaboratively, to validate results with different input data, to understand the process used to solve a particular problem, and to reuse the knowledge involved in the data analysis process. In addition, it greatly limits the longevity of the data products–without precise and sufficient information about how the data product was generated, its value is greatly diminished. SDA systems aimed at the scientific domain need to provide a flexible framework that not only enables scientists to perform complex analyses over large data but also captures detailed provenance of the analysis process.

4. VisTrails: provenance and data exploration

Data exploration is an inherently creative process that requires users to locate relevant data, to visualize these data and discover relationships, to collaborate with peers while exploring different solutions, and to disseminate results. Given the size of data and complexity of analyses that are common in scientific exploration, new tools are needed and existing tools should be extended to better support creativity [3, 4].

The ability to systematically capture provenance is a key requirement for these tools. The provenance (also referred to as the audit trail, lineage, and pedigree) of a data product contains information about the process and data used to derive the data product. The importance of keeping provenance information for data products is well recognized in the scientific community [20–27]. Provenance provides important documentation that is key to preserving the data, to determining the data's quality and authorship, and to reproducing as well as validating the results. In addition, the availability of provenance supports reflective reasoning, allowing users to store temporary results, to make inferences from stored knowledge, and to follow chains of reasoning backward and forward [6].

VisTrails (`www.vistrails.org`) is an open-source system that was designed to support exploratory computational tasks such as visualization, data mining, and integration. VisTrails provides comprehensive provenance management infrastructure and can be easily combined with existing tools and libraries. A new concept we introduced with VisTrails is the notion of *provenance of workflow evolution* [28]. In contrast to previous workflow and visualization systems that maintain provenance only for derived data products, VisTrails treats the workflows (or pipelines) as first-class data items and keeps their provenance. In Section 4.1, we discuss

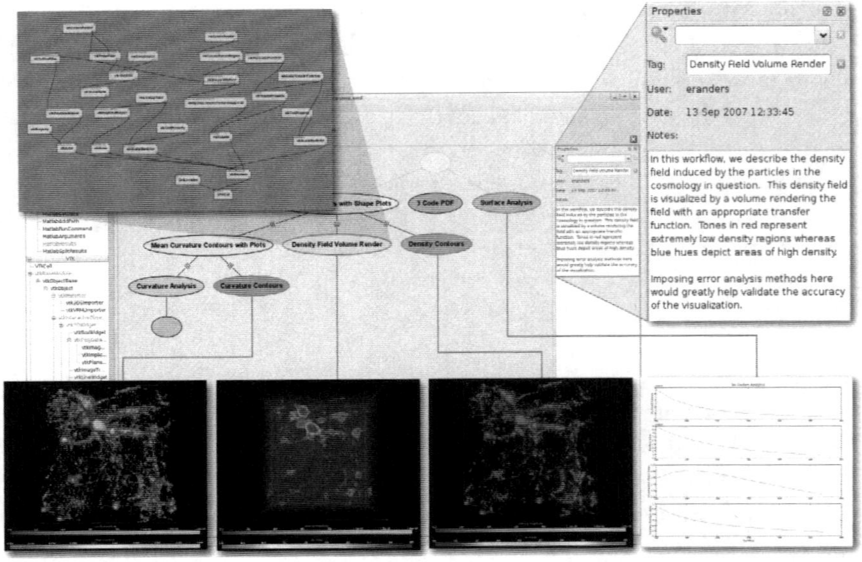

Figure 3. Example of an exploratory visualization for studying cellestial structures derived from cosmological simulations using VisTrails [30]. Complete provenance of the exploration process is displayed as a vistrail (history tree) with each node representing a workflow that generates a unique visualization. Detailed metadata is also stored including free-text notes made by the scientist, the date and time the workflow was created or modied, optional descriptive tags, and the user that created it.

how workflow evolution provenance enables a series of operations which simplify exploratory processes and foster reflective reasoning. For example, scientists can easily navigate through the space of workflows created for a given exploration task, visually compare workflows and their results, and explore large parameter spaces.

As part of its provenance infrastructure, VisTrails provides usable interfaces for exploring the provenance information and supporting knowledge reuse. In Section 4.2, we describe three components of the system that serve this purpose: a query-by-example interface, a mechanism for refining workflows by analogy, and a recommendation system that aids users in the design of workflows.

VisTrails is an extensible system. Like workflow systems, it allows pipelines to be created that combine multiple libraries. In addition, the VisTrails provenance infrastructure can be integrated with interactive tools, which cannot be easily wrapped in a workflow system. We give a brief overview of how new packages can be added to VisTrails in Section 4.3, where we also describe a VisTrails-based provenance plug-in developed for ParaView [29].

4.1. Using provenance to streamline data exploration

Compared both to visualization [14, 29] and scientific workflow systems [31–34], a key distinguishing feature of VisTrails is that, besides recording provenance for data products, it captures information about how workflows evolve over time. VisTrails uses a new change-based provenance mechanism [28, 35] that uniformly captures changes to parameter values as well as to workflow definitions. By maintaining *detailed provenance of the exploration process*, it not only ensures reproducibility but also allows scientists to easily navigate through the space of workflows created for a given exploration task.

Interacting with History. Figure 3 shows an example of an exploratory visualization using

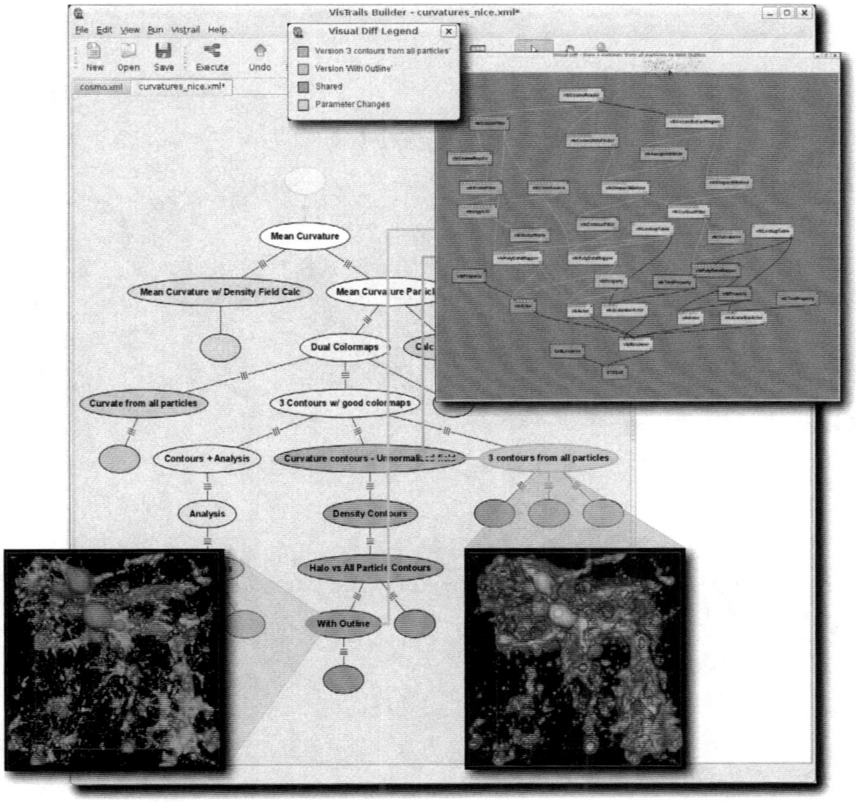

Figure 4. By representing provenance as a series of actions that modify a pipeline, visualizing the differences between two workflows becomes possible. The difference between workflows is described in a meaningful way, as an aggregation of the two workflows forming it. This representation of the difference is both informative and intuitive, reducing the time it takes to understand how two workflows are functionally different.

VisTrails. In the center, the vistrail (the visual trail) captures all modifications users apply to the visualizations. Each node in the vistrail tree corresponds to a workflow (or pipeline) and edges between two nodes correspond to changes applied to transform the parent pipeline into the child (e.g., through the addition of a module or a change to a parameter value). The tree-based representation allows a scientist to return to a previous version in an intuitive way, to undo bad changes, to compare different workflows, and to be reminded of the actions that led to a particular result.

Comparing Workflows and their Results. Another important benefit of the change-based provenance model is that it enables a series of operations that simplify the exploration process including the ability to compare workflows and a scalable mechanism for exploring large parameter spaces [28]. The discovery process requires many trial-and-error steps, and it is not uncommon for tens to hundreds of different workflows to be generated in the course of a study. In such a scenario it becomes important to understand what the differences between workflows are, especially if multiple people are collaboratively exploring the data. As illustrated in figure 4, VisTrails provides a visual difference that allows structural comparison of workflows. This figure depicts two similar workows that yield drastically different results. By analyzing the difference between the two workflows, the salient aspects of each visualization can be more conveniently explored. The visual difference shows in blue modules unique to the workflow on the left; in

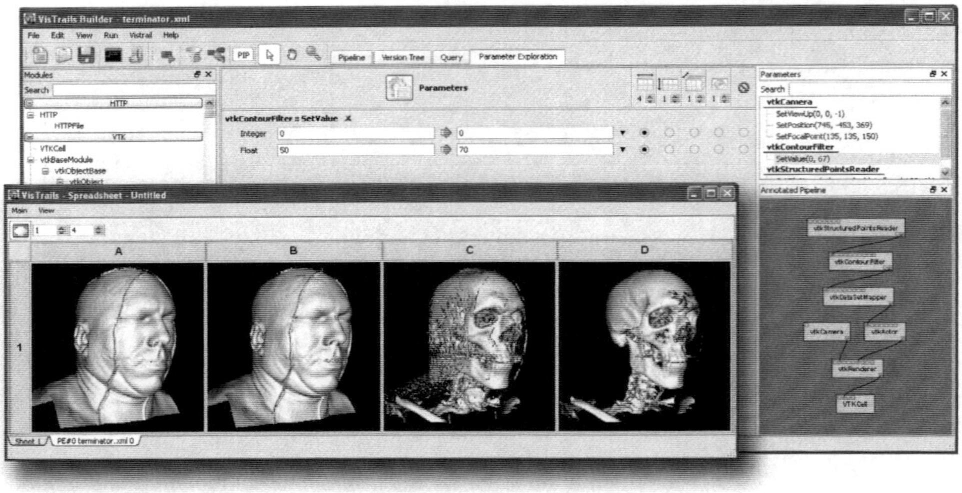

Figure 5. Parameter exploration is performed in VisTrails using a simple interface. The results are computed efficiently by avoiding redundant computation and displayed in the spreadsheet for interactive comparative visualization.

orange modules unique to the workflow on the right; and in gray, the modules shared by the two workflows.

Exploring Parameter Spaces. The parameter exploration mechanism allows the user to specify a set of parameters to be explored, as well as how to explore them and display the results. As a simple one dimensional example, we show an exploration of the isosurface for the visible human head value as four steps between 50 and 70 and display them horizontally in the VisTrails Spreadsheet. Figure 5 shows the interface as well as the results of this exploration. The VisTrails Spreadsheet provides the means to compare visualizations in different dimensions (row, column, sheet, time). The cells in the spreadsheet can be linked to synchronize the interactions between visualizations. We note that VisTrails leverages the dataflow specifications to identify and avoid redundant operations. By using the same cache for different cells in the spreadsheet, VisTrails allows the efficient exploration of a large number of related visualizations [36].

4.2. Exploring and reusing provenance

Although the benefits of using workflow and workflow-based visualization systems are well known, the fact that workflows are hard to create and maintain has been a major barrier to wider adoption of the technology in the scientific domain [1, 37, 38]. VisTrails provides infrastructure that allows users to reuse provenance information to simplify the creation and refinement of workflows. This includes the ability to query provenance; to refine workflows by analogy, using previous workflow refinements as examples; and a recommendation system that guides users through the process of workflow design by using a database of previously created workflows.

Auto-Completion for Workflows. Scientific workflows and visualization pipelines are specified as graphs, where nodes represent processes (or modules) and edges capture the flow of data between the processes. Assembling these workflows requires programming expertise as well as detailed knowledge of multiple tools and libraries. Often, it also entails tightly coupled collaborative efforts. Even for systems that have sophisticated visual programming interfaces, such as DX, AVS and SCIRun, the path from the raw data to insightful visualizations is quite laborious and error-prone. Significant effort is required to build new workflows and to manually

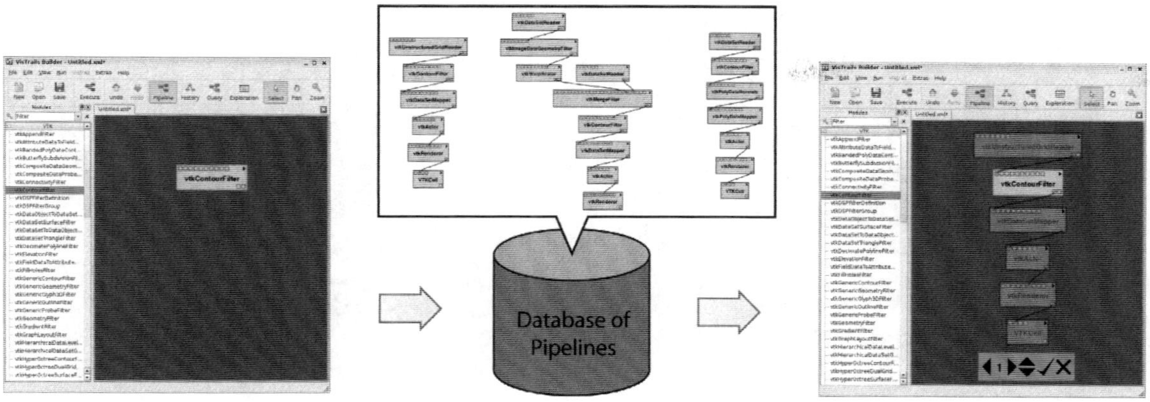

(a) (b) (c)

Figure 6. Workflow auto-completion: (a) A user starts by adding a module to the pipeline. (b) The most likely completions are generated using indexed paths computed from a database of pipelines. (c) A suggested completion is presented to the user as semi-transparent modules and connections. The user can browse through suggestions using the interface and choose to accept or reject the completion.

modify existing workflows to cater to new requirements (e.g., to use different parameters or different visualization techniques). The difficulty of these tasks makes the use of workflows impossible for domain scientists with little or no programming expertise.

By learning common paths used in existing workflows, VisTrails predicts a set of likely module sequences that can be presented to the user as suggestions during the design process. The system acts as an auto-completion mechanism for workflows, suggesting potential modules and connections in a manner similar to a Web browser suggesting URLs. The workflow completions are presented graphically in a way that allows the user to easily explore and accept suggestions or disregard them and continue working as they were. Figure 6 gives an overview of how the auto-completion mechanism works. For more details, see [39].

Querying Workflows. The ability to query both the specification and provenance of workflows enables users to better understand exploratory processes and their results. For example, users can identify workflows that are suitable for and can be reused for a given task or can identify workflow instances that have been found to contain anomalies. Provenance information can also be associated with data products (e.g., images, graphs), allowing structured queries to be posed over these unstructured data.

Besides supporting simple, keyword-based queries as well as structured queries, VisTrails also allows queries to be specified by example, using an interface that mirrors the interface for building a workflow [40]. As figure 7 illustrates, a user can construct (or copy and paste) a pipeline fragment into the VisTrails query tab to identify in the history tree all nodes that contain that fragment. The user can then browse through the highlighted nodes. When a highlighted node is clicked on, the workflow is displayed and the modules that match the query are highlighted. The user can then click on the individual modules to view execution log records associated with modules.

Refining Workflows by Analogy. While the query interface allows users to identify workflows (and subworkflows) that are relevant for a particular task, the *analogy operation* provides a mechanism for reusing these pipelines to construct new data products in a semi-automated manner—without requiring users to directly manipulate or edit the dataflow specifications [40].

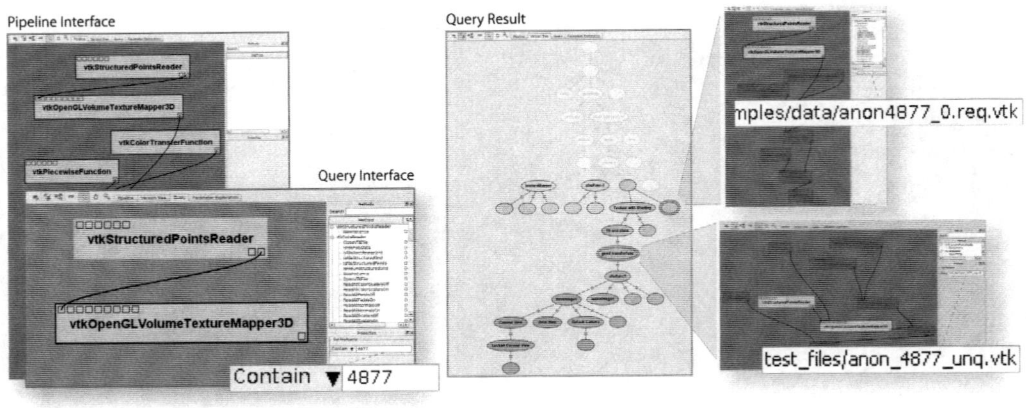

Figure 7. Querying by example. The interface for building a query over an ensemble of pipelines is essentially the same as the one for constructing and updating pipelines. In fact, they work together: portions of a pipeline can become query templates by directly pasting them onto the Query Canvas. In this figure, the user is looking for a volume-rendered image of a file whose name contains the string "4877." The system highlights the matches both at the visualization level (version tree, shown in the middle) and at the module level (shown in the right insets).

Figure 8 illustrates the use of analogies for creating visualizations. To apply an analogy, first the difference between a source pair of analogous visualizations is computed. This difference is then transferred to a third visualization. Analogies can be used as the basis for scalable updates: the user does not need to have knowledge of the exact details of the three visualization workflows to perform the operation.

4.3. Provenance for workflows and interactive applications

Integrating Tools and Libraries. VisTrails provides an infrastructure that allows users to integrate user-defined functions and libraries. Specifically, users can incorporate their own visualization and simulation codes into pipelines by defining custom modules (or wrappers) which are bundled in *packages*. A VisTrails package is a collection of Python classes, where each class corresponds to a module. These wrappers specify the input and output parameters for the modules as well as how the underlying function is invoked. Detailed instructions on how to add packages are given in [41]. The VisTrails distribution provides support for several libraries, including VTK, Web Services, and SciPy.

Provenance-Enabling Existing Applications. The initial focus of the VisTrails project was to develop a provenance infrastructure for computational tasks specified as workflows. But although the change-based provenance model was originally designed to capture changes applied to a workflow definition (Section 4.1), it is extensible and can be applied in different environments. In particular, it can capture change actions that are applied by a user through an interactive user interface (e.g., slicing a volume or editing a parameter in a scientific visualization system). The change-based representation of provenance is easily incorporated into existing applications that provide a mechanism for controlling the actions that are being performed by a user via a graphical interface. And by doing so, the interfaces and mechanisms provided by VisTrails for interacting and exploring provenance can be reused. An important advantage of this approach is that, instead of forcing users to learn to use a new environment, it allows users to capture and leverage provenance using the applications and environments that they are used

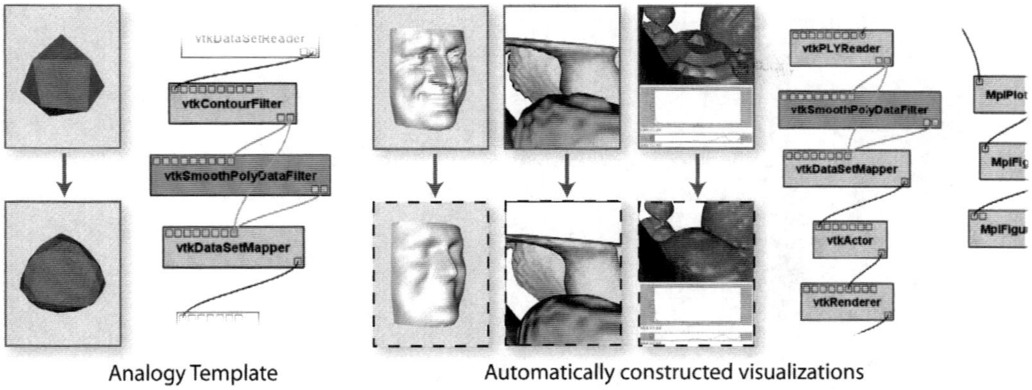

Figure 8. Visualization by analogy. The user chooses a pair of visualizations to serve as an analogy template. In this case, the pair represents a change where a file downloaded from the WWW is smoothed. Then, the user chooses a set of other visualizations that will be used to derive new visualizations, with the same change. These new visualizations are derived automatically. The pipeline on the left reflects the original changes, and the one on the right reflects the changes when translated to the last visualization on the right. The pipeline pieces to be removed are portrayed in orange, and the ones to be added, in blue. Note that the surrounding modules do not match exactly: the system figures out the most likely match.

to.

In general, applications that take advantage of the model-view-controller paradigm [42] have a mechanism for storing and reusing actions: the undo and redo operations. In a scientific visualization system, for instance, with *undo* a user is able to walk through the steps they took to create an image, albeit backwards. Although undo does not capture the complete exploration process nor does it persist across sessions, it provides valuable context for granularity of actions. The designers of the software have already determined the granularity of actions by designing the undo stack. The undo stack of an application may individually capture single mouse events or keyboard strokes if they are needed to recreate the state. Furthermore, interactions performed by the user may cause multiple actions to be performed, which the undo stack will store as one step. We can capture actions at the same granularity in which the undo stack does. In fact, in practice it is simpler to capture actions as they are being added to the undo stack instead of where they are handled by the controller. Obviously, this depends on the completeness and availability of the undo/redo mechanism in the application.

Figure 9 shows ParaView together with the VisTrails Provenance Explorer, transparently capturing the complete exploration process. The provenance capture mechanism was implemented by inserting monitoring code in ParaView's undo/redo mechanism, which captures changes to the underlying pipeline specification. Essentially, the action on top of the undo stack is added to the vistrail in the appropriate place, and undo is reinterpreted to mean "move up the version tree." Note that the change-based representation is both simple and compact—it uses substantially less space than the alternative of storing multiple instances or *versions* of the state.

5. Conclusions and future directions

In this paper, we discussed important issues that must be considered in the design of software environments for exploratory data analysis and visualization. The end goal is clear: build tools that aid scientists and streamline scientific discoveries. But to achieve this we need

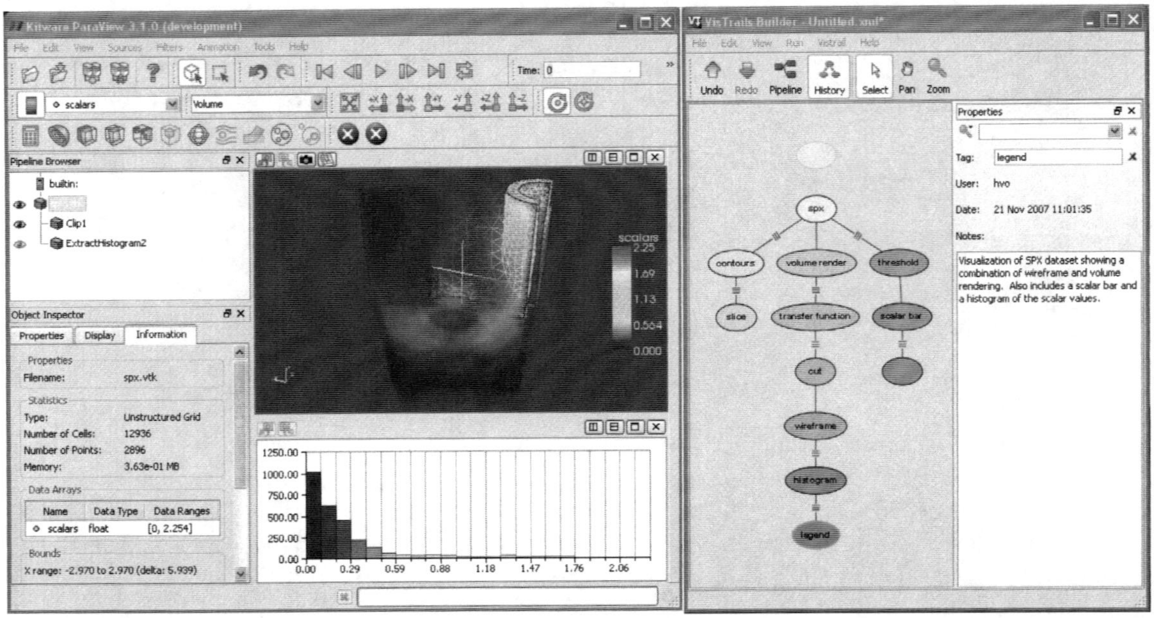

Figure 9. Screenshot of ParaView (left) with the provenance captured by VisTrails and displayed as a version tree in a separate window (right). This preliminary prototype taps into ParaView undo/redo mechanism to capture the exploration process.

creativity support tools that enable scientists to be more efficient and that provide effective means for collaboration. We view provenance management and scientific workflows as key enabling technologies that can serve as the basis for these tools.

There is still much work left for the visualization community to make this vision, of having visualization as an integral part of the scientific process, a reality. Below, we describe some of the directions we are currently pursuing. The ability to support and integrate multiple tools is of paramount importance in practice. Users often have a set of tools that they have been using and/or developing, and on which they have made substantial investments. For most, replacing those tools is not an option. We believe that any "scientific discovery environment" needs to be built in such a way as to interoperate with existing tools: it needs to be modular, and to have components that can be easily incorporated into or combined with custom tools. An early example of work toward enhancing the capabilities of an existing powerful tool is our provenance plugin for ParaView. We plan to work on similar capabilities for VisIt. On a more ambitious plan, we are also working with Dean Williams and his group on extensions to CDAT. Although CDAT has been developed with climate applications in mind, most of the concepts of the system could be reused for developing a more general Data Analysis Tool (DAT) by merging CDAT and VisTrails functionality.

Publications are an essential component of the scientific process. But although the way science is done has changed substantially in recent years, the same cannot be said of scientific publications. We have been exploring new tools and ways to extend existing tools (e.g., LaTeX, wiki, MS Word) to support the creation of articles that are more dynamic, reproducible, and verifiable. By allowing readers to replicate the experiments in a paper and to maintain detailed provenance of artifacts used across papers, the publications of the future have the potential to accelerate scientific discoveries and to improve the quality of science.

As provenance-enabled systems (including workflow systems) are deployed, large volumes of provenance information are being collected. Having these large collections creates opportunities to explore provenance in novel ways. But it also presents several challenges. Just collecting provenance information does not bring major benefits. For the data to be truly useful, we need effective and efficient ways to analyze it. As provenance collections grow, new mechanisms are needed to deal with a potential information overload. The problem of mining and extracting knowledge from provenance data, however, has been largely unexplored. We are interested in "provenance analytics," which we define to be a new area aimed at investigating new techniques for mining, visualizing and reusing provenance of computational tasks.

Acknowledgments

We thank our students and collaborators for contributing to this work, and the anonymous reviewers for comments that helped us to improve our manuscript. Our research has been funded the Department of Energy SciDAC (VACET and SDM centers), the National Science Foundation (grants IIS-0746500, CNS-0751152, IIS-0713637, OCE-0424602, IIS-0534628, CNS-0514485, IIS-0513692, CNS-0524096, CCF-0401498, OISE-0405402, CCF-0528201, CNS-0551724), and IBM Faculty Awards (2005, 2006, and 2007).

References

[1] Johnson C R, Moorhead R, Munzner T, Pfister H, Rheingans P and Yoo T S 2006 *NIH-NSF Visualization Research Challenges* (IEEE Computer Society) `http://tab.computer.org/vgtc/vrc/index.html`
[2] van Wijk J 2005 *Proceedings of IEEE Visualization*
[3] Shneiderman B 2007 *Commun. ACM* **50** 20–32
[4] Shneiderman B 2002 *Commun. ACM* **45** 31–34
[5] Kniss J, Kindlmann G and Hansen C 2002 *IEEE Transactions on Visualization and Computer Graphics* **8** 270–285
[6] Norman D A 1994 *Things That Make Us Smart: Defending Human Attributes in the Age of the Machine* (Addison Wesley)
[7] Tufte E R 1997 *Visual Explanations: Images and Quantities, Evidence and Narrative* (Graphics Press)
[8] Plaisant C 2004 *Proceedings of the working conference on Advanced visual interfaces* (ACM) pp 109–116
[9] Upson et al C 1989 *IEEE Computer Graphics and Applications* **9** 30–42
[10] IBM OpenDX `http://www.research.ibm.com/dx`
[11] Schroeder W, Martin K and Lorensen B 2003 *The Visualization Toolkit An Object-Oriented Approach To 3D Graphics* (Kitware)
[12] Law C C, Henderson A and Ahrens J 2001 *IEEE Symposium on Parallel and Large-data Visualization and Graphics 2001* pp 125–128
[13] Childs H, Brugger E S, Bonnell K S, Meredith J S, Miller M, Whitlock B J and Max N 2005 *IEEE Visualization 2005* pp 190–198
[14] Parker S G and Johnson C R 1995 *Proceedings of Supercomputing* p 52
[15] Brodlie K, Poon A, Wright H, Brankin L, Banecki G and Gay A 1993 *IEEE Visualization 1993* pp 102–109
[16] Shrinivasan Y B and van Wijk J J 2008 *ACM SIGCHI 2008* pp 1237–1246
[17] Pirolli P and Card S 1999 *Psychological Review* **106** 643–675
[18] Pirolli P and Card S 2005 *International Conference on Intelligence Analysis*
[19] Freire J and Silva C 2008 *CHI Social Data Analysis Workshop*
[20] Freire J, Koop D, Santos E and Silva C T 2008 *Computing in Science and Engineering* **10** 11–21
[21] Davidson S B and Freire J 2008 *Proceedings of ACM SIGMOD* pp 1345–1350
[22] Davidson S B, Boulakia S C, Eyal A, Ludäscher B, McPhillips T M, Bowers S, Anand M K and Freire J 2007 *IEEE Data Eng. Bull.* **30** 44–50
[23] Miles S, Groth P, Branco M and Moreau L 2006 The requirements of using provenance in e-science experiments Tech. rep. ECS, University of Southampton
[24] Bose R and Frew J 2005 *ACM Computing Surveys* **37** 1–28
[25] Simmhan Y L, Plale B and Gannon D 2005 *SIGMOD Record* **34** 31–36
[26] Moreau L, Ludäscher B, Altintas I, Barga R S, Bowers S, Callahan S, Chin Jr G, Clifford B, Cohen S, Cohen-Boulakia S, Davidson S, Deelman E, Digiampietri L, Foster I, Freire J, Frew J, Futrelle J, Gibson T, Gil Y, Goble C, Golbeck J, Groth P, Holland D A, Jiang S, Kim J, Koop D, Krenek A, McPhillips T, Mehta G, Miles S, Metzger D, Munroe S, Myers J, Plale B, Podhorszki N, Ratnakar V, Santos E, Scheidegger C, Schuchardt K, Seltzer M, Simmhan Y L, Silva C, Slaughter P, Stephan E, Stevens R, Turi D, Vo H, Wilde M, Zhao J and Zhao Y 2007 *Concurrency and Computation: Practice and Experience*

[27] Freire J, Miles S and Moreau L 2007 Second provenance challenge `http://twiki.ipaw.info/bin/view/Challenge/SecondProvenanceChallenge`

[28] J Freire, C T Silva, S P Callahan, Santos E, Scheidegger C E and Vo H T 2006 *International Provenance and Annotation Workshop (IPAW)* LNCS 4145 pp 10–18

[29] Kitware Paraview `http://www.paraview.org`

[30] Anderson E W, Ahrens J, Heitmann K, Habib S and Silva C 2008 *Computing in Science and Engineering* **10** 30–37

[31] The Kepler Project `http://kepler-project.org`

[32] The Taverna Project `http://taverna.sourceforge.net`

[33] The Triana Project `http://www.trianacode.org`

[34] Ailamaki A, Ioannidis Y E and Livny M 1998 *Proc. of the Conference on Scientific and Statistical Database Management* pp 190–199

[35] Callahan S, Freire J, Santos E, Scheidegger C, Silva C and Vo H 2006 *IEEE Workshop on Workflow and Data Flow for Scientific Applications (SciFlow)*

[36] Bavoil L, Callahan S, Crossno P, Freire J, Scheidegger C, Silva C and Vo H 2005 *Proceedings of IEEE Visualization* pp 135–142

[37] Maechling P, Chalupsky H, Dougherty M, Deelman E, Gil Y, Gullapalli S, Gupta V, Kesselman C, Kim J, Mehta G, Mendenhall B, Russ T, Singh G, Spraragen M, Staples G and Vahi K 2005 *SIGMOD Rec.* **34** 24–30

[38] Schopf J, Coleman I, Procter R and Voss A 2006 Report of the user requirements and web based access for eresearch workshop Tech. Rep. UKeS-2006-07 UK National e-Science Centre

[39] Koop D, Scheidegger C, Callahan S, Freire J and Silva C 2008 *IEEE Transactions on Visualization and Computer Graphics* Accepted with minor revisions.

[40] Scheidegger C, Koop D, Vo H, Freire J and Silva C 2007 *IEEE Transactions on Visualization and Computer Graphics*

[41] VisTrails Packages `http://www.vistrails.org/index.php/UsersGuideVisTrailsPackages`

[42] Krasner G E and Pope S T 1988 *Journal of Object-Oriented Programming* **1**(3) 26–49

AUTHOR INDEX

AUTHOR INDEX

AUTHOR INDEX

AUTHOR INDEX

AUTHOR INDEX

AUTHOR INDEX

AUTHOR INDEX

AUTHOR INDEX

AUTHOR INDEX

Institute of Physics Publishing
Dirac House, Temple Back
Bristol BS1 6BE UK

ISBN: 978-1-60560-608-8